ENVIRONMENTAL CHEMISTRY

NINTH EDITION

ENVIRONMENTAL CHEMISTRY

NINTH EDITION

STANLEY E. MANAHAN

CRC Press is an imprint of the
Taylor & Francis Group, an **informa** business

The cover illustration depicts the pathway of solar energy captured by photosynthesis to convert atmospheric carbon dioxide to biomass that can be used as a feedstock for the thermochemical production of methane. This synthetic natural gas product is the cleanest burning of all the carbon-based fuels. Methane made by the biomass pathway is neutral with respect to the production of greenhouse gas carbon dioxide because any of that gas generated by combustion of biomass-based synthetic natural gas was removed from the atmosphere by photosynthesis in producing the biomass feedstock. Biomass-based synthetic fuel production and other sustainable energy technologies are discussed in detail in Chapter 19, "Sustainable Energy: The Key to Everything."

MIX
Paper from responsible sources
FSC® C014174

CRC Press
Taylor & Francis Group
6000 Broken Sound Parkway NW, Suite 300
Boca Raton, FL 33487-2742

Printed in the United States of America on acid-free paper
10 9 8

International Standard Book Number: 978-1-4200-5920-5 (Hardback)

Library of Congress Cataloging-in-Publication Data

Manahan, Stanley E.
Environmental chemistry / Stanley E. Manahan. -- 9th ed.
p. cm.
Includes bibliographical references and index.
ISBN 978-1-4200-5920-5 (hardcover : alk. paper)
1. Environmental chemistry. I. Title.

TD193.M36 2010
628--dc22 2009040939

Visit the Taylor & Francis Web site at
http://www.taylorandfrancis.com

and the CRC Press Web site at
http://www.crcpress.com

Contents

Preface

Environmental Chemistry, ninth edition, maintains much the same organizational structure, level, and emphasis that have been developed through preceding editions, with updates in keeping with the emerging face of the dynamic science of environmental chemistry. Therefore, rather than entering into an immediate discussion of a specific environmental problem, such as stratospheric ozone depletion, the book systematically develops the concept of environmental chemistry so that, when specific pollution problems are discussed, the reader has the background to understand such problems. Chapters 1 and 2 have been significantly changed from the eighth edition to provide a better perspective on sustainability, environmental science as a whole, chemical fate and transport, cycles of matter, the nature of environmental chemistry, and green chemistry. The chapter on terrorism in the eighth edition has been removed, but specific aspects of this topic, such as the potential role of toxic substances in terrorist attacks, have been placed in other chapters. Because of the importance of energy to the environment and sustainability, an extensive new chapter on this topic has been added to the book. Separate chapters on basic chemistry and organic chemistry that were in the seventh and earlier editions have not been included, but may be obtained from the author or publisher in pdf format upon request.

The book views the environment as consisting of five spheres: (1) hydrosphere, (2) atmosphere, (3) geosphere, (4) biosphere, and (5) anthrosphere. It emphasizes the importance of the anthrosphere—that part of the environment made and operated by humans and their technologies. This environmental sphere has so much influence on the Earth and its environmental systems that, according to Nobel Prize winner Paul Crutzen, the Earth is leaving the Holocene epoch, throughout which humankind has existed on Earth until now and is entering the Anthropocene epoch, in which human influences, such as emissions of gases that significantly affect the warming and protective functions of the atmosphere, will have a dominant influence on conditions under which humankind exists on Earth. Since technology will in fact be used to attempt to support humankind on the planet, it is important that the anthrosphere be designed and operated in a manner that is compatible with sustainability and interacts constructively with the other environmental spheres. In this endeavor, environmental chemistry has a key role to play.

Environmental chemistry has evolved significantly since the first edition of *Environmental Chemistry* was published in 1972. (One interesting footnote to this evolution has been that with each new edition the calculation of the pH of ordinary rainwater has had to be revised because levels of atmospheric carbon dioxide had increased enough since each preceding edition to affect the result.) Whereas in the early 1970s environmental chemistry dealt largely with pollution and its effects, it now has a current emphasis upon sustainability. During the lifetime of the book, problems with organochlorine pesticides and detergent phosphates that cause water eutrophication have largely disappeared as the manufacture and sale of these substances have essentially ceased. When the book was first published, it was not known with certainty what happened to large quantities of carbon monoxide emitted to the atmosphere by automobiles; it was suspected that soil microorganisms metabolized this pollutant, but it is now known that the ubiquitous hydroxyl radical scavenges CO from the atmosphere. In 1972, the potential for stratospheric ozone depletion was just emerging as a major issue, but it was not known that refrigerant chlorofluorocarbons (freon compounds) were predominantly responsible for this threat. As the book progressed through various editions, the threat of these materials was revealed; the southern hemisphere springtime Antarctic ozone hole was discovered, which grew ominously year by year; the manufacture of chlorofluorocarbons was banned as a consequence; and Molina, Rowland, and Crutzen shared a well-deserved Nobel Prize, the first ever in environmental chemistry, for their work in this field. The potential for greenhouse warming due to growing emissions of infrared-capturing carbon dioxide, methane, and other gases was shown to be a potentially huge problem for Earth and one that has not yet been resolved. In 1972, the terms green chemistry and industrial ecology had not yet been coined, but these disciplines emerged from the 1990s as crucial elements of environmental chemistry.

Chapter 1 provides an overview and background in environmental and sustainability science. The chapter is introduced with a brief discussion of the central issue of our time—energy, "From the Sun to Fossil Fuels and Back Again." This chapter introduces chemical fate and transport, environmental terrorism, and environmental forensics.

Chapter 2 defines environmental chemistry and green chemistry in some detail. The chapter discusses the important concept of cycles of matter. It introduces the anthrosphere, how it integrates with the other environmental spheres, and its effects on Earth. Components of the anthrosphere that influence the environment are discussed with emphasis on the all-important infrastructure that is part of the anthrosphere.

Chapters 3 through 8 deal with the hydrosphere. Chapter 3 introduces the special characteristics of water and the environmental chemistry of water. The remaining Chapters 4 through 8 discuss specific aspects of aquatic chemistry, aquatic biochemistry, and water sustainability and treatment.

Chapters 9 through 14 discuss atmospheric chemistry. Chapter 14 emphasizes the greatest success story of environmental chemistry to date—the study of ozone-depleting chlorofluorocarbons, which resulted in the first Nobel Prize being awarded in environmental chemistry as mentioned above. It also emphasizes the greenhouse effect, which may be the greatest of all threats to the global environment as we know it.

Chapters 15 and 16 deal with the geosphere, the latter chapter emphasizing soil and agricultural chemistry. Included in the discussion of agricultural chemistry is the important and controversial new area of transgenic crops. Another area discussed is conservation tillage, which makes limited use of herbicides to grow crops with minimum soil disturbance.

Chapter 17 goes into detail on the topic of green chemistry and the closely related area of industrial ecology. Chapter 18 discusses resources and sustainable materials.

Chapter 19, on energy, is new as a separate chapter in the ninth edition. Entitled "Sustainable Energy: The Key to Everything," it covers key topics on sustainable energy including conservation and renewable sources. The chapter ends with a proposed system of industrial ecology designed to produce methane from renewable biofuels and hydrogen generated from the electrolysis of water using renewable wind or solar energy.

The nature and environmental chemistry of hazardous wastes are covered in Chapter 20 and industrial ecology for waste minimization, utilization, and treatment in Chapter 21.

Chapters 22 and 23 cover the biosphere. Chapter 22 is an overview of biochemistry with emphasis on environmental aspects. Chapter 23 introduces and outlines the topic of toxicological chemistry. Chapter 24 discusses the toxicological chemistry of various classes of chemical substances.

Chapters 25 through 28 deal with environmental chemical analysis, including water, wastes, air, and xenobiotics in biological materials.

As noted above, two chapters on basic chemistry and organic chemistry that were present in the seventh and earlier editions have been removed from this edition for the sake of brevity. Readers who need this material can obtain files containing these chapters from the publisher or the author.

I welcome comments and questions from readers. I can be reached by e-mail at manahans@missouri.edu.

Stanley E. Manahan

Author

Stanley E. Manahan is professor emeritus of chemistry at the University of Missouri–Columbia, where he has been on the faculty since 1965. He received his AB in chemistry from Emporia State University in 1960 and his PhD in analytical chemistry from the University of Kansas in 1965.

Since 1968 his primary research and professional activities have been in environmental chemistry, toxicological chemistry, and waste treatment. His classic textbook, *Environmental Chemistry*, has been in print continuously in various editions since 1972 and is the longest standing title on this subject in the world. His other books are *Fundamentals of Environmental Chemistry,* 3rd ed. (Taylor & Francis/CRC Press, 2009), *Fundamentals of Sustainable Chemical Science* (Taylor & Francis/ CRC Press, 2009), *Environmental Science and Technology*, 2nd ed. (Taylor & Francis, 2006), *Green Chemistry and the Ten Commandments of Sustainability*, 2nd ed. (ChemChar Research, Inc, 2006), *Toxicological Chemistry and Biochemistry*, 3rd ed. (CRC Press/Lewis Publishers, 2001), *Industrial Ecology: Environmental Chemistry and Hazardous Waste* (CRC Press/Lewis Publishers, 1999), *Environmental Science and Technology* (CRC Press/ Lewis Publishers, 1997), *Hazardous Waste Chemistry, Toxicology and Treatment* (Lewis Publishers, 1992), *Quantitative Chemical Analysis* (Brooks/Cole, 1986), and *General Applied Chemistry*, 2nd ed. (Willard Grant Press, 1982).

Dr. Manahan has lectured on the topics of environmental chemistry, toxicological chemistry, waste treatment, and green chemistry throughout the United States as an American Chemical Society Local Section Tour Speaker and has presented plenary lectures on these topics in international meetings in Puerto Rico; the University of the Andes in Mérida, Venezuela; Hokkaido University in Japan; the National Autonomous University in Mexico City; France; and Italy. He was the recipient of the Year 2000 Award of the Environmental Chemistry Division of the Italian Chemical Society. His research specialty is gasification of hazardous wastes.

1 The Environment and Sustainability Science

1.1 FROM THE SUN TO FOSSIL FUELS AND BACK AGAIN

An old Chinese proverb states that, "If we do not change direction, we are likely to end up where we are headed." The New Millennium has brought with it evidence that we have been on a course that, if not altered, will have profound, unpleasant effects on humankind and on the Earth upon which this species and all other living things depend for their existence. The World Trade Center attacks of September 11, 2001, as well as later attacks on the London subway system, trains in Madrid, hotels in Mumbai, and other places around the world showed the vulnerability of our civilization to malevolent actions of those who feel compelled to commit evil acts and raised concerns about the possibilities for even more damaging attacks with chemical, biological, or radioactive agents. The first part of 2008 saw runaway prices for key commodities including petroleum, metals such as copper, and grain. Crude oil prices reached almost $150 per barrel in July 2008 and predictions were made that gasoline prices in the United States would exceed $5.00 per gallon for the foreseeable future. These trends were reversed in the latter part of 2008 with the occurrence of the greatest economic collapse the world had seen since the Great Depression of the 1930s. Housing prices, which had reached levels unsupportable by people with ordinary incomes, collapsed, and the prices of some commodities plummeted. In early 2009, world leaders were struggling to find solutions to severe economic problems.

As humans and their governments struggle with economic challenges, evidence has continued to gather that their activities are degrading the Earth's life support system upon which they depend for their existence. The emission to the atmosphere of carbon dioxide and other greenhouse gases is almost certainly causing global warming. During the early 2000s, the Arctic ice cap diminished to a level never before observed in recorded history. Discharge of pollutants has degraded the atmosphere, hydrosphere, and geosphere in industrialized areas. Natural resources including minerals, fossil fuels, freshwater, and biomass have become stressed and depleted. The productivity of agricultural land has been diminished by water and soil erosion, deforestation, desertification, contamination, and conversion to nonagricultural uses. Wildlife habitats including woodlands, grasslands, estuaries, and wetlands have been destroyed or damaged. About 3 billion people (half of the world's population) live in dire poverty on less than the equivalent of U.S. $2 per day. The majority of these people lack access to sanitary sewers and the conditions under which they live give rise to debilitating viral, bacterial, and protozoal diseases including malaria and diarrhea. At the other end of the standard of living scale, a relatively small fraction of the world's population consumes an inordinate amount of resources with lifestyles that involve living too far from where they work, in energy-wasting houses that are far larger than they need, commuting long distances in large "sport utility vehicles" that consume far too much fuel, and overeating to the point of unhealthy obesity with accompanying problems of heart disease, diabetes, and other obesity-related maladies.

In a sense, the history of humankind and its relationship to Planet Earth is a story of "from the sun to fossil fuels and back again." During virtually all their existence on Earth, humans have been dependent upon the bounty provided by the sun. Solar radiation provided the warmth required for humans to exist and was augmented by fire from burning of biomass generated by photosynthesis and by garments made of skins of animals that had fed upon photosynthetically produced biomass. Food consumed by humans came from plants that converted solar energy into biomass chemical energy and from meat produced by plant-eating animals. As human societies developed, indirect means of utilizing solar energy were harnessed as well. Wind generated by solar heating of the atmosphere was utilized to drive windmills and to propel sailboats used for transportation. Humans learned to impound water and to convert the energy of flowing water to mechanical energy with waterwheels. The circulation of this water was part of the solar-powered hydrological cycle. Essentially everything that humans used and depended upon for their existence came ultimately from the sun.

1.1.1 The Brief but Spectacular Era of Fossil Fuels

As civilizations developed, humans discovered the uses of fossil fuels for energy. Although coal in those few locations where it was readily accessible from the surface had been used as a source of heat for centuries, the development of this energy source really took off around 1800, especially with the development of the steam engine as a practical power source. This began a massive shift from solar and biomass energy sources to fossil fuels, progressing from coal to petroleum and eventually natural gas. The result was an enormous revolution in human society with the development of huge heavy industries; transportation systems including rail, motor vehicles, and aircraft; and means of greatly increased food production. In Germany in the early 1900s, Carl Bosch and Fritz Haber developed the process for converting atmospheric elemental nitrogen to ammonia (NH_3), a high-pressure, energy-intensive process that required large amounts of fossil fuels. This discovery enabled the production of huge quantities of relatively inexpensive nitrogen fertilizer and the resulting increase in agricultural production may well have saved Europe, with a rapidly developing population at the time, from widespread starvation. So it was that the era of fossil fuel, which has been described as "fossiled sunshine,"[1] dating from about 1800 enabled humankind to enjoy unprecedented material prosperity and to increase in numbers from around 1 billion to over 6 billion.

Now, however, it is apparent that the fossil fuel era, if not ending, will no longer be sustainable as the bedrock of industrial society. Approximately half of the world's total petroleum resource has already been consumed and, despite periods of reduced demand such as occurred during the global economic downturn in 2009, petroleum will continue to become more scarce and expensive and can last for only a few more decades as the dominant fuel and organic chemicals source for humankind. Coal is much more abundant, but its utilization has troubling environmental implications, especially as the major source of greenhouse gas carbon dioxide. Natural gas, an ideal, clean-burning fossil fuel that produces the least amount of carbon dioxide per unit energy generated, is relatively abundant and means have now been developed to extract it from previously inaccessible tight shale formations; it can serve as a "bridging fuel" for several decades until other sources can be developed. Nuclear energy, properly used with nuclear fuel reprocessing, can take on a greater share of energy production, especially for base load electricity generation.

1.1.2 Back to the Sun

Given the fact that humans cannot rely upon fossil hydrocarbons for future fuel and raw materials, it is back to the sun for meeting basic human needs. The most direct use of the sun is for solar heating and for photovoltaic power generation. But there are probably even greater indirect uses of the sun for the production of energy and material. Arguably the fastest growing energy source in the world is electricity generated by wind. Wind is solar powered; basically, the sun heats masses of air, the air expands, and the wind blows. Biomass generated by solar-powered photosynthesis can be used as a raw material to replace petroleum in petrochemicals manufacture. Furthermore, biomass can be converted to any of the hydrocarbon fuels including methane, diesel fuel, and gasoline as discussed in Chapter 19. The use of biomass for making liquid fuels has not gotten off to a particularly auspicious start in the United States and some other countries. This is because the two major synthetic fuels made from biomass, ethanol from fermentation of sugar and biodiesel fuel made from plant lipid oils, largely use the most valuable parts of food plants—corn grain to get sugar for fermentation to ethanol and soybeans to get oil for the synthesis of biodiesel fuel. The yields from these pathways are relatively low and almost as much energy is required to grow and process grain to make fuel as is obtained from the fuel itself. The demand for corn and soybeans to make synthetic fuels has caused disruptions in the agricultural grain markets and resulted in inflated prices that have created hardships for people dependent upon these crops for food. Sugarcane grown in more tropical regions, especially Brazil, produces very high yields of fermentable carbohydrate and is a practical energy source. Oil palm trees produce fruits and seeds with high contents of oils used to make biodiesel fuel, but the intensive cultivation of oil palm trees in countries such as Malaysia has resulted in environmentally harmful destruction of rain forests. The diversion of palm oil to make fuel has resulted in less of it being available for food.

Fortunately, means exist to produce the biomass needed for fuel and raw material without seriously disrupting the world's food supply. The main pathway for doing so is through the thermochemical conversion of biomass to synthesis gas, a mixture of CO and H_2, followed by chemical synthesis of methane and other hydrocarbons by long-established technology discussed in Chapter 19. The raw material for doing so may come from a number of renewable sources including crop by-products, dedicated crops, and algae. Crop by-products generated in huge quantities in agricultural areas include corn stover (stalks, leaves, husks, and cobs of the plant), wheat straw, and rice straw. Although a certain amount of these materials needs to be returned to soil to maintain soil quality, a significant fraction can be removed for fuel and chemical synthesis (until the practice was prohibited because of air pollution, rice straw was commonly burned in the fields to prevent excess accumulation of residues on soil). Dedicated crops such as hybrid poplar trees and sawgrass can be grown in large quantities for biomass. Microscopic aquatic algae can be vastly more productive of biomass than plants grown on soil and can thrive in brackish (somewhat saline) water in containments in desert areas. In all these cases mineral nutrients from the residues of thermochemical processing, especially potassium, can be returned to the soil. Clearly, photosynthetically produced biomass will have a large role to play in "re-entering the solar age."

1.2 THE SCIENCE OF SUSTAINABILITY

Environmentalists, including the practitioners of environmental chemistry, are often accused of being gloomy and pessimistic. A close look at the state of the world can certainly give credence to such a view. However, the will and ingenuity of humans that have been directed toward exploiting world resources, giving rise to conditions leading to deterioration of Planet Earth, can be—indeed, are being—harnessed to preserve the planet, its resources, and its characteristics that are conducive to healthy and productive human life. The key is *sustainability* or *sustainable development* defined by the Bruntland Commission in 1987 as industrial progress that meets the needs of the present without compromising the ability of future generations to meet their own needs.[2] A key aspect of sustainability is the maintenance of the Earth's carrying capacity, that is, its ability to maintain an acceptable level of human activity and consumption over a sustained period of time.[3]

Interviewed in February 2009, Dr. Steven Chu, a Nobel Prize-winning physicist who had just been appointed Secretary of Energy in U.S. President Barack Obama's new administration, listed three main areas that require Nobel-level breakthroughs in the achievement of sustainability: solar power, electric batteries, and the development of new crops that can be turned into fuel. He contended that the efficiency of solar energy capture and conversion to electricity needed to improve several-fold. Improved electric batteries are needed to store electrical energy generated by renewable means and to enable practical driving ranges in electric vehicles. Improved crops are needed that convert solar energy to chemical energy in biomass at higher efficiencies than current crops. In this case the potential for improvement is enormous because most plants convert <1% of the solar energy falling on them to chemical energy through photosynthesis. Through genetic engineering, it is likely that this efficiency could be improved several-fold leading to vastly increased generation of biomass. Clearly, the achievement of sustainability employing high-level scientific developments will be an exciting development in decades to come.

1.2.1 Environmental Science

This book is about environmental chemistry. To understand that topic, it is important to have some appreciation of environmental science and sustainability science as a whole. *Environmental science* in its broadest sense is the science of the complex interactions that occur among the terrestrial, atmospheric, aquatic, living, and anthropological systems that compose Earth and the surroundings that may affect living things. It includes all the disciplines, such as chemistry, biology, ecology, sociology, and government, that affect or describe these interactions.[4] For the purposes of this book, environmental science will be defined as the study of the earth, air, water, and living environments, and the effects of technology thereon. To a significant degree, environmental science has evolved from investigations of the ways by which, and places in which, living organisms carry out their life cycles. This discipline used to be known as natural history, which later evolved into ecology, the study of environmental factors that affect organisms and how organisms interact with these factors and with each other.

1.2.2 Green Science and Technology

In recent years the environmental movement has seen a shift away from emphasis primarily upon pollution, its effects, and how to overcome these ill effects toward a broader view of sustainability. The more modern orientation is often labeled as "green." As applied to chemistry, the practice of inherently safer and more environmentally friendly chemical science is called *green chemistry*,[5] a topic that is discussed in more detail later in this book. A spin-off of green chemistry to engineering—especially chemical engineering—is known as *green engineering*.[6] In the most general sense, the practice of sustainable science and technology can be called *green science and technology*. As humankind attempts to provide for numbers of people that are already very large for a planet of limited resources, the practice of green science and technology has become a matter of crucial importance.

1.3 CHEMISTRY AND THE ENVIRONMENT

As the science of all matter, chemistry obviously has a large role in understanding the environment and preserving its quality. In the past, grievous damage was done to the environment by misguided and ignorant practices of chemical science and engineering. Wastes from chemical processes were discarded by the cheapest, most convenient routes, which usually meant up the stack, down the drain, or onto the ground (Figure 1.1). Biologists observed the effects of these practices in the form of fish kills, declining bird populations, and deformed animals. Physicians noticed maladies caused by air and water pollution, such as respiratory problems from breathing polluted air. And ordinary citizens without any special scientific knowledge could observe obscured visibility in polluted atmospheres and waterways choked with excessive plant growth from nutrients discharged into the water; eyes and nose alone were often sufficient to detect significant pollution problems.

But, as the science of matter, chemistry has an essential role to play in environmental protection and improvement. Increasingly, chemists have become familiar with chemical processes that occur in the environment and have developed means of directing chemical science toward environmental improvement. Since about 1970, the science of chemistry in the environment—*environmental chemistry*, the topic of this book—has emerged as a strong and dynamic science that has made enormous contributions to understanding the environment and the chemical and biochemical processes that occur in it. Toxicological chemistry has developed as a discipline that relates the chemical nature of substances to their toxic effects.[7]

But to understand environmental problems is not enough. Measures need to be taken to not only alleviate such problems, but to prevent their developing in the first place. Toward those ends, other

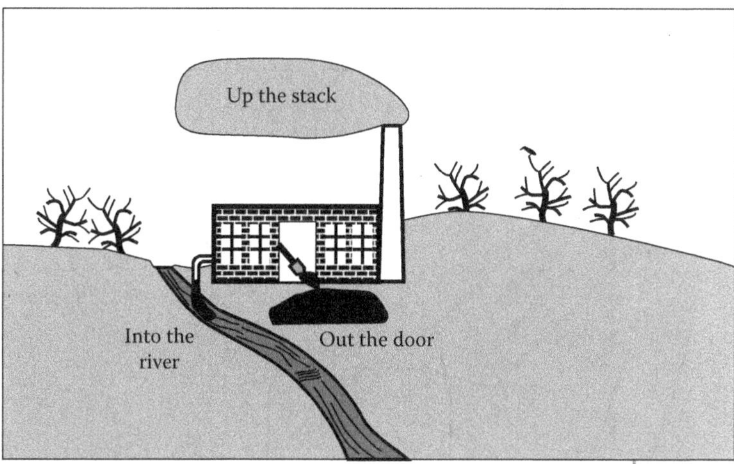

FIGURE 1.1 In the "bad old way" of doing things that prevailed until the mid-1900s, and even today in some places, wastes and by-products of chemical operations were commonly discarded to waterways, sent up stacks, or dumped in the geosphere, resulting in enormous environmental problems.

disciplines are developing that point the way to more environmentally friendly ways of doing things. Sustainable development, the practice of industrial ecology, and the practice of green chemistry are all directed toward enabling human societies and industrial systems to exist in better harmony with the Earth's support systems upon which all living beings ultimately depend for their existence. These areas, all dependent upon environmental chemistry, are developed in detail later in this book.

1.4 WATER, AIR, EARTH, LIFE, AND TECHNOLOGY

Water, air, earth, life, and technology are strongly interconnected as shown in Figure 1.2, which, in a sense, summarizes and outlines the theme of the rest of this book. Traditionally, environmental science has been divided into the study of the atmosphere, the hydrosphere, the geosphere, and the biosphere. However, for better or for worse, the environment in which all humans must live has been affected irreversibly by technology. Therefore, technology is considered strongly in this book within a separate environmental sphere called the anthrosphere in terms of how it affects the environment and in the ways by which, applied intelligently by those knowledgeable of environmental science, it can serve, rather than damage, this Earth upon which all living beings depend for their welfare and existence.

The strong interactions among living organisms and the various spheres of the abiotic (nonliving) environment are best described by cycles of matter that involve biological, chemical, and geological processes and phenomena. Such cycles are called biogeochemical cycles, and are discussed in more detail in Chapter 2 and elsewhere in this book.

In light of the above definitions, it is now possible to consider environmental chemistry from the viewpoint of the interactions among water, air, earth, life, and the anthrosphere as outlined in Figure 1.2. These five environmental "spheres" and the interrelationships among them are summarized in this section. In addition, the chapters in which each of these topics is discussed in greater detail are designated here.

1.4.1 Water and the Hydrosphere

The *hydrosphere* contains the Earth's water, a vitally important substance in all parts of the environment. Water, the environmental chemistry of which is discussed in detail in Chapters 3 through 8, is an essential part of all living systems and is the medium from which life evolved and in which life exists. Water covers about 70% of the Earth's surface. Over 97% of the Earth's water is in oceans, and

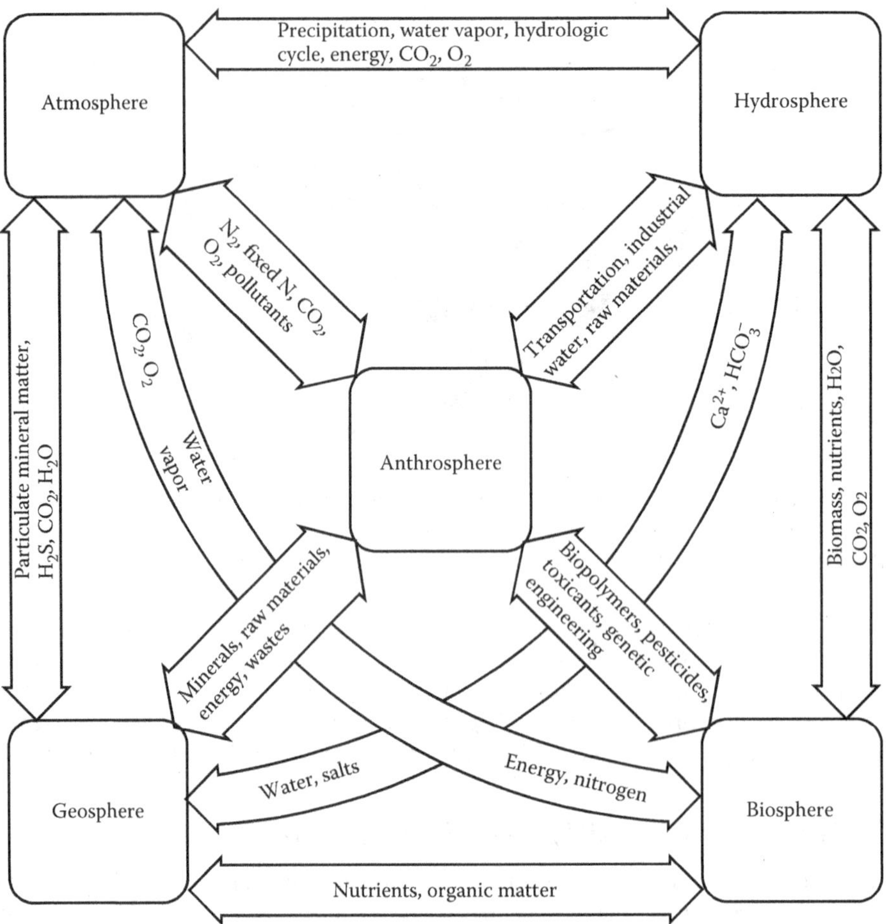

FIGURE 1.2 Illustration of the close relationships among the air, water, and earth environments with each other and with living systems, as well as the tie-in with technology (the anthrosphere).

most of the remaining freshwater is in the form of ice. Therefore, only a relatively small percentage of the total water on Earth is actually involved with terrestrial, atmospheric, and biological processes.

Water plays a key role in many segments of the anthrosphere, such as in boilers or municipal water distribution systems. Energy and matter are carried through various spheres of the environment by water. Water leaches soluble constituents from mineral matter and carries them to the ocean or leaves them as mineral deposits some distance from their sources. Water carries plant nutrients from soil into the bodies of plants by way of plant roots. Solar energy absorbed in the evaporation of ocean water is carried as latent heat and released inland. This process provides a large fraction of the energy that is transported from equatorial regions toward the Earth's poles and that powers massive storms in the atmosphere.

1.4.2 Air and the Atmosphere

The *atmosphere* is a thin protective blanket that nurtures life on Earth and protects it from the hostile environment of outer space by absorbing energy and damaging ultraviolet radiation from the sun and by moderating the Earth's temperature to within a range conducive to life. It is the source of carbon dioxide for plant photosynthesis and of oxygen for respiration. It provides the elemental nitrogen that nitrogen-fixing bacteria and ammonia-manufacturing industrial plants use to produce chemically bound nitrogen, an essential component of life molecules. As a basic part of the hydrologic cycle (Chapter 3, Figure 3.1), the atmosphere transports water from the oceans to land.

Atmospheric science deals with the movement of air masses in the atmosphere, atmospheric heat balance, and atmospheric chemical composition and reactions. Atmospheric chemistry is covered in Chapters 9 through 14 of this book.

1.4.3 Earth, the Geosphere

The *geosphere*, discussed in general in Chapter 15, consists of the solid earth, including soil, which supports most plant life (Chapter 16). The geosphere is composed of a solid, iron-rich inner core, a molten outer core, mantle, and the crust. The crust, only 5–40 km thick, is a thin outer skin composed largely of lighter silicate-based minerals and is the most important part of the geosphere insofar as interactions with the other spheres of the environment are concerned. It is that part of the Earth upon which humans live and from which they extract most of their food, minerals, and fuels.

Geology is the science of the geosphere and is very important in considerations of the environment. It pertains mostly to the solid mineral portions of the Earth's crust. But it must also consider water, which is involved in weathering rocks and in producing mineral formations; the atmosphere and climate, which have profound effects on the geosphere and interchange matter and energy with it; and living systems, which largely exist on the geosphere and in turn have significant effects on it. Modern technology, for example, the ability to move massive quantities of dirt and rock around, has a profound influence on the geosphere.

1.4.4 Life, the Biosphere

All living entities on the Earth compose the *biosphere*. Living organisms and the aspects of the environment pertaining directly to them are called biotic, and other portions of the environment are abiotic. Biology is the science of life. It is based on biologically synthesized chemical species, many of which exist as large molecules called macromolecules. As living beings, the ultimate concern of humans with their environment is the interaction of the environment with life. Therefore, biological science is a key component of environmental science and environmental chemistry.

The role of life in environmental science is discussed in the following section and in numerous other parts of this book. The crucial effects of microorganisms on aquatic chemistry are covered in Chapter 6. Chapter 22 addresses biochemistry as it applies to the environment. The effects on living beings of toxic substances, many of which are environmental pollutants, are addressed in Chapters 23 and 24. Other chapters discuss aspects of the interaction of living systems with various parts of the environment.

1.4.5 Technology and the Environment

Technology refers to the ways in which humans do and make things with materials and energy, that is, how they construct and operate the anthrosphere. Technology is the product of engineering based on science, which explains interrelated natural phenomena of energy, matter, time, and space. Engineering applies science to provide the plans and means to achieve specific practical objectives. Technology uses these plans to carry out desired objectives.

It is essential to consider technology, engineering, and industrial activities in studying environmental science because of the enormous influence that they have on the environment. Humans will use technology to provide the food, shelter, goods, and transportation that they need for their well-being and survival. The challenge is to integrate technology with considerations of the environment and ecology such that the two are mutually advantageous rather than in opposition to each other.

Technology, properly applied, is an enormously positive influence for environmental protection. The most obvious such application is in air and water pollution control. As necessary as "end-of-pipe" measures are for the control of air and water pollution, it is much better to use technology in manufacturing processes to prevent the formation of pollutants. Technology is being used increasingly to develop highly efficient processes of energy conversion, renewable energy resource utilization, and conversion of raw materials to finished goods with minimum generation of hazardous waste by-products. In the transportation area, properly applied technology in areas such as high-speed train transport can enormously increase the speed, energy efficiency, and safety of means by which people and goods are moved.

Until very recently, technological advances were made largely without heed to environmental impacts. Now, however, the greatest technological challenge is to reconcile technology with environmental consequences. The survival of humankind and of the planet that supports it now requires that the established two-way interaction between science and technology become a three-way relationship including environmental protection and emphasizing sustainability.

1.5 ECOLOGY, ECOTOXICOLOGY, AND THE BIOSPHERE

1.5.1 The Biosphere

Virtually the entire biosphere composed of living organisms is contained by the geosphere and hydrosphere in the very thin layer where these environmental spheres interface with the atmosphere. There are some specialized life forms at extreme depths in the ocean, but these are still relatively close to the atmospheric interface.

The biosphere strongly influences, and in turn is strongly influenced by, the other parts of the environment. It is believed that organisms were responsible for converting the Earth's original reducing atmosphere to an oxygen-rich one. Photosynthetic organisms remove CO_2 from the atmosphere, thus preventing runaway greenhouse warming of the Earth's surface. Organisms largely determine aquatic chemistry in bodies of water, and are strongly involved in weathering processes that break down rocks in the geosphere and convert rock matter to soil.

The biosphere is based upon *plant photosynthesis*, which fixes solar energy ($h\nu$) and carbon from atmospheric CO_2 in the form of high-energy biomass, represented as $\{CH_2O\}$:

$$CO_2 + H_2O + h\nu \rightarrow \{CH_2O\} + O_2\,(g) \qquad (1.1)$$

In so doing, plants and algae function as autotrophic organisms, those that utilize solar or chemical energy to fix elements from simple, nonliving inorganic material into complex life molecules that compose living organisms. Carbon that was originally fixed photosynthetically forms the basis of all fossil fuels in the geosphere. The opposite process of photosynthesis, namely biodegradation, breaks down biomass, preventing its accumulation in the environment, and releasing carbon dioxide to the atmosphere:

$$\{CH_2O\} + O_2\,(g) \rightarrow CO_2 + H_2O \qquad (1.2)$$

There is a strong interconnection between the biosphere and the anthrosphere. Humans depend upon the biosphere for food, fuel, and raw materials. Sustainability in the face of diminishing supplies of petroleum will increasingly rely on the biosphere for the production of raw materials and fuel in the future. Human influence on the biosphere continues to change it drastically. Fertilizers, pesticides, and cultivation practices have vastly increased yields of biomass, grains, and food. Destruction of habitat is resulting in the extinction of many species, in some cases even before they are discovered. Bioengineering of organisms with recombinant DNA technology and older techniques of selection and hybridization are causing great changes in the characteristics of organisms and promise to result in even more striking alterations in the future. It is the responsibility of humankind to make such changes intelligently and to protect and nurture the biosphere.

1.5.2 Ecology

Ecology is the science that deals with the relationships between living organisms with their physical environment and with each other. An *ecosystem* consists of an assembly of mutually interacting organisms (a community) and their environment in which materials are interchanged in a largely cyclical manner. An ecosystem has physical, chemical, and biological components along with energy sources and pathways of energy and materials interchange. The environment in which a particular organism lives is called its *habitat*. The role of an organism in a habitat is its *niche*. A major community of organisms based on predominant producers of biomass and adaptations of various organisms in the community to their environment is called a *biome*.

To study ecology it is often convenient to divide the environment into general categories. The terrestrial environment is based on land and consists of biomes, such as grasslands, savannas, deserts, or one of several kinds of forests. The freshwater environment can be further subdivided into standing-water habitats (lakes and reservoirs) and running-water habitats (streams and rivers). The oceanic marine environment is characterized by saltwater and may be divided broadly into the shallow waters of the continental shelf composing the neritic zone and the deeper waters of the ocean that constitute the oceanic region. An environment in which two or more kinds of organisms exist to their mutual benefit is termed a symbiotic environment.

A particularly important factor in describing ecosystems is that of *populations* consisting of numbers of a specific species occupying a specific habitat. Populations may be stable, or they may grow exponentially as a population explosion. A population explosion that is unchecked results in resource depletion, waste accumulation, and predation, culminating in an abrupt decline called a population crash. Behavior in areas such as hierarchies, territoriality, social stress, and feeding patterns plays a strong role in determining the fates of populations. All of these aspects, including the possibility of a population crash, apply to human populations as well.

1.5.3 Ecotoxicology

As discussed in detail in Chapters 23 and 24, *toxicology* refers to the detrimental effects of substances on organisms. Substances with such effects are called toxic substances, toxicants, or poisons. Whether or not a substance is toxic depends upon the amount of the substance an organism is exposed to and the manner of exposure. Some substances that are harmless or even beneficial at low levels are toxic at higher levels of exposure.

Toxic substances have a strong influence on ecosystems and the organisms in ecosystems, so the interactions between ecology and toxicology are very important. These interactions may be complex and involve a number of organisms. They may involve food chains and complex food webs. For example, persistent organohalide compounds may become more highly concentrated through food chains and exert their most adverse effects on organisms such as birds of prey at the top of the food chain. The combination of ecology and toxicology—the study of the effects of toxic substances upon ecosystems—has come to be known as *ecotoxicology*, which has been developed into an important discipline in environmental science.

Ecotoxicological effects may be considered through several different organizational levels. The first is introduction of a toxicant or pollutant into the system. This may result in biochemical changes at the molecular level. As a result, physiological changes may occur in tissues and organs. These may result in detrimental alterations of organisms. As a result, the organisms affected may undergo population changes, such as those that occurred in the 1950s and 1960s with decreased populations of hawks exposed to DDT. Such changes can alter communities; for example, decreased numbers of hawks may allow increased numbers of rodents accompanied by greater destruction of grain crops. Finally, whole ecosystems may be altered significantly.

Whereas toxicology usually deals with the effects of toxic substances on individuals, ecotoxicology emphasizes populations. Large numbers of individuals may be poisoned by toxic substances, whereas the population may survive. Dealing with populations as it does, ecotoxicology is significantly more complex than toxicology. In reducing risk, emphasis is usually placed upon the protection of species composition since that automatically protects ecosystem processes.

1.6 ENERGY AND CYCLES OF ENERGY

Biogeochemical cycles and virtually all other processes on the Earth are driven by energy from the sun. The sun acts as a so-called blackbody radiator with an effective surface temperature of 5780 K (absolute temperature in which each unit is the same as a Celsius degree, but with zero taken at absolute zero). It transmits energy to the Earth as electromagnetic radiation (see below) with a maximum energy flux at about 500 nm, which is in the visible region of the spectrum. A 1 m^2 area perpendicular to the line of solar flux at the top of the atmosphere receives energy at a rate of 1340 W, sufficient, for example, to power an electric iron. This is called the solar flux (see Chapter 9, Figure 9.3).

1.6.1 Light and Electromagnetic Radiation

Electromagnetic radiation, particularly light, is of utmost importance in considering energy in environmental systems. Therefore, the following important points related to electromagnetic radiation should be noted:

- Energy can be carried through space at the speed of light (c), 3.00×10^8 meters per second (m/s) in a vacuum, by electromagnetic radiation, which includes visible light, ultraviolet radiation, infrared radiation, microwaves, radio waves, γ-rays, and x-rays.
- Electromagnetic radiation has the character of a wave. The waves move at the speed of light, c, and have the characteristics of wavelength (λ), amplitude, and frequency (ν, Greek "nu") as illustrated below:

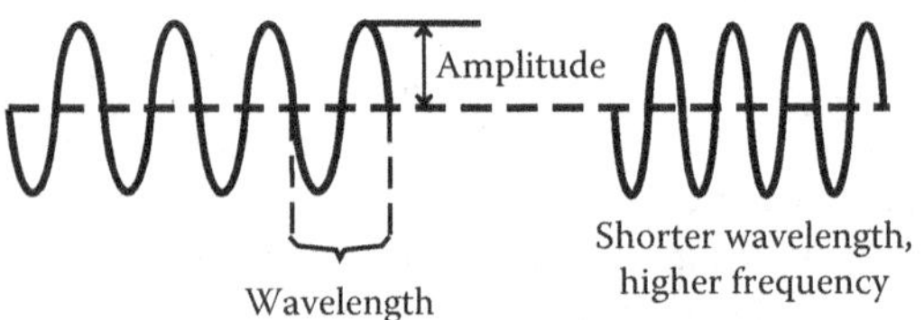

- The wavelength is the distance required for one complete cycle, and the frequency is the number of cycles per unit time. They are related by the following equation:

$$\nu\lambda = c$$

  where ν is expressed in units of cycles per second [s^{-1}, a unit called the hertz (Hz)] and λ is in meters (m).
- In addition to behaving as a wave, electromagnetic radiation has the characteristics of particles.
- The dual wave/particle nature of electromagnetic radiation is the basis of the quantum theory of electromagnetic radiation, which states that radiant energy may be absorbed or emitted only in discrete packets called quanta or photons. The energy, E, of each photon is given by

$$E = h\nu$$

  where h is Planck's constant, 6.63×10^{-34} J s.
- From the preceding, it is seen that the energy of a photon is higher when the frequency of the associated wave is higher (and the wavelength shorter).

1.6.2 Energy Flow and Photosynthesis in Living Systems

Whereas materials are recycled through ecosystems, the flow of useful energy is essentially a one-way process. Incoming solar energy can be regarded as high-grade energy because it can cause useful reactions to occur, such as production of electricity in photovoltaic cells or photosynthesis in plants. As shown in Figure 1.3, solar energy captured by green plants energizes chlorophyll, which in turn powers metabolic processes that produce carbohydrates from water and carbon dioxide. These carbohydrates are repositories of stored chemical energy that can be converted to heat and work by metabolic reactions with oxygen in organisms. Ultimately, most of the energy is converted to low-grade heat, which is eventually reradiated away from the Earth by infrared radiation.

1.6.3 Energy Utilization

During the last two centuries, the growing, enormous human impact on energy utilization has resulted in many of the environmental problems now facing humankind. This time period has seen a transition from the almost exclusive use of energy captured by photosynthesis and utilized as biomass (food to provide muscle power, wood for heat) to the use of fossil fuel petroleum, natural gas, and coal for about 90%, and nuclear energy for about 5%, of all energy employed commercially. By 2008, it had become painfully obvious that resources of the most desirable fossil fuels (petroleum and natural gas) were becoming depleted and falling behind in meeting demand. Of particular importance is the fact that all fossil fuels produce carbon dioxide, a greenhouse gas that is almost certainly causing global warming. Therefore, as noted at the beginning of this chapter, it will be necessary to move toward the utilization of alternate renewable energy sources, including solar energy and biomass. The study of energy utilization is crucial in sustainability science, and it is discussed in greater detail in Chapter 19.

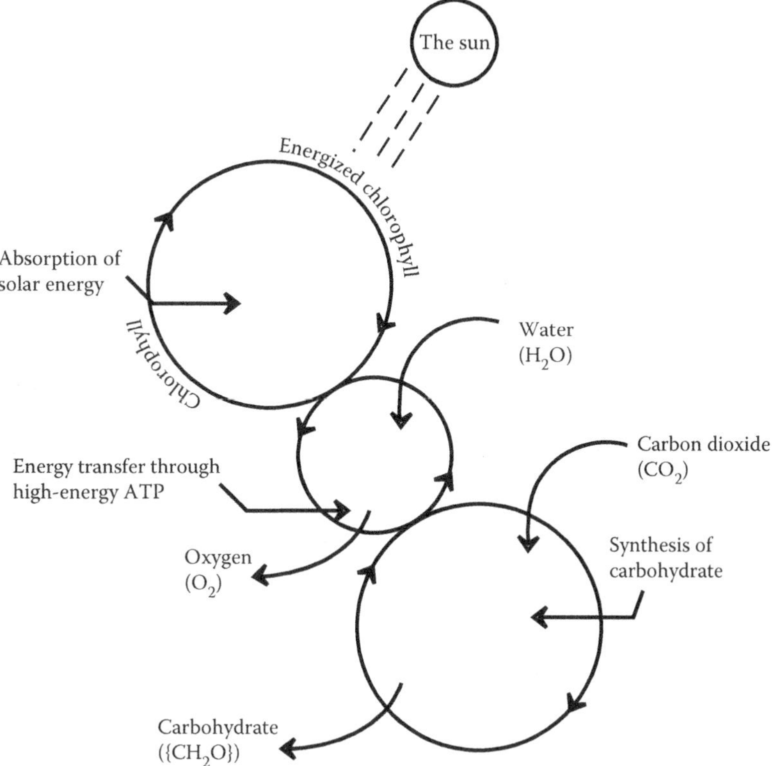

FIGURE 1.3 Energy conversion and transfer by photosynthesis.

1.7 HUMAN IMPACT AND POLLUTION

The demands of increasing population coupled with the desire of most people for a higher material standard of living are resulting in worldwide *pollution* on a massive scale. All five of the major environmental spheres can suffer from pollution and all are linked with respect to pollution phenomena. For example, some gases emitted to the atmosphere can be converted to strong acids by atmospheric chemical processes, fall to the Earth as acid rain, and pollute water with acidity. Improperly discarded hazardous wastes can leach into groundwater that is eventually released as polluted water into streams.

1.7.1 Some Definitions Pertaining to Pollution

In some cases, pollution is a clear-cut phenomenon, while in other cases it lies largely in the eyes of the beholder. Frequently, time and place determine what may be called a pollutant. The phosphate that the sewage treatment plant operator has to remove from wastewater is chemically the same as the phosphate that the farmer a few miles away has to buy at high prices for fertilizer. Most pollutants are, in fact, resources gone to waste; as resources become scarcer and more expensive, economic pressures can provide impetus for solutions to many pollution problems. A key aspect of sustainability is the diversion of matter in pollutants to useful applications.

A reasonable definition of a *pollutant* is a substance present in greater than natural concentration as a result of human activity that has a net detrimental effect upon its environment or upon something of value in that environment. *Contaminants*, which are not classified as pollutants unless they have some detrimental effect, cause deviations from the normal composition of an environment.

Every pollutant originates from a *source*. The source is particularly important because it is generally the logical place to eliminate pollution. After a pollutant is released from a source, it may act upon a receptor. The *receptor* is anything that is affected by the pollutant. Humans whose eyes smart from oxidants in the atmosphere are receptors. Trout fingerlings that may die after exposure to acid from acid rain in water are also receptors. Eventually, if the pollutant is long-lived, it may be deposited in a *sink*, a long-time repository of the pollutant. Here it will remain for a long time, though not necessarily permanently. Thus, a limestone wall may be a sink for atmospheric sulfuric acid by conversion of $CaCO_3$ in limestone into $CaSO_4$.

1.7.2 Pollution of Various Spheres of the Environment

Pollution of surface water and groundwater is discussed in some detail in Chapter 7. Particulate air pollutants are covered in Chapter 10, gaseous inorganic air pollutants in Chapter 11, and organic air pollutants and associated photochemical smog in Chapters 12 and 13. Some air pollutants, particularly those that may result in irreversible global warming or destruction of the protective stratospheric ozone layer, are of such a magnitude that they have the potential to threaten life on the Earth. These are discussed in Chapter 14. Hazardous wastes are discussed in Chapters 20 and 21. Toxic pollutants are addressed in Chapters 23 and 24.

1.8 CHEMICAL FATE AND TRANSPORT

The movement and fate of environmental pollutants is a key consideration in determining their impacts. This concern is addressed by the discipline of *chemical fate and transport* or *environmental fate and transport*. Figure 1.4 illustrates the major pathways involved in chemical fate and transport. Substances regarded as pollutants almost always originate from the anthrosphere (although substances such as sulfur-containing volcanic gases can act as pollutants as well). They may move into the air, onto the ground, into water (surface water or groundwater), into sediments, and into biota (plants and animals). Where such substances go and what they do depend upon their properties

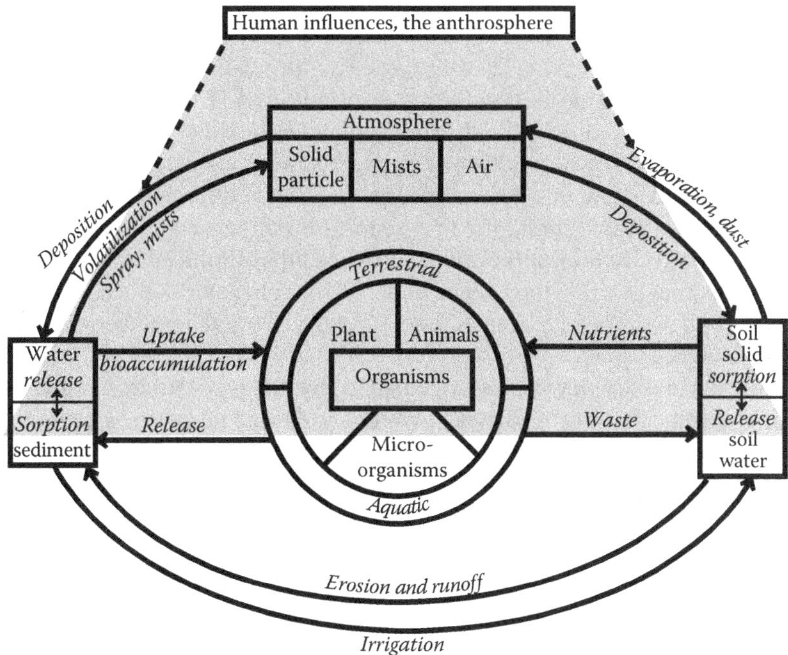

FIGURE 1.4 Interchanges of contaminants released from the anthrosphere among various segments of the other environmental spheres and illustrations of pathways involved in chemical fate and transport.

and the conditions of the environment into which they are introduced. As a general rule, the fate and transport of contaminants are controlled by their *physical transport* (movement without reacting or interacting with other phases) and their *reactivity*, including chemical or biochemical reactions or physical interactions with other phases.

It is convenient to view environmental fate and transport in terms of three major environmental compartments: (1) the atmosphere, (2) surface waters, and (3) the terrestrial or subsurface including soil, mineral strata, and groundwater. These compartments are illustrated in Figure 1.5.

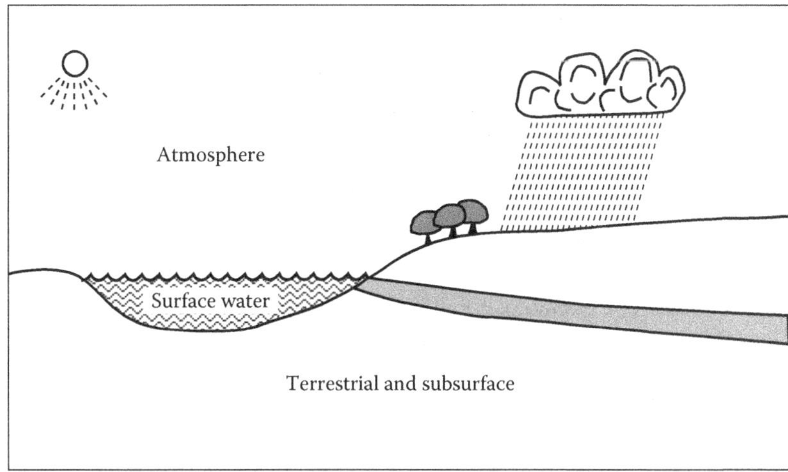

FIGURE 1.5 The three major environmental compartments considered in environmental fate and transport.

1.8.1 Physical Transport

Although there are numerous physical transport processes depending upon the medium in which the contaminants are found, they may be divided into two categories. The first of these is *advection* due to movement of masses of fluids that simply carry pollutants with them. Vertical advection of air or water is called *convection*. The second type of movement of chemical species is *diffusive transport* or *Fickian transport*, most commonly regarded as molecular diffusion, the natural tendency of molecules to move from regions of higher to regions of lower concentration by the random movement of molecules. Diffusive transport is also approximated by turbulent mixing. Turbulent mixing can be seen in the eddies of a flowing stream, and a similar phenomenon occurs in air. The mixing that occurs as water flows around and among small particles as it flows underground is also treated as a form of diffusive transport.

Air contaminants move with wind and air currents, by diffusion processes, and by settling and suspension of atmospheric aerosol particles. Water contaminants move by currents in water, by turbulent mixing processes, by diffusion, and by settling and suspension of particles. In solids, such as mineral formations, contaminants move by the action of groundwater or by diffusion in the vapor phase. The mobility of contaminants depends upon the fluidity of the medium in which they are contained. Contaminants in the atmosphere may move by as much as several kilometers per hour in wind, whereas their movement in soil and solids may be almost imperceptible.

1.8.2 Reactivity

Reactivity includes chemical reactions, biological uptake, and binding to and release from surfaces. Reactivity processes include the two broad categories of chemical reactions and interphase exchange. In water, interphase exchange could include the binding of soluble species to suspended particles in water and in air it could include evaporation and condensation of species. Biological processes fall into the broad category of interphase exchange, but any biochemical reactions that pollutants taken up by organisms undergo are obviously chemical changes.

1.8.3 Mass Balance Expression

Like all matter, pollutants are governed by mass conservation, which simply accounts for all the matter in a pollutant wherever it is moved and whatever reactions it might undergo. In considering mass conservation in the environment, it is useful to define a *control volume* as a part of the environment within which all sources and sinks of a pollutant can be accounted for and across the boundaries of which movement of pollutants may be known. In relationship to such a portion of the environment, a pollutant is described by the *mass balance relationship* as shown in Figure 1.6. A special case of the mass balance relationship that often simplifies the study of environmental fate and transport is when there is no net change in the mass of pollutant within the control volume, a condition of *steady state*.

A typical control volume might be the water in a lake, not including the sediment layer. Uptake of a substance by the sediment would depend upon the relative affinity of the substance for water and sediment. For example, a relatively hydrophobic substance would have a strong tendency to leave the water phase and enter the organic phase in the sediment. A substance that is volatile would tend to evaporate at the surface. The substance might be changed by biodegradation processes mediated by microorganisms suspended in the lake water. Water might flow into the lake from a stream source and out of the lake through an outlet stream. From these considerations it should be obvious that although the concept of the mass balance relationship is straightforward, its calculation for a substance in a particular control volume can become quite complicated.

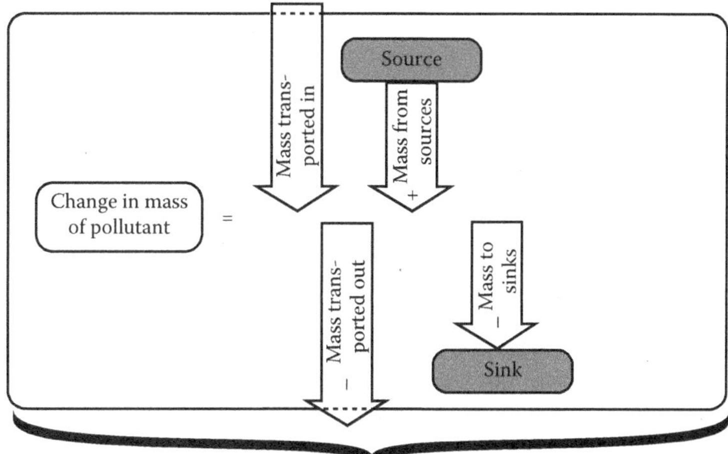

FIGURE 1.6 Illustration of the mass balance relationship for a pollutant with respect to a specified compartment of the environment.

1.8.4 Distribution among Phases

Distribution among phases is very important in environmental fate and transport. This may involve movement between major environmental compartments shown in Figure 1.5. Or it may entail partitioning between phases within an environmental compartment. For example, a gaseous air pollutant in the atmosphere might be taken up by plant leaves and move from the atmospheric to the terrestrial compartment. Or it might absorb onto particulate matter in the atmosphere. A chemical species in the subsurface may be absorbed to soil particles, dissolve in soil water, or even partition into air spaces in the Earth.

The various factors involved in partitioning environmental species between phases are addressed in more detail in Chapters 5 (water), 9 (the atmosphere), and 15 (terrestrial); several examples are mentioned here. The tendency of a substance to partition between water and solids in contact with the water depends upon the substance's water solubility or hydrophilic tendencies. A substance in water that has a high vapor pressure will tend to evaporate into the atmosphere. Species in the atmosphere that are organophilic may be taken up by the wax-like surfaces of plant structures, such as pine needles.

There are two major ways in which a substance may be sequestered by another phase. It may bind with the surface, such as occurs when some organic contaminants in water are taken up by surface interactions with solids, a process called *adsorption*. Or it may actually be incorporated into the body of material to which it is bound, a process called *absorption*. A more general term that applies to both processes is *sorption*. A substance is said to be *sorbed*; it is called the *sorbate* and the material to which it is bound is called the *sorbent*.

1.9 CHEMICAL FATE AND TRANSPORT IN THE ATMOSPHERE, HYDROSPHERE, AND GEOSPHERE

This section briefly addresses chemical fate and transport in the atmosphere, hydrosphere, and geosphere. It is important to have a rudimentary perspective of chemical fate and transport in these spheres to understand chemical fate and transport in later parts of the book. These aspects are discussed in more detail in later chapters dealing with specific spheres of the environment.

1.9.1 Pollutants in the Atmosphere

Substances that tend to be transported to the atmosphere are those that are relatively volatile. Such substances include those that are gases under normal ambient conditions, including compounds such as nitric oxide (NO) or carbon monoxide (CO). A number of organic compounds including those in gasoline or chlorofluorocarbons are called *volatile organic compounds* (VOC). Organic compounds that are less volatile nevertheless get into air are classified as *semivolatile organic compounds.*

An important consideration regarding substances that get into the atmosphere is the degree to which they are *hydrophobic*. Hydrophobic substances, including a number of the VOC and semivolatile organic compounds, are those that are repelled by water. Their opposites, *hydrophilic compounds*, readily dissolve in water and are removed by precipitation (rain). For example, both methanol and dichloromethane are VOC, but methanol is highly water soluble and is rapidly removed from the atmosphere with rainfall, whereas dichloromethane is hydrophobic and tends to stay in the air.

A number of important air pollutants are in the form of particulate matter, popularly called *particulates*. Particles may be emitted directly to the atmosphere from sources such as power plants emitting fly ash. Or they may be formed by the reactions of gases as occurs when particles characteristic of photochemical smog are formed from hydrocarbon vapors, nitrogen oxides, and oxygen in the atmosphere. Heavy particles tend to settle rapidly and deposit on soil or on the surfaces of plant leaves. Lighter particles tend to stay in the atmosphere longer and travel farther from their sources. Particles of water-soluble materials or with hydrophilic surfaces are readily removed by rainfall and may get into water.

1.9.2 Pollutants in the Hydrosphere

Pollutants may be discharged directly to water or may get into water from the atmosphere or from soil runoff. Both surface water and water under the ground—groundwater—must be considered. Groundwater may pick up pollutants as contaminated surface water that flows from the surface into underground aquifers as part of the process of recharge. Another pollutant source of major concern is leachate from wastes that have been improperly disposed on the surface or in landfills. Particularly, soluble (hydrophilic) chemicals have a strong tendency to stay dissolved in water and to move with the flow of surface water or groundwater. More hydrophobic species in water will have a greater tendency to be held on mineral surfaces (in groundwater) or in sediments (in surface water).

The tendency of a solute in water to be retained by soil, sediment, or aquifer mineral is expressed by a parameter K_d, commonly called the *soil–water partition coefficient* and expressed as

$$K_d = \frac{C_s}{C_w} \tag{1.3}$$

where C_s is the equilibrium concentration of the contaminant in the solid and C_w is the equilibrium concentration of the contaminant in water. The organic fraction of the solid material in soil, sediments, and even minerals is generally the material with the predominant affinity for organic contaminants. Designating this fraction as f_{oc} and the partition coefficient of the organic contaminant for the pure organic solid as K_{oc}, the following holds:

$$K_d = f_{oc} \times K_{oc} \tag{1.4}$$

This relationship shows that as the fraction of organic sorbent in the solid increases, the effective value of K_d increases as well.

1.9.3 Pollutants in the Geosphere

The transport of contaminants in the geosphere is largely by movement of groundwater through rock formations composing aquifers. In cases where the aquifer mineral material is composed of finely divided matter such as sand, the contaminants in the groundwater are continuously exposed to the mineral surface and may be absorbed by it. Very commonly, however, the aquifer consists of solid rock fragmented by fractures through which groundwater may move rapidly and over long distances. In such a case, contaminants do not have a good opportunity to be taken up by the mineral surface. This can result in the rapid appearance of contaminants from an infiltrating water source in well water supplies of drinking water. Numerous cases have occurred in which surface pollution, such as by heavy metals or organic solvents, have rapidly infiltrated groundwater and moved to a well used as a water source. An interesting case consists of dense nonaqueous phase liquids (DNAPLs) that sink to the bottom of the aquifer and accumulate as pools of material that might get into a water source.

1.10 ENVIRONMENTAL MISCHIEF AND TERRORISM

Unfortunately, chemistry has a great deal of potential for wreaking havoc in the hands of those who would do evil deeds. The potential of chemistry to cause harm in the wrong hands comes to mind in considering terrorism. The terrorists' favorite weapons, explosives, act by means of very rapid chemical reactions that very rapidly release large amounts of energy and heat. Another potentially deadly terrorist tool consists of toxic substances made by chemists. Some of these, such as hydrogen cyanide, are used industrially. Others, such as deadly nerve gases, are made specifically for military poisons. But chemistry also works to thwart terrorism, for example, by the use of very sensitive analytical techniques that can detect threats, such as explosives or toxic chemicals.

Environmental chemistry certainly has a strong connection to understanding and combating terrorism. It is not much of an exaggeration to consider widespread environmental harm, such as that inflicted by the improper use of chemicals and chemical processes, as a form of terrorism. It takes no great stretch of the imagination to believe that a crowded urban area afflicted with smoke, gasoline fumes, nitrogen oxides, and ozone is under a form of terrorist attack. Some of the great environmental catastrophes, such as the release of methyl isocyanate in Bhopal, India, in 1984 that killed 3500 people, or the catastrophic explosion and fire of the Chernobyl nuclear power plant in the former Soviet Union in 1986, have resembled terrorist attacks in their development and effects. So did a release of natural gas containing deadly hydrogen sulfide and accompanying fire from a natural gas well in China in December 2003, killing almost 200 people. Such attacks can take the form of environmental damage to water supplies or soil. The study of unfortunate environmental catastrophes of the past can provide guidance in dealing with terrorist attacks in the future.

1.10.1 Protection through Green Chemistry and Engineering

The single best way that chemical science can avoid terrorist threats is to follow the precepts of green chemistry, a topic that is developed further in Chapter 2 and later in this book. This is so because green chemistry is *safe chemistry* and *sustainable chemistry*. Green chemistry avoids the hazards that might be turned to malevolent ends. Sustainable chemistry means that the chemical industry and other enterprises that depend upon it are resistant to disruptive incidents such as those that could cause problems when supplies of essential raw materials are curtailed.

Green chemistry avoids the use or generation of substances that pose hazards to humans and the environment. When such substances are not made or used, they are simply unavailable for theft or diversion by criminal elements. Green chemical products are as effective as possible for their designated purpose, but with minimum toxicity. Chemical products that do what they are supposed to do when used in minimum quantities reduce the hazards posed by potential improper use. Green chemistry minimizes or avoids the use of auxiliary substances. These may include materials, such as flammable solvents, that may be hazardous, so reducing or minimizing their use can increase safety. Green chemistry avoids the use or generation of substances that are likely to react violently, burn, build up excessive pressures, or otherwise cause unforeseen incidents in the manufacturing process; the safety aspects of avoiding such substances are obvious. Green chemistry attempts to minimize the severity of conditions, particularly those of temperature and pressure, under which chemical processing is carried out. This can have the effect of greatly reducing damage that would occur through damage to chemical reactors and other processing equipment in the event of a malfunction.

The practice of green chemistry minimizes energy consumption. One way in which this is done is to use biological processes, which, because of the conditions under which organisms grow and their enzymes function, must occur at moderate temperatures and in the absence of toxic substances, thereby enhancing safety. Any measures that reduce the dependence on potentially hostile foreign sources of energy indirectly reduce vulnerability to terrorism. The same principle applies to reduced use of critical raw materials by the application of green chemistry.

The successful practice of green chemistry requires real-time, in-process monitoring techniques coupled with process control. Such a control is consistent with minimizing hazards including those caused by sabotage. It should be noted that passive systems of safety should be employed that will work by default if intricate control systems fail or are damaged (e.g., passive cooling of nuclear reactors by water flowing under gravity in the event of cooling system malfunction).

It should be noted that the chemical industry and related enterprises are taking steps to implement the practice of green chemistry to reduce vulnerabilities to attack. For example, manufacturing operations that once stored highly toxic methyl isocyanate, the agent of the catastrophic 1984 Bhopal, India, disaster, now make this material on demand as needed at the plant. Some water treatment plants that used to store reactive, toxic liquid chlorine disinfectant under pressure in 90-ton rail cars located on site have replaced it with much safer solid sodium hypochlorite.

1.11 ENVIRONMENTAL FORENSICS

Environmental forensics is the science that deals with the legal and medical aspects of environmental pollution.[8] It is an important area because of the health effects of pollutants and because often large sums of money are at stake in lawsuits aimed at determining parties responsible for environmental contamination, such as hazardous waste sites. Furthermore, environmental forensics can be used for determining parties responsible for terrorist attacks that had used chemical agents. This discipline studies the sources, transport, and effects of pollutants with the goal of determining parties responsible for pollution and harmful environmental events. Important aspects are the source, timing, or extent of an environmental incident. Typically, in cases where hazardous chemical wastes are improperly disposed, soil and groundwater are examined to determine the site history through studies of groundwater flow, groundwater chemical and physical analyses, and modeling. The chemical analysis of groundwater can provide a fingerprint of the source of pollution leading to knowledge of its location and extent, often pointing the way to parties responsible for it. In addition to chemical analysis, other disciplines that may be involved include chemical use history (which can help establish a time frame for when the contamination occurred), geology and forensic geochemistry, hydrogeology, and historical aerial photo interpretation. Computerized contaminant transport modeling of complex subsurface environments can also be very useful.

As discussed later in this book, the renovation of land areas damaged by chemical pollution—the restoration of so-called brownfields—is an important aspect of sustainability. Before a brownfield land area that has been damaged by dumping of wastes can be restored and used for commercial or residential applications, the nature and extent of contamination must be assessed and measures must be taken to remove or neutralize any hazardous material, in some cases with charges addressed to parties responsible for the wastes. Environmental forensics has a role to play in these activities. The findings of environmental forensics can reduce the time and costs of legal proceedings leading to negotiated settlements between the parties involved.

LITERATURE CITED

1. Baum, R. M., Sustainability: Learning to live off the sun in real time, *Chemical and Engineering News*, **86**, 42–46, 2008.
2. World Commission on Environment and Development, *Our Common Future*, Oxford University Press, New York, 1987.
3. Manahan, S. E., *Environmental Science and Technology: A Sustainable Approach to Green Science and Technology*, Taylor & Francis/CRC Press, Boca Raton, FL, 2007.
4. Cunningham, W. P., M. A. Cunningham, and B. W. Saigo, *Environmental Science, a Global Concern*, 9th ed., McGraw-Hill Higher Education, Boston, 2007.
5. Manahan, S. E., *Green Chemistry and the Ten Commandments of Sustainability*, 2nd ed., ChemChar Research, Inc., Columbia, MO, 2006.
6. Doble, M. and A. Kumar, *Green Chemistry and Engineering*, Elsevier, Amsterdam, 2007.
7. Manahan, S. E., *Toxicological Chemistry and Biochemistry*, 3rd ed., Taylor & Francis/CRC Press, Boca Raton, FL, 2002.
8. Sklash, M., J. Bolin, and M. Schroeder, Who done it? The ABCs of environmental forensics, *Chemical Engineering Progress*, **102**, 40–45, 2006.

SUPPLEMENTARY REFERENCES

Botkin, D. B. and E. A. Keller, *Environmental Science: Earth as a Living Planet*, 6th ed., Wiley, Hoboken, NJ, 2007.

Connell, D. W., *Basic Concepts of Environmental Chemistry*, 2nd ed., Taylor & Francis/CRC Press, Boca Raton, FL, 2005.

Lichtfouse, E., J. Schwarzbauer, and R. Didier, Eds, *Environmental Chemistry: Green Chemistry and Pollutants in Ecosystems*, Springer, Berlin, 2005.

Mackenzie, F. T., *Our Changing Planet: An Introduction to Earth System Science and Global Environmental Change*, 3rd ed., Prentice-Hall, Upper Saddle River, NJ, 2002.

Miller, G. Tyler and S. Spoolman, *Environmental Science: Problems, Connections and Solutions*, 12th ed., Brooks Cole, Belmont, CA, 2008.

Newman, M. C. and W. H. Clements, *Ecotoxicology: A Comprehensive Treatment,* Taylor & Francis, Boca Raton, FL, 2008.

Newman, M. C. and M. A. Unger, *Fundamentals of Ecotoxicology*, 2nd ed., Taylor & Francis/CRC Press, Boca Raton, FL, 2003.

Raven, P. H., L. R. Berg, and D. M. Hassenzahl, *Environment*, 6th ed., Wiley, Hoboken, NJ, 2008.

Walker, C. H., Ed., *Principles of Ecotoxicology*, 3rd ed., Taylor & Francis/CRC Press, Boca Raton, FL, 2006.

Weiner, E. R., *Applications of Environmental Aquatic Chemistry: A Practical Guide*, 2nd ed., Taylor & Francis/CRC Press, Boca Raton, FL, 2008.

Withgott, J. and S. Brennan, *Essential Environment: The Science Behind the Stories*, 3rd ed., Pearson Benjamin Cummings, San Francisco, 2009.

Worldwatch Institute, *State of the World 2009: Into a Warming World*, Worldwatch Institute, Williamsport, PA, 2003.

Worldwatch Institute, *Vital Signs 2007–2008: The Trends that are Shaping Our Future*, Worldwatch Institute, Williamsport, PA, 2007.

Yen, T. F., *Chemical Processes for Environmental Engineering*, Imperial College Press, London, 2007.

QUESTIONS AND PROBLEMS

The use of internet resources is assumed in answering any of the questions. These would include such things as constants and conversion factors as well as additional information needed to complete an answer.

1. Under what circumstances does a contaminant become a pollutant?
2. Chemical plants and the chemical industry have been the subject of much concern regarding the potential for terrorist attack. Look up this subject in the Internet and attempt to

discern which products and processes pose the greater risks. What is being done to reduce these risks? Can you find reports of any major incidents in which the products of chemical manufacturing have been used for attacks?

3. The argument has been made that nuclear power generation and spent nuclear fuel transport and processing pose less danger of terrorist attacks than do corresponding operations in the chemical industry. Suggest the rationale for such an argument. If criminals were to obtain highly radioactive materials, explain why it might be very difficult for them to do anything with them.
4. Explain how toxicological chemistry differs from environmental biochemistry.
5. Distinguish between the geosphere, lithosphere, and crust of the Earth. Which science deals with these parts of the environment?
6. Define ecology and explain how it relates to at least one of the major biogeochemical cycles, such as the carbon cycle.
7. Although energy is not destroyed, why is it true that the flow of *useful* energy through an environmental system is essentially a one-way process?
8. Describe some ways in which patterns of energy use have "resulted in many of the environmental problems now facing humankind."
9. Compare nuclear energy to fossil fuel energy sources and defend or refute the statement, "Nuclear energy, with modern, safe, and efficient reactors, is gaining increasing attention as a reliable, environmentally friendly energy source."
10. The sun powers a cycle of matter that is key to the indirect utilization of solar energy through the medium of water. What is this cycle? How does it enable solar energy to be converted to mechanical energy and ultimately to electrical energy?
11. Using Internet resources, look up the percentages of carbon and hydrogen in natural gas (hint: What is the major hydrocarbon compound in natural gas?), gasoline, and typical bituminous coal and come up with a chemical formula in the form of CH_h for each of these fuels, where h will not necessarily be a whole number. Write a chemical reaction for the combustion of each in the form of $CH_h + h/4O_2 \rightarrow h/2H_2O$. Noting that the combustion of both hydrocarbon-bound C and H produces heat, rank these three kinds of fuels in order from the one that produces the least greenhouse gas CO_2 per molecule burned to the one that produces the most.
12. A former industrial site ("brownfield") has been chosen for an urban renewal project. It is known to be contaminated with potentially harmful materials released from previous manufacturing operations. Suggest an environmental forensics program that could be used as part of the effort to renovate the site.
13. In China in 2003, the release of gas from a gas well killed about 200 people. To help remedy the problem, the gas was deliberately set on fire. What pollutant species (other than CO_2) was produced by the fire? Despite generation of this pollutant, why was setting the gas on fire the best emergency measure to take until the well could be capped?
14. In 1995, people in Tokyo were the victims of a terrorist attack using a chemical agent. What were the effects of this attack? What agent was used?
15. What did Nobel Prize winner Paul Crutzen have to say about the anthrosphere? Name and describe the term that Dr. Crutzen coined pertaining to the anthrosphere.
16. Suppose that each of the molecules in a mole of "compound *X*" absorbed a photon of ultraviolet radiation from the sun of wavelength 300 nm to become an energetically "excited" species (see Chapter 9). Calculate the energy in joules absorbed.
17. As a general rule, the fate and transport of contaminants are controlled by their physical transport and their reactivity. Can these two factors be considered to be entirely separate? Explain.

18. Conventionally, environmental fate and transport are considered in terms of three major environmental compartments: (1) the atmosphere, (2) surface waters, and (3) the terrestrial or subsurface environment. Should the biosphere and anthrosphere be considered as separate compartments? Give arguments for and against doing so.
19. The fate and transport of contaminants is controlled in part by their reactivity. Suggest examples of chemical, biological, and physical reactivity processes.
20. Explain the significance of mass conservation, control volume, and steady state in the mass balance relationship.
21. How may volatility, hydrophobic tendency, and hydrophilic tendency be involved in environmental fate and transport in the atmosphere?
22. A pollutant, "*X*," was found to be in equilibrium between soil and water in contact with the soil. Its concentration in the soil was determined to be 0.44 mg/kg of dry soil and its concentration in water was measured as 0.037 mg/L. What is the value of K_d and what are its units?
23. If the soil in the preceding question was 5.2% organic matter and the entire affinity of the soil for "*X*" was due to organic matter, what is the value of the partition coefficient of the organic contaminant for the pure organic solid, K_{oc}?
24. Even though the ability of contaminants to move through solid soil is usually very low, transport of a contaminant from one terrestrial location to another often occurs rather quickly. Explain.
25. Using Internet resources, find at least one case in which criminal penalties have been assessed for environmental pollution and discuss how environmental forensics were used in the case.

2 Chemistry and the Anthrosphere: Environmental Chemistry and Green Chemistry

2.1 ENVIRONMENTAL CHEMISTRY

In Chapter 1, the environment was defined as consisting of five spheres: the hydrosphere, the atmosphere, the geosphere, the biosphere, and the anthrosphere; that is, water, air, the Earth, life, and those parts of the environment consisting of human constructs and activities. The chemistry of the environment, *environmental chemistry*, may be defined as *the study of the sources, reactions, transport, effects, and fates of chemical species in the hydrosphere, the atmosphere, the geosphere, and the anthrosphere and the effects of human activities thereon*. This definition is illustrated for a typical environmental pollutant in Figure 2.1. The pollutant sulfur dioxide is generated during the combustion of sulfur in coal, transported to the atmosphere with flue gas, and oxidized by chemical and photochemical processes to sulfuric acid. Sulfuric acid, in turn, falls as acidic precipitation, where it may have detrimental effects such as toxic effects on trees and other plants. Eventually, sulfuric acid is carried by stream runoff to a lake or ocean where its ultimate fate is to be stored in solution in the water or precipitated as solid sulfates.

Environmental chemistry is complicated by the continuous and variable interchange of chemical species among various environmental spheres. This complexity is illustrated for sulfur species in Figure 2.1. The sulfur in coal is taken from the geosphere, converted to gaseous sulfur dioxide by an anthrospheric process (combustion), transported and undergoes chemical reactions in the atmosphere, may affect plants in the biosphere, and ends up in a sink in the hydrosphere or back in the geosphere. Throughout this sequence, sulfur takes on several forms including organically bound sulfur or pyrite (FeS_2) in coal, sulfur dioxide produced in the combustion of coal, sulfuric acid produced by the oxidation of sulfur dioxide in the atmosphere, and sulfate salts produced from sulfuric acid when it reaches the geosphere. Throughout an environmental system there are variations in temperature, mixing, intensity of solar radiation, input of materials, and various other factors that strongly influence chemical conditions and behavior. Because of its complexity, environmental chemistry must be approached with simplified models.

Green chemistry, the practice of chemical science and technology in a nonpolluting, safe, and sustainable manner, and industrial ecology, which treats industrial systems in a manner analogous to natural ecosystems, are discussed in detail later in this chapter and in Chapter 17. Environmental chemistry has a strong connection to both of these disciplines. A major goal of green chemistry is to avoid environmental pollution, an endeavor that requires knowledge of environmental chemistry. The design of an integrated system of industrial ecology must consider the principles and processes of environmental chemistry. Environmental chemistry must be considered in the extraction of materials from the geosphere and other environmental spheres to provide the materials required by

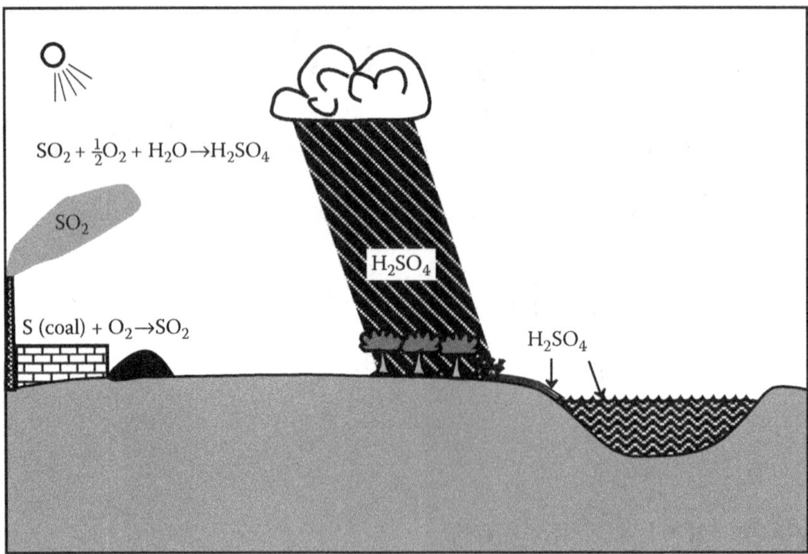

FIGURE 2.1 Illustration of the definition of environmental chemistry by the example of the pollutant sulfuric acid formed by the oxidation of sulfur dioxide generated during the combustion of coal.

industrial systems in a manner consistent with minimum environmental impact. The facilities and processes of an industrial ecology system can be sited and operated for minimal adverse environmental impact if environmental chemistry is considered in their planning and operation. Environmental chemistry clearly points the way to minimize the environmental impacts of the emissions and by-products of industrial systems, and is very helpful in reaching the ultimate goal of a system of industrial ecology, which is to reduce these emissions and by-products to zero.

Environmental chemistry can be divided into several categories. The first of these addressed in this book is *aquatic chemistry*, which deals with chemical phenomena in water. Various aspects of aquatic chemistry are discussed in Chapters 3 through 8. As the name implies, *atmospheric chemistry* deals with chemical phenomena in the atmosphere. Atmospheric chemistry is discussed in Chapters 9 through 14. General aspects of the environmental chemistry of the geosphere are discussed in Chapter 15, and soil chemistry is the topic of Chapter 16. The anthrosphere and environmental chemistry are discussed in Chapters 17 through 21. The biosphere related to environmental chemistry is mentioned in various contexts throughout the book as it relates to environmental chemical processes in water and soil. It is the main topic of discussion of Chapters 22 through 24. Toxicological chemistry defined as *the chemistry of toxic substances with emphasis on their interactions with biologic tissue and living organisms*[1] is addressed specifically in Chapters 23 and 24. *Analytical chemistry* is uniquely important in environmental chemistry. Environmental chemical analysis is discussed in Chapters 25 through 28.

2.2 MATTER AND CYCLES OF MATTER

Very much connected with environmental chemistry, *cycles of matter*, often based on elemental cycles, are of utmost importance in the environment. Global geochemical cycles can be regarded from the viewpoint of various reservoirs, such as oceans, sediments, and the atmosphere, connected by conduits through which matter moves continuously among the hydrosphere, atmosphere, geosphere, biosphere, and, increasingly, the anthrosphere. The movement of a specific kind of matter between two particular reservoirs may be reversible or irreversible. The fluxes of movement for particular kinds of matter vary greatly as do the contents of such matter in a specified reservoir. Most cycles of matter have a strong biotic component, especially through the biochemical processes

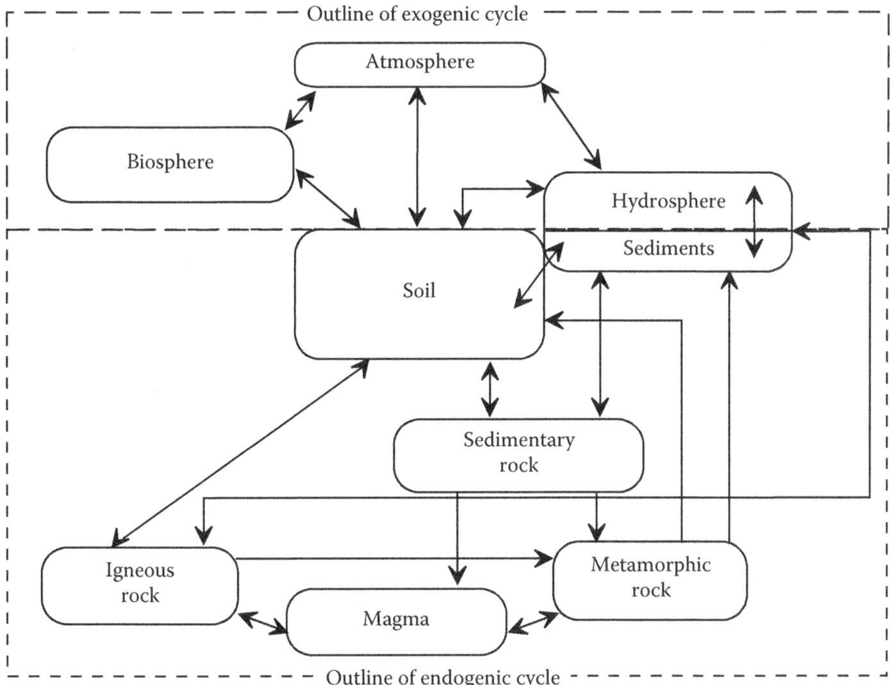

FIGURE 2.2 General outline of exogenic and endogenic cycles of matter.

of plants and microorganisms. The cycles in which organisms participate are called *biogeochemical cycles*, which describe the circulation of matter, particularly plant and animal nutrients, through ecosystems. Most biogeochemical cycles can be described as elemental cycles involving nutrient elements such as carbon, nitrogen, oxygen, phosphorus, and sulfur. As part of the carbon cycle, atmospheric carbon in CO_2 is fixed as biomass; as part of the nitrogen cycle, atmospheric N_2 is fixed in organic matter. The reverse of these kinds of processes is mineralization, in which biologically bound elements are returned to inorganic states. Biogeochemical cycles are ultimately powered by solar energy, which is fine-tuned and directed by energy expended by organisms. In a sense, the solar-energy-powered hydrological cycle (Figure 3.1) acts as an endless conveyer belt to move materials essential for life through ecosystems. Cycles of matter are also very much involved with the chemical fate and transport of pollutants discussed in Section 1.8.

As shown in Figure 2.2, cycles of matter may be divided into two major categories. *Exogenic cycles* occur largely on the Earth's surface and are those in which the element in question spends part of the cycle in the atmosphere—O_2 for oxygen, N_2 for nitrogen, and CO_2 for carbon. *Endogenic cycles*, notably the sulfur cycle, predominantly involve subsurface rocks of various kinds and are without a gaseous component. In general, sediment and soil can be viewed as being shared between the two cycles and constitute the predominant interface between them. All sedimentary cycles involve salt solutions or soil solutions (see Section 16.2) that contain dissolved substances leached from weathered minerals; these substances may be deposited as mineral formations, or they may be taken up by organisms as nutrients.

2.2.1 Carbon Cycle

Carbon circulates through the *carbon cycle* shown in Figure 2.3. It shows that carbon may be present as gaseous atmospheric CO_2, constituting a relatively small but highly significant portion of global carbon. Some of the carbon is dissolved in surface water and groundwater as HCO_3^- or

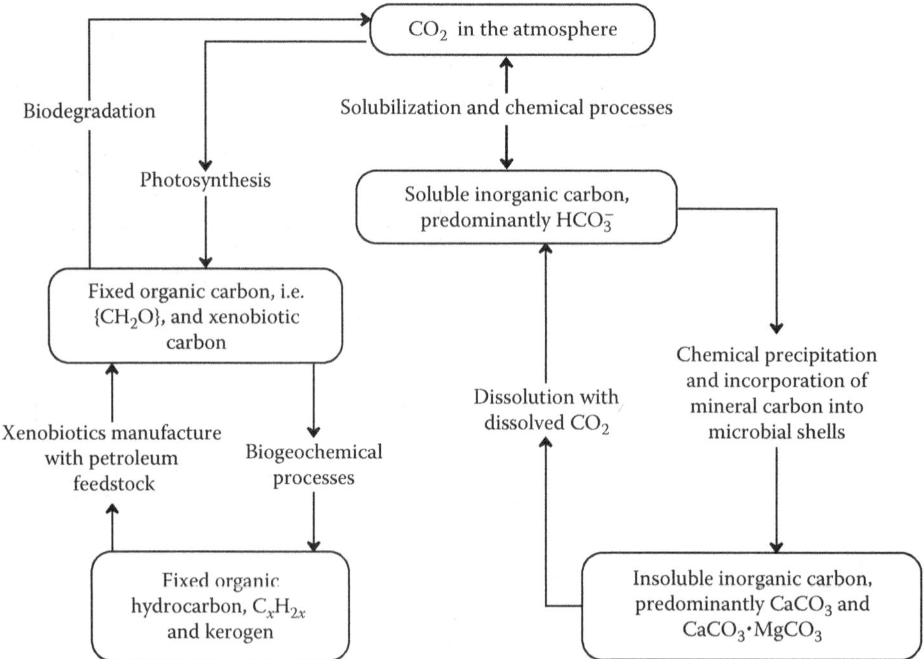

FIGURE 2.3 The carbon cycle. Mineral carbon is held in a reservoir of limestone, $CaCO_3$, from which it may enter a mineral solution as dissolved hydrogen carbonate ion, HCO_3^-, formed when dissolved CO_2 (*aq*) reacts with $CaCO_3$. In the atmosphere carbon is present as carbon dioxide, CO_2. Atmospheric carbon dioxide is fixed as organic matter by photosynthesis, and organic carbon is released as CO_2 by microbial decay of organic matter.

molecular CO_2 (*aq*). A very large amount of carbon is present in minerals, particularly magnesium and calcium carbonates such as $CaCO_3$. Photosynthesis fixes inorganic C as *biological carbon*, represented as {CH_2O}, which is a constituent of all life molecules. Another fraction of carbon is fixed as petroleum and natural gas, with a much larger amount as hydrocarbonaceous kerogen (the organic matter in oil shale), coal, and lignite. Manufacturing processes are used to convert hydrocarbons to xenobiotic compounds with functional groups containing halogens, oxygen, nitrogen, phosphorus, or sulfur. Although a very small amount of total environmental carbon, these compounds are particularly significant because of their toxicological chemical effects.

An important aspect of the carbon cycle is that it is the one by which solar energy is transferred to biological systems and ultimately to the geosphere and anthrosphere as fossil carbon and fossil fuels. Organic, or biological, carbon, {CH_2O}, is contained in energy-rich molecules that can react biochemically with O_2, to regenerate carbon dioxide and to produce energy. This can occur biochemically in an organism through aerobic respiration as shown in Equation 1.2, or it may occur as combustion, such as when wood or fossil fuels are burned.

Microorganisms are strongly involved in the carbon cycle, mediating crucial biochemical reactions discussed later in this section. Photosynthetic algae are the predominant carbon-fixing agents in water; as they consume CO_2 to produce biomass, the pH of the water increases, enabling precipitation of $CaCO_3$ and $CaCO_3 \cdot MgCO_3$. Biogeochemical processes transform organic carbon fixed by microorganisms to fossil petroleum, kerogen, coal, and lignite. Microorganisms degrade organic carbon from biomass, petroleum, and xenobiotic sources, ultimately returning it to the atmosphere as CO_2. Hydrocarbons such as those in crude oil and some synthetic hydrocarbons are degraded by microorganisms. This is an important mechanism for eliminating pollutant hydrocarbons, such as those that are accidentally spilled on soil or in water. Biodegradation also acts to destroy carbon-containing compounds in hazardous wastes.

2.2.2 Nitrogen Cycle

As shown in Figure 2.4, nitrogen occurs prominently in all the spheres of the environment. The atmosphere is 78% elemental nitrogen, N_2, by volume and comprises an inexhaustible reservoir of this essential element. Nitrogen, although constituting much less of biomass than carbon or oxygen, is an essential constituent of proteins. The N_2 molecule is very stable so that breaking it down into atoms that can be incorporated with inorganic and organic chemical forms of nitrogen is the limiting step in the nitrogen cycle. This does occur by highly energetic processes in lightning discharges that produce nitrogen oxides. Elemental nitrogen is also incorporated into chemically bound forms, or fixed by biochemical processes mediated by microorganisms. The biological nitrogen is mineralized to the inorganic form during the decay of biomass. Large quantities of nitrogen are fixed synthetically under high-temperature and high-pressure conditions according to the following overall reaction:

$$N_2 + 3H_2 \rightarrow 2NH_3 \quad (2.1)$$

The production of gaseous N_2 and N_2O by microorganisms and the evolution of these gases to the atmosphere complete the nitrogen cycle through a process called *denitrification*. The nitrogen cycle is discussed from the viewpoint of microbial processes in Section 6.11.

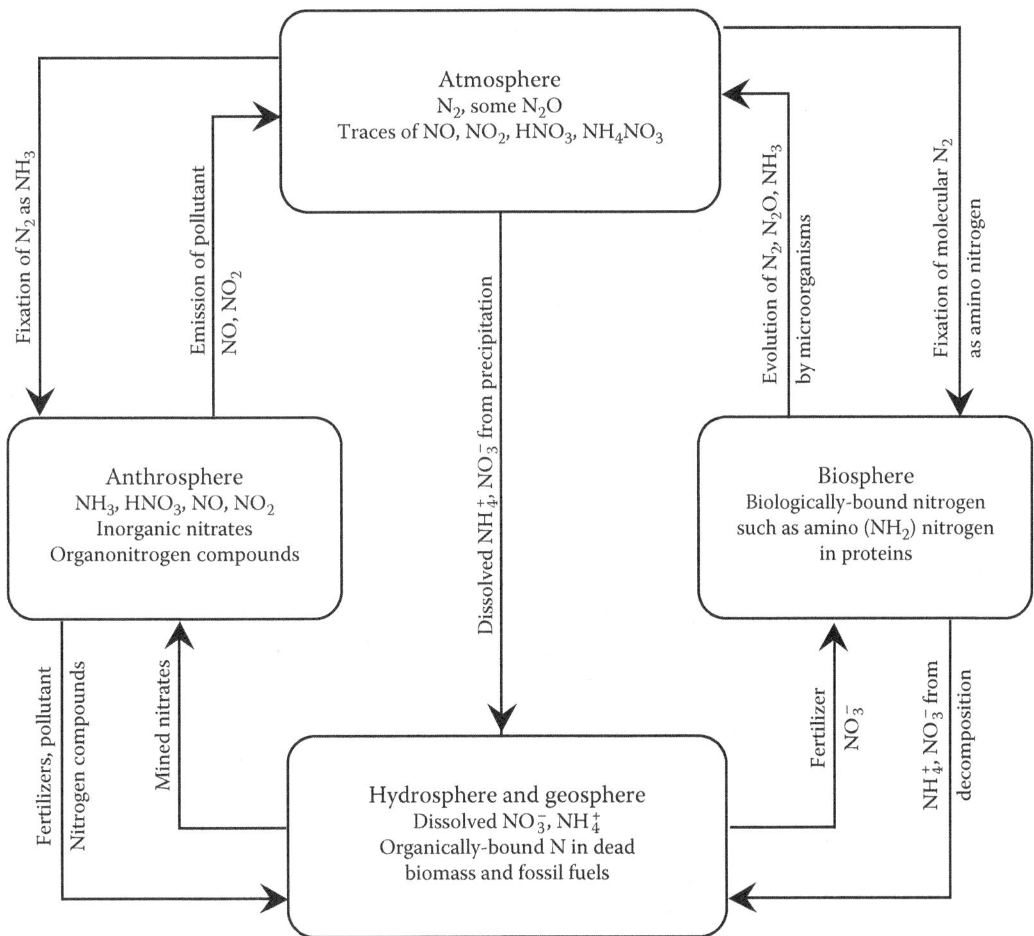

FIGURE 2.4 The nitrogen cycle.

2.2.3 Oxygen Cycle

The *oxygen cycle* is discussed in Chapter 9 and is illustrated in Figure 9.11. It involves the interchange of oxygen between the elemental form of gaseous O_2, contained in a huge reservoir in the atmosphere, and chemically bound O in CO_2, H_2O, minerals, and organic matter. It is strongly tied with other elemental cycles, particularly the carbon cycle. Elemental oxygen becomes chemically bound by various energy-yielding processes, particularly combustion and metabolic processes in organisms. It is released in photosynthesis. This element readily combines with and oxidizes other species such as carbon in the process of aerobic respiration (Reaction 1.2), or carbon and hydrogen in the combustion of fossil fuels such as methane:

$$CH_4 + 2O_2 \rightarrow CO_2 + 2H_2O \tag{2.2}$$

Elemental oxygen also oxidizes inorganic substances such as iron(II) in minerals:

$$4FeO + O_2 \rightarrow 2Fe_2O_3 \tag{2.3}$$

A particularly important aspect of the oxygen cycle is stratospheric ozone, O_3. As discussed in Chapter 9, Section 9.10, a relatively small concentration of ozone in the stratosphere, more than 10 km high in the atmosphere, filters out ultraviolet radiation in the wavelength range of 220–330 nm, thus protecting life on the Earth from the highly damaging effects of this radiation.

The oxygen cycle is completed by the return of elemental O_2 to the atmosphere. The only significant way in which this is done is through photosynthesis mediated by plants. The overall reaction for photosynthesis is given in Reaction 1.1.

2.2.4 Phosphorus Cycle

The *phosphorus cycle*, Figure 2.5, is crucial because phosphorus is usually the limiting nutrient in ecosystems. There are no common stable gaseous forms of phosphorus, so the phosphorus cycle is endogenic. In the geosphere, phosphorus is held largely in poorly soluble minerals, such as hydroxyapatite, a calcium salt, deposits of which constitute the major reservoir of environmental phosphate. Soluble phosphorus from phosphate minerals and other sources such as fertilizers is taken up by plants and incorporated into nucleic acids that make up the genetic material of organisms. Mineralization of biomass by microbial decay returns phosphorus to the salt solution from which it may precipitate as mineral matter.

The anthrosphere is a major reservoir of phosphorus in the environment with large quantities of phosphates extracted from phosphate minerals for fertilizer, chemicals, and food additives. Phosphorus is a constituent of some extremely toxic compounds, especially organophosphate insecticides and military poison nerve gases including the infamous Sarin and VX (see Chapter 24).

2.2.5 Sulfur Cycle

The *sulfur cycle*, which is illustrated in Figure 2.6, is relatively complex in that it involves several gaseous species, poorly soluble minerals, and several species in solution. It is tied with the oxygen cycle in that sulfur combines with oxygen to form gaseous sulfur dioxide, SO_2, an atmospheric pollutant, and soluble sulfate ion, SO_4^{2-}. Among the significant species involved in the sulfur cycle are gaseous hydrogen sulfide, H_2S; volatile dimethyl sulfide, $(CH_3)_2S$, released to the atmosphere by biological processes in the ocean; mineral sulfides, such as PbS; sulfuric acid, H_2SO_4, the main constituent of acid rain; and biologically bound sulfur in sulfur-containing proteins.

Insofar as pollution is concerned, the most significant part of the sulfur cycle is the presence of pollutant SO_2 gas and H_2SO_4 in the atmosphere. The former is a somewhat toxic gaseous air pollutant

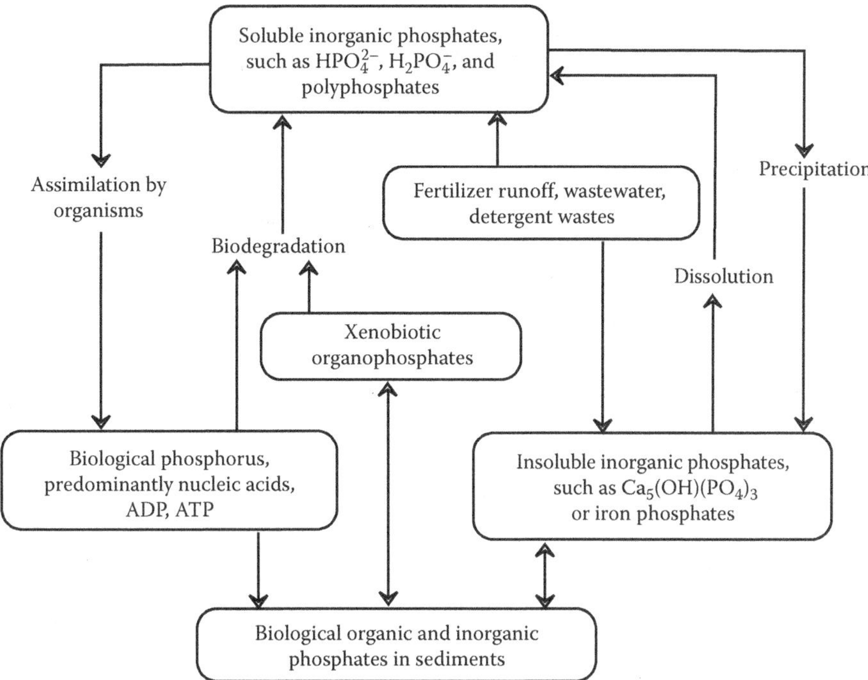

FIGURE 2.5 The phosphorus cycle.

evolved in the combustion of sulfur-containing fossil fuels. Sulfur dioxide is discussed further as an air pollutant in Chapter 11, and its toxicological chemistry is covered in Chapter 24. The major detrimental effect of sulfur dioxide in the atmosphere is its tendency to oxidize in the atmosphere to produce sulfuric acid. This species is responsible for acidic precipitation, "acid rain," discussed as a major atmospheric pollutant in Chapter 14.

2.3 ANTHROSPHERE AND ENVIRONMENTAL CHEMISTRY

As the source of most environmental pollution, the anthrosphere has a strong connection with environmental chemistry. The *anthrosphere* may be defined as that part of the environment made or modified by humans and used for their activities. Of course, there are some ambiguities associated with that definition. Clearly, a factory building used for manufacture is part of the anthrosphere as is an ocean-going ship used to ship goods made in the factory. The ocean on which the ship moves belongs to the hydrosphere, but it is clearly used by humans. A pier constructed on the ocean shore and used to load the ship is part of the anthrosphere, but it is closely associated with the hydrosphere and anchored to the geosphere.

During most of its time on the Earth, humankind made comparatively little impact on the planet. Simple huts or tents for dwellings, narrow trails worn across the land for movement, and food gathered largely from nature's sources affected the environment minimally. Even so, there is evidence to suggest that prehistoric humans were beginning to have an impact on nature, probably hunting some species to extinction and burning forests to provide grazing land to attract wild game. However, with increasing effect as the industrial revolution developed, and especially during the last century, humans have built structures and modified the other environmental spheres, especially the geosphere, such that it is necessary to consider the anthrosphere as a separate area with a pronounced, sometimes overwhelming influence on the environment as a whole. The influence of the anthrosphere and human activities is so great that some authorities on the environment assert that a

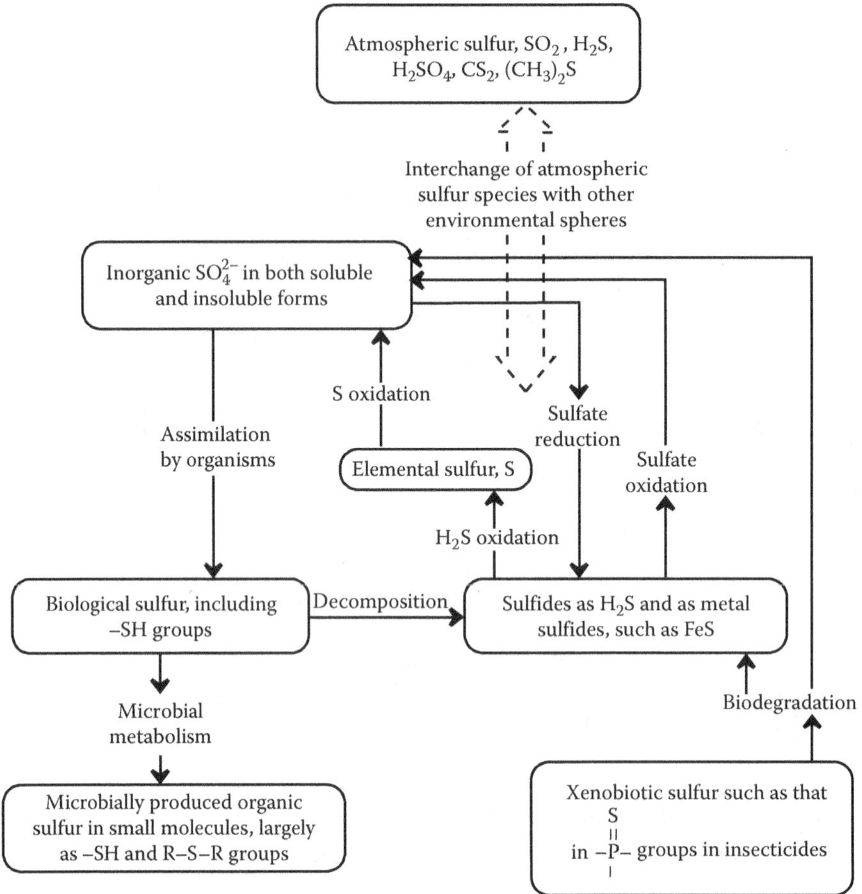

FIGURE 2.6 The sulfur cycle.

transition is under way to a new era, the *anthropocene*, in which the nature of the Earth's environment is largely determined by human activities in the anthrosphere.

2.3.1 Components of the Anthrosphere

Like other spheres of the environment, the anthrosphere consists of a number of different parts. These may be categorized by considering where humans live; how they move; how they make or provide the things or services they need or want; how they produce food, fiber, and wood; how they obtain, distribute, and use energy; how they communicate; how they extract and process nonrenewable minerals; and how they collect, treat, and dispose of wastes. With these factors in mind, it is possible to divide the anthrosphere into the following categories (Figure 2.7):

- Structures used for dwellings
- Structures used for manufacturing, commerce, education, and other activities
- Utilities, including water, fuel, and electricity distribution systems, and waste distribution systems, such as sewers
- Structures used for transportation, including roads, railroads, airports, and waterways constructed or modified for water transport
- Structures and other parts of the environment modified for food production, such as fields used for growing crops and water systems used to irrigate the fields

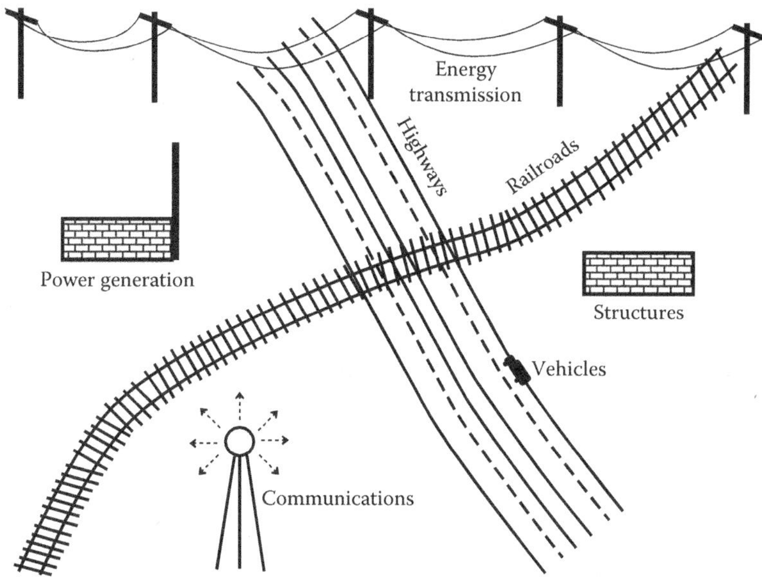

FIGURE 2.7 Key components of the anthrosphere.

- Machines of various kinds, including automobiles, farm machinery, and airplanes
- Structures and devices used for communications, such as telephone lines or radio transmitter towers
- Structures, such as mines or oil wells, associated with extractive industries

From the list given above it is obvious that the anthrosphere is very complex with an enormous potential to affect the environment. Prior to addressing these environmental effects, several aspects of the anthrosphere will be discussed in more detail.

2.4 TECHNOLOGY AND THE ANTHROSPHERE

Since the anthrosphere is the result of technology, it is appropriate to discuss technology at this point. *Technology* refers to the ways in which humans do and make things with materials and energy. In the modern era, technology is to a large extent the product of engineering based on scientific principles. Science deals with the discovery, explanation, and development of theories pertaining to interrelated natural phenomena of energy, matter, time, and space. Based on the fundamental knowledge of science, engineering provides the plans and means to achieve specific practical objectives. Technology uses these plans to carry out the desired objectives.

Technology has a long history and, indeed, goes back into prehistory to times when humans used primitive tools made from stone, wood, and bone. As humans settled in cities, human and material resources became concentrated and focused such that technology began to develop at an accelerating pace. Technological advances predating the Roman era include the development of *metallurgy*, beginning with native copper around 4000 BC, domestication of the horse, discovery of the wheel, architecture to enable construction of substantial buildings, control of water for canals and irrigation, and writing for communication. The Greek and Roman eras saw the development of *machines*, including the windlass, pulley, inclined plane, screw, catapult for throwing missiles in warfare, and water screw for moving water. Later, the water wheel was developed for power transmitted by wooden gears. Many technological innovations, such as printing with wood blocks starting around the year 740 and gunpowder about a century later, originated in China.

The 1800s saw an explosion in technology. Among the major advances during this century were widespread use of steam power, steam-powered railroads, the telegraph, telephone, electricity as a power source, textiles, the use of iron and steel in building and bridge construction, cement, photography, and the invention of the internal combustion engine, which revolutionized transportation in the following century.

Since about 1900, advancing technology has been characterized by vastly increased uses of energy; much higher speed in manufacturing processes, information transfer, computation, transportation, and communication; automated control; a vast new variety of chemicals; new and improved materials for new applications; and, more recently, the widespread application of computers to manufacturing, communication, and transportation. In transportation, the development of passenger-carrying airplanes has enabled an astounding change in the ways in which people get around and how high-priority freight is moved. Rapid advances in biotechnology now promise to revolutionize food production and medical care.

The technological advances of the 1900s were largely attributable to two factors. The first of these was the application of electronics, now based on solid state devices, to technology in areas such as communications, sensors, and computers for manufacturing control. The second area largely responsible for modern technological innovations has been based on improved materials. For example, special strong light alloys of aluminum were used in the construction of airliners before World War II and more recently these alloys have been supplanted to a degree by even more advanced composites. Synthetic materials with a significant impact on modern technology include plastics, fiber-reinforced materials, composites, and ceramics.

Until fairly recently, technological advances were made largely without heed to environmental impacts. Now, however, the greatest technological challenge is to reconcile technology with environmental consequences. The survival of humankind and of the planet that supports it now requires that the established two-way interaction between science and technology become a three-way relationship including environmental protection and sustainability.

2.4.1 Engineering

Engineering uses fundamental knowledge acquired through science to provide the plans and means to achieve specific objectives in areas such as manufacturing, communication, and transportation. There are many engineering specialties including chemical, electrical, environmental, aerospace, agricultural, biomedical, computer-aided design and computer-aided manufacturing (CAD/CAM), ceramic, industrial, materials, metallurgical, mining, and petroleum engineering. Some other categories of engineering are defined below.

- *Mechanical engineering*, which deals with machines and the manner in which they handle forces, motion, and power
- *Electrical engineering* dealing with the generation, transmission, and utilization of electrical energy
- *Electronics engineering* dealing with phenomena based on the behavior of electrons in vacuum tubes and other devices
- *Chemical engineering*, which uses the principles of chemical science, physics, and mathematics to design and operate processes that generate products and materials
- *Civil engineering* dealing largely with infrastructure components, such as highways, airports, and water distribution systems
- *Environmental engineering* involving pollution control, waste treatment, and pollution prevention

The role of engineering in constructing and operating the various components of the anthrosphere is obvious. In the past, engineering was often applied without much, if any, consideration of environmental factors. As examples, huge machines designed by mechanical engineers were used to dig up and rearrange the Earth's surface without regard for the environmental consequences, and chemical engineering was used to make a broad range of products without consideration of the wastes produced. Fortunately, that approach is changing rapidly. Examples of environmentally friendly engineering include machinery designed to minimize noise, much improved energy efficiency in machines, and the use of earth-moving equipments for environmentally beneficial purposes, such as restoration of strip-mined lands and construction of wetlands. Efficient generation, distribution, and utilization of electrical energy based on the principles of electrical engineering constitute one of the most promising avenues of endeavor leading to environmental improvement. Automated factories developed through applications of electronic engineering and controlled by computers can turn out goods with the lowest possible consumption of energy and materials while minimizing air and water pollutants, production of hazardous wastes, and worker exposure to hazards. Chemical factories can be engineered to maximize the most efficient utilization of energy and materials while minimizing waste production. In recent years, environmentally friendly and sustainable engineering have been formalized as *green engineering*.[2]

2.5 INFRASTRUCTURE

The *infrastructure* consists of the utilities, facilities, and systems used in common by members of a society and upon which the society depends for its normal function. The infrastructure includes both physical components—roads, bridges, and pipelines—and the instructions—laws, regulations, and operational procedures—under which the physical infrastructure operates. Parts of the infrastructure may be publicly owned, such as the U.S. Interstate Highway System and some European railroads, or privately owned, as is the case with virtually all railroads in the United States. Some of the major components of the infrastructure of a modern society are the following.

- Transportation systems, including railroads, highways, and air transport systems
- Energy-generating and distribution systems
- Buildings
- Telecommunications systems
- Water supply and distribution systems
- Waste treatment and disposal systems, including those for municipal wastewater, municipal solid refuse, and industrial wastes

In general, the infrastructure refers to the facilities that large segments of a population must use in common in order for a society to function. In a sense, the infrastructure is analogous to the operating system of a computer. A computer operating system determines how individual applications operate and the manner in which they distribute and store the documents, spreadsheets, illustrations, and communications created by the applications. Similarly, the infrastructure is used to move raw materials and power to factories and to distribute and store their output. Just as an outdated, cumbersome computer operating system with a tendency to crash is detrimental to the efficient operation of a computer, an outdated, cumbersome, broken-down infrastructure, which has become all to common in many countries including the United States, causes society to operate in a very inefficient manner and is subject to catastrophic failure.

For a society to be successful, it is of utmost importance to maintain a modern, viable infrastructure. Such an infrastructure is consistent with environmental protection. Properly designed utilities and other infrastructural elements, such as water supply systems and wastewater treatment systems, minimize pollution and environmental damage.

Components of the infrastructure are subject to deterioration. To a large extent this is due to natural aging processes. Fortunately, many of these processes can be slowed or even reversed. Corrosion of steel structures, such as bridges, is a big problem for infrastructures; however, use of corrosion-resistant materials and maintenance with corrosion-resistant coatings can virtually stop this deterioration process. The infrastructure is subject to human insult, such as vandalism, misuse, and neglect. Often the problem begins with the design and basic concept of a particular component of the infrastructure. For example, many river dikes destroyed by flooding should never have been built because they attempt to thwart to an impossible extent the natural tendency of rivers to flood periodically.

Technology plays a major role in building and maintaining a successful infrastructure. Many of the most notable technological advances applied to the infrastructure were made from 150 to 100 years ago. By the year 1900, railroads, electric utilities, telephones, and steel building skeletons had been developed. The net effect of most of these technological innovations was to enable humankind to "conquer" or at least temporarily subdue nature. The telegraph and telephone helped to overcome isolation, high-speed rail transport and later air transport conquered distance, and dams were used to control rivers and water flow. The development of new and improved materials, such as stronger, lighter steel structural elements, continues to have a significant influence on the infrastructure.

The major challenge in designing and operating the infrastructure in the future will be to use it to work with the environment and to enhance environmental quality to the benefit of humankind. Obvious examples of environmentally friendly infrastructures are state-of-the-art sewage treatment systems, high-speed rail systems that can replace inefficient highway transport, and stack gas emission control systems in power plants. More subtle approaches with a tremendous potential for making the infrastructure more environmentally friendly include employment of workers at computer terminals in their homes so that they do not need to commute, instantaneous electronic mail that avoids the necessity of moving letters physically, and solar electric-powered installations to operate remote signals and relay stations, so that electric power lines do not need to be run to them.

Developments in electronics and computers are enabling significant advances in infrastructure. One of the areas in which the influence of modern electronics and computers is most visible is in telecommunications. Dial telephones and mechanical relays were satisfactory in their time, but have been made totally obsolete by innovations in electronics, computer control, and fiber optics. Air transport controlled by a truly modern, state-of-the-art computerized control system can enable present airports to handle many more airplanes safely and efficiently, thus reducing the need for airport construction. Sensors for monitoring strain, temperature, movement, and other parameters can be imbedded in the structural members of bridges and other structures. Information from these sensors can be processed by computers to warn of failure and to aid in proper maintenance. Many similar examples can be cited.

2.5.1 Vulnerable Infrastructure

Almost any part of the infrastructure is vulnerable to attack and disruption. In the modern era, an action that disables a segment of infrastructure may cause severe disruptions throughout the system. This is because of *vulnerability due to interconnectivity*, which is a characteristic of modern electrical, transportation, communications, and other infrastructure systems, the components of which are interconnected and mutually dependent to an extreme degree. This was illustrated dramatically on August 14, 2003, when electrical power was abruptly terminated to tens of millions of people in the Northeastern United States and Ontario including the major cities of New York City, Detroit, Cleveland, and Toronto. About 100 powerplants, an astounding 68,100 MW of generating capacity, and dozens of transmission lines were shut down. All of these facilities stopped operations within about 5 min, although the core of the actual event probably occurred within only about 10 s, after which the shutdown was irreversible.

The incident that caused the great blackout of August 2003 was so minor that even 2 days later experts were not certain of the actual precipitating event. Although it was quickly established that the event was not a deliberate incident of terrorism or sabotage, it well could have been and it stands as an ominous symbol of the potential of terrorist attacks to disrupt essential infrastructure services. This power failure was an illustration of *cascading failures* on complex networks. In the case of electrical power, hundreds of generating plants are interconnected by power lines and the load is continually redistributed to meet fluctuating demand. This provides adequate power with much less total capacity and hence lower cost than would be the case if individual power plants had to have extra capacity to provide for fluctuating local demand. The computer Internet works by means of many interconnected computing facilities. Routers on the Internet are programmed to redirect traffic around malfunctioning routers. Modern manufacturing operations depend on "just-in-time" delivery of dozens of components. Occasional power outages, Internet failures, and assembly line shutdowns due to the lack of an essential part serve as reminders of the vulnerability of such networks to cascading failures and the ever-present possibility of sabotage in causing such failures.

As the complexity of urban infrastructure systems has grown, so has the importance of emergency operations centers in charge of managing damages to the infrastructure that can result in disruption. Historically, these centers have dealt with natural disasters such as tornadoes, floods, and earthquakes and they have also handled major fires, explosions, and building collapses caused by human error and design flaws. With increased concern over terrorist threats, however, emergency operations centers have had to become equipped to deal with attacks. Terrorist attacks that may entail highly toxic chemicals, pathogens, or nuclear materials present a special challenge to first responders, who may be put in serious personal danger due to the event.

Fortunately, fire and other emergency personnel have long been trained to handle spills and releases of toxic and otherwise hazardous substances. Toxic vapors are almost always associated with fires so that fire-fighting personnel possess protective and breathing gear that would protect them in case of a chemical attack. Military personnel have been trained to respond to attack with military poisons and have appropriate gear to protect them from such attack.

The various elements of the infrastructure are highly connected and interdependent. Perhaps no component is more central to the correct functioning of the other elements of the infrastructure than the electrical supply. Without electricity, water pumping systems fail and water may become unavailable for fighting fires. Larger cities with subway systems depend on electricity for moving subway trains. Traffic light systems require electricity and, without them, traffic in urban areas quickly disintegrates into chaos and gridlock. Electricity is required to run elevators in large buildings.

Buildings can also be regarded as part of the essential infrastructure. And buildings are vulnerable as illustrated most vividly by the September 11, 2001, attack on the World Trade Center in New York in which two huge buildings collapsed due to fires from the fuel of impacting aircraft. Other examples illustrating the vulnerability of buildings to attack include the 1995 attack on the Murrah Federal Building in Oklahoma City, and terrorist attacks on hotels in Mumbai, India, in 2008.

Chemistry certainly has a role to play in protecting infrastructure. Materials that resist attack, such as flame-resistant building structural materials, can greatly reduce damage from attack. In general those design characteristics that are consistent with the best practices of environmental chemistry and green chemistry likewise increase the resistance of elements of the infrastructure to attack. Sensitive, portable instruments for the detection of toxic substances and explosives can reduce the hazards of these materials to the infrastructure.

2.6 COMPONENTS OF THE ANTHROSPHERE THAT INFLUENCE THE ENVIRONMENT

Several components of the anthrosphere have especially strong influences on the environment. They also have enormous potential for modification and improvement to achieve the goal of sustainability. Several of these are mentioned briefly here.

- *Dwellings and buildings*: Although much of the world dwells in substandard housing, a large part of the current housing in the United States and other industrialized countries is not in keeping with the best practices of sustainability and has a very high environmental impact. There is a need to locate dwellings closer to where people work and to commercial areas. Homes should be of reasonable size, efficiently laid out, and as energy efficient as possible. Buildings should be versatile so that they can be converted to alternate uses without destroying the entire structure. Technological advances can be used to make buildings much more environmentally friendly. Advanced window design and effective insulating materials can significantly reduce energy consumption. Modern heating and air conditioning systems operate with a high degree of efficiency. Automated and computerized control of building utilities, particularly those used for cooling and heating, can significantly reduce energy consumption by regulating temperatures and lighting to the desired levels at specific locations and times as needed in the building.

- *Transportation*: The widespread use of automobiles, trucks, and buses has enormous effects on the environment. Entire landscapes have been entirely rearranged to construct highways, interchanges, and parking lots. Emissions from the internal combustion engines used in automobiles are the major source of air pollution in many urban areas. The automobile has made possible the "urban sprawl" that is characteristic of residential and commercial patterns of development in the United States, and in many other industrialized countries as well. Huge new suburban housing tracts and the commercial developments, streets, and parking lots constructed to support them continue to consume productive farmland at a frightening rate. Applications of advanced engineering and technology to transportation can be of tremendous benefit to the environment. Modern rail and subway transportation systems, concentrated in urban areas and carefully connected to airports for longer distance travel, can enable the movement of people rapidly, conveniently, and safely with minimum environmental damage. The *telecommuter society*, composed of workers who do their work at home and "commute" through their computers, modems, fax machines, and the Internet connected by way of high-speed communication lines, can significantly relieve the strain on transportation systems and reduce their adverse environmental effects.
- *Communications*: The major areas to consider with respect to information are its acquisition, recording, computing, storing, displaying, and transmission. All these have been tremendously augmented by recent technological advances. Perhaps the greatest such advance has been that of silicon-integrated circuits. Optical memory consisting of information recorded and read by microscopic beams of laser light has enabled the storage of astounding quantities of information on a single compact disk. The use of optical fibers to transmit information digitally by light has resulted in a comparable advance in the communication of information. The central characteristic of communication in the modern age is the combination of telecommunications with computers called *telematics,* such as is employed in automatic bank teller machines. An important capability in modern communications is the ability to rapidly and accurately acquire, analyze, and communicate information about the environment.
- *Food and agriculture*: The environmental impact of agriculture is enormous. One of the most rapid and profound changes in the environment that has ever taken place was the conversion of vast areas of the North American continent from forests and grasslands to cropland, which occurred predominantly during the 1800s. This enabled the production of huge amounts of food but also resulted in damaging water and wind erosion. Recognition of these problems led to intense soil conservation efforts starting around 1900 and continuing to the present. In recent decades, valuable farmland has faced the threat posed by the urbanization of rural areas—urban sprawl—as prime agricultural land has been turned into subdivisions and paved over to create parking lots and streets. Increasing population and higher living standards will put more pressure on the agricultural resources. It is likely that production of synthetic fuels from biomass will also increase the demands on the agricultural sector. Global warming and resulting climate change will likely have profound effects on agriculture.
- *Manufacturing*: The manufacture of goods carries with it the potential to cause significant air and water pollution and production of hazardous wastes. The earlier in the design

and development process that environmental considerations are taken into account, the more "environmentally friendly" a manufacturing process will be. Application of the principles of industrial ecology and green chemistry can greatly increase the sustainability of manufacturing. Three relatively new developments that have revolutionized manufacturing and that continue to do so are automation, robotics, and computers. *Automation* uses automatic devices to perform repetitive tasks such as assembly line operations. *Robotics* refers to the use of machines to simulate human movements and activities. CAD is employed to convert an idea to a manufactured product, replacing innumerable sketches, engineering drawings, and physical mockups that used to be employed. CAM uses computers to plan and control manufacturing operations, for quality control, and to manage entire manufacturing plants. This increases the efficiency of manufacturing and contributes to sustainability.

2.7 EFFECTS OF THE ANTHROSPHERE ON EARTH

The effects of the anthrosphere on Earth have been many and profound. Persistent and potentially harmful products of human activities have been widely dispersed and concentrated in specific locations in the anthrosphere as well as other spheres of the environment as the result of human activities. Among the most troublesome of these are toxic heavy metals and organochlorine compounds. Such materials have accumulated in the anthrosphere in painted and coated surfaces, such as organotin-containing paints used to prevent biofouling on boats; under and adjacent to airport runways; under and along highway paving; buried in old factory sites; in landfills; and in materials dredged from waterways and harbors that are sometimes used as landfill on which buildings, airport runways, and other structures have been placed. In many cases, productive topsoil used to grow food has been contaminated with discarded industrial wastes, phosphate fertilizers, and dried sewage sludge containing levels of metals harmful to crops.

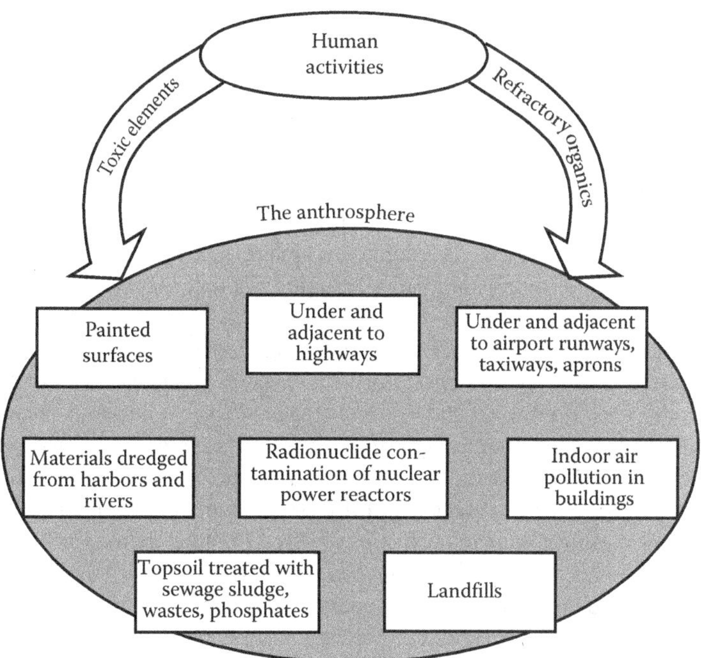

FIGURE 2.8 The anthrosphere is a repository of many of the pollutant by-products of human activities.

Some of the portions of the anthrosphere that may be severely contaminated by human activities are shown in Figure 2.8. In some cases the contamination has been so pervasive and persistent that the effects will remain for centuries. Some of the most vexsome environmental and waste problems are due to contamination of various parts of the anthrosphere by persistent and toxic waste materials.

Potentially harmful wastes and pollutants of anthrospheric origin have found their way into water, air, soil, and living organisms. For example, chlorofluorocarbons (CFCs) (Freons) have been released to the atmosphere in such quantities and are so stable that they are now constituents of "normal" atmospheric air and pose a threat to the protective ozone layer in the stratosphere. Lake sediments, stream beds, and deltas deposited by flowing rivers are contaminated with heavy metals and refractory organic compounds of anthrospheric origin. The most troubling repository of wastes in the hydrosphere is groundwater. Some organisms have accumulated high enough levels of persistent organic compounds or heavy metals to do harm to themselves or to humans who use them as a food source.

2.8 INTEGRATION OF THE ANTHROSPHERE INTO THE TOTAL ENVIRONMENT

Over the eons of the Earth's existence, natural processes free from sudden, catastrophic disturbances (such as those that have occurred from massive asteroid impacts, for example) have resulted in a finely tuned balance among the systems composing the Earth's natural environment. Fortuitously, these conditions—adequate water, moderate temperatures, and an atmosphere that serves as a shield against damaging solar radiation—have resulted in conditions amenable to various life forms. Indeed, these life forms have had a strong impact in changing their own environments. According to the *Gaia hypothesis* advanced by the British chemist James Lovelock, organisms on the Earth have modified the Earth's climate and other environmental conditions, such as by regulating the CO_2/O_2 balance in the atmosphere, in a manner conducive to the existence and reproduction of the organisms.

To a degree, the early anthrosphere created by preindustrial humans caused minimal environmental degradation. The relatively harmonious relationship between the anthrosphere and the rest of the environment existed so long as the energy required to alter the environment came mainly from human and animal muscle power. This situation began to change markedly with the introduction of machines, particularly power sources, beginning with the steam engine, that greatly multiplied the capabilities of humans to alter their surroundings. As humans developed their use of machines and other attributes of industrialized civilization, they did so with little consideration of the environment and in a way that was out of synchronization with the other environmental spheres. A massive environmental imbalance has resulted, the magnitude of which has been realized only in recent decades. The most commonly cited manifestation of this imbalance has been pollution of air or water.

Because of the detrimental effects of human activities undertaken without due consideration of environmental consequences, significant efforts have been made to reduce the environmental impacts of these activities. Figure 2.9 shows three stages of the evolution of the anthrosphere from an unintegrated appendage to the natural environment to a system more attuned to its surroundings. The first approach to dealing with the pollutants and wastes produced by industrial activities—particulate matter from power plant stacks, sulfur dioxide from copper smelters, and mercury-contaminated wastes from chlor-alkali manufacture—was to ignore them. However, as smoke from uncontrolled factory furnaces, raw sewage, and other by-products of human activities became more troublesome, "end-of-pipe" measures were adopted to prevent the release of pollutants after they were generated. Such measures have included electrostatic precipitators and flue gas desulfurization to remove particulate matter and sulfur dioxide from flue gas; physical processes used in primary sewage treatment; microbial processes used for secondary sewage treatment; and physical, chemical,

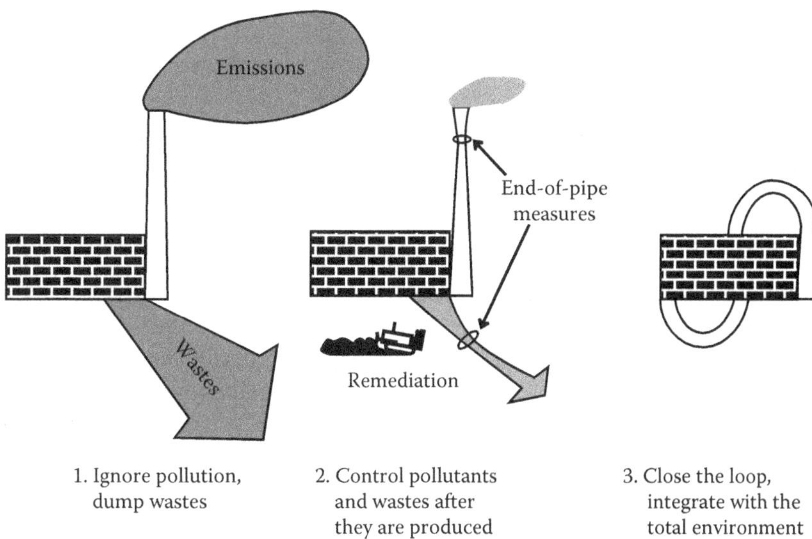

FIGURE 2.9 Steps in evolution of the anthrosphere to a more environmentally compatible form.

and biological processes for advanced (tertiary) sewage treatment. Such treatment measures are often very sophisticated and effective. Another kind of end-of-pipe treatment is the disposal of wastes in a supposedly safe place. In some cases, such as municipal solid wastes, radioactive materials, hazardous chemicals, power plant ash, and contaminated soil, the disposal of sequestered wastes in a secure location is practiced as a direct treatment process. In other cases, including flue-gas desulfurization sludge, sewage sludge, and sludge from chemical treatment of industrial wastewater, disposal is practiced as an adjunct to other end-of-pipe measures. Waste disposal practices later found to be inadequate have spawned an entirely separate end-of-pipe treatment activity called *remediation* in which discarded wastes are dug up, sometimes subjected to additional treatment, and then placed in a more secure disposal site.

Although sometimes unavoidable, the production of pollutants followed by measures taken to control or remediate them to reduce the quantities and potential harmfulness of wastes are not very desirable. Such measures do not usually eliminate wastes and may, in fact, transfer a waste problem from one part of the environment to another. An example of this is the removal of air pollutants from stack gas and their disposal as wastes on land, where they have the potential to cause groundwater pollution. Clearly, it is now unacceptable to ignore pollution and to dump wastes, and the control of pollutants and wastes after they are produced is not a good permanent solution to waste problems. Therefore, it has become accepted practice to "close the loop" on industrial processes, recycling materials as much as possible and allowing only benign waste products to be released to the environment.

2.8.1 Anthrosphere and Industrial Ecology

"Closing the loop" on industrial processes is the basis of industrial ecology and one of the most effective ways of integrating the anthrosphere into the total environment. Discussed in detail in Chapter 17, *industrial ecology* is practiced when industrial enterprises interact in a mutually advantageous manner to produce goods and services with maximum efficiency and with minimum environmental impact. Although practiced in rudimentary forms ever since industrial production began, modern practice of industrial ecology dates from a 1989 article by Frosch and Gallopoulos.[3] Industrial ecology is carried out in *industrial ecosystems* analogous to natural ecosystems. The processing of matter and energy through industrial ecosystems is known as *industrial metabolism*. The efficient practice of industrial ecology is a crucial aspect of modern environmental chemistry.

2.9 GREEN CHEMISTRY

As the modern environmental movement developed since around 1970, enormous progress has been made in environmental quality. This has been done largely through a *command and control* approach based on laws and regulations. Many of the measures taken to reduce pollution have been "end-of-pipe" measures in which water and air pollutants were produced but removed before discharge to the environment. However, in those countries with good, well-enforced pollution control regulations, most of the easy measures in pollution control have been taken and small additional reductions in pollutant emissions require high expenditures. Furthermore, enforcement is a continuous, expensive, and often litigious challenge.

It has become obvious that, to the extent possible, systems are needed that are inherently nonpolluting and sustainable. Starting in the 1990s, this need has been approached through a system known as green chemistry. *Green chemistry* can be defined as *the sustainable, safe, and nonpolluting practice of chemical science and manufacturing in a manner that consumes minimum amounts of materials and energy while producing little or no waste material.* In a word, green chemistry is *sustainable chemistry.*[4] Green chemistry is based on "the 12 principles of green chemistry"[5] addressed below.

1. Minimize or eliminate the need for waste cleanup by emphasizing waste prevention.
2. To the extent possible, all materials involved in making a product should be incorporated into it. This rule involves the key concept of atom economy discussed below.
3. Avoid the use and generation of hazardous substances that may harm humans or the environment.
4. Design and use chemical products that have minimum toxicity.
5. Minimize or eliminate the use of auxiliary substances that do not become part of the final product. Solvents are examples of such substances that should be avoided if possible.
6. Minimize energy consumption.
7. Use renewable raw materials instead of depletable feedstocks. For example, biomass raw material, which can be produced renewably by plants, is preferable to petroleum, of which there is a finite supply.
8. In organic synthesis, the use of protecting groups should be avoided because the material used in such groups does not become part of the final product.
9. Reagents should be chosen for maximum selectivity of function.
10. Products that will be released to the environment or discarded as wastes should break down rapidly to innocuous materials.
11. Manufacturing processes should be monitored and controlled in-process and in real time with appropriate computerized systems.
12. Processes and materials that are likely to cause extreme temperatures or pressures or unforeseen incidents such as explosions, runaway reactions, and fires should be avoided.

2.9.1 GREEN SYNTHETIC CHEMISTRY

Much of what is known as green chemistry to date has been applied to *synthetic chemistry*, which is involved with the production of new and existing chemical products. In recent times, the costs of making chemical products have come to include substantial expenses beyond those associated with raw materials, energy, production, and marketing. Among these added costs are those of regulatory compliance, treatment and disposal of waste by-products, liability, and, more recently, operation of security measures to provide protection from the threats of terrorism. Ideally, the practice of green chemistry greatly reduces these added costs by avoiding hazardous feedstocks and catalysts, eliminating the generation of dangerous intermediates and by-products, and avoiding severe conditions that might lead to hazards.

A key concept of green synthetic chemistry is that of high *atom economy*. In conventional synthesis, a high yield is obtained when the reaction making the product goes to completion. But even with 100% yield, by-products inherent to the synthesis may be generated posing disposal problems. *Atom economy* refers to the fraction of all reagents that go into the final product. Ideally, all materials that go into a synthetic process would be incorporated into the product, resulting in 100% atom economy.

2.9.2 Risk Reduction

Risk reduction is an important goal of any manufacturing process and the products that are generated and marketed. In all cases risks should be minimized to workers, the surrounding community, customers, and the environment as a whole. Risk is a function of *hazard* and *exposure*:

$$\text{Risk} = F\{\text{hazard} \times \text{exposure}\} \quad (2.4)$$

In past times, emphasis has been placed on reduction of exposure. For example, hazardous still bottoms from the distillation of a toxic insecticide have been placed in "secure" landfills to prevent exposure to the environment. The practice of green chemistry reduces the hazard by substituting insecticides whose synthesis does not generate such hazardous by-products.

2.9.3 Specific Aspects of Green Chemistry

During its brief lifetime as a recognized discipline, green chemistry has seen the development of several specific kinds of processes and products that can fairly be deemed to be "green." Since the discipline is still developing, new aspects are recognized each year. Several established specific aspects of green chemistry are those listed below.

1. *Chemical transformations under mild conditions*: Many chemical processes are carried out under conditions of high temperature and high pressure using potentially hazardous catalysts and media (solvents) that may present hazards. Enzymatic transformations using biological catalysts (enzymes) must occur under mild conditions and, wherever possible, are being substituted for more established harsh conditions.
2. *Green catalysts*: These are catalysts with maximum specificity and efficiency that do not themselves pose major hazards. Appropriate catalysts reduce energy and material requirements, decrease the need for separations because of their higher selectivity, and permit the use of less toxic reagents, such as H_2O_2, in place of more hazardous reagents, such as those containing heavy metals.
3. *Solventless processes*: Since many of the hazards and environmental problems associated with chemical processes arise from the use of solvents, processes that do not use solvents are favored.
4. *Less dangerous and polluting solvents*: Benign solvents, especially water, are favored for operations that require solvents.
5. *Supercritical fluids*: Supercritical carbon dioxide often provides good reaction media and makes product separation easier.
6. *Process intensification*: Small reactors such as channeled microreactors, spinning disc reactors, and tube-in-tube reactors can concentrate a chemical process in a small volume reducing the hazard posed by reacting large quantities of material in large-volume containers.
7. *Electricity*: In some cases, addition of electrons (reduction) and removal of electrons (oxidation) can be accomplished with electricity, a mass-less reagent that is cheap (only a few cents per mole) and easy to transport.

8. *Renewable feedstocks*: The chemical industry is based largely on petroleum feedstocks, which inevitably will become much less available and in short supply. Biomass sources from plants are being developed to substitute for petroleum and natural gas raw materials.
9. *Design for degradability*: Products that will be released into the environment are being designed for maximum degradability so that any hazard that they might present is transient and eliminated by natural processes.
10. *Biodegradable polymers*: Polymers used in textiles, plastic bags, and a vast number of other applications are likely to be discarded into the environment. Therefore, these polymers should be biodegradable. Biochemically synthesized polymeric materials including those from carbohydrate feedstocks are especially amenable to biodegradation.

2.9.4 Three Undesirable Characteristics of Chemicals: Persistence, Bioaccumulation, and Toxicity

Green chemistry seeks to avoid substances that are most likely to cause problems in the environment, including those that (1) are persistent, (2) have a strong tendency to undergo bioaccumulation, and (3) are toxic. Basically, these are materials that stay around for a long time; accumulate in organisms, particularly the lipid tissues of animals; and once inside an organism are toxic to the host. Some of the once commonly used organohalide pesticides meet all three of these criteria. Certainly, DDT is persistent in the environment, undergoes biomagnification in the food chain, and is toxic, through interference with reproductive processes, to birds of prey at the top of the food chain. Insecticidal parathion is quite toxic but its danger to the environment is relatively lower because it is readily biodegraded.

Through years of careful study and often unfortunate experience, the persistence/bioaccumulation/toxicity (PBT) characteristics of a number of common chemicals have become well established. The body of knowledge gathered from studies of such chemicals combined with increasingly sophisticated tools relating chemical structures to characteristics of substances has led to new ways to predict which chemicals may be troublesome. The U.S. Environmental Protection Agency has published a computerized PBT Profiler designed to predict undesirable PBT behavior.[6] Although not a substitute for careful experimental study, this tool is helpful in warning of possible problems with new chemicals that may be introduced into the environment.

2.9.5 Green Chemistry and Environmental Chemistry

Green chemistry and environmental chemistry are closely linked. In order to develop green chemical processes, environmental chemistry must be carefully considered to know how these processes and their products will affect the environment. Environmental chemistry's goal of environmental protection can best be accomplished by the implementation of green chemistry. Green chemistry is closely linked with the practice of industrial ecology; both of these topics are discussed in more detail in Chapter 17.

LITERATURE CITED

1. Manahan, S. E., *Toxicological Chemistry and Biochemistry*, Lewis Publishers/CRC Press, Boca Raton, FL, 2002.
2. Anastas, P. T. and J. B. Zimmerman, Design through the twelve principles of green engineering, *Environmental Science and Technology*, **37**, 95A–101A, 2003.
3. Frosch, R. A. and N. E. Gallopoulos, Strategies for manufacturing, *Scientific American*, **261**, 94–102, 1989.
4. Lankey, R. and P. Anastas, *Advancing Sustainability through Green Chemistry and Engineering*, Oxford University Press, Oxford, 2003.
5. Anastas, P. T. and J. C. Warner, *Green Chemistry Theory and Practice*, Oxford University Press, Oxford, 1998.
6. The profiler is available on the web at http://www.epa.gov/oppt/sf/tools/pbtprofiler.htm

SUPPLEMENTARY REFERENCES

Anastas, P., Ed., *Handbook of Green Chemistry*, Wiley-VCH, New York, 2010.

Baird, C. and M. Cann, *Environmental Chemistry*, 4th ed., W. H. Freeman, New York, 2008.

Boehnke, D. N. and R. D. Delumyea, *Laboratory Experiments in Environmental Chemistry*, Prentice Hall, Upper Saddle River, NJ, 1999.

Chen, W.-K., *Computer Aided Design and Design Automation*, CRC Press, Boca Raton, FL, 2009.

Clark, J. H. and D. Macquarrie, Eds, *Handbook of Green Chemistry and Technology*, Blackwell Science Inc, Malden, MA, 2002.

Connell, D. W., *Basic Concepts of Environmental Chemistry*, 2nd ed., Taylor & Francis/CRC Press, Boca Raton, FL, 2005.

Doble, M. and A. K. Kruthiventi, *Green Chemistry and Processes*, Academic Press, Amsterdam, 2007.

Doxsee, K. and J. Hutchison, *Green Organic Chemistry: Strategies, Tools, and Laboratory Experiments*, Brooks Cole, Monterey, CA, 2003.

Graedel, T. E. and B. R. Allenby, *Industrial Ecology*, 2nd ed., Prentice Hall, Upper Saddle River, NJ, 2002.

Hanrahan, G., *Environmental Chemometrics: Principles and Modern Applications*, Taylor & Francis/CRC Press, Boca Raton, FL, 2009.

Harrison, R. M., *Principles of Environmental Chemistry*, Royal Society of Chemistry, Cambridge, UK, 2007.

Hawken, P., A. Lovins, and L. H. Lovins, *Natural Capitalism: Creating the Next Industrial Revolution*, Back Bay Books, Boston, 2008.

Hites, R. A., *Elements of Environmental Chemistry*, Wiley, Hoboken, NJ, 2007.

Holtzapple, M. T. and W. D. Reece, *Concepts in Engineering*, 2nd ed., McGraw-Hill, Dubuque, IA, New York, 2008.

Ibanez, J. G., *Environmental Chemistry: Fundamentals*, Springer, Berlin, 2007.

Lancaster, M., *Green Chemistry*, Springer-Verlag, Berlin, 2002.

Lankey, R. L. and P. T. Anastas, Eds, *Advancing Sustainability Through Green Chemistry and Engineering*, American Chemical Society, Washington, DC, 2002.

Lichtfouse, E., J. Schwarzbauer, and R. Didier, Eds, *Environmental Chemistry: Green Chemistry and Pollutants in Ecosystems*, Springer, Berlin, 2005.

Matlack, A. S., *Introduction to Green Chemistry*, Marcel Dekker, New York, 2001.

Pond, R. J. and J. L. Rankinen, *Introduction to Engineering Technology*, 7th ed., Prentice Hall, Upper Saddle River, NJ, 2009.

Proakis, J. G. and M. Salehi, *Digital Communications*, 5th ed., McGraw-Hill, Boston, 2008.

Raven, P. H., L. R. Berg, and D. M. Hassenzahl, *Environment*, 6th ed., Wiley, Hoboken, NJ, 2008.

Robinson, P. and R. Gallo, Eds, *Environmental Chemistry Research Progress*, Nova Science Publishers, Hauppauge, NY, 2009.

Roukes, M. L. and S. Fritz, Eds, *Understanding Nanotechnology*, Warner Books, New York, 2002.

Sawyer, C. N., P. L. McCarty, and G. F. Parkin, *Chemistry for Environmental Engineering and Science*, 5th ed., McGraw-Hill, New York, 2002.

Schlesinger, W. H., *Biogeochemistry: An Analysis of Global Change*, 2nd ed., Academic Press, San Diego, 1997.

Schwarzenbach, R. P., P. M. Gschwend, and D. M. Imboden, *Environmental Organic Chemistry*, 2nd ed., Wiley, New York, 2002.

Sellers, K., *Nanotechnology and the Environment*, CRC Press/Taylor & Francis, Boca Raton, FL, 2008.

Van Loon, G. W. and S. J. Duffy, *Environmental Chemistry: A Global Perspective*, 2nd ed., Oxford University Press, Oxford, UK, 2005.

Warren, D., *Green Chemistry: A Teaching Resource*, Royal Society of Chemistry, London, 2002.

Withgott, J., *Essential Environment: The Science Behind the Stories*, 2nd ed., Pearson Benjamin Cummings, San Francisco, 2007.

Yen, T. F., *Chemical Processes for Environmental Engineering*, Imperial College Press, London, 2007.

Yen, T. F., *Environmental Chemistry: Chemistry of Major Environmental Cycles*, Imperial College Press, London, 2005.

Zeid, I., *Mastering CAD/CAM*, McGraw-Hill Higher Education, Boston, 2005.

QUESTIONS AND PROBLEMS

The use of Internet resources is assumed in answering any of the questions. These would include such things as constants and conversion factors as well as additional information needed to complete an answer.

1. Much of the Netherlands consists of land reclaimed from the sea that is actually below sea level as a result of dredging and dike construction. Discuss how this may relate to the anthrosphere and the other spheres of the environment.
2. With a knowledge of the chemical behavior of iron and copper, explain why copper was used as a metal long before iron, even though iron has some superior qualities for a number of applications.
3. How does engineering relate to basic science and to technology?
4. Suggest ways in which an inadequate infrastructure of a city may contribute to environmental degradation.
5. In what sense are automobiles not part of the infrastructure whereas trains are?
6. Discuss how the application of computers can make an existing infrastructure run more efficiently.
7. Although synthetic materials require relatively more energy and nonrenewable resources for their fabrication, how may it be argued that they are often the best choice from an environmental viewpoint for the construction of buildings?
8. What is a telecommuter society and what are its favorable environmental characteristics?
9. What are the major areas to consider with respect to information?
10. What was the greatest threat to farmland in the United States during the 1930s, and what was done to alleviate that threat? What is currently the greatest threat, and what can be done to alleviate it?
11. What is shown by the reaction below?

$$2\{CH_2O\} \rightarrow CO_2(g) + CH_4(g)$$

How is this process related to aerobic respiration?
12. Define cycles of matter and explain how the definition given relates to the definition of environmental chemistry.
13. What are the main features of the carbon cycle?
14. What is the CAD/CAM combination?
15. What are some of the parts of the anthrosphere that may be severely contaminated by human activities?
16. What largely caused or marked the change between the "relatively harmonious relationship between the anthrosphere and the rest of the environment," which characterized most of human existence on Earth, and the current situation in which the anthrosphere is a highly perturbing, potentially damaging influence? What does this have to do with the anthropocene, a term that may be looked up on the Internet?
17. What are the three major stages in the evolution of industry with respect to how it relates to the environment
18. How is an industrial facility based on the principles of industrial ecology similar to a natural ecological system?
19. Describe, with an example, if possible, what is meant by "end-of-pipe" measures for pollution control. Why are such measures sometimes necessary? Why are they relatively less desirable? What are the alternatives?
20. Discuss how at least one kind of air pollutant might become a water pollutant.
21. Suggest one or two examples of how technology, properly applied, can be "environmentally friendly."
22. Suggest a definition that incorporates both industrial ecology and green chemistry.
23. In what respects are biological transformations, such as those using bacteria, inherently green? Are there any circumstances under which such transformations might not be in keeping with the practice of green chemistry?

24. Consider the reaction scheme below in which stoichiometric amounts of reagents A and B were reacted to yield product and by-product that is generated as part of the synthesis reaction, with some of the reactants left over. The mass in kilograms of each material is given in parentheses. From the information given, calculate the percent yield and the percent atom economy.

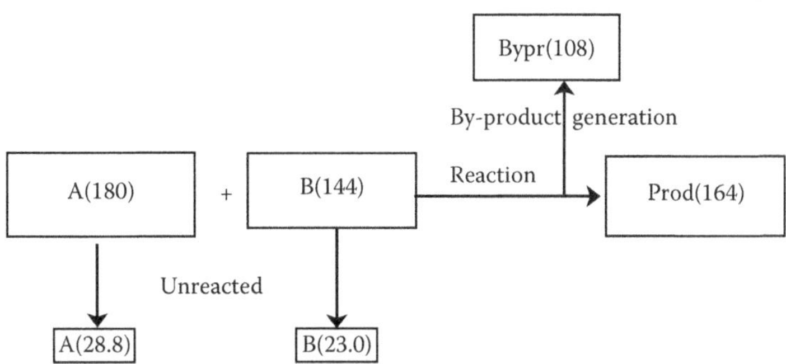

25. Equation 2.4 expresses risk as a function of hazard and exposure. Argue why reduction of hazard is inherently more desirable and foolproof than reduction of exposure. Suggest circumstances under which reduction of exposure might be the only viable alternative.
26. Describe the role of organisms in the nitrogen cycle.
27. Describe how the oxygen cycle is closely related to the carbon cycle.
28. In what important respect does the phosphorus cycle differ from cycles of other similar elements such as nitrogen and sulfur?
29. Suppose that each of the molecules in a mole of "compound *X*" absorbed a photon of ultraviolet radiation from the sun of wavelength 300 nm to become an energetically "excited" species (see Chapter 9). Calculate the energy in joules absorbed.

3 Fundamentals of Aquatic Chemistry

3.1 IMPORTANCE OF WATER

Throughout history, the quality and quantity of water available to humans have been vital factors in determining their well-being. Whole civilizations have disappeared because of water shortages resulting from changes in climate. Even in temperate climates, fluctuations in precipitation cause problems. Devastating droughts and destructive floods are frequent occurrences in many areas of the world.

Waterborne diseases such as cholera and typhoid killed millions of people in the past and still cause great misery in less developed countries. Ambitious programs of dam and dike construction have reduced flood damage, but they have had a number of undesirable side effects in some areas, such as inundation of farmland by reservoirs and failure of unsafe dams. Globally, problems with quantity and quality of water supply remain and in some respects are becoming more serious. These problems include increased water use due to population growth, contamination of drinking water by improperly discarded hazardous wastes (see Chapter 20), and destruction of wildlife by water pollution.

Aquatic chemistry, the subject of this chapter, must consider water in rivers, lakes, estuaries, oceans, and underground, as well as the phenomena that determine the distribution and circulation of chemical species in natural waters. Its study requires some understanding of the sources, transport, characteristics, and composition of water. The chemical reactions that occur in water and the chemical species found in it are strongly influenced by the environment in which the water is found. The chemistry of water exposed to the atmosphere is quite different from that of water at the bottom of a lake. Microorganisms play an essential role in determining the chemical composition of water. Thus, in discussing water chemistry, it is necessary to consider the many general factors that influence this chemistry.

The study of water is known as *hydrology* and is divided into a number of subcategories. *Limnology* is the branch of the science dealing with the characteristics of freshwater including biological properties, as well as chemical and physical properties. *Oceanography* is the science of the ocean and its physical and chemical characteristics. The chemistry and biology of the Earth's vast oceans are unique because of the ocean's high salt content, great depth, and other factors.

Water and the hydrosphere are crucial in the processes involved with chemical fate and transport in the environment. These include the physical processes of volatilization, dissolution, precipitation, and uptake and release by sediments. The chemical processes involved with chemical fate and transport in water are chemical reactions that result in dissolution or precipitation, hydrolysis, complexation, oxidation–reduction, and photochemical reactions. These processes are very much influenced by biochemical phenomena such as bioaccumulation and magnification in food chains and biodegradation. The fate and transport of pollutants in the hydrosphere are very important (Chapter 7).

3.2 WATER: FROM MOLECULES TO OCEANS

3.2.1 SOURCES AND USES OF WATER: THE HYDROLOGICAL CYCLE

The world's water supply is found in the five parts of the hydrologic cycle (Figure 3.1). About 97% of the Earth's water is found in the oceans. Another fraction is present as water vapor in the

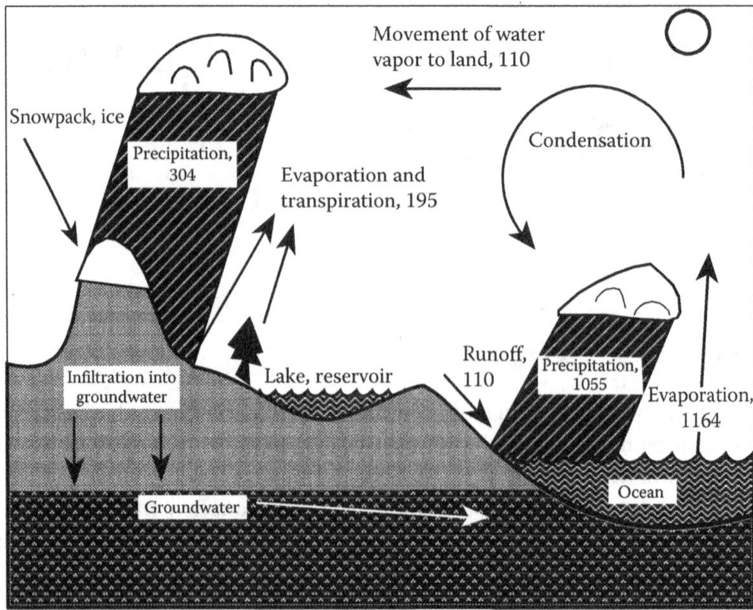

FIGURE 3.1 Hydrological cycle with quantities of water in trillions of liters per day.

atmosphere (clouds). Some water is contained in the solid state as ice and snow in snowpacks, glaciers, and the polar ice caps. Surface water is found in lakes, streams, and reservoirs. Groundwater is located in aquifers underground.

There is a strong connection between the hydrosphere, where water is found, and the lithosphere, which is that part of the geosphere accessible to water. Human activities affect both. For example, disturbance of land by conversion of grasslands or forests to agricultural land or intensification of agricultural production may reduce vegetation cover, decreasing transpiration (loss of water vapor by plants) and affecting the microclimate. The result is increased rain runoff, erosion, and accumulation of silt in bodies of water. The nutrient cycles may be accelerated, leading to nutrient enrichment of surface waters. This, in turn, can profoundly affect the chemical and biological characteristics of bodies of water.

The water that humans use is primarily fresh surface water and groundwater, the sources of which may differ from each other significantly. In arid regions, a small fraction of the water supply comes from the ocean, a source that will continue to increase as the world's supply of freshwater dwindles relative to demand. Saline or brackish groundwaters can also be utilized in some areas.

In the continental United States, an average of approximately 1.48×10^{13} L of water fall as precipitation each day, which translates to 76 cm/yr. Of that amount, approximately 1.02×10^{13} L/day, or 53 cm/yr, are lost by evaporation and transpiration. Thus, the water theoretically available for use is approximately 4.6×10^{12} L/day, or only 23 cm/yr. At present, the United States uses 1.6×10^{12} L/day, or 8 cm of the average annual precipitation. This amounts to an almost 10-fold increase from a usage of 1.66×10^{11} L/day in 1900. Even more striking is the per capita increase from about 40 L/day in 1900 to around 600 L/day now. Much of this increase is accounted for by high agricultural and industrial use, which each account for approximately 46% of total consumption. Municipal use consumes the remaining 8%.

Starting around 1980, however, the increase in water use in the United States slowed significantly. This trend, which is illustrated in Figure 3.2, has been attributed to the success of efforts to conserve water, especially in the industrial (including power generation) and agricultural sectors. Conservation and recycling have accounted for much of the decreased use in the industrial sector. Irrigation water has been used much more efficiently by replacing spray irrigators, which lose large quantities of water to the action of wind and to evaporation, with irrigation systems that apply water

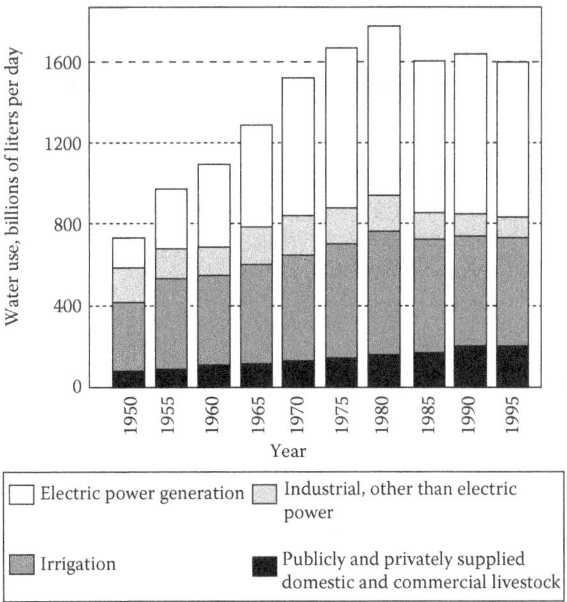

FIGURE 3.2 Trends in water use in the United States. (Data from U.S. Geological Survey.)

directly to soil. Trickle irrigation systems that apply just the amount of water needed directly to plant roots are especially efficient.

A major problem with water supply is its nonuniform distribution with location and time. As shown in Figure 3.3, precipitation falls unevenly in the continental United States. This causes difficulties because people in areas with low precipitation often consume more water than people in regions with more rainfall. Rapid population growth in the more arid southwestern states of the United States during recent decades has further aggravated the problem. Water shortages are becoming more acute in this region, which contains six of the nation's 11 largest cities (Los Angeles, Houston, Dallas, San Diego, Phoenix, and San Antonio). Other problem areas include Florida, where overdevelopment of coastal areas threatens Lake Okeechobee; the Northeast, plagued by deteriorating water systems; and the High Plains, ranging from the Texas panhandle to Nebraska,

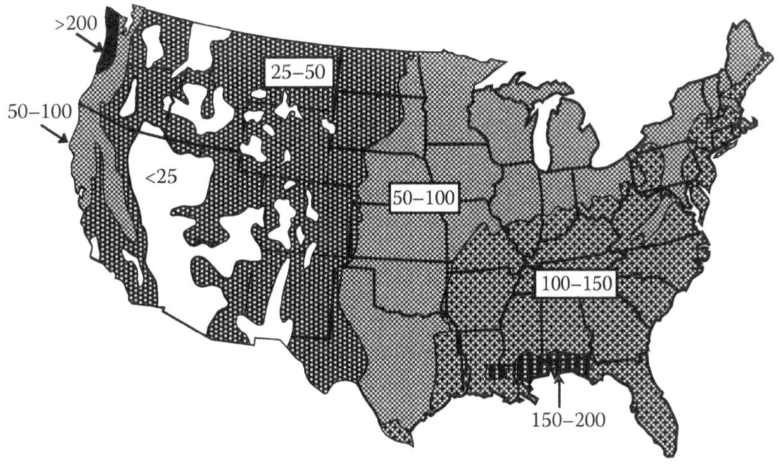

FIGURE 3.3 Distribution of precipitation in the continental United States, showing average annual rainfall in centimeters.

where irrigation demands on the Ogallala aquifer are causing a drop in the water table steadily with no hope of recharge. These problems are minor, however, in comparison to those in some parts of Africa where water shortages are contributing to real famine conditions.

3.2.2 Properties of Water, a Unique Substance

Water has a number of unique properties that are essential to life. Some of the special characteristics of water include its polar character, tendency to form hydrogen bonds, and ability to hydrate metal ions. These properties are listed in Table 3.1.

3.2.3 Water Molecule

Water's properties can best be understood by considering the structure and bonding of the water molecule shown in Figure 3.4. The water molecule is made up of two hydrogen atoms bonded to an oxygen atom. The three atoms are not in a straight line; instead, as shown above, they form an angle of 105°. Because of water's bent structure and the fact that the oxygen atom attracts the negative electrons more strongly than do the hydrogen atoms, the water molecule behaves like a dipole having opposite electrical charges at either end. The water dipole may be attracted to either positively or negatively charged ions. For example, when NaCl dissolves in water as positive Na^+ ions and negative Cl^- ions, the positive sodium ions are surrounded by water molecules with their negative ends pointed at the ions, and the chloride ions are surrounded by water molecules with their positive ends pointing at the negative ions, as shown in Figure 3.4. This kind of attraction for ions is the reason why water dissolves many ionic compounds and salts that do not dissolve in other liquids.

A second important characteristic of the water molecule is its ability to form hydrogen bonds. *Hydrogen bonds* are a special type of bond that can form between the hydrogen in one water molecule and the oxygen in another water molecule. This bonding takes place because the oxygen has a partial negative charge and the hydrogen a partial positive charge. Hydrogen bonds, shown in Figure 3.5 as dashed lines, hold the water molecules together in large groups.

Hydrogen bonds also help to hold some solute molecules or ions in solution. This happens when hydrogen bonds form between the water molecules and hydrogen, nitrogen, or oxygen atoms on the solute molecule (see Figure 3.5). Hydrogen bonding also aids in retaining extremely small particles called colloidal particles in suspension in water (see Section 5.4).

TABLE 3.1
Important Properties of Water

Property	Effects and Significance
Excellent solvent	Transport of nutrients and waste products, making biological processes possible in aqueous medium
Highest dielectric constant of any common liquid	High solubility of ionic substances and their ionization in solution
Higher surface tension than any other liquid	Controlling factor in physiology; governs drop and surface phenomena
Transparent to visible and longer wavelength fraction of ultraviolet light	Colorless so that light required for photosynthesis reaches considerable depths in bodies of water
Maximum density as a liquid at 4°C	Ice floats; vertical circulation in bodies of water
Higher heat of evaporation than any other material	Determines transfer of heat and water molecules between the atmosphere and bodies of water
Higher latent heat of fusion of any common liquid	Temperature stabilized at the freezing point of water
Higher heat capacity than any other liquid except liquid ammonia	Stabilization of temperatures of organisms and geographical regions

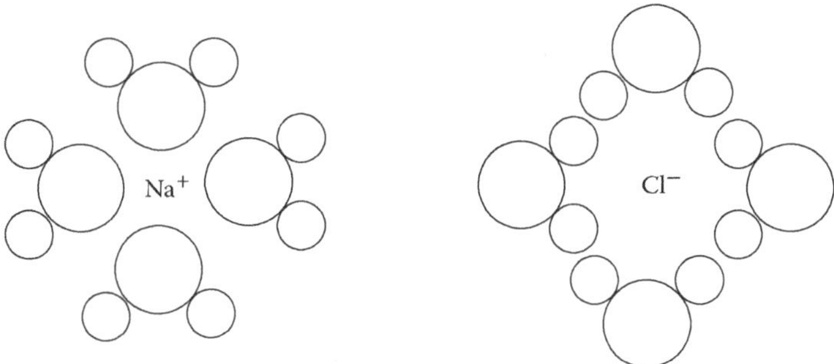

FIGURE 3.4 Polar water molecules surrounding Na^+ ion (left) and Cl^- ion (right).

Water is an excellent solvent for many materials; thus it is the basic transport medium for nutrients and waste products in life processes. The extremely high dielectric constant of water relative to other liquids has a profound effect upon its solvent properties in that most ionic materials are dissociated in water. With the exception of liquid ammonia, water has the highest heat capacity of any liquid or solid, 4.186 J/g/deg (1 cal/g/deg). Because of this high heat capacity, a relatively large amount of heat is required to change appreciably the temperature of a mass of water; hence, a body of water can have a stabilizing effect upon the temperature of nearby geographic regions. In addition, this property prevents sudden large changes of temperature in large bodies of water and thereby protects aquatic organisms from the shock of abrupt temperature variations. The extremely high heat of vaporization of water, 2446 J/g at 25°C, likewise stabilizes the temperature of bodies of water and the surrounding geographic regions. It also influences the transfer of heat and water vapor between bodies of water and the atmosphere. Water has its maximum density at 4°C, a temperature above its freezing point. The fortunate consequence of this fact is that ice floats, so that few large bodies of water ever freeze solid. Furthermore, the pattern of vertical circulation of water in lakes, a determining factor in their chemistry and biology, is governed largely by the unique temperature–density relationship of water.

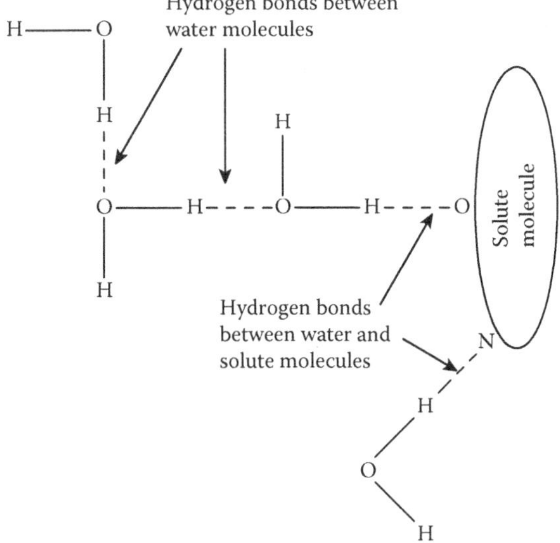

FIGURE 3.5 Hydrogen bonding between water molecules and between water molecules and a solute molecule in solution.

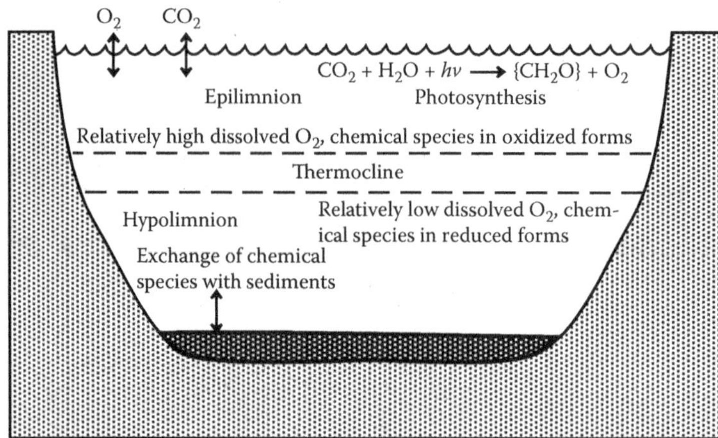

FIGURE 3.6 Stratification of a lake.

3.3 CHARACTERISTICS OF BODIES OF WATER

The physical condition of a body of water strongly influences the chemical and biological processes that occur in water. *Surface water* occurs primarily in streams, lakes, and reservoirs. *Wetlands* are flooded areas in which the water is shallow enough to enable growth of bottom-rooted plants. *Estuaries* are arms of the ocean into which streams flow. The mixing of fresh and salt water gives estuaries unique chemical and biological properties. Estuaries are the breeding grounds of much marine life, which makes their preservation very important.

Water's unique temperature–density relationship results in the formation of distinct layers within nonflowing bodies of water, as shown in Figure 3.6. During the summer a surface layer (*epilimnion*) is heated by solar radiation and, because of its lower density, floats upon the bottom layer, or *hypolimnion*. This phenomenon is called *thermal stratification*. When an appreciable temperature difference exists between the two layers, they do not mix but behave independently and have very different chemical and biological properties. Exposed to sunlight, the epilimnion may have a heavy growth of algae. As a result of exposure to the atmosphere and (during daylight hours) because of the photosynthetic activity of algae, the epilimnion contains relatively higher levels of dissolved oxygen (DO) and, generally, is aerobic. In the hypolimnion, bacterial action on biodegradable organic material may cause the water to become anaerobic (lacking DO). As a consequence, chemical species in a relatively reduced form tend to predominate in the hypolimnion.

The shear plane, or layer between epilimnion and hypolimnion, is called the *metalimnion* or *thermocline*. During autumn, when the epilimnion cools, a point is reached at which the temperatures of the epilimnion and hypolimnion are equal. This disappearance of thermal stratification causes the entire body of water to behave as a hydrological unit, and the resultant mixing is known as *overturn*. An overturn also generally occurs in the spring. During the overturn, the chemical and physical characteristics of the body of water become much more uniform, and a number of chemical, physical, and biological changes may result. Biological activity may increase from the mixing of nutrients. Changes in water composition during overturn may cause disruption in water treatment processes.

3.4 AQUATIC LIFE

The living organisms (*biota*) in an aquatic ecosystem may be classified as either autotrophic or heterotrophic. *Autotrophic* organisms utilize solar or chemical energy to fix elements from simple, nonliving inorganic material into complex life molecules that compose living organisms. Algae are the most important autotrophic aquatic organisms because they are *producers* that utilize solar energy to generate biomass from CO_2 and other simple inorganic species.

Heterotrophic organisms utilize the organic substances produced by autotrophic organisms as energy sources and as the raw materials for the synthesis of their own biomass. *Decomposers* (or *reducers*) are a subclass of the heterotrophic organisms and consist of chiefly bacteria and fungi, which ultimately break down material of biological origin to the simple compounds originally fixed by the autotrophic organisms.

The ability of a body of water to produce living material is known as its *productivity*. Productivity results from a combination of physical and chemical factors. High productivity requires an adequate supply of carbon (CO_2), nitrogen (nitrate), phosphorus (orthophosphate), and trace elements such as iron. Water of low productivity generally is desirable for water supply or for swimming. Relatively high productivity is required for the support of fish and to serve as the basis of the food chain in an aquatic ecosystem. Excessive productivity results in decay of the biomass produced, consumption of DO, and odor production, a condition called *eutrophication*.

Life forms higher than algae and bacteria—fish, for example—comprise a comparatively small fraction of the biomass in most aquatic systems. The influence of these higher life forms upon aquatic chemistry is minimal. However, aquatic life is strongly influenced by the physical and chemical properties of the body of water in which it lives. *Temperature*, *transparency*, and *turbulence* are the three main physical properties affecting aquatic life. Very low water temperatures result in very slow biological processes, whereas very high temperatures are fatal to most organisms. The transparency of water is particularly important in determining the growth of algae. Turbulence is an important factor in mixing processes and transport of nutrients and waste products in water. Some small organisms (*plankton*) depend upon water currents for their own mobility.

DO frequently is the key substance in determining the extent and kinds of life in a body of water. Oxygen deficiency is fatal to many aquatic animals such as fish. The presence of oxygen can be equally fatal to many kinds of anaerobic bacteria. *Biochemical oxygen demand*, *BOD*, discussed as a water pollutant in Section 7.9, refers to the amount of oxygen utilized when the organic matter in a given volume of water is degraded biologically.

Carbon dioxide is produced by respiratory processes in water and sediments and can also enter water from the atmosphere. Carbon dioxide is required for the photosynthetic production of biomass by algae and, in some cases, is a limiting factor. High levels of carbon dioxide produced by the degradation of organic matter in water can cause excessive algal growth and biomass productivity.

The salinity of water also determines the kinds of life forms present. Irrigation waters may pick up harmful levels of salt. Marine life obviously requires or tolerates salt water, whereas many freshwater organisms are intolerant of salt.

3.5 INTRODUCTION TO AQUATIC CHEMISTRY

To understand water pollution, it is first necessary to have an appreciation of chemical phenomena that occur in water. The remaining sections of this chapter discuss aquatic acid–base and complexation phenomena. Oxidation–reduction reactions and equilibria are discussed in Chapter 4, and the details of solubility calculations and interactions between liquid water and other phases are given in Chapter 5. The main categories of aquatic chemical phenomena are illustrated in Figure 3.7.

Aquatic environmental chemical phenomena involve processes familiar to chemists, including acid–base, solubility, oxidation–reduction, and complexation reactions.[1] Although most aquatic chemical phenomena are discussed here from the thermodynamic (equilibrium) viewpoint, it is important to keep in mind that reaction rates (kinetics) are very important in aquatic chemistry. Biological processes play a key role in aquatic chemistry. For example, algae undergoing photosynthesis can raise the pH of water by removing aqueous CO_2, thereby converting an HCO_3^- ion to a CO_3^{2-} ion; this ion in turn reacts with Ca^{2+} in water to precipitate $CaCO_3$.

Compared with the carefully controlled conditions of the laboratory, it is much more difficult to describe chemical phenomena in natural water systems. Such systems are very complex and a description of their chemistry must take many variables into consideration. In addition to water, these systems contain

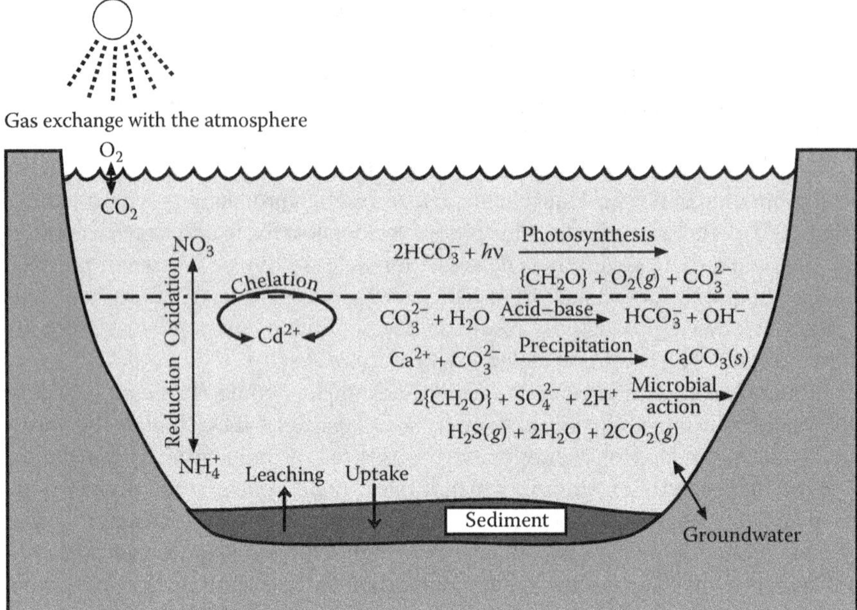

FIGURE 3.7 Major aquatic chemical processes.

mineral phases, gas phases, and organisms. As open, dynamic systems, they have variable inputs and outputs of energy and mass. Therefore, except under unusual circumstances, a true equilibrium condition is not obtained, although an approximately steady-state aquatic system frequently exists. Most metals found in natural waters do not exist as simple hydrated cations in the water, and oxyanions often are found as polynuclear species, rather than as simple monomers. The nature of chemical species in water containing bacteria or algae is strongly influenced by the action of these organisms. Thus, an exact description of the chemistry of a natural water system based upon acid–base, solubility, and complexation equilibrium constants, redox potential, pH, and other chemical parameters is not possible. Therefore, the systems must be described by simplified *models*, often based around equilibrium chemical concepts. Though not exact, nor entirely realistic, such models can yield useful generalizations and insights pertaining to the nature of aquatic chemical processes, and provide guidelines for the description and measurement of natural water systems. Though greatly simplified, such models are very helpful in visualizing the conditions that determine chemical species and their reactions in natural waters and wastewaters.

3.6 GASES IN WATER

Dissolved gases—O_2 for fish and CO_2 for photosynthetic algae—are crucial to the welfare of living species in water. Some gases in water can also cause problems, such as the death of fish from bubbles of nitrogen formed in the blood caused by exposure to water supersaturated with N_2. Volcanic carbon dioxide evolved from supersaturated dissolved CO_2 in the waters of Lake Nyos in the African country of Cameroon asphyxiated 1700 people in 1986.

The solubilities of gases in water are calculated with *Henry's law*, which states that *the solubility of a gas in a liquid is proportional to the partial pressure of that gas in contact with the liquid.* These calculations are discussed in some detail in Chapter 5.

3.6.1 Oxygen in Water

Without an appreciable level of DO, many kinds of aquatic organisms cannot exist in water. DO is consumed by the degradation of organic matter in water. Many fish kills are caused not from the direct toxicity of pollutants but from a deficiency of oxygen because of its consumption in the biodegradation of pollutants.

Most elemental oxygen comes from the atmosphere, which is 20.95% oxygen by volume of dry air. Therefore, the ability of a body of water to re-oxygenate itself by contact with the atmosphere is an important characteristic. Oxygen is produced by the photosynthetic action of algae, but this process is really not an efficient means of oxygenating water because some of the oxygen formed by photosynthesis during the daylight hours is lost at night when the algae consume oxygen as part of their metabolic processes. When the algae die, the degradation of their biomass also consumes oxygen.

The solubility of oxygen in water depends upon water temperature, the partial pressure of oxygen in the atmosphere, and the salt content of the water. It is important to distinguish between oxygen *solubility*, the maximum dissolved O_2 concentration at equilibrium, and DO *concentration*, which is generally not the equilibrium concentration and is limited by the rate at which oxygen dissolves. The calculation of oxygen solubility as a function of partial pressure is discussed in Section 5.3, where it is shown that the concentration of oxygen in water at 25°C in equilibrium with air at atmospheric pressure is only 8.32 mg/L. Thus, water in equilibrium with air cannot contain a high level of DO compared with many other solute species. If oxygen-consuming processes are occurring in the water, the DO level may rapidly approach zero unless some efficient mechanism for the reaeration of water is operative, such as turbulent flow in a shallow stream or air pumped into the aeration tank of an activated sludge secondary waste treatment facility (see Chapter 8). The problem becomes largely one of kinetics, in which there is a limit to the rate at which oxygen is transferred across the air–water interface. This rate depends upon turbulence, air bubble size, temperature, and other factors.

If organic matter of biological origin is represented by the formula $\{CH_2O\}$, the consumption of oxygen in water by the degradation of organic matter may be expressed by the following biochemical reaction:

$$\{CH_2O\} + O_2 \rightarrow CO_2 + H_2O \tag{3.1}$$

The mass of organic material required to consume the 8.3 mg of O_2 in a liter of water in equilibrium with the atmosphere at 25°C is given by a simple stoichiometric calculation based on Equation 3.1, which yields a value of 7.8 mg of $\{CH_2O\}$. Thus, the microorganism-mediated degradation of only 7 or 8 mg of organic material can completely consume the O_2 in 1 L of water initially saturated with air at 25°C. The depletion of oxygen to levels below those that sustain oxic organisms requires the degradation of even less organic matter at higher temperatures (where the solubility of oxygen is less) or in water not initially saturated with atmospheric oxygen. Furthermore, there are no common aquatic chemical reactions that replenish DO; except for oxygen provided by photosynthesis, it must come from the atmosphere.

The temperature effect on the solubility of gases in water is especially important in the case of oxygen. The solubility of oxygen in water in equilibrium with atmospheric air decreases from 14.74 mg/L at 0°C to 7.03 mg/L at 35°C. At higher temperatures, the decreased solubility of oxygen, combined with the increased respiration rate of aquatic organisms, frequently causes a condition in which a higher demand for oxygen accompanied by lower solubility of the gas in water results in severe oxygen depletion.

3.7 WATER ACIDITY AND CARBON DIOXIDE IN WATER

Acid–base phenomena in water involve loss and acceptance of H^+ ion. Many species act as *acids* in water by releasing H^+ ion, while others act as *bases* by accepting H^+, and the water molecule itself does both. An important species in the acid–base chemistry of water is bicarbonate ion, HCO_3^-, which may act as either an acid or a base:

$$HCO_3^- \rightleftarrows CO_3^{2-} + H^+ \tag{3.2}$$

$$HCO_3^- + H^+ \rightleftarrows CO_2(aq) + H_2O \tag{3.3}$$

Acidity as applied to natural water and wastewater is the capacity of the water to neutralize OH^-; it is analogous to alkalinity, the capacity to neutralize H^+, which is discussed in the next section. Although virtually all water has some alkalinity, acidic water is not frequently encountered, except in cases of severe pollution. Acidity generally results from the presence of weak acids, particularly CO_2, but sometimes includes others such as $H_2PO_4^-$, H_2S, proteins, and fatty acids. Acidic metal ions, particularly Fe^{3+}, may also contribute to acidity.

Strong acids are the most important contributors to water pollutant acidity. The term *free mineral acid* is applied to strong acids such as H_2SO_4 and HCl in water. Acid mine water is a common water pollutant that contains an appreciable concentration of free mineral acid. Whereas total acidity is determined by titration with base to the phenolphthalein end point (pH 8.2), free mineral acid is determined by titration with base to the methyl orange end point (pH 4.3).

As shown by the hydrolysis of hydrated Al^{3+} below, the acidic character of some hydrated metal ions may contribute to acidity. Some industrial wastes, such as spent steel pickling liquor, contain acidic metal ions and often some excess strong acid. The acidity of such wastes must be measured in calculating the amount of lime or other chemicals required to neutralize the acid.

$$Al(H_2O)_6^{3+} \rightleftarrows Al(H_2O)_5OH^{2+} + H^+ \tag{3.4}$$

3.7.1 Carbon Dioxide in Water

The most important weak acid in water is carbon dioxide, CO_2. Because of the presence of carbon dioxide in air and its production from microbial decay of organic matter, dissolved CO_2 is present in virtually all natural waters and wastewaters. Rainfall from even an absolutely unpolluted atmosphere is slightly acidic due to the presence of dissolved CO_2. Dissolution in seawater is an important mechanism for the reduction of atmospheric carbon dioxide, the "greenhouse" gas that makes the greatest contribution to climate warming. Proposals have been made to pump CO_2 from combustion into places in the ocean where seawater is sinking to sequester it from the atmosphere for thousands of years.[2]

Carbon dioxide, and its ionization products, bicarbonate ion (HCO_3^-), and carbonate ion (CO_3^{2-}), have an extremely important influence upon the chemistry of water. Many minerals are deposited as salts of the carbonate ion. Algae in water utilize dissolved CO_2 in the synthesis of biomass. The equilibrium of dissolved CO_2 with gaseous carbon dioxide in the atmosphere,

$$CO_2(\text{water}) \rightleftarrows CO_2(\text{atmosphere}) \tag{3.5}$$

and the equilibrium of CO_3^{2-} ion between aquatic solution and solid carbonate minerals,

$$MCO_3(\text{slightly soluble carbonate salt}) \rightleftarrows M^{2+} + CO_3^{2-} \tag{3.6}$$

have a strong buffering effect upon the pH of water.

Carbon dioxide is only about 0.039% by volume of normal dry air. As a consequence of the low level of atmospheric CO_2, water totally lacking in alkalinity (capacity to neutralize H^+, see Section 3.8) in equilibrium with the atmosphere contains only a very low level of carbon dioxide. However, the formation of HCO_3^- and CO_3^{2-} greatly increases the solubility of carbon dioxide. High concentrations of free carbon dioxide in water may adversely affect respiration and gas exchange of aquatic animals. It may even cause death and should not exceed levels of 25 mg/L in water.

A large share of the carbon dioxide found in water is a product of the breakdown of organic matter by bacteria. Even algae, which utilize CO_2 in photosynthesis, produce it through their metabolic processes in the absence of light. As water seeps through layers of decaying organic matter while infiltrating the ground, it may dissolve a great deal of CO_2 produced by the respiration of organisms in the soil. Later, as water goes through limestone formations, it dissolves calcium carbonate because of the presence of the dissolved CO_2:

$$CaCO_3(s) + CO_2(aq) + H_2O \rightleftarrows Ca^{2+} + 2HCO_3^- \tag{3.7}$$

This process is the one by which limestone caves are formed. The implications of the above reaction for aquatic chemistry are discussed in greater detail in Section 3.9.

The concentration of gaseous CO_2 in the atmosphere varies with location and season; it is increasing by about one part per million (ppm) by volume per year. For purposes of calculation here, the concentration of atmospheric CO_2 will be taken as 390 ppm (0.0390%) in dry air. At 25°C, water in equilibrium with unpolluted air containing 390 ppm carbon dioxide has a $CO_2(aq)$ concentration of 1.276×10^{-5} M (see Henry's law calculation of gas solubility in Section 5.3), and this value will be used for subsequent calculations.

Although CO_2 in water is often represented as H_2CO_3, the equilibrium constant for the reaction

$$CO_2(aq) + H_2O \rightleftarrows H_2CO_3 \tag{3.8}$$

is only around 2×10^{-3} at 25°C, so just a small fraction of the dissolved carbon dioxide is actually present as H_2CO_3. In this text, nonionized carbon dioxide in water will be designated simply as CO_2, which in subsequent discussions will stand for the total of dissolved molecular CO_2 and undissociated H_2CO_3.

The CO_2–HCO_3^-–CO_3^{2-} system in water may be described by the equations,

$$CO_2(aq) + H_2O \rightleftarrows H^+ + HCO_3^- \tag{3.9}$$

$$K_{a1} = \frac{[H^+][HCO_3^-]}{[CO_2]} = 4.45 \times 10^{-7}, \quad pK_{a1} = 6.35 \tag{3.10}$$

$$HCO_3^- \rightleftarrows H^+ + CO_3^{2-} \tag{3.11}$$

$$K_{a2} = \frac{[H^+][CO_3^{2-}]}{[HCO_3^-]} = 4.69 \times 10^{-11}, \quad pK_{a2} = 10.33 \tag{3.12}$$

where $pK_a = -\log K_a$. The predominant species formed by CO_2 dissolved in water depends upon pH. This is best shown by a distribution of species diagram with pH as a master variable as illustrated in Figure 3.8. Such a diagram shows the major species present in solution as a function of pH. For CO_2 in aqueous solution, the diagram is a series of plots of the fractions present as CO_2, HCO_3^-, and CO_3^{2-} as a function of pH. These fractions, designated as α_x, are given by the following expressions:

$$\alpha_{CO_2} = \frac{[CO_2]}{[CO_2] + [HCO_3^-] + [CO_3^{2-}]} \tag{3.13}$$

$$\alpha_{HCO_3^-} = \frac{[HCO_3^-]}{[CO_2] + [HCO_3^-] + [CO_3^{2-}]} \tag{3.14}$$

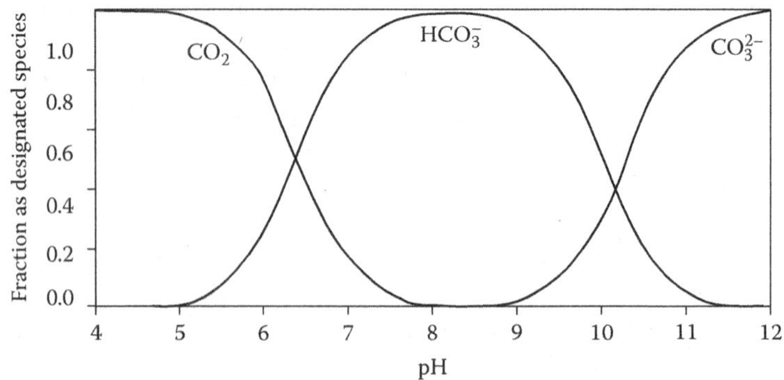

FIGURE 3.8 Distribution of species diagram for the CO_2–HCO_3^-–CO_3^{2-} system in water.

$$\alpha_{CO_3^{2-}} = \frac{[CO_3^{2-}]}{[CO_2] + [HCO_3^-] + [CO_3^{2-}]} \tag{3.15}$$

Substitution of the expressions for K_{a1} and K_{a2} into the α expressions gives the fractions of species as a function of acid dissociation constants and hydrogen ion concentration:

$$\alpha_{CO_2} = \frac{[H^+]^2}{[H^+]^2 + K_{a1}[H^+] + K_{a1}K_{a2}} \tag{3.16}$$

$$\alpha_{HCO_3^-} = \frac{K_{a1}[H^+]}{[H^+]^2 + K_{a1}[H^+] + K_{a1}K_{a2}} \tag{3.17}$$

$$\alpha_{CO_3^{2-}} = \frac{K_{a1}K_{a2}}{[H^+]^2 + K_{a1}[H^+] + K_{a1}K_{a2}} \tag{3.18}$$

Calculations from these expressions show the following:

- For pH significantly below pK_{a1}, α_{CO_2} is essentially 1
- When $pH = pK_{a1}$, $\alpha_{CO_2} = \alpha_{HCO_3^-}$
- When $pH = \frac{1}{2}(pK_{a1} + pK_{a2})$, $\alpha_{HCO_3^-}$ is at its maximum value of 0.98
- When $pH = pK_{a2}$, $\alpha_{HCO_3^-} = \alpha_{CO_3^{2-}}$
- For pH significantly above pK_{a2}, $\alpha_{CO_3^{2-}}$ is essentially 1

The distribution of species diagram in Figure 3.8 shows that hydrogen carbonate (bicarbonate) ion (HCO_3^-) is the predominant species in the pH range found in most waters, with CO_2 predominating in more acidic waters.

As mentioned above, the value of [CO_2 (*aq*)] in water at 25°C in equilibrium with air that is 390 ppm CO_2 is 1.276×10^{-5} M. The carbon dioxide dissociates partially in water to produce equal concentrations of H^+ and HCO_3^-:

$$CO_2 + H_2O \rightleftarrows HCO_3^- + H^+ \tag{3.19}$$

The concentrations of H^+ and HCO_3^- are calculated from K_{a1}:

$$K_{a1} = \frac{[H^+][HCO_3^-]}{[CO_2]} = \frac{[H^+]^2}{1.276 \times 10^{-5}} = 4.45 \times 10^{-7} \tag{3.20}$$

$$[H^+] = [HCO_3^-] = (1.276 \times 10^{-5} \times 4.45 \times 10^{-7})^{1/2} = 2.38 \times 10^{-6}$$
$$pH = 5.62$$

This calculation explains why pure water that has equilibrated with the unpolluted atmosphere is slightly acidic with a pH somewhat below 7.

3.8 ALKALINITY

The capacity of water to accept H^+ ions (protons) is called *alkalinity*. Alkalinity is important in water treatment and in the chemistry and biology of natural waters. Frequently, the alkalinity of water must be known to calculate the quantities of chemicals to be added in treating the water. Highly alkaline water often has a high pH and generally contains elevated levels of dissolved solids. These characteristics may be detrimental for water to be used in boilers, food processing, and municipal water systems. Alkalinity serves as a pH buffer and reservoir for inorganic carbon, thus helping us to determine the ability of a water to support algal growth and other aquatic life, so it can be used as a measure of water fertility. Generally, the basic species responsible for alkalinity in water are bicarbonate ion, carbonate ion, and hydroxide ion:

$$HCO_3^- + H^+ \rightleftarrows CO_2 + H_2O \tag{3.21}$$

$$CO_3^{2-} + H^+ \rightleftarrows HCO_3^- \tag{3.22}$$

$$OH^- + H^+ \rightleftarrows H_2O \tag{3.23}$$

Other, usually minor, contributors to alkalinity are ammonia and the conjugate bases of phosphoric, silicic, boric, and organic acids.

At pH values below 7, $[H^+]$ in water detracts significantly from alkalinity, and its concentration must be subtracted in computing the total alkalinity. Therefore, the following equation is the complete equation for alkalinity in a medium where the only contributors to it are HCO_3^-, CO_3^{2-}, and OH^-:

$$[alk] = [HCO_3^-] + 2[CO_3^{2-}] + [OH^-] - [H^+] \tag{3.24}$$

Alkalinity generally is expressed as *phenolphthalein alkalinity*, corresponding to titration with acid to the pH at which HCO_3^- is the predominant carbonate species (pH 8.3), or *total alkalinity*, corresponding to titration with acid to the methyl orange end point (pH 4.3), where both bicarbonate and carbonate species have been converted to CO_2.

It is important to distinguish between high *basicity*, manifested by an elevated pH, and high *alkalinity*, the capacity to accept H^+. Whereas pH is an *intensity* factor, alkalinity is a *capacity* factor. This may be illustrated by comparing a solution of 1.00×10^{-3} M NaOH with a solution of 0.100 M $NaHCO_3$. The sodium hydroxide solution is very basic, pH 11, but a liter of it will neutralize only 1.00×10^{-3} mol of acid. The pH of the $NaHCO_3$ solution is 8.34, much lower than that of the NaOH. However, a liter of the sodium bicarbonate solution will neutralize 0.100 mol of acid; therefore, its alkalinity is 100 times that of the more basic NaOH solution.

In engineering terms, alkalinity frequently is expressed in units of mg/L of $CaCO_3$, based upon the following acid-neutralizing reaction:

$$CaCO_3 + 2H^+ \rightleftarrows Ca^{2+} + CO_2 + H_2O \tag{3.25}$$

The equivalent mass of calcium carbonate is one-half its formula mass. Expressing alkalinity in terms of mg/L of $CaCO_3$ can, however, lead to confusion, and the preferable notation for the chemist is equivalents/L (eq/L), the number of moles of H^+ neutralized by the alkalinity in a liter of solution.

3.8.1 Contributors to Alkalinity at Different pH Values

Natural water typically has an alkalinity ("[alk]") of 1.00×10^{-3} eq/L, meaning that the alkaline solutes in 1 L of the water will neutralize 1.00×10^{-3} mol of acid. The contributions made by different species to alkalinity depend upon pH. This is shown here by calculation of the relative contributions to alkalinity of HCO_3^-, CO_3^{2-}, and OH^- at pH 7.0 and 10.0. First, for water at pH 7.0, $[OH^-]$ is too low to make any significant contribution to the alkalinity. Furthermore, as shown by Figure 3.8, at pH 7.0, $[HCO_3^-] \gg [CO_3^{2-}]$. Therefore, the alkalinity is due to HCO_3^- and $[HCO_3^-] = 1.00 \times 10^{-3}$ M. Substitution into the expression for K_{a1} shows that at pH 7.0 and $[HCO_3^-] = 1.00 \times 10^{-3}$ M, the value of $[CO_2\,(aq)]$ is 2.25×10^{-4} M, higher than the value that arises from water in equilibrium with atmospheric air, but readily reached due to the presence of carbon dioxide from bacterial decay in water and sediments.

Consider next the case of water with the same alkalinity, 1.00×10^{-3} eq/L but with a pH of 10.0. At this higher pH both OH^- and CO_3^{2-} are present at significant concentrations compared with HCO_3^- and the following may be calculated:

$$[\text{alk}] = [HCO_3^-] + 2[CO_3^{2-}] + [OH^-] = 1.00 \times 10^{-3} \tag{3.26}$$

The concentration of CO_3^{2-} is multiplied by 2 because each CO_3^{2-} ion can neutralize two H^+ ions. The other two equations that must be solved to get the concentrations of HCO_3, CO_3^{2-}, and OH^- are

$$[OH^-] = \frac{K_W}{[H^+]} = \frac{1.00 \times 10^{-14}}{1.00 \times 10^{-10}} = 1.00 \times 10^{-4} \tag{3.27}$$

and

$$[CO_3^{2-}] = \frac{K_{a2}[HCO_3^-]}{[H^+]} \tag{3.28}$$

Solving these three equations gives $[HCO_3^-] = 4.64 \times 10^{-4}$ M and $[CO_3^{2-}] = 2.18 \times 10^{-4}$ M, so the contributions to the alkalinity of this solution are the following:

$$\begin{aligned} & 4.64 \times 10^{-4} \text{ eq/L from } HCO_3^- \\ 2 \times 2.18 \times 10^{-4} = {} & 4.36 \times 10^{-4} \text{ eq/L from } CO_3^{2-} \\ & \underline{1.00 \times 10^{-4} \text{ eq/L from}\qquad\quad} \\ \text{alk} = {} & 1.00 \times 10^{-3} \text{ eq/L} \end{aligned}$$

3.8.2 Dissolved Inorganic Carbon and Alkalinity

The values given above can be used to show that at the same alkalinity value the concentration of total dissolved inorganic carbon, [C],

$$[C] = [CO_2] + [HCO_3^-] + [CO_3^{2-}] \tag{3.29}$$

varies with pH. At pH 7.0,

$$[C]_{pH\,7} = 2.25 \times 10^{-4} + 1.00 \times 10^{-3} + 0 = 1.22 \times 10^{-3} \quad (3.30)$$

whereas at pH 10.0,

$$[C]_{pH\,10} = 0 + 4.64 \times 10^{-4} + 2.18 \times 10^{-4} = 6.82 \times 10^{-4} \quad (3.31)$$

The calculation above shows that the dissolved inorganic carbon concentration at pH 10.0 is only about half that at pH 7.0. This is because at pH 10.0 major contributions to alkalinity are made by CO_3^{2-} ion, each of which has twice the alkalinity of each HCO_3^- ion, and by OH^-, which does not contain any carbon. The lower inorganic carbon concentration at pH 10.0 shows that an initially relatively low pH aquatic system (pH 7.0) can donate dissolved inorganic carbon for use in photosynthesis with a change in pH but none in alkalinity. This pH-dependent difference in dissolved inorganic carbon concentration represents a significant potential source of carbon for algae growing in water, which fix carbon by the overall reactions

$$CO_2 + H_2O + h\nu \rightleftarrows \{CH_2O\} + O_2 \quad (3.32)$$

and

$$HCO_3^- + H_2O + h\nu \rightleftarrows \{CH_2O\} + OH^- + O_2 \quad (3.33)$$

As dissolved inorganic carbon is used up to synthesize biomass, $\{CH_2O\}$, the water becomes more basic. The amount of inorganic carbon that can be consumed before the water becomes too basic to allow algal reproduction is proportional to the alkalinity. In going from pH 7.0 to pH 10.0, the amount of inorganic carbon consumed from 1.00 L of water having an alkalinity of 1.00×10^{-3} eq/L is

$$[C]_{pH\,7} \times 1\text{ L} - [C]_{pH\,10} \times 1\text{ L} = 1.22 \times 10^{-3}\text{ mol} - 6.82 \times 10^{-4}\text{ mol} = 5.4 \times 10^{-4}\text{ mol} \quad (3.34)$$

This translates to an increase of 5.4×10^{-4} mol/L of biomass. Since the formula mass of $\{CH_2O\}$ is 30, the weight of biomass produced amounts to 16 mg/L. Assuming no input of additional CO_2, at higher alkalinity more biomass is produced for the same change in pH, whereas at lower alkalinity less is produced. Because of this effect, biologists use alkalinity as a measure of water fertility.

3.8.3 Influence of Alkalinity on CO_2 Solubility

The increased solubility of carbon dioxide in water with an elevated alkalinity can be illustrated by comparing its solubility in pure water (alkalinity 0) to its solubility in water initially containing 1.00×10^{-3} M NaOH (alkalinity 1.00×10^{-3} eq/L). The number of moles of CO_2 that will dissolve in a liter of pure water from the atmosphere containing 390 ppm carbon dioxide is

$$\text{Solubility} = [CO_2\,(aq)] + [HCO_3^-] \quad (3.35)$$

Substituting values calculated in Section 3.7 gives

$$\text{Solubility} = 1.276 \times 10^{-5} + 2.38 \times 10^{-6} = 1.514 \times 10^{-5}\text{ M}$$

The solubility of CO_2 in water, initially 1.00×10^{-3} M in NaOH, is about 100-fold higher because of uptake of CO_2 by the reaction

$$CO_2\,(aq) + OH^- \rightarrow HCO_3^- \tag{3.36}$$

so that

$$\text{Solubility} = [CO_2\,(aq)] + [HCO_3^-] = 1.276 \times 10^{-5} + 1.00 \times 10^{-3} = 1.01 \times 10^{-3}\ \text{M} \tag{3.37}$$

3.9 CALCIUM AND OTHER METALS IN WATER

The kinds and concentrations of metal ions in water are determined largely by the rocks that the water contacts. Metal ions in water, commonly denoted M^{n+}, exist in numerous forms. A bare metal ion, Ca^{2+} for example, cannot exist as a separate entity in water. In order to secure the highest stability of their outer electron shells, metal ions in water are bonded, or *coordinated*, to other species. These may be water molecules or other stronger bases (electron-donor partners) that might be present. Therefore, metal ions in water solution are present in forms such as the *hydrated* metal cation $M(H_2O)_x^{n+}$. Metal ions in aqueous solution seek to reach a state of maximum stability through chemical reactions including acid–base,

$$Fe(H_2O)_6^{3+} \rightleftarrows Fe(H_2O)_5OH^{2+} + H^+ \tag{3.38}$$

precipitation,

$$Fe(H_2O)_6^{3+} \rightleftarrows Fe(OH)_3(s) + 3H_2O + 3H^+ \tag{3.39}$$

and oxidation–reduction reactions:

$$Fe(H_2O)_6^{2+} \rightleftarrows Fe(OH)_3(s) + 3H_2O + e^- + 3H^+ \tag{3.40}$$

These all provide a means through which metal ions in water are transformed to more stable forms. Because of reactions such as these and the formation of dimeric species, such as $Fe_2(OH)_2^{4+}$, the concentration of simple hydrated $Fe(H_2O)_6^{3+}$ ions in water is vanishingly small; the same holds true for many other hydrated metal ions dissolved in water.

3.9.1 Hydrated Metal Ions as Acids

Hydrated metal ions, particularly those with a charge of +3 or more, tend to lose H^+ ions from the water molecules bound to them in aqueous solution, and fit the definition of Brönsted acids, according to which acids are H^+ donors and bases are H^+ acceptors. The acidity of a metal ion increases with charge and decreases with increasing radius. As shown by the reaction,

$$Fe(H_2O)_6^{3+} \rightleftarrows Fe(H_2O)_5OH^{2+} + H^+ \tag{3.41}$$

hydrated iron(III) ion is an acid, a relatively strong one with a K_{a1} of 8.9×10^{-4}, so that solutions of iron(III) tend to have low pH values. Hydrated trivalent metal ions, such as iron(III), generally are minus at least one hydrogen ion at neutral pH values or above. For tetravalent metal ions, the completely protonated forms, $M(H_2O)_x^{4+}$, are rare even at very low pH values. Commonly, O^{2-} is coordinated to tetravalent metal ions; an example is the vanadium(IV) species, VO^{2+}. Generally, divalent metal ions do not lose a hydrogen ion at pH values below 6, whereas monovalent metal ions such as Na^+ do not act as acids at all, and exist in water solution as simple hydrated ions.

The tendency of hydrated metal ions to behave as acids may have a profound effect upon the aquatic environment. A good example is *acid mine water* (see Chapter 7), which derives part of its acidic character from the acidic nature of hydrated iron(III):

$$Fe(H_2O)_6^{3+} \rightleftarrows Fe(OH)_3(s) + 3H^+ + 3H_2O \quad (3.42)$$

Hydroxide, OH^-, bonded to a metal ion, may function as a bridging group to join two or more metals together as shown by the following dehydration–dimerization process:

$$2Fe(H_2O)_5OH^{2+} \rightarrow (H_2O)_5Fe\overset{\displaystyle H}{\underset{\displaystyle H}{\overset{O}{\underset{O}{\diamond}}}}Fe(H_2O)_4^{4+} + 2(H_2O) \quad (3.43)$$

Among the metals other than iron(III) forming polymeric species with OH^- as a bridging group are Al(III), Be(II), Bi(III), Ce(IV), Co(III), Cu(II), Ga(III), Mo(V), Pb(II), Sc(II), Sn(IV), and U(VI). Additional hydrogen ions may be lost from water molecules bonded to the dimers, furnishing OH^- groups for further bonding and leading to the formation of polymeric hydrolytic species. If the process continues, colloidal hydroxy polymers are formed and, finally, precipitates are produced. This process is thought to be the general one by which hydrated iron(III) oxide, $Fe_2O_3 \cdot x(H_2O)$ [also called iron(II) hydroxide, $Fe(OH)_3$], is precipitated from solutions containing iron(III).

3.9.2 Calcium in Water

Of the cations found in most freshwater systems, calcium generally has the highest concentration. The chemistry of calcium, although complicated enough, is simpler than that of the transition metal ions found in water. Calcium is a key element in many geochemical processes, and minerals constitute the primary sources of calcium ion in waters. Among the primary contributing minerals are gypsum, $CaSO_4 \cdot 2H_2O$; anhydrite, $CaSO_4$; dolomite, $CaMg(CO_3)_2$; and calcite and aragonite, which are different mineral forms of $CaCO_3$.

Calcium ion, along with magnesium and sometimes iron(II) ion, accounts for *water hardness*. The most common manifestation of water hardness is the curdy precipitate formed by soap in hard water. *Temporary hardness* is due to the presence of calcium and bicarbonate ions in water and may be eliminated by boiling the water:

$$Ca^{2+} + 2HCO_3^- \rightleftarrows CaCO_3(s) + CO_2(g) + H_2O \quad (3.44)$$

Increased temperature may force this reaction to the right by evolving CO_2 gas, and a white precipitate of calcium carbonate may form in boiling water having temporary hardness.

Water containing a high level of carbon dioxide readily dissolves calcium from its carbonate minerals:

$$CaCO_3(s) + CO_2(aq) + H_2O \rightleftarrows Ca^{2+} + 2HCO_3^- \quad (3.45)$$

When this reaction is reversed and CO_2 is lost from the water, calcium carbonate deposits are formed. The concentration of CO_2 in water determines the extent of dissolution of calcium carbonate. The carbon dioxide that water may gain by equilibration with the atmosphere is not sufficient to account for the levels of calcium dissolved in natural waters, especially groundwaters.

Rather, the respiration of microorganisms degrading organic matter in water, sediments, and soil,

$$\{CH_2O\} + O_2 \rightleftarrows CO_2 + H_2O \tag{3.46}$$

accounts for the very high levels of CO_2 and HCO_3^- observed in water and is very important in aquatic chemical processes and geochemical transformations.

3.9.3 Dissolved Carbon Dioxide and Calcium Carbonate Minerals

The equilibrium between dissolved carbon dioxide and calcium carbonate minerals is important in determining several natural water chemistry parameters such as alkalinity, pH, and dissolved calcium concentration (Figure 3.9). For freshwater, the typical figures quoted for the concentrations of both HCO_3^- and Ca^{2+} are 1.00×10^{-3} M. It may be shown that these are reasonable values when the water is in equilibrium with limestone, $CaCO_3$, and with atmospheric CO_2. The concentration of CO_2 in water in equilibrium with air has already been calculated as 1.276×10^{-5} M. The other constants needed to calculate $[HCO_3^-]$ and $[Ca^{2+}]$ are the acid dissociation constant for CO_2:

$$K_{a1} = \frac{[H^+][HCO_3^-]}{[CO_2]} = 4.45 \times 10^{-7} \tag{3.47}$$

the acid dissociation constant of HCO_3^-:

$$K_{a2} = \frac{[H^+][CO_3^{2-}]}{[HCO_3^-]} = 4.69 \times 10^{-11} \tag{3.48}$$

and the solubility product of calcium carbonate (calcite):

$$K_{sp} = [Ca^{2+}][CO_3^{2-}] = 4.47 \times 10^{-9} \tag{3.49}$$

The reaction between calcium carbonate and dissolved CO_2 is

$$CaCO_3(s) + CO_2(aq) + H_2O \rightleftarrows Ca^{2+} + 2HCO_3^- \tag{3.50}$$

for which the equilibrium expression is the following:

$$K' = \frac{[Ca^{2+}][HCO_3^-]^2}{[CO_2]} = \frac{K_{sp}K_{a1}}{K_{a2}} = 4.24 \times 10^{-5} \tag{3.51}$$

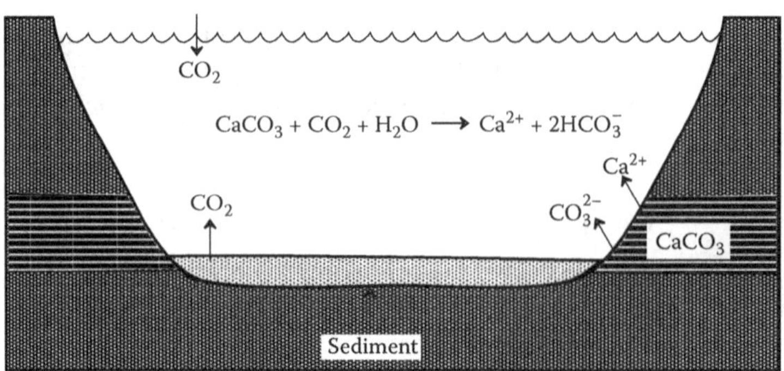

FIGURE 3.9 Carbon dioxide–calcium carbonate equilibria.

The stoichiometry of Reaction 3.50 gives a bicarbonate ion concentration that is twice that of calcium. Substitution of the value of CO_2 concentration into the expression for K' yields values of 5.14×10^{-4} M for $[Ca^{2+}]$ and 1.03×10^{-3} for $[HCO_3^-]$. Substitution into the expression for K_{sp} yields 8.70×10^{-6} M for $[CO_3^{2-}]$. When known concentrations are substituted into the product $K_{a1}K_{a2}$,

$$K_{a1}K_{a2} = \frac{[H^+]^2[CO_3^{2-}]}{[CO_2]} = 2.09 \times 10^{-17} \tag{3.52}$$

a value of 5.54×10^{-9} M is obtained for $[H^+]$ (pH 8.26). The alkalinity is essentially equal to $[HCO_3^-]$, which is much higher than $[CO_3^{2-}]$ or $[OH^-]$.

To summarize, for water in equilibrium with solid calcium carbonate and atmospheric CO_2, the following concentrations are calculated:

$$[CO_2] = 1.276 \times 10^{-5}\ \text{M}, \quad [Ca^{2+}] = 5.14 \times 10^{-4}\ \text{M}$$

$$[HCO_3^-] = 1.03 \times 10^{-3}\ \text{M}, \quad [H^+] = 5.54 \times 10^{-9}\ \text{M}$$

$$[CO_3^{2-}] = 8.70 \times 10^{-6}\ \text{M}, \quad \text{pH} = 8.26$$

Factors such as nonequilibrium conditions, high CO_2 concentrations in bottom regions, and increased pH due to algal uptake of CO_2 cause deviations from these values. Nevertheless, they are close to the values found in a large number of natural water bodies.

3.10 COMPLEXATION AND CHELATION

The properties of metals dissolved in water depend largely upon the nature of metal species dissolved in the water. Therefore, *speciation* of metals plays a crucial role in their environmental chemistry in natural waters and wastewaters. In addition to the hydrated metal ions, for example, $Fe(H_2O)_6^{3+}$ and hydroxy species such as $FeOH(H_2O)_5^{2+}$ discussed in the preceding section, metals may exist in water reversibly bound to inorganic anions or to organic compounds as *metal complexes*. For example, a cyanide ion can bond to dissolved iron(II):

$$Fe(H_2O)_6^{2+} + CN^- \rightleftarrows FeCN(H_2O)_5^+ + H_2O \tag{3.53}$$

Additional cyanide ions may bond to the iron to form $Fe(CN)_2$, $Fe(CN)_3^-$, $Fe(CN)_4^{2-}$, $Fe(CN)_5^{3-}$, and $Fe(CN)_6^{4-}$, where the water molecules still bound to the iron(II) are omitted for simplicity. This phenomenon is called *complexation*; the species that binds with the metal ion, CN^-, in the example above, is called a *ligand*, and the product in which the ligand is bound with the metal ion is a *complex*, *complex ion*, or *coordination compound*. A special case of complexation in which a ligand bonds in two or more places to a metal ion is called *chelation*. In addition to being present as metal complexes, metals may occur in water as *organometallic* compounds containing carbon-to-metal bonds. The solubilities, transport properties, and biological effects of such species are often vastly different from those of the metal ions themselves. Subsequent sections of this chapter consider metal species with an emphasis upon metal complexation, especially chelation, in which particularly strong metal complexes are formed.

In the example above, the cyanide ion is a *unidentate ligand*, which means that it possesses only one site that bonds to a metal ion. Complexes of unidentate ligands are of relatively little importance in solution in natural waters. Of considerably more importance are complexes with *chelating agents*. A chelating agent has more than one atom that may be bonded to a central metal ion at one time to form a ring structure. Thus, pyrophosphate ion, $P_2O_7^{4-}$, bonds to two sites on a calcium ion to form a chelate:

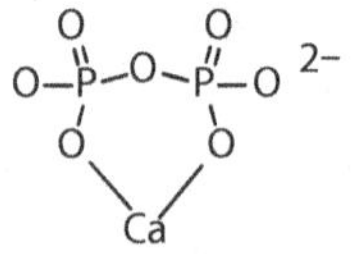

In general, since a chelating agent may bond to a metal ion in more than one place simultaneously (Figure 3.10), chelates are more stable than complexes with unidentate ligands. Stability tends to increase with the number of chelating sites available on the ligand. Structures of metal chelates take a number of different forms, all characterized by rings in various configurations. The structure of a tetrahedrally coordinated chelate of nitrilotriacetate (NTA) ion is shown in Figure 3.10.

The ligands found in natural waters and wastewaters contain a variety of functional groups which can donate the electrons required to bond the ligand to a metal ion. Among the most common of these groups are:

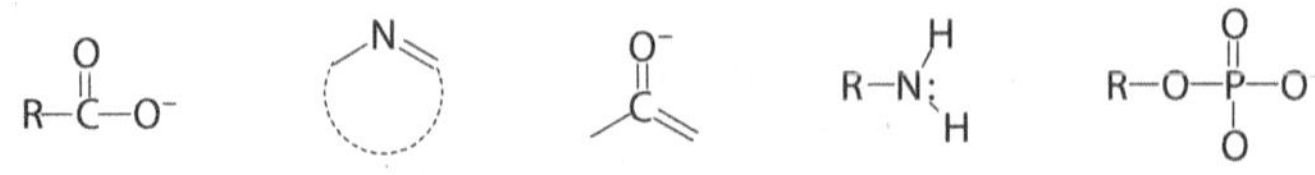

Carboxylate Heterocyclic nitrogen Phenoxide Aliphatic and aromatic amino Phosphate

These ligands complex most metal ions found in unpolluted waters and biological systems (Mg^{2+}, Ca^{2+}, Mn^{2+}, Fe^{2+}, Fe^{3+}, Cu^{2+}, Zn^{2+}, and VO^{2+}). They also bind to contaminant metal ions such as Co^{2+}, Ni^{2+}, Sr^{2+}, Cd^{2+}, and Ba^{2+}.

Complexation may have a number of effects, including reactions of both ligands and metals. Among the ligand reactions are oxidation–reduction, decarboxylation, hydrolysis, and biodegradation. Complexation may cause changes in oxidation state of the metal and may result in a metal becoming solubilized from an insoluble compound. The formation of insoluble complex compounds removes metal ions from solution. Complexation may strongly influence metals' adsorption, distribution, transport, and fate as well as biochemical effects including bioavailability, toxicity, and plant uptake.[3]

Complex compounds of metals such as iron (in hemoglobin) and magnesium (in chlorophyll) are vital to life processes. Naturally occurring chelating agents, such as humic substances and amino acids, are found in water and soil. The high concentration of chloride ion in seawater results in the formation of some chloro complexes. Synthetic chelating agents such as sodium tripolyphosphate, sodium ethylenediaminetetraacetate (EDTA), sodium NTA, and sodium citrate are produced in large quantities for use in metal-plating baths, industrial water treatment, detergent formulations, and food preparation. These compounds may enter aquatic systems through waste discharges.

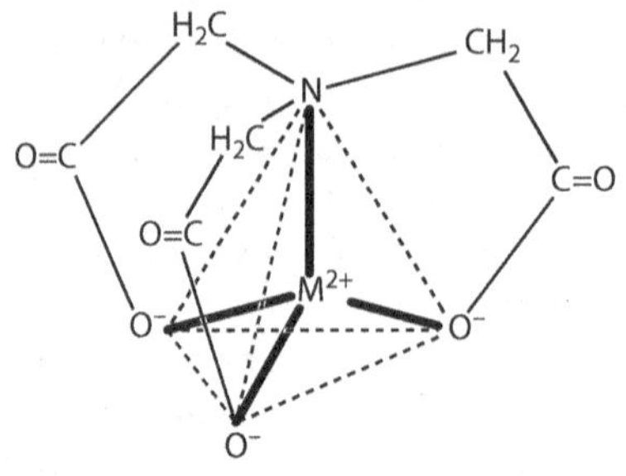

FIGURE 3.10 Nitrilotriacetate chelate of a divalent metal ion in a tetrahedral configuration.

3.10.1 Occurrence and Importance of Chelating Agents in Water

Chelating agents are common potential water pollutants. These substances can occur in sewage effluent and industrial wastewater such as metal plating wastewater. In addition to pollutant sources, there are natural sources of chelating agents. One such chelating agent is ethylenediaminedisuccinic acid,

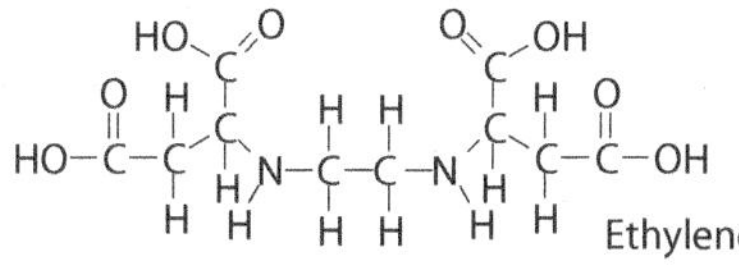

Ethylenediaminedisuccinic acid

a metabolite of the soil actinomycete, *Amycolatopsis orientalis*. Unlike a number of common synthetic chelating agents, such as EDTA (see below), ethylenediaminedisuccinic acid is biodegradable and has been used as an extractant for phytoremediation (see Chapter 9, Section 21.9) of sites contaminated with heavy metals.

The most important pollutant chelating agents are aminopolycarboxylates, of which the most common examples are NTA (Figure 3.10) and EDTA (structure illustrated at the beginning of Section 3.13). Both EDTA and NTA are common water pollutants.[4] Because of their strong bonding to metal ions, the aminopolycarboxylate chelating agents are almost always encountered in the bound form, which has a strong effect on their chemistry and environmental fate and transport. To the extent that EDTA is present in wastewater, it prevents some metals from binding to and settling out with biomass sludge in biological wastewater treatment processes. Therefore, EDTA chelates compose most of the copper, nickel, and zinc in wastewater effluents (most of these metals in excess of strong chelating agents present are incorporated into the sludge).

Chelation by EDTA has been shown to greatly increase the migration rates of radioactive ^{60}Co from pits and trenches used by the Oak Ridge National Laboratory in Oak Ridge, Tennessee, for disposal of intermediate-level radioactive waste. EDTA was used as a cleaning and solubilizing agent for the decontamination of hot cells, equipment, and reactor components. Analysis of water from sample wells in the disposal pits showed EDTA concentrations of 3.4×10^{-7} M. The presence of EDTA 12–15 years after its burial attests to its low rate of biodegradation. In addition to cobalt, EDTA strongly chelates radioactive plutonium and radioisotopes of Am^{3+}, Cm^{3+}, and Th^{4+}. Such chelates with negative charges are much less strongly sorbed by mineral matter and are vastly more mobile than the unchelated metal ions.

Contrary to the above findings, only very low concentrations of chelatable radioactive plutonium were observed in groundwater near the Idaho Chemical Processing Plant's low-level waste disposal well. No plutonium was observed in wells at any significant distance from the disposal well. The waste processing procedure used was designed to destroy any chelating agents in the waste prior to disposal, and no chelating agents were found in the water pumped from the test wells.

The fate of radionuclide metal chelates that have been discarded in soil is obviously important. If some mechanism exists to destroy the chelating agents, the radioactive metals will be much less mobile. Although EDTA is only poorly biodegradable, NTA is degraded by the action of *Chlatobacter heintzii* bacteria. In addition to uncomplexed NTA, these bacteria have been shown to degrade NTA that is chelated to metals, including cobalt, iron, zinc, aluminum, copper, and nickel.

Complexing agents in wastewater are of concern primarily because of their ability to solubilize heavy metals from plumbing, and from deposits containing heavy metals. Complexation may increase the leaching of heavy metals from waste disposal sites and reduce the efficiency with which heavy metals are removed with sludge in conventional biological waste treatment. Removal of chelated iron is difficult with conventional municipal water treatment processes. Iron(III) and perhaps several other essential micronutrient metal ions are kept in solution by chelation in algal cultures. The availability of chelating agents may be a factor in determining growth of algae and plants in water.[5] The yellow-brown color of some natural waters is due to naturally occurring chelates of iron.

3.11 BONDING AND STRUCTURE OF METAL COMPLEXES

This section discusses some of the fundamentals of complexation in water. A complex consists of a central metal atom to which neutral or negatively charged ligands possessing electron-donor properties are bonded. The complex may be neutral or may have a positive or negative charge. The ligands are contained within the *coordination sphere* of the central metal atom. Depending upon the type of bonding involved, the ligands within the coordination sphere are held in a definite structural pattern. However, in solution, ligands of many complexes exchange rapidly between solution and the coordination sphere of the central metal ion.

The *coordination number* of a metal atom, or ion, is the number of ligand electron-donor groups that are bonded to it. The most common coordination numbers are 2, 4, and 6. Polynuclear complexes contain two or more metal atoms joined together through bridging ligands, frequently OH, as shown for iron(III) in Reaction 3.43.

3.11.1 Selectivity and Specificity in Chelation

Although chelating agents are never entirely specific for a particular metal ion, some complicated chelating agents of biological origin approach almost complete specificity for certain metal ions. One example of such a chelating agent is ferrichrome, synthesized by and extracted from fungi, which forms extremely stable chelates with iron(III). It has been observed that cyanobacteria of the *Anabaena* species secrete appreciable quantities of iron-selective hydroxamate chelating agents during periods of heavy algal bloom. These photosynthetic organisms readily take up iron chelated by hydroxamate-chelated iron, whereas some competing green algae, such as *Scenedesmus*, do not. Thus, the chelating agent serves a dual function of promoting the growth of certain cyanobacteria while suppressing the growth of competing species, allowing the cyanobacteria to exist as the predominant species. The production of chelating agents selective for iron(III) has been observed in cyanobacterial *Plectonema* and *Spirulina* as well as *Chlorella*, *Scenedesmus*, and *Porphyrium* algae.

3.12 CALCULATIONS OF SPECIES CONCENTRATIONS

The stability of complex ions in solution is expressed in terms of *formation constants*. These can be *stepwise formation constants* (*K* expressions) representing the bonding of individual ligands to a metal ion, or *overall formation constants* (β expressions) representing the binding of two or more ligands to a metal ion. These concepts are illustrated for complexes of zinc ion with ammonia by the following:

$$Zn^{2+} + NH_3 \rightleftarrows ZnNH_3^{2+} \quad (3.54)$$

$$K_1 = \frac{[ZnNH_3^{2+}]}{[Zn^{2+}][NH_3]} = 3.9 \times 10^2 \text{ (Stepwise formation constant)} \quad (3.55)$$

$$ZnNH_3^{2+} + NH_3 \rightleftarrows Zn(NH_3)_2^{2+} \quad (3.56)$$

$$K_2 = \frac{[Zn(NH_3)_2^{2+}]}{[ZnNH_3^{2+}][NH_3]} = 2.1 \times 10^2 \quad (3.57)$$

$$Zn^{2+} + 2NH_3 \rightleftarrows Zn(NH_3)_2^{2+} \tag{3.58}$$

$$\beta_2 = \frac{[Zn(NH_3)_2^{2+}]}{[Zn^{2+}][NH_3]^2} = K_1K_2 = 8.2 \times 10^4 \text{ (Overall formation constant)} \tag{3.59}$$

(For $Zn(NH_3)_3^{2+}$, $\beta_3 = K_1K_2K_3$ and for $Zn(NH_3)_4^{2+}$, $\beta_4 = K_1K_2K_3K_4$.)

The following sections show some calculations involving chelated metal ions in aquatic systems. Because of their complexity, the details of these calculations may be beyond the needs of some readers, who may choose to simply consider the results. In addition to the complexation itself, consideration must be given to competition of H^+ for ligands, competition among metal ions for ligands, competition among different ligands for metal ions, and precipitation of metal ions by various precipitants. Not the least of the problems involved in such calculations is the lack of accurately known values of equilibrium constants to be used under the conditions being considered, a factor that can yield questionable results from even the most elegant computerized calculations. Furthermore, kinetic factors are often quite important. Such calculations, however, can be quite useful to provide an overall view of aquatic systems in which complexation is important, and as general guidelines to determine areas in which more data should be obtained.

3.13 COMPLEXATION BY DEPROTONATED LIGANDS

Generally, calculation of complex species concentrations is complicated by competition between metal ions and H^+ for ligands. First consider an example in which the ligand has lost all ionizable hydrogen. At pH values of 11 or above, EDTA is essentially all in the completely ionized tetranegative form, Y^{4-}, illustrated below:

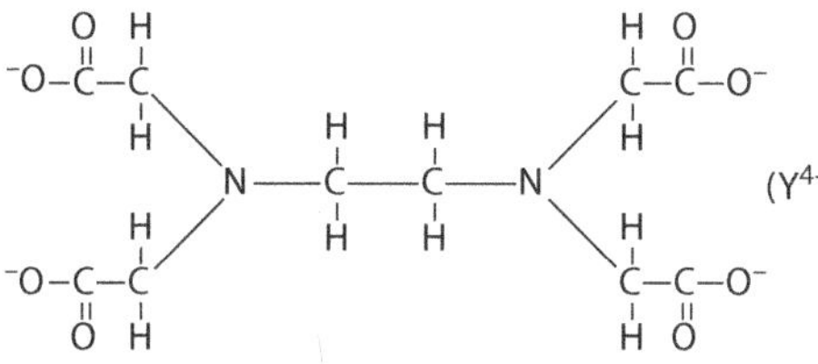

```
    O H                         H O
    ‖ |                         | ‖
 ⁻O–C–C                         C–C–O⁻
       |                       /|
       H \     H   H         /  H
          N——C——C——N
       H /     |   |         \  H
       |       H   H           \|
 ⁻O–C–C                         C–C–O⁻
    ‖ |                         | ‖
    O H                         H O
```

(Y^{4-})

Consider a wastewater with an alkaline pH of 11 containing copper(II) at a total level of 5.0 mg/L and excess uncomplexed EDTA at a level of 200 mg/L (expressed as the disodium salt, $Na_2H_2C_{10}H_{12}O_8N_2 \cdot 2H_2O$, formula mass 372). At this pH, uncomplexed EDTA is present as ionized Y^{4-}. The questions to be asked are: Will most of the copper be present as the EDTA complex? If so, what will be the equilibrium concentration of the hydrated copper(II) ion, Cu^{2+}? To answer the former question it is first necessary to calculate the molar concentration of uncomplexed excess EDTA, Y^{4-}. Since disodium EDTA with a formula mass of 372 is present at 200 mg/L (ppm), the total molar concentration of EDTA as Y^{4-} is 5.4×10^{-4} M. The formation constant K_1 of the copper–EDTA complex CuY^{2-} is

$$K_1 = \frac{[CuY^{2-}]}{[Cu^{2+}][Y^{4-}]} = 6.3 \times 10^{18} \tag{3.60}$$

The ratio of complexed to uncomplexed copper is

$$\frac{[CuY^{2-}]}{[Cu^{2+}]} = [Y^{4-}]K_1 = 5.4 \times 10^{-4} \times 6.3 \times 10^{18} = 3.3 \times 10^{15} \tag{3.61}$$

and, therefore, essentially all of the copper is present as the complex ion. The molar concentration of total copper(II) in a solution containing 5.0 mg/L copper(II) is 7.9×10^{-5} M, which in this case is essentially all in the form of the EDTA complex. The very low concentration of uncomplexed, hydrated copper(II) ion is given by

$$[\mathrm{Cu}^{2+}] = \frac{[\mathrm{CuY}^{2-}]}{K_1[\mathrm{Y}^{4-}]} = \frac{7.9\times10^{-5}}{6.3\times10^{18}\times5.4\times10^{-4}} = 2.3\times10^{-20}\ \mathrm{M} \tag{3.62}$$

In the medium described, the concentration of hydrated Cu^{2+} ion is extremely low compared to total copper(II). Solution phenomena that depend upon the concentration of the hydrated Cu^{2+} ion (such as a physiological effect or an electrode response) would differ greatly in the medium described, compared to the effect observed if all of the copper at a level of 5.0 mg/L were present as Cu^{2+} in a more acidic solution and in the absence of complexing agent. The phenomenon of reducing the concentration of hydrated metal ion to very low values through the action of strong chelating agents is one of the most important effects of complexation in natural aquatic systems.

3.14 COMPLEXATION BY PROTONATED LIGANDS

Generally, complexing agents, particularly chelating compounds, are conjugate bases of Brönsted acids; for example, glycinate anion, $H_2NCH_2CO_2^-$, is the conjugate base of glycine, $^+H_3NCH_2CO_2^-$. Therefore, in many cases hydrogen ion competes with metal ions for a ligand, so that the strength of chelation depends upon pH. In the nearly neutral pH range usually encountered in natural waters, most organic ligands are present in a conjugated acid form.

In order to understand the competition between hydrogen ion and metal ion for a ligand, it is useful to know the distribution of ligand species as a function of pH. Consider nitrilotriacetic acid (NTA), commonly designated H_3T, as an example. The trisodium salt of this compound is a strong chelating agent that can be used in metal plating, as a detergent phosphate substitute, and in other applications in which strong chelating capability is required. Biological processes are required for NTA degradation, and under some conditions it persists for long times in water. Given the ability of NTA to solubilize and transport heavy metal ions, this material is of potential environmental concern.

Nitrilotriacetic acid, H_3T, loses hydrogen ion in three steps to form the NTA anion, T^{3-}, the structural formula of which is

$$^{-}\mathrm{O{-}C({=}O){-}CH_2{-}N(CH_2{-}CO_2^-)_2} \quad (\mathrm{T}^{3-})$$

The T^{3-} species may coordinate through three $-CO_2^-$ groups and through the nitrogen atom, as shown in Figure 3.10. Note the similarity of the NTA structure to that of EDTA, discussed in Section 3.13. The stepwise ionization of H_3T is described by the following equilibria:

$$\mathrm{H_3T} \rightleftarrows \mathrm{H^+ + H_2T^-} \tag{3.63}$$

$$K_{a1} = \frac{[\mathrm{H^+}][\mathrm{H_2T^-}]}{[\mathrm{H_3T}]} = 2.18\times10^{-2},\quad \mathrm{p}K_{a1} = 1.66 \tag{3.64}$$

TABLE 3.2
Fractions of NTA Species at Selected pH Values

pH Value	α_{H_3T}	$\alpha_{H_2T^-}$	$\alpha_{HT^{2-}}$	$\alpha_{T^{3-}}$
pH below 1.0	1.00	0.00	0.00	0.00
$pH = pK_{a1}$	0.49	0.49	0.02	0.00
$pH = \frac{1}{2}(pK_{a1} + pK_{a2})$	0.16	0.68	0.16	0.00
$pH = pK_{a2}$	0.02	0.49	0.49	0.00
$pH = \frac{1}{2}(pK_{a1} + pK_{a3})$	0.00	0.00	1.00	0.00
$pH = pK_{a3}$	0.00	0.00	0.50	0.50
pH above 12	0.00	0.00	0.00	1.00

$$H_2T^- \rightleftarrows H^+ + HT^{2-} \tag{3.65}$$

$$K_{a2} = \frac{[H^+][HT^{2-}]}{[H_2T^-]} = 1.12 \times 10^{-3}, \quad pK_{a2} = 2.95 \tag{3.66}$$

$$HT^{2-} \rightleftarrows H^+ + T^{3-} \tag{3.67}$$

$$K_{a3} = \frac{[H^+][T^{3-}]}{[HT^{2-}]} = 5.25 \times 10^{-11}, \quad pK_{a3} = 10.28 \tag{3.68}$$

NTA may exist in solution as any one of the four species, H_3T, H_2T^-, HT^{2-}, or T^{3-}, depending upon the pH of the solution. As was shown for the $CO_2/HCO_3^-/CO_3^{2-}$ system in Section 3.7 and Figure 3.8, fractions of NTA species can be illustrated graphically by a diagram of the distribution-of-species with pH as a master (independent) variable. The key points used to plot such a diagram for NTA are given in Table 3.2, and the plot of fractions of species (α values) as a function of pH is shown in Figure 3.11. Examination of the plot shows that the complexing anion T^{3-} is the predominant species only at relatively high pH values, much higher than would be usually encountered in natural waters. The HT^{2-} species has an extremely wide range of predominance, however, spanning the entire normal pH range of ordinary freshwaters.

3.15 SOLUBILIZATION OF LEAD ION FROM SOLIDS BY NTA

A major concern regarding the widespread introduction of strong chelating agents such as NTA into aquatic ecosystems from sources such as detergents or electroplating wastes is that of possible solubilization of toxic heavy metals from solids through the action of chelating agents. Experimentation is required to determine whether this may be a problem, but calculations are helpful in predicting probable effects. The extent of solubilization of heavy metals depends upon a number of factors, including the stability of the metal chelates, the concentration of the complexing agent in the water, pH, and the nature of the insoluble metal deposit. Several example calculations are given here.

Consider first the solubilization of lead from solid $Pb(OH)_2$ by NTA at pH 8.0. As illustrated in Figure 3.11, essentially all uncomplexed NTA is present as HT^{2-} ion at pH 8.0. Therefore, the solubilization reaction is

$$Pb(OH)_2\,(s) + HT^{2-} \rightleftarrows PbT^- + OH^- + H_2O \tag{3.69}$$

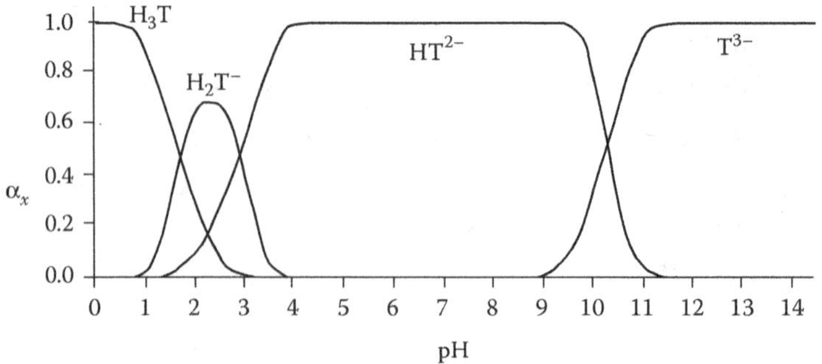

FIGURE 3.11 Plot of fraction of species α_x as a function of pH for NTA species in water.

which may be obtained by adding the following reactions:

$$Pb(OH)_2(s) \rightleftarrows Pb^{2+} + 2OH^- \tag{3.70}$$

$$K_{sp} = [Pb^{2+}][OH^-]^2 = 1.61 \times 10^{-20} \tag{3.71}$$

$$HT^{2-} \rightleftarrows H^+ + T^{3-} \tag{3.67}$$

$$K_{a3} = \frac{[H^+][T^{3-}]}{[HT^{2-}]} = 5.25 \times 10^{-11} \tag{3.68}$$

$$Pb^{2+} + T^{3-} \rightleftarrows PbT^- \tag{3.72}$$

$$K_f = \frac{[PbT^-]}{[Pb^{2+}][T^{3-}]} = 2.45 \times 10^{11} \tag{3.73}$$

$$H^+ + OH^- \rightleftarrows H_2O \tag{3.74}$$

$$\frac{1}{K_W} = \frac{1}{[H^+][OH^-]} = \frac{1}{1.00 \times 10^{-14}} \tag{3.75}$$

$$Pb(OH)_2(s) + HT^{2-} \rightleftarrows PbT^- + OH^- + H_2O \tag{3.69}$$

$$K = \frac{[PbT^-][OH^-]}{[HT^{2-}]} = \frac{K_{sp}K_{a3}K_f}{K_W} = 2.07 \times 10^{-5} \tag{3.76}$$

Assume that a sample of water contains 25 mg/L of $N(CH_2CO_2Na)_3$, the trisodium NTA salt, with a formula mass of 257. The total concentration of both complexed and uncomplexed NTAs is 9.7×10^{-5} mmol/mL. Assuming a system in which NTA at pH 8.0 is in equilibrium with solid $Pb(OH)_2$, the NTA may be primarily in the uncomplexed form, HT^{2-}, or in the lead complex, PbT^-. The predominant species may be determined by calculating the $[PbT^-]/[HT^{2-}]$ ratio from the expression for K, noting that at pH 8.0, $[OH^-] = 1.00 \times 10^{-6}$ M.

$$\frac{[PbT^-]}{[HT^{2-}]} = \frac{K}{[OH^-]} = \frac{2.07 \times 10^{-5}}{1.00 \times 10^{-6}} = 20.7 \tag{3.77}$$

Since $[PbT^-]/[HT^{2-}]$ is approximately from 20 to 1, most of the NTA in solution is present as the lead chelate. The molar concentration of PbT^- is just slightly less than the 9.7×10^{-5} mmol/mL total NTA present. The atomic mass of lead is 207, so the concentration of lead in solution is approximately 20 mg/L. This reaction is pH dependent such that the fraction of NTA chelated decreases with increasing pH.

3.15.1 Reaction of NTA with Metal Carbonate

Carbonates are common forms of heavy metal ion solids. Solid lead carbonate, $PbCO_3$, is stable within the pH region and alkalinity conditions often found in natural waters and wastewaters. An example similar to the one in the preceding section may be worked, assuming that equilibrium is established with $PbCO_3$ rather than with solid $Pb(OH)_2$. In this example it is assumed that 25 mg/L of trisodium NTA is in equilibrium with $PbCO_3$ at pH 7.0 and a calculation is made to determine whether the lead will be complexed appreciably by the NTA. The carbonate ion, CO_3^{2-}, reacts with H^+ to form HCO_3^-. As discussed in Section 3.7, the acid–base equilibrium reactions for the $CO_2/HCO_3^-/CO_3^{2-}$ system are

$$CO_2(aq) + H_2O \rightleftarrows H^+ + HCO_3^- \tag{3.9}$$

$$K'_{a1} = \frac{[H^+][HCO_3^-]}{[CO_2]} = 4.45 \times 10^{-7},\ pK_{a1} = 6.35 \tag{3.10}$$

$$HCO_3^- \rightleftarrows H^+ + CO_3^{2-} \tag{3.11}$$

$$K'_{a2} = \frac{[H^+][CO_3^{2-}]}{[HCO_3^-]} = 4.69 \times 10^{-11} \tag{3.12}$$

where the acid dissociation constants of the carbonate species are designated with " ′ " to distinguish them from the acid dissociation constants of NTA. Figure 3.8 shows that within a pH range of about 7–10 the predominant carbonic species is HCO_3^-; therefore, the CO_3^{2-} released by the reaction of NTA with $PbCO_3$ will go into solution as HCO_3^-:

$$PbCO_3(s) + HT^{2-} \rightleftarrows PbT^- + HCO_3^- \tag{3.78}$$

This reaction and its equilibrium constant are obtained as follows:

$$PbCO_3(s) \rightleftarrows Pb^{2+} + CO_3^{2-} \tag{3.79}$$

$$K_{sp}(s) = [Pb^{2+}][CO_3^{2-}] = 1.48 \times 10^{-13} \tag{3.80}$$

$$Pb^{2+} + T^{3-} \rightleftarrows PbT^- \tag{3.72}$$

$$K_f = \frac{[PbT^-]}{[Pb^{2+}][T^{3-}]} = 2.45 \times 10^{11} \tag{3.73}$$

$$HT^{2-} \rightleftarrows H^+ + T^{3-} \tag{3.67}$$

$$K_{a3} = \frac{[H^+][T^{3-}]}{[HT^{2-}]} = 5.25 \times 10^{-11} \tag{3.68}$$

$$H^+ + CO_3^{2-} \rightleftarrows HCO_3^- \tag{3.81}$$

$$\frac{1}{K'_{a2}} = \frac{[HCO_3^-]}{[H^+][CO_3^{2-}]} = \frac{1}{4.69 \times 10^{-11}} \tag{3.82}$$

$$PbCO_3(s) + HT^{2-} \rightleftarrows PbT^- + HCO_3^- \tag{3.78}$$

$$K = \frac{[PbT^-][HCO_3^-]}{[HT^{2-}]} = \frac{K_{sp}K_{a3}K_f}{K'_{a2}} = 4.06 \times 10^{-2} \tag{3.83}$$

From the expression for K, Equation 3.83, it may be seen that the degree to which $PbCO_3$ is solubilized as PbT^- depends upon the concentration of HCO_3^-. Although this concentration will vary appreciably, the figure commonly used to describe natural waters is a bicarbonate ion concentration of 1.00×10^{-3}, as shown in Section 3.9. Using this value the following may be calculated:

$$\frac{[PbT^-]}{[HT^{2-}]} = \frac{K}{[HCO_3^-]} = \frac{4.06 \times 10^{-2}}{1.00 \times 10^{-3}} = 40.6 \tag{3.84}$$

Thus, under the given conditions, most of the NTA in equilibrium with solid $PbCO_3$ would be present as the lead complex. As in the previous example, at a trisodium NTA level of 25 mg/L, the concentration of soluble lead(II) would be approximately 20 mg/L. At relatively higher concentrations of HCO_3^-, the tendency to solubilize lead would be diminished, whereas at lower concentrations of HCO_3^-, NTA would be more effective in solubilizing lead.

3.15.2 Effect of Calcium Ion upon the Reaction of Chelating Agents with Slightly Soluble Salts

Chelatable calcium ion, Ca^{2+}, which is generally present in natural waters and wastewaters, competes for the chelating agent with a metal in a slightly soluble salt, such as $PbCO_3$. At pH 7.0, the reaction between calcium ion and NTA is

$$Ca^{2+} + HT^{2-} \rightleftarrows CaT^- + H^+ \tag{3.85}$$

described by the following equilibrium expression:

$$K' = \frac{[CaT^-][H^+]}{[Ca^{2+}][HT^{2-}]} = 1.48 \times 10^8 \times 5.25 \times 10^{-11} = 7.75 \times 10^{-3} \tag{3.86}$$

The value of K' is the product of the formation constant of CaT^-, (1.48×10^8), and K_{a3} of NTA, 5.25×10^{-11}. The fraction of NTA bound as CaT^- depends upon the concentration of Ca^{2+} and the pH. Typically, $[Ca^{2+}]$ in water is 1.00×10^{-3} M. Assuming this value and pH 7.0, the ratio of NTA present in solution as the calcium complex to that present as HT^{2-} is

$$\frac{[CaT^-]}{[HT^{2-}]} = \frac{[Ca^{2+}]}{[H^+]}K' = \frac{1.00 \times 10^{-3}}{1.00 \times 10^{-7}} \times 7.75 \times 10^{-3} = 77.5 \tag{3.87}$$

Therefore, most of the NTA in equilibrium with 1.00×10^{-3} M Ca^{2+} would be present as the calcium complex, CaT^-, which would react with $PbCO_3$ as follows:

$$PbCO_3(s) + CaT^- + H^+ \rightleftarrows Ca^{2+} + HCO_3^- + PbT^- \tag{3.88}$$

$$K'' = \frac{[Ca^{2+}][HCO_3^-][PbT^-]}{[CaT^-][H^+]} \tag{3.89}$$

Reaction 3.88 may be obtained by subtracting Reaction 3.85 from Reaction 3.78, and its equilibrium constant may be obtained by dividing the equilibrium constant of Reaction 3.85 by that of Reaction 3.78:

$$PbCO_3(s) + HT^{2-} \rightleftarrows PbT^- + HCO_3^- \tag{3.78}$$

$$-(Ca^{2+} + HT^{2-} \rightleftarrows CaT^- + H^+) \tag{3.85}$$

$$K' = \frac{[CaT^-][H^+]}{[Ca^{2+}][HT^{2-}]} = 7.75 \times 10^{-3} \tag{3.86}$$

$$PbCO_3(s) + CaT^- + H^+ \rightleftarrows Ca^{2+} + HCO_3^- + PbT^- \tag{3.88}$$

$$K'' = \frac{K}{K'} = \frac{4.06 \times 10^{-2}}{7.75 \times 10^{-3}} = 5.24 \tag{3.90}$$

Having obtained the value of K'', it is now possible to determine the distribution of NTA between PbT^- and CaT^-. Thus, for water containing NTA chelated to calcium at pH 7.0, a concentration of HCO_3^- of 1.00×10^{-3}, a concentration of Ca^{2+} of 1.00×10^{-3}, and in equilibrium with solid $PbCO_3$, the distribution of NTA between the lead complex and the calcium complex is

$$\frac{[PbT^-]}{[CaT^-]} = \frac{[H^+]K''}{[Ca^{2+}][HCO_3^-]} = 0.524$$

It may be seen that only about one-third of the NTA would be present as the lead chelate, whereas under the identical conditions, but in the absence of Ca^{2+}, approximately all of the NTA in equilibrium with solid $PbCO_3$ was chelated to NTA. Since the fraction of NTA present as the lead chelate is directly proportional to the solubilization of $PbCO_3$, differences in calcium concentration will affect the degree to which NTA solubilizes lead from lead carbonate.

3.16 POLYPHOSPHATES AND PHOSPHONATES IN WATER

Phosphorus occurs as many oxoanions, anionic forms in combination with oxygen. Some of these are strong complexing agents. Since about 1930, salts of polymeric phosphorus oxoanions have been used for water treatment, for water softening, and as detergent builders. When used for water treatment, polyphosphates "sequester" calcium ion in a soluble or suspended form. The effect is to reduce the equilibrium concentration of calcium ion and prevent the precipitation of calcium carbonate in installations such as water pipes and boilers. Furthermore, when water is softened properly with polyphosphates, calcium does not form precipitates with soaps or interact detrimentally with detergents.

The simplest form of phosphate is orthophosphate, PO_4^{3-}:

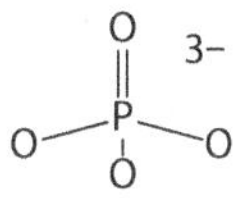

```
     O   3-
     ||
 O — P — O
     |
     O
```

Orthophosphate ion has three sites for attachment of H^+. Orthophosphoric acid, H_3PO_4, has a pK_{a1} of 2.17, a pK_{a2} of 7.31, and a pK_{a3} of 12.36. Because the third hydrogen ion is so difficult to remove from orthophosphate, as evidenced by the very high value of pK_{a3}, very basic conditions are required for PO_4^{3-} to be present at significant levels in water. It is possible for orthophosphate in natural waters to originate from the hydrolysis of polymeric phosphate species.

3.16.1 Polyphosphates

Pyrophosphate ion, $P_2O_7^{4-}$, is the first of a series of unbranched chain polyphosphates produced by the condensation of orthophosphate:

$$2PO_4^{3-} + H_2O \rightleftarrows P_2O_7^{4-} + 2OH^- \tag{3.91}$$

A long series of linear polyphosphates may be formed, the second of which is triphosphate ion, $P_3O_{10}^{5-}$. These species consist of PO_4 tetrahedra with adjacent tetrahedra sharing a common oxygen atom at one corner. The structural formulas of the acidic forms, $H_4P_2O_7$ and $H_5P_3O_{10}$, are given in Figure 3.12.

It is easy to visualize the longer chains composing the higher linear polyphosphates. *Vitreous sodium phosphates* are mixtures consisting of linear phosphate chains with from 4 to approximately 18 phosphorus atoms each. Those with intermediate chain lengths comprise the majority of the species present.

3.16.2 Hydrolysis of Polyphosphates

All of the polymeric phosphates hydrolyze to simpler products in water. The rate of hydrolysis depends upon a number of factors, including pH, and the ultimate product is always some form of orthophosphate. The simplest hydrolytic reaction of a polyphosphate is that of pyrophosphoric acid to produce orthophosphoric acid:

$$H_4P_2O_7 + H_2O \rightleftarrows 2H_3PO_4 \tag{3.92}$$

Researchers have found evidence that algae and other microorganisms catalyze the hydrolysis of polyphosphates. Even in the absence of biological activity, polyphosphates hydrolyze chemically at a significant rate in water. Therefore, there is much less concern about the possibility of polyphosphates binding to heavy metal ions and transporting them than is the case with organic chelating agents such as NTA or EDTA, which must depend upon microbial degradation for their decomposition.

3.16.3 Complexation by Polyphosphates

In general, chain phosphates are good complexing agents and even form complexes with alkali-metal ions. Ring phosphates form much weaker complexes than do chain species. The different

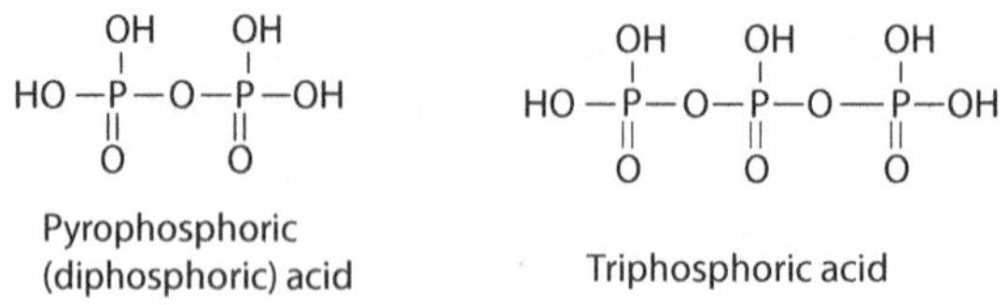

```
      OH   OH                    OH    OH    OH
      |    |                     |     |     |
  HO—P—O—P—OH               HO—P—O—P—O—P—OH
      ||   ||                    ||    ||    ||
      O    O                     O     O     O
```

Pyrophosphoric (diphosphoric) acid   Triphosphoric acid

FIGURE 3.12 Structural formulas of diphosphoric acid and triphosphoric acid. For diphosphoric acid, pK_{a1} is quite small (relatively strong acid), whereas pK_{a2} is 2.64, pK_{a3} is 6.76, and pK_{a4} is 9.42. For triphosphoric acid, pK_{a1} and pK_{a2} values are low, pK_{a3} is 2.30, pK_{a4} is 6.50, and pK_{a5} is 9.24. These values reflect the relative ease of removing H^+ from OH groups on P atoms that have no other O^- groups compared with those that already have one O^- group.

chelating abilities of chain and ring phosphates are due to structural hindrance of bonding by the ring polyphosphates.

3.16.4 Phosphonates

Phosphonate chelating agents are organic compounds structurally analogous to the aminopolycarboxylate chelating agents such as NTA and EDTA discussed earlier in this chapter. The structural formula of the undissociated acid form of a typical phosphonate chelating agent, nitrilotris(methylenephosphonic) acid, NTMP, is the following:

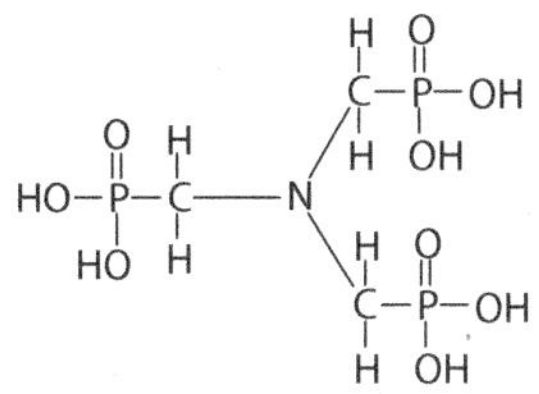

```
                     H  O
                     |  ||
                     C—P—OH
       O  H         /|  |
       ||  |       / H  OH
HO—P—C——N                    NTMP
       |   |       \ H  O
      HO  H         \|  ||
                     C—P—OH
                     |  |
                     H  OH
```

Note the similarity of the structural formula of this compound to NTA discussed earlier in this chapter.

The uses of phosphonate chelating agents are increasing steadily for applications that include inhibition of scale and corrosion, metal finishing, cleaning and laundry agents, ore recovery, and petroleum drilling. They are used in agricultural applications and in pulp, paper, and textile production. In part because of the difficulty in determining low levels of these substances in water, their environmental chemistry is not very well known. Although they are not very biodegradable, they interact strongly with surfaces and are removed with the sludge from biological waste treatment.

3.17 COMPLEXATION BY HUMIC SUBSTANCES

The most important class of complexing agents that occur naturally are the *humic substances*. These are degradation-resistant materials formed during the decomposition of vegetation that occur as deposits in soil, marsh sediments, peat, coal, lignite, or in almost any location where large quantities of vegetation have decayed. Humic substances abound in natural waters and soil.[6] They are commonly classified on the basis of solubility. If a material containing humic substances is extracted with strong base, and the resulting solution is acidified, the products are (a) a nonextractable plant residue called *humin*; (b) a material that precipitates from the acidified extract, called *humic acid*; and (c) an organic material that remains in the acidified solution, called *fulvic acid*. Because of their acid–base, sorptive, and complexing properties, both the soluble and insoluble humic substances have a strong effect upon the properties of water. In general, fulvic acid dissolves in water and exerts its effects as the soluble species. Humin and humic acid remain insoluble and affect water quality through exchange of species, such as cations or organic materials, with water.

Humic substances are high-molecular-mass, polyelectrolytic macromolecules. Molecular masses range from a few hundred for fulvic acid to tens of thousands for the humic acid and humin fractions. These substances contain a carbon skeleton with a high degree of aromatic character and with a large percentage of the molecular mass incorporated in functional groups, most of which contain oxygen. The elementary composition of most humic substances is within the following ranges: C, 45–55%; O, 30–45%; H, 3–6%; N, 1–5%; and S, 0–1%. The terms *humin*, *humic acid*, and *fulvic acid* do not refer to single compounds but to a wide range of compounds of generally similar origin with many properties in common. Humic substances have been known since before 1800, but their structural and chemical characteristics are still being explained.

Some feeling for the nature of humic substances may be obtained by considering the structure of a hypothetical molecule of fulvic acid shown:

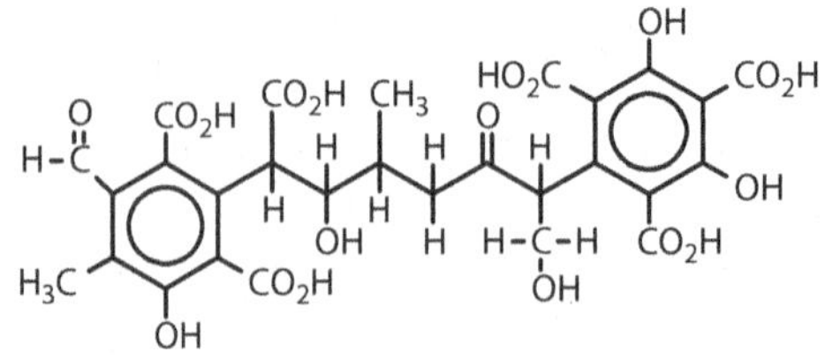

This structure is typical of the type of compound composing fulvic acid. The compound has a formula mass of 666, and its chemical formula may be represented as $C_{20}H_{15}(CO_2H)_6(OH)_5(CO)_2$. As shown in the hypothetical compound, the functional groups that may be present in fulvic acid are carboxyl, phenolic hydroxyl, alcoholic hydroxyl, and carbonyl. The functional groups vary with the particular acid sample. Approximate ranges in units of milliequivalents per gram of acid are total acidity, 12–14; carboxyl, 8–9; phenolic hydroxyl, 3–6; alcoholic hydroxyl, 3–5; and carbonyl, 1–3. In addition, some methoxyl groups, $-OCH_3$, may be encountered at low levels.

The binding of metal ions by humic substances is one of the most important environmental qualities of humic substances. This binding can occur as chelation between a carboxyl group and a phenolic hydroxyl group, as chelation between two carboxyl groups, or as complexation with a carboxyl group (see Figure 3.13).

One of the most significant characteristics of humic substances is their ability to bind with metal cations. Iron and aluminum are very strongly bound to humic substances, whereas magnesium is rather weakly bound. Other common ions, such as Ni^{2+}, Pb^{2+}, Ca^{2+}, and Zn^{2+}, are intermediate in their binding to humic substances.

The role played by soluble fulvic acid complexes of metals in natural waters is not well known. They probably keep some of the biologically important transition-metal ions in solution, and are particularly involved in iron solubilization and transport. Yellow fulvic acid-type compounds called *Gelbstoffe*, and frequently encountered along with soluble iron, are associated with color in water.

Insoluble humins and humic acids, effectively exchange cations with water and may accumulate large quantities of metals. Lignite coal, which is largely a humic acid material, tends to remove some metal ions from water.

Special attention has been given to humic substances since about 1970, following the discovery of *trihalomethanes* (THMs, such as chloroform and dibromochloromethane) in water supplies. It is now generally believed that these suspected carcinogens can be formed in the presence of humic substances during the disinfection of raw municipal drinking water by chlorination (see Chapter 8). The humic substances produce THMs by reaction with chlorine. The formation of THMs can be reduced by removing as much of the humic material as possible prior to chlorination.

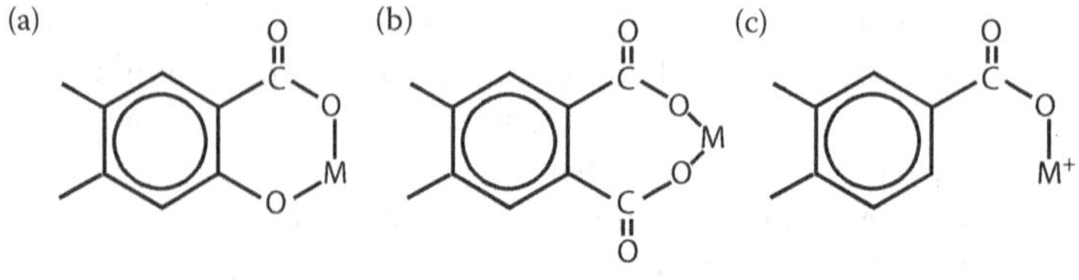

FIGURE 3.13 Binding of a metal ion, M^{2+}, by humic substances (a) by chelation between carboxyl and phenolic hydroxyl, (b) by chelation between two carboxyl groups, and (c) by complexation with a carboxyl group.

3.18 COMPLEXATION AND REDOX PROCESSES

Complexation may have a strong effect upon oxidation–reduction equilibria by shifting reactions, such as that for the oxidation of lead,

$$Pb \rightleftarrows Pb^{2+} + 2e^- \quad (3.93)$$

strongly to the right by binding to the product ion, thus cutting its concentration down to very low levels. Perhaps more important is the fact that upon oxidation,

$$M + \tfrac{1}{2}O_2 \rightleftarrows MO \quad (3.94)$$

many metals form self-protective coatings of oxides, carbonates, or other insoluble species that prevent further chemical reaction. Copper and aluminum roofing and structural iron are examples of materials that are thus self-protecting. A chelating agent in contact with such metals can result in continual dissolution of the protective coating so that the exposed metal corrodes readily. For example, chelating agents in wastewater may increase the corrosion of metal plumbing, thus adding heavy metals to effluents. Solutions of chelating agents employed to clean metal surfaces in metal plating operations have a similar effect.

LITERATURE CITED

1. Graham, M. C. and J. G. Farmer, Chemistry of freshwaters, in *Principles of Environmental Chemistry*, R.M. Harrison, Ed., pp. 80–169, Royal Society of Chemistry, Cambridge, UK, 2007.
2. Anon, Carbon conveyer, *Nature Geoscience*, **2**, 1, 2009.
3. Nowack, B. and J. M. VanBriesen, Eds, *Biogeochemistry of Chelating Agents*, American Chemical Society, Washington, DC, 2005.
4. Yuan, Z. and J. M. VanBriesen, The formation of intermediates in EDTA and NTA biodegradation, *Environmental Engineering Science*, **23**, 533–544, 2006.
5. Bell, P. R. and I. Elmetri, Some chemical factors regulating the growth of *Lyngbya majuscula* in Moreton Bay, Australia: Importance of sewage discharges, *Hydrobiologia*, **592**, 359–371, 2007.
6. Van Zomeren, A., A. Costa, J. P. Pinheiro, and R. N. J. Comans, Proton binding properties of humic substances originating from natural and contaminated materials, *Environmental Science and Technology*, **43**, 1393–1399, 2009.

SUPPLEMENTARY REFERENCES

Benjamin, M. M., *Water Chemistry*, McGraw-Hill, New York, 2002.

Berk, Z., *Water Science for Food, Health, Agriculture and Environment*, Technomic Publishing, Lancaster, PA, 2001.

Bianchi, T., *Biogeochemistry of Estuaries*, Oxford University Press, Oxford, 2007.

Brightwell, C., *Marine Chemistry*, T.F.H. Publications, Neptune City, NJ, 2007.

Brownlow, A. H., *Geochemistry*, 2nd ed., Prentice-Hall, Upper Saddle River, NJ, 1996.

Buffle, J. D., *Complexation Reactions in Aquatic Systems: An Analytical Approach*, Ellis Horwood, Chichester, 1988.

Butler, J. N., *Ionic Equilibrium: Solubility and pH Calculations*, Wiley, New York, 1998.

Closs, G., B. Downes, and A. Boulton, *Freshwater Ecology: A Scientific Introduction*, Blackwell Publishing Company, Malden, MA, 2004.

Dodds, W. K., *Freshwater Ecology: Concepts and Environmental Applications*, Academic Press, San Diego, CA, 2002.

Dodson, S. I., *Introduction to Limnology*, McGraw-Hill, New York, 2005.

Drever, J. I., *The Geochemistry of Natural Waters: Surface and Groundwater Environments*, 3rd ed., Prentice-Hall, Upper Saddle River, NJ, 1997.

Eby, G. N., *Principles of Environmental Geochemistry*, Thomson-Brooks/Cole, Pacific Grove, GA, 2004.

Essington, M. E., *Soil and Water Chemistry: An Integrative Approach*, Taylor & Francis/CRC Press, Boca Raton 2004.

Faure, G., *Principles and Applications of Geochemistry: A Comprehensive Textbook for Geology Students*, 2nd ed., Prentice-Hall, Upper Saddle River, NJ, 1998.

Findlay, S. E. G. and R. L. Sinsabaugh, Eds, *Aquatic Ecosystems: Interactivity of Dissolved Organic Matter*, Academic Press, San Diego, CA, 2003.

Ghabbour, E. A. and G. Davis, Eds, *Humic Substances: Nature's Most Versatile Materials*, Garland Publishing, New York, 2003.

Hem, J. D., *Study and Interpretation of the Chemical Characteristics of Natural Water*, 2nd ed., U.S. Geological Survey Paper 1473, Washington, DC, 1970.

Hessen, D. O. and L. J. Tranvik, Eds, *Aquatic Humic Substances: Ecology and Biogeochemistry*, Springer Verlag, Berlin, 1998.

Holland, H. D. and K. K. Turekian, Eds, *Treatise on Geochemistry*, Elsevier/Pergamon, Amsterdam, 2004.

Howard, A. G., *Aquatic Environmental Chemistry*, Oxford University Press, Oxford, UK, 1998.

Hynes, H. B. N., *The Ecology of Running Waters*, Blackburn Press, Caldwell, NJ, 2001.

Jensen, J. N., *A Problem-Solving Approach to Aquatic Chemistry*, Wiley, New York, 2003.

Jones, J. B. and P. J. Mulholland, Eds, *Streams and Ground Waters*, Academic Press, San Diego, CA, 2000.

Kalff, J., *Limnology: Inland Water Ecosystems*, Prentice-Hall, Upper Saddle River, NJ, 2002.

Kumar, A., Ed., *Aquatic Ecosystems*, A.P.H. Publishing Corporation, New Delhi, 2003.

Langmuir, D., *Aqueous Environmental Geochemistry*, Prentice-Hall, Upper Saddle River, NJ, 1997.

Morel, F. M. M. and J. G. Hering, *Principles and Applications of Aquatic Chemistry*, Wiley-Interscience, New York, 1993.

Polevoy, S., *Water Science and Engineering*, Krieger Publishing, Melbourne, FL, 2003.

Spellman, F. R., *The Science of Water: Concepts and Applications*, 2nd ed., Taylor & Francis/CRC Press, Boca Raton, FL, 2008.

Steinberg, C., *Ecology of Humic Substances in Freshwaters: Determinants from Geochemistry to Ecological Niches*, Springer, Berlin, 2003.

Stewart, B. A. and T. A. Howell, Eds, *Encyclopedia of Water Science*, Marcel Dekker, New York, 2003.

Stober, I. and K. Bucher, Eds, *Water–Rock Interaction*, Kluwer Academic Publishers, Hingham, MA, 2002.

Stumm, W. and J. J. Morgan, *Aquatic Chemistry: Chemical Equilibria and Rates in Natural Waters*, 3rd ed., Wiley, New York, 1995.

Stumm, W., *Chemistry of the Solid–Water Interface: Processes at the Mineral-Water and Particle-Water Interface in Natural Systems*, Wiley, New York, 1992.

Trimble, S. W., *Encyclopedia of Water Science*, 2nd ed., Taylor & Francis/CRC Press, Boca Raton, FL, 2008.

Weiner, E. R., *Applications of Environmental Aquatic Chemistry: A Practical Guide*, 2nd ed., Taylor & Francis/CRC Press, Boca Raton, FL, 2008.

Welch, E. B. and J. Jacoby, *Pollutant Effects in Fresh Waters: Applied Limnology*, Taylor & Francis, London, 2007.

Wetzel, R. G., *Limnology: Lake and River Ecosystems*, 3rd ed., Academic Press, San Diego, CA, 2001.

Wiener, E., R., *Applications of Environmental Aquatic Chemistry: A Practical Guide*, 2nd ed., Taylor & Francis/CRC Press, Boca Raton, FL, 2008.

QUESTIONS AND PROBLEMS

1. Alkalinity is determined by titration with standard acid. The alkalinity is often expressed as mg/L of $CaCO_3$. If V_p mL of acid of normality N are required to titrate V_s mL of sample to the phenolphthalein end point, what is the formula for the phenolphthalein alkalinity as mg/L of $CaCO_3$?
2. Exactly 100 pounds of cane sugar (dextrose), $C_{12}H_{22}O_{11}$, were accidentally discharged into a small stream saturated with oxygen from the air at 25°C. How many liters of this water could be contaminated to the extent of removing all the DO by biodegradation?
3. Water with an alkalinity of 2.00×10^{-3} eq/L has a pH of 7.0. Calculate $[CO_2]$, $[HCO_3^-]$, $[CO_3^{2-}]$, and $[OH^-]$.
4. Through the photosynthetic activity of algae, the pH of the water in Problem 3 was changed to 10.0. Calculate all the preceding concentrations and the weight of biomass, $\{CH_2O\}$, produced. Assume no input of atmospheric CO_2.

5. Calcium chloride is quite soluble, whereas the solubility product of calcium fluoride, CaF_2, is only 3.9×10^{-11}. A waste stream of 1.00×10^{-3} M HCl is injected into a formation of limestone, $CaCO_3$, where it comes into equilibrium. Give the chemical reaction that occurs and calculate the hardness and alkalinity of the water at equilibrium. Do the same for a waste stream of 1.00×10^{-3} M HF.
6. For a solution having 1.00×10^{-3} eq/L total alkalinity (contributions from HCO_3^-, CO_3^{2-}, and OH^-) at $[H^+] = 4.69 \times 10^{-11}$, what is the percentage contribution to alkalinity from CO_3^{2-}?
7. A wastewater disposal well for carrying various wastes at different times is drilled into a formation of limestone ($CaCO_3$), and the wastewater has time to come to complete equilibrium with the calcium carbonate before leaving the formation through an underground aquifer. Of the following components in the wastewater, which of the following would not cause an increase in alkalinity due either to the component itself or to its reaction with limestone: (a) NaOH, (b) CO_2, (c) HF, (d) HCl, or (e) all of these.
8. Calculate the ratio $[PbT^-]/[HT^{2-}]$ for NTA in equilibrium with $PbCO_3$ in a medium having $[HCO_3^-] = 3.00 \times 10^{-3}$ M.
9. If the medium in Problem 8 contained excess calcium such that the concentration of uncomplexed calcium, $[Ca^{2+}]$, were 5.00×10^{-3} M, what would be the ratio $[PbT^-]/[CaT^-]$ at pH 7?
10. A wastewater stream containing 1.00×10^{-3} M disodium NTA, Na_2HT, as the only solute is injected into a limestone ($CaCO_3$) formation through a waste disposal well. After going through this aquifer for some distance and reaching equilibrium, the water is sampled through a sampling well. What is the reaction between NTA species and $CaCO_3$? What is the equilibrium constant for the reaction? What are the equilibrium concentrations of CaT^-, HCO_3^-, and HT^{2-}? (The appropriate constants may be looked up in this chapter.)
11. If the wastewater stream in Problem 10 were 0.100 M in NTA and contained other solutes that exerted a buffering action such that the final pH were 9.0, what would be the equilibrium value of HT^{2-} concentration in mol/L?
12. Exactly 1.00×10^{-3} mol of $CaCl_2$, 0.100 mol of NaOH, and 0.100 mol of Na_3T were mixed and then diluted to 1.00 L. What was the concentration of Ca^{2+} in the resulting mixture?
13. How does chelation influence corrosion?
14. The following ligand has more than one site for binding to a metal ion. How many such sites does it have?

```
    O  H     H  O
    ||  |     |  ||
 ⁻O–C–C–N–C–C–O⁻
        |     |
        H     H
```

15. If a solution containing initially 25 mg/L trisodium NTA is allowed to come to equilibrium with solid $PbCO_3$ at pH 8.5 in a medium that contains 1.76×10^{-3} M HCO_3^- at equilibrium, what is the value of the ratio of the concentration of NTA bound with lead to the concentration of unbound NTA, $[PbT^-]/[HT^{2-}]$?
16. After a low concentration of NTA has equilibrated with $PbCO_3$ at pH 7.0 in a medium having $[HCO_3^-] = 7.50 \times 10^{-4}$ M, what is the ratio of $[PbT]/[HT^{2-}]$?
17. What detrimental effect may dissolved chelating agents have on conventional biological waste treatment?
18. Why is chelating agent usually added to artificial algal growth media?
19. What common complex compound of magnesium is essential to certain life processes?
20. What is always the ultimate product of polyphosphate hydrolysis?
21. A solution containing initially 1.00×10^{-5} M CaT^- is brought to equilibrium with solid $PbCO_3$. At equilibrium, pH = 7.0, $[Ca^{2+}] = 1.50 \times 10^{-3}$ M and $[HCO_3^-] = 1.10 \times 10^{-3}$ M. At equilibrium, what is the fraction of total NTA in solution as PbT^-?

22. What is the fraction of NTA present after HT^{2-} has been brought to equilibrium with solid $PbCO_3$ at pH 7.0 in a medium in which $[HCO_3^-] = 1.25 \times 10^{-3}$ M.
23. Describe ways in which measures taken to alleviate water supply and flooding problems might actually aggravate such problems.
24. The study of water is known as ______________________, ______________________ is the branch of the science dealing with the characteristics of freshwater, and the science that deals with about 97% of all the Earth's water is called ______________________.
25. Consider the hydrological cycle in Figure 3.1. List or discuss the kinds or classes of environmental chemistry that might apply to each major part of this cycle.
26. Consider the unique and important properties of water. What molecular or bonding characteristics of the water molecules are largely responsible for these properties? List or describe one of the following unique properties of water: (a) thermal characteristics, (b) transmission of light, (c) surface tension, and (d) solvent properties.
27. Discuss how thermal stratification of a body of water may affect its chemistry.
28. Relate aquatic life to aquatic chemistry. In so doing, consider the following: autotrophic organisms, producers, heterotrophic organisms, decomposers, eutrophication, dissolved oxygen, and biochemical oxygen demand.
29. Assuming levels of atmospheric CO_2 are 390 ppm CO_2, what is the pH of rainwater due to the presence of carbon dioxide? Some estimates are for atmospheric carbon dioxide levels to double in the future. What would be the pH of rainwater if this happens?
30. Assume a sewage treatment plant processing 1 million liters of wastewater per day containing 200 mg/L of degradable biomass, $\{CH_2O\}$. Calculate the volume of dry air at 25°C that must be pumped into the wastewater per day to provide the oxygen required to degrade the biomass (Reaction 3.1).
31. Anaerobic bacteria growing in a lake sediment produced equal molar amounts of carbon dioxide and carbon monoxide according to the biochemical reaction $2\{CH_2O\} \rightarrow CO_2 + CH_4$, so that the water in the lake was saturated with both CO_2 gas and CH_4 gas. In units of mol/L/atm the Henry's law constant for CO_2 is 3.38×10^{-2} and that of CH_4 has a value of 1.34×10^{-3}. At the depth at which the gas was being evolved, the total pressure was 1.10 atm and the temperature was 25°C, so the vapor pressure of water was 0.0313 atm. Calculate the concentrations of dissolved CO_2 and dissolved CH_4.

4 Oxidation–Reduction in Aquatic Chemistry

4.1 THE SIGNIFICANCE OF OXIDATION–REDUCTION

Oxidation–reduction (*redox*) reactions are those involving changes of oxidation states of reactants. Such reactions are easiest to visualize as the transfer of electrons from one species to another. For example, soluble cadmium ion, Cd^{2+}, is removed from wastewater by reaction with metallic iron. The overall reaction is

$$Cd^{2+} + Fe \rightarrow Cd + Fe^{2+} \tag{4.1}$$

This reaction is the sum of two *half-reactions*, a reduction half-reaction in which cadmium ion accepts two electrons and is reduced,

$$Cd^{2+} + 2e^- \rightarrow Cd \tag{4.2}$$

and an oxidation half-reaction in which elemental iron is oxidized:

$$Fe \rightarrow Fe^{2+} + 2e^- \tag{4.3}$$

When these two half-reactions are added algebraically, the electrons cancel on both sides and the result is the overall reaction given in Equation 4.1.

Oxidation–reduction phenomena are highly significant in the environmental chemistry of natural waters and wastewaters. In a lake, for example, the reduction of oxygen (O_2) by organic matter (represented by $\{CH_2O\}$),

$$\{CH_2O\} + O_2 \rightarrow CO_2 + H_2O \tag{4.4}$$

results in oxygen depletion, which can kill fish. The rate at which sewage is oxidized is crucial to the operation of a waste treatment plant. Reduction of insoluble iron(III) to soluble iron(II),

$$Fe(OH)_3(s) + 3H^+ + e^- \rightarrow Fe^{2+} + 3H_2O \tag{4.5}$$

in a reservoir contaminates the water with dissolved iron, which is hard to remove in the water treatment plant. Oxidation of NH_4^+ to NO_3^- in water

$$NH_4^+ + 2O_2 \rightarrow NO_3^- + 2H^+ + H_2O \tag{4.6}$$

converts ammonium nitrogen to nitrate, a form more assimilable by algae in the water. Many other examples can be cited of the ways in which the types, rates, and equilibria of redox reactions largely determine the nature of important solute species in water.

This chapter discusses redox processes and equilibria in water. In so doing, it emphasizes the concept of p*E*, analogous to pH and defined as the negative log of electron activity. Low p*E* values are indicative of reducing conditions; high p*E* values reflect oxidizing conditions.

Two important points should be stressed regarding redox reactions in natural waters and wastewaters. First, as is discussed in Chapter 6, "Aquatic Microbial Biochemistry," many of the most important redox reactions are catalyzed by microorganisms. Bacteria are the catalysts by which molecular oxygen reacts with organic matter, iron(III) is reduced to iron(II), and ammonia is oxidized to nitrate ion.

The second important point regarding redox reactions in the hydrosphere is their close relationship to acid–base reactions. Whereas the activity of the H^+ ion is used to express the extent to which water is acidic or basic, the activity of the electron, e^-, is used to express the degree to which an aquatic medium is oxidizing or reducing. Water with a high hydrogen-ion activity, such as runoff from "acid rain," is *acidic*. By analogy, water with a high *electron* activity, such as that in the anoxic digester of a sewage treatment plant, is said to be *reducing*. Water with a low H^+ ion activity (high concentration of OH^-)—such as landfill leachate contaminated with waste sodium hydroxide—is *basic*, whereas water with a low electron activity—highly chlorinated water, for example—is said to be *oxidizing*. Actually, neither free electrons nor free H^+ ions as such are found dissolved in aquatic solution; they are always strongly associated with solvent or solute species. However, the concept of electron activity, like that of hydrogen-ion activity, remains a very useful one to the aquatic chemist.

Many species in water undergo exchange of both electrons and H^+ ions. For example, acid mine water contains the hydrated iron(III) ion, $Fe(H_2O)_6^{3+}$, which readily loses H^+ ion

$$Fe(H_2O)_6^{3+} \rightleftarrows Fe(H_2O)_5OH^{2+} + H^+ \tag{4.7}$$

to contribute acidity to the medium. The same ion accepts an electron

$$Fe(H_2O)_6^{3+} + e^- \rightleftarrows Fe(H_2O)_6^{2+} \tag{4.8}$$

to give iron(II).

Generally, the transfer of electrons in a redox reaction is accompanied by H^+ ion transfer, and there is a close relationship between redox and acid–base processes. For example, if iron(II) loses an electron at pH 7, three hydrogen ions are also lost to form highly insoluble iron(II) hydroxide,

$$Fe(H_2O)_6^{2+} \rightleftarrows e^- + Fe(OH)_3(s) + 3H_2O + 3H^+ \tag{4.9}$$

an insoluble, gelatinous solid.

The stratified body of water shown in Figure 4.1 can be used to illustrate redox phenomena and relationships in an aquatic system. The anoxic sediment layer is so reducing that carbon can be reduced to its lowest possible oxidation state, –4 in CH_4. If the lake becomes anoxic, the hypolimnion may contain elements in their reduced states: NH_4^+ for nitrogen, H_2S for sulfur, and soluble $Fe(H_2O)_6^{2+}$ for iron. Saturation with atmospheric oxygen makes the surface layer a relatively oxidizing medium. If allowed to reach thermodynamic equilibrium, it is characterized by the more oxidized forms of the elements present: CO_2 for carbon, NO_3^- for nitrogen, iron as insoluble $Fe(OH)_3$, and sulfur as SO_4^{2-}. Substantial changes in the distribution of chemical species in water resulting from redox reactions are vitally important to aquatic organisms and have a tremendous influence on water quality.

It should be pointed out that the systems presented in this chapter are assumed to be at equilibrium, a state almost never achieved in any real natural water or wastewater system. Most real

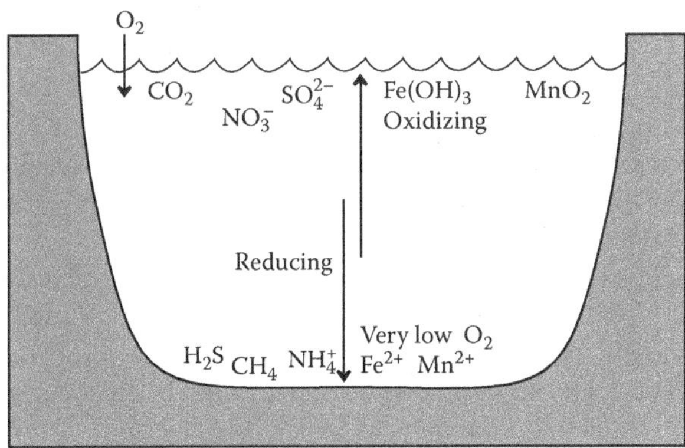

FIGURE 4.1 Predominance of various chemical species in a stratified body of water that has a high oxygen concentration (oxidizing, high pE) near the surface and a low oxygen concentration (reducing, low pE) near the bottom.

aquatic systems are dynamic systems that may approach a steady state, rather than true equilibrium. Nevertheless, the picture of a system at equilibrium is very useful in visualizing trends in natural water and wastewater systems, yet the model is still simple enough to comprehend. It is important to realize the limitations of such a model, however, especially in trying to make measurements of the redox status of water.

4.2 ELECTRON AND REDOX REACTIONS

In order to explain redox processes in natural waters, it is necessary to have an understanding of redox reactions. In a formal sense such reactions can be viewed as the transfer of electrons between species. This section considers such reactions in a simple system. All redox reactions involve changes in the oxidation states of some of the species that take part in the reaction. Consider, for example, a solution containing iron(II) and iron(III) that is sufficiently acidic to prevent precipitation of solid $Fe(OH)_3$; such a medium might be acid mine water or a steel pickling liquor waste. Suppose that the solution is treated with elemental hydrogen gas over a suitable catalyst to bring about the reduction of iron(III) to iron(II). The overall reaction can be represented as

$$2Fe^{3+} + H_2 \rightleftarrows 2Fe^{2+} + 2H^+ \tag{4.10}$$

The reaction is written with a double arrow, indicating that it is *reversible* and could proceed in either direction; for normal concentrations of reaction participants, this reaction goes to the right. As the reaction goes to the right, the hydrogen is *oxidized* as it changes from an *oxidation state* (number) of 0 in elemental H_2 to a higher oxidation number of +1 in H^+. The oxidation state of iron goes from +3 in Fe^{3+} to +2 in Fe^{2+}; the oxidation number of iron decreases, which means that it is *reduced.*

All redox reactions such as this one can be broken down into a reduction *half-reaction*, in this case

$$2Fe^{3+} + 2e^- \rightleftarrows 2Fe^{2+} \tag{4.11}$$

(for one electron, $Fe^{3+} + e^- \rightleftarrows Fe^{3+}$), and an oxidation half-reaction, in this case

$$H_2 \rightleftarrows 2H^+ + 2e^- \tag{4.12}$$

Note that adding these two half-reactions together gives the overall reaction. *The addition of an oxidation half-reaction and a reduction half-reaction, each expressed for the same number of electrons so that the electrons cancel on both sides of the arrows, gives a whole redox reaction.*

The equilibrium of a redox reaction, that is, the degree to which the reaction as written tends to lie to the right or left, can be deduced from information about its constituent half-reactions. To visualize this, assume that the two half-reactions can be separated into two half-cells of an electrochemical cell as shown for Reaction 4.10 in Figure 4.2.

If the initial activities of H^+, Fe^{2+}, and Fe^{3+} were of the order of 1 (concentrations of 1 M), and if the pressure of H_2 were 1 atm, H_2 would be oxidized to H^+ in the left half-cell, Fe^{3+} would be reduced to Fe^{2+} in the right half-cell, and ions would migrate through the salt bridge to maintain electroneutrality in both half-cells. The net reaction occurring would be Reaction 4.10.

If a voltmeter were inserted in the circuit between the two electrodes, no significant current could flow and the two half-reactions could not take place. However, the voltage registered by the voltmeter would be a measure of the relative tendencies of the two half-reactions to occur. In the left half-cell the oxidation half-reaction

$$H_2 \rightleftarrows 2H^+ + 2e^- \tag{4.12}$$

will tend to go to the right, releasing electrons to the platinum electrode in the half-cell and giving that electrode a relatively negative (–) potential. In the right half-cell, the reduction half-reaction

$$2Fe^{3+} + e^- \rightleftarrows Fe^{2+} \tag{4.11}$$

will tend to go to the right, taking electrons from the platinum electrode in the half-cell and giving that electrode a relatively positive (+) potential. The difference in these potentials is a measure of the "driving force" of the overall reaction. If each of the reaction participants were at unit activity, the potential difference would be 0.77 V.

The left electrode shown in Figure 4.2 is the standard electrode against which all other electrode potentials are compared. It is called the *standard hydrogen electrode* (*SHE*). This electrode has

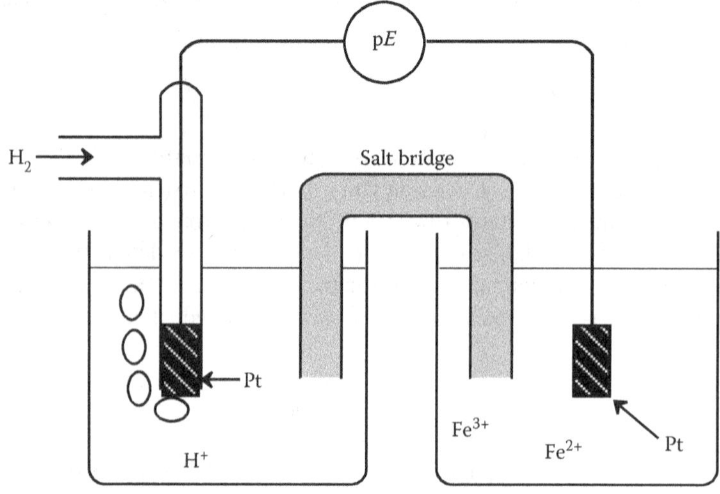

FIGURE 4.2 Electrochemical cell in which the reaction $2Fe^{3+} + 2H^+ \rightleftarrows 2Fe^{2+} + H^+$ can be carried out in two half-cells.

been assigned a potential value of exactly 0 V by convention, and its half-reaction is written as the following:

$$2H^+ + 2e^- \rightleftarrows H_2, \quad E^0 = 0.00\,V \tag{4.13}$$

The measured potential of the right-hand electrode in Figure 4.2 versus the SHE is called the *electrode potential*, *E*. If the Fe^{2+} and Fe^{3+} ions are both at unit activity, the potential is the *standard electrode potential* (according to IUPAC convention, *the standard reduction potential*), E^0. The standard electrode potential for the Fe^{3+}/Fe^{2+} couple is 0.77 V expressed conventionally as follows:

$$Fe^{3+} + e^- \rightleftarrows Fe^{2+}, \quad E^0 = +0.77\,V \tag{4.14}$$

4.3 ELECTRON ACTIVITY AND p*E*

In this book, for the most part, p*E* and pE^0 are used instead of *E* and E^0 to more clearly illustrate redox equilibria in aquatic systems over many orders of magnitude of electron activity in a manner analogous to pH. Numerically, p*E* and pE^0 are simply the following:

$$pE = \frac{E}{2.303RT/F} = \frac{E}{0.0591} \quad \text{(at 25°C)} \tag{4.15}$$

$$pE^0 = \frac{E^0}{2.303RT/F} = \frac{E^0}{0.0591} \quad \text{(at 25°C)} \tag{4.16}$$

where *R* is the molar gas constant, *T* is the absolute temperature, and *F* is the Faraday constant. The "p*E* concept" is explained below.

Just as pH is defined as

$$pH = -\log(a_H^+) \tag{4.17}$$

where a_H^+ is the activity of hydrogen ion in solution, p*E* is defined as

$$pE = -\log(a_{e^-}) \tag{4.18}$$

where a_{e^-} is the activity of the electron in solution. Since hydrogen-ion concentration may vary over many orders of magnitude, pH is a convenient way of expressing a_H^+ in terms of manageable numbers. Similarly, electron activities in water may vary over more than 20 orders of magnitude so that it is convenient to express a_{e^-} as p*E*.

Values of p*E* are defined in terms of the following half-reaction for which pE^0 is defined as exactly zero:*

$$2H^+(aq) + 2e^- \rightleftarrows H_2(g), \quad E^0 = +0.00\,V, \quad pE^0 = 0.00 \tag{4.19}$$

* Thermodynamically, the free-energy change for this reaction is defined as exactly zero when all reaction participants are at unit activity. For ionic solutes, activity—the effective concentration in a sense—approaches concentration at low concentrations and low ionic strengths. The activity of a gas is equal to its partial pressure. Furthermore, the free energy, *G*, decreases for spontaneous processes occurring at constant temperature and pressure. Processes for which the free-energy change, Δ*G*, is zero have no tendency toward spontaneous change and are in a state of equilibrium. Reaction 4.13 is the one upon which free energies of formation of all ions in aqueous solution are based. It also forms the basis for defining free-energy changes for oxidation–reduction processes in water.

It is relatively easy to visualize the activities of ions in terms of concentration, whereas it is harder to visualize the activity of the electron, and therefore pE, in similar terms. For example, at 25°C in pure water, a medium of zero ionic strength, the hydrogen-ion concentration is 1.0×10^{-7} M, the hydrogen-ion *activity* is 1.0×10^{-7}, and the pH is 7.0. The electron activity, however, must be defined in terms of Equation 4.19. When H^+ (*aq*) at unit activity is in equilibrium with hydrogen gas at 1 atm pressure (and likewise at unit activity), the activity of the electron in the medium is exactly 1.00 and the pE is 0.0. If the electron activity were increased by a factor of 10 (as would be the case if H^+ (*aq*) at an activity of 0.100 were in equilibrium with H_2 at an activity of 1.00), the electron activity would be 10 and the pE value would be –1.0.

4.4 THE NERNST EQUATION

The *Nernst equation* is used to account for the effect of different activities upon electrode potential. Referring to Figure 4.2, if the Fe^{3+} ion concentration is increased relative to the Fe^{2+} ion concentration, it is readily visualized that the potential and the pE of the right electrode will become more positive because the higher concentration of electron-deficient Fe^{3+} ions clustered around it tends to draw electrons from the electrode. Decreased Fe^{3+} ion or increased Fe^{2+} ion concentration has the opposite effect. Such concentration effects upon E and pE are expressed by the *Nernst equation*. As applied to the half-reaction

$$\mathrm{Fe}^{3+} + e^- \rightleftarrows \mathrm{Fe}^{2+}, \quad E^0 = +0.77\,\mathrm{V}, \quad \mathrm{p}E^0 = 13.2 \tag{4.20}$$

the Nernst equation is the following, where $2.303RT/F = 0.0591$ at 25°C:

$$E = E^0 + \frac{2.303RT}{nF}\log\frac{[\mathrm{Fe}^{3+}]}{[\mathrm{Fe}^{2+}]} = E^0 + \frac{0.0591}{n}\log\frac{[\mathrm{Fe}^{3+}]}{[\mathrm{Fe}^{2+}]} \tag{4.21}$$

where n is the number of electrons involved in the half-reaction (1 in this case), and the activities of Fe^{3+} and Fe^{2+} ions have been taken as their concentrations (a simplification valid for more dilute solutions, which will be made throughout this chapter). Considering that

$$\mathrm{p}E = \frac{E}{2.303RT/F} \quad \text{and} \quad \mathrm{p}E^0 = \frac{E^0}{2.303RT/F}$$

the Nernst equation can be expressed in terms of pE and pE^0

$$\mathrm{p}E = \mathrm{p}E^0 + \frac{1}{n}\log\frac{\mathrm{Fe}^{3+}}{\mathrm{Fe}^{2+}} \quad (\text{in this case } n = 1) \tag{4.22}$$

The Nernst equation in this form is quite simple and offers some advantages in calculating redox relationships.

If, for example, the value of $[Fe^{3+}]$ is 2.35×10^{-3} M and $[Fe^{2+}] = 7.85 \times 10^{-5}$ M, the value of pE is

$$\mathrm{p}E = 13.2 + \log\frac{2.35 \times 10^{-3}}{7.85 \times 10^{-5}} = 14.7 \tag{4.23}$$

As the concentration of Fe^{3+} increases relative to the concentration of Fe^{2+}, the value of pE becomes higher (more positive) and as the concentration of Fe^{2+} increases relative to the concentration of Fe^{3+}, the value of pE becomes lower (more negative).

4.5 REACTION TENDENCY: WHOLE REACTION FROM HALF-REACTIONS

This section discusses how half-reactions can be combined to give whole reactions and how the pE^0 values of the half-reactions can be used to predict the directions in which reactions will go. The half-reactions discussed here are the following:

$$Hg^{2+} + 2e^- \rightleftarrows Hg, \quad pE^0 = 13.35 \tag{4.24}$$

$$Fe^{3+} + e^- \rightleftarrows Fe^{2+}, \quad pE^0 = 13.2 \tag{4.25}$$

$$Cu^{2+} + 2e^- \rightleftarrows Cu, \quad pE^0 = 5.71 \tag{4.26}$$

$$2H^+ + 2e^- \rightleftarrows H_2, \quad pE^0 = 0.00 \tag{4.27}$$

$$Pb^{2+} + 2e^- \rightleftarrows Pb, \quad pE^0 = -2.13 \tag{4.28}$$

Such half-reactions and their pE^0 values can be used to explain observations such as the following: A solution of Cu^{2+} flows through a lead pipe and the lead acquires a layer of copper metal through the reaction

$$Cu^{2+} + Pb \rightarrow Cu + Pb^{2+} \tag{4.29}$$

This reaction occurs because the copper(II) ion has a greater tendency to acquire electrons than the lead ion has to retain them. This reaction can be obtained by subtracting the lead half-reaction, Equation 4.28, from the copper half-reaction, Equation 4.26:

$$Cu^{2+} + 2e^- \rightleftarrows Cu, \qquad pE^0 = 5.71 \tag{4.26}$$

$$-(Pb^{2+} + 2e^- \rightleftarrows Pb), \qquad pE^0 = -2.13 \tag{4.28}$$

$$Cu^{2+} + Pb \rightleftarrows Cu + Pb^{2+}, \qquad pE^0 = 7.84 \tag{4.30}$$

The positive value of pE^0 for this reaction, 7.84, indicates that the reaction tends to go to the right as written. This occurs when lead metal directly contacts a solution of copper(II) ion. Therefore, if a waste solution containing copper(II) ion, a relatively innocuous pollutant, comes into contact with lead in plumbing, toxic lead may go into solution.

In principle, half-reactions may be allowed to occur in separate electrochemical half-cells, as could occur for Reaction 4.30 in the cell shown in Figure 4.3 if the meter (pE) were bypassed by an electrical conductor; they are therefore called *cell reactions.*

If the activities of Cu^{2+} and Pb^{2+} are not unity, the direction of the reaction and value of pE are deduced from the Nernst equation. For Reaction 4.30 the Nernst equation is

$$pE = pE^0 + \frac{1}{n}\log\frac{[Cu^{2+}]}{[Pb^{2+}]} = 7.84 + \frac{1}{2}\log\frac{[Cu^{2+}]}{[Pb^{2+}]} \tag{4.31}$$

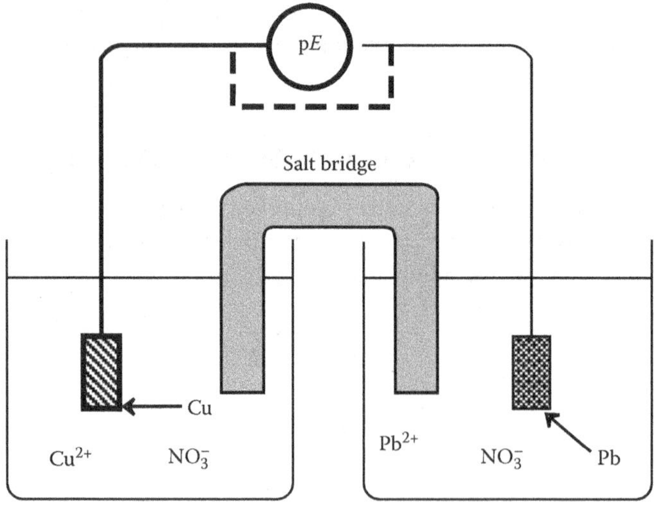

FIGURE 4.3 Cell for the measurement of pE between a lead half-cell and a copper half-cell. In this configuration the "pE" meter has a very high resistance and current cannot flow.

By combining the appropriate half-reactions, it can be shown that copper metal will not cause hydrogen gas to be evolved from solutions of strong acid [hydrogen ion has less attraction for electrons than does copper(II) ion], whereas lead metal, in contrast, will displace hydrogen gas from acidic solutions.

4.6 THE NERNST EQUATION AND CHEMICAL EQUILIBRIUM

Refer again to Figure 4.3. Imagine that instead of the cell being set up to measure the potential between the copper and lead electrodes, the voltmeter used to measure pE was removed and the electrodes directly connected with a wire so that the current might flow between them. The reaction

$$\mathrm{Cu^{2+} + Pb \rightleftarrows Cu + Pb^{2+}}, \quad \mathrm{p}E^0 = 7.84 \tag{4.30}$$

will occur until the concentration of lead ion becomes so high, and that of copper ion so low, that the reaction stops. The system is at equilibrium and, since current no longer flows, pE is exactly zero. The equilibrium constant K for the reaction is given by the expression

$$K = \frac{[\mathrm{Pb^{2+}}]}{[\mathrm{Cu^{2+}}]} \tag{4.32}$$

The equilibrium constant can be calculated from the Nernst equation, noting that under equilibrium conditions pE is zero and $[Cu^{2+}]$ and $[Pb^{2+}]$ are at equilibrium concentrations:

$$\begin{aligned} \mathrm{p}E &= \mathrm{p}E^0 + \frac{1}{n}\log\frac{[\mathrm{Cu^{2+}}]}{[\mathrm{Pb^{2+}}]} = 7.84 + \frac{1}{2}\log\frac{[\mathrm{Cu^{2+}}]}{[\mathrm{Pb^{2+}}]} \\ \mathrm{p}E &= 0.00 = 7.84 - \frac{1}{2}\log\frac{[\mathrm{Pb^{2+}}]}{[\mathrm{Cu^{2+}}]} = 7.84 - \frac{1}{2}\log K \end{aligned} \tag{4.33}$$

Note that the reaction products are placed over reactants in the log term, and a minus sign is placed in front to put the equilibrium constant in the correct form (a purely mathematical operation). The value of $\log K$ obtained from solving the above equation is 15.7.

The equilibrium constant for a redox reaction involving n electrons is given in terms of pE simply by

$$\log K = n(\mathrm{p}E^0) \tag{4.34}$$

4.7 THE RELATIONSHIP OF pE TO FREE ENERGY

Aquatic systems and the organisms that inhabit them—just like the steam engine or students hoping to pass physical chemistry—must obey the laws of thermodynamics. Bacteria, fungi, and human beings derive their energy from acting as mediators (catalysts) of chemical reactions and extracting a certain percentage of useful energy from them. In predicting or explaining the behavior of an aquatic system, it is helpful to be able to predict the useful energy that can be extracted from chemical reactions in the system, such as microbially mediated oxidation of organic matter to CO_2 and water, or the fermentation of organic matter to methane by anoxic bacteria in the absence of oxygen. Such information may be obtained by knowing the free-energy change, ΔG, for the redox reaction; ΔG, in turn, may be obtained from pE for the reaction. The free-energy change for a redox reaction involving n electrons at an absolute temperature of T is given by

$$\Delta G = -2.303nRT(\mathrm{p}E) \tag{4.35}$$

where R is the gas constant. When all reaction participants are in their standard states (pure liquids, pure solids, solutes at an activity of 1.00, gas pressures at 1 atm), ΔG is the standard free-energy change, ΔG^0, given by

$$\Delta G^0 = -2.303nRT(\mathrm{p}E^0) \tag{4.36}$$

4.8 REACTIONS IN TERMS OF ONE ELECTRON-MOLE

For comparing free-energy changes between different redox reactions, it is most meaningful to consider the reactions in terms of the transfer of exactly 1 mol of electrons. This concept may be understood by considering two typical and important redox reactions that occur in aquatic systems—nitrification

$$\mathrm{NH^{4+} + 2O_2 \rightleftarrows NO_3^- + 2H^+ + H_2O}, \quad \mathrm{p}E^0 = 5.85 \tag{4.37}$$

and oxidation of iron(II) to iron(III):

$$\mathrm{4Fe^{2+} + O_2 + 10H_2O \rightleftarrows 4Fe(OH)_3}(s)\mathrm{+8H^+}, \quad \mathrm{p}E^0 = 7.6 \tag{4.38}$$

What do reactions written in this way really mean? If any thermodynamic calculations are to be made involving the reactions, Reaction 4.37 means that 1 mol of ammonium ion reacts with 2 mol of elemental oxygen to yield 1 mol of nitrate ion, 2 mol of hydrogen ion, and 1 mol of water. Reaction 4.38 is taken to mean that 4 mol of iron(II) ion react with 1 mol of oxygen and 10 mol of water to produce 4 mol of $Fe(OH)_3$ and 8 mol of hydrogen ions. The free-energy changes calculated for these quantities of reaction participants do not enable meaningful comparisons of their free-energy changes. Such comparisons may be made, though, on the common basis of the transfer of 1 mol of electrons, writing each reaction in terms of one electron-mole. The advantage of this approach is illustrated by considering Reaction 4.37, which involves an eight-electron change, and Reaction 4.38, which involves a four-electron change. Rewriting Equation 4.37 for one electron-mole yields

$$\tfrac{1}{8}\mathrm{NH_4^+} + \tfrac{1}{4}\mathrm{O_2} \rightleftarrows \tfrac{1}{8}\mathrm{NO_3^-} + \tfrac{1}{4}\mathrm{H^+} + \tfrac{1}{8}\mathrm{H_2O}, \quad \mathrm{p}E^0 = 5.85 \tag{4.39}$$

whereas Reaction 4.38, when rewritten for one electron-mole rather than four, yields

$$Fe^{2+} + \tfrac{1}{4}O_2 + \tfrac{5}{2}H_2O \rightleftarrows Fe(OH)_3(s) + 2H^+, \quad pE^0 = 7.6 \tag{4.40}$$

From Equation 4.36, the standard free-energy change for a reaction is

$$\Delta G^0 = -2.303nRT(pE^0) \tag{4.36},$$

which, for a one electron-mole reaction is simply

$$\Delta G^0 = -2.303RT(pE^0) \tag{4.36}.$$

Therefore, for reactions written in terms of one electron-mole, a comparison of pE^0 values provides a direct comparison of ΔG^0 values.

As shown in Equation 4.34 for a redox reaction involving n electrons, pE^0 is related to the equilibrium constant by

$$\log K = n(pE^0) \tag{4.41}$$

which for a one electron-mole reaction becomes simply

$$\log K = pE^0 \tag{4.42}$$

Reaction 4.39, the nitrification reaction written in terms of one electron-mole, has a pE^0 value of +5.85. The equilibrium-constant expression for this reaction is

$$K = \frac{[NO_3^-]^{1/8}[H^+]^{1/4}}{[NH_4^+]^{1/8}PO_2^{1/4}} \tag{4.43}$$

a computationally cumbersome form for which log K is given very simply as the following:

$$\log K = pE^0 = 5.85 \quad \text{or} \quad K = 7.08 \times 10^5 \tag{4.44}$$

Table 4.1 is a compilation of pE^0 values for redox reactions that are especially important in aquatic systems. Most of these values are calculated from thermodynamic data rather than from direct potentiometric measurements in an electrochemical cell, as shown in Figure 4.2. Most electrode systems that might be devised do not give potential responses corresponding to the Nernst equation; that is, they do not behave *reversibly.* It is true that one may place a platinum electrode and a reference electrode in water and measure a potential. This potential, referred to the SHE, is the so-called E_H value. Furthermore, the measured potential will be more positive (more oxidizing) in an oxidizing medium, such as the oxic surface layers of a lake, than in a reducing medium, such as the anaerobic bottom regions of a body of water. However, attaching any quantitative significance to the E_H value measured directly with an electrode is a very dubious practice. Acid mine waters containing relatively high levels of sulfuric acid and dissolved iron give reasonably accurate E_H values by direct measurement, but most aquatic systems do not yield meaningful values of E_H. The most accurate method of assessing the redox status of water is through calculation of *oxidative capacity*, a parameter analogous to buffer capacity for acids and bases and consisting of a single descriptive parameter derived from the sum of the species that can be oxidized or reduced.

TABLE 4.1
pE^0 Values of Redox Reactions Important in Natural Waters (at 25°C)

	Reaction	pE^0	pE^0 (W)[a]
1.	$\frac{1}{4}O_2(g) + H^+(W) + e^- \rightleftarrows \frac{1}{2}H_2O$	+20.75	+13.75
2.	$\frac{1}{5}NO_3^- + \frac{6}{5}H^+ + e^- \rightleftarrows \frac{1}{10}N_2 + \frac{3}{5}H_2O$	+12.65	+21.05
3.	$\frac{1}{2}MnO_2 + \frac{1}{2}HCO_3^-(1\times10^{-3}\ M) + \frac{3}{2}H^+(W) + e^- \rightleftarrows \frac{1}{2}MnCO_3(s) + H_2O$	—	+8.5[b]
4.	$\frac{1}{2}NO_3^- + H^+(W) + e^- \rightleftarrows \frac{1}{2}NO_2^- + \frac{1}{2}H_2O$	+14.15	+7.15
5.	$\frac{1}{8}NO_3^- + \frac{5}{4}H^+(W) + e^- \rightleftarrows \frac{1}{8}NH_4^+ + \frac{3}{8}H_2O$	+14.90	+6.15
6.	$\frac{1}{6}NO_2^- + \frac{4}{3}H^+(W) + e^- \rightleftarrows \frac{1}{6}NH_4^+ + \frac{1}{3}H_2O$	+15.14	+5.82
7.	$\frac{1}{2}CH_3OH + H^+(W) + e^- \rightleftarrows \frac{1}{2}CH_4(g) + \frac{1}{2}H_2O$	+9.88	+2.88
8.	$\frac{1}{4}CH_2O + H^+(W) + e^- \rightleftarrows \frac{1}{4}CH_4(g) + \frac{1}{4}H_2O$	+6.94	−0.06
9.	$FeOOH(g) + HCO_3^-(1\times10^{-3}\ M) + 2H^+(W) + e^- \rightleftarrows FeCO_3(s) + 2H_2O$	—	−1.67[b]
10.	$\frac{1}{2}CH_2O + H^+(W) + e^- \rightleftarrows +\frac{1}{2}CH_3OH$	+3.99	−3.01
11.	$\frac{1}{6}SO_4^{2-} + \frac{4}{3}H^+(W) + e^- \rightleftarrows \frac{1}{6}S(s) + \frac{2}{3}H_2O$	+6.03	−3.30
12.	$\frac{1}{8}SO_4^{2-} + \frac{5}{4}H^+(W) + e^- \rightleftarrows \frac{1}{8}H_2S(g) + \frac{1}{2}H_2O$	+5.75	−3.50
13.	$\frac{1}{8}SO_4^{2-} + \frac{9}{8}H^+(W) + e^- \rightleftarrows \frac{1}{8}HS^- + \frac{1}{2}H_2O$	+4.13	−3.75
14.	$\frac{1}{2}S(s) + H^+(W) + e^- \rightleftarrows \frac{1}{2}H_2S(g)$	+2.89	−4.11
15.	$\frac{1}{8}CO_2 + H^+ + e^- \rightleftarrows \frac{1}{8}CH_4 + \frac{1}{4}H_2O$	+2.87	−4.13
16.	$\frac{1}{6}N_2 + \frac{4}{3}H^+(W) + e^- \rightleftarrows \frac{1}{3}NH_4^+$	+4.68	−4.65
17.	$H^+(W) + e^- \rightleftarrows \frac{1}{2}H_2(g)$	0.00	−7.00
18.	$\frac{1}{4}CO_2(g) + H^+(W) + e^- \rightleftarrows \frac{1}{4}CH_2O + \frac{1}{4}H_2O$	−1.20	−8.20

Source: Stumm, Werner and James J. Morgan, *Aquatic Chemistry*, John Wiley and Sons, New York, 1970, p. 318. Reproduced by permission of John Wiley & Sons, Inc.

[a] (W) indicates $a_{H^+} = 1.00 \times 10^{-7}$ M and pE^0(W) is a pE^0 at $a_{H^+} = 1.00 \times 10^{-7}$ M.

[b] These data correspond to $a_{HCO_3^-} = 1.00 \times 10^{-3}$ M rather than unity and, hence, are not exactly pE^0(W); they represent typical aquatic conditions more exactly than pE^0 values do.

4.9 THE LIMITS OF p*E* IN WATER

There are pH-dependent limits to the p*E* values at which water is thermodynamically stable. Water may be either oxidized as

$$2H_2O \rightleftarrows O_2 + 4H^+ + 4e^- \tag{4.45}$$

or reduced as

$$2H_2O + 2e^- \rightleftarrows H_2 + 2OH^- \tag{4.46}$$

These two reactions determine the limits of pE in water. On the oxidizing side (relatively more positive pE values), the pE value is limited by the oxidation of water, half-reaction 4.45. The evolution of hydrogen, Half-reaction 4.46, limits the pE value on the reducing side.

The condition under which oxygen from the oxidation of water has a pressure of 1.00 atm can be regarded as the oxidizing limit of water, whereas a hydrogen pressure of 1.00 atm may be regarded as the reducing limit of water. These are *boundary conditions* that enable calculation of the stability boundaries of water. Writing the reverse of Reaction 4.45 for one electron and setting $P_{O_2} = 100$ yield

$$\tfrac{1}{4}O_2 + H^+ + e^- \rightleftarrows \tfrac{1}{2}H_2O, \quad pE^0 = 20.75 \quad \text{(from Table 4.1)} \tag{4.47}$$

Thus, Equation 4.49 defines the pH-dependent oxidizing limit of water. At a specified pH, pE values more positive than the one given by Equation 4.49 cannot exist at equilibrium in water in contact with the atmosphere.

$$pE = pE^0 + \log(P_{O_2}^{1/4}[H^+]) \tag{4.48}$$

$$pE = 20.75 - pH \tag{4.49}$$

The pE–pH relationship for the reducing limit of water taken at $P_{H_2} = 1$ atm is given by the following derivation:

$$H^+ + e^- \rightleftarrows \tfrac{1}{2}H_2, \quad pE^0 = 0.00 \tag{4.50}$$

$$pE = pE^0 + \log[H^+] \tag{4.51}$$

$$pE = -pH \tag{4.52}$$

For neutral water (pH = 7.00), substitution into Equations 4.52 and 4.49 yields –7.00 to 13.75 for the pE range of water. The pE–pH boundaries of stability for water are shown by the dashed lines in Figure 4.4 of Section 4.11.

The decomposition of water is very slow in the absence of a suitable catalyst. Therefore, water may have temporary nonequilibrium pE values more negative than the reducing limit or more positive than the oxidizing limit. A common example of the latter is a solution of chlorine in water.

4.10 p*E* VALUES IN NATURAL WATER SYSTEMS

Although it is not generally possible to obtain accurate pE values by direct potentiometric measurements in natural aquatic systems, in principle, pE values may be calculated from the species present in water at equilibrium. An obviously significant pE value is that of neutral water in thermodynamic equilibrium with the atmosphere. In water under these conditions, $P_{O_2} = 0.21$ atm and $[H^+] = 1.00 \times 10^{-7}$ M. Substitution into Equation 4.48 yields

$$pE = 20.75 + \log\{(0.21)^{1/4} \times 1.00 \times 10^{-7}\} = 13.8 \tag{4.53}$$

According to this calculation, a pE value of around +13 is to be expected for water in equilibrium with the atmosphere, that is, an oxic water. At the other extreme, consider anoxic water in which

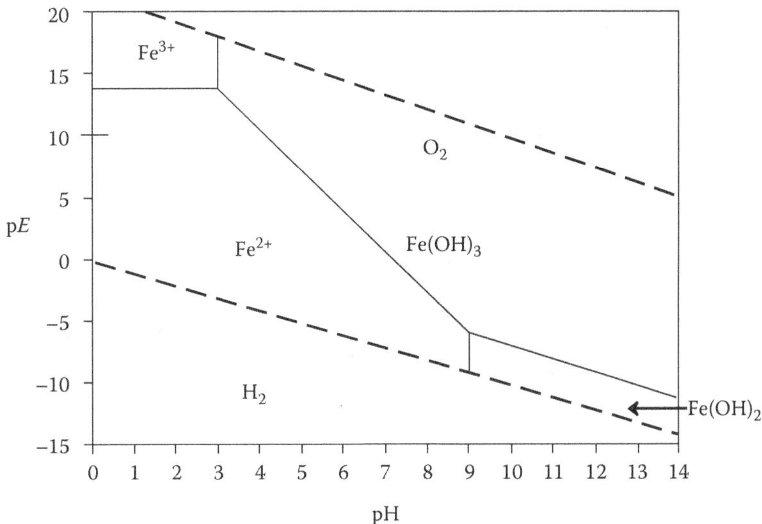

FIGURE 4.4 Simplified pE–pH diagram for iron in water. The maximum soluble iron concentration is 1.00×10^{-5} M.

methane and CO_2 are being produced by microorganisms. Assume $P_{CO_2} = P_{CH_4}$ and that pH = 7.00. The relevant half-reaction is

$$\tfrac{1}{8}CO_2 + H^+ + e^- \rightleftarrows \tfrac{1}{8}CH_4 + \tfrac{1}{4}H_2O \tag{4.54}$$

for which the Nernst equation is

$$pE = 2.87 + \log \frac{P_{CO_2}^{1/8}\,[H^+]}{P_{CH_2}^{1/8}} = 2.87 + \log[H^+] = 2.87 - 7.00 = -4.13 \tag{4.55}$$

Note that the pE value of −4.13 does not exceed the reducing limit of water at pH 7.00, which from Equation 4.52 is −7.00. It is of interest to calculate the pressure of oxygen in neutral water at this low pE value of −4.13. Substitution into Equation 4.48 yields

$$-4.13 = 20.75 + \log(P_{O_2}^{1/4} \times 1.00 \times 10^{-7}) \tag{4.56}$$

from which the pressure of oxygen is calculated to be 3.0×10^{-72} atm. This impossibly low figure for the pressure of oxygen means that equilibrium with respect to oxygen partial pressure is not achieved under these conditions. Certainly, under any condition approaching equilibrium between comparable levels of CO_2 and CH_4, the partial pressure of oxygen must be extremely low.

4.11 p*E*–pH DIAGRAMS

The examples cited so far have shown the close relationships between pE and pH in water. This relationship may be expressed graphically in the form of a pE–pH diagram. Such diagrams show the regions of stability and the boundary lines for various species in water. Because of the numerous species that may be formed, such diagrams may become extremely complicated. For example, if a metal is being considered, several different oxidation states of the metal, hydroxy complexes, and different forms of the solid metal oxide or hydroxide may exist in different regions described by the pE–pH diagram. Most waters contain carbonate, and many contain sulfates and sulfides, so that various metal carbonates, sulfates, and sulfides may predominate in different regions of the diagram. In order to illustrate the principles involved, however, a simplified pE–pH diagram is considered here. The reader is referred to more advanced works on geochemistry and aquatic chemistry for more complicated (and more realistic) pE–pH diagrams.[1,2]

A pE–pH diagram for iron may be constructed assuming a maximum concentration of iron in solution, in this case 1.0×10^{-5} M. The following equilibria will be considered:

$$Fe^{3+} + e^- \rightleftarrows Fe^{2+}, \quad pE^0 = +13.2 \tag{4.57}$$

$$Fe(OH)_2(s) + 2H^+ \rightleftarrows Fe^{2+} + 2H_2O \tag{4.58}$$

$$K_{sp} = \frac{[Fe^{2+}]}{[H^+]^2} = 8.0 \times 10^{12} \tag{4.59}$$

$$Fe(OH)_3(s) + 3H^+ \rightleftarrows Fe^{3+} + 3H_2O \tag{4.60}$$

$$K'_{sp} = \frac{[Fe^{3+}]}{[H^+]^3} = 9.1 \times 10^3 \tag{4.61}$$

[The constants K_{sp} and K'_{sp} are derived from the solubility products of $Fe(OH)_2$ and $Fe(OH)_3$, respectively, and are expressed in terms of $[H^+]$ to facilitate the calculations.] Note that the formation of species such as $Fe(OH)^{2+}$, $Fe(OH)_2^+$, and solid $FeCO_3$ or FeS, all of which might be of significance in a natural water system, is not considered. Rapid hydrolysis of iron(III) in solution yields solid ferrihydrite, a hydrated iron(III) hydroxide/oxide with a high surface area and strong affinity to coprecipitate metals other than iron.

In constructing the pE–pH diagram, several boundaries must be considered. The first two of these are the oxidizing and reducing limits of water (see Section 4.9). At the high pE end, the stability limit of water is defined by Equation 4.49 derived previously:

$$pE = 20.75 - pH \tag{4.49}$$

The low pE limit is defined by Equation 4.52:

$$pE = -pH \tag{4.52}$$

The pE–pH diagram constructed for the iron system must fall between the boundaries defined by these two equations.

Below pH 3, Fe^{3+} may exist in equilibrium with Fe^{2+}. The boundary line that separates these two species, where $[Fe^{3+}] = [Fe^{2+}]$, is given by the following calculation:

$$pE = 13.2 + \log\frac{[Fe^{3+}]}{[Fe^{2+}]} \tag{4.62}$$

$$[Fe^{3+}] = [Fe^{2+}] \tag{4.63}$$

$$pE = 13.2 \quad \text{(independent of pH)} \tag{4.64}$$

At pE exceeding 13.2, as the pH increases from very low values, $Fe(OH)_3$ precipitates from a solution of Fe^{3+}. The pH at which precipitation occurs depends, of course, upon the concentration of Fe^{3+}. In this example, a maximum soluble iron concentration of 1.00×10^{-5} M has been chosen so that at the $Fe^{3+}/Fe(OH)_3$ boundary, $[Fe^{3+}] = 1.00 \times 10^{-5}$ M. Substitution in Equation 4.61 yields

$$[H^+]^3 = \frac{[Fe^{3+}]}{K'_{sp}} = \frac{1.00 \times 10^{-5}}{9.1 \times 10^3} \tag{4.65}$$

$$pH = 2.99 \tag{4.66}$$

In a similar manner, the boundary between Fe^{2+} and solid $Fe(OH)_2$ may be defined, assuming $[Fe^{2+}] = 1.00 \times 10^{-5}$ M (the maximum soluble iron concentration specified at the beginning of this exercise) at the boundary:

$$[H^+]^2 = \frac{[Fe^{2+}]}{K_{sp}} = \frac{1.00 \times 10^{-5}}{8.0 \times 10^{12}} \tag{4.67}$$

$$pH = 8.95 \tag{4.68}$$

Throughout a wide pE–pH range, Fe^{2+} is the predominant soluble iron species in equilibrium with the solid hydrated iron(III) oxide, $Fe(OH)_3$. The boundary between these two species depends upon both pE and pH. Substituting Equation 4.61 into Equation 4.62 yields

$$pE = 13.2 + \frac{K'_{sp}[H^+]^3}{[Fe^{2+}]} \tag{4.69}$$

$$pE = 13.2 + \log 9.1 \times 10^3 - \log 1.00 \times 10^{-5} + 3 \times \log[H^+]$$

$$pE = 22.2 - 3\ pH \tag{4.70}$$

The boundary between the solid phases $Fe(OH)_2$ and $Fe(OH)_3$ likewise depends upon both pE and pH, but it does not depend upon an assumed value for total soluble iron. The required relationship is derived from substituting both Equations 4.59 and 4.61 into Equation 4.62:

$$pE = 13.2 + \log \frac{K'_{sp}}{K_{sp}[H^+]^2}[H^+]^3$$

$$pE = 13.2 + \log \frac{9.1 \times 10^3}{8.0 \times 10^{12}} + \log[H^+] \tag{4.71}$$

$$pE = 4.3 - pH \tag{4.72}$$

All of the equations needed to prepare the pE–pH diagram for iron in water have now been derived. To summarize, the equations are (Equation 4.49), O_2–H_2O boundary; (Equation 4.52), H_2–H_2O boundary; (Equation 4.64), Fe^{3+}–Fe^{2+} boundary; (Equation 4.66), Fe^{3+}–$Fe(OH)_3$ boundary; (Equation 4.68), Fe^{2+}–$Fe(OH)_2$ boundary; (Equation 4.70), Fe^{2+}–$Fe(OH)_3$ boundary; and (Equation 4.72), $Fe(OH)_2$–$Fe(OH)_3$ boundary.

The pE–pH diagram for the iron system in water is shown in Figure 4.4. In this system, at a relatively high hydrogen-ion activity and high electron activity (an acidic reducing medium), iron(II) ion, Fe^{2+}, is the predominant iron species; some groundwaters contain appreciable levels of iron(II) under these conditions. (In most natural water systems the solubility range of Fe^{2+} is relatively narrow because of the precipitation of FeS or $FeCO_3$.) At a very high hydrogen-ion activity and low electron activity (an acidic oxidizing medium), Fe^{3+} ion predominates. In an oxidizing medium at lower acidity, solid $Fe(OH)_3$ is the primary iron species present. Finally, in a basic reducing medium, with low hydrogen-ion activity and high electron activity, solid $Fe(OH)_2$ can be stable.

Note that within the pH regions normally encountered in a natural aquatic system (approximately pH 5–9) $Fe(OH)_3$ or Fe^{2+} is the predominant stable iron species. In fact, it is observed that in waters containing dissolved oxygen at any appreciable level (a relatively high pE), hydrated iron(III) oxide $Fe(OH)_3$ is essentially the only inorganic iron species found. Such waters contain a high level of suspended iron, but any truly soluble iron must be in the form of a complex (see Chapter 3).

In highly anoxic, low-pE water, appreciable levels of Fe^{2+} may be present. When such water is exposed to atmospheric oxygen, the pE rises and $Fe(OH)_3$ precipitates. The resulting deposits of hydrated iron(III) oxide can stain laundry and bathroom fixtures with a refractory red/brown stain. This phenomenon also explains why red iron oxide deposits are found near pumps and springs that bring deep, anaerobic water to the surface. In shallow wells, where the water may become oxic, solid $Fe(OH)_3$ may precipitate on the well walls, clogging the aquifer outlet. This usually occurs through bacterially mediated reactions, which are discussed in Chapter 6.

One species not yet considered is elemental iron. For the half-reaction

$$Fe^{2+} + 2e^- \rightleftarrows Fe, \quad pE^0 = -7.45 \tag{4.73}$$

the Nernst equation gives pE as a function of $[Fe^{2+}]$

$$pE = -7.45 + \tfrac{1}{2}\log[Fe^{2+}] \tag{4.74}$$

For iron metal in equilibrium with 1.00×10^{-5} M Fe^{2+}, the following pE value is obtained:

$$pE = -7.45 + \tfrac{1}{2}\log 1.00 \times 10^{-5} = -9.95 \tag{4.75}$$

Examination of Figure 4.4 shows that the pE values for elemental iron in contact with Fe^{2+} are below the reducing limit of water. This shows that iron metal in contact with water is thermodynamically unstable with respect to reducing water and going into solution as Fe^{2+}, a factor that contributes to the tendency of iron to undergo corrosion.

4.12 HUMIC SUBSTANCES AS NATURAL REDUCTANTS

Humic substances (see Section 3.17) are oxidation–reduction-active species that may be significant as reducing species in chemical and biochemical processes that occur in natural water and wastewater systems. The capability of humic substances to act as reductants is due primarily to the presence of the quinone/hydroquinone group that acts as an oxidation–reduction couple as follows:

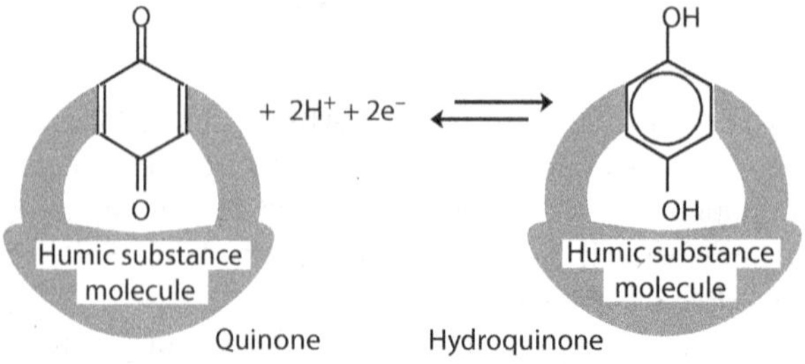

O O Humic substance molecule (Quinone) $+ 2H^+ + 2e^- \rightleftarrows$ OH OH Humic substance molecule (Hydroquinone) (4.76)

There is some evidence suggesting that soluble humic substances can act as intermediates in the reduction of species in water by solid reductants, such as solid iron(II) species.

It is believed that in addition to being reducing agents, humic substances act as electron shuttles to transfer electrons to iron(III) in the microorganism-mediated bioreduction of iron(III) to iron(II) in water.[3] The process is faster in the presence of high carboxyl ($-CO_2H$) group content in the humic substances presence indicative of the importance of complexation of iron ions by the humic substance.

4.13 PHOTOCHEMICAL PROCESSES IN OXIDATION–REDUCTION

As discussed in Chapter 9, absorption of a light photon can introduce a high level of energy into chemical species that can become involved in oxidation–reduction processes in water. An important oxidizing agent in water exposed to sunlight is believed to be *superoxide ion*, $O_2^{\bullet -}$, produced by the action of photochemically excited natural organic matter (humic substance) upon dissolved O_2. This superoxide species is a *free radical* (see Chapter 9) with an unpaired electron denoted by the dot in the formula $O_2^{\bullet -}$. Superoxide ion is capable of oxidizing inorganically complexed metals such as iron.

An important intermediate involved with photochemically induced oxidation processes in water is hydrogen peroxide, H_2O_2, generated when $O_2^{\bullet -}$ reacts with water. In the presence of iron(II) the Fenton reaction,

$$Fe(II) + H_2O_2 \rightarrow Fe(III) + OH^- + HO^\bullet \quad (4.77)$$

generates hydroxyl radical, $HO^\bullet$. Hydroxyl radical is a very reactive species that reacts with even refractory organic species, leading to their oxidation. As a consequence of reactions involving superoxide radical ion and hydroxyl radical, photochemical processes are important in promoting oxidation of oxidizable species in water.

Aqueous suspensions of TiO_2 are photochemically active. When photons of ultraviolet radiation of a wavelength <318 nm ($h\nu$) strike a TiO_2 surface,

$$TiO_2 + h\nu \rightarrow e^-_{cb} + h^+_{vb} \quad (4.78)$$

conduction band electrons (e^-_{cb}) capable of initiating photoreduction and valence band holes (h^+_{vb}) capable of initiating photooxidation are produced. This has led to interest in the use of suspensions of TiO_2 exposed to sunlight to photoreduce and photooxidize pollutant and hazardous waste species.

4.14 CORROSION

One of the most damaging redox phenomena is *corrosion*, defined as the destructive alteration of metal through interactions with its surroundings. In addition to its multibillion dollar annual costs due to destruction of equipment and structures, corrosion introduces metals into water systems and destroys pollution control equipment and waste disposal pipes; it is aggravated by water and air pollutants and some kinds of hazardous wastes (see corrosive wastes in Chapter 20, Section 20.6).

Thermodynamically, all commonly used metals are unstable relative to their environments. Elemental metals tend to undergo chemical changes to produce the more stable forms of ions, salts, oxides, and hydroxides. Fortunately, the rates of corrosion are normally slow, so that metals exposed to air and water may endure for long periods of time. However, protective measures are necessary. Sometimes these measures fail; for example, witness the gaping holes in automobile bodies exposed to salt used to control road ice.

Corrosion normally occurs when an electrochemical cell is set up on a metal surface. The area corroded is the anode, where the following oxidation reaction occurs, illustrated for the formation of a divalent metal ion from a metal, M:

$$M \rightarrow M^{2+} + 2e^- \quad (4.79)$$

Several cathodic reactions are possible. One of the most common of these is the reduction of H^+ ion:

$$2H^+ + 2e^- \rightarrow H_2 \tag{4.80}$$

Oxygen may also be involved in cathodic reactions, including reduction to hydroxide, reduction to water, and reduction to hydrogen peroxide:

$$O_2 + 2H_2O + 4e^- \rightarrow 4OH^- \tag{4.81}$$

$$O_2 + 4H^+ + 4e^- \rightarrow 2H_2O \tag{4.82}$$

$$O_2 + 2H_2O + 2e^- \rightarrow 2OH^- + H_2O_2 \tag{4.83}$$

Oxygen may either accelerate corrosion processes by participating in reactions, such as these, or retard them by forming protective oxide films. As discussed in Chapter 6, bacteria are often involved with corrosion.

Corrosion is a very important consideration in domestic water distribution systems. Metals that may be corroded in such systems include iron, copper, lead, and red brass. Corrosion of these materials involves the redox potential of the water as determined by the solutes in it and the solid materials with which it is in contact. Important factors in corrosion include type and concentration of oxidant (usually from added Cl_2), metal release rate, and corrosion scale properties of the corroding metals.

LITERATURE CITED

1. Stumm, W. and J. J. Morgan, *Aquatic Chemistry: Chemical Equilibria and Rates in Natural Waters*, 3rd ed., Wiley, New York, 1995.
2. Garrels, R. M. and C. M. Christ, *Solutions, Minerals, and Equilibria*, Harper and Row, New York, 1965.
3. Rakshit, S., M. Uchimiya, and G. Sposito, Iron(III) bioreduction in soil in the presence of added humic substances, *Soil Science Society of America Journal*, **73**, 65–71, 2009.

SUPPLEMENTARY REFERENCES

Albarè de, F., *Geochemistry*, Cambridge University Press, New York, 2004.

Amatya, S. R. and S. Mika, Influence of Eh/pH—barriers on releasing/accumulation of manganese and iron at sediment–water interface, *Research Journal of Chemistry and Environment*, **12**, 7–13, 2008.

Auque, L., M. J. Gimeno, J. Gomez, and A. C. Nilsson, Potentiometrically measured Eh in groundwaters from the Scandinavian shield, *Applied Geochemistry*, 1820–1833, 2008.

Baas Becking, L. G. M., I. R. Kaplan, and D. Moore, Limits of the natural environment in terms of pH and oxidation–reduction potentials in natural waters, *Journal of Geology*, **68**, 243–284, 1960.

Bates, R. G., The modern meaning of pH, *Critical Reviews in Analytical Chemistry*, **10**, 247–278, 1981.

Chapelle, F. H., *Ground-Water Microbiology and Geochemistry*, 2nd ed., Wiley, New York, 2000.

Chester, R., *Marine Geochemistry*, Blackwell Science, Oxford, UK, 2000.

Faure, G., *Principles and Applications of Geochemistry: A Comprehensive Textbook for Geology Students*, 2nd ed., Prentice-Hall, Upper Saddle River, NJ, 1998.

Faust, B. C., A review of the photochemical redox reactions of Iron(III) species in atmospheric, oceanic, and surface waters: Influences on geochemical cycles and oxidant formation, in *Aquatic and Surface Photochemistry*, George Helz, R. G. Zepp, and D. G. Crosby, Eds, CRC Press/Lewis Publishers, Boca Raton, FL, pp. 3–37, 1994.

Garrison S., Oxidation reduction reactions, in *The Chemistry of Soils*, 2nd edition, Oxford University Press, New York, pp. 144–173, 2008.

Holland, H. D. and K. K. Turekian, Eds, *Treatise on Geochemistry*, Elsevier/Pergamon, Amsterdam, 2004.

McSween, H. Y., S. M. Richardson, and M. E. Uhle, *Geochemistry: Pathways and Processes*, 2nd ed., Columbia University Press, New York, 2003.
Pankow, J. F., *Aquatic Chemistry Concepts*, Lewis Publishers/CRC Press, Boca Raton, FL, 1991.
Pankow, J. F., *Aquatic Chemistry Problems*, Titan Press, OR, Portland, OR, 1992.
Rowell, D. L., Oxidation and reduction, in *The Chemistry of Soil Processes*, D. J. Greenland and M. H. B. Hayes, Eds, Wiley, Chichester, 1981.
Spellman, F. R., *The Science of Water: Concepts and Applications*, 2nd ed., Taylor & Francis/CRC Press, Boca Raton, FL, 2008.
Stewart, B. A., *Encyclopedia of Water Science*, Marcel Dekker, New York, 2003.
Stumm, W., *Redox Potential as an Environmental Parameter: Conceptual Significance and Operational Limitation*, Third International Conference on Water Pollution Research (Munich, Germany), Water Pollution Control Federation, Washington, DC, 1966.
Sun, L., S. C. Su, L. Chao, and T. Sun, Effects of flooding on changes in Eh, pH and speciation of cadmium and lead in contaminated soil, *Bulletin of Environmental Contamination and Toxicology*, **79**, 2007.
Trimble, S. W., *Encyclopedia of Water Science*, 2nd ed., Taylor & Francis, CRC Press, Boca Raton, FL, 2008.

QUESTIONS AND PROBLEMS

The use of Internet resources is assumed in answering any of the questions. These resources would include such things as constants and conversion factors as well as additional information needed to complete an answer.

1. The acid–base reaction for the dissociation of acetic acid is

$$HOAc + H_2O \rightarrow H_3O^+ + OAc^-$$

with $K_a = 1.75 \times 10^{-5}$. Break this reaction down into two half-reactions involving H^+ ion. Break down the redox reaction

$$Fe^{2+} + H^+ \rightarrow Fe^{3+} + \tfrac{1}{2}H_2$$

into two half-reactions involving the electron. Discuss the analogies between the acid–base and redox processes.

2. Assuming a bicarbonate ion concentration $[HCO_3^-]$ of 1.00×10^{-3} M and a value of 3.5×10^{-11} for the solubility product of $FeCO_3$, what would you expect to be the stable iron species at pH 9.5 and pE −8.0, as shown in Figure 4.4?
3. Assuming that the partial pressure of oxygen in water is that of atmospheric O_2, 0.21 atm, rather than the 1.00 atm assumed in deriving Equation 4.49, derive an equation describing the oxidizing pE limit of water as a function of pH.
4. Plot $\log P_{O_2}$ as a function of pE at pH 7.00.
5. Calculate the pressure of oxygen for a system in equilibrium in which $[NH_4^+] = [NO_3^-]$ at pH 7.00.
6. Calculate the values of $[Fe^{3+}]$, pE, and pH at the point in Figure 4.4 where Fe^{2+} is at a concentration of 1.00×10^{-5} M, and $Fe(OH)_2$ and $Fe(OH)_3$ are in equilibrium.
7. What is the pE value in a solution in equilibrium with air (21% O_2 by volume) at pH 6.00?
8. What is the pE value at the point on the Fe^{2+}–$Fe(OH)_3$ boundary line (see Figure 4.4) in a solution with a soluble iron concentration of 1.00×10^{-4} M at pH 6.00?
9. What is the pE value in an acid mine water sample having $[Fe^{3+}] = 7.03 \times 10^{-3}$ M and $[Fe^{2+}] = 3.71 \times 10^{-4}$ M?
10. At pH 6.00 and pE 2.58, what is the concentration of Fe^{2+} in equilibrium with $Fe(OH)_2$?
11. What is the calculated value of the partial pressure of O_2 in acid mine water of pH 2.00, in which $[Fe^{3+}] = [Fe^{2+}]$?

12. What is the major advantage of expressing redox reactions and half-reactions in terms of exactly one electron-mole?
13. Why are p*E* values that are determined by reading the potential of a platinum electrode versus a reference electrode generally not very meaningful?
14. What determines the oxidizing and reducing limits, respectively, for the thermodynamic stability of water?
15. How would you expect p*E* to vary with depth in a stratified lake?
16. Upon what half-reaction is the rigorous definition of p*E* based?
17. Analysis of water in a sediment sample at equilibrium at pH 7.00 showed $[SO_4^{2-}] = 2.00 \times 10^{-5}$ M and a partial pressure of H_2S of 0.100 atm. Show with appropriate calculations if methane, CH_4, would be expected in the sediment.
18. Choose the correct answer of the following, explain why it is true, and explain why the other choices are untrue: (A) high p*E* is associated with species such as CH_4, NH_4^+, and Fe^{2+}; (B) low p*E* is associated with species such as CO_2, O_2, and NO_3^-; (C) values of p*E* in bodies of water range from about 1×10^{-7} to about 1×10^7; (D) p*E* is a number, but cannot be related to anything real, such as is the case with pH; (E) p*E* uses convenient numbers to express electron activity over many orders of magnitude.
19. Match the following:

A. For 1 electron-mole	1. $Fe(H_2O)_6^{2+} \rightleftarrows e^- + Fe(OH)_3(s) + 3H_2O + 3H^+$
B. Reaction for standard electrode	2. $H_2 \rightleftarrows 2H^+ + 2e^-$
C. At the upper p*E* limit of water	3. $\frac{1}{8}NH_4^+ + \frac{1}{4}O_2 \rightleftarrows \frac{1}{8}NO_3^- + \frac{1}{4}H^+ + H_2O$
D. Formation of a pollutant when anoxic water is brought to the surface	4. $2H_2O \rightleftarrows O_2 + 4H^+ + 4e^-$

20. Of the following, the true statement regarding oxidation–reduction reactions and phenomena in natural water systems is: (A) At a p*E* higher than the oxidizing limit of stability, water decomposes to evolve H_2; (B) the production of CH_4 at a very low p*E* is caused by the action of bacteria; (C) in the p*E*–pH diagram for iron, the region of greatest area is occupied by solid $Fe(OH)_2$; (D) it is easy to accurately measure the p*E* of water with a platinum electrode; (E) there are no p*E*–pH limits for the regions of stability of H_2O.

5 Phase Interactions in Aquatic Chemistry

5.1 CHEMICAL INTERACTIONS INVOLVING SOLIDS, GASES, AND WATER

Homogeneous chemical reactions occurring entirely in aqueous solution are rather rare in natural waters and wastewaters. Instead, most significant chemical and biochemical phenomena in water involve interactions between species in water and another phase. Some of these important interactions are illustrated in Figure 5.1. Several examples of phase interactions in water illustrated by the figure are the following: production of solid biomass through the photosynthetic activity of algae occurs within a suspended algal cell and involves exchange of dissolved solids and gases between the surrounding water and the cell. Similar exchanges occur when bacteria degrade organic matter (often in the form of small particles) in water. Chemical reactions occur that produce solids or gases in water. Iron and many important trace-level elements are transported through aquatic systems as colloidal chemical compounds or are sorbed to solid particles. Pollutant hydrocarbons and some pesticides may be present on the water surface as an immiscible liquid film. Sediment can be washed physically into a body of water.

This chapter discusses the importance of interactions among different phases in aquatic chemical processes. In a general sense, in addition to water, these phases may be divided between *sediments* (bulk solids) and *suspended colloidal material*. The ways in which sediments are formed and the significance of sediments as repositories and sources of aquatic solutes are discussed. Mentioned in earlier chapters, solubilities of solids and gases (Henry's law) are covered here in some detail.

Much of this chapter deals with the behavior of colloidal material that consists of very fine particles of solids, gases, or immiscible liquids suspended in water. Colloidal material is involved with many significant aquatic chemical phenomena. It is very reactive because of its high surface-area-to-volume ratio.

5.2 IMPORTANCE AND FORMATION OF SEDIMENTS

Sediments are the layers of relatively finely divided matter covering the bottoms of rivers, streams, lakes, reservoirs, bays, estuaries, and oceans. Sediments typically consist of mixtures of fine-, medium-, and coarse-grained minerals, including clay, silt, and sand, mixed with organic matter. They may vary in composition from pure mineral matter to predominantly organic matter. Sediments are reservoirs of a variety of biological, chemical, and pollutant detritus in bodies of water and are repositories of pollutants, such as heavy metals and toxic organic compounds. Of particular concern is the transfer of chemical species from sediments into aquatic food chains via organisms that spend significant parts of their life cycles in contact with or living in sediments. Among the sediment-dwelling organisms are various kinds of shellfish (shrimp, crayfish, crab, and clams) and a variety of worms, insects, amphipods, bivalves, and other smaller organisms that are of particular concern because they are located near the base of the food chain.

Pollutant transfer from sediments to organisms may involve an intermediate stage in water solution as well as direct transfer from sediments to organisms. This is probably particularly important

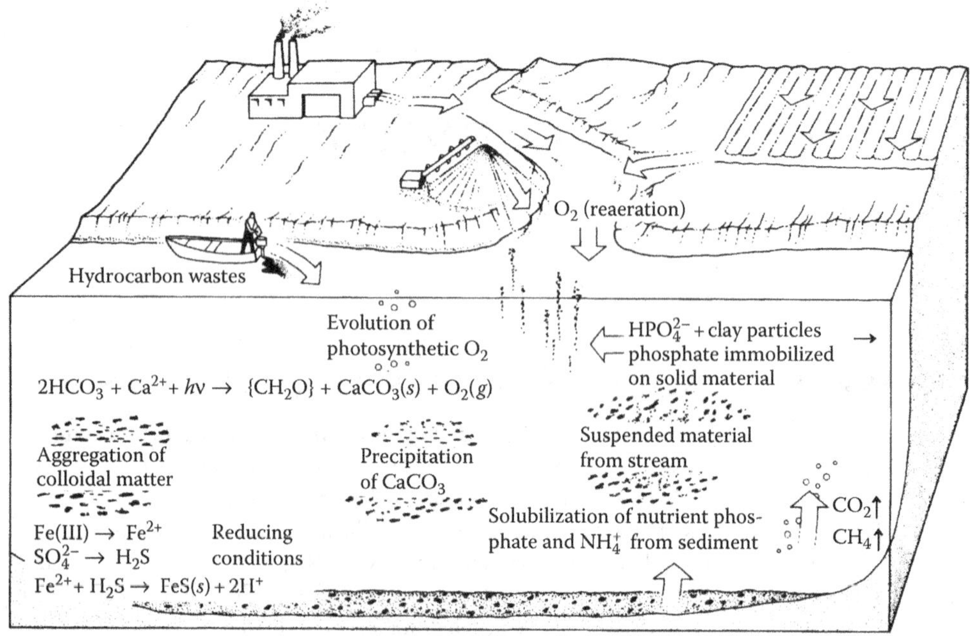

FIGURE 5.1 Most important environmental chemical processes in water involve interactions between water and another phase.

for poorly water-soluble organophilic pollutants, such as organohalide pesticides. The portion of substances held in sediments that is most available to organisms is that contained in *pore water*, obtained in microscopic pores within the sediment mass. Pore water is commonly extracted from sediments for measurements of toxicity to aquatic test organisms.

5.2.1 Formation of Sediments

Physical, chemical, and biological processes may all result in the deposition of sediments in the bottom regions of bodies of water. These sediments may eventually be covered and produce sedimentary minerals. Sedimentary material may be simply carried into a body of water by erosion or through sloughing (caving in) of the shore. Thus, clay, sand, organic matter, and other materials may be washed into a lake and settle out as layers of sediment.

Sediments may be formed by simple precipitation reactions, several of which are discussed below. When a phosphate-rich wastewater enters a body of water containing a high concentration of calcium ions, the following reaction occurs to produce solid hydroxyapatite:

$$5Ca^{2+} + H_2O + 3HPO_4^{2-} \rightarrow Ca_5OH(PO_4)_3(s) + 4H^+ \tag{5.1}$$

Calcium carbonate sediment may form when water rich in carbon dioxide and containing a high level of calcium as temporary hardness (see Section 3.5) loses carbon dioxide to the atmosphere:

$$Ca^{2+} + 2HCO_3^- \rightarrow CaCO_3(s) + CO_2(g) + H_2O \tag{5.2}$$

or when the pH is raised by a photosynthetic reaction:

$$Ca^{2+} + 2HCO_3^- + h\nu \rightarrow \{CH_2O\} + CaCO_3(s) + O_2(g) \tag{5.3}$$

Oxidation of reduced forms of an element can result in its transformation to an insoluble species, such as occurs when iron(II) is oxidized to iron(III) to produce a precipitate of insoluble iron(III) hydroxide:

$$4Fe^{2+} + 10H_2O + O_2 \rightarrow 4Fe(OH)_3(s) + 8H^+ \tag{5.4}$$

A decrease in pH can result in the production of an insoluble humic acid sediment from base-soluble organic humic substances in solution (see Section 3.17).

Biological activity is responsible for the formation of some aquatic sediments. Some bacterial species produce large quantities of iron(III) oxide (see Section 6.14) as part of their energy-extracting mediation of the oxidation of iron(II) to iron(III). In anoxic (oxygen-deficient) bottom regions of bodies of water, some bacteria use sulfate ion as an electron receptor:

$$SO_4^{2-} \rightarrow H_2S \tag{5.5}$$

whereas other bacteria reduce iron(III) to iron(II):

$$Fe(OH)_3(s) \rightarrow Fe^{2+} \tag{5.6}$$

These two products can react to give a black layer of iron(II) sulfide sediment:

$$Fe^{2+} + H_2S \rightarrow FeS(s) + 2H^+ \tag{5.7}$$

This frequently occurs during the winter, alternating with the production of calcium carbonate by-product from photosynthesis (Reaction 5.3) during the summer. Under such conditions, a layered bottom sediment is produced composed of alternate layers of black FeS and white $CaCO_3$ as shown in Figure 5.2.

5.2.2 Organic and Carbonaceous Sedimentary Materials

Carbonaceous sediments from organic materials are particularly important because of their affinity for poorly soluble organic water pollutants. (As noted in Section 1.9, in chemical fate and transport calculations involving the uptake of organic materials from water by sediments containing organic solids, the sediment–water partition coefficient for the organic substance partitioning between water and soil may be expressed as a function of the fraction of organic matter in the sediment, f_{oc}, and the partition coefficient of the organic contaminant for the pure organic solid, K_{oc}.) Sediment organic carbon comes from biological sources and from fossil fuels. Biological sources may consist of plant, animal, and microbial biomass including materials such as cellulose, lignin, collagen, and cuticle, and the degradation products thereof, especially humic substances. Fossil fuel sources include coal tar, petroleum residues (such as asphalt), soot, coke, charcoal, and coal.

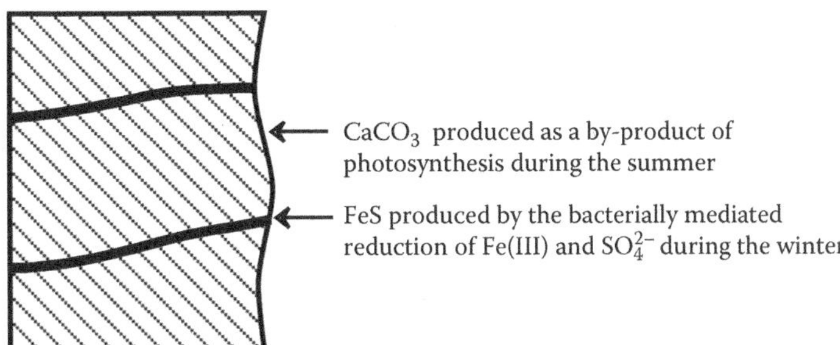
$CaCO_3$ produced as a by-product of photosynthesis during the summer
FeS produced by the bacterially mediated reduction of Fe(III) and SO_4^{2-} during the winter

FIGURE 5.2 Alternate layers of FeS and $CaCO_3$ in a lake sediment. This phenomenon has been observed in Lake Zürich in Switzerland.

Black carbon is the name given to small particles of carbon left over from the combustion of fossil fuels and biomass. Significant quantities of black carbon are produced by combustion processes and it is found in atmospheric particulate matter, in soil, and in sediments. Elemental carbon has an affinity for organic matter (see the discussion of activated carbon in Chapter 8) and is a significant sink for hydrophobic organic compounds in sediments.[1]

Hydrophobic organic compounds are bound preferentially to sediment organic carbon in sediments. The two most prominent examples of such compounds are polycyclic aromatic hydrocarbons (PAHs) and polychlorinated biphenyls (PCBs). From 60% to 90% of these hydrophobic organic compounds are bound with sediment organic carbon even though typically the sediment organic carbon comprises only 5–7% of the sediments. Sediment organic carbon provides a reservoir for the storage of hydrophobic organic compounds that may persist for many years after the source of pollution is removed. However, compounds held by these solids are relatively less bioavailable and less readily biodegraded than compounds that are in solution or are bound to mineral sedimentary materials.

5.3 SOLUBILITIES

The formation and stabilities of nonaqueous phases in water are strongly dependent upon solubilities. Calculations of the solubilities of solids and gases are addressed in this section.

5.3.1 Solubilities of Solids

Generally, the solubility of a solid in water is of concern when the solid is slightly soluble, often having such a low solubility that it is called "insoluble." In Section 3.15 the solubility of lead carbonate was considered. This salt can introduce toxic lead ion into water by reactions such as

$$PbCO_3(s) \rightleftarrows Pb^{2+} + CO_3^{2-} \tag{3.79}$$

A relatively straightforward calculation of the solubility of an ionic solid can be performed on barium sulfate, which dissolves according to the reaction

$$BaSO_4(s) \rightleftarrows Ba^{2+} + SO_4^{2-} \tag{5.8}$$

for which the equilibrium constant is the following:

$$K_{sp} = [Ba^{2+}][SO_4^{2-}] = 1.23 \times 10^{-10} \tag{5.9}$$

An equilibrium constant in this form that expresses the solubility of a solid that forms ions in water is a *solubility product* and is designated K_{sp}. In the simplest cases, a solubility product can be used alone to calculate the solubility of a slightly soluble salt in water. The solubility (S, mol/L) of barium sulfate is calculated as follows:

$$[Ba^{2+}] = [SO_4^{2-}] = S \tag{5.10}$$

$$[Ba^{2+}][SO_4^{2-}] = S \times S = K_{sp} = 1.23 \times 10^{-10} \tag{5.11}$$

$$S = (K_{sp})^{1/2} = (1.23 \times 10^{-10})^{1/2} = 1.11 \times 10^{-5} \tag{5.12}$$

Even such a simple calculation may be complicated by variations in activity coefficients resulting from differences in ionic strength.

Intrinsic solubilities account for the fact that a significant portion of the solubility of an ionic solid is due to the dissolution of the neutral form of the salt and must be added to the solubility calculated from K_{sp} to obtain the total solubility, as illustrated below for the calculation of the solubility of calcium sulfate. When calcium sulfate dissolves in water that does not have any other sources of calcium or sulfate ions, the two major reactions are

$$CaSO_4(s) \rightleftarrows CaSO_4(aq) \tag{5.13}$$

$$[CaSO_4(aq)] = 5.0 \times 10^{-3} \text{ M } (25°C) \tag{5.14}$$

(intrinsic solubility of $CaSO_4$)

$$CaSO_4(s) \rightleftarrows Ca^{2+} + SO_4^{2-} \tag{5.15}$$

$$[Ca^{2+}][SO_4^{2-}] = K_{sp} = 2.6 \times 10^{-5} \ (25°C) \tag{5.16}$$

and the total solubility of $CaSO_4$ is calculated as follows:

$$S = \underbrace{[Ca^{2+}]}_{\text{Contribution to solubility from solubility product}} + \underbrace{[CaSO_4(aq)]}_{\text{Contribution to solubility from intrinsic solubility}}$$

$$\begin{aligned} S &= (K_{sp})^{1/2} + [CaSO_4(aq)] = (2.6 \times 10^{-5})^{1/2} + 5.0 \times 10^{-3} \\ &= 5.1 \times 10^{-3} + 5.0 \times 10^{-3} = 1.01 \times 10^{-2} \text{ M} \end{aligned} \tag{5.18}$$

It is seen that, in this case, the intrinsic solubility accounts for half of the solubility of the salt.

In Section 3.15, it was seen that solubilities of ionic solids can be very much affected by reactions of cations and anions. It was shown that the solubility of $PbCO_3$ is increased by the chelation of lead ion by NTA,

$$Pb^{2+} + T^{3-} \rightleftarrows PbT^- \tag{3.72}$$

increased by reaction of carbonate ion with H^+,

$$H^+ + CO_3^{2-} \rightleftarrows HCO_3^- \tag{5.19}$$

and decreased by the presence of carbonate ion from water alkalinity:

$$CO_3^{2-}(\text{from dissociation of } HCO_3^-) + Pb^{2+} \rightleftarrows PbCO_3(s) \tag{5.20}$$

These examples illustrate that reactions of both cations and anions must often be considered in calculating the solubilities of ionic solids.

5.3.2 Solubilities of Gases

The solubilities of gases in water are described by Henry's law, which states that *at constant temperature the solubility of a gas in a liquid is proportional to the partial pressure of the gas in contact with the liquid.* For a gas, "X," this law applies to equilibria of the type

$$X(g) \rightleftarrows X(aq) \tag{5.21}$$

and does not account for additional reactions of the gas species in water such as

$$NH_3 + H_2O \rightleftarrows NH_4^+ + OH^- \tag{5.22}$$

$$SO_2 + HCO_3^-(\text{from water alkalinity}) \rightleftarrows CO_2 + HSO_3^- \tag{5.23}$$

TABLE 5.1
Henry's Law Constants for Some Gases in Water at 25°C

Gas	K (mol/L/atm)
O_2	1.28×10^{-3}
CO_2	3.38×10^{-2}
H_2	7.90×10^{-4}
CH_4	1.34×10^{-3}
N_2	6.48×10^{-4}
NO	2.0×10^{-4}

which may result in much higher solubilities than predicted by Henry's law alone. Mathematically, Henry's law is expressed as

$$[X(aq)] = KP_X \tag{5.24}$$

where $[X(aq)]$ is the aqueous concentration of the gas, P_X is the partial pressure of the gas, and K is the Henry's law constant applicable to a particular gas at a specified temperature. For gas concentrations in units of moles per liter and gas pressures in atmospheres, the units of K are moles per liter per atmosphere. Some values of K for dissolved gases that are significant in water are given in Table 5.1.

In calculating the solubility of a gas in water, a correction must be made for the partial pressure of water by subtracting it from the total pressure of the gas. At 25°C, the partial pressure of water is 0.0313 atm; values at other temperatures are readily obtained from standard handbooks. The concentration of oxygen in water saturated with air at 1.00 atm and 25°C may be calculated as an example of a simple gas solubility calculation. Considering that dry air is 20.95% by volume oxygen, factoring in the partial pressure of water gives the following:

$$P_{O_2} = (1.0000 \text{ atm} - 0.0313 \text{ atm}) \times 0.2095 = 0.2029 \text{ atm} \tag{5.25}$$

$$[O_2(aq)] = K \times P_{O_2} = 1.28 \times 10^{-3} \text{ mol / L / atm} \times 0.2029 \text{ atm} = 2.60 \times 10^{-4} \text{ mol/L} \tag{5.26}$$

The molecular mass of oxygen is 32, so the concentration of dissolved O_2 in water in equilibrium with air under the conditions given above is 8.32 mg/L or 8.32 parts per million (ppm).

The solubilities of gases decrease with increasing temperature. Account is taken of this factor with the *Clausius–Clapeyron* equation,

$$\log \frac{C_2}{C_1} = \frac{\Delta H}{2.303R}\left[\frac{1}{T_1} - \frac{1}{T_2}\right] \tag{5.27}$$

where C_1 and C_2 denote the gas concentration in water at absolute temperatures of T_1 and T_2, respectively; ΔH is the heat of the solution; and R is the gas constant. The value of R is 1.987 cal/deg/mol, which gives ΔH in units of cal/mol.

5.4 COLLOIDAL PARTICLES IN WATER

Many minerals, some organic pollutants, proteinaceous materials, some algae, and some bacteria are suspended in water as very small particles. Such particles, which have some characteristics of both species in solution and larger particles in suspension, which range in diameter from about 0.001 micrometer (μm) to about 1 μm, and which scatter white light as a light blue hue observed at right angles to the incident light, are classified as *colloidal particles.* The characteristic light-scattering phenomenon of colloids results from their being the same order of size as the wavelength of light and is called the *Tyndall effect.* The unique properties and behavior of colloidal particles are strongly influenced by their physical–chemical characteristics, including high specific area, high interfacial energy, and high surface/charge density ratio. Colloids play a very important role in determining the properties and behavior of natural waters and wastewaters.

5.4.1 Contaminant Transport by Colloids in Water

An important influence of colloids in aquatic chemistry is their ability to transport various kinds of organic and inorganic contaminants. *Colloid-facilitated transport* in which contaminants bound to the surface of colloidal particles can be an important means of moving substances that would otherwise be sorbed to sediments or, in the case of groundwater transport, to aquifer rocks. This mechanism is of concern with respect to circumvention of natural and artificial barriers in the long-term subsurface disposal of some kinds of wastes, such as high-level nuclear wastes including plutonium.[2]

5.4.2 Occurrence of Colloids in Water

Colloids composed of a variety of organic substances (including humic substances), inorganic materials (especially clays), and pollutants occur in natural water and wastewater. These substances have a number of effects, including those on organisms and pollutant transport. The characterization of colloidal materials in water is obviously very important, and a variety of means are used to isolate and characterize those materials. The two most widely used methods are filtration and centrifugation, although other techniques including voltammetry and field-flow fractionation can be used.

5.4.3 Kinds of Colloidal Particles

Colloids may be classified as *hydrophilic colloids*, *hydrophobic colloids*, or *association colloids.* These three classes are briefly summarized below.

Hydrophilic colloids generally consist of macromolecules, such as proteins and synthetic polymers that are characterized by strong interaction with water resulting in spontaneous formation of colloids when they are placed in water. In a sense, hydrophilic colloids are solutions of very large molecules or ions. Suspensions of hydrophilic colloids are less affected by the addition of salts to water than are suspensions of hydrophobic colloids.

Hydrophobic colloids interact to a lesser extent with water and are stable because of their positive or negative electrical charges as shown in Figure 5.3. The charged surface of the colloidal particle and the *counterions* that surround it compose an *electrical double layer,* which causes the particles to repel each other.

Hydrophobic colloids are usually caused to settle from suspension by the addition of salts. Examples of hydrophobic colloids are clay particles, petroleum droplets, and very small gold particles.

Association colloids consist of special aggregates of ions and molecules called *micelles.* To understand how this occurs, consider sodium stearate, a typical soap with the structural formula shown below:

```
  H H H H H H H H H H H H H H H H H O
H-C-C-C-C-C-C-C-C-C-C-C-C-C-C-C-C-C-C-O⁻ Na⁺
  H H H H H H H H H H H H H H H H H
```

Represented as ~~~~~~~~~~~~~~~~(-) (Na+)

The stearate ion has both a hydrophilic $-CO_2^-$ head and a long organophilic tail, $CH_3(CH_2)_{16}^-$. As a result, stearate anions in water tend to form clusters consisting of as many as 100 anions clustered

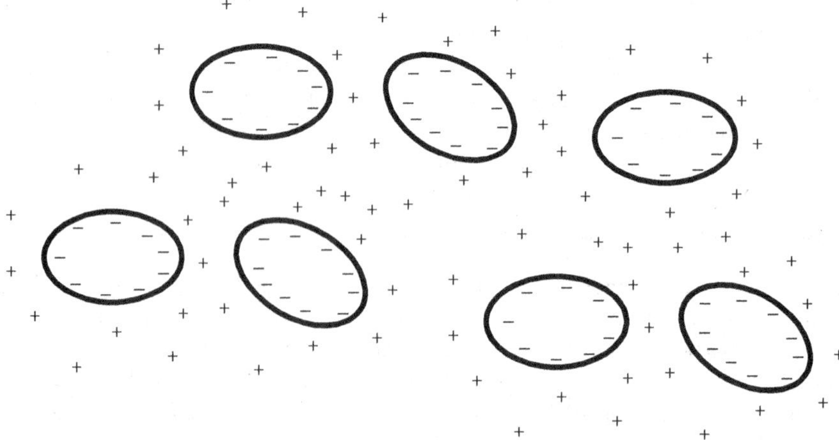

FIGURE 5.3 Representation of negatively charged hydrophobic colloidal particles surrounded in solution by positively charged counter-ions, forming an electrical double layer. (Colloidal particles suspended in water may have either a negative or positive charge.)

together with their hydrocarbon "tails" on the inside of a spherical colloidal particle and their ionic "heads" on the surface in contact with water and with Na^+ counter-ions. Such *micelles* (Figure 5.4) can be visualized as droplets of oil about 3–4 nanometers (nm) in diameter and covered with ions or polar groups. According to this model, micelles form when a certain concentration of surfactant species, typically around 1×10^{-3} M, is reached. The concentration at which this occurs is called the *critical micelle concentration.*

5.4.4 Colloid Stability

The stability of colloids is a prime consideration in determining their behavior. It is involved in important aquatic chemical phenomena including the formation of sediments, dispersion and agglomeration of bacterial cells, and dispersion and removal of pollutants (such as crude oil from an oil spill).

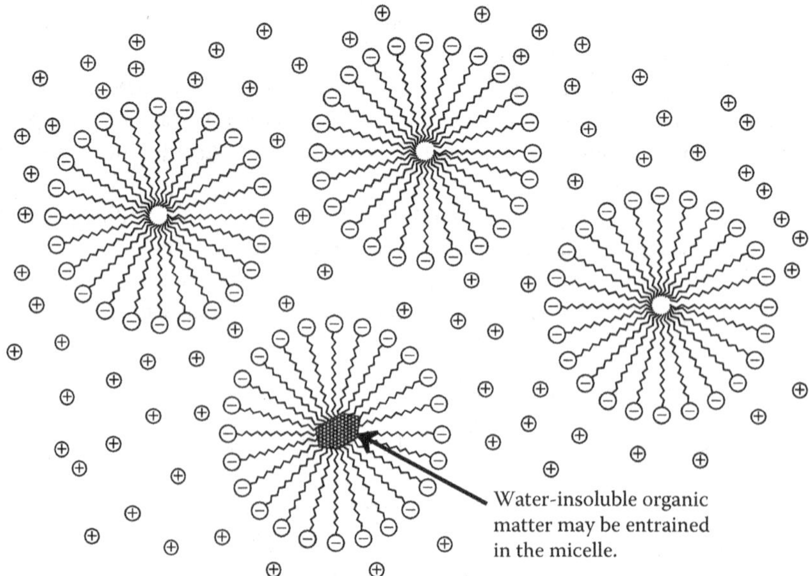
Water-insoluble organic matter may be entrained in the micelle.

FIGURE 5.4 Representation of colloidal soap micelle particles.

As discussed above, the two main phenomena contributing to the stabilization of colloids are *hydration* and *surface charge*. The layer of water on the surface of hydrated colloidal particles prevents contact, which would result in the formation of larger units. A surface charge on colloidal particles may prevent aggregation, since like-charged particles repel each other. The surface charge is frequently pH dependent; at around pH 7, most colloidal particles in natural waters are negatively charged. Negatively charged aquatic colloids include algal cells, bacterial cells, proteins, and colloidal petroleum droplets. Natural organic matter in water tends to bind with colloidal particle surfaces and, because of the negatively charged functional groups characteristic of this matter, gives colloidal particles in water a predominantly negative charge.

One of the three major ways in which a particle may acquire a surface charge is by *chemical reaction at the particle surface*. This phenomenon, which frequently involves hydrogen ion and is pH dependent, is typical of hydroxides and oxides and is illustrated for manganese dioxide, MnO_2, in Figure 5.5.

As an illustration of pH-dependent charge on colloidal particle surfaces, consider the effects of pH on the surface charge of hydrated manganese oxide, represented by the chemical formula $MnO_2(H_2O)(s)$. In a relatively acidic medium, the reaction

$$MnO_2(H_2O)(s) + H^+ \rightarrow MnO_2(H_3O)^+(s) \quad (5.28)$$

may occur on the surface giving the particle a net positive charge. In a more basic medium, hydrogen ions may be lost from the hydrated oxide surface to yield negatively charged particles:

$$MnO_2(H_2O)(s) \rightarrow MnO_2(OH)^-(s) + H^+ \quad (5.29)$$

At some intermediate pH value, called the *zero point of charge* (ZPC), colloidal particles of a given hydroxide will have a net charge of zero, which favors aggregation of particles and precipitation of a bulk solid:

$$\text{Number of } MnO_2(H_3O)^+ \text{ sites} = \text{Number of } MnO_2(OH)^- \text{ sites} \quad (5.30)$$

Individual cells of microorganisms that behave as colloidal particles have a charge that is pH dependent. The charge is acquired through the loss and gain of H^+ ions by carboxyl and amino groups on the cell surface:

$$\underset{\text{low pH}}{{}^+H_3N(+\text{cell})CO_2H} \quad \underset{\text{intermediate pH}}{{}^+H_3N(\text{neutral cell})CO_2^-} \quad \underset{\text{high pH}}{H_2N(-\text{cell})CO_2^-}$$

Ion absorption is a second way in which colloidal particles become charged. This phenomenon involves attachment of ions onto the colloidal particle surface by means other than conventional covalent bonding, including hydrogen bonding and London (van der Waal) interactions.

Ion replacement is a third way in which a colloidal particle may gain a net charge; for example, replacement of some of the Si(IV) with Al(III) in the basic SiO_2 chemical unit in the crystalline lattice of some clay minerals as shown in Equation 5.31,

$$[SiO_2] + Al(III) \rightarrow [AlO_2^-] + Si(IV) \quad (5.31)$$

yields sites with a net negative charge. Similarly, replacement of Al(III) by a divalent metal ion such as Mg(II) in the clay crystalline lattice produces a net negative charge.

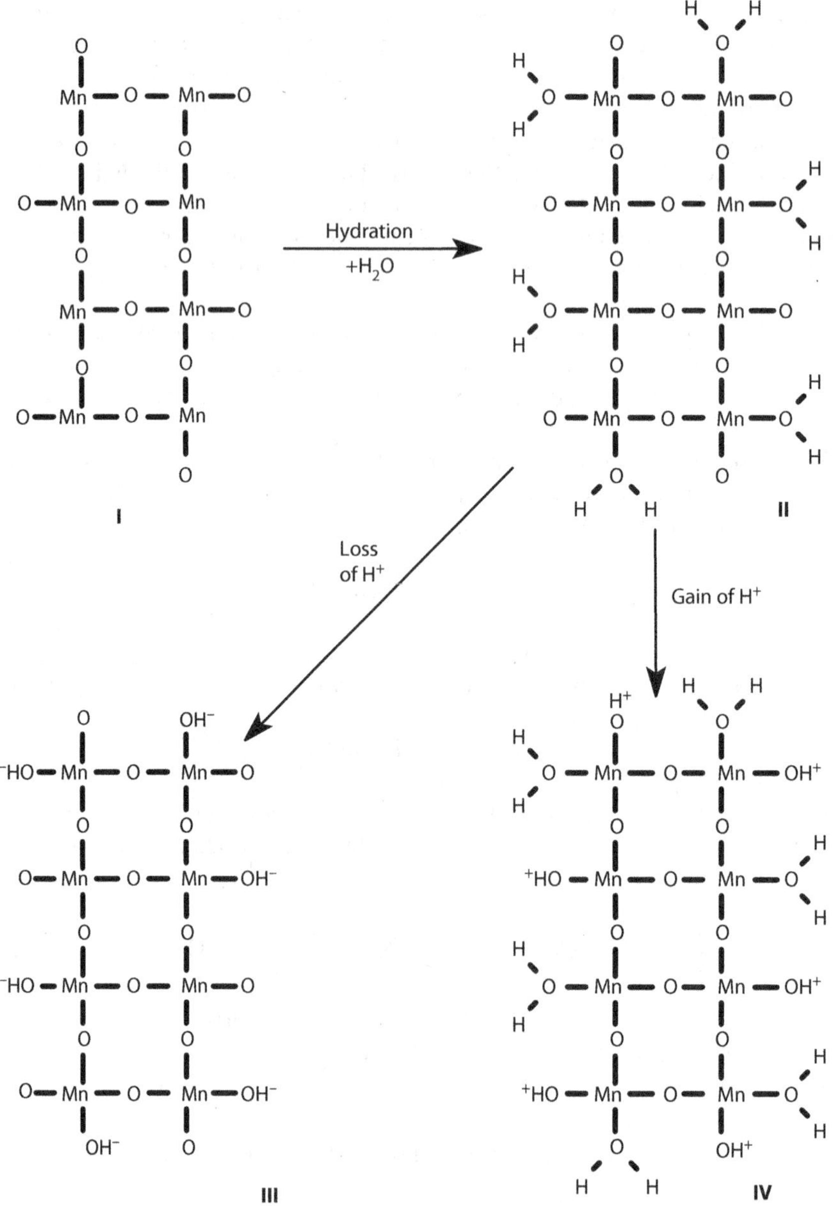

FIGURE 5.5 Acquisition of surface charge by colloidal MnO_2 in water. (I) Anhydrous MnO_2 has two O atoms per Mn atom. (II) Suspended in water as a colloid, it binds to water molecules to form hydrated MnO_2. (III) Loss of H^+ from the bound H_2O yields a negatively charged colloidal particle. (IV) Gain of H^+ by surface O atoms yields a positively charged particle. The former process (loss of H^+ ion) predominates for metal oxides.

5.5 COLLOIDAL PROPERTIES OF CLAYS

Clays constitute the most important class of common minerals occurring as colloidal matter in water. The composition and properties of clays are discussed in some detail in Section 15.7 (as solid terrestrial minerals) and are briefly summarized here. *Clays* consist largely of hydrated aluminum and silicon oxides and are *secondary minerals*, which are formed by weathering and other processes acting on primary rocks (see Sections 15.2 and 15.8). The general formulas of some common clays are given below:

- Kaolinite: $Al_2(OH)_4Si_2O_5$
- Montmorillonite: $Al_2(OH)_2Si_4O_{10}$
- Nontronite: $Fe_2(OH)_2Si_4O_{10}$
- Hydrous mica: $KAl_2(OH)_2(AlSi_3)O_{10}$

The most common clay minerals are illites, montmorillonites, chlorites, and kaolinites. These clay minerals are distinguished from each other by general chemical formula, structure, and chemical and physical properties. Iron and manganese are commonly associated with clay minerals.

Clays are characterized by layered structures consisting of sheets of silicon oxide alternating with sheets of aluminum oxide. Units of two or three sheets make up *unit layers*. Some clays, particularly the montmorillonites, may absorb large quantities of water between unit layers, a process accompanied by swelling of the clay.

As described in Section 5.4, clay minerals may attain a net negative charge by ion replacement, in which Si(IV) and Al(III) ions are replaced by metal ions of similar size but lesser charge. This negative charge must be compensated for by association of cations with the clay layer surfaces. Since these cations need not fit specific sites in the crystalline lattice of the clay, they may be relatively large ions, such as K^+, Na^+, or NH_4^+. These cations are called *exchangeable cations* and are exchangeable for other cations in water. The amount of exchangeable cations, expressed as milliequivalents (of monovalent cations) per 100 g of dry clay, is called the *cation-exchange capacity* (CEC) of the clay and is a very important characteristic of colloids and sediments that have cation-exchange capabilities.

Because of their structure and high surface area per unit mass, clays have a strong tendency to sorb chemical species from water. Thus, clays play a role in the transport and reactions of biological wastes, organic chemicals, gases, and other pollutant species in water. However, clay minerals may also effectively immobilize dissolved chemicals in water and so exert a purifying action. Some microbial processes occur at clay particle surfaces and, in some cases, sorption of organics by clay inhibits biodegradation. Thus, clay may play a role in the microbial degradation or nondegradation of organic wastes.

5.6 AGGREGATION OF PARTICLES

The processes by which particles aggregate and precipitate from colloidal suspension are quite important in the aquatic environment. For example, the settling of biomass during biological waste treatment depends upon the aggregation of bacterial cells. Other processes involving the aggregation of colloidal particles are the formation of bottom sediments and the clarification of turbid water for domestic or industrial use. Particle aggregation is complicated and may be divided into the two general classes of *coagulation* and *flocculation*. These are discussed below.

Colloidal particles are prevented from aggregating by the electrostatic repulsion of the electrical double layers (adsorbed-ion layer and counter-ion layer). *Coagulation* involves the reduction of this electrostatic repulsion such that colloidal particles of identical materials may aggregate. *Flocculation* uses *bridging compounds*, which form chemically bonded links between colloidal particles and enmesh the particles in relatively large masses called *floc networks*.

Hydrophobic colloids are often readily coagulated by the addition of small quantities of salts that contribute ions to solution. Such colloids are stabilized by electrostatic repulsion. Therefore, the simple explanation of coagulation by ions in solution is that the ions reduce the electrostatic repulsion between particles to such an extent that the particles aggregate. Because of the double layer of electrical charge surrounding a charged particle, this aggregation mechanism is sometimes called *double-layer compression*. It is particularly noticeable in estuaries where sediment-laden

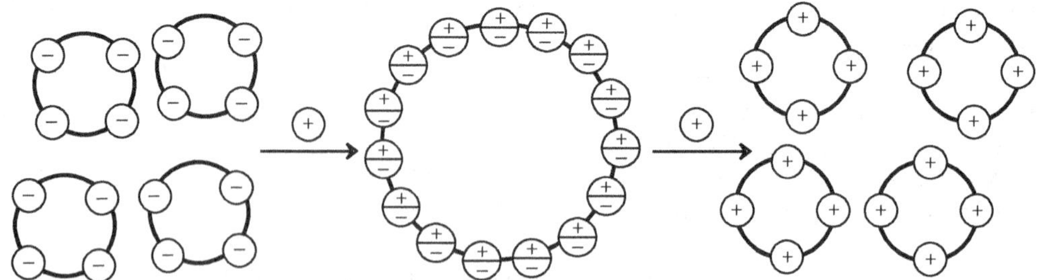

FIGURE 5.6 Aggregation of negatively charged colloidal particles by reaction with positive ions, followed by restabilization as a positively charged colloid.

fresh water flows into the sea, and is largely responsible for deltas formed where large rivers enter oceans.

The binding of positive ions to the surface of an initially negatively charged colloid can result in precipitation followed by colloid restabilization as shown in Figure 5.6. This kind of behavior is explained by an initial neutralization of the negative surface charge on the particles by sorption of positive ions, allowing coagulation to occur. As more of the source of positive ions is added, their sorption results in the formation of positive colloidal particles.

5.6.1 Flocculation of Colloids by Polyelectrolytes

Polyelectrolytes of both natural and synthetic origin may cause colloids to flocculate. Polyelectrolytes are polymers with a high formula mass that normally contain ionizable functional groups. Typical examples of synthetic polyelectrolytes are shown in Table 5.2.

It can be seen from Table 5.2 that anionic polyelectrolytes have negatively charged functional groups such as $-SO_3^-$ and $-CO_2^-$. Cationic polyelectrolytes have positively charged functional groups, normally H^+ bonded to N. Nonionic polymers that serve as flocculants normally do not have charged functional groups.

Somewhat paradoxically, *anionic* polyelectrolytes may flocculate *negatively charged* colloidal particles, a process that is particularly important in biological systems, for example, in the cohesion of tissue cells, clumping of bacterial cells, and antibody–antigen reactions. The mechanism by which this occurs involves bridging between the colloidal particles by way of the polyelectrolyte anions. It is facilitated by the presence of a low concentration of a metal ion capable of serving as a bridge between the negatively charged anionic polyelectrolytes and negatively charged functional groups on the colloidal particle surface.

5.6.2 Flocculation of Bacteria by Polymeric Materials

The aggregation and settling of microorganism cells is a very important process in aquatic systems and is essential to the function of biological waste treatment systems. In these processes, using bacteria to degrade organic wastes, a significant fraction of the carbon in the waste is removed as *bacterial floc*, consisting of aggregated bacterial cells that have settled from the water. The formation of this floc is obviously an important phenomenon in biological waste treatment. Polymeric substances, including polyelectrolytes, that are formed by the bacteria induce bacterial flocculation.

Within the pH range of normal natural waters (pH 5–9), bacterial cells are negatively charged. The ZPC of most bacteria is within the pH range 2–3. However, even at the ZPC, stable bacterial suspensions may exist. Therefore, surface charge is not necessarily required to maintain bacterial cells in suspension in water, and it is likely that bacterial cells remain in suspension because of the hydrophilic character of their surfaces. As a consequence, some sort of chemical interaction involving bridging species must be involved in bacterial flocculation.

TABLE 5.2
Synthetic Polyelectrolytes and Neutral Polymers Used as Flocculants

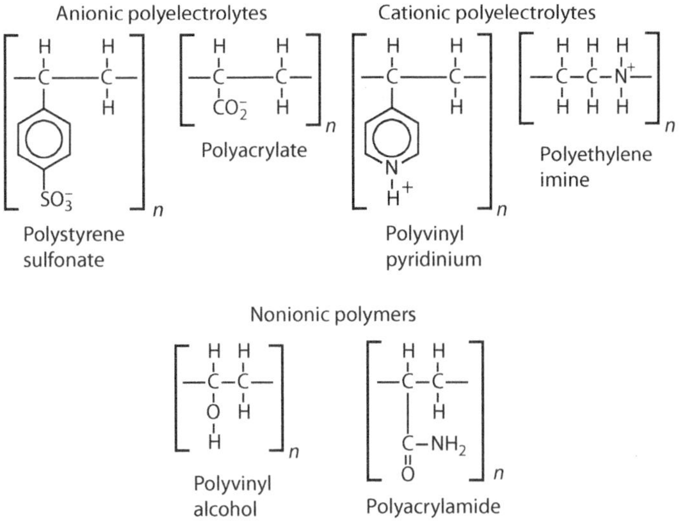

Anionic polyelectrolytes

Polystyrene sulfonate: $\left[-\mathrm{CH}-\mathrm{CH_2}-\right]_n$ with the ring carrying SO_3^-

Polyacrylate: $\left[-\mathrm{CH}(CO_2^-)-\mathrm{CH_2}-\right]_n$

Cationic polyelectrolytes

Polyvinyl pyridinium: $\left[-\mathrm{CH}-\mathrm{CH_2}-\right]_n$ with the pyridine ring protonated at N ($\mathrm{N}-\mathrm{H}^+$)

Polyethylene imine: $\left[-\mathrm{CH_2}-\mathrm{CH_2}-\mathrm{NH_2^+}-\right]_n$

Nonionic polymers

Polyvinyl alcohol: $\left[-\mathrm{CH}(\mathrm{OH})-\mathrm{CH_2}-\right]_n$

Polyacrylamide: $\left[-\mathrm{CH}(\mathrm{C(=O)NH_2})-\mathrm{CH_2}-\right]_n$

5.7 SURFACE SORPTION BY SOLIDS

Many of the properties and effects of solids in contact with water have to do with the sorption of solutes by solid surfaces. Surfaces in finely divided solids tend to have excess surface energy because of an imbalance of chemical forces among surface atoms, ions, and molecules. Surface energy level may be lowered by a reduction in surface area. Normally this reduction is accomplished by aggregation of particles or by sorption of solute species.

Some kinds of surface interactions can be illustrated with metal oxide surfaces binding with metal ions in water. (Such a surface, its reaction with water, and its subsequent acquisition of a charge by loss or gain of H^+ ion were shown in Figure 5.5 for MnO_2.) Other inorganic solids, such as clays, probably behave much like solid metal oxides. Soluble metal ions, such as Cd^{2+}, Cu^{2+}, Pb^{2+}, or Zn^{2+}, may be bound with metal oxides such as $MnO_2 \cdot xH_2O$ by nonspecific ion-exchange adsorption, complexation with surface –OH groups, coprecipitation in solid solution with the metal oxide, or as a discrete oxide or hydroxide of the sorbed metal. Sorption of metal ions, Mt^{z+}, by complexation to the surface is illustrated by the reaction

$$\mathrm{M{-}OH} + \mathrm{Mt}^{z+} \rightleftarrows \mathrm{M{-}OMt}^{z-1} + \mathrm{H}^+ \tag{5.32}$$

and chelation by the following process:

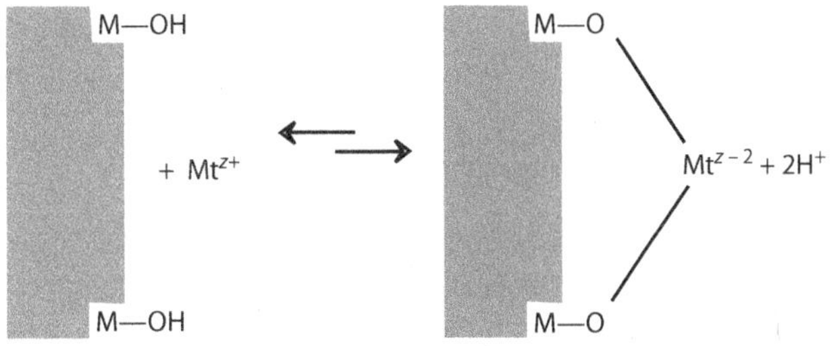

$$\begin{matrix}\mathrm{M{-}OH}\\ \\ \mathrm{M{-}OH}\end{matrix} + \mathrm{Mt}^{z+} \rightleftarrows \begin{matrix}\mathrm{M{-}O}\\ \\ \mathrm{M{-}O}\end{matrix}\!\!>\mathrm{Mt}^{z-2} + 2\mathrm{H}^+ \tag{5.33}$$

A metal ion complexed with a ligand, L, may bond by displacement of either H^+ or OH^-:

$$\text{M–OH} + \text{MtL}^{z+} \rightleftarrows \text{M–OMtL}^{(z-1)} + \text{H}^+ \tag{5.34}$$

$$\text{M–OH} + \text{MtL}^{z+} \rightleftarrows \text{M–(MtL)}^{(z+1)} + \text{OH}^- \tag{5.35}$$

Furthermore, in the presence of a ligand, dissociation of the complex and sorption of the metal complex and ligand must be considered as shown by the scheme below in which "(sorbed)" represents sorbed species and "(*aq*)" represents dissolved species:

$$\begin{array}{ccc} \text{Mt}^{z+}(\text{sorbed}) & \rightleftarrows & \text{Mt}^{z+}(aq) \\ \uparrow\downarrow & & \uparrow\downarrow \\ \text{MtL}^{z+}(\text{sorbed}) & \rightleftarrows & \text{MtL}^{z+}(aq) \\ \uparrow\downarrow & & \uparrow\downarrow \\ \text{L(sorbed)} & \rightleftarrows & \text{L}(aq) \end{array}$$

Some hydrated metal oxides, such as manganese(IV) oxide and iron(III) oxide, are especially effective in sorbing various species from aquatic solution. The sorption ability is especially pronounced for relatively fresh metal hydroxides or hydrated oxides such as colloidal MnO_2. This oxide usually is produced in natural waters by the oxidation of Mn(II) present in natural waters placed there by the bacterially mediated reduction of manganese oxides in anoxic bottom sediments. Colloidal hydrated manganese(II) oxide can also be produced by the reduction of manganese(VII), which can be deliberately added to water as an oxidant in the form of permanganate salts to diminish taste and odor or to oxidize iron(II).

Freshly precipitated MnO_2 may have a surface area as large as several hundred square meters per gram. The hydrated oxide acquires a charge by loss and gain of H^+ ion and has a ZPC in an acidic pH range between 2.8 and 4.5. Since the pH of most normal natural waters exceeds 4.5, hydrous MnO_2 colloids are usually negatively charged.

The sorption of anions by solid surfaces is harder to explain than the sorption of cations. Phosphates may be sorbed on hydroxylated surfaces by displacement of hydroxides (ion exchange):

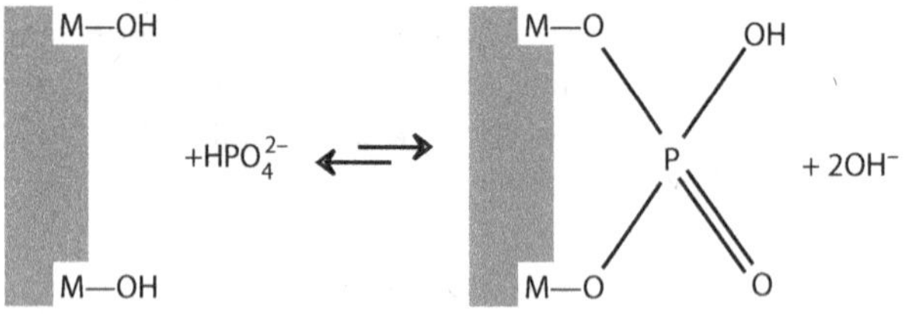

$$\begin{array}{l} \text{M—OH} \\ \\ \text{M—OH} \end{array} + \text{HPO}_4^{2-} \rightleftarrows \begin{array}{l} \text{M—O} \\ \\ \text{M—O} \end{array} \!\!>\!\! \text{P}(\text{OH})(=\text{O}) + 2\text{OH}^- \tag{5.36}$$

The degree of anion sorption varies. As with phosphate, sulfate may be sorbed by chemical bonding, usually at a pH < 7. Chloride and nitrate are sorbed by electrostatic attraction, such as occurs with positively charged colloidal particles in soil at a low pH. More specific bonding mechanisms may be involved in the sorption of fluoride, molybdate, selenate, selenite, arsenate, and arsenite anions.

5.8 SOLUTE EXCHANGE WITH BOTTOM SEDIMENTS

Bottom sediments are important sources and sinks of inorganic and organic matter in streams, freshwater impoundments, estuaries, and oceans. It is incorrect to consider bottom sediments simply as wet soil. Normal soils are in contact with the atmosphere and are oxic, whereas the environment around bottom sediments is usually anoxic, so sediments are subjected to reducing conditions. Bottom sediments undergo continuous leaching, whereas soils do not. The level of organic matter in sediments is generally higher than that in soils.

One of the most important characteristics of bottom sediments is their ability to exchange cations with the surrounding aquatic medium. CEC measures the capacity of a solid, such as a sediment, to sorb cations. It varies with pH and salt concentration. Another parameter, *exchangeable cation status* (ECS), refers to the amounts of specific ions bonded to a given amount of sediment. Generally, both CEC and ECS are expressed as milliequivalents per 100 g of solid.

Because of the generally anoxic nature of bottom sediments, special care must be exercised in their collection and treatment. Particularly, contact with atmospheric oxygen rapidly oxidizes exchangeable Fe^{2+} and Mn^{2+} to nonexchangeable oxides containing the metals in higher oxidation states as Fe_2O_3 and MnO_2. Therefore, sediment samples must be sealed and frozen as soon as possible after they are collected.

A common method for the determination of CEC consists of (1) treating the sediment with a solution of an ammonium salt so that all exchangeable sites are occupied by NH_4^+ ion; (2) displacing the ammonium ion with a solution of NaCl; and (3) determining the quantity of displaced ammonium ion. The CEC values may then be expressed as the number of milliequivalents of ammonium ion exchanged per 100 g of dried sample. Note that the sample must be dried *after* exchange.

The basic method for the determination of ECS consists of stripping all of the exchangeable metal cations from the sediment sample with ammonium acetate. Metal cations, including Fe^{2+}, Mn^{2+}, Zn^{2+}, Cu^{2+}, Ni^{2+}, Na^+, K^+, Ca^{2+}, and Mg^{2+}, are then determined in the leachate. Exchangeable hydrogen ion is very difficult to determine by direct methods. It is generally assumed that the total CEC minus the sum of all exchangeable cations except hydrogen ion is equal to the exchangeable hydrogen ion.

Freshwater sediments typically have CEC values of 20–30 milliequivalents/100 g. The ECS values for individual cations typically range from <1 to 10–20 milliequivalents/100 g. Sediments are important repositories of metal ions that may be exchanged with surrounding waters. Furthermore, because of their capacity to sorb and release hydrogen ions, sediments have an important pH-buffering effect in some waters.

5.8.1 Trace-Level Metals in Suspended Matter and Sediments

Sediments and suspended particles are important repositories for trace amounts of metals such as chromium, cadmium, copper, molybdenum, nickel, cobalt, and manganese. These metals may be present as discrete compounds, ions held by cation-exchanging clays, bound to hydrated oxides of iron or manganese, or chelated by insoluble humic substances. The form of the metals depends upon p*E*. Examples of specific trace-metal-containing compounds that may be stable in natural waters under oxidizing and reducing conditions are given in Table 5.3. Solubilization of metals from sedimentary or suspended matter is often a function of the complexing agents present. These include amino acids, such as histidine, tyrosine, or cysteine; citrate ion; and, in the presence of seawater, chloride ion. Suspended particles containing trace elements may be in the submicrometer size range. Although less available than metals in true solution, metals held by very small particles are more accessible than those in sediments. Among the factors involved in metal availability are the identity of the metal, its chemical form (type of binding, oxidation state), the nature of the suspended material, the type of organism taking up the metal, and the physical and chemical conditions in the water. The pattern of trace-metal occurrence in suspended matter in relatively unpolluted water tends to correlate well with that of the parent minerals from which the suspended solids originated; anomalies appear in polluted waters where industrial sources add to the metal content of the stream.

Two inorganic species are particularly important in the sequestration of heavy metals in sediments. One of these is hydrated iron(III) oxide, commonly represented as $Fe_2O_3 \cdot x(H_2O)$, $Fe(OH)_3$, or FeOOH, which coprecipitates heavy metals. This material forms when soluble iron(II) is exposed to oxidizing conditions. Under anoxic conditions the material is reduced to soluble iron(II) (Reaction 5.6), which

TABLE 5.3
Inorganic Trace Metal Compounds That May Be Stable under Oxidizing and Reducing Conditions

	Discrete Compound That May Be Present	
Metal	**Oxidizing Conditions**	**Reducing Conditions**
Cadmium	$CdCO_3$	CdS
Copper	$Cu_2(OH)_2CO_3$	CuS
Iron	$Fe_2O_3 \cdot x(H_2O)$	FeS, FeS_2
Mercury	HgO	HgS
Manganese	$MnO_2 \cdot x(H_2O)$	MnS, $MnCO_3$
Nickel	$Ni(OH)_2$, $NiCO_3$	NiS
Lead	$2PbCO_3 \cdot Pb(OH)_2$, $PbCO_3$	PbS
Zinc	$ZnCO_3$, $ZnSiO_3$	ZnS

releases bound heavy metals. The second important solid that binds heavy metals consists of *acid-volatile sulfides*, especially FeS. Most heavy metals have a stronger affinity for sulfide than does iron(II), so heavy metals tend to displace Fe(II) from this compound. The conditions conducive to reduction of hydrated iron(III) oxide leading to release of sequestered heavy metal ions are also conducive to formation of FeS, which may release sulfide to bind such ions in an insoluble form.

The toxicities of heavy metals in sediments and their availability to organisms are very important in determining the environmental effects of heavy metals in aquatic systems. Many sediments are anoxic, so that microbial reduction of sulfate to sulfide leads to a preponderance of metal sulfides in sediments. The very low solubilities of sulfides tend to limit bioavailability of metals in anoxic sediments. However, exposure of such sediments to air, and subsequent oxidation of sulfide to sulfate, can release significant amounts of heavy metals. Dredging operations can expose anoxic sediments to air, leading to oxidation of sulfides and release of metals such as lead, mercury, cadmium, zinc, and copper.

5.8.2 Phosphorus Exchange with Bottom Sediments

Phosphorus is one of the key elements in aquatic chemistry and is the limiting nutrient in the growth of algae under many conditions. Exchange with sediments plays a role in making phosphorus available for algae and contributes, therefore, to eutrophication. Sedimentary phosphorus may be classified into the following types:

- *Phosphate minerals*, particularly hydroxyapatite, $Ca_5OH(PO_4)_3$.
- *Nonoccluded phosphorus*, such as orthophosphate ion bound to the surface of SiO_2 or $CaCO_3$. Such phosphorus is generally more soluble and more available than occluded phosphorus (below).
- *Occluded phosphorus* consisting of orthophosphate ions contained within the matrix structures of amorphous hydrated oxides of iron and aluminum and amorphous aluminosilicates. Such phosphorus is not as readily available as nonoccluded phosphorus.
- *Organic phosphorus* incorporated within aquatic biomass, usually of algal or bacterial origin.

In some waters receiving heavy loads of domestic or industrial wastes, inorganic polyphosphates (e.g., from detergents) may be present in sediments. Runoff from fields where liquid polyphosphate fertilizers have been used can result in polyphosphates sorbed on sediments. The action of microorganisms on phosphate minerals, especially iron phosphates, can release phosphate to water.[3]

5.8.3 Organic Compounds on Sediments and Suspended Matter

Many organic compounds interact with suspended material and sediments in bodies of water. Colloids can play a significant role in the transport of organic pollutants in surface waters, through treatment processes, and even, to a limited extent, in groundwater. Settling of suspended material containing sorbed organic matter carries organic compounds into the sediment of a stream or lake. For example, this phenomenon is largely responsible for the presence of herbicides in sediments containing contaminated soil particles eroded from cropland. Some organics are carried into sediments by the remains of organisms or by fecal pellets from zooplankton that have accumulated organic contaminants.

Suspended particulate matter affects the mobility of organic compounds sorbed to particles. Furthermore, sorbed organic matter undergoes chemical degradation and biodegradation at different rates and by different pathways compared to organic matter in solution. There is, of course, a vast variety of organic compounds that get into water and they react with sediments in different ways, the type and strength of binding varying with the type of compound.

The most common types of sediments considered for their organic binding abilities are clays, organic (humic) substances, and complexes between clay and humic substances. Both clays and humic substances act as cation exchangers. Therefore, these materials sorb cationic organic compounds through ion exchange. This is a relatively strong sorption mechanism, greatly reducing the mobility and biological activity of the organic compound. When sorbed by clays, cationic organic compounds are generally held between the layers of the clay mineral structure where their biological activity is essentially zero.

Since most sediments lack strong anion-exchange sites, negatively charged organics are not held strongly at all. Thus, these compounds are relatively mobile and biodegradable in water despite the presence of solids.

The degree of sorption of organic compounds is generally inversely proportional to their water solubility. The more water-insoluble compounds tend to be taken up strongly by lipophilic ("fat-loving") solid materials, such as humic substances (see Section 3.17). Compounds having a relatively high vapor pressure can be lost from water or solids by evaporation. When this happens, photochemical processes (see Chapter 9) can play an important role in their degradation.

The herbicide 2,4-dichlorophenoxyacetic acid (2,4-D) has been studied extensively in regard to sorption reactions. Most of these studies have dealt with pure clay minerals, however, whereas soils and sediments are likely to have a strong clay–fulvic acid complex component. The sorption of 2,4-D by such a complex can be described using an equation of the Freundlich isotherm type

$$X = KC^n \tag{5.37}$$

where X is the amount sorbed per unit weight of solid, C is the concentration of 2,4-D in water solution at equilibrium, and n and K are constants. These values are determined by plotting log X versus log C. If a Freundlich-type equation is obeyed, the plot will be linear with a slope of n and an intercept of log K.

Sorption of comparatively nonvolatile hydrocarbons by sediments not only removes these materials from contact with aquatic organisms but also greatly retards their biodegradation. Aquatic plants produce some of the hydrocarbons that are found in sediments. Photosynthetic organisms, for example, produce quantities of *n*-heptadecane. The distribution of hydrocarbon types in sediments can be used to distinguish natural from pollutant sources.

The sorption of neutral species like petroleum obviously cannot be explained by ion-exchange processes. It probably involves phenomena such as van der Waal's forces (a term sometimes invoked when the true nature of an attractive force is not understood, but generally regarded as consisting of induced dipole–dipole interaction involving a neutral molecule), hydrogen bonding, charge-transfer complexation, and hydrophobic interactions.

In some cases, pollutant compounds become covalently bounded as *bound residues* to humic substances in sediments and soil. Among the kinds of organic compounds for which there is evidence of bound residues in sediments are pesticides and their metabolites, plasticizers, halogenated aromatic compounds, and nitro compounds.[4] It is very difficult to remove such residues from humic substances thermally, biochemically, or by exposure to acid or base (hydrolysis). Therefore, the formation of bound residues by persistent organic pollutants during humification of organic matter is an immobilization and detoxification process that is significant compared with biodegradation and mineralization in effectively removing such pollutants from the environment. The binding is thought to occur through the action of enzymes from some organisms. These enzymes are extracellular enzymes (those acting outside the cell) that act as oxidoreductases, which catalyze oxidation–reduction reactions. Microbial phenoloxidase enzymes act upon aromatic substrates from lignin decomposition, which are present in humic substances produced from lignin precursors, and also act upon aromatic xenobiotic aromatic compounds, such as anilines and phenols. This can result in the binding of xenobiotic aromatic compounds with humic substances as illustrated below for the coupling of the pollutant 2,4-dichlorophenol to an aryl ring on a humic substance molecule:

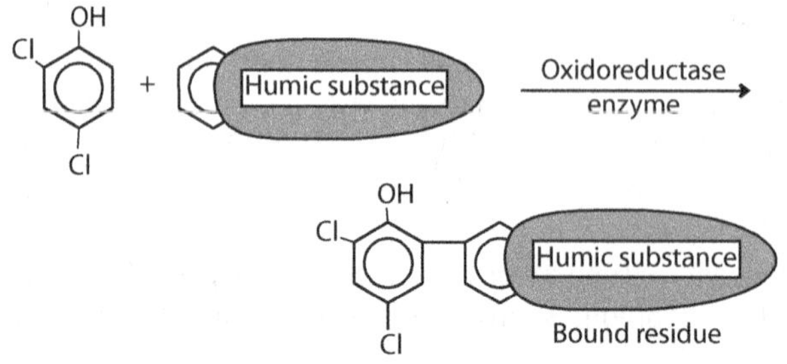

OH, Cl, Cl + Humic substance —Oxidoreductase enzyme→ (5.38)

OH, Cl, Cl — Humic substance

Bound residue

Obviously, uptake of organic matter by suspended and sedimentary material in water is an important phenomenon. This phenomenon significantly reduces the overall toxicity of pesticides in water. Biodegradation is generally slowed down appreciably, however, by binding of a substance to a solid. In certain intensively farmed areas, there is a very high accumulation of pesticides in the sediments of streams, lakes, and reservoirs. The sorption of pesticides by solids and the resulting influence on their biodegradation is an important consideration in the licensing of new pesticides.

The transfer of surface water to groundwater often results in sorption of some water contaminants by soil and mineral material. To take advantage of this purification effect, some municipal water supplies are drawn from beneath the surface of natural or artificial river banks as a first step in water treatment. The movement of water from waste landfills to aquifers is also an important process (see Chapter 20) in which pollutants in the landfill leachate may be sorbed by the solid material through which the water passes.

Surfactant alkylphenolpolyethoxylates and their biodegradation product alkylphenols (see Section 7.10), which have many industrial and domestic uses, are discharged in significant quantities in wastewater treatment plant effluents. These compounds have the potential to disrupt the endocrine systems of aquatic organisms and even humans, tend to accumulate in sediments, and may be poorly biodegradable under anoxic conditions. Levels of these compounds up to as high as 14 mg/kg have been found in sediments in the United States and some fish from the United Kingdom have shown tissue levels approaching 1 μg/kg.

5.8.4 Bioavailability of Sediment Contaminants

An important consideration with respect to contaminants in sediments and soils is *bioavailability*, defined generally as the degree to which a substance can be absorbed into the system of an organism. Bioavailability is an especially important consideration in assessing the hazards posed to organisms by pollutants in sediments and soils. In some cases, it may be best to not attempt to

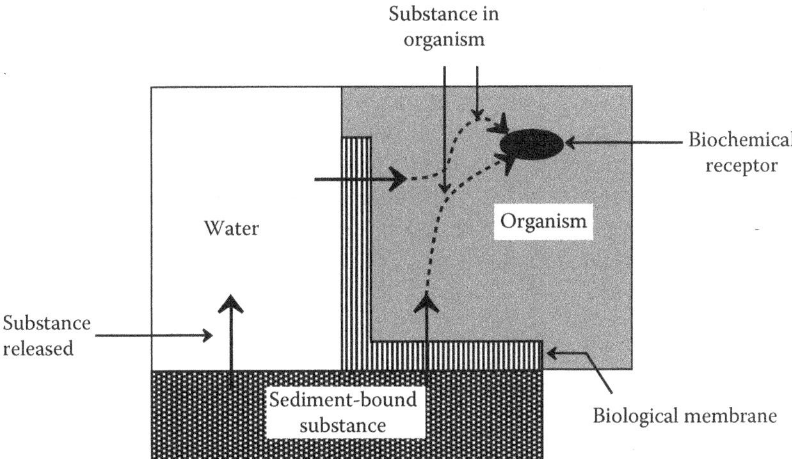

FIGURE 5.7 Transfer of contaminants in sediment to organisms where they may have an effect by acting on a biochemical receptor.

remediate these media if the substances of concern have a low bioavailability. In other cases, very high bioavailability of toxic substances may make them particularly dangerous.

Figure 5.7 shows the overall process by which a contaminant may get into a microorganism from a sediment. A substance bound in a sediment may be released into water, then passed across a biological membrane into an organism, or it may transfer directly across a membrane from sediment to an organism. Once inside the organism, a potentially toxic substance may migrate to a biochemical receptor in the organism (see the discussion of receptors in Chapter 23) to exert a toxic effect. These processes may actually occur within an organism when contaminated soil or sediment material is ingested and traverses the gut wall.

5.9 INTERSTITIAL WATER

Interstitial water or *pore water* consists of water held by sediments. The solutes in interstitial water reflect the chemical and biochemical conditions in the sediments. These solutes include species capable of undergoing oxidation/reduction, reduced metal ions, nutrients such as NH_4^+, and soluble organic compounds. These species include all of the products expected from the decomposition and mineralization of planktonic biomass, largely through the activity of anoxic bacteria in the sediments. Circulation is strongly curtailed in sediments so that there is a pronounced vertical gradient in the species present. At sediment surfaces that may contact water with some oxygen dissolved in it, some more oxidized species are encountered, whereas deeper in the sediments reduced species predominate.

Interstitial water is an important reservoir for gases in natural water systems. Generally, the gas concentrations in interstitial waters are different from those in the overlying water. Typically, interstitial water at the surface of the sediment contains significant amounts of N_2 and relatively little CH_4, which requires anoxic conditions for its production and biodegrades in contact with oxygen. At depths of around 1 m in sediments, CH_4 levels are high and N_2 concentrations are low because nitrogen is stripped from the interstitial water by microbially produced methane and carbon dioxide, which are produced by the anoxic fermentation of organic matter:

$$2\{CH_2O\} \rightarrow CH_4(g) + CO_2(g) \tag{5.39}$$

The concentrations of argon and nitrogen are much lower at a depth of 1 m than they are at the sediment surface. This finding may be explained by the stripping action of the fermentation-produced methane rising to the sediment surface.

5.10 PHASE INTERACTIONS IN CHEMICAL FATE AND TRANSPORT

In Section 1.8, the topic of environmental chemical fate and transport was introduced. The hydrosphere is arguably the most important environmental sphere with respect to chemical fate and transport processes, many of which involve distribution of species between phases. It is appropriate at this point to consider chemical fate and transport in the hydrosphere. This is an environmentally important topic that increasingly utilizes sophisticated mathematical models and calculations that are beyond the scope of this book. For more details and example calculations, the reader is referred to a comprehensive reference work on the subject.[5]

5.10.1 Rivers

The physical movement of chemical species in a river is largely by advection (see Section 1.8) due to the gravitational movement of masses of water downstream. This results in relatively rapid mixing and dilution because of *velocity shear* in which parcels of water move at different rates in a river where water in contact with the banks and bottom moves more slowly than water in the middle of the stream. Also contributing to this mixing are turbulent mixing and diffusive transport of dissolved species and colloidal particles. The net result of these processes is that a pollutant introduced into the river in a relatively concentrated "plug" becomes spread out as it flows downstream.

5.10.2 Lakes and Reservoirs

Lakes and reservoirs are relatively quiet compared with rivers, but there is still movement within such bodies of water. Water enters by stream inflow, inflow from underground springs and aquifers, and directly by rainfall. It exits by outflow, such as over a spillway in a reservoir, infiltration into ground, and evaporation. The balance between input and loss of water in a body of water means that the water has a *hydraulic residence time*. A major factor in mixing processes in lakes is the influence of wind. This occurs because windblown surface water moves at a rate that is 2–3% of the wind velocity, a phenomenon called *wind drift*. Water moved toward one side of a lake by wind drift must return some distance below the water surface as a *return current*. In a shallow, unstratified body of water, the return current flows along the bottom where it contacts and may agitate bottom sediments. In stratified lakes (see Figure 3.6) the circulation of water occurs in the upper epilimnion layer as shown in Figure 5.8.

5.10.3 Exchange with the Atmosphere

Exchange of chemical species between water and overlying air is an important process. It is the means by which atmospheric oxygen enters water to provide oxygen needed by fish. Carbon dioxide required for algal growth may come from the air. Air pollutants such as acid gases may enter water from the

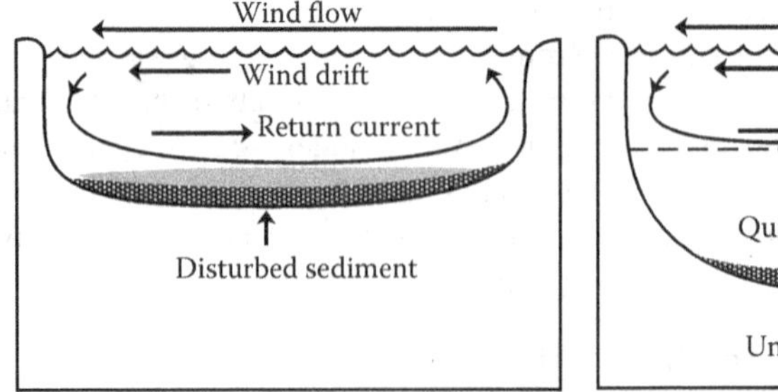

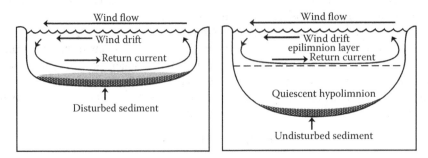

FIGURE 5.8 Mixing due to wind in a shallow unstratified lake (left) and in a stratified lake (right). In the former case in which the return current contacts the sediment, the sediment may be stirred, enabling release of substances to the water.

atmosphere. Under circumstances of high algal photosynthetic activity, oxygen produced by algae is released to the air. Decay of organic matter can oversaturate water with carbon dioxide, requiring its release to the air. Anoxic microbial processes in sediments can produce hydrogen sulfide and methane that are released to air. Volatile organic water pollutants may move from water to the atmosphere.

The air–water interface is the boundary across which species move and is crucial in determining the rate of exchange of materials. Current models of these processes assume that there is a thin, stationary liquid film on the surface of the water in direct contact with a thin, stationary layer of air. Both of these layers are thin—as little as a few micrometers—and molecular diffusion is the only mechanism for movement of solute species in them. Immediately below the thin surface film of water, turbulent diffusion mixes water solutes, and immediately above the thin air film, turbulent diffusion mixes the species in air.

5.10.4 Exchange with Sediments

Sediments are very important in chemical fate and transport in the hydrosphere. This is because substances, including pollutants such as heavy metals or hydrophobic organics, bind with particles as they settle through the water column and are incorporated into sediments. The settling flux density, *J*, is equal to the mass of a substance transported across an area through which the settling occurs per unit time and is given by the product of the rate at which sedimentary materials settle and the concentration of the substance in question in the settling materials:

$$J = (\text{sediment deposition rate}) \times (\text{substance concentration in particles}) \tag{5.40}$$

Incorporation of pollutants by settling from the water column into sediments is a particularly important mode of chemical fate and transport under quiescent conditions during which very small (colloidal) particles are aggregating together (flocculation). If undisturbed, the layers of sediments can provide a record of pollution. A hypothetical, though typical, such plot is shown for lead in Figure 5.9. Although sediments are normally repositories of pollutants and reduce their environmental harm, they can also provide sources of pollutants that can be mobilized by physical, chemical, or biochemical processes. For example, mercury precipitated in sediments can be mobilized as soluble methylmercury species by the action of anoxic bacteria in the oxygen-deficient sediments (see Section 7.3).

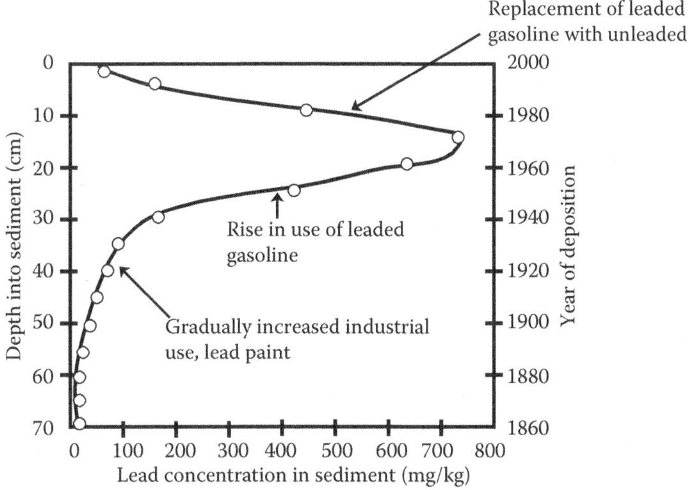

FIGURE 5.9 Typical sediment record for lead deposition reflecting the sharp increase in lead from gasoline and a drop of lead after phaseout of leaded gasoline.

LITERATURE CITED

1. Ghosh, U., The role of black carbon in influencing availability of PAHs in sediments, *Human and Ecological Risk Assessment*, **13**, 276–285, 2007.
2. Massoudieh, A. and T. R. Ginn, Modeling colloid-facilitated transport of multi-species contaminants in unsaturated porous media, *Journal of Contaminant Hydrology*, **92**, 162–183, 2007.
3. Huang, T.-L., X.-C. Ma, C. Hai-Bing, and B.-B. Chai, Microbial effects on phosphorus release in aquatic sediments, *Water Science and Technology*, **58**, 1285–1289, 2008.
4. Schwarzbauer, J., M. Ricking, B. Gieren, and R. Keller, Anthropogenic organic contaminants incorporated into the non-extractable particulate matter of riverine sediments from the Teltow Canal (Berlin). In Eric Lichtfouse, Jan Schwarzbauer, and Robert Didier, Eds, *Environmental Chemistry*, Springer, Berlin, 329–352, 2005.
5. Gulliver, J. S., *Introduction to Chemical Transport in the Environment*, Cambridge University Press, New York, 2007.

SUPPLEMENTARY REFERENCES

Allen, H. E., Ed., *Metal Contaminated Aquatic Sediments*, Ann Arbor Press, Chelsea, MI, 1995.

Barnes, G. and I. Gentle, *Interfacial Science: An Introduction*, Oxford University Press, New York, 2005.

Beckett, R., Ed., *Surface and Colloid Chemistry in Natural Waters and Water Treatment*, Plenum, New York, 1990.

Berkowitz, B., *Contaminant Geochemistry*, Springer, New York, 2007.

Bianchi, T. S., *Biogeochemistry of Estuaries*, Oxford University Press, New York, 2007.

Birdi, K. S., Ed., *Handbook of Surface and Colloid Chemistry*, 2nd ed., CRC Press, Boca Raton, FL, 2003.

Burdige, D. J., *Geochemistry of Marine Sediments*, Princeton University Press, Princeton, NJ, 2006.

Calabrese, E. J., P. T. Kostecki, and J. Dragun, Eds, *Contaminated Soils, Sediments, and Water, Volume 10, Successes and Challenges*, Springer, New York, 2006.

Eby, G. N., *Principles of Environmental Geochemistry*, Thomson-Brooks/Cole, Pacific Grove, CA, 2004.

Evans, R. D., J. Wisniewski, and J. R. Wisniewski, Eds, *The Interactions Between Sediments and Water*, Kluwer, Dordrecht, The Netherlands, 1997.

Golterman, H. L., Ed., *Sediment–Water Interaction 6*, Kluwer, Dordrecht, The Netherlands, 1996.

Gustafsson, O. and P. M. Gschwend, Aquatic colloids: Concepts, definitions, and current challenges, *Limnology and Oceanography*, **42**, 519–528, 1997.

Holland, H. D., and K K. Turekian, Eds, *Treatise on Geochemistry*, Elsevier/Pergamon, Amsterdam, 2004.

Holmberg, K., D. O. Shah, and M. J. Schwuger, Eds, *Handbook of Applied Surface and Colloid Chemistry*, Wiley, New York, 2002.

Hunter, R. J., *Foundations of Colloid Science*, 2nd ed., Oxford University Press, New York, 2001.

John V. W., *Essentials of Geochemistry*, 2nd ed., Jones and Bartlett Publishers, Sudbury, MA, 2009.

Jones, M. N. and N. D. Bryan, Colloidal properties of humic substances, *Advances in Colloid and Interfacial Science*, **78**, 1–48, 1998.

Jones, S. J. and L. E. Frostick, Eds, *Sediment Flux to Basins: Causes, Controls and Consequences*, Geological Society, London, 2002.

McSween, H. Y., S. M. Richardson, and M. E. Uhle, *Geochemistry: Pathways and Processes*, 2nd ed., Columbia University Press, New York, 2003.

Mudroch, A., J. M. Mudroch, and P. Mudroch, Eds, *Manual of Physico-Chemical Analysis of Aquatic Sediments*, CRC Press, Boca Raton, FL, 1997.

Myers, D., *Surfaces, Interfaces, and Colloids: Principles and Applications*, 2nd ed., Wiley, New York, 1999.

Shchukin, E. D., *Colloid and Surface Chemistry*, Elsevier, Amsterdam, 2001.

Stumm, W., L. Sigg, and B. Sulzberger, *Chemistry of the Solid–Water Interface*, Wiley, New York, 1992.

Stumm, W. and J. J. Morgan, *Aquatic Chemistry: Chemical Equilibria and Rates in Natural Waters*, 3rd ed., Wiley, New York, 1995.

U.S. Environmental Protection Agency, website on contaminated sediments, 2003, available at http://www.epa.gov/waterscience/cs/

Wilkinson, K. J. and J. R. Lead, Eds, *Environmental Colloids and Particles: Behaviour, Separation and Characterisation*, Wiley, Hoboken, NJ, 2007.

QUESTIONS AND PROBLEMS

The use of internet resources is assumed in answering any of the questions. These resources would include such things as constants and conversion factors as well as additional information needed to complete an answer.

1. A sediment sample was taken from a lignite strip-mine pit containing highly alkaline (pH 10) water. Cations were displaced from the sediment by treatment with HCl. A total analysis of cations in the leachate yielded, on the basis of millimoles per 100 g of dry sediment, 150 mmol of Na^+, 5 mmol of K^+, 20 mmol of Mg^{2+}, and 75 mmol of Ca^{2+}. What is the CEC of the sediment in milliequivalents per 100 g of dry sediment? Why does H^+ not have to be considered in this case?
2. What is the value of $[O_2(aq)]$ for water saturated with a mixture of 50% O_2, and 50% N_2 by volume at 25°C and a total pressure of 1.00 atm?
3. Of the following, the least likely mode of transport of iron(III) in a normal stream is: (a) bound to suspended humic material, (b) bound to clay particles by cation-exchange processes, (c) as suspended Fe_2O_3, (d) as soluble Fe^{3+} ion, and (e) bound to colloidal clay–humic substance complexes.
4. How does freshly precipitated colloidal iron(III) hydroxide interact with many divalent metal ions in solution?
5. What stabilizes colloids composed of bacterial cells in water?
6. The solubility of oxygen in water is 14.74 mg/L at 0°C and 7.03 mg/L at 35°C. Estimate the solubility at 50°C.
7. What is thought to be the mechanism by which bacterial cells aggregate?
8. What is a good method for the production of freshly precipitated MnO_2?
9. A sediment sample was equilibrated with a solution of NH_4^+ ion, and the NH_4^+ was later displaced by Na^+ for analysis. A total of 33.8 milliequivalents of NH_4^+ were bound to the sediment and later displaced by Na^+. After drying, the sediment weighed 87.2 g. What was its CEC in milliequivalents per 100 g?
10. A sediment sample with a CEC of 67.4 milliequivalents per 100 g was found to contain the following exchangeable cations in milliequivalents per 100 g: Ca^{2+}, 21.3; Mg^{2+}, 5.2; Na^+, 4.4; K^+, 0.7. The quantity of hydrogen ion, H^+, was not measured directly. What was the ECS of H^+ in milliequivalents per 100 g?
11. What is the meaning of ZPC as applied to colloids? Is the surface of a colloidal particle totally without charged groups at the ZPC?
12. The concentration of methane in an interstitial water sample was found to be 150 mL/L at standard temperature and pressure of 0°C and 1 atm (STP). Assuming that the methane was produced by the fermentation of organic matter, $\{CH_2O\}$, what mass of organic matter was required to produce the methane in a liter of the interstitial water?
13. What is the difference between CEC and ECS?
14. Match the sedimentary mineral on the left with its conditions of formation on the right:

a. $FeS(s)$	1. May be formed when anoxic water is exposed to O_2.
b. $Ca_5OH(PO_4)_3$	2. May be formed when oxic water becomes anoxic.
c. $Fe(OH)_3$	3. Photosynthesis by-product.
d. $CaCO_3$	4. May be formed when wastewater containing a particular kind of contaminant flows into a body of very hard water.

15. In terms of their potential for reactions with species in solution, how might metal atoms, M, on the surface of a metal oxide, MO, be described?
16. Air is 20.95% oxygen by volume. If air at 1.0000 atm pressure is bubbled through water at 25°C, what is the partial pressure of O_2 in the water?

17. The volume percentage of CO_2 in a mixture of that gas with N_2 was determined by bubbling the mixture at 1.00 atm and 25°C through a solution of 0.0100 M $NaHCO_3$ and measuring the pH. If the equilibrium pH was 6.50, what was the volume percentage of CO_2?
18. For what purpose is a polymer with the following general formula used?

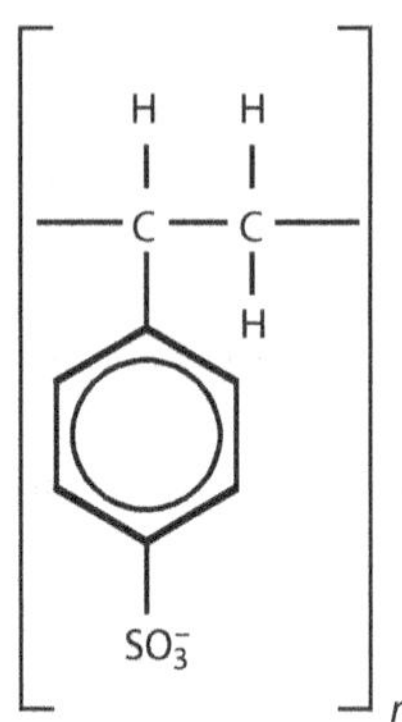

19. Of the following statements, the one that is true regarding colloids is: (A) Hydrophilic colloids consist of aggregates of relatively small molecules; (B) hydrophobic colloids do not have electrical charges; (C) hydrophilic colloids are those formed by clusters of species, such as $H_3C(CH_2)_{16}CO_2^-$; (D) association colloids form micelles; (E) the electrical charges of hydrophobic colloids are insignificant.
20. For a slightly soluble divalent metal sulfate, MSO_4, $K_{sp} = 9.00 \times 10^{-14}$. An excess of pure solid MSO_4 was equilibrated with pure water to give a solution that contained 6.45×10^{-7} mol/L of dissolved M. Considering these observations the true statement is: (A) MSO_4 has a significant degree of intrinsic solubility; (B) the solubility product, alone, accurately predicts solubility; (C) the value of the solubility product is in error; (D) the concentration of "M" in water must have been in error; (E) the only explanation for the observations is formation of HSO_4^{2-}.
21. Of the following, the *incorrect* statement regarding sediments and their formation is: (A) Physical, chemical, and, biological processes may all result in the deposition of sediments in the bottom regions of bodies of water; (B) indirectly, photosynthesis can result in formation of $CaCO_3$ sediment; (C) oxidation of Fe^{2+} ion can result in formation of an insoluble species that can be incorporated into sediment; (D) sediments typically consist of mixtures of clay, silt, sand, organic matter, and various minerals, and may vary in composition from pure mineral matter to predominantly organic matter; (E) FeS that gets into sediment tends to form at the surface of water in contact with O_2.
22. Given that at 25°C, the Henry's law constant for oxygen is 1.28×10^{-3} mol/L/atm and the partial pressure of water vapor is 0.0313 atm, what is the value of $[O_2(aq)]$ for water saturated with a mixture of 33.3% O_2 and 66.7% N_2 by volume at 25°C and a total pressure of 1.00 atm in units of mol/L?
23. Match the following regarding colloids:

A. Hydrophilic colloids	1. $CH_3CO_2^-Na^+$
B. Association colloids	2. Macromolecular proteins
C. Hydrophobic colloids	3. Often removed by addition of salt
D. Noncolloidal	4. $CH_3(CH_2)_{16}CO_2^-Na^+$

6 Aquatic Microbial Biochemistry

6.1 AQUATIC BIOCHEMICAL PROCESSES

Microorganisms—*bacteria, fungi, protozoa*, and *algae*—are living catalysts that enable a vast number of chemical processes to occur in water and soil. A majority of the important chemical reactions that take place in water, particularly those involving organic matter and oxidation–reduction processes, occur through bacterial intermediaries. Algae are the primary producers of biological organic matter (biomass) in water. Microorganisms are responsible for the formation of many sediment and mineral deposits; they also play the dominant role in secondary waste treatment. Some of the effects of microorganisms on the chemistry of water in nature are illustrated in Figure 6.1.

Pathogenic microorganisms must be eliminated from water purified for domestic use. In the past, major epidemics of typhoid, cholera, and other water-borne diseases resulted from pathogenic microorganisms in water supplies. Even today, constant vigilance is required to ensure that water for domestic use is free of pathogens.

Most of this chapter is concerned with aquatic chemical transformations mediated by microorganisms. Although they are not involved in such transformations, special mention should be made of viruses in water. Viruses cannot grow by themselves, but reproduce in the cells of host organisms. They are only about 1/30–1/20 the size of bacterial cells, and they cause a number of diseases, such as polio, viral hepatitis, and perhaps cancer. It is thought that many of these diseases are waterborne.

Because of their small size (0.025–0.100 μm) and biological characteristics, viruses are difficult to isolate and culture. They often survive municipal water treatment, including chlorination. Thus, although viruses have no effect on the overall environmental chemistry of water, they are an important consideration in the treatment and use of water.

Microorganisms are divided into the two broad categories of *prokaryotes* and *eukaryotes*; the latter have well-defined cell nuclei enclosed by a nuclear membrane, whereas the former lack a nuclear membrane and the nuclear genetic material is more diffuse in the cell. Other differences between these two classes of organisms include location of cell respiration, means of photosynthesis, means of motility, and reproductive processes. All classes of microorganisms produce *spores*, metabolically inactive bodies that form and survive under adverse conditions in a "resting" state until conditions favorable for growth occur.

Fungi, protozoa, and bacteria (with the exception of photosynthetic bacteria and protozoa) are classified as *reducers*, which break down chemical compounds to more simple species and thereby extract the energy needed for their growth and metabolism. Algae are classified as *producers* because they utilize light energy and store it as chemical energy. In the absence of sunlight, however, algae utilize chemical energy for their metabolic needs. In a sense, therefore, bacteria, protozoa, and fungi may be looked upon as environmental catalysts, whereas algae function as aquatic solar fuel cells.

All microorganisms can be put into one of the four following classifications based on the sources of energy and carbon that they utilize: chemoheterotrophs, chemoautotrophs, photoheterotrophs, and photoautotrophs. These classifications are based on (1) the energy source and (2) the carbon source utilized by the organism. *Chemotrophs* use chemical energy derived from oxidation–reduction

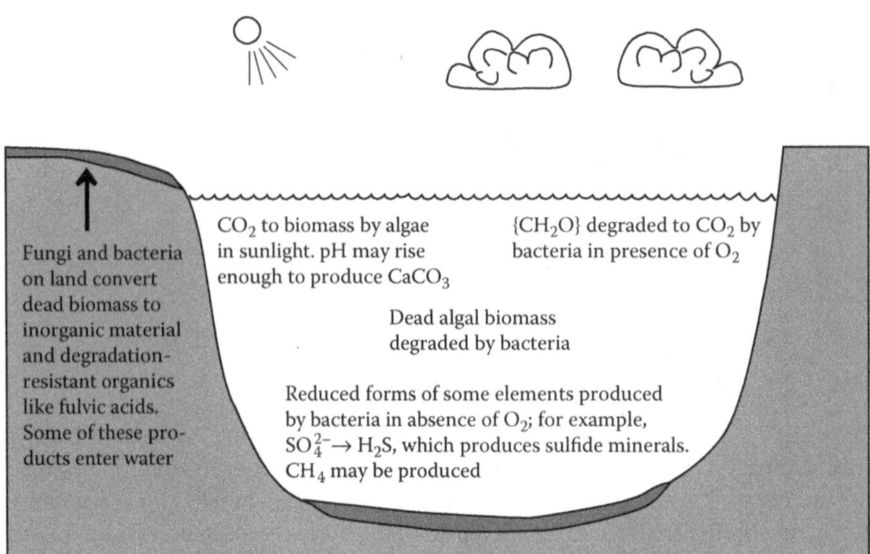

FIGURE 6.1 Effects of microorganisms on the chemistry of water in nature.

reactions of simple inorganic chemical species for their energy needs. *Phototrophs* utilize light energy from photosynthesis. *Heterotrophs* obtain their carbon from other organisms; *autotrophs* use carbon dioxide and ionic carbonates for the C that they require. Figure 6.2 summarizes the classifications into which microorganisms may be placed with these definitions.

6.1.1 Microorganisms at Interfaces

Aquatic microorganisms tend to grow at interfaces. Many such microorganisms grow on solids that are suspended in water or are present in sediments. Large populations of aquatic bacteria typically reside on the surface of water at the air–water interface. In addition to being in contact with air that aerobic microorganisms need for their metabolic processes, this interface also accumulates food in the form of lipids (oils, fats), polysaccharides, and proteins. Bacteria at this interface are generally

Energy source → / Carbon source ↓	Chemical	Photochemical (light)
Organic matter	**Chemoheterotrophs** All fungi and protozoans, most bacteria. Chemoheterotrophs use organic sources for both energy and carbon	**Photoheterotrophs** A few specialized bacteria that use photoenergy, but are dependent on organic matter for a carbon source
Inorganic carbon (CO_2, HCO_3^-)	**Chemoautotrophs** Use CO_2 for biomass and oxidize substances such as H_2 (*Pseudomonas*), NH_4^+ (*Nitrosomonas*), S (*Thiobacillus*) for energy	**Photoautotrophs** Algae, and photosynthetic bacteria including cyanobacteria that use light energy to convert CO_2 (HCO_3^-) to biomass by photosynthesis

FIGURE 6.2 Classification of microorganisms among chemoheterotrophs, chemoautotrophs, photoheterotrophs, and photoautotrophs.

different from those in the body of water and may have a hydrophobic cell character. When surface bubbles burst, bacteria at the air–water interface can be incorporated into aerosol water droplets and carried by wind. This is a matter of some concern with respect to sewage treatment plants as a possible vector for spreading disease-causing microorganisms.

6.2 ALGAE

For the purposes of discussion here, *algae* may be considered as generally microscopic organisms that subsist on inorganic nutrients and produce organic matter from carbon dioxide by photosynthesis. In addition to single cells, algae grow as filaments, sheets, and colonies. Some algae, particularly the marine kelps, are huge multicellular organisms. The study of algae is called *phycology.*

The four main classes of unicellular algae of importance in environmental chemistry are the following.

- *Chrysophyta*, which contain pigments that give these organisms a yellow-green or golden-brown color. Chrysophyta are found in both freshwater and marine systems. They store food as carbohydrate or oil. The most well known of these algae are *diatoms*, characterized by silica-containing cell walls.
- *Chlorophyta*, commonly known as green algae, are responsible for most of the primary productivity in freshwaters.
- *Pyrrophyta*, commonly known as dinoflagellates, are motile with structures that enable them to move about in water, thus giving them characteristics of protozoa (Section 6.4). Pyrrophyta occur in both marine and freshwater environments. "Blooms" of *Gymnodinium* and *Gonyaulax* species release toxins that cause harmful "red tides."
- *Euglenophyta* likewise exhibit characteristics of both plants and animals. Although capable of photosynthesis, these algae are not exclusively photoautotrophic (see Figure 6.2), and they utilize biomass from other sources for at least part of their carbon needs.

The general nutrient requirements of algae are carbon (obtained from CO_2 or HCO_3^-), nitrogen (generally as NO_3^-), phosphorus (as some form of orthophosphate), sulfur (as SO_4^{2-}), and trace elements including sodium, potassium, calcium, magnesium, iron, cobalt, and molybdenum.

In a highly simplified form, the production of organic matter by algal photosynthesis is described by the reaction

$$CO_2 + H_2O + h\nu \rightarrow \{CH_2O\} + O_2(g) \tag{6.1}$$

where $\{CH_2O\}$ represents a unit of carbohydrate and $h\nu$ stands for the energy of a quantum of light. A reasonably accurate representation of the overall formula of the algae *Chlorella* is $C_{5.7}H_{9.8}O_{2.3}NP_{0.06}$. Using this formula for algal biomass exclusive of the phosphorus, the overall reaction for photosynthesis is

$$5.7CO_2 + 3.4H_2O + NH_3 + h\nu \rightarrow C_{5.7}H_{9.8}O_{2.3}N + 6.25O_2(g) \tag{6.2}$$

In the absence of light, algae metabolize organic matter in the same manner as do nonphotosynthetic organisms. Thus, algae may satisfy their metabolic demands by utilizing chemical energy from the degradation of stored starches or oils, or from the consumption of algal protoplasm itself. In the absence of photosynthesis, the metabolic process consumes oxygen, so during the hours of darkness an aquatic system with a heavy growth of algae may become depleted in oxygen.

Symbiotic relationships of algae with other organisms are common. There are even reports of unicellular green algae growing inside hairs on polar bears, which are hollow for purposes of insulation; the sight of a green polar bear is alleged to have driven more than one Arctic explorer to the brink of madness. The most common symbiotic relationship involving algae is that of *lichen* in which algae coexist with fungi; both kinds of organisms are woven into the same thallus (tubular vegetative unit). The fungus provides moisture and nutrients required by the algae, which generates food photosynthetically. Lichen are involved in weathering processes of rocks (see Section 15.2).

The main role of algae in aquatic systems is the production of biomass. This occurs through photosynthesis, which fixes carbon dioxide and inorganic carbon from dissolved carbonate species as organic matter, thus providing the basis of the food chain for other organisms in the system. The production of moderate amounts of biomass is beneficial to other organisms in the aquatic system. Under some conditions, the growth of algae can produce metabolites that are responsible for odor and even toxicity in water.

An interesting aspect of the growth of oceanic plankton including algae is their uptake of micronutrient metals resulting in extremely low concentrations of essential metals in surface seawater. In some cases the resultant low availability of nutrient metals limits photosynthetic rates in seawater. Some microorganisms in seawater increase bioavailability of nutrient metals and accelerate their cycling in water by releasing strong chelating agents and catalyzing oxidation/reduction reactions that put nutrient metals in a more soluble and available form.

6.3 FUNGI

Fungi are nonphotosynthetic, often filamentous, organisms exhibiting a wide range of morphology (structure). The study of fungi is called *mycology*. Some fungi are as simple as the microscopic unicellular yeasts, whereas other fungi form large, intricate toadstools. The microscopic filamentous structures of fungi generally are much larger than bacteria, and usually are 5–10 μm in width. Fungi are aerobic or oxic (oxygen-requiring) organisms and generally can thrive in more acidic media than can bacteria. They are also more tolerant of higher concentrations of heavy metal ions than are most bacteria.

Perhaps the most important function of fungi in the environment is the breakdown of cellulose in wood and other plant materials. To accomplish this, fungal cells secrete an extracellular enzyme (exoenzyme), *cellulase*, that hydrolyzes insoluble cellulose to soluble carbohydrates that can be absorbed by the fungal cell.

Fungi do not grow well in water. However, they play an important role in determining the composition of natural waters and wastewaters because of the large amount of their decomposition products that enter water. An example of such a product is humic material, which interacts with hydrogen ions and metals (see Section 3.17).

6.4 PROTOZOA

Protozoa are microscopic animals consisting of single eukaryotic cells. The numerous kinds of protozoa are classified on the bases of morphology (physical structure), means of locomotion (flagella, cilia, pseudopodia), presence or absence of chloroplasts, presence or absence of shells, ability to form cysts (consisting of a reduced-size cell encapsulated in a relatively thick skin that can be carried in the air or by animals in the absence of water), and ability to form spores. Protozoa occur in a wide variety of shapes and their movement in the field of a microscope is especially fascinating to watch. Some protozoa contain chloroplasts and are photosynthetic.

Protozoa play a relatively small role in environmental biochemical processes, but are nevertheless significant in the aquatic and soil environment for the following reasons.

- Several devastating human diseases, including malaria, sleeping sickness, and some kinds of dysentery, are caused by protozoa that are parasitic to the human body
- Parasitic protozoa can cause debilitating, even fatal, diseases in livestock and wildlife
- Vast limestone ($CaCO_3$) deposits have been formed by the deposition of shells from the *foramifera* group of protozoa

- Protozoa are active in the oxidation of degradable biomass, particularly in sewage treatment
- Protozoa may affect bacteria active in degrading biodegradable substances by "grazing" on bacterial cells

Although they are single celled, protozoa have a fascinating variety of structures that enable them to function. The protozoal cell membrane is protected and supported by a relatively thick pellicle, or by a mineral shell that may act as an exoskeleton. Food is ingested through a structure called a cytosome from which it is concentrated in a cytopharynx or oral groove, and then digested by enzymatic action in a food vacuole. Residue from food digestion is expelled through a cytopyge and soluble metabolic products, such as urea or ammonia, are eliminated by a contractile vacuole, which also expels water from the cell interior.

6.5 BACTERIA

Bacteria are single-celled prokaryotic microorganisms that may be shaped as rods (*bacillus*), spheres (*coccus*), or spirals (*vibrios, spirilla, spirochetes*). Bacteria cells may occur individually or grow as groups ranging from two to millions of individual cells. Most bacteria fall into the size range of 0.5–3.0 μm. However, considering all species, a size range of 0.3–50 μm is observed. Characteristics of most bacteria include a semirigid cell wall, motility with flagella for those capable of movement, unicellular nature (although clusters of cloned bacterial cells are common), and multiplication by binary fission in which each of two daughter cells is genetically identical to the parent cell. Like other microorganisms, bacteria produce spores.

The metabolic activity of bacteria is greatly influenced by their small size. Their surface-to-volume ratio is extremely large, so that the inside of a bacterial cell is highly accessible to a chemical substance in the surrounding medium. Thus, for the same reason that a finely divided catalyst is more efficient than a more coarse one, bacteria may bring about very rapid chemical reactions compared to those mediated by larger organisms. Bacteria excrete exoenzymes that break down solid food material to soluble components which can penetrate bacterial cell walls, where the digestion process is completed.

Although individual bacteria cells cannot be seen by the naked eye, bacterial colonies arising from individual cells are readily visible. A common method of counting individual bacterial cells in water consists of spreading a measured volume of an appropriately diluted water sample on a plate of agar gel containing bacterial nutrients. Wherever a viable bacterial cell adheres to the plate, a bacterial colony consisting of many cells will grow. These visible colonies are counted and related to the number of cells present initially. Because bacterial cells may already be present in groups, and because individual cells may not live to form colonies or even have the ability to form colonies on a plate, plate counts tend to underestimate the number of viable bacteria.

6.5.1 Autotrophic and Heterotrophic Bacteria

Bacteria may be divided into two main categories, autotrophic and heterotrophic. Autotrophic bacteria are not dependent on organic matter for growth and thrive in a completely inorganic medium; they use carbon dioxide or other carbonate species as a carbon source. A number of sources of energy may be used, depending on the species of bacteria; however, a biologically mediated chemical reaction always supplies the energy.

An example of autotrophic bacteria is *Gallionella*. In the presence of oxygen, these bacteria are grown in a medium consisting of NH_4Cl, phosphates, mineral salts, CO_2 (as a carbon source), and solid FeS (as an energy source). It is believed that the following is the energy-yielding reaction for this species:

$$4FeS(s) + 9O_2 + 10H_2O \rightarrow 4Fe(OH)_3(s) + 4SO_4^{2-} + 8H^+ \tag{6.3}$$

Starting with the simplest inorganic materials, autotrophic bacteria must synthesize all of the complicated proteins, enzymes, and other materials needed for their life processes. It follows, therefore, that the biochemistry of autotrophic bacteria is quite sophisticated. Because of their consumption and production of a wide range of minerals, autotrophic bacteria are involved in many geochemical transformations.

Heterotrophic bacteria depend on organic compounds, both for energy and for carbon to build their biomass. They are much more common in occurrence than autotrophic bacteria. Heterotrophic bacteria are the microorganisms primarily responsible for the breakdown of pollutant organic matter in water, and of organic wastes in biological waste-treatment processes.

Some bacteria are capable of photosynthesis to supply their energy and carbon. The most common such bacteria in water are the cyanobacteria. These organisms were once thought to be algae and were called blue-green algae. They can grow prolifically under some conditions and may impart such disagreeable tastes and odors to water that they may make it unfit for domestic consumption during cyanobacteria blooms.

A type of cyanobacteria of unique importance consists of the marine cyanobacteria from the *Prochlorococcus* genus.[1] Only discovered in 1988, these bacteria are the smallest photosynthetic organisms known, only about 0.5 μm in size. They are the most abundant photosynthetic organisms in the oceans (therefore, on the entire planet), constituting 40–50% of phytoplankton biomass in ocean waters in the region between 40° north and 40° south. One of the two major strains of *Prochlorococcus* lives near the surface where light is abundant, whereas a second strain conducts photosynthesis to depths of 200 m reached by only a fraction of 1% of incident light. Despite their small cell size, *Prochlorococcus* produce a significant fraction of photosynthetically generated biomass and are very important in marine food chains. Because they fix vast quantities of carbon dioxide, they have a potentially significant role in the alleviation of greenhouse warming effects. They are noted for their rapid genetic adaptation abilities, which hopefully will enable them to continue functioning efficiently under altered conditions, such as reduced pH in ocean waters, caused by elevated carbon dioxide levels in the atmosphere.

6.5.2 Oxic and Anoxic Bacteria

Another classification system for bacteria depends on their requirement for molecular oxygen. Oxic (aerobic bacteria) require oxygen as an electron receptor:

$$O_2 + 4H^+ + 4e^- \rightarrow 2H_2O \tag{6.4}$$

Anaerobic bacteria, or anoxic bacteria, function only in the complete absence of molecular oxygen. Frequently, molecular oxygen is quite toxic to anoxic bacteria. Anoxic bacteria have gained increasing attention for their ability to degrade organic wastes.

A third class of bacteria, facultative bacteria, utilize free oxygen when it is available and use other substances as electron receptors (oxidants) when molecular oxygen is not available. Common oxygen substitutes in water are nitrate ion (see Section 6.11) and sulfate ion (see Section 6.12).

6.5.3 Marine Bacteria

Most of the attention that has been given to bacteria in water has been focused on bacteria found in freshwater. Recently, more attention has been given to marine bacteria including those in ocean sediments. An example of such bacteria is *Salinospora*, a type of actinomycete bacteria that thrive in ocean sediments under dark, cold, high pressure, saline conditions. Known freshwater and terrestrial actinomycetes have been the major source of antibiotics such as streptomycin and vancomycin and the possibility of finding new antibiotics and even anticancer drugs in marine actinomycetes has generated intense interest in these organisms.

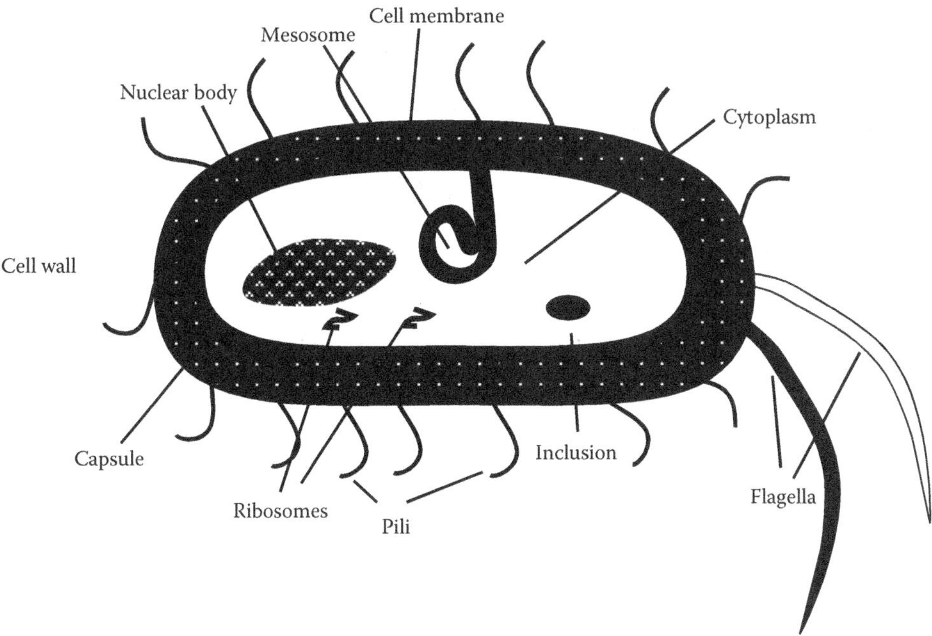

FIGURE 6.3 Generic prokaryotic bacterial cell illustrating major cell features.

6.6 THE PROKARYOTIC BACTERIAL CELL

Figure 6.3 illustrates a generic prokaryotic bacterial cell. Bacterial cells are enclosed in a *cell wall*, which holds the contents of the bacterial cell and determines the shape of the cell. The cell wall in many bacteria is frequently surrounded by a *slime layer* (capsule). This layer protects the bacteria and helps the bacterial cells to adhere to surfaces.

The *cell membrane* or *cytoplasmic membrane* composed of protein and phospholipid occurs as a thin layer only about 7 nm in thickness on the inner surface of the cell wall enclosing the cellular cytoplasm. The cytoplasmic membrane is of crucial importance to cell function in that it controls the nature and quantity of materials transported into and out of the cell. It is also very susceptible to damage from some toxic substances.

Folds in the cytoplasmic membrane called *mesosomes* serve several functions. One of these is to increase the surface area of the membrane to enhance transport of materials through it. Another function is to act as a site for division of the cell during reproduction. Bacterial DNA is separated at the mesosome during cell division.

Hairlike *pili* on the surface of a bacterial cell enable the cell to stick to surfaces. Specialized *sex pili* enable nucleic acid transfer between bacterial cells during an exchange of genetic material. Somewhat similar to pili—but larger, more complex, and fewer in number—are *flagella*, moveable appendages that cause bacterial cells to move by their whipping action. Bacteria with flagella are termed *motile*.

Bacterial cells are filled with an aqueous solution and suspension containing proteins, lipids, carbohydrates, nucleic acids, ions, and other materials. Collectively, these materials are referred to as *cytoplasm*, the medium in which the cell's metabolic processes are carried out. The major cell bodies suspended in cytoplasm are the following.

- *Nuclear body* consisting of a single DNA macromolecule that controls metabolic processes and reproduction
- *Inclusions* of reserve food material consisting of fats, carbohydrates, and even elemental sulfur
- *Ribosomes*, which are sites of protein synthesis and which contain protein and RNA

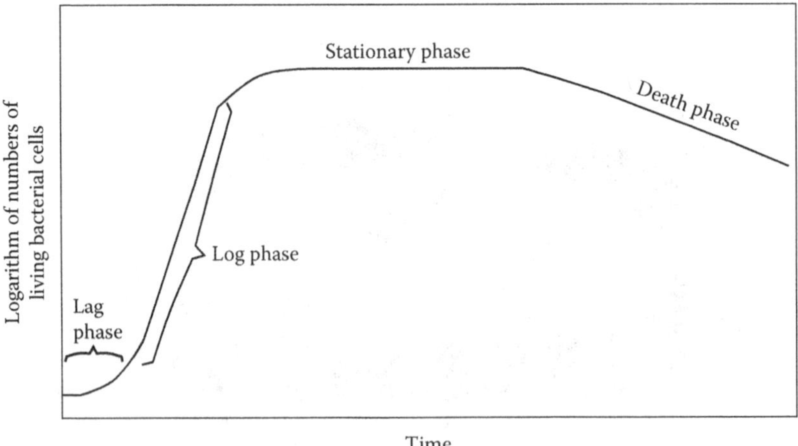

FIGURE 6.4 Population curve for a bacterial culture.

6.7 KINETICS OF BACTERIAL GROWTH

The population size of bacteria and unicellular algae as a function of time in a growth culture is illustrated by Figure 6.4, which shows a *population curve* for a bacterial culture. Such a culture is started by inoculating a rich nutrient medium with a small number of bacterial cells. The population curve consists of four regions. The first region is characterized by little bacterial reproduction and is called the *lag phase*. The lag phase occurs because the bacteria must become acclimated to the new medium. Following the lag phase comes a period of very rapid bacterial growth. This is the *log phase*, or exponential phase, during which the population doubles over a regular time interval called the *generation time*. This behavior can be described by a mathematical expression that applies when the growth rate is proportional to the number of individuals present and there are no limiting factors such as death or lack of food:

$$\frac{dN}{dt} = kN \tag{6.5}$$

$$\ln\left(\frac{N}{N_0}\right) = kt \quad \text{or} \quad N = N_0 e^{kt} \tag{6.6}$$

where N is the population at time t and N_0 is the population at time $t = 0$. Thus, another way of describing population growth during the log phase is to say that the logarithm of bacterial population increases linearly with time. The generation time, or doubling time, is $(\ln 2)/k$, analogous to the half-life of radioactive decay. Fast growth during the log phase can cause very rapid microbial transformations of chemical species in water.

The log phase terminates and the *stationary phase* begins when a limiting factor is encountered. Typical factors limiting growth are depletion of an essential nutrient, buildup of toxic material, and exhaustion of oxygen. During the stationary phase, the number of viable cells remains virtually constant. After the stationary phase, the bacteria begin to die faster than they reproduce, and the population enters the *death phase*.

6.8 BACTERIAL METABOLISM

Bacteria obtain the energy and raw materials needed for their metabolic processes and reproduction by mediating chemical reactions. Nature provides a large number of such reactions, and

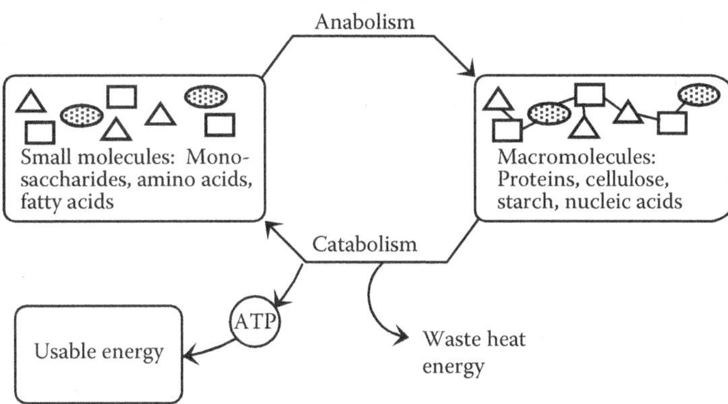

FIGURE 6.5 Bacterial metabolism and energy production.

bacterial species have evolved that utilize many of these. As a consequence of their participation in such reactions, bacteria are involved in many biogeochemical processes in water and soil. Bacteria are essential participants in many important elemental cycles in nature, including those of nitrogen, carbon, and sulfur. They are responsible for the formation of many mineral deposits, including some of iron and manganese. On a smaller scale, some of these deposits form through bacterial action in natural water systems and even in pipes used to transport water.

Bacterial metabolism addresses the biochemical processes by which chemical species are modified in bacterial cells. It is basically a means of deriving energy and cellular material from nutrient substances. Figure 6.5 summarizes the essential features of bacterial metabolism. The two major divisions of bacterial metabolism are catabolism, energy-yielding degradative metabolism that breaks macromolecules down to their small monomeric constituents, and anabolism, synthetic metabolism in which small molecules are assembled into large ones. A key distinction among bacteria has to do with the terminal electron acceptor in the electron transport chain involved in the process by which bacteria gain energy by oxidizing food materials. If the terminal electron acceptor is molecular O_2, the process is *oxic respiration*. If it is another reducible species, commonly including SO_4^{2-}, NO_3^-, HCO_3^-, or iron(III), the process is called *anoxic respiration*. As examples, *Desulfovibrio* bacteria convert SO_4^{2-} to H_2S, *Methanobacterium* reduce HCO_3^- to CH_4, and assorted bacteria reduce NO_3^- to NO_2^-, N_2O, N_2, or NH_4^+.

6.8.1 Factors Affecting Bacterial Metabolism

Bacterial metabolic reactions are mediated by enzymes, biochemical catalysts endogenous to living organisms that are discussed in detail in Chapter 22. Enzymatic processes in bacteria are essentially the same as those in other organisms. At this point, however, it is useful to review several factors that influence bacterial enzyme activity and, therefore, bacterial growth.

Figure 6.6 illustrates the effect of *substrate concentration* on enzyme activity, where a substrate is a substance on which an enzyme acts. It is seen that enzyme activity increases in a linear fashion up to a value that represents saturation of the enzyme activity. Beyond this concentration, increasing substrate levels do not result in increased enzyme activity. This kind of behavior is reflected in bacterial activity, which increases with available nutrients up to a saturation value. Superimposed on this plot in a bacterial system is increased bacterial population which, in effect, increases the amount of available enzyme.

Figure 6.7 shows the effect of *temperature* on enzyme activity and on bacterial growth and metabolism. Over a relatively short range of temperature in which the bacteria thrive, enzyme activity increases with temperature. The curve shows a maximum growth rate with an optimum temperature that is skewed toward the high-temperature end of the curve, and exhibits an abrupt drop-off beyond

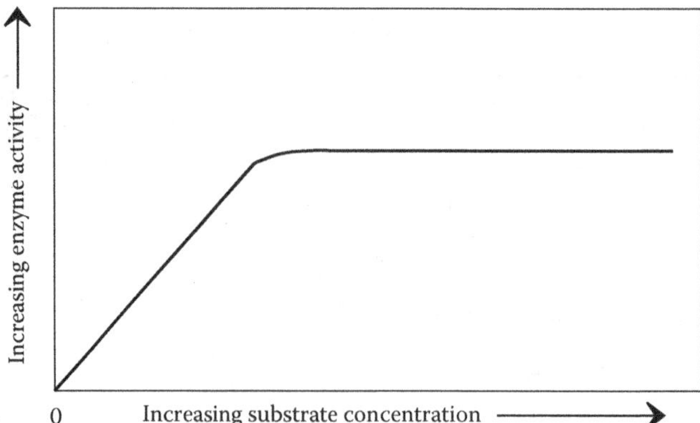

FIGURE 6.6 Effect of increasing substrate concentration on enzyme activity. Bacterial metabolism parallels such a plot.

the temperature maximum. This occurs because enzymes are destroyed by being denatured at temperatures not far above the optimum. Bacteria show different temperature optima. *Psychrophilic bacteria* are bacteria having temperature optima below approximately 20°C. The temperature optima of *mesophilic bacteria* lie between 20°C and 45°C. Bacteria having temperature optima above 45°C are called *thermophilic bacteria.* The temperature range for optimum growth of different kinds of bacteria is remarkably wide, with some bacteria being able to grow at 0°C, and some thermophilic bacteria existing in boiling hot water.

There has been much interest in the adaptation of thermophilic bacteria and their enzymes to industrial applications because, as shown by Figure 6.7, enzyme activity generally increases with increasing temperature.[2] An example of such an application is the potential use of a heat-resistant catalase enzyme from a thermophilic bacterium isolated from a hot spring pool in Yellowstone National Park to catalyze the breakdown of hydrogen peroxide in industrial bleaching wastewaters. The application of hydrogen peroxide as a bleaching agent has increased greatly in recent years because it is generally regarded as more of a "green chemical" than chlorine and hypochlorite that it has been replacing. It is important to break down spent hydrogen peroxide to oxygen and water before it is discharged, but catalase enzymes from conventional sources are inhibited or destroyed

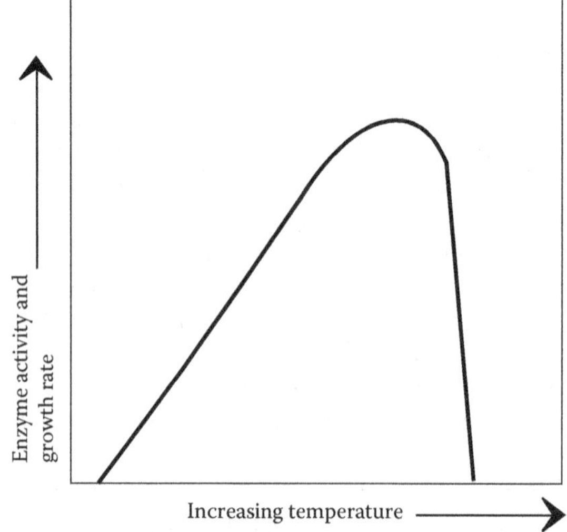

FIGURE 6.7 Enzyme activity as a function of temperature. A plot of bacterial growth versus temperature has the same shape.

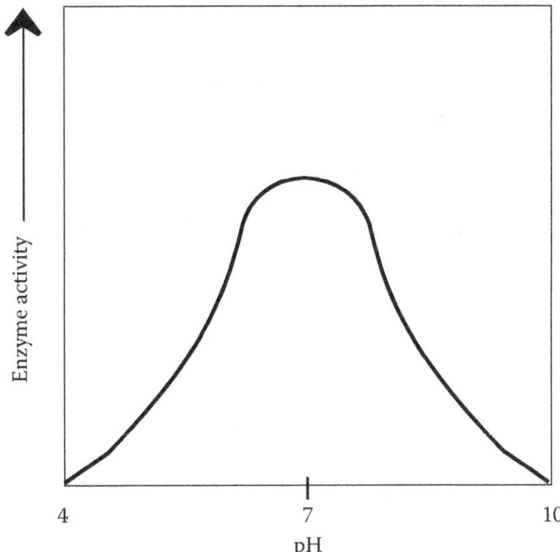

FIGURE 6.8 Enzyme activity and bacterial growth rate as a function of pH.

at the elevated temperatures and pH values typical of the bleaching solutions. However, the catalase enzyme from thermophilic bacteria can function well at 70°C and a pH up to 10.

Figure 6.8 is a plot of pH versus bacterial enzyme activity. Although the optimum pH will vary somewhat, enzymes typically have a pH optimum around neutrality. Enzymes tend to become denatured at pH extremes. For some bacteria, such as those that generate sulfuric acid by the oxidation of sulfide or that produce organic acids by fermentation of organic matter, the pH optimum may be quite acidic.

6.8.2 Microbial Oxidation and Reduction

The metabolic processes by which bacteria obtain their energy involve mediation of oxidation–reduction reactions. The most environmentally important oxidation–reduction reactions occurring in water and soil through the action of bacteria are summarized in Table 6.1. Much of the remainder of this chapter is devoted to a discussion of important redox reactions mediated by bacteria, particularly those summarized in Table 6.1.

TABLE 6.1
Principal Microbially Mediated Oxidation and Reduction Reactions

Oxidation	$pE^0(w)$[a]
(1) $\frac{1}{4}\{CH_2O\}+\frac{1}{4}H_2O \rightleftarrows \frac{1}{4}CO_2+H^+(w)+e^-$	−8.20
(1a) $\frac{1}{2}HCOO^- \rightleftarrows \frac{1}{2}CO_2(g)+\frac{1}{2}H^+(w)+e^-$	−8.73
(1b) $\frac{1}{2}\{CH_2O\}+\frac{1}{2}H_2O \rightleftarrows \frac{1}{2}HCOO^-+\frac{3}{2}H^+(w)+e^-$	−7.68
(1c) $\frac{1}{2}CH_3OH \rightleftarrows \frac{1}{2}\{CH_2O\}+H^+(w)+e^-$	−3.01
(1d) $\frac{1}{2}CH_4(g)+\frac{1}{2}H_2O \rightleftarrows \frac{1}{2}CH_3OH+H^+(w)+e^-$	−2.88
(2) $\frac{1}{8}HS^-+\frac{1}{2}H_2O \rightleftarrows \frac{1}{8}SO_4^{2-}+\frac{9}{8}H^+(w)+e^-$	−3.75
(3) $\frac{1}{8}NH_4^++\frac{3}{8}H_2O \rightleftarrows \frac{1}{8}NO_3^-+\frac{5}{4}H^+(w)+e^-$	+6.16

continued

TABLE 6.1 (continued)
Principal Microbially Mediated Oxidation and Reduction Reactions

Oxidation	**p$E^0(w)$[a]**
(4)[a] $FeCO_3(s) + 2H_2O \rightleftarrows FeOOH(s) + HCO_3^-(10^{-3}) + 2H^+(w) + e^-$	−1.67
(5)[a] $\frac{1}{2}MnCO_3(s) + H_2O \rightleftarrows \frac{1}{2}MnO_2 + \frac{1}{2}HCO_3^-(10^{-3}) + \frac{3}{2}H^+(w) + e^-$	−8.5
Reduction	
(A) $\frac{1}{4}O_2(g) + H^+(w) + e^- \rightleftarrows \frac{1}{2}H_2O$	+13.75
(B) $\frac{1}{5}NO_3^- + \frac{6}{5}H^+(w) + e^- \rightleftarrows \frac{1}{10}N_2 + \frac{3}{5}H_2O$	+12.65
(C) $\frac{1}{8}NO_3^- + \frac{5}{4}H^+(w) + e^- \rightleftarrows \frac{1}{8}NH_4^+ + \frac{3}{8}H_2O$	+6.15
(D) $\frac{1}{2}\{CH_2O\} + H^+(w) + e^- \rightleftarrows \frac{1}{2}CH_3OH$	−3.01
(E) $\frac{1}{8}SO_4^{2-} + \frac{9}{8}H^+(w) + e^- \rightleftarrows \frac{1}{8}HS^- + \frac{1}{2}H_2O$	−3.75
(F) $\frac{1}{8}CO_2(g) + H^+(w) + e^- \rightleftarrows \frac{1}{8}CH_4(g) + \frac{1}{4}H_2O$	−4.13
(G) $\frac{1}{6}N_2 + \frac{4}{3}H^+(w) + e^- \rightleftarrows \frac{1}{3}NH_4^+$	−4.68

Sequence of Microbial Mediation

Model 1: Excess of organic material (water initially contains O_2, NO_3^-, SO_4^{2-}, and HCO_3^-)
Examples: Hypolimnion of a eutrophic lake, sediments, sewage treatment plant digester.

	Combination	**p$E^0(w)$[b]**	**$\Delta G^0(w)$ (kcal)**
Oxic respiration	(1) + (A)	21.95	−29.9
Denitrification	(1) + (B)	20.85	−28.4
Nitrate reduction	(1) + (C)	14.36	−19.6
Fermentation[c]	(1b) + (D)	4.67	−6.4
Sulfate reduction	(1) + (E)	4.45	−5.9
Methane fermentation	(1) + (F)	4.07	−5.6
N-fixation	(1) + (G)	3.52	−4.8

Model 2: Excess O_2 [water initially contains organic matter SH^-, NH_4^+, and possibly Fe(II) and Mn(II)]
Examples: Oxic waste treatment, self-purification in streams, epilimnion of lake.

	Combination	**p$E^0(w)$[b]**	**$\Delta G^0(w)$ (kcal)**
Oxic respiration	(A) + (1)	21.95	−29.9
Sulfide oxidation	(A) + (2)	17.50	−23.8
Nitrification	(A) + (3)	7.59	−10.3
Iron(II) oxidation[d]	(A) + (4)	15.42	21.0
Manganese(II) oxidation[d]	(A) + (5)	5.75	−7.2

Source: Stumm, Werner, and James J. Morgan, *Aquatic Chemistry*, pp. 336–337, Wiley-Interscience, New York, 1970. Reproduced with permission of John Wiley & Sons, Inc.

[a] These pE^0 values are at H^+ ion activity of 1.00×10^{-7}; $H^+(w)$ designates water in which $[H^+] = 1.00 \times 10^{-7}$. p$E^0$ values for half-reactions (1) through (5) are given for reduction, although the reaction is written as an oxidation.

[b] pE^0 values = log K(w) for a reaction written for a one-electron transfer. The term K(w) is the equilibrium constant for the reaction in which the activity of the hydrogen ion has been set at 1.00×10^{-7} and incorporated into the equilibrium constant.

[c] Fermentation is interpreted as an organic redox reaction where one organic substance is reduced by oxidizing another organic substance (for example, alcoholic fermentation; the products are metastable thermodynamically with respect to CO_2 and CH_4).

[d] The data for pE^0(w) or ΔG^0(w) of these reactions correspond to an activity of HCO_3^- ion of 1.00×10^{-3} rather than unity.

6.9 MICROBIAL TRANSFORMATIONS OF CARBON

Carbon is an essential life element and composes a high percentage of the dry mass of microorganisms. For most microorganisms, the bulk of net energy-yielding or energy-consuming metabolic processes involve changes in the oxidation state of carbon. These chemical transformations of carbon have important environmental implications. For example, when algae and other plants fix CO_2 as carbohydrate, represented as $\{CH_2O\}$,

$$CO_2 + H_2O + h\nu \rightarrow \{CH_2O\} + O_2(g) \tag{6.7}$$

carbon changes from the +4 to the 0 oxidation state. Energy from sunlight is stored as chemical energy in organic compounds. However, when the algae die, bacterial decomposition occurs through oxic respiration in the reverse of the biochemical process represented by the above reaction for photosynthesis, energy is released, and oxygen is consumed.

In the presence of oxygen, the principal energy-yielding reaction of bacteria is the oxidation of organic matter. Since it is generally more meaningful to compare reactions on the basis of the reaction of one electron-mole, the oxic degradation of organic matter is conveniently written as

$$\tfrac{1}{4}\{CH_2O\} + \tfrac{1}{4}O_2(g) \rightarrow \tfrac{1}{4}CO_2 + \tfrac{1}{4}H_2O \tag{6.8}$$

for which the free-energy change is −29.9 kcal (see oxic respiration, Table 6.1). From this general type of reaction, bacteria and other microorganisms extract the energy needed to carry out their metabolic processes, to synthesize new cell material, for reproduction, and for locomotion.

Partial microbial decomposition of organic matter is a major step in the production of peat, lignite, coal, oil shale, and petroleum. Under reducing conditions, particularly below water, the oxygen content of the original plant material (approximate empirical formula, $\{CH_2O\}$) is lowered, leaving materials with relatively higher carbon contents.

6.9.1 Methane-Forming Bacteria

The production of methane in anoxic (oxygenless) sediments is favored by high organic levels and low nitrate and sulfate levels. Methane production plays a key role in local and global carbon cycles as the final step in the anoxic decomposition of organic matter. This process is the source of about 80% of the methane entering the atmosphere.

The carbon from microbially produced methane can come from either the reduction of CO_2 or the fermentation of organic matter, particularly acetate. The anoxic production of methane can be represented in the following simplified manner. When carbon dioxide acts as an electron receptor in the absence of oxygen, methane gas is produced:

$$\tfrac{1}{8}CO_2 + H^+ + e^- \rightarrow \tfrac{1}{8}CH_4 + \tfrac{1}{4}H_2O \tag{6.9}$$

This reaction is mediated by methane-forming bacteria. When organic matter is degraded microbially, the half-reaction for one electron-mole of $\{CH_2O\}$ is

$$\tfrac{1}{4}\{CH_2O\} + \tfrac{1}{4}H_2O \rightarrow \tfrac{1}{4}CO_2 + H^+ + e^- \tag{6.10}$$

Adding half-reactions 6.9 and 6.10 yields the overall reaction for the anoxic degradation of organic matter by methane-forming bacteria, which involves a free-energy change of −5.55 kcal per electron-mole:

$$\tfrac{1}{4}\{CH_2O\} \rightarrow \tfrac{1}{8}CH_4 + \tfrac{1}{8}CO_2 \tag{6.11}$$

This reaction, in reality a series of complicated processes, is a *fermentation reaction*, defined as a redox process in which both the oxidizing agent and reducing agent are organic substances. It may be seen that only about one-fifth as much free energy is obtained from one electron-mole of methane formation as from a one electron-mole reaction involving complete oxidation of the organic matter, Reaction 6.8.

Methane formation is a valuable process responsible for the degradation of large quantities of organic wastes, both in biological waste-treatment processes (see Chapter 8) and in nature. Methane production is used in biological waste treatment plants to further degrade excess sludge from the activated sludge process. In the bottom regions of natural waters, methane-forming bacteria degrade organic matter in the absence of oxygen. This eliminates organic matter, which would otherwise require oxygen for its biodegradation. If this organic matter were transported to oxic water containing dissolved O_2, it would exert a BOD. Methane production is a very efficient means for the removal of BOD. The reaction

$$CH_4 + 2O_2 \rightarrow CO_2 + 2H_2O \tag{6.12}$$

shows that 1 mol of methane requires 2 mols of oxygen for its oxidation to CO_2. Therefore, the production of 1 mol of methane and its subsequent evolution from water are equivalent to the removal of 2 mols of oxygen demand. In a sense, therefore, the removal of 16 g (1 mol) of methane is equivalent to the addition of 64 g (2 mols) of available oxygen to the water.

In favorable cases, methane fuel can be produced cost-effectively as a renewable resource from anoxic digestion of organic wastes. Some installations use cattle feedlot wastes. Methane is routinely generated by the action of anoxic bacteria and is used for heat and engine fuel at sewage treatment plants (see Chapter 8). Methane produced underground in municipal landfills is being tapped by some municipalities; however, methane seeping into basements of buildings constructed on landfill containing garbage has caused serious explosions and fires.

6.9.2 Bacterial Utilization of Hydrocarbons

Methane is oxidized under oxic conditions by a number of strains of bacteria. One of these, *Methanomonas*, is a highly specialized organism that cannot use any material other than methane as an energy source. Methanol, formaldehyde, and formic acid are intermediates in the microbial oxidation of methane to carbon dioxide. As discussed in Section 6.10, several types of bacteria can degrade higher hydrocarbons and use them as energy and carbon sources.

6.9.3 Microbial Utilization of Carbon Monoxide

Carbon monoxide is removed from the atmosphere by contact with soil. Since neither sterilized soil nor green plants grown under sterile conditions show any capacity to remove carbon monoxide from air, this ability must be because of microorganisms in the soil. Fungi capable of CO metabolism include some commonly occurring strains of the ubiquitous *Penicillium* and *Aspergillus*. It is also possible that some bacteria are involved in CO removal. However some microorganisms metabolize CO, other aquatic and terrestrial organisms produce this gas.

6.10 BIODEGRADATION OF ORGANIC MATTER

The biodegradation of organic matter in the aquatic and terrestrial environments is a crucial environmental process. Some organic pollutants are biocidal; for example, effective fungicides must be antimicrobial in action. Therefore, in addition to killing harmful fungi, fungicides frequently harm beneficial saprophytic fungi (fungi that decompose dead organic matter) and bacteria. Herbicides, which are designed for plant control, and insecticides, which are used to control insects, generally do not have any detrimental effect on microorganisms.

The biodegradation of organic matter by microorganisms occurs by way of a number of stepwise, microbially catalyzed reactions. These reactions will be discussed individually with examples.

6.10.1 Oxidation

Oxidation occurs by the action of oxygenase enzymes (see Chapter 22 for a discussion of biochemical terms). *Epoxidation* is a major step in many oxidation mechanisms and consists of adding an oxygen atom between two C atoms in an unsaturated system as shown for benzene below:

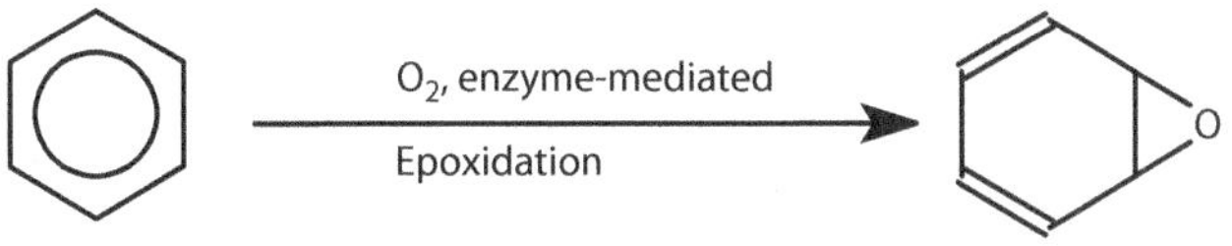

$$\text{C}_6\text{H}_6 \xrightarrow[\text{Epoxidation}]{O_2\text{, enzyme-mediated}} \text{benzene epoxide} \qquad (6.13)$$

Epoxidation is a particularly important means of metabolic attack upon aromatic rings that abound in many xenobiotic compounds and is part of the process by which potentially carcinogenic PAH compounds such as benzo(a)pyrene are biodegraded.[3] Epoxidation of an aromatic ring may be followed by *ring cleavage*, an important step in the biodegradation of aromatic compounds:

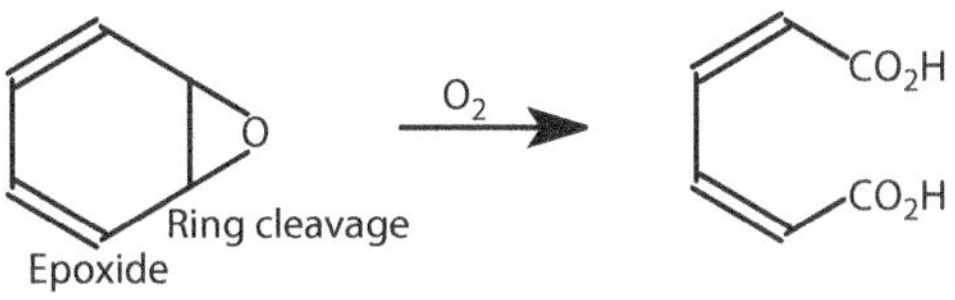

$$\text{Epoxide} \xrightarrow[\text{Ring cleavage}]{O_2} \text{HO}_2\text{C-CH=CH-CH=CH-CO}_2\text{H} \qquad (6.14)$$

6.10.1.1 Microbial Oxidation of Hydrocarbons

The degradation of hydrocarbons by microbial oxidation is an important environmental process because it is the primary means by which petroleum wastes are eliminated from water and soil. Bacteria capable of degrading hydrocarbons include *Micrococcus*, *Pseudomonas*, *Mycobacterium*, and *Nocardia*.

The most common initial step in the microbial oxidation of alkanes involves conversion of a terminal $-CH_3$ group to a $-CO_2H$ group. After formation of a carboxylic acid from the alkane, further oxidation normally occurs by β-oxidation:

$$CH_3CH_2CH_2CH_2CO_2H + 3O_2 \rightarrow CH_3CH_2CO_2H + 2CO_2 + 2H_2O \qquad (6.15)$$

A complicated cycle with a number of steps is involved in β-oxidation. The residue at the end of each cycle is an organic acid with two fewer carbon atoms than its precursor at the beginning of the cycle.

Hydrocarbons vary significantly in their biodegradability, and microorganisms show a strong preference for straight-chain hydrocarbons. A major reason for this preference is that branching inhibits β-oxidation at the site of the branch. The presence of a quaternary carbon (below) particularly inhibits alkane degradation.

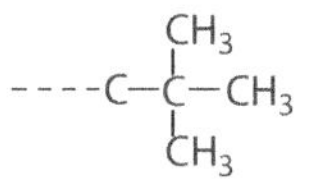

$$\text{----C-C(CH}_3)_2\text{-CH}_3$$

The biodegradation of petroleum is essential for the elimination of oil spills (of the order of a million metric tons per year). This oil is degraded by both marine bacteria and filamentous fungi. In some cases, the rate of degradation is limited by available nitrate and phosphate and it has been observed that the microbial degradation of hydrocarbons from oil spills is enhanced by providing nitrogen, phosphorus, and potassium nutrients. The physical form of crude oil makes a large difference in its degradability. Degradation in water occurs at the water–oil interface and in layers of oil floating on top of water and exposed to the atmosphere. Therefore, thick layers of crude oil prevent contact with bacterial enzymes and O_2.

Hydroxylation often accompanies microbial oxidation. It is the attachment of –OH groups to hydrocarbon chains or rings. In the biodegradation of foreign compounds, hydroxylation often follows epoxidation as shown by the following rearrangement reaction for benzene epoxide:

$$\text{Benzene epoxide (H, O, H)} \xrightarrow{\text{Rearrangement}} \text{C}_6\text{H}_5\text{OH}$$

Hydroxylation can consist of the addition of more than one hydroxide group. An example of epoxidation and hydroxylation is the metabolic production of carcinogenic 7,8-diol-9,10-epoxide from benzo(a)pyrene discussed in Section 24.4.

6.10.2 Other Biochemical Processes in Biodegradation of Organics

Hydrolysis, which involves the addition of H_2O to a molecule accompanied by cleavage of the molecule into two products, is a major step in microbial degradation of many pollutant compounds, especially pesticidal esters, amides, and organophosphate esters. The types of enzymes that bring about hydrolysis are *hydrolase enzymes*; those that enable the hydrolysis of esters are called *esterases*, whereas those that hydrolyze amides are *amidases*. At least one species of *Pseudomonas* hydrolyzes malathion in a type of hydrolysis reaction typical of those by which pesticides are degraded:

$$\underset{\text{Malathion}}{(CH_3O)_2P(=S)-S-CH(C(=O)-O-C_2H_5)-CH_2-C(=O)-O-C_2H_5} \xrightarrow{H_2O} (CH_3O)_2P(=S)-SH + HO-CH(C(=O)-O-C_2H_5)-CH_2-C(=O)-O-C_2H_5 \quad (6.16)$$

Reductions are carried out by *reductase enzymes*; for example, nitroreductase enzyme catalyzes the reduction of the nitro group. Table 6.2 gives the major kinds of functional groups reduced by microorganisms.

Dehalogenation reactions of organohalide compounds involve the bacterially mediated replacement of a covalently bound halogen atom (F, Cl, Br, I) with –OH, and are discussed in more detail in Section 6.13.

Many environmentally significant organic compounds contain alkyl groups, such as the methyl ($-CH_3$) group, attached to atoms of O, N, and S. An important step in the microbial metabolism of many of these compounds is *dealkylation*, replacement of alkyl groups by H as shown in Figure 6.9. Examples of these kinds of reactions include O-dealkylation of methoxychlor insecticides, N-dealkylation of carbaryl insecticide, and S-dealkylation of dimethyl sulfide. Alkyl groups removed by dealkylation usually are attached to oxygen, sulfur, or nitrogen atoms; those attached to carbon are normally not removed directly by microbial processes.

TABLE 6.2
Functional Groups That Undergo Microbial Reduction

Reactant	Process	Product
R—C(=O)—H	Aldehyde reduction	R—CH₂—OH (R—C(H)(H)—OH)
R—C(=O)—R′	Ketone reduction	R—C(OH)(H)—R′
R—S(=O)—R′	Sulfoxide reduction	R—S—R′
R—SS—R′	Disulfide reduction	R—SH, R′—SH
R(H)C=C(H)R′	Alkene reduction	R—C(H)(H)—C(H)(H)—R′
R—NO_2	Nitro group reduction	R—NO, R—NH_2, R—N(H)(OH)

6.11 MICROBIAL TRANSFORMATIONS OF NITROGEN

Some of the most important microorganism-mediated chemical reactions in aquatic and soil environments are those involving nitrogen compounds.[4] They are summarized in the *nitrogen cycle* shown in Figure 6.10. This cycle describes the dynamic processes through which nitrogen is interchanged among the atmosphere, organic matter, and inorganic compounds. It is one of nature's most vital dynamic processes.

Among the biochemical transformations in the nitrogen cycle are nitrogen fixation, whereby molecular nitrogen is fixed as organic nitrogen; nitrification, the process of oxidizing ammonia to nitrate; nitrate reduction, the process by which nitrogen in nitrate ion is reduced to form compounds having nitrogen in a lower oxidation state; and denitrification, the reduction of nitrate and nitrite to N_2, with a resultant net loss of nitrogen gas to the atmosphere. Each of these important chemical processes will be discussed separately.

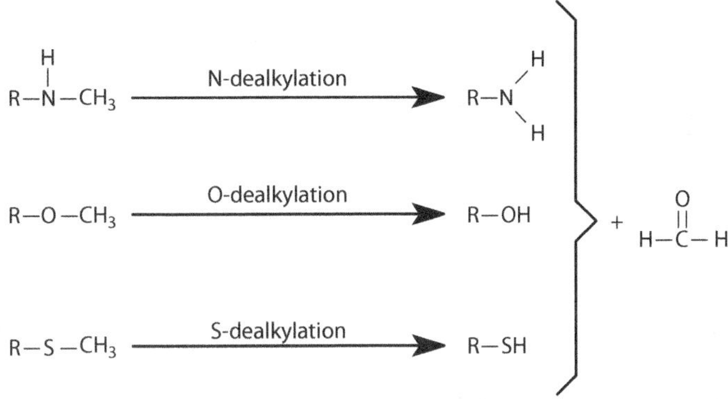

FIGURE 6.9 Metabolic dealkylation reactions shown for the removal of CH_3 from N, O, and S atoms in organic compounds.

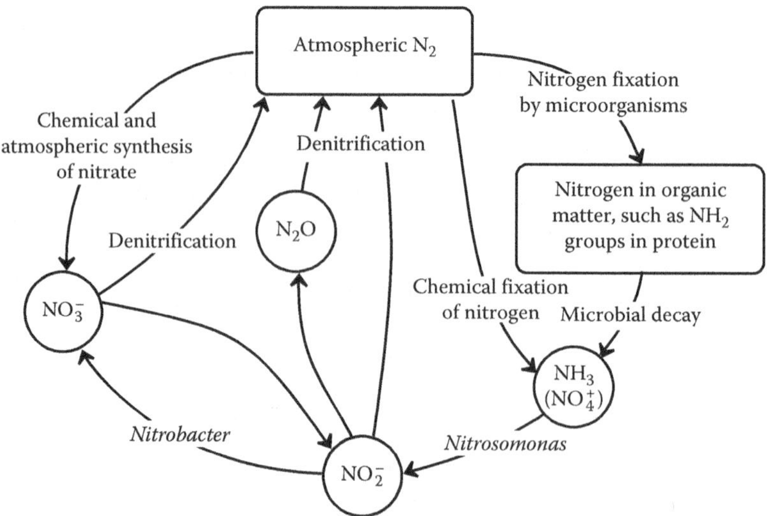

FIGURE 6.10 The nitrogen cycle.

6.11.1 Nitrogen Fixation

The overall microbial process for *nitrogen fixation*, the binding of atmospheric nitrogen in a chemically combined form

$$3\{CH_2O\} + 2N_2 + 3H_2O + 4H^+ \rightarrow 3CO_2 + 4NH_4^+ \qquad (6.17)$$

is actually quite a complicated process that is the subject of extensive investigation. Biological nitrogen fixation is a key biochemical process in the environment and is essential for plant growth in the absence of synthetic fertilizers.

Only a few species of aquatic microorganisms have the ability to fix atmospheric nitrogen. Among the aquatic bacteria having this capability are photosynthetic bacteria, *Azotobacter*, several species of *Clostridium*, and cyanobacteria. In most natural freshwater systems, however, the fraction of nitrogen fixed by organisms in the water relative to that originating from the decay of organic material, fertilizer runoff, and other external sources is relatively low.

The best-known and most important form of nitrogen-fixing bacteria is *Rhizobium*, which enjoys a symbiotic (mutually advantageous) relationship with leguminous plants such as clover or alfalfa. The *Rhizobium* bacteria are found in root nodules, special structures bound to the roots of legumes (see Figure 16.2) and connected directly to the vascular (circulatory) system of the plant, enabling the bacteria to derive photosynthetically produced energy directly from the plant. Thus, the plant provides the energy required to break the strong triple bonds in the dinitrogen molecule, converting the nitrogen to a reduced form which is directly assimilated by the plant. When the legumes die and decay, NH_4^+ ion is released and is converted by microorganisms to nitrate ion, which is assimilable by other plants. Some of the ammonium ion and nitrate released may be carried into natural water systems.

Some nonlegume angiosperms fix nitrogen through the action of actinomycetes bacteria contained in root nodules. Shrubs and trees in the nitrogen-fixing category are abundant in fields, forests, and wetlands throughout the world. Their rate of nitrogen fixation is comparable to that of legumes.

Free-living bacteria associated with some grasses are stimulated by the grasses to fix nitrogen. One such bacterium is *Spirillumlipoferum*. In tropical surroundings, the amount of reduced nitrogen fixed by such bacteria can amount to the order of 100 kg per hectare per year.

Because of the cost of energy required to fix nitrogen synthetically, efforts are underway to increase the efficiency of natural means of nitrogen fixation. One approach uses recombinant DNA methodologies in attempts to transfer the nitrogen-fixing capabilities of nitrogen-fixing bacteria directly to plant cells. Although a fascinating possibility, this transfer has not yet been achieved on a practical basis. The other approach uses more conventional plant breeding and biological techniques in attempts to increase the range and effectiveness of the symbiotic relationship existing between some plants and nitrogen-fixing bacteria.

One matter of concern is that successful efforts to increase nitrogen fixation may upset the global nitrogen balance. Total annual global fixation of nitrogen is now more than 50% higher than the preindustrial level of 150 million metric tons estimated for 1850. Potential accumulation of excess fixed nitrogen is the subject of some concern because of aquatic nitrate pollution and microbial production of N_2O gas. Some atmospheric scientists fear that excess N_2O gas may be involved in depletion of the protective atmospheric ozone layer (see Chapter 14).

6.11.2 Nitrification

Nitrification, the conversion of N(–III) to N(V), is a very common and extremely important process in water and in soil. Aquatic nitrogen in thermodynamic equilibrium with air is in the +5 oxidation state as NO_3^-, whereas in most biological compounds, nitrogen is present as N(–III), such as $-NH_2$ in amino acids. The equilibrium constant of the overall nitrification reaction, written for one electron-mole

$$\tfrac{1}{4}O_2 + \tfrac{1}{8}NH_4^+ \rightarrow \tfrac{1}{8}NO_3^- + \tfrac{1}{4}H^+ + \tfrac{1}{8}H_2O \tag{6.18}$$

is $10^{7.59}$ (Table 6.1), showing that the reaction is highly favored from a thermodynamic viewpoint.

Nitrification is especially important in nature because nitrogen is absorbed by plants primarily as nitrate. When fertilizers are applied in the form of ammonium salts or anhydrous ammonia, a microbial transformation to nitrate enables maximum assimilation of nitrogen by the plants.

In nature, nitrification is catalyzed by two groups of bacteria, *Nitrosomonas* and *Nitrobacter*. *Nitrosomonas* bacteria bring about the transition of ammonia to nitrite

$$NH_3 + \tfrac{3}{2}O_2 \rightarrow H^+ + NO_2^- + H_2O \tag{6.19}$$

whereas *Nitrobacter* mediates the oxidation of nitrite to nitrate:

$$NO_2^- + \tfrac{1}{2}O_2 \rightarrow NO_3^- \tag{6.20}$$

Both of these highly specialized types of bacteria are *obligate aerobes*; that is, they function only in the presence of molecular O_2. These bacteria are also *chemolithotrophic*, meaning that they can utilize oxidizable inorganic materials as electron donors in oxidation reactions to yield needed energy for metabolic processes.

For the oxic conversion of one electron-mole of ammoniacal nitrogen to nitrite ion at pH 7.00,

$$\tfrac{1}{4}O_2 + \tfrac{1}{6}NH_4^+ \rightarrow \tfrac{1}{6}NO_2^- + \tfrac{1}{3}H^+ + \tfrac{1}{6}H_2O \tag{6.21}$$

the free-energy change is –10.8 kcal. The free-energy change for the oxic oxidation of one electron-mole of nitrite ion to nitrate ion:

$$\tfrac{1}{4}O_2 + \tfrac{1}{2}NO_2^- \rightarrow \tfrac{1}{2}NO_3^- \tag{6.22}$$

is –9.0 kcal. Both steps of the nitrification process involve an appreciable yield of free energy.

6.11.3 Nitrate Reduction

As a general term, *nitrate reduction* refers to microbial processes by which nitrogen in chemical compounds is reduced to lower oxidation states. In the absence of free oxygen, nitrate may be used by some bacteria as an alternate electron receptor. The most complete possible reduction of nitrogen in nitrate ion involves the acceptance of eight electrons by the nitrogen atom, with the consequent conversion of nitrate to ammonia (+V to −III oxidation state). Nitrogen is an essential component of protein, and any organism that utilizes nitrogen from nitrate for the synthesis of protein must first reduce the nitrogen to the −III oxidation state (ammoniacal form). However, incorporation of nitrogen into protein generally is a relatively minor use of the nitrate undergoing microbially mediated reactions and is more properly termed nitrate assimilation.

Nitrate ion functioning as an electron receptor usually produces NO_2^-:

$$\tfrac{1}{2}NO_3^- + \tfrac{1}{4}\{CH_2O\} \rightarrow \tfrac{1}{2}NO_2^- + \tfrac{1}{4}H_2O + \tfrac{1}{4}CO_2 \tag{6.23}$$

The free-energy yield per electron-mole is only about two-third of the yield when oxygen is the oxidant; however, nitrate ion is a good electron receptor in the absence of O_2. One of the factors limiting the use of nitrate ion in this function is its generally low concentration in most waters. Furthermore, nitrite, NO_2^-, is relatively toxic and tends to inhibit the growth of many bacteria after building up to a certain level. Sodium nitrate has been used as a "first-aid" treatment in sewage lagoons that have become oxygen deficient. It provides an emergency source of oxygen to reestablish normal bacterial growth.

Nitrate ion can be an effective oxidizing agent for a number of species in water that are oxidized by the action of microorganisms. Therefore, nitrate salts are sometimes added as an alternate source of oxygen in the biological treatment of oxidizable wastes.

6.11.4 Denitrification

An important special case of nitrate reduction is *denitrification*, in which the reduced nitrogen product is a nitrogen-containing gas, usually N_2. At pH 7.00, the free-energy change per electron-mole of reaction

$$\tfrac{1}{5}NO_3^- + \tfrac{1}{4}\{CH_2O\} + \tfrac{1}{5}H^+ \rightarrow \tfrac{1}{10}N_2 + \tfrac{1}{4}CO_2 + \tfrac{7}{20}H_2O \tag{6.24}$$

is −2.84 kcal. The free-energy yield per mole of nitrate reduced to N_2 (five electron-moles) is lower than that for the reduction of the same quantity of nitrate to nitrite. More important, however, the reduction of a nitrate ion to N_2 gas consumes five electrons, compared to only two electrons for the reduction of NO_3^- to NO_2^-.

Denitrification is an important process in nature. It is the mechanism by which fixed nitrogen is returned to the atmosphere. Denitrification is also used in advanced water treatment for the removal of nutrient nitrogen (see Chapter 8). Because nitrogen gas is a nontoxic volatile substance that does not inhibit microbial growth, and since nitrate ion is a very efficient electron acceptor, denitrification allows the extensive growth of bacteria under anoxic conditions.

Loss of nitrogen to the atmosphere may also occur through the formation of N_2O and NO by bacterial action on nitrate and nitrite catalyzed by the action of several types of bacteria. Production of N_2O relative to N_2 is enhanced during denitrification in soils by increased concentrations of NO_3^-, NO_2^-, and O_2.

6.11.5 Competitive Oxidation of Organic Matter by Nitrate Ion and Other Oxidizing Agents

The successive oxidation of organic matter by dissolved O_2, NO_3^-, and SO_4^{2-} brings about an interesting sequence of nitrate-ion levels in sediments and hypolimnion waters initially containing O_2 but

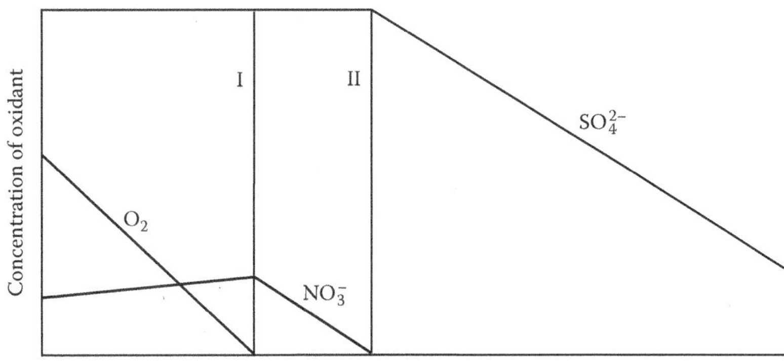

FIGURE 6.11 Oxidation of organic matter by O_2, NO_3^-, and SO_4^{2-}.

lacking a mechanism for reaeration. This is shown in Figure 6.11, where concentrations of dissolved O_2, NO_3^-, and SO_4^{2-} are plotted as a function of total organic matter metabolized. This behavior can be explained by the following sequence of biochemical processes:

$$O_2 + \text{organic matter} \rightarrow \text{products} \tag{6.25}$$

$$NO_3^- + \text{organic matter} \rightarrow \text{products} \tag{6.26}$$

$$SO_4^{2-} + \text{organic matter} \rightarrow \text{products} \tag{6.27}$$

So long as some O_2 is present, some nitrate may be produced from nitrogenous organic matter. After exhaustion of molecular oxygen, nitrate is the favored oxidizing agent, and its concentration falls from a maximum value (I) to zero (II). Sulfate, which is usually present in a large excess over the other two oxidants, then becomes the favored electron receptor, enabling biodegradation of organic matter to continue.

6.12 MICROBIAL TRANSFORMATIONS OF PHOSPHORUS AND SULFUR

6.12.1 Phosphorus Compounds

Biodegradation of phosphorus compounds is important in the environment for two reasons. The first of these is that it provides a source of algal nutrient orthophosphate from the hydrolysis of polyphosphates (see Section 3.16). Secondly, biodegradation deactivates toxic organophosphate compounds, such as the organophosphate insecticides.

The organophosphorus compounds of greatest environmental concern tend to be sulfur-containing *phosphorothionate* and *phosphordithioate* ester insecticides. These are discussed in more detail in Chapter 7, Section 7.11, and structural formulas of several representative examples are shown in Figure 7.8. The biodegradation of these compounds is an important environmental chemical process. Fortunately, unlike the organohalide insecticides that they largely displaced until less toxic classes of compounds came into general use, the organophosphates readily undergo biodegradation and do not bioaccumulate.

Hydrolysis is an important step in the biodegradation of phosphorothionate, phosphorodithioate, and phosphate ester insecticides leading to their detoxification and loss of insecticidal activity. Hydrolysis of these insecticides is shown by the following general reactions where *R* is an alkyl group, Ar is a substituent group that is frequently aromatic, and X is either S or O:

$$\text{R—O—}\underset{\substack{|\\ \text{O}\\ |\\ \text{R}}}{\overset{\substack{\text{X}\\ \|}}{\text{P}}}\text{—OAr} \xrightarrow{H_2O} \text{R—O—}\underset{\substack{|\\ \text{O}\\ |\\ \text{R}}}{\overset{\substack{\text{X}\\ \|}}{\text{P}}}\text{—OH} + \text{HOAr} \tag{6.28}$$

$$\text{R—O—}\underset{\substack{|\\ \text{O}\\ |\\ \text{R}}}{\overset{\substack{\text{X}\\ \|}}{\text{P}}}\text{—SAr} \xrightarrow{H_2O} \text{R—O—}\underset{\substack{|\\ \text{O}\\ |\\ \text{R}}}{\overset{\substack{\text{X}\\ \|}}{\text{P}}}\text{—OH} + \text{HSAr} \tag{6.29}$$

6.12.2 Sulfur Compounds

Sulfur compounds are very common in water. Sulfate ion, SO_4^{2-}, is found in varying concentrations in practically all natural waters. Organic sulfur compounds, both those of natural origin and pollutant species, are very common in natural aquatic systems, and the degradation of these compounds is an important microbial process. Sometimes the degradation products, such as odiferous and toxic H_2S, cause serious problems with water quality.

Sulfur in living material is present primarily in its most reduced state, for example, as the hydrosulfide group, –SH. When organic sulfur compounds are decomposed by bacteria, the initial sulfur product is generally the reduced form, H_2S. Some bacteria produce and store elemental sulfur from sulfur compounds. In the presence of oxygen, some bacteria convert reduced forms of sulfur to the oxidized form in SO_4^{2-} ion.

6.12.2.1 Oxidation of H_2S and Reduction of Sulfate by Bacteria

In the presence of oxygen some bacteria can oxidize sulfide (H_2S) to sulfate and in the absence of oxygen other bacteria can reduce sulfate to sulfide.[5] The bacteria *Desulfovibrio* can reduce inorganic sulfate ion to H_2S. In so doing, they utilize sulfate as an electron acceptor in the oxidation of organic matter. The overall reaction for the microbially mediated oxidation of biomass with sulfate is

$$SO_4^{2-} + 2\{CH_2O\} + 2H^+ \rightarrow H_2S + 2CO_2 + 2H_2O \tag{6.30}$$

and it requires other bacteria besides *Desulfovibrio* to oxidize organic matter completely to CO_2. Because of the high concentration of sulfate ion in seawater, bacterially mediated formation of H_2S causes pollution problems in some coastal areas and is a major source of atmospheric sulfur. In waters where sulfide formation occurs, the sediment is often black in color due to the formation of FeS.

Some bacteria such as purple sulfur bacteria and green sulfur bacteria can oxidize sulfur in hydrogen sulfide to higher oxidation states. The oxic colorless sulfur bacteria may use molecular oxygen to oxidize H_2S to elemental sulfur and can oxidize elemental S and thiosulfate ($S_2O_3^{2-}$) to sulfate.

Oxidation of sulfur in a low oxidation state to sulfate ion produces sulfuric acid, a strong acid. One of the colorless sulfur bacteria, *Thiobacillus thiooxidans*, is tolerant of one normal acid solutions; a remarkable acid tolerance. When elemental sulfur is added to excessively alkaline soils, the acidity is increased because of a microorganism-mediated reaction, which produces sulfuric acid. Elemental sulfur may be deposited as granules in the cells of purple sulfur bacteria and colorless sulfur bacteria. Such processes are important sources of elemental sulfur deposits.

6.12.3 Microorganism-Mediated Degradation of Organic Sulfur Compounds

Sulfur occurs in many types of biological compounds. As a consequence, organic sulfur compounds of natural and pollutant origin are very common in water. The degradation of these compounds is an important microbial process having a strong effect upon water quality.

Among some of the common sulfur-containing functional groups found in aquatic organic compounds are hydrosulfide (–SH), disulfide (–SS–), sulfide (–S–), sulfoxide ($-\overset{O}{\overset{\|}{S}}-$), sulfonic acid ($-SO_2OH$), thioketone ($-\overset{S}{\overset{\|}{C}}-$), and thiazole (a heterocyclic sulfur group). Protein contains some amino acids with sulfur functional groups—cysteine,

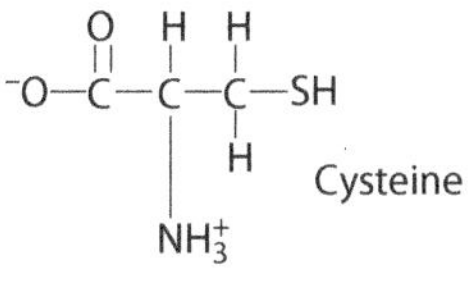

$$^{-}O-\overset{O}{\overset{\|}{C}}-\underset{NH_3^+}{\overset{H}{\overset{|}{C}}}-\underset{H}{\overset{H}{\overset{|}{\underset{|}{C}}}}-SH \quad \text{Cysteine}$$

cystine, and methionine—whose breakdown is important in natural waters. The amino acids are readily degraded by bacteria and fungi. The biodegradation of sulfur-containing amino acids can result in production of volatile organic sulfur compounds such as methane thiol, CH_3SH, and dimethyl disulfide, CH_3SSCH_3. These compounds have strong, unpleasant odors. Their formation, in addition to that of H_2S, accounts for much of the odor associated with the biodegradation of sulfur-containing organic compounds.

Hydrogen sulfide is formed from a large variety of organic compounds through the action of a number of different kinds of microorganisms. A typical sulfur-cleavage reaction producing H_2S is the conversion of cysteine to pyruvic acid through the action of cysteine desulfhydrase enzyme in bacteria:

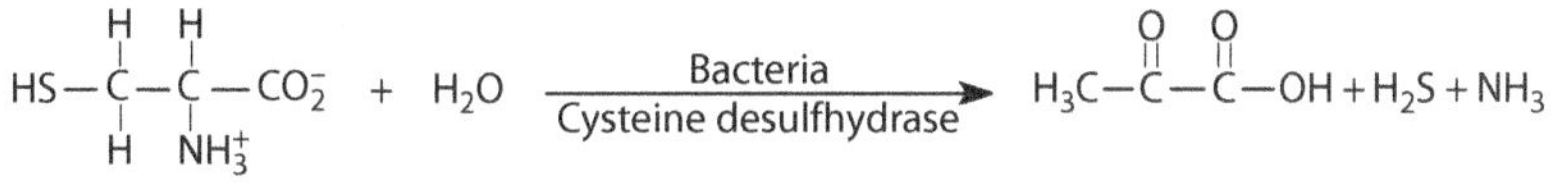

$$HS-\underset{H}{\overset{H}{\overset{|}{\underset{|}{C}}}}-\underset{NH_3^+}{\overset{H}{\overset{|}{\underset{|}{C}}}}-CO_2^- + H_2O \xrightarrow[\text{Cysteine desulfhydrase}]{\text{Bacteria}} H_3C-\overset{O}{\overset{\|}{C}}-\overset{O}{\overset{\|}{C}}-OH + H_2S + NH_3 \tag{6.31}$$

Because of the numerous forms in which organic sulfur may exist, a variety of sulfur products and biochemical reaction paths must be associated with the biodegradation of organic sulfur compounds.

6.13 MICROBIAL TRANSFORMATIONS OF HALOGENS AND ORGANOHALIDES

Microbially mediated reactions are important in the environmental degradation of pollutant organohalide compounds.[6] *Dehalogenation* reactions involving the replacement of a halogen atom, for example,

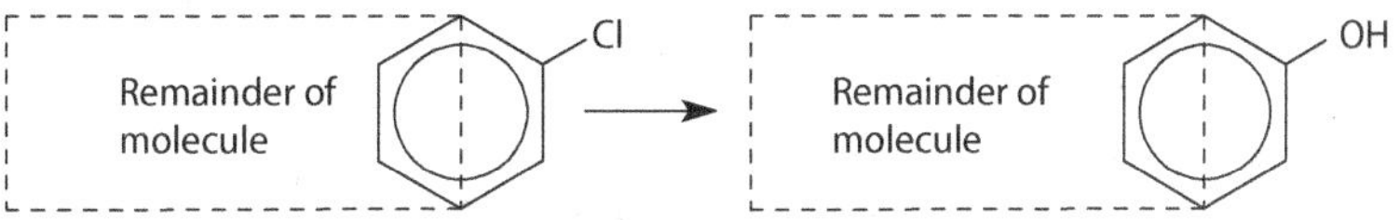

[Remainder of molecule]–C_6H_4–Cl → [Remainder of molecule]–C_6H_4–OH

represent a major pathway for the biodegradation of organohalide hydrocarbons. In some cases, organohalide compounds serve as sole carbon sources, sole energy sources, or electron acceptors for anoxic bacteria. Microorganisms need not utilize a particular organohalide compound as a sole carbon source in order to cause its degradation. This is due to the phenomenon of *cometabolism*, which results from a lack of specificity in the microbial degradation processes. Thus, bacterial degradation of small amounts of an organohalide compound may occur while the microorganism involved is metabolizing much larger quantities of another substance.

Some anoxic bacteria can reductively dechlorinate chlorinated aliphatic and aromatic compounds using relatively highly chlorinated compounds as electron acceptors as illustrated by the reaction

$$\{CH_2O\} + H_2O + 2Cl\text{–}R \rightarrow CO_2 + 2H^+ + 2Cl^- + 2H\text{–}R \quad (6.32)$$

where Cl–R represents a site of chlorine substitution on a chlorinated hydrocarbon molecule and H–R represents a site of hydrogen substitution. The net result of this process called *dehalorespiration* is the replacement of Cl by H on chlorinated hydrocarbons.

Microbially mediated dichloroelimination from 1,1,2,2-tetrachloroethane in the anoxic biodegradation of this compound can produce one of three possible isomers of dichloroethylene; the process followed by successive hydrogenolysis reactions can produce vinyl chloride and ethene (ethylene):

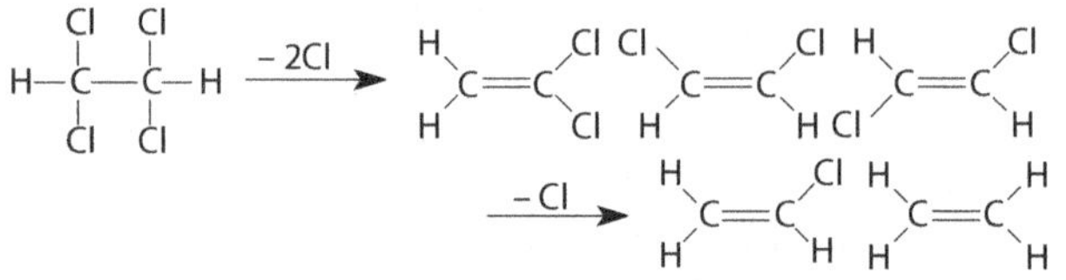

- 2Cl
- Cl
Vinyl chloride
Ethene (ethylene)
(6.33)

Successive hydrogenolysis reactions of 1,1,2,2-tetrachloroethane can produce ethane derivatives with 3, 2, 1, and 0 chlorine atoms

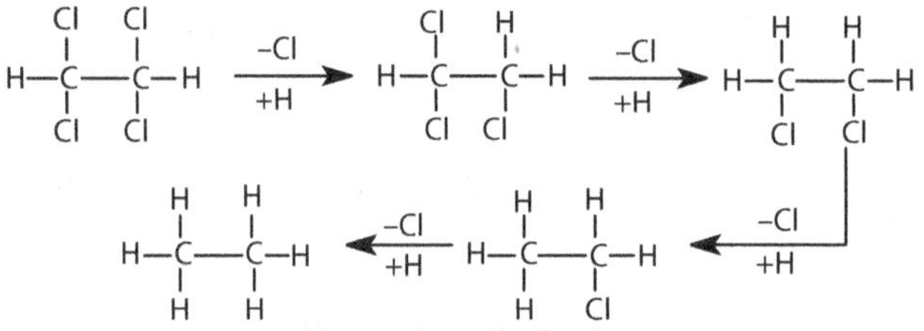

-Cl +H
(6.34)

Bioconversion of DDT to replace Cl with H yields DDD:

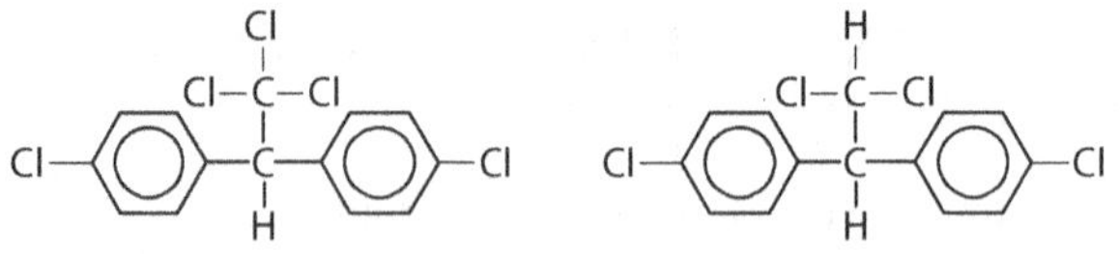

DDT
DDD

The latter compound is more toxic to some insects than DDT and has even been manufactured as a pesticide.

6.14 MICROBIAL TRANSFORMATIONS OF METALS AND METALLOIDS

Some bacteria, including *Ferrobacillus*, *Gallionella*, and some forms of *Sphaerotilus*, utilize iron compounds to obtain energy for their metabolic needs. These bacteria catalyze the oxidation of iron(II) to iron(III) by molecular oxygen:

$$4Fe(II) + 4H^+ + O_2 \rightarrow 4Fe(III) + 2H_2O \quad (6.35)$$

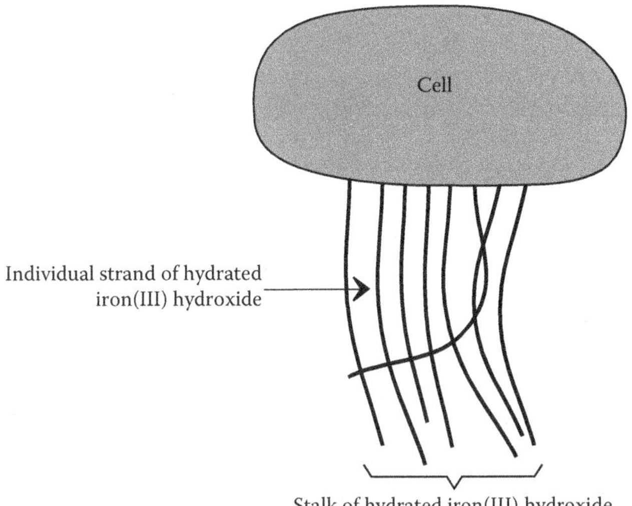

FIGURE 6.12 Sketch of a cell of *Gallionella* showing iron(III) oxide secretion.

The carbon source for some of these bacteria is CO_2. Since they do not require organic matter for carbon, and they derive energy from the oxidation of inorganic matter, these bacteria may thrive in environments where organic matter is absent.

The microorganism-mediated oxidation of iron(II) is not a particularly efficient means of obtaining energy for metabolic processes. For the reaction

$$FeCO_3(s)+\tfrac{1}{4}O_2+\tfrac{3}{2}H_2O \rightarrow Fe(OH)_3(s)+CO_2 \quad (6.36)$$

the change in free energy is approximately 10 kcal/electron-mole. Approximately 220 g of iron(II) must be oxidized to produce 1.0 g of cell carbon. The calculation assumes CO_2 as a carbon source and a biological efficiency of 5%. The production of only 1.0 g of cell carbon would produce approximately 430 g of solid $Fe(OH)_3$. It follows that large deposits of hydrated iron(III) oxide form in areas where iron-oxidizing bacteria thrive.

Some of the iron bacteria, notably *Gallionella*, secrete large quantities of hydrated iron(III) oxide in intricately branched structures. The bacterial cell grows at the end of a twisted stalk of the iron oxide. Individual cells of *Gallionella*, photographed through an electron microscope, have shown that the stalks consist of a number of strands of iron oxide secreted from one side of the cell (Figure 6.12).

At nearly neutral pH values, bacteria deriving energy by mediating the air oxidation of iron(II) must compete with direct chemical oxidation of iron(II) by O_2. The latter process is relatively rapid at pH 7. As a consequence, these bacteria tend to grow in a narrow layer in the region between the oxygen source and the source of iron(II). Therefore, iron bacteria are sometimes called *gradient organisms*, and they grow at intermediate p*E* values.

Bacteria are strongly involved in the oceanic manganese cycle. Manganese nodules, a potentially important source of manganese, copper, nickel, and cobalt that occur on ocean floors, yield different species of bacteria which enzymatically mediate both the oxidation and reduction of manganese.

6.14.1 Acid Mine Waters

One consequence of bacterial action on metal compounds is acid mine drainage, one of the most common and damaging problems in the aquatic environment. Many waters flowing from coal mines and draining from the "gob piles" left over from coal processing and washing are practically sterile due to high acidity.

Acid mine water results from the presence of sulfuric acid produced by the oxidation of pyrite, FeS_2. Microorganisms are closely involved in the overall process, which consists of several reactions and involves a number of microbial species that can exist under acidic conditions. The first of these reactions is the oxidation of pyrite, which is followed by the oxidation of iron(II) ion to iron(III) ion:

$$2FeS_2(s) + 2H_2O + 7O_2 \rightarrow 4H^+ + 4SO_4^{2-} + 2Fe^{2+} \tag{6.37}$$

$$4Fe^{2+} + O_2 + 4H^+ \rightarrow 4Fe^{3+} + 2H_2O \tag{6.38}$$

a process that occurs very slowly at low pH values found in acid mine waters. Below pH 3.5, the iron oxidation is catalyzed by the iron bacterium *Thiobacillus ferrooxidans*, and in the pH range 3.5–4.5 it may be catalyzed by a variety of *Metallogenium*, a filamentous iron bacterium. Other bacteria that may be involved in acid mine water formation are *Thiobacillus thiooxidans* and *Ferrobacillus ferrooxidans*. The Fe^{3+} ion further dissolves pyrite

$$FeS_2(s) + 14Fe^{3+} + 8H_2O \rightarrow 15Fe^{2+} + 2SO_4^{2-} + 16H^+ \tag{6.39}$$

which in conjunction with Reaction 6.38 constitutes a cycle for the dissolution of pyrite. $Fe(H_2O)_6^{3+}$ is an acidic ion and at pH values much above 3, the iron(III) precipitates as the hydrated iron(III) oxide:

$$Fe^{3+} + 3H_2O \rightarrow Fe(OH)_3(s) + 3H^+ \tag{6.40}$$

The beds of streams afflicted with acid mine drainage often are covered with "yellowboy," an unsightly deposit of amorphous, semigelatinous $Fe(OH)_3$. The most damaging component of acid mine water, however, is sulfuric acid. It is directly toxic and has other undesirable effects such as excessive weathering of minerals that it contacts.

Limestone, $CaCO_3$, is commonly used to treat acid mine water. When acid mine water is treated with limestone, the following reaction occurs:

$$CaCO_3(s) + 2H^+ + SO_4^{2-} \rightarrow Ca^{2+} + SO_4^{2-} + H_2O + CO_2(g) \tag{6.41}$$

Unfortunately, because iron(III) is generally present, $Fe(OH)_3$ precipitates as the pH is raised (Reaction 6.40). The hydrated iron(III) oxide product covers the particles of carbonate rock with a relatively impermeable layer. This armoring effect prevents further neutralization of the acid.

6.14.2 Microbial Transitions of Selenium

Directly below sulfur in the periodic table, selenium is subject to bacterial oxidation and reduction. These transitions are important because selenium is a crucial element in nutrition, particularly of livestock. Diseases related to either selenium excesses or deficiency have been reported in at least half of the states of the United States and in 20 other countries, including the major livestock producing countries. Livestock in New Zealand, in particular, suffer from selenium deficiency.

Microorganisms are closely involved with the selenium cycle, and microbial reduction of oxidized forms of selenium has been known for some time. Reductive processes under anoxic conditions can reduce both SeO_3^{2-} and SeO_4^{2-} ions to elemental selenium, which can accumulate as a sink for selenium in anoxic sediments. Some bacteria such as selected strains of *Thiobacillus* and *Leptothrix* can oxidize elemental selenium to selenite, SeO_3^{2-}, thus remobilizing this element from deposits of Se(0).

The major volatile selenium species emitted to the atmosphere by microbial processes in water and soil is dimethylselenide, $(CH_3)_2Se$. (The single largest source of natural sulfur discharged to the

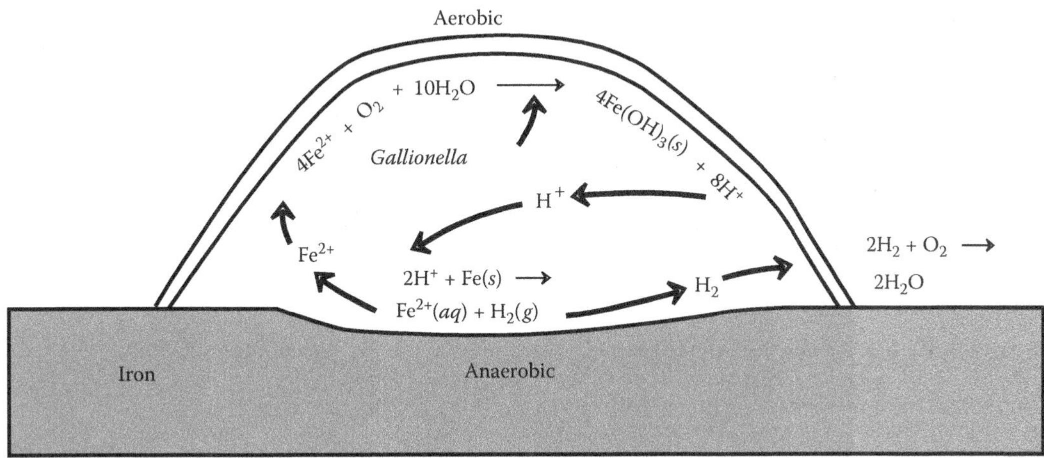

FIGURE 6.13 Tubercle in which the bacterially mediated corrosion of iron occurs through the action of *Gallionella*.

atmosphere is the analogous sulfur compound, biogenic dimethyl sulfide, $(CH_3)_2S$, from marine sources; see Section 11.4.) The major biological selenium compound that is the precursor to dimethylselenide formation is selenomethionine:

$$H_3C-Se-\underset{H}{\overset{H}{C}}-\underset{H}{\overset{H}{C}}-\underset{NH_2}{\overset{H}{C}}-\overset{O}{\overset{\|}{C}}-OH \quad \text{Selenomethionine}$$

6.14.3 Microbial Corrosion

Corrosion involving the deterioration of iron as well as other metals is a redox phenomenon and was discussed in Section 4.14. Much corrosion is bacterial in nature.[7] Bacteria involved with corrosion set up their own electrochemical cells in which a portion of the surface of the metal being corroded forms the anode of the cell and is oxidized. Structures called *tubercles* form in which bacteria pit and corrode metals as shown in Figure 6.13.

LITERATURE CITED

1. Vaulot, D., W. Eikrem, M. Viprey, and H. Moreau, The diversity of small eukaryotic phytoplankton (≤3 μm) in marine ecosystems, *FEMS Microbiology Reviews*, **32**, 795–820, 2008.
2. Koskinen, P. E. P., S. R. Beck, J. Orlygsson, and J. A. Puhakka, Ethanol and hydrogen production by two thermophilic, anaerobic bacteria isolated from icelandic geothermal areas, *Biotechnology and Bioengineering*, **101**(4), 679–690, 2008.
3. Diab, E. A., Phytoremediation of polycyclic aromatic hydrocarbons (PAHs) in a polluted desert soil, with special reference to the biodegradation of the carcinogenic PAHs, *Australian Journal of Basic and Applied Sciences*, **2**, 757–762, 2008.
4. Jetten, M. S. M., The microbial nitrogen cycle, *Environmental Microbiology*, **10**, 2903–2909, 2008.
5. Pereyra, L. P., S. R. Hiibel, A. Pruden, and K. F. Reardon, Comparison of microbial community composition and activity in sulfate-reducing batch systems remediating mine drainage, *Biotechnology and Bioengineering*, **101**, 702–713, 2008.
6. Scheutz, C., N. D. Durant, P. Dennis, M. H. Hansen, T. Joergensen, R. Jakobsen, E. E. Cox, and P. L. Bjerg, Concurrent ethene generation and growth of dehalococcoides containing vinyl chloride reductive dehalogenase genes during an enhanced reductive dechlorination field demonstration, *Environmental Science and Technology*, **42**, 9302–9309, 2008.

7. Nizhegorodov, S. Yu., S. A. Voloskov, V. A. Trusov, L. M. Kaputkina, and T. A. Syur, Corrosion of steels due to the action of microorganisms, *Metal Science and Heat Treatment*, **50**, 191–195, 2008.

SUPPLEMENTARY REFERENCES

Amsler, C. D., Ed., *Algal Chemical Ecology*, Springer, Berlin, 2008.
Bitton, G., *Wastewater Microbiology*, 3rd ed., Wiley-Liss, New York, 2005.
Bitton, G., *Encyclopedia of Environmental Microbiology*, Wiley, New York, 2002.
Chapelle, F. H., *Ground-Water Microbiology and Geochemistry*, 2nd ed., Wiley, New York, 2001.
Crawford, R. L. and C. J. Hurst, Eds, *Manual of Environmental Microbiology*, ASM Press, Washington, DC, 2002.
Deacon, J. W., *Introduction to Modern Mycology*, Blackwell Science Inc., Cambridge, MA, 1997.
Gerardi, M. H., *Microscopic Examination of the Activated Sludge Process*, Wiley, Hoboken, NJ, 2008.
Gerardi, M. H., *Wastewater Bacteria*, Wiley-Interscience, Hoboken, NJ, 2006.
Glymph, T., *Wastewater Microbiology: A Handbook for Operators*, American Water Works Association, Denver, 2005.
Graham, L. E., L. W. Wilcox, and J. Graham, *Algae*, 2nd ed., Pearson/Benjamin Cummings, San Francisco, 2009.
Granéli, E. and J. T. Turner, Eds, *Ecology of Harmful Algae*, Springer, Berlin, 2008.
Howard, A. D., Ed., *Algal Modelling: Processes and Management*, Kluwer Academic Publishing, the Netherlands, 1999.
Kim, M.-B., Ed., *Progress in Environmental Microbiology*, Nova Biomedical Books, New York, 2008.
Leadbetter, J. R., Ed., *Environmental Microbiology*, Elsevier Academic Press, Amsterdam, 2005.
Lee, R. E., *Phycology*, 4th ed., Cambridge University Press, New York, 2008.
León, R., A. Galván, and E. Fernández, *Transgenic Microalgae as Green Cell Factories*, Springer, New York, 2007.
Madigan, M. T. and J. M. Martinko, *Brock Biology of Microorganisms*, 11th ed., Prentice Hall, Upper Saddle River, NJ, 2006.
Mara, D., Ed., *Handbook of Water and Wastewater Microbiology*, Academic Press, San Diego, CA, 2003.
McKinney, R. E., *Microbiology for Sanitary Engineers*, McGraw-Hill, New York, 1962.
Mitchell, R., *Water Pollution Microbiology*, Vol. 1, Wiley-Interscience, New York, 1970.
Mitchell, R., *Water Pollution Microbiology*, Vol. 2, Wiley-Interscience, New York, 1978.
Mohandas, A. and I. S. Bright Singh, Eds, *Frontiers in Applied Environmental Microbiology*, A.P.H. Publishing Corporation, New Delhi, 2002.
Nester, E. W., *Microbiology: A Human Perspective*, 4th ed., McGraw-Hill, Boston, 2004.
Sigee, D. C., *Freshwater Microbiology: Biodiversity and Dynamic Interactions of Microorganisms in the Aquatic Environment*, Wiley, Hoboken, NJ, 2005.
Sutton, B., Ed., *A Century of Mycology*, Cambridge University Press, New York, 1996.
Talaro, K. P., *Foundations in Microbiology: Basic Principles*, 7th ed., McGraw-Hill, Dubuque, IA, 2009.
Varnam, A. H. and M. G. Evans, *Environmental Microbiology*, Manson, London, 2000.
Wheelis, M. L., *Principles of Modern Microbiology*, Jones and Bartlett Publishers, Sudbury, MA, 2008.

QUESTIONS AND PROBLEMS

The use of internet resources is assumed in answering any of the questions. These would include such things as constants and conversion factors as well as additional information needed to complete an answer.

1. As $CH_3CH_2CH_2CH_2CO_2H$ biodegrades in several steps to carbon dioxide and water, various chemical species are observed. What stable chemical species would be observed as a result of the first step of this degradation process?
2. Which of the following statements is true regarding the production of methane in water: (a) it occurs in the presence of oxygen, (b) it consumes oxygen, (c) it removes biological oxygen demand from the water, (d) it is accomplished by oxic bacteria, (e) it produces more energy per electron-mole than does oxic respiration.

3. At the time zero, the cell count of a bacterial species mediating oxic respiration of wastes was 1×10^6 cells/L. At 30 min it was 2×10^6; at 60 min it was 4×10^6; at 90 min, 7×10^6; at 120 min, 10×10^6; and at 150 min, 13×10^6. From these data, which of the following logical conclusions would you draw? (a) The culture was entering the log phase at the end of the 150-min period, (b) the culture was in the log phase throughout the 150-min period, (c) the culture was leaving the log phase at the end of the 150-min period, (d) the culture was in the lag phase throughout the 150-min period, (e) the culture was in the death phase throughout the 150-min period.
4. What may be said about the biodegradability of a hydrocarbon containing the following structure?

$$----C-\overset{\displaystyle CH_3}{\underset{\displaystyle CH_3}{\overset{|}{\underset{|}{C}}}}-CH_3$$

5. Suppose that the anoxic fermentation of organic matter, $\{CH_2O\}$, in water yields 15.0 L of CH_4 (at standard temperature and pressure). How many grams of oxygen would be consumed by the oxic respiration of the same quantity of $\{CH_2O\}$? (Recall the significance of 22.4 L in chemical reaction of gases.)
6. What mass of $FeCO_3(s)$, using Reaction (A) + (4) in Table 6.1, gives the same free energy yield as 1.00 g of organic matter, using Reaction (A) + (1), when oxidized by oxygen at pH 7.00?
7. How many bacteria would be produced after 10 h by one bacterial cell, assuming exponential growth with a generation time of 20 min?
8. Referring to Reaction 6.18, calculate the concentration of ammonium ion in equilibrium with oxygen in the atmosphere and 1.00×10^{-5} M NO_3^- at pH 7.00.
9. When a bacterial nutrient medium is inoculated with bacteria grown in a markedly different medium, the lag phase (Figure 6.4) often is quite long, even if the bacteria eventually grow well in the new medium. Can you explain this behavior?
10. Most plants assimilate nitrogen as nitrate ion. However, ammonia (NH_3) is a popular and economical fertilizer. What essential role do bacteria play when ammonia is used as a fertilizer? Do you think any problems might occur when using ammonia in a waterlogged soil lacking oxygen?
11. Why is the growth rate of bacteria as a function of temperature (Figure 6.7) not a symmetrical curve?
12. Discuss the analogies between bacteria and a finely divided chemical catalyst.
13. Would you expect autotrophic bacteria to be more complex physiologically and biochemically than heterotrophic bacteria? Why?
14. Wastewater containing 8 mg/L O_2 (atomic mass O = 16), 1.00×10^{-3} M NO_3^-, and 1.00×10^{-2} M soluble organic matter, $\{CH_2O\}$, is stored isolated from the atmosphere in a container richly seeded with a variety of bacteria. Assume that denitrification is one of the processes that will occur during storage. After the bacteria have had a chance to do their work, which of the following statements will be true? (a) No $\{CH_2O\}$ will remain, (b) some O_2 will remain, (c) some NO_3^- will remain, (d) denitrification will have consumed more of the organic matter than oxic respiration, (e) the composition of the water will remain unchanged.
15. Of the four classes of microorganisms—algae, fungi, bacteria, and virus—which has the least influence on water chemistry?
16. Figure 6.3 shows the main structural features of a bacterial cell. Which of these do you think might cause the most trouble in water-treatment processes such as filtration or ion exchange, where the maintenance of a clean, unfouled surface is critical? Explain.

17. A bacterium capable of degrading 2,4-D herbicide was found to have its maximum growth rate at 32°C. Its growth rate at 12°C was only 10% of the maximum. Do you think there is another temperature at which the growth rate would also be 10% of the maximum? If you believe this to be the case, of the following temperatures, choose the one at which it is most plausible for the bacterium to also have a growth rate of 10% of the maximum: 52°C, 37°C, 8°C, and 20°C.
18. The day after a heavy rain washed a great deal of cattle feedlot waste into a farm pond, the following counts of bacteria were obtained:

Time	Thousands of Viable Cells/mL	Time	Thousands of Viable Cells/mL
6:00 a.m.	0.10	11:00 a.m.	0.40
7:00 a.m.	0.11	12:00 Noon	0.80
8:00 a.m.	0.13	1:00 p.m.	1.60
9:00 a.m.	0.16	2:00 p.m.	3.20
10:00 a.m.	0.20		

To which portion of the bacterial growth curve, Figure 6.3, does this time span correspond?
19. Addition of which two half-reactions in Table 6.1 is responsible for (a) elimination of an algal nutrient in secondary sewage effluent using methanol as a carbon source, (b) a process responsible for a bad-smelling pollutant when bacteria grow in the absence of oxygen, (c) a process that converts a common form of commercial fertilizer to a form that most crop plants can absorb, (d) a process responsible for the elimination of organic matter from wastewater in the aeration tank of an activated sludge sewage-treatment plant, (e) a characteristic process that occurs in the anoxic digester of a sewage treatment plant.
20. What is the surface area in square meters of 1.00 g of spherical bacterial cells, 1.00 μm in diameter, having a density of 1.00 g/cm^3?
21. What is the purpose of exoenzymes in bacteria?
22. Match each species of bacteria listed in the left column with its function on the right.

a. *Spirillumlipoferum*	1. Reduces sulfate to H_2S
b. *Rhizobium*	2. Catalyzes oxidation of Fe^{2+} to Fe^{3+}
c. *Thiobacillusferrooxidans*	3. Fixes nitrogen in grasses
d. *Desulfovibrio*	4. On legume roots

23. What factors favor the production of methane in anoxic surroundings?
24. Below are listed three kinds of microorganisms, and below that are listed with numbers several chemical species or energy sources. In parentheses to the *left* of each kind of microorganism, place the numbers corresponding to at least *two* things that the microorganism might *need* or *use*. In parentheses to the *right* of each kind of microorganism, place the numbers corresponding to at least *two* things that the microorganism might produce.
( ) Algae ( )
( ) Aerobic, nonphotosynthetic, autotrophic *Gallionella* bacteria ( )
( ) Anaerobic, heterotrophic bacteria ( )

1. CO_2, 2. $h\nu$, 3. O_2, 4. $\{CH_2O\}$, 5. CH_4, 6. $Fe(OH)_3$, 7. electron acceptor other than O_2.

25. Bacteria growing in the log phase on a waste in water were assayed at a particular time, t_0, and found to number 3.01×10^5 cells/mL. At 90 min after t_0 the count was 2.41×10^6 cells/mL. The best estimate of the following of the bacterial population at 60 min after t_0 is (A) 4.20×10^5, (B) 8.25×10^5, (C) 6.48×10^5, (D) 1.20×10^6, and (E) 3.21×10^6.
26. Consider a type of bacteria that derive energy by mediating the oxidation of sulfides, such as H_2S, FeS, or FeS_2, with molecular oxygen, O_2. Such bacteria (A) are most likely

heterotrophic, (B) cannot be autotrophic, (C) should be thermophilic, (D) should be acid-tolerant, (E) cannot exist.

27. The growth rate of bacteria, "*G*," was plotted versus an unidentified parameter, "*X*," giving the following values in which both *G* and *X* are in arbitrary units:

X	*G*	*X*	*G*
5	100	30	600
10	200	35	700
15	300	40	700
20	400	45	350
25	500	50	25

Based on the above data *X* is most likely (A) time, showing the bacteria with a generation time of 5 units of *X*, (B) nutrient concentration, (C) waste product, (D) pH, (E) temperature.

28. Of the following, the *untrue* statement pertaining to biodegradation of organic matter is (A) epoxidation consists of adding an oxygen atom between two carbon atoms, (B) oxidation of hydrocarbon chains tends to occur two carbon atoms at a time, (C) esterases are a specific category of hydrolase ions, (D) carbon atoms bonded to three or four other carbon atoms are especially susceptible to microbial epoxidation, (E) exoenzymes are involved in the biodegradation of cellulose.

29. Of the following, the *untrue* statement is (A) all fungi and protozoans are chemoheterotrophs, (B) photoheterotrophs that use photoenergy, but are dependent on organic matter as a carbon source, are especially abundant and widespread, (C) chemoautotrophs use CO_2 for biomass and oxidize substances such as NH_4^+ for energy, (D) algae are photoautotrophs, (E) some bacteria, such as cyanobacteria, perform photosynthesis.

30. The reaction, $4FeS(s) + 9O_2 + 10H_2O \rightarrow 4Fe(OH)_3(s) + 4SO_4^{2-} + 8H^+$, (A) illustrates sulfate reduction, (B) is a means for autotrophic *Gallionella* bacteria to obtain energy, (C) illustrates the action of heterotrophic bacteria, (D) is not mediated by bacteria, (E) is written for one electron-mole.

7 Water Pollution

7.1 NATURE AND TYPES OF WATER POLLUTANTS

Throughout history, the quality of drinking water has been a factor in determining human welfare. Fecal pollution of drinking water has frequently caused waterborne diseases that have decimated the populations of whole cities. Unwholesome water polluted by sewage has caused great hardship for people forced to drink it or use it for irrigation. Although waterborne diseases are now well controlled in technologically advanced countries, the shortage of safe drinking water is still a major problem in regions afflicted by strife and poverty.

An ongoing concern with water safety now is the potential presence of chemical pollutants. These may include organic chemicals, inorganics, and heavy metals from industrial, urban runoff, and agricultural sources. Water pollutants can be divided into some general categories, as summarized in Table 7.1. Most of these categories of pollutants, and several subcategories, are discussed in this chapter.

7.1.1 Markers of Water Pollution

Markers of water pollution are substances that show the presence of pollution sources. These include herbicides indicative of agricultural runoff, fecal coliform bacteria that are characteristic of pollution from sewage, and pharmaceuticals, pharmaceutical metabolites, and even caffeine that show contamination by domestic wastewater.

Biomarkers of water pollution are organisms that live in or are closely associated with bodies of water and provide evidence of pollution either from accumulation of water pollutants or their metabolites or from effects on the organism due to pollutant exposure. Fish are the most common bioindicators of water pollution, and fish lipid (fat) tissue is commonly analyzed for persistent organic water pollutants.

An organism that has been described as "a worldwide sentinel species to assess and monitor environmental pollution in rivers, lakes, reservoirs, and estuaries"[1] is the osprey (*Pandionhaliaetus*), a large raptor bird with a wingspan that can exceed 1.5 m and a mass up to 2 kg. Found in all continents throughout the world except for Antarctica, the osprey feeds almost exclusively on fish. In addition to these characteristics, other aspects of the osprey that make it a good indicator species are that it is well adapted to human landscapes where pollution is most likely to occur, it is at the top of the aquatic food web through which persistent pollutants undergo bioaccumulation and biomagnification, it is sensitive to many pollutants, and it has a relatively long life span. This bird builds highly visible nests that are evenly dispersed over wide areas, usually remains with a single nest, and tolerates disturbance of nests for brief periods. The osprey is very sensitive to some pollutants, and although populations are now comfortably high, they were almost wiped out by the effects of DDT before that insecticide was banned. Chemical and biochemical analyses of osprey feathers, eggs, blood, and organs along with observations of behavior, nesting habits, and populations have been used to assess water pollution.

7.2 ELEMENTAL POLLUTANTS

Table 7.2 lists the more important *trace elements* (those encountered at levels of a few ppm or less) found in natural waters. Typical of the behavior of many substances in the aquatic environment,

TABLE 7.1
General Types of Water Pollutants

Class of Pollutant	Significance
Trace elements	Health, aquatic biota, toxicity
Heavy metals	Health, aquatic biota, toxicity
Organically-bound metals	Metal transport
Radionuclides	Toxicity
Inorganic pollutants	Toxicity, aquatic biota
Asbestos	Human health
Algal nutrients	Eutrophication
Acidity, alkalinity, salinity (in excess)	Water quality, aquatic life
Trace organic pollutants	Toxicity
Polychlorinated biphenyls	Possible biological effects
Pesticides	Toxicity, aquatic biota, wildlife
Petroleum wastes	Effect on wildlife, esthetics
Sewage, human, and animal wastes	Water quality, oxygen levels
Biochemical oxygen demand	Water quality, oxygen levels
Pathogens	Health effects
Detergents	Eutrophication, wildlife, esthetics
Chemical carcinogens	Incidence of cancer
Sediments	Water quality, aquatic biota, wildlife
Taste, odor, and color	Esthetics

TABLE 7.2
Important Trace Elements in Natural Waters

Element	Sources	Effects and Significance
Arsenic	Mining by-product, chemical waste	Toxic[a], possibly carcinogenic
Beryllium	Coal, industrial wastes	Toxic
Boron	Coal, detergents, wastes	Toxic
Chromium	Metal plating	Essential as Cr(III), toxic as Cr(VI)
Copper	Metal plating, mining, industrial waste	Essential trace element, toxic to plants and algae at higher levels
Fluorine (F^-)	Natural geological sources, wastes, water additive	Prevents tooth decay at around 1 mg/L, toxic at higher levels
Iodine (I^-)	Industrial wastes, natural brines, seawater intrusion	Prevents goiter
Iron	Industrial wastes, corrosion, acid mine water, microbial action	Essential nutrient, damages fixtures by staining
Lead	Industrial waste, mining, fuels	Toxic, harmful to wildlife
Manganese	Industrial wastes, acid mine water microbial action	Toxic to plants, damages fixtures, by staining
Mercury	Industrial waste, mining, coal	Toxic, mobilized as methyl mercury compounds by anaerobic bacteria
Molybdenum	Industrial wastes, natural sources	Essential to plants, toxic to animals
Selenium	Natural sources, coal	Essential at lower levels, toxic at higher levels
Zinc	Industrial waste, metal plating, plumbing	Essential element, toxic to plants at higher levels

[a] Toxicities of these elements are discussed in Chapter 23.

some of these are essential plant and animal nutrients at low levels but toxic at higher levels. Several of them, such as lead or mercury, have such toxicological and environmental significance that they are discussed in detail in separate sections.

A few of the *heavy metals* are among the most harmful of the elemental pollutants and are of particular concern because of their toxicities to humans. These elements are in general the transition metals, and some of the representative elements, such as lead and tin, are placed in the lower right-hand corner of the periodic table. Heavy metals include essential elements like iron as well as toxic metals like cadmium and mercury. Most of them have a tremendous affinity for sulfur and disrupt enzyme function by forming bonds with sulfur groups in enzymes. Protein carboxylic acid ($-CO_2H$) and amino ($-NH_2$) groups are also chemically bound by heavy metals. Cadmium, copper, lead, and mercury ions bind to cell membranes, hindering transport processes through the cell wall. Heavy metals may also precipitate phosphate biocompounds or catalyze their decomposition. The biochemical effects of metals are discussed in Chapter 24.

Some of the *metalloids*, elements on the borderline between metals and nonmetals, are significant water pollutants. Arsenic, selenium, and antimony are of particular interest.

Inorganic chemicals' manufacture has the potential to contaminate water with trace elements. Among the industries regulated for potential trace element pollution of water are those producing chlor-alkali, hydrofluoric acid, sodium dichromate (sulfate process and chloride ilmenite process), aluminum fluoride, chrome pigments, copper sulfate, nickel sulfate, sodium bisulfate, sodium hydrosulfate, sodium bisulfite, titanium dioxide, and hydrogen cyanide.

7.3 HEAVY METALS

7.3.1 Cadmium

Pollutant *cadmium* in water may arise from mining wastes and industrial discharges, especially from metal plating. Chemically, cadmium is very similar to zinc, and these two metals frequently undergo geochemical processes together. Both metals are found in water in the +2 oxidation state.

The effects of acute cadmium poisoning in humans include high blood pressure, kidney damage, damage to testicular tissue, and destruction of red blood cells. Much of the physiological action of cadmium is due to its chemical similarity to zinc. Cadmium may replace zinc in some enzymes, thereby altering the stereostructure of the enzyme and impairing its catalytic activity, causing disease symptoms.

Cadmium and zinc are common water and sediment pollutants in harbors surrounded by industrial installations. Concentrations of more than 100 ppm dry weight sediment have been found in harbor sediments. Typically, during periods of calm in the summer when the water stagnates, the anaerobic bottom layer of harbor water has a low soluble Cd concentration because microbial reduction of sulfate by organic matter, $\{CH_2O\}$, produces sulfide, which precipitates cadmium as insoluble cadmium sulfide:

$$2\{CH_2O\} + SO_4^{2-} + H^+ \rightarrow 2CO_2 + HS^- + 2H_2O \quad (7.1)$$

$$CdCl^+(\text{chloro complex in seawater}) + HS^- \rightarrow CdS(s) + H^+ + Cl^- \quad (7.2)$$

Mixing of bay water from outside the harbor and harbor water by high winds during the winter results in desorption of cadmium from harbor sediments by aerobic bay water. This dissolved cadmium is carried out into the bay where it is absorbed by suspended solid materials, which then becomes incorporated with the bay sediments. This is an example of the sort of complicated interaction of hydraulic, chemical solution–solid, and microbiological factors involved in the transport and distribution of a pollutant in an aquatic system.

7.3.2 Lead

Inorganic *lead* arising from a number of industrial and mining sources and formerly from leaded gasoline occurs in water in the +2 oxidation state. In addition to pollutant sources, lead-bearing limestone and galena (PbS) contribute lead to natural waters in some locations. Evidence from hair samples and other sources indicates that body burdens of this toxic metal have decreased during recent decades, largely the result of less lead used in plumbing and other products that come in contact with food or drink.

Acute lead poisoning in humans causes severe dysfunction in the kidneys, reproductive system, liver, and the brain and central nervous system and results in sickness or death. Lead poisoning from environmental exposure is thought to have caused mental retardation in many children. Mild lead poisoning causes anemia. The victim may have headaches and sore muscles, and may feel generally fatigued and irritable.

Except in isolated cases, lead is probably not a major problem in drinking water, although the potential exists in cases where old lead pipes are still in use. Lead used to be a constituent of solder and some pipe-joint formulations, such that household water does have some contact with lead. Water that has stood in household plumbing for some time may have elevated levels of lead (along with zinc, cadmium, and copper) and should be drained for a while before use.

7.3.3 Mercury

Because of its toxicity, mobilization as methylated forms by anaerobic bacteria, and other pollution factors, mercury generates a great deal of concern as a heavy-metal pollutant. Mercury is found as a trace component of many minerals, with continental rocks containing an average of around 80 parts per billion, or slightly less, of this element. Fossil fuel coal and lignite contain mercury, often at levels of 100 parts per billion or even higher, and emissions from the combustion of these fuels are a major source of environmental mercury.

Metallic mercury once was commonly used as an electrode in the electrolytic generation of chlorine gas, in laboratory vacuum apparatus, and in other applications, and significant quantities of inorganic mercury(I) and mercury(II) compounds were used annually. Organic mercury compounds used to be widely applied as pesticides, particularly fungicides. These mercury compounds included aryl mercurials such as phenyl mercuric dimethyldithiocarbamate,

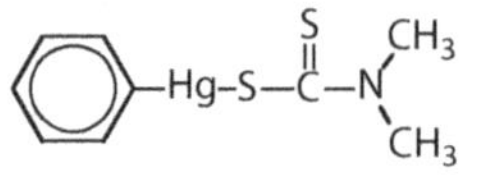

$$C_6H_5\text{—}Hg\text{—}S\text{—}\overset{\overset{\displaystyle S}{\|}}{C}\text{—}N(CH_3)_2$$

(formerly used in paper mills as a slimicide and as a mold retardant for paper) and alkyl-mercurials such as ethylmercuric chloride, C_2H_5HgCl, that was used as a seed fungicide. Because of concern over the health and environmental effects of mercury, these uses have been severely curtailed in recent years. As of 2008, approximately 3800 metric tons of mercury were traded worldwide each year.

Globally, one of the greatest sources of mercury pollution has been mercury used to extract gold from gold-bearing ores. It is estimated that each year 15 million miners in approximately 40 developing countries use 650–1000 metric tons of mercury to extract gold. This use subjects many local areas to mercury contamination and contributes significantly to the global environmental burden of mercury as well as exposing the miners, many of whom are children, to toxic mercury. The problem has been made worse in recent years as industrialized countries have reduced mercury use with the result that surplus mercury has been made available on the world market for gold extraction. In recognition of this problem, in 2008 the European Union adopted a mercury export ban to take effect in 2011 and the United States adopted a similar ban to be effective in 2013.

The toxicity of mercury was tragically illustrated in the Minamata Bay area of Japan during the period 1953–1960. A total of 111 cases of mercury poisoning and 43 deaths were reported among people who had consumed seafood from the bay that had been contaminated with mercury waste from a chemical plant that drained into Minamata Bay. Congenital defects were observed in

19 babies whose mothers had consumed seafood contaminated with mercury. The level of metal in the contaminated seafood was 5–20 ppm.

Among the toxicological effects of mercury are neurological damage, including irritability, paralysis, blindness, or insanity; chromosome breakage; and birth defects. The milder symptoms of mercury poisoning such as depression and irritability have a psychopathological character. Because of the resemblance of these symptoms to common human behavioral problems, mild mercury poisoning may escape detection. Some forms of mercury are relatively nontoxic and were formerly used as medicines, for example, in the treatment of syphilis. Other forms of mercury, particularly organic compounds, are highly toxic.

Because there are few major natural sources of mercury, and since most inorganic compounds of this element are relatively insoluble, it was assumed for some time that mercury was not a serious water pollutant. However, in 1970, alarming mercury levels were discovered in fish in Lake Saint Clair located between Michigan and Ontario, Canada. A subsequent survey by the U.S. Federal Water Quality Administration revealed a number of other waters contaminated with mercury. It was found that several chemical plants, particularly caustic chemical manufacturing operations, were each releasing up to 14 or more kilograms of mercury in wastewaters each day.

The unexpectedly high concentrations of mercury found in water and in fish tissues result from the formation of soluble monomethylmercury ion, CH_3Hg^+, and volatile dimethylmercury, $(CH_3)_2Hg$, by anaerobic bacteria in sediments. Mercury from these compounds becomes concentrated in fish lipid (fat) tissue and the concentration factor from water to fish may exceed 10^3. The methylating agent by which inorganic mercury is converted to methylmercury compounds is methylcobalamin, a vitamin B_{12} analog:

$$HgCl_2 \xrightarrow{\text{Methylcobalamin}} CH_3HgCl + Cl^- \tag{7.3}$$

It is believed that the bacteria that synthesize methane produce methylcobalamin as an intermediate in the synthesis. Thus, waters and sediments in which anaerobic decay is occurring provide the conditions under which methylmercury production occurs. In neutral or alkaline waters, volatile dimethylmercury, $(CH_3)_2Hg$, may be formed.

7.4 METALLOIDS

The most significant water pollutant metalloid element is arsenic, a toxic element that has been the chemical villain of more than a few murder plots. Acute arsenic poisoning can result from the ingestion of more than about 100 mg of the element. Chronic poisoning occurs with the ingestion of small amounts of arsenic over a long period of time. There is some evidence that this element is also carcinogenic.

Arsenic occurs in the Earth's crust at an average level of 2–5 ppm. The combustion of fossil fuels, particularly coal, introduces large quantities of arsenic into the environment, much of it reaching natural waters. Arsenic occurs with phosphate minerals and enters into the environment along with some phosphorus compounds. Some formerly used pesticides, particularly those from before World War II, contain highly toxic arsenic compounds. The most common of these are lead arsenate, $Pb_3(AsO_4)_2$; sodium arsenite, Na_3AsO_3; and Paris green, $Cu_3(AsO_3)_2$. Another major source of arsenic is mine tailings. Arsenic produced as a by-product of copper, gold, and lead refining exceeds the commercial demand for arsenic, and it accumulates as waste material.

Like mercury, arsenic may be converted to more mobile and toxic methyl derivatives by bacteria, first by reduction of H_3AsO_4 to H_3AsO_3 then by methylation to produce $CH_3AsO(OH)_2$ (methylarsinic acid), $(CH_3)_2AsO(OH)$ (dimethyarsinic acid), and $(CH_3)_2AsH$ (dimethylarsine).

In what may have been the largest mass poisoning of a human population in history, between 35 million and 77 million people of the 125 million inhabitants of Bangladesh have been exposed to potentially toxic levels of arsenic in drinking water. This catastrophic public health problem has resulted from well-intentioned programs funded initially by the United Nations Children's Fund to install shallow tube wells that provided a source of drinking water free from disease-causing pathogens. By 1987, numerous cases of arsenic-induced skin lesions characterized by pigmentation changes, predominantly on the upper chest, arms, and legs, and keratoses of the palms of the hands and soles of the feet were being observed. These effects were characteristic of arsenic poisoning, leading to the discovery that arsenic-contaminated drinking water from the tube wells was responsible for this. Since the initial discovery of arsenic poisoning, huge numbers of new cases have been revealed and it is possible that tens of thousands of people in Bangladesh will die prematurely because of exposure to arsenic in drinking water. Other countries with acute problems from arsenic in drinking water include Vietnam (where large numbers of tube wells were drilled more recently than those in Bangladesh), Argentina, Chile, China, Mexico, Taiwan, and Thailand.

The geochemical conditions that result in arsenic contamination of water are often associated with the presence of iron, sulfur, and organic matter in (alluvial) deposits produced by water. Iron released from rocks eroded by river water forms iron oxide deposits on rock particle surfaces. The iron oxide accumulates arsenic and concentrates it from the river water. These particles then get buried along with degradable organic matter in sediments, and the insoluble iron(III) in the iron oxides is converted to soluble iron(II) by the anaerobic reducing conditions under which the organic matter biodegrades. This releases the bound arsenic that can get into well water.

7.5 ORGANICALLY BOUND METALS AND METALLOIDS

An appreciation of the strong influence of complexation and chelation on heavy metals' behavior in natural waters and wastewaters may be gained by reading Sections 3.10 through 3.17, which deal with that subject. Methylmercury formation is discussed in Section 7.3. Both topics involve the combination of metals and organic entities in water. It must be stressed that the interaction of metals with organic compounds is of utmost importance in determining the role played by the metal in an aquatic system.

There are two major types of metal–organic interactions to be considered in an aquatic system. The first of these is complexation, usually chelation when organic ligands are involved. As discussed in Chapter 3, complexation and chelation involve reversible binding of a metal ion with a ligand species.

Organometallic compounds, on the other hand, contain metals bound to organic entities by way of a carbon atom and do not dissociate reversibly at lower pH or greater dilution. Furthermore, the organic component, and sometimes the particular oxidation state of the metal involved, may not be stable apart from the organometallic compound. The major categories of organometallic compounds encountered in the environment are (1) those in which the organic group is an alkyl group such as ethyl in tetraethyl lead, $Pb(C_2H_5)_4$; (2) carbonyls, some of which are quite volatile and toxic, having carbon monoxide bonded to metals; and (3) those in which the organic group is a π electron donor, such as ethylene (C_2H_4) or benzene (C_6H_6).

Combinations of the three general types of compounds outlined above exist, the most prominent of which are arene carbonyl species, in which a metal atom is bonded to both an aryl entity such as benzene and to several carbon monoxide molecules.

A large number of compounds exist that have at least one bond between the metal and a C atom on an organic group, as well as other covalent or ionic bonds between the metal and atoms other than carbon. An example is monomethylmercury chloride, CH_3HgCl, in which the organometallic CH_3Hg^+ ion is ionically bonded to the chloride anion. Another class of compounds with organometallic character are those that have organic groups bonded to a metal atom through atoms other than carbon. An example of such a compound is isopropyl titanate, $Ti(i\text{-}OC_3H_7)_4$, in which the hydrocarbon groups are bonded to the metal by way of oxygen atoms.

The interaction of trace metals with organic compounds in natural waters is too vast an area to cover in detail in this chapter; however, it may be noted that metal–organic interactions may involve organic species of both pollutant (such as EDTA) and natural (such as fulvic acids) origin. These interactions are influenced by, and sometimes play a role in, redox equilibria, formation and dissolution of precipitates, colloid formation and stability, acid–base reactions, and microorganism-mediated reactions in water. Metal–organic interactions may increase or decrease the toxicity of metals in aquatic ecosystems, and they have a strong influence on the growth of algae in water.

7.5.1 Organotin Compounds

Of all the metals, tin has had the greatest number of organometallic compounds in commercial use, with global production that reached 40,000 metric tons per year before uses were restricted because of concerns over water pollution. In addition to synthetic organotin compounds, methylated tin species can be produced biologically in the environment. Figure 7.1 gives some examples of the many known organotin compounds.

Major industrial uses of organotin compounds in the past have included applications of tin compounds in fungicides, acaricides, disinfectants, antifouling paints, stabilizers to lessen the effects of heat and light in polyvinylchloride (PVC) plastics, catalysts, and precursors for the formation of films of SnO_2 on glass. Tributyl tin (TBT) chloride and related TBT compounds have bactericidal, fungicidal, and insecticidal properties and formerly were of particular environmental significance because of their use as industrial biocides. In addition to TBT chloride, other TBT compounds used as biocides include the hydroxide, the naphthenate, bis(tributyltin) oxide, and tris(tributylstannyl) phosphate. TBT was once widely used in boat and ship hull coatings to prevent the growth of fouling organisms. Other applications have included preservation of wood, leather, paper, and textiles. Antifungal TBT compounds have been used as slimicides in cooling tower water.

Obviously the many applications of organotin compounds for a variety of uses have posed a significant potential for environmental pollution. Because of their applications near or in contact with bodies of water, organotin compounds are potentially significant water pollutants and have been linked to endocrine disruption in shellfish, oysters, and snails. Because of such concerns, several countries, including the United States, England, and France, prohibited TBT application on vessels smaller than 25 m in length during the 1980s. In response to concerns over water pollution, in 2001 the International Maritime Organization agreed to ban organotin antifouling paints on ships and yachts. Provisions of this agreement came into effect in September 2008.

7.6 INORGANIC SPECIES

Some important inorganic water pollutants were mentioned in Sections 7.2 through 7.4 as part of the discussion of pollutant trace elements. Inorganic pollutants that contribute acidity, alkalinity, or salinity to water are considered separately in this chapter. Still another class is that of algal nutrients. This leaves unclassified, however, some important inorganic pollutant species, of which cyanide ion, CN^-, is probably the most important. Others include ammonia, carbon dioxide, hydrogen sulfide, nitrite, and sulfite.

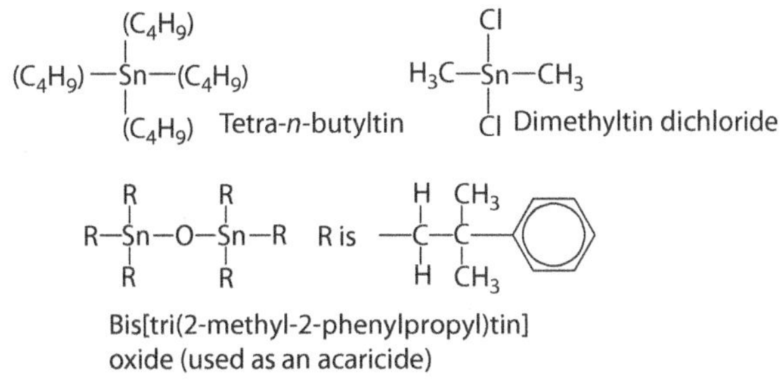

FIGURE 7.1 Examples of organotin compounds.

7.6.1 Cyanide

Cyanide, a deadly poisonous substance, exists in water as HCN, a weak acid, K_a of 6×10^{-10}. The cyanide ion has a strong affinity for many metal ions, forming relatively less toxic ferrocyanide, $Fe(CN)_6^{4-}$, with iron(II), for example. Volatile HCN is very toxic and has been used in gas chamber executions in the United States.

Cyanide is widely used in industry, especially for metal cleaning and electroplating. It is also one of the main gas and coke scrubber effluent pollutants from gas works and coke ovens. Cyanide is also used in certain mineral-processing operations. Numerous fish kills have resulted from the discharge of cyanide from mineral processing operations into waterways.

7.6.2 Ammonia and Other Inorganic Pollutants

Excessive levels of ammoniacal nitrogen cause water-quality problems. *Ammonia* is the initial product of the decay of nitrogenous organic wastes, and its presence frequently indicates the presence of such wastes. It is a normal constituent of low-pE groundwaters and is sometimes added to drinking water as an aid to disinfection, where it reacts with chlorine to provide residual chlorine (see Section 8.11). Since the pK_a of ammonium ion, NH_4^+, is 9.26, most ammonia in water is present as NH_4^+ rather than as NH_3.

Hydrogen sulfide, H_2S, is a product of the anaerobic decay of organic matter containing sulfur. It is also produced in the anaerobic reduction of sulfate by microorganisms (see Chapter 6) and is evolved as a gaseous pollutant from geothermal waters. Wastes from chemical plants, paper mills, textile mills, and tanneries may also contain H_2S. Its presence is easily detected by its characteristic rotten-egg odor. In water, H_2S is a weak diprotic acid with pK_{a1} of 6.99 and pK_{a2} of 12.92; S^{2-} is not present in normal natural waters. The sulfide ion has tremendous affinity for many heavy metals, and precipitation of metallic sulfides often accompanies production of H_2S.

Free *carbon dioxide*, CO_2, is frequently present in water at high levels due to decay of organic matter. It is also added to softened water during water treatment as part of a recarbonation process (see Chapter 8). Excessive carbon dioxide levels may make water more corrosive and may be harmful to aquatic life.

Nitrite ion, NO_2^-, occurs in water as an intermediate oxidation state of nitrogen over a relatively narrow pE range. Nitrite is added to some industrial process water as a corrosion inhibitor. However, it rarely occurs in drinking water at levels over 0.1 mg/L.

Sulfite ion, SO_3^{2-}, is found in some industrial wastewaters. Sodium sulfite is commonly added to boiler feedwaters as an oxygen scavenger:

$$2SO_3^{2-} + O_2 \rightarrow 2SO_4^{2-} \tag{7.4}$$

Since pK_{a1} of sulfurous acid is 1.76 and pK_{a2} is 7.20, sulfite exists as either HSO_3^- or SO_3^{2-} in natural waters, depending on the pH. It may be noted that hydrazine, N_2H_4, also functions as an oxygen scavenger:

$$N_2H_4 + O_2 \rightarrow 2H_2O + N_2(g) \tag{7.5}$$

Perchlorate ion, ClO_4^-, emerged as a water pollution problem in some areas in the 1990s when advances in ion chromatography enabled its detection in the low parts per billion range of concentrations. Ammonium perchlorate, NH_4ClO_4, has been widely manufactured as an oxidizer in solid rocket propellants and contamination from ammonium perchlorate manufacturing facilities has been regarded as the major source of contamination. Perchlorate in water is very unreactive and all common perchlorate salts other than $KClO_4$ are soluble, so it is difficult to remove. Physiologically, it competes with iodide ion and diminishes essential iodide uptake by the thyroid. The U.S. Environmental Protection Agency has recommended a drinking water standard for perchlorate of around 1 part per billion.

7.6.3 Asbestos in Water

The toxicity of inhaled asbestos is well established. The fibers scar lung tissue and cancer eventually develops, often 20 or 30 years after exposure. It is not known for sure whether asbestos is toxic in drinking water. This has been a matter of considerable concern because of the dumping of taconite (iron ore tailings) containing asbestos-like fibers into Lake Superior. The fibers have been found in drinking waters of cities around the lake. After having dumped the tailings into Lake Superior since 1952, the Reserve Mining Company at Silver Bay on Lake Superior solved the problem in 1980 by constructing a 6-square-mile containment basin inland from the lake. This $370-million facility keeps the taconite tailings covered with a 3-m layer of water to prevent the escape of fiber dust.

7.7 ALGAL NUTRIENTS AND EUTROPHICATION

The term *eutrophication*, derived from the Greek word meaning "well-nourished," describes a condition of lakes or reservoirs involving excess algal growth. Although some algal productivity is necessary to support the food chain in an aquatic ecosystem, excess growth under eutrophic conditions may eventually lead to severe deterioration of the body of water. The first step in eutrophication of a body of water is an input of plant nutrients (Table 7.3) from watershed runoff or sewage. The nutrient-rich body of water then produces a great deal of plant biomass by photosynthesis, along with a smaller amount of animal biomass. Dead biomass accumulates in the bottom of the lake, where it partially decays, recycling nutrient carbon dioxide, phosphorus, nitrogen, and potassium. If the lake is not too deep, bottom-rooted plants begin to grow, accelerating the accumulation of solid material in the basin. Eventually a marsh is formed, which finally fills in to produce a meadow or forest.

Eutrophication is often a natural phenomenon; for instance, it is basically responsible for the formation of huge deposits of coal and peat. However, human activity can greatly accelerate the process. To understand why this is so, consider that most of the nutrients required for plant and algae growth shown in Table 7.3 are available in adequate amounts from natural sources. The nutrients most likely to be limiting are the "fertilizer" elements: nitrogen, phosphorus, and potassium. All these are present in sewage and are, of course, found in runoff from heavily fertilized fields. They are also constituents of various kinds of industrial wastes. Each of these elements can also come

TABLE 7.3
Essential Plant Nutrients: Sources and Functions

Nutrient	Source	Function
	Macronutrients	
Carbon (CO_2)	Atmosphere, decay	Biomass constituent
Hydrogen	Water	Biomass constituent
Oxygen	Water	Biomass constituent
Nitrogen (NO_3^-)	Decay, pollutants, atmosphere	Protein constituent (from nitrogen-fixing organisms)
Phosphorus	Decay, minerals (phosphate)	DNA/RNA constituent pollutants
Potassium	Minerals, pollutants	Metabolic function
Sulfur (sulfate)	Minerals	Proteins, enzymes
Magnesium	Minerals	Metabolic function
Calcium	Minerals	Metabolic function
	Micronutrients	
B, Cl, Co, Cu, Fe, Mo, Mn, Na, Si, V, Zn	Minerals, pollutants	Metabolic function and/or constituent of enzymes

from natural sources—phosphorus and potassium from mineral formations, and nitrogen fixed by bacteria, cyanobacteria, or discharge of lightning in the atmosphere.

In some cases, nitrogen or even carbon may be the limiting nutrients, the presence of which determines the rate of algal growth. This is particularly true of nitrogen in seawater. In seawater, micronutrients, particularly iron, may be limiting.

In most cases in freshwaters, the single plant nutrient most likely to be limiting is phosphorus, and it is generally regarded as the culprit in excessive eutrophication. Household detergents were once a common source of phosphate in wastewater, and eutrophication control has concentrated on eliminating phosphates from detergents, removing phosphate at the sewage treatment plant, and preventing phosphate-laden sewage effluents from entering bodies of water enabling the excessive growth of algae that can cause eutrophication.

7.8 ACIDITY, ALKALINITY, AND SALINITY

Aquatic biota are sensitive to extremes of pH. Largely because of osmotic effects, they cannot live in a medium having a salinity to which they are not adapted. Thus, a freshwater fish soon succumbs in the ocean, and sea fish normally cannot live in freshwater. Excess salinity soon kills plants not adapted to it. There are, of course, ranges in salinity and pH in which organisms live. As shown in Figure 7.2, these ranges frequently may be represented by a reasonably symmetrical curve, along the fringes of which an organism may live without really thriving. These curves do not generally exhibit a sharp cutoff at one end or the other, as does the high-temperature end of the curve representing the growth of bacteria as a function of temperature (Figure 6.7).

The most common source of *pollutant acid* in water is acid mine drainage. The sulfuric acid in such drainage arises from the microbial oxidation of pyrite or other sulfide minerals as described in Chapter 6. The values of pH encountered in acid-polluted water may fall below 3, a condition deadly to most forms of aquatic life except the culprit bacteria mediating the pyrite and iron(II) oxidation, which thrive under very low pH conditions. Industrial wastes frequently have the potential to contribute strong acid to water. Sulfuric acid produced by the air oxidation of pollutant sulfur dioxide (see Chapter 11) enters natural waters as acidic rainfall. In cases where the water does not have contact with a basic mineral, such as limestone, the water pH may become dangerously low. This condition occurs in some Canadian lakes, for example.

Excess *alkalinity*, and frequently accompanying high pH, generally are not introduced directly into water from anthropogenic sources. However, in many geographic areas, the soil and mineral strata are alkaline and impart a high alkalinity to water. Human activity can aggravate the situation, for example, by exposure of alkaline overburden from strip mining to surface water or groundwater. Excess alkalinity in water is manifested by a characteristic fringe of white salts at the edges of a body of water or on the banks of a stream.

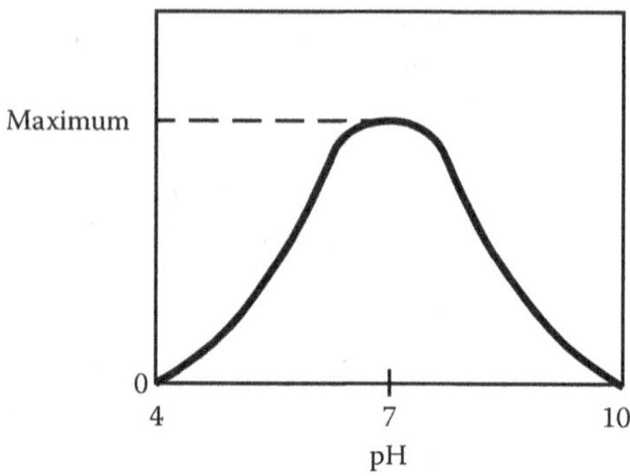

FIGURE 7.2 A generalized plot of the growth of an aquatic organism as a function of pH.

Water *salinity* may be increased by a number of human activities. Water passing through a municipal water system inevitably picks up salt from sources such as recharging water softeners with sodium chloride. Salts can leach from spoil piles. One of the major environmental constraints on the production of shale oil, for example, is the high percentage of leachable sodium sulfate in piles of spent shale. Careful control of these wastes is necessary to prevent further saline pollution of water in areas where salinity is already a problem. Irrigation adds a great deal of salt to water, a phenomenon responsible for the Salton Sea in California, and is a source of conflict between the United States and Mexico over saline contamination of the Rio Grande and Colorado rivers. Irrigation and intensive agricultural production have caused saline seeps in some of the Western states. These occur when water seeps into a slight depression in tilled, sometimes irrigated, fertilized land, carrying salts (particularly sodium, magnesium, and calcium sulfates) along with it. The water evaporates in the dry summer heat, leaving behind a salt-laden area that no longer supports much plant growth. With time, these areas spread, destroying the productivity of crop land.

7.9 OXYGEN, OXIDANTS, AND REDUCTANTS

Oxygen is a vitally important species in water (see Chapter 2). In water, oxygen is consumed rapidly by the oxidation of organic matter, $\{CH_2O\}$:

$$\{CH_2O\} + O_2 \xrightarrow{\text{Microorganisms}} CO_2 + H_2O \tag{7.6}$$

Unless the water is reaerated efficiently, as by turbulent flow in a shallow stream, it rapidly loses oxygen and will not support higher forms of aquatic life.

In addition to the microorganism-mediated oxidation of organic matter, oxygen in water may be consumed by the biooxidation of nitrogenous material:

$$NH_4^+ + 2O_2 \rightarrow 2H^+ + NO_3^- + H_2O \tag{7.7}$$

and by the chemical or biochemical oxidation of chemical reducing agents:

$$4Fe^{2+} + O_2 + 10H_2O \rightarrow 4Fe(OH)_3(s) + 8H^+ \tag{7.8}$$

$$2SO_3^{2-} + O_2 \rightarrow 2SO_4^{2-} \tag{7.9}$$

All these processes contribute to the deoxygenation of water.

The degree of oxygen consumption by microbially mediated oxidation of contaminants in water is called the BOD (biochemical or biological oxygen demand). This parameter is commonly measured by determining the quantity of oxygen utilized by suitable aquatic microorganisms during a 5-day period.

The addition of oxidizable pollutants to streams produces a typical oxygen sag curve as shown in Figure 7.3. Initially, a well-aerated, unpolluted stream is relatively free of oxidizable material; the oxygen level is high; and the bacterial population is relatively low. With the addition of oxidizable pollutants, the oxygen level drops because re-aeration cannot keep up with oxygen consumption. In the decomposition zone, the bacterial population rises. The septic zone is characterized by a high bacterial population and very low oxygen levels. The septic zone terminates when the oxidizable pollutant is exhausted, and then the recovery zone begins. In the recovery zone, the bacterial population decreases and the dissolved oxygen level increases until the water regains its original condition.

Although BOD is a reasonably realistic measure of water quality insofar as oxygen is concerned, the test for determining it is time consuming and cumbersome to perform. Total organic carbon (TOC) is frequently measured by catalytically oxidizing carbon in the water and measuring the CO_2 that is evolved. It has become popular because TOC is readily determined instrumentally.

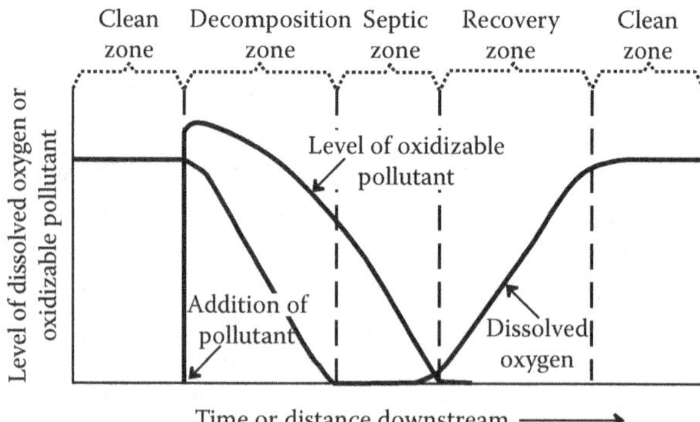

FIGURE 7.3 Oxygen sag curve resulting from the addition of oxidizable pollutant material to a stream.

7.10 ORGANIC POLLUTANTS

7.10.1 Bioaccumulation of Organic Pollutants

An important characteristic of organic water pollutants, especially those that have an affinity for lipid (fat) tissue and that resist biodegradation, is the *bioconcentration factor* (BCF), which is defined as the ratio of a substance's concentration in the tissue of an aquatic organism to the concentration of the substance in the water where the organism lives. It assumes that exposure is through water only and that the concentration in water is stable over a long period of time. A related parameter, the *bioaccumulation factor* (BAF) is defined in the same way, except that it assumes that both the organism and its food are exposed similarly to a pollutant over a long period of time. Although literally thousands of such factors involving hundreds of species and substances that are taken up by organisms from water have been measured or estimated, there is considerable uncertainty in these numbers; however, they remain indicative of the pollution potential of persistent organic compounds.[2]

7.10.2 Sewage

As shown in Table 7.4, sewage from domestic, commercial, food-processing, and industrial sources contains a wide variety of pollutants, including organic pollutants. Some of these pollutants,

TABLE 7.4
Some of the Primary Constituents of Sewage from a City Sewage System

Constituent	Potential Sources	Effects in Water
Oxygen-demanding substances	Mostly organic materials, particularly human feces and urine	Consume dissolved oxygen
Refractory organics	Industrial wastes, household products	Toxic to aquatic life
Viruses	Human wastes	Cause disease (possibly cancer); major deterrent to sewage recycle through water systems
Detergents	Household detergents	Esthetics, prevent grease and oil removal, toxic to aquatic life
Phosphates	Detergents	Algal nutrients
Grease and oil	Cooking, food processing, industrial wastes	Esthetics, harmful to some aquatic life
Salts	Human wastes, water softeners, industrial wastes	Increase water salinity
Heavy metals	Industrial wastes, chemical laboratories	Toxicity
Chelating agents	Some detergents, industrial wastes	Heavy metal ion solubilization and transport
Solids	All sources	Esthetics, harmful to aquatic life

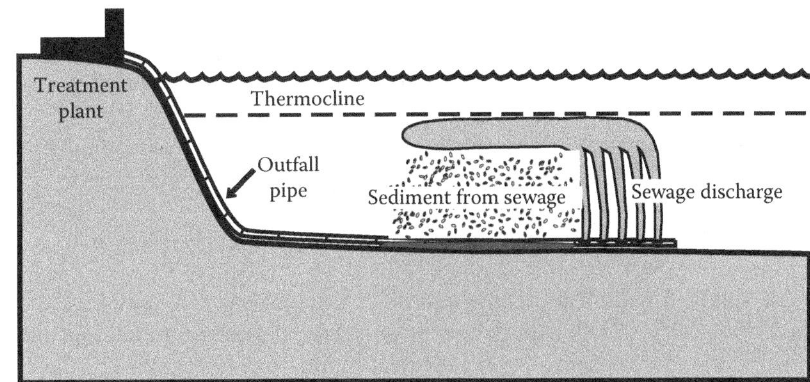

FIGURE 7.4 Settling of solids from an ocean-floor sewage effluent discharge.

particularly oxygen-demanding substances (see Section 7.9)—oil, grease, and solids—are removed by primary and secondary sewage-treatment processes. Others, such as salts, heavy metals, and refractory (degradation-resistant) organics, are not efficiently removed.

Disposal of inadequately treated sewage can cause severe problems. For example, offshore disposal of sewage, once commonly practiced by coastal cities, results in the formation of beds of sewage residues. Municipal sewage typically contains about 0.1% solids, even after treatment, and these settle out in the ocean in a typical pattern, illustrated in Figure 7.4. The warm sewage water rises in the cold hypolimnion and is carried in one direction or another by tides or currents. It does not rise above the thermocline (metalimnion); instead, it spreads out as a cloud from which the solids rain down on the ocean floor. Aggregation of sewage colloids is aided by dissolved salts in seawater (see Chapter 5), thus promoting the formation of sludge-containing sediments.

Another major disposal problem with sewage is the sludge produced as a product of the sewage treatment process (see Chapter 8). This sludge contains organic material, which continues to degrade slowly; refractory organics; and heavy metals. The amounts of sludge produced are truly staggering. For example, the city of Chicago produces about 3 million tons of sludge each year. A major consideration in the safe disposal of such amounts of sludge is the presence of potentially dangerous components such as heavy metals.

Careful control of sewage sources is needed to minimize sewage pollution problems. Particularly, heavy metals and refractory organic compounds need to be controlled at the source to enable use of sewage, or treated sewage effluents, for irrigation, recycling to the water system, or groundwater recharge.

Soaps, detergents, and associated chemicals are potential sources of organic pollutants. These pollutants are discussed briefly here.

7.10.3 Soaps, Detergents, and Detergent Builders

7.10.3.1 Soaps

Soaps are salts of higher fatty acids, such as sodium stearate, $C_{17}H_{35}COO^{-}Na^{+}$. The cleaning action of soap results largely from its emulsifying power and its ability to lower the surface tension of water. This concept may be understood by considering the dual nature of the soap anion. An examination of its structure shows that the stearate ion consists of an ionic carboxyl "head" and a long hydrocarbon "tail":

$$\sim\sim\sim\sim\sim\sim\sim\sim\sim\overset{\overset{\displaystyle O}{\|}}{C}{-}O^{-}Na^{+}$$

In the presence of oils, fats, and other water-insoluble organic materials, the tendency is for the "tail" of the anion to dissolve in the organic matter, whereas the "head" remains in aquatic solution. Thus, the soap emulsifies, or suspends, organic material in water. In the process, the anions form colloidal soap micelles, as shown in Figure 5.4.

The primary disadvantage of soap as a cleaning agent comes from its reaction with divalent cations to form insoluble salts of fatty acids:

$$2C_{17}H_{35}COO^{-}Na^{+} + Ca^{2+} \rightarrow Ca(C_{17}H_{35}CO_2)_2(s) + 2Na^{+} \quad (7.10)$$

These insoluble solids, usually salts of magnesium or calcium, are not at all effective as cleaning agents. In addition, the insoluble "curds" form unsightly deposits on clothing and in washing machines. If sufficient soap is used, all of the divalent cations may be removed by their reaction with soap, and the water containing excess soap will have good cleaning qualities. This is the approach commonly used when soap is employed with unsoftened water in the bathtub or wash basin, where the insoluble calcium and magnesium salts can be tolerated. However, in applications such as washing clothing, the water must be softened by the removal of calcium and magnesium or their complexation by substances such as polyphosphates (see Section 3.16).

Although the formation of insoluble calcium and magnesium salts has resulted in the essential elimination of soap as a cleaning agent for clothing, dishes, and most other materials, it has distinct advantages from the environmental standpoint. As soon as soap gets into sewage or an aquatic system, it generally precipitates as calcium and magnesium salts. Hence, any effects that soap might have in solution are eliminated. With eventual biodegradation, the soap is completely eliminated from the environment. Therefore, aside from the occasional formation of unsightly scum, soap does not cause any substantial pollution problems.

7.10.3.2 Detergents

Synthetic *detergents* have good cleaning properties and do not form insoluble salts with "hardness ions" such as calcium and magnesium. Such synthetic detergents have the additional advantage of being the salts of relatively strong acids and, therefore, they do not precipitate out of acidic waters as insoluble acids, an undesirable characteristic of soaps. The potential of detergents to contaminate water is high because of their heavy use throughout the consumer, institutional, and industrial markets. Over 1 billion pounds of detergent surfactants are consumed annually in the U.S. household market alone, with slightly more consumed in Europe. Most of this material, along with the other ingredients associated with detergent formulations, is discarded with wastewater.

The key ingredient of detergents is the *surfactant* or surface-active agent, which acts in effect to make water "wetter" and a better cleaning agent. Surfactants concentrate at interfaces of water with gases (air), solids (dirt), and immiscible liquids (oil). They do so because of their *amphiphilic structure*, meaning that one part of the molecule is a polar or ionic group (head) with a strong affinity for water, and the other part is a hydrocarbon group (tail) with an aversion to water. This kind of structure is illustrated below for the molecule of alkyl benzene sulfonate (ABS) surfactant:

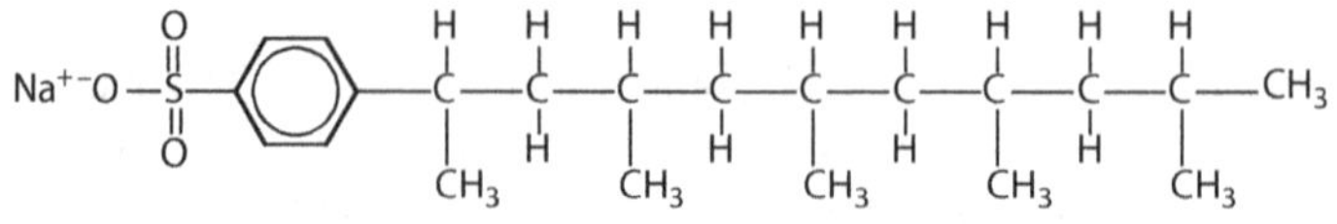

Until the early 1960s, ABS was the most common surfactant used in detergent formulations. However, it suffered the distinct disadvantage of being only very slowly biodegradable because of its branched-chain structure (see Section 6.10). An objectionable manifestation of the nonbiodegradable detergents was the "head" of foam that began to appear in glasses of drinking water in areas where sewage was recycled through the domestic water supply. Spectacular beds of foam appeared near sewage outflows and in sewage treatment plants and at least one fatality was reported when an individual fell in a bed of foam at a sewage treatment plant and was asphyxiated by the gases in the foam. Occasionally, the entire aeration tank of an activated sludge plant would be smothered by a blanket of foam. Among the other undesirable effects of persistent detergents upon waste-treatment processes were lowered surface tension of water, deflocculation of colloids, flotation of solids, emulsification of grease and oil, and destruction of useful bacteria. Consequently, ABS was replaced by a biodegradable surfactant known as linear alkyl sulfonate (LAS).

LAS, α-benzenesulfonate, has the general structural formula illustrated below where the benzene ring may be attached at any point on the alkyl chain except at the ends:

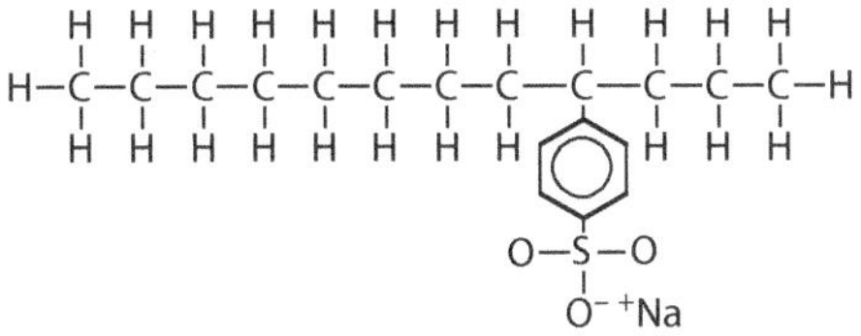

```
    H   H   H   H   H   H   H   H    H    H   H   H
    |   |   |   |   |   |   |   |    |    |   |   |
H - C - C - C - C - C - C - C - C  - C  - C - C - C - H
    |   |   |   |   |   |   |   |    |    |   |   |
    H   H   H   H   H   H   H   H  (C6H4) H   H   H
                                     |
                                 O - S - O
                                     |
                                   O- +Na
```

LAS

LAS is more biodegradable than ABS because the alkyl portion of LAS is not branched and does not contain the tertiary carbon which is so detrimental to biodegradability. Since LAS has replaced ABS in detergents, the problems arising from the surface-active agent in the detergents (such as toxicity to fish fingerlings) have greatly diminished and the levels of surface-active agents found in water have decreased markedly.

Some detergent surfactants are nonionic. One type that has proven troublesome consists of the alkyl phenol polyethoxylates:

```
     ⎡ H   H      ⎤          H   H   H   H   H   H   H   H   H
     ⎢ |   |      ⎥          |   |   |   |   |   |   |   |   |
HO - ⎢ C - C - O  ⎥ -(C6H4)- C - C - C - C - C - C - C - C - C - H
     ⎢ |   |      ⎥          |   |   |   |   |   |   |   |   |
     ⎣ H   H      ⎦n         H   H   H   H   H   H   H   H   H
```

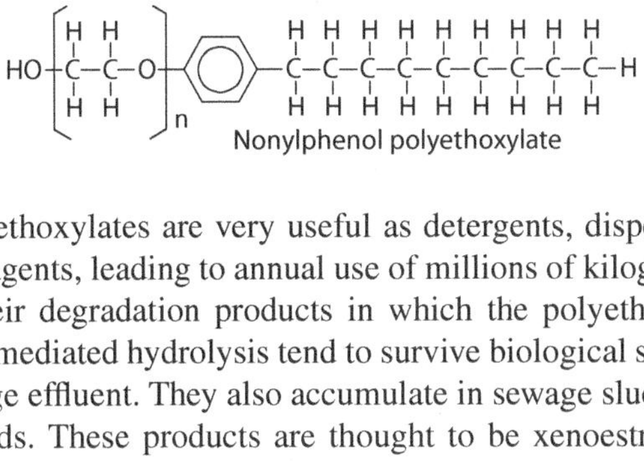

Nonylphenol polyethoxylate

The alkyl phenol polyethoxylates are very useful as detergents, dispersing agents, emulsifiers, solubilizers, and wetting agents, leading to annual use of millions of kilograms in the United States. These substances and their degradation products in which the polyethoxylate chains have been shortened by microbially mediated hydrolysis tend to survive biological sewage treatment and to be discharged with the sewage effluent. They also accumulate in sewage sludge, much of which is disposed on agricultural lands. These products are thought to be xenoestrogens (estrogen hormone mimickers) and their potential entry into the food chain from sludge-treated soil is of concern. Because of these concerns, the uses of these compounds have been severely restricted in some European countries.

Detergent *builders* added to detergents to bind to hardness ions, making the detergent solution alkaline and greatly improving the action of the detergent surfactant, can cause environmental problems. A commercial solid detergent contains only 10–30% surfactant. Detergent formulations also contain complexing agents added to complex calcium and to function as builders. Other ingredients include ion exchangers, alkalies (sodium carbonate), anticorrosive sodium silicates, amide foam stabilizers, soil-suspending carboxymethylcellulose, bleaches, fabric softeners, enzymes, optical brighteners, fragrances, dyes, and diluent sodium sulfate. The polyphosphates formerly used in builders have caused the most concern as environmental pollutants, although these problems have largely been resolved as polyphosphates have been phased out.

Increasing demands on the performance of detergents have led to the use of enzymes in detergent formulations destined for both domestic and commercial applications. To a degree, enzymes can take the place of chlorine and phosphates, both of which can have detrimental environmental consequences. Lipases and cellulases are the most useful enzymes for detergent applications.

7.10.4 Naturally Occurring Chlorinated and Brominated Compounds

Although halogenated organic compounds in water, such as those discussed as pesticides in Section 7.11, are normally considered to be from anthropogenic sources, more than 2000 such compounds have been identified from natural sources.[3] These are produced largely by marine species, especially some kinds of red algae, probably as chemical defense agents. Some marine microorganisms, worms, sponges, and tunicates are also known to produce organochlorine and organobromine compounds. Various ones of these compounds have been detected in samples in Arctic regions including air, fish, seabird eggs, marine mammals, and Eskimo women's milk. An example of such a compound commonly encountered in the marine environment is 1,2′-bi-1H-pyrrole,2,3,3′,4,4′,5,5′-heptachloro-1′-methyl-($C_9H_3N_2Cl_7$), and its analogous bromine compound:

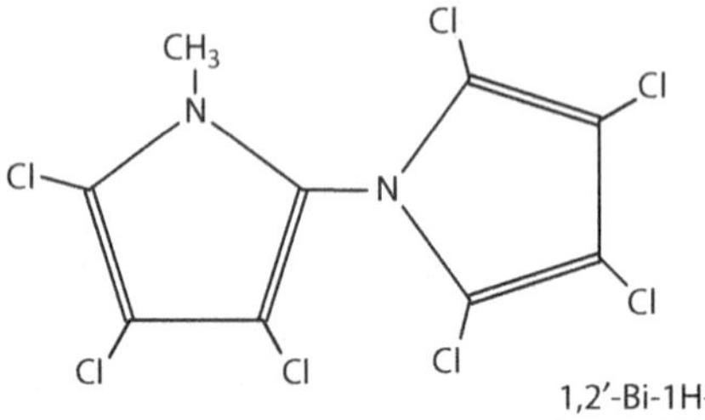

CH3 N Cl Cl Cl N Cl Cl Cl Cl

1,2′-Bi-1H-pyrrole,2,3,3′,4,4′,5,5′-heptachloro-1′-methyl

7.10.5 Microbial Toxins

Bacteria and protozoa in water can produce toxins that can cause illness or even death. Toxins produced in rivers, lakes, and reservoirs by cyanobacteria including *Anabaena*, *Microcystis*, and *Nodularia* have caused adverse health effects in Australia, Brazil, England, and elsewhere in the world. There are about 40 species of cyanobacteria that produce toxins from six chemical groups. *Cylindrospermopsin* toxin (below) produced by cyanobacteria has poisoned people who have drunk water contaminated by the toxin.

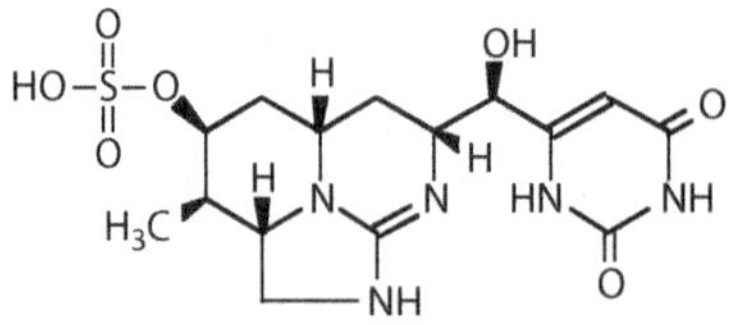

HO–S(=O)(=O)–O H OH H H N N HN NH O O H3C NH

Cylindrospermopsin

Most of the protozoans that produce toxins belong to the order *dinoflagellata*, which are predominantly marine species. The cells of these organisms are enclosed in cellulose envelopes, which often have beautiful patterns on them. Among the effects caused by toxins from these organisms are gastrointestinal, respiratory, and skin disorders in humans; mass kills of various marine animals; and paralytic conditions caused by eating infested shellfish.

The marine growth of dinoflagellates is characterized by occasional incidents in which they multiply at such an explosive rate that they color the water yellow, olive-green, or red by their vast numbers. In 1946, some sections of the Florida coast became so afflicted by "red tide" that the water became viscous and for many miles the beaches were littered with the remains of dead fish, shellfish, turtles, and other marine organisms. The sea spray in these areas became so irritating that coastal schools and resorts were closed.

The greatest danger to humans from dinoflagellata toxins comes from the ingestion of shellfish, such as mussels and clams, that have accumulated the protozoa from sea water. In this form, the toxic material is called paralytic shellfish poison. As little as 4 mg of this toxin, the amount found in several severely infested mussels or clams, can be fatal to a human. The toxin depresses respiration and affects the heart, resulting in complete cardiac arrest in extreme cases.

7.11 PESTICIDES IN WATER

The introduction of DDT during World War II marked the beginning of a period of very rapid growth in pesticide use. Pesticides are employed for many different purposes. Chemicals used in the control of invertebrates include *insecticides*, *molluscicides* for the control of snails and slugs, and *nematicides* for the control of microscopic roundworms. Vertebrates are controlled by *rodenticides* that kill rodents, *avicides* are used to repel birds, and *piscicides* are used in fish control. *Herbicides* are used to kill plants, particularly weeds in agricultural crops. Plant *growth regulators*, *defoliants*, and *plant desiccants* are used for various purposes in the cultivation of plants. *Fungicides* are used against fungi, *bactericides* against bacteria, *slimicides* against slime-causing organisms in water, and *algicides* against algae. As of the mid-1990s, U.S. agriculture used about 365 million kg of pesticides per year, whereas about 900 million kg of insecticides were used in nonagricultural applications including forestry, landscaping, gardening, food distribution, and home pest control. Insecticide production has remained about level during the last three or four decades. However, insecticides and fungicides are the most important pesticides with respect to human exposure in food because they are applied shortly before or even after harvesting. Herbicide production has increased as chemicals have increasingly replaced cultivation of land in the control of weeds and now accounts for the majority of agricultural pesticides. The potential exists for large quantities of pesticides to enter water either directly, in applications such as mosquito control or indirectly, primarily from drainage of agricultural lands.

Several classes of pesticides and other chemicals are of particular concern as water pollutants because of their potential effects. These are (1) highly biodegradation-resistant compounds, (2) known or probable carcinogens, (3) toxicants with adverse reproductive or developmental effects, (4) neurotoxins including cholinesterase inhibitors, (5) substances with high acute toxicities, and (6) known groundwater contaminants. Table 7.5 lists some of the most commonly used pesticides that may be of concern as water pollutants.

Degradation (often hydrolysis) products of some pesticides are encountered in water at levels similar to, or even higher than, the parent pesticides. In some cases, the degradation products are more toxic than the parent pesticides. An example of a pesticide degradation product commonly found in water is aminomethyl phosphonic acid produced from herbicidal glyphosate (Figure 7.13) which, under the brand name of Roundup, is the most widely produced pesticide in the world.

7.11.1 Natural Product Insecticides, Pyrethrins, and Pyrethroids

Plants provide several significant classes of insecticides including *nicotine* from tobacco, *rotenone* from certain legume roots, and *pyrethrins* (see structural formulas in Figure 7.5). Because of the ways in which they are applied and their biodegradabilities, these substances are unlikely to be significant water pollutants.

Pyrethrins and their synthetic analogs represent both the oldest and newest of insecticides. Extracts of dried chrysanthemum or pyrethrum flowers, which contain pyrethrin I and related compounds, have been known for their insecticidal properties for a long time, and may have even been used as botanical insecticides in China almost 2000 years ago. The most important commercial sources of insecticidal pyrethrins are chrysanthemum varieties grown in Kenya. Pyrethrins have several advantages as insecticides, including facile enzymatic degradation, which makes them relatively safe for mammals; ability to rapidly paralyze ("knock down") flying insects; and good biodegradability characteristics.

Synthetic analogs of the pyrethrins, *pyrethroids*, have been widely produced as insecticides during recent years. The first of these was allethrin, and another common example is fenvalerate

TABLE 7.5
Pesticides That May Be Found as Water Pollutants[a]

Pesticide	Use	Compound Type
Alachlor	Herbicide	Chloroacetanilide
Aldicarb	Insecticide	Carbamate
Allethrin	Insecticide	Pyrethroid
Atrazine	Herbicide	Triazine
Azadirachtin	Insecticide, nematicide	Complex botanical compound
Azinphos-methyl	Insecticide	Organophosphate
Azoxystrobin	Fungicide	Strobin
Captan	Fungicide	Thiophthalimide
Carbaryl	Insecticide, plant growth regulator, nematicide	Carbamate
Carbofuran	Insecticide, nematicide	Carbamate
Chlorothalonil	Fungicide	Substituted benzene
Chlorpyrifos	Insecticide, nematicide	Organophosphate
Cypermethrin	Insecticide	Pyrethroid
Deltamethrin	Insecticide	Pyrethroid
Diazinon	Insecticide	Organophosphorus
Diclofop-methyl	Herbicide	Chlorophenoxy acid/ester
Diuron	Herbicide	Urea
EPTC	Herbicide	Thiocarbamate
Ethephon	Plant growth regulator	Organophosphate
Fenvalerate	Insecticide	Pyrethroid
Fluquinconazole	Fungicide	Azole
Glyphosate	Herbicide	Phosphonoglycine
Iprodione	Fungicide	Dicarboximide
Kresoxim-methyl	Fungicide	Strobin
Linuron	Herbicide	Urea
Malathion	Insecticide	Organophosphate
Mecoprop	Herbicide	Chlorphenoxy compound
Metam-sodium	Fumigant, herbicide, microbiocide, algacide	Dithiocarbamate
Methiocarb	Insecticide, molluscicide	Carbamate
Metolachlor	Herbicide	Chloroacetanilide
Metribuzin	Herbicide	Triazinone
Phosmet	Insecticide	Organophosphate
Pirimicarb	Insecticide	Carbamate
Prometon	Herbicide	Triazine
Propachlor	Herbicide	Chloroacetanilide
Propanil	Herbicide	Anilide
Propiconazole	Fungicide	Azole
Simazine	Herbicide	Triazine
Spinosad	Insecticide	Compound from bacteria
Tebuconazole	Fungicide	Azole
Tebuthiuron	Herbicide	Urea
Terbacil	Herbicide	Uracil
Terbuthylazine	Algicide, herbicide, microbiocide	Triazine
Thifensulfuron-methyl	Herbicide	Sulfonylurea
Thiophanate-methyl	Fungicide	Benzimidazole
Tri-allate	Herbicide	Thiocarbamate
Triclopyr	Herbicide	Chloropyridinyl
Trifloxystrobin	Fungicide	Strobin
Trifluralin	Herbicide	2,6-Dinitroaniline

[a] For additional information see Pesticide Action Network: http://www.pesticideinfo.org/

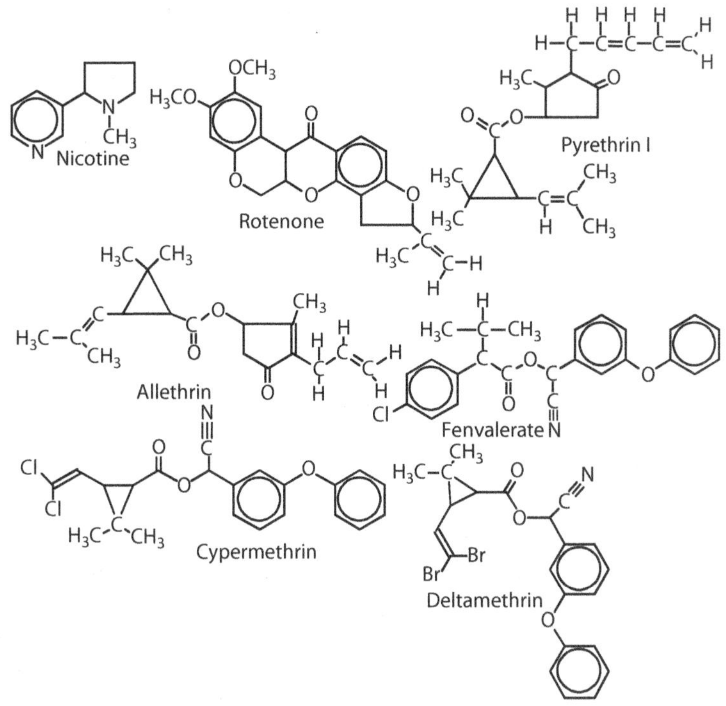

FIGURE 7.5 Common botanical insecticides and synthetic analogs of the pyrethrins.

(see structural formulas in Figure 7.5). Other examples of insecticidal pyrethroids that have water pollution potential include cypermethrin and deltamethrin.

7.11.2 DDT and Organochlorine Insecticides

Chlorinated hydrocarbon or organochlorine insecticides are hydrocarbon compounds in which various numbers of hydrogen atoms have been replaced by Cl atoms (Figure 7.6). Of the organochlorine insecticides, the most notable has been DDT, which was used in massive quantities following World War II. It has a low acute toxicity to mammals, although there is some evidence that it might be

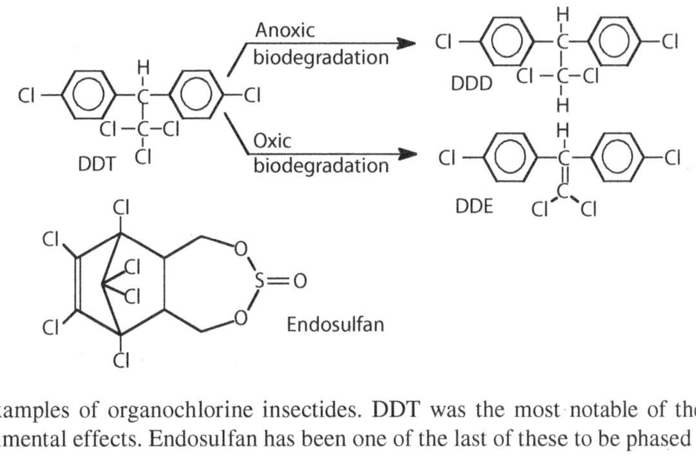

FIGURE 7.6 Examples of organochlorine insectides. DDT was the most notable of these because of its history of environmental effects. Endosulfan has been one of the last of these to be phased out of use.

carcinogenic. It is a very persistent insecticide and accumulates in food chains. It has been banned in the United States since 1972.

Many organochlorine insecticides were widely used in past decades but are now banned because of their toxicities, and particularly their accumulation and persistence in food chains. Now largely of historical interest, these include methoxychlor (once a popular replacement for DDT), dieldrin, endrin, chlordane, aldrin, dieldrin/endrin, heptachlor, toxaphene, lindane, and endosulfan (one of the last to be phased out of general use).

7.11.3 Organophosphate Insecticides

Organophosphate insecticides are insecticidal organic compounds that contain phosphorus, some of which are organic esters of orthophosphoric acid, such as paraoxon:

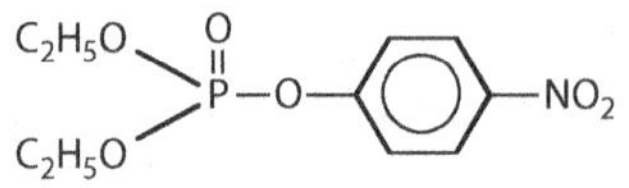

$$(C_2H_5O)_2P(=O)-O-C_6H_4-NO_2$$

More commonly, insecticidal phosphorus compounds are phosphorothionate and phosphorodithioate compounds, such as those shown in Figure 7.7, which have an =S group rather than an =O group bonded to P.

The toxicities of organophosphate insecticides vary a great deal. Their major toxic effect is the inhibition of acetylcholinesterase, an enzyme essential for nerve function. For example, as little as 120 mg of parathion has been known to kill an adult human, and a dose of 2 mg has killed a child. Most accidental poisonings have occurred by absorption through the skin. Since its use began, several hundred people have been killed by parathion. In contrast, *malathion* shows how differences in structural formula can cause pronounced differences in the properties of organophosphate pesticides. Malathion has two carboxyester linkages, which are hydrolyzable by carboxylase enzymes to relatively nontoxic products as shown by the following reaction:

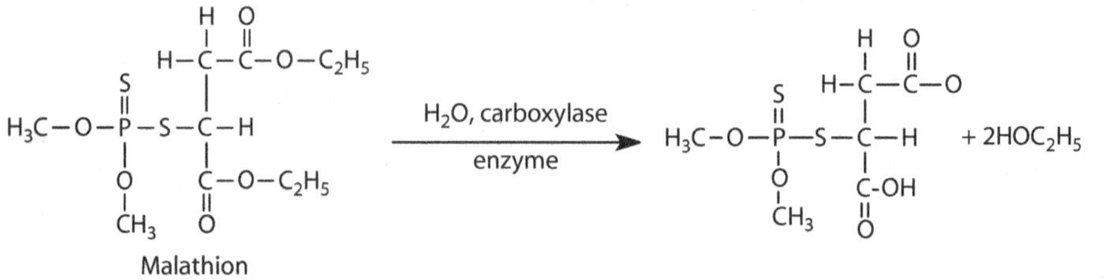

$$\underset{\text{Malathion}}{H_3C-O-P(=S)(-O-CH_3)-S-CH(-C(=O)-O-C_2H_5)-CH_2-C(=O)-O-C_2H_5} \xrightarrow[\text{enzyme}]{H_2O,\ \text{carboxylase}} H_3C-O-P(=S)(-O-CH_3)-S-CH(-C(=O)-OH)-CH_2-C(=O)-O + 2HOC_2H_5 \quad (7.11)$$

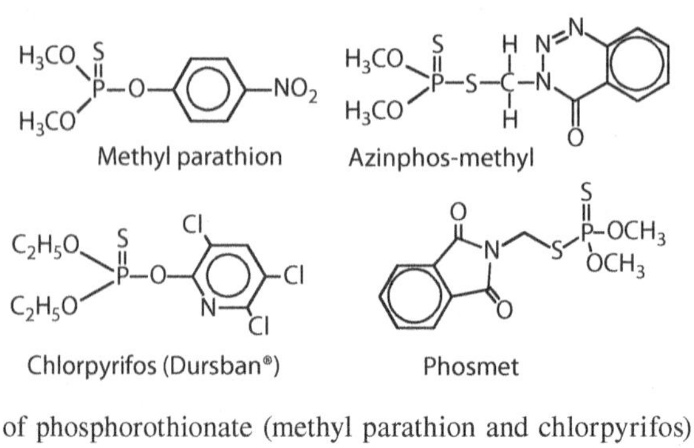

FIGURE 7.7 Examples of phosphorothionate (methyl parathion and chlorpyrifos) and phosphorodithioate (azinphos-methylphosmet) insecticides.

The enzymes that accomplish malathion hydrolysis are possessed by mammals, but not by insects, so mammals can detoxify malathion, whereas insects cannot. The result is that malathion has selective insecticidal activity. For example, although malathion is a very effective insecticide, its LD_{50} (dose required to kill 50% of test subjects) for adult male rats is about 100 times that of parathion, reflecting the much lower mammalian toxicity of malathion compared to some of the more toxic organophosphate insecticides, such as parathion.

Unlike the organohalide compounds they largely displaced, the organophosphates readily undergo biodegradation and do not bioaccumulate. Because of their high biodegradability and restricted use, organophosphates are of comparatively little significance as water pollutants.

7.11.4 Carbamates

Pesticidal organic derivatives of carbamic acid, for which the formula is shown in Figure 7.8, are known collectively as *carbamates*. Carbamate pesticides have been widely used because some are more biodegradable than the formerly popular organochlorine insecticides, and have lower dermal toxicities than most common organophosphate pesticides.

Carbaryl has been widely used as an insecticide on lawns or gardens. It has a low toxicity to mammals. *Carbofuran* has a high water solubility and acts as a plant systemic insecticide. As a plant systemic insecticide, it is taken up by the roots and leaves of plants so that insects are poisoned by the plant material on which they feed. *Pirimicarb* has been widely used in agriculture as a systemic aphicide. Unlike many carbamates, it is rather persistent, with a strong tendency to bind to soil.

Carbamates are toxic to animals because they inhibit acetylcholinesterase. Unlike some of the organophosphate insecticides, they do so without the need for undergoing a prior biotransformation and are therefore classified as direct inhibitors. Their inhibition of acetylcholinesterase is relatively reversible. Loss of acetylcholinesterase inhibition activity may result from hydrolysis of the carbamate ester, which can occur metabolically.

7.11.5 Fungicides

Fungicides are widely applied to cereal and food crops to prevent fungal infections of these crops. Because of this, fungicides have the potential to contaminate water. Structural formulas of three commonly used fungicides are shown in Figure 7.9. Of the ones shown, chlorothalonil has been used for more than 30 years with annual applications in the United States of more than 5 million kg. It is typically applied at a rate of 1 kg per hectare per application with 4–9 applications per year. The strobilurin fungicides exemplified by azoxystrobin and the triazole fungicides exemplified by propiconazole came into use during the 1990s and have been effective, although some problems with resistance developed by target organisms have been encountered.

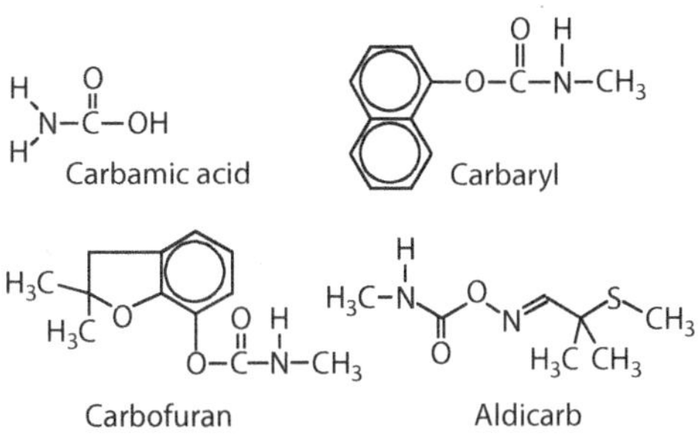

Carbamic acid

Carbaryl

Carbofuran

Aldicarb

FIGURE 7.8 Carbamic acid and three insecticidal carbamates.

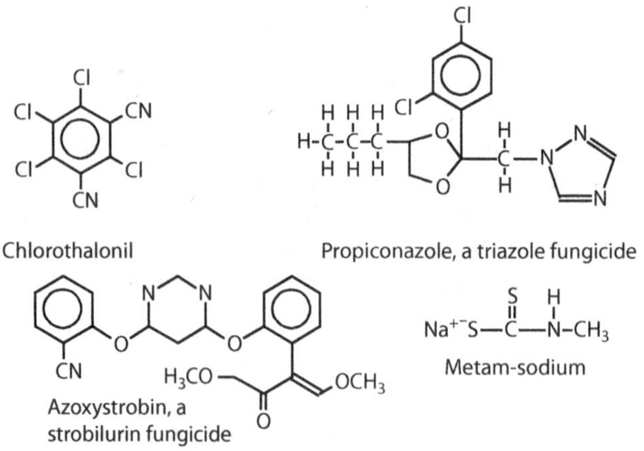

FIGURE 7.9 Four examples of widely used fungicides that may be of concern as potential water pollutants.

7.11.6 Herbicides

Herbicides are applied over millions of acres of farmland worldwide and are widespread water pollutants as a result of this intensive use. Herbicides are commonly found in surface water and groundwater. Among those commonly encountered in surface water and groundwater are atrazine, simazine, and cyanazine widely used to control weeds on corn and soybeans in the "Corn Belt" states of Kansas, Nebraska, Iowa, Illinois, and Missouri as well as agricultural regions throughout the world. Other herbicides found in water include prometon, metolachlor, metribuzin, tebuthiuron, trifluran, alachlor, and the atrazine metabolites deisopropylatrazine and deethylatrazine. Although glyphosate is a very widely used herbicide to control weeds on crops genetically engineered to resist its effects, it is rarely found at levels of concern in water because of its very strong affinity for soil solids.

7.11.6.1 Bipyridilium Compounds

As shown by the structures in Figure 7.10, a bipyridilium compound contains two pyridine rings per molecule. The two important pesticidal compounds of this type are the herbicides *diquat* and *paraquat*, the structural formulas of which are illustrated in Figure 7.11.

Other members of this class of herbicides include chlormequat, morfamquat, and difenzoquat. Applied directly to the plant tissue, these compounds rapidly destroy plant cells and give the plant a frostbitten appearance. However, they bind tenaciously to soil, especially the clay mineral fraction, which results in rapid loss of herbicidal activity so that sprayed fields can be planted within a day or two of herbicide application.

Paraquat, which was registered for use in 1965, has been one of the most used of the bipyridilium herbicides. Highly toxic, it is reputed to have "been responsible for hundreds of human deaths."[4] Exposure to fatal or dangerous levels of paraquat can occur by all pathways, including inhalation of spray, skin contact, ingestion, and even suicidal hypodermic injections. Despite these possibilities and its widespread application, paraquat is used safely without ill effects when proper procedures are followed.

Because of its widespread use as a herbicide, the possibility exists of substantial paraquat contamination of food. Drinking water contamination by paraquat has also been observed.

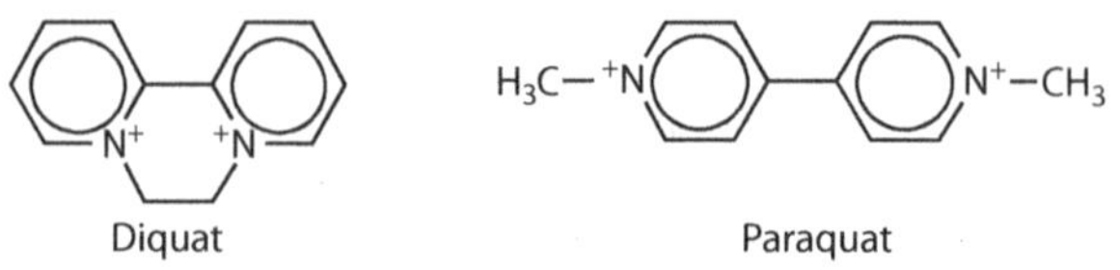

FIGURE 7.10 The two major bipyridilium herbicides (cation forms).

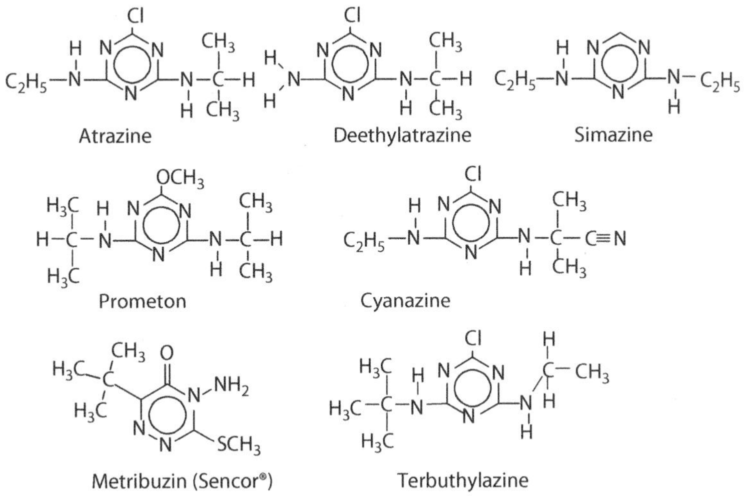

Atrazine Deethylatrazine Simazine

Prometon Cyanazine

Metribuzin (Sencor®) Terbuthylazine

FIGURE 7.11 Triazine herbicides. These compounds are commonly encountered as water pollutants in agricultural areas where they are widely used.

7.11.6.2 Herbicidal Heterocyclic Nitrogen Compounds

A number of important herbicides contain three heterocyclic nitrogen atoms in ring structures and are therefore called *triazines* (Figure 7.11). Triazine herbicides inhibit photosynthesis. Selectivity is gained by the inability of target plants to metabolize and detoxify the herbicide. The most long established and common example of this class is atrazine, commonly applied to kill weeds in corn, and a widespread water pollutant in corn-growing regions. Another member of this class is metribuzin, which is widely used on soybeans, sugarcane, and wheat.

7.11.6.3 Chlorophenoxy Herbicides

The chlorophenoxy herbicides, including 2,4-D and 2,4,5-trichlorophenoxyacetic acid (2,4,5-T) shown in Figure 7.12, were manufactured on a large scale for weed and brush control and as military defoliants. At one time the latter was of particular concern because of contaminant 2,3,7,8-tetrachlorodibenzo-*p*-dioxin (TCDD) (see figures) present as a manufacturing by-product.

7.11.6.4 Miscellaneous Herbicides

Many herbicides that are environmentally significant do not fall into the classifications described above. The most commonly used of these that are likely to be encountered as water pollutants are shown in Figure 7.13.

Nitroaniline herbicides are characterized by the presence of NO_2 and a substituted $-NH_2$ group on a benzene ring as shown for trifluralin. This class of herbicides is widely represented in agricultural

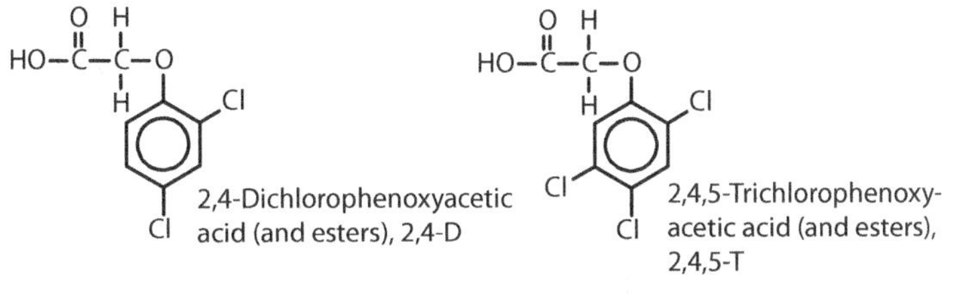

2,4-Dichlorophenoxyacetic acid (and esters), 2,4-D

2,4,5-Trichlorophenoxyacetic acid (and esters), 2,4,5-T

FIGURE 7.12 Chlorophenoxy herbicides.

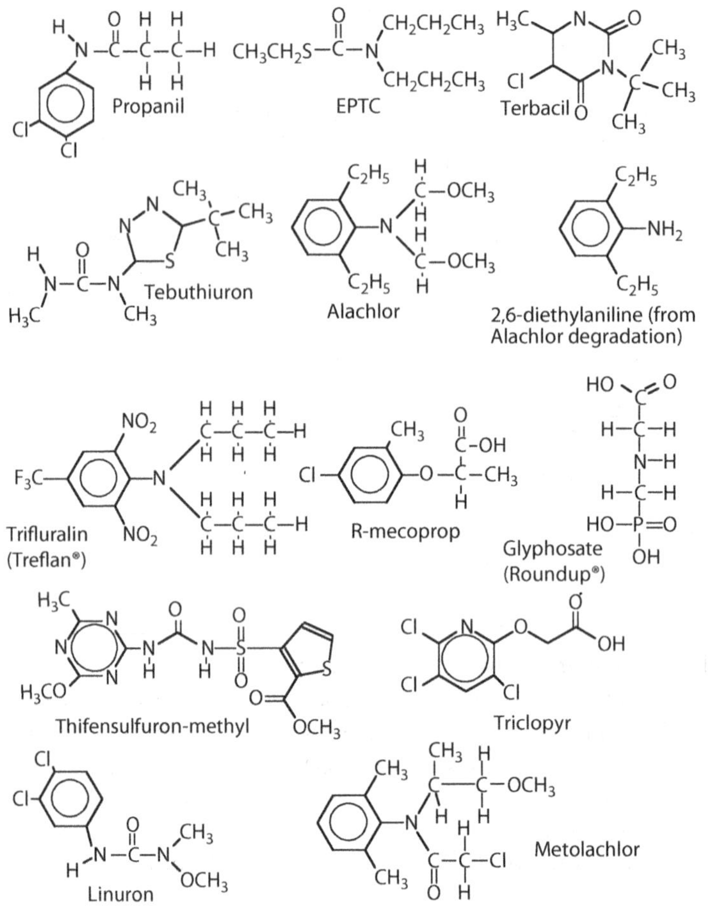

FIGURE 7.13 Miscellaneous herbicides, some of which are commonly found as water pollutants.

applications and includes benefin (Balan®), oryzalin (Surflan®), pendimethalin (Prowl®), and fluchoralin (Basalin®).

A wide variety of chemicals have been used as herbicides, and have been potential water pollutants. One such compound is R-mecoprop (Figure 7.13). Other types of herbicides include substituted ureas, carbamates, and thiocarbamates.

Until about 1960, arsenic trioxide and other inorganic arsenic compounds (see Section 7.4) were employed to kill weeds. Because of the incredibly high rates of application of up to several hundred kilograms per acre, and because arsenic is nonbiodegradable, the potential still exists for arsenic pollution of surface water and groundwater from fields formerly dosed with inorganic arsenic. Organic arsenicals, such as cacodylic acid, have also been widely applied to kill weeds:

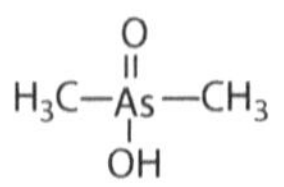

$$H_3C-\overset{\overset{\displaystyle O}{\|}}{\underset{\underset{\displaystyle OH}{|}}{As}}-CH_3$$

Hydroxydimethylarsine oxide, cacodylic acid

7.11.7 By-Products of Pesticide Manufacture

A number of water pollution and health problems have been associated with the manufacture of organochlorine pesticides. The most notorious by-products of pesticide manufacture are *polychlorinated*

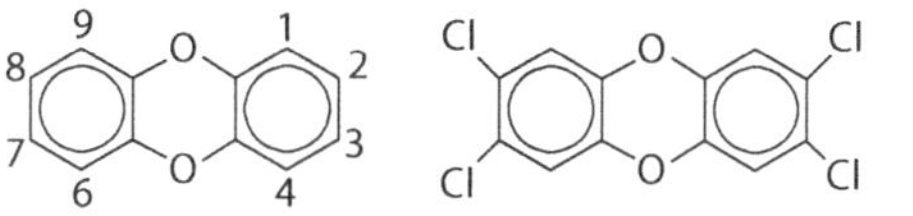

Dibenzo-*p*-dioxin   2,3,7,8-tetrachlorodibenzo-*p*-dioxin

FIGURE 7.14 Dibenzo-*p*-dioxin and TCDD, often called simply "dioxin." In the structure of dibenzo-*p*-dioxin, each number refers to a numbered carbon atom to which an H atom is bound, and the names of derivatives are based upon the carbon atoms where another group has been substituted for the H atoms, as is seen by the structure and name of 2,3,7,8-tetrachlorodibenzo-*p*-dioxin.

dibenzodioxins. From 1 to 8 Cl atoms may be substituted for H atoms on dibenzo-*p*-dioxin (Figure 7.14), giving a total of 75 possible chlorinated derivatives. Commonly referred to as "dioxins," these species have a high environmental and toxicological significance. Of the dioxins, the most notable pollutant and hazardous waste compound is TCDD, often referred to simply as "*dioxin*." This compound was produced as a low-level contaminant in the manufacture of some aryl, oxygen-containing organohalide compounds such as chlorophenoxy herbicides (mentioned previously in this section) synthesized by processes used until the 1960s.

TCDD has a very low vapor pressure, a high melting point of 305°C, and a water solubility of only 0.2 μg/L. It is chemically unreactive, stable thermally up to about 700°C, and is poorly biodegradable. It is very toxic to some animals, with an LD_{50} of only about 0.6 μg/kg body mass in male guinea pigs. (The type and degree of its toxicity to humans are largely unknown; it is known to cause a severe skin condition called chloracne.) Because of its properties, TCDD is a stable, persistent environmental pollutant and a hazardous waste constituent of considerable concern. It has been identified in some municipal incineration emissions, in which it is believed to form when chlorine from the combustion of organochlorine compounds reacts with carbon in the incinerator.

TCDD contamination has resulted from improper waste disposal, the most notable case of which resulted from the spraying of waste oil mixed with TCDD on roads and horse arenas in Missouri in the early 1970s. Contamination of the soil in Times Beach, Missouri, resulted in the whole town being bought out and its topsoil dug up and incinerated at a cost exceeding $100 million.

One of the greater environmental disasters ever to result from pesticide manufacture involved the production of Kepone, structural formula

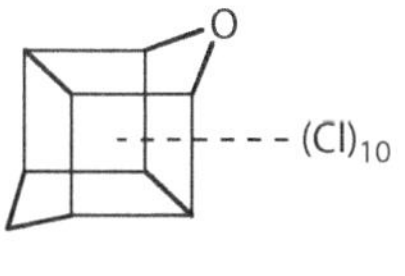

$(Cl)_{10}$

This pesticide has been used for the control of banana-root borer, tobacco wireworm, ants, and cockroaches. Kepone exhibits acute, delayed, and cumulative toxicity in birds, rodents, and humans, and it causes cancer in rodents. It was manufactured in Hopewell, Virginia, during the mid-1970s. During this time, workers were exposed to Kepone and are alleged to have suffered health problems as a result. As much as 53,000 kg of kepone may have been dumped through the sewage system of Hopewell and into the James River. The cost of dredging the river to remove this waste was estimated at a prohibitive cost of several billion dollars.

7.12 POLYCHLORINATED BIPHENYLS

First discovered as environmental pollutants in 1966, PCB compounds have been found throughout the world in water, sediments, bird tissue, and fish tissue. These compounds constitute an important

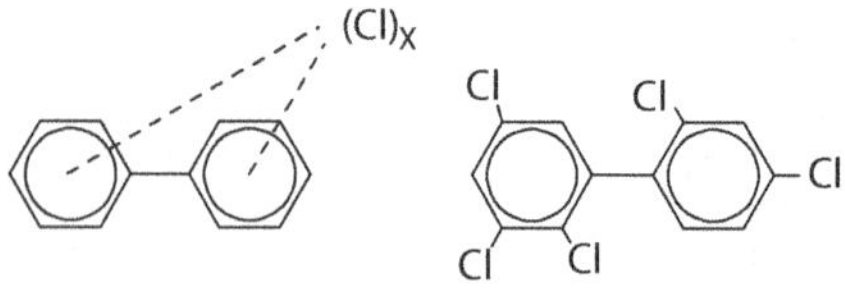

FIGURE 7.15 General formula of polychlorinated biphenyls (left, where X may range from 1 to 10) and a specific 5-chlorine congener (right).

class of special wastes. They are made by substituting from 1 to 10 Cl atoms onto the biphenyl aryl structure as shown on the left in Figure 7.15. This substitution can produce 209 different compounds (congeners), of which one example is shown on the right in Figure 7.15.

Polychlorinated biphenyls have very high chemical, thermal, and biological stability; low vapor pressure; and high dielectric constants. These properties have led to the use of PCBs as coolant-insulation fluids in transformers and capacitors; for the impregnation of cotton and asbestos; as plasticizers; and as additives to some epoxy paints. The same properties that made extraordinarily stable PCBs so useful also contributed to the widespread dispersion and accumulation of these substances in the environment. By regulations issued in the United States under the authority of the Toxic Substances Control Act passed in 1976, the manufacture of PCBs was discontinued in the United States, and their uses and disposal were strictly controlled. Some degree of biodegradation of PCBs in the environment does occur.[5]

Substitutes for PCBs for electrical applications have been developed. Disposal of PCBs from discarded electrical equipment and other sources have caused problems, particularly since PCBs can survive ordinary incineration by escaping as vapors through the smokestack. However, they can be destroyed by special incineration processes.

PCBs are especially prominent pollutants in the sediments of the Hudson River as a result of waste discharges from two capacitor manufacturing plants that operated about 60 km upstream from the southernmost dam on the river from 1950 to 1976. The river sediments downstream from the plants exhibit PCB levels of about 10 ppm, 1–2 orders of magnitude higher than levels commonly encountered in river and estuary sediments. In 2002, General Electric Co. was ordered to dredge and decontaminate sections of the Hudson River polluted with PCBs at a cost exceeding $100 million. As of 2009, the actual cleanup was barely underway.

7.13 EMERGING WATER POLLUTANTS, PHARMACEUTICALS, AND HOUSEHOLD WASTES

The continued development of new products used for various purposes has led to interest in *emerging pollutants* of various kinds which may be of concern in water. Prominent among these are *nanomaterials* consisting of very small entities in the 1–100 nm size range. Nanomaterials of various kinds have unique properties of high thermal stability, low permeability, high strength, and high conductivity. These and other properties are leading to uses in electronics, automobiles, apparel, sunscreens, cosmetics, water purification, and other products. It is anticipated that the use of nanomaterials in drug delivery will increase rapidly in the future. The commercial uses of nanomaterials are in their infancy, but very rapid growth is underway. Little is known about the potential pollution effects and toxicities of nanomaterials, so their potential effects as water pollutants are of significant concern.

Another class of emerging pollutants consists of *siloxanes* (commonly called silicones) including octamethylcyclotetrasiloxane, decamethylcyclopentasiloxane, and dodecamethylcyclohexasiloxane. Siloxanes are thermally and chemically very stable leading to their uses as coolants in transformers, protective encapsulating materials in semiconductors, lubricants, coatings, and sealants. Siloxanes are widely used in personal care products including deodorants, cosmetics, soaps, hair conditioners, and hair dyes and other products such as water repellent windshield coatings, detergent antifoaming agents, and even food additives. Siloxanes are resistant to biodegradation and as a result are encountered in water that has received wastewater.

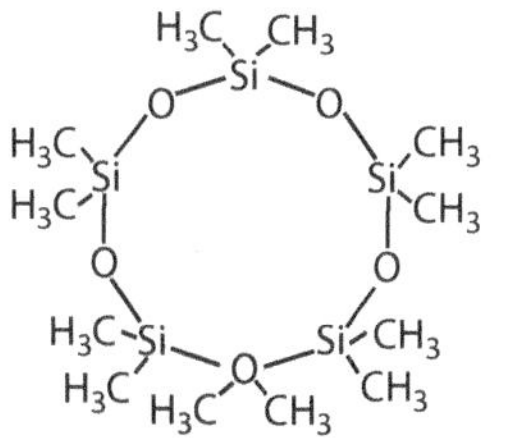

Decamethylcyclopentasiloxane, a cyclic siloxane

Disinfection by-products are of some concern as water pollutants. These are compounds containing halogens and nitrogen that result from reactions of water disinfectants including chlorine, hypochlorite, and chlorine dioxide. In addition to exposure in drinking water, humans may be exposed to these substances through skin contact in bathing or swimming and as vapors emitted from water during showers. Brominated and iodated compounds are formed in chlorinated water by reactions of bromide or iodide in the water, usually present in areas of seawater or saline groundwater intrusion.

The most common disinfection by-products are the *trihalomethanes*—chloroform ($CHCl_3$), dibromochloromethane ($CHClBr_2$), bromodichloromethane ($CHCl_2Br$), and tribromomethane ($CHBr_3$). These are all by-products of water chlorination and are Group B carcinogens (shown to cause cancer in laboratory animals). By far the most abundant of these in water systems is trichloromethane (chloroform). Dibromochloromethane is regarded as posing about 10 times the risk of cancer as chloroform, and tribromomethane is thought to pose just a slightly greater cancer risk than chloroform. As of 2009, the maximum allowable limit for total trihalomethanes in drinking water was 100 μg/L.

Various substances associated with household wastes are found in water, especially in treated sewage discharges. These materials include steroids, surfactants, flame retardants, fragrances, plasticizers, and pharmaceuticals and their metabolites. It should be noted that, although significant numbers of such compounds are found, they are generally at subpart-per-billion levels and are detectable only by the remarkable capability of modern analytical instrumentation. A study of "exotic" organics found in groundwater and water supplies[6] detected a variety of substances including cholesterol, nicotine metabolite cotinine, β-sitosterol (a natural plant sterol), 1,7-dimethylxanthine (caffeine metabolite), bisphenol-A plasticizer, and fire retardant (2-chloroethyl) phosphate.

Pharmaceutical compounds and their partial degradation products are discharged with sewage as wastes from human ingestion and from being discarded with wastewater.[7] The quantities of these substances in sewage in developed countries can reach of the order of 100 metric tons per year. Levels of common pharmaceuticals of around 1 μg/L have been observed in river water. Although these are relatively low values, they are of some concern because of the biological activity inherent to pharmaceutical products, which are increasingly designed for higher potency, bioavailability, and resistance to degradation. Figure 7.16 shows some of the most common pharmaceuticals and their degradation products that have been observed in water.

The most obvious effect of pharmaceuticals and their metabolic products in water has been the feminization of male fish observed downstream from treated sewage discharges resulting from estrogenic compounds in wastewater. First noted in England and later in the United States and Europe, these male fish have been observed to produce proteins associated with egg production by female fish and to produce early-stage eggs in their testes. These effects are largely attributed to residues of synthetic 17α-ethinylestradiol and the natural hormone 17β-estradiol, used in oral contraceptives. The glucuronide and sulfate conjugates of 17α-ethinylestradiol (see Chapter 23 for a discussion of conjugates formed metabolically) are excreted with urine and cleaved by bacteria in water to regenerate the original compound.

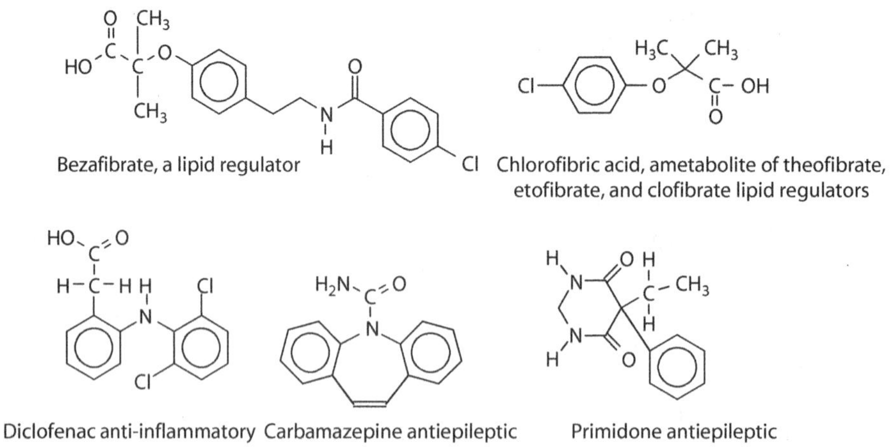

FIGURE 7.16 Pharmaceuticals and metabolites of concern as water pollutants.

7.13.1 Bactericides

Bactericides used in cleaning and consumer products may be encountered in water. One of the most common of these is triclosan,

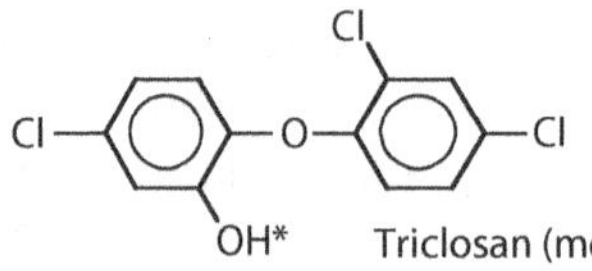

Triclosan (methyl triclosan has a $-CH_3$ group substituted for the H designated H*)

widely used in antibacterial soaps as well as other consumer items such as shampoos, deodorants, lotions, toothpastes, sportswear, shoes, carpets, and even refuse containers. This compound and its methyl derivative have been found in natural waters in Switzerland.[8]

7.13.2 Estrogenic Substances in Wastewater Effluents

A class of water pollutants of particular concern commonly found in sewage and even treated sewage effluent are *estrogenic substances* that can disrupt the crucial endocrine gland activities that regulate the metabolism and reproductive functions of organisms (Section 22.8). Aquatic organisms including fish, frogs, and reptiles such as alligators exposed to such substances may exhibit reproductive dysfunction, alterations in secondary sex characteristics, and abnormal serum steroid levels. Such substances include exogenous estrogenic substances among which are 17α-ethynyl estradiol, diethylstilbestrol, mestranol, levonorgestrel, and norethindrone used in oral contraceptives, treatment of hormonal disorders, and cancer treatment. Some synthetic substances also act as estrogen disruptors. Of prime concern as water pollutants are nonionic surfactant polyethoxylates, mentioned in the discussion of detergents above, and their major degradation product, persistent nonylphenol. Although these substances are orders of magnitude less potent than hormonal substances, annual usage of millions of kilograms of nonionic surfactants make them a significant factor as water pollutants.

7.13.3 Biorefractory Organic Pollutants

Millions of tons of organic compounds are manufactured globally each year. Significant quantities of several thousand such compounds appear as water pollutants. Most of these compounds, particularly the less biodegradable ones, are substances to which living organisms have not been exposed until recent years. Often, their effects on organisms are not known, particularly for long-term exposures at very low levels. The potential exists for synthetic organics to cause genetic damage, cancer, or other ill effects. On the positive side, organic pesticides enable a level of agricultural productivity without which millions would starve. Synthetic organic chemicals are increasingly taking the place of natural products in short supply. Thus it is seen that organic chemicals are essential for the operation of a modern society. Because of their potential danger, however, acquisition of knowledge about their environmental chemistry must have a high priority.

Biorefractory organics are the organic compounds of most concern in wastewater, particularly when they are found in sources of drinking water. These are poorly biodegradable substances and are sometimes referred to as *persistent organic pollutants* (POP), prominent among which are aryl or chlorinated hydrocarbons. Biorefractory compounds that may be found in water include benzene, chloroform, methyl chloride, styrene, tetrachloroethylene, trichloroethane, and toluene. In addition to their potential toxicity, biorefractive compounds can cause taste and odor problems in water. They are not completely removed by biological treatment, and water contaminated with these compounds must be treated by physical and chemical means, including air stripping, solvent extraction, ozonation, and carbon adsorption.

Methyl *tert*-butyl ether (MTBE),

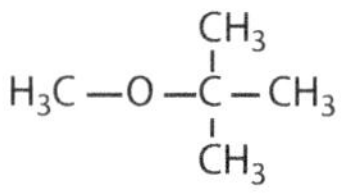

$$H_3C-O-\underset{CH_3}{\overset{CH_3}{\overset{|}{\underset{|}{C}}}}-CH_3$$

was once used as a gasoline octane booster, but was phased out after it appeared as a low-level water pollutant in the United States. Levels of this chemical in recreational lakes and reservoirs were attributed largely to emissions of unburned fuel from recreational motorboats and personal watercraft having two-cycle engines that discharge their exhausts directly to the water.

Perfluorinated organic compounds constitute a unique class of POPs. They occur as completely fluorinated hydrocarbon derivatives, such as CF_4, in the atmosphere, where they are regarded as atmospheric pollutants and potential greenhouse gases (see Section 12.7). Other perfluorinated compounds that are organic acids or their salts have been encountered as water pollutants. Most commonly cited are salts of perfluorooctane sulfonic acid; others include salts of perfluorinated carboxylic acids, such as perfluorohexanoic acid:

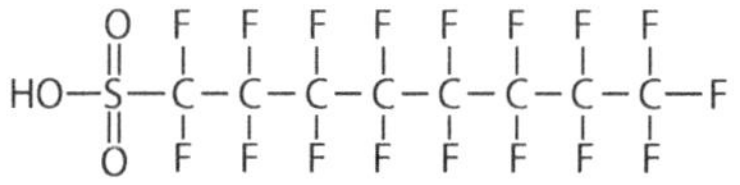

$$HO-S(=O)_2-(CF_2)_7-CF_3$$

Perfluorooctane sulfonic acid

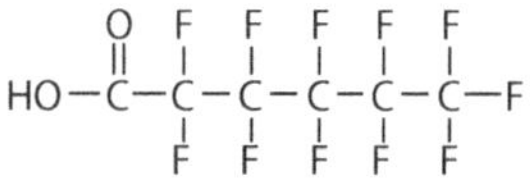

$$HO-C(=O)-(CF_2)_4-CF_3$$

Perfluorohexanoic acid

Perfluorocarbons have been used commercially since the 1950s, primarily as coatings to resist soil and grease in paper products, fabrics, carpet material, and leather. Scotchgard fabric protector, once manufactured by 3M Corporation, contained perfluorooctane sulfonates, but this use has been discontinued. Perfluorocarbons have also been used as surfactants in oil drilling fluids and firefighting foams. Other applications have included alkaline cleaners, floor polish formulations, etching baths, and even denture cleaners. Perfluorocarbons have been detected in water, fish blood and liver, and human blood.

Brominated compounds have been recognized as significant environmental and water pollutants in recent years and have even been found in mothers' milk in some countries. These compounds have been manufactured as flame retardants, largely for use in polymers and textiles. The most common of the brominated compounds likely to be encountered as pollutants are polybrominated diphenyl ethers and tetrabromobisphenol:

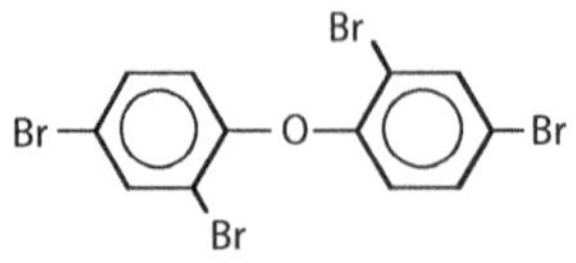

2,2′,4,4′ tetrabromodiphenyl ether 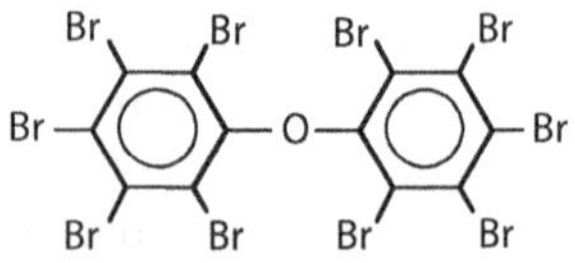 Decabromodiphenyl ether

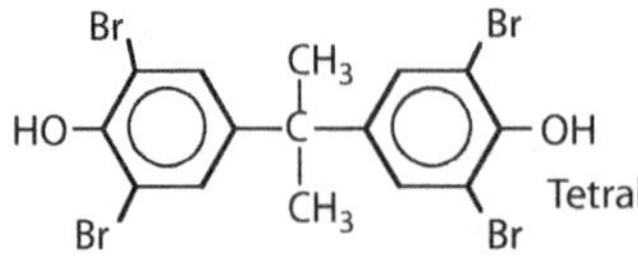

Tetrabromobisphenol A

Benzotriazole and tolyltriazoles (structural formulas below) are complexing agents for metals that are widely used as anticorrosive additives by forming a thin complexing film on metal surfaces, thereby protecting the metal from corrosion. These compounds are used in a variety of products including hydraulic fluids, cooling fluids, antifreeze formulations, and aircraft deicer fluids, and for silver protection in dishwasher detergents. Because of these uses, the triazoles are commonly encountered as "down-the-drain" chemicals that get into wastewaters. Their widespread use, high water solubility, and poor biodegradability make them among the most widely encountered chemicals in waters receiving treated wastewater.[9]

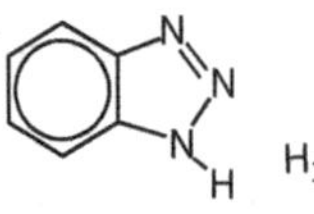

Benzotriazole 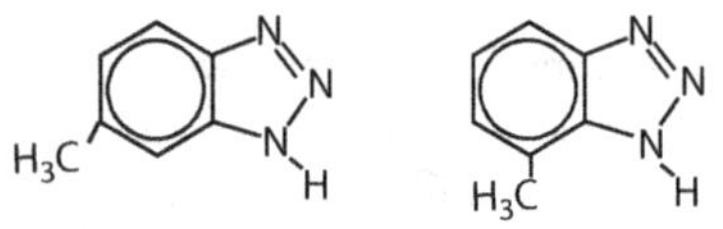 Tolyltriazoles

Naphthenic acid is a complex and variable mixture of carboxylic acids of molar masses in the approximate range of 180–350 and typically containing one $-CO_2H$ group and one 5- or 6-membered ring per molecule that is a by-product of petroleum refining. Naphthenic acids recovered from petroleum refining are used for solvents, lubricants, corrosion inhibitors, metal naphthenate synthesis, fuel additives, deicing, dust control, wood preservation, and road stabilization. Water pollution from naphthenic acids is most severe in the tar sands processing area of Alberta, Canada, where caustic hot water is used to wash heavy hydrocarbon crude oil from sand, leaving huge quantities of tailings of clay, sand, and water contaminated with 80–120 mg/L naphthenic acids. The acids are toxic to aquatic organisms and are endocrine disrupting substances.

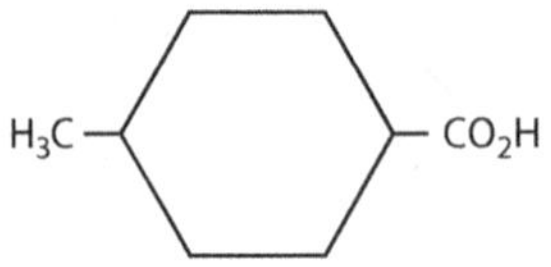

4-Methyl-1-cyclohexane carboxylic acid, a typical low-molar-mass naphthenic acid

7.14 RADIONUCLIDES IN THE AQUATIC ENVIRONMENT

The massive production of *radionuclides* (radioactive isotopes) by weapons and nuclear reactors since World War II has been accompanied by increasing concern about the effects of radioactivity on health and the environment. Radionuclides are produced as fission products of heavy nuclei of elements such as uranium or plutonium. They are also produced by the reaction of neutrons with stable nuclei. These

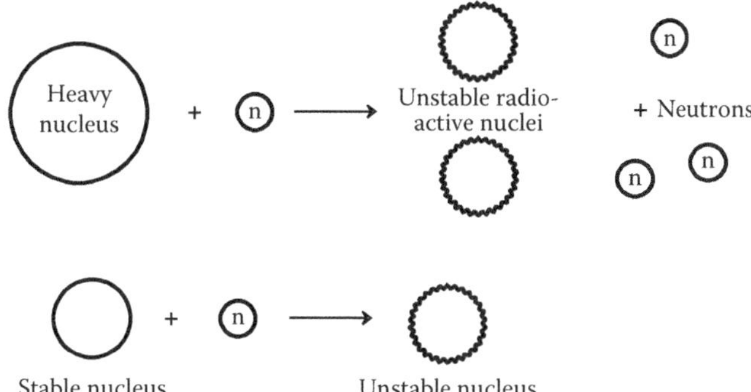

FIGURE 7.17 A heavy nucleus, such as that of ^{235}U, may absorb a neutron and break up (undergo fission), yielding lighter radioactive nuclei. A stable nucleus may absorb a neutron to produce a radioactive nucleus.

phenomena are illustrated in Figure 7.17 and specific examples are given in Table 7.6. Radionuclides are formed in large quantities as waste products in nuclear power generation. Their ultimate disposal is a problem that has caused much controversy regarding the widespread use of nuclear power. Artificially produced radionuclides are also widely used in industrial and medical applications, particularly as "tracers." With so many possible sources of radionuclides, it is impossible to entirely

TABLE 7.6
Radionuclides in Water

Radionuclide	Half-Life	Nuclear Reaction, Description, Source
		Naturally Occurring and from Cosmic Reactions
Carbon-14	5730 y[a]	$^{14}N(n,p)^{14}C$,[b] thermal neutrons from cosmic or nuclear-weapon sources reacting with N_2
Silicon-32	~300 y	$^{40}Ar(p,x)^{32}Si$, nuclear spallation (splitting of the nucleus) of atmospheric argon by cosmic-ray protons
Potassium-40	$\sim 1.4 \times 10^9$ y	0.0119% of natural potassium including potassium in the body
		Naturally Occurring from ^{238}U Series
Radium-226	1620 y	Diffusion from sediments, atmosphere
Lead-210	21 y	$^{226}Ra \rightarrow$ 6 steps $\rightarrow {}^{210}Pb$
Thorium-230	75,200 y	$^{238}U \rightarrow$ 3 steps $\rightarrow {}^{230}Th$ produced *in situ*
Thorium-234	24 d	$^{238}U \rightarrow {}^{234}Th$ produced *in situ*
		From Reactor and Weapons Fission[c]
Strontium-90 (28 y) Iodine-131 (8 d) Cesium-137 (30 y)		
Barium-140 (13 d) > Zirconium-95 (65 d) > Cerium-141 (33 d) > Strontium-89 (51 d) > Ruthenium-103 (40 d) > Krypton-85 (10.3 y)		
		From Nonfission Sources
Cobalt-60	5.25 y	From nonfission neutron reactions in reactors
Manganese-54	310 d	From nonfission neutron reactions in reactors
Iron-55	2.7 y	$^{56}Fe(n,2n)^{55}Fe$, from high-energy neutrons acting on iron in weapons hardware
Plutonium-239	24,300 y	$^{238}U(n,g)^{239}Pu$, neutron capture by uranium

[a] Abbreviations: y, years; d, days.
[b] This notation shows the isotope nitrogen-14 reacting with a neutron, *n*, giving off a proton, *p*, and forming the carbon-14 isotope; other nuclear reactions can be deduced from this notation where *x* represents nuclear fragments from spallation.
[c] The first three fission-product radioisotopes listed below as products of reactor and weapons fission are of most significance because of their high yields and biological activity. The other fission products are listed in generally decreasing order of yield.

eliminate radioactive contamination of aquatic systems. Furthermore, radionuclides may enter aquatic systems from natural sources. Therefore, the transport, reactions and biological concentration of radionuclides in aquatic ecosystems are of great importance to the environmental chemist.

Radionuclides differ from other nuclei in that they emit *ionizing radiation*—alpha particles, beta particles, and gamma rays. The most massive of these emissions is the *alpha particle*, a helium nucleus of atomic mass 4, consisting of two neutrons and two protons. The symbol for an alpha particle is ${}^{4}_{2}\alpha$. An example of alpha production is found in the radioactive decay of uranium-238:

$$ {}^{238}_{92}\text{U} \rightarrow {}^{234}_{90}\text{Th} + {}^{4}_{2}\alpha \tag{7.12} $$

This transformation occurs when a uranium nucleus, atomic number 92 and atomic mass 238, loses an alpha particle, atomic number 2 and atomic mass 4, to yield a thorium nucleus, atomic number 90 and atomic mass 234.

Beta radiation consists of either highly energetic, negative electrons, which are designated ${}^{0}_{-1}\beta$, or positive electrons, called positrons, and designated ${}^{0}_{1}\beta$. A typical beta emitter, chlorine-38, may be produced by irradiating chlorine with neutrons. The chlorine-37 nucleus, natural abundance 24.5%, absorbs a neutron to produce chlorine-38 and gamma radiation:

$$ {}^{37}_{17}\text{Cl} + {}^{1}_{0}\text{n} \rightarrow {}^{38}_{17}\text{Cl} + \gamma \tag{7.13} $$

The chlorine-38 nucleus is radioactive and loses a negative *beta particle* to become an argon-38 nucleus:

$$ {}^{38}_{17}\text{Cl} \rightarrow {}^{38}_{18}\text{Ar} + {}^{0}_{-1}\beta \tag{7.14} $$

Since the negative beta particle has essentially no mass and a –1 charge, the stable product isotope, argon-38, has the same mass and a charge 1 greater than chlorine-38.

Gamma rays are electromagnetic radiation similar to x-rays, though more energetic. Since the energy of gamma radiation is often a well-defined property of the emitting nucleus, it may be used in some cases for the qualitative and quantitative analysis of radionuclides.

The primary effect of alpha particles, beta particles, and gamma rays on materials is the production of ions; therefore, they are called *ionizing radiation*. Due to their large size, alpha particles do not penetrate matter deeply, but cause an enormous amount of ionization along their short path of penetration. Therefore, alpha particles present little hazard outside the body, but are very dangerous when ingested. Although beta particles are more penetrating than alpha particles, they produce much less ionization per unit path length. Gamma rays are much more penetrating than particulate radiation, but cause much less ionization. Their degree of penetration is proportional to their energy.

The *decay* of a specific radionuclide follows first-order kinetics; that is, the number of nuclei disintegrating in a short time interval is directly proportional to the number of radioactive nuclei present. The rate of decay, $-\mathrm{d}N/\mathrm{d}t$, is given by the equation

$$ \text{Decay rate} = \frac{\mathrm{d}N}{\mathrm{d}t} = \lambda N \tag{7.15} $$

where N is the number of radioactive nuclei present and λ is the rate constant, which has units of reciprocal time. Since the exact number of disintegrations per second is difficult to determine in the laboratory, radioactive decay is often described in terms of the measured *activity*, A, which is proportional to the absolute rate of decay. The first-order decay equation may be expressed in terms of A:

$$A = A_0 e^{-\lambda t} \tag{7.16}$$

where A is the activity at time t, A_0 is the activity when t is zero, and e is the natural logarithm base. The *half-life*, $t_{1/2}$, is generally used instead of λ to characterize a radionuclide:

$$t_{1/2} = \frac{0.693}{\lambda} \tag{7.17}$$

As the term implies, a half-life is the period of time during which half of a given number of atoms of a specific kind of radionuclide decay. Ten half-lives are required for the loss of 99.9% of the activity of a radionuclide.

Radiation damages living organisms by initiating harmful chemical reactions in tissues. For example, bonds are broken in the macromolecules that carry out life processes. In cases of acute radiation poisoning, bone marrow, which produces red blood cells, is destroyed and the concentration of red blood cells is diminished. Radiation-induced genetic damage is of great concern. Such damage may not become apparent until many years after exposure. As humans have learned more about the effects of ionizing radiation, the dosage level considered to be safe has steadily diminished. For example, the U.S. Nuclear Regulatory Commission has dropped the maximum permissible concentration of some radioisotopes to levels of less than one ten-thousandth of those considered safe in the early 1950s. Although it is possible that even the slightest exposure to ionizing radiation entails some damage, some radiation is unavoidably received from natural sources including the radioactive ^{40}K found in all humans. For the majority of the population, exposure to natural radiation exceeds that from artificial sources.

The study of the ecological and health effects of radionuclides involves consideration of many factors. Among these are the type and energy of radiation emitter and the half-life of the source. In addition, the degree to which the particular element is absorbed by living species and the chemical interactions and transport of the element in aquatic ecosystems are important factors. Radionuclides having very short half-lives may be hazardous when produced but decay too rapidly to affect the environment into which they are introduced. Radionuclides with very long half-lives may be quite persistent in the environment but of such low activity that little environmental damage is caused. Therefore, in general, radionuclides with intermediate half-lives are the most dangerous. They persist long enough to enter living systems while still retaining a high activity. Because they may be incorporated within living tissue, radionuclides of "life elements" are particularly dangerous. Much concern has been expressed over strontium-90, a common waste product of nuclear testing. This element is interchangeable with calcium in bone. Strontium-90 fallout drops onto pasture and cropland and is ingested by cattle. Eventually, it enters the bodies of infants and children by way of cow's milk.

Some radionuclides found in water, primarily radium and potassium-40, originate from natural sources, particularly leaching from minerals. Others come from pollutant sources, primarily nuclear power plants and testing of nuclear weapons. The levels of radionuclides found in water typically are measured in units of pCi/L (picocuries per liter), where a curie is 3.7×10^{10} disintegrations per second, and a pCi is 1×10^{-12} that amount, or 3.7×10^{-2} disintegrations per second (2.2 disintegrations per minute).

The radionuclide of most concern in drinking water is *radium*, Ra. Areas in the United States where significant radium contamination of water has been observed include the uranium-producing regions of the western United States, Iowa, Illinois, Wisconsin, Missouri, Minnesota, Florida, North Carolina, Virginia, and the New England states.

The U.S. Environmental Protection Agency specifies maximum contaminant levels (MCL) for total radium ($^{226}Ra + ^{228}Ra$) in drinking water in units of pCi/L. In the past, perhaps as many as several hundred municipal water supplies in the United States have exceeded permissible levels, which has required finding alternative sources or additional treatment to remove radium. Fortunately, conventional water softening processes, which are designed to take out excessive levels of calcium, are relatively efficient in removing radium from water.

The possible contamination of water by fission-product radioisotopes from nuclear power production is of some concern. (If nations continue to refrain from testing nuclear weapons above ground, it is hoped that radioisotopes from this source will contribute only minor amounts of radioactivity to water.) Table 7.6 summarizes the major natural and artificial radionuclides likely to be encountered in water.

Transuranic elements are also of concern in the oceanic environment. These alpha emitters are long-lived and highly toxic. Included among these elements are various isotopes of neptunium, plutonium, americium, and curium. Specific isotopes, with half-lives in years given in parentheses, are Np-237 (2.14×10^6), Pu-236 (2.85), Pu-238 (87.8), Pu-239 (2.44×10^4), Pu-240 (6.54×10^3), Pu-241 (15), Pu-242 (3.87×10^5), Am-241 (433), Am-243 (7.37×10^6), Cm-242 (0.22), and Cm-244 (17.9).

LITERATURE CITED

1. Grove, R. A., C. J. Henny, and J. L. Kaiser, Worldwide sentinel species for assessing and monitoring environmental contamination in rivers, lakes, reservoirs, and estuaries, *Journal of Toxicology and Environmental Health, Part B: Critical Reviews*, **12**, 25–44, 2008.
2. Arnot, J. A. and Frank A. P. C. Gobas, A review of bioconcentration factor (BCF) and bioaccumulation factor (BAF) assessments for organic chemicals in aquatic organisms, *Environmental Reviews*, **14**, 257–297, 2008.
3. Vetter, W., Marine halogenated natural products of environmental relevance, *Reviews of Environmental Contamination and Toxicology*, **188**, 1–57, 2006.
4. Gosselin, R. E., R. P. Smith, and H. C. Hodge, Paraquat, *Clinical Toxicology of Commercial Products*, 5th ed., pp. III-328–III-336, Williams and Wilkins, Baltimore/London, 1984.
5. Pieper, D. H. and M. Seeger, Bacterial metabolism of polychlorinated biphenyls, *Journal of Molecular Microbiology and Biotechnology*, **15**, 121–138, 2008.
6. Focazio, M. J., D. W. Kolpin, K. K. Barnes, E. T. Furlong, M. T. Meyer, S. D. Zaugg, L. B. Barber, and M. E. Thurman, A national reconnaissance of pharmaceuticals and other organic wastewater contaminants in the United States, *Science of the Total Environment*, **402**, 192–216, 2008.
7. Khetan, S. K. and T. J. Collins, Human pharmaceuticals in the aquatic environment: A challenge to green chemistry, *Chemical Reviews*, **107**, 2319–2364, 2007.
8. Lindström, A., I. J. Buerge, T. Poiger, P.-A. Bergqvist, M. D. Müller, and H.-R. Buser, Occurrence and environmental behavior of the bactericide triclosan and its methyl derivative in surface waters and in wastewater, *Environmental Science and Technology*, **36**, 2322–2329, 2002.
9. Giger, W., Christian S., and H.-P. Kohler, Benzotriazole and tolyltriazole as aquatic contaminants. 1. Input and occurrence in rivers and lakes, *Environmental Science and Technology*, **40**, 7186–7192, 2006.

SUPPLEMENTARY REFERENCES

Alley, E. R., *Water Quality Control Handbook*, 2nd ed., McGraw-Hill, New York, 2007.

Burk, A. R., Ed., *Water Pollution: New Research*, Nova Science Publishers, New York, 2005.

Calhoun, Y., Ed., *Water Pollution*, Chelsea House Publishers, Philadelphia, 2005.

Eckenfelder, W. W., D. L. Ford, and A. J. Englande, *Industrial Water Quality*, 4th ed., McGraw-Hill, New York, 2009.

Fellenberg, G. and A. Wier, *The Chemistry of Pollution*, 3rd ed., Wiley, Hoboken, NJ, 2000.

Gilliom, R. J., *Pesticides in the Nation's Streams and Ground Water, 1992–2001: The Quality of our Nation's Waters*, U.S. Geological Survey, Reston, VA, 2006.

Hamilton, D. and S. Crossley, Eds, *Pesticide Residues in Food and Drinking Water: Human Exposure and Risks*, Wiley, New York, 2004.

Howd, R. A. and A. M. Fan, *Risk Assessment for Chemicals in Drinking Water*, Wiley, Hoboken, NJ, 2007.

Kaluarachchi, J. J., *Groundwater Contamination by Organic Pollutants: Analysis and Remediation*, American Society of Civil Engineers, Reston, VA, 2001.

Knepper, T. P., D. Barcelo, and P. de Voogt, *Analysis and Fate of Surfactants in the Aquatic Environment*, Elsevier, Amsterdam, 2003.

Laws, E. A. *Aquatic Pollution: An Introductory Text*, 3rd ed., Wiley, New York, 2000.

Lewinsky, A. A., Ed., *Hazardous Materials and Wastewater: Treatment, Removal and Analysis*, Nova Science Publishers, New York, 2007.

Lipnick, R. L., D. C. G. Muir, K. C. Jones, J. L. M. Hermens, and D. Mackay, Eds, *Persistent, Bioaccumulative, and Toxic Chemicals II: Assessment and Emerging New Chemicals*, American Chemical Society, Washington, DC, 2000.

Livingston, J. V., Ed., *Focus on Water Pollution Research*, Nova Science Publishers, New York, 2006.

Mason, C. F., *Biology of Freshwater Pollution*, 4th ed., Prentice Hall College Division, Upper Saddle River, NJ, 2002.

Raven, P. H., L. R. Berg, and D. M. Hassenzahl, *Environment*, 6th ed., Wiley, Hoboken, NJ, 2008.

Ravenscroft, P., H. Brammer, and K. Richards, *Arsenic Pollution: A Global Synthesis*, Blackwell, Malden, MA, 2009.

Research Council Committee on Drinking Water Contaminants, *Classifying Drinking Water Contaminants for Regulatory Consideration*, National Academy Press, Washington, DC, 2001.

Rico, D. P., C. A. Brebbia, and Y. Villacampa Esteve, Eds, *Water Pollution IX*, WIT Press, Southampton, UK, 2008.

Ritter, W. F. and A. Shirmohammadi, Eds, *Agricultural Nonpoint Source Pollution: Watershed Management and Hydrology*, CRC Press/Lewis Publishers, Boca Raton, FL, 2000.

Stollenwerk, K. G. and A. H. Welch, Eds, *Arsenic in Ground Water: Geochemistry and Occurrence*, Kluwer Academic Publishers, Hingham, MA, 2003.

Sullivan, P. J., F. J. Agardy, and J. J. Clark, *The Environmental Science of Drinking Water*, Elsevier Butterworth-Heinemann, Burlington, MA, 2005.

Viessman, W. and M. J. Hammer, *Water Supply and Pollution Control*, 7th ed., Pearson Prentice Hall, Upper Saddle River, NJ, 2005.

Water Environment Research, a research publication of the Water Environment Federation, Water Environment Federation, Alexandria, VA. This journal contains may articles of interest to water science; the annual reviews are especially informative.

Wheeler, W. B., *Pesticides in Agriculture and the Environment*, Marcel Dekker, New York, 2002.

Whitacre, D. M., Ed., *Reviews of Environmental Contamination and Toxicology*, **196**, Springer-Verlag, New York, 2008 (published annually).

Xie, Y., *Disinfection Byproducts in Drinking Water: Form, Analysis, and Control*, CRC Press, Boca Raton, FL, 2002.

QUESTIONS AND PROBLEMS

1. Which of the following statements are true regarding chromium in water: (a) chromium(III) is suspected of being carcinogenic, (b) chromium(III) is less likely to be found in a soluble form than chromium(VI), (c) the toxicity of chromium(III) in electroplating wastewaters is decreased by oxidation to chromium(VI), (d) chromium is not an essential trace element, (e) chromium is known to form methylated species analogous to methylmercury compounds.
2. What do mercury and arsenic have in common in regard to their interactions with bacteria in sediments?
3. What are some characteristics of radionuclides that make them especially hazardous to humans?
4. To what class do pesticides containing the following group belong?

```
  H  O
  |  ||
—N—C—
```

5. Consider the following compound:

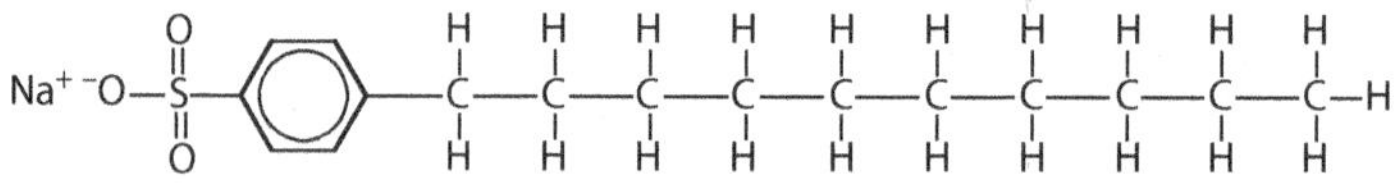

Which of the following characteristics is not possessed by the compound: (a) one end of the molecule is hydrophilic and the other end in hydrophobic, (b) surface-active qualities, (c) the ability to lower surface tension of water, (d) good biodegradability, (e) tendency to cause foaming in sewage treatment plants.
6. A certain pesticide is fatal to fish fingerlings at a level of 0.50 ppm in water. A leaking metal can containing 5.00 kg of the pesticide was dumped into a stream with a flow of 10.0 L/s moving at 1 km/h. The container leaks pesticide at a constant rate of 5 mg/s. For what distance (in km) downstream is the water contaminated by fatal levels of the pesticide by the time the container is empty?
7. Give a reason why Na_3PO_4 would not function well as a detergent builder, whereas $Na_3P_3O_{10}$ is satisfactory, though it is a source of pollutant phosphate.
8. Of the compounds $CH_3(CH_2)_{10}CO_2H$, $(CH_3)_3C(CH_2)_2CO_2H$, $CH_3(CH_2)_{10}CH_3$, and ϕ-$(CH_2)_{10}CH_3$ (where ϕ represents a benzene ring), which is the most readily biodegradable?
9. A pesticide sprayer got stuck while trying to ford a stream flowing at a rate of 136 L/s. Pesticide leaked into the stream for exactly 1 h and at a rate that contaminated the stream at a uniform 0.25 ppm of methoxychlor. How much pesticide was lost from the sprayer during this time?
10. A sample of water contaminated by the accidental discharge of a radionuclide used for medicinal purposes showed an activity of 12,436 counts per second at the time of sampling and 8966 counts per second exactly 30 days later. What is the half-life of the radionuclide?
11. What are the two reasons that soap is environmentally less harmful than ABS surfactant used in detergents?
12. What is the exact chemical formula of the specific compound designated as PCB?
13. Match each compound designated by a letter with the description corresponding to it designated by a number.

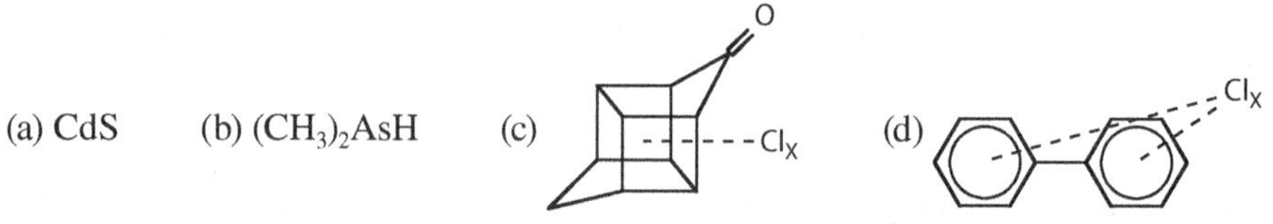

   1. Pollutant released to a U.S. stream by a poorly controlled manufacturing process.
   2. Insoluble form of a toxic trace element likely to be found in anaerobic sediments.
   3. Common environmental pollutant formerly used as a transformer coolant.
   4. Chemical species thought to be produced by bacterial action.
14. A radioisotope has a nuclear half-life of 24 h and a biological half-life of 16 h (half of the element is eliminated from the body in 16 h). A person accidentally swallowed sufficient quantities of this isotope to give an initial "whole body" count rate of 1000 counts per minute. What was the count rate after 16 h?
15. What is the primary detrimental effect on organisms of salinity in water arising from dissolved NaCl and Na_2SO_4?
16. Give a specific example of each of the following general classes of water pollutants: (a) trace elements, (b) metal–organic combinations, and (c) pesticides.
17. A polluted water sample is suspected of being contaminated with one of the following: soap, ABS surfactant, or LAS surfactant. The sample has a very low BOD (biochemical oxygen demand) relative to its TOC (total organic carbon content). Which is the contaminant?

18. Of the following, the one that is *not* a cause of or associated with eutrophication is (A) eventual depletion of oxygen in the water, (B) excessive phosphate, (C) excessive algal growth, (D) excessive nutrients, and (E) excessive O_2.
19. Match the pollutants on the left with effects or other significant aspects on the right, below:

A. Salinity	1. Excessive productivity
B. Alkalinity	2. Can enter water from pyrite or from the atmosphere
C. Acidity	3. Osmotic effects on organisms
D. Nitrate	4. From soil and mineral strata

20. Of the following heavy metals, choose the one most likely to have microorganisms involved in its mobilization in water and explain why this is so: (A) lead, (B) mercury, (C) cadmium, (D) chromium, and (E) zinc.
21. Of the following, choose the true statement: (A) eutrophication results from the direct discharge of toxic pollutants into water, (B) treatment of a lake with phosphates is a process used to deter eutrophication, (C) alkalinity is the most frequent limiting nutrient in eutrophication, (D) eutrophication results from excessive plant or algal growth, (E) eutrophication is generally a beneficial phenomenon because it produces oxygen.
22. Of the following, the statement that is *untrue* regarding radionuclides in the aquatic environment is (A) they emit ionizing radiation, (B) they invariably come from human activities, (C) radionuclides of "life elements," such as iodine-131, are particularly dangerous, (D) normally the radionuclide of the most concern in drinking water is radium, (E) they may originate from the fission of uranium nuclei.
23. From the formulas below match the following: (A) lowers surface tension of water, (B) a carbamate, (C) a herbicide, (D) a noncarbamate insecticide

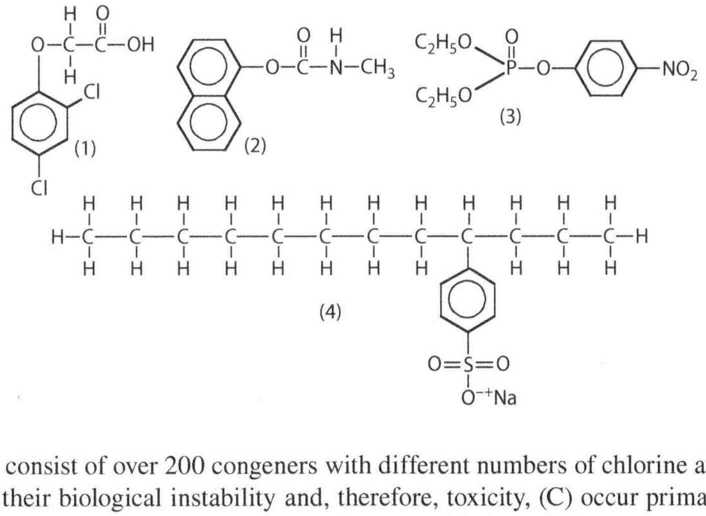

24. PCBs (A) consist of over 200 congeners with different numbers of chlorine atoms, (B) are noted for their biological instability and, therefore, toxicity, (C) occur primarily as localized pollutants, (D) are not known to undergo any biodegradation processes, (E) had no common uses, but were produced as manufacturing by-products.

8 Water Treatment

8.1 WATER TREATMENT AND WATER USE

The treatment of water may be divided into three major categories:

- Purification for domestic use
- Treatment for specialized industrial applications
- Treatment of wastewater to make it acceptable for release or reuse

The type and degree of treatment are strongly dependent upon the source and intended use of the water. Water for domestic use must be thoroughly disinfected to eliminate disease-causing microorganisms, but may contain appreciable levels of dissolved calcium and magnesium (hardness). Water to be used in boilers may contain bacteria but must be quite soft to prevent scale formation. Wastewater being discharged into a large river may require less rigorous treatment than water to be reused in an arid region. As world demand for limited water resources grows, more sophisticated and extensive means will have to be employed to treat water.

Most physical and chemical processes used to treat water involve similar phenomena, regardless of their application to the three main categories of water treatment listed above. Therefore, after introductions to water treatment for municipal use, industrial use, and disposal, each major kind of treatment process is discussed as it applies to all of these applications.

8.2 MUNICIPAL WATER TREATMENT

The modern water treatment plant is often called upon to perform wonders with the water fed to it. The clear, safe, even tasteful water that comes from a faucet may have started as a murky liquid pumped from a polluted river laden with mud and swarming with bacteria. Or, its source may have been well water, much too hard for domestic use and containing high levels of stain-producing dissolved iron and manganese. The water treatment plant operator's job is to make sure that the water plant product presents no hazards to the consumer.

A schematic diagram of a typical municipal water treatment plant is shown in Figure 8.1. This particular facility treats water containing excessive hardness and a high level of iron. The raw water taken from wells first goes to an aerator. Contact of the water with air removes volatile solutes such as hydrogen sulfide, carbon dioxide, methane, and volatile odorous substances such as methane thiol (CH_3SH) and bacterial metabolites. Contact with oxygen also aids iron removal by oxidizing soluble iron(II) to insoluble iron(III). The addition of lime as CaO or $Ca(OH)_2$ after aeration raises the pH and results in the formation of precipitates containing the hardness ions Ca^{2+} and Mg^{2+}. These precipitates settle from the water in a primary basin. Much of the solid material remains in suspension and requires the addition of coagulants [such as iron(III) and aluminum sulfates, which form gelatinous metal hydroxides] to settle the colloidal particles. Activated silica or synthetic polyelectrolytes may also be added to stimulate coagulation or flocculation. The settling occurs in a secondary basin after the addition of carbon dioxide to lower the pH. Sludge from both the primary and secondary basins is pumped to a sludge lagoon. The water is finally chlorinated, filtered, and pumped to the city water mains.

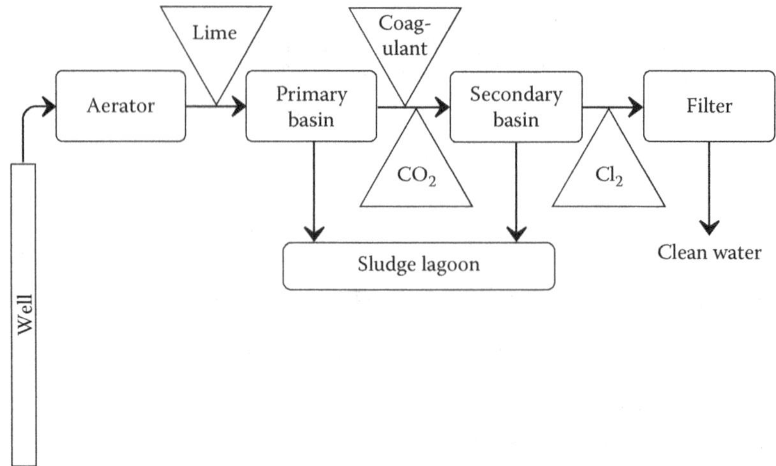

FIGURE 8.1 Schematic of a municipal water treatment plant.

8.3 TREATMENT OF WATER FOR INDUSTRIAL USE

Water is widely used in various process applications in industry. Other major industrial uses are boiler feedwater and cooling water. The kind and degree of treatment of water in these applications depends upon the end use. As examples, cooling water may require only minimal treatment, removal of corrosive substances and scale-forming solutes is essential for boiler feedwater, and water used in food processing must be free of pathogens and toxic substances. Improper treatment of water for industrial use can cause problems such as corrosion, scale formation, reduced heat transfer in heat exchangers, reduced water flow, and product contamination. These effects may cause reduced equipment performance or equipment failure, increased energy costs due to inefficient heat utilization or cooling, increased costs for pumping water, and product deterioration. Obviously, the effective treatment of water at minimum cost for industrial use is a very important area of water treatment.

Numerous factors must be taken into consideration in designing and operating an industrial water treatment facility. These include the following:

- Water requirement
- Quantity and quality of available water sources
- Sequential use of water (successive uses for applications requiring progressively lower water quality)
- Water recycle
- Discharge standards

The various specific processes employed to treat water for industrial use are discussed in later sections of this chapter. *External treatment*, usually applied to the plant's entire water supply, uses processes such as aeration, filtration, and clarification to remove material that may cause problems from water. Such substances include suspended or dissolved solids, hardness, and dissolved gases. Following this basic treatment, the water may be divided into different streams, some to be used without further treatment, and the rest to be treated for specific applications.

Internal treatment is designed to modify the properties of water for specific applications. Examples of internal treatment include the following:

- Reaction of dissolved oxygen (DO) with hydrazine or sulfite
- Addition of chelating agents to react with dissolved Ca^{2+} and prevent formation of calcium deposits

- Addition of precipitants, such as phosphate used for calcium removal
- Treatment with dispersants to inhibit scale
- Addition of inhibitors to prevent corrosion
- Adjustment of pH
- Disinfection for food processing uses or to prevent bacterial growth in cooling water

An important aspect of industrial process water treatment is the use of antiscalants and dispersants. The former prevent scale from materials such as $CaCO_3$ from building up and dispersants keep scale particles from adhering to surfaces by keeping them dispersed in water. One of the most effective agents for this purpose is polyacrylate polymer formed by the polymerization of acrylic acid and treatment with base. This polymer binds with scale-forming substances and keeps them dispersed in water by virtue of the negative charge of the polyacrylate. This same dispersant quality is useful in detergent formulations, some of which are about 5% polyacrylate. Polyacrylate is not biodegradable and accumulates with sludge residues from water treatment processes.

$$\left[-CH_2-CH(COO^-)- \right]_n$$

Polyacrylate polymer

8.4 SEWAGE TREATMENT

Typical municipal sewage contains oxygen-demanding materials, sediments, grease, oil, scum, pathogenic bacteria, viruses, salts, algal nutrients, pesticides, refractory organic compounds, heavy metals, and an astonishing variety of flotsam ranging from children's socks to sponges. It is the job of the waste treatment plant to remove as much of this material as possible.

Several characteristics are used to describe sewage. These include turbidity (international turbidity units), suspended solids (ppm), total dissolved solids (ppm), acidity (H^+ ion concentration or pH), and DO (in ppm O_2). BOD is used as a measure of oxygen-demanding substances.

Current processes for the treatment of wastewater may be divided into three main categories of primary treatment, secondary treatment, and tertiary treatment, each of which is discussed separately. Also discussed are total wastewater treatment systems, based largely upon physical and chemical processes.

Waste from a municipal water system is normally treated in a *publicly owned treatment works, POTW*. In the United States, these systems are allowed to discharge only effluents that have attained a certain level of treatment, as mandated by Federal law.

8.4.1 PRIMARY WASTE TREATMENT

Primary treatment of wastewater consists of the removal of insoluble matter such as grit, grease, and scum from water. The first step in primary treatment normally is screening. Screening removes or reduces the size of trash and large solids that get into the sewage system. These solids are collected on screens and scraped off for subsequent disposal. Most screens are cleaned with power rakes. Comminuting devices shred and grind solids in the sewage. Particle size may be reduced to the extent that the particles can be returned to the sewage flow.

Grit in wastewater consists of such materials as sand and coffee grounds which do not biodegrade well and generally have a high settling velocity. *Grit removal* is practiced to prevent its accumulation in other parts of the treatment system, to reduce clogging of pipes and other parts, and to protect moving parts from abrasion and wear. Grit is normally allowed to settle in a tank under conditions of low flow velocity, and it is then scraped mechanically from the bottom of the tank.

Primary sedimentation removes both settleable and floatable solids. During primary sedimentation there is a tendency for flocculant particles to aggregate for better settling, a process that may be aided by the addition of chemicals. The material that floats in the primary settling basin is known collectively as grease. In addition to fatty substances, the grease consists of oils, waxes, free fatty acids, and insoluble soaps containing calcium and magnesium. Normally, some of the grease settles with the sludge and some floats to the surface, where it may be removed by a skimming device.

8.4.2 Secondary Waste Treatment by Biological Processes

The most obvious harmful effect of biodegradable organic matter in wastewater is BOD, consisting of a biochemical oxygen demand for dissolved oxygen by microorganism-mediated degradation of the organic matter. *Secondary wastewater treatment* is designed to remove BOD, usually by taking advantage of the same kind of biological processes that would otherwise consume oxygen in water receiving the wastewater. Secondary treatment by biological processes takes many forms but consists basically of the action of microorganisms provided with added oxygen degrading organic material in solution or in suspension until the BOD of the waste has been reduced to acceptable levels.[1] The waste is oxidized biologically under conditions controlled for optimum bacterial growth, and at a site where this growth does not influence the environment.

One of the simplest biological waste treatment processes is the *trickling filter* (Figure 8.2) in which wastewater is sprayed over rocks or other solid support material covered with microorganisms. The structure of the trickling filter is such that contact of the wastewater with air is allowed and degradation of organic matter occurs by the action of the microorganisms.

Rotating biological reactors (*contactors*), another type of treatment system, consist of groups of large plastic discs mounted close together on a rotating shaft. The device is positioned such that at any particular instant half of each disc is immersed in wastewater and half exposed to air. The shaft rotates constantly, so that the submerged portion of the discs is always changing. The discs, usually made of high-density polyethylene or polystyrene, accumulate thin layers of attached biomass, which degrades organic matter in the sewage. Oxygen is absorbed by the biomass and by the layer of wastewater adhering to it during the time that the biomass is exposed to air.

Both trickling filters and rotating biological reactors are examples of fixed-film biological (FFB) or attached growth processes. The greatest advantage of these processes is their low energy consumption. The energy consumption is minimal because it is not necessary to pump air or oxygen into the water, as is the case with the popular activated sludge process described below. The trickling filter has long been a standard means of wastewater treatment, and a number of wastewater treatment plants use trickling filters at present.

The *activated sludge process*, Figure 8.3, is probably the most versatile and effective of all wastewater treatment processes. Microorganisms in the aeration tank convert organic material in wastewater

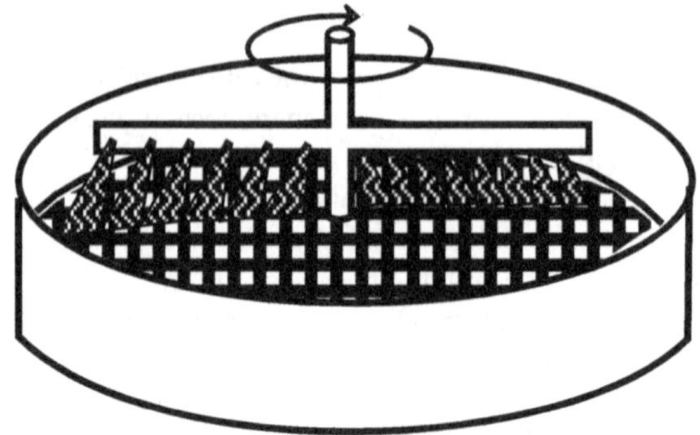

FIGURE 8.2 Trickling filter for secondary waste treatment.

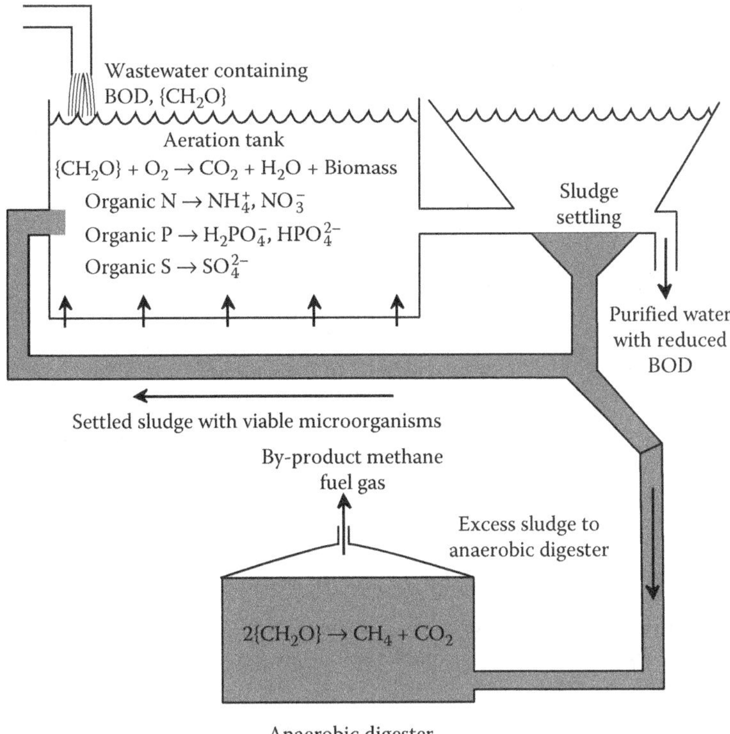

FIGURE 8.3 Activated sludge process.

to microbial biomass and CO_2. Organic nitrogen is converted to ammonium ion or nitrate. Organic phosphorus is converted to orthophosphate. The microbial cell matter formed as part of the waste degradation processes is normally kept in the aeration tank until the microorganisms are past the log phase of growth (Section 6.7), at which point the cells flocculate relatively well to form settleable solids. These solids settle out in a settler and a fraction of them is discarded. Part of the solids, the return sludge, is recycled to the head of the aeration tank and comes into contact with fresh sewage. The combination of a high concentration of "hungry" cells in the return sludge and a rich food source in the influent sewage provides optimum conditions for the rapid degradation of organic matter. The degradation of organic matter that occurs in an activated sludge facility also occurs in streams and other aquatic environments. However, in general, when a degradable waste is put into a stream, it encounters only a relatively small population of microorganisms capable of carrying out the degradation process. Thus, several days may be required for the buildup of a sufficient population of organisms to degrade the waste. In the activated sludge process, continual recycling of active organisms provides the optimum conditions for waste degradation, and a waste may be degraded within the very few hours that it is present in the aeration tank.

The activated sludge process provides two pathways for the removal of BOD, as illustrated schematically in Figure 8.4. BOD may be removed by (1) oxidation of organic matter to provide energy for the metabolic processes of the microorganisms and (2) synthesis, incorporation of the organic matter into cell mass. In the first pathway, carbon is removed in the gaseous form as CO_2. The second pathway provides for removal of carbon as a solid in biomass. That portion of the carbon converted to CO_2 is vented to the atmosphere and does not present a disposal problem. The disposal of waste sludge, however, is a problem, primarily because it is only about 1% solids and contains many undesirable components. Normally, partial water removal is accomplished by drying on sand filters, vacuum filtration, or centrifugation. The dewatered sludge may be incinerated or used as landfill.

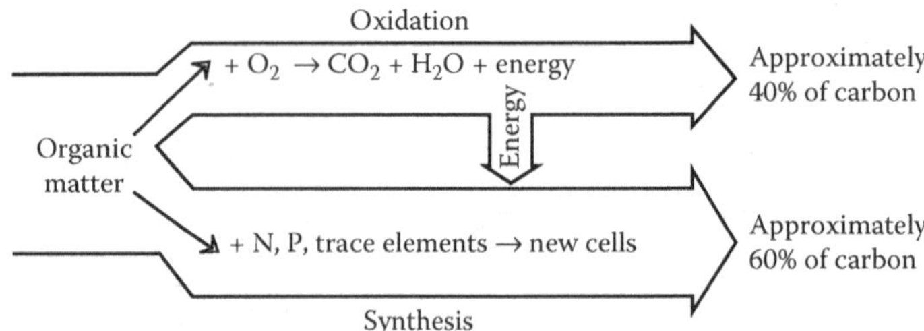

FIGURE 8.4 Pathways for the removal of BOD in biological wastewater treatment.

To a certain extent, sewage sludge may be digested in the absence of oxygen by methane-producing anaerobic bacteria to produce methane and carbon dioxide, a process that reduces both the volatile-matter content and the volume of the sludge by about 60%. A carefully designed plant may produce enough methane to provide for all of its power needs:

$$2\{CH_2O\} \rightarrow CH_4 + CO_2 \tag{8.1}$$

One of the most desirable means of sludge disposal is to use it to fertilize and condition soil. However, care has to be taken that excessive levels of heavy metals are not applied to the soil as sludge contaminants. Problems with various kinds of sludges resulting from water treatment are discussed further in Section 8.10.

Activated sludge wastewater treatment is the most common example of an aerobic suspended culture process. Many factors must be considered in the design and operation of an activated sludge wastewater treatment system. These include parameters involved with the process modeling and kinetics. The microbiology of the system must be considered. In addition to BOD removal, phosphorus and nitrogen removal must also be taken into account. Oxygen transfer and solids separation are important. Industrial wastes and the fates and effects of industrial chemicals (xenobiotics) must also be considered.

Nitrification (the microbially mediated conversion of ammonium nitrogen to nitrate; see Section 6.11) is a significant process that occurs during biological waste treatment. Ammonium ion is normally the first inorganic nitrogen species produced in the biodegradation of nitrogenous organic compounds. It is oxidized, under the appropriate conditions, first to nitrite by *Nitrosomonas* bacteria, then to nitrate by *Nitrobacter*:

$$2NH_4^+ + 3O_2 \rightarrow 4H^+ + 2NO_2^- + 2H_2O \tag{8.2}$$

$$2NO_2^- + O_2 \rightarrow 2NO_3^- \tag{8.3}$$

These reactions occur in the aeration tank of the activated sludge plant and are favored in general by long retention times, low organic loadings, large amounts of suspended solids, and high temperatures. Nitrification can reduce sludge settling efficiency because the denitrification reaction

$$4NO_3^- + 5\{CH_2O\} + 4H^+ \rightarrow 2N_2(g) + 5CO_2(g) + 7H_2O \tag{8.4}$$

occurring in the oxygen-deficient settler causes bubbles of N_2 to form on the sludge floc (aggregated sludge particles), making it so buoyant that it floats to the top. This prevents settling of the sludge and increases the organic load in the receiving waters. Under the appropriate conditions, however, advantage can be taken of this phenomenon to remove nutrient nitrogen from water (see Section 8.9).

8.4.3 Membrane Bioreactor

A problem with the activated sludge process is the difficulty encountered in settling the suspended biomass. Incomplete separation of the suspended solids in the sludge settling unit can result in solids contamination of the effluent and a dilute sludge that lacks sufficient biomass of active organisms required for effective waste biodegradation. These problems can be overcome with a *membrane bioreactor* in which a suspension of active biomass is maintained in an aeration tank and treated water is withdrawn through a membrane filter (see Section 8.6). In some configurations the membrane is immersed in the aeration chamber and the treated effluent drawn through the membrane filter under vacuum whereas in others the treated effluent is pumped through the filter under pressure.

8.4.4 Tertiary Waste Treatment

Unpleasant as the thought may be, many people drink used water—water that has been discharged from a municipal sewage treatment plant or from some industrial process. This raises serious questions about the presence of pathogenic organisms or toxic substances in such water. Because of high population density and heavy industrial development, the problem is especially acute in Europe where some municipalities process 50% or more of their water from "used" sources. Obviously, there is a great need to treat wastewater in a manner that makes it amenable to reuse. This requires treatment beyond the secondary processes.

Tertiary waste treatment (sometimes called *advanced waste treatment*) is a term used to describe a variety of processes performed on the effluent from secondary waste treatment.[1] The contaminants removed by tertiary waste treatment fall into the general categories of (1) suspended solids, (2) dissolved inorganic materials, and (3) dissolved organic compounds, including the important class of algal nutrients. Low levels of substances and their metabolites, such as pharmaceuticals, synthetic and natural hormones, and personal care products that are discharged in sewage, pose challenges in advanced wastewater treatment. Suspended solids are primarily responsible for residual biological oxygen demand in secondary sewage effluent waters. The dissolved organics are the most hazardous from the standpoint of potential toxicity. The major problem with dissolved inorganic materials is that presented by algal nutrients, primarily nitrates, and phosphates. In addition, potentially hazardous toxic metals may be found among the dissolved inorganics.

In addition to these chemical contaminants, secondary sewage effluent often contains a number of disease-causing microorganisms, requiring disinfection in cases where humans may later come into contact with the water. Among the bacteria that may be found in secondary sewage effluent are organisms causing tuberculosis, dysenteric bacteria (*Bacillus dysenteriae, Shigella dysenteriae, Shigella paradysenteriae, Proteus vulgaris*), cholera bacteria (*Vibriocholerae*), bacteria causing mud fever (*Leptospiraicterohemorrhagiae*), and bacteria causing typhoid fever (*Salmonella typhosa, Salmonella paratyphi*). In addition, viruses causing diarrhea, eye infections, infectious hepatitis, and polio may be encountered. Ingestion of sewage still causes disease, even in more developed nations.

8.4.5 Physical–Chemical Treatment of Municipal Wastewater

Complete physical–chemical wastewater treatment systems offer both advantages and disadvantages relative to biological treatment systems. The capital costs of physical–chemical facilities can be less than those of biological treatment facilities, and they usually require less land. They are better able to cope with toxic materials and overloads. However, they require careful operator control and consume relatively large amounts of energy.

Basically, a physical–chemical treatment process involves:

- Removal of scum and solid objects
- Clarification, generally with addition of a coagulant, and frequently with the addition of other chemicals (such as lime for phosphorus removal)

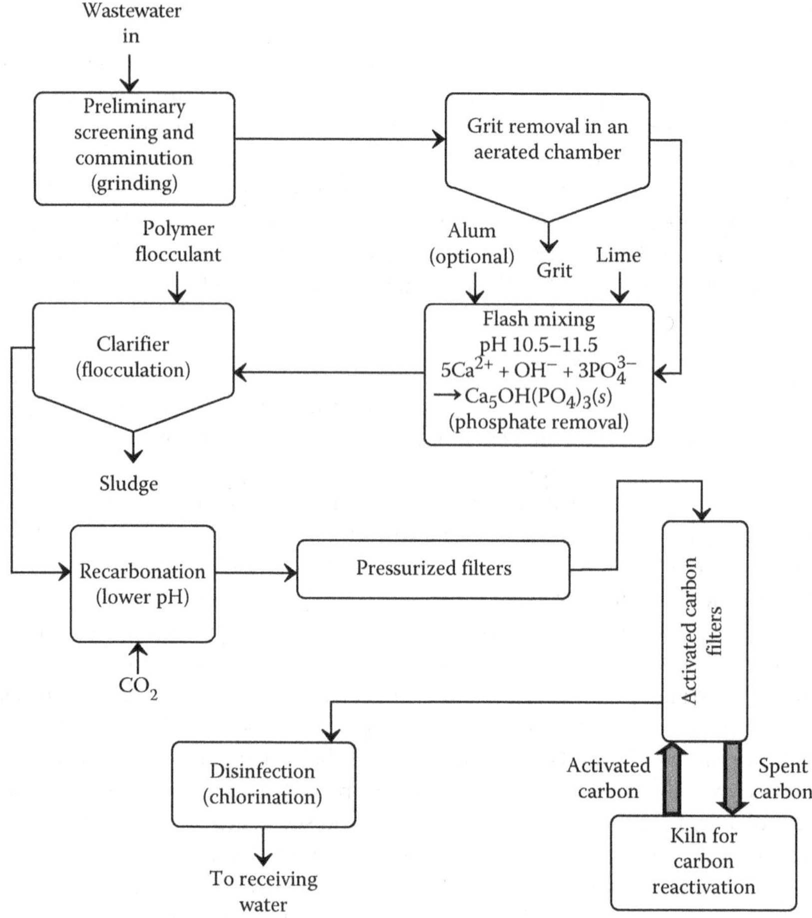

FIGURE 8.5 Major components of a complete physical–chemical treatment facility for municipal wastewater.

- Filtration to remove filterable solids
- Activated carbon adsorption
- Disinfection

The basic steps of a complete physical–chemical wastewater treatment facility are shown in Figure 8.5.

During the early 1970s, it appeared likely that physical–chemical treatment would largely replace biological treatment. However, higher chemical and energy costs since then have slowed the development of physical–chemical facilities.

8.5 INDUSTRIAL WASTEWATER TREATMENT

Before treatment, industrial wastewater should be characterized fully and the biodegradability of wastewater constituents determined. The options available for the treatment of wastewater are summarized briefly in this section and discussed in greater detail in later sections.

One of two major ways of removing organic wastes is biological treatment by an activated sludge or related process (see Section 8.4 and Figure 8.3). It may be necessary to acclimate microorganisms to the degradation of constituents that are not normally biodegradable. Consideration needs to be given to possible hazards of biotreatment sludges, such as those containing excessive levels of heavy metal ions. The other major process for the removal of organics from wastewater is sorption by activated carbon (see Section 8.8), usually in columns of granular activated carbon. Activated carbon and biological treatment can be combined with the use of powdered activated carbon in the activated sludge process. The powdered activated carbon sorbs some constituents that may be toxic to microorganisms and is collected with the sludge. A major consideration with the use of activated carbon to treat wastewater is the hazard that spent activated carbon may present from the wastes it retains. These hazards may include those of toxicity or reactivity, such as those hazards posed by wastes from the manufacture of explosives sorbed to activated carbon. Regeneration of the carbon is expensive and can be hazardous in some cases.

Wastewater can be treated by a variety of chemical processes, including acid/base neutralization, precipitation, and oxidation/reduction. Sometimes these steps must precede biological treatment; for example, acidic or alkaline wastewater must be neutralized in order for microorganisms to thrive in it. Cyanide in the wastewater may be oxidized with chlorine and organics with ozone, hydrogen peroxide promoted with ultraviolet radiation, or DO at high temperatures and pressures. Heavy metals may be precipitated with base, carbonate, or sulfide.

Wastewater can be treated by several physical processes. In some cases, simple density separation and sedimentation can be used to remove water-immiscible liquids and solids. Filtration is frequently required, and flotation by gas bubbles generated on particle surfaces may be useful. Wastewater solutes can be concentrated by evaporation, distillation, and membrane processes, including reverse osmosis, hyperfiltration, and ultrafiltration. Organic constituents can be removed by solvent extraction, air stripping, or steam stripping.

Synthetic resins are useful for removing some pollutant solutes from wastewater. Organophilic resins have proven useful for the removal of alcohols; aldehydes; ketones; hydrocarbons; chlorinated alkanes, alkenes, and aryl compounds; esters, including phthalate esters; and pesticides. Cation exchange resins are effective for the removal of heavy metals.

8.6 REMOVAL OF SOLIDS

Relatively large solid particles are removed from water by simple *settling* and *filtration*. The removal of colloidal solids from water usually requires *coagulation*.[2] Salts of aluminum and iron are the coagulants most often used in water treatment. Of these, alum or filter alum is the most commonly used. This substance is a hydrated aluminum sulfate, $Al_2(SO_4)_3 \cdot 18H_2O$. When this salt is added to water, the aluminum ion hydrolyzes by reactions that consume alkalinity in the water, such as

$$Al(H_2O)_6^{3+} + 3HCO_3^- \rightarrow Al(OH)_3(s) + 3CO_2 + 6H_2O \tag{8.5}$$

The gelatinous hydroxide thus formed carries suspended material with it as it settles. Furthermore, it is likely that positively charged hydroxyl-bridged dimers such as

$$(H_2O)_4Al\begin{matrix} \overset{H}{|} \\ O \\ / \quad \backslash \\ \backslash \quad / \\ O \\ \underset{H}{|} \end{matrix}Al(H_2O)_4^{4+}$$

and higher polymers are formed which interact specifically with colloidal particles, bringing about coagulation. Sodium silicate partially neutralized by acid aids coagulation, particularly when used with alum. Metal ions in coagulants also react with virus proteins and destroy viruses in water.

Anhydrous iron(III) sulfate added to water forms iron(III) hydroxide in a reaction analogous to Reaction 8.5. An advantage of iron(III) sulfate is that it works over a wide pH range of approximately

4–11. Hydrated iron(II) sulfate, or copperas, $FeSO_4 \cdot 7H_2O$, is also commonly used as a coagulant. It forms a gelatinous precipitate of hydrated iron(III) oxide; in order to function, it must be oxidized to iron(III) by DO in the water at a pH higher than 8.5, or by chlorine, which can oxidize iron(II) at lower pH values.

Natural and synthetic polyelectrolytes are used in flocculating particles. Among the natural compounds so used are starch and cellulose derivatives, proteinaceous materials, and gums composed of polysaccharides. More recently, selected synthetic polymers, including neutral polymers and both anionic and cationic polyelectrolytes, that are effective flocculants have come into use.

Coagulation–filtration is a much more effective procedure than filtration alone for the removal of suspended material from water. As the term implies, the process consists of the addition of coagulants that aggregate the particles into larger size particles, followed by filtration. Either alum or lime, often with added polyelectrolytes, is most commonly employed for coagulation.

The filtration step of coagulation–filtration is usually performed on a medium such as sand or anthracite coal. Often, to reduce clogging, several media with progressively smaller interstitial spaces are used. One example is the *rapid sand filter*, which consists of a layer of sand supported by layers of gravel particles, the particles becoming progressively larger with increasing depth. The substance that actually filters the water is coagulated material that collects in the sand. As more material is removed, the buildup of coagulated material eventually clogs the filter and must be removed by back-flushing.

An important class of solids that must be removed from wastewater consists of suspended solids in secondary sewage effluent that arise primarily from sludge that was not removed in the settling process. These solids account for a large part of the BOD in the effluent and may interfere with other aspects of tertiary waste treatment, such as by clogging membranes in reverse osmosis water treatment processes. The quantity of material involved may be rather high. Processes designed to remove suspended solids often will remove 10–20 mg/L of organic material from secondary sewage effluent. In addition, a small amount of the inorganic material is removed.

An important class of solids that must be removed from wastewater consists of suspended solids in secondary sewage effluent that arise primarily from sludge that was not removed in the settling process. These solids account for a large part of the BOD in the effluent and may interfere with other aspects of tertiary waste treatment, such as by clogging membranes in reverse osmosis water treatment processes. The quantity of material involved may be rather high. Processes designed to remove suspended solids often will remove 10–20 mg/L of organic material from secondary sewage effluent. In addition, a small amount of the inorganic material is removed.

8.6.1 Dissolved Air Flotation

Many of the particles found in water have low densities close to or even less than that of water. Particles less dense than water have a tendency to rise to the surface from which they can be skimmed off, but this is often a slow and incomplete process. The removal of such particles can be aided by *dissolved air flotation* in which small air bubbles are formed that attach to particles causing them to float. As shown in Figure 8.6, flotation of particles with air can be accompanied by coagulation of the particles as aided by a coagulant. Water supersaturated with air under pressure is released to the bottom of a tank where bubbles are formed in a layer of milky (white) water. Bubble formation accompanied by flocculation of the particles entrains bubbles in the floc, which floats to the surface where it is skimmed off.

8.6.2 Membrane Filtration Processes

Filtration through membranes under pressure is an especially effective means of removing solids and impurities from water.[3] The purified water that goes through a membrane is the *permeate* and the smaller amount of material that does not go through the membrane is a *retentate*. Membranes

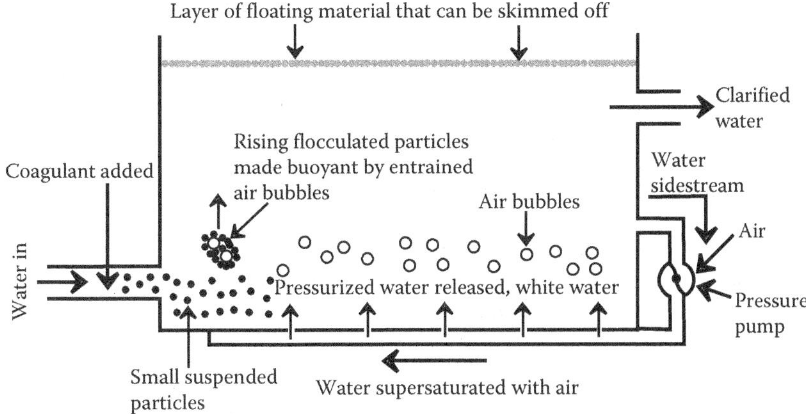

FIGURE 8.6 Illustration of dissolved air flotation in which pressurized water supersaturated with air is released into the bottom of a tank of water producing an abundance of small bubbles giving the water a white or milky appearance. Flocculated particles that entrain air bubbles are buoyant and rise to the top where the flocculated material may be skimmed off.

normally operate by size exclusion. As membranes with smaller openings are used, smaller particles and even molecules and ions are excluded, but higher pressures are required and more energy is consumed. In order of decreasing pore size, the common membrane processes are *microfiltration* > *ultrafiltration* > *nanofiltration* > and *hyperfiltration*. Microfiltration membranes have pores of 0.1–2 μm in size, and the other processes use membranes with progressively smaller pores. The main types of membrane processes and their uses are summarized in Table 8.1 and reverse osmosis is discussed in Section 8.9.

A problem common to all membrane processes is that posed by the retentate, which is concentrated in the substances that are removed from water. In some cases, this material may be discharged with wastewater, and retentate from the reverse osmosis desalination of seawater may be returned to the sea. Other options, depending upon the source of water treated, include evaporation of water and incineration of the residue, reclamation of materials from some industrial wastewaters, and disposal in deep saline water aquifers.

8.7 REMOVAL OF CALCIUM AND OTHER METALS

Calcium and magnesium salts, which generally are present in water as bicarbonates or sulfates, cause water hardness. One of the most common manifestations of water hardness is the insoluble "curd" formed by the reaction of soap with calcium or magnesium ions. The formation of these insoluble soap salts is discussed in Section 7.10. Although ions that cause water hardness do not

TABLE 8.1
Major Membrane Processes Used for Water Treatment

Process	Pressure (atm)	Contaminants Removed
Microfiltration	<5	Suspended solids, emulsified components, bacteria, protozoa
Ultrafiltration	2–8	Macromolecules above 5000–100,000 molecular mass (a function of pore size)
Nanofiltration	5–15	Molecules above 200–500 molecular mass (a function of pore size)
Hyperfiltration	15–100	Most solutes and ions; brackish water requires reverse osmosis pressures up to 15 bar; seawater desalination requires pressures up to 100 bar

form insoluble products with detergents, they do adversely affect detergent performance. Therefore, calcium and magnesium must be complexed or removed from water in order for detergents to function properly.

Another problem caused by hard water is the formation of mineral deposits. For example, when water containing calcium and bicarbonate ions is heated, insoluble calcium carbonate is formed:

$$Ca^{2+} + 2HCO_3^- \rightarrow CaCO_3(s) + CO_2(g) + H_2O \tag{8.6}$$

This product coats the surfaces of hot water systems, clogging pipes, and reducing heating efficiency. Dissolved salts such as calcium and magnesium bicarbonates and sulfates can be especially damaging in boiler feedwater. Clearly, the removal of water hardness is essential for many uses of water.

Several processes are used for softening water. On a large scale, such as in community water-softening operations, the lime-soda process is used. This process involves the treatment of water with lime, $Ca(OH)_2$, and soda ash, Na_2CO_3. Calcium is precipitated as $CaCO_3$ and magnesium as $Mg(OH)_2$. When the calcium is present primarily as "bicarbonate hardness," it can be removed by the addition of $Ca(OH)_2$ alone:

$$Ca^{2+} + 2HCO_3^- + Ca(OH)_2 \rightarrow 2CaCO_3(s) + 2H_2O \tag{8.7}$$

When bicarbonate ion is not present at substantial levels, a source of CO_3^{2-} must be provided at a high enough pH to prevent conversion of most of the CO_3^{2-} to HCO_3^-. These conditions are obtained by the addition of Na_2CO_3. For example, calcium present as the chloride can be removed from water by the addition of soda ash:

$$Ca^{2+} + 2Cl^- + 2Na^+ + CO_3^{2-} \rightarrow CaCO_3(s) + 2Cl^- + 2Na^+ \tag{8.8}$$

Note that the removal of bicarbonate hardness results in a net removal of soluble salts from solution, whereas removal of nonbicarbonate hardness involves the addition of at least as many equivalents of ionic material as are removed.

The precipitation of magnesium as the hydroxide requires a higher pH than the precipitation of calcium as the carbonate:

$$Mg^{2+} + 2OH^- \rightarrow Mg(OH)_2(s) \tag{8.9}$$

The high pH required may be provided by the basic carbonate ion from soda ash:

$$CO_3^{2-} + H_2O \rightarrow HCO_3^- + OH^- \tag{8.10}$$

Some large-scale lime-soda softening plants make use of the precipitated calcium carbonate product as a source of additional lime. The calcium carbonate is first heated to at least 825°C to produce quicklime, CaO:

$$CaCO_3 + \text{heat} \rightarrow CaO + CO_2(g) \tag{8.11}$$

The quicklime is then slaked with water to produce calcium hydroxide:

$$CaO + H_2O \rightarrow Ca(OH)_2 \tag{8.12}$$

The water softened by lime-soda softening plants usually suffers from two defects. First, because of supersaturation effects, some $CaCO_3$ and $Mg(OH)_2$ usually remain in solution. If not removed, these compounds will precipitate at a later time and cause harmful deposits or undesirable cloudiness in water. The second problem results from the use of highly basic sodium carbonate, which gives the product water an excessively high pH, up to pH 11. To overcome these problems, the water is recarbonated by bubbling CO_2 into it. The carbon dioxide converts the slightly soluble calcium carbonate and magnesium hydroxide to their soluble bicarbonate forms:

$$CaCO_3(s) + CO_2 + H_2O \rightarrow Ca^{2+} + 2HCO_3^- \quad (8.13)$$

$$Mg(OH)_2(s) + 2CO_2 \rightarrow Mg^{2+} + 2HCO_3^- \quad (8.14)$$

The CO_2 also neutralizes excess hydroxide ion:

$$OH^- + CO_2 \rightarrow HCO_3^- \quad (8.15)$$

The pH generally is brought within the range 7.5–8.5 by recarbonation. The source of CO_2 used in the recarbonation process may be from the combustion of carbonaceous fuel. Scrubbed stack gas from a power plant frequently is utilized. Water adjusted to a pH, alkalinity, and Ca^{2+} concentration very close to $CaCO_3$ saturation is labeled *chemically stabilized*. It neither precipitates $CaCO_3$ in water mains, which can clog the pipes, nor dissolves protective $CaCO_3$ coatings from the pipe surfaces. Water with Ca^{2+} concentration much below $CaCO_3$ saturation is called *aggressive* water.

Calcium may be removed from water very efficiently by the addition of orthophosphate:

$$5Ca^{2+} + 3PO_4^{3-} + OH^- \rightarrow Ca_5OH(PO_4)_3(s) \quad (8.16)$$

It should be pointed out that the chemical formation of a slightly soluble product for the removal of undesired solutes such as hardness ions, phosphate, iron, and manganese must be followed by sedimentation in a suitable apparatus. Frequently, coagulants must be added, and filtration employed for complete removal of these sediments.

Water may be purified by ion exchange, the reversible transfer of ions between aquatic solution and a solid material capable of bonding ions. The removal of NaCl from solution by two ion exchange reactions is a good illustration of this process. First the water is passed over a solid cation exchanger in the hydrogen form, represented by $H^{+-}\{Cat(s)\}$:

$$H^{+-}\{Cat(s)\} + Na^+ + Cl^- \rightarrow Na^{+-}\{Cat(s)\} + H^+ + Cl^- \quad (8.17)$$

Next, the water is passed over an anion exchanger in the hydroxide ion form, represented by $OH^{-+}\{An(s)\}$:

$$OH^{-+}\{An(s)\} + H^+ + Cl^- \rightarrow Cl^{-+}\{An(s)\} + H_2O \quad (8.18)$$

Thus, the cations in solution are replaced by hydrogen ion and the anions by hydroxide ion, yielding water as the product.

The softening of water by ion exchange does not require the removal of all ionic solutes, just those cations responsible for water hardness. Generally, therefore, only a cation exchanger is necessary. Furthermore, the sodium rather than the hydrogen form of the cation exchanger is used, and the divalent cations are replaced by sodium ion. Sodium ion at low concentrations is harmless in water to be used for most purposes, and sodium chloride is a cheap and convenient substance with which to recharge the cation exchangers.

A number of materials have ion-exchanging properties. Among the minerals especially noted for their ion exchange properties are the aluminum silicate minerals, or *zeolites*. An example of a zeolite which has been used commercially in water softening is glauconite, $K_2(MgFe)_2Al_6(Si_4O_{10})_3(OH)_{12}$. Synthetic zeolites have been prepared by drying and crushing the white gel produced by mixing solutions of sodium silicate and sodium aluminate.

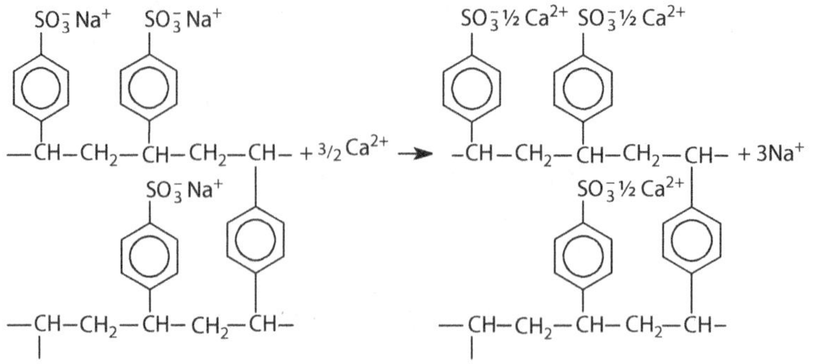

FIGURE 8.7 Strongly acidic cation exchanger. Sodium exchange for calcium in water is shown.

The discovery in the mid-1930s of synthetic ion exchange resins composed of organic polymers with attached functional groups marked the beginning of modern ion exchange technology. Structural formulas of typical synthetic ion exchangers are shown in Figures 8.7 and 8.8. The cation exchanger shown in Figure 8.7 is called a *strongly acidic cation exchanger* because the parent $-SO_3^-H^+$ group is a strong acid. When the functional group binding the cation is the $-CO_2^-$ group, the exchange resin is called a *weakly acidic cation exchanger,* because the $-CO_2H$ group is a weak acid. Figure 8.8 shows a *strongly basic anion exchanger* in which the functional group is a quaternary ammonium group, $-N^+(CH_3)_3$. In the hydroxide form, $-N^+(CH_3)_3OH^-$, the hydroxide ion is readily released, so the exchanger is classified as *strongly basic.*

The water-softening capability of a cation exchanger is shown in Figure 8.7, where sodium ion on the exchanger is exchanged for calcium ion in solution. The same reaction occurs with magnesium ion. Water softening by cation exchange is a widely used, effective, and economical process. In many areas having a low water flow, however, widespread home water softening by ion exchange causes deterioration of water quality arising from the contamination of wastewater by sodium chloride. Such contamination results from the periodic need to regenerate a water softener with sodium chloride in order to displace calcium and magnesium ions from the resin and replace these hardness ions with sodium ions:

$$Ca^{2+-}\{Cat(s)\}_2 + 2Na^+ + 2Cl^- \rightarrow 2Na^{+-}\{Cat(s)\} + Ca^{2+} + 2Cl^- \quad (8.19)$$

During the regeneration process, a large excess of sodium chloride must be used—several pounds for a home water softener. Appreciable amounts of dissolved sodium chloride can be introduced into sewage by this route.

Strongly acidic cation exchangers are used for the removal of water hardness. Weakly acidic cation exchangers having the $-CO_2H$ group as a functional group are useful for removing alkalinity. Alkalinity generally is manifested by bicarbonate ion, a species that is a sufficiently strong base to neutralize the acid of a weak acid cation exchanger:

$$2R\text{-}CO_2H + Ca^{2+} + 2HCO_3^- \rightarrow [R\text{-}CO_2^-]_2Ca^{2+} + 2H_2O + 2CO_2. \quad (8.20)$$

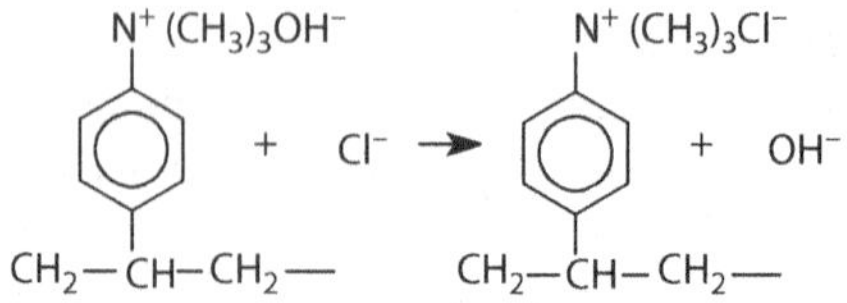

FIGURE 8.8 Strongly basic anion exchanger. Chloride exchange for hydroxide ion is shown.

However, weak bases such as sulfate ion or chloride ion are not strong enough to remove hydrogen ion from the carboxylic acid exchanger. An additional advantage of these exchangers is that they may be regenerated almost stoichiometrically with dilute strong acids, thus avoiding the potential pollution problem caused by the use of excess sodium chloride to regenerate strongly acidic cation changers.

Chelation or, as it is sometimes known, *sequestration*, is an effective method of softening water without actually having to remove calcium and magnesium from solution. A complexing agent is added which greatly reduces the concentrations of free hydrated cations, as shown by some of the example calculations in Chapter 3. For example, chelating calcium ion with excess EDTA anion (Y^{4-}),

$$Ca^{2+} + Y^{4-} \rightarrow CaY^{2-} \tag{8.21}$$

reduces the concentration of hydrated calcium ion, preventing the precipitation of calcium carbonate:

$$Ca^{2+} + CO_3^{2-} \rightarrow CaCO_3(s) \tag{8.22}$$

Polyphosphate salts, EDTA, and NTA (see Chapter 3) are chelating agents commonly used for water softening. Polysilicates are used to complex iron.

8.7.1 Removal of Iron and Manganese

Soluble iron and manganese are found in many groundwaters because of reducing conditions which favor the soluble +2 oxidation state of these metals (see Chapter 4). Iron is the more commonly encountered of the two metals. In groundwater, the level of iron seldom exceeds 10 mg/L, and that of manganese is rarely higher than 2 mg/L. The basic method for removing both of these metals depends upon oxidation to higher insoluble oxidation states. The oxidation is generally accomplished by aeration. The rate of oxidation is pH-dependent in both cases, with a high pH favoring more rapid oxidation. The oxidation of soluble Mn(II) to insoluble MnO_2 is a complicated process. It appears to be catalyzed by solid MnO_2, which is known to adsorb Mn(II). This adsorbed Mn(II) is slowly oxidized on the MnO_2 surface.

Chlorine and potassium permanganate are sometimes employed as oxidizing agents for iron and manganese. There is some evidence that organic chelating agents with reducing properties hold iron(II) in a soluble form in water. In such cases, chlorine is effective because it destroys the organic compounds and enables the oxidation of iron(II).

In water with a high level of carbonate, $FeCO_3$ and $MnCO_3$ may be precipitated directly by raising the pH above 8.5 by the addition of sodium carbonate or lime. This approach is less popular than oxidation, however.

Relatively high levels of insoluble iron(III) and manganese(IV) frequently are found in water as colloidal material which is difficult to remove. These metals may be associated with humic colloids or "peptizing" organic material that binds to colloidal metal oxides, stabilizing the colloid.

Heavy metals such as copper, cadmium, mercury, and lead are found in wastewaters from a number of industrial processes. Because of the toxicity of many heavy metals, their concentrations must be reduced to very low levels prior to release of the wastewater. A number of approaches are used in heavy metals removal.

Lime treatment, discussed earlier in this section for calcium removal, precipitates heavy metals as insoluble hydroxides, basic salts, or coprecipitated with calcium carbonate or iron(III) hydroxide. This process does not completely remove mercury, cadmium, or lead, so their removal is aided by addition of sulfide (most heavy metals are sulfide-seekers):

$$Cd^{2+} + S^{2-} \rightarrow CdS(s) \tag{8.23}$$

Heavy chlorination is frequently necessary to break down metal-solubilizing ligands (see Chapter 3). Lime precipitation does not normally permit recovery of metals and is sometimes undesirable from the economic viewpoint.

Electrodeposition (reduction of metal ions to metal by electrons at an electrode), *reverse osmosis* (see Section 8.9), and *ion exchange* are frequently employed for metal removal. Solvent extraction using organic-soluble chelating substances is also effective in removing many metals. *Cementation*, a process by which a metal deposits by reaction of its ion with a more readily oxidized metal, may be employed:

$$Cu^{2+} + \text{Fe (iron scrap)} \rightarrow Fe^{2+} + Cu \tag{8.24}$$

Activated carbon adsorption effectively removes some metals from water at the part per million level. Sometimes a chelating agent is sorbed to the charcoal to increase metal removal.

Even when not specifically designed for the removal of heavy metals, most waste treatment processes remove appreciable quantities of the more troublesome heavy metals encountered in wastewater. Biological waste treatment effectively removes metals from water. These metals accumulate in the sludge from biological treatment, so sludge disposal must be given careful consideration.

Various physical–chemical treatment processes effectively remove heavy metals from wastewaters. One such treatment is lime precipitation followed by activated-carbon filtration. Activated-carbon filtration may also be preceded by treatment with iron(III) chloride to form an iron(III) hydroxide floc, which is an effective heavy metals scavenger. Similarly, alum, which forms aluminum hydroxide, may be added prior to activated-carbon filtration.

The form of the heavy metal has a strong effect upon the efficiency of metal removal. For instance, chromium(VI) is normally more difficult to remove than chromium(III). Chelation may prevent metal removal by solubilizing metals (see Chapter 3).

In the past, removal of heavy metals has been largely a fringe benefit of wastewater treatment processes. Currently, however, more consideration is being given to design and operating parameters that specifically enhance heavy-metals removal as part of wastewater treatment.

8.8 REMOVAL OF DISSOLVED ORGANICS

Very low levels of exotic organic compounds in drinking water are suspected of contributing to cancer and other maladies. Water disinfection processes, which by their nature involve chemically rather severe conditions, particularly of oxidation, have a tendency to produce *disinfection by-products*. Some of these are chlorinated organic compounds produced by chlorination of organics in water, especially humic substances. Removal of organics to very low levels prior to chlorination has been found to be effective in preventing THM formation. Another major class of disinfection by-products consists of organooxygen compounds such as aldehydes, carboxylic acids, and oxoacids.

A variety of organic compounds survive, or are produced by, secondary wastewater treatment and should be considered as factors in discharge or reuse of the treated water. Almost half of these are humic substances (see Section 3.17) with a molecular-mass range of 1000–5000. Among the remainder are found ether-extractable materials, carbohydrates, proteins, detergents, tannins, and lignins. Because of their high molecular mass and anionic character, the humic compounds influence some of the physical and chemical aspects of waste treatment. The ether extractables contain many of the compounds that are resistant to biodegradation and are of particular concern regarding potential toxicity, carcinogenicity, and mutagenicity. In the ether extract are found many fatty acids, hydrocarbons of the *n*-alkane class, naphthalene, diphenylmethane, diphenyl, methylnaphthalene, isopropylbenzene, dodecylbenzene, phenol, phthalates, and triethylphosphate.

The standard method for the removal of dissolved organic material is adsorption on activated carbon, a product that is produced from a variety of carbonaceous materials including wood, pulpmill char, peat, and lignite. The carbon is produced by charring the raw material anaerobically below 600°C, followed by an activation step consisting of partial oxidation. Carbon dioxide may be employed as an oxidizing agent at 600–700°C.

$$CO_2 + C \rightarrow 2CO \tag{8.25}$$

or the carbon may be oxidized by steam at 800–900°C:

$$H_2O + C \rightarrow H_2 + CO \tag{8.26}$$

These processes develop porosity, increase the surface area, and leave the C atoms in arrangements that have affinities for organic compounds.

Activated carbon comes in two general types: granulated activated carbon, consisting of particles 0.1–1 mm in diameter, and powdered activated carbon, in which most of the particles are 50–100 μm in diameter.

The exact mechanism by which activated carbon holds organic materials is not known. However, one reason for the effectiveness of this material as an adsorbent is its tremendous surface area. A solid cubic foot of carbon particles may have a combined pore and surface area of approximately 10 square miles!

Although interest is increasing in the use of powdered activated carbon for water treatment, currently granular carbon is more widely used. It may be employed in a fixed bed, through which water flows downward. Accumulation of particulate matter requires periodic backwashing. An expanded bed in which particles are kept slightly separated by water flowing upward may be used with less chance of clogging.

Economics require regeneration of the carbon, which is accomplished by heating it to 950°C in a steam–air atmosphere. This process oxidizes adsorbed organics and regenerates the carbon surface, with an approximately 10% loss of carbon.

Removal of organics may also be accomplished by adsorbent synthetic polymers. Such polymers as Amberlite XAD-4 have hydrophobic surfaces and strongly attract relatively insoluble organic compounds, such as chlorinated pesticides. The porosity of these polymers is up to 50% by volume, and the surface area may be as high as 850 m^2/g. They are readily regenerated by solvents such as isopropanol and acetone. Under appropriate operating conditions, these polymers remove virtually all nonionic organic solutes; for example, phenol at 250 mg/L is reduced to less than 0.1 mg/L by appropriate treatment with Amberlite XAD-4.

Oxidation of dissolved organics holds some promise for their removal. Ozone, hydrogen peroxide, molecular oxygen (with or without catalysts), chlorine and its derivatives, permanganate, or ferrate [iron(VI)] can be used. Electrochemical oxidation may be possible in some cases. High-energy electron beams produced by high-voltage electron accelerators also have the potential to destroy organic compounds.

8.8.1 Removal of Herbicides

Because of their widespread application and persistence, herbicides have proven to be particularly troublesome in drinking water sources. Their levels vary with season related to times that they are applied to control weeds. The more soluble ones are most likely to enter drinking water sources. One of the most troublesome is atrazine, which is often manifested by its metabolite desethylatrazine. Activated carbon treatment is the best means of removing herbicides and their metabolites from drinking water sources. A problem with activated carbon is that of *preloading*, in which natural organic matter in the water loads up the carbon and hinders uptake of pollutant organics such as herbicides. Pretreatment to remove such organic matter, such as flocculation and precipitation of humic substances, can significantly increase the efficacy of activated carbon for the removal of herbicides and other organics.

8.9 REMOVAL OF DISSOLVED INORGANICS

In order for complete water recycling to be feasible, inorganic-solute removal is essential. The effluent from secondary waste treatment generally contains 300–400 mg/L more dissolved inorganic material than does the municipal water supply. It is obvious, therefore, that 100% water recycling without removal of inorganics would cause the accumulation of an intolerable level of dissolved material. Even when water is not destined for immediate reuse, the removal of the inorganic nutrients phosphorus and nitrogen is highly desirable to reduce eutrophication downstream. In some cases the removal of toxic trace metals is needed.

One of the most obvious methods for removing inorganics from water is distillation. However, the energy required for distillation is generally quite high, so that distillation is not generally economically feasible. Furthermore, volatile materials such as ammonia and odorous compounds are carried over to a large extent in the distillation process unless special preventative measures are taken. Freezing produces a very pure water, but is considered uneconomical with present technology. Ion exchange and membrane processes are the most cost-effective means of removing inorganic materials from water. These processes are discussed below.

8.9.1 Ion Exchange

The ion exchange method for softening water is described in detail in Section 8.7. The ion exchange process used for removal of inorganics consists of passing the water successively over a solid cation exchanger and a solid anion exchanger, which replace cations and anions by hydrogen ion and hydroxide ion, respectively, so that each equivalent of salt is replaced by a mole of water. For the hypothetical ionic salt MX, the reactions are the following where $^{-}\{\mathrm{Cat}(s)\}$ represents the solid cation exchanger and $^{+}\{\mathrm{An}(s)\}$ represents the solid anion exchanger:

$$H^{+-}\{\mathrm{Cat}(s)\} + M^{+} + X^{-} \rightarrow M^{+-}\{\mathrm{Cat}(s)\} + H^{+} + X^{-} \quad (8.27)$$

$$OH^{-+}\{\mathrm{An}(s)\} + H^{+} + X^{-} \rightarrow X^{-+}\{\mathrm{An}(s)\} + H_2O \quad (8.28)$$

The cation exchanger is regenerated with strong acid and the anion exchanger with strong base.

Demineralization by ion exchange generally produces water of a very high quality. Unfortunately, some organic compounds in wastewater foul ion exchangers, and microbial growth on the exchangers can diminish their efficiency. In addition, regeneration of the resins is expensive, and the concentrated wastes from regeneration require disposal in a manner that will not damage the environment.

8.9.2 Electrodialysis

Electrodialysis consists of applying a direct current across a body of water separated into layers by membranes alternately permeable to cations and anions. Cations migrate toward the cathode and anions toward the anode. Cations and anions both enter one layer of water, and both leave the adjacent layer. Thus, layers of water enriched in salts alternate with those from which salts have been removed. The water in the brine-enriched layers is recirculated to a certain extent to prevent excessive accumulation of brine. The principles involved in electrodialysis treatment are shown in Figure 8.9.

Fouling caused by various materials can cause problems with reverse osmosis treatment of water. Although the relatively small ions constituting the salts dissolved in wastewater readily pass through the membranes, large organic ions (proteins, for example) and charged colloids migrate to the membrane surfaces, often fouling or plugging the membranes and reducing efficiency. In addition, growth of microorganisms on the membranes can cause fouling.

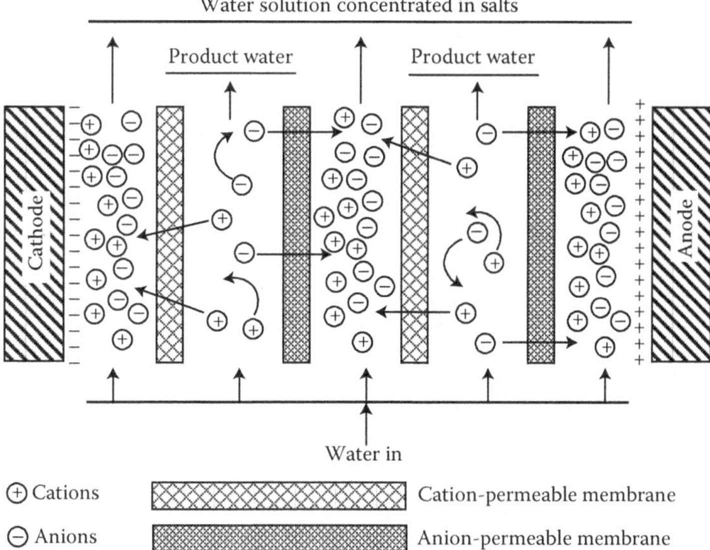

FIGURE 8.9 Electrodialysis apparatus for the removal of ionic material from water.

Experience with pilot plants indicates that electrodialysis has the potential to be a practical and economical method to remove up to 50% of the dissolved inorganics from secondary sewage effluent after pretreatment to eliminate fouling substances. Such a level of efficiency would permit repeated recycling of water without dissolved inorganic materials reaching unacceptably high levels.

8.9.3 Reverse Osmosis

Reverse osmosis is one of several pressure-driven membrane processes for water purification which also include nonfiltration, ultrafiltration, and microfiltration (Section 8.6). Reverse osmosis is a very useful and well-developed technique for the purification and desalination of water. As illustrated in Figure 8.10, it consists of forcing pure water through a semipermeable membrane that allows the passage of water but not of other material. This process, which is not simply sieve separation or ultrafiltration, depends on the preferential sorption of water on the surface of a porous cellulose acetate or polyamide membrane. Pure water from the sorbed layer is forced through pores in the membrane under pressure. If the thickness of the sorbed water layer is d, the pore diameter for optimum separation

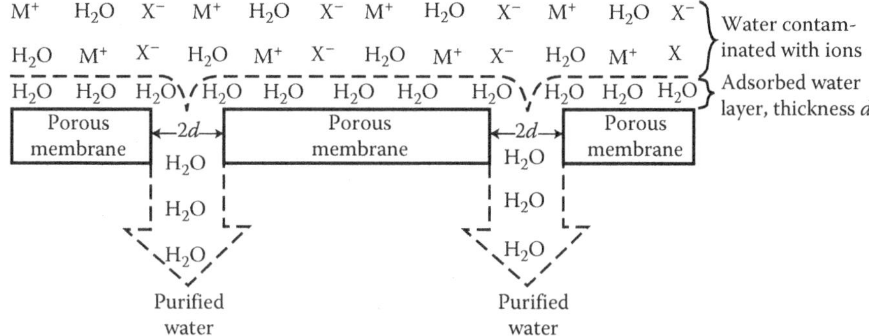

FIGURE 8.10 Solute removal from water by reverse osmosis.

should be $2d$. The optimum pore diameter depends upon the thickness of the sorbed pure water layer and may be several times the diameters of the solute and solvent molecules.

In an example of recycling that is in keeping with the practice of green chemistry, spent reverse osmosis membranes have been treated with potassium permanganate, which removes substances that clog the membranes, but also removes the salt-rejecting surface layer, drastically reducing their salt rejection properties. The treated membranes are then used for filtration to remove as much as 94% of suspended solids from water.

Nanofiltration is a pressurized membrane filtration process that does not remove singly charged salt ions that are removed by osmosis, but may be effective in removing hardness (Ca^{2+}). Nanofiltration operates at lower pressures than reverse osmosis, thus requiring less energy and less expense. Recently it has gained popularity as a drinking water treatment process.

8.9.4 Phosphorus Removal

Advanced waste treatment normally requires removal of phosphorus to reduce algal growth. Algae may grow at PO_4^{3-} levels as low as 0.05 mg/L. Growth inhibition requires levels well below 0.5 mg/L. Since municipal wastes typically contain approximately 25 mg/L of phosphate (as orthophosphates, polyphosphates, and insoluble phosphates), the efficiency of phosphate removal must be quite high to prevent algal growth. This removal may occur in the sewage treatment process (1) in the primary settler, (2) in the aeration chamber of the activated sludge unit, or (3) after secondary waste treatment.

Activated sludge treatment removes about 20% of the phosphorus from sewage. Thus, an appreciable fraction of largely biological phosphorus is removed with the sludge. Detergents and other sources contribute significant amounts of phosphorus to domestic sewage and considerable phosphate ion remains in the effluent. However, some wastes, such as carbohydrate wastes from sugar refineries, are so deficient in phosphorus that supplementation of the waste with inorganic phosphorus is required for proper growth of the microorganisms degrading the wastes.

Under some sewage plant operating conditions, much greater than normal phosphorus removal has been observed. In such plants, characterized by high DO and high pH levels in the aeration tank, removal of 60–90% of the phosphorus has been attained, yielding two or three times the normal level of phosphorus in the sludge. In a conventionally operated aeration tank of an activated sludge plant, the CO_2 level is relatively high because of the release of the gas by the degradation of organic material. A high CO_2 level results in a relatively low pH, because of the weakly acidic nature of dissolved CO_2 in water. The aeration rate is generally not maintained at a very high level because oxygen is transferred relatively more efficiently from air when the DO levels in water are relatively low. Therefore, the aeration rate normally is not high enough to sweep out sufficient dissolved carbon dioxide to bring its concentration down to low levels. Thus, the pH generally is low enough that phosphate is maintained primarily in the form of the $H_2PO_4^-$ ion. However, at a higher rate of aeration in a relatively hard water, the CO_2 is swept out, the pH rises, and reactions such as the following occur:

$$5Ca^{2+} + 3HPO_4^{2-} + H_2O \rightarrow Ca_5OH(PO_4)_3(s) + 4H^+ \qquad (8.29)$$

The precipitated hydroxyapatite or other form of calcium phosphate is incorporated in the sludge floc. Reaction 8.29 is strongly hydrogen ion-dependent, and an increase in the hydrogen ion concentration drives the equilibrium back to the left. Thus, under anaerobic conditions when the sludge medium becomes more acidic due to higher CO_2 levels, the phosphate returns to solution.

Chemically, phosphate is most commonly removed by precipitation. Some common precipitants and their products are shown in Table 8.2. Precipitation processes are capable of at least 90–95% phosphorus removal at reasonable cost. Lime has the advantages of low cost and ease of regeneration. The efficiency with which phosphorus is removed by lime is not as high as would be predicted by the low solubility of hydroxyapatite, $Ca_5OH(PO_4)_3$. Some of the possible reasons for this are slow precipitation of $Ca_5OH(PO_4)_3$, formation of nonsettling colloids; precipitation of calcium as $CaCO_3$

TABLE 8.2
Chemical Precipitants for Phosphate and Their Products

Precipitant(s)	Product
$Ca(OH)_2$	$Ca_5OH(PO_4)_3$ (hydroxyapatite)
$Ca(OH)_2 + NaF$	$Ca_5F(PO_4)_3$ (fluorapatite)
$Al_2(SO_4)_3$	$AlPO_4$
$FeCl_3$	$FePO_4$
$MgSO_4$	$MgNH_4PO_4$

in certain pH ranges, and the fact that phosphate may be present as condensed phosphates (polyphosphates) which form soluble complexes with calcium ion.

Lime, $Ca(OH)_2$, is the chemical most commonly used for phosphorus removal:

$$5Ca(OH)_2 + 3HPO_4^{2-} \rightarrow Ca_5OH(PO_4)_3(s) + 3H_2O + 6OH^- \quad (8.30)$$

Phosphate can be removed from solution by adsorption on some solids, particularly activated alumina, Al_2O_3. Removals of up to 99.9% of orthophosphate have been achieved with this method.

8.9.5 Nitrogen Removal

Next to phosphorus, nitrogen is the algal nutrient most commonly removed as part of advanced wastewater treatment. Numerous chemical, biological, ion exchange, and membrane techniques can be used to remove nitrogen from water. Mineralization of organic nitrogen in wastewater treatment first produces ammonium ion, NH_4^+. The most common chemical process for nitrogen removal from wastewater is reaction of nitrogen in the form of ammonium ion with hypochlorite (from chlorine) to produce elemental nitrogen that is evolved as a gas:

$$NH_4^+ + HOCl \rightarrow NH_2Cl + H_2O + H^+ \quad (8.31)$$

$$2NH_2Cl + HOCl \rightarrow N_2(g) + 3H^+ + 3Cl^- H_2O \quad (8.32)$$

Nitrification followed by denitrification is the most common biological means of removing inorganic nitrogen from wastewater and arguably the most effective technique. The first step is an essentially complete conversion of ammonia and organic nitrogen to nitrate under strongly aerobic conditions, achieved by more extensive than normal aeration of the sewage:

$$NH_4^+ + 2O_2 \text{ (nitrifying bacteria)} \rightarrow NO_3^- + 2H^+ + H_2O \quad (8.33)$$

The second step is the reduction of nitrate to nitrogen gas. This reaction is also bacterially catalyzed and requires a carbon source and a reducing agent such as methanol, CH_3OH.

$$6NO_3^- + 5CH_3OH + 6H^+ \text{ (denitrifying bacteria)} \rightarrow 3N_2(g) + 5CO_2 + 13H_2O \quad (8.34)$$

The denitrification process may be carried out either in a tank or on a carbon column. In pilot plant operation, conversions of 95% of the ammonia to nitrate and 86% of the nitrate to nitrogen have been achieved.

8.10 SLUDGE

Perhaps the most pressing water treatment problem at this time has to do with sludge collected or produced during water treatment. Finding a safe place to put the sludge or a use for it has proven troublesome, and the problem is aggravated by the large number of water treatment systems.

Some sludge is present in wastewater prior to treatment and may be collected from it. Such sludge includes human wastes, garbage grindings, organic wastes, and inorganic silt and grit from storm water runoff, and organic and inorganic wastes from commercial and industrial sources. There are two major kinds of sludge generated in a waste treatment plant. The first of these is organic sludge from activated sludge, trickling filter, or rotating biological reactors. The second is inorganic sludge from the addition of chemicals, such as in phosphorus removal (see Section 8.9).

Most commonly, sewage sludge is subjected to anaerobic digestion in a digester designed to allow bacterial action to occur in the absence of air. This reduces the mass and volume of sludge and ideally results in the formation of a stabilized humus material. Disease agents are also destroyed in the process.

Following digestion, sludge is generally conditioned and thickened to concentrate and stabilize it and make it more amenable to removal of water. Relatively inexpensive processes, such as gravity thickening, may be employed to get the moisture content down to about 95%. Sludge may be further conditioned chemically by the addition of iron or aluminum salts, lime, or polymers.

Sludge dewatering is employed to convert the sludge from an essentially liquid material to a damp solid containing not more than about 85% water. This may be accomplished on sludge drying beds consisting of layers of sand and gravel. Mechanical devices may also be employed, including vacuum filtration, centrifugation, and filter presses. Heat may be used to aid the drying process.

Ultimately, disposal of the sludge is required. Until banned by the U.S. Congress in the late 1980s, ocean disposal of sewage sludge was commonly practiced. Currently, the two main alternatives for sludge disposal are land spreading and incineration.

Rich in nutrients, waste sewage sludge contains around 5% N, 3% P, and 0.5% K on a dry-mass basis and can be used to fertilize and condition soil. The humic material in the sludge improves the physical properties and CEC of the soil. Possible accumulation of heavy metals is of some concern insofar as the use of sludge on cropland is concerned. Sewage sludge is an efficient heavy metals scavenger and may contain elevated levels of zinc, copper, nickel, and cadmium. These and other metals tend to remain immobilized in soil by chelation with organic matter, adsorption on clay minerals, and precipitation as insoluble compounds such as oxides or carbonates. However, increased application of sludge on cropland has caused distinctly elevated levels of zinc and cadmium in both leaves and grain of corn. Therefore, caution has been advised in heavy or prolonged application of sewage sludge to soil. Prior control of heavy metal contamination from industrial sources has greatly reduced the heavy metal content of sludge and enabled it to be used more extensively on soil.

A significant problem in sewage treatment can arise from sludge sidestreams. These consist of water removed from sludge by various treatment processes. Sewage treatment processes can be divided into mainstream treatment processes (primary clarification, trickling filter, activated sludge, and rotating biological reactor) and sidestream processes. During sidestream treatment, sludge is dewatered, degraded, and disinfected by a variety of processes, including gravity thickening, dissolved air flotation, anaerobic (anoxic) digestion, oxic digestion, vacuum filtration, centrifugation, belt-filter press filtration, sand-drying-bed treatment, sludge-lagoon settling, wet air oxidation, and pressure filtration. Each of these produces a liquid by-product sidestream which is circulated back to the mainstream. These add to the BOD and suspended solids of the mainstream.

A variety of chemical sludges are produced by various water treatment and industrial processes. Among the most abundant of such sludges is alum sludge produced by the hydrolysis of Al(III) salts used in the treatment of water, which creates gelatinous aluminum hydroxide:

$$Al^{3+} + 3OH^- (aq) \rightarrow Al(OH)_3(s) \tag{8.35}$$

Alum sludges containing aluminum hydroxide normally are 98% or more water and are very difficult to dewater.

Both iron(II) and iron(III) compounds are used for the removal of impurities from wastewater by precipitation of $Fe(OH)_3$. The sludge contains $Fe(OH)_3$ in the form of soft, fluffy precipitates that are difficult to dewater beyond 10% or 12% solids.

The addition of either lime, $Ca(OH)_2$, or quicklime, CaO, to water is used to raise the pH to about 11.5 and cause the precipitation of $CaCO_3$, along with metal hydroxides and phosphates. Calcium carbonate is readily recovered from lime sludges and can be recalcined to produce CaO, which can be recycled through the system.

Metal hydroxide sludges are produced in the removal of metals such as lead, chromium, nickel, and zinc from wastewater by raising the pH to such a level that the corresponding hydroxides or hydrated metal oxides are precipitated. The disposal of these sludges is a substantial problem because of their toxic heavy metal content. Reclamation of the metals is an attractive alternative for these sludges.

Pathogenic (disease-causing) microorganisms may persist in the sludge left from the treatment of sewage. These include bacteria such as *Enterobacter* and *Shigella*, viruses including hepatitis and enteroviruses, protozoa such as *Entamoeba* and *Giardia*, and helminth worms such as *Ascaris* and *Toxocara*. Many of these organisms present potential health hazards, and there is risk of public exposure when the sludge is applied to soil. Therefore, it is necessary both to be aware of pathogenic microorganisms in municipal wastewater treatment sludge and to find a means of reducing the hazards caused by their presence.

The most significant organisms in municipal sewage sludge include (1) indicators of fecal pollution, including fecal and total coliform, (2) pathogenic bacteria, including *Salmonellae* and *Shigellae*, (3) enteric (intestinal) viruses, including enterovirus and poliovirus, and (4) parasites, such as *Entamoeba histolytica* and *Ascaris lumbricoides*.

Several ways are recommended to significantly reduce levels of pathogens in sewage sludge. Aerobic digestion involves aerobic agitation of the sludge for periods of 40–60 days (longer times are employed with low sludge temperatures). Air drying involves draining and/or drying of the liquid sludge for at least 3 months in a layer 20–25 cm thick. This operation may be performed on underdrained sand beds or in basins. Anaerobic digestion involves maintenance of the sludge in an anaerobic state for periods of time ranging from 60 days at 20°C to 15 days at temperatures exceeding 35°C. Composting involves mixing dewatered sludge cake with bulking agents subject to decay, such as wood chips or shredded municipal refuse, and allowing the action of bacteria to promote decay at temperatures ranging up to 45–65°C. The higher temperatures tend to kill pathogenic bacteria. Finally, pathogenic organisms may be destroyed by lime stabilization in which sufficient lime is added to raise the pH of the sludge to 12 or higher.

8.11 WATER DISINFECTION

Chlorine is the most commonly used disinfectant employed for killing bacteria in water. When chlorine is added to water, it rapidly hydrolyzes according to the reaction

$$Cl_2 + H_2O \rightarrow H^+ + Cl^- + HOCl \tag{8.36}$$

which has the following equilibrium constant:

$$K = \frac{[H^+][Cl^-][HOCl]}{[Cl_2]} = 4.5 \times 10^{-4} \tag{8.37}$$

Hypochlorous acid, HOCl, is a weak acid that dissociates according to the reaction,

$$HOCl \rightleftarrows H^+ + OCl^- \tag{8.38}$$

with an ionization constant of 2.7×10^{-8}. From the above it can be calculated that the concentration of elemental Cl_2 is negligible at equilibrium above pH 3 when chlorine is added to water at levels below 1.0 g/L.

Sometimes, hypochlorite salts are substituted for chlorine gas as a disinfectant. Calcium hypochlorite, $Ca(OCl)_2$, is commonly used. The hypochlorites are safer to handle than gaseous chlorine.

The two chemical species formed by chlorine in water, HOCl and OCl^-, are known as *free available chlorine*. Free available chlorine is very effective in killing bacteria. In the presence of ammonia, monochloramine, dichloramine, and trichloramine are formed:

$$NH_4^+ + HOCl \rightarrow NH_2Cl \text{ (monochloramine)} + H_2O + H^+ \quad (8.39)$$

$$NH_2Cl + HOCl \rightarrow NHCl_2 \text{ (dichloramine)} + H_2O \quad (8.40)$$

$$NHCl_2 + HOCl \rightarrow NCl_3 \text{ (trichloramine)} + H_2O \quad (8.41)$$

The chloramines are called *combined available chlorine*. Chlorination practice frequently provides for formation of combined available chlorine which, although a weaker disinfectant than free available chlorine, is more readily retained as a disinfectant throughout the water distribution system. Too much ammonia in water is considered undesirable because it exerts excess demand for chlorine.

At sufficiently high Cl:N ratios in water containing ammonia, some HOCl and OCl^- remain unreacted, and a small quantity of NCl_3 is formed. The ratio at which this occurs is called the *breakpoint*. Chlorination beyond the breakpoint ensures disinfection. It has the additional advantage of destroying the more common materials that cause odor and taste in water.

At moderate levels of NH_3-N (approximately 20 mg/L), when the pH is between 5.0 and 8.0, chlorination with a minimum 8:1 mass ratio of Cl to NH_3-nitrogen produces efficient denitrification:

$$NH_4^+ + HOCl \rightarrow NH_2Cl + H_2O + H^+ \quad (8.39)$$

$$2NH_2Cl + HOCl \rightarrow N_2(g) + 3H^+ + 3Cl^- + H_2O \quad (8.42)$$

This reaction is used to remove pollutant ammonia from wastewater. However, problems can arise from chlorination of organic wastes. Typical of such by-products is chloroform, produced by the chlorination of humic substances in water.

Chlorine is used to treat water other than drinking water. It is employed to disinfect effluent from sewage treatment plants, as an additive to the water in electric power plant cooling towers, and to control microorganisms in food processing.

8.11.1 Chlorine Dioxide

Chlorine dioxide, ClO_2, is an effective water disinfectant that is of particular interest because, in the absence of impurity Cl_2, it does not produce impurity THMs in water treatment. In acidic and neutral water, respectively, the two half-reactions for ClO_2 acting as an oxidant are the following:

$$ClO_2 + 4H^+ + 5e^- \rightleftarrows Cl^- + 2H_2O \quad (8.43)$$

$$ClO_2 + e^- \rightleftarrows ClO_2^- \quad (8.44)$$

In the neutral pH range, chlorine dioxide in water remains largely as molecular ClO_2 until it contacts a reducing agent with which to react. Chlorine dioxide is a gas that is violently reactive with organic matter and explosive when exposed to light. For these reasons, it is not shipped, but is generated on-site by processes such as the reaction of chlorine gas with solid sodium hypochlorite:

$$2NaClO_2\,(s) + Cl_2\,(g) \rightleftarrows 2ClO_2\,(g) + 2NaCl\,(s) \qquad (8.45)$$

A high content of elemental chlorine in the product may require its purification to prevent unwanted side reactions from Cl_2.

As a water disinfectant, chlorine dioxide does not chlorinate or oxidize ammonia or other nitrogen-containing compounds. Some concern has been raised over possible health effects of its main degradation by-products, ClO_2^- and ClO_3^-.

8.11.2 Ozone and Other Oxidants

Ozone is sometimes used as a disinfectant in place of chlorine, particularly in Europe. Figure 8.11 shows the main components of an ozone water treatment system. Basically, air is filtered, cooled, dried, and pressurized, then subjected to an electrical discharge of approximately 20,000 V. The ozone produced is then pumped into a contact chamber where water contacts the ozone for 10–15 min. Concern over possible production of toxic organochlorine compounds by water chlorination processes has increased interest in ozonation. Furthermore, ozone is more destructive to viruses than is chlorine. Unfortunately, the solubility of ozone in water is relatively low, which limits its disinfective power.

Ozone oxidizes water contaminants directly through the reaction of O_3 and indirectly by generating hydroxyl radical, $HO^{\bullet}$, a reactive strong oxidant. A major consideration with ozone is the rate at which it decomposes spontaneously in water, according to the overall reaction,

$$2O_3 \rightarrow 3O_2\,(g) \qquad (8.46)$$

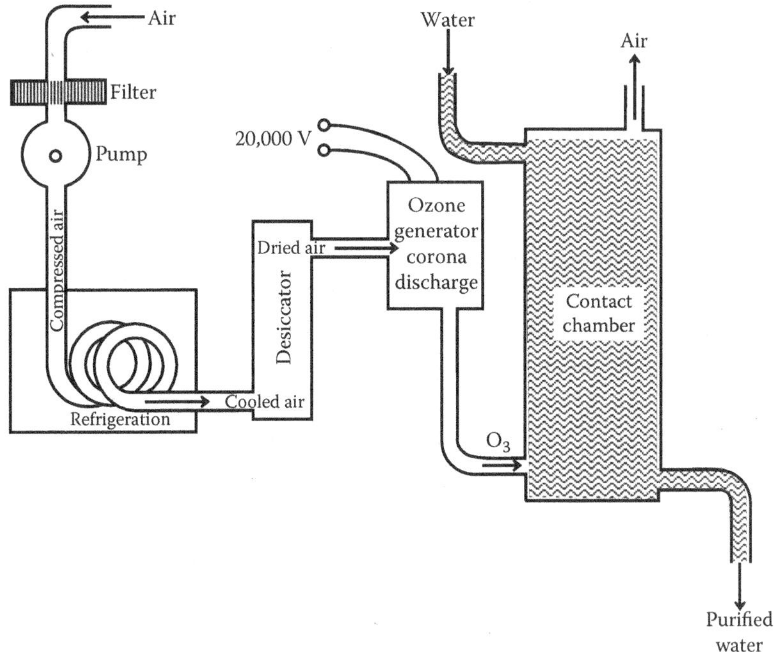

FIGURE 8.11 Schematic diagram of a typical ozone water-treatment system.

Because of the decomposition of ozone in water, some chlorine must be added to maintain disinfectant throughout the water distribution system.

Iron(VI) in the form of ferrate ion, FeO_4^{2-}, is a strong oxidizing agent with excellent disinfectant properties. It has the additional advantage of removing heavy metals, viruses, and phosphate. It may well find increased application for disinfection in the future.[4]

Another oxidant that is becoming widely used as a cleaning and bleaching agent in detergents and that, because of its oxidizing ability, probably has disinfecting abilities as well, is sodium percarbonate. This solid, commonly called *percarbonate*, is produced by mixing hydrogen peroxide, H_2O_2, with sodium carbonate, Na_2CO_3, along with $MgSO_4$ and Na_2SiO_3 stabilizers to produce a stable solid with an approximate empirical formula of $Na_2CO_3{\cdot}3H_2O_2$ in which the H_2O_2 may be stabilized by hydrogen bonding to CO_3^{2-} ion. In water, this material releases oxygen and acts as a bleaching agent, detergent booster, and probably bactericide as well.

8.11.3 Disinfection with Ultraviolet Radiation

Ultraviolet radiation, commonly represented as $h\nu$, is an effective disinfection agent that acts by direct breakage of chemical bonds in the biomolecules of pathogenic organisms and by generating reactive ions and radicals that destroy microorganisms. Ultraviolet radiation may be generated by high-powered mercury vapor lamps and solar radiation can be used as a source of ultraviolet radiation. Because it is reagent-free, ultraviolet radiation is a particularly green means of water disinfection. It works well in combination with other disinfection agents, such as chlorine or ozone, and reduces the quantities of such agents required. Studies are underway on the use for water disinfection of titanium dioxide, TiO_2, a photocatalytic material that can destroy pathogens when activated by ultraviolet radiation.

8.12 NATURAL WATER PURIFICATION PROCESSES

Virtually all of the materials that waste treatment processes are designed to eliminate may be absorbed by soil or degraded in soil.[5] In fact, most of these materials are essential for soil fertility. Wastewater may provide the water that is essential to plant growth, in addition to the nutrients—phosphorus, nitrogen, and potassium—usually provided by fertilizers. Wastewater also contains essential trace elements and vitamins. Stretching the point a bit, the degradation of organic wastes provides the CO_2 essential for photosynthetic production of plant biomass.

Soil may be viewed as a natural filter for wastes. Most organic matter is readily degraded in soil and, in principle, soil constitutes an excellent primary, secondary, and tertiary treatment system for water. Soil has physical, chemical, and biological characteristics that can enable wastewater detoxification, biodegradation, chemical decomposition, and physical and chemical fixation. A number of soil characteristics are important in determining its use for land treatment of wastes. These characteristics include physical form, ability to retain water, aeration, organic content, acid–base characteristics, and oxidation–reduction behavior. Soil is a natural medium for a number of living organisms that may have an effect upon biodegradation of wastewaters, including those that contain industrial wastes. Of these, the most important are bacteria, including those from the genera *Agrobacterium*, *Arthrobacteri*, *Bacillus*, *Flavobacterium*, and *Pseudomonas*. Actinomycetes and fungi are important in decay of vegetable matter and may be involved in biodegradation of wastes. Other unicellular organisms that may be present in or on soil are protozoa and algae. Soil animals, such as earthworms, affect soil parameters such as soil texture. The growth of plants in soil may have an influence on its waste treatment potential in such aspects as uptake of soluble wastes and erosion control.

Early civilizations, such as the Chinese, used human organic wastes to increase soil fertility, and the practice continues today. The ability of soil to purify water was noted well over a century ago. In 1850 and 1852, J. Thomas Way, a consulting chemist to the Royal Agricultural Society in England, presented two papers to the Society entitled "Power of Soils to Absorb Manure." Way's experiments showed that soil is an ion exchanger. Much practical and theoretical information on the ion exchange process resulted from his work.

If soil treatment systems are not properly designed and operated, odor can become an overpowering problem. The author of this book is reminded of driving into a small town, recalled from some years before as a very pleasant place, and being assaulted with a virtually intolerable odor. The disgruntled residents pointed to a large spray irrigation system on a field in the distance—unfortunately upwind—spraying liquified pig manure as part of an experimental feedlot waste treatment operation. The experiment was not deemed a success and was discontinued by the investigators, presumably before they met with violence from the local residents.

8.12.1 Industrial Wastewater Treatment by Soil

Wastes that are amenable to land treatment are biodegradable organic substances, particularly those contained in municipal sewage and in wastewater from some industrial operations, such as food processing. However, through acclimation over a long period of time, soil bacterial cultures may develop that are effective in degrading normally recalcitrant compounds that occur in industrial wastewater. Acclimated microorganisms are found particularly at contaminated sites, such as those where soil has been exposed to crude oil for many years.

Soil used for land treatment must meet a number of criteria. It should be at least a meter thick because wastes that reach underlying rock strata tend to degrade poorly. The soil slope should be moderate to reduce runoff. The water table should lie at a sufficient depth to avoid leaching into groundwater. Treatment sites should be located at appropriate distances from dwellings or other anthrospheric constructs that could be adversely affected by exposure to wastes. Shallow cultivation of soil to which wastes have been applied promotes degradation by exposure to atmospheric oxygen.

Land treatment is most used for petroleum refining wastes and is applicable to the treatment of fuels and wastes from leaking underground storage tanks. It can also be applied to biodegradable organic chemical wastes, including some organohalide compounds. Land treatment is not suitable for the treatment of wastes containing acids, bases, toxic inorganic compounds, salts, heavy metals, and organic compounds that are excessively soluble, volatile, or flammable.

8.13 GREEN WATER

Much of the world including parts of the United States suffers from chronic shortages of water. Throughout history, water used by humans has come from freshwater groundwater and surface sources, relying on nature's hydrologic cycle for purification to remove dissolved salt. Largely untapped is the more than 97% of earth's water that is too saline to use, most of which is present in the oceans, but including significant amounts in underground saline aquifers. Assuming sufficient availability of affordable energy, these waters can be subjected to *desalination* to provide freshwater suitable for human consumption and irrigation.

The two main means of water desalination are distillation and reverse osmosis. Distillation normally is carried out in several stages in which brine feedstock is heated to progressively higher temperatures as it is used to condense steam from each stage, the highest of which is heated by an external source. At each step, the heated brine is subjected to a lower pressure enabling more water to evaporate or "flash" to produce steam that condenses in a heat exchanger cooled by incoming brine feed. This approach enables maximum efficiency of energy use. Because of the energy intensity of this multistage distillation, it is usually combined with a power plant using otherwise waste heat from the power plant steam turbine outlet. An interesting possibility for small-scale water desalination is the use of solar power to heat the desalination units.

The world's largest seawater distillation plant is the multistage flash distillation unit in Shoaibi, Saudi Arabia, a nation that gets about 70% of its municipal water supplies from water desalination. This operation is combined with an oil-fired steam power plant, which uses the desalination plant for cooling and provides the heat needed for desalination.

In addition to distillation, the other significant alternative for providing freshwater is reverse osmosis discussed in Section 8.9. Some communities in arid lands close to a seacoast get their freshwater from this source.

Both distillation and reverse osmosis produce concentrated brine by-products (retentate) that requires disposal. This material can be pumped back into the ocean or, in some cases, to deep underground saltwater aquifers. Some of it can also be evaporated, in some cases using solar energy, to yield a salt product with commercial value.

Some inland areas have large resources of saline groundwater (brackish water) that can be desalinated. An advantage of this source is that the salt level is often less than that of seawater, which makes the energy requirements relatively lower. Residue from purification of saline groundwater can be pumped back into the aquifer at a depth or location that does not contaminate the source. Utilization of saline groundwater could vastly increase the availability of water in many water-deficient areas.

8.13.1 Reuse and Recycling of Wastewater

The most readily available water source is wastewater discharged by municipalities, industrial concerns, and even irrigation. Wastewater from municipal and industrial use usually requires treatment prior to discharge and it is relatively straightforward and inexpensive to remove inorganic and organic solutes from the water to make it suitable for use. It has an advantage over saline seawater or groundwater in that its dissolved salt content is usually much less, lowering the cost of desalination.

Water reuse and recycling are becoming much more common as demands for water exceed supply. *Unplanned reuse* occurs as the result of waste effluents entering receiving waters or groundwater and subsequently being taken into a water distribution system. A typical example of unplanned water reuse occurs in London, which withdraws water from the Thames River that may have been through other water systems at least once, and which uses groundwater sources unintentionally recharged with sewage effluents from a number of municipalities. *Planned reuse* utilizes wastewater treatment systems deliberately designed to bring water up to standards required for subsequent applications. The term *direct reuse* refers to water that has retained its identity from a previous application; reuse of water that has lost its identity is termed *indirect reuse*. The distinction also needs to be made between recycling and reuse. *Recycling* occurs internally before water is ever discharged. An example is condensation of steam in a steam power plant followed by return of the steam to boilers. *Reuse* occurs, for example, when water discharged by one user is taken as a water source by another user.

Reuse of water continues to grow because of two major factors. The first of these is lack of supply of water. The second is that widespread deployment of modern water treatment processes significantly enhances the quality of water available for reuse. These two factors come into play in semi-arid regions in countries with advanced technological bases. For example, Israel, which is dependent upon irrigation for essentially all its agriculture, reuses about two-third of the country's sewage effluent for irrigation, whereas the United States, where water is relatively more available, uses only about 2–3% of its water for this purpose.

Since drinking water and water used for food processing require the highest quality of all large applications, intentional reuse for potable water is relatively less desirable, though widely practiced unintentionally or out of necessity. This leaves three applications with the greatest potential for reuse:

- Irrigation for cropland, golf courses, and other applications requiring water for plant and grass growth. This is the largest potential application for reused water and one that can take advantage of plant nutrients, particularly nitrogen and phosphorus, in water.
- Cooling and process water in industrial applications. For some industrial applications, relatively low quality water can be used and secondary sewage effluent is a suitable source.
- *Groundwater recharge*. Groundwater can be recharged with reused water either by direct injection into an aquifer or by applying the water to land, followed by percolation into the

aquifer. The latter, especially, takes advantage of biodegradation and chemical sorption processes to further purify the water.

The major concern with recycled water is the potential presence of harmful pollutants. Water is readily disinfected to eliminate disease-causing bacteria. Viruses are more persistent, but can also be eliminated. Organic contaminants at very low levels may cause harm and can be very difficult to eliminate. As discussed in Section 7.13, these contaminants can include pharmaceuticals and their metabolic products such as synthetic 17α-ethinylestradiol and the natural hormone 17β-estradiol, used in oral contraceptives; other endocrine disrupting species may be present as well.

It is inevitable that water recycling and reuse will continue to grow. This trend will increase the demand for water treatment, both qualitatively and quantitatively. In addition, it will require more careful consideration of the original uses of water to minimize water deterioration and enhance its suitability for reuse.

Figure 8.12 shows a system for reuse of water. Most of the water treatment processes discussed in this chapter may be utilized in purifying water for reuse and employed in an operation such as the one shown (see especially the physical/chemical wastewater purification system shown in Figure 8.5). The system shown in Figure 8.12 shows some of the main operations employed in a system for total water recycle and the applications of the purified water product. It takes secondary wastewater effluent from which most of the BOD has been removed by the activated sludge process and discharges it into wetlands where plants and algae grow profusely in the nutrient-rich water. This biomass has the potential to be harvested and utilized as a fuel or raw material using thermochemical gasification or related measures. Depending upon geological conditions, significant amounts of water in the wetlands may infiltrate into groundwater, where it is purified further by natural processes underground. Water exiting the wetlands may be run over a bed of activated

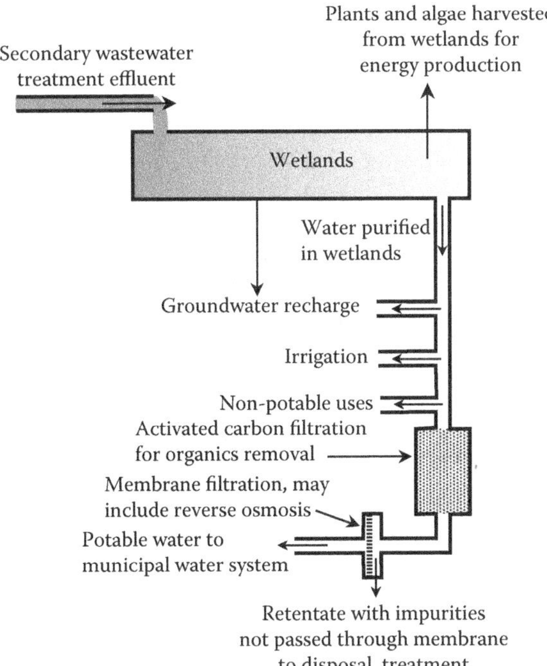

FIGURE 8.12 System in which wastewater effluent from secondary sewage treatment is upgraded to water that can be used for a number of purposes, including even recirculation back into the municipal water system.

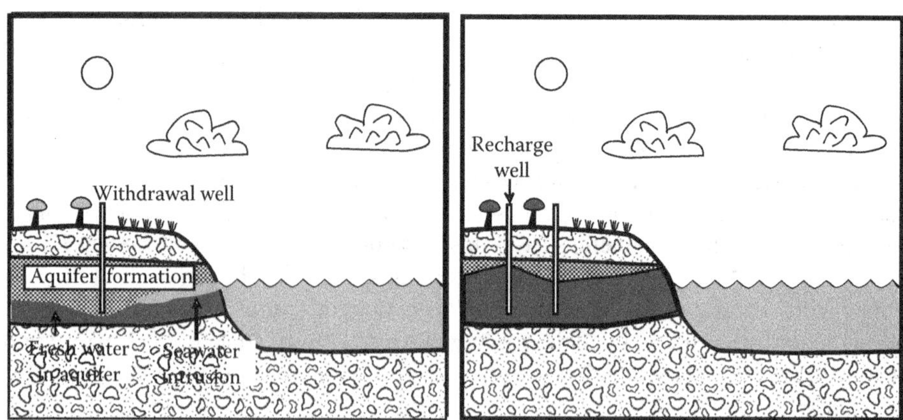

FIGURE 8.13 Injection of reclaimed water into underground aquifers in coastal areas can prevent seawater intrusion into the aquifer formation.

carbon to remove organics. The water can also be subjected to highly efficient membrane filtration processes including even reverse osmosis for removal of all salts. The solute-enriched retentate from treatment of water by membrane processes can be pumped into deep underground saline aquifers in some areas, discarded to the ocean in coastal regions (often with mixing with wastewater or cooling water to dilute the impurities in the retentate), or evaporated to produce a solid or sludge residue that can be disposed.

The world's largest water plant devoted to bringing sewage effluent up to a standard permitting potable uses began operation in the Orange County Water District in southern California in late 2007. Known as the Groundwater Replenishment System this $481 million project is fed with sewage treated by secondary wastewater treatment processes and pumps its clean water product into underground aquifers to serve as future municipal water sources and to prevent saltwater intrusion from the nearby ocean. Around 275 million liters per day of sewage effluent is first forced through microfilters to remove solids, then treated by reverse osmosis to remove dissolved salts, and finally subjected to intense ultraviolet radiation to break down organic contaminants, especially pharmaceuticals and their metabolites before being discharged to underground aquifers. It is expected to serve as a model for similar installations around the world.

In coastal areas and on islands, a major advantage of upgrading wastewater to a standard that it can be injected into underground aquifers is the prevention of seawater intrusion into the aquifers (Figure 8.13). As aquifers are pumped down, the lowered water level in the aquifer can enable infiltration of saline water, thus ruining the groundwater quality. Injection of treated wastewater into the aquifer can raise the water table in the aquifer to a level that prevents seawater intrusion.

8.14 WATER CONSERVATION

Water conservation, simply using less water, is one of the most effective, and clearly the fastest, means to ensure water supply. Water conservation falls into several major categories:

1. Indoor and household water conservation practices
2. Water-conserving devices and appliances
3. Outdoor and landscaping water conservation
4. Efficient irrigation practices
5. Conservation in agricultural operations other than irrigation
6. Water-efficient manufacturing

Each of these categories is summarized briefly below.

Indoor and household water conservation practices are generally those simple measures that water consumers may implement to reduce use of water. Examples include sequential uses of water such as saving water from rinsing dishes to use for household cleaning, ensuring that there are no leaking faucets or toilet appliances, taking showers by first lathering one's wet body without water flowing followed by rinsing with water for the minimum length of time, minimizing the number and water volume of toilet flushes, minimizing use of kitchen sink disposal devices for food waste disposal, operating clothes washers and dishwashers only with full loads, and other common sense measures to reduce water use. Implementation of these measures is largely a matter of education and good environmental citizenship. In cases of real water shortage, force may be applied through the painful mechanism of sharply higher water utility rates.

A number of water-saving devices and appliances are available. These can be quite simple and cheap, such as aerators with flow constrictors on faucets or showerheads, or they may be large and expensive such as low-water-consumption clothes washers that function well with cold water. Toilets that use minimal water per flush are now mandated by law, although the less effective of these may actually increase water consumption when multiple flushes are required. In public facilities, automatically flushing toilets and urinals are now commonplace, although some have the counterproductive tendency of performing multiple automatic flushes for each instance of use. Instant in-line water heaters installed near faucet outlets can reduce water use by making it unnecessary to run the hot water faucet for some time to obtain hot water.

Water used for lawns and landscaping often is one of the greatest consumers of water. In some areas, such as public parks, purified wastewater is used for this purpose. In principle, water used in some household applications can be recycled to water yard plants or grass. Some of the greater savings can be achieved in this area by growing ground cover and plants that require minimal water. This may mean planting native grass species that can withstand drought without much watering. In extreme cases, yards have been converted to gravel cover. Mulching with materials such as tree bark or grass clippings can reduce water use significantly.

Irrigation of agricultural crops is one of the largest consumers of water and one that is amenable to conservation. Spray irrigation devices lose large quantities of water to evaporation and should be replaced by systems that do not spray the water into the air. The ultimate in irrigation efficiency is drip irrigation that applies minimum quantities of water directly to the plant roots. Application of excess water in irrigation should be minimized. However, particularly in areas that rarely receive enough rain to flush the soil, application of minimal amounts of irrigation water can result in harmful accumulation of salts in the soil.

Livestock production and processing agricultural products use large amounts of water. The production of 1 kg of beef requires around 20,000 L of water, including the water required to grow the grain that the animal eats. About 6000 L of water are required to produce 1 kg of chicken meat. Significantly less water is needed to produce an equivalent amount of vegetable protein. The production of fuel ethanol from the fermentation of sugar from corn requires relatively large amounts of water. Choices made in the kinds of agricultural products produced for market have important implications for water conservation.

Manufacturing can be very water-intensive. The manufacture of a single automobile requires around 150,000 L of water. Refining crude oil to petroleum requires approximately 44 times the volume of water as the gasoline produced. The conservation of water in manufacturing and commercial activities is an important aspect of industrial ecology (see Chapter 17).

8.15 PROTECTING WATER SUPPLIES FROM ATTACK

The fact that drinking water is distributed centrally to large numbers of people makes water systems a concern to terrorist threats. History abounds with examples of large populations devastated by contaminated water supplies that caused diseases such as cholera and typhoid. In modern times, contaminated water in developing countries causes outbreaks of debilitating and often fatal dysentery, a leading cause of mortality in infants and children in some areas. Even in industrialized countries in modern times, drinking water contamination can cause illness and even fatalities. For example, in 1993 contamination of the Milwaukee water system by the protozoan *Cryptosporidium parvum* caused illness in over 400,000 people, more than 50 of whom died as a result.

Deliberate chemical contamination of an entire water system is an unlikely, but not impossible, event. Incidents have been recorded in which terrorist groups attempted to obtain cyanide salts, presumably with the objective of poisoning water supplies. Toxins from microorganisms are likely weapons of groups wishing to poison water. The most toxic of these is the botulinus toxin, an astoundingly poisonous substance produced by *Botulinus* bacteria. Direct contamination by bacteria is possible. The most likely candidates are *Bacillus anthracis*, *Shigella dysenteriae*, *Vibriocholerae*, and *Yersiniapestis*. Waterborne strains of *Escherichia coli* bacteria that produce the shiga toxin (the toxin produced by *Shigelladysenteriae*) continue to cause fatal infections in the United States from time to time. Shiga-producing *E. coli* in the municipal water supply of Walkerton, Ontario, Canada, caused approximately 3000 people to become ill and seven to die in May 2000. In an incident suggestive of the potential for terrorist attack, 12 staff of a laboratory of a large medical center developed acute diarrheal illness requiring hospitalization of four of the victims due to infection by *Shigelladysenteriae*. Cultures taken from the victims were found to be identical to those found on doughnuts or muffins that they had eaten in the facility break room and to a culture of *Shigelladysenteriae* maintained in the laboratory, a portion of which was missing. This led to the conclusion that the contamination was deliberate.[6]

Fortunately, water supplies are not particularly vulnerable to attack. It would be rather difficult to infiltrate a water treatment system, and injecting a toxic substance into the water distribution system beyond the treatment plant is unlikely. Vulnerability to infectious microbial agents is greatly reduced by maintaining a disinfectant chlorine residual in the system. Some of the likely chemical agents hydrolyze in water or are destroyed by the action of chlorine, and their effects are reduced by the great dilution provided by water systems that distribute huge quantities of water. Even in the event of contamination of a water supply, provision of relatively small quantities of bottled water for drinking would reduce the hazard to very low levels. Prudence dictates that water supply, treatment, and distribution systems be maintained in a secure state and monitored for likely agents of attack, but water systems need not be the subject of undue concern.

LITERATURE CITED

1. Wiesmann, U., I. S. Choi, and E.-M. Dombrowski, *Fundamentals of Biological Wastewater Treatment*, Wiley-VCH, Hoboken, NJ, 2007.
2. Bratby, J., *Coagulation and Flocculation in Water and Wastewater Treatment*, 2nd ed., IWA Publishing, Seattle, 2006.
3. Hager, L. S., *Membrane Systems for Wastewater Treatment*, McGraw-Hill, New York, 2006.
4. Sharma, V., Ed., *Ferrates: Synthesis, Properties, and Applications in Water and Wastewater Treatment*, American Chemical Society, Washington, DC, 2008.
5. Crites, R. W., E. J. Middlebrooks, and S. C. Reed, *Natural Wastewater Treatment Systems*, CRC/Taylor & Francis, Boca Raton, FL, 2006.
6. Kolavic, S. A., A. Kimura, S. L. Simons, L. Slutsker, S. Barth, and C. E. Haley, An outbreak of *Shigelladysenteriae* type 2 among laboratory workers due to intentional food contamination, *Journal of the American Medical Association*, **278**, 396–398, 1997.

SUPPLEMENTARY REFERENCES

American Water Works Association, *Reverse Osmosis and Nanofiltration*, 2nd ed., American Water Works Association, Denver, CO, 2007.

American Water Works Association, *Water Chlorination/Chloramination Practices and Principles*, 2nd ed., American Water Works Association, Denver, CO, 2006.

Ash, M. and I. Ash, *Handbook of Water Treatment Chemicals*, Gower, Aldershot, England, 1996.
Brennan, Waterworks: Research accelerates on advanced water-treatment technologies as their use in purification grows, *Chemical and Engineering News*, 79, 32–38, 2001.
Crittenden, J. C., *Water Treatment Principles and Design*, 2nd ed., Wiley, New York, 2005.
Faust, S. D. and O. M. Aly, Eds., *Chemistry of Water Treatment*, 2nd ed., American Water Works Association, Denver, CO, 1997.
Hammer, M. J., *Water and Wastewater Technology*, 6th ed., Prentice Hall, Upper Saddle River, NJ, 2008.
Rakness, K. L., *Ozone in Drinking Water Treatment: Process Design, Operation, and Optimization*, American Water Works Association, Denver, 2005.
Rimer, A. E., *The Impacts of Membrane Process Residuals on Wastewater Treatment: A Guidance Manual*, WateReuse Foundation, Alexandria, VA, 2008.
Russell, D. L., *Practical Wastewater Treatment*, Wiley-Interscience, Hoboken, NJ, 2006.
Spellman, F. R., Handbook of Water and Wastewater Treatment Plant Operations, 2nd ed., Taylor & Francis, Boca Raton, FL, 2008.
Sullivan, P. J., F. J. Agardy, and J. J. J. Clark, *The Environmental Science of Drinking Water*, Elsevier Butterworth-Heinemann, Burlington, MA, 2005.
Water Environment Federation, *Operation of Municipal Wastewater Treatment Plants*, 6th ed., McGraw-Hill, New York, 2008.

QUESTIONS AND PROBLEMS

The use of internet resources is assumed in answering any of the questions. These would include such things as constants and conversion factors as well as additional information needed to complete an answer.

1. Consider the equilibrium reactions and expressions covered in Chapter 3. How many moles of NTA should be added to 1000 L of water having a pH of 9 and containing CO_3^{2-} at 1.00×10^{-4} M to prevent precipitation of $CaCO_3$? Assume a total calcium level of 40 mg/L.
2. What is the purpose of the return sludge step in the activated sludge process?
3. What are the two processes by which the activated sludge process removes soluble carbonaceous material from sewage?
4. Why might hard water be desirable as a medium if phosphorus is to be removed by an activated sludge plant operated under conditions of high aeration?
5. How does reverse osmosis differ from a simple sieve separation or ultrafiltration process?
6. How many liters of methanol would be required daily to remove the nitrogen from a 200,000-L/day sewage treatment plant producing an effluent containing 50 mg/L of nitrogen? Assume that the nitrogen has been converted to NO_3^- in the plant. The denitrifying reaction is Reaction 8.34.
7. Discuss some of the advantages of physical–chemical treatment of sewage as opposed to biological wastewater treatment. What are some disadvantages?
8. Why is recarbonation necessary when water is softened by the lime-soda process?
9. Assume that a waste contains 300 mg/L of biodegradable $\{CH_2O\}$ and is processed through a 200,000-L/day sewage treatment plant which converts 40% of the waste to CO_2 and H_2O. Calculate the volume of air (at 25°C, 1 atm) required for this conversion. Assume that the O_2 is transferred to the water with 20% efficiency.
10. If all of the $\{CH_2O\}$ in the plant described in Question 9 could be converted to methane by anaerobic digestion, how many liters of methane (STP) could be produced daily?
11. Assuming that aeration of water does not result in the precipitation of calcium carbonate, of the following, which one that would not be removed by aeration: hydrogen sulfide, carbon dioxide, volatile odorous bacterial metabolites, alkalinity, iron?
12. In which of the following water supplies would moderately high water hardness be most detrimental: municipal water, irrigation water, boiler feedwater, drinking water (in regard to potential toxicity).

13. Which solute in water is commonly removed by the addition of sulfite or hydrazine?
14. A wastewater containing dissolved Cu^{2+} ion is to be treated to remove copper. Which of the following processes would *not* remove copper in an insoluble form: lime precipitation, cementation, treatment with NTA, ion exchange, reaction with metallic Fe.
15. Match each water contaminant in the left column with its preferred method of removal in the right column.

A. Mn^{2+}	1. Activated carbon
B. Ca^{2+} and HCO_3^-	2. Raise pH by addition of Na_2CO_3
C. THM compounds	3. Addition of lime
D. Mg^{2+}	4. Oxidation

16. A cementation reaction employs iron to remove Cd^{2+} present at a level of 350 mg/L from a wastewater stream. Given that the atomic mass of Cd is 112.4 and that of Fe is 55.8, how many kilograms of Fe are consumed in removing all the Cd from 4.50×10^6 L of water?
17. Consider municipal drinking water from two different kinds of sources, one a flowing, well-aerated stream with a heavy load of particulate matter, and the other an anaerobic groundwater. Describe possible differences in the water treatment strategies for these two sources of water.
18. In treating water for industrial use, consideration is often given to "sequential use of the water." What is meant by this term? Give some plausible examples of sequential use of water.
19. Active biomass is used in the secondary treatment of municipal wastewater. Describe three ways of supporting a growth of the biomass, contacting it with wastewater, and exposing it to air.
20. Using appropriate chemical reactions for illustration, show how calcium present as the dissolved HCO_3^- salt in water is easier to remove than other forms of hardness, such as dissolved $CaCl_2$.
21. Match the pollutant or impurity on the left with a reagent used to treat it from the list on the right, below:

A. Bacteria	1. Ozone
B. PCB	2. $Al_2(SO_4)_3$
C. H_2S	3. Activated carbon
D. Colloidal matter	4. Air

22. Match the water treatment step or process on the right, below, with the substance that it treats or removes on the left.

A. Oxygen	1. Live microorganisms
B. Calcium, Ca^{2+}	2. Phosphate, PO_4^{3-}
C. $\{CH_2O\}$	3. Hydrazine (H_4N_2) or sulfite (SO_3^{2-})
D. Colloidal solids	4. $Al_2(SO_4)_3{\cdot}18H_2O$

23. Regarding secondary wastewater treatment, the true statement of the following is (A) the activated sludge process is predominantly a physical/chemical process, (B) trickling filters make use of a mass of biological sludge that is continuously pumped over the filter, (C) excess sludge from activated sludge treatment is likely to undergo the process represented by $2\{CH_2O\} \rightarrow CH_4 + CO_2$, (D) the activated sludge process gets rid of all of the sludge as soon as it is made, (E) the trickling filter is an anaerobic (oxygen-free) treatment process.
24. Of the following, the true statement pertaining to water disinfection is (A) disinfection with ozone is particularly desirable because the ozone persists throughout the water distribution system, (B) disinfection with ozone is noted for producing toxic organic

by-products, (C) organic substances are left in water to be treated by chlorine to retain disinfecting capacity in the water, (D) the main disadvantage of chlorination of water is that Cl_2 gas is the only effective chlorinating agent, (E) chlorine dioxide, ClO_2, is an effective water disinfectant that does not produce impurity THMs in water treatment.

25. Of the following, designate the technique *least likely* to significantly lower the concentration of heavy metals in water and explain your choice: (A) cementation, (B) anion exchange, (C) addition of S^{2-}, (D) biological waste treatment, (E) lime.
26. Of the following, designate the *least likely* to be effective in reducing total levels of dissolved ions (salts) from water and explain your choice: (A) reverse osmosis, (B) cation exchange followed by anion exchange, (C) distillation, (D) ion exchange as practiced for water softening, (E) electrodialysis.
27. Using internet resources, attempt to find instances in which deliberate poisoning of water supplies has been used as a tactic for warfare or terrorism.
28. Bottled water is a product with enormous commercial value. What are the advantages of using this source of drinking water? What are the environmental disadvantages of this product?

9 The Atmosphere and Atmospheric Chemistry

9.1 INTRODUCTION

The *atmosphere* consists of the thin layer of mixed gases covering the Earth's surface. Exclusive of water, atmospheric air is 78.1% (by volume) nitrogen, 21.0% oxygen, 0.9% argon, and 0.04% carbon dioxide. Normally, air contains 1–3% water vapor by volume. In addition, air contains a large variety of trace level gases below 0.002%, including neon, helium, methane, krypton, nitrous oxide, hydrogen, xenon, sulfur dioxide, ozone, nitrogen dioxide, ammonia, and carbon monoxide. The atmosphere behaves as it does as a consequence of the gases in it from both natural and anthropogenic sources and the physical forces acting on it.

The atmosphere is divided into several layers on the basis of temperature. Of these, the most significant are the troposphere extending in altitude from the Earth's surface to approximately 11 km, and the stratosphere from about 11 km to approximately 50 km. Atmospheric stratification is discussed in more detail in Section 9.3.

9.1.1 Photochemistry and Some Important Terms

Various aspects of the environmental chemistry of the atmosphere are discussed in Chapters 9 through 14. Important aspects of atmospheric chemistry include effects of solar radiation leading to photolysis of trace gases and photooxidation of oxidizable trace gases in the troposphere. The most significant feature of atmospheric chemistry is the occurrence of *photochemical reactions* resulting from the absorption by molecules of *photons* of electromagnetic radiation from the sun, mostly in the ultraviolet region of the spectrum.[1] Photochemical reactions and photochemistry are discussed in more detail in Section 9.8. However, it is important to define several key aspects of photochemistry at this point to enable understanding of the rest of the material in this chapter.

- The meaning of $h\nu$: The energy, E, of a photon of visible or ultraviolet light is given by the equation, $E = h\nu$, where h is Planck's constant and ν is the frequency of electromagnetic inversely proportional to its wavelength, λ. The involvement of a photon in a photochemical reaction is shown by the following reaction with ozone, O_3:

$$O_3 + h\nu\ (\lambda < 420\ \text{nm}) \rightarrow O^* + O_2 \qquad (9.1)$$

  Ultraviolet radiation has a higher frequency than visible light and is, therefore, more energetic and more likely to break chemical bonds in molecules that absorb it.
- *Excited state*, *: The product of a photochemical reaction may be electronically energized in an excited state commonly denoted with an asterisk, *, as shown for the excited oxygen atom, O*, above. This excess electronic energy may make the excited species chemically more reactive.

- *Free radicals* containing an unpaired electron denoted with a dot, •: Reaction of an excited oxygen atom with a molecule of water vapor,

$$O^* + H_2O \rightarrow HO^\bullet + HO^\bullet \tag{9.2}$$

produces two hydroxyl radicals, $HO^\bullet$. Because of the strong pairing tendency of electrons, free radicals are normally highly reactive and cause most of the important chemical processes that occur in the atmosphere.

- *Third body, M*: An important reaction in the stratosphere is that between an oxygen atom and an O_2 molecule

$$O + O_2 + M \rightarrow O_3 + M \tag{9.3}$$

to produce stratospheric ozone. In this reaction, M is a third body, almost always a molecule of O_2 or N_2, that absorbs the energy released in the reaction, which otherwise would cause the product molecule to fall apart.

9.1.2 Atmospheric Composition

Dry air up to several kilometers altitude consists of two *major components*

- Nitrogen 78.08% (by volume)
- Oxygen 20.95%

and two *minor components*

- Argon 0.934%
- Carbon dioxide 0.039%

in addition to argon, four more noble *gases*,

- Neon $1.818 \times 10^{-3}\%$
- Krypton $1.14 \times 10^{-4}\%$
- Helium $5.24 \times 10^{-4}\%$
- Xenon $8.7 \times 10^{-6}\%$

and *trace gases* as given in Table 9.1. Atmospheric air may contain 0.1–5% water by volume, with a normal range of 1–3%.

9.1.3 Gaseous Oxides in the Atmosphere

Oxides of carbon, sulfur, and nitrogen are important constituents of the atmosphere and are pollutants at higher levels. Of these, carbon dioxide, CO_2, is the most abundant. It is a natural atmospheric constituent, and it is required for plant growth. However, the level of carbon dioxide in the atmosphere, now at about 390 ppm by volume, is increasing by about 2 ppm per year. As discussed in Chapter 14, this increase in atmospheric CO_2 may well cause general atmospheric warming—the "greenhouse effect," with potentially very serious consequences for the global atmosphere and for life on Earth. Although not a global threat, carbon monoxide, CO, can be a serious health threat because it prevents blood from transporting oxygen to body tissues.

The two most serious nitrogen oxide air pollutants are nitric oxide, NO, and nitrogen dioxide, NO_2, collectively denoted as "NO_x." These tend to enter the atmosphere as NO, and photochemical processes in the atmosphere can convert NO to NO_2. Further reactions can result in the formation of corrosive nitrate salts or nitric acid, HNO_3. Nitrogen dioxide is particularly significant in atmospheric chemistry because of its photochemical dissociation by light with a wavelength $<$430 nm to produce highly reactive O atoms. This is the first step in the formation of photochemical smog (see below). Sulfur dioxide, SO_2, is a reaction product of the combustion of sulfur-containing fuels such

TABLE 9.1
Atmospheric Trace Gases in Dry Air Near Ground Level

Gas or Species	Volume Percent[a]	Major Sources	Process for Removal from the Atmosphere
CH_4	1.8×10^{-4}	Biogenic[b]	Photochemical[c]
CO	$\sim 1.2 \times 10^{-5}$	Photochemical, anthropogenic[d]	Photochemical
N_2O	3×10^{-5}	Biogenic	Photochemical
NOx[e]	10^{-10}–10^{-6}	Photochemical, lightning, anthropogenic	Photochemical
HNO_3	10^{-9}–10^{-7}	Photochemical	Washed out by precipitation
NH_3	10^{-8}–10^{-7}	Biogenic	Photochemical, washed out by precipitation
H_2	5×10^{-5}	Biogenic, photochemical	Photochemical
H_2O_2	10^{-8}–10^{-6}	Photochemical	Washed out by precipitation
$HO\bullet$[f]	10^{-13}–10^{-10}	Photochemical	Photochemical
$HO_2\bullet$[f]	10^{-11}–10^{-9}	Photochemical	Photochemical
H_2CO	10^{-8}–10^{-7}	Photochemical	Photochemical
CS_2	10^{-9}–10^{-8}	Anthropogenic, biogenic	Photochemical
OCS	10^{-8}	Anthropogenic, biogenic, photochemical	Photochemical
SO_2	$\sim 2 \times 10^{-8}$	Anthropogenic, photochemical, volcanic	Photochemical
I_2	0-trace	—	—
CCl_2F_2[g]	2.8×10^{-5}	Anthropogenic	Photochemical
H_3CCCl_3[h]	$\sim 1 \times 10^{-8}$	Anthropogenic	Photochemical

[a] Levels in the absence of gross pollution.
[b] From biological sources.
[c] Reactions induced by the absorption of light energy as described later in this chapter and in Chapters 13 and 14.
[d] Sources arising from human activities.
[e] Sum of NO, NO_2, and NO_3, of which NO_3 is a major reactive species in the atmosphere at night.
[f] Reactive free radical species with one unpaired electron; these are transient species whose concentrations become much lower at night.
[g] A CFC, Freon F-12.
[h] Methyl chloroform.

as high-sulfur coal. Part of this sulfur dioxide is converted in the atmosphere to sulfuric acid, H_2SO_4, normally the predominant contributor to acid precipitation.

9.1.4 Atmospheric Methane

The most abundant hydrocarbon in the atmosphere is methane, CH_4, released from underground sources as natural gas and produced by the fermentation of organic matter. Methane is one of the least reactive atmospheric hydrocarbons and is produced by diffuse sources, so that its role in producing intense regionalized incidents of air pollution is limited. However, because of its presence throughout the atmosphere and despite its relatively low reactivity, it is a major factor in atmospheric chemical processes. Information gathered from ice cores has shown that during the last 250 years levels of atmospheric methane have more than doubled as the result of fossil fuel utilization, agricultural practices (particularly cultivation of rice in which methane is evolved from anoxic bacteria growing in waterlogged soil), and waste fermentation. Per molecule, methane is a much more effective greenhouse gas than is carbon dioxide. Methane affects both tropospheric and stratospheric chemistry, particularly by influencing levels of hydroxyl radical, ozone, and stratospheric water vapor.

9.1.5 Hydrocarbons and Photochemical Smog

The most significant atmospheric pollutant hydrocarbons are the reactive ones produced as automobile exhaust emissions. In the presence of NO, under conditions of temperature inversion (see Chapter 11), low humidity, and sunlight, these hydrocarbons produce undesirable *photochemical smog* manifested by the presence of visibility-obscuring particulate matter, oxidants such as ozone, and noxious organic species such as aldehydes.

9.1.6 Particulate Matter

Particles ranging from aggregates of a few molecules to pieces of dust readily visible to the naked eye are commonly found in the atmosphere and are discussed in detail in Chapter 10. Some atmospheric particles, such as sea salt formed by the evaporation of water from droplets of sea spray, are natural and even beneficial atmospheric constituents. Very small particles called *condensation nuclei* serve as bodies for atmospheric water vapor to condense upon and are essential for the formation of rain drops. Colloidal-sized particles in the atmosphere are called *aerosols*. Those formed by grinding up bulk matter are known as *dispersion aerosols*, whereas particles formed from chemical reactions of gases are *condensation aerosols*; the latter tend to be smaller. Smaller particles are in general the most harmful because they have a greater tendency to scatter light and are the most respirable-(tendency to be inhaled into the lungs).

Much of the mineral particulate matter in a polluted atmosphere is in the form of oxides and other compounds produced during the combustion of high-ash fossil fuel. Smaller particles of *fly ash* enter furnace flues and are efficiently collected in a properly equipped stack system. However, some fly ash escapes through the stack and enters the atmosphere. Unfortunately, the fly ash thus released tends to consist of smaller particles that do the most damage to human health, plants, and visibility.

9.1.7 Primary and Secondary Pollutants

Primary pollutants in the atmosphere are those that are emitted directly. An example of a primary pollutant is sulfur dioxide, SO_2, which directly harms vegetation and is a lung irritant. Of greater importance in most cases are *secondary pollutants* that are formed by atmospheric chemical processes acting upon primary pollutants and even nonpollutant species in the atmosphere. Secondary pollutants are generally produced by the natural tendency of the atmosphere to oxidize trace gases in it. Secondary pollutant sulfuric acid, H_2SO_4, is generated by the oxidation of primary pollutant SO_2 and secondary pollutant NO_2 is produced when the primary pollutant NO is oxidized. One of the most important secondary pollutants in the troposphere is ozone, O_3, for which the primary raw material is atmospheric oxygen, O_2. Pollutant levels of ozone are produced in the troposphere by photochemical processes in the presence of hydrocarbons and NO_x ($NO + NO_2$) as discussed in Chapter 13. Another important kind of secondary pollutant consists of particulate matter generated by atmospheric chemical reactions operating on gaseous primary pollutants.

9.2 IMPORTANCE OF THE ATMOSPHERE

The atmosphere is a protective blanket that nurtures life on Earth and protects it from the hostile environment of outer space. The atmosphere is the source of carbon dioxide for plant photosynthesis and of oxygen for respiration. It provides the nitrogen that nitrogen-fixing bacteria and ammonia-manufacturing plants use to produce chemically bound nitrogen, an essential component of life molecules. As a basic part of the hydrologic cycle (Figure 3.1) the atmosphere transports water from the oceans to land, thus acting as the condenser in a vast solar-powered still. Unfortunately, the atmosphere also has been used as a dumping ground for many pollutant materials—ranging from sulfur dioxide to refrigerant Freon—a practice that causes damage to vegetation and materials, shortens human life, and alters the characteristics of the atmosphere itself.

In its essential role as a protective shield, the atmosphere absorbs most of the cosmic rays from outer space and protects organisms from their effects. It also absorbs most of the electromagnetic radiation from the sun, allowing transmission of significant amounts of radiation only in the regions of 300–2500 nm (near-ultraviolet, visible, and near-infrared radiation) and 0.01–40 m (radio waves). By absorbing electromagnetic radiation below 300 nm, the atmosphere filters out damaging ultraviolet radiation that would otherwise be very harmful to living organisms. Furthermore, because it reabsorbs much of the infrared radiation by which absorbed solar energy is re-emitted to space, the atmosphere stabilizes the Earth's temperature, preventing the tremendous temperature extremes that occur on planets and moons lacking substantial atmospheres.

9.3 PHYSICAL CHARACTERISTICS OF THE ATMOSPHERE

Atmospheric science deals with the movement of air masses in the atmosphere, atmospheric heat balance, and atmospheric chemical composition and reactions. In order to understand atmospheric chemistry and air pollution, it is important to have an overall appreciation of the atmosphere, its composition, and physical characteristics as discussed in the first parts of this chapter.

9.3.1 Variation of Pressure and Density with Altitude

As anyone who has exercised at high altitudes well knows, the density of the atmosphere decreases sharply with increasing altitude as a consequence of the gas laws and gravity. More than 99% of the total mass of the atmosphere is found within approximately 30 km (about 20 mile) of the Earth's surface. Such an altitude is miniscule compared to the Earth's diameter, so it is not an exaggeration to characterize the atmosphere as a "tissue-thin" protective layer. In fact, if Earth were a globe the size of one typically seen in a geography classroom, the greater part of the atmosphere upon which humankind is absolutely dependent for its existence would be of a thickness only about that of the coat of varnish on the globe! Although the total mass of the global atmosphere is huge, approximately 5.14×10^{15} metric tons, it is still only about one-millionth of the Earth's total mass.

The fact that atmospheric pressure decreases as an approximately exponential function of altitude largely determines the characteristics of the atmosphere. Ideally, in the absence of mixing and at a constant absolute temperature, T, the pressure at any given height, P_h, is given in the exponential form

$$P_h = P_0 e^{-Mgh/RT} \tag{9.4}$$

where P_0 is the pressure at zero altitude (sea level), M is the average molar mass of air (28.97 g/mol in the troposphere), g is the acceleration of gravity (981 cm/s^2 at sea level), h is the altitude in cm, and R is the gas constant (8.314×10^7 erg/deg/mol). These units are given in the cgs (centimeter-gram-second) system for consistency; altitude can be converted to meter or kilometer as appropriate.

The factor RT/Mg is defined as the *scale height*, which represents the increase in altitude by which the pressure drops by e^{-1}. At an average sea-level temperature of 288 K, the scale height is 8×10^5 cm or 8 km; at an altitude of 8 km, the pressure is only about 39% of that at sea level.

Conversion of Equation 9.4 to the logarithmic (base 10) form and expression of h in km yields

$$\log P_h = \log P_0 - \frac{Mgh \times 10^5}{2.303RT} \tag{9.5}$$

and taking the pressure at sea level to be exactly 1 atm gives the following expression:

$$\log P_h = -\frac{Mgh \times 10^5}{2.303RT} \tag{9.6}$$

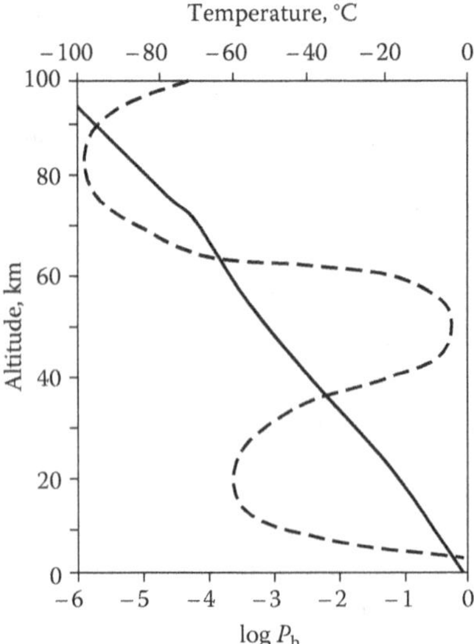

FIGURE 9.1 Variation of pressure (solid line) and temperature (dashed line) with altitude.

Plots of P_h and temperature *versus* altitude are shown in Figure 9.1. The plot of P_h is nonlinear because of variations arising from nonlinear variations in temperature with altitude that are discussed later in this section and from the mixing of air masses.

The characteristics of the atmosphere vary widely with altitude, time (season), location (latitude), and even solar activity. Extremes of pressure and temperature are illustrated in Figure 9.1. At very high altitudes, normally reactive species such as atomic oxygen, O, persist for long periods of time. That occurs because the pressure is very low at these altitudes such that the distance traveled by a reactive species before it collides with a potential reactant—its mean free path—is quite high. A particle with a mean free path of 1×10^{-6} cm at sea level has a mean free path greater than 1×10^{6} cm at an altitude of 500 km, where the pressure is lower by many orders of magnitude.

9.3.2 Stratification of the Atmosphere

As shown in Figure 9.2, the atmosphere is stratified on the basis of the temperature/density relationships resulting from interactions between physical and photochemical processes in air. The lowest layer of the atmosphere extending from sea level to an altitude of 10–16 km is the *troposphere*, characterized by a generally homogeneous composition of major gases other than water and decreasing temperature with increasing altitude. To understand why the temperature decreases with increasing altitude in the troposphere, consider a hypothetical mass of air at the surface rising to higher altitudes in the troposphere. As the air rises, it expands, doing work on its surroundings so that its temperature must fall. The extent of the temperature decrease *for dry air* with increasing altitude is known as the *adiabatic lapse rate*, which has a value of 9.8 K/km. However, air contains water vapor that condenses as the air mass rises, releasing heat of vaporization and lowering the lapse rate to an average of about 6.5 km^{-1}. The upper limit of the troposphere, which has a temperature minimum of about −56°C, varies in altitude by a kilometer or more with atmospheric temperature, underlying terrestrial surface, latitude, and time. The homogeneous composition of the troposphere results from constant mixing by convection currents in air masses, driven by the unstable situation with colder air above warmer air (the name of the troposphere is based on the Greek *tropos* for mixing). However, the water vapor content of the troposphere is extremely variable because of cloud formation, precipitation, and evaporation of water from terrestrial water bodies.

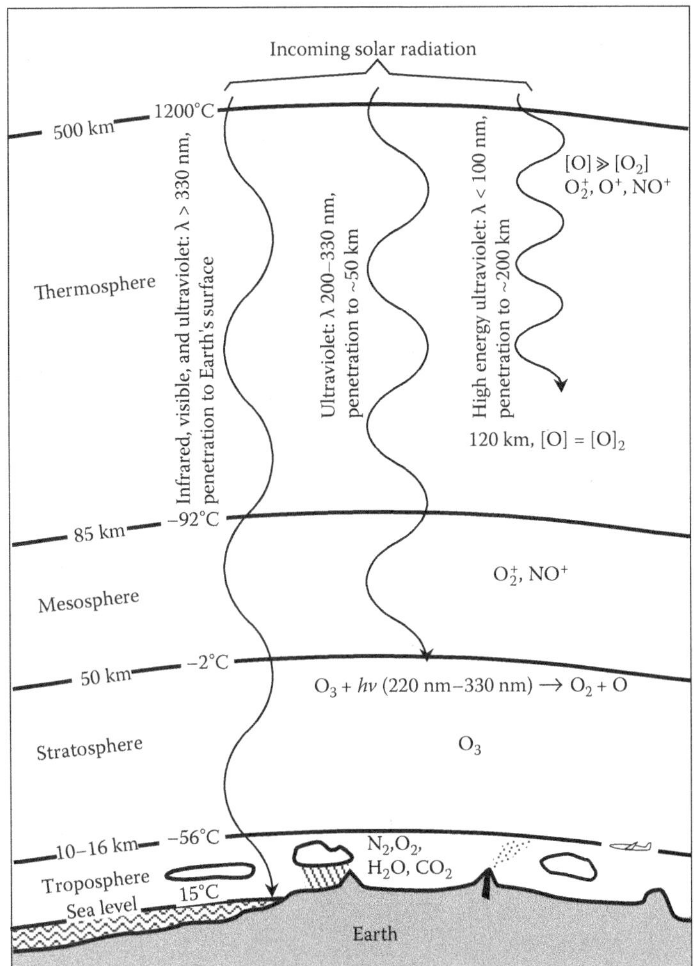

FIGURE 9.2 Major regions of the atmosphere (not to scale).

The very cold *tropopause* layer at the top of the troposphere serves as a barrier that causes water vapor to condense to ice so that it cannot reach altitudes at which it would photodissociate through the action of intense high-energy ultraviolet radiation. If this happened, the hydrogen produced would escape the Earth's atmosphere and be lost. (Much of the hydrogen and helium gases originally present in the Earth's atmosphere were lost by this process.)

The atmospheric layer directly above the troposphere is the *stratosphere*, in which the temperature rises to a maximum of about –2°C with increasing altitude. The increasing temperature with higher altitude in this region results in very little vertical mixing (the name stratosphere is based on the Greek *stratus* for mixing). This phenomenon is due to the presence of ozone, O_3, which may reach a level of around 10 ppm by volume in the mid-range of the stratosphere. The heating effect is caused by the absorption of ultraviolet radiation energy by ozone, a phenomenon discussed later in this chapter.

The absence of high levels of radiation-absorbing species in the *mesosphere* immediately above the stratosphere results in a further temperature decrease to about –92°C at an altitude around 85 km. The upper regions of the mesosphere and higher define a region called the exosphere from which molecules and ions can completely escape the atmosphere. Extending to the far outer reaches of the atmosphere is the *thermosphere*, in which the highly rarified gas reaches temperatures as high as 1200°C by the absorption of very energetic radiation of wavelengths less than approximately 200 nm by gas species in this region.

9.4 ENERGY TRANSFER IN THE ATMOSPHERE

The physical and chemical characteristics of the atmosphere and the critical heat balance of Earth are determined by energy and mass transfer processes in the atmosphere. Energy transfer phenomena are addressed in this section and mass transfer in Section 9.4. Incoming solar energy is largely in the visible region of the spectrum. The shorter wavelength blue solar light is scattered relatively more strongly by molecules and particles in the upper atmosphere, which is why the sky appears blue as it is viewed by scattered light and appears red by transmitted light, particularly around sunset and sunrise and when the atmosphere contains a high level of particles. The solar energy flux reaching the atmosphere is huge, amounting to 1.34×10^3 W/m^2 (19.2 kcal/min/m^2) perpendicular to the line of solar flux at the top of the atmosphere, as illustrated in Figure 9.3. This value is the *solar constant*, and may be termed *insolation*, which stands for "incoming solar radiation." This energy must be radiated back into space, and a delicate energy balance is involved in maintaining the Earth's temperature within very narrow limits that enable conditions of climate that support present levels of life on Earth. The great climate changes that resulted in past ice ages lasting thousands of years alternating with long periods of tropical conditions were caused by variations of only a few degrees in average temperature. Marked climate changes within recorded history have been marked by much smaller average temperature changes. The mechanisms by which the Earth's average temperature is retained within its present narrow range are complex and still the topic of intensive study, but the main features are explained here.[2]

About half of the solar radiation entering the atmosphere reaches the Earth's surface either directly or after scattering by clouds, atmospheric gases, or particles. The remaining half of the radiation is either reflected directly back or absorbed in the atmosphere, and its energy radiated back into space at a later time as infrared radiation. Most of the solar energy reaching the surface is absorbed and is returned to space to maintain heat balance. In addition, a very small amount of

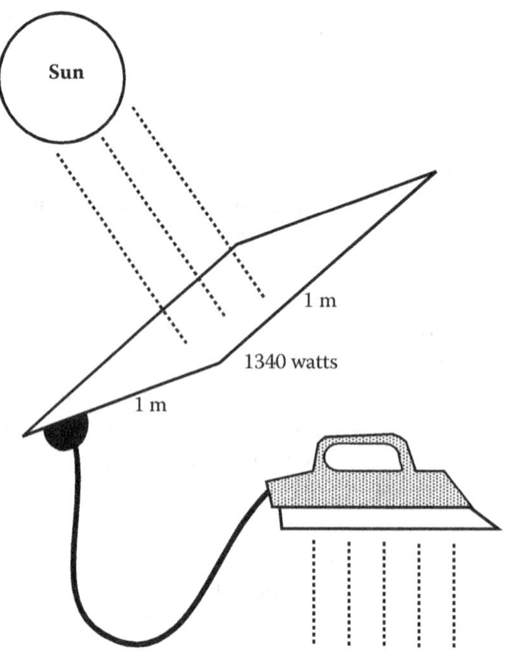

FIGURE 9.3 The solar flux at the distance of Earth from the sun is 1.34×10^3 W/m^2.

energy (<1% of that received from the sun) reaches the Earth's surface by convection and conduction processes from the Earth's hot mantle, and this, too, must be lost.

Energy transport, which is crucial to eventual reradiation of energy from Earth, is accomplished by three major mechanisms. These are conduction, convection, and radiation. *Conduction* of energy occurs through the interaction of adjacent atoms or molecules without the bulk movement of matter and is a relatively slow means of transferring energy in the atmosphere. *Convection* involves the movement of whole masses of air, which may be either relatively warm or cold. It is the mechanism by which abrupt temperature variations occur when large masses of air move across an area. Besides carrying *sensible heat* due to the kinetic energy of molecules, convection carries *latent heat* in the form of water vapor, which releases heat as it condenses. An appreciable fraction of the Earth's surface heat is transported to clouds in the atmosphere by conduction and convection before being lost ultimately by radiation.

Radiation of energy in the Earth's atmosphere occurs through electromagnetic radiation. Electromagnetic radiation is the only way in which energy is transmitted through a vacuum; therefore, it is the means by which all of the energy that must be lost from the planet to maintain its heat balance is ultimately returned to space. The maximum intensity of incoming radiation is at 0.5 μm (500 nm) in the visible region, with essentially none outside the range of 0.2–3 μm. This range encompasses the whole visible region and small parts of the ultraviolet and infrared adjacent to it. Outgoing radiation is in the infrared region, with maximum intensity at about 10 μm, primarily between 2 and 40 μm. Thus Earth loses energy by electromagnetic radiation of a much longer wavelength (lower energy per photon) than the radiation by which it receives energy, a crucial factor in maintaining the Earth's heat balance, and one susceptible to upset by human activities.

9.4.1 The Earth's Radiation Budget

The Earth's radiation budget is illustrated in Figure 9.4. The average surface temperature is maintained at a relatively comfortable 15°C because of an atmospheric "greenhouse effect" in which water vapor and, to a lesser extent, carbon dioxide reabsorb much of the outgoing radiation and reradiate about half of it back to the surface. Were this not the case, the surface temperature would average around −18°C. Most of the absorption of infrared radiation is done by water molecules in the atmosphere. Absorption is weak in the regions 7–8.5 and 11–14 μm, and nonexistent between 8.5 and 11 μm, leaving a "hole" in the infrared absorption spectrum through which radiation may escape. Carbon dioxide, though present at a much lower concentration than water vapor, absorbs strongly between 12 and 16.3 μm, and plays a key role in maintaining the heat balance. There is concern that an increase in the carbon dioxide level in the atmosphere could prevent sufficient energy loss to cause a perceptible and damaging increase in the Earth's temperature. This phenomenon, discussed in more detail in Section 9.7 and Chapter 14, is popularly known as the *greenhouse effect* and may occur from elevated CO_2 levels caused by increased use of fossil fuels and the destruction of photosynthesizing trees in forests.

An important aspect of solar radiation that reaches the Earth's surface is the percentage reflected from the surface, described as *albedo*. Albedo is important in determining the Earth's heat balance in that absorbed radiation heats the surface and reflected radiation does not. Albedo varies spectacularly with the surface. At the two extremes, freshly fallen snow has an albedo of 90% because it reflects 9/10th of incoming radiation, whereas freshly plowed black topsoil has an albedo of only about 2.5%.

9.5 ATMOSPHERIC MASS TRANSFER, METEOROLOGY, AND WEATHER

Meteorology is the science of atmospheric phenomena, encompassing the study of the movement of air masses as well as physical forces in the atmosphere—heat, wind, and transitions of water, primarily liquid to vapor or *vice versa*.[3] Meteorological phenomena affect, and in turn are affected by, the chemical properties of the atmosphere. Meteorological phenomena determine whether or not

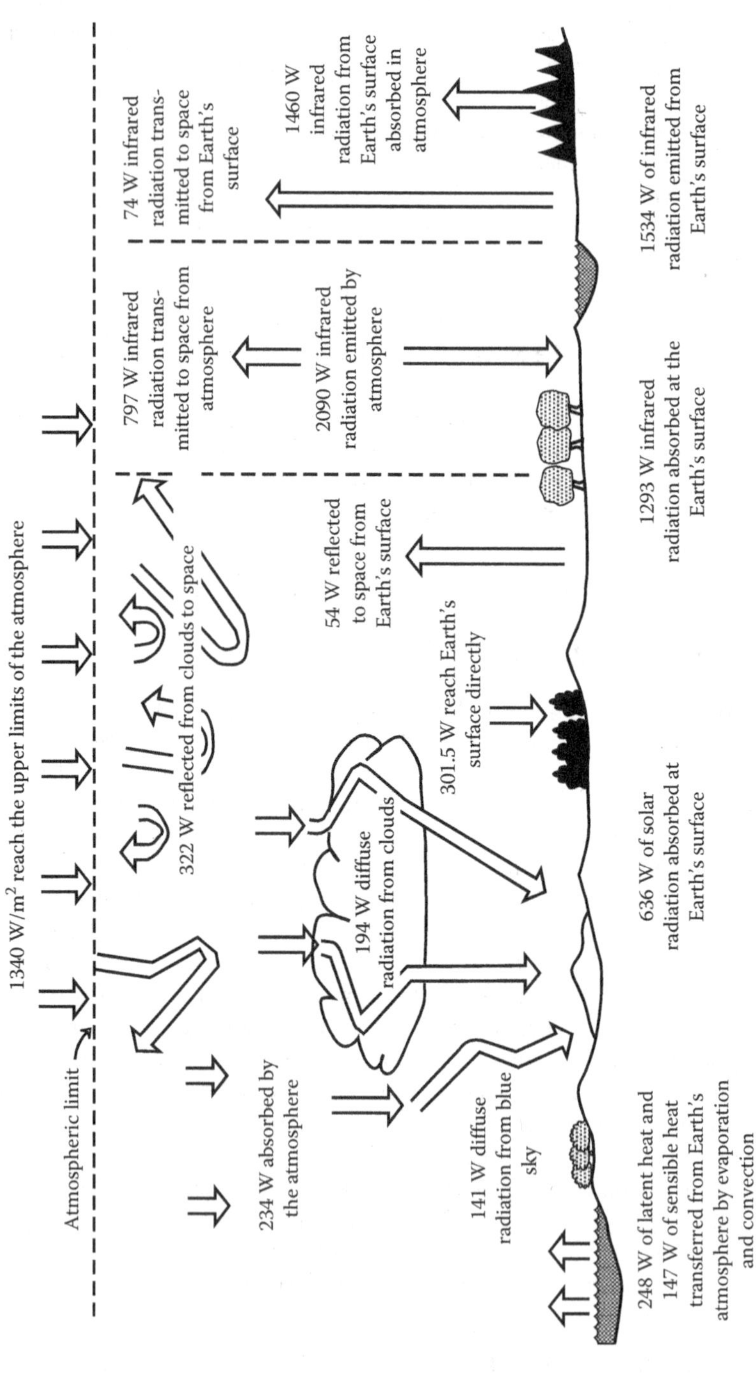

FIGURE 9.4 The Earth's radiation budget expressed on the basis of portions of the 1340 W/m^2 composing the solar flux.

pollutants emitted from a point source such as a power plant stack rise in the atmosphere and are dispersed or settle near the source where they may do maximum harm. Los Angeles largely owes its susceptibility to smog to the meteorology of the Los Angeles basin, which holds hydrocarbons and nitrogen oxides long enough to cook up an unpleasant brew of photochemical smog (Chapter 13). Short-term variations in the state of the atmosphere constitute *weather* defined in terms of seven closely interrelated major factors: temperature, clouds, winds, humidity, horizontal visibility (as affected by fog, etc.), type and quantity of precipitation, and atmospheric pressure. Longer-term variations and trends within a particular geographical region in those factors that compose weather are described as *climate*, a term defined and discussed in Section 9.6.

9.5.1 Atmospheric Water in Energy and Mass Transfer

The driving force behind weather and climate is the distribution and ultimate reradiation to space of solar energy. A large fraction of solar energy is converted to latent heat by evaporation of water into the atmosphere. As water condenses from atmospheric air, large quantities of heat are released. This is a particularly significant means for transferring energy from the ocean to land. Solar energy falling on the ocean is converted to latent heat by the evaporation of water, then the water vapor moves inland where it condenses. The latent heat released when the water condenses warms the surrounding land mass.

Atmospheric water can be present as vapor, liquid, or ice. The water vapor content of air can be expressed as *humidity*. *Relative humidity*, expressed as a percentage, describes the amount of water vapor in the air as a ratio of the maximum amount that the air can hold at that temperature. Air with a given relative humidity can undergo any of several processes to reach the saturation point at which water vapor condenses in the form of rain or snow. For this condensation to happen, air must be cooled below a temperature called the *dew point*, and *condensation nuclei* must be present. These nuclei are hygroscopic substances such as salts, sulfuric acid droplets, and some organic materials, including bacterial cells. Air pollution in some forms is an important source of condensation nuclei.

The liquid water in the atmosphere is present largely in *clouds*. Clouds normally form when rising, adiabatically cooling air can no longer hold water in the vapor form and the water forms very small aerosol droplets. Clouds may be classified into three major forms. Cirrus clouds occur at great altitudes and have a thin feathery appearance. Cumulus clouds are detached masses with a flat base and frequently a "bumpy" upper structure. Stratus clouds occur in large sheets and may cover all of the sky visible from a given point as overcast. Clouds are important absorbers and reflectors of radiation (heat). Their formation is affected by the products of human activities, especially particulate matter pollution and emission of deliquescent gases such as SO_2 and HCl. Some atmospheric chemical processes occur in solution in cloud droplets and crystalline ice particles in stratospheric clouds act as reservoirs for ozone-destroying chlorine species (see Section 14.5).

The formation of precipitation from the very small droplets of water that compose clouds is a complicated and important process. Cloud droplets normally take somewhat longer than a minute to form by condensation. They average about 0.04 mm across and do not exceed 0.2 mm in diameter. Raindrops range from 0.5 to 4 mm in diameter. Condensation processes do not form particles large enough to fall as precipitation (rain, snow, sleet, or hail). The small condensation droplets must collide and coalesce to form precipitation-size particles. When droplets reach a threshold diameter of about 0.04 mm, they grow more rapidly by coalescence with other particles than by condensation of water vapor.

9.5.2 Air Masses

Distinct air masses are a major feature of the troposphere. These air masses are uniform and horizontally homogeneous. Their temperature and water vapor content are particularly uniform. These characteristics are determined by the nature of the surface over which a large air mass forms. Polar continental air masses form over cold land regions; polar maritime air masses form over polar oceans. Air masses originating in the tropics may be similarly classified as tropical continental air masses or tropical maritime air masses. The movement of air masses and the conditions in them may have important effects on pollutant reactions, effects, and dispersal.

Solar energy received by Earth is largely redistributed by the movement of huge masses of air with different pressures, temperatures, and moisture contents separated by boundaries called *fronts*. Horizontally moving air is called *wind*, whereas vertically moving air is referred to as an *air current*. Atmospheric air moves constantly, with behavior and effects that reflect the laws governing the behavior of gases. First of all, gases will move horizontally and/or vertically from regions of *high atmospheric pressure* to those of *low atmospheric pressure*. Furthermore, expansion of gases causes cooling, whereas compression causes warming. A mass of warm air tends to move from the Earth's surface to higher altitudes where the pressure is lower; in so doing, it expands *adiabatically* (i.e., without exchanging energy with its surroundings) and becomes cooler. If there is no condensation of moisture from the air, the cooling effect is about 10°C/1000 m of altitude, a figure known as the *dry adiabatic lapse rate*. A cold mass of air at a higher altitude does the opposite; it sinks and becomes warmer at about 10°C/1000 m. Often, however, when there is sufficient moisture in rising air, water condenses from it, releasing latent heat. This partially counteracts the cooling effect of the expanding air, giving a *moist adiabatic lapse rate* of about 6°C/1000 m. Parcels of air do not rise and fall, or even move horizontally in a completely uniform way, but exhibit eddies, currents, and various degrees of turbulence.

As noted above, *wind* is air moving horizontally, whereas *air currents* are created by air moving up or down. Wind occurs because of differences in air pressure from high-pressure regions to low-pressure areas. Air currents are largely *convection currents* formed by differential heating of air masses. Air that is over a solar heated land mass is warmed, becomes less dense, and therefore rises and is replaced by cooler and more dense air. Wind and air currents are strongly involved with air pollution phenomena. Wind carries and disperses air pollutants. In some cases the absence of wind can enable pollutants to collect in a region and undergo processes that lead to even more (secondary) pollutants. Prevailing wind direction is an important factor in determining the areas most affected by an air pollution source. Wind is an important renewable energy resource (see Chapter 18). Furthermore, wind plays an important role in the propagation of life by dispersing spores, seeds, and organisms, such as spiders.

9.5.3 Topographical Effects

Topography, the surface configuration and relief features of the Earth's surface may strongly affect winds and air currents. Differential heating and cooling of land surfaces and bodies of water can result in *local convective winds*, including land breezes and sea breezes at different times of the day along the seashore, as well as breezes associated with large bodies of water inland. Mountain topography causes complex and variable localized winds. The masses of air in mountain valleys heat up during the day causing upslope winds, and cool off at night causing downslope winds. Upslope winds flow over ridge tops in mountainous regions. The blocking of wind and of masses of air by mountain formations some distance inland from seashores can trap bodies of air, particularly when temperature inversion conditions occur (see Section 9.6).

9.5.4 Movement of Air Masses

Basically, weather is the result of the interactive effects of (1) redistribution of solar energy, (2) horizontal and vertical movement of air masses with varying moisture contents, and (3) evaporation and condensation of water, accompanied by uptake and release of heat. To see how these factors determine weather—and ultimately climate—on a global scale, first consider the cycle illustrated in Figure 9.5. This figure shows solar energy being absorbed by a body of water and causing some water to evaporate. The warm, moist mass of air thus produced moves from a region of high pressure to one of low pressure, and cools by expansion as it rises in what is called a *convection column*. As the air cools, water condenses from it and energy is released; this is a major pathway by which energy is transferred from the Earth's surface to high in the atmosphere. As a result of condensation

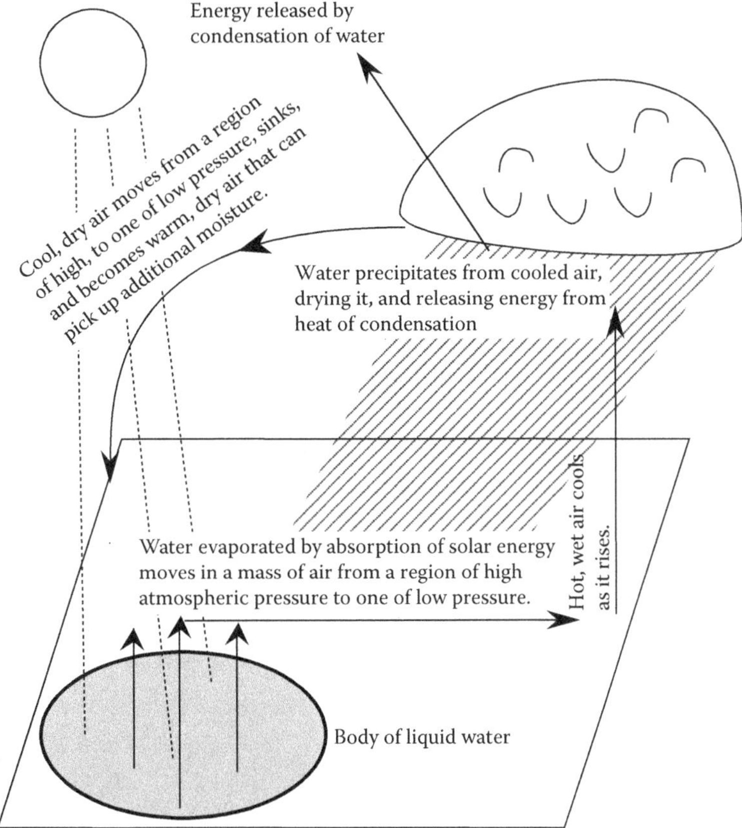
Energy released by
condensation of water
Cool, dry air moves from a region
of high, to one of low pressure, sinks,
and becomes warm, dry air that can
pick up additional moisture.
Water precipitates from cooled air,
drying it, and releasing energy from
heat of condensation
Water evaporated by absorption of solar energy
moves in a mass of air from a region of high
atmospheric pressure to one of low pressure.
Hot, wet air cools
as it rises.
Body of liquid water

FIGURE 9.5 Circulation patterns involved with movement of air masses and water; uptake and release of solar energy as latent heat in water vapor.

of water and loss of energy, the air is converted from warm, moist air to cool, dry air. Furthermore, the movement of the parcel of air to high altitudes results in a degree of "crowding" of air molecules and creates a zone of relatively high pressure high in the troposphere at the top of the convection column. This air mass, in turn, moves from the upper-level region of high pressure to one of low pressure; in so doing, it subsides, thus creating an upper-level low-pressure zone, and becomes warm, dry air in the process. The pileup of this air at the surface creates a surface high-pressure zone where the cycle described above began. The warm, dry air in this surface high-pressure zone again picks up moisture, and the cycle begins again.

9.5.5 Global Weather

The factors discussed above that determine and describe the movement of air masses are involved in the massive movement of air, moisture, and energy that occurs globally. The central feature of global weather is the redistribution of solar energy that falls unequally on Earth at different latitudes (relative distances from the equator and poles). Consider Figure 9.6. Sunlight, and the energy flux from it, is most intense at the equator because, averaged over the seasons, solar radiation comes in perpendicular to the Earth's surface at the equator. With increasing distance from the equator (higher latitudes) the angle is increasingly oblique and more of the energy-absorbing atmosphere must be traversed, so that progressively less energy is received per unit area of the Earth's surface. The net result is that equatorial regions receive a much greater share of solar radiation, progressively less is

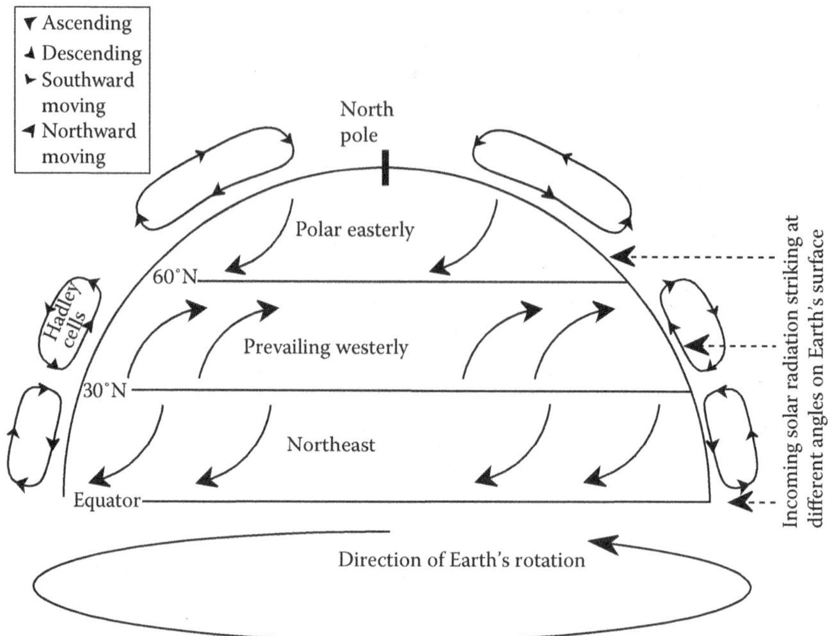

FIGURE 9.6 Global circulation of air in the northern hemisphere.

received farther from the equator, and the poles receive a comparatively miniscule amount. The excess heat energy in the equatorial regions causes the air to rise. The air ceases to rise when it reaches the stratosphere because in the stratosphere the air becomes warmer with higher elevation. As the hot equatorial air rises in the troposphere, it cools by expansion and loss of water, and then sinks again. The air circulation patterns in which this occurs are called *Hadley cells.* As shown in Figure 9.6, there are three major groupings of these cells, which result in very distinct climatic regions on the Earth's surface. The air in the Hadley cells does not move straight north and south, but is deflected by the Earth's rotation and by contact with the rotating Earth; this is the *Coriolis effect*, which results in spiral-shaped air circulation patterns called cyclonic or anticyclonic, depending on the direction of rotation. These give rise to different directions of prevailing winds, depending on latitude. The boundaries between the massive bodies of circulating air shift markedly over time and season, resulting in significant weather instability.

The movement of air in Hadley cells combined with other atmospheric phenomena results in the development of massive *jet streams* that are, in a sense, shifting rivers of air that may be several kilometers deep and several tens of kilometers wide. Jet streams move through discontinuities in the tropopause (see Section 9.3), generally from west to east at velocities around 200 km/h (well over 100 mph); in so doing, they redistribute huge amounts of air and have a strong influence on weather patterns.

The air and wind circulation patterns described above shift massive amounts of energy over long distances on Earth. If it were not for this effect, the equatorial regions would be unbearably hot, and the regions closer to the poles intolerably cold. About half of the heat that is redistributed is carried as sensible heat by air circulation, almost one-third is carried by water vapor as latent heat, and the remaining approximately 20% by ocean currents.

9.5.6 Weather Fronts and Storms

As noted earlier, the interface between two masses of air that differ in temperature, density, and water content is called a front. A mass of cold air moving such that it displaces one of warm air is a *cold front*, and a mass of warm air displacing one of cold air is a *warm front*. Since cold air is more dense than warm air, the air in a cold mass of air along a cold front pushes under warmer air. This causes the warm, moist air to rise such that water condenses from it. The condensation of water releases energy, so the air rises further. The net effect can be the formation of massive cloud formations (thunderheads) that may reach stratospheric levels. These spectacular thunderheads may produce heavy rainfall and even hail, and sometimes violent storms with strong winds, including tornadoes. Warm fronts cause somewhat similar effects such as warm, moist air pushes over colder air. However, the front is usually much broader, and the weather effects milder, typically resulting in widespread drizzle rather than intense rainstorms.

Swirling *cyclonic storms*, such as typhoons, hurricanes, and tornadoes, are created in low-pressure areas by rising masses of warm, moist air. As such air cools, water vapor condenses, and the latent heat released warms the air more, sustaining and intensifying its movement upward in the atmosphere. Air rising from surface level creates a low-pressure zone into which surrounding air moves. The movement of the incoming air assumes a spiral pattern, thus causing a cyclonic storm.

9.6 INVERSIONS AND AIR POLLUTION

The complicated movement of air across the Earth's surface is a crucial factor in the creation and dispersal of air pollution phenomena. When air movement ceases, stagnation can occur with a resultant buildup of atmospheric pollutants in localized regions. Although the temperature of air relatively near the Earth's surface normally decreases with increasing altitude, certain atmospheric conditions can result in the opposite condition—increasing temperature with increasing altitude. Such conditions are characterized by high atmospheric stability and are known as *temperature inversions*. Because they limit the vertical circulation of air, temperature inversions result in air stagnation and the trapping of air pollutants in localized areas.

Inversions can occur in several ways. An inversion can form when a warm air mass overrides a cold air mass. *Radiation inversions* are likely to form in still air at night when Earth is no longer receiving solar radiation. The air closest to Earth cools faster than the air higher in the atmosphere, which remains warm, thus less dense. Furthermore, cooler surface air tends to flow into valleys at night, where it is overlain by warmer, less dense air. *Subsidence inversions*, often widespread and accompanied by radiation inversions, can form in the vicinity of a surface high-pressure area when high-level air subsides to take the place of surface air blowing out of the high-pressure zone. The subsiding air is warmed as it compresses and can remain as a warm layer several hundred meters above ground level. A *marine inversion* is produced during the summer months when cool air laden with moisture from the ocean blows onshore and under warm, dry inland air.

As noted above, inversions contribute significantly to the effects of air pollution because, as shown in Figure 9.7, they prevent mixing of air pollutants, thus keeping the pollutants in one area. This not only prevents the pollutants from escaping, but also acts like a container in which additional pollutants accumulate. Furthermore, in the case of secondary pollutants formed by atmospheric chemical processes, such as photochemical smog (see Chapter 13), the pollutants may be kept together such that they react with each other and with sunlight to produce even more noxious products.

9.7 GLOBAL CLIMATE AND MICROCLIMATE

Perhaps the single most important influence on the Earth's environment is *climate*, consisting of long-term weather patterns over large geographical areas. As a general rule, climatic conditions are characteristic of a particular region. This does not mean that climate remains the same throughout the year, of course, because it varies with season. One important example of such variation is the *monsoon*, seasonal variations in wind patterns between oceans and continents accompanied by alternate periods of high and low rainfall. The climates of Africa and the Indian subcontinent are particularly influenced by monsoons. Summer monsoon rains are responsible for tropical rain

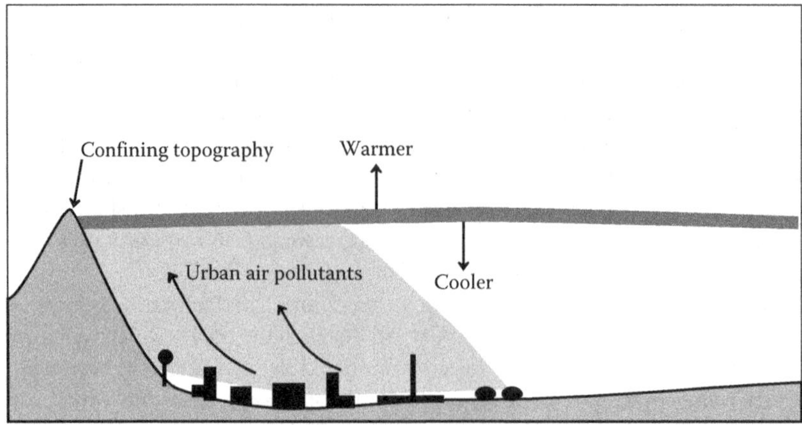

FIGURE 9.7 Illustration of pollutants trapped in a temperature inversion.

forests in Central Africa. The interface between this region and the Sahara Desert varies over time. When the boundary is relatively far north, rain falls on the Sahel desert region at the interface, crops grow, and the people do relatively well. When the boundary is more to the south, a condition that may last for several years, devastating droughts and even starvation may occur.

It is known that there are fluctuations, cycles, and cycles imposed on cycles in climate. The causes of these variations are not completely understood, but they are known to be substantial, and even devastating to civilization. The last *ice age*, which ended only about 10,000 years ago and which was preceded by several similar ice ages, produced conditions under which much of the present land mass of the northern hemisphere was buried under thick layers of ice and were uninhabitable. A "mini-ice age" occurred during the 1300s, causing crop failures and severe hardship in northern Europe. In modern times the El-Niño-Southern oscillation occurs with a period of several years when a large, semipermanent tropical low-pressure area shifts into the Central Pacific region from its more common location in the vicinity of Indonesia. This shift modifies prevailing winds, changes the pattern of ocean currents, and affects upwelling of ocean nutrients with profound effects on weather, rainfall, and fish and bird life over a vast area of the Pacific from Australia to the west coasts of South and North America.

9.7.1 Atmospheric Carbon Dioxide and Human Modifications of Climate

Although the Earth's atmosphere is huge and has an enormous ability to resist and correct for detrimental change, it is likely that human activities are reaching a point at which they are significantly affecting climate. One such way is by emission of large quantities of carbon dioxide and other greenhouse gases into the atmosphere, with the potential to cause global warming and substantial climatic change. It is known that atmospheric carbon dioxide levels are increasing by about 2 ppm volume per year and are on track to more than double from preindustrial levels during the current century. Furthermore, global temperatures during recent decades have increased measurably, consistent with models of global warming from atmospheric greenhouse gases. The influence of atmospheric carbon dioxide on global temperatures and possible impacts on climate are discussed in more detail in Section 14.2.

9.7.2 Microclimate

The preceding section described climate on a large scale, ranging up to global dimensions. The climate that organisms and objects on the surface are exposed to close to the ground, under rocks, and surrounded by vegetation is often quite different from the surrounding macroclimate. Such highly localized climatic conditions are termed the *microclimate*. Microclimate effects are largely determined by the uptake and loss of solar energy very close to the Earth's surface, and by the fact that air circulation due to wind is much lower at the surface. During the day, solar energy absorbed by relatively bare soil heats the surface, but is lost only slowly because of very limited air circulation at the surface. This provides a warm blanket of surface air several centimeters thick, and an even thinner layer of warm soil. At night, radiative loss of heat from the surface of soil and vegetation can result in surface temperatures several degrees colder than the air about 2 m above ground level. These lower temperatures result in condensation of *dew* on vegetation and the soil surface, thus providing a relatively more moist microclimate near ground level. Heat absorbed during early morning evaporation of the dew tends to prolong the period of cold experienced right at the surface.

Vegetation substantially affects microclimate. In relatively dense growths, circulation may be virtually zero at the surface because vegetation severely limits convection and diffusion. The crown surface of the vegetation intercepts most of the solar energy, so that maximum solar heating may be a significant distance up from the Earth's surface. The region below the crown surface of vegetation thus becomes one of relatively stable temperature. In addition, in a dense growth of vegetation, most of the moisture loss is not from evaporation from the soil surface, but rather from transpiration from plant leaves. The net result is the creation of temperature and humidity conditions that provide a favorable living environment for a number of organisms, such as insects and rodents.

Another factor influencing microclimate is the degree to which the slope of land faces north or south. South-facing slopes of land in the northern hemisphere receive greater solar energy. Advantage has been taken of this phenomenon in restoring land strip-mined for brown coal in Germany by terracing the land such that the terraces have broad south slopes and very narrow north slopes. On the south-sloping portions of the terrace, the net effect has been to extend the short summer growing season by several days, thereby significantly increasing crop productivity. In areas where the growing season is longer, better growing conditions may exist on a north slope because it is less subject to temperature extremes and to loss of water by evaporation and transpiration.

9.7.3 Effects of Urbanization on Microclimate

A particularly marked effect on microclimate is that induced by urbanization. In a rural setting, vegetation and bodies of water have a moderating effect, absorbing modest amounts of solar energy and releasing it slowly. The stone, concrete, and asphalt pavement of cities have an opposite effect, strongly absorbing solar energy, and reradiating heat back to the urban microclimate. Rainfall is not allowed to accumulate in ponds, but is drained away as rapidly and efficiently as possible. Human activities generate significant amounts of heat, and produce large quantities of CO_2 and other greenhouse gases that retain heat. The net result of these effects is that a city is capped by a *heat dome* in which the temperature is as much as 5°C warmer than in the surrounding rural areas, such that large cities have been described as "heat islands." The rising warmer air over a city brings in a breeze from the surrounding area and causes a local greenhouse effect that probably is largely counterbalanced by reflection of incoming solar energy by particulate matter above cities. Overall, compared to climatic conditions in nearby rural surroundings, the city microclimate is warmer, foggier, and overlain with more cloud cover a greater percentage of the time, and is subject to more precipitation, though generally less humid.

9.8 CHEMICAL AND PHOTOCHEMICAL REACTIONS IN THE ATMOSPHERE

Figure 9.8 represents some of the major atmospheric chemical processes, which are discussed under the topic of *atmospheric chemistry*. Several key concepts of atmospheric chemistry that are important for this discussion were defined at the beginning of the chapter. These include the energy of a

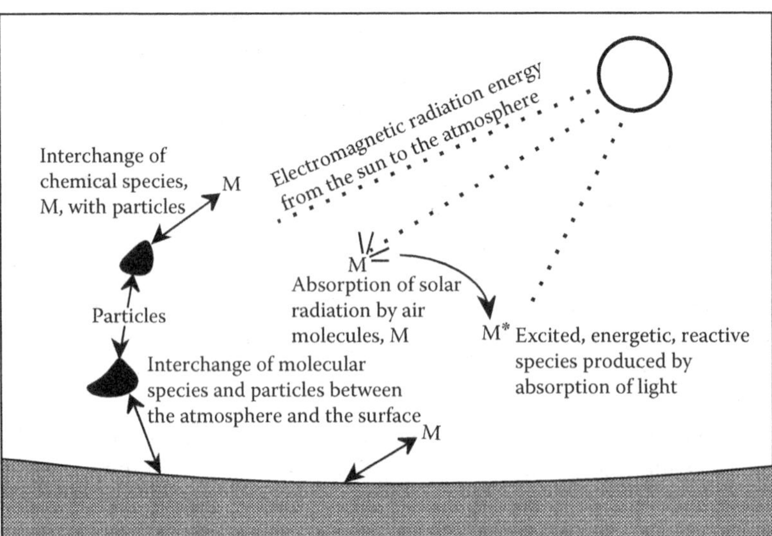

FIGURE 9.8 Representation of major atmospheric chemical processes.

photon, $h\nu$, the excited state, often designated with an asterisk, *, free radicals such as the hydroxyl radical, $HO^{\bullet}$, and energy absorbing third bodies denoted M.

The atmosphere is a huge, variable "laboratory" in which the study of chemical processes is very challenging. One of the primary obstacles encountered in studying atmospheric chemistry is that the chemist generally must deal with incredibly low concentrations, so that the detection and analysis of reaction products is quite difficult (see Chapter 26). Simulating high-altitude conditions in the laboratory can be extremely hard because of interferences, such as those from species given off from container walls under conditions of very low pressure. Many chemical reactions that require a third body to absorb excess energy occur very slowly in the upper atmosphere where there is a sparse concentration of third bodies, but occur readily in a container whose walls effectively absorb energy. Container walls may serve as catalysts for some important reactions, or they may absorb important species and react chemically with the more reactive ones.

Atmospheric chemistry involves the unpolluted atmosphere, highly polluted atmospheres, and a wide range of gradations in between. The same general phenomena govern all and produce one huge atmospheric cycle in which there are numerous subcycles. Gaseous atmospheric chemical species fall into the following somewhat arbitrary and overlapping classifications: *inorganic oxides* (CO, CO_2, NO_2, SO_2), *oxidants* (O_3, H_2O_2, $HO^{\bullet}$ radical, $HO_2^{\bullet}$ radical, $ROO^{\bullet}$ radicals, NO_3), *reductants* (CO, SO_2, H_2S), *organics* (also reductants; in the unpolluted atmosphere, CH_4 is the predominant organic species, whereas alkanes, alkenes, and aryl compounds are common around sources of organic pollution), *oxidized organic species* (carbonyls, organic nitrates), *photochemically active species* (NO_2, formaldehyde), acids (H_2SO_4), bases (NH_3), salts (NH_4HSO_4), and *unstable reactive species* (electronically excited NO_2^*, $HO^{\bullet}$ radical). In addition, both solid and liquid particles in atmospheric aerosols and clouds play a strong role in atmospheric chemistry as sources and sinks for gas-phase species, as sites for surface reactions (solid particles) and as bodies for aqueous-phase reactions (liquid droplets). Two constituents of utmost importance in atmospheric chemistry are radiant energy from the sun, predominantly in the ultraviolet region of the spectrum, and the hydroxyl radical, $HO^{\bullet}$. The former provides a way to pump a high level of energy into a single gas molecule to start a series of atmospheric chemical reactions, and the latter is the most important reactive intermediate and "currency" of daytime atmospheric chemical phenomena; NO_3 radicals are important intermediates in nighttime atmospheric chemistry. These are addressed in more detail in this chapter and Chapters 10 through 14.

Of particular importance in atmospheric chemistry is the discipline of *chemical kinetics*, which deals with rates of reactions. A detailed discussion of the kinetics of atmospheric chemical processes is beyond the scope of this book and the reader is referred to other sources, such as those listed in the supplementary references at the end of this chapter.

The rate constants, which describe chemical reaction rates, vary widely and are important tools for the atmospheric chemist to describe and explain atmospheric chemical processes. Uncertainties still exist regarding the values of particular rate constants.

9.8.1 Photochemical Processes

The absorption by chemical species of light, broadly defined here to include ultraviolet radiation from the sun, can bring about reactions, called *photochemical reactions*, which do not otherwise occur under the conditions (particularly the temperature) of the medium in the absence of light. Thus, photochemical reactions, even in the absence of a chemical catalyst, occur at temperatures much lower than those which otherwise would be required. Photochemical reactions, which are induced by intense solar radiation, play a very important role in determining the nature and ultimate fate of a chemical species in the atmosphere.

Nitrogen dioxide, NO_2, is one of the most photochemically active species found in a polluted atmosphere and is an essential participant in the smog-formation process. A species such as NO_2 may absorb light of energy $h\nu$, producing an *electronically excited molecule*,

$$NO_2 + h\nu \rightarrow NO_2^* \tag{9.7}$$

designated in the reaction above by an asterisk, *. The photochemistry of nitrogen dioxide is discussed in greater detail in Chapters 11 and 13.

Electronically excited molecules compose one of the three kinds of relatively reactive and unstable species that are encountered in the atmosphere and are strongly involved with atmospheric chemical processes. The other two species are atoms or molecular fragments with unshared electrons, called *free radicals*, and *ions* consisting of electrically charged atoms or molecular fragments.

Electronically excited molecules are produced when stable molecules absorb energetic electromagnetic radiation in the ultraviolet or visible regions of the spectrum. A molecule may possess several possible excited states, but generally ultraviolet or visible radiation is energetic enough to excite molecules only to several of the lowest energy levels. The nature of the excited state may be understood by considering the disposition of electrons in a molecule. Most molecules have an even number of electrons. The electrons occupy orbitals, with a maximum of two electrons with opposite spin occupying the same orbital. The absorption of light may promote one of these electrons to a vacant orbital of higher energy. In some cases the electron thus promoted retains a spin opposite to that of its former partner, giving rise to an *excited singlet state*. In other cases the spin of the promoted electron is reversed, such that it has the same spin as its former partner; this gives rise to an *excited triplet state*.

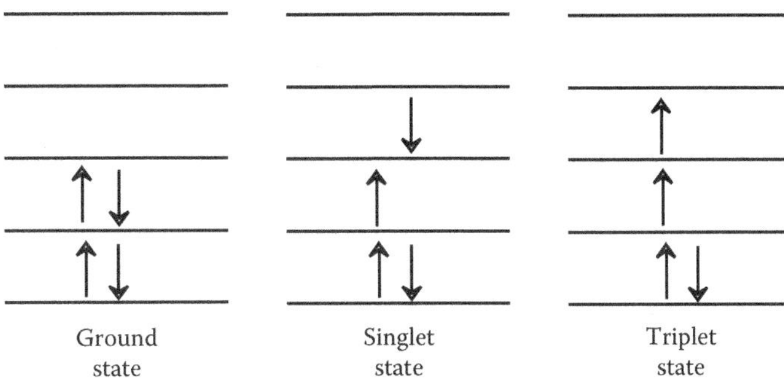

The chemical species in excited states are relatively energized compared to the ground state and are chemically reactive species. Their participation in atmospheric chemical reactions, such as those involved in smog formation, will be discussed later in detail.

In order for a photochemical reaction to occur, light must be absorbed by the reacting species. If the absorbed light is in the visible region of the sun's spectrum, the absorbing species is colored. Colored NO_2 is a common example of such a species in the atmosphere. Normally, the first step in a photochemical process is the activation of the molecule by the absorption of a single unit of photochemical energy characteristic of the frequency of the light called a *quantum* of light. The energy of one quantum is equal to the product $h\nu$, where h is Planck's constant, 6.63×10^{-34} J s (6.63×10^{-27} erg s), and ν is the frequency of the absorbed light in s^{-1} (inversely proportional to its wavelength, λ).

The reactions that occur following absorption of a photon of light to produce an electronically excited species are largely determined by the way in which the excited species loses its excess energy. This may occur by one of the following processes.

- Loss of energy to another molecule or atom (M) by *physical quenching*, followed by dissipation of the energy as heat

$$O_2^* + M \rightarrow O_2 + M \text{ (higher translational energy)} \tag{9.8}$$

- *Dissociation* of the excited molecule (the process responsible for the predominance of atomic oxygen in the upper atmosphere)

$$O_2^* \rightarrow O + O \tag{9.9}$$

- *Direct reaction* with another species

$$O_2^* + O_3 \rightarrow 2O_2 + O \tag{9.10}$$

- *Luminescence* consisting of loss of energy by the emission of electromagnetic radiation

$$NO_2^* \rightarrow NO_2 + h\nu \tag{9.11}$$

  If the re-emission of light is almost instantaneous, luminescence is called *fluorescence*, and if it is significantly delayed, the phenomenon is *phosphorescence*. *Chemiluminescence* is said to occur when the excited species (such as NO_2^* below) is formed by a chemical process:

$$O_3 + NO \rightarrow NO_2^* + O_2 \text{ (higher energy)} \tag{9.12}$$

- *Intermolecular energy transfer* in which an excited species transfers energy to another species which then becomes excited

$$O_2^* + Na \rightarrow O_2 + Na^* \tag{9.13}$$

  A subsequent reaction by the second species is called a *photosensitized* reaction.

- *Intramolecular transfer* in which energy is transferred within a molecule

$$XY^* \rightarrow XY^{\dagger} \text{ (}^{\dagger}\text{ denotes another excited state of the same molecule)} \tag{9.14}$$

- *Photoionization* through loss of an electron

$$N_2^* \rightarrow N_2^+ + e^- \tag{9.15}$$

Electromagnetic radiation absorbed in the infrared region lacks the energy to break chemical bonds, but does cause the receptor molecules to gain vibrational and rotational energy. The energy absorbed as infrared radiation ultimately is dissipated as heat and raises the temperature of the whole atmosphere. As noted in Section 9.3, the absorption of infrared radiation is very important in the Earth's acquiring heat from the sun and in the retention of energy radiated from the Earth's surface.

9.8.2 Ions and Radicals in the Atmosphere

One of the characteristics of the upper atmosphere which is difficult to duplicate under laboratory conditions is the presence of significant levels of electrons and positive ions. Because of the rarefied conditions, these ions may exist in the upper atmosphere for long periods before recombining to form neutral species.

At altitudes of approximately 50 km and up, ions are so prevalent that the region is called the *ionosphere*. The presence of the ionosphere has been known since about 1901, when it was discovered that radio waves could be transmitted over long distances where the curvature of Earth makes line-of-sight transmission impossible. These radio waves bounce off the ionosphere.

Ultraviolet light is the primary producer of ions in the ionosphere. In darkness, the positive ions slowly recombine with free electrons. The process is more rapid in the lower regions of the ionosphere where the concentration of species is relatively high. Thus, the lower limit of the ionosphere lifts at night and makes possible the transmission of radio waves over much greater distances.

The Earth's magnetic field has a strong influence on the ions in the upper atmosphere. Probably the best-known manifestation of this phenomenon is found in the Van Allen belts, discovered in 1958 consisting of two belts of ionized particles encircling the Earth. They can be visualized as two doughnuts with the axis of the Earth's magnetic field extending through the holes in the doughnuts. The inner belt consists of protons and the outer belt consists of electrons. A schematic diagram of the Van Allen belts is shown in Figure 9.9.

Although ions are produced in the upper atmosphere primarily by the action of energetic electromagnetic radiation, they may also be produced in the troposphere by the shearing of water droplets

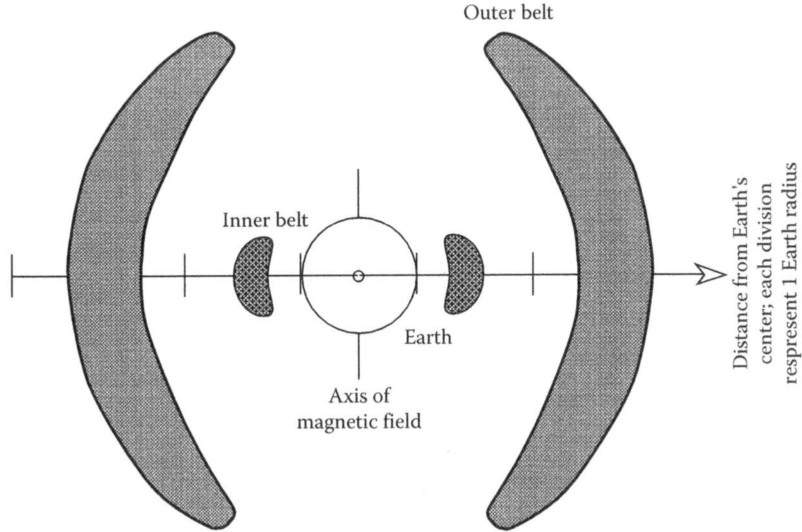

FIGURE 9.9 Cross section of the Van Allen belts encircling Earth.

during precipitation. The shearing may be caused by the compression of descending masses of cold air or by strong winds over hot, dry land masses. The last phenomenon is known as the foehn, sharav (in the Near East), or Santa Ana (in southern California). These hot, dry winds cause severe discomfort. The ions they produce consist of electrons and positively charged molecular species.

9.8.2.1 Free Radicals

In addition to forming ions by photoionization, energetic electromagnetic radiation in the atmosphere may produce atoms or groups of atoms with unpaired electrons called *free radicals*:

$$H_3C{-}\overset{\overset{\displaystyle O}{\|}}{C}{-}H + h\nu \rightarrow H_3C\bullet + \bullet\overset{\overset{\displaystyle O}{\|}}{C}{-}H \tag{9.16}$$

Free radicals are involved with most significant atmospheric chemical phenomena and are of utmost importance in the atmosphere. Because of their unpaired electrons and the strong pairing tendencies of electrons under most circumstances, free radicals are highly reactive. The upper atmosphere is so rarefied, however, that at very high altitudes radicals may have half-lives of several minutes, or even longer. Radicals can take part in chain reactions in which one of the products of each reaction is a radical. Eventually, through processes such as reaction with another radical, one of the radicals in a chain is destroyed and the chain ends:

$$H_3C^\bullet + H_3C^\bullet \rightarrow C_2H_6 \tag{9.17}$$

This process is a *chain-terminating reaction*. Reactions involving free radicals are responsible for smog formation, discussed in Chapter 13.

Free radicals are quite reactive; therefore, they generally have short lifetimes. It is important to distinguish between high reactivity and instability. A totally isolated free radical or atom would be quite stable. Therefore, free radicals and single atoms from diatomic gases tend to persist under the rarefied conditions of very high altitudes because they can travel long distances before colliding with another reactive species. However, electronically excited species have a finite, generally very short lifetime because they can lose energy through radiation without having to react with another species.

9.8.3 Hydroxyl and Hydroperoxyl Radicals in the Atmosphere

As illustrated in Figure 9.10, the hydroxyl radical, $HO^\bullet$, is the single most important reactive intermediate species in atmospheric chemical processes. It is formed by several mechanisms. At higher altitudes it is produced by photolysis of water:

$$H_2O + h\nu \rightarrow HO^\bullet + H \tag{9.18}$$

In the presence of organic matter, hydroxyl radical is produced in abundant quantities as an intermediate in the formation of photochemical smog (see Chapter 13). To a certain extent in the atmosphere, and for laboratory experimentation, $HO^\bullet$ is made by the photolysis of nitrous acid vapor:

$$HONO + h\nu \rightarrow HO^\bullet + NO \tag{9.19}$$

Hydroxyl radical is also generated by photodissociation of hydrogen peroxide, H_2O_2, the most important oxidant in solution in atmospheric particles of fog, cloud, or rain:

$$H_2O_2 + h\nu \rightarrow HO^\bullet + HO^\bullet \tag{9.20}$$

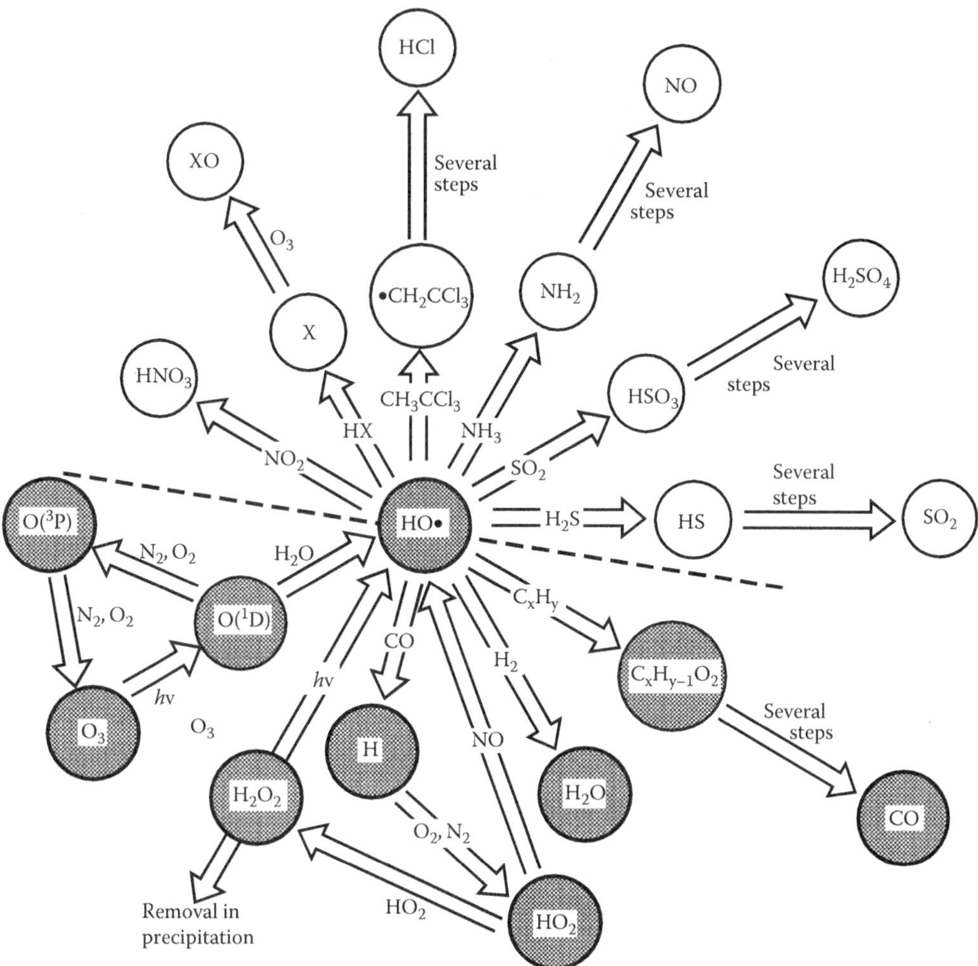

FIGURE 9.10 Control of trace gas concentrations by $HO^\bullet$ radical in the troposphere. Processes below the dashed line are those largely involved in controlling the concentrations of $HO^\bullet$ in the troposphere; those above the line control the concentrations of the associated reactants and products. Reservoirs of atmospheric species are shown in circles, reactions denoting conversion of one species to another are shown by arrows, and the reactants or photons needed to bring about a particular conversion are shown along the arrows. Hydrogen halides are denoted by HX and hydrocarbons by $HxYy$. (Reprinted from D.D. Davis and W.L. Chameides, Chemistry in the troposphere, *Chemical and Engineering News*, October 4, 1982, pp. 39–52. With permission.)

In the relatively unpolluted troposphere, hydroxyl radical is produced as a result of photolysis of ozone:

$$O_3 + h\nu(\lambda < 315 \text{ nm}) \rightarrow O^* + O_2 \tag{9.21}$$

followed by the reaction of a fraction of the excited oxygen atoms with water molecules:

$$O^* + H_2O \rightarrow 2HO^\bullet \tag{9.22}$$

Involvement of the hydroxyl radical in chemical transformations of a number of trace species in the atmosphere is summarized in Figure 9.10, and some of the pathways illustrated are discussed in later chapters. Among the important atmospheric trace species that react with hydroxyl radical are carbon monoxide, sulfur dioxide, hydrogen sulfide, methane, and nitric oxide.

Hydroxyl radical is most frequently removed from the troposphere by reaction with methane or carbon monoxide:

$$CH_4 + HO^\bullet \rightarrow H_3C^\bullet + H_2O \tag{9.23}$$

$$CO + HO^\bullet \rightarrow CO_2 + H \tag{9.24}$$

The highly reactive methyl radical, $H_3C^\bullet$, reacts with O_2,

$$H_3C^\bullet + O_2 \rightarrow H_3COO^\bullet \tag{9.25}$$

to form *methylperoxyl radical*, $H_3COO^\bullet$. (Further reactions of this species are discussed in Chapter 13.) The hydrogen atom produced in Reaction 9.24 reacts with O_2 to produce *hydroperoxyl radical*:

$$H + O_2 \rightarrow HOO^\bullet \tag{9.26}$$

The hydroperoxyl radical can undergo chain termination reactions, such as

$$HOO^\bullet + HO^\bullet \rightarrow H_2O + O_2 \tag{9.27}$$

$$HOO^\bullet + HOO^\bullet \rightarrow H_2O_2 + O_2 \tag{9.28}$$

or reactions that regenerate hydroxyl radical:

$$HOO^\bullet + NO \rightarrow NO_2 + HO^\bullet \tag{9.29}$$

$$HOO^\bullet + O_3 \rightarrow 2O_2 + HO^\bullet \tag{9.30}$$

The global concentration of hydroxyl radical, averaged diurnally and seasonally, is estimated to range from 2×10^5 to 1×10^6 radicals per cm^3 in the troposphere. Because of the higher humidity and higher incident sunlight which result in elevated O* levels, the concentration of $HO^\bullet$ is higher in tropical regions. The southern hemisphere probably has about a 20% higher level of $HO^\bullet$ than does the northern hemisphere because of greater production of anthropogenic, $HO^\bullet$-consuming CO in the northern hemisphere.

The hydroperoxyl radical, $HOO^\bullet$, is an intermediate in some important chemical reactions. In addition to its production by the reactions discussed above, in polluted atmospheres, hydroperoxyl radical is made by the following two reactions, starting with photolytic dissociation of formaldehyde to produce a reactive formyl radical:

$$HCHO + h\nu \rightarrow H + H\dot{C}O \tag{9.31}$$

$$H\dot{C}O + O_2 \rightarrow HOO^\bullet + CO \tag{9.32}$$

The hydroperoxyl radical reacts more slowly with other species than does the hydroxyl radical. The kinetics and mechanisms of hydroperoxyl radical reactions are difficult to study because it is hard to retain these radicals free of hydroxyl radicals.

9.8.4 Chemical and Biochemical Processes in Evolution of the Atmosphere

It is now widely believed that the Earth's atmosphere originally was very different from its present state and that the changes were brought about by biological activity and accompanying chemical changes. Approximately 3.5 billion years ago, when the first primitive life molecules were formed, the atmosphere was probably free of oxygen and consisted of a variety of gases such as carbon dioxide, water vapor, and perhaps even methane, ammonia, and hydrogen. The atmosphere was bombarded by intense, bond-breaking ultraviolet light which, along with lightning and radiation from radionuclides, provided the energy to bring about chemical reactions that resulted in the production of relatively complicated molecules, including even amino acids and sugars. From the rich chemical mixture in the sea, life molecules evolved. Initially, these very primitive life forms derived their energy from fermentation of organic matter formed by chemical and photochemical processes, but eventually they gained the capability to produce organic matter, "$\{CH_2O\}$," by photosynthesis:

$$CO_2 + H_2O + h\nu \rightarrow \{CH_2O\} + O_2(g) \tag{9.33}$$

Photosynthesis released oxygen, thereby setting the stage for the massive biochemical transformation that resulted in the production of almost all the atmosphere's O_2.

The oxygen initially produced by photosynthesis was probably quite toxic to primitive life forms. However, much of this oxygen was converted to iron oxides by reaction with soluble iron(II):

$$4Fe^{2+} + O_2 + 4H_2O \rightarrow 2Fe_2O_3 + 8H^+ \tag{9.34}$$

This resulted in the formation of enormous deposits of iron oxides, the existence of which provides major evidence for the liberation of free oxygen in the primitive atmosphere.

Eventually, enzyme systems developed that enabled organisms to mediate the reaction of waste-product oxygen with oxidizable organic matter in the sea. Later, this mode of waste-product disposal was utilized by organisms to produce energy by respiration, which is now the mechanism by which nonphotosynthetic organisms obtain energy.

In time, O_2 accumulated in the atmosphere, providing an abundant source of oxygen for respiration. It had an additional benefit in that it enabled the formation of an ozone shield (see Section 9.9). The ozone shield absorbs bond-rupturing ultraviolet light. With the ozone shield protecting tissue from destruction by high-energy ultraviolet radiation, Earth became a much more hospitable environment for life, and life forms were enabled to move from the sea to land.

9.9 ACID–BASE REACTIONS IN THE ATMOSPHERE

Acid–base reactions occur between acidic and basic species in the atmosphere. The atmosphere is normally at least slightly acidic because of the presence of a low level of carbon dioxide, which dissolves in atmospheric water droplets and dissociates slightly:

$$CO_2(g) \xrightarrow{\text{Water}} CO_2(aq) \tag{9.35}$$

$$CO_2(aq) + H_2O \rightarrow H^+ + HCO_3^- \tag{9.36}$$

Atmospheric sulfur dioxide forms a somewhat stronger acid than carbon dioxide when it dissolves in water:

$$SO_2(g) + H_2O \rightarrow H^+ + HSO_3^- \tag{9.37}$$

In terms of pollution, however, strongly acidic HNO_3 and H_2SO_4 formed by the atmospheric oxidation of N oxides, SO_2, and H_2S are much more important because they lead to the formation of damaging acid rain (see Chapter 14).

As reflected by the generally acidic pH of rainwater, basic species are relatively less common in the atmosphere. Particulate calcium oxide, hydroxide, and carbonate can get into the atmosphere from ash and ground rock, and can react with acids such as in the following reaction:

$$Ca(OH)_2(s) + H_2SO_4(aq) \rightarrow CaSO_4(s) + 2H_2O \tag{9.38}$$

The most important basic species in the atmosphere is gas-phase ammonia, NH_3. The major source of atmospheric ammonia is from biodegradation of nitrogen-containing biological matter and from bacterial reduction of nitrate:

$$NO_3^-(aq) + 2\{CH_2O\}(\text{biomass}) + H^+ \rightarrow NH_3(g) + 2CO_2 + H_2O \tag{9.39}$$

Ammonia is particularly important as a base in the air because it is the only water-soluble base present at significant levels in the atmosphere. Dissolved in atmospheric water droplets, it plays a strong role in neutralizing atmospheric acids:

$$NH_3(aq) + HNO_3(aq) \rightarrow NH_4NO_3(aq) \tag{9.40}$$

$$NH_3(aq) + H_2SO_4(aq) \rightarrow NH_4HSO_4(aq) \tag{9.41}$$

These reactions have three effects: (1) they result in the presence of NH_4^+ ion in the atmosphere as dissolved or solid salts, (2) they serve in part to neutralize acidic constituents of the atmosphere, and (3) they produce relatively corrosive ammonium salts.

9.10 REACTIONS OF ATMOSPHERIC OXYGEN

Some of the primary features of the exchange of oxygen among the atmosphere, geosphere, hydrosphere, biosphere, and anthrosphere are summarized in Figure 9.11. The oxygen cycle is critically important in atmospheric chemistry, geochemical transformations, and life processes.

Oxygen in the troposphere plays a strong role in processes that occur on the Earth's surface. Atmospheric oxygen takes part in energy-producing reactions, such as the burning of fossil fuels:

$$CH_4 \text{ (in natural gas)} + 2O_2 \rightarrow CO_2 + 2H_2O \tag{9.42}$$

Atmospheric oxygen is utilized by aerobic organisms in the degradation of organic material. Some oxidative weathering processes of minerals (see Section 15.2) consume oxygen, such as

$$4FeO + O_2 \rightarrow 2Fe_2O_3 \tag{9.43}$$

Oxygen is returned to the atmosphere through plant photosynthesis:

$$CO_2 + H_2O + h\nu \rightarrow \{CH_2O\} + O_2 \tag{9.44}$$

All molecular oxygen now in the atmosphere is thought to have originated through the action of photosynthetic organisms, which shows the importance of photosynthesis in the oxygen balance of the atmosphere. It can be shown that most of the carbon fixed by these photosynthetic processes is dispersed in mineral formations as humic material (Section 3.17); only a very small fraction is deposited in fossil fuel beds. Therefore, although combustion of fossil fuels consumes large amounts of O_2, there is no danger of running out of atmospheric oxygen.

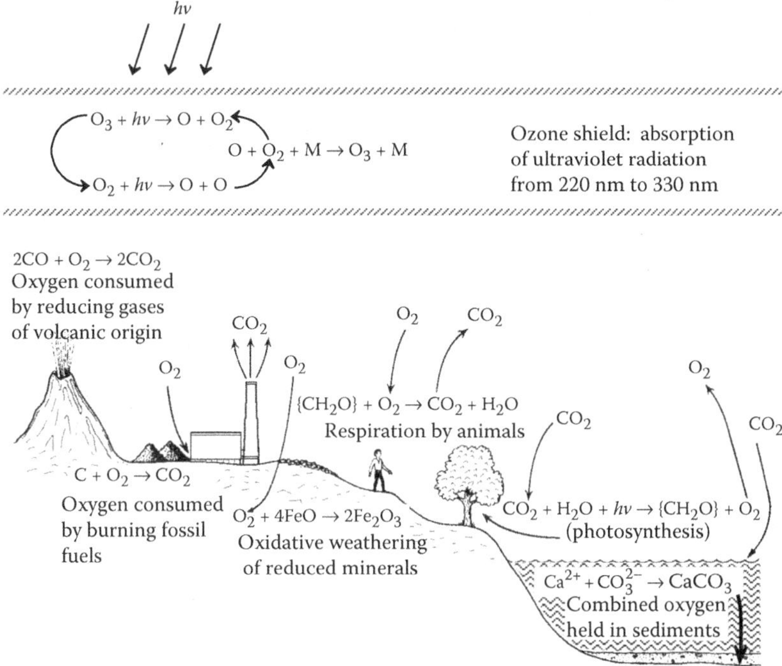

FIGURE 9.11 Oxygen exchange among the atmosphere, geosphere, hydrosphere, and biosphere.

9.11 REACTIONS OF ATMOSPHERIC NITROGEN

The 78% by volume of nitrogen contained in the atmosphere constitutes an inexhaustible reservoir of that essential element. The nitrogen cycle and nitrogen fixation by microorganisms were discussed in Chapter 6. A small amount of nitrogen is fixed in the atmosphere by lightning, and some is also fixed by combustion processes, particularly in internal combustion and turbine engines.

Unlike oxygen, which is almost completely dissociated to the monatomic form in higher regions of the thermosphere, molecular nitrogen is not readily dissociated by ultraviolet radiation. However, at altitudes exceeding approximately 100 km, atomic nitrogen is produced by photochemical reactions:

$$N_2 + h\nu \rightarrow N + N \tag{9.45}$$

Several reactions of ionic species in the ionosphere may generate N atoms as well.

The N_2^+ ion is generated by photoionization in the atmosphere:

$$N_2 + h\nu \rightarrow N_2^+ + e^- \tag{9.46}$$

and may react to form other ions. The NO^+ ion is one of the predominant ionic species in the so-called E region of the ionosphere.

Pollutant oxides of nitrogen, particularly NO_2, are key species involved in air pollution and the formation of photochemical smog. For example, NO_2 is readily dissociated photochemically to NO and reactive atomic oxygen:

$$NO_2 + h\nu \rightarrow NO + O \tag{9.47}$$

This reaction is the most important primary photochemical process involved in smog formation. The roles played by nitrogen oxides in smog formation and other forms of air pollution are discussed in Chapters 11 through 14.

9.12 ATMOSPHERIC WATER

The water vapor content of the troposphere is normally within a range of 1–3% by volume with a global average of about 1%. However, air can contain as little as 0.1% or as much as 5% water. The percentage of water in the atmosphere decreases rapidly with increasing altitude. Water circulates through the atmosphere in the hydrologic cycle as shown in Figure 3.1.

Water vapor absorbs infrared radiation even more strongly than does carbon dioxide, thus greatly influencing the Earth's heat balance. Clouds formed from water vapor reflect light from the sun and have a temperature-lowering effect. On the other hand, water vapor in the atmosphere acts as a kind of "blanket" at night, retaining heat from the Earth's surface by absorption of infrared radiation.

As discussed in Section 9.5, water vapor and the heat released and absorbed by transitions of water between the vapor state and liquid or solid are strongly involved in atmospheric energy transfer. Condensed water vapor in the form of very small droplets is of considerable concern in atmospheric chemistry. The harmful effects of some air pollutants—for instance, the corrosion of metals by acid-forming gases—requires the presence of water which may come from the atmosphere. Atmospheric water vapor has an important influence on pollution-induced fog formation under some circumstances. Water vapor interacting with pollutant particulate matter in the atmosphere may reduce visibility to undesirable levels through the formation of very small atmospheric aerosol particles.

As noted in Section 9.2, the cold tropopause serves as a barrier to the movement of water into the stratosphere. Thus, little water is transferred from the troposphere to the stratosphere, and the main source of water in the stratosphere is the photochemical oxidation of methane:

$$CH_4 + 2O_2 + h\nu \xrightarrow[\text{several steps}]{} CO_2 + 2H_2O \tag{9.48}$$

The water produced by this reaction serves as a source of stratospheric hydroxyl radical as shown by the following reaction:

$$H_2O + h\nu \rightarrow HO^\bullet + H \tag{9.49}$$

9.13 INFLUENCE OF THE ANTHROSPHERE

Human activities have had an enormous influence on the atmosphere and atmospheric chemistry. Agricultural, industrial, and transportation activities have substantially altered the composition of trace gases in the atmosphere. These effects are particularly pronounced in the area of greenhouse gases (carbon dioxide and methane in particular) that may lead to global warming and in alteration of levels of ozone in both the troposphere and stratosphere. Emissions of NO from industrial and transportation sources, and perhaps most importantly from biomass burning, have led to significant increases in tropospheric ozone in mid- and low latitudes, where it is an undesirable air pollutant. Emissions of CFCs (freons) have caused decreases of ozone in the stratosphere, particularly in the Antarctic, where it serves a vital protective function against ultraviolet radiation.

Although the atmosphere has a substantial ability to cleanse itself of harmful pollutants by oxidizing them, this ability is being overtaxed in important respects. Of particular concern for the future are growing emissions from sources in developing countries.

9.14 CHEMICAL FATE AND TRANSPORT IN THE ATMOSPHERE

The atmosphere is very much involved in fate and transport processes of environmental chemicals. To understand these processes, it is necessary to consider sources, transport, dispersal, and fluxes of airborne contaminants. Interactions at the atmosphere/surface boundary are important, including flow and dispersal of material in the atmosphere over complex terrain, and around surface obstacles, such as trees and buildings. Interactions and interchange with surface media including rock and soil, water, and vegetation must be considered. Transport and dispersal by advection due to movement of masses of air and diffusive and Fickian transport (Section 1.8) are important considerations as are dispersion, and degradation half-life.[1,2] Pollutants in the atmosphere may be viewed on local, long-range, and global scales.

Local-scale chemical fate and transport may be viewed with respect to a smokestack point source. As Figure 9.12 illustrates, gases and particles are emitted from a stack and carried and dispersed by wind and air currents while undergoing mixing and dilution. Because stack gases are carried upward by a rising current warmer than the surrounding atmosphere, the effective height of a stack is always greater than its actual height. The farther from the stack source before pollutants reach ground level, the more dilute they are. The dispersion of pollutants is strongly influenced by atmospheric conditions such as wind, air turbulence, and the occurrence of temperature inversions (Figure 9.7). High stacks reduce the immediate impact of air pollutants and illustrate the once-prevailing philosophy that "The solution to pollution is dilution."

Long-range transport of species in the atmosphere is an important aspect of air pollution. One illustration of long-range transport of air pollutants was the contamination of much of Europe including northern reaches of Scandinavia by radionuclides emitted in the Chernobyl nuclear reactor meltdown and fire. New England and southeastern Canada are affected by acid rainfall originating from sulfur dioxide emitted by power plants in the U.S. Ohio River Valley hundreds of kilometers distant.

Modeling long-range environmental fate and transport is an important exercise in determining sources of pollutants and mitigating their effects. Such models using sophisticated computer programs and high computing capacity must consider advective transport and mixing phenomena. Very large areas must be considered and averages taken over long time intervals. Weather conditions are important factors.

Some important atmospheric pollutants must be considered on a global scale. Such pollutants have very long lifetimes so that they persist long enough to mix with and spread throughout the global atmosphere and they are produced from a variety of widely dispersed sources. An example of such a substance is greenhouse gas carbon dioxide emitted to the atmosphere by billions of heating and cooking stoves, millions of automobiles, and thousands of power plants throughout the globe.

The Earth's atmosphere cannot be considered as a single large mixing bowl for contaminants on a global scale. Prevailing winds cause relatively rapid mixing within the northern and southern

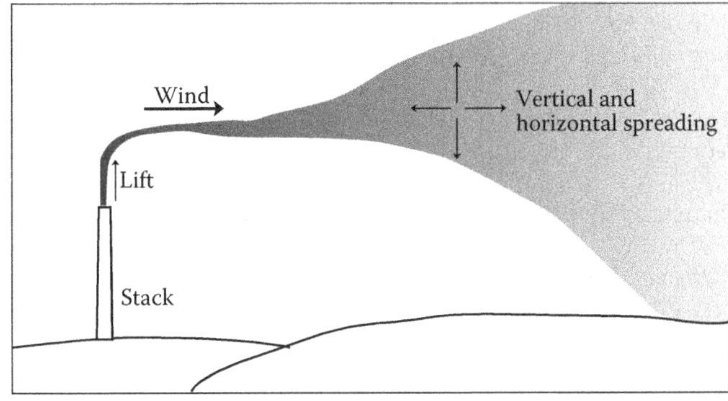

FIGURE 9.12 Illustration of localized chemical fate and transport with air pollutants from a point source (smokestack).

hemisphere, whereas transport of atmospheric constituents across the equator is relatively slow. This phenomenon is illustrated by the discussion in Section 14.2 of atmospheric greenhouse gas carbon dioxide. In the northern hemisphere, which has an abundance of photosynthesizing plants, there is a pronounced annual fluctuation of several ppm in the levels of atmospheric carbon dioxide produced by the annual seasonal cycle of photosynthesis, whereas the fluctuation is much less pronounced in the southern hemisphere, which has much less photosynthetic activity. Mixing between the two hemispheres over a year's period of time is insufficient to dampen the fluctuation, whereas the average concentration of carbon dioxide is essentially the same in both hemispheres reflecting mixing over several years' periods of time.

An interesting aspect of fate and transport involving the atmosphere is provided by the accumulation of semivolatile POP in polar regions. This phenomenon occurs by a distillation effect in which such pollutants are evaporated in warmer latitudes, carried by air currents toward the poles, and condensed in colder polar regions. As a result, surprisingly high levels of some semivolatile POP (e.g., PCBs in Arctic polar bear fat) have been found in samples from polar regions.

Environmental fate and transport involving the atmosphere are obviously important in environmental chemistry and a detailed discussion of this area is beyond the scope of this book. For additional material regarding this topic the reader is referred to reference works on the subject.[4–6]

LITERATURE CITED

1. Pandis, S. N. and J. H. Seinfeld, *Atmospheric Chemistry and Physics: From Air Pollution to Climate Change*, 2nd ed., Wiley, Hoboken, NJ, 2006.
2. Ackerman, S. A. and J. A. Knox, *Meteorology: Understanding the Atmosphere*, 2nd ed., Thomson Brooks/Cole, Belmont, CA, 2007.
3. Lutgens, F. K. and E. J. Tarbuck, *The Atmosphere: An Introduction to Meteorology*, 10th ed., Pearson Prentice Hall, Upper Saddle River, NJ, 2007.
4. Gulliver, J. S., *Introduction to Chemical Transport in the Environment*, Cambridge University Press, New York, 2007.
5. Dunnivant, F. M. and E. Anders, *A Basic Introduction to Pollutant Fate and Transport: An Integrated Approach with Chemistry, Modeling, Risk Assessment, and Environmental Legislation*, Wiley-Interscience, Hoboken, NJ, 2006.
6. Ramaswami, A., J. B. Milford, and M. J. Small, *Integrated Environmental Modeling: Pollutant Transport, Fate, and Risk in the Environment*, Wiley, Hoboken, NJ, 2005.

SUPPLEMENTARY REFERENCES

Aguado, E. and J. E. Burt, *Understanding Weather and Climate*, 4th ed., Pearson Education, Upper Saddle River, NJ, 2007.

Ahrens, C. D., *Meteorology Today: An Introduction to Weather, Climate, and the Environment,* 9th ed., Thomson Brooks/Cole, Belmont, CA, 2009.

Ahrens, C. D., *Essentials of Meteorology Today: An Invitation to the Atmosphere*, 5th ed., Thomson Brooks/Cole, Belmont, CA, 2008.

Allaby, M., *Atmosphere: A Scientific History of Air, Weather, and Climate*, Facts on File, New York, 2009.

Austin, J., P. Brimblecombe, W. Sturges, Eds, *Air Pollution Science for the 21st Century*, Elsevier Science, New York, 2002.

Barker, J. R., A brief introduction to atmospheric chemistry, *Advances Series in Physical Chemistry*, **3**, 1–33, 1995.

Brasseur, G. P., J. J. Orlando, and G. S. Tyndall, Eds, *Atmospheric Chemistry and Global Change*, Oxford University Press, New York, 1999.

Desonie, D., *Atmosphere: Air Pollution and its Effects*, Chelsea House Publishers, New York, 2007.

Garratt, R., *Atmosphere: A Scientific History of Air, Weather, and Climate*, Facts on File, New York, 2009.

Hewitt, C. N. and A. Jackson, *Atmospheric Science for Environmental Scientists*, Wiley-Blackwell, Hoboken, NJ, 2009.

Hewitt, N. and A. Jackson, Eds, *Handbook of Atmospheric Science,* Blackwell Publishing, Malden, MA, 2003.

Hobbs, P. V., *Introduction to Atmospheric Chemistry*, Cambridge University Press, New York, 2000.
Jacob, D. J., *Introduction to Atmospheric Chemistry*, Princeton University Press, Princeton, NJ, 1999.
Nalwa, H. S., Ed., *Handbook of Photochemistry and Photobiology*, American Scientific Publishers, Stevenson Ranch, CA, 2003.
Oliver, J. E. and J. J. Hidore, *Climatology: An Atmospheric Science*, Prentice Hall, Upper Saddle River, NJ, 2002.
Spellman, F. R., *The Science of Air: Concepts and Applications*, 2nd ed., Taylor & Francis, Boca Raton, FL, 2009.
Wallace, J. M., *Atmospheric Science: An Introductory Survey*, 2nd ed., Elsevier Academic Press, Amsterdam, 2006.

QUESTIONS AND PROBLEMS

The use of internet resources is assumed in answering any of the questions. These would include things such as constants and conversion factors as well as additional information needed to complete an answer.

1. What phenomenon is responsible for the temperature maximum at the boundary of the stratosphere and the mesosphere?
2. What function does a third body serve in an atmospheric chemical reaction?
3. Why does the lower boundary of the ionosphere lift at night?
4. Considering the total number of electrons in NO_2, why might it be expected that the reaction of a free radical with NO_2 is a chain-terminating reaction?
5. It may be argued that wind energy, which is now used by growing numbers of large turbines to generate renewable electricity, is actually a form of solar energy. Explain on the basis of meteorological phenomena the rationale for this argument.
6. Suppose that 22.4 L of air at STP is used to burn 1.50 g of carbon to form CO_2, and that the gaseous product is adjusted to STP. What is the volume and the average molar mass of the resulting mixture?
7. If the pressure is 0.01 atm at an altitude of 38 km and 0.001 at 57 km, what is it at 19 km (ignoring temperature variations)?
8. Measured in μm, what are the lower wavelength limits of solar radiation reaching Earth; the wavelength at which maximum solar radiation reaches Earth; and the wavelength at which maximum energy is radiated back into space?
9. Of the species O, $HO^{\bullet}$, NO_2^*, $H_3C^{\bullet}$, and N^+, which could most readily revert to a nonreactive, "normal" species in total isolation?
10. Of the gases neon, sulfur dioxide, helium, oxygen, and nitrogen, which shows the most variation in its atmospheric concentration?
11. A 12.0-L sample of air at 25°C and 1.00 atm pressure was collected and dried. After drying, the volume of the sample was exactly 11.50 L. What was the percentage *by mass* of water in the original air sample?
12. The sunlight incident upon a 1 m^2 area perpendicular to the line of transmission of the solar flux just above the Earth's atmosphere provides energy at a rate most closely equivalent to (A) that required to power a pocket calculator, (B) that required to provide a moderate level of lighting for a 40-person capacity classroom illuminated with fluorescent lights, (C) that required to propel a 2500 pound automobile at 55 mph, (D) that required to power a 100-W incandescent light bulb, (E) that required to heat a 40-person classroom to 70°F when the outside temperature is −10°F.
13. At an altitude of 50 km, the average atmospheric temperature is essentially 0°C. What is the average number of air molecules per cubic centimeter of air at this altitude?
14. What is the distinction between chemiluminescence and luminescence caused when light is absorbed by a molecule or atom?

15. State two factors that make the stratosphere particularly important in terms of acting as a region where atmospheric trace contaminants are converted to other, chemically less reactive, forms.
16. What two chemical species are most generally responsible for the removal of hydroxyl radical from the unpolluted troposphere?
17. What is the distinction between the symbols * and • in discussing chemically active species in the atmosphere?
18. Of the following the true statement is (A) incoming solar energy is primarily in the form of infrared radiation, (B) the very cold tropopause layer at the top of the troposphere is the major absorber of harmful ultraviolet radiation from the sun, (C) the stratosphere is defined as a region of the atmosphere in which temperature decreases with increasing altitude, (D) a large fraction of solar energy is converted to latent heat by evaporation of water in the atmosphere, (E) temperature inversions are most useful because they cause air pollutants to disperse.
19. Of the following the true statement is (A) chemiluminescence refers to a chemical reaction that results from a molecule having absorbed a photon of light, (B) O* denotes an excited oxygen atom, (C) O_2^* denotes a free radical, (D) HO• is an insignificant species in the atmosphere, (E) the longer the wavelength of incoming solar radiation, the more likely it is to cause a photochemical reaction to occur.
20. Match the following pertaining to classes of atmospheric chemical species:

A. NO_2 | 1. Reductant
B. H_2S | 2. Corrosive substance
C. NH_4HSO_4 | 3. Photochemically active species
D. O_3^* | 4. Of the species shown, most likely to dissociate without additional outside input

21. Free radicals *do not* or *are not* (A) have unpaired electrons, (B) normally highly reactive, (C) last longer in the stratosphere than in the troposphere, (D) take part in chain reactions, (E) lose their energy spontaneously, reverting to a stable species by themselves.
22. Of the following the true statement is (A) the central feature of global weather is the redistribution of moisture from equatorial areas where it falls to polar areas where it freezes, (B) cyclonic storms are caused by temperature inversions, (C) temperature inversions limit the vertical circulation of air, (D) albedo refers to the percentage of infrared radiation that is reabsorbed as energy is emitted from Earth, (E) the troposphere has a homogeneous composition of all gases and vapors including water.
23. Using numbers ranging from 1 to 4, put the following in order of their anticipated lifetime in the troposphere from the shortest-lived (1) to the longest-lived (4) and explain: CH_4, CCl_2F_2, NO_2^*, and SO_2.
24. The Earth's atmosphere is stratified into layers. Of the following, the true statement regarding this stratification, the characteristics of the layers, and the characteristics of species in the layers is (A) the stratosphere and troposphere have essentially the same composition, (B) the upper boundary of the stratosphere is colder than the upper boundary of the troposphere because the former is higher, (C) ozone is most desirable near the Earth's surface in the troposphere, (D) the composition of the troposphere is characterized by both its high and uniform content of water vapor, (E) the boundary between the troposphere and the stratosphere serves as a barrier to the movement of one of the important constituents of tropospheric air.

10 Particles in the Atmosphere

10.1 INTRODUCTION

Particles abound in the atmosphere, ranging in size from about 1.5 mm (the size of sand or drizzle) down to molecular dimensions. Atmospheric particles are made up of an amazing variety of materials and discrete objects that may consist of either solids or liquid droplets. A number of terms are commonly used to describe atmospheric particles; the more important of these are summarized in Table 10.1. *Particulates* is a term that has come to stand for particles in the atmosphere, although *particulate matter* or simply *particles* is preferred usage.

Particulate matter makes up the most visible and obvious form of air pollution. Atmospheric *aerosols* are solid or liquid particles smaller than 100 μm in diameter. Pollutant particles in the 0.001–10 μm range are commonly suspended in the air near sources of pollution such as the urban atmosphere, industrial plants, highways, and power plants.

Very small, solid particles include carbon black, silver iodide, combustion nuclei, and sea-salt nuclei (Figure 10.1). Larger particles include cement dust, wind-blown soil dust, foundry dust, and pulverized coal. Liquid particulate matter, *mist*, includes raindrops, fog, and sulfuric acid mist. Particulate matter may be organic or inorganic; both types are very important atmospheric contaminants.

There are several major sources of particulate matter in the industrialized urban atmosphere. These include particles from coal combustion, secondary sulfate, secondary nitrate associated with upwind and local sources of NO_x and NH_3, secondary organic aerosols produced by chemical processes operating on organic pollutants from a number of sources, and direct emissions from motor vehicle traffic, such as particles from diesel engine exhausts.

Some particles are of biological origin, such as viruses, bacteria, bacterial spores, fungal spores, and pollen. In addition to organic materials, organisms may contribute to sulfate particulate matter in the atmosphere. Marine biological sources may contribute significantly to atmospheric aerosols. Biogenic materials reacting in and on the surface of sea-salt aerosols produce some significant atmospheric chemical species, such as halogen radicals, and in so doing influence cycles involving atmospheric sulfur, nitrogen, and oxidants.

As discussed later in this chapter, particulate matter originates from a wide variety of sources and processes, ranging from simple grinding of bulk matter to complicated chemical or biochemical syntheses. The effects of particulate matter are also widely varied. Possible effects on climate are discussed in Chapter 14. Either by itself or in combination with gaseous pollutants, particulate matter may be detrimental to human health. Atmospheric particles may damage materials, reduce visibility, and cause undesirable esthetic effects. It is now recognized that very small particles have a particularly high potential for harm, including adverse health effects, and specific regulations now apply to particles with a diameter of 2.5 μm or less.

For the most part, aerosols consist of carbonaceous material, metal oxides and glasses, dissolved ionic species (electrolytes), and ionic solids. The predominant constituents are carbonaceous material, water, sulfate, nitrate, ammonium nitrogen, and silicon. The composition of aerosol particles varies significantly with size. The very small particles tend to be acidic and often originate from gases, such as from the conversion of SO_2 to H_2SO_4. Larger particles tend to consist of materials generated mechanically, such as by the grinding of limestone, and have a greater tendency to be basic.

TABLE 10.1
Important Terms Describing Atmospheric Particles

Term	Meaning
Aerosol	Colloidal-sized atmospheric particle
Condensation aerosol	Formed by condensation of vapors or reactions of gases
Dispersion aerosol	Formed by grinding of solids, atomization of liquids, or dispersion of dusts
Fog	Term denoting high level of water droplets
Haze	Denotes decreased visibility due to the presence of particles
Mists	Liquid particles
Smoke	Particles formed by incomplete combustion of fuel

10.2 PHYSICAL BEHAVIOR OF PARTICLES IN THE ATMOSPHERE

As shown in Figure 10.2, atmospheric particles undergo a number of processes in the atmosphere. Small colloidal particles are subject to *diffusion processes.* Smaller particles *coagulate* together to form larger particles. *Sedimentation* or *dry deposition* of particles, which have often reached sufficient size to settle by coagulation, is one of two major mechanisms for particle removal from the atmosphere. In many areas, dry deposition on vegetation is a significant mechanism for particle removal.[1] In addition to sedimentation, the other major pathway for particle removal from the atmosphere is *scavenging* by raindrops and other forms of precipitation. Particles also react with atmospheric gases.

Particle size usually expresses the diameter of a particle, though sometimes it is used to denote the radius. The sizes of atmospheric particles cover several orders of magnitude from <0.01 μm to around 100 μm. Volume and mass of particles are a function of d^3, where d is the particle diameter. As a consequence, in general, the total mass of atmospheric particles is concentrated in the larger size range, whereas the total number and surface area of atmospheric particles are in the smaller fraction.

The rate at which a particle settles is a function of particle diameter and density. The settling rate is important in determining the effect of the particle on the atmosphere. For spherical particles greater than approximately 1 μm in diameter, Stokes' law applies,

$$v = \frac{gd^2(\rho_1 - \rho_2)}{18\eta} \tag{10.1}$$

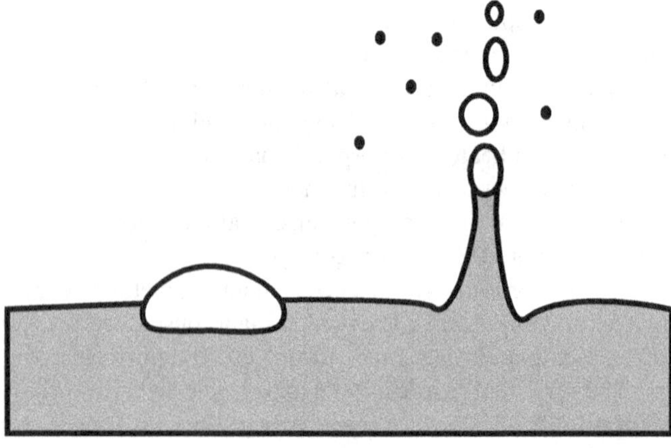

FIGURE 10.1 Bursting bubbles in seawater form small liquid aerosol particles. Evaporation of water from aerosol particles results in the formation of small solid particles of sea-salt nuclei.

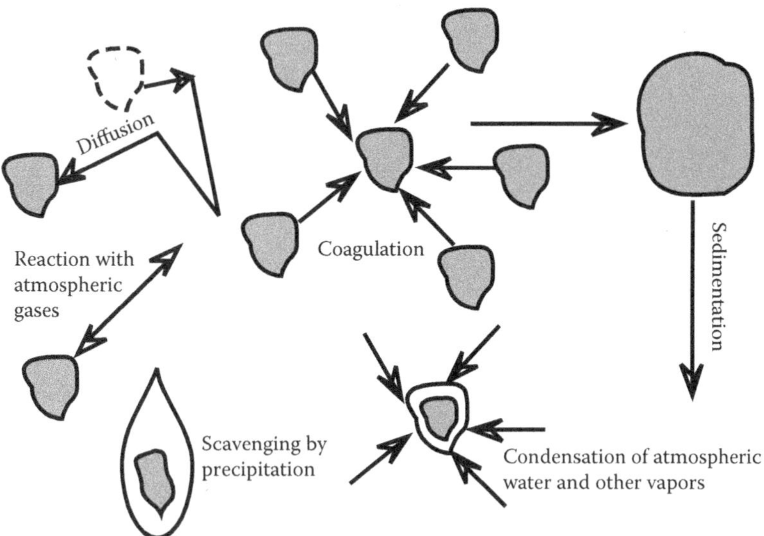

FIGURE 10.2 Processes that particles undergo in the atmosphere.

where v is the settling velocity in cm/s, g is the acceleration of gravity in cm/s^2, ρ_1 is the density of the particle in g/cm^3, ρ_2 is the density of air in g/cm^3, and η is the viscosity of air in poise. Stokes' law can also be used to express the effective diameter of an irregular nonspherical particle. These are called *Stokes diameters* (aerodynamic diameters) and are normally the ones given when particle diameters are expressed. Furthermore, since the density of a particle is often not known, an arbitrary density of 1 g/cm^3 is conventionally assigned to ρ_1; when this is done, the diameter calculated from Equation 10.1 is called the *reduced sedimentation diameter.*

10.2.1 Size and Settling of Atmospheric Particles

Most kinds of aerosol particles have unknown diameters and densities and occur over a range of sizes. For such particles, the term *mass median diameter* (MMD) may be used to describe aerodynamically equivalent spheres having an assigned density of 1 g/cm^3 at a 50% mass collection efficiency, as determined in sampling devices calibrated with spherical aerosol particles having a known, uniform size. (Polystyrene latex is commonly used as a material for the preparation of such standard aerosols.) The determination of MMD is accomplished by plotting the log of particle size as a function of the percentage of particles smaller than the given size on a probability scale. Two such plots are shown in Figure 10.3. It is seen from the plot that particles of aerosol X have a MMD of 2.0 μm (ordinate corresponding to 50% on the abscissa). In the case of aerosol Y, linear extrapolation to sizes below the lower measurable size limit of about 0.7 μm gives an estimated value of 0.5 μm for the MMD.

The settling characteristics of particles smaller than about 1 μm in diameter deviate from Stokes' law because the settling particles "slip between" air molecules. Extremely small particles are subject to *Brownian motion* resulting from random movement due to collisions with air molecules and do not obey Stokes' law. Deviations are also observed for particles above 10 μm in diameter because they settle rapidly and generate turbulence as they fall.

10.3 PHYSICAL PROCESSES FOR PARTICLE FORMATION

Dispersion aerosols, such as dusts, formed from the disintegration of larger particles are usually above 1 μm in size. Typical processes for forming dispersion aerosols include evolution of dust from coal grinding, formation of spray in cooling towers, and blowing of dirt from dry soil.

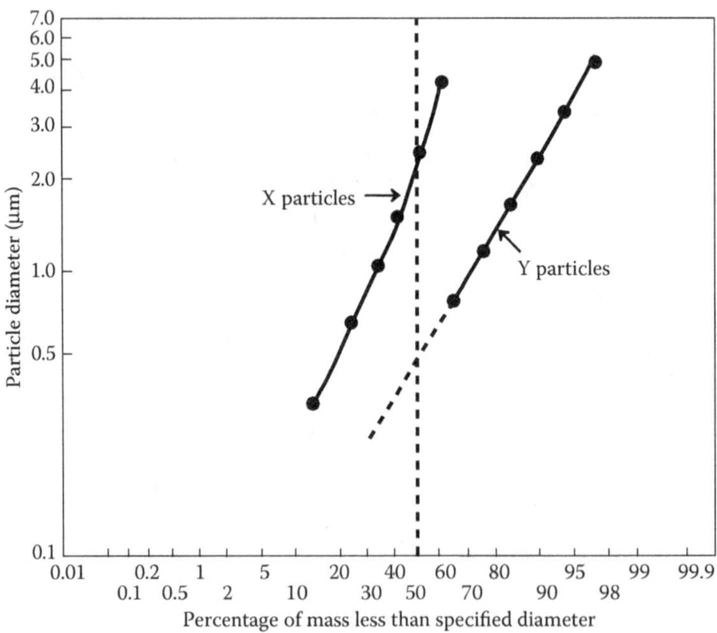

FIGURE 10.3 Particle size distribution for particles of X (MMD = 2.0 µm) and Y (MMD = 0.5 µm).

Many dispersion aerosols originate from natural sources such as sea spray, windblown dust, and volcanic dust. However, a vast variety of human activities break up material and disperse it to the atmosphere. "All terrain" vehicles churn across desert lands, coating fragile desert plants with layers of dispersed dust. Quarries and rock crushers spew out plumes of ground rock. Cultivation of land has made it much more susceptible to dust-producing wind erosion. Areas of North America are now sometimes afflicted by plumes of particles stirred up by windstorms in Asia that disturb soil converted to desert by global warming, improper cultivation, and overgrazing.

However, since much more energy is required to break material down into small particles than is required for or released by the synthesis of particles through chemical synthesis or the adhesion of smaller particles, most dispersion aerosols are relatively large. Larger particles tend to have fewer harmful effects than smaller ones. As examples, larger particles are less *respirable* in that they do not penetrate so far into the lungs as smaller ones, and larger particles are relatively easier to remove from air pollution effluent sources.

Huge volcanic eruptions can cause highly elevated levels of particles in the atmosphere. These can be from the physical process of simply blowing as much as several cubic kilometers of volcanic ash as high as the stratosphere. As noted below, volcanic gases can produce secondary particles by chemical processes.

10.4 CHEMICAL PROCESSES FOR PARTICLE FORMATION

Chemical processes in the atmosphere convert large quantities of atmospheric gases to particulate matter.[2] Among the chemical species most responsible for this conversion are the organic pollutants and nitrogen oxides that cause formation of ozone and photochemical smog (see Chapter 13) in the troposphere. Smaller particles formed by chemical processes tend to have higher contents of organic matter than do coarser particles. To an extent, therefore, control of hydrocarbon and NO_x emissions to reduce smog also curtails atmospheric particulate matter pollution.

A major fraction of ambient particulate matter arises from atmospheric gas-to-particle conversion. Attempts to reduce particulate matter levels require control of the same organic and nitrogen oxide (NO_x) emissions that are precursors to urban and regional ozone formation.

Most chemical processes that produce particles are combustion processes, including fossil-fuel-fired power plants; incinerators; home furnaces, fireplaces, and stoves; cement kilns; internal combustion engines; forest, brush, and grass fires; and active volcanoes. Particles from combustion sources tend to occur in a size range below 1 μm. Such very small particles are particularly important because they are most readily carried into the alveoli of lungs (see pulmonary route of exposure to toxicants in Chapter 22) and they are likely to be enriched in more hazardous constituents, such as toxic heavy metals and arsenic. The pattern of occurrence of such trace elements can enable the use of small particle analysis for tracking sources of particulate pollutants.

10.4.1 Inorganic Particles

Metal oxides constitute a major class of inorganic particles in the atmosphere. These are formed whenever fuels containing metals are burned. For example, particulate iron oxide is formed during the combustion of pyrite-containing coal:

$$3FeS_2 + 8O_2 \rightarrow Fe_3O_4 + 6SO_2 \tag{10.2}$$

Organic vanadium in residual fuel oil is converted to particulate vanadium oxide. Part of the calcium carbonate in the ash fraction of coal is converted to calcium oxide and is emitted into the atmosphere through the stack:

$$CaCO_3 + \text{heat} \rightarrow CaO + CO_2 \tag{10.3}$$

A common process for the formation of aerosol mists involves the oxidation of atmospheric sulfur dioxide to sulfuric acid, a hygroscopic substance that accumulates atmospheric water to form small liquid droplets:

$$2SO_2 + O_2 + 2H_2O \rightarrow 2H_2SO_4 \tag{10.4}$$

In the presence of basic air pollutants, such as ammonia or calcium oxide, the sulfuric acid reacts to form salts:

$$H_2SO_4(\mathit{droplet}) + 2NH_3(g) \rightarrow (NH_4)_2SO_4(\mathit{droplet}) \tag{10.5}$$

$$H_2SO_4(\mathit{droplet}) + CaO(s) \rightarrow CaSO_4(\mathit{droplet}) + H_2O \tag{10.6}$$

Under low-humidity conditions, water is lost from these droplets and a solid aerosol is formed.

Volcanic SO_2 and H_2S gases can be precursors to large quantities of atmospheric particulate sulfuric acid and sulfates. A study of the 1982 eruption of El Chichón volcano in Mexico showed that volcanic glass, sodium chloride, and sulfate from the volcano were deposited in snow in Greenland. The June 15, 1991 eruption of Mount Pinatubo in the Philippines caused perceptible perturbations in the earth's atmospheric solar and infrared radiation transmission.

Nitrogen in ammonium and nitrate salts is a common constituent of inorganic particulate matter. Particulate ammonium nitrate and ammonium chloride are produced by the following reversible reactions:

$$NH_3(g) + HNO_3(g) \rightleftarrows NH_4NO_3(s, \mathit{particulate}) \tag{10.7}$$

$$NH_3(g) + HCl(g) \rightleftarrows NH_4Cl(s, \mathit{particulate}) \tag{10.8}$$

Ammonium nitrate, chloride, and sulfate salts in the atmosphere are corrosive to metals such as the metal contacts in electrical relays.

Biogenic sources provide the ingredients for the production of significant amounts of particulate ammonium, nitrate, and sulfate salts. Ammonia is evolved from the decay of organic matter. Gaseous sulfur enters the atmosphere as H_2S gas from the decay of sulfur-containing organic matter and from anoxic microbial processes using sulfate as an electron receptor (oxidant). Marine organisms evolve large quantities of dimethyl sulfide, $(CH_3)_2S$, from oceans. Hydrogen sulfide and dimethyl sulfide are oxidized to sulfate in the atmosphere by atmospheric chemical processes. Microorganisms evolve significant quantities of gaseous N_2O, which becomes oxidized in the atmosphere to nitrate. Some atmospheric ammonia is also oxidized to nitrate.

The preceding examples show several ways in which solid or liquid inorganic aerosols are formed by chemical reactions. Such reactions constitute an important general process for the formation of aerosols, particularly the smaller particles.

10.4.2 Organic Particles

A significant portion of organic particulate matter is produced by internal combustion engines in complicated processes that involve pyrolysis and pyrosynthesis processes. These products may include nitrogen-containing compounds and oxidized hydrocarbon polymers. Engine lubricating oil and its additives may also contribute to organic particulate matter.

10.4.3 Polycyclic Aromatic Hydrocarbon Synthesis

The organic particles of greatest concern are polycyclic aromatic hydrocarbon (PAH), which consist of condensed ring aromatic (aryl) molecules. The most often cited example of a PAH compound is benzo(a)pyrene, a compound that the body can metabolize to a carcinogenic form:

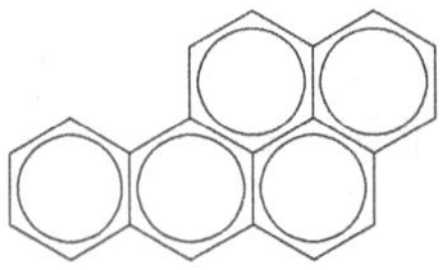

Benzo(a)pyrene

PAHs and derivatives from them are formed during incomplete combustion of hydrocarbons. Although natural combustion processes, such as forest and grass fires, produce PAHs, most of the especially troublesome pollutant PAHs come from anthrospheric processes.[3] PAHs may be synthesized from saturated hydrocarbons under oxygen-deficient conditions. Hydrocarbons with very low molecular masses, including even methane, may act as precursors for the polycyclic aromatic compounds. Low-molar-mass hydrocarbons form PAHs by *pyrosynthesis*. This happens at temperatures exceeding approximately 500°C at which carbon–hydrogen and carbon–carbon bonds are broken to form free radicals. These radicals undergo dehydrogenation and combine chemically to form aryl ring structures, which are resistant to thermal degradation. The basic process for the formation of such rings from pyrosynthesis starting with ethane is,

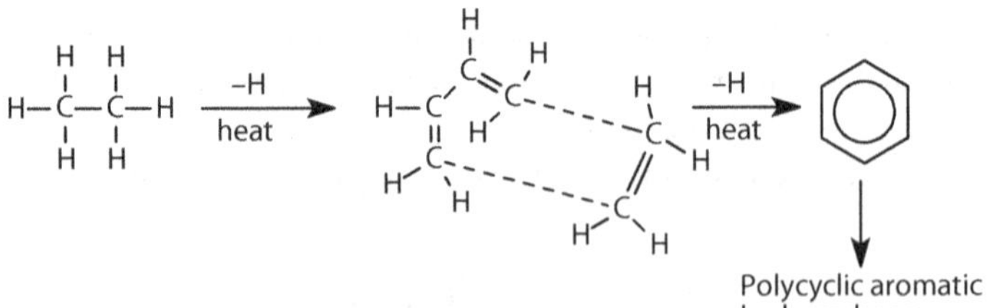

-H heat → -H heat →

Polycyclic aromatic hydrocarbons

which results in the formation of stable PAH structures. The tendency of hydrocarbons to form PAHs by pyrosynthesis varies in the order aromatics > cycloolefins > olefins > paraffin's. The existing ring structure of cyclic compounds is conducive to PAH formation. Unsaturated compounds are especially susceptible to the addition reactions involved in PAH formation.

Polycyclic aromatic compounds may be formed from higher alkanes present in fuels and plant materials by the process of *pyrolysis*, the "cracking" of organic compounds to form smaller and less stable molecules and radicals.

10.5 THE COMPOSITION OF INORGANIC PARTICLES

Figure 10.4 illustrates the basic factors responsible for the composition of inorganic particulate matter. In general, the proportions of elements in atmospheric particulate matter reflect relative abundances of elements in the parent material.

The source of particulate matter is reflected in its elemental composition, taking into consideration chemical reactions that may change the composition. For example, particulate matter largely from an ocean spray origin in a coastal area receiving sulfur dioxide pollution may show anomalously high sulfate and corresponding low chloride content. The sulfate comes from atmospheric oxidation of sulfur dioxide to form nonvolatile ionic sulfate, whereas some chloride originally from the NaCl in the seawater may be lost from the solid aerosol as volatile HCl:

$$2SO_2 + O_2 + 2H_2O \rightarrow 2H_2SO_4 \tag{10.9}$$

$$H_2SO_4 + 2NaCl\ (particulate) \rightarrow Na_2SO_4\ (particulate) + 2HCl \tag{10.10}$$

Acids other than sulfuric acid can also be involved in the modification of sea-salt particles. The most common such acid is nitric acid formed by reactions of nitrogen oxides in the atmosphere. Traces of nitrate salts may be found among sea-salt particles.

Among the constituents of inorganic particulate matter found in polluted atmospheres are salts, oxides, nitrogen compounds, sulfur compounds, various metals, and radionuclides. In coastal areas, sodium and chlorine get into atmospheric particles as sodium chloride from sea spray. The major trace elements that typically occur at levels above 1 μg/m³ in particulate matter are aluminum, calcium, carbon, iron, potassium, sodium, and silicon; note that most of these tend to originate from

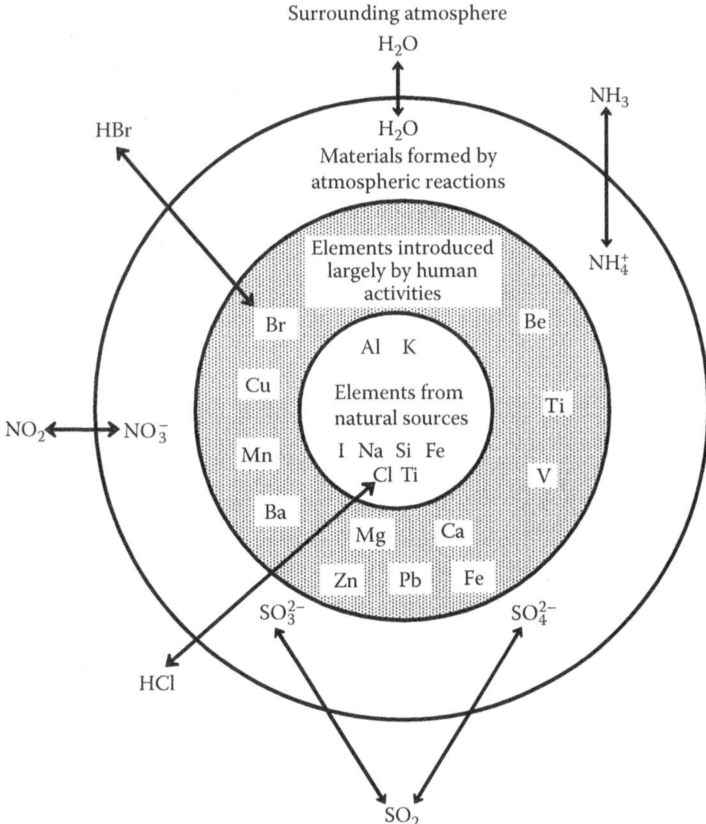

FIGURE 10.4 Some of the components of inorganic particulate matter and their origins.

terrestrial sources. Lesser quantities of copper, lead, titanium, and zinc and even lower levels of antimony, beryllium, bismuth, cadmium, cobalt, chromium, cesium, lithium, manganese, nickel, rubidium, selenium, strontium, and vanadium are commonly observed. The likely sources of some of these elements are given below:

- *Al, Fe, Ca, Si*: Soil erosion, rock dust, coal combustion
- *C*: Incomplete combustion of carbonaceous fuels
- *Na, Cl*: Marine aerosols, chloride from incineration of organohalide polymer wastes
- *Sb, Se*: Very volatile elements, possibly from the combustion of oil, coal, or refuse
- *V*: Combustion of residual petroleum (present at very high levels in residues from Venezuelan crude oil)
- *Zn*: Tends to occur in small particles, probably from combustion
- *Pb*: Combustion of fuels and wastes containing lead

Particulate carbon as soot, carbon black, coke, and graphite originates from auto and truck exhausts, heating furnaces, incinerators, power plants, and steel and foundry operations, and composes one of the more visible and troublesome particulate air pollutants. Because of its good adsorbent properties, carbon can be a carrier of gaseous and other particulate pollutants. Both nitrogen and sulfur compounds in exhaust gases are adsorbed onto particulate carbon that is emitted by poorly controlled diesel engines. Particulate carbon surfaces may catalyze some heterogeneous atmospheric reactions, including the important conversion of SO_2 into sulfate.

10.5.1 Fly Ash

Much of the mineral particulate matter in a polluted atmosphere is in the form of oxides and other compounds produced during the combustion of high-ash fossil fuel. Some of the mineral matter in fossil fuels such as coal or lignite is converted during combustion into a fused, glassy bottom ash that presents no air pollution problems. Smaller particles of *fly ash* enter furnace flues and are efficiently collected in a properly equipped stack system. However, some fly ash escapes through the stack and enters the atmosphere. Unfortunately, the fly ash thus released tends to consist of smaller particles that do the most damage to human health, plants, and visibility.

The composition of fly ash varies widely, depending upon the fuel. The predominant constituents are oxides of aluminum, calcium, iron, and silicon. Other elements that occur in fly ash are magnesium, sulfur, titanium, phosphorus, potassium, and sodium. Elemental carbon (soot and carbon black) is a significant fly ash constituent.

The size of fly ash particles is a very important factor in determining their removal from stack gas and their ability to enter the body through the respiratory tract. Fly ash from coal-fired utility boilers has shown a bimodal (two peak) distribution of size, with a peak at about 0.1 μm as illustrated in Figure 10.5. Although only about 1–2% of the total fly ash mass is in the smaller size fraction, it includes the vast majority of the total number of particles and particle surface area. Submicrometer particles probably result from a volatilization–condensation process during combustion, as reflected in a higher concentration of more volatile elements such as As, Sb, Hg, and Zn. In addition to their being relatively much more respirable and potentially toxic, the very small particles are the most difficult to remove by electrostatic precipitators and bag houses (see Section 10.12).

10.5.2 Asbestos

Asbestos is the name given to a group of fibrous silicate minerals, typically those of the serpentine group, approximate formula $Mg_3P(Si_2O_5)(OH)_4$. The tensile strength, flexibility, and nonflammability of asbestos have led to many uses of this material. In 1973, annual U.S. consumption of asbestos

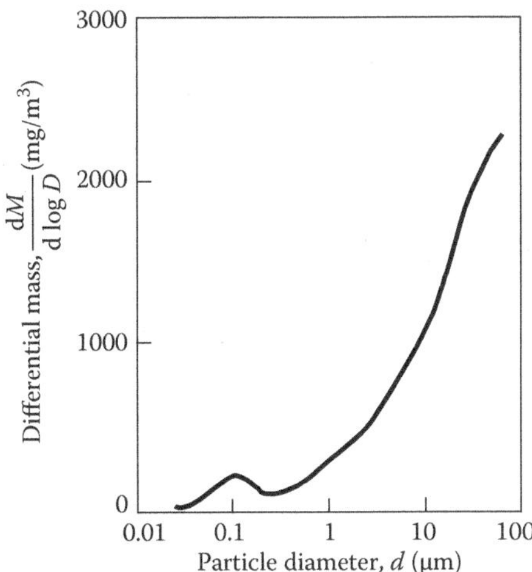

FIGURE 10.5 General appearance of particle-size distribution in coal-fired power plant ash. The data are given on differential mass coordinates, where M is the mass, so that the area under the curve in a given size range is the mass of the particles in that size range.

peaked at 652,000 metric tons for applications including brake linings and pads, roofing products, structural materials, cement/asbestos pipe, gaskets, heat-resistant packing, and specialty papers. By 1988, because of findings regarding the adverse health effects of inhaled asbestos, annual consumption had dropped to 85,000 metric tons. In 1989, the U.S. Environmental Protection Agency (EPA) announced regulations that phased out most uses of asbestos by 1996, and now virtually no asbestos is used in the United States.

Asbestos is of concern as an air pollutant because when inhaled it may cause asbestosis (a pneumonia condition), mesothelioma (tumor of the mesothelial tissue lining the chest cavity adjacent to the lungs), and bronchogenic carcinoma (cancer originating with the air passages in the lungs). Therefore, uses of asbestos have been severely curtailed and widespread programs have been undertaken to remove the material from buildings.

10.6 TOXIC METALS IN THE ATMOSPHERE

Some of the metals found predominantly as particulate matter in polluted atmospheres are known to be hazardous to human health.[4] All of these except beryllium are so-called "heavy metals." Lead is the toxic metal of greatest concern in the urban atmosphere because it comes closest to being present at a toxic level; mercury ranks second. Others include beryllium, cadmium, chromium, vanadium, nickel, and arsenic (a metalloid).

10.6.1 Atmospheric Mercury

Atmospheric mercury is of concern because of its toxicity, volatility, and mobility. Some atmospheric mercury is associated with particulate matter. Much of the mercury entering the atmosphere does so as volatile elemental mercury from coal combustion and volcanoes. Volatile organomercury compounds such as dimethylmercury, $(CH_3)_2Hg$, and monomethylmercury salts, such as CH_3HgBr, are also encountered in the atmosphere.

10.6.2 Atmospheric Lead

Lead is one of six priority pollutants regulated by the U.S. EPA (see Section 10.9). With the reduction of leaded fuels, atmospheric lead is of less concern than it used to be. However, during the decades when leaded gasoline containing tetraethyllead ($Pb(C_2H_5)_4$) was the predominant automotive fuel, particulate lead halides were emitted in large quantities. This occurred through the action of dichloroethane and dibromoethane added as halogenated scavengers to prevent the accumulation of lead oxides inside engines. The lead halides formed by the chemical process represented by Equation 10.11 are volatile enough to exit through the exhaust system but condense in the air to form particles. During the period of peak usage of leaded gasoline in the early 1970s, about 200,000 tons of lead were entering the atmosphere each year by this route in the United States.

$$Pb(C_2H_5)_4 + O_2 + \text{halogenated scavengers} \rightarrow CO_2 + H_2O + PbCl_2 + PbClBr + PbBr_2 \text{ (unbalanced)} \tag{10.11}$$

10.6.3 Atmospheric Beryllium

Only about 200 metric tons of beryllium are consumed each year in the United States for the formulation of specialty alloys used in electrical equipment, electronic instrumentation, space gear, and nuclear reactor components. Therefore, distribution of beryllium is quite limited compared to other toxic substances produced in larger quantities, such as lead.

During the 1940s and 1950s, the toxicity of beryllium and beryllium compounds became widely recognized; it has the lowest allowable limit in the atmosphere of all the elements. One of the main results of the recognition of beryllium toxicity hazards was the elimination of this element from phosphors (coatings that produce visible light from ultraviolet light) in fluorescent lamps.

10.7 RADIOACTIVE PARTICLES

Some of the radioactivity detected in atmospheric particles is of natural origin. This activity includes that produced when cosmic rays act on nuclei in the atmosphere to produce radionuclides, including ^{7}Be, ^{10}Be, ^{14}C, ^{39}Cl, ^{3}H, ^{22}Na, ^{32}P, and ^{33}P. A significant natural source of radionuclides in the atmosphere is *radon*, a noble gas product of radium decay. Radon may enter the atmosphere as either of two isotopes, ^{222}Rn (half-life: 3.8 days) and ^{220}Rn (half-life: 54.5 s). Both are alpha emitters in decay chains that terminate with stable isotopes of lead. The initial decay products, ^{218}Po and ^{216}Po, are nongaseous and adhere readily to atmospheric particulate matter.

The catastrophic 1986 meltdown and fire at the Chernobyl nuclear reactor in the former Soviet Union spread large quantities of radioactive materials over a wide area of Europe. Much of this radioactivity was in the form of particles.

One of the more serious problems in connection with radon is that of radioactivity originating from uranium mine tailings that have been used in some areas as backfill, soil conditioner, and a base for building foundations. Radon produced by the decay of radium exudes from foundations and walls constructed on tailings. Higher than normal levels of radioactivity have been found in some structures in the city of Grand Junction, Colorado, where uranium mill tailings have been used extensively in construction. Some medical authorities have suggested that the rates of birth defects and infant cancer in areas where uranium mill tailings have been used in residential construction are significantly higher than normal. The combustion of fossil fuels introduces radioactivity into the atmosphere in the form of radionuclides contained in fly ash. Large coal-fired power plants lacking ash-control equipment may introduce up to several hundred millicuries of radionuclides into the atmosphere each year, far more than either an equivalent nuclear or oil-fired power plant.

The radioactive noble gas ^{85}Kr (half-life: 10.3 years) is emitted into the atmosphere by the operation of nuclear reactors and the processing of spent reactor fuels. In general, other radionuclides produced by reactor operation are either chemically reactive and can be removed from the reactor effluent, or have such short half-lives that a short time delay prior to emission prevents their leaving the reactor. Although ^{85}Kr is largely contained in spent reactor fuel during reactor operation, nuclear fuel reprocessing releases most of this gas from the fuel elements. Fortunately, biota cannot concentrate this chemically unreactive element.

The above-ground detonation of nuclear weapons can add large amounts of radioactive particulate matter to the atmosphere. Among the radioisotopes that have been detected in rainfall collected after atmospheric nuclear weapon detonation are ^{91}Y, ^{141}Ce, ^{144}Ce, ^{147}Nd, ^{147}Pm, ^{149}Pm, ^{151}Sm, ^{153}Sm, ^{155}Eu, ^{156}Eu, ^{89}Sr, ^{90}Sr, ^{115m}Cd, ^{129m}Te, ^{131}I, ^{132}Te, and ^{140}Ba. (Note that "m" denotes a metastable state that decays by gamma-ray emission to an isotope of the same element.) The rate of travel of radioactive particles through the atmosphere is a function of particle size. Appreciable fractionation of nuclear debris is observed because of differences in the rates at which the various debris constituents move through the atmosphere.

10.8 THE COMPOSITION OF ORGANIC PARTICLES

The composition of organic particulate matter reflects its origins. Much organic particulate matter, such as the organic particles characteristic of photochemical smog (see Chapter 13), is formed as secondary material that results from photochemical processes operating on volatile and semivolatile organic compounds emitted to the atmosphere. The compounds emitted to the atmosphere are predominantly hydrocarbon in nature and the incorporation of oxygen and/or nitrogen through atmospheric chemical processes gives less volatile material in the form of organic particles.

Organic atmospheric particles occur in a wide variety of compounds. For analysis, such particles can be collected onto a filter; extracted with organic solvents; fractionated into neutral, acid, and basic groups; and analyzed for specific constituents by chromatography and mass spectrometry. The neutral group contains predominantly hydrocarbons, including aliphatic, aromatic, and oxygenated fractions. The aliphatic fraction of the neutral group contains a high percentage of long-chain hydrocarbons, predominantly those with 16–28 carbon atoms. These relatively unreactive compounds are not particularly toxic and do not participate strongly in atmospheric chemical reactions. The aromatic fraction, however, contains carcinogenic PAHs, which are discussed below. Aldehydes, ketones, epoxides, peroxides, esters, quinones, and lactones are found among the oxygenated neutral components, some of which may be mutagenic or carcinogenic. The acidic group contains long-chain fatty acids and nonvolatile phenols. Among the acids recovered from air-pollutant particulate matter are lauric, myristic, palmitic, stearic, behenic, oleic, and linoleic acids. The basic group consists largely of alkaline N-heterocyclic hydrocarbons such as acridine:

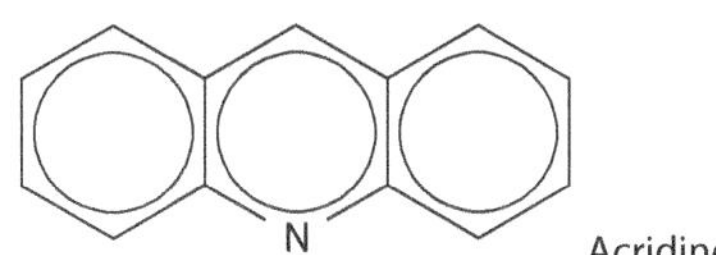

10.8.1 PAHs

PAHs in atmospheric particles have received a great deal of attention because of the known carcinogenic effects of some of these compounds, which are discussed in greater detail in Chapter 23. Prominent among these compounds are benzo(a)pyrene, benz(a)anthracene, chrysene, benzo(e)pyrene, benz(e)acephenanthrylene, benzo(j)fluoranthene, and indenol. Some representative structures of PAH compounds are given below:

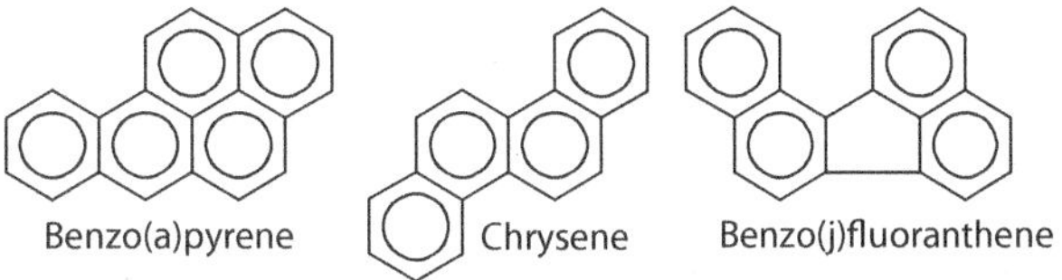

Elevated levels of PAH compounds of up to about 20 μg/m^3 are found in the atmosphere. Elevated levels of PAHs are most likely to be encountered in polluted urban atmospheres, and in the vicinity of natural fires such as forest and prairie fires. Coal furnace stack gas may contain over 1000 μg/m^3 of PAH compounds, and cigarette smoke may contain almost 100 μg/m^3.

Atmospheric PAHs are found almost exclusively in the solid phase, largely sorbed to soot particles. Soot itself is a highly condensed product of PAHs. Soot contains 1–3% hydrogen and 5–10% oxygen, the latter due to partial surface oxidation. Benzo(a)pyrene adsorbed on soot disappears very rapidly in the presence of light, yielding oxygenated products; the large surface area of the particle contributes to the high rate of reaction. Oxidation products of benzo(a)pyrene include epoxides, quinones, phenols, aldehydes, and carboxylic acids as illustrated by the composite structures shown below:

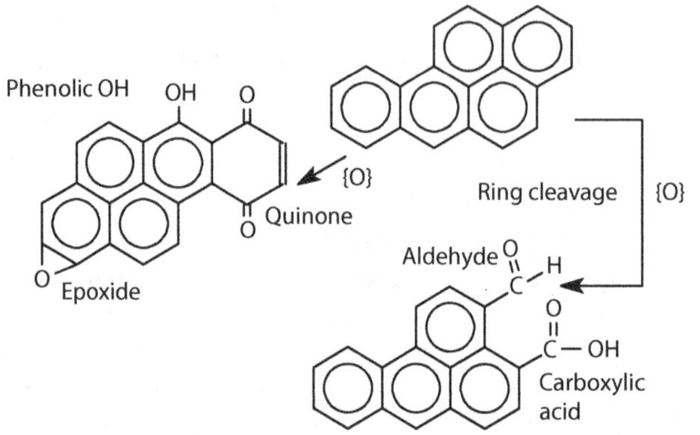

Phenolic OH OH O
{O}
Quinone
O
Epoxide
Ring cleavage {O}
Aldehyde O H C
O C—OH
Carboxylic acid

10.8.2 Carbonaceous Particles from Diesel Engines

Diesel engines emit significant levels of carbonaceous particles. Although an appreciable fraction of these particles have aerodynamic diameters <1 μm, they may exist as aggregates of several thousand smaller particles in clusters up to 30 μm in diameter. This particulate matter is composed largely of elemental carbon, although as much as 40% of the particle mass consists of organic-extractable hydrocarbons and hydrocarbon derivatives including organosulfur and organonitrogen compounds. As noted in Section 10.12, modern diesel engines employ filters to remove exhaust particles, which are periodically burned off of the filter surfaces.

10.9 EFFECTS OF PARTICLES

Atmospheric particles have numerous effects. Because of their pollutant effects, particles are one of six so-called *criteria pollutants* for which the U.S. EPA is required to issue standards. (The other five priority pollutants are sulfur dioxide, carbon monoxide, ozone, nitrogen dioxide, and lead.) Standards for atmospheric particulate matter were first issued by EPA in 1971 with revisions in 1987, 1997, and 2006, most recently with special attention given to particles of a size of 2.5 μm and smaller ($PM_{2.5}$).

The most obvious effect of atmospheric particles is reduction and distortion of visibility. They provide active surfaces upon which heterogeneous atmospheric chemical reactions can occur thereby strongly influencing air pollution phenomena. The ability of particles to act as nucleation bodies for the condensation of atmospheric water vapor can influence precipitation and weather.[5]

The most visible effects of aerosol particles upon air quality result from their optical effects. Particles smaller than about 0.1 μm in diameter scatter light much like molecules, that is, Rayleigh scattering. Generally, such particles have an insignificant effect upon visibility in the atmosphere. The light-scattering and intercepting properties of particles larger than 1 μm are approximately proportional to the particles' cross-sectional areas. Particles of 0.1–1 μm cause interference phenomena because they are about the same dimensions as the wavelengths of visible light, so their light-scattering effects are especially pronounced.

Atmospheric particles inhaled through the respiratory tract may damage health, and exposure to particles in the atmosphere has been linked to a number of health effects including aggravated asthma and premature death from heart and lung diseases. Relatively large particles are likely to be retained in the nasal cavity and in the pharynx, whereas very small particles below 2.5 μm in size are *respirable particles* that are likely to reach the lungs and be retained by them. The respiratory system possesses mechanisms for the expulsion of inhaled particles. In the ciliated region of the respiratory system, particles are carried as far as the entrance to the gastrointestinal tract by a flow of mucus. Macrophages in the nonciliated pulmonary regions carry particles to the ciliated region.

The respiratory system may be damaged directly by particulate matter. In addition, the particulate material or soluble components of it may enter the blood system or lymph system through the lungs to be transported to organs some distance from the lungs and have a detrimental effect on these organs. Particles cleared from the respiratory tract are to a large extent swallowed into the gastrointestinal tract.

A strong correlation has been found between increases in the daily mortality rate and acute episodes of air pollution including particulate pollution. In such cases, high levels of particulate matter are accompanied by elevated concentrations of SO_2 and other pollutants that may have adverse health effects in combination with particles.

A classic case of adverse health effects associated with high levels of atmospheric particles occurred in London during an incident in a 5-day period in 1952 in which a temperature inversion stabilized a mass of air laden with fog, coal smoke, and other particles. Epidemiological data from this period showed approximately 4000 excess deaths. Ultrafine particles <0.1 μm in size and acid-forming constituents have been suspected of contributing to the deaths. Lung and respiratory tract samples archived from victims of this event and subjected to electron microscopic examination 50 years later showed carbonaceous soot predominant in the retained particles in the respiratory tract. Particles bearing metals including lead, zinc, and tin were also found.

10.9.1 Partitioning of Semivolatile Organic Substances between Air and Particles

An effect of atmospheric particles is the partitioning of semivolatile organic compounds such as PCBs between air and particles. As shown in Figure 10.6, these particles can act as carriers to deposit such compounds from the atmosphere onto surfaces in the other environmental spheres. Binding with particles can influence the reactivity of organic compounds, particularly with respect to oxidation.

10.10 WATER AS PARTICULATE MATTER

Droplets of water are very widespread in the atmosphere. Although a natural phenomenon, such droplets can have significant and sometimes harmful effects. The most important such consequence is reduction of visibility, with accompanying detrimental effects on driving, flying, and boat navigation. Water droplets in fog act as carriers of pollutants. The most important of these are solutions of corrosive salts, particularly ammonium nitrates and sulfates, and solutions of strong acids. As discussed in Chapter 14, Section 14.4, the pH of water in acidic mist droplets collected during a Los Angeles acidic fog has been as low as 1.7, far below that of acidic precipitation. Such acidic mist can be especially damaging to the respiratory tract because it is very penetrating.

Arguably the most significant effect of water droplets in the atmosphere is as aquatic media in which important atmospheric chemical processes occur. The single most significant such process may well be the oxidation of S(IV) species to sulfuric acid and sulfate salts, a process that may be facilitated by the presence of iron. The S(IV) species so oxidized include SO_2 (*aq*), HSO_3^-, and SO_3^{2-}.

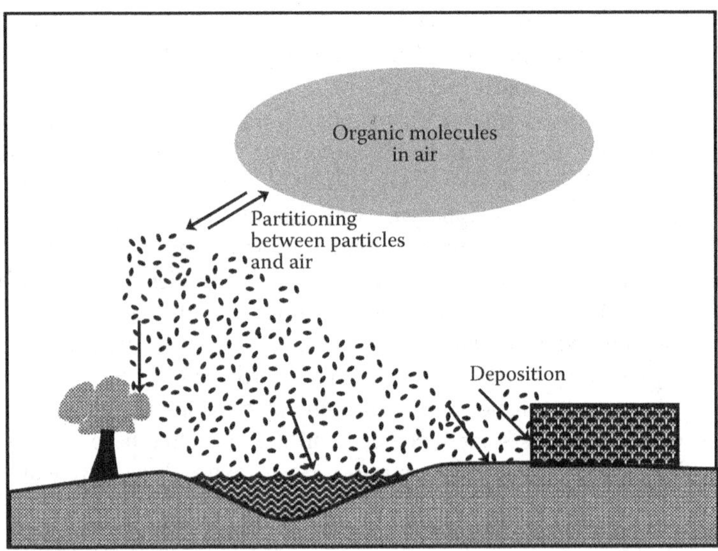

FIGURE 10.6 Semivolatile organic compounds in the atmosphere may be partitioned between air and particles, which in turn may deposit along with the organic compounds on vegetation, water, soil, and structures in the anthrosphere.

Another important oxidation that takes place in atmospheric water droplets is the oxidation of aldehydes to organic carboxylic acids.

The hydroxyl radical, $HO^{\bullet}$, is very important in initiating atmospheric oxidation reactions such as those noted above. Hydroxyl radical as $HO^{\bullet}$ can enter water droplets from the gas-phase atmosphere, it can be produced in water droplets photochemically, or it can be generated from H_2O_2 and $^{\bullet}O_2^-$ radical ion, which dissolve in water from the gas phase and then produce $HO^{\bullet}$ by solution chemical reaction:

$$H_2O_2 + {}^{\bullet}O_2^- \rightarrow HO^{\bullet} + O_2 + OH^- \tag{10.12}$$

Several solutes can react photochemically in aqueous solution (as opposed to the gas phase) to produce hydroxyl radical. One of these is hydrogen peroxide:

$$H_2O_2(aq) + h\nu \rightarrow 2HO^{\bullet}(aq) \tag{10.13}$$

Nitrite as NO_2^- or HNO_2, nitrate (NO_3^-), and iron(III) as $Fe(OH)^{2+}(aq)$ can also react photochemically in aqueous solution to produce $HO^{\bullet}$. It has been observed that ultraviolet radiation at 313 nm and simulated sunlight can react to produce $HO^{\bullet}$ radical in authentic samples of water collected from cloud and fog sources.[6] Based on the results of this study and related investigations, it may be concluded that the aqueous-phase formation of hydroxyl radical is an important, and in some cases dominant, means by which this key atmospheric oxidant is introduced into atmospheric water droplets.

10.11 ATMOSPHERIC CHEMICAL REACTIONS INVOLVING PARTICLES

In recent years there has been an increasing recognition of the importance in atmospheric chemistry of chemical processes that occur on particle surfaces and in solution in liquid particles (Figure 10.7). Challenging as it is, gas-phase atmospheric chemistry is relatively straightforward compared to the heterogeneous chemistry that involves particles. Particles may serve as sources

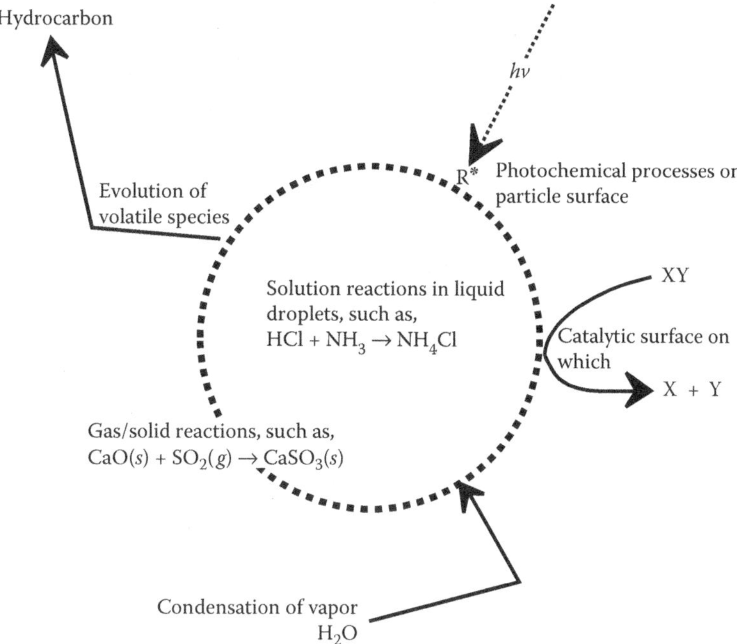

FIGURE 10.7 Particles provide sites for many important atmospheric chemical processes.

and sinks of atmospheric chemical reaction participant species. Solid particle surfaces may adsorb reactants and products, serve a catalytic function, exchange electrical charge, and absorb photons of electromagnetic radiation, thus acting as photocatalytic surfaces. Liquid water droplets may act as media for solution reactions including photochemical reactions that occur in solution.

Reactions on particle surfaces are very difficult to study because of factors such as variability in atmospheric particulate matter, the virtual impossibility of duplicating conditions that occur with suspended particles in the atmosphere, and the effects of water vapor and water condensed on particle surfaces. Prominent among solid particles that serve as reaction sites are soot and elemental carbon, oxides, carbonates, silica, and mineral dust. Particles may be liquid aerosols, dry solids, or solids with deliquescent surfaces. They exhibit wide variations in diameter, surface area, and chemical composition. Some of the atmospheric chemical processes that likely occur on particle surfaces are N_2O_5 hydrolysis, surface aging of soot particles by surface oxidation, generation of HONO (a precursor to $HO^{\bullet}$) by reaction of nitrogen oxides and water vapor on soot and silica particle surfaces, reactions of $HO^{\bullet}$ with nonvolatile chemical species sorbed to particle surfaces, uptake and reactions of carbonyl compounds such as acetone on particulate oxides and mineral dusts, and processes involving particles that influence residence times of atmospheric chemicals.

An interesting example of chemical processes on particle surfaces is the accumulation of sulfate on the surfaces of sodium chloride particles produced by evaporation of water from seawater spray droplets. This phenomenon has been attributed in part to a process that begins with the reaction of deliquesced (moist) sodium chloride with hydroxyl radical[7]:

$$2NaCl + 2HO^{\bullet} \rightarrow 2NaOH + Cl_2 \qquad (10.14)$$

Part of the Cl_2 on the surface reacts with NaOH,

$$Cl_2 + 2NaOH \rightarrow NaOCl + NaCl + H_2O \qquad (10.15)$$

to produce sodium hypochlorite, which has been observed on the particle surfaces. The NaOH on the surface reacts with atmospheric sulfuric acid,

$$2NaOH + H_2SO_4 \rightarrow Na_2SO_4 + H_2O \quad (10.16)$$

to produce sodium sulfate. The basic sodium hydroxide on the surface also facilitates oxidation of atmospheric SO_2:

$$SO_2 + 2NaOH + \frac{1}{2}O_2 \rightarrow Na_2SO_4 + H_2O \quad (10.17)$$

The net result is that the oxidation of atmospheric sulfur dioxide is promoted and particulate sodium chloride contains significant amounts of sodium sulfate.

10.12 CONTROL OF PARTICULATE EMISSIONS

The removal of particulate matter from gas streams is the most widely practiced means of air pollution control. A number of devices have been developed for this purpose that differ widely in effectiveness, complexity, and cost. The selection of a particle removal system for a gaseous waste stream depends upon the particle loading, nature of particles (size distribution), and type of gas-scrubbing system used.

10.12.1 Particle Removal by Sedimentation and Inertia

The simplest means of particulate matter removal is *sedimentation*, a phenomenon that occurs continuously in nature. Gravitational settling chambers may be employed for the removal of particles from gas streams by simply settling under the influence of gravity. These chambers take up large amounts of space and have low collection efficiencies, particularly for small particles.

Gravitational settling of particles is enhanced by increased particle size, which occurs spontaneously by coagulation. Thus, over time, the sizes of particles increase and the number of particles decreases in a mass of air that contains particles. Brownian motion of particles less than about 0.1 μm in size is primarily responsible for their contact, enabling coagulation to occur. Particles greater than about 0.3 μm in radius do not diffuse appreciably and serve primarily as receptors of smaller particles.

Inertial mechanisms are effective for particle removal. These depend upon the fact that the radius of the path of a particle in a rapidly moving, curving air stream is larger than the path of the stream as a whole. Therefore, when a gas stream is spun by vanes, a fan, or a tangential gas inlet, the particulate matter may be collected on a separator wall because the particles are forced outward by centrifugal force. Devices utilizing this mode of operation are called *dry centrifugal collectors* (cyclones).

10.12.2 Particle Filtration

Fabric filters, as their name implies, consist of fabrics that allow the passage of gas but retain particulate matter. These are used to collect dust in bags contained in structures called *baghouses*. Periodically, the fabric composing the filter is shaken to remove the particles and to reduce back-pressure to acceptable levels. Typically, the bag is in a tubular configuration as shown in Figure 10.8. Numerous other configurations are possible. Collected particulate matter is removed from bags by mechanical agitation, blowing air on the fabric, or rapid expansion and contraction of the bags.

Although simple, baghouses are generally effective in removing particles from exhaust gas. Particles as small as 0.01 μm in diameter are removed, and removal efficiency is relatively high for particles down to 0.5 μm in diameter. Aided by the development of mechanically strong, heat-resistant fabrics from which the bags are fabricated, baghouse installations have increased significantly in the effort to control particulate emissions.

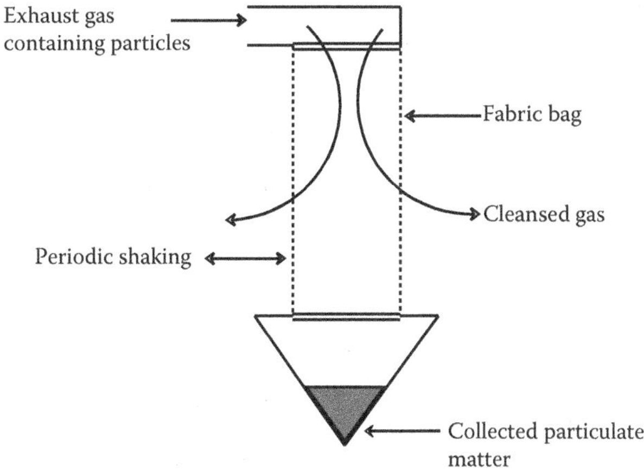

FIGURE 10.8 Baghouse collection of particulate emissions.

Diesel engines, especially those in heavy-duty trucks and buses, are major sources of particulate matter in urban areas and along highways. Although exhaust particulate filters were developed for use on diesel-powered highway vehicles in the latter 1970s, because of advances in engine design and control these devices were deemed unnecessary for a number of years. Later findings regarding suspected health effects of diesel particulate matter subsequently led to an increase in interest in diesel particulate filters. Devices for controlling diesel emissions, such as those that trap carbonaceous particles on ceramic filters followed by cycles in which the accumulated material is burned off the filter, have now reached a high level of sophistication and effectiveness. Given the attractiveness of the diesel engine for its high fuel economy, diesel particulate filters are becoming common equipment on diesel-powered vehicles.

10.12.3 Scrubbers

A venturi scrubber passes gas through a device that leads the gas stream through a converging section, throat, and diverging section as shown in Figure 10.9. Injection of the scrubbing liquid at right angles to incoming gas breaks the liquid into very small droplets, which are ideal for scavenging

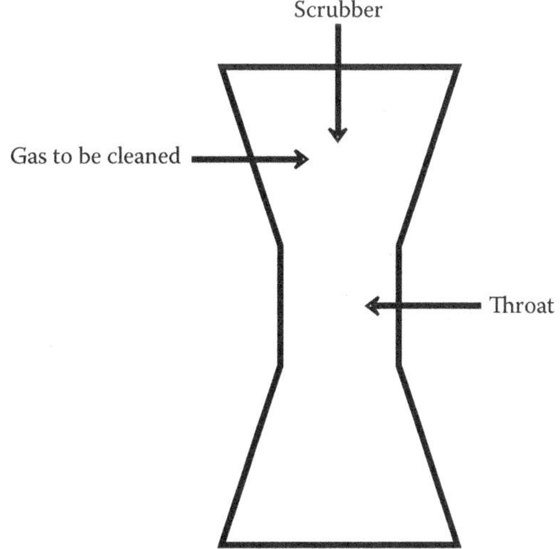

FIGURE 10.9 Venturi scrubber.

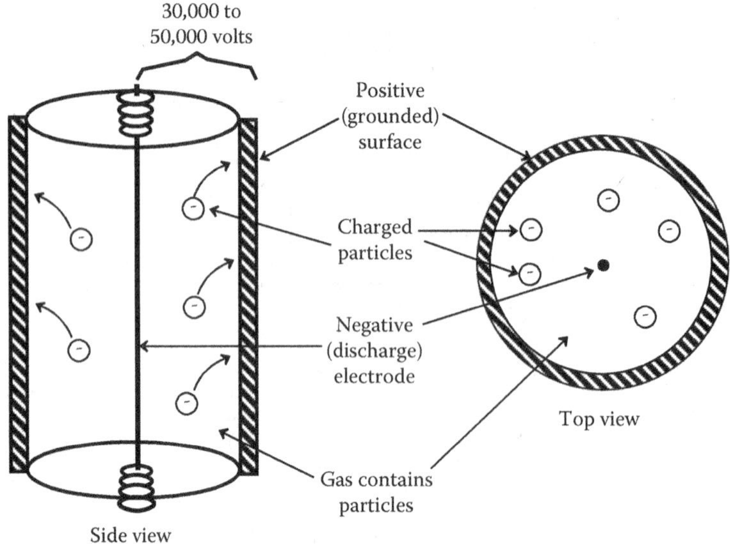

FIGURE 10.10 Schematic diagram of an electrostatic precipitator.

particles from the gas stream. In the reduced-pressure (expanding and, therefore, cooling) region of the venturi, some condensation can occur of vapor from liquid initially evaporated in the generally hot waste gas, adding to the scrubbing efficiency. In addition to removing particles, venturis may serve as quenchers to cool exhaust gas, and as scrubbers for pollutant gases.

Ionizing wet scrubbers place an electrical charge on particles upstream from a wet scrubber. Larger particles and some gaseous contaminants are removed by scrubbing action. Smaller particles tend to induce opposite charges in water droplets in the scrubber and in its packing material and are removed by attraction of the opposite charges.

10.12.4 Electrostatic Removal

Aerosol particles may acquire electrical charges. In an electric field, such particles are subjected to a force, F, given by

$$F = Eq \tag{10.18}$$

where E is the electrical potential gradient between the oppositely charged electrodes in the space in which the particle is suspended and q is the electrostatic charge on the particle. This phenomenon has been widely used in highly efficient *electrostatic precipitators*, as shown in Figure 10.10. The particles acquire a charge when the gas stream is passed through a high-voltage, direct current corona. Because of the charge, the particles are attracted to a grounded surface from which they may be later removed. Ozone may be produced by the corona discharge.

LITERATURE CITED

1. Petroff, A., A. Mailliat, M. Amielh, and F. Anselmet, Aerosol dry deposition on vegetative canopies. Part I: Review of present knowledge, *Atmospheric Environment*, **42**, 3625–3653, 2008.
2. Clement, C. F., Mass transfer to aerosols, in *Environmental Chemistry of Aerosols*, Ian Colbeck, Ed., pp. 49–89, Blackwell Publishing Ltd., Oxford, UK, 2008.
3. Ravindra, K., R. Sokhi, and R. V. Grieken, Atmospheric polycyclic aromatic hydrocarbons: Source attribution, emission factors and regulation, *Atmospheric Environment*, **42**, 2895–2921, 2008.

4. Chen, L. C. and M. Lippmann, Effects of metals within ambient air particulate matter (PM) on human health, *Inhalation Toxicology*, **21**(1), 1–31, 2009.
5. Rosenfeld, D., U. Lohmann, G. B. Raga, C. D. O'Dowd, M. Kulmala, S. Fuzzi, A. Reissell, and M. O. Andreae, Flood or drought: How do aerosols affect precipitation? *Science*, **321**, 1309–1313, 2008.
6. Zuo, Y., Light-induced formation of hydroxyl radicals in fog waters determined by an authentic fog constituent, hydroxymethanesulfonate, *Chemosphere*, **51**, 175–179, 2003.
7. Laskin, A., D. J. Gaspar, W. Wang, S. W. Hunt, J. P. Cowin, S. D. Colson, and B. J. Finlayson-Pitts, Reactions at interfaces as a source of sulfate formation in sea-salt particles, *Science*, **301**(5631), 340–344, 2003.

SUPPLEMENTARY REFERENCES

Austin, J., P. Brimblecombe, and W. Sturges, Eds, *Air Pollution Science for the 21st Century*, Elsevier Science, New York, 2002.

Barker, J. R., A brief introduction to atmospheric chemistry, *Advances Series in Physical Chemistry*, **3**, 1–33, 1995.

Baron, P. A. and K. Willeke, Eds, *Aerosol Measurements*, Wiley, New York, 2001.

Brasseur, G. P., J. J. Orlando, and G. S. Tyndall, Eds, *Atmospheric Chemistry and Global Change*, Oxford University Press, New York, 1999.

Colbeck, I., Ed., *Environmental Chemistry of Aerosols*, Blackwell Publishing, Oxford, UK, 2008.

Desonie, D., *Atmosphere: Air Pollution and its Effects*, Chelsea House Publishers, New York, 2007.

Donaldson, K. and P. Borm, Eds, *Particle Toxicology*, Taylor & Francis/CRC Press, Boca Raton, FL, 2007.

Hewitt, C. N. and A. Jackson, Eds, *Atmospheric Science for Environmental Scientists*, Wiley-Blackwell, Hoboken, NJ, 2009.

Hewitt, N. and A. Jackson, Eds, *Handbook of Atmospheric Science*, Blackwell Publishing, Malden, MA, 2003.

Hobbs, P. V., *Introduction to Atmospheric Chemistry*, Cambridge University Press, New York, 2000.

Jacob, D. J., *Introduction to Atmospheric Chemistry*, Princeton University Press, Princeton, NJ, 1999.

Kidd, J. S. and R. A. Kidd, *Air Pollution: Problems and Solutions*, Chelsea House Publishers, New York, 2006.

Lewis, E. R. and S. E. Schwartz, *Sea Salt Aerosol Production: Mechanisms, Methods, Measurements and Models: A Critical Review*, American Geophysical Union, Washington, DC, 2004.

Nalwa, H. S., Ed., *Handbook of Photochemistry and Photobiology*, American Scientific Publishers, Stevenson Ranch, CA, 2003.

Pandis, S. N. and J. H. Seinfeld, *Atmospheric Chemistry and Physics: From Air Pollution to Climate Change*, 2nd ed., Wiley, Hoboken, NJ, 2006.

Seinfeld, J. H. and S. N. Pandis, *Atmospheric Chemistry and Physics: From Air Pollution to Climate Change*, 2nd ed., Wiley, Hoboken, NJ, 2006.

Sokhi, R. S., Ed., *World Atlas of Atmospheric Pollution*, Anthem Press, New York, 2007.

Spellman, F. R., *The Science of Air: Concepts and Applications*, 2nd ed., Taylor & Francis, Boca Raton, FL, 2009.

Vallero, D. A., *Fundamentals of Air Pollution*, 4th ed., Elsevier, Amsterdam, 2008.

QUESTIONS AND PROBLEMS

The use of internet resources is assumed in answering any of the questions. These would include such things as constants and conversion factors as well as additional information needed to complete an answer.

1. In 2006 the U.S. EPA proposed lowering the allowable $PM_{2.5}$ level to 35 $\mu g/m^3$. How many particles would this be in a cubic meter of air assuming that all the particles were spheres of a diameter of 2.5 μm and had a density of exactly 1 g/cm^3?
2. For small charged particles, those that are 0.1 μm or less in size, an average charge of 4.77×10^{-10} esu, is normally assumed for the whole particle. What is the surface charge in esu/cm^2 for a charged spherical particle with a radius of 0.1 μm?

3. What is the settling velocity of a particle having a Stokes diameter of 10 μm and a density of 1 g/cm^3 in air at 1.00 atm pressure and 0°C temperature? (The viscosity of air at 0°C is 170.8 micropoise. The density of air under these conditions is 1.29 g/L.)
4. A freight train that included a tank car containing anhydrous NH_3 and one containing concentrated HCl was wrecked, causing both of the tank cars to leak. In the region between the cars, a white aerosol formed. What was it, and how was it produced?
5. Examination of aerosol fume particles produced by a welding process showed that 2% of the particles were >7 μm in diameter and only 2% were <0.5 μm. What is the MMD of the particles?
6. What two vapor forms of mercury might be found in the atmosphere?
7. Analysis of particulate matter collected in the atmosphere near a seashore shows considerably more Na than Cl on a molar basis. What does this indicate?
8. What type of process results in the formation of very small aerosol particles?
9. Which size range encompasses most of the particulate matter mass in the atmosphere?
10. Why are aerosols in the 0.1–1 μm size range especially effective in scattering light?
11. Per unit mass, why are smaller particles relatively more effective catalysts for atmospheric chemical reactions?
12. In terms of origin, what are the three major categories of elements found in atmospheric particles?
13. What are the five major classes of material making up the composition of atmospheric aerosol particles?
14. The size distribution of particles emitted from coal-fired power plants is bimodal. What are some of the properties of the smaller fraction in terms of potential environmental implications?
15. Of the following, the statement that is *untrue* regarding particles in the atmosphere is (explain): (A) Dispersion aerosol particles formed by grinding up bulk matter are typically relatively large; (B) very small particles tend to be acidic and often originate from gases; (C) Al, Fe, Ca, and Si in particles often come from soil erosion; (D) carcinogenic PAHs may be synthesized from saturated hydrocarbons under oxygen-deficient conditions; (E) larger particles are more harmful because they contain more matter.
16. Of the following, the species that is *least likely* to be a constituent of solid or liquid atmospheric particulate matter is (explain): (A) C, (B) O_3, (C) H_2SO_4, (D) NaCl, (E) benzo(a)pyrene.
17. Of the following, the one that is *not* a characteristic of dispersion aerosols is (explain): (A) They are most readily carried into the alveoli of lungs; (B) they are usually above 1 μm in size; (C) they are relatively easier to remove; (D) they are generally less respirable; (E) they are produced when bulk materials (larger particles) are ground up or subdivided.
18. Match the constituent of particulate matter from the left with its most likely source from the right:

A. Si	1. Natural sources, soil erosion
B. PAH	2. Incomplete combustion of hydrocarbons
C. SO_4^{2-}	3. Element largely introduced by human activities
D. Pb	4. Reaction of a gas in the atmosphere

19. Of the following, the most likely to be formed by pyrosynthesis is (explain): (A) sulfate particles, (B) ammonium particles, (C) sulfuric acid mist, (D) PAHs, (E) ozone in smog.
20. Match each particle constituent below, left, with its likely source:

A. Si	1. From gases in the surrounding atmosphere
B. V	2. From natural sources
C. Benzo(a)pyrene	3. Combustion of certain kinds of fuel oil
D. Sulfuric acid droplets	4. From incomplete combustion

11 Gaseous Inorganic Air Pollutants

11.1 INORGANIC POLLUTANT GASES

A number of gaseous inorganic pollutants enter the atmosphere as the result of human activities.[1] Those added in the greatest quantities are CO, SO_2, NO, and NO_2. (These quantities are relatively small compared to the amount of CO_2 in the atmosphere. The possible environmental effects of increased atmospheric CO_2 levels are discussed in Chapter 14.) Other inorganic pollutant gases include NH_3, N_2O, N_2O_5, H_2S, Cl_2, HCl, and HF. Substantial quantities of some of these gases are added to the atmosphere each year by human activities. Globally, atmospheric emissions of carbon monoxide, sulfur oxides, and nitrogen oxides are of the order of one to several hundred million tons per year.

11.2 PRODUCTION AND CONTROL OF CARBON MONOXIDE

Carbon monoxide, CO, is a natural constituent of the atmosphere and a pollutant when it is present above normal background concentrations. It causes problems in cases of locally high concentrations because of its toxicity (see Chapter 24). The overall atmospheric concentration of carbon monoxide is about 0.1 ppm, corresponding to a burden in the Earth's atmosphere of approximately 500 million metric tons of CO with an average residence time ranging from 36 to 110 days. Much of this CO is present as an intermediate in the oxidation of methane by hydroxyl radical. From Table 9.1 it may be seen that the methane content of the atmosphere is about 1.6 ppm, more than 10 times the concentration of CO. Therefore, any oxidation process for methane that produces carbon monoxide as an intermediate is certain to contribute substantially to the overall carbon monoxide burden, probably around two-thirds of the total CO.

Degradation of chlorophyll during the autumn months releases CO, amounting to perhaps as much as 20% of the total annual release. Anthropogenic sources account for about 6% of CO emissions. The remainder of atmospheric CO comes from largely unknown sources. These include some plants and marine organisms known as siphonophores, an order of *Hydrozoa*. Carbon monoxide is also produced by decay of plant matter other than chlorophyll.

Because of carbon monoxide emissions from internal combustion engines, the highest levels of this toxic gas tend to occur in congested urban areas at times when the maximum number of people are exposed, such as during rush hours. At such times, carbon monoxide levels in the atmosphere have become as high as 50–100 ppm, definitely hazardous to human health.

Atmospheric levels of carbon monoxide in urban areas show a positive correlation with the density of vehicular traffic and a negative correlation with wind speed. Urban atmospheres may show average carbon monoxide levels of the order of several parts per million (ppm), much higher than those in remote areas.

11.2.1 Control of Carbon Monoxide Emissions

Since the internal combustion engine is the primary source of localized pollutant carbon monoxide emissions, control measures have been concentrated on the automobile. Carbon monoxide emissions may be lowered by employing a leaner air–fuel mixture, that is, one in which the mass ratio of air to fuel is relatively high. At air–fuel (mass:mass) ratios exceeding approximately 16:1, an internal combustion engine emits very little carbon monoxide.

Modern automobiles use computerized control of engines with catalytic exhaust reactors to cut down on carbon monoxide emissions. Excess air is pumped into the exhaust gas, and the mixture is passed through a catalytic converter in the exhaust system, resulting in the oxidation of CO to CO_2.

11.3 FATE OF ATMOSPHERIC CO

It is generally agreed that carbon monoxide is removed from the atmosphere by reaction with hydroxyl radical, $HO^{\bullet}$:

$$CO + HO^{\bullet} \rightarrow CO_2 + H \tag{11.1}$$

The reaction produces hydroperoxyl radical as a product:

$$O_2 + H + M \rightarrow HOO^{\bullet} + M \tag{11.2}$$

$HO^{\bullet}$ is regenerated from $HOO^{\bullet}$ by the following reactions:

$$HOO^{\bullet} + NO \rightarrow HO^{\bullet} + NO_2 \tag{11.3}$$

$$HOO^{\bullet} + HOO^{\bullet} \rightarrow H_2O_2 + O_2 \tag{11.4}$$

The latter reaction is followed by photochemical dissociation of H_2O_2 to regenerate $HO^{\bullet}$:

$$H_2O_2 + h\nu \rightarrow 2HO^{\bullet} \tag{11.5}$$

Methane is also involved through the atmospheric $CO/HO^{\bullet}/CH_4$ cycle.

Soil microorganisms act to remove CO from the atmosphere. Therefore, soil is a sink for carbon monoxide.

11.4 SULFUR DIOXIDE SOURCES AND THE SULFUR CYCLE

Figure 11.1 shows the main aspects of the global sulfur cycle. This cycle involves primarily H_2S, $(CH_3)_2S$, SO_2, SO_3, and sulfates. There are many uncertainties regarding the sources, reactions, and fates of these atmospheric sulfur species. On a global basis, sulfur compounds enter the atmosphere to a very large extent through human activities. Approximately 100 million metric tons of sulfur per year enter the global atmosphere through anthropogenic activities, primarily as SO_2 from the combustion of coal and residual fuel oil. In the United States, anthropogenic emissions of sulfur dioxide peaked at 28.8 Tg (teragrams, millions of metric tons) in 1990 and have been reduced by more than 40% since then. As the result of air pollution control measures, sulfur dioxide emissions in the area of Europe covered by the United Nations Economic Commission for Europe Environmental Monitoring and Evaluation Program fell from 59 Tg in 1980 to 27 Tg in 1997 and emissions in the United Kingdom dropped from 6.4 Tg in 1970 to 1.2 Tg in 1999. Emissions of sulfur dioxide as measured directly and inferred by analysis of atmospheric sulfate have continued to drop in Europe.[2]

The greatest uncertainties in the sulfur cycle have to do with nonanthropogenic sulfur, which enters the atmosphere largely as SO_2 and H_2S from volcanoes, and as $(CH_3)_2S$ and H_2S from the biological decay of organic matter and the reduction of sulfate. The single largest source of natural sulfur discharged to the atmosphere is now believed to be biogenic dimethyl sulfide, $(CH_3)_2S$, from marine sources. Any H_2S that does get into the atmosphere is converted rapidly to SO_2 by the following overall process:

$$H_2S + \tfrac{3}{2}O_2 \rightarrow SO_2 + H_2O \tag{11.6}$$

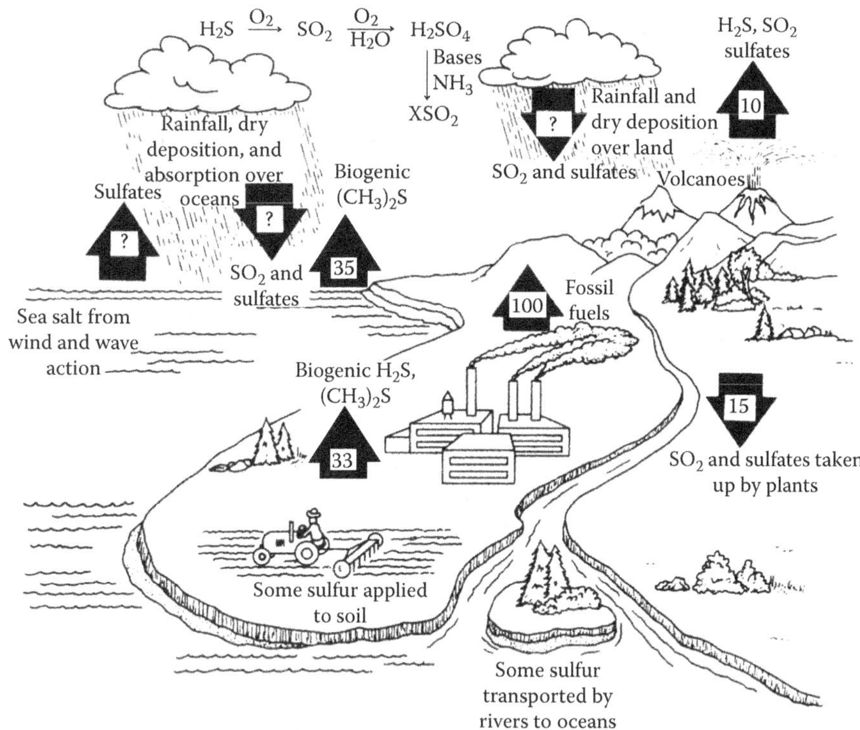

FIGURE 11.1 Global atmospheric sulfur cycle. Fluxes of sulfur represented by the arrows are in millions of metric tons per year. Those marked with a question mark are uncertain, but large, probably of the order of 100 million metric tons per year.

The initial reaction is hydrogen ion abstraction by a hydroxyl radical,

$$H_2S + HO^\bullet \rightarrow HS^\bullet + H_2O \tag{11.7}$$

followed by the following two reactions to give SO_2:

$$HS^\bullet + O_2 \rightarrow HO^\bullet + SO \tag{11.8}$$

$$SO + O_2 \rightarrow SO_2 + O \tag{11.9}$$

The primary source of anthropogenic sulfur dioxide is coal, from which sulfur must be removed at considerable expense to keep sulfur dioxide emissions at acceptable levels. Approximately half of the sulfur in coal is in some form of pyrite, FeS_2, and the other half is organic sulfur. The production of sulfur dioxide by the combustion of pyrite is given by the following reaction:

$$4FeS_2 + 11O_2 \rightarrow 2Fe_2O_3 + 8SO_2 \tag{11.10}$$

Essentially all of the sulfur is converted to SO_2 and only 1% or 2% to SO_3.

11.5 SULFUR DIOXIDE REACTIONS IN THE ATMOSPHERE

Many factors, including temperature, humidity, light intensity, atmospheric transport, and surface characteristics of particulate matter, may influence the atmospheric chemical reactions of sulfur dioxide. Like many other gaseous pollutants, sulfur dioxide reacts to form particulate matter, which then settles or is scavenged from the atmosphere by rainfall or other processes. It is known that high levels of air pollution normally are accompanied by a marked increase in aerosol particles and a consequent reduction in visibility. Reaction products of sulfur dioxide are thought to be responsible for some aerosol formation. Whatever the processes involved, much of the sulfur dioxide in the atmosphere is ultimately oxidized to sulfuric acid and sulfate salts, particularly ammonium sulfate and ammonium hydrogen sulfate. In fact, it is likely that these sulfates account for the turbid haze that covers much of the eastern part of the United States under all atmospheric conditions except those characterized by massive intrusions of Arctic air masses during the winter months. The potential of sulfates to induce climatic change is high and must be taken into account when considering control of sulfur dioxide.

Some of the possible ways in which sulfur dioxide may react in the atmosphere are (1) photochemical reactions; (2) photochemical and chemical reactions in the presence of nitrogen oxides and/or hydrocarbons, particularly alkenes; (3) chemical processes in water droplets, particularly those containing metal salts and ammonia; and (4) reactions on solid particles in the atmosphere. Since the atmosphere is a highly dynamic system with great variations in temperature, composition, humidity, and intensity of sunlight, different processes may predominate under various atmospheric conditions.

Photochemical reactions are probably involved in some of the processes resulting in the atmospheric oxidation of SO_2. Light with wavelengths above 218 nm is not sufficiently energetic to bring about the photodissociation of SO_2, so direct photochemical reactions in the troposphere are of no significance. The oxidation of sulfur dioxide at the parts-per-million level in an otherwise unpolluted atmosphere is a slow process. Therefore, other pollutant species must be involved in the process in atmospheres polluted with SO_2.

The presence of hydrocarbons and nitrogen oxides greatly increases the oxidation rate of atmospheric SO_2. As discussed in Chapter 13, hydrocarbons, nitrogen oxides, and ultraviolet light are the ingredients necessary for the formation of photochemical smog. This disagreeable condition is characterized by high levels of various oxidizing species (photochemical oxidants) capable of oxidizing SO_2. In the smog-prone Los Angeles area, the oxidation of SO_2 ranges up to 5–10% per hour. Among the oxidizing species present that could bring about this fast reaction are $HO^\bullet$, $HOO^\bullet$, O, O_3, NO_3, N_2O_5, $ROO^\bullet$, and $RO^\bullet$. As discussed in Chapters 12 and 13, the latter two species are reactive organic free radicals containing oxygen. Although ozone, O_3, is an important product of photochemical smog, it is believed that the oxidation of SO_2 by ozone in the gas phase is too slow to be appreciable, but oxidation by ozone and hydrogen peroxide is probably significant in water droplets.[3]

The most important gas-phase reaction leading to the oxidation of SO_2 is the addition of an $HO^\bullet$ radical,

$$HO^\bullet + SO_2 \rightarrow HOSO_2^\bullet \tag{11.11}$$

forming a reactive free radical that is eventually converted to a form of sulfate.

In all but relatively dry atmospheres, it is probable that sulfur dioxide is oxidized by reactions occurring inside water aerosol droplets. The overall process of sulfur dioxide oxidation in the aqueous phase is rather complicated. It involves the transport of gaseous SO_2 and oxidant to the aqueous phase, diffusion of species in the aqueous droplet, hydrolysis and ionization of SO_2, and oxidation of SO_2 by the following overall process, where {O} represents an oxidizing agent such as H_2O_2, $HO^\bullet$, or O_3 and S(IV) is $SO_2(aq)$, $HSO_3^-(aq)$, and $SO_3^{2-}(aq)$:

$$\{O\}(aq) + S(IV)(aq) \rightarrow 2H^+ + SO_4^{2-}(\text{unbalanced}) \tag{11.12}$$

In the absence of catalytic species, the reaction with dissolved molecular O_2,

$$\tfrac{1}{2}O_2(aq) + SO_2(aq) + H_2O \rightarrow H_2SO_4(aq) \tag{11.13}$$

is too slow to be significant. Hydrogen peroxide is an important oxidizing agent in the atmosphere. It reacts with dissolved sulfur dioxide through the overall reaction

$$SO_2(aq) + H_2O_2(aq) \rightarrow H_2SO_4(aq) \quad (11.14)$$

to produce sulfuric acid. The major reaction is thought to be between hydrogen peroxide and HSO_3^- ion with peroxymonosulfurous acid, $HOOSO_2^-$, as an intermediate.

Ozone, O_3, oxidizes sulfur dioxide in water. The fastest reaction is with sulfite ion:

$$SO_3^{2-}(aq) + O_3(aq) \rightarrow SO_4^{2-}(aq) + O_2 \quad (11.15)$$

Reactions are slower with $HSO_3^-(aq)$ and $SO_2(aq)$ and the rate of oxidation of aqueous SO_2 species by ozone increases with increasing pH. The oxidation of sulfur dioxide in water droplets is faster in the presence of ammonia, which reacts with sulfur dioxide to produce bisulfite ion and sulfite ion in solution:

$$NH_3 + SO_2 + H_2O \rightarrow NH_4^+ + HSO_3^- \quad (11.16)$$

Some solutes dissolved in water catalyze the oxidation of aqueous SO_2. Both iron(III) and Mn(II) have this effect. The reactions catalyzed by these two ions are faster with increasing pH. Dissolved nitrogen species, NO_2 and HNO_2, oxidize aqueous sulfur dioxide in the laboratory. As noted in Section 10.10, nitrite dissolved in water droplets may react photochemically to produce an $HO^\bullet$ radical, and this species in turn could act to oxidize dissolved sulfite.

Heterogeneous reactions on solid particles may also play a role in the removal of sulfur dioxide from the atmosphere. In atmospheric photochemical reactions, such particles may function as nucleation centers. Thus, they act as catalysts and grow in size by accumulating reaction products. The final result would be the production of an aerosol with a composition unlike that of the original particle. Soot particles, which consist of elemental carbon contaminated with polynuclear aromatic hydrocarbons (see Chapter 10, Section 10.4) produced in the incomplete combustion of carbonaceous fuels, can catalyze the oxidation of sulfur dioxide to sulfate as indicated by the presence of sulfate on the soot particles. Soot particles are very common in polluted atmospheres, so it is very likely that they are strongly involved in catalyzing the oxidation of sulfur dioxide.

Oxides of metals such as aluminum, calcium, chromium, iron, lead, and vanadium may also be catalysts for the heterogenous oxidation of sulfur dioxide. These oxides may also adsorb sulfur dioxide. However, the total surface area of oxide particulate matter in the atmosphere is very low, so that the fraction of sulfur dioxide oxidized on metal oxide surfaces is relatively small.

11.5.1 Effects of Atmospheric Sulfur Dioxide

Though not terribly toxic to most people, low levels of sulfur dioxide in air do have some health effects. Its primary effect is upon the respiratory tract, producing irritation and increasing airway resistance, especially to people with respiratory weaknesses and sensitized asthmatics. Therefore, exposure to the gas may increase the effort required to breathe. Mucus secretion is also stimulated by exposure to air contaminated by sulfur dioxide. Although SO_2 causes death in humans at 500 ppm, it has not been found to harm laboratory animals at 5 ppm.

Sulfur dioxide has been at least partially implicated in several acute incidents of air pollution. In December 1930, a thermal inversion trapped waste products from a number of industrial sources in the narrow Meuse River Valley of Belgium. Sulfur dioxide levels reached 38 ppm. Approximately 60 people died in the episode, and some cattle were killed. In October 1948, a similar incident caused illness in over 40% of the population of Donora, Pennsylvania, and 20 people died. Sulfur dioxide concentrations of 2 ppm were recorded. During a 5-day period marked by a temperature inversion and fog in London in December 1952, approximately 3500–4000 deaths in excess of normal occurred. Levels of SO_2 reached 1.3 ppm. Autopsies revealed irritation of the respiratory tract, and high levels of sulfur dioxide in combination with inhaled particles were suspected of contributing to excess mortality.

Atmospheric sulfur dioxide is harmful to plants, some species of which are affected more than others. Acute exposure to high levels of the gas kills leaf tissue, a condition called leaf necrosis. The edges of the leaves and the areas between the leaf veins show characteristic damage. Chronic exposure of plants to sulfur dioxide causes chlorosis, a bleaching or yellowing of the normally green portions of the leaf. Plant injury increases with increasing relative humidity. Plants incur most injury from sulfur dioxide when their stomata (small openings in plant surface tissue that allow interchange of gases with the atmosphere) are open. For most plants, the stomata are open during the daylight hours, and most damage from sulfur dioxide occurs then. Long-term, low-level exposure to sulfur dioxide can reduce the yields of grain crops such as wheat or barley. In areas with high levels of sulfur dioxide pollution, plants may be damaged by sulfuric acid aerosols formed by oxidation of SO_2. Such damage appears as small spots where sulfuric acid droplets have impinged on leaves.

One of the more costly effects of sulfur dioxide pollution is deterioration of building materials. Limestone, marble, and dolomite are calcium and/or magnesium carbonate minerals that are attacked by atmospheric sulfur dioxide to form products that are either water-soluble or composed of poorly adherent solid crusts on the rock's surface, adversely affecting the appearance, structural integrity, and life of the building. Although both SO_2 and NO_x attack such stone, chemical analysis of the crusts shows predominantly sulfate salts. Dolomite, a calcium/magnesium carbonate mineral, reacts with atmospheric sulfur dioxide as follows:

$$CaCO_3 \cdot MgCO_3 + 2SO_2 + O_2 + 9H_2O \rightarrow CaSO_4 \cdot 2H_2O + MgSO_4 \cdot 7H_2O + 2CO_2 \quad (11.17)$$

11.5.2 Sulfur Dioxide Removal

Several kinds of processes are being used to remove sulfur and sulfur oxides from fuel before combustion and from stack gas after combustion. Most of these efforts concentrate on coal, since it is the major source of sulfur oxides pollution. Physical separation techniques may be used to remove discrete particles of pyritic sulfur from coal. Chemical methods may also be employed for removing sulfur from coal. *Fluidized bed combustion* of coal can largely eliminate SO_2 emissions at the point of combustion. The process consists of burning granular coal in a bed of finely divided limestone or dolomite maintained in a fluid-like condition by air injection. Heat calcines the limestone,

$$CaCO_3 \rightarrow CaO + CO_2 \quad (11.18)$$

and the lime produced absorbs SO_2:

$$CaO + SO_2 \rightarrow CaSO_3 \text{ (which may be oxidized to } CaSO_4\text{)} \quad (11.19)$$

Many processes have been proposed or studied for the removal of sulfur dioxide from stack gas. These vary by the nature of the adsorbent, the means of contacting flue gas with the adsorbent, and whether or not the final product is dry. Sorbents include $CaCO_3$ (limestone), $CaCO_3 \cdot MgCO_3$ (dolomite), $Ca(OH)_2$ (lime), alkaline fly ash from coal combustion, sodium sulfite solution, sodium carbonate and soda ash, soda liquor waste from production of trona (a sodium carbonate mineral), and magnesium oxide. Flue gas may be contacted with adsorbents by spray processes, spray dry processes in which the water in the absorbent solution is evaporated and dry solid residue collected, venturi systems, packed beds, bubbling reactors, and trays. Wet processes are the most

TABLE 11.1
Major Stack Gas Scrubbing Systems

Process	Reaction	Significant Advantages or Disadvantages
Lime slurry scrubbing[a]	$Ca(OH)_2 + SO_2 \rightarrow CaSO_3 + H_2O$	Up to 200 kg of lime are needed per metric ton of coal, producing large amounts of wastes
Limestone slurry scrubbing[a]	$CaCO_3 + SO_2 \rightarrow CaSO_3 + CO_2(g)$	Less basic than lime slurry, not so efficient
Magnesium oxide scrubbing	$Mg(OH)_2(slurry) + SO_2 \rightarrow MgSO_3 + H_2O$	The sorbent can be regenerated, which can be done off site
Sodium-base scrubbing	$Na_2SO_3 + H_2O + SO_2 \rightarrow 2NaHSO_3$	No major technological limitations
	$2NaHSO_3 + heat \rightarrow Na_2SO_3 + H_2O + SO_2$ (regeneration)	Relatively high annual costs
Double alkali[a]	$2NaOH + SO_2 \rightarrow Na_2SO_3 + H_2O$ $Ca(OH)_2 + Na_2SO_3 \rightarrow CaSO_3(s) + 2NaOH$ (regeneration of NaOH)	Allows for regeneration of expensive sodium alkali solution with less costly lime

Source: Lunt, Richard R. and John D. Cunic, *Profiles in Flue Gas Desulfurization*, American Institute of Chemical Engineers, New York, 2000.

[a] These processes have also been adapted to produce a gypsum product by oxidation of $CaSO_3$ in the spent scrubber medium: $CaSO_3 + \frac{1}{2}O_2 + 2H_2O \rightarrow CaSO_4{\cdot}2H_2O(s)$. Gypsum has some commercial value, such as in the manufacture of plasterboard, and makes a relatively settleable waste product.

commonly used. Table 11.1 summarizes some stack gas scrubbing systems including throwaway and recovery systems.

The wet systems commonly employed to remove sulfur dioxide from flue gas provide a number of challenges including scaling, corrosion, general messiness of slurries, and cooling of the gas that requires subsequent heating of the gas so that it will rise in the stack. Therefore, dry systems would be more desirable, but they have not proven to be particularly effective. A dry throwaway system used with only limited success involves the injection of dry limestone or dolomite into the boiler followed by recovery of dry lime, sulfites, and sulfates. The overall reaction, shown here for dolomite, is the following:

$$CaCO_3 \cdot MgCO_3 + SO_2 + \tfrac{1}{2}O_2 \rightarrow CaSO_4 + MgO + 2CO_2 \tag{11.20}$$

The solid sulfate and oxide products are removed by electrostatic precipitators or cyclone separators. The process has an efficiency of 50% or less for the removal of sulfur oxides.

As seen from the reactions shown in Table 11.1, all wet sulfur dioxide removal processes, except for catalytic oxidation, depend upon the absorption of SO_2 by reaction with a base. Most scrubbing processes produce a solution of calcium sulfite, $CaSO_3$, and solid calcium sulfite hemihydrate, $CaSO_3 \cdot \frac{1}{2}H_2O(s)$ may form as well. Gypsum is formed in the scrubbing process by the oxidation of sulfite,

$$SO_3^{2-} + \tfrac{1}{2}O_2 \rightarrow SO_4^{2-} \tag{11.21}$$

followed by a reaction of sulfate ion with a calcium ion:

$$Ca^{2+} + SO_4^{2-} + 2H_2O \rightleftarrows CaSO_4 \cdot 2H_2O(s) \tag{11.22}$$

Gypsum is the desired final product and is commonly used to produce wallboard as a commercial product (mentioned as an example of the practice of industrial ecology in Section 17.15). Some systems blow air into the calcium sulfite product to oxidize it to calcium sulfate in a forced oxidation system.

Recovery systems in which sulfur dioxide or elemental sulfur are removed from the spent sorbing material, which is recycled, are much more desirable from an environmental viewpoint than are throwaway systems. Many kinds of recovery processes have been investigated, including those that involve scrubbing with magnesium oxide slurry, sodium hydroxide solution, sodium sulfite solution, ammonia solution, or sodium citrate solution.

Sulfur dioxide recovered from a stack-gas-scrubbing process can be converted to hydrogen sulfide by reaction with synthesis gas (H_2, CO, CH_4)

$$SO_2 + (H_2, CO, CH_4) \rightarrow H_2S + CO_2(H_2O) \tag{11.23}$$

The Claus reaction is then employed to produce elemental sulfur:

$$2H_2S + SO_2 \rightarrow 2H_2O + 3S \tag{11.24}$$

The elemental sulfur product is a commercially valuable material used to make sulfuric acid. The recovery of sulfur by this means is a good example of a green chemical process.

11.6 NITROGEN OXIDES IN THE ATMOSPHERE

The three oxides of nitrogen normally encountered in the atmosphere are nitrous oxide (N_2O), nitric oxide (NO), and nitrogen dioxide (NO_2). In addition, nitrate radical, NO_3, is an important species involved in the nightime chemistry of photochemical smog (see Section 13.5; this species is not important in daytime chemistry because it undergoes photodissociation very rapidly in sunlight). The chemistry of nitrogen oxides and other reactive inorganic nitrogen species is very important in the atmosphere in areas such as formation of photochemical smog, production of acid rain, and depletion of stratospheric ozone.

Nitrous oxide, a commonly used anesthetic known as "laughing gas," is produced by microbiological processes and is a component of the unpolluted atmosphere at a level of approximately 0.3 ppm (see Table 9.1). This gas is relatively unreactive and probably does not significantly influence important chemical reactions in the lower atmosphere. Its concentration decreases rapidly with altitude in the stratosphere due to the photochemical reaction

$$N_2O + h\nu \rightarrow N_2 + O \tag{11.25}$$

and some reaction with singlet atomic oxygen:

$$N_2O + O \rightarrow N_2 + O_2 \tag{11.26}$$

$$N_2O + O \rightarrow 2NO \tag{11.27}$$

These reactions are significant in terms of depletion of the ozone layer. Increased global fixation of nitrogen, accompanied by increased microbial production of N_2O, could contribute to ozone layer depletion.

Colorless, odorless nitric oxide (NO) and pungent red-brown nitrogen dioxide (NO_2) are very important in polluted air. Collectively designated NO_x, these gases enter the atmosphere from natural sources, such as lightning and biological processes, and from pollutant sources. The latter are much more significant because of regionally high NO_2 concentrations which can cause severe air quality deterioration. Estimates of the quantities of NO_x entering the atmosphere vary widely, but generally range from a few tens of millions of metric tons per year to somewhat more than 100 million. The biggest share of anthropogenic NO_x amounting to around 20 million metric tons per year enters the atmosphere from combustion of fossil fuels in both stationary and mobile sources. A similar amount of NO_x is emitted from soil, much of it from the action of microorganisms on nitrogen fertilizer. Other natural sources are biomass burning, lightning, and, to a lesser extent, atmospheric oxidation of NH_3. There is a relatively small flux of NO_x from the stratosphere to the troposphere. The contribution of automobiles to nitric oxide production in the United States has become somewhat lower in the last decade as newer automobiles have replaced older models.

Most NO_x entering the atmosphere from pollution sources does so as NO generated from internal combustion engines. At very high temperatures, the following overall reaction occurs with intermediate steps:

$$N_2 + O_2 \rightarrow 2NO \tag{11.28}$$

The speed with which this reaction takes place increases steeply with temperature. The equilibrium concentration of NO in a mixture of 3% O_2 and 75% N_2, typical of that which occurs in the combustion chamber of an internal combustion engine, is shown as a function of temperature in Figure 11.2. At room temperature (27°C) the equilibrium concentration of NO is only 1.1×10^{-10} ppm, whereas at high temperatures it is much higher. Therefore, high temperatures favor both a high equilibrium concentration and a rapid rate of formation of NO. Rapid cooling of the exhaust gas from combustion "freezes" NO at a relatively high concentration because equilibrium is not maintained. Thus, by its very nature, the combustion process both in the internal combustion engine and in furnaces produces high levels of NO in the combustion products. The mechanism for the formation of nitrogen oxides from N_2 and O_2 during combustion is a complicated process. Both oxygen and nitrogen atoms are formed at the very high combustion temperatures by the reactions

$$O_2 + M \rightarrow O + O + M \tag{11.29}$$

$$N_2 + M \rightarrow N^{\bullet} + N^{\bullet} + M \tag{11.30}$$

where M is a third body highly energized by heat that imparts enough energy to the molecular N_2 and O_2 to break their chemical bonds. The energies required for these reactions are quite high because breakage of the oxygen bond requires 118 kcal/mol and breakage of the nitrogen bond requires 225 kcal/mol. Because of its relatively weaker bond, dissociation of O_2 predominates over

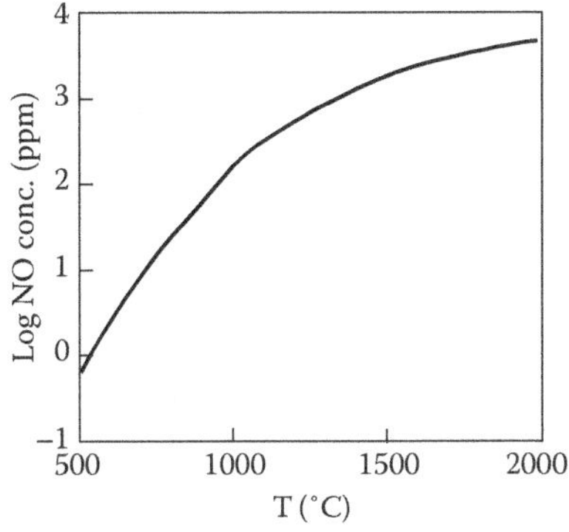

FIGURE 11.2 Log of equilibrium NO concentration as a function of temperature in a mixture containing 75% N_2 and 3% O_2.

that of N_2. Once formed, O and N atoms participate in the following chain reaction for the formation of nitric oxide from nitrogen and oxygen:

$$N_2 + O \rightarrow NO + N \tag{11.31}$$

$$N + O_2 \rightarrow NO + O \tag{11.32}$$

Leading to the net reaction:

$$N_2 + O_2 \rightarrow 2NO \tag{11.33}$$

There are, of course, many other species present in the combustion mixture besides those shown. The oxygen atoms are especially reactive toward hydrocarbon fragments by reactions such as the following:

$$RH + O \rightarrow R^\bullet + HO^\bullet \tag{11.34}$$

where $R^\bullet$ represents a hydrocarbon fragment of a molecule from which a hydrogen atom has been extracted. These fragments compete with N_2 for oxygen atoms. It is partly for this reason that the formation of NO is appreciably higher at air/fuel ratios exceeding the stoichiometric ratio (lean mixture), as shown in Figure 13.3.

The hydroxyl radical itself can participate in the formation of NO. The reaction is

$$N + HO^\bullet \rightarrow NO + H^\bullet \tag{11.35}$$

Nitric oxide, NO, is a product of the combustion of coal and petroleum containing chemically bound nitrogen. Production of NO by this route occurs at much lower temperatures than those required for "thermal" NO, discussed previously.

11.6.1 Atmospheric Reactions of NO_x

Atmospheric chemical reactions convert NO_x to nitric acid, inorganic nitrate salts, organic nitrates, and peroxyacetyl nitrate (see Chapter 13). The principal reactive nitrogen oxide species in the troposphere are NO, NO_2, and HNO_3. These species cycle among each other, as shown in Figure 11.3. Although NO is the primary form in which NO_x is released to the atmosphere, the conversion of NO to NO_2 is relatively rapid in the troposphere.

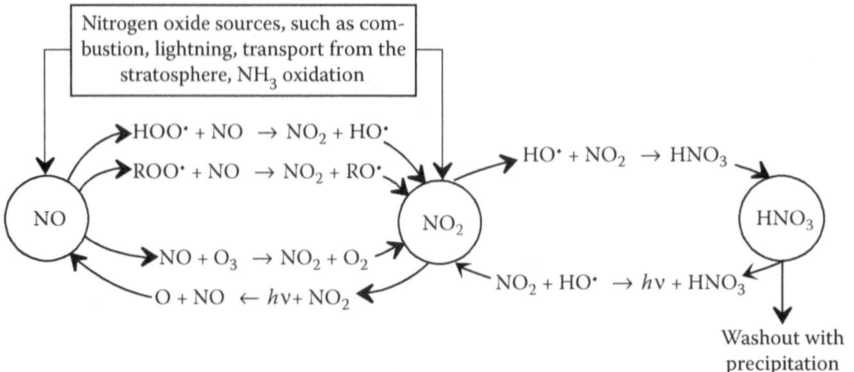

FIGURE 11.3 Principal reactions among NO, NO_2, and HNO_3 in the atmosphere. $ROO^\bullet$ represents an organic peroxyl radical, such as the methylperoxyl radical, $CH_3OO^\bullet$.

Nitrogen dioxide is a very reactive and significant species in the atmosphere. It absorbs light throughout the ultraviolet and visible spectrum penetrating the troposphere. At wavelengths below 398 nm, photodissociation occurs,

$$NO_2 + h\nu \rightarrow NO + O \tag{11.36}$$

to produce ground state oxygen atoms. Above 430 nm, only excited molecules of NO_2 are formed,

$$NO_2 + h\nu \rightarrow NO_2^* \tag{11.37}$$

whereas at wavelengths between 398 and 430 nm, either process may occur. Photodissociation at these wavelengths requires input of rotational energy from rotation of the NO_2 molecule. The tendency of NO_2 to photodissociate is shown clearly by the fact that in direct sunlight the half-life of NO_2 is much shorter than that of any other common molecular atmospheric species. (The steady-state level of NO_2 may remain relatively high because it rapidly forms again from NO through the action of oxidant-free radical species.) The photodissociation of nitrogen dioxide can give rise to the following significant inorganic reactions in addition to a host of atmospheric reactions involving organic species:

$$O + O_2 + M \text{ (third body)} \rightarrow O_3 + M \tag{11.38}$$

$$NO + O_3 \rightarrow NO_2 + O_2 \tag{11.39}$$

This reaction results in rapid conversion of NO to NO_2 and during the daytime NO_2 is rapidly converted back to NO by Reaction 11.36. In the absence of photodissociation at night, NO_2 predominates over NO.

$$NO_2 + O_3 \rightarrow NO_3 + O_2 \tag{11.40}$$

The NO_3 product undergoes rapid photodissociation in daytime, but builds up at night. As discussed in Chapter 13, NO_3 plays a role in atmospheric chemical processes at night, including those involved in photochemical smog formation.

$$O + NO_2 \rightarrow NO + O_2 \tag{11.41}$$

$$O + NO_2 + M \rightarrow NO_3 + M \tag{11.42}$$

$$NO_2 + NO_3 \rightarrow N_2O_5 \tag{11.43}$$

$$NO + NO_3 \rightarrow 2NO_2 \tag{11.44}$$

$$O + NO + M \rightarrow NO_2 + M \tag{11.45}$$

Nitrogen dioxide ultimately is removed from the atmosphere as nitric acid, nitrates, or (in atmospheres where photochemical smog is formed) as organic nitrogen. Dinitrogen pentoxide formed in Reaction 11.43 is the anhydride of nitric acid, which it forms by reacting with water:

$$N_2O_5 + H_2O \rightarrow 2HNO_3 \tag{11.46}$$

In the stratosphere, nitrogen dioxide reacts with hydroxyl radicals to produce nitric acid:

$$HO^{\bullet} + NO_2 \rightarrow HNO_3 \tag{11.47}$$

In this region, the nitric acid can also be destroyed by hydroxyl radicals,

$$HO^{\bullet} + HNO_3 \rightarrow H_2O + NO_3 \tag{11.48}$$

or by a photochemical reaction,

$$HNO_3 + h\nu \rightarrow HO^{\bullet} + NO_2 \quad (11.49)$$

so that HNO_3 serves as a temporary sink for NO_2 in the stratosphere. Nitric acid produced from NO_2 is removed as precipitation, or reacts with bases (ammonia, particulate lime) to produce particulate nitrates.

Reactions of both nitrogen oxides and sulfur dioxide in power plant plumes are important in the environmental fate and transport of NO_x and SO_2. The presence of water vapor and droplets as well as particles may facilitate reactions of NO_x and SO_2 in plumes.

11.6.2 Harmful Effects of Nitrogen Oxides

Nitric oxide, NO, is less toxic than NO_2. Like carbon monoxide and nitrite, NO attaches to hemoglobin and reduces oxygen transport efficiency. However, in a polluted atmosphere, the concentration of nitric oxide normally is much lower than that of carbon monoxide so that the effect on hemoglobin is much less.

Acute exposure to NO_2 can be quite harmful to human health. For exposures ranging from several minutes to one hour, a level of 50–100 ppm of NO_2 causes inflammation of lung tissue for a period of 6–8 weeks, after which time the subject normally recovers. Exposure of the subject to 150–200 ppm of NO_2 causes *bronchiolitis fibrosa obliterans*, a condition fatal within 3–5 weeks after exposure. Death generally results within 2–10 days after exposure to 500 ppm or more of NO_2. "Silo-filler's disease," caused by NO_2 generated by the fermentation of ensilage (moist chopped stalks of corn or sorghum used for cattle feed) containing nitrate, is a particularly striking example of nitrogen dioxide poisoning. Deaths have resulted from the inhalation of NO_2-containing gases from burning celluloid and nitrocellulose film, and from spillage of NO_2 oxidant (used with liquid hydrazine fuel) from missile rocket motors.

Although extensive damage to plants is observed in areas receiving heavy exposure to NO_2, most of this damage probably comes from secondary products of nitrogen oxides, such as peroxyacetyl nitrate (PAN) formed in smog (see Chapter 14). Exposure of plants to several ppm of NO_2 in the laboratory causes leaf spotting and break down of plant tissue. Exposure to 10 ppm of NO causes a reversible decrease in the rate of photosynthesis. The effect on plants of long-term exposure to a few tenths of a ppm of NO_2 is less certain.

Nitrogen oxides are known to cause fading of dyes and inks used in some textiles. This has been observed in gas clothes dryers and is due to NO_x formed in the dryer flame. Much of the damage to materials caused by NO_x comes from secondary nitrates and nitric acid. For example, stress-corrosion cracking of springs once widely used in telephone relays occurs far below the yield strength of the nickel-brass spring metal because of the action of particulate nitrates and aerosol nitric acid formed from NO_x.

Concern has been expressed about the possibility that NO_x emitted to the atmosphere by supersonic transport planes could catalyze the partial destruction of the stratospheric ozone layer that absorbs damaging short-wavelength (240–300 nm) ultraviolet radiation. This possibility raised initial concerns regarding anthropogenic damage to the stratospheric ozone layer around 1970. Detailed consideration of this effect is quite complicated, and only the main features are considered here.

In the upper stratosphere and in the mesosphere, molecular oxygen is photodissociated by ultraviolet light of less than 242-nm wavelength:

$$O_2 + h\nu \rightarrow O + O \quad (11.50)$$

In the presence of energy-absorbing third bodies, the atomic oxygen reacts with molecular oxygen to produce ozone:

$$O_2 + O + M \rightarrow O_3 + M \quad (11.51)$$

Ozone can be destroyed by reaction with atomic oxygen,

$$O_3 + O \rightarrow O_2 + O_2 \tag{11.52}$$

and its formation can be prevented by recombination of oxygen atoms:

$$O + O + M \rightarrow O_2 + M \tag{11.53}$$

Addition of the reaction of nitric oxide with ozone,

$$NO + O_3 \rightarrow NO_2 + O_2 \tag{11.54}$$

to the reaction of nitrogen dioxide with atomic oxygen,

$$NO_2 + O \rightarrow NO + O_2 \tag{11.55}$$

results in a net reaction for the destruction of ozone:

$$O + O_3 \rightarrow O_2 + O_2 \tag{11.56}$$

Along with NO_x, water vapor is also emitted into the atmosphere by aircraft exhausts, which could accelerate ozone depletion by the following two reactions:

$$O + H_2O \rightarrow HO^\bullet + HO^\bullet \tag{11.57}$$

$$HO^\bullet + O_3 \rightarrow HOO^\bullet + O_2 \tag{11.58}$$

However, there are many natural stratospheric buffering reactions, which tend to mitigate the potential ozone destruction from those reactions outlined above. Atomic oxygen capable of regenerating ozone is produced by the photochemical reaction

$$NO_2 + h\nu \rightarrow NO + O\ (\lambda < 420\ \text{nm}) \tag{11.59}$$

A competing reaction removing catalytic NO is

$$NO + HOO^\bullet \rightarrow NO_2 + HO^\bullet \tag{11.60}$$

Current belief is that emissions from even relatively large fleets of supersonic aircraft (unlikely in light of the 2003 demise of the Concorde aircraft) will not cause nearly as much damage to the ozone layer as CFCs.

11.6.3 Control of Nitrogen Oxides

As noted in Section 10.9, nitrogen oxides compose a class of the six criteria pollutants regulated by the U.S. EPA. Furthermore, they are involved in the formation of tropospheric ozone, O_3, another of the regulated criteria pollutants. As discussed in Chapter 13, both hydrocarbons and nitrogen oxides are the required ingredients for the production of ozone and other noxious species in the photochemical smog forming process. Initially, efforts to control such production of ozone were concentrated on reducing emissions of volatile organic hydrocarbons. However, in most cases nitrogen oxides are the limiting reactants in the production of photochemical smog and ozone so that control of nitrogen oxides emissions is very important in reducing tropospheric air pollution.

There are two general approaches to the control of NO_x emissions. The first of these is modification of combustion conditions to prevent formation of NO and the second is treatment of flue gas to remove NO_x before it is released to the atmosphere.[4]

The level of NO_x emitted from stationary sources such as power plant furnaces generally falls within the range of 50–1000 ppm. NO production is favored both kinetically and thermodynamically by high temperatures and by high excess oxygen concentrations and increases with increasing time of exposure to these conditions. Reduction of these three conditions during combustion lowers the amount of NO formed. Reduction of flame temperature to prevent NO formation is accomplished by adding recirculated exhaust gas, cool air, inert gases, or water, although this decreases the efficiency of energy conversion as calculated by the Carnot equation (see Chapter 19).

Low-excess-air firing is effective in reducing NO_x emissions during the combustion of fossil fuels. As the term implies, low-excess-air firing uses the minimum amount of excess air required for oxidation of the fuel, so that less oxygen is available for the reaction

$$N_2 + O_2 \rightarrow 2NO \tag{11.61}$$

in the high-temperature region of the flame. Incomplete fuel burnout with the emission of hydrocarbons, soot, and CO is an obvious problem with low-excess-air firing. This may be overcome by a two-stage combustion process consisting of the following steps:

1. A first stage in which the fuel is fired at a relatively high temperature with a substoichiometric amount of air, for example, 90–95% of the stoichiometric requirement. NO formation is limited by the absence of excess oxygen.
2. A second stage in which fuel burnout is completed at a relatively low temperature in excess air. The low temperature prevents formation of NO.

In some gas-fired power plants, the emission of NO has been reduced by as much as 90% by a two-stage combustion process.

Removal of NO_x from stack gas presents some formidable problems. Possible approaches to NO_x removal are catalytic decomposition of nitrogen oxides, catalytic reduction of nitrogen oxides, and sorption of NO_x by liquids or solids; sorption is limited by the low water solubility of NO, the predominant nitrogen oxide species in stack gas.

Commonly used NO_x reduction processes use a reducing agent such as methane or ammonia to reduce the N oxide to elemental N_2 and H_2O over a honeycombed catalyst of vanadium and tungsten oxides on an alumina (Al_2O_3) support as shown by the following reaction in which ammonia is the reducing agent

$$4NH_3 + 4NO + O_2 \rightarrow 4N_2 + 6H_2O \tag{11.62}$$

Denoted as selective catalytic reduction (SCR) the catalytic reduction of NO_x is carried out over a catalyst at 300–400°C with about 80% efficiency. It can also be carried out without a catalyst at a much higher temperature of 900–1000°C, but that is only 40–60% efficient.

Sorption processes for SO_2 removal using bases are relatively ineffective in removing nitric oxide because of the low acidity and low water solubility of NO. Proposals have been made for oxidation of NO to NO_2 and N_2O_3, which are more efficiently absorbed by base. The NO X SO regenerable dry sorbent process uses sodium carbonate on high-surface-area alumina beads to absorb SO_2 and NO_x in a fluidized bed through which the flue gas is passed. The beads are regenerated to restore their sorptive properties.

An interesting possibility for NO_x control is the use of *biofilters*.[5] A relatively new development in air pollution control, biofilters employ microorganisms in a fixed or fluidized bed configuration to contact gases and absorb pollutants. The microorganisms degrade pollutants retained on the filtration media and produce innocuous products (in the case of sorbed N compounds, N_2 is the favored product). Ideally, biofilters operate at relatively low cost and with low maintenance. However, the challenge of maintaining a viable biofilter in the exhaust stream of a large power plant is daunting.

11.7 ACID RAIN

As discussed in this chapter, much of the sulfur and nitrogen oxides entering the atmosphere are converted to sulfuric and nitric acids, respectively. When combined with hydrochloric acid arising from hydrogen chloride emissions, these acids cause acidic precipitation (acid rain) that is now a major pollution problem in some areas.

Headwater streams and high-altitude lakes are especially susceptible to the effects of acid rain and may sustain loss of fish and other aquatic life. Other effects include reductions in forest and crop productivity; leaching of nutrient cations and heavy metals from soils, rocks, and the sediments of lakes and streams; dissolution of metals such as lead and copper from water distribution pipes; corrosion of exposed metal; and dissolution of the surfaces of limestone buildings and monuments. Dissolution of phytotoxic aluminum from soil by acidic precipitation is especially detrimental to plant and forest growth.

As a result of its widespread distribution and effects, acid rain is an air pollutant that may pose a threat to the global atmosphere. Therefore, it is discussed in greater detail in Chapter 14.

11.8 AMMONIA IN THE ATMOSPHERE

Ammonia, one of the most abundant nitrogen-containing species in the atmosphere, is present even in unpolluted air as a result of natural biochemical and chemical processes. Among the various sources of atmospheric ammonia are microorganisms in soil, decay of animal wastes, ammonia fertilizer, sewage treatment, coke manufacture, ammonia manufacture, and leakage from ammonia-based refrigeration systems. Livestock and the feedlots in which they are kept are the largest source category of ammonia in the United States with production of probably more than a billion kilogram per year.[6] High concentrations of ammonia gas in the atmosphere are generally indicative of accidental release of the gas.

Ammonia is removed from the atmosphere by its affinity for water and by its action as a base; it is the major atmospheric base. It is a key species in the formation and neutralization of nitrate and sulfate aerosols in polluted atmospheres. Ammonia reacts with these acidic aerosols to form ammonium salts:

$$NH_3 + HNO_3 \rightarrow NH_4NO_3 \tag{11.63}$$

$$NH_3 + H_2SO_4 \rightarrow NH_4HSO_4 \tag{11.64}$$

Atmospheric ammonia may have a number of effects. Ammonium salts are among the more corrosive salts in atmospheric aerosols. Roughly half of the very small particulate matter ($PM_{2.5}$) in the atmosphere of the eastern United States consists of ammonium sulfate, significantly affecting visibility. Ammonia may affect vegetation with acute exposures causing visible foliar injury.

11.9 FLUORINE, CHLORINE, AND THEIR GASEOUS COMPOUNDS

Fluorine, hydrogen fluoride, and other volatile fluorides are produced in the manufacture of aluminum, and hydrogen fluoride is a byproduct in the conversion of fluorapatite (rock phosphate) to phosphoric acid, superphosphate fertilizers, and other phosphorus products. The wet process for the production of phosphoric acid involves the reaction of fluorapatite, $Ca_5F(PO_4)_3$, with sulfuric acid:

$$Ca_5F(PO_4)_3 + 5H_2SO_4 + 10H_2O \rightarrow 5CaSO_4 \cdot 2H_2O + HF + 3H_3PO_4 \tag{11.65}$$

It is necessary to recover most of the by-product fluorine from rock phosphate processing to avoid severe pollution problems. Recovery as fluorosilicic acid, H_2SiF_6, is normally practiced.

Hydrogen fluoride gas is a dangerous substance that is so corrosive it even reacts with glass. It is irritating to body tissues, and the respiratory tract is very sensitive to it. Brief exposure to HF vapors at the part-per-thousand level may be fatal. The acute toxicity of F_2 is even higher than that of HF. Chronic exposure to high levels of fluorides causes fluorosis, the symptoms of which include mottled teeth and pathological bone conditions.

Plants are particularly susceptible to the effects of gaseous fluorides, which appear to enter the leaf tissue through the stomata. Fluoride is a cumulative poison in plants, and exposure of sensitive plants to even very low levels of fluorides for prolonged periods results in damage. Characteristic symptoms of fluoride poisoning are chlorosis (fading of green color due to conditions other than the absence of light), edge burn, and tip burn. Conifers (such as pine trees) afflicted with fluoride poisoning may have reddish-brown, necrotic needle tips. The sensitivity of some conifers to fluoride poisoning is illustrated by the fact that fluorine produced by aluminum plants in Norway has destroyed forests of *Pinus sylvestris* up to 8 miles distant; trees were damaged at distances as great as 20 miles.

Silicon tetrafluoride gas, SiF_4, is another gaseous fluoride pollutant produced during some steel and metal smelting operations that employ CaF_2, fluorspar. Fluorspar reacts with silicon dioxide (sand), releasing SiF_4 gas:

$$2CaF_2 + 3SiO_2 \rightarrow 2CaSiO_3 + SiF_4 \tag{11.66}$$

Another gaseous fluorine compound, sulfur hexafluoride, SF_6, occurs in the atmosphere at levels of about 0.3 parts per trillion. It is extremely unreactive with an atmospheric lifetime estimated at 3200 years, and is used as an atmospheric tracer. It does not absorb ultraviolet light in either the troposphere or the stratosphere, and is probably destroyed above 60 km by reactions beginning with its capture of free electrons. Current atmospheric levels of SF_6 are significantly higher than the estimated background level of 0.04 parts per trillion in 1953 when commercial production of it began. The compound is very useful in specialized applications including gas-insulated electrical equipment and inert blanketing/degassing of molten aluminum and magnesium. Increasing uses of sulfur hexafluoride have caused concern because it is the most powerful greenhouse gas known, with a global warming potential (per molecule added to the atmosphere) about 23,900 times that of carbon dioxide.

11.9.1 Chlorine and Hydrogen Chloride

Chlorine gas, Cl_2, does not occur as an air pollutant on a large scale but can be quite damaging on a local scale. Chlorine was the first poisonous gas deployed in World War I. It is widely used as a manufacturing chemical in the plastics industry, for example, as well as for water treatment and as a bleach. Therefore, possibilities for its release exist in a number of locations. Chlorine is quite toxic and is a mucous-membrane irritant. It is very reactive and a powerful oxidizing agent. Chlorine dissolves in atmospheric water droplets, yielding hydrochloric acid and hypochlorous acid, an oxidizing agent:

$$H_2O + Cl_2 \rightarrow H^+ + Cl^- + HOCl \tag{11.67}$$

Spills of chlorine gas have caused fatalities among exposed persons. For example, the rupture of a derailed chlorine tank car at Youngstown, Florida, on February 25, 1978, resulted in the death of eight people who inhaled the deadly gas, and a total of 89 people were injured. Massive chlorine releases from sabotage are a major concern with respect to terrorism.

Hydrogen chloride, HCl, is emitted from a number of sources. Incineration of chlorinated plastics, such as PVC, a polymer of vinyl chloride with an empirical formula of C_2H_3Cl, releases HCl as a combustion product.

Some compounds released to the atmosphere as air pollutants hydrolyze to form HCl. Accidents have occurred in which compounds such as silicon tetrachloride, $SiCl_4$, and aluminum chloride, $AlCl_3$, have been released and reacted with atmospheric water vapor:

$$SiCl_4 + 2H_2O \rightarrow SiO_2 + 4HCl \tag{11.68}$$

$$AlCl_3 + 3H_2O \rightarrow Al(OH)_3 + 3HCl \tag{11.69}$$

Hydrogen chloride may be evolved from inorganic chloride salts, such as NaCl, KCl, and $CaCl_2$, at high temperatures in the presence of SO_2, O_2, and H_2O by the process of sulfation. The overall reaction of sulfation is illustrated below for $CaCl_2$:

$$CaCl_2 + \tfrac{1}{2}O_2 + SO_2 + H_2O \rightarrow CaSO_4 + 2HCl \tag{11.70}$$

As discussed in Section 10.11, it is now believed that deliquesced (moist) chloride particles can react with hydroxyl radical:

$$2NaCl + 2HO^{\bullet} \rightarrow 2NaOH + Cl_2 \tag{10.14}$$

This leaves a basic surface conducive to the retention and oxidation of SO_2. Hydrolysis of released Cl_2 can produce HCl, contributing to acidic material in the atmosphere:

$$Cl_2 + H_2O \rightarrow HCl + HOCl \tag{11.71}$$

11.10 REDUCED SULFUR GASES

Hydrogen sulfide, carbonyl sulfide (OCS), carbon disulfide (CS_2), and dimethyl sulfide ($S(CH_3)_2$) are important gaseous compounds in the atmosphere in which the sulfur is in a lower oxidation state. These gases are oxidized in the atmosphere to sulfate, in some cases with production of intermediate SO_2, and are important sources of atmospheric sulfur.

Hydrogen sulfide is produced by microbial processes including the decay of sulfur compounds and bacterial reduction of sulfate (see Chapter 6). Hydrogen sulfide is also released from geothermal steam, from wood pulping, and from a number of miscellaneous natural and anthropogenic sources. Approximately 8×10^9 kg (8 Tg) of H_2S are released to the global atmosphere annually. Because it is so readily oxidized, most atmospheric hydrogen sulfide is rapidly converted to SO_2. The organic homologs of hydrogen sulfide, the mercaptans, enter the atmosphere from decaying organic matter and have particularly objectionable odors.

Hydrogen sulfide pollution from artificial sources is not nearly so widespread as sulfur dioxide pollution. However, there have been several acute incidents of hydrogen sulfide emissions resulting in damage to human health and even fatalities. One of the most notorious such episodes occurred in Poza Rica, Mexico, in 1950. Accidental release of hydrogen sulfide from a plant used for the recovery of sulfur from natural gas caused the deaths of 22 people and the hospitalization of more than 300, although some unofficial reports have put the number of fatalities much higher. In December 2003, a blowout of natural gas contaminated with hydrogen sulfide killed 242 people and seriously injured more than 2000 in China. As an emergency measure, the gas was set on fire, which produced copious quantities of sulfur dioxide, still an air pollutant, but not nearly so deadly as hydrogen sulfide. Efforts to tap very deep natural gas formations have increased the danger of accidental hydrogen sulfide release.

Hydrogen sulfide in "sour" natural gas has posed an increasing problem as larger amounts of relatively abundant sour natural gas are tapped for energy. The problem is especially acute in Alberta, Canada, which has some of the world's most abundant sour natural gas. Recovery of the hydrogen sulfide and conversion to elemental sulfur has caused the supply of sulfur to exceed demand. Now, increasing amounts of hydrogen sulfide accompanied by carbon dioxide, another acidic component of sour natural gas are being injected into underground formations.[7]

Hydrogen sulfide at levels well above ambient concentrations destroys immature plant tissue. This type of plant injury is readily distinguished from that due to other phytotoxins. More sensitive species are killed by continuous exposure to around 3000 ppb (parts per billion by volume) H_2S, whereas other species exhibit reduced growth, leaf lesions, and defoliation.

Damage to certain kinds of materials is a very expensive effect of hydrogen sulfide pollution. Paints containing lead pigments, $2PbCO_3 \cdot Pb(OH)_2$ (no longer used), are particularly susceptible to darkening by H_2S. Darkening results from exposure over several hours to as little as 50 ppb H_2S. The lead sulfide originally produced by reaction of the lead pigment with hydrogen sulfide eventually may be converted to white lead sulfate by atmospheric oxygen after removal of the source of H_2S, thus partially reversing the damage.

A black layer of copper sulfide forms on copper metal exposed to H_2S. Eventually, this layer is replaced by a green coating of basic copper sulfate, $CuSO_4 \cdot 3Cu(OH)_2$. The green "patina," as it is called, is very resistant to further corrosion. Such layers of corrosion can seriously impair the function of copper contacts on electrical equipment. Hydrogen sulfide also forms a black sulfide coating on silver.

Carbonyl sulfide, OCS, is now recognized as a component of the atmosphere at a tropospheric concentration of approximately 500 parts per trillion by volume, corresponding to a global burden of about 2.4 Tg. Approximately 1.3 Tg of carbonyl sulfide is released to the atmosphere annually. About half that amount of carbon disulfide, CS_2, is released to the atmosphere annually. Carbonyl sulfide has an atmospheric lifetime of approximately 18 months and is the longest-lived of the common reduced sulfur gases in the atmosphere.

Both OCS and CS_2 are oxidized in the atmosphere by reactions initiated by the hydroxyl radical. The initial reactions are

$$HO^{\bullet} + OCS \rightarrow CO_2 + HS^{\bullet} \quad (11.72)$$

$$HO^{\bullet} + CS_2 \rightarrow OCS + HS^{\bullet} \quad (11.73)$$

These reactions with hydroxyl radical initiate oxidation processes that occur through a series of atmospheric chemical reactions. The sulfur-containing products that are initially formed as shown by Reactions 11.72 and 11.73 undergo further reactions to sulfur dioxide and, eventually, to sulfate species. The reaction of carbonyl sulfide with hydroxyl radical is so slow that it is probably a minor pathway for OCS loss in the troposphere. This gas is metabolized by higher plant carbonic anhydrase, which normally metabolizes carbon dioxide, and uptake by higher plants is believed to be a significant sink for carbonyl sulfide.

Carbonyl sulfide is so long-lived that significant amounts of it reach the stratosphere. Here it can undergo photolysis, leading to the following sequence of reactions:

$$OCS + h\nu \rightarrow CO + S \quad (11.74)$$

$$O + OCS \rightarrow CO + SO \quad (11.75)$$

$$S + O_2 \rightarrow SO + O \quad (11.76)$$

$$SO + O_2 \rightarrow SO_2 + O \quad (11.77)$$

$$SO + NO_2 \rightarrow SO_2 + NO \quad (11.78)$$

The sulfur dioxide produced by these reactions is eventually oxidized to sulfate and sulfuric acid aerosol. This produces a stratospheric aerosol layer extending from the bottom of the stratosphere to an altitude of approximately 30 km. Named after Christian Junge, who discovered it in 1960 while searching for cosmic dust and debris from nuclear weapons testing, this layer is called the Junge layer.

LITERATURE CITED

1. Livingston, J. V., *Air Pollution: New Research*, Nova Science Publishers, New York, 2007.
2. Berglen, T. F., G. Myhre, I. Isaksen, V. Vestreng, S. J. Smith, Sulphate trends in Europe: Are we able to model the recent observed decrease?, *Tellus, Series B: Chemical and Physical Meteorology*, **59B**, 773–786, 2007.
3. Penkett, S. A., B. M. R. Jones, K. A. Brice, and A. E. J. Eggleton, The importance of atmospheric ozone and hydrogen peroxide in oxidising sulphur dioxide in cloud and rainwater, *Atmospheric Environment*, **41**, S154–S168, 2008.
4. Basu, S., Chemical and biochemical processes for NO_x control from combustion off-gases, *Chemical Engineering Communications*, **194**, 1374–1395, 2007.
5. Zhang, S.-H., L.-L. Cai, X.-H. Mi, J.-L. Jiang, and W. Li, NO_x removal from simulated flue gas by chemical absorption-biological reduction integrated approach in a biofilter, *Environmental Science and Technology*, **42**, 3814–3820, 2008.
6. Faulkner, W. B. and B. W. Shaw, Review of ammonia emission factors for United States animal agriculture, *Atmospheric Environment*, **42**, 6567–6574, 2008.
7. Bachu, S. and W. D. Gunter, Acid-gas injection in the Alberta Basin, Canada: A CO_2-storage experiment, *Geological Society Special Publication*, **233**, 225–234, 2004.

SUPPLEMENTARY REFERENCES

Ashmore, M. R., L. Emberson, and F. Murray, Eds., *Air Pollution Impacts on Crops and Forests,* Imperial College Press, London, 2003.

Balduino, S. P., Ed., *Progress in Air Pollution Research,* Nova Science Publishers, New York, 2007.

Baukal, C. E., *Industrial Combustion Pollution and Control,* Marcel Dekker, New York, 2004.

Bodine, C. G., Ed., *Air Pollution Research Advances,* Nova Science Publishers, New York, 2007.

Cheremisinoff, N. P., *Handbook of Air Pollution Prevention and Control,* Butterworth-Heinemann, Boston, 2002.

Colls, J., *Air Pollution,* 2nd ed., Spon Press, New York, 2003.

Cooper, C. D. and F. C. Alley, *Air Pollution Control: A Design Approach,* Waveland Press, Prospect Heights, IL, 2002.

Davis, W. T., Ed., *Air Pollution Engineering Manual,* 2nd ed., Wiley, New York, 2000.

Heck, R. M., R. J. Farrauto, and S. T. Gulati, *Catalytic Air Pollution Control: Commercial Technology,* 3rd ed., Wiley, New York, 2009.

Heumann, W. L., Ed., *Industrial Air Pollution Control Systems,* McGraw-Hill, New York, 1997.

Kidd, J. S. and R. A. Kidd, *Air Pollution: Problems and Solutions,* Chelsea House, New York, 2006.

Lunt, R. R., A. D. Little, and J. D. Cunic, *Profiles in Flue Gas Desulfurization,* American Institute of Chemical Engineers, New York, 2000.

Metcalfe, S., and D. Derwent, *Atmospheric Pollution and Environmental Change,* Oxford University Press, New York, 2005.

Romano, G. C. and A. G. Conti, Eds., *Air Quality in the 21st Century,* Nova Science Publishers, New York, 2008.

Schnelle, K. B., C. A. Brown, and F. Kreith, Eds., *Air Pollution Control Technology Handbook,* CRC Press, Boca Raton, FL, 2002.

Stevens, L. B., W. L. Cleland, and E. Roberts Alley, *Air Quality Control Handbook,* McGraw-Hill, New York, 1998.

Thad, G., *Air Quality,* 4th ed., Taylor & Francis/CRC Press, Boca Raton, FL, 2004.

Tomita, A., *Emissions Reduction: NO_x/SO_x Suppression,* Elsevier, New York, 2001.

Vallero, D. A., *Fundamentals of Air Pollution,* 4th ed., Elsevier, Amsterdam, 2008.

Wang, L. K., N. C. Pereira, and Y.-T. Hung, Eds., *Air Pollution Control Engineering,* Humana Press, Totowa, NJ, 2004.

Warner, C. F., W. T. Davis, and K. Wark, *Air Pollution: Its Origin and Control,* 3rd ed., Addison-Wesley, Reading, MA, 1997.

QUESTIONS AND PROBLEMS

The use of internet resources is assumed in answering any of the questions. These would include such things as constants and conversion factors as well as additional information needed to complete an answer.

1. Why is it that "highest levels of carbon monoxide tend to occur in congested urban areas at times when the maximum number of people are exposed?"
2. Which unstable, reactive species is responsible for the removal of CO from the atmosphere?
3. Which of the following fluxes in the atmospheric sulfur cycle is smallest: (a) sulfur species washed out in rainfall over land, (b) sulfates entering the atmosphere as "sea salt," (c) sulfur species entering the atmosphere from volcanoes, (d) sulfur species entering the atmosphere from fossil fuels, (e) hydrogen sulfide entering the atmosphere from biological processes in coastal areas and on land?
4. Of the following agents, the one that would not favor conversion of sulfur dioxide to sulfate species in the atmosphere is: (a) ammonia, (b) water, (c) contaminant reducing agents, (d) ions of transition metals such as manganese, (e) sunlight.
5. Of the stack gas scrubber processes discussed in this chapter, which is the least efficient for the removal of SO_2?
6. The air inside a garage was found to contain 10 ppm CO by volume at standard temperature and pressure (STP). What is the concentration of CO in mg/L and in ppm by mass?
7. How many metric tons of 5%-S coal would be needed to yield the H_2SO_4 required to produce a 3.00-cm rainfall of pH 2.00 over a 100-km^2 area?
8. In what major respect is NO_2 a more significant species than SO_2 in terms of participation in atmospheric chemical reactions?
9. Assume that an incorrectly adjusted lawn mower is operated in a garage such that the combustion reaction in the engine is

$$C_8H_{18} + \tfrac{17}{2}O_2 \rightarrow 8CO + 9H_2O$$

If the dimensions of the garage are $5 \times 3 \times 3$ m, how many grams of gasoline must be burned to raise the level of CO in the air to 1000 ppm by volume at STP?
10. A 12.0-L sample of waste air from a smelter process was collected at 25°C and 1.00 atm pressure, and the sulfur dioxide was removed. After SO_2 removal, the volume of the air sample was 11.50 L. What was the percentage by mass of SO_2 in the original sample?
11. What is the oxidant in the Claus reaction? What is the commercial product of this reaction?
12. Carbon monoxide is present at a level of 10 ppm by volume in an air sample taken at 15°C and 1.00 atm pressure. At what temperature (at 1.00 atm pressure) would the sample also contain 10 mg/m^3 of CO?
13. How many metric tons of coal containing an average of 2% S are required to produce the SO_2 emitted by fossil fuel combustion shown in Figure 11.1? (Note that the values given in the figure are in terms of elemental sulfur, S.) How many metric tons of SO_2 are emitted? How does this amount of coal compare to that currently used in the world?
14. Assume that the wet limestone process requires 1 metric ton of $CaCO_3$ to remove 90% of the sulfur from 4 metric tons of coal containing 2% S. Assume that the sulfur product is $CaSO_4$. Calculate the percentage of the limestone converted to calcium sulfate.
15. Referring to the two preceding problems, calculate the number of metric tons of $CaCO_3$ required each year to remove 90% of the sulfur from 1 billion metric tons of coal (approximate annual U.S. consumption), assuming an average of 2% sulfur in the coal.

16. If a power plant burning 10,000 metric tons of coal per day with 10% excess air emits stack gas containing 100 ppm by volume of NO, what is the daily output of NO? Assume the coal is pure carbon.
17. How many cubic kilometers of air at 25°C and 1 atm pressure would be contaminated to a level of 0.5 ppm NO_x from the power plant discussed in the preceding question?
18. Match the following pertaining to gaseous inorganic air pollutants:

A. CO	1. Produced in internal combustion engines as a precursor to photochemical smog formation
B. O_3	2. Formed in connection with photochemical smog
C. SO_2	3. Not associated particularly with smog or acid rain formation, but of concern because of its direct toxic effects
D. NO	4. Does not cause smog to form, but is a precursor to acid rain

19. Of the following, the one that is *not* an inorganic pollutant gas is (explain): (A) Benzo(a)-pyrene, (B) SO_2, (C) NO, (D) NO_2, (E) H_2S.
20. Of the following, the statement that is *not* true regarding carbon monoxide in the atmosphere is (explain): (A) It is produced in the stratosphere by a process that starts with H abstraction from CH_4 by $HO^•$, (B) it is removed from the atmosphere largely by reaction with hydroxyl radical, (C) it is removed from the atmosphere in part by its being metabolized by soil microorganisms, (D) it has some natural, as well as pollutant, sources, (E) at its average concentration in the global atmosphere, it is probably a threat to human health.

12 Organic Air Pollutants

12.1 ORGANIC COMPOUNDS IN THE ATMOSPHERE

Organic pollutants may have a strong effect upon atmospheric quality. The effects of organic pollutants in the atmosphere may be divided into two major categories. The first consists of *direct effects*, such as cancer caused by exposure to vinyl chloride. The second is the formation of *secondary pollutants*, especially photochemical smog (discussed in detail in Chapter 13). In the case of pollutant hydrocarbons in the atmosphere, the latter is the more important effect. In some localized situations, particularly the workplace, direct effects of organic air pollutants may be equally important.

This chapter discusses the nature and distribution of organic compounds in the atmosphere. Chapter 13 deals with photochemical smog and addresses the mechanisms by which organic compounds undergo photochemical reactions in the atmosphere.

12.1.1 Loss of Organic Substances from the Atmosphere

Organic contaminants are lost from the atmosphere by a number of routes. These include dissolution in precipitation (rainwater), dry deposition, photochemical reactions, formation of and incorporation into particulate matter, and uptake by plants. Reactions of organic atmospheric contaminants are particularly important in determining their manner and rates of loss from the atmosphere. Such reactions are discussed in this chapter.

Forest trees present a large surface area to the atmosphere and are particularly important in filtering organic contaminants from air. Forest trees and plants contact the atmosphere through plant cuticle layers, the biopolymer "skin" on the leaves, and needles of the plants. The cuticle layer is lipophilic, meaning that it has a particular affinity for organic substances, including those in the atmosphere. Uptake increases with increasing lipophilicity of the compounds and with increasing surface area of the leaves. This phenomenon points to the importance of forests in atmospheric purification and illustrates an important kind of interaction between the atmosphere and the biosphere.

12.1.2 Global Distillation and Fractionation of Persistent Organic Pollutants (POP)

Persistent organic pollutants (POP), compounds that are resistant to chemical and biochemical degradation, are significant atmospheric contaminants.[1] On a global scale, it is likely that POP undergo a cycle of distillation and fractionation in which they are vaporized into the atmosphere in warmer regions of the Earth and condense and are deposited in colder regions. The theory of this phenomenon holds that the distribution of such pollutants is governed by their physicochemical properties and the temperature conditions to which they are exposed. As a result, the least volatile POP are deposited near their sources, those of relatively high volatility are distilled into polar regions, and those of intermediate volatility are deposited predominantly at mid-latitudes. This phenomenon has some important implications regarding the accumulation of POP in environmentally fragile polar regions far from industrial sources.

12.2 BIOGENIC ORGANIC COMPOUNDS

Biogenic organic compounds in the atmosphere are those produced by organisms. Biogenic compounds are abundant in the atmosphere of forest regions and are significant participants in the atmospheric chemistry of these regions.[2] Natural sources are the most important contributors of organics to the atmosphere, and hydrocarbons generated and released by human activities constitute only about one-seventh of the total hydrocarbons in the atmosphere. The release to the atmosphere of organic compounds by organisms is a very important kind of interaction between the atmosphere and the biosphere. Other than methane released primarily by bacteria (see below), the greatest source of biogenic organic compounds in the atmosphere is vegetation. Various plants release hydrocarbons, including isoprene, $C_{10}H_{16}$ monoterpenes, and $C_{15}H_{24}$ sesquiterpenes. Oxygenated compounds are released in smaller quantities, but in a great variety, including alcohols, such as methanol and 2-methyl-3-buten-2-ol; ketones, such as 6-methyl-5-hepten-2-one; and hexene derivatives.

The heavy preponderance of biogenic organic compounds in the atmosphere is in large part the result of the huge quantities of methane produced by anoxic bacteria in the decomposition of organic matter in water, sediments, and soil:

$$2\{CH_2O\}(\text{bacterial action}) \rightarrow CO_2(g) + CH_4(g) \tag{12.1}$$

Flatulent emissions from domesticated animals, arising from bacterial decomposition of food in their digestive tracts, add about 85 million metric tons of methane to the atmosphere each year. Anoxic conditions in intensively cultivated rice fields produce large amounts of methane, perhaps as much as 100 million metric tons per year. Methane is a natural constituent of the atmosphere and is present at a level of about 1.8 ppm in the troposphere.

Methane in the troposphere contributes to the photochemical production of carbon monoxide and ozone. The photochemical oxidation of methane is a major source of water vapor in the stratosphere.

As the most important natural source of nonmethane biogenic compounds, it is possible that vegetation emits thousands of different organic compounds to the atmosphere. Other natural sources include microorganisms, forest fires, animal wastes, and volcanoes.

One of the simplest organic compounds given off by plants is ethylene, C_2H_4. This compound is produced by a variety of plants and released to the atmosphere in its role as a messenger species regulating plant growth. Because of its double bond, ethylene is highly reactive with hydroxyl radical, $HO^\bullet$, and with oxidizing species in the atmosphere. Ethylene from vegetation sources is an active participant in atmospheric chemical processes.

Most of the hydrocarbons emitted by plants are *terpenes*, which constitute a large class of organic compounds found in essential oils. Essential oils are obtained when parts of some types of plants are subjected to steam distillation. Most of the plants that produce terpenes are conifers (evergreen trees and shrubs such as pine and cypress), plants of the genus *Myrtus*, and trees and shrubs of the genus *Citrus*. One of the most common terpenes emitted by trees is α-pinene, a principal component of turpentine. The terpene limonene, found in citrus fruit and pine needles, is encountered in the atmosphere around these sources. Isoprene (2-methyl-1,3-butadiene), a hemiterpene, has been identified in the emissions from cottonwood, eucalyptus, oak, sweetgum, and white spruce trees. Linalool is a terpene with the chemical formula $(CH_3)_2C{=}CHCH_2CH_2C(CH_3)(OH)CH{=}CH_2$, which is given off by some plant species common to Italy and Austria, including the pine *Pinuspinea* and orange blossoms. Other terpenes that are given off by trees include β-pinene, myrcene, ocimene, and α-terpinene.

As exemplified by the structural formulas of α-pinene, β-pinene, D^3-carene, isoprene, and limonene, shown in Figure 12.1, terpenes contain alkenyl (olefinic) bonds, in some cases, two or more per molecule. Because of these and other structural features, terpenes are among the most reactive compounds in the atmosphere. The reaction of terpenes with hydroxyl radical, $HO^\bullet$, is very rapid, and terpenes also react with other oxidizing agents in the atmosphere, particularly ozone, O_3, and nitrate radical, NO_3. Turpentine, a mixture of terpenes, has been widely used in paint because it reacts with atmospheric oxygen to form a peroxide, then a hard resin. Terpenes such as α-pinene

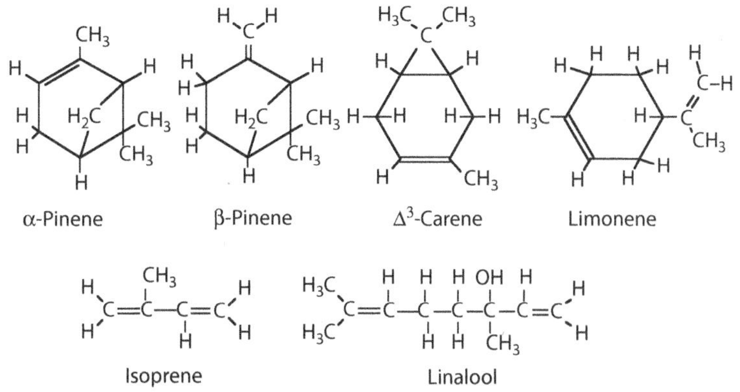

FIGURE 12.1 Some common terpenes emitted to the atmosphere by vegetation, primarily trees such as pine and citrus trees. These reactive compounds are involved in the formation of much of the small particulate matter encountered in the atmosphere.

and isoprene undergo similar reactions in the atmosphere to form particulate matter. Identified as *secondary organic aerosol* (SOA), the resulting Aitken nuclei aerosols (see Chapter 10) are a cause of the blue haze in the atmosphere above some heavy growths of vegetation. These SOA particles are a large part of atmospheric fine particulate matter, especially in forested areas.[3]

Sesquiterpenes consisting of three isoprene units and having the molecular formula $C_{15}H_{24}$ comprise an important class of terpenes. A common example of a sesquiterpene is δ-cadinene:

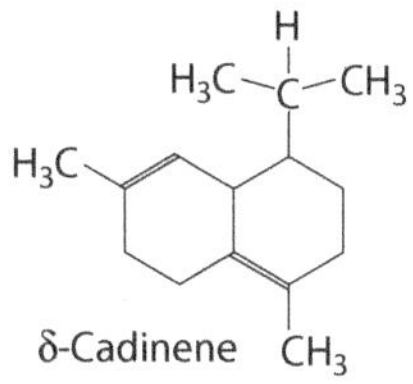

Laboratory and smog-chamber experiments have been performed in an effort to determine the fates of atmospheric terpenes. Oxidation initiated by reaction with NO_3 of the four cyclic monoterpenes listed above, α-pinene, β-pinene, Δ^3-carene, and limonene, gives products containing carbonyl (C=O) functionality and organically bound nitrogen as organic nitrate. When a mixture of α-pinene with NO and NO_2 in air is irradiated with ultraviolet light, pinonic acid is formed:

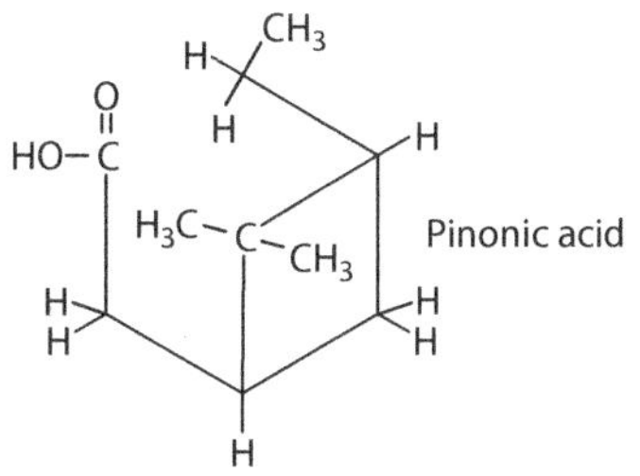

Found in forest aerosol particles, this compound is produced by photochemical processes acting upon α-pinene.

In addition to products of reactions with NO_3 and hydroxyl radical, a significant fraction of the atmospheric aerosol formed as the result of reactions of unsaturated biogenic hydrocarbons results from reactions between them and ozone. Pinonic acid (see above) is produced by the reaction of α-pinene with ozone. Two of the products of the reaction of limonene with ozone are formaldehyde and 4-acetyl-1-methylcyclohexene:

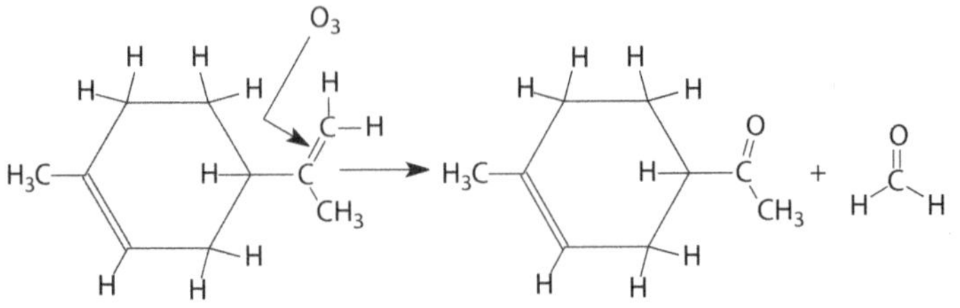

Limonene 4-Acetyl-1-methylcyclohexene

Perhaps the greatest variety of compounds emitted by plants consists of *esters*. However, they are released in such small quantities that they have little influence upon atmospheric chemistry. Esters are primarily responsible for the fragrances associated with much vegetation. Some typical esters that are released by plants to the atmosphere are shown below:

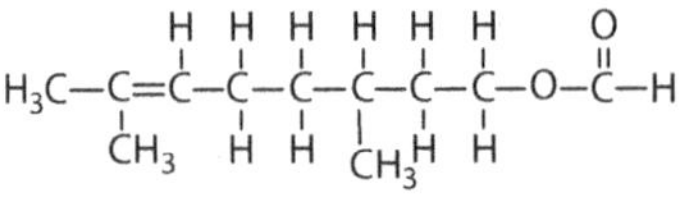

Citronellyl formate

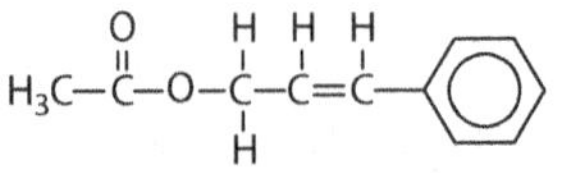

Cinnamyl acetate

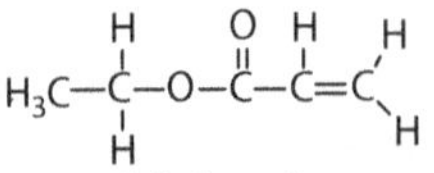

Ethyl acrylate

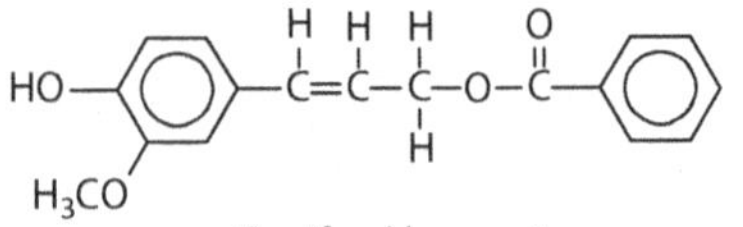

Coniferyl benzoate

12.2.1 Removal of Atmospheric Organic Compounds by Plants

In addition to being sources of atmospheric organic compounds as discussed above, plants are also repositories of POP thus playing a significant role in their environmental fate and transport, as noted in Section 12.1. The leaves, needles (on pine trees), and stems of higher green plants are covered with an epicuticular wax that is organophilic and thus has an affinity for organic compounds in air. The most important plants in this respect are those in the evergreen boreal coniferous forests in the northern temperate zone. The importance of these forests is due to the heavy forestation of the northern temperate zone and the high amount of leaf surface per unit area of such forests.

12.3 POLLUTANT HYDROCARBONS

Ethylene and terpenes, which were discussed in the preceding section, are *hydrocarbons*, organic compounds containing only hydrogen and carbon. The major classes of hydrocarbons are *alkanes*, such as 2,2,3-trimethylbutane:

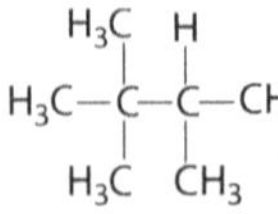

2,2,3-Trimethylbutane

alkenes (compounds with double bonds between adjacent carbon atoms), such as ethylene, *alkynes* (compounds with triple bonds), such as acetylene:

$$H{-}C{\equiv}C{-}H$$

and *aromatic* (aryl) *compounds*, such as naphthalene:

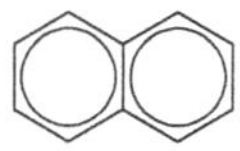

Because of their widespread use in fuels, hydrocarbons predominate among organic atmospheric pollutants. Petroleum products, primarily gasoline, are the source of most of the anthropogenic (originating through human activities) pollutant hydrocarbons found in the atmosphere. Hydrocarbons may enter the atmosphere either directly or as by-products of the partial combustion of other hydrocarbons. The latter are particularly important because they tend to be unsaturated and relatively reactive (see Chapter 13 for a discussion of hydrocarbon reactivity in photochemical smog formation). Most hydrocarbon pollutant sources produce about 15% reactive hydrocarbons, whereas those from incomplete combustion of gasoline are about 45% reactive. The hydrocarbons in uncontrolled automobile exhausts are only about 1/3 alkanes, with the remainder divided approximately equally between more reactive alkenes and aromatic hydrocarbons, thus accounting for the relatively high reactivity of automotive exhaust hydrocarbons.

Alkanes are among the more stable hydrocarbons in the atmosphere. Straight-chain alkanes with one to more than 30 carbon atoms, and branched-chain alkanes with six or fewer carbon atoms, are commonly present in polluted atmospheres. Because of their high vapor pressures, alkanes with six or fewer carbon atoms are normally present as gases, alkanes with 20 or more carbon atoms are present as aerosols or sorbed to atmospheric particles, and alkanes with 6–20 carbon atoms per molecule may be present either as vapor or particles, depending upon conditions.

In the atmosphere, alkanes (general formula C_xH_{2x+2}) are attacked primarily by hydroxyl radical, $HO^\bullet$, resulting in the loss of a hydrogen atom and formation of an *alkyl radical*:

$$C_xH_{2x+1}^\bullet$$

Subsequent reaction with O_2 causes formation of *alkylperoxyl radical*:

$$C_xH_{2x+1}O_2^\bullet$$

These radicals may act as oxidants, losing oxygen (usually to NO forming NO_2) to produce *alkoxyl radicals*:

$$C_xH_{2x+1}O^\bullet$$

As a result of these and subsequent reactions, lower-molecular-mass alkanes are eventually oxidized to species that can be precipitated from the atmosphere with particulate matter to ultimately undergo biodegradation in soil.

Alkenes enter the atmosphere from a variety of processes, including emissions from internal combustion engines and turbines, foundry operations, and petroleum refining. Several alkenes, including the ones shown below, are among the top 50 chemicals produced each year, with annual worldwide production of several billion kilograms:

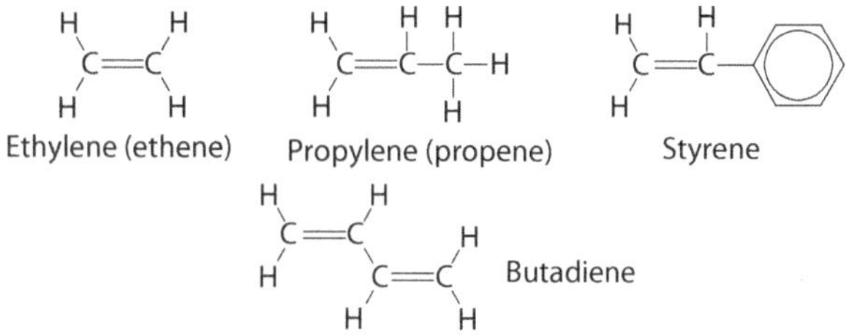

Ethylene (ethene) Propylene (propene) Styrene

Butadiene

These compounds are used primarily as monomers, which are polymerized to create polymers for plastics (polyethylene, polypropylene, and polystyrene), synthetic rubber (styrenebutadiene and polybutadiene), latex paints (styrenebutadiene), and other applications. All of these compounds, as well as others manufactured in lesser quantities, are released into the atmosphere. In addition to the direct release of alkenes, these hydrocarbons are commonly produced by the partial combustion and "cracking" at high temperatures of alkanes, particularly in the internal combustion engine.

Alkynes occur much less commonly in the atmosphere than do alkenes. Detectable levels are sometimes found of acetylene used as a fuel for welding torches and 1-butyne used in synthetic rubber manufacture:

$$\underset{\text{Acetylene}}{\text{H}-\text{C}\equiv\text{C}-\text{H}} \qquad \underset{\text{1-Butyne}}{\text{H}-\text{C}\equiv\text{C}-\text{CH}_2-\text{CH}_3}$$

Unlike alkanes, alkenes are highly reactive in the atmosphere, especially in the presence of NO_x and sunlight. Hydroxyl radical reacts readily with alkenes, adding to the double bond and (rarely) by abstracting a hydrogen atom. If hydroxyl radical adds to the double bond in propylene, for example, the product is:

$$\text{HO}-\text{CH}_2-\dot{\text{C}}\text{H}-\text{CH}_3$$

Addition of molecular O_2 to this radical results in the formation of a peroxyl radical:

$$\text{HO}-\text{CH}_2-\text{CH}(\text{OO}^\bullet)-\text{CH}_3$$

These radicals then react with HOO•, alkylperoxyl radicals (ROO•), or NO, depending upon the availability of these species, leading to the formation of aldehydes and other reactive species that can participate in reaction chains, such as those discussed for the formation of photochemical smog in Chapter 13.

Ozone, O_3, adds across double bonds and is rather reactive with alkenes. As shown for the reaction with ozone of the natural alkene limonene in Section 12.2, aldehydes are among the products of reactions between alkenes and ozone.

Although the reaction of alkenes with NO_3 is much slower than that with HO•, the much higher levels of NO_3 relative to HO•, especially at night, make it a significant reactant for atmospheric alkenes. (The NO_3 radical is present in the atmosphere at appreciable levels only at night because it undergoes photolysis very rapidly when exposed to sunlight.) The initial reaction with NO_3 is addition across the alkene double bond which, because NO_3 is a free radical species, forms a radical species. A typical reaction sequence is the following:

$$\begin{aligned}&\text{CH}_3-\text{CH}=\text{CH}_2 + \text{NO}_3 \longrightarrow \text{CH}_3-\dot{\text{C}}\text{H}-\text{CH}_2-\text{ONO}_2 \xrightarrow{\text{O}_2} \text{CH}_3-\text{CH}(\text{OO}^\bullet)-\text{CH}_2-\text{ONO}_2\\ &\xrightarrow{\text{NO}} \text{CH}_3-\text{CH}(\text{O}^\bullet)-\text{CH}_2-\text{ONO}_2 \longrightarrow \text{CH}_3-\text{C}(=\text{O})-\text{H} + \text{H}-\text{C}(=\text{O})-\text{H} + \text{NO}_2\end{aligned} \qquad (12.2)$$

The reaction of alkenes with hydroxyl radical in the presence of nitrogen oxides can produce β-hydroxynitrates and dihydroxynitrates.[4] An example of the formation of a β-hydroxynitrate from a 1-alkene is shown in Reaction 12.3.

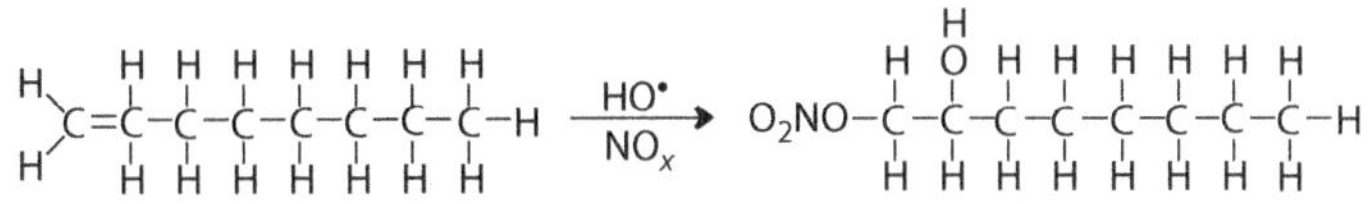

$$H_2C{=}CH{-}(CH_2)_6{-}H \xrightarrow[NO_x]{HO^\bullet} O_2NO{-}CH_2{-}CH(OH){-}(CH_2)_6{-}H \tag{12.3}$$

12.3.1 Aromatic Hydrocarbons

Aromatic (aryl) hydrocarbons may be divided into the two major classes of those that have only one benzene ring and those with multiple rings. As discussed in Chapter 10, the latter are PAH. Aromatic hydrocarbons with two rings, such as naphthalene, are intermediate in their behavior. Some typical aromatic hydrocarbons are:

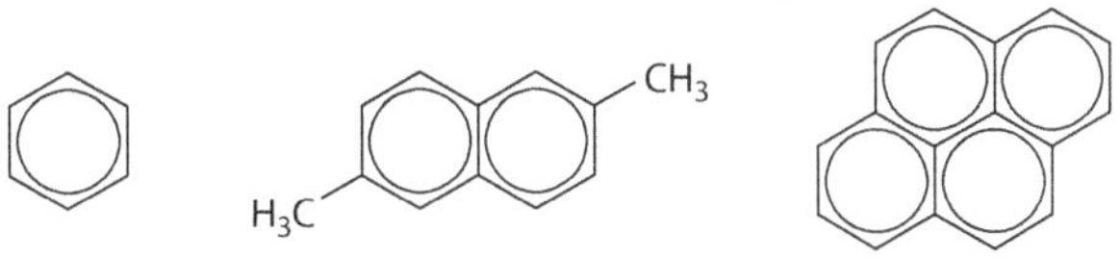

Benzene 2,6-Dimethylnaphthalene Pyrene

The following aromatic hydrocarbons are among the top 50 chemicals manufactured each year:

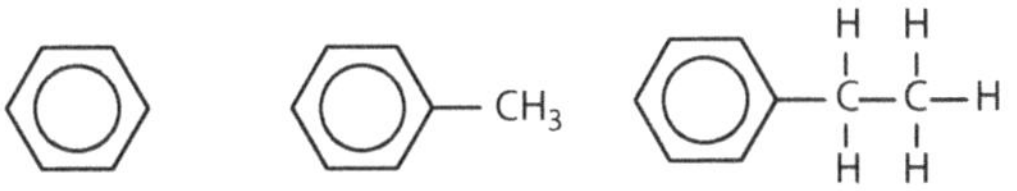

Benzene Toluene Ethylbenzene

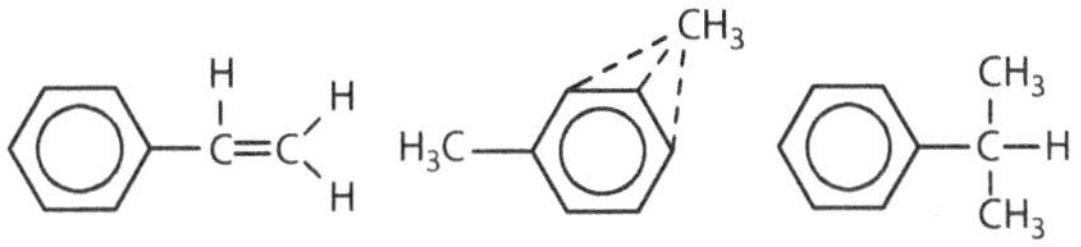

Styrene Xylene (3 isomers) Cumene

Single-ring aromatic compounds are important constituents of gasoline, although the benzene content of gasoline is restricted because of its possible health effects. Aromatic hydrocarbons are raw materials for the manufacture of monomers and plasticizers in polymers. Styrene is a monomer used in the manufacture of plastics and synthetic rubber. Cumene is oxidized to produce phenol and acetone, which are valuable by-products. Because of these applications, plus production of these compounds as combustion by-products, aromatic compounds are common atmospheric pollutants. A group of single-ring aromatic hydrocarbons that are found as urban air pollutants are those in the group known as BTEX consisting of benzene, toluene, ethylbenzene, *o*-xylene, *m*-xylene, *p*-xylene.[5]

Many hydrocarbons containing a single benzene ring and a number of hydrocarbon derivatives of naphthalene have been found as atmospheric pollutants. In addition, several compounds containing two or more *unconjugated* rings (not sharing the same π electron cloud between rings) have been detected as atmospheric pollutants. These compounds have been detected in tobacco smoke and biphenyl has been found in diesel engine smoke.

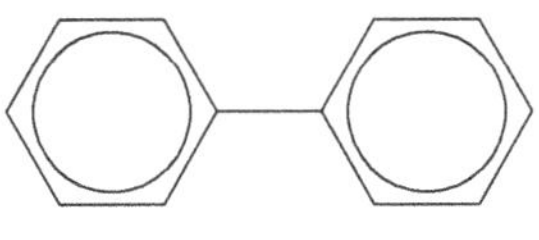

Biphenyl

As discussed in Section 10.8, PAHs are present as aerosols in the atmosphere because of their extremely low vapor pressures. These compounds are the most stable forms of hydrocarbons having low hydrogen-to-carbon ratios and are formed by the combustion of hydrocarbons under oxygen-deficient conditions. The partial combustion of coal, which has a hydrogen-to-carbon ratio <1, is a major source of PAH compounds. In addition to being formed in the anthrosphere by incomplete combustion of carbonaceous fuels, PAHs are produced and emitted to the atmosphere by grass and forest fires.[6]

12.3.2 Reactions of Atmospheric Aromatic Hydrocarbons

As with most atmospheric hydrocarbons, the most likely reaction of benzene and its derivatives is with hydroxyl radical. Addition of $HO^{\bullet}$ to the benzene ring results in the formation of an unstable radical species,

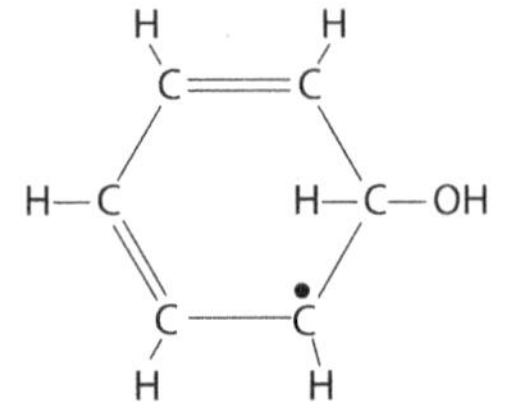

H H
C=C
H—C H—C—OH
C—Ċ
H H

where the dot denotes an unpaired electron. The electron is not confined to one atom; therefore, it is *delocalized* and may be represented in the aromatic radical structure by a half-circle with a dot in the middle. Using this notation for the radical above, its reaction with O_2 is

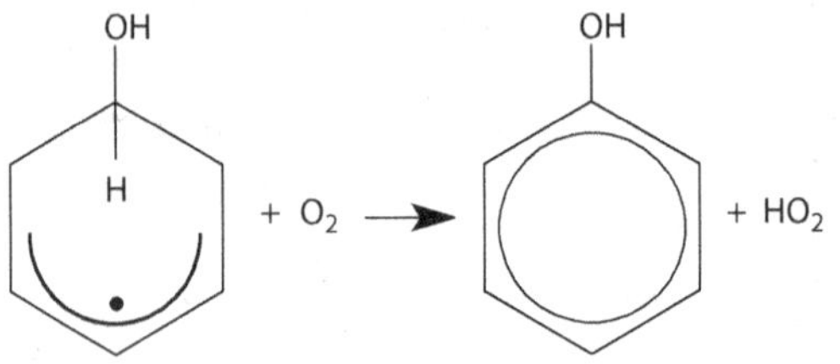

OH OH
H + O_2 → + HO_2 (12.4)

to form stable phenol and reactive hydroperoxyl radical, $HOO^{\bullet}$. Alkyl-substituted aromatics may undergo reactions involving the alkyl group. For example, abstraction of alkyl H by $HO^{\bullet}$ from a compound such as *p*-xylene can result in the formation of a radical

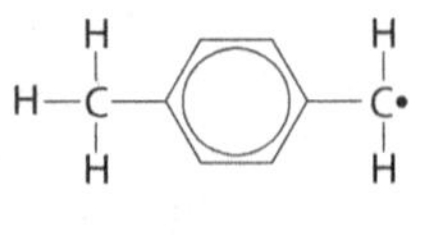

H H
H—C—(benzene ring)—C•
H H

which can react further with O_2 to form a peroxyl radical, and then enter chain reactions involved in the formation of photochemical smog (Chapter 13).

Although reaction with hydroxyl radical is the most common fate of aromatic compounds during daylight, they react with NO_3 at night. This oxide of nitrogen is formed by the reaction of ozone with NO_2:

$$NO_2 + O_3 \rightarrow NO_3 + O_2 \quad (12.5)$$

and can remain in the atmosphere for some time as its addition product with NO_2:

$$NO_2 + NO_3 + M \rightarrow N_2O_5 + M \quad (12.6)$$

12.4 CARBONYL COMPOUNDS: ALDEHYDES AND KETONES

Carbonyl compounds, consisting of aldehydes and ketones that have a carbonyl moiety, C=O, are often the first species formed after unstable reaction intermediates in the photochemical oxidation of atmospheric hydrocarbons. Carbonyls are very important in atmospheric chemistry because (1) they are formed by the photochemical oxidation of almost all hydrocarbons; (2) they lead to production of very reactive and harmful free radicals, ozone, and peroxyacylnitrates; and (3) some carbonyls, especially formaldehyde, acetaldehyde, and acrolein, are toxic mutagens, potential carcinogens, and eye irritants. The general formulas of aldehydes and ketones are represented by the following, where R and R′ represent the hydrocarbon *moieties* (portions), such as the $-CH_3$ group.

$$R{-}\overset{\overset{\Large O}{\|}}{C}{-}H \qquad R{-}\overset{\overset{\Large O}{\|}}{C}{-}R' \qquad {-}\overset{\overset{\Large O}{\|}}{C}{-}$$

Aldehyde   Ketone   Carbonyl moiety

Carbonyl compounds are by-products of the generation of hydroperoxyl radicals from organic alkoxyl radicals (see Section 12.3) by reactions such as the following:

$$H_3C{-}CH(O^{\bullet}){-}CH_3 + O_2 \rightarrow H_3C{-}\overset{\overset{\Large O}{\|}}{C}{-}CH_3 + HO_2^{\bullet} \tag{12.7}$$

$$H_3C{-}CH_2{-}CH_2O^{\bullet} + O_2 \rightarrow H_3C{-}CH_2{-}\overset{\overset{\Large O}{\|}}{C}{-}H + HO_2^{\bullet} \tag{12.8}$$

The simplest and most widely produced of the carbonyl compounds is the lowest aldehyde, *formaldehyde*:

$$H{-}\overset{\overset{\Large O}{\|}}{C}{-}H \quad \text{Formaldehyde}$$

Formaldehyde is produced in the atmosphere as a product of the reaction of atmospheric hydrocarbons beginning with their reactions with hydroxyl radical. For example, formaldehyde is the product of the reaction of methoxyl radical with O_2:

$$H_3CO^{\bullet} + O_2 \rightarrow H{-}\overset{\overset{\Large O}{\|}}{C}{-}H + HOO^{\bullet} \tag{12.9}$$

With annual global industrial production exceeding 1 billion kg, formaldehyde is used in the manufacture of plastics, resins, lacquers, dyes, and explosives. It is uniquely important because of its widespread distribution and toxicity. Humans may be exposed to formaldehyde in the manufacture and use of phenol, urea, and melamine resin plastics, and from formaldehyde-containing adhesives in pressed wood products such as particle board, used in especially large quantities in mobile home construction. However, significantly improved manufacturing processes have greatly reduced formaldehyde emissions from these synthetic building materials. Formaldehyde occurs in the atmosphere primarily in the gas phase.

The structural formulas of some important aldehydes and ketones are shown below:

```
    H  O      H     H  O         H  O  H         H  O  H  H
    |  ||      \    |  ||        |  || |         |  || |  |
 H—C—C—H      C=C—C—H        H—C—C—C—H       H—C—C—C—C—H
    |         /                  |     |         |     |  |
    H        H                   H     H         H     H  H
 Acetaldehyde   Acrolein         Acetone       Methylethyl ketone
```

Acetaldehyde is a widely produced organic chemical used in the manufacture of acetic acid, plastics, and raw materials. Approximately 1 billion kg of acetone are produced each year as a solvent and for applications in the rubber, leather, and plastics industries. Methylethyl ketone is employed as a low-boiling solvent for coatings and adhesives, and for the synthesis of other chemicals.

In addition to their production from hydrocarbons by photochemical oxidation, carbonyl compounds enter the atmosphere from a large number of sources and processes. These include direct emissions from internal combustion engine exhausts, incinerator emissions, spray painting, polymer manufacture, printing, petrochemicals manufacture, and lacquer manufacture. Formaldehyde and acetaldehyde are produced by microorganisms, and acetaldehyde is emitted by some kinds of vegetation.

Aldehydes are second only to NO_2 as atmospheric sources of free radicals produced by the absorption of light. This is because the carbonyl group is a *chromophore*, a molecular group that readily absorbs light. It absorbs well in the near-ultraviolet region of the spectrum. The activated compound produced when a photon is absorbed by an aldehyde dissociates into a formyl radical,

$$H\dot{C}O$$

and an alkyl radical. The photodissociation of acetaldehyde illustrates this two-step process:

```
    H  O                  H  O                     H        O
    |  ||                 |  || *                  |        ||
 H—C—C—H + hv  ⟶   H—C—C—H          ⟶   H—C•  +  •C—H        (12.10)
    |                     |  (Photoexcited)        |
    H                     H                        H
```

Photolytically excited formaldehyde, CH_2O^*, may dissociate in two ways. The first of these produces an H atom and HCO radical; the second produces chemically stable H_2 and CO.

As a result of their reaction with $HO^\bullet$ followed by O_2 then NO_2, aldehydes are precursors to the production of strongly oxidizing peroxyacyl nitrates (PANs) such as peroxyacetyl nitrate and peroxypropionyl nitrate. This process is discussed in Chapter 13, Section 13.5.

Because of the presence of both double bonds and carbonyl groups, alkenyl aldehydes are especially reactive in the atmosphere. The most common of these found in the atmosphere is acrolein,

```
 H     H  O
  \    |  ||
   C=C—C—H
  /
 H          Acrolein
```

a powerful lachrymator (tear producer) used as an industrial chemical and produced as a combustion by-product. The presence of significant amounts of acrolein in the atmosphere has been attributed to atmospheric reactions operating on 2-furaldehyde, a biomass burning indicator that occurs in wood smoke.[7]

The most abundant atmospheric ketone is acetone, $CH_3C(O)CH_3$. About half of the acetone in the atmosphere is generated as a product of the atmospheric oxidation of propane, isobutane, isobutene, and other hydrocarbons. Most of the remainder comes about equally from direct biogenic emissions and biomass burning, with about 3% from direct anthropogenic emissions.

Acetone photolyzes in the atmosphere,

$$H_3C\text{—}C(=O)\text{—}CH_3 + h\nu \rightarrow H_3C\text{—}\dot{C}(=O) + \cdot CH_3 \qquad (12.11)$$

to produce the PAN precursor acetyl radical. It is believed that the mechanism for the removal of the higher ketones from the atmosphere involves an initial reaction with $HO^\bullet$ radical.

Carbonyls are commonly found in ambient air associated with severe incidents of photochemical smog formation. For the most part, these compounds are produced as secondary pollutants from the photochemical oxidation of hydrocarbons.

12.5 MISCELLANEOUS OXYGEN-CONTAINING COMPOUNDS

Oxygen-containing aldehydes, ketones, and esters in the atmosphere were covered in preceding sections. This section discusses the oxygen-containing organic compounds consisting of *aliphatic alcohols*, *phenols*, *ethers*, and *carboxylic acids*. These compounds have the general formulas given below, where R and R′ represent hydrocarbon moieties, and Ar stands specifically for an aryl moiety, such as the phenyl group (benzene less an H atom):

R—OH	Ar—OH	R—O—R′	R—C(=O)—OH
Aliphatic alcohols	Phenols	Ethers	Carboxylic acids

These classes of compounds include many important organic chemicals.

12.5.1 ALCOHOLS

Of the alcohols, methanol, ethanol, isopropanol, and ethylene glycol rank among the top 50 chemicals with annual worldwide production of the order of 1 billion kg or more. The most common of the many uses of these chemicals is for the manufacture of other chemicals. Methanol is widely used in the manufacture of formaldehyde (see Section 12.4) as a solvent, and mixed with water as an antifreeze formulation. Ethanol is used as a solvent and as the starting material for the manufacture of acetaldehyde, acetic acid, ethyl ether, ethyl chloride, ethyl bromide, and several important esters. Both methanol and ethanol can be used as motor vehicle fuels, usually in mixtures with gasoline. Ethylene glycol is a common antifreeze compound.

Numerous aliphatic alcohols have been reported in the atmosphere. Because of their volatility, the lower alcohols, especially methanol and ethanol, predominate as atmospheric pollutants. Among the other alcohols released into the atmosphere are 1-propanol, 2-propanol, propylene glycol, 1-butanol, and even octadecanol [chemical formula: $CH_3(CH_2)_{16}CH_2OH$], which is evolved by plants. Alcohols can undergo photochemical reactions, beginning with abstraction of hydrogen by hydroxyl radical. Mechanisms for scavenging alcohols from the atmosphere are relatively efficient because the lower alcohols are quite water soluble and the higher ones have low vapor pressures.

Some alkenyl alcohols have been found in the atmosphere, largely as by-products of combustion. Typical of these is 2-buten-1-ol,

$$H_3C\text{—}CH\text{=}CH\text{—}CH_2\text{—}OH$$

which has been detected in automobile exhausts. Some alkenyl alcohols are emitted by plants. One of these, *cis*-3-hexen-1-ol, $CH_3CH_2CH=CHCH_2CH_2OH$, is emitted from grass, trees, and crop plants and is known as "leaf alcohol." In addition to reacting with $HO^•$ radical, alkenyl radicals react strongly with atmospheric ozone, which adds across the double bond.

12.5.2 Phenols

Phenols are aromatic alcohols that have an –OH group bonded to an aryl ring. They are more noted as water pollutants than as air pollutants. Some typical phenols that have been reported as atmospheric contaminants are the following:

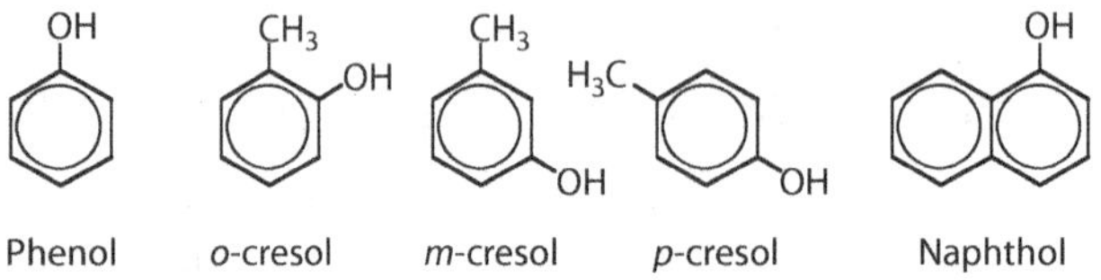

Phenol *o*-cresol *m*-cresol *p*-cresol Naphthol

The simplest of these compounds, phenol, is among the top 50 chemicals produced annually. It is most commonly used in the manufacture of resins and polymers, such as bakelite, a phenol–formaldehyde copolymer. Phenols are produced by the pyrolysis of coal and are major by-products of coking. Thus, in local situations involving coal coking and similar operations, phenols can be troublesome air pollutants.

12.5.3 Ethers

Ethers are relatively uncommon atmospheric pollutants; however, the flammability hazard of diethyl ether vapor in an enclosed work space is well known. In addition to aliphatic ethers, such as dimethyl ether and diethyl ether, several alkenyl ethers, including vinylethyl ether, are produced by internal combustion engines. A cyclic ether and important industrial solvent, tetrahydrofuran, occurs as an air contaminant. Methyltertiarybutyl ether (MTBE) became the octane booster of choice to replace tetraethyllead in gasoline. Because of its widespread distribution, MTBE has the potential to be an air pollutant, although its hazard is limited by its low vapor pressure. Largely because of its potential to contaminate water, MTBE has been largely replaced by ethanol as the oxygenated octane booster of choice in gasoline. Another possible air contaminant because of its potential uses as an octane booster is diisopropyl ether (DIPE). The structural formulas of the ethers mentioned above are given below:

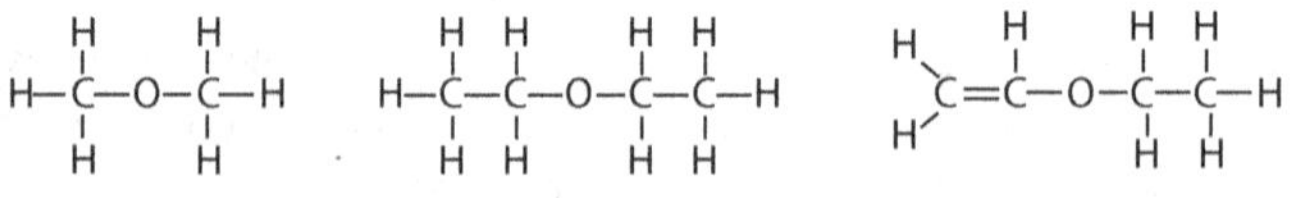

Dimethyl ether Diethyl ether Vinylethyl ether

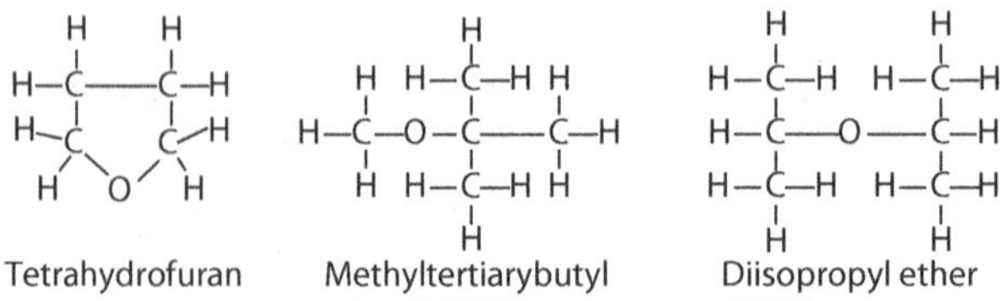

Tetrahydrofuran Methyltertiarybutyl ether (MTBE) Diisopropyl ether (DIPE)

Ethers are relatively unreactive and not as water-soluble as the lower alcohols or carboxylic acids. The predominant process for their atmospheric removal begins with hydroxyl radical attack.

12.5.4 Oxides

Ethylene oxide and propylene oxide,

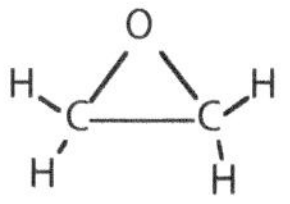

Ethylene oxide 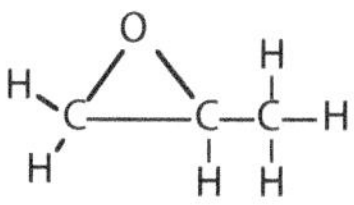 Propylene oxide

rank among the 50 most widely produced industrial chemicals and have a limited potential to enter the atmosphere as pollutants. Ethylene oxide is a moderately to highly toxic, sweet-smelling, colorless, flammable, explosive gas used as a chemical intermediate, sterilant, and fumigant. It is a mutagen and a carcinogen to experimental animals. It is classified as hazardous because of both its toxicity and ignitability.

12.5.5 Carboxylic Acids

Carboxylic acids have one or more of the functional group,

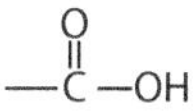

$$-\overset{\displaystyle O}{\overset{\|}{C}}-OH$$

attached to an alkane, alkene, or aryl hydrocarbon moiety. A carboxylic acid, pinonic acid, produced by the photochemical oxidation of naturally produced α-pinene, was discussed in Section 12.2. Many of the carboxylic acids found in the atmosphere probably result from the photochemical oxidation of other organic compounds through gas-phase reactions or by reactions of other organic compounds dissolved in aqueous aerosols. These acids are often the end products of photochemical oxidation because their low vapor pressures and relatively high water solubilities make them susceptible to scavenging from the atmosphere, although they are among the most stable organic compounds while in the atmosphere. They are removed from the atmosphere by both wet and dry deposition processes.

Low-molecular-mass formic acid, HCOOH, and acetic acid, H_3CCOOH, occur in the gas phase and partition into aqueous aerosol droplets in fog and clouds, playing a significant role in the chemistry that occurs in these droplets. The heavier organic acids are the most abundant constituents of small particles in the urban atmosphere (SOA), examples of which are shown in Figure 12.2.

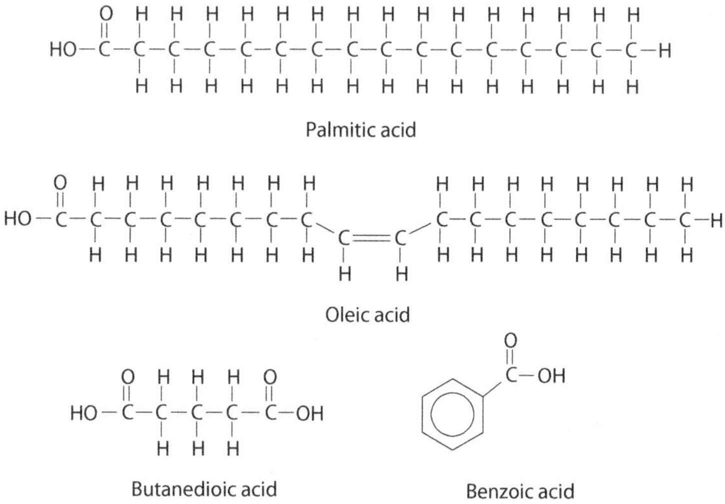

Palmitic acid

Oleic acid

Butanedioic acid Benzoic acid

FIGURE 12.2 Organic acids. Typical examples of more than 50 organic acids typically found in polluted and unpolluted atmospheres.

12.6 ORGANONITROGEN COMPOUNDS

Organonitrogen compounds formed largely by atmospheric chemical processes operating on hydrocarbons and oxygenated species in the atmosphere are abundant and are involved in the transfer of some atmospheric nitrogen to the hydrosphere and geosphere. These compounds in the atmosphere may be divided between reduced nitrogen compounds such as amines and oxidized organic compounds such as nitrates. The reduced species are predominantly those emitted directly to the atmosphere whereas the oxidized species are normally produced by oxidizing photochemical processes involving $HO^{\bullet}$, O_3, NO_x, and NO_3 radical.

Organic nitrogen compounds that may be found as atmospheric contaminants may be classified as *amines*, *amides*, *nitriles*, *nitro compounds*, or *heterocyclic nitrogen compounds*. Structural formulas of common examples of each of these five classes of compounds reported as atmospheric contaminants are:

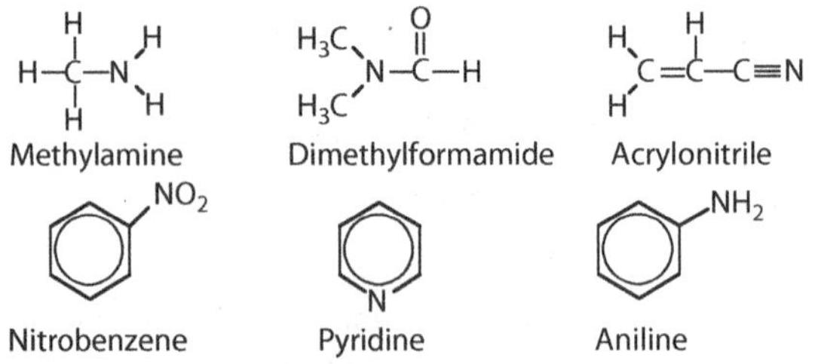

Methylamine Dimethylformamide Acrylonitrile

Nitrobenzene Pyridine Aniline

The organonitrogen compounds shown above can come from anthropogenic pollution sources. Significant amounts of anthropogenic atmospheric nitrogen may also come from reactions of inorganic nitrogen with reactive organic species. Examples include nitrates produced by the reaction of atmospheric NO_3.

Amines consist of compounds in which one or more of the hydrogen atoms in NH_3 has been replaced by a hydrocarbon moiety. Lower-molecular-mass amines are volatile. These are prominent among the compounds giving rotten fish their characteristic odor—an obvious reason why air contamination by amines is undesirable. The simplest and most important aryl amine is aniline, used in the manufacture of dyes, amides, photographic chemicals, and drugs. A number of amines are widely used industrial chemicals and solvents, so industrial sources have the potential to contaminate the atmosphere with these chemicals. Decaying organic matter, especially protein wastes, produce amines, so rendering plants, packing houses, and sewage treatment plants are important sources of these substances.

Aromatic amines are of special concern as air pollutants, especially in the workplace, because some cause urethral tract cancer (particularly of the bladder) in exposed individuals. Aromatic amines are widely used as chemical intermediates, antioxidants, and curing agents in the manufacture of polymers (rubber and plastics), drugs, pesticides, dyes, pigments, and inks. In addition to aniline, some aromatic amines of potential concern are the following:

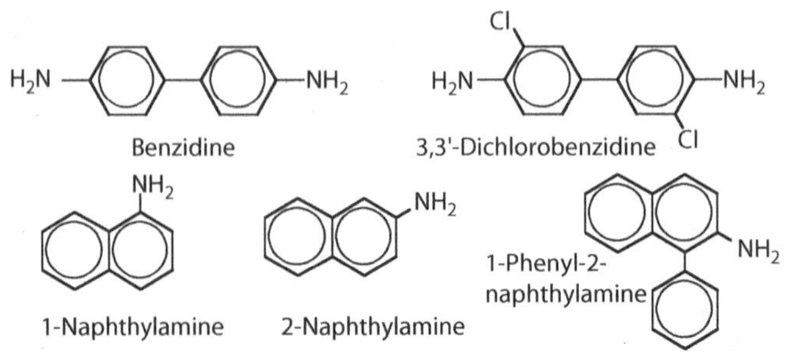

Benzidine 3,3'-Dichlorobenzidine

1-Naphthylamine 2-Naphthylamine 1-Phenyl-2-naphthylamine

In the atmosphere, amines can be attacked by hydroxyl radicals and undergo further reactions. Amines are bases (electron-pair donors). Therefore, their acid–base chemistry in the atmosphere may be important, particularly in the presence of acids in acidic precipitation.

The amide most likely to be encountered as an atmospheric pollutant is dimethylformamide. It is widely used commercially as a solvent for the synthetic polymer, polyacrylonitrile (Orlon, Dacron). Most amides have relatively low vapor pressures, which limit their entry into the atmosphere.

Nitriles, which are characterized by the –C≡N group, have been reported as air contaminants, particularly from industrial sources. Both acrylonitrile and acetonitrile (CH_3CN) have been reported in the atmosphere as a result of synthetic rubber manufacture. As expected from their volatilities and levels of industrial production, most of the nitriles reported as atmospheric contaminants are low-molecular-mass aliphatic or alkenyl nitriles or aryl nitriles with only one benzene ring. Acrylonitrile, used to make polyacrylonitrile polymer, is the only nitrogen-containing organic chemical among the top 50 chemicals with annual worldwide production exceeding 1 billion kg.

Among the nitro compounds, RNO_2, reported as air contaminants are nitromethane, nitroethane, and nitrobenzene. These compounds are produced from industrial sources. Highly oxygenated compounds containing the NO_2 group, particularly peroxyacetyl nitrate (PAN) and peroxypropionyl nitrate, have no known emission sources and are secondary pollutant end products of the photochemical oxidation of hydrocarbons in urban atmospheres. These compounds are intense eye irritants responsible for much of the disagreeable qualities of photochemical smog. They are mutagens and phytotoxins and are suspected of causing skin cancer.

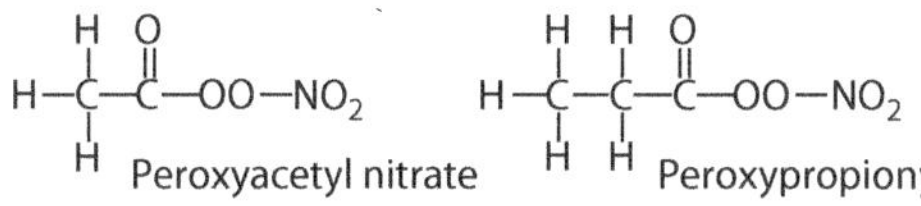

H–C(H)(H)–C(=O)–OO–NO_2 Peroxyacetyl nitrate

H–C(H)(H)–C(H)(H)–C(=O)–OO–NO_2 Peroxypropionyl nitrate

A large number of *heterocyclic nitrogen compounds* have been reported in tobacco smoke, and it is inferred that many of these compounds can enter the atmosphere from burning vegetation. Coke ovens are another major source of these compounds. In addition to the derivatives of pyridine, some of the heterocyclic nitrogen compounds are derivatives of pyrrole:

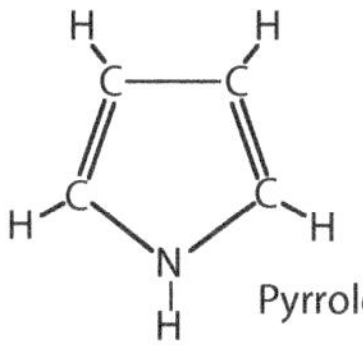

Pyrrole

Heterocyclic nitrogen compounds in the atmosphere occur almost entirely in association with aerosols.

Nitrosamines, which contain the N–N=O group, and having, therefore, the general formula

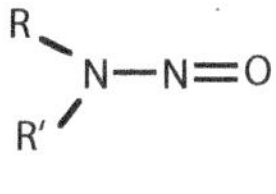

$$\begin{matrix} R \\ & N{-}N{=}O \\ R' \end{matrix}$$

deserve special mention as atmospheric contaminants because some are known carcinogens. As discussed in Chapter 22, nitrosamines include compounds that can attach alkyl groups to DNA resulting in modified DNA, which can lead to cancer. Both *N*-nitrosodimethylamine and *N*-nitrosodiethylamine have been detected in the atmosphere.

12.7 ORGANOHALIDE COMPOUNDS

Organohalides consist of halogen-substituted hydrocarbon molecules, each of which contains at least one atom of F, Cl, Br, or I. They may be saturated (*alkyl halides*), unsaturated (*alkenyl halides*), or aromatic (*aryl halides*). The organohalides of environmental and toxicological concern exhibit a wide range of physical and chemical properties. Although most organohalide compounds of pollution concern are from anthropogenic sources, it is now known that a large variety of such compounds are generated by organisms, particularly those in marine environments.

Structural formulas of several alkyl halides commonly encountered in the atmosphere are given below:

Chloromethane (bp –24°C)	Dichloromethane (methylene chloride, bp 40°C)	Dichlorodifluoromethane, Freon-12, bp –29°C)	1,1,1-Trichloroethane, (methyl chloroform, bp 74°C)

Dichloromethane is a volatile liquid with excellent solvent properties for nonpolar organic solutes. It has been used as a solvent for the decaffeination of coffee, in paint strippers, as a blowing agent in urethane polymer manufacture, and to depress vapor pressure in aerosol formulations. *Dichlorodifluoromethane* is one of the CFC compounds once widely manufactured as a refrigerant and involved in stratospheric ozone depletion. One of the more common industrial chlorinated solvents is *1,1,1-trichloroethane.*

Viewed as halogen-substituted derivatives of alkenes, the *alkenyl organohalides* contain at least one halogen atom and at least one carbon–carbon double bond. The most significant of these are the lighter chlorinated compounds.

Vinyl chloride is consumed in large quantities as a raw material to manufacture pipe, hose, wrapping, and other products fabricated from polyvinyl chloride plastic. This highly flammable, volatile, sweet-smelling gas is known to cause angiosarcoma, a rare form of liver cancer. *Trichloroethylene* is a clear, colorless, nonflammable, volatile liquid. It is an excellent degreasing and dry-cleaning solvent, and has been used as a household solvent and for food extraction (e.g., in decaffeination of coffee). *Allyl chloride*, 3-chloropropene, is an intermediate in the manufacture of allyl alcohol and other allyl compounds, including pharmaceuticals, insecticides, and thermosetting varnish and plastic resins.

Some commonly used aryl halide derivatives of benzene and toluene are shown below:

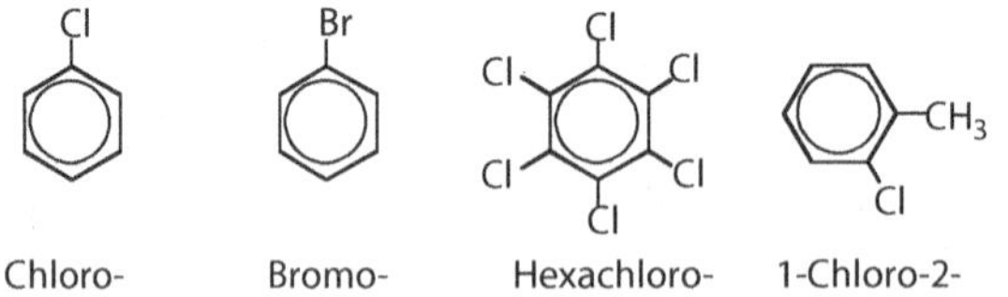

Chlorobenzene	Bromobenzene	Hexachlorobenzene	1-Chloro-2-methylbenzene

Aryl halide compounds have many uses. The inevitable result of all these uses has been widespread occurrences of human exposure and environmental contamination. PCBs, a group of compounds formed by the chlorination of biphenyl,

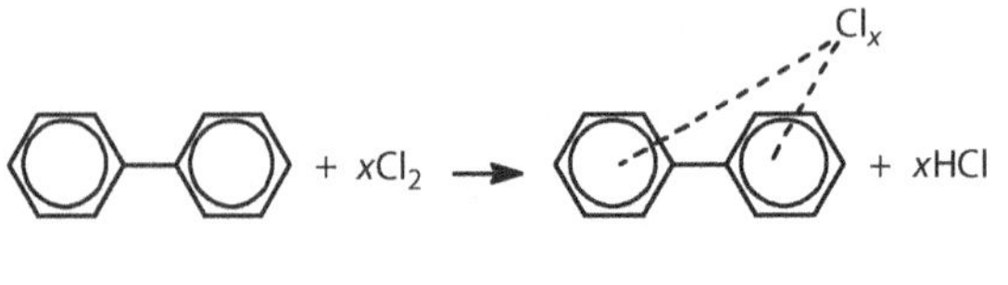

$$\text{Biphenyl} + x\text{Cl}_2 \rightarrow \text{Cl}_x\text{-biphenyl} + x\text{HCl} \quad (12.12)$$

have extremely high physical and chemical stabilities and other qualities that led to their being used in many applications, including heat transfer fluids, hydraulic fluids, and dielectrics until their manufacture and use were banned in the 1970s because of their pollution potential.

As expected from their high vapor pressures and volatilities, the lighter organohalide compounds are the most likely to be found at detectable levels in the atmosphere. On a global basis, the most abundant organochlorine compound in the atmosphere is methyl chloride, CH_3Cl, with atmospheric concentrations of the order of tenths of a part per billion largely from natural sources in the ocean and tropical coastal regions. Methyl chloroform, CH_3CCl_3, is relatively persistent in the atmosphere, with residence times of several years. Therefore, it may pose a threat to the stratospheric ozone layer in the same way as CFCs. Among the other lighter organohalides that have been found in the atmosphere are methylene chloride; methyl bromide, CH_3Br; bromoform, $CHBr_3$; assorted CFCs; and halogen-substituted ethylene compounds such as trichloroethylene, vinyl chloride, perchloroethylene, ($CCl_2{=}CCl_2$), and solvent ethylene dibromide ($CHBr{=}CHBr$).

12.7.1 Chlorofluorocarbons

CFCs, such as dichlorodifluoromethane, commonly called freons, are volatile 1- and 2-carbon compounds that contain Cl and F bonded to carbon. These compounds are notably stable and nontoxic. They have been widely used in recent decades in the fabrication of flexible and rigid foams and as fluids for refrigeration and air conditioning. Until their uses were banned (see below) the most widely manufactured of these compounds were CCl_3F (CFC-11, bp 24°C), CCl_2F_2(CFC-12, bp –28°C), $C_2Cl_3F_3$ (CFC-113), $C_2Cl_2F_4$ (CFC-114), and C_2ClF_5 (CFC-115).

Halons are related compounds that contain bromine and are used in fire extinguisher systems. The major commercial halons are $CBrClF_2$ (halon-1211), $CBrF_3$ (halon-1301), and $C_2Br_2F_4$ (halon-2402), where the sequence of numbers denotes the number of carbon, fluorine, chlorine, and bromine atoms, respectively, per molecule. Halons are particularly effective fire extinguishing agents because of the way in which they stop combustion. Halons act by chain reactions that destroy hydrogen atoms that sustain combustion. The basic sequence of reactions involved is outlined below:

$$CBrClF_2 + H^\bullet \rightarrow CClF_2^\bullet + HBr \tag{12.13}$$

$$HBr + H^\bullet \rightarrow Br^\bullet + H_2 \tag{12.14}$$

$$H^\bullet + Br^\bullet \rightarrow HBr \tag{12.15}$$

Halons are used in automatic fire extinguishing systems, particularly those located in flammable solvent storage areas, and in specialty fire extinguishers, particularly those on aircraft. Because of their potential to destroy stratospheric ozone discussed below, the use of halons in fire extinguishers was severely curtailed in a ban imposed in developed nations on January 1, 1994. The ban on halons has caused concern because of the favorable properties of halons in fire extinguishers, particularly on aircraft. It is possible that hydrogen-containing analogs of halons may be effective as fire extinguishers without posing a threat to ozone.

The nonreactivity of CFC compounds, combined with worldwide production that once reached approximately one-half million metric tons per year and deliberate or accidental release to the atmosphere, has resulted in CFCs becoming homogeneous components of the global atmosphere. In 1974, it was convincingly suggested, in a classic work that earned the authors a Nobel Prize, that chlorofluoromethanes could catalyze the destruction of stratospheric ozone that filters out cancer-causing ultraviolet radiation from the sun.[8] Subsequent data on ozone levels in the stratosphere and on increased ultraviolet radiation at earth's surface showed that the threat to stratospheric ozone posed by CFCs is real. Although quite inert in the lower atmosphere, CFCs undergo photodecomposition by the action of high-energy ultraviolet radiation in the stratosphere, which is energetic enough to break their very strong C–Cl bonds through reactions such as

$$Cl_2CF_2 + h\nu \rightarrow Cl^{\bullet} + ClCF_2^{\bullet} \quad (12.16)$$

thereby releasing Cl atoms. The Cl atoms are very reactive species that initiate chain reactions that destroy stratospheric ozone as discussed in detail in Section 14.4, "Ozone Layer Destruction."

The U.S. EPA regulations, imposed in accordance with the 1986 Montreal Protocol on Substances that Deplete the Ozone Layer, curtailed the production of CFCs and halocarbons in the United States starting from 1989. The substitutes for these halocarbons are hydrogen-containing chlorofluorocarbons (HCFCs), hydrogen-containing fluorocarbons (HFCs), and some volatile hydrocarbon formulations. These have included CH_2FCF_3 (HFC-134a, 1,1,1,2-tetrafluoroethane, which became the standard substitute for CFC-12 in automobile air conditioners and refrigeration equipment), $CHCl_2CF_3$ (HCFC-123), CH_3CCl_2F (HCFC-141b), $CHClF_2$ (HCFC-22), and CH_2F_2 (HFC-152a, a flammable material). Because of the more readily broken H–C bonds they contain, these compounds are more easily destroyed by atmospheric chemical reactions (particularly with hydroxyl radical) before they reach the stratosphere. The HFC compounds are favored as CFC substitutes because they contain only fluorine and hydrogen bound to carbon so that they cannot generate any ozone-destroying chlorine atoms.

Some of the HFC compounds, particularly HFC-134a, have been criticized for their global warming potential. Among the newer alternatives with much less global warming potential are HFO-1234yf, chemical formula $CF_3CF{=}CH_2$. Concerned about the global warming potential of HFCs, European automakers are developing automobile air conditioning systems that use CO_2 as a refrigerant fluid. Although carbon dioxide contributes to global warming, the incremental effects from release of this gas from automobile air conditioners would be minuscule because the quantities used in automobile air conditioners would be insignificant compared to emissions from fossil fuels combustion.

12.7.2 Atmospheric Reactions of Hydrofluorocarbons and Hydrochlorofluorocarbons

The atmospheric chemistry of hydrofluorocarbons (HFCs) and hydrochlorofluorocarbons (HCFCs) is important, even though these compounds do not pose much danger to the ozone layer. Of particular importance is the photooxidation of these compounds and the fates and effects of their photooxidation products. Initial attack on these compounds leading to their eventual destruction can be from hydroxyl radical or chlorine atoms.[9] HFC 134a, CF_3CH_2F, reacts as follows with hydroxyl radical in the troposphere:

$$CF_3CH_2F + HO^{\bullet} \rightarrow CF_3CHF^{\bullet} + H_2O \quad (12.17)$$

The alkyl radical produced by this reaction forms a peroxy radical with molecular oxygen:

$$CF_3CHF^{\bullet} + O_2 + M \rightarrow CF_3CHFO_2^{\bullet} + M \quad (12.18)$$

and the peroxy radical reacts with NO:

$$CF_3CHFO_2^{\bullet} + NO \rightarrow CF_3CHFO^{\bullet} + NO_2 \quad (12.19)$$

The product of that reaction can either decompose:

$$CF_3CHFO^\bullet \rightarrow CF_3^\bullet + HC(O)F \tag{12.20}$$

or react with molecular O_2:

$$CF_3CHFO^\bullet + O_2 \rightarrow CF_3C(O)F + HO_2^\bullet \tag{12.21}$$

These latter two processes are thought to occur to about equal extents. Ultimately, products are formed that are purged from the atmosphere.

12.7.3 Perfluorocarbons

Perfluorocarbons are completely fluorinated organic compounds, the simplest examples of which are carbon tetrafluoride (CF_4) and hexafluoroethane (C_2F_6). Several hundred metric tons of these compounds are produced annually as etching agents in the electronics industry. However, about 30,000 metric tons of CF_4 and about 10% that amount of C_2F_6 are emitted to the global atmosphere each year from aluminum production.

Nontoxic perfluorocarbons do not react with hydroxyl radical, ozone, or other reactive substances in the atmosphere, and the only known significant mechanism by which they are destroyed in the atmosphere is photolysis by radiation less than 130 nm in wavelength. Because of their extreme lack of reactivity, they are involved in neither photochemical smog formation nor ozone layer depletion. As a result of this stability, perfluorocarbons are very long lived in the atmosphere; the lifetime of CF_4 is estimated to be an astoundingly long 50,000 years! The major atmospheric concern with these compounds is their potential to cause greenhouse warming (see Chapter 14). Taking into account their nonreactivity and ability to absorb infrared radiation, perfluorocarbons have a potential to cause global warming over a very long time span with an aggregate effect per molecule several thousand times that of carbon dioxide.

12.7.4 Marine Sources of Organohalogen Compounds

The predominant natural source of volatile atmospheric organohalogens is from marine organisms. This is not surprising considering the high concentrations of chloride ion and lesser levels of bromide and iodide in seawater. Various marine algae are major sources of atmospheric organohalogens.

12.7.5 Chlorinated Dibenzo-*p*-dioxins and Dibenzofurans

Polychlorinated dibenzo-*p*-dioxins (PCDDs) and polychlorinated dibenzofurans (PCDFs) are pollutant compounds with the general formulas shown below:

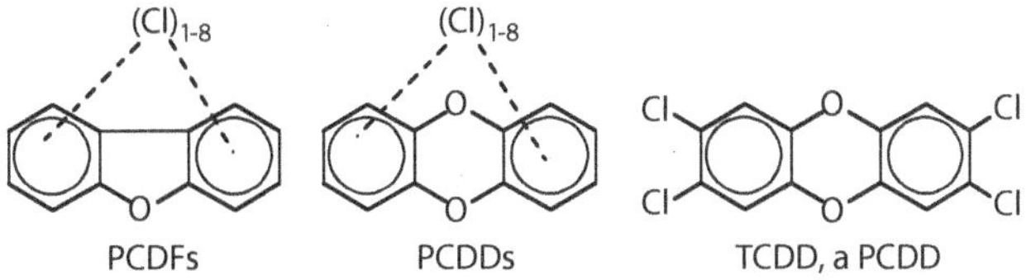

$(Cl)_{1-8}$ PCDFs

$(Cl)_{1-8}$ PCDDs

TCDD, a PCDD

As discussed in Chapters 7 and 24, these compounds are of considerable concern because of their toxicities. One of the more infamous environmental pollutant chemicals is 2,3,7,8-TCDD, often known simply as "dioxin."

PCDDs and PCDFs enter the air from numerous sources, including automobile engines, waste incinerators, and steel and other metal production. A particularly important source may well be municipal solid waste incinerators. The formation of PCDDs and PCDFs in such incinerators results in part because of the presence of both chlorine (such as from PVC plastic in municipal waste) and catalytic metals. Furthermore, PCDDs and PCDFs are produced by *de novo* synthesis on carbonaceous fly ash surfaces in the postcombustion region of an incinerator at relatively low temperatures of around 300°C in the presence of oxygen and sources of chlorine and hydrogen.

Atmospheric levels of PCDDs and PCDFs are quite low, in the range of 0.4–100 pg/m^3 of air. Because of their lower volatilities, the more highly chlorinated congeners of these compounds tend to occur in atmospheric particulate matter, in which they are relatively protected from photolysis and reaction with hydroxyl radical, which are the two main mechanisms by which PCDDs and PCDFs are eliminated from the atmosphere. Furthermore, the less highly chlorinated congeners are more reactive because of their C–H bonds, which are susceptible to attack from hydroxyl radical.

12.8 ORGANOSULFUR COMPOUNDS

Substitution of alkyl or aryl hydrocarbon groups such as phenyl and methyl for H on hydrogen sulfide, H_2S, leads to a number of different organosulfurthiols (mercaptans, R–SH) and sulfides, also called thioethers (R–S–R). Structural formulas of examples of these compounds are shown below.

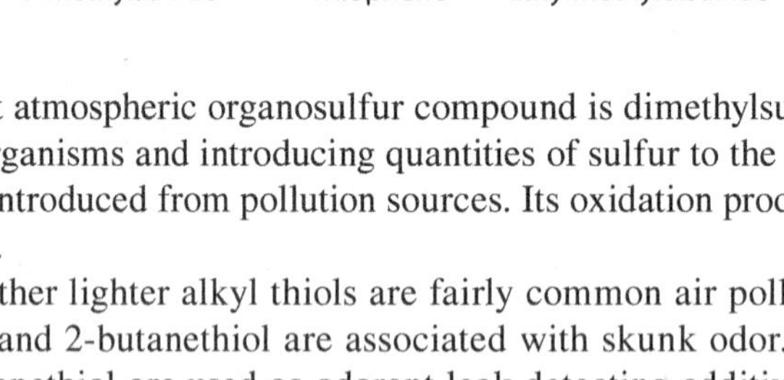

Methanethiol 2-Propene-1-thiol Benzenethiol

Dimethylsulfide Thiophene Ethylmethyldisulfide

The most significant atmospheric organosulfur compound is dimethylsulfide, produced in large quantities by marine organisms and introducing quantities of sulfur to the atmosphere comparable in magnitude to those introduced from pollution sources. Its oxidation produces most of the SO_2 in the marine atmosphere.

Methanethiol and other lighter alkyl thiols are fairly common air pollutants that have "ultragarlic" odors; both 1- and 2-butanethiol are associated with skunk odor. Gaseous methanethiol and volatile liquid ethanethiol are used as odorant leak-detecting additives for natural gas, propane, and butane, and are also employed as intermediates in pesticide synthesis. Allylmercaptan (2-propene-1-thiol) is a toxic, irritating volatile liquid with a strong garlic odor. Benzenethiol (phenyl mercaptan) is the simplest of the aryl thiols. It is a toxic liquid with a severely "repulsive" odor.

Alkyl sulfides or thioethers contain the C–S–C functional group. The lightest of these compounds is dimethyl sulfide, a volatile liquid (bp 38°C) that is moderately toxic by ingestion. Cyclic sulfides contain the C–S–C group in a ring structure. The most common of these compounds is thiophene, a heat-stable liquid (bp 84°C) with a solvent action much like that of benzene that is used in the manufacture of pharmaceuticals, dyes, and resins.

Although not highly significant as atmospheric contaminants on a large scale, organic sulfur compounds can cause local air pollution problems because of their bad odors. Major sources of organosulfur compounds in the atmosphere include microbial degradation, wood pulping, volatile matter evolved from plants, animal wastes, packing house and rendering plant wastes, starch manufacture, sewage treatment, and petroleum refining.

Although the impact of organosulfur compounds on atmospheric chemistry is minimal in areas such as aerosol formation or production of acid precipitation components, these compounds are the worst of all in producing odor. Therefore, it is important to prevent their release into the atmosphere.

As with all hydrogen-containing organic species in the atmosphere, reaction of organosulfur compounds with hydroxyl radical is a first step in their atmospheric photochemical reactions. The sulfur from both mercaptans and sulfides ends up as SO_2. In both cases, there is thought to be a readily oxidized SO intermediate, and $HS^{\bullet}$ radical may also be an intermediate in the oxidation of mercaptans. Another possibility is the addition of O atoms to S, resulting in the formation of free radicals as shown below for methyl mercaptan:

$$CH_3SH + O \rightarrow H_3C^{\bullet} + HSO^{\bullet} \quad (12.22)$$

The $HSO^{\bullet}$ radical is readily oxidized by atmospheric O_2 to SO_2.

12.9 ORGANIC PARTICULATE MATTER

Organic species are important constituents of atmospheric particulate matter. Some particles consist almost entirely of organic matter. Others have significant amounts of organic compounds adsorbed to the surfaces of nonorganic material. The visibility-obscuring particles characteristic of photochemical smog (Chapter 13) are composed largely of oxygenated organic material that is the end product of the photochemical smog process. Particles of elemental carbon and of highly condensed PAHs produced as products of incomplete combustion of hydrocarbons from sources such as diesel engines have a strong affinity for organic vapors in the atmosphere.

Organic particulate matter may be emitted directly from sources as primary pollutants or formed as secondary pollutants produced by atmospheric chemical processes operating on organic vapors. Through the action of reactive atmospheric species, especially $HO^{\bullet}$ radical, O_3, NO_x, and NO_3 radical, oxygen and nitrogen are added to vaporous organic molecules producing much less volatile species that condense and form particles.

12.10 HAZARDOUS AIR POLLUTANT ORGANIC COMPOUNDS

Hazardous air pollutants were designated in the U.S. Clean Air Act of 1970 as those likely to cause adverse health effects. These substances are generally regarded as ones that may come from specific sources, such as release from a particular factory, as distinguished from *criteria air pollutants*, such as SO_2 and NO_x, that are widespread and from a number of sources. The Clean Air Amendments of 1990 specified a list of hazardous air pollutants, which has been modified slightly since then. Table 12.1 lists organic compounds, most of which are on the list of hazardous air pollutants (several have been delisted since the original publication was published). Most of the substances in the list of hazardous air pollutants are organic compounds produced industrially. Although space does not permit a discussion of these substances, the CAS number given for each compound will enable the reader to look up its formula, properties, and literature related to atmospheric pollution on databases such as SciFinder.

TABLE 12.1
Organic Compounds from EPA List of Hazardous Air Pollutants

CAS Number	Chemical Name	CAS Number	Chemical Name
75070	Acetaldehyde	108394	*m*-Cresol
60355	Acetamide	106445	*p*-Cresol
75058	Acetonitrile	98828	Cumene
98862	Acetophenone	94757	2,4-D, salts and esters
53963	2-Acetylaminofluorene	3547044	DDE
107028	Acrolein	334883	Diazomethane
79061	Acrylamide	132649	Dibenzofurans
79107	Acrylic acid	96128	1,2-Dibromo-3-chloropropane
107131	Acrylonitrile	84742	Dibutylphthalate
107051	Allyl chloride	106467	1,4-Dichlorobenzene(p)
92671	4-Aminobiphenyl	91941	3,3-Dichlorobenzidene
62533	Aniline	111444	Dichloroethyl ether [bis(2-chloroethyl)ether]
90040	*o*-Anisidine	542756	1,3-Dichloropropene
71432	Benzene (including benzene from gasoline)	62737	Dichlorvos
92875	Benzidine	121697	*N,N*-Diethyl aniline (*N,N*-dimethylaniline)
98077	Benzotrichloride	64675	Diethyl sulfate
100447	Benzyl chloride	119904	3,3-Dimethoxybenzidine
92524	Biphenyl	60117	Dimethyl aminoazobenzene
117817	Bis(2-ethylhexyl)phthalate (DEHP)	119937	3,3′-Dimethyl benzidine
542881	Bis(chloromethyl)ether	79447	Dimethyl carbamoyl chloride
75252	Bromoform	68122	Dimethyl formamide
106990	1,3-Butadiene	57147	1,1-Dimethyl hydrazine
133062	Captan	131113	Dimethyl phthalate
63252	Carbaryl	77781	Dimethyl sulfate
56235	Carbon tetrachloride	534521	4,6-Dinitro-*o*-cresol, and salts
120809	Catechol	51285	2,4-Dinitrophenol
133904	Chloramben	121142	2,4-Dinitrotoluene
57749	Chlordane	123911	1,4-Dioxane (1,4-diethyleneoxide)
7782505	Chlorine	122667	1,2-Diphenylhydrazine
79118	Chloroacetic acid	106898	Epichlorohydrin (1-chloro-2,3-epoxypropane)
532274	2-Chloroacetophenone	106887	1,2-Epoxybutane
108907	Chlorobenzene	140885	Ethyl acrylate
510156	Chlorobenzilate	100414	Ethyl benzene
67663	Chloroform	51796	Ethyl carbamate (urethane)
107302	Chloromethyl methyl ether	75003	Ethyl chloride (chloroethane)
126998	Chloroprene	106934	Ethylene dibromide (dibromoethane)
1319773	Cresols/Cresylic acid (isomers and mixture)	98953	Nitrobenzene
95487	*o*-Cresol	92933	4-Nitrobiphenyl
107062	Ethylene dichloride (1,2-dichloroethane)	100027	4-Nitrophenol
107211	Ethylene glycol	79469	2-Nitropropane
151564	Ethylene imine (aziridine)	684935	*N*-Nitroso-*N*-methylurea
75218	Ethylene oxide	62759	*N*-Nitrosodimethylamine
96457	Ethylene thiourea	59892	*N*-Nitrosomorpholine
75343	Ethylidene dichloride (1,1-dichloroethane)	56382	Parathion
50000	Formaldehyde	82688	Pentachloronitrobenzene
76448	Heptachlor	87865	Pentachlorophenol
118741	Hexachlorobenzene	108952	Phenol

continued

TABLE 12.1 (continued)
Organic Compounds from EPA List of Hazardous Air Pollutants

CAS Number	Chemical Name	CAS Number	Chemical Name
87683	Hexachlorobutadiene	95476	*o*-Xylenes
77474	Hexachlorocyclopentadiene	108383	*m*-Xylenes
67721	Hexachloroethane	106423	*p*-Xylenes
822060	Hexamethylene-1,6-diisocyanate	106503	*p*-Phenylenediamine
680319	Hexamethylphosphoramide	85449	Phthalic anhydride
110543	Hexane	1336363	PCBs (aroclors)
123319	Hydroquinone	1120714	1,3-Propane sultone
78591	Isophorone	57578	β-Propiolactone
58899	Lindane (all isomers)	123386	Propionaldehyde
108316	Maleic anhydride	114261	Propoxur (baygon)
67561	Methanol	78875	Propylene dichloride (1,2-dichloropropane)
72435	Methoxychlor	75569	Propylene oxide
74839	Methyl bromide (bromomethane)	75558	1,2-Propylenimine (2-methyl aziridine)
74873	Methyl chloride (chloromethane)	91225	Quinoline
71556	Methyl chloroform (1,1,1-trichloroethane)	106514	Quinone
78933	Methyl ethyl ketone (2-butanone)	100425	Styrene
60344	Methyl hydrazine	96093	Styrene oxide
74884	Methyl iodide (iodomethane)	1746016	2,3,7,8-Tetrachlor-odibenzo-*p*-dioxin
108101	Methyl isobutyl ketone	79345	1,1,2,2-Tetrachloroethane
624839	Methyl isocyanate	127184	Tetrachloroethylene (perchloroethylene)
80626	Methyl methacrylate diisocyanate (MDI)	7550450	Titanium tetrachloride
1634044	Methyl *tert*-butyl ether	108883	Toluene
101144	4,4-methylene bis (2-chloroaniline)	95807	2,4-Toluene diamine
75092	Methylene chloride (dichloromethane)	584849	2,4-Toluene diisocyanate
101688	Methylene diphenyl	88062	2,4,5-Trichlorophenol
101779	4,4-Methylenedianiline	95534	*o*-Toluidine
91203	Naphthalene	8001352	Toxaphene (chlorinated camphene)
540841	2,2,4-Trimethylpentane	120821	1,2,4-Trichlorobenzene
108054	Vinyl acetate	79005	1,1,2-Trichloroethane
593602	Vinyl bromide	79016	Trichloroethylene
75014	Vinyl chloride	95954	2,4,6-Trichlorophenol
75354	Vinylidene chloride (1,1-dichloroethylene)	121448	Triethylamine
1330207	Xylenes (isomers and mixture)	1582098	Trifluralin

LITERATURE CITED

1. Pozo, K., T. Harner, S. C. Lee, F. Wania, D. C. G. Muir, and K. C. Jones, Seasonally resolved concentrations of persistent organic pollutants in the global atmosphere from the first year of the GAPS study, *Environmental Science and Technology*, **43**, 796–803, 2009.
2. Karl, T., A. Guenther, R. J. Yokelson, J. Greenberg, M. Potosnak, D. Blake, and P. Artaxo, The tropical forest and fire emissions experiment: Emission, chemistry, and transport of biogenic volatile organic compounds in the lower atmosphere over Amazonia, *Journal of Geophysical Research*, **112**, D18302/1–D18302/17, 2007.
3. Sakulyanontvittaya, T., A. Guenther, D. Helmig, J. Milford, and C. Wiedinmyer, Secondary organic aerosol from sesquiterpene and monoterpene emissions in the United States, *Environmental Science and Technology*, **42**, 8784–8790, 2008.

4. Matsunaga, A. and P. J. Ziemann, Yields of β-hydroxynitrates and dihydroxynitrates in aerosol formed from OH radical-initiated reactions of linear alkenes in the presence of NO_x, *Journal of Physical Chemistry A*, **113**, 599–606, 2009.
5. Iovino, P., R. Polverino, S. Salvestrini, and S. Capasso, Temporal and spatial distribution of BTEX pollutants in the atmosphere of metropolitan areas and neighbouring towns, *Environmental Monitoring and Assessment*, **150**, 437–444, 2009.
6. Genualdi, S. A., R. K. Killin, J. Woods, G. Wilson, D. Schmedding, and S. L. Simonich, Trans-pacific and regional atmospheric transport of polycyclic aromatic hydrocarbons and pesticides in biomass burning emissions to Western North America, *Environmental Science and Technology*, **43**, 1061–1066, 2009.
7. Spada, N., E. Fujii, and T. M. Cahill, Diurnal cycles of acrolein and other small aldehydes in regions impacted by vehicle emissions, *Environmental Science and Technology*, **42**, 7084–7090, 2008.
8. Molina, M. J. and F. S. Rowland, Stratospheric sink for chlorofluoromethanes, *Nature*, **249**, 810–812, 1974.
9. Inoue, Y., M. Kawasaki, T. J. Wallington, and M. D. Hurley, Atmospheric chemistry of $CF_3CH_2CF_2CH_3$ (HFC-365mfc): Kinetics and mechanism of chlorine atom initiated oxidation, infrared spectrum, and global warming potential, *Chemical Physics Letters*, **462**, 164–168, 2008.

SUPPLEMENTARY REFERENCES

Austin, J., P. Brimblecombe, and W. Sturges, Eds, *Air Pollution Science for the 21st Century*, Elsevier Science, New York, 2002.

Balduino, S. P., Ed., *Progress in Air Pollution Research*, Nova Science Publishers, New York, 2007.

Bodine, C. G., Ed., *Air Pollution Research Advances*, Nova Science Publishers, New York, 2007.

Brimblecombe, P., *Air Composition and Chemistry*, 2nd ed., Cambridge University Press, Cambridge, UK, 1996.

Desonie, D., *Atmosphere: Air Pollution and Its Effects*, Chelsea House Publishers, New York, 2007.

Granier, C., P. Artaxo, and C. E. Reeves, *Emissions of Atmospheric Trace Compounds*, Kluwer Academic Publishers, Boston, 2004.

Hewitt, C. N., Ed., *Reactive Hydrocarbons in the Atmosphere*, Academic Press, San Diego, CA, 1999.

Kidd, J. S. and R. A. Kidd, *Air Pollution: Problems and Solutions*, Chelsea House Publishers, New York, 2006.

Livingston, J. V., *Air Pollution: New Research*, Nova Science Publishers, New York, 2007.

Seinfeld, J. H. and S. N. Pandis, *Atmospheric Chemistry and Physics: From Air Pollution to Climate Change*, 2nd ed., Wiley, Hoboken, NJ, 2006.

Sokhi, R. S., Ed., *World Atlas of Atmospheric Pollution*, Anthem Press, New York, 2007.

Vallero, D. A., *Fundamentals of Air Pollution*, 4th ed., Elsevier, Amsterdam, 2008.

QUESTIONS AND PROBLEMS

The use of internet resources is assumed in answering any of the questions. These would include such things as constants and conversion factors as well as additional information needed to complete an answer.

1. Match each organic pollutant in the left column with its expected effect in the right column, below:

a. CH_3SH	1. Most likely to have a secondary effect in the atmosphere
b. $CH_3CH_2CH_2CH_3$	2. Most likely to have a direct effect
c. $H_2C{=}CH{-}CH_2{-}CH_3$	3. Should have the least effect of these three

2. Why are hydrocarbon emissions from uncontrolled automobile exhaust particularly reactive?
3. Assume an accidental release of a mixture of gaseous alkanes and alkenes into an urban atmosphere early in the morning. Assume that the mass of air is subjected to intense sunlight during the day and is kept in a stagnant condition by a thermal inversion. If the atmosphere at the release site is monitored for these compounds, what can be said about their total and relative concentrations at the end of the day? Explain.
4. Match each radical in the left column with its type in the right column, below:

a. $H_3C\bullet$	1. Formyl radical
b. $CH_3CH_2O^{\bullet}$	2. Alkylperoxyl radical
c. $H\dot{C}O$	3. Alkyl radical
d. $CH_xCH_{2x+1}O_2^{\bullet}$	4. Alkoxyl radical

5. When reacting with hydroxyl radical, alkenes have a reaction mechanism not available to alkanes, which makes the alkenes much more reactive. What is this mechanism?
6. What is the most stable type of hydrocarbon that has a very low hydrogen-to-carbon ratio?
7. In the sequence of reactions leading to the oxidation of hydrocarbons in the atmosphere, what is the first stable class of compounds generally produced?
8. Give a sequence of reactions leading to the formation of acetaldehyde from ethane starting with the reaction of hydroxyl radical.
9. What important photochemical property do carbonyl compounds share with NO_2?
10. Of the following, the statement that is *untrue* regarding air pollutant hydrocarbons is (explain): (A) although methane, CH_4, is normally considered as coming from natural sources, and may be thought of as a nonpollutant, human activities have increased atmospheric methane levels, with the potential for doing harm, (B) some organic species from trees can result in the formation of secondary pollutants in the atmosphere, (C) alkenyl hydrocarbons containing the C=C group have a means of reacting with hydroxyl radical that is not available for alkanes, (D) the reactivities of individual hydrocarbons as commonly measured for their potential to form smog, vary only about ±25%, (E) most nonmethane hydrocarbons in the atmosphere are of concern because of their ability to produce secondary pollutants.
11. Of the following regarding organic air pollutants, the true statement is (explain): (A) carbonyl compounds (aldehydes and ketones) are usually the last organic species formed during the photochemical oxidation of hydrocarbons, (B) carboxylic acids (containing the $-CO_2H$ group) are especially long lived and persistent in the atmosphere, (C) CFCs, such as CCl_2F_2, are secondary pollutants, (D) PAN is a primary pollutant, (E) HFCs pose a greater danger to the stratospheric ozone layer than do CFCs.
12. Of the following, the true statement regarding atmospheric hydrocarbons is (explain): (A) alkanes readily undergo addition reactions with hydroxyl radicals, (B) alkenes undergo addition reactions with hydroxyl radical, (C) ozone tends to add across C–H bonds in alkanes, (D) hydrocarbons tend to be formed by the chemical reduction of esters evolved by plants, (E) unsaturated alkenes tend to be evolved from evaporation of gasoline, whereas alkanes are produced as automotive exhaust products.
13. Of the following, the *untrue* statement pertaining to hydrocarbons in the atmosphere is (explain): (A) hydrocarbons generated and released by human activities constitute only about one-seventh of the total hydrocarbons in the atmosphere, (B) natural sources are the most important contributors of organics in the atmosphere, (C) the reaction $2\{CH_2O\}$(bacterial action) $\rightarrow CO_2(g) + CH_4(g)$ is a huge contributor to atmospheric hydrocarbons, (D) methane, CH_4, is produced by a variety of plants and released to the atmosphere, (E) a number of plants evolve a simple hydrocarbon that is highly reactive with hydroxyl radical, $HO^{\bullet}$, and with oxidizing species in the atmosphere.

14. An important characteristic of atmospheric carbonyl compounds is (explain): (A) aldehydes are second only to NO_2 as atmospheric sources of free radicals produced by the absorption of light because the carbonyl group is a *chromophore*, (B) they are normally the final products of oxidation of atmospheric hydrocarbons and are relatively harmless in the atmosphere, (C) they are free radicals with unpaired electrons, (D) they are the predominant organic compounds emitted from auto exhausts, (E) alkenyl aldehydes such as acrolein are especially stable and unreactive in the atmosphere.

13 Photochemical Smog

13.1 INTRODUCTION

This chapter discusses the *oxidizing smog* or *photochemical smog* that permeates atmospheres in Los Angeles, Mexico City, Zurich, and many other urban areas. Although *smog* is the term used in this book to denote a photochemically oxidizing atmosphere, the word originally was used to describe the unpleasant combination of smoke and fog laced with sulfur dioxide which was formerly prevalent in London when high-sulfur coal was the primary fuel used in that city. This mixture is characterized by the presence of sulfur dioxide, a reducing compound; therefore it is a *reducing smog* or *sulfurous smog*. In fact, sulfur dioxide is readily oxidized and has a short lifetime in an atmosphere where oxidizing photochemical smog is present.

Smog has a long history. Exploring what is now southern California in 1542, Juan Rodriguez Cabrillo named San Pedro Bay "The Bay of Smokes" because of the heavy haze that covered the area. Complaints of eye irritation from anthropogenically polluted air in Los Angeles were recorded as far back as 1868. Characterized by reduced visibility, eye irritation, cracking of rubber, and deterioration of materials, smog became a serious nuisance in the Los Angeles area during the 1940s. It is now recognized as a major air pollution problem in many areas of the world.

Smoggy conditions are manifested by moderate to severe eye irritation or visibility below 3 miles when the relative humidity is below 60%. The formation of oxidants in the air, particularly ozone, is indicative of smog formation. Serious levels of photochemical smog may be assumed to be present when the oxidant level exceeds 0.15 ppm for more than 1 h. The three ingredients required to generate photochemical smog are ultraviolet light, hydrocarbons, and nitrogen oxides. Advanced techniques of analysis have shown a large variety of hydrocarbon precursors to smog formation in the atmosphere.

The importance of ozone as an atmospheric pollutant in atmospheres contaminated with photochemical smog have been recognized by changing regulations to lower allowable ozone concentrations in the United States. Allowable ozone levels were further reduced in 2008.

From the time it was recognized as a significant air pollution problem in the 1940s, the photochemical smog problem has been the subject of intense studies by chemists, efforts that were largely responsible for the evolution of the discipline of atmospheric chemistry. Much of the progress made in this area was due to advances in the study of chemical kinetics in the gas phase, the power of computers to make complex calculations, and advances in instrumentation to measure low levels of chemical species in polluted atmospheres. This chapter discusses the chemistry of photochemical smog.

Photochemical smog forms in the troposphere and is very much influenced by tropospheric conditions. The troposphere can be viewed as being divided into two major regions. The lowest layer, typically around 1 km thick, contains the *planetary boundary layer* in which there is maximum interaction between tropospheric air and Earth's surface. It is the region in which temperature inversions (see Figure 9.7) form and hold smog-forming chemicals with minimal mixing and dispersion so that they can interact with sunlight and each other to produce smog. Above this lower layer and extending up to the tropopause where the stratosphere begins is the *free troposphere*.

In August 2003, Europe experienced a devastating heat wave that killed thousands of people. In addition to the intense heat, the incident was characterized by a stagnant boundary layer, strong emissions of hydrocarbons and nitrogen oxides from the anthrosphere, and widespread forest fires that pumped large quantities of smog-forming emissions into the atmosphere. The result was a long period of photochemical smog formation adding to the misery caused by the prolonged period of high temperatures.[1]

13.2 SMOG-FORMING EMISSIONS

Internal combustion engines used in automobiles and trucks produce reactive hydrocarbons and nitrogen oxides, two of the three key ingredients required for smog to form. Therefore, automotive air emissions are discussed next.

The production of nitrogen oxides was discussed in Section 11.6. At the high temperature and pressure conditions in an internal combustion engine, products of incompletely burned gasoline undergo chemical reactions, which produce several hundred different hydrocarbons. Many of these are highly reactive in forming photochemical smog. As shown in Figure 13.1, the automobile has several potential sources of hydrocarbon emissions other than the exhaust. The first of these to be controlled was the mist of hydrocarbons composed of lubricating oil and "blowby" emanating from the engine crankcase. The latter consists of exhaust gas and unoxidized fuel/air mixture that enters the crankcase from the combustion chambers around the pistons. This mist is destroyed by recirculating it through the engine intake manifold by way of the positive crankcase ventilation (PCV) valve.

A second major source of automotive hydrocarbon emissions is the fuel system, from which hydrocarbons are emitted from the fuel tank and vents on carburetors, which used to be the primary means of introducing fuel/air mixtures into automobile engines. When the engine is shut off and the engine heat warms up the fuel system, gasoline may be evaporated and emitted to the atmosphere. In addition, heating during the daytime and cooling at night causes the fuel tank to breathe and emit gasoline fumes. Such emissions are reduced by fuel formulated to reduce volatility. Automobiles are equipped with canisters of carbon, which collect evaporated fuel from the fuel tank and fuel system, to be purged and burned when the engine is operating. Modern automobile engines with fuel injection systems emit much less hydrocarbon vapor than earlier models equipped with carburetors.

13.2.1 Control of Exhaust Hydrocarbons

In order to understand the production and control of automotive hydrocarbon exhaust products, it is helpful to understand the basic principles of the internal combustion engine. As shown in Figure 13.2, the four steps involved in one complete cycle of the four-cycle engine used in most vehicles are the following:

1. *Intake*: Air is drawn into the cylinder through the open intake valve. Gasoline is either injected with the intake air or injected separately into the cylinder.
2. *Compression*: The combustible mixture is compressed at a ratio of about 7:1. Higher compression ratios favor thermal efficiency and complete combustion of hydrocarbons. However, higher temperatures, premature combustion ("pinging"), and high production of nitrogen oxides also result from higher compression ratios.

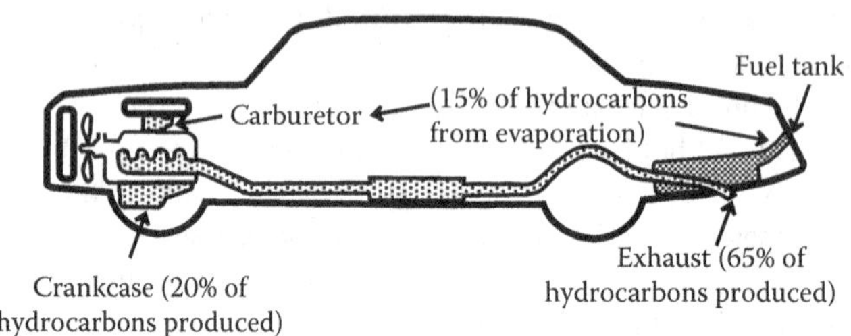

FIGURE 13.1 Potential sources of pollutant hydrocarbons from an automobile built in years before automotive air pollutants were controlled.

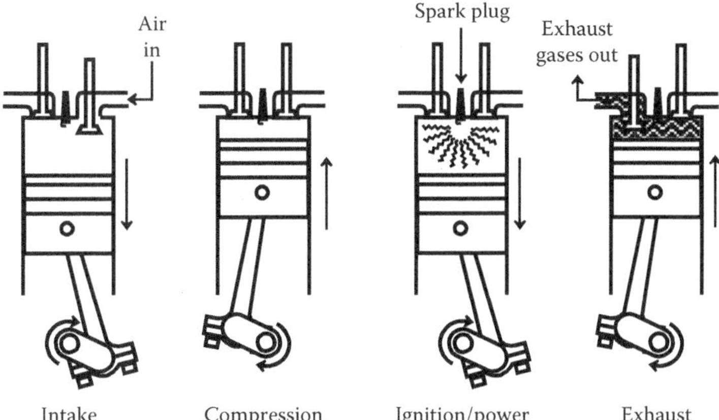

FIGURE 13.2 Steps in one complete cycle of a four-cycle internal combustion engine. Fuel is mixed with the intake air or injected separately into each cylinder.

3. *Ignition and power stroke*: As the fuel–air mixture normally produced by injecting fuel into the cylinder is ignited by the spark plug near top-dead-center, a temperature of about 2500°C is reached very rapidly at pressures up to 40 atm. As the gas volume increases with downward movement of the piston, the temperature decreases in a few milliseconds. This rapid cooling "freezes" nitric oxide in the form of NO without allowing it to dissociate to N_2 and O_2, which are thermodynamically favored at the normal temperatures and pressures of the atmosphere.
4. *Exhaust*: Exhaust gases consisting largely of N_2 and CO_2, with traces of CO, NO, hydrocarbons, and O_2, are pushed out through the open exhaust valve, thus completing the cycle.

The primary cause of unburned hydrocarbons in the engine cylinder is wall quench, wherein the relatively cool wall in the combustion chamber of the internal combustion engine causes the flame to be extinguished within several thousandths of a centimeter from the wall. Part of the remaining hydrocarbons may be retained as residual gas in the cylinder, and part may be oxidized in the exhaust system. The remainder is emitted to the atmosphere as pollutant hydrocarbons. Engine misfire due to improper adjustment and deceleration greatly increases the emission of hydrocarbons. Turbine engines are not subject to the wall quench phenomenon because their surfaces are always hot.

Several engine design characteristics favor lower exhaust hydrocarbon emissions. Wall quench, which is mentioned above, is diminished by design that decreases the combustion chamber surface/volume ratio through reduction of compression ratio, more nearly spherical combustion chamber shape, increased displacement per engine cylinder, and increased ratio of stroke relative to bore.

Spark retard also reduces exhaust hydrocarbon emissions. For optimum engine power and economy, the spark should be set to fire appreciably before the piston reaches the top of the compression stroke and begins the power stroke. Retarding the spark to a point closer to top-dead-center reduces the hydrocarbon emissions markedly. One reason for this reduction is that the effective surface/volume ratio of the combustion chamber is reduced, thus cutting down on wall quench. Second, when the spark is retarded, the combustion products are purged from the cylinders sooner after combustion. Therefore, the exhaust gas is hotter, and reactions consuming hydrocarbons are promoted in the exhaust system.

As shown in Figure 13.3, the air/fuel ratio in the internal combustion engine has a marked effect upon the emission of hydrocarbons. As the air/fuel ratio becomes richer in fuel than the stoichiometric ratio, the emission of hydrocarbons increases significantly. There is a moderate decrease in hydrocarbon emissions when the mixture becomes appreciably leaner in fuel than the stoichiometric ratio requires. The lowest level of hydrocarbon emissions occurs at an air/fuel ratio somewhat leaner in fuel than the stoichiometric ratio. This behavior is the result of a combination of factors, including minimum quench layer thickness at an air/fuel ratio somewhat richer in fuel than the

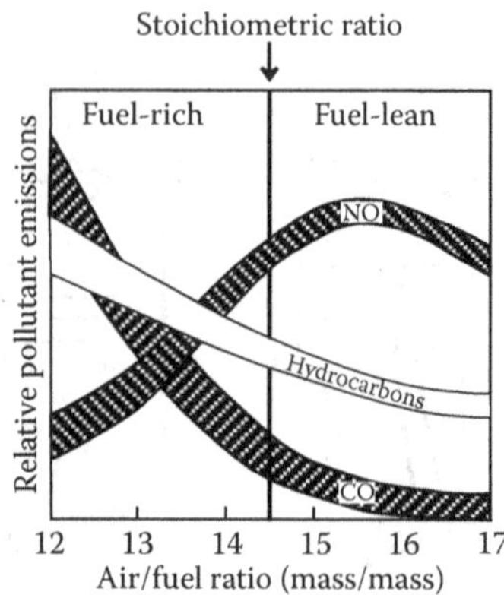

FIGURE 13.3 Effects of air/fuel ratio on pollutant emissions from an internal combustion piston engine.

stoichiometric ratio, decreasing hydrocarbon concentration in the quench layer with a leaner mixture, increasing oxygen concentration in the exhaust with a leaner mixture, and a peak exhaust temperature at a ratio slightly leaner in fuel than the stoichiometric ratio.

Catalytic converters are now used to destroy pollutants in exhaust gases. Currently, the most commonly used automotive catalytic converter is the three-way conversion catalyst, so called because a single catalytic unit destroys all three of the main class of automobile exhaust pollutants—hydrocarbons, carbon monoxide, and nitrogen oxides. This catalyst depends upon accurate sensing of oxygen levels in the exhaust combined with computerized engine control, which cycles the air/fuel mixture several times per second back and forth between slightly lean and slightly rich relative to the stoichiometric ratio. Under these conditions carbon monoxide, hydrogen, and hydrocarbons (C_cH_h) are oxidized.

$$CO + \tfrac{1}{2}O_2 \rightarrow CO_2 \tag{13.1}$$

$$H_2 + \tfrac{1}{2}O_2 \rightarrow H_2O \tag{13.2}$$

$$C_cH_h + (c + \tfrac{h}{4})O_2 \rightarrow cCO_2 + \tfrac{h}{2}H_2O \tag{13.3}$$

Nitrogen oxides are reduced on the catalyst to N_2 by carbon monoxide, hydrocarbons, or hydrogen as shown by the following reduction with CO:

$$CO + NO \rightarrow \tfrac{1}{2}N_2 + CO_2 \tag{13.4}$$

Automotive exhaust catalysts are dispersed on a high surface area substrate, most commonly consisting of cordierite, a ceramic composed of alumina (Al_2O_3), silica, and magnesium oxide. The substrate is formed as a honeycomb-type structure providing maximum surface area to contact exhaust gases. The support needs to be mechanically strong to withstand vibrational stresses from the automobile, and it must resist severe thermal stresses in which the temperature may rise from ambient temperatures to approximately 900°C over an approximately 2-min period during "light-off" when the engine is started. The catalytic material, which composes only about 0.10–0.15% of the catalyst body, consists of a mixture of precious metals. Platinum and palladium catalyze the oxidation of hydrocarbons and carbon monoxide, and rhodium acts as a catalyst for the reduction of nitrogen oxides; presently, palladium is the most common precious metal in exhaust catalysts.

Since lead can poison auto exhaust catalysts, automobiles equipped with catalytic exhaust-control devices require lead-free gasoline, which has now displaced gasoline containing antiknock tetraethyl lead, the dominant automobile engine fuel that was standard until the 1970s. Sulfur in gasoline is also detrimental to catalyst performance, and the sulfur contents of gasoline and, more recently, diesel fuel have been greatly reduced in recent years.

The internal combustion automobile engine has been developed to a remarkably high degree of sophistication in terms of its emissions. Increased use of hybrid automobiles combining an internal combustion engine with an electric motor/generator that enable the internal combustion engine to run evenly under optimum operating conditions are lowering emissions even further.

The 1990 U.S. Clean Air Act called for reformulating gasoline by adding more oxygenated compounds to reduce emissions of hydrocarbons and carbon monoxide. However, this measure became rather controversial and problems were encountered with one of the major oxygenated additives, MTBE, which was detected as a common water pollutant in some areas. Because of these concerns, MTBE has been largely eliminated from gasoline and replaced by ethanol as an oxygenated additive.

Ethanol in gasoline poses some environmental and sustainability problems. Ethanol is regarded as a renewable fuel source made by fermentation of sugars, primarily from corn in the United States and abundant sugarcane in Brazil. Some studies have suggested that the life cycle of corn-based ethanol in gasoline will increase photochemical smog compared to gasoline that is entirely from petroleum sources.[2] Emissions to the atmosphere of volatile ethanol from fuel that is 85% ethanol and 15% gasoline (E85) could contribute to elevated atmospheric levels of photochemically produced acetaldehyde, a noxious smog component.

13.2.2 Automotive Emission Standards

Federal law and California state law mandate automotive emission standards. The allowable emissions have been on a downward trend since the first standards were imposed in the mid-1960s. Table 13.1 gives Federal emissions prior to controls and those for various time intervals since 1970.

13.2.3 Polluting Green Plants

In some areas, biogenic hydrocarbons emitted to the atmosphere by plants are significant—even dominant—sources of hydrocarbons contributing to smog. The hydrocarbons from plants that contribute the

TABLE 13.1
Exhaust Emission Standards for Light-Duty Motor Vehicles in the United States[a]

Model Year	HCs[b]	CO[b]	NO_x[b]
Before controls[c]	10.60	84.0	4.1
1970	4.1	34.0	—
1975	1.5	15.0	3.1
1980	0.41	7.0	2.0
1985	0.41	3.4	1.0
1990	0.41	3.4	1.0
1998	0.41 (0.25)[d]	3.4	0.4
2008	0.41 (0.25)[d]	3.4	0.4

[a] Standards for gasoline-fueled vehicles.
[b] HCs, hydrocarbons from exhaust; CO, carbon monoxide; NO_x, sum of NO and NO_2; all values in g/mile.
[c] Estimated average emissions per vehicle before implementation of controls.
[d] Values in parentheses are for nonmethane hydrocarbons.

most to smog formation are terpenes, highly reactive alkenes. Some of the most common biogenic terpenes are shown in Figure 12.1. These compounds include α-pinene from pine trees and limonene from citrus trees. The most widely emitted terpene from plants is isoprene,[3] a monomer in natural rubber:

$$H_2C{=}C(CH_3){-}CH{=}CH_2 \quad \text{Isoprene}$$

Photochemical oxidation of isoprene results in the formation of much of the aerosol encountered in forest regions. The primary oxidation products of isoprene under smog-forming conditions are the carbonyls methacrolein and methylvinyl ketone:

$$H_2C{=}C(CH_3){-}CH{=}O \qquad O{=}C(CH_3){-}CH{=}CH_2$$

Methacrolein Methylvinyl ketone

A significant product of the atmospheric oxidation of isoprene consists of isoprene nitrates, which can be formed by the reaction of isoprene with hydroxyl radical, $HO^{\bullet}$, in the presence of nitrogen oxides and by reaction of isoprene with nitrate radical, NO_3.[4]

13.3 SMOG-FORMING REACTIONS OF ORGANIC COMPOUNDS IN THE ATMOSPHERE

Hydrocarbons are eliminated from the atmosphere by a number of chemical and photochemical reactions. These reactions are responsible for the formation of many noxious secondary pollutant products and intermediates from relatively innocuous hydrocarbon precursors. These pollutant products and intermediates make up photochemical smog.

Hydrocarbons and most other organic compounds in the atmosphere are thermodynamically unstable toward oxidation and tend to be oxidized through a series of steps. The oxidation process terminates with the formation of CO_2, solid organic particulate matter which settles from the atmosphere, or water-soluble products (e.g., acids, aldehydes) which are removed by rain. Inorganic species such as ozone or nitric acid are byproducts of these reactions.

13.3.1 Photochemical Reactions of Methane

Some of the major reactions involved in the oxidation of atmospheric hydrocarbons may be understood by considering the oxidation of methane, the most common and widely dispersed atmospheric hydrocarbon (though also the least reactive in the atmosphere). Like other hydrocarbons, methane reacts with oxygen atoms (generally produced by the photochemical dissociation of NO_2 to O and NO) to generate the all-important hydroxyl radical and an alkyl (methyl) radical

$$CH_4 + O \rightarrow H_3C\bullet + HO\bullet \qquad (13.5)$$

The methyl radical produced reacts rapidly with molecular oxygen to form very reactive peroxyl radicals,

$$H_3C\bullet + O_2 + M \text{ (energy-absorbing third body, usually a molecule of } N_2 \text{ or } O_2) \rightarrow H_3COO\bullet + M \qquad (13.6)$$

in this case, methyl peroxyl radical, $H_3COO^{\bullet}$. Such radicals participate in a variety of subsequent chain reactions, including those leading to smog formation. The hydroxyl radical reacts rapidly with hydrocarbons to form reactive hydrocarbon radicals,

$$CH_4 + HO^\bullet \rightarrow H_3C^\bullet + H_2O \tag{13.7}$$

in this case, the methyl radical, $H_3C^\bullet$. The following are more reactions involved in the overall oxidation of methane:

$$H_3COO^\bullet + NO \rightarrow H_3CO^\bullet + NO_2 \tag{13.8}$$

(This is a very important kind of reaction in smog formation because the oxidation of NO by peroxyl radicals is the predominant means of regenerating NO_2 in the atmosphere after it has been photochemically dissociated to NO.)

$$H_3CO^\bullet + O_3 \rightarrow \text{various products} \tag{13.9}$$

$$H_3CO^\bullet + O_2 \rightarrow CH_2O + HOO^\bullet \tag{13.10}$$

$$H_3COO^\bullet + NO_2 + M \rightarrow CH_3OONO_2 + M \tag{13.11}$$

$$\underset{\text{(photodissociation of formaldehyde)}}{CH_2O + h\nu \rightarrow \text{photodissociation products}} \tag{13.12}$$

As noted throughout this chapter, hydroxyl radical, $HO^\bullet$, and hydroperoxyl radical, $HOO^\bullet$, are ubiquitous intermediates in photochemical chain-reaction processes. These two species are known collectively as *odd hydrogen radicals.*

Reactions such as 13.5 and 13.7 are *abstraction reactions* involving the removal of an atom, usually hydrogen, by reaction with an active species. *Addition reactions* of organic compounds are also common. Typically, hydroxyl radical reacts with an alkene such as propylene to form another reactive free radical:

$$HO^\bullet + H_3C\text{—}CH{=}CH_2 \rightarrow H_3C\text{—}\dot{C}H\text{—}CH_2\text{—}OH \tag{13.13}$$

Ozone adds to unsaturated compounds to form reactive ozonides:

$$H_3C\text{—}CH{=}CH_2 + O_3 \rightarrow H_3C\text{—}\underset{\diagdown O \diagup}{\overset{O\text{—}O}{CH\quad CH_2}} \tag{13.14}$$

Organic compounds (in the troposphere, almost exclusively carbonyls) can undergo primary photochemical reactions resulting in the direct formation of free radicals. By far the most important of these is the photochemical dissociation of aldehydes:

$$H_3C\text{—}C({=}O)\text{—}H + h\nu \rightarrow H_3C^\bullet + H\dot{C}O \tag{13.15}$$

Organic free radicals undergo a number of chemical reactions. Hydroxyl radicals may be generated from organic peroxyl reactions such as

$$H_3C\text{—}CH(OO^\bullet)\text{—}CH_3 \rightarrow H_3C\text{—}C({=}O)\text{—}CH_3 + HO^\bullet \tag{13.16}$$

leaving an aldehyde or ketone. The hydroxyl radical may react with other organic compounds, maintaining the chain reaction. Gas-phase reaction chains commonly have many steps. Furthermore, chain-branching reactions take place in which a free radical reacts with an excited molecule causing it to produce two new radicals. Chain termination may occur in several ways, including reaction of two free radicals,

$$2HO^\bullet \rightarrow H_2O_2 \tag{13.17}$$

adduct formation with nitric oxide or nitrogen dioxide (which, because of their odd numbers of electrons, are themselves stable free radicals),

$$HO^\bullet + NO_2 + M \rightarrow HNO_3 + M \tag{13.18}$$

or reaction of the radical with a solid particle surface.

Hydrocarbons may undergo heterogeneous reactions on particles in the atmosphere. Dusts composed of metal oxides or charcoal have a catalytic effect upon the oxidation of organic compounds. Metal oxides may enter into photochemical reactions. For example, zinc oxide photosensitized by exposure to light promotes oxidation of organic compounds.

The kinds of reactions just discussed are involved in the formation of photochemical smog in the atmosphere. Much of what is known about the reactions that take place in a smog-forming atmosphere has been learned from studies in large chambers containing a stationary mass of air that is subjected to conditions conducive to the formation of photochemical smog including exposure to ultraviolet radiation, low humidity, and presence of reactive hydrocarbons and nitrogen oxides.[5] Next, consideration is given to the smog-forming process.

13.4 OVERVIEW OF SMOG FORMATION

This section addresses the conditions that are characteristic of a smoggy atmosphere and the overall processes involved in smog formation. In atmospheres that receive hydrocarbon and NO pollution accompanied by intense sunlight and stagnant air masses, oxidants tend to form. In air-pollution parlance, *gross photochemical oxidant* is a substance in the atmosphere capable of oxidizing iodide ion to elemental iodine. Sometimes other reducing agents are used to measure oxidants. The primary oxidant in the atmosphere is ozone. Other atmospheric oxidants include H_2O_2, organic peroxides (ROOR′), organic hydroperoxides (ROOH), and peroxyacyl nitrates such as peroxyacetyl nitrate (PAN).

$$H_3C-C(=O)-OO-NO_2$$

Peroxyacetyl nitrate (PAN)

Nitrogen dioxide, NO_2, is not regarded as a gross photochemical oxidant. However, it is about 15% as efficient as O_3 in oxidizing iodide to iodine(0), and a correction is made in measurements for the positive interference of NO_2. Sulfur dioxide is oxidized by O_3 and produces a negative interference for which a measurement correction must also be made.

PAN and related compounds containing the $-C(O)OONO_2$ moiety, such as peroxybenzoyl nitrate (PBN),

$$C_6H_5-C(=O)-OO-NO_2$$

Peroxybenzoyl nitrate (PBN)

a powerful eye irritant and lachrymator, are produced photochemically in atmospheres containing alkenes and NO_x. PAN, especially, is a notorious organic oxidant. It has several adverse effects including eye irritation, phytotoxicity, and mutagenicity and is perhaps the best single indicator of photochemical smog conditions. In addition to PAN and PBN, some other specific organic oxidants that may be important in polluted atmospheres are peroxypropionyl nitrate (PPN); peracetic acid, $CH_3(CO)OOH$; acetylperoxide, $CH_3(CO)OO(CO)CH_3$; butyl hydroperoxide, $CH_3CH_2CH_2CH_2OOH$; and *tert*-butylhydroperoxide, $(CH_3)_3COOH$. Fortunately, levels of PAN, PPN, and other organic oxidants as well have decreased significantly in smog-prone areas such as southern California from the 1960s to the present, the result of emission control measures that have been implemented.

As shown in Figure 13.4, smoggy atmospheres show characteristic variations with time of day in levels of NO, NO_2, hydrocarbons, aldehydes, and oxidants. Examination of the figure shows that shortly after sunrise the level of NO in the atmosphere decreases markedly, a decrease that is accompanied by a peak in the concentration of NO_2. During midday (significantly, after the concentration of NO has fallen to a very low level), the levels of aldehydes and oxidants become relatively high. The concentration of total hydrocarbons in the atmosphere peaks sharply in the morning, then decreases during the remaining daylight hours.

An overview of the processes responsible for the behavior just discussed is summarized in Figure 13.5. The chemical bases for the processes illustrated in this figure are explained in the following section.

13.5 MECHANISMS OF SMOG FORMATION

Here are discussed some of the primary aspects of photochemical smog formation. For more details the reader is referred to books on atmospheric chemistry and atmospheric chemistry and physics listed in the Supplementary References at the end of this chapter. Since the exact chemistry of photochemical smog formation is very complex, many of the reactions are given as plausible illustrative examples rather than proven mechanisms.

The kind of behavior summarized in Figure 13.4 contains several apparent anomalies, which puzzled scientists for many years. The first of these was the rapid increase in NO_2 concentration and decrease in NO concentration under conditions where it was known that photodissociation of NO_2 to O and NO was occurring. Furthermore, it could be shown that the disappearance of alkenes and other hydrocarbons was much more rapid than could be explained by their relatively slow reactions with O_3 and O. These anomalies are now explained by chain reactions involving the interconversion of NO and NO_2, the oxidation of hydrocarbons, and the generation of reactive intermediates, particularly hydroxyl radical ($HO^\bullet$).

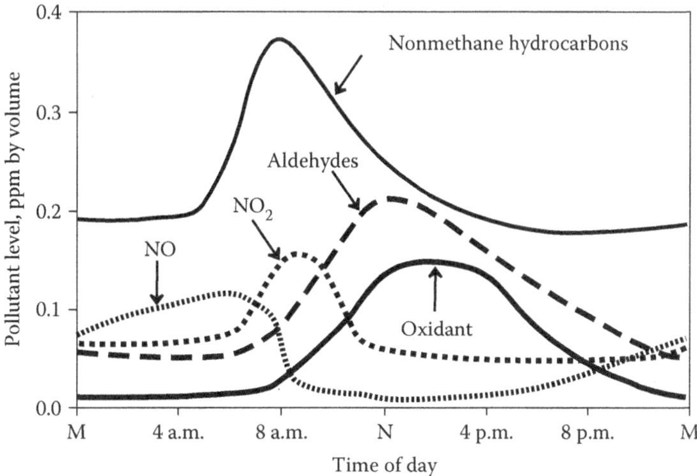

FIGURE 13.4 Generalized plot of atmospheric concentrations of species involved in smog formation as a function of time of day.

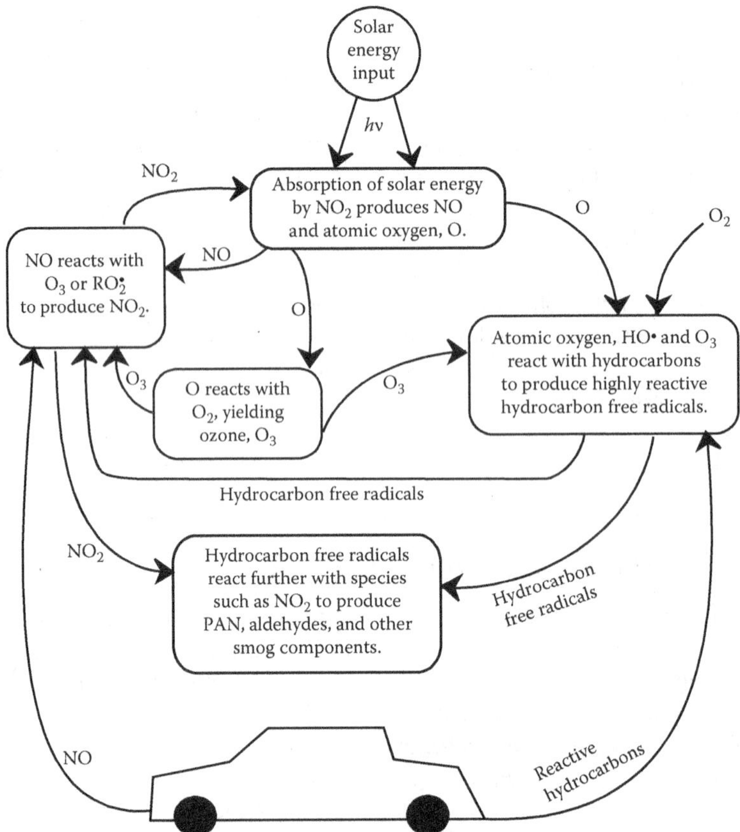

FIGURE 13.5 Generalized scheme for the formation of photochemical smog.

Figure 13.5 shows the overall reaction scheme for smog formation, which is based upon the photochemically initiated reactions that occur in an atmosphere containing nitrogen oxides, reactive hydrocarbons, and oxygen. The time variations in levels of hydrocarbons, ozone, NO, and NO_2 are explained by the following overall reactions:

1. Primary photochemical reaction producing oxygen atoms:

$$NO_2 + h\nu\ (\lambda < 394\text{ nm}) \rightarrow NO + O \tag{13.19}$$

2. Reactions involving oxygen species (M is an energy-absorbing third body):

$$O_2 + O + M \rightarrow O_3 + M \tag{13.20}$$

$$O_3 + NO \rightarrow NO_2 + O_2 \tag{13.21}$$

Because the latter reaction is rapid, the concentration of O_3 remains low until that of NO falls to a low value. Automotive emissions of NO tend to keep O_3 concentrations lower along freeways.

3. Production of organic free radicals from hydrocarbons, RH:

$$O + RH \rightarrow R^{\bullet} + \text{other products} \tag{13.22}$$

$$O_3 + RH \rightarrow R^\bullet + \text{and/or other products} \quad (13.23)$$

($R^\bullet$ is a free radical in which the dot denotes an unpaired electron, which may or may not contain oxygen.)

4. Chain propagation, branching, and termination by a variety of reactions such as the following:

$$NO + ROO^\bullet \rightarrow NO_2 + \text{and/or other products} \quad (13.24)$$

$$NO_2 + R^\bullet \rightarrow \text{products (e.g., PAN)} \quad (13.25)$$

The latter kind of reaction is the most common chain-terminating process in smog because NO_2 is a stable free radical present at high concentrations. Chains may terminate also by reaction of free radicals with NO or by reaction of two $R^\bullet$ radicals, although the latter is uncommon because of the relatively low concentrations of radicals compared to molecular species. Chain termination by radical sorption on a particle surface is also possible and may contribute to aerosol particle growth.

A large number of specific reactions are involved in the overall scheme for the formation of photochemical smog. The formation of atomic oxygen by a primary photochemical reaction (Reaction 13.19) leads to several reactions involving oxygen and nitrogen oxide species:

$$O + O_2 + M \rightarrow O_3 + M \quad (13.26)$$

$$O + NO + M \rightarrow NO_2 + M \quad (13.27)$$

$$O + NO_2 \rightarrow NO + O_2 \quad (13.28)$$

$$O_3 + NO \rightarrow NO_2 + O_2 \quad (13.29)$$

$$O + NO_2 + M \rightarrow NO_3 + M \quad (13.30)$$

$$O_3 + NO_2 \rightarrow NO_3 + O_2 \quad (13.31)$$

There are a number of significant atmospheric reactions involving nitrogen oxides, water, nitrous acid, and nitric acid:

$$NO_3 + NO_2 \rightarrow N_2O_5 \quad (13.32)$$

$$N_2O_5 \rightarrow NO_3 + NO_2 \quad (13.33)$$

$$NO_3 + NO \rightarrow 2NO_2 \quad (13.34)$$

$$N_2O_5 + H_2O \rightarrow 2HNO_3 \quad (13.35)$$

(This reaction is slow in the gas phase but may be fast on surfaces.)

Very reactive $HO^\bullet$ radicals can be formed by the reaction of excited atomic oxygen with water,

$$O^* + H_2O \rightarrow 2HO^\bullet \quad (13.36)$$

by photodissociation of hydrogen peroxide,

$$H_2O_2 + h\nu\ (\lambda < 350\text{ nm}) \rightarrow 2HO^\bullet \quad (13.37)$$

or by the photolysis of nitrous acid,

$$HNO_2 + h\nu \rightarrow HO^\bullet + NO \quad (13.38)$$

This reaction may be the main source of hydroxyl radical during early morning hours when sources deriving from the presence of ozone are minimal. Among the inorganic species with which the hydroxyl radical reacts are oxides of nitrogen,

$$HO^{\bullet} + NO_2 \rightarrow HNO_3 \tag{13.39}$$

$$HO^{\bullet} + NO + M \rightarrow HNO_2 + M \tag{13.40}$$

and carbon monoxide,

$$CO + HO^{\bullet} + O_2 \rightarrow CO_2 + HOO^{\bullet} \tag{13.41}$$

The last reaction is significant in that it is responsible for the disappearance of much atmospheric CO (see Section 11.3) and because it produces the hydroperoxyl radical $HOO^{\bullet}$. One of the major inorganic reactions of the hydroperoxyl radical is the oxidation of NO:

$$HOO^{\bullet} + NO \rightarrow HO^{\bullet} + NO_2 \tag{13.42}$$

For purely inorganic systems, kinetic calculations and experimental measurements cannot explain the rapid transformation of NO to NO_2 that occurs in an atmosphere undergoing photochemical smog formation and predict that the concentration of NO_2 should remain very low. However, in the presence of reactive hydrocarbons, NO_2 accumulates very rapidly by a reaction process beginning with its photodissociation! It may be concluded, therefore, that the organic compounds form species that react with NO directly rather than with NO_2.

A number of chain reactions have been shown to result in the general type of species behavior with time shown in Figure 13.4. When alkane hydrocarbons, RH, react with O, O_3, or $HO^{\bullet}$ radical,

$$RH + O + O_2 \rightarrow ROO^{\bullet} + HO^{\bullet} \tag{13.43}$$

$$RH + HO^{\bullet} + O_2 \rightarrow ROO^{\bullet} + H_2O \tag{13.44}$$

reactive oxygenated organic radicals, $ROO^{\bullet}$, are produced. Alkenes are much more reactive, undergoing primarily addition reactions with hydroxyl radical,

$$R_2C{=}CR_2 + HO^{\bullet} \xrightarrow{\text{Very rapid}} \underset{\text{Radical adduct}}{HO{-}CR_2{-}CR_2^{\bullet}} \rightarrow \text{Oxidation products} \tag{13.45}$$

(where R may be one of a number of hydrocarbon moieties or an H atom), with oxygen atoms,

$$R_2C{=}CR_2 + O \rightarrow \underset{\text{Biradical}}{R{-}C(O^{\bullet})(R){-}CR_2^{\bullet}} \rightarrow \text{Oxidation products} \tag{13.46}$$

or with ozone:

$$R_2C{=}CR_2 + O_3 \rightarrow R{-}\overset{O-O-O}{C(R){-}C(R)}{-}R \rightarrow \text{Oxidation products} \tag{13.47}$$

Aromatic hydrocarbons, Ar-H, may also react with O and HO$^\bullet$. Addition reactions of aromatics with HO$^\bullet$ are favored. The product of the reaction of benzene with HO$^\bullet$ is phenol, as shown by the following reaction sequence in which the half-circle and dot inside the hexagon structure represents a free radical formed by adding HO$^\bullet$ to the aromatic ring:

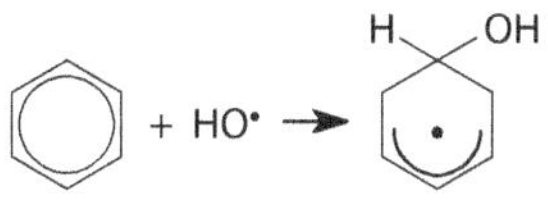

$$C_6H_6 + HO^\bullet \rightarrow C_6H_6(OH)^\bullet \quad (13.48)$$

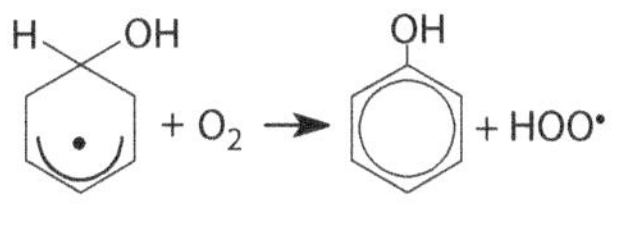

$$C_6H_6(OH)^\bullet + O_2 \rightarrow C_6H_5OH + HOO^\bullet \quad (13.49)$$

In the case of alkyl benzenes, such as toluene, the hydroxyl radical attack may occur on the alkyl group, leading to reaction sequences such as those of alkanes.

Aldehydes react with HO$^\bullet$,

$$R{-}\overset{\overset{O}{\|}}{C}{-}H + HO^\bullet + O_2 \rightarrow R{-}\overset{\overset{O}{\|}}{C}{-}OO^\bullet + H_2O \quad (13.50)$$

$$H_2C{=}O + HO^\bullet + \tfrac{3}{2}O_2 \rightarrow CO_2 + HOO^\bullet + H_2O \quad (13.51)$$

and undergo photochemical reactions:

$$R{-}\overset{\overset{O}{\|}}{C}{-}H + h\nu + 2O_2 \rightarrow ROO^\bullet + CO + HOO^\bullet \quad (13.52)$$

$$H_2C{=}O + h\nu + 2O_2 \rightarrow CO + HOO^\bullet \quad (13.53)$$

Hydroxyl radical (HO$^\bullet$), which reacts with some hydrocarbons at rates that are almost diffusion-controlled, is the predominant reactant in early stages of smog formation. Significant contributions are made by hydroperoxyl radical (HOO$^\bullet$), O_3, and (at night) NO_3 after smog formation is well underway.

One of the most important reaction sequences in the smog-formation process begins with the abstraction by HO$^\bullet$ of a hydrogen atom from a hydrocarbon and leads to the oxidation of NO to NO_2 as follows:

$$RH + HO^\bullet \rightarrow R^\bullet + H_2O \quad (13.54)$$

The alkyl radical, R$^\bullet$, reacts with O_2 to produce a peroxyl radical, ROO$^\bullet$:

$$R^\bullet + O_2 \rightarrow ROO^\bullet \quad (13.55)$$

This strongly oxidizing species very effectively oxidizes NO to NO_2,

$$ROO^\bullet + NO \rightarrow RO^\bullet + NO_2 \quad (13.56)$$

thus explaining the once-puzzling rapid conversion of NO to NO_2 in an atmosphere in which the latter is undergoing photodissociation. The alkoxyl radical product, RO$^\bullet$, is not so stable as ROO$^\bullet$.

In cases where the oxygen atom is attached to a carbon atom that is also bonded to H, a carbonyl compound is likely to be formed by the following type of reaction:

$$H_3CO^\bullet + O_2 \rightarrow H{-}\overset{\displaystyle O}{\overset{\|}{C}}{-}H + HOO^\bullet \tag{13.57}$$

The rapid production of photosensitive carbonyl compounds from alkoxyl radicals is an important stimulant for further atmospheric photochemical reactions. In the absence of extractable hydrogen, cleavage of a radical containing the carbonyl group occurs:

$$H_3C{-}\overset{\displaystyle O}{\overset{\|}{C}}{-}O^\bullet \rightarrow H_3C^\bullet + CO_2 \tag{13.58}$$

Carbonyl compounds, which are discussed in more detail in Section 12.4, are important initiators, intermediates and end products of smog formation. More than 50 carbonyls may be present in highly polluted air. The most abundant carbonyls in the polluted urban atmosphere are usually formaldehyde and acetaldehyde; acetone is usually the most abundant ketone.

The oxidation of NO to NO_2 in the atmosphere was noted above. Another reaction that can lead to the oxidation of NO is of the following type:

$$R{-}\overset{\displaystyle O}{\overset{\|}{C}}{-}OO^\bullet + NO + O_2 \rightarrow ROO^\bullet + NO_2 + CO_2 \tag{13.59}$$

Peroxyacyl nitrates are formed by an addition reaction with NO_2:

$$R{-}\overset{\displaystyle O}{\overset{\|}{C}}{-}OO^\bullet + NO_2 \rightarrow R{-}\overset{\displaystyle O}{\overset{\|}{C}}{-}OO{-}NO_2 \tag{13.60}$$

When R is the methyl group, the product is PAN, mentioned in Section 13.4. Although it is thermally unstable, PAN does not undergo photolysis rapidly, reacts only slowly with $HO^\bullet$ radical, and has a low water solubility. Therefore, the major pathway by which it is lost from the atmosphere is thermal decomposition, the reverse of Reaction 13.60.

Peroxyacyl nitrates are highly significant air pollutants that are symptomatic of photochemical smog. These compounds are eye irritants and mutagens. As potent phytotoxins, they are damaging to plants and they are agents for the atmospheric transport of reactive nitrogen. There is concern that the increasing use of ethanol as a gasoline antiknock additive will increase levels of PAN. This is because ethanol is photochemically oxidized to acetaldehyde, the photochemical oxidation of which leads to the formation of PAN.[6]

Alkyl nitrates and alkyl nitrites may be formed by the reaction of alkoxyl radicals ($RO^\bullet$) with nitrogen dioxide and nitric oxide, respectively:

$$RO^\bullet + NO_2 \rightarrow RONO_2 \tag{13.61}$$

$$RO^\bullet + NO \rightarrow RONO \tag{13.62}$$

Addition reactions with NO_2 and NO such as those shown in Reactions 13.61 and 13.62 are important in terminating the reaction chains involved in smog formation. The fact that NO_2 is involved in a chain termination step (Reaction 13.61) as well as a chain initiation step (Reaction 13.19) may limit the effectiveness of moderate reductions in NO_x emissions in curtailing smog formation. Nitrates and peroxynitrates are important constituents of smoggy atmospheres.

As shown in Reaction 13.57, the reaction of oxygen with alkoxyl radicals produces hydroperoxyl radical. Peroxyl radicals can react with one another to produce reactive hydrogen peroxide, alkoxyl radicals, and hydroxyl radicals:

$$HOO^\bullet + HOO^\bullet \rightarrow H_2O_2 + O_2 \tag{13.63}$$

$$HOO^{\bullet} + ROO^{\bullet} \rightarrow RO^{\bullet} + HO^{\bullet} + O_2 \quad (13.64)$$

$$ROO^{\bullet} + ROO^{\bullet} \rightarrow 2RO^{\bullet} + O_2 \quad (13.65)$$

13.5.1 Nitrate Radical

First observed in the troposphere in 1980, nitrate radical, NO_3, is now recognized as being an important atmospheric chemical species, especially at night. This species is formed by the reaction

$$NO_2 + O_3 \rightarrow NO_3 + O_2 \quad (13.66)$$

and exists in equilibrium with NO_2:

$$NO_2 + NO_3 + M \rightleftarrows N_2O_5 + M \quad \text{(energy-absorbing third body)} \quad (13.67)$$

Levels of NO_3 remain low during daylight, typically with a lifetime at noon of only about 5 s, because of the following two dissociation reactions:

$$NO_3 + h\nu\ (\lambda < 700\text{ nm}) \rightarrow NO + O_2 \quad (13.68)$$

$$NO_3 + h\nu\ (\lambda < 580\text{ nm}) \rightarrow NO_2 + O \quad (13.69)$$

It is likely that NO_3 levels become high enough in the hour before sunset to begin to have an effect on tropospheric chemistry. At night the levels of NO_3 become much higher, typically reaching values of around 8×10^7 molecules/cm^3 compared to only about 1×10^6 molecules/cm^3 for hydroxyl radical. Although the hydroxyl radical reacts 10–1000 times faster than NO_3, the much higher concentration of the latter means that it is responsible for much of the atmospheric chemistry that occurs at night. The nitrate radical adds across the double bonds in alkenes leading to the formation of reactive radical species that participate in smog formation.

13.5.2 Photolyzable Compounds in the Atmosphere

It may be useful at this time to review the types of compounds capable of undergoing photolysis in the troposphere and thus initiating chain reaction. Under most tropospheric conditions, the most important of these is NO_2:

$$NO_2 + h\nu\ (\lambda < 394\text{ nm}) \rightarrow NO + O \quad (13.19)$$

In relatively polluted atmospheres, the next most important photodissociation reaction is that of carbonyl compounds, particularly formaldehyde:

$$CH_2O + h\nu\ (\lambda < 335\text{ nm}) \rightarrow H^{\bullet} + H\dot{C}O \quad (13.70)$$

Hydrogen peroxide photodissociates to produce two hydroxyl radicals:

$$HOOH + h\nu\ (\lambda < 350\text{ nm}) \rightarrow 2HO^{\bullet} \quad (13.71)$$

Finally, organic peroxides may be formed and subsequently dissociate by the following reactions, starting with a peroxyl radical:

$$H_3COO^{\bullet} + HOO^{\bullet} \rightarrow H_3COOH + O_2 \quad (13.72)$$

$$H_3COOH + h\nu\ (\lambda < 350\text{ nm}) \rightarrow H_3CO^{\bullet} + HO^{\bullet} \quad (13.73)$$

It should be noted that Reactions 13.70, 13.71, and 13.73 each gives rise to two free radical species per photon absorbed. Ozone undergoes photochemical dissociation to produce excited oxygen atoms at wavelengths less than 315 nm. These atoms may react with H_2O to produce hydroxyl radicals.

13.6 REACTIVITY OF HYDROCARBONS

The reactivity of hydrocarbons in the smog formation process is an important consideration in understanding the process and in developing control strategies. It is useful to know which are the most reactive hydrocarbons so that their release can be minimized. Less reactive hydrocarbons, of which propane is a good example, may cause smog formation far downwind from the point of release.

Hydrocarbon reactivity is best based upon the interaction of hydrocarbons with hydroxyl radical. Methane, the least reactive common gas-phase hydrocarbon with an atmospheric half-life exceeding 10 days, is assigned a reactivity of 1.0. (Despite its low reactivity, methane is so abundant in the atmosphere that it accounts for a significant fraction of total hydroxyl radical reactions.) In contrast, β-pinene produced by conifer trees and other vegetation, is almost 9000 times as reactive as methane, and *d*-limonene, produced by orange rind, is almost 19,000 times as reactive. Relative to their rates of reaction with hydroxyl radical, hydrocarbon reactivities may be classified from I through V as shown in Table 13.2.

13.7 INORGANIC PRODUCTS FROM SMOG

Two major classes of inorganic products from smog are sulfates and nitrates. Inorganic sulfates and nitrates, along with sulfur and nitrogen oxides, can contribute to acidic precipitation, corrosion, reduced visibility, and adverse health effects.

TABLE 13.2
Relative Reactivities of Hydrocarbons and CO with HO• Radical

Reactivity Class	Reactivity Range[a]	Approximate Half-Life in the Atmosphere	Compounds in Increasing Order of Reactivity
I	<10	>10 days	Methane
II	10–100	24 h–10 days	CO, acetylene, ethane
III	100–1000	2.4–24 h	Benzene, propane, *n*-butane, isopentane, methylethyl ketone, 2-methylpentane, toluene, *n*-propylbenzene, isopropylbenzene, ethene, *n*-hexane, 3-methylpentane, ethylbenzene
IV	1000–10,000	15 min–2.4 h	*p*-xylene, *p*-ethyltoluene, *o*-ethyl-toluene, *o*-xylene, Methylisobutyl ketone, *m*-ethyltoluene, *m*-xylene, 1,2,3-trimethylbenzene, propene, 1,2,4-trimethyl-benzene, 1,3,5-trimethylbenzene, *cis*-2-butene, β-pinene, 1,3-butadiene
V	>10,000	<15 min	2-methyl-2-butene, 2,4-dimethyl-2-butene, *d*-limonene

Source: Based on data from Darnall, K. R. et al., reactivity scale for atmospheric hydrocarbons based on reaction with hydroxyl radical, *Environmental Science and Technology*, **10**, 692–696, 1976.

[a] Based on an assigned reactivity of 1.0 for methane reacting with hydroxyl radical.

Although slow in a clean atmosphere, the oxidation of SO_2 to sulfate species is relatively rapid under smoggy conditions. During severe photochemical smog conditions, oxidation rates of 5–10% per hour may occur, as compared to only a fraction of a percent per hour under normal atmospheric conditions. Thus, sulfur dioxide exposed to smog can produce very high local concentrations of sulfate, which can aggravate already bad atmospheric conditions.

Several oxidant species in smog can oxidize SO_2. Among the oxidants are compounds, including O_3, NO_3, and N_2O_5, as well as reactive radical species, particularly $HO^{\bullet}$, $HOO^{\bullet}$, O, $RO^{\bullet}$, and $ROO^{\bullet}$. The two major primary reactions are oxygen transfer,

$$SO_2 + O \text{ (from O, RO}^{\bullet}\text{, ROO}^{\bullet}\text{)} \rightarrow SO_3 \rightarrow H_2SO_4\text{, sulfates} \quad (13.74)$$

or addition of other species. As an example of the latter, $HO^{\bullet}$ adds to SO_2 to form a reactive species which can further react with oxygen, nitrogen oxides, or other species to yield sulfates, other sulfur compounds, or compounds of nitrogen:

$$HO^{\bullet} + SO_2 \rightarrow HOSOO^{\bullet} \quad (13.75)$$

The presence of $HO^{\bullet}$ (typically at a level of 3×10^6 radicals/cm^3, but appreciably higher in a smoggy atmosphere) makes this a likely route. Addition of SO_2 to $RO^{\bullet}$ or $ROO^{\bullet}$ can yield organic sulfur compounds.

It should be noted that the reaction of H_2S with $HO^{\bullet}$ is quite rapid. As a result, the normal atmospheric half-life of H_2S of about one-half day becomes much shorter in the presence of photochemical smog.

Inorganic nitrates or nitric acid are formed by several reactions in smog. Among the important reactions forming nitric acid are the reaction of N_2O_5 with water (Reaction 13.35) and the addition of hydroxyl radical to NO_2 (Reaction 13.39). The oxidation of NO or NO_2 to nitrate species may occur after absorption of gas by an aerosol droplet. Nitric acid formed by these reactions reacts with ammonia in the atmosphere to form ammonium nitrate:

$$NH_3 + HNO_3 \rightarrow NH_4NO_3 \quad (13.76)$$

Other nitrate salts may also be formed.

Nitric acid and nitrates are among the more damaging end products of smog. In addition to possible adverse effects on plants and animals, they cause severe corrosion problems. Electrical relay contacts and small springs associated with electrical switches are especially susceptible to damage from nitrate-induced corrosion.

13.8 EFFECTS OF SMOG

The harmful effects of smog occur mainly in the areas of (1) human health and comfort, (2) damage to materials, (3) effects on the atmosphere, and (4) toxicity to plants. The exact degree to which exposure to smog affects human health is not known, although substantial adverse effects are suspected. Pungent-smelling, smog-produced ozone is known to be toxic. Ozone at 0.15 ppm causes coughing, wheezing, bronchial constriction, and irritation to the respiratory mucous system in healthy, exercising individuals. In March 2008, the U.S. EPA released a revised eight-hour national ambient air quality standard for ground-level ozone of 0.075 ppm ozone based upon likely health effects of this pollutant. At the same time, the EPA also revised the secondary standard for ground-level ozone to the same 0.075 ppm level. The secondary standard is based upon evidence of ozone damage to plants, trees, and crops during the growing season. In addition to ozone, oxidant peroxyacyl nitrates and aldehydes found in smog are eye irritants.

Materials are adversely affected by some smog components. Rubber has a high affinity for ozone and is cracked and aged by it. Indeed, the cracking of rubber used to be employed as a test for the presence of ozone. Ozone attacks natural rubber and similar materials by oxidizing and breaking double bonds in the polymer according to the following reaction:

$$\left[-\overset{H}{\underset{H}{C}}-\overset{}{\underset{H}{C}}=\overset{CH_3}{C}-\overset{H}{\underset{H}{C}}- \right]_n + O_3 \rightarrow R-\overset{H}{C}-C\langle O-O, O\rangle C(CH_3)-R' \rightarrow R-\overset{O}{\overset{\|}{C}}-OH + H_3C-\overset{O}{\overset{\|}{C}}-R' \qquad (13.77)$$

Rubber polymer

This oxidative scission type of reaction causes bonds in the polymer structure to break and results in deterioration of the polymer.

Aerosol particles that reduce visibility are formed by the polymerization of the smaller molecules produced in smog-forming reactions.[7] Since these reactions largely involve the oxidation of hydrocarbons, it is not surprising that oxygen-containing organics make up the bulk of the particulate matter produced from smog. Ether-soluble aerosols collected from the Los Angeles atmosphere have shown an empirical formula of approximately CH_2O. Among the specific kinds of compounds identified in organic smog aerosols are alcohols, aldehydes, ketones, organic acids, esters, and organic nitrates. Hydrocarbons of plant origin are prominent among the precursors to particle formation in photochemical smog.

Smog aerosols likely form by condensation on existing nuclei rather than by self-nucleation of reaction product molecules. In support of this view are electron micrographs of these aerosols showing that smog aerosol particles in the micrometer-size region consist of liquid droplets with an inorganic electron-opaque core (Figure 13.6). Thus, particulate matter from a source other than smog may have some influence on the formation and properties of smog aerosols.

In view of worldwide shortages of food, the known harmful effects of smog on plants are of particular concern. These effects are largely due to oxidants in the smoggy atmosphere. The three major oxidants involved are ozone, PAN, and nitrogen oxides. Of these, PAN has the highest toxicity to plants, attacking younger leaves and causing "bronzing" and "glazing" of their surfaces. Exposure for several hours to an atmosphere containing PAN at a level of only 0.02–0.05 ppm will damage vegetation. The sulfhydryl group of proteins in organisms is susceptible to damage by PAN, which reacts with such groups as both an oxidizing agent and an acetylating agent. Fortunately, PAN is usually present at only low levels. Nitrogen oxides occur at relatively high concentrations during smoggy conditions, but their toxicity to plants is relatively low.

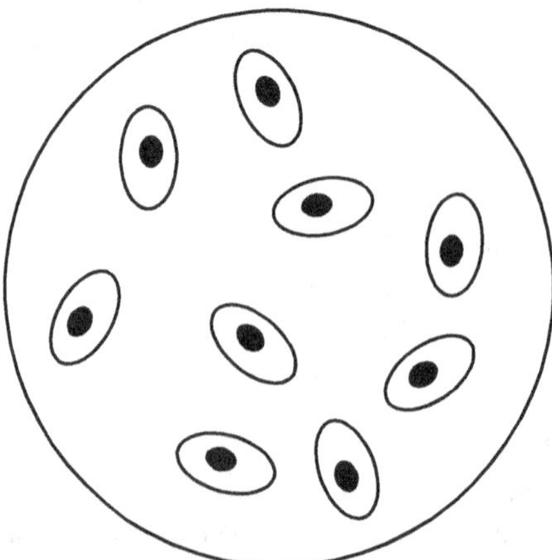

FIGURE 13.6 Representation of an electron micrograph of smog aerosol particles collected by a jet inertial impactor, showing electron-opaque nuclei in the centers of the impacted droplets.

Short-chain alkyl hydroperoxides, which were mentioned in Section 13.4, occur at low levels under smoggy conditions, and even in remote atmospheres. It is possible that these species can oxidize DNA bases, causing adverse genetic effects. Alkyl hydroperoxides are formed under smoggy conditions by the reaction of alkyl peroxy radicals with hydroperoxy radical, $HO_2^{\bullet}$, as shown for the formation of methyl hydroperoxide below:

$$H_3CO_2^{\bullet} + HO_2^{\bullet} \rightarrow H_3COOH + O_2 \tag{13.78}$$

Ames assays of methyl, ethyl, *n*-propyl, and *n*-butyl hydroperoxides (see Chapter 22) have shown some tendency toward mutagenicity on select strains of *Salmonella typhimurium*, although any conclusions drawn from such studies on human health should be made with caution.

The low toxicity of nitrogen oxides and the usually low levels of PAN, hydroperoxides, and other oxidants present in smog leave ozone as the greatest smog-produced threat to plant life. Some plant species, including sword-leaf lettuce, black nightshade, quickweed, and double-fortune tomato, are so susceptible to the effects of ozone and other photochemical oxidants that they are used as bioindicators of the presence of smog. Typical of the phytotoxicity of O_3, ozone damage to a lemon leaf is manifested by chlorotic stippling (characteristic yellow spots on a green leaf), as represented in Figure 13.7. Ponderosa and Jeffrey pines exposed to ozone and smog in California's San Bernardino Mountains have suffered chlorotic mottle and premature needle death. Reduction in plant growth may occur without visible lesions on the plant.

Brief exposure to approximately 0.06 ppm of ozone may temporarily cut photosynthesis rates in some plants in half. Crop damage from ozone and other photochemical air pollutants in California alone is estimated to cost millions of dollars each year. The geographic distribution of damage to plants in California is illustrated in Figure 13.8.

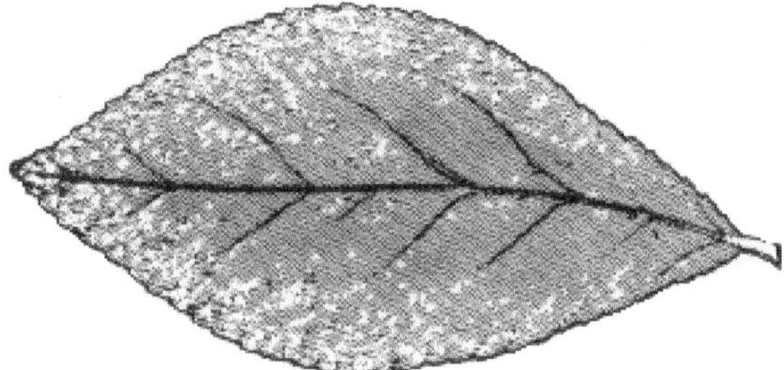

FIGURE 13.7 Representation of ozone damage to a lemon leaf. In color, the spots appear as yellow chlorotic stippling on the green upper surface caused by ozone exposure.

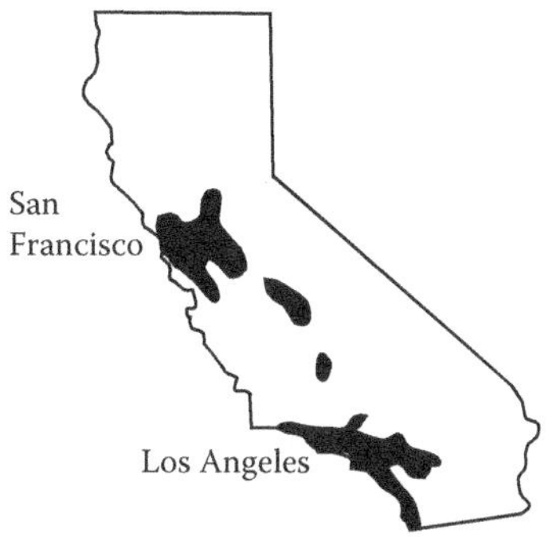

FIGURE 13.8 Geographic distribution of plant damage from smog in California.

LITERATURE CITED

1. Guerova, G. and N. Jones, A global model study of ozone enhancement during the August 2003 heat wave in Europe, *Environmental Chemistry*, **4**, 285–292, 2007.
2. Kim, S. and B. Dale, Life cycle assessment of fuel ethanol derived from corn grain via dry milling, *Bioresource Technology*, **99**, 5250–5260, 2008.
3. Sharkey, T. D., A. E. Wiberley, and A. R. Donohue, Isoprene emission from plants: Why and how, *Annals of Botany*, **101**, 5–18, 2008.
4. Horowitz, L. W., A. M. Fiore, G. P. Milly, R. C. Cohen, A. Perring, P. J. Wooldridge, P. G. Hess, L. K. Emmons, and J.-F. Lamarque, Observational constraints on the chemistry of isoprene nitrates over the eastern United States, *Journal of Geophysical Research (Atmospheres)*, **112**, D12S08/1–D12S08/13, 2007.
5. Wu, S., Z. Lu, J. Hao, Z. Zhao, J. Li, H. Takekawa, H. Minoura, and A. Yasuda, Construction and characterization of an atmospheric simulation smog chamber, *Advances in Atmospheric Sciences*, **24**, 250–258, 2007.
6. Jacobson, M. Z., Effects of ethanol (E85) versus gasoline vehicles on cancer and mortality in the United States, *Environmental Science and Technology*, **41**, 4150–4157, 2007.
7. Tsigaridis, K. and M. Kanakidou, Secondary organic aerosol importance in the future atmosphere, *Atmospheric Environment*, **41**, 4682–4692, 2007.

SUPPLEMENTARY REFERENCES

Allaby, M., *Fog, Smog, and Poisoned Rain*, Facts on File, New York, 2003.

Balduino, S. P., Ed., *Progress in Air Pollution Research*, Nova Science Publishers, New York, 2007.

Barnes, I., *Global Atmospheric Change and Its Impact on Regional Air*, Kluwer Academic Publishers, Norwell, MA, 2002.

Bodine, C. G., Ed., *Air Pollution Research Advances*, Nova Science Publishers, New York, 2007.

Brasseur, G. P., R. G. Prinn, and A. A. P. Pszenny, *The Changing Atmosphere: An Integration and Synthesis of a Decade of Tropospheric Chemistry Research*, Springer-Verlag, New York, 2003.

Chapman, M. and R. Bowden, *Air Pollution: Our Impact on the Planet*, Raintree Publishers, Austin, TX, 2002.

Colls, J. and A. Tiwary, *Air Pollution: Measurement, Modeling, and Mitigation*, 3rd ed., Spon Press, London, 2009.

Desonie, D., *Atmosphere: Air Pollution and its Effects*, Chelsea House Publishers, New York, 2007.

Finlayson-Pitts, B. J. and J. N. Pitts, *Chemistry of the Upper and Lower Atmosphere: Theory, Experiments, and Applications*, Academic Press, San Diego, CA, 2000.

Heck, R. M., R. J. Farrauto, and S. T. Gulati, *Catalytic Air Pollution Control: Commercial Technology*, 3rd ed., Wiley, Hoboken, NJ, 2009.

Harrington, W. and V. McConnell, Eds., *Controlling Automobile Air Pollution*, Ashgate, Burlington, VT, 2007.

Hewitt, C. N., Ed., *Reactive Hydrocarbons in the Atmosphere*, Academic Press, San Diego, CA, 1999.

Kidd, J. S. and R. A. Kidd, *Air Pollution: Problems and Solutions*, Chelsea House Publishers, New York, 2006.

Klán, P. and J. Wirz, *Photochemistry of Organic Compounds: From Concepts to Practice*, Wiley, Hoboken, NJ, 2009.

Montalti, M., *Handbook of Photochemistry*, 3rd ed., CRC Press/Taylor & Francis, Boca Raton, FL, 2006.

Livingston, J. V., *Air Pollution: New Research*, Nova Science Publishers, New York, 2007.

Seinfeld, J. H. and S. N. Pandis, *Atmospheric Chemistry and Physics: From Air Pollution to Climate Change*, 2nd ed., Wiley, Hoboken, NJ, 2006.

Sokhi, R. S., Ed., *World Atlas of Atmospheric Pollution*, Anthem Press, New York, 2007.

Turro, N. J., V. Ramamuthy, and J. C. Scaiano, *Principles of Molecular Photochemistry: An Introduction*, University Science Books, Sausalito, CA, 2009.

Vallero, D. A., *Fundamentals of Air Pollution*, 4th ed., Elsevier, Amsterdam, 2008.

Wayne, R. P., *Chemistry of the Atmospheres*, 3rd ed., Oxford University Press, Oxford, UK, 2000.

QUESTIONS AND PROBLEMS

1. Of the following species, the one which is the least likely product of the absorption of a photon of light by a molecule of NO_2 is: (a) O, (b) a free radical species, (c) NO, (d) NO_2^*, (e) N atoms.

2. Which of the following statements is true: (a) RO• reacts with NO to form alkyl nitrates, (b) RO• is a free radical, (c) RO• is not a very reactive species, (d) RO• is readily formed by the action of stable hydrocarbons and ground-state NO_2, (e) RO• is not thought to be an intermediate in the smog-forming process.
3. Of the following species, the one most likely to be found in reducing smog is: ozone, relatively high levels of atomic oxygen, SO_2, PAN, PBN.
4. Why are automotive exhaust pollutant hydrocarbons even more damaging to the environment than their quantities would indicate?
5. At what point in the smog-producing chain reaction is PAN formed?
6. Of the smog products mentioned in this chapter, which particularly irritating product is likely to be formed in the laboratory by the irradiation of a mixture of benzaldelhyde and NO_2 with ultraviolet light?
7. Which of the following species reaches its peak value last on a smog-forming day: NO, oxidants, hydrocarbons, NO_2?
8. What is the main species responsible for the oxidation of NO to NO_2 in a smoggy atmosphere?
9. Give two reasons why a turbine engine should have lower hydrocarbon emissions than an internal combustion engine.
10. What pollution problem does a lean mixture aggravate when employed to control hydrocarbon emissions from an internal combustion engine?
11. Why is a modern automotive catalytic converter called a "three-way conversion catalyst"?
12. What is the distinction between *reactivity* and *instability* as applied to some of the chemically active species in a smog-forming atmosphere?
13. Why might carbon monoxide be chosen as a standard against which to compare automotive hydrocarbon emissions in atmospheres where smog is formed? What are some pitfalls created by this choice?
14. What is the purpose of alumina in an automotive exhaust catalyst? What kind of material actually catalyzes the destruction of pollutants in the catalyst?
15. Some atmospheric chemical reactions are abstraction reactions and others are addition reactions. Which of these applies to the reaction of hydroxyl radical with propane? With propene (propylene)?
16. How might oxidants be detected in the atmosphere?
17. Each of the following occurs during smog formation. Place each in order from the one that occurs first (denoted 1) to the one that occurs last (denoted 5) and explain your choice: (A) An alkyl peroxyl radical, ROO•, is produced, (B) particles in the atmosphere obscure visibility, (C) NO and another product are produced from NO_2, (D) NO reacts to produce NO_2, (E) an alkyl radical, R•, is produced from a hydrocarbon.
18. Why is ozone especially damaging to rubber?
19. Show how hydroxyl radical, HO•, might react differently with ethylene, $H_2C{=}CH_2$, and methane, CH_4.
20. Name the stable product that results from an initial addition reaction of hydroxyl radical, HO•, with benzene.
21. Of the following, the true statement is (explain): (A) NO_2 is *not* involved in the processes that initiate smog formation, only in those that tend to stop it, (B) NO undergoes photodissociation to start the process of smog formation, (C) NO_2 can react with free radical species to terminate chain reactions involved in smog formation, (D) once NO_2 has undergone photodissociation, there is not a mechanism in a smog-forming atmosphere by which it can be regenerated, (E) NO_2 is the most phytotoxic (toxic to plants) species present in a smoggy atmosphere.

22. Of the following, the statement that is *untrue* regarding air pollutant hydrocarbons is (explain): (A) Although methane, CH_4, is normally considered as coming from natural sources, and may be thought of as a nonpollutant, human activities have increased atmospheric methane levels, with the potential for doing harm, (B) some organic species from trees can result in the formation of secondary pollutants in the atmosphere, (C) alkenyl hydrocarbons containing the C=C group have a means of reacting with hydroxyl radical that is not available for alkanes, (D) the reactivities of individual hydrocarbons as commonly measured for their potential to form smog, vary only about ±25%, (E) most nonmethane hydrocarbons in the atmosphere are of concern because of their ability to produce secondary pollutants.
23. Of the following, the true statement pertaining to hydrocarbon reactivity in smog formation is (explain): (A) Alkanes are more reactive than alkenes (olefins), (B) reactivity is based on reaction with HO, (C) methane, CH_4, is the most reactive hydrocarbon, (D) terpenes, such as *d*-limonene, are unreactive, (E) hydroxyl radical is classified as an unreactive hydrocarbon.
24. Smog aerosol droplets are composed of organic matter surrounding a small inorganic (mineral matter) core. Suggest what this shows regarding the process by which these aerosols are formed. Is the organic portion of the aerosol likely to be pure hydrocarbon (explain)?

14 The Endangered Global Atmosphere

14.1 CLIMATE CHANGE AND ANTHROPOGENIC EFFECTS

Several pieces of evidence are used to infer the long record of the earth's climate. These include fossil records, isotopic abundances in polar ice, and air entrained in it. For conditions going back to several centuries, the size and trace element content of tree rings are particularly useful and reflect abundance of water, temperatures, air composition, and presence of pollutants under which each ring was formed.[1]

There is a very strong connection between life forms on Earth and the nature of the earth's climate, which determines its suitability for life. As proposed by James Lovelock, a British chemist, this forms the basis of the *Gaia hypothesis*, which contends that the atmospheric O_2/CO_2 balance established and sustained by organisms determines and maintains the earth's climate and other environmental conditions. For about 3.5 billion years, stabilizing feedback mechanisms have maintained the earth/atmosphere boundary region within narrow liquid water conditions in which life can exist. It is incumbent upon humankind to avoid upsetting this delicate balance within just a few years' period of time.

Ever since life first appeared on Earth, the atmosphere has been influenced by the metabolic processes of living organisms. When the first primitive life molecules were formed approximately 3.5 billion years ago, the atmosphere was very different from its present state. At that time, it was chemically reducing and was thought to contain nitrogen, methane, ammonia, water vapor, and hydrogen, but no elemental oxygen. These gases and water in the sea were bombarded by intense, bond-breaking ultraviolet radiation which, along with lightning and radiation from radionuclides, provided the energy to bring about chemical reactions that resulted in the production of relatively complicated molecules, including even amino acids and sugars. From this rich chemical mixture, life molecules evolved. Initially, these very primitive life forms derived their energy from fermentation of organic matter formed by chemical and photochemical processes, but eventually they gained the capability to produce organic matter, "$\{CH_2O\}$," by photosynthesis utilizing solar light energy ($h\nu$),

$$CO_2 + H_2O + h\nu \rightarrow \{CH_2O\} + O_2(g) \tag{14.1}$$

and the stage was set for the massive biochemical transformation that resulted in the production of almost all the atmosphere's oxygen.

The oxygen initially produced by photosynthesis was probably quite toxic to primitive life forms. However, much of this oxygen was converted to iron oxides by reaction with soluble iron(II):

$$4Fe^{2+} + O_2 + 4H_2O \rightarrow 2Fe_2O_3 + 8H^+ \tag{14.2}$$

The enormous deposits of iron oxides thus formed provide convincing evidence for the liberation of free oxygen into the primitive atmosphere.

Eventually enzyme systems developed that enabled organisms to mediate the reaction of waste-product oxygen with oxidizable organic matter in the sea. Later this mode of waste-product disposal was utilized by organisms to produce energy for respiration, which is now the mechanism by which nonphotosynthetic organisms obtain energy.

In time, oxygen accumulated in the atmosphere, providing an abundant source of O_2 for respiration. It had an additional benefit in that it enabled the formation of an ozone shield against solar ultraviolet radiation in the stratosphere. With this shield in place, Earth became a much more hospitable environment for life, and life forms were enabled to move from the protective surroundings of the sea to the more exposed environment of the land.

Other instances of climatic change and regulation induced by organisms can be cited. An important example is the maintenance of atmospheric carbon dioxide at low levels through the action of photosynthetic organisms (note from Reaction 14.1 that photosynthesis removes CO_2 from the atmosphere). But, at an ever-accelerating pace during the last 200 years, another organism, humankind, has engaged in a number of activities that are altering the atmosphere profoundly. As noted in Chapter 1, human influences are so strong that it is useful to invoke a fifth sphere of the environment, the anthrosphere.

The atmosphere receives a number of contaminants from the anthrosphere. These substances may have pronounced effects far out of proportion to their fraction of the total mass of the atmosphere, especially in the following areas: (1) absorption of outgoing infrared radiation, thereby warming the atmosphere, (2) scattering and reflection of sunlight, (3) formation of photochemically reactive species such as NO_2 that are activated by absorption of ultraviolet radiation, (4) formation of catalytic species such as ozone-destroying Cl atoms produced by the photodissociation of CFCs in the stratosphere.

Human activities that have a strong influence on the atmosphere include industrial activities that emit particles and pollutant gases; fossil fuel combustion that emits particles and oxides of carbon, sulfur, and nitrogen; modes of transportation reliant on fossil fuels that emit air pollutants; alteration of land surfaces, including deforestation and desertification; burning of biomass and vegetation that emits soot and carbon and nitrogen oxides; and agricultural practices such as the cultivation of rice, which emits large quantities of methane. Major effects of these processes include increased acidity of the atmosphere, elevated levels of atmospheric oxidants, increased global warming, increased levels of gases that threaten the stratospheric ozone layer, and increased corrosivity of the atmosphere.

In 1957, photochemical smog was only beginning to be recognized as a serious problem, acid rain and the greenhouse effect were scientific curiosities, and the ozone-destroying potential of CFCs had not even been imagined. In that year, Revelle and Suess[2] prophetically referred to human perturbations of Earth and its climate as a massive "geophysical experiment." The effects that this experiment may have on the global atmosphere are discussed in this chapter.

14.1.1 Changes in Climate

Ample evidence exists of massive changes in the earth's climate in times past. Indeed, humankind exists now in an approximately 10,000-year interglacial era called the *holocene*. Evidence from the past suggests that major changes in climate may occur very rapidly, within a few years' period of time. These may occur through positive feedback mechanisms in which, once a certain threshold is passed, the change feeds on itself and proceeds rapidly and irreversibly. One analogy is that of a canoe. Leaning gently to one side will cause the canoe to tip slightly such that it rights itself once the occupant has stopped leaning. However, beyond a certain point, the canoe tips over completely and irreversibly. Cooling of the climate may result in more coverage of the earth's surface with ice and snow, which reflects solar energy and results in more cooling and more ice and snow. Drought can destroy vegetation, without which there is less transpiration of moisture into the atmosphere, causing less rainfall and even greater loss of vegetation.

Fluctuations in climate have significant ecological effects, both direct and indirect. In recent years, attention has shifted from short-term, localized weather phenomena (rainfall, snow coverage, temperature) to climate phenomena on a larger scale and over longer time periods. Global scale phenomena, especially the El Niño Southern Oscillation and the North Atlantic Oscillation can have very significant ecological effects lasting for several years over huge areas of the globe. Effects on terrestrial plants and their productivity can cause changes in animal populations and the relationships between herbivores and carnivores. Upwelling of nutrients and ocean temperatures can cause variations in photosynthetic activity in marine environments, thereby affecting fish populations and other marine biota.

14.2 GLOBAL WARMING

This section deals with infrared-absorbing trace gases (other than water vapor) in the atmosphere that contribute to global warming and with the influence of particles on temperature. Figure 14.1 shows global temperature trends since 1880 and illustrates a steady warming trend during recent decades. In addition to being a scientific issue, greenhouse warming of the atmosphere has also become a major policy, political, and economic issue.

Carbon dioxide and other infrared-absorbing trace gases in the atmosphere contribute to global warming—the "greenhouse effect"—by allowing incoming solar radiant energy to penetrate into the earth's surface while reabsorbing infrared radiation emanating from it. As shown in Figure 14.2, atmospheric levels of "greenhouse gas" carbon dioxide have increased at a rapid rate during recent decades and are continuing to do so. Concern over this phenomenon has intensified since about 1980. Adding to that concern is that, according to the Goddard Institute of Space Science, the eight warmest years on record have occurred since 1998 and the 14 warmest years recorded have taken place since 1990. The warmest of these was 2005. The year 2007 was tied with 1998 as the second warmest year on record. The near record warmth of 2007 is all the more remarkable because the year was at a minimum of solar irradiance and the natural El Niño–La Niña cycle of the equatorial Pacific Ocean was in its cool phase.

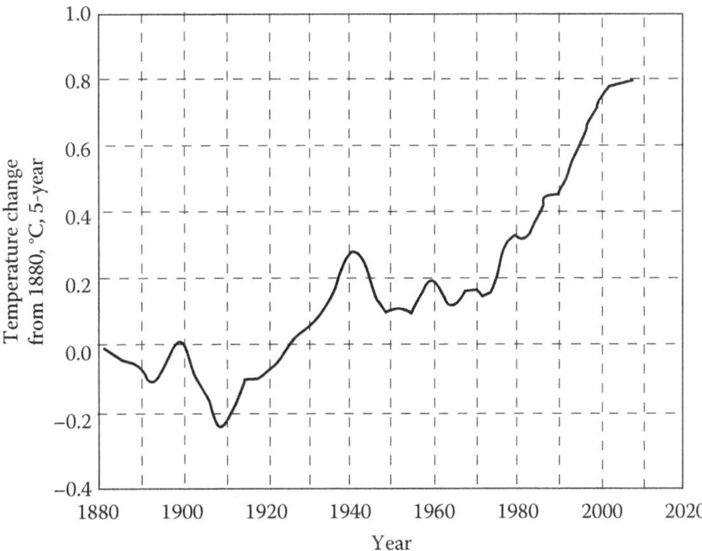

FIGURE 14.1 Global temperature trend since 1880. Earlier values are less certain because of the lack of sophisticated means of measuring temperature. More recent values are very accurate because of the use of satellite-based technologies for measuring temperature.

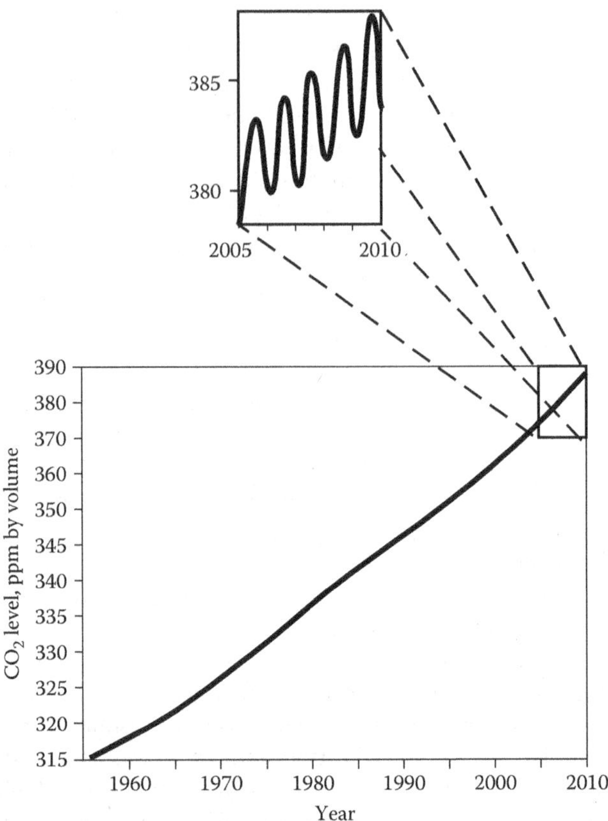

FIGURE 14.2 Increases in atmospheric CO_2 levels in recent years. The inset illustrates seasonal variations in the northern hemisphere.

Atmospheric carbon dioxide is the gas believed to be contributing most to global warming. Chemically and photochemically, carbon dioxide is a comparatively insignificant species because of its relatively low concentrations and low photochemical reactivity. The one significant photochemical reaction that carbon dioxide undergoes, and a major source of stratospheric CO, is the photodissociation of CO_2 by energetic solar ultraviolet radiation in the stratosphere:

$$CO_2 + h\nu \rightarrow CO + O \tag{14.3}$$

The most obvious factor in increased atmospheric carbon dioxide is consumption of carbon-containing fossil fuels. In addition, release of CO_2 from the biodegradation of biomass and uptake by photosynthesis are important factors determining overall CO_2 levels in the atmosphere. The role of photosynthesis is illustrated in Figure 14.2, which shows a seasonal cycle in carbon dioxide levels in the northern hemisphere. Maximum values occur in April and minimum values in late September or October. These oscillations are due to the "photosynthetic pulse," influenced most strongly by forests in middle latitudes and located predominantly in the Northern Hemisphere. Forests have a much greater influence than other vegetation because trees carry out more photosynthesis. Furthermore, forests store enough fixed but readily oxidizable carbon in the form of wood and humus to have a marked influence on atmospheric CO_2 content. Thus, during the summer months, forest trees carry out enough photosynthesis to reduce the atmospheric carbon dioxide content markedly. During the winter, metabolism of biota, such as bacterial decay of humus, releases a significant amount of CO_2. Therefore, the current worldwide destruction of forests and conversion of forest lands into agricultural uses contributes substantially to a greater overall increase in atmospheric CO_2 levels.

With current trends, it is likely that global CO_2 levels will double from preindustrial levels later this century, which may well raise the earth's mean surface temperature by 1.5–4.5°C. Such a change might have the potential to cause more massive irreversible environmental changes than any other disaster short of global nuclear war or asteroid impact.

A consistent annual increase of atmospheric carbon dioxide of about 1 ppm/year was characteristic of the trends shown in Figure 14.2 until relatively recently, but now the annual increase is closer to 2 ppm. As shown in Figure 14.3, per capita carbon dioxide emissions are highest for industrialized countries and development of countries with high populations, such as China and India, can be expected to add large quantities of carbon dioxide to the atmosphere in the future. CFCs, which also are greenhouse gases, were not even introduced into the atmosphere until the 1930s. Although trends in levels of these gases are well known, their effects on global temperature and climate are much less certain. The phenomenon has been the subject of much computer modeling. Most models predict global warming of at least 3.0°C and up to 5.5°C occurring over a period of just a few decades. These estimates are sobering because they correspond to the approximate temperature increase since the last ice age 18,000 years ago, which took place at a much slower pace of only about 1°C or 2°C per 1000 years. Such warming would have profound effects on rainfall, plant growth, and sea levels, which might rise as much as 0.5–1.5 m.

Both positive and negative feedback mechanisms may be involved in determining the rates at which carbon dioxide and methane (discussed below) build up in the atmosphere. Laboratory studies indicate that increased CO_2 levels in the atmosphere cause accelerated uptake of this gas by plants undergoing photosynthesis, which tends to slow the buildup of atmospheric CO_2. Given adequate rainfall, plants living in a warmer climate that would result from the greenhouse effect would grow faster and take up more CO_2. This could be an especially significant effect of forests, which have a high CO_2-fixing ability. However, the projected rate of increase in carbon dioxide levels is so rapid that forests would lag behind in their ability to fix additional CO_2. Similarly, higher atmospheric

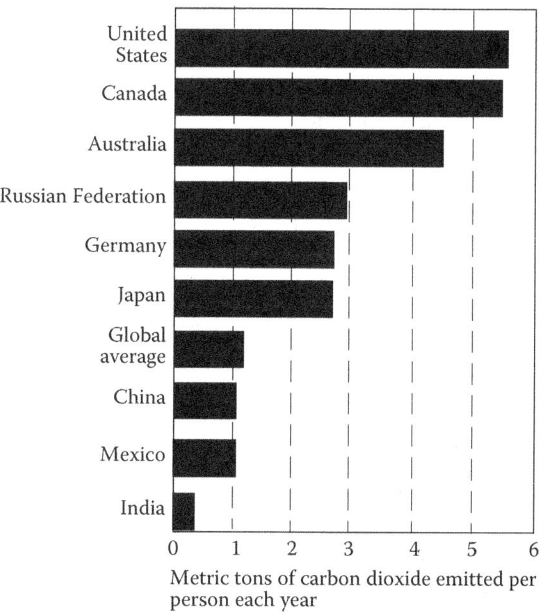

FIGURE 14.3 Per capita emissions of carbon dioxide. Rapid industrialization of high-population countries, particularly China and India, can be expected to add much larger quantities of carbon dioxide to the atmosphere in the future. Because of its huge population, China is almost equal to the United States in total annual carbon dioxide emissions for the entire country.

CO_2 concentrations will result in accelerated sorption of the gas by oceans. The amount of dissolved CO_2 in the oceans is about 60 times the amount of CO_2 gas in the atmosphere. However, the times for the transfer of carbon dioxide from the atmosphere to the ocean are of the order of years. Because of low mixing rates, the times for the transfer of carbon dioxide from the upper approximately 100-m layer of the oceans to ocean depths is much longer, of the order of decades. Therefore, like the uptake of CO_2 by forests, increased absorption by oceans will lag behind the emissions of CO_2. A concern with increased levels of CO_2 in the oceans is the lowering of ocean water pH that will result. Even though such an effect will be slight, of the order of one tenth to several tenths of a pH unit, it has the potential to strongly impact organisms that live in ocean water. Severe drought conditions resulting from climatic warming could cut down substantially on CO_2 uptake by plants. Warmer conditions would accelerate the release of both CO_2 and CH_4 by microbial degradation of organic matter. (It is important to realize that about twice as much carbon is held in soil in dead organic matter—necrocarbon—potentially degradable to CO_2 and CH_4 as is present in the atmosphere.) Global warming might speed up the rates at which biodegradation adds these gases to the atmosphere.

14.2.1 Methane and Other Greenhouse Gases

Among the gases other than carbon dioxide that contribute to global warming are CFCs, fluorocarbons, HCFC, HFCs, N_2O, and, especially, methane, CH_4. Now at a level of around 1.8 ppm in the atmosphere, methane is going up at a rate of almost 0.02 ppm/year. The comparatively very rapid increase in methane levels is attributed to a number of factors resulting from human activities. Among these are direct leakage of natural gas, by-product emissions from coal mining and petroleum recovery, and release from the burning of savannas and tropical forests. Biogenic sources resulting from human activities produce large amounts of atmospheric methane. These include methane from bacteria degrading organic matter such as municipal refuse in landfills; methane evolved from anaerobic biodegradation of organic matter in rice paddies; and methane emitted as the result of bacterial action in the digestive tracts of ruminant animals.

In addition to acting as a greenhouse gas, methane has significant effects upon atmospheric chemistry. It produces atmospheric CO as an intermediate oxidation product and influences concentrations of atmospheric hydroxyl radicals and ozone. In the stratosphere, it produces hydrogen and H_2O, but acts to remove ozone-destroying chlorine.

A term called *radiative forcing* is used to describe the reduction in infrared radiation penetrating outward through the atmosphere per unit increase in the level of gas in the atmosphere. The radiative forcing of CH_4 is about 25 times that of CO_2. Increases in the concentration of methane and several other greenhouse gases have such a disproportionate effect on retention of infrared radiation because their infrared absorption spectra fill gaps in the overall spectrum of outbound radiation not covered by much more abundant carbon dioxide and water vapor. Therefore, whereas an increase in carbon dioxide concentration has a comparatively small incremental effect because the gas is already absorbing such a high fraction of infrared radiation in regions of the spectrum where it absorbs, an increase in the concentration of methane, CFC, or other greenhouse gases has comparatively a much larger effect.

14.2.2 Particles and Global Warming

Whereas the effects of carbon dioxide and other gases on temperature are relatively easy to calculate, the effects of particles are much more complicated. Atmospheric particles have both direct effects exerted by scattering and absorbing radiation and indirect effects in changing the microphysical structure, lifetimes, and quantities of clouds (see condensation nuclei in Section 9.5). Particles interact with and scatter the radiation that is of a wavelength similar to the size of the particles most strongly. Most of the incoming solar energy is at wavelengths <4 μm and most particles are smaller than 4 μm, so the major effect of atmospheric particles is to scatter radiation from incoming solar energy, which has a cooling effect on the atmosphere. Some kinds of particles, such as those composed of black carbon and soot, absorb incoming solar radiation, thereby warming the atmosphere.

Liquid water droplets composing clouds can both scatter incoming radiation and absorb outbound infrared radiation. Clouds at lower altitudes act mainly to lower atmospheric temperature by scattering lower wavelength radiation, whereas clouds at higher altitudes tend to absorb outbound infrared causing temperature increases. Aerosol particles, such as sulfate salts that act as cloud condensation nuclei upon which atmospheric water vapor condenses, tend to increase the number of particles, which are therefore more numerous. In general, this has a cooling effect.

Overall, the effects of particles on global temperature are variable and not particularly well understood. Both cooling and warming effects may occur. Modeling these effects are much more challenging than modeling the effects of gaseous atmospheric constituents such as carbon dioxide.

14.2.3 The Outlook for Global Warming and Associated Effects

It is certain that atmospheric CO_2 levels will continue to increase significantly. The degree to which this occurs depends upon future levels of CO_2 production and the fraction of that production that remains in the atmosphere. Given the plausible projections of CO_2 production and a reasonable estimate that half of that amount will remain in the atmosphere, projections can be made that indicate that sometime approximately during the next 100 years, the concentration of this gas will reach 600 ppm in the atmosphere. This is well over twice the levels estimated for preindustrial times. Much less certain are the effects that this change will have on climate. It is virtually impossible for the elaborate computer models used to estimate these effects to accurately take account of all variables, such as the degree and nature of cloud cover. Clouds both reflect incoming light radiation and absorb outgoing infrared radiation, with the former effect tending to predominate. The magnitudes of these effects depend upon the degree of cloud cover, brightness, altitude, and thickness. In the case of clouds, too, feedback phenomena occur; for example, warming induces formation of more clouds, which reflect more incoming energy.

Drought is one of the most serious problems that could arise from major climatic change resulting from greenhouse warming. Typically, a three-degree warming would be accompanied by a 10% decrease in precipitation. Water shortages would be aggravated, not just from decreased rainfall, but from increased evaporation as well. Increased evaporation results in decreased runoff, thereby reducing water available for agricultural, municipal, and industrial use. Water shortages, in turn, lead to increased demand for irrigation and to the production of lower quality, higher salinity runoff water and wastewater. In the United States, such a problem would be especially intense in the Colorado River basin, which supplies much of the water used in the rapidly growing U.S. Southwest.

A variety of other problems, some of them unforeseen as of now, could result from global warming. An example is the effect of warming on plant and animal pests—insects, weeds, diseases, and rodents. Many of these would certainly thrive much better under warmer conditions.

Interestingly, another air pollutant, acid-rain-forming sulfur dioxide (see Section 14.4), may have a counteracting effect on greenhouse gases. This is because sulfur dioxide is oxidized in the atmosphere to sulfuric acid, forming a light-reflecting haze. Furthermore, the sulfuric acid and resulting sulfates act as condensation nuclei upon which atmospheric water vapor condenses, thereby increasing the extent, density, and brightness of light-reflecting cloud cover. Sulfate aerosols are particularly effective in counteracting greenhouse warming in central Europe and the eastern United States during the summer.

14.3 GREEN SCIENCE AND TECHNOLOGY TO ALLEVIATE GLOBAL WARMING

Although there are still "global-warming denyers" who attempt to discredit those who have concern over global warming, the overwhelming consensus of reputable scientists is that global warming is taking place and that greenhouse gases, particularly carbon dioxide, are the main cause. The question for responsible people then becomes what to do about the problem. The possibilities can be divided into three categories: (1) minimization, (2) counteracting measures, and (3) adaptation.

14.3.1 Minimization

Minimization refers to measures taken to reduce emissions of greenhouse gases. Minimization is tied inextricably to energy production and utilization, since most greenhouse gas emissions are from the burning of fossil fuels. Fortunately, fossil fuel may be consumed much more efficiently utilizing existing and developing energy conservation measures. Efficient utilization of fossil energy can be accomplished without major economic disruption. A prime example would be the conversion of the U.S. private vehicle fleet to hybrid internal combustion/electric propulsion. Hybrid vehicles can be manufactured with the capability to charge their batteries from house electrical current to provide sufficient charge for approximately 30 km of travel before the internal combustion engine even has to be turned on, which would take care of as much as half of routine commuting and travel. As a green fringe benefit without additional charge, employers could provide charging stations in parking facilities so that vehicles could be recharged during the work day. Such a system could readily reduce by 50% the amount of fuel and hence the emissions of greenhouse gases of current private automotive transport systems.

Conversion of the current U.S. freight transportation system from truck to electrified rail to the maximum extent possible would make a further major reduction in greenhouse gas emissions. Since about 1990, there has been a significant shift in the amount of freight transported by rail in the form of shipping containers and truck trailers to the point that some rail freight lines are nearing their maximum capacity. Unlike Europe, most railways in the United States are not electrified, although that certainly could be done and the electricity could be generated by means that do not require burning of fossil fuels. Interstate highway rights of way can serve as routes for new rail lines in some cases.

Utilization of fossil fuels for heating and air conditioning can be done much more efficiently. The best fuel for home heating is natural gas (methane, CH_4). This fuel emits the least amount of greenhouse gas carbon dioxide per unit heat delivered because of its relatively high content of hydrogen. Rather than burning methane in a furnace, it can be used as a fuel in a small internal combustion engine connected to a heat pump to pump heat from the outside. The exhaust gases from the engine can be cooled and the water vapor in them condensed to capture additional heat.

One way to reduce the release of carbon dioxide is by using biomass as fuel or raw material for the manufacture of various products. Burning a biomass fuel does release carbon dioxide to the atmosphere, but an exactly equal amount of carbon dioxide is removed from the atmosphere in the photosynthetic process by which the biomass is made, so there is no net addition of CO_2. Unless and until biomass-derived materials used in feedstocks are burned, their use represents a net loss of carbon dioxide from the atmosphere.

Another potential use of green chemistry to prevent addition of carbon dioxide to the atmosphere is through *carbon sequestration* in which carbon dioxide is produced, but is bound in a form such that it is not released to the atmosphere.[3] This approach has the greatest potential in applications where the carbon dioxide is produced in a concentrated form. Carbon from coal can be reacted with oxygen and water to produce elemental hydrogen and carbon dioxide in a coal gasification process. The net reaction for this production is the following:

$$2C + O_2 + 2H_2O \rightarrow 2CO_2 + 2H_2 \tag{14.4}$$

The hydrogen generated can be used as a pollution-free fuel in fuel cells or combustion engines. The carbon dioxide can be pumped into deep ocean waters, although this has the potential to lower ocean pH slightly, which would be detrimental to marine organisms. Another option is to pump the carbon dioxide deep underground. A side benefit of the latter approach is that in some areas carbon dioxide pumped underground can be used to recover additional crude oil from depleted oil-bearing formations.

Carbon dioxide can actually be converted to fuel if there is an abundant and cheap source of elemental H_2, such as might be the case by electrolyzing water with excess electrical power generated by wind or by abundant geothermal energy, a process used in Iceland. The way in which this can be done is by the following reverse water gas shift reaction:

$$CO_2 + H_2 \rightarrow CO + H_2O \tag{14.5}$$

Fischer–Tropsch synthesis can then be used to make hydrocarbons and other organic fuels and chemicals from the CO and additional elemental hydrogen.

The best way to reduce greenhouse gas emissions from home heating is to avoid the use of fossil fuels entirely. Substantial progress has been made in solar heating systems that heat homes without using fossil fuels. Another good measure is the use of electricity for heating and air conditioning, generating the electricity from energy sources other than fossil fuels. Additional options for generating electricity without adding to the burden of greenhouse gases are discussed in Chapter 19.

A green technological approach to the reduction of carbon dioxide emissions is to develop alternative methods of energy production. One thing that would be very beneficial is the development of more efficient photovoltaic cells. These devices have become marginally competitive for the generation of electricity, and even relatively small improvements in efficiency would enable their much wider use, replacing fossil fuel sources of electricity generation. Another device that would be extremely useful is a system for the direct photochemical dissociation of water to produce elemental hydrogen and oxygen, which could be used in fuel cells. An application of green biochemistry that would reduce carbon dioxide emissions is the development of plants with much higher efficiencies for photosynthesis. Plants now are only about 0.5% efficient in converting light energy to chemical energy. Raising this value to only 1% would make a vast difference in the economics of producing biomass as a substitute for fossil carbon.

Regardless of the technologies used to reduce emissions of greenhouse gases, political and economic pressures must be applied to ensure that they are implemented and sustained. One of the most directly effective measures consists of *mileage standards* for vehicles. These were imposed in the United States during the earlier years of the "energy crisis" but inexcusably were allowed to languish during the latter 1900s and early 2000s. Finally, in 2007, laws were passed requiring gradual imposition of higher mileage for new vehicles. Another useful device is that of *carbon taxes* on fuels that emit carbon dioxide. Very high taxes on gasoline and diesel fuel have for years resulted in the use of very efficient vehicles in Europe. Inherently more efficient diesel-powered vehicles are popular in Europe, but diesel automobiles are rare in the United States where gasoline has been relatively cheap. Another effective means for reducing carbon dioxide emissions is the *cap and trade* system in which concerns are allocated specific amounts of carbon that they can emit. If they emit less than the limit, they may sell the right to emit carbon to a concern that is exceeding its limit. All of these measures have the advantage of not dictating technologies to be used, leaving that aspect to the ingenuity of the private sector.

One of the more contentious issues related to greenhouse gas emissions has been the Kyoto treaty and the refusal of the United States to ratify that agreement. This treaty evolved from a 1997 meeting of 160 nations in Kyoto, Japan. The agreement called for stabilization of greenhouse gas emissions to 1990 levels during the period 2008–2012, which would have led to 23% less of the emissions below levels projected during the 1990s without remedial action. The United States has refused to ratify the agreement because it exempted large developing countries, especially India and China, from participating for economic reasons. Such countries produce only a small fraction of as much greenhouse gas per capita as is the case with the United States and several other industrial countries. During the early 2000s, as the economies of India and China have developed, their emissions of greenhouse gases have accelerated significantly.

14.3.2 Counteracting Measures

Counteracting measures for reducing global warming consist of schemes such as the injection of light-reflecting particles into the upper atmosphere. The scale required of such measures is so great that they are probably impossible to implement in a meaningful way. One possibility is to increase (or at least not reduce) the amount of sulfur gases emitted to the atmosphere that oxidize to sulfuric acid which serves to generate condensation nuclei that produce light-reflecting clouds. Another far-out possibility would be to use high-flying tanker aircraft to separately inject hydrogen chloride and ammonia vapors into the atmosphere. Any chemistry student who has had separate beakers of hydrochloric acid and ammonia solution close together knows that HCl and NH_3 react, as shown in the following equation,

$$HCl(g) + NH_3(g) \rightarrow NH_4Cl(s) \tag{14.6}$$

to produce a dense fog of ammonium chloride particles that might serve as cloud condensation nuclei. In addition to the obvious questions regarding air pollution that this remedy raises, it might be difficult to find aircrew willing to fly Airbus A380 jumbo jet aircraft into the lower stratosphere laden with pressurized tanks of corrosive hydrogen chloride and ammonia.

A possibly significant counteracting measure is modification of the earth's surface in a manner so as to reflect light. This can potentially be done with appropriate kinds of vegetation in the form of forests and grasslands and by agricultural practices that minimize exposure of light-absorbing freshly cultivated soil. Again, the scale required to make any significant difference would be huge. Another small effect might be obtained by designing surfaces (roofs and parking lot surfaces) of anthrospheric structures to maximize reflection of solar radiation. Such a measure would favor reflective aluminum roofs over dark roofing and light-colored concrete over black asphalt for parking lot surfaces.

14.3.3 Adaptation

Since global warming will in fact occur and neither minimization nor counteracting measures will be sufficient to stop it, adaptation to climate warming will be required. Other than increases in global temperature, there are many effects of global warming suggesting a variety of adaptations. It may be anticipated that adaptation to global warming will be one of the most significant activities of green science and technology in the future.

Water shortages and drought will constitute perhaps the most troublesome aspects of climate warming. Water, already in short supply in many parts of the world, will become scarcer. It will be necessary to implement more efficient irrigation practices and to grow crops that require less irrigation. One approach with significant promise is to grow crops on arid coastal lands irrigated with seawater. Plants that can grow in seawater are called halophytes and can produce 1–2 kg of dry biomass per square meter of field area, which is about the same production as conventional crops such as alfalfa. Some of the most productive plants that grow in seawater have relatively unattractive names including glasswort, sea blite, saltbush, and salt grass. Though not producing grain, some of these plants produce abundant forage that can be eaten by animals. One small problem is that because of the high salt content of the forage, animals that consume it must drink significantly greater amounts of fresh water. One saltwater plant that has does produce abundant seeds is *Salicornia bigelovii* that rapidly colonizes new mud flats. With a salt content of $<3\%$, the seeds are 35% protein and 30% highly polyunsaturated oil, similar to safflower oil in composition. The seeds contain bitter saponins, which somewhat limit the amount of seed or meal left after extracting oil that can be fed to animals. Especially for oil production, production of oil-producing algae in saltwater ponds, such as those filled with brackish groundwater, is a promising approach.

One of the greatest adverse effects of global warming results from the effects of high temperatures on people, particularly the vulnerable elderly. This was illustrated tragically in Europe in

August 2003. The highest temperatures ever recorded in the United Kingdom to that date occurred on August 10, 2003, with temperatures of 38°C (100.4°F) at London's Heathrow airport and 38.1°C at Gravesend, Kent. Over 1000 people died from the heat wave in the United Kingdom. However, the greatest toll was in France where about 15,000 people, mostly elderly, died from the heat in August 2003. The problem was exacerbated by the fact that France is not used to extreme heat and the custom is for many people, including government ministers and physicians who would have been involved in remedial measures, to take August off for vacation. Funeral homes were overwhelmed and refrigerated warehouses had to be used to store bodies until they could be identified and buried. A major concern was that the nuclear reactors that provide much of France's electrical power could not be adequately cooled and in some cases their cooling had to be augmented by spraying with hoses. Other countries were adversely affected. Portugal lost 10% of its forests in forest fires. On the positive side, the hot dry summer yielded grapes with very high sugar contents resulting in one of the best years ever for French wines. In the hot, dry summer of 2007, much of the forests in Greece were consumed by fire.

As global warming occurs, a major adaptation will need to be the installation of air conditioning and other cooling measures in regions of the world where air conditioning in homes and commercial buildings has been uncommon. This is particularly true in Europe where periods of hot weather will become more common, though of shorter duration than in much of the United States, for example. In addition to the installation of air conditioning, provision will need to be made to provide sustainable power for it. This could include a need for fuel turbine peaking facilities for electrical power generation and for greater reserves of cooling water for nuclear power reactors.

14.4 ACID RAIN

Precipitation made acidic by the presence of acids stronger than $CO_2(aq)$ is commonly called *acid rain*; the term applies to all kinds of acidic aqueous precipitation, including fog, dew, snow, and sleet.[4] In a more general sense, *acid deposition* refers to the deposition on the earth's surface of aqueous acids, acid gases (such as SO_2), and acidic salts (such as NH_4HSO_4). According to this definition, deposition in solution form is *acid precipitation*, and deposition of dry gases and compounds is *dry deposition*. Although carbon dioxide is present at higher levels in the atmosphere, sulfur dioxide, SO_2, contributes more to the acidity of precipitation for two reasons. The first of these is that sulfur dioxide is significantly more soluble in water than is carbon dioxide, as indicated by its Henry's law constant (Section 5.3) of 1.2 mol/L/atm compared to a value of 3.38×10^{-2} mol/L/atm for CO_2. Secondly, the value of K_{a1} for $SO_2(aq)$,

$$SO_2(aq) + H_2O \rightleftarrows H^+ + HSO_3^- \tag{14.7}$$

$$K_{a1} = \frac{[H^+][HSO_3^-]}{[SO_2]} = 1.7 \times 10^{-2} \tag{14.8}$$

is more than four orders of magnitude higher than the value of 4.45×10^{-7} for CO_2.

Although acid rain can originate from the direct emission of strong acids, such as HCl gas or sulfuric acid mist, most of it is a secondary air pollutant produced by the atmospheric oxidation of acid-forming gases such as the following, both shown as overall reactions consisting of several steps:

$$SO_2 + \tfrac{1}{2}O_2 + H_2O \rightarrow \{2H^+ + SO_4^{2-}\}(aq) \tag{14.9}$$

$$2NO_2 + \tfrac{1}{2}O_2 + H_2O \rightarrow 2\{H^+ + NO_3^-\}(aq) \tag{14.10}$$

Chemical reactions such as these play a dominant role in determining the nature, transport, and fate of acid precipitation. As the result of such reactions the chemical properties (acidity, ability to react with other substances) and physical properties (volatility, solubility) of acidic atmospheric pollutants are altered drastically. For example, even the small fraction of NO that does dissolve in water does not react significantly. However, its ultimate oxidation product, HNO_3, though volatile, is highly water-soluble, strongly acidic, and very reactive with other materials. Therefore, it tends to be removed readily from the atmosphere and to do a great deal of harm to plants, corrodible materials, and other things that it contacts.

Although emissions from industrial operations and fossil fuel combustion are the major sources of acid-forming gases, acid rain has also been encountered in areas far from such sources. This is due in part to the fact that acid-forming gases are oxidized to acidic constituents and deposited over several days, during which time the air mass containing the gas may have moved as much as several thousand kilometers. It is likely that the burning of biomass, such as is employed in "slash-and-burn" agriculture, evolves the gases that lead to acid formation in more remote areas. In arid regions, dry acid gases or acids sorbed to particles may be deposited with effects similar to those of acid rain deposition.

Acid rain spreads out over areas of several hundred to several thousand kilometers. This classifies it as a *regional* air pollution problem compared to a *local* air pollution problem for smog and a *global* one for ozone-destroying CFCs and greenhouse gases. Other examples of regional air pollution problems are those caused by soot, smoke, and fly ash from combustion sources and fires (forest fires). Nuclear fallout from weapons testing or from reactor fires (of which, fortunately, there has been only one major one to date—the one at Chernobyl in the former Soviet Union) may also be regarded as a regional phenomenon.

Acid precipitation shows a strong geographic dependence, as illustrated in Figure 14.4, representing the pH of precipitation in the continental United States. The preponderance of acidic rainfall in the northeastern United States, which also affects southeastern Canada, is obvious. Analyses of the movements of air masses have shown a correlation between acid precipitation and prior movement of an air mass over major sources of anthropogenic sulfur and nitrogen oxides emissions. This is particularly obvious in southern Scandinavia, which receives a heavy burden of air pollution from densely populated, heavily industrialized areas in Europe.

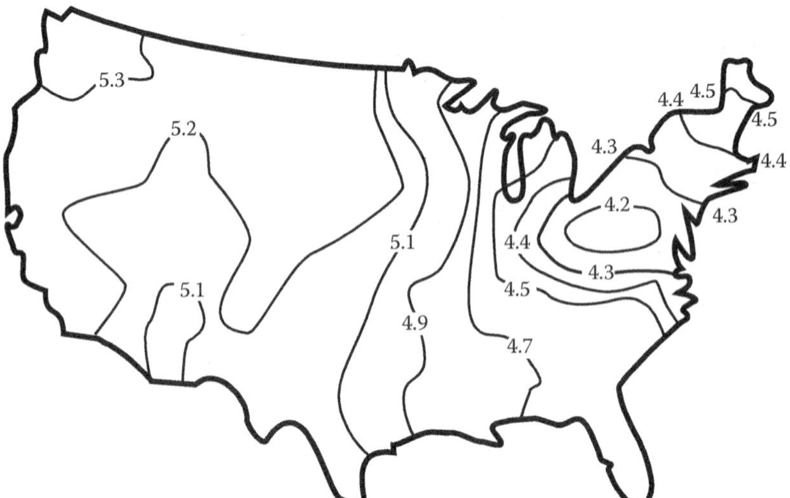

FIGURE 14.4 Isopleths of pH illustrating a hypothetical, but typical, precipitation-pH pattern in the lower 48 continental United States. Actual values found may vary with the time of year and climatic conditions.

Acid rain has been observed for well over a century, with many of the older observations from Great Britain. The first manifestations of this phenomenon were elevated levels of SO_4^{2-} in precipitation collected in industrialized areas. More modern evidence was obtained from analyses of precipitation in Sweden in the 1950s, and of U.S. precipitation a decade or so later. A vast research effort on acid rain was conducted in North America by the National Acid Precipitation Assessment Program, which resulted from the U.S. Acid Precipitation Act of 1980 and data are still being collected as an outgrowth of this program.

Table 14.1 shows typical major cations and anions in pH-4.25 precipitation. Although actual values encountered vary greatly with time and location of collection, this table does show some major features of ionic solutes in precipitation. From the predominance of sulfate anion, it is apparent that sulfuric acid is the major contributor to acid precipitation. Nitric acid makes a smaller but growing contribution to the acid present. Hydrochloric acid ranks third.

Ample evidence exists of the damaging effects of acid rain. The major effects are the following:

- Direct phytotoxicity to plants from excessive acid concentrations. (Evidence of direct or indirect phytoxicity of acid rain is provided by the declining health of Eastern United States and Scandinavian forests and especially by damage to Germany's Black Forest.)
- Phytotoxicity from acid-forming gases, particularly SO_2 and NO_2, which accompany acid rain.
- Indirect phytotoxicity. One of the most harmful effects of acidic precipitation is the dissolution of Al^{3+} from soil at levels that are harmful to plants.
- Destruction of sensitive forests.
- Respiratory effects on humans and other animals.
- Acidification of lake water with toxic effects to Lake Flora and Fauna, especially fish fingerlings.
- Corrosion of exposed structures, electrical relays, equipment, and ornamental materials. Because of the effect of hydrogen ion:

$$2H^+ + CaCO_3(s) \rightarrow Ca^{2+} + CO_2(g) + H_2O$$

  Limestone, $CaCO_3$, is especially susceptible to damage from acid rain.
- Associated effects, such as reduction of visibility by sulfate aerosols and the influence of sulfate aerosols on physical and optical properties of clouds. (As mentioned in Section 14.2, intensification of cloud cover and changes in the optical properties of cloud

TABLE 14.1
Typical Values of Ion Concentrations in Acidic Precipitation

Cations		Anions	
Ion	Concentration Equivalents/L × 10^6	Ion	Concentration Equivalents/L × 10^6
H^+	56	SO_4^{2-}	51
NH_4^+	10	NO_3^-	20
Ca^{2+}	7	Cl^-	12
Na^+	5	Total	83
Mg^{2+}	3		
K^+	2		
Total	83		

droplets—specifically, increased reflectance of light—resulting from acid sulfate in the atmosphere may even have a mitigating effect on greenhouse warming of the atmosphere.) A significant association exists between acidic sulfate in the atmosphere and haziness.

Soil sensitivity to acid precipitation can be estimated from CEC (see Chapter 5). Soil is generally insensitive if free carbonates are present or if it is flooded frequently. Soils with a CEC above 15.4 meq/100 g (on the basis of the dry mass of soil) are also insensitive because the soil acts as a buffer by taking up H^+ ion. Soils with cation exchange capacities between 6.2 meq/100 g and 15.4 meq/100 g are slightly sensitive. Soils with cation exchange capacities below 6.2 meq/100 g normally are sensitive if free carbonates are absent and the soil is not frequently flooded.

Forms of precipitation other than rainfall may contain excess acidity. Acidic fog can be especially damaging because it is very penetrating. In early December 1982, Los Angeles experienced a severe, 2-day episode of acid fog. This fog consisted of a heavy concentration of acidic mist particles at ground level which reduced visibility and were very irritating to breathe. The pH of the water in these particles was 1.7, much lower than ever before recorded for acid precipitation. Another source of precipitation heavy in the ammonium, sulfate, and nitrate ions associated with atmospheric acid is *acid rime*. Rime is frozen cloudwater which may condense on snowflakes or exposed surfaces. Rime constitutes up to 60% of the snowpack in some mountainous areas, and the deposition of acidic constituents with rime may be a significant vector for the transfer of acidic atmospheric constituents to the Earth's surface in some cases.

14.5 STRATOSPHERIC OZONE DESTRUCTION

Recall from Section 9.9 that stratospheric ozone, O_3, serves as a shield to absorb harmful ultraviolet radiation in the stratosphere, protecting living beings on Earth from the effects of excessive amounts of such radiation. The two reactions by which stratospheric ozone is produced are

$$O_2 + h\nu \rightarrow O + O \quad (\lambda < 242.4 \text{ nm}) \tag{14.11}$$

$$O + O_2 + M \rightarrow O_3 + M \quad (\text{energy-absorbing } N_2 \text{ or } O_2) \tag{14.12}$$

and it is destroyed by photodissociation

$$O_3 + h\nu \rightarrow O_2 + O \quad (\lambda < 325 \text{ nm}) \tag{14.13}$$

and a series of reactions from which the net result is the following:

$$O + O_3 \rightarrow 2O_2 \tag{14.14}$$

The concentration of ozone in the stratosphere is a steady-state concentration resulting from the balance of ozone production and destruction by the above processes. The region of the stratosphere in which significant amounts of ozone occur is called the *ozone layer*. The boundaries of the ozone layer vary with altitude, but it is generally regarded as extending from 15 to 35 km. At higher altitudes above 35 km, the level of atmospheric gases is very low, most of the oxygen is dissociated to O atoms, and there is very little molecular O_2 with which the O atoms can combine to produce O_3. At lower altitudes below about 15 km, most of the short-wavelength ultraviolet radiation capable of producing O atoms required for ozone production has been filtered from sunlight by the atmosphere above.

A total of about 350,000 metric tons of ozone are formed and destroyed daily in the ozone layer. Ozone never makes up more than a small fraction of the gases in the ozone layer. In fact, if the entire atmosphere's ozone were averaged in a uniform layer at sea level at a temperature of 25°C and 1 atm pressure, it would be only 3 mm thick! This would represent an ozone layer of 0.3 atm-cm. The total amount of ozone in the atmosphere above a particular location is measured in Dobson units (DU). The DU is 0.001 atm-cm so the average ozone layer is about 300 DU.

14.5.1 Shielding Effect of the Ozone Layer

Ozone absorbs ultraviolet radiation very strongly in the region 220–330 nm. Therefore, it is effective in filtering out dangerous UV-B radiation, 290 nm $<\lambda<$320 nm. (UV-A radiation, 320–400 nm, is relatively less harmful and UV-C radiation, <290 nm does not penetrate into the troposphere.) If UV-B were not absorbed by ozone, severe damage would result to exposed forms of life on Earth. Absorption of electromagnetic radiation by ozone converts the radiation's energy to heat and is responsible for the temperature maximum encountered at the boundary between the stratosphere and the mesosphere at an altitude of approximately 50 km. The reason that the temperature maximum occurs at a higher altitude than that of the maximum ozone concentration arises from the fact that ozone is such an effective absorber of ultraviolet light, so that most of this radiation is absorbed in the upper stratosphere where it generates heat and only a small fraction reaches the lower altitudes, which remain relatively cool.

Increased intensities of ground-level ultraviolet radiation caused by stratospheric ozone destruction would have some significant adverse consequences. One major effect would be on plants, including crops used for food. The destruction of microscopic plants that are the basis of the ocean's food chain (phytoplankton) could severely reduce the productivity of the world's seas. Human exposure would result in an increased incidence of cataracts. The effect of most concern to humans is the elevated occurrence of skin cancer in individuals exposed to ultraviolet radiation. This is because of photochemical reactions in cellular DNA (Chapter 22) that has absorbed UV-B radiation such that the genetic code is improperly translated during cell division leading to uncontrolled cell division and skin cancer. People with light complexions lack protective melanin, which absorbs UV-B radiation, and are especially susceptible to its effects. The most common type of skin cancer resulting from ultraviolet exposure is squamous cell carcinoma, which forms lesions that are readily removed and has little tendency to spread (metastasize). Readily metastasized malignant melanoma caused by absorption of UV-B radiation is often fatal. Fortunately, this form of skin cancer is relatively uncommon.

14.5.2 Ozone Layer Destruction

One of the major threats to the environment is destruction of stratospheric ozone by substances released to the atmosphere that catalyze conversion of O_3 back to O_2. The major culprit in ozone depletion consists of CFC compounds, commonly known as "Freons." In 1974 it was convincingly suggested by Mario Molina and F. Sherwood Rowland that chlorofluoromethanes could catalyze the destruction of stratospheric ozone. Subsequent data on ozone levels in the stratosphere and on increased ultraviolet radiation at the earth's surface showed that the threat to stratospheric ozone posed by CFCs is real. Along with atmospheric scientist Paul Crutzen, these investigators received a richly deserved Nobel Prize in 1995 for this classic work.

Developed in the 1930s as substitutes for hazardous sulfur dioxide and ammonia refrigerant fluids, CFCs were used and released to a very large extent over several decades. In addition to their major use as refrigerant fluids, other applications have included solvents, aerosol propellants, and blowing agents in the fabrication of foam plastics. The same extreme chemical stability that makes CFCs nontoxic enables them to persist for years in the atmosphere and to enter the stratosphere. In the stratosphere, as discussed in Section 12.7, the photochemical dissociation of CFCs by intense ultraviolet radiation,

$$CF_2Cl_2 + h\nu \rightarrow Cl^{\bullet} + CClF_2^{\bullet} \quad (14.15)$$

yields chlorine atoms, each of which can go through chain reactions involving first the reaction of atomic chlorine with ozone:

$$Cl^{\bullet} + O_3 \rightarrow ClO\bullet + O_2 \quad (14.16)$$

In the most common sequence of reactions involved with stratospheric ozone destruction, the $ClO^{\bullet}$ radicals react to form a dimer, which then reacts to regenerate Cl atoms, which in turn react with ozone to regenerate $ClO^{\bullet}$ in the following cyclical reaction sequence (where M is an energy-absorbing third body, such as an N_2 molecule):

$$ClO^{\bullet} + ClO^{\bullet} \rightarrow ClOOCl \quad (14.17)$$

$$ClOOCl + h\nu \rightarrow ClOO^{\bullet} + Cl^{\bullet} \quad (14.18)$$

$$ClOO^{\bullet} + M \rightarrow Cl^{\bullet} + O_2 + M \quad (14.19)$$

$$2Cl^{\bullet} + 2O_3 \rightarrow 2ClO^{\bullet} + 2O_2 \quad (14.20)$$

$$2O_3 \rightarrow 3O_2 \text{ (net reaction)} \quad (14.21)$$

The net effect of these reactions is catalysis of the destruction of several thousand molecules of O_3 for each Cl atom produced. Because of their widespread use and persistency, the two CFCs of most concern in ozone destruction are CFC-11 and CFC-12, $CFCl_3$, and CF_2Cl_2, respectively. Even in the intense ultraviolet radiation of the stratosphere, the most persistent CFCs have lifetimes of the order of 100 years.

The most prominent instance of ozone layer destruction is the so-called "Antarctic ozone hole" that was first noted in 1985 by the British Antarctic Survey and observed with great alarm in subsequent years. (Re-examination of earlier data showed that the ozone hole had been ongoing for several years prior to 1985.) This phenomenon is manifested by the appearance of severely depleted stratospheric ozone (up to 50%) over the polar region during the Antarctic's late winter and early spring months of September and October. The ozone hole is defined by a boundary of 220 DU chosen because prior to 1979, no ozone measurements below 220 DU were observed.

Figure 14.5 shows the size and thickness of the Antarctic ozone hole in recent years. In the Southern Hemisphere spring of 2008, the hole in the stratospheric ozone layer reached the fifth largest on record, according to the U.S. National Oceanic and Atmospheric Administration (NOAA).[5] The 2008 ozone hole reached its maximum size on September 12, covering an area of 27.2 million km^2 and extending 6.5 km deep, according to NOAA. The record ozone hole was measured in 2006, peaking at a size of 29.5 km^2 million sq miles. Although the production of CFCs responsible for the Antarctic ozone hole has been officially phased out, they persist for decades in the atmosphere and may take several years to rise high in the atmosphere where their effects are felt. NOAA scientists estimate that full recovery of the Antarctic ozone hole may not occur until after 2050.

The reasons why the Antarctic ozone hole occurs is due to the formation of a unique cloud in the lower stratosphere when temperatures drop well below –70°C for several months in the winter (a similar thing happens in the Arctic, but to a much lesser extent). In the absence of sunlight in the winter, this cloud persists as a rapidly whirling vortex. It is composed largely of ice crystals and supercooled ternary mixtures of HNO_3, $HNO_3{\cdot}3H_2O$, H_2SO_4, and H_2O. Also present is some HCl generated when CFCs undergo photochemical dissociation and abstract H from stratospheric CH_4. This HCl and chlorine nitrate produced by the following reaction are unreactive in ozone destruction:

$$ClO + NO_2 \rightarrow ClONO_2 \quad (14.22)$$

During the darkness of the Antarctic winter, Cl sequestered in the stratospheric cloud cover above Antarctica undergoes sequences of reactions that produce accumulations of photochemically reactive Cl_2 and HOCl. During the winter, these species have no effect on ozone because of the absence

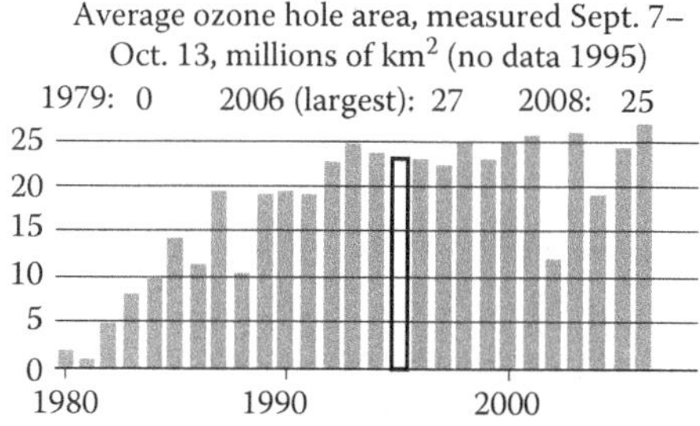

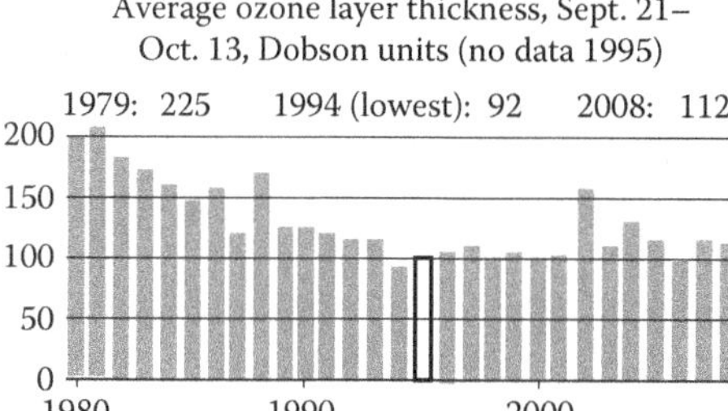

FIGURE 14.5 The area and ozone density of the Antarctic ozone hole during the September/October period when it reaches its maximum extent. The area increased steadily until the mid-1990s and has been relatively level since then reflecting a worldwide ban on CFC manufacture. The lower plot illustrates the thickness of the ozone layer in DU showing a decrease until the mid-1990s and generally level values since then. Data from NASA.

of ultraviolet solar radiation from the sun required to produce $Cl^\bullet$ atoms. Several important chemical processes are initiated on the surface of the stratospheric cloud particles which may be covered with a thin liquid aqueous layer. Chlorine nitrate may release HOCl:

$$ClONO_2 + H_2O \rightarrow HOCl + HNO_3 \quad (14.23)$$

Chlorine nitrate reacts in several steps with gaseous HCl to produce elemental chlorine and leads to an accumulation of photochemically reactive Cl_2:

$$ClONO_2 + HCl \rightarrow Cl_2 + HNO_3 \quad (14.24)$$

Some Cl_2 reacts with water to produce HOCl,

$$Cl_2 + H_2O \rightarrow HOCl + H^+ + Cl^- \quad (14.25)$$

And some chloride from HCl reacts with HOCl to generate Cl_2:

$$Cl^- + HOCl \rightarrow Cl_2 + OH^- \quad (14.26)$$

As sunlight returns to the lower stratosphere above Antarctica in spring, the Cl_2 and the HOCl produced by the reactions above undergo photodissociation,

$$Cl_2 + h\nu \rightarrow 2Cl^\bullet \quad (14.27)$$

$$HOCl + h\nu \rightarrow HO^\bullet + Cl^\bullet \quad (14.28)$$

to produce Cl atoms that can undergo the sequence of chain reactions (Reactions 14.16 through 14.20) leading to the net reaction for ozone destruction (Reaction 14.21). Therefore, over the winter months, photoreactive Cl_2 and HOCl accumulate in the Antarctic stratospheric region in the absence of sunlight and then undergo a burst of photochemical activity when spring arrives, leading to stratospheric ozone destruction and the formation of the Antarctic ozone hole.

14.5.3 Green Chemistry Solutions to Stratospheric Ozone Depletion

As discussed in Section 12.7, the U.S. EPA regulations, imposed in accordance with the 1986 Montreal Protocol on Substances that Deplete the Ozone Layer, curtailed the production of CFCs and halocarbons in the United States, starting in 1989. One of the most active areas of endeavor in green chemistry has been to find environmentally acceptable substitutes for ozone-depleting CFCs.

One of the easiest measures for CFC replacement was the application of carbon dioxide in producing foams for polystyrene plastic foams used in packing, food containers, and insulation. Global production of polystyrene foam amounts to about 5 million metric tons each year. The volume (and waste disposal problems) with this material can be appreciated considering that the foam is about 95% gas used as a blowing agent in making it. The Dow Chemical Company won the 1996 Presidential Green Challenge Award for their development of a process to use carbon dioxide as a complete replacement for CFCs in making polystyrene foam. (The incremental addition of greenhouse-gas carbon dioxide from this application is miniscule and CO_2 by-product of processes such as fermentation to produce ethanol can be used for foam blowing.)

As noted in Chapter 12, the substitutes for CFCs have been hydrogen-containing chlorofluorocarbons (HCFCs), hydrogen-containing fluorocarbons (HFCs), and some volatile hydrocarbon formulations. The presence of hydrogen in an H–C bond in these compounds provides a point of attack for hydroxyl radical to initiate compound decomposition in the troposphere. The first class of substitutes to be widely used was the HCFCs, which generally have only about 5% of the ozone-depleting potential of CFCs. One of the most popular of these has been HCFC-22, chemical formula $CHClF_2$, that was widely used in air conditioners, refrigerators, and as a foam blowing agent. Other popular HCFCs have included HCFC-142b (CH_3CClF_2, used as a refrigerant, often in a blend with HCFC-22), $CHCl_2CF_3$ (HCFC-123), CF_3CHClF (HCFC-124), and CH_3CCl_2F (HCFC-141b).

Although the HCFCs deplete stratospheric ozone much less than the CFCs, they still do so and are being phased out of general use. Ozone depletion potentials of CFC substitutes have been compiled that depend upon several factors including numbers of H–C bonds susceptible to attack and rate of reaction with $HO^•$ radical. Ozone depletion potentials are expressed relative to a value of 1.0 for CFC-11, a non-hydrogen-containing CFC with a formula of $CFCl_3$, which is particularly destructive to ozone because of its long atmospheric lifetime and high Cl content. Low ozone depletion potential correlates with short tropospheric lifetime, which means that the compound is destroyed in the troposphere before migrating to the stratosphere. The ozone-depletion potentials of some HCFC compounds that have been used as substitutes for CFCs are HCFC-22, 0.030; HCFC-123, 0.013; HCFC-141b, 0.10; HCFC-124, 0.035; and HCFC-142b, 0.038.

A kind of CFC substitute that does not harm stratospheric ozone consists of HFCs. These have included CH_2FCF_3 (HFC-134a, 1,1,1,2-tetrafluoroethane, which became the standard substitute for CFC-12 in automobile air conditioners and refrigeration equipment) and CH_2F_2 (HFC-152a, a flammable material). These compounds not only contain the more readily broken H–C bonds, they are chlorine-free and cannot generate the Cl atoms that attack stratospheric ozone. HFCs are favored as CFC replacements in North America. There is some concern over their global warming potential and, particularly in the case of popular HFC-134a, their potential to generate toxic fluoroacetic acid degradation products.

Arguably the most environmentally benign CFC substitutes are low-molecular-mass hydrocarbons including cyclopentane and isobutane. First used as refrigerants in 1867 and favored along with ammonia for this purpose until CFCs were developed in the 1930s, volatile hydrocarbons are now banned from use as home refrigerants in the United States because of their flammability, but are favored in Europe and other parts of the world.

14.6 ATMOSPHERIC BROWN CLOUDS

In Chapter 13, photochemical smog was discussed largely as a regional problem affecting large urban areas such as the one in the Los Angeles Basin. Smog is actually part of a larger problem that the United Nations Environment Programme defined in November 2008 as the Atmospheric Brown Cloud.[6] Generated by sources including automobiles, coal-fired power plants, slash-and burn agriculture, and cooking over fires fueled by wood or dung, the brown cloud consists of a layer of polluted air approximately 3 km thick that extends from the Arabian Peninsula across China and the western Pacific Ocean, at times even touching on the west coast of the United States Laden with soot and other particles, noxious air pollutants including ozone and greenhouse gases, the cloud is now recognized to have many adverse effects including the darkening of megacities such as New Delhi and Beijing, increased melting of Himalayan glaciers, decreased agricultural productivity, and greater extremes in weather events. Increased melting of the Hindu Kush–Himalayan–Tibetan glaciers is a particular concern because they provide head-waters for most Asian rivers that provide water to hundreds of millions of people as well as irrigation water essential to the production of rice and other crops. Because of its content of sunlight-reflecting sulfates and other particles, the cloud may be masking greenhouse warming of affected areas by 20–80%.

A major cause of the brown cloud phenomenon is inefficient, unsustainable burning of fossil fuels, particularly coal, but also including biomass. Deforestation, including burning of forest residues, also contributes.

The brown cloud has been studied most intensively in Asia because the region contains about half of the world's population, has undergone very rapid economic growth in recent years, and has a variable climate including annual monsoons. However, similar brown clouds occur in parts of Europe, North America, southern Africa, and the Amazon Basin. The severities of brown cloud events that occur in the North American eastern seaboard and Europe are reduced by winter precipitation, which wash pollutants out of the sky. The United Nations report identifies the following several regional brown cloud hotspots of most crucial concern:

- East Asia, a region covering eastern China
- The southeast Asian countries of Cambodia, Indonesia, Thailand, and Vietnam
- The Indo-Gangetic plains extending from eastern Pakistan across India to Bangladesh and Myanmar
- The southeast Asian region of Cambodia, Indonesia, Thailand, and Vietnam
- The sub-Saharan region of Africa extending into Angola, Zambia, and Zimbabwe
- The South American Amazon Basin region
- At least 13 megacities in Asia and Africa: Bangkok, Beijing, Cairo, Dhaka, Karachi, Kolkata, Lagos, Mumbai, New Delhi, Seoul, Shanghai, Shenzhen, and Tehran

The particles in brown clouds have two major effects. Sulfates, including sulfuric acid droplets and ammonium sulfates, largely from burning coal, scatter sunlight and reflect it back into space. Black soot and carbon particles from incomplete combustion of fossil fuels absorb incoming sunlight. These phenomena have a light-dimming effect, particularly in urban areas. India as a whole has noted light dimming of about 2% per decade since 1960 and China has experienced dimming of 3–4% since the 1950s; in both countries the effect has become more pronounced since 1980. Dimming of 10–25% has been noted in Beijing, Shanghai, Guangzhou, Karachi, and New Delhi. Brown cloud particles may also contribute to increased cloud cover from condensation of atmospheric water vapor to small cloud droplets, further increasing the dimming effect.

The dimming effects of brown clouds result in cooling of affected areas. Some climatologists believe that this masks the greenhouse gas warming of affected areas by 20–80%. This leads to the interesting conclusion that abrupt elimination of brown cloud pollution (a very unlikely event) could trigger warming of affected areas by about 2°C, which would adversely affect climate. Overall, temperature patterns in India, China, and surrounding areas have exhibited complex behavior during recent decades with warming in some places and cooling in others, which may well be the result of the influence of the brown clouds.

A particular concern regarding the potential effects of the brown cloud has to do with its influence on Asian monsoons, which are crucial to food production in the region. Several trends in Asian monsoons and rainfall have been observed in recent years which may be influenced by the brown cloud phenomenon. Since the 1950s, monsoon precipitation over India and southeast Asia has decreased 5–7% and the Indian summer monsoon has experienced fewer rainy days. Extreme rain events of more than 10 cm have increased in both China and India and very heavy rainfalls of more than 15 cm per day have nearly doubled.

A particular concern has been the potential influence of the brown cloud on glaciers. The most important of these in Asia has been the Hindu Kush–Himalayan–Tibetan glaciers that compose the head-waters of the Brahmaputra, Ganges, Mekong, and Yangtze rivers. With a population of more than 400 million people, India's Ganges basin contains 40% of India's irrigated farmland, so water input to the Ganges river from glacial sources is crucially important. About 70% of Ganges river water comes from the Gangotri glacier. China's 47,000 glaciers have decreased by about 5% and 3000 km^2 since 1980. Deposition of black carbon on glacier surfaces has increased light absorption and accelerated melting.

The brown cloud is likely having significant effects on agriculture and food production including the key food crops of rice, wheat, corn, and soybeans. Ground-level ozone above 40 parts per billion is believed to reduce crop production. Such levels are reached in parts of Asia, peaking during February to June and again between September and November, reducing yields of wheat, rice, and legumes. Other adverse brown cloud effects on crop production come from deposition of acidic and toxic particles on plants and reduced photosynthesis due to dimming of sunlight that reaches the earth's surface.

Several important health effects could be caused by brown clouds especially by particles smaller than 2.5 μm ($PM_{2.5}$ discussed in Section 10.9). Toxic substances including carcinogens are inhaled with air in brown clouds. Brown cloud air pollutants are linked to respiratory disease and cardiovascular disease. Hundreds of thousands of premature deaths in countries affected by the brown cloud may be due to inhalation of smoke particles from indoor cooking fires fueled by wood, coal, and even dried animal dung.

14.6.1 Yellow Dust

Closely tied to the brown cloud, *yellow dust* consists of huge masses of windblown dust and sand that afflict parts of the world each year. The most common yellow dust phenomenon begins with winds over deserts in Mongolia and China spreading east across China and affecting the Korean peninsula, Japan, and even Russia's Pacific port of Vladivostok. In its movement across China's industrial areas, the yellow dust mixes with the brown cloud of air pollutants to produce a highly noxious brew of very unpleasant air. Yellow dust is most prevalent in China from March until May. Deforestation and desertification, probably aggravated by global warming are contributing to the yellow dust phenomenon.

The economic damage from yellow dust amounts to millions of dollars each year. Manufacturers of high-tech goods that require clean environments are especially afflicted. The human costs are quite high with increased occurrence of asthma, lung disease, and immune diseases.

In contrast to the damaging dust storms that afflict southeast Asia, dust storms that originate in Africa's Sahara desert are thought to be important in sustaining life in some other parts of the world. About half of the Saharan dust comes from a small region called the Bodele Depression where a gap between the Tibesti and Ennedi mountains forms a natural wind tunnel through which winds blow across the Bodele Depression picking up an average of 700,000 metric tons of dust in a day. Easterly trade winds blow this dust across the Atlantic Ocean, depositing about 40 million tons of dust each year onto the Amazon region of South America. This dust is a significant source of nutrients for the Amazon forests helping to counteract the strong leaching effect of heavy rainfall in these regions. Much of the Saharan dust is deposited in the Atlantic Ocean as it is carried toward South America where it fertilizes the phytoplankton organisms that form the base of the oceanic food chain. Ocean water tends to be iron-deficient and it is believed that the Saharan dust replenishes nutrient iron in regions of the Atlantic where it is deposited.

14.7 ATMOSPHERIC DAMAGE BY PHOTOCHEMICAL SMOG

Photochemical smog, an air pollution phenomenon discussed in Chapter 13, is a significant contributor to the brown cloud discussed above. It occurs in urban areas where the combination of pollution-forming emissions and appropriate atmospheric conditions are right for its formation. In order for high levels of smog to form, relatively stagnant air must be subjected to sunlight under low humidity conditions in the presence of pollutant nitrogen oxides and hydrocarbons. The automobile is a major source of these pollutants, but hydrocarbons may come from biogenic sources, of which α-pinene and isoprene from trees are the most abundant (see Section 12.2). Under smog-forming conditions, the urban atmosphere acts as a huge chemical reactor in which hydrocarbons, nitrogen oxides, sulfur oxides, and oxygen undergo reactions driven by sunlight to produce ozone, organic oxidants, aldehydes, organic particles, nitrates, sulfates, and other noxious products. Smog does pose significant hazards to living things and materials in local urban areas in which millions of people are exposed and the oxidants generated by it have detrimental effects on crop production.

Ironically, ozone, which serves an essential protective function in the stratosphere, is the major culprit in tropospheric smog. In fact, surface ozone levels are used as a measure of smog. Ozone's phytotoxicity raises particular concern with respect to trees and crops. Ozone is the smog constituent responsible for most of the respiratory system distress and eye irritation, characteristic of human exposure to smog. Breathing is impaired at ozone levels approaching only about 0.1 ppm. Ozone is the "criterion" air pollutant that has been most resistant to control measures. Because of its strongly oxidizing nature, ozone attacks unsaturated bonds in fatty acid constituents of cell membranes. Other oxidants, such as PAN (Section 13.4), also contribute to the toxicity of smog, as do aldehydes produced as reactive intermediates in smog formation.

Smog is a secondary air pollutant that forms some time after and some distance from the injection into the atmosphere of the primary pollutant nitrogen oxides and reactive hydrocarbons required for its formation. The U.S. EPA's Empirical Kinetic Modeling Approach uses the concept of an *air parcel* to model smog formation. This model utilizes the concept of a "parcel" of relatively unpolluted air moving across an urban area in which it becomes contaminated with smog-forming gases. When the upper boundary of this parcel is restricted to about 1000 m by a temperature inversion and subjected to sunlight, the primary pollutants react to form smog in a system that involves photochemical reaction processes, transport, mixing, and dilution. As the hydrocarbons are consumed by photochemical oxidation processes in the air, and as nitrogen oxides are removed as nitrates and nitric acid (especially at nighttime), ozone levels reach a peak concentration at a time and place that may be some distance removed from the source of pollutants.

The most visible manifestation of smog is the *urban aerosol*, which greatly reduces visibility in smoggy urban atmospheres and contributes to the brown cloud phenomenon. Many of the particles composing this aerosol are condensation aerosols made from gases by chemical processes (see Chapter 10) and are therefore quite small, usually less than 2 μm. Particles of such a size are especially harmful because they scatter light most efficiently and are the most respirable. Aerosol particles formed from smog often contain toxic constituents, such as respiratory tract irritants and mutagens. The urban aerosol also contains particle constituents that originate from processes other than smog formation. Oxidation of pollutant sulfur dioxide by the strongly oxidizing conditions of photochemical smog,

$$SO_2 + \tfrac{1}{2}O_2 + H_2O \rightarrow H_2SO_4 \quad \text{(overall process)} \tag{14.29}$$

produces sulfuric acid and sulfate particles. Nitric acid and nitrates are produced at night when sunlight is absent, a process that involves intermediate NO_3 radical:

$$O_3 + NO_2 \rightarrow O_2 + NO_3 \tag{14.30}$$

$$NO_3 + NO_2 + \text{ M (energy-absorbing third body)} \rightarrow N_2O_5 + M \tag{14.31}$$

$$N_2O_5 + H_2O \rightarrow 2HNO_3 \tag{14.32}$$

$$HNO_3 + NH_3 \rightarrow NH_4NO_3 \tag{14.33}$$

As indicated by the last reaction above, ammonium salts are common constituents of urban aerosol particles; they tend to be particularly corrosive. Metals, which may contribute to the toxicity of urban aerosol particles and which may catalyze reactions on their surfaces, occur in the particles. Water is always present, even in low humidity atmospheres, and is usually present in urban aerosol particles. Carbon and PAHs from partial combustion and diesel engine emissions are usually abundant constituents; elemental carbon is usually the particulate constituent most responsible for absorbing light in the urban aerosol. If the air parcel originates over the ocean, it contains sea salt particles consisting largely of NaCl, from which some of the chloride may be lost as volatile HCl by the action of less volatile strong acids produced by smog. This phenomenon is responsible for Na_2SO_4 and $NaNO_3$ found in the urban aerosol.

PAHs (see Section 10.8), are among the urban aerosol particle constituents of most concern, particularly because metabolites of some of these compounds (see the 7,8-diol-9,10-epoxide of benzo(a)pyrene in Figure 23.3) are carcinogenic. PAHs include unsubstituted compounds as well as those with alkyl, oxygen, or nitrogen substituents, or O or N hetero atoms:

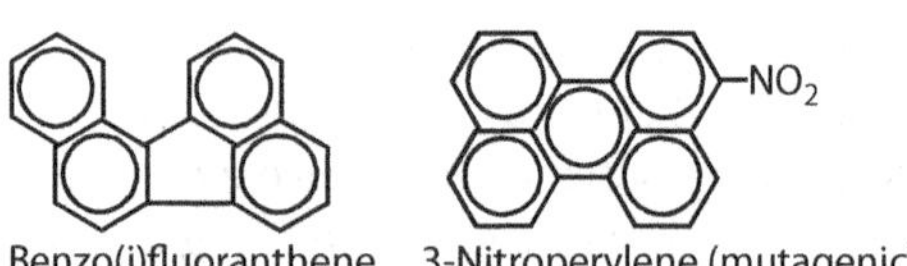

These kinds of compounds are emitted by internal combustion engine exhausts and occur in both the gas and particulate phases. Numerous mechanisms exist for their destruction and chemical alteration, particularly reaction with oxidant species—$HO^{\bullet}$, O_3, NO_2, N_2O_5, and HNO_3. Direct photolysis is also possible. PAH compounds in the vapor phase are destroyed relatively rapidly by these means, whereas PAHs sorbed to particles are much more resistant to reaction.

Another kind of urban aerosol particulate matter of considerable concern is *acid fog*, which may have pH values below 2 due to the presence of H_2SO_4 or HNO_3. Acid fog formation covers a wide range of atmospheric chemical and physical phenomena. The gas-phase oxidation of SO_2 and NO produces strong acids, which form very small aerosol particles. These, in turn, act as condensation nuclei for water vapor. Acid–base phenomena occur in the droplets, and the droplets act as scavengers to remove ionic species from air. Because fog aerosol particles form in areas of intense acid gas pollution near the surface, the concentrations of acids and ionic species in fog aerosol droplets tend to be much higher than in cloud aerosol droplets at higher altitudes.

In addition to health effects and damage to materials, one of the greater problems caused by smog is the destruction of crops and the reduction of crop yields. The annual cost of these effects in California alone is several billion dollars.

Even lightly populated nonindustrial areas are subject to the effects of smog brought about by human activities. Particularly, the practice of burning savanna grasses for agricultural purposes causes smog. This burning produces NO_x and reactive hydrocarbons that are required for smog formation. Furthermore, these grasses grow in tropical regions which have the intense sunlight required for smog formation. The net result is the rapid development of smoggy conditions as manifested by ozone levels several times normal background values.

Since about 1970, significant progress has been made in eliminating the emissions of organic compounds and NO_x that cause smog formation. These efforts resulted in a less-than-anticipated reduction in smog; for example, urban ozone in the United States declined only about 8% during the 1980s. In an effort to reduce ozone and other manifestations of air pollution more rapidly, the U.S. Congress passed a new, more rigorous set of Clean Air Act Amendments in 1990. The Amendments were written to further reduce automotive exhaust emissions, require changes in automotive fuel formulations, reduce emissions from stationary sources, and mandate other changes designed to reduce photochemical smog and other forms of air pollution. As a result of the enforcement of regulations under the Clean Air Act and technological advances, particularly in the area of automobile emissions, substantial progress has been made in controlling smog in the United States since the early 1990s.

14.8 NUCLEAR WINTER

Nuclear winter is a term used to describe a catastrophic atmospheric effect that might occur after a massive exchange of nuclear firepower between major powers as well as natural phenomena, particularly huge volcanic eruptions and asteroid impact, which would have much the same effect. The heat from the nuclear blasts and from resulting fires would cause powerful updrafts carrying sooty combustion products to stratospheric regions. This would result in several years of much lower temperatures and freezing temperatures even during summertime. There are several reasons for such an effect. First of all, the highly absorbent, largely black particulate matter would absorb solar radiation high in the atmosphere so that it would not reach the earth's surface. Cooling would also occur from a phenomenon opposite to that of the greenhouse effect. That is because outgoing infrared radiation from particles high in the atmosphere would have to penetrate relatively much less into the atmosphere and, therefore, would be exposed to much less infrared-absorbing water vapor and carbon dioxide gas. This would deprive the lower atmosphere of the warming effect of outgoing infrared radiation and would mean that less infrared would be re-radiated from the atmosphere back to the earth's surface. The cooling would also inhibit the evaporation of water, thereby reducing the amount of infrared-absorbing water vapor in the atmosphere and slowing the process by which particulate matter is scavenged from the atmosphere by rain. In addition to the direct suffering caused, starvation of millions of people would result from crop failures accompanying years of nuclear winter.

Conditions similar to those of a nuclear winter have resulted from huge volcanic eruptions. One of these occurred in 1816, "the year without a summer," following the astoundingly enormous Tambora, the Indonesian volcanic explosion of 1815. Brutally cold years around 210 BC that followed a similar volcanic incident in Iceland were recorded in ancient China. The June 1991 explosion of the Philippine volcano Pinatubo, which blasted millions of tons of material into the atmosphere, resulted in an approximately 0.5°C cooling the following year. The greatest climatic effects of volcanoes result from emission of millions of tons of sulfur gases into the atmosphere, which produce sunlight-reflecting sulfuric acid aerosols and sulfate particles. The pronounced cooling effect of the Pinatubo volcano was mainly due to the 15–30 million tons of sulfur dioxide that it emitted to the atmosphere. Eruptions near the equator cause global events whereas those at higher latitudes affect the hemisphere in which they occur. The volcanic incidents cited above clearly illustrate the climatic effects of huge quantities of particulate matter ejected high into the atmosphere. And it will happen again, perhaps on a scale that will dwarf even the Tambora eruption. Massive destructive eruptions can occur with supervolcanoes, one of the largest of which is the Yellowstone volcano in the United States. With a caldera about 50 km in diameter, another major eruption of this volcano could well occur again. An article entitled "Is the 'Beast' Building to a Violent Tantrum?" states the following: "When the volcano in Yellowstone National Park blew 6,400 centuries ago, it obliterated a mountain range, felled herds of prehistoric camels hundreds of miles away, and left a smoking hole in the ground the size of the Los Angeles Basin."[7] Such an event could well extinguish civilization on Earth.

Evidence exists to suggest that military explosives can result in the introduction of large quantities of particulate matter into the atmosphere. For example, carpet bombings of cities, such as the tragic bombing of Dresden, Germany, near the end of World War II, have produced huge firestorms that created their own wind causing a particle-laden updraft into the atmosphere. Of course, the effect of a full-scale nuclear exchange would be manyfold higher.

An idea of the potential climatic effect resulting from a full-scale nuclear exchange may be obtained by considering the magnitude of the blasts that might be involved. Only two nuclear bombs have been used in warfare, both dropped on cities in Japan in 1945. The Hiroshima fission bomb had the explosive force of 12 kilotons of Trinitrotoluene (TNT) explosive. Its blast, fireball, and instantaneous emissions of neutrons and gamma radiation, followed by fires and exposure to radioactive fission products, killed about 100,000 people and destroyed the city on which it was dropped. In comparison with this 12-kiloton bomb, modern fusion bombs are typically rated at 500 kilotons, and 10-megaton weapons are common. A full-scale nuclear exchange might involve a total of the order of 5000 megatons of nuclear explosives. As a result, unimaginable quantities of soot from the partial combustion of wood, plastics, paving asphalt, petroleum, forests, and other combustibles would be carried to the stratosphere. At such high altitudes, tropospheric removal mechanisms for particles are not effective because there is not enough water in the stratosphere to produce rainfall to wash particles from the air, and convection processes are very limited. Much of the particulate matter would be in the micrometer size range in which light is reflected, scattered, and absorbed most effectively and settling is very slow. Therefore, vast areas of Earth would be overlain by a stable cloud of particles and the fraction of sunlight reaching the earth's surface would be drastically reduced, resulting in a dramatic cooling effect. There would be other effects as well. The extreme heat and pressure in the fireball would result in the fixation of nitrogen by the following reaction:

$$O_2 + N_2 \rightarrow 2NO \tag{14.34}$$

The large quantities of NO generated would catalyze the destruction of protective stratospheric ozone.

The timing and location of nuclear blasts are very important in determining their climatic effects. Atmospheric testing of nuclear weapons, including a 58-megaton monster detonated by the Soviet Union, has had little atmospheric effect. Such tests were carried out at widely spaced intervals on deserts, small tropical islands, and other locations with minimal combustible matter. In contrast, military use of nuclear weapons would involve a high concentration of firepower, both in time and in space, on industrial and military targets consisting largely of combustibles. Furthermore, destruction of hardened military sites requires blasts that disrupt large quantities of soil, rock, and concrete, which are pulverized, vaporized, and blown into the atmosphere.

On a hopeful note, the East–West conflict that dominated world politics and threatened nuclear war from the mid-1900s until 1990 has now abated and the probability of nuclear warfare seems to have diminished. However, continued armed conflict in the Middle East, tense relations between nuclear-armed India and Pakistan, efforts by potentially unstable countries to acquire nuclear weapons, racial hatred accompanied by a determination to perform "ethnic cleansing," and a "trigger-happy" state of mind among even educated people who should know better still give cause for concern with respect to the prospect of "nuclear winter."

14.8.1 Doomsday Visitors from Space

Of all the possible atmospheric catastrophes that can occur, arguably the most threatening would be one caused by the collision of a large asteroid with Earth. Convincing evidence now exists that mass extinctions of species in the past have resulted from Earth being hit by asteroids several kilometers in diameter. It is well established that the impact of a huge asteroid caused the extinction of dinosaurs and most other animal species 66 million years ago. Fossil evidence suggests that even greater extinctions that occurred millions of years earlier, including one that destroyed about 90% of all living species 251 million years ago, were probably the result of asteroid impact. Such an impact would cause much the same effects as those from "nuclear winter," though with a large asteroid, the effects would be much more pronounced.

Around 13,000 years ago, Earth was just emerging from the most recent Ice Age when it reverted to a cold period lasting for 1300 years called the Younger Dryas or "Big Freeze."[8] As a result, about 35 animal species including several large mammals such as the short-faced bear, American lion, and mastodons became extinct. Geologically, the beginning of the Younger Dryas is marked by a thin black layer of carbon indicative of massive fires, probably from burning forests. It is possible that this catastrophic event was caused by the impact of a large comet that rained large numbers of meteorites over the earth's surface. The discovery of very small diamonds ("nanodiamonds") in the black layer is indicative of such an event.

14.9 WHAT IS TO BE DONE?

Of all environmental hazards, there is little doubt that major disruptions in the atmosphere and climate have the greatest potential for catastrophic and irreversible environmental damage. If levels of greenhouse gases and reactive trace gases continue to increase at present rates, major environmental effects are virtually certain. On a hopeful note, the bulk of these emissions arise from industrialized nations which, in principle, can apply the resources needed to reduce them substantially. The best example to date has been the 1987 "Montreal Protocol on Substances that Deplete the Ozone Layer," an international treaty through which a large number of nations agreed to cut CFC emissions by 50% by the year 2000. This agreement and subsequent ones, particularly the Copenhagen Amendment of 1992, may pave the way for more encompassing agreements covering carbon dioxide and other trace gases.

More ominous, however, is the combination of population pressure and desire for better living standards on a global basis. Consider, for example, the demand that these two factors place on energy resources and the environmental disruption that may result. In many highly populated developing nations, high-sulfur coal is the most readily available, cheapest source of energy. It is understandably difficult to persuade populations faced with real hunger to forego short-term economic gain for the sake of long-term environmental quality. Destruction of rain forests by "slash-and-burn" agricultural methods does seem to make economic sense to those engaged in subsistence farming to obtain badly needed hard currency, which can be earned by converting forest to pasture land and exporting fast-food-hamburger beef to wealthier nations.

What is to be done? First of all, it is important to keep in mind that the atmosphere has a strong ability to cleanse itself of pollutant species. Water-soluble gases, including greenhouse-gas CO_2, acid-gas SO_2, and fine particulate matter are removed with precipitation. For most gaseous contaminants, oxidation precedes or accompanies removal processes. To a degree, oxidation is carried out by O_3. To a larger extent, the most active atmospheric oxidant is hydroxyl radical $HO^{\bullet}$. As illustrated in Figure 9.10, this atmospheric scavenger species reacts with all important trace gas species except for CO_2 and CFCs. It is now generally recognized that $HO^{\bullet}$ is an almost universal atmospheric cleansing agent. Given this crucial role of $HO^{\bullet}$ radical, any pollutants that substantially reduce its concentration in the atmosphere are potentially troublesome. One concern over carbon monoxide emissions to the atmosphere is the reactivity of $HO^{\bullet}$ with CO,

$$CO + HO^{\bullet} \rightarrow CO_2 + H \tag{14.35}$$

which could result in removal of $HO^{\bullet}$ from the atmosphere.

Of all the major threats to the global climate, it is virtually certain that humankind will have to try to cope with greenhouse warming and the climatic effects thereof. The measures to be taken in dealing with this problem fall into the three following categories:

- *Minimization* by reducing emissions of greenhouse gases, switching to alternate energy sources, increasing energy conservation, and reversing deforestation. It is especially sensible to use measures that have major benefits in addition to reduction of greenhouse warming. Such measures include, as examples, reforestation, restoration of grasslands, increased energy conservation, and a massive shift to solar and wind energy sources.
- *Counteracting measures*, such as injecting light-reflecting particles into the upper atmosphere.
- *Adaptation*, particularly through increased efficiency and flexibility of the distribution and use of water, which might be in very short supply in many parts of the world as a consequence of greenhouse warming. Important examples are implementation of more efficient irrigation practices and changes in agriculture to grow crops that require less irrigation. Emphasis on adaptation is favored by those who contend that not enough is known about the types and severity of global warming to justify massive expenditures on minimization and counteractive measures. In any case, adaptation will certainly have to be employed as a means of coping with global warming.

Potentially, tax strategy can be very effective in reducing the use of carbonaceous fuels and greenhouse CO_2 emissions. This is the rationale behind the *carbon tax*, which is tied with the carbon content of various fuels. Another option is to dispose off carbon dioxide to a sink other than the atmosphere. The most obvious such sink is the ocean; other possibilities are deep subterranean aquifers and exhausted oil and gas wells.

A common measure taken against the effects of another atmospheric hazard, ultraviolet radiation, provides an example of adaptation. This measure is the use of sunscreens placed on the skin as lotions to filter out UV-B and in some formulations UV-A radiation, thus reducing the likelihood of skin cancer. The active ingredient of sunscreen must absorb or reflect ultraviolet light effectively. Suspensions of small particles of zinc oxide or titanium oxide are used in sunscreen formulations as physical blockers of ultraviolet radiation. Water-insoluble suspensions of organic octinoxate (octyl methoxycinnamate) absorb UV-B radiation and oxybenzone is effective against UV-A radiation (see structural formulas below). These and related compounds act as chemical absorbers, which are generally aromatic compounds conjugated with a carbonyl group. Chemical absorbers normally act by absorbing ultraviolet radiation to reach an excited state and then dissipating the absorbed energy by reverting to the ground state to regenerate the absorbing species.

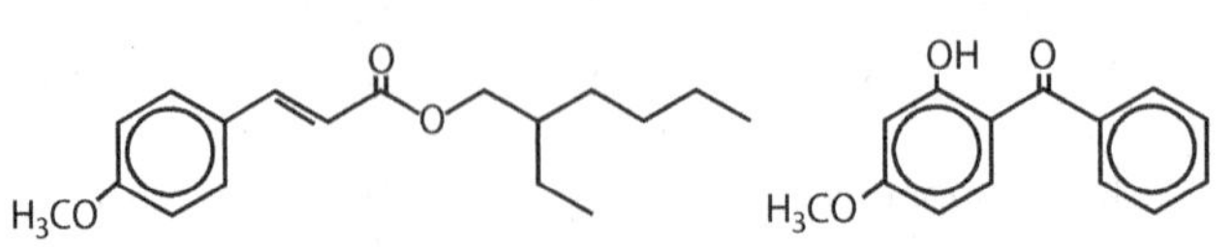

H_3CO O OH O H_3CO

Octinoxate (octylmethoxycinnamate) Oxybenzone (2-hydroxy-4-methoxyphenyl)phenylmetanone

The "*tie-in strategy*," sometimes called the "no-regrets" policy, has been proposed as a sensible approach to dealing with the kinds of global environmental problems discussed in this chapter. This approach was first enunciated in 1980.[9] It advocates taking measures consisting of "high-leverage actions," which are designed to prevent problems from occurring, which improve resiliency and adaptability, and which have substantial merit even if the major problems that they are designed to avoid do not materialize. An example is implementation of environmentally sound substitutes for fossil fuels to lower atmospheric CO_2 output and prevent greenhouse warming. Even if it turns out that the greenhouse effect is exaggerated, such substitutes would save Earth from other kinds of environmental damage, such as disruption of land by strip mining coal or preventing oil spills from petroleum transport. Definite economic and political benefits would also accrue from lessened dependence on uncertain, volatile petroleum supplies. Increased energy efficiency would diminish both greenhouse gas and acid rain production, while lowering costs of production and reducing the need for expensive and environmentally disruptive new power plants.

Even when changes such as in climate are inevitable, measures that slow the changes are desirable because adaptation is usually much more feasible over longer time periods. The implementation of tie-in strategies requires some degree of incentive beyond normal market forces and, therefore, is opposed by some on ideological grounds. A good example is opposition to mandatory fuel mileage standards for automobiles. Although free market purists tend to oppose measures such as fuel mileage standards, a market that does not factor in environmental costs is not truly a free market.

LITERATURE CITED

1. Petkewich, R., Trees testify to pollution: Tree rings help scientists trace the source and timing of chemical leaks and spills in the environment, *Chemical and Engineering News*, **86**, 37–38 (2008).
2. Revelle, R. and H. E. Suess, Carbon dioxide exchange between atmosphere and ocean and the question of an increase of atmospheric CO_2 during the past decades, *Tellus*, **9**, 18–27 (1957).
3. Wilson, E. J. and D. Gerard, *Carbon Capture and Sequestration: Integrating Technology, Monitoring and Regulation*, Blackwell Publishing, Ames, IA, 2007.
4. Visgilio, G. R. and D. M. Whitelaw, *Acid in the Environment: Lessons Learned and Future Prospects*, Springer, Berlin, 2007.
5. Bakker, S. H., Ed., *Ozone Depletion, Chemistry, and Impacts,* Nova Science Publishers, Hauppauge, NY, 2008.
6. The summary of a 2008 report entitled "Atmospheric Brown Clouds" is available on the United Nations Environment Programme for Development on the UNEP website: http://www.unep.org/
7. Breining, G., *Super Volcano: The Ticking Time Bomb Beneath Yellowstone National Park*, MBI Publishing, Minneapolis, 2010.
8. Kennett, D. J., J. P. Kennett, A. West, C. Mercer, S. S. Que Hee, L. Bement, T. E. Bunch, M. Sellers, and W. S. Wolbach, Nanodiamonds in the Younger Dryas boundary sediment layer, *Science*, **323**, 94 (2009).
9. E. Boulding, in *Carbon Dioxide Effects, Research and Assessment Program: Workshop on Environmental and Societal Consequence of a Possible CO_2–Induced Climatic Change*, Report 009, CONF-7904143, U.S. Department of Energy, U.S. Government Printing Office, Washington, DC, October 1980, pp. 79–10.

SUPPLEMENTARY REFERENCES

Aguado, E. and J. E. Burt, *Understanding Weather and Climate*, 4th ed., Pearson Education, Upper Saddle River, NJ, 2007.

Ahrens, C. D., *Meteorology Today: An Introduction to Weather, Climate, and the Environment*, 9th ed., Thomson Brooks/Cole, Belmont, CA, 2009.

Allaby, M., *Atmosphere: A Scientific History of Air, Weather, and Climate*, Facts on File, New York, 2009.

Austin, J., P. Brimblecombe, and W. Sturges, Eds, *Air Pollution Science for the 2lst Century*, Elsevier Science, New York, 2002.

Bowen, M., *Censoring Science: Inside the Political Attack on Dr. James Hansen and the Truth of Global Warming*, Dutton, New York, 2008.
Casper, J. K., *Greenhouse Gases: Worldwide Impacts*, Facts on File, New York, 2009.
Desonie, D., *Climate: Causes and Effects of Climate Change*, Chelsea House Publishers, New York, 2008.
Dewet, A., *Whole Earth: Earth System Science and Global Change*, W.H. Freeman & Company, New York, 2004.
Graedel, T. E. and P. J. Crutzen, *Atmospheric Change: An Earth System Perspective*, W.H. Freeman & Company, New York, 1993.
Hardy, J. T., *Climate Change: Causes, Effects and Solutions*, Wiley, New York, 2003.
Hewitt, C. N. and A. Jackson, *Atmospheric Science for Environmental Scientists*, Wiley-Blackwell, Hoboken, NJ, 2009.
Houghton, J. T., *Global Warming*, Cambridge University Press, New York, 2004.
Kidd, J. S. and R. A. Kidd, *Air Pollution: Problems and Solutions*, Chelsea House Publishers, New York, 2006.
Koppes, S. N., *Killer Rocks from Outer Space: Asteroids, Comets, and Meteorites*, Lerner Publications Company, Minneapolis, 2004.
Kushner, J. A., *Global Climate Change and the Road to Extinction*, Carolina Academic Press, Durham, NC, 2008.
Livingston, J. V., *Air Pollution: New Research*, Nova Science Publishers, New York, 2007.
Mathez, E. A., *Climate Change: The Science of Global Warming and Our Energy Future*, Columbia University Press, New York, 2009.
Peretz, L. N., *Climate Change Research Progress*, Nova Science Publishers, New York, 2008.
Rohli, R. V. and A. J. Vega, *Climatology*, Jones and Bartlett Publishers, Sudbury, MA, 2008.
Spalding, F., *Under a Black Cloud: Our Atmosphere under Attack*, Rosen Publishing, New York, 2009.
Vallero, D. A., *Fundamentals of Air Pollution*, 4th ed., Elsevier, Amsterdam, 2008.
Zedillo, E., Ed., *Global Warming: Looking Beyond Kyoto*, Brookings Institution Press, Washington, DC, 2008.

QUESTIONS AND PROBLEMS

1. How do modern transportation problems contribute to the kinds of atmospheric problems discussed in this chapter?
2. What is the rationale for classifying most acid rain as a secondary pollutant?
3. Distinguish among UV-A, UV-B, and UV-C radiation. Why does UV-B pose the greatest danger in the troposphere?
4. How does the extreme cold of stratospheric clouds in Antarctic regions contribute to the Antarctic ozone hole?
5. How does the oxidizing nature of ozone from smog contribute to the damage that it does to cell membranes?
6. What may be said about the time and place of the occurrence of maximum ozone levels from smog with respect to the origin of the primary pollutants that result in smog formation?
7. What is the basis for "nuclear winter"?
8. Discuss the analogies between the effects of a large asteroid hit on Earth with "nuclear winter."
9. What is meant by a "tie-in strategy"?
10. Of the following, the statement that is *untrue* is (explain): (A) acid rain is denoted by any precipitation with a pH less than neutral (7.00), (B) acid may be deposited as acidic salts and acid gases, in addition to aqueous acids, (C) acid rain is a regional air pollution problem as distinguished from local or global problems, (D) carbon dioxide makes rainfall slightly acidic, (E) acid rain is often associated with elevated levels of sulfate ion, SO_4^{2-}.
11. Of the following related to greenhouse gases and global warming, the true statement is (explain): (A) levels of greenhouse gas methane are increasing by about 1 ppm per year in the atmosphere, (B) per molecule, methane has a greater effect on greenhouse warming

than does carbon dioxide, (C) radiative forcing of CO_2 is about 25 times that of CH_4, (D) carbon dioxide is the only gas considered significant as a cause of greenhouse warming, (E) although models predict greenhouse warming, there is no evidence in recent years that it may have in fact begun.

12. Of the following, the true statement pertaining to the "Antarctic ozone hole" is (explain): (A) it reaches its peak during the Antarctic summer, (B) it does not involve chlorine species, (C) it involves only species that occur in the gas phase, (D) it does not involve ClO radical, (E) it is related to species that are frozen in stratospheric cloud particles at very low temperature.
13. Of the following, the one that is *not* an effect of acid rain is (explain): (A) direct phytotoxicity (plant toxicity) from H^+, (B) phytotoxicity from acid-forming gases, such as SO_2, (C) phytotoxicity from liberated Al^{3+}, (D) toxicity to fish fingerlings from acid accumulated in lakes, (E) all the above are effects.
14. Of the following, the one that is *not* a measure for decreasing adverse effects on global climate is (explain): (A) minimization, such as reducing emissions of greenhouse gases, (B) counteracting measures, such as injection of light-reflecting particles into the upper atmosphere, (C) replacement of nuclear energy with fossil energy (D) adaptation, such as more efficient irrigation, (E) "tie-in" strategy.
15. Of the following, the true statement pertaining to "the Endangered Global Atmosphere" is (explain): (A) atmospheric SO_2 may indirectly help reduce the greenhouse effect, (B) atmospheric carbon dioxide levels are projected to decrease after the year 2010, (C) "nuclear winter" is of concern primarily for the greenhouse effect, (D) the major effect of volcanic eruptions is greenhouse warming, (E) photochemical smog is primarily a global problem, not regional or local.
16. Using the internet, look up the most recent near collision of a large, potentially very destructive, asteroid with Earth. When did it happen? How close did the asteroid come to Earth? How large was it? What would have been the likely effects of a collision?

15 The Geosphere and Geochemistry

15.1 INTRODUCTION

The *geosphere*, or solid earth, is that part of Earth upon which humans live and from which humans extract most of their food, minerals, and fuels. Once thought to have an almost unlimited buffering capacity against the perturbations of humankind, the geosphere is now known to be rather fragile and subject to harm from human activities. For example, some billions of tons of earth material are mined or otherwise disturbed each year in the extraction of minerals and coal. Two atmospheric pollutant phenomena—excess carbon dioxide and acid rain (see Chapter 14)—have the potential to cause major changes in the geosphere. Too much carbon dioxide in the atmosphere may cause global heating ("greenhouse effect"), which could significantly alter rainfall patterns and turn currently productive areas of Earth into desert regions. The low pH characteristic of acid rain can bring about drastic changes in the solubilities and oxidation–reduction rates of minerals. Erosion caused by intensive cultivation of land is washing away vast quantities of topsoil from fertile farmlands each year. In some areas of industrialized countries, the geosphere has been the dumping ground for toxic chemicals. Ultimately, the geosphere must provide disposal sites for the nuclear wastes of the more than 400 nuclear reactors that have operated worldwide. It may be readily seen that the preservation of the geosphere in a form suitable for human habitation is one of the greatest challenges facing humankind.

The interface between the geosphere and the atmosphere at the Earth's surface is very important to the environment. Human activities on the Earth's surface may affect climate, most directly through the change of surface albedo, defined as the percentage of incident solar radiation reflected by a land or water surface. For example, if the sun radiates 100 units of energy per minute to the outer limits of the atmosphere, and the Earth's surface receives 60 units per minute of the total, then reflects 30 units upward, the albedo is 50%. Some typical albedo values for different areas on the Earth's surface are: evergreen forests, 7–15%; dry, plowed fields, 10–15%; deserts, 25–35%; fresh snow, 85–90%; asphalt, 8%. In some heavily developed areas, anthropogenic (human-produced) heat release is comparable to the solar input. The anthropogenic energy release over the 60 km^2 of Manhattan Island averages about four times the solar energy falling on the area; over the 3500 km^2 of Los Angeles the anthropogenic energy release is about 13% of the solar flux.

One of the greatest impacts of humans on the geosphere is the creation of desert areas through abuse of land with marginal amounts of rainfall. This process, called *desertification*, is manifested by declining groundwater tables, salinization of topsoil and water, reduction of surface waters, unnaturally high soil erosion, and desolation of native vegetation. The problem is severe in some parts of the world, particularly Africa's Sahel (southern rim of the Sahara), where the Sahara advanced southward at a particularly rapid rate during the period 1968–1973, contributing to widespread starvation in Africa during the 1980s. Large, arid areas of the western United States are experiencing at least some desertification as the result of human activities and severe droughts.

As the populations of the western United States increase, one of the greatest challenges facing the residents is to prevent additional conversion of land into desert.

The most important part of the geosphere for life on Earth is soil. It is the medium upon which plants grow, and virtually all terrestrial organisms depend upon it for their existence. The productivity of soil is strongly affected by environmental conditions and pollutants. Because of the importance of soil, all of Chapter 16 is devoted to its environmental chemistry.

With increasing population and industrialization, one of the more important aspects of human use of the geosphere has to do with the protection of water sources. Mining, agricultural, chemical, and radioactive wastes all have the potential for contaminating both surface water and groundwater. Sewage sludge spread on land may contaminate water by release of nitrate and heavy metals. Landfills may likewise be sources of contamination. Leachates from unlined pits and lagoons containing hazardous liquids or sludges may pollute drinking water.

It should be noted, however, that many soils have the ability to assimilate and neutralize pollutants. Various chemical and biochemical phenomena in soils operate to reduce the harmful nature of pollutants. These phenomena include oxidation–reduction processes, hydrolysis, acid–base reactions, precipitation, sorption, and biochemical degradation. Some hazardous organic chemicals may be degraded to harmless products on soil, and heavy metals may be sorbed by it. In general, however, extreme care should be exercised in disposing of chemicals, sludges, and other potentially hazardous materials on soil, particularly where there is a possibility of water contamination.

15.2 THE NATURE OF SOLIDS IN THE GEOSPHERE

Earth is divided into layers, including the solid iron-rich inner core, molten outer core, mantle, and crust. Environmental chemistry is most concerned with the *lithosphere*, which consists of the outer mantle and the *crust*. The crust is the Earth's outer skin that is accessible to humans. It is extremely thin compared to the Earth's diameter, ranging from 5 to 40 km thick.

Most of the solid earth crust consists of rocks. Rocks are composed of minerals, where a *mineral* is a naturally occurring inorganic solid with a definite internal crystal structure and chemical composition. A *rock* is a solid, cohesive mass of pure mineral or an aggregate of two or more minerals.

15.2.1 Structure and Properties of Minerals

The combination of two characteristics is unique to a particular mineral. These characteristics are a defined chemical composition, as expressed by the mineral's chemical formula, and a specific crystal structure. The *crystal structure* of a mineral refers to the way in which the atoms are arranged relative to each other. It cannot be determined from the appearance of visible crystals of the mineral, but requires structural methods such as x-ray structure determination. Different minerals may have the same chemical composition, or they may have the same crystal structure, but may not be identical for truly different minerals.

Physical properties of minerals can be used to classify them. The characteristic external appearance of a pure crystalline mineral is its *crystal form*. Because of space constraints on the ways that minerals grow, the pure crystal form of a mineral is often not expressed. *Color* is an obvious characteristic that can vary widely due to the presence of impurities. The appearance of a mineral surface in reflected light describes its *luster*, which may be metallic, partially metallic (submetallic), vitreous (like glass), dull or earthy, resinous, or pearly. The color observed when a mineral is rubbed across an unglazed porcelain plate is known as *streak*. *Hardness* is expressed on the Mohs scale, which ranges from 1 to 10 and is based upon 10 minerals that vary from talc, hardness 1, to diamond, hardness 10. *Cleavage* denotes the manner in which minerals break along planes and the angles in which these planes intersect. For example, mica cleaves to form thin sheets. Most minerals *fracture* irregularly, although some fracture along smooth curved surfaces or into fibers or splinters. *Specific gravity*—density relative to that of water—is another important physical characteristic of minerals.

15.2.2 Kinds of Minerals

Although over 2000 minerals are known, only about 25 *rock-forming minerals* make up most of the Earth's crust.[1] The nature of these minerals may be better understood with a knowledge of the elemental composition of the crust. Oxygen and silicon make up 49.5% and 25.7% by mass of the Earth's crust, respectively. Therefore, most minerals are *silicates* such as quartz, SiO_2, or orthoclase, $KAlSi_3O_8$. In descending order of abundance; the other elements in the Earth's crust are aluminum (7.4%), iron (4.7%), calcium (3.6%), sodium (2.8%), potassium (2.6%), magnesium (2.1%), and other (1.6%). Table 15.1 summarizes the major kinds of minerals in the Earth's crust.

Secondary minerals are formed by alteration of parent mineral matter. *Clays* are silicate minerals, usually containing aluminum, that constitute one of the most significant classes of secondary minerals. Olivine, augite, hornblende, and feldspars all form clays. Clays are discussed in detail in Section 15.7.

15.2.3 Evaporites

Evaporites are soluble salts that precipitate from a solution under special arid conditions, commonly as the result of the evaporation of seawater. The most common evaporite is *halite*, NaCl. Other simple evaporite minerals are sylvite (KCl), thenardite (Na_2SO_4), and anhydrite ($CaSO_4$). Many evaporites are hydrates, including bischofite ($MgCl_2 \cdot 6H_2O$), gypsum ($CaSO_4 \cdot 2H_2O$), kieserite ($MgSO_4 \cdot H_2O$), and epsomite ($MgSO_4 \cdot 7H_2O$). Double salts, such as carnallite ($KMgCl_3 \cdot 6H_2O$), kainite ($KMgClSO_4 \cdot \frac{11}{4}H_2O$), glaserite ($K_3Na(SO_4)_2$), polyhalite ($K_2MgCa_2(SO_4)_4 \cdot 2H_2O$), and loeweite ($Na_{12}Mg_7(SO_4)_{13} \cdot 15H_2O$), are very common in evaporites.

The precipitation of evaporites from marine and brine sources depends upon a number of factors. Prominent among these are the concentrations of the evaporite ions in the water and the solubility products of the evaporite salts. The presence of a common ion decreases solubility; for example, $CaSO_4$ is precipitated more readily from brine containing Na_2SO_4 than from a solution containing no other source of sulfate. The presence of other salts that do not have a common ion increases solubility because it decreases activity coefficients. Differences in temperature result in significant differences in solubility.

TABLE 15.1
Major Mineral Groups in the Earth's Crust

Mineral Group	Examples	Formula
Silicates	Quartz	SiO_2
	Olivine	$(Mg,Fe)_2SiO_4$
	Potassium feldspar	$KAlSi_3O_8$
Oxides	Corundum	Al_2O_3
	Magnetite	Fe_3O_4
Carbonates	Calcite	$CaCO_3$
	Dolomite	$CaCO_3 \cdot MgCO_3$
Sulfides	Pyrite	FeS_2
	Galena	PbS
Sulfates	Gypsum	$CaSO_4 \cdot 2H_2O$
Halides	Halite	NaCl
	Fluorite	CaF_2
Native elements	Copper	Cu
	Sulfur	S

The nitrate deposits that occur in the hot and extraordinarily dry regions of northern Chile are chemically unique because of the stability of highly oxidized nitrate salts. The dominant salt, which has been mined for its nitrate content for use in explosives and fertilizers, is Chile saltpeter, $NaNO_3$. Traces of highly oxidized $CaCrO_4$ and $Ca(ClO_4)_2$ are also encountered in these deposits, and some regions contain enough $Ca(IO_3)_2$ to serve as a commercial source of iodine.

15.2.4 Volcanic Sublimates

A number of mineral substances are gaseous at the magmatic temperatures of volcanoes and are mobilized with volcanic gases. These kinds of substances condense near the mouths of volcanic fumaroles and are called *sublimates.* Elemental sulfur is a common sublimate. Some oxides, particularly of iron and silicon, are deposited as sublimates. Most other sublimates consist of chloride and sulfate salts. The cations most commonly involved are monovalent cations of ammonium ion, sodium, and potassium; magnesium; calcium; aluminum; and iron. Fluoride and chloride sublimates are sources of gaseous HF and HCl formed by their reactions at high temperatures with water, such as the following:

$$2H_2O + SiF_4 \rightarrow 4HF + SiO_2 \tag{15.1}$$

15.2.5 Igneous, Sedimentary, and Metamorphic Rock

At elevated temperatures deep beneath the Earth's surface, rocks and mineral matter melt to produce a molten substance called *magma.* Cooling and solidification of magma produces *igneous rock.* Common igneous rocks include granite, basalt, quartz (SiO_2), pyroxene ($(Mg,Fe)SiO_3$), feldspar ($(Ca,Na,K)AlSi_3O_8$), olivine ($(Mg,Fe)_2SiO_4$), and magnetite (Fe_3O_4). Igneous rocks are formed under water-deficient, chemically reducing conditions of high temperature and high pressure. Exposed igneous rocks are under wet, oxidizing, low-temperature, and low-pressure conditions. Since such conditions are the opposite of those conditions under which igneous rocks were formed, they are not in chemical equilibrium with their surroundings when they become exposed. As a result, such rocks disintegrate by a process called *weathering.* Weathering tends to be slow because igneous rocks are often hard, nonporous, and of low reactivity. Erosion from wind, water, or glaciers picks up materials from weathering rocks and deposits it as *sediments* or *soil.* A process called *lithification* describes the conversion of sediments to *sedimentary rocks.* In contrast to the parent igneous rocks, sediments and sedimentary rocks are porous, soft, and chemically reactive. Heat and pressure convert sedimentary rock into *metamorphic rock.*

Sedimentary rocks may be *detrital rocks* consisting of solid particles eroded from igneous rocks as a consequence of weathering; quartz is the most likely to survive weathering and transport from its original location chemically intact. A second kind of sedimentary rocks consists of *chemical sedimentary rocks* produced by the precipitation or coagulation of dissolved or colloidal weathering products. *Organic sedimentary rocks* contain residues of plant and animal remains. Carbonate minerals of calcium and magnesium—*limestone or dolomite*—are especially abundant in sedimentary rocks. Important examples of sedimentary rocks are the following:

- Sandstone produced from sand-sized particles of minerals such as quartz
- Conglomerates made up of relatively larger particles of variable size
- Shale formed from very fine particles of silt or clay
- Limestone, $CaCO_3$, produced by the chemical or biochemical precipitation of calcium carbonate:

$$Ca^{2+} + CO_3^{2-} \rightarrow CaCO_3(s)$$

$$Ca^{2+} + 2HCO_3 + h\nu \text{ (algal photosynthesis)} \rightarrow \{CH_2O\} \text{ (biomass)} + CaCO_3(s) + O_2(g)$$

- Chert consisting of microcrystalline SiO_2

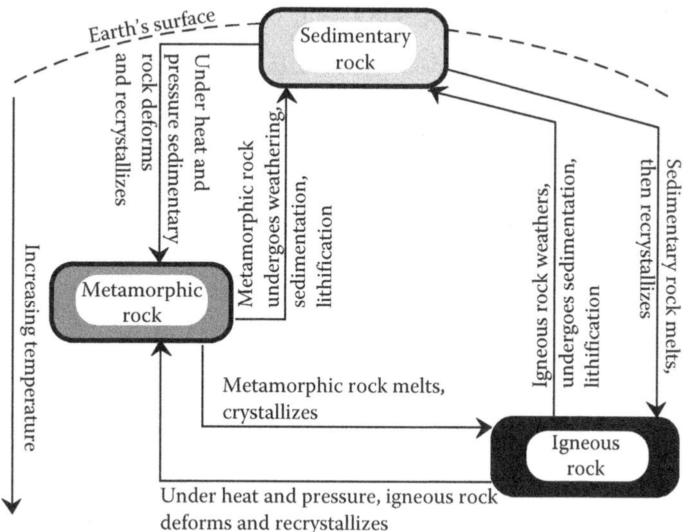

FIGURE 15.1 The rock cycle.

15.2.5.1 Rock Cycle

The interchanges and conversions among igneous, sedimentary, and metamorphic rocks, as well as the processes involved therein, are described by the *rock cycle*. A rock of any of these three types may be changed to a rock of any other type. Or a rock of any of these three kinds may be changed to a different rock of the same general type in the rock cycle. The rock cycle is illustrated in Figure 15.1.

15.2.5.2 Stages of Weathering

Weathering can be classified into *early*, *intermediate*, and *advanced stages*. The stage of weathering to which a mineral is exposed depends upon time; chemical conditions, including exposure to air, carbon dioxide, and water; and physical conditions such as temperature and mixing with water and air.

Reactive and soluble minerals such as carbonates, gypsum, olivine, feldspars, and iron(II)-rich substances can survive only early weathering. This stage is characterized by dry conditions, low leaching, absence of organic matter, reducing conditions, and limited time of exposure. Quartz, vermiculite, and smectites can survive the intermediate stage of weathering manifested by retention of silica, sodium, potassium, magnesium, calcium, and iron(II) not present in iron(II) oxides. These substances are mobilized in advanced-stage weathering, other characteristics of which are intense leaching by freshwater, low pH, oxidizing conditions [iron(II) $\rightarrow$ iron(III)], presence of hydroxy polymers of aluminum, and dispersion of silica.

15.3 PHYSICAL FORM OF THE GEOSPHERE

The most fundamental aspect of the physical form of the geosphere has to do with the Earth's shape and dimensions. Earth is shaped as a *geoid* defined by a surface corresponding to the average sea level of the oceans and continuing as hypothetical sea levels under the continents. This shape is not a perfect sphere because of variations in the attraction of gravity at various places on the Earth's surface. This slight irregularity in shape is important in surveying to precisely determine the locations of points on the Earth's surface according to longitude, latitude, and elevation above sea level. Of more direct concern to humans is the nature of landforms and the processes that occur on them. This area of study is classified as *geomorphology*.

15.3.1 Plate Tectonics and Continental Drift

The geosphere has a highly varied, constantly changing physical form. Most of the Earth's land mass is contained in several massive continents separated by vast oceans. Towering mountain ranges span across the continents, and in some places the ocean bottom is at extreme depths. Earthquakes, which often cause great destruction and loss of life, and volcanic eruptions, which sometimes throw enough material into the atmosphere to cause temporary changes in climate, serve as reminders that Earth is a dynamic, living body that continues to change. There is convincing evidence, such as the close fit between the western coast of Africa and the eastern coast of South America, that suggest that widely separated continents were once joined and have moved relative to each other. This ongoing phenomenon is known as *continental drift*. It is now believed that 200 million years ago much of the Earth's land mass was all part of a supercontinent, now called Gowandaland. This continent split apart to form the present-day continents of Antarctica, Australia, Africa, and South America, as well as Madagascar, the Seychelle Islands, and India.

The observations described above are explained by the theory of *plate tectonics*.[2] This theory views the Earth's solid surface as consisting of several rigid plates that move relative to each other. These plates drift at an average rate of several centimeters per year atop a relatively weak, partially molten layer that is part of the Earth's upper mantle called the *asthenosphere*. The science of plate tectonics explains the large-scale phenomena that affect the geosphere, including the creation and enlargement of oceans as the ocean floors open up and spread, the collision and breaking apart of continents, the formation of mountain chains, volcanic activities, the creation of islands of volcanic origin, and earthquakes.

The boundaries between these plates are where most geological activity such as earthquakes and volcanic activity occur. These boundaries are of the three following types:

- *Divergent boundaries*, where the plates are moving away from each other. Occurring on ocean floors, these are regions in which hot magma flows upward and cools to produce new solid lithosphere. This new solid material creates *ocean ridges*.
- *Convergent boundaries*, where plates move toward each other. One plate may be pushed beneath the other in a *subduction zone* in which matter is buried in the asthenosphere and eventually remelted to form new magma. When this does not occur, the lithosphere is pushed up to form mountain ranges along a collision boundary.
- *Transform fault boundaries* in which two plates slide past each other. These boundaries create faults that result in earthquakes.

The phenomena described above are parts of the *tectonic cycle*, a geological cycle that describes how tectonic plates move relative to each other, magma rises to form new solid rocks, and solid lithospheric rocks sink to become melted thus forming new magma. The tectonic cycle is illustrated in Figure 15.2.

15.3.2 Structural Geology

The Earth's surface is constantly being reshaped by geological processes. The movement of rock masses during processes such as the formation of mountains results in substantial deformation of rock. At the opposite extreme of the size scale are defects in crystals at a microscopic level. *Structural geology* addresses the geometric forms of geological structures over a wide range of sizes, the nature of structures formed by geological processes, and the formation of folds, faults, and other geological structures.

Primary structures are those that have resulted from the formation of a rock mass from its parent materials. Primary structures are modified and deformed to produce *secondary structures*. A basic premise of structural geology is that most layered rock formations were deposited in a horizontal configuration. Cracking of such a formation without displacement of the separate parts

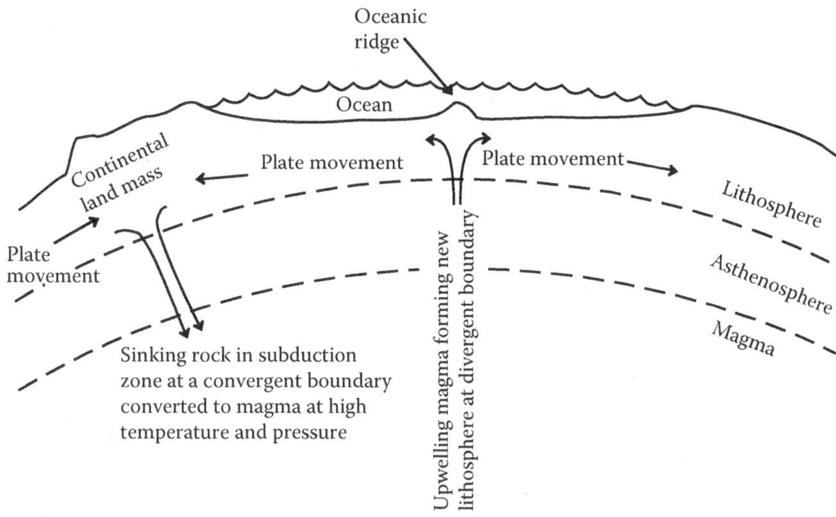

FIGURE 15.2 Illustration of the tectonic cycle in which upwelling magma along a boundary where two plates diverge creates new lithosphere on the ocean floor, and sinking rock in a subduction zone is melted to form magma.

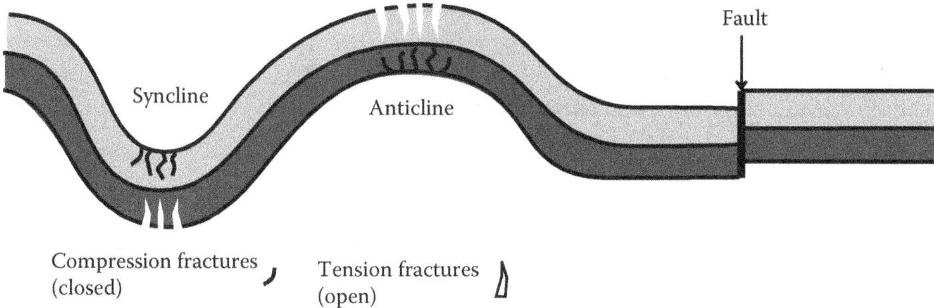

FIGURE 15.3 Folds (syncline and anticline) are formed by the bending of rock formations. Faults are produced by rock formations moving vertically or laterally with respect to each other.

of the formation relative to each other produces a *joint*, whereas displacement produces a *fault* (Figure 15.3).

A key relationship in structural geology is that between the force or *stress* placed upon a geological formation or object and the deformation resulting from it, called the *strain*. An important aspect of structural geology, therefore, is *rheology*, which deals with the deformation and flow of solids and semisolids. Whereas rocks tend to be strong, rigid, and brittle under the conditions at the Earth's surface, their rheology changes such that they may become weak and pliable under the extreme conditions of temperature and pressure at significant depths below the Earth's surface.

15.4 INTERNAL PROCESSES

The preceding section addressed the physical form of the geosphere. Related to the physical configuration of the geosphere are several major kinds of processes that occur that change this configuration and that have the potential to cause damage and even catastrophic effects.[3] These can be divided into the two main categories: *internal processes* that arise from phenomena located significantly below the Earth's surface and *surface processes* that occur on the surface. Internal processes are addressed in this section, and surface processes in Section 15.5.

15.4.1 Earthquakes

Earthquakes usually arise from plate tectonic processes and originate along plate boundaries occurring as motion of ground resulting from the release of energy that accompanies an abrupt slippage of rock formations subjected to stress along a fault. Basically, two huge masses of rock tend to move relative to each other, but are locked together along a fault line. This causes deformation of the rock formations, which increases with increasing stress. Eventually, the friction between the two moving bodies is insufficient to keep them locked in place, and movement occurs along an existing fault, or a new fault is formed. Freed from constraints on their movement, the rocks undergo elastic rebound, causing Earth to shake. The serious earthquake damage that may ensue is discussed further in Section 15.10.

In addition to shaking of ground, which can be quite violent, earthquakes can cause the ground to rupture, subside, or rise. *Liquefaction* is an important phenomenon that occurs during earthquakes with ground that is poorly consolidated and in which the water table may be high. Liquefaction results from separation of soil particles accompanied by water infiltration. When this occurs, the ground behaves like a fluid.

The location of the initial movement along a fault that causes an earthquake to occur is called the *focus* of the earthquake. The surface location directly above the focus is the *epicenter.* Energy is transmitted from the focus by *seismic waves.* Seismic waves that travel through the interior of Earth are called *body waves* and those that traverse the surface are *surface waves.* Body waves are further categorized into *P-waves*, compressional vibrations that result from the alternate compression and expansion of geospheric material, and *S-waves*, consisting of shear waves manifested by sideways oscillations of material. The motions of these waves are detected by a *seismograph*, often at great distances from the epicenter. The two types of waves move at different rates, with P-waves moving faster. From the arrival times of the two kinds of waves at different seismographic locations, it is possible to locate the epicenter of an earthquake.

15.4.2 Volcanoes

In addition to earthquakes, the other major subsurface process that has the potential to massively affect the environment consists of emissions of molten rock (lava), gases, steam, ash, and particles due to the presence of magma near the Earth's surface. This phenomenon is called a *volcano* (Figure 15.4). Volcanoes can be very destructive and damaging to the environment. Aspects of the potential harm from volcanoes are discussed in Section 15.11.

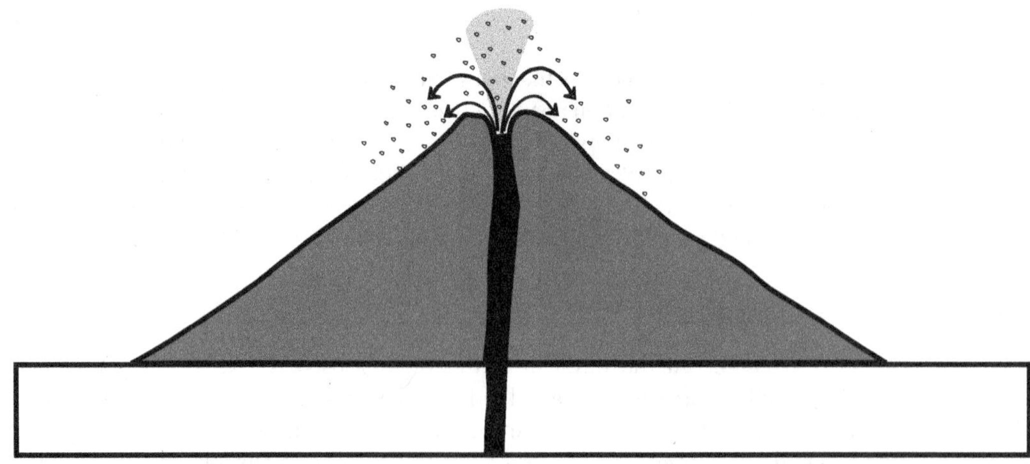

FIGURE 15.4 Volcanoes come in many shapes and forms. A classically shaped volcano may be a cinder cone formed by ejection of rock and lava, called pyroclastics, from the volcano to produce a relatively uniform cone.

Volcanoes take on a variety of forms, which are beyond the scope of this chapter to cover in detail. Basically, they are formed when magma rises to the surface. This frequently occurs in subduction zones created where one plate is pushed beneath another (see Figure 15.2). The downward movement of solid lithospheric material subjects it to high temperatures and pressures that cause the rock in it to melt and raise to the surface as magma. Molten magma issuing from a volcano at temperatures usually in excess of 500°C and often as high as 1400°C, is called *lava*, and is one of the most common manifestations of volcanic activity.

15.4.3 Surface Processes

Surface geological features are formed by upward movement of materials from the Earth's crust. With exposure to water, oxygen, freeze–thaw cycles, organisms, and other influences on the surface, surface features are subject to two processes that largely determine the landscape: weathering and erosion. As noted earlier in this chapter, weathering consists of the physical and chemical breakdown of rock and erosion is the removal and movement of weathered products by the action of wind, liquid water, and ice. Weathering and erosion work together in that one augments the other in breaking down rock and moving the products. Weathered products removed by erosion are eventually deposited as sediments and may undergo diagenesis and lithification to form sedimentary rocks.

One of the most common surface processes that can adversely affect humans consists of *landslides* that occur when soil or other unconsolidated materials slide down a slope. Related phenomena include rockfalls, mudflows, and snow avalanches. Destructive surface processes are discussed further in Section 15.12.

15.5 SEDIMENTS

Vast areas of land, as well as lake and stream sediments, are formed from sedimentary rocks. The properties of these masses of material depend strongly upon their origins and transport. Water is the main vehicle of sediment transport, although wind can also be significant. Hundreds of millions of tons of sediment are carried by major rivers each year.

The action of flowing water in streams cuts away stream banks and carries sedimentary materials for great distances. Sedimentary materials such as the following may be carried by flowing water in streams:

- *Dissolved load* from sediment-forming minerals in solution
- *Suspended load* from solid sedimentary materials carried along in suspension
- *Bed load* dragged along the bottom of the stream channel

The transport of calcium carbonate as dissolved calcium bicarbonate provides a straightforward example of dissolved load and is the most prevalent type of such a load. Water with a high dissolved carbon dioxide content (usually present as the result of bacterial action) in contact with calcium carbonate formations contains Ca^{2+} and HCO_3^- ions. Flowing water containing calcium as the HCO_3^- salt has *temporary hardness* but may become more basic by loss of CO_2 to the atmosphere, consumption of CO_2 by algal growth, or contact with dissolved base, resulting in the deposition of insoluble $CaCO_3$:

$$Ca^{2+} + 2HCO_3^- \rightarrow CaCO_3(s) + CO_2(g) + H_2O \tag{15.2}$$

Most flowing water that contains dissolved load originates underground, where the water has had the opportunity to dissolve minerals from the rock strata that it has passed through.

Most sediments are transported by streams as suspended load, obvious in the observation of "mud" in the flowing water of rivers draining agricultural areas or finely divided rock in Alpine streams fed by melting glaciers. Under normal conditions, finely divided silt, clay, or sand makes up most of the suspended load, although larger particles are transported in rapidly flowing water. The degree and rate of movement of suspended sedimentary material in streams are functions of the velocity of water flow and the settling velocity of the particles in suspension.

Bed load is moved along the bottom of a stream by the action of water "pushing" particles along. Particles carried as bed load do not move continuously. The grinding action of such particles is an important factor in stream erosion.

Typically, about two-third of the sediment carried by a stream is transported in suspension, about one-fourth in solution, and the remaining relatively small fraction as bed load. The ability of a stream to carry sediment increases with both the overall rate of flow of the water (mass per unit time) and the velocity of the water. Both of these are higher under flood conditions, so floods are particularly important in the transport of sediments.

Streams mobilize sedimentary materials through *erosion*, *transport* the materials along with stream flow, and release them in a solid form during *deposition*. Deposits of stream-borne sediments are called *alluvium*. As conditions such as lowered stream velocity begin to favor deposition, larger, more settleable particles are released first. This results in *sorting* such that particles of a similar size and type tend to occur together in alluvial deposits. Much sediment is deposited in flood plains where streams overflow their banks.

15.6 CLAYS

Clays are extremely common and important in mineralogy. Furthermore, in general, clays predominate in the inorganic components of most soils (see Chapter 16) and are very important in holding water and in plant nutrient cation exchange. All clays contain silicate and most contain aluminum and water. Physically, clays consist of very fine grains having sheet-like structures. For the purposes of the discussion here, *clay* is defined as a group of microcrystalline secondary minerals consisting of hydrous aluminum silicates that have sheet-like structures. Clay minerals are distinguished from each other by general chemical formula, structure, and chemical and physical properties. The three major groups of clay minerals are the following:

- *Montmorillonite*, $Al_2(OH)_2Si_4O_{10}$
- *Illite*, $K_{0-2}Al_4(Si_{8-6}Al_{0-2})O_{20}(OH)_4$
- *Kaolinite*, $Al_2Si_2O_5(OH)_4$

Many clays contain large amounts of sodium, potassium, magnesium, calcium, and iron, as well as trace quantities of other metals. Clays bind cations such as Ca^{2+}, Mg^{2+}, K^+, Na^+, and NH_4^+, which protects the cations from leaching by water but keeps them available in soil as plant nutrients. Since many clays are readily suspended in water as colloidal particles, they may be leached from soil or carried to lower soil layers.

Olivine, augite, hornblende, and feldspars are all parent minerals that form clays. An example is the formation of kaolinite ($Al_2Si_2O_5(OH)_4$) from potassium feldspar rock ($KAlSi_3O_8$):

$$2KAlSi_3O_8(s) + 2H^+ + 9H_2O \rightarrow Al_2Si_2O_5(OH)_4(s) + 2K^+(aq) + 4H_4SiO_4(aq) \quad (15.3)$$

The layered structures of clays consist of sheets of silicon oxide alternating with sheets of aluminum oxide. The silicon oxide sheets are made up of tetrahedra in which each silicon atom is surrounded by four oxygen atoms. Of the four oxygen atoms in each tetrahedron, three are shared with other silicon atoms that are components of other tetrahedra. This sheet is called the *tetrahedral sheet*. The aluminum oxide is contained in an *octahedral sheet*, so named because each aluminum atom is surrounded by six oxygen atoms in an octahedral configuration. The structure is such that some of the oxygen atoms are shared between aluminum atoms and some are shared with the tetrahedral sheet.

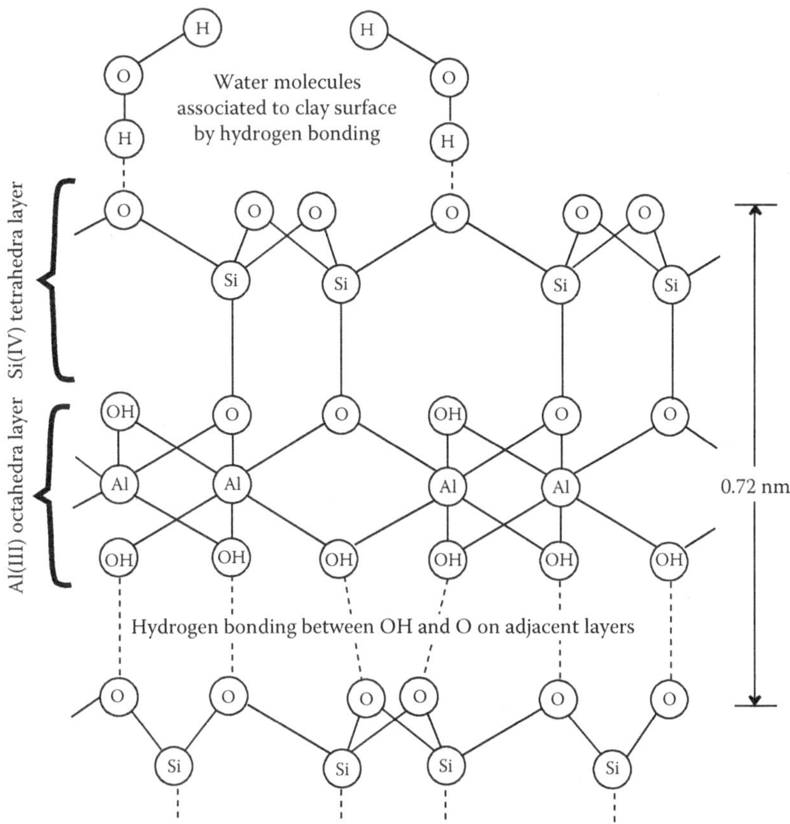

FIGURE 15.5 Representation of the structure of kaolinite, a two-layer clay.

Structurally, clays may be classified as either *two-layer clays* in which oxygen atoms are shared between a tetrahedral sheet and an adjacent octahedral sheet, or *three-layer clays* in which an octahedral sheet shares oxygen atoms with tetrahedral sheets on either side. These layers composed of either two or three sheets are called *unit layers*. A unit layer of a two-layer clay typically is around 0.7 nm thick, whereas that of a three-layer clay exceeds 0.9 nm in thickness. The structure of the two-layer clay kaolinite is represented in Figure 15.5. Some clays, particularly the montmorillonites, may absorb large quantities of water between unit layers, a process accompanied by swelling of the clay.

As described in Section 5.4, clay minerals may attain a net negative charge by *ion replacement*, in which Si(IV) and Al(III) ions are replaced by metal ions of similar size but lesser charge. Compensation must be made for this negative charge by association of cations with the clay layer surfaces. Since these cations need not fit specific sites in the crystalline lattice of the clay, they may be relatively large ions, such as K^+, Na^+, or NH_4^+. These cations are called *exchangeable cations* and are exchangeable for other cations in water. The amount of exchangeable cations, expressed as milliequivalents (of monovalent cations) per 100 g of dry clay, is called the CEC of the clay and is a very important characteristic of colloids and sediments that have cation-exchange capabilities.

15.7 GEOCHEMISTRY

Geochemistry deals with chemical species, reactions, and processes in the lithosphere and their interactions with the atmosphere and hydrosphere. The branch of geochemistry that explores the complex interactions among the rock/water/air/life systems that determine the chemical characteristics of the surface environment is *environmental geochemistry*. Obviously, geochemistry and its environmental subdiscipline are very important areas of environmental chemistry with many applications related to the environment.

15.7.1 Physical Aspects of Weathering

Defined in Section 15.2, *weathering* is discussed here as a geochemical phenomenon. Rocks tend to weather more rapidly when there are pronounced differences in physical conditions—alternate freezing and thawing and wet periods alternating with severe drying. Other mechanical aspects are swelling and shrinking of minerals with hydration and dehydration, as well as growth of roots through cracks in rocks. Temperature is involved in that the rates of chemical weathering (below) increase with increasing temperature.

15.7.2 Chemical Weathering

As a chemical phenomenon, weathering can be viewed as the result of the tendency of the rock/water/mineral system to attain equilibrium. This occurs through the usual chemical mechanisms of dissolution/precipitation, acid–base reactions, complexation, hydrolysis, and oxidation–reduction.

Weathering occurs extremely slowly in dry air but is many orders of magnitude faster in the presence of water. Water itself is a chemically active weathering substance and it holds weathering agents in solution such that they are transported to chemically active sites on rock minerals and contact the mineral surfaces at the molecular and ionic levels. Prominent among such weathering agents are CO_2, O_2, organic acids (including humic and fulvic acids, see Section 3.17), sulfur acids ($SO_2(aq)$ and H_2SO_4), and nitrogen acids (HNO_3 and HNO_2). Water provides the source of H^+ ion needed for acid-forming gases to act as acids as shown by the following:

$$CO_2 + H_2O \rightarrow H^+ + HCO_3^- \tag{15.4}$$

$$SO_2 + H_2O \rightarrow H^+ + HSO_3^- \tag{15.5}$$

Rainwater is essentially free of mineral solutes. It is usually slightly acidic due to the presence of dissolved carbon dioxide or more highly acidic because of acid-rain-forming constituents. As a result of its slight acidity and lack of alkalinity and dissolved calcium salts, rainwater is *chemically aggressive* (see Section 8.7) toward some kinds of mineral matter, which it breaks down by chemical weathering processes. Because of this process, river water has a higher concentration of dissolved inorganic solids than does rainwater.

The processes involved in chemical weathering may be divided into the following major categories:

- *Hydration/dehydration*, for example:

$$CaSO_4(s) + 2H_2O \rightarrow CaSO_4 \cdot 2H_2O(s)$$

$$2Fe(OH)_3 \cdot xH_2O(s) \rightarrow Fe_2O_3(s) + (3 + 2x)H_2O$$

- *Dissolution*, for example:

$$CaSO_4 \cdot 2H_2O(s)\ (\text{water}) \rightarrow Ca^{2+}(aq) + SO_4^{2-}(aq) + 2H_2O$$

- *Oxidation*, such as occurs in the dissolution of pyrite:

$$4FeS_2(s) + 15O_2(g) + (8 + 2x)H_2O \rightarrow 2Fe_2O_3 \cdot xH_2O + 8SO_4^{2-}(aq) + 16H^+(aq)$$

or in the following example in which dissolution of an iron(II) mineral is followed by oxidation of iron(II) to iron(III):

$$Fe_2SiO_4(s) + 4CO_2(aq) + 4H_2O \rightarrow 2Fe^{2+} + 4HCO_3^- + H_4SiO_4$$

$$4Fe^{2+} + 8HCO_3^- + O_2(g) \rightarrow 2Fe_2O_3(s) + 8CO_2 + 4H_2O$$

The second of these two reactions may occur at some distance from the first, resulting in net transport of iron from its original location. Iron, manganese, and sulfur are the major elements that undergo oxidation as part of the weathering process.

- *Dissolution with hydrolysis* as occurs with the hydrolysis of carbonate ion when mineral carbonates dissolve:

$$CaCO_3(s) + H_2O \rightarrow Ca^{2+}(aq) + HCO_3^-(aq) + OH^-(aq)$$

Hydrolysis is the major means by which silicates undergo weathering as shown by the following reaction of forsterite:

$$Mg_2SiO_4(s) + 4CO_2 + 4H_2O \rightarrow 2Mg^{2+} + 4HCO_3^- + H_4SiO_4$$

The weathering of silicates yields soluble silicon as species such as H_4SiO_4, and residual silicon-containing minerals (clay minerals).

- *Acid hydrolysis*, which accounts for the dissolution of significant amounts of $CaCO_3$ and $CaCO_3 \cdot MgCO_3$ in the presence CO_2-rich water:

$$CaCO_3(s) + H_2O + CO_2(aq) \rightarrow Ca^{2+}(aq) + 2HCO_3^-(aq)$$

- *Complexation*, as exemplified by the reaction of oxalate ion, $C_2O_4^{2-}$, with aluminum in muscovite, $K_2(Si_6Al_2)Al_4O_{20}(OH)_4$:

$$K_2(Si_6Al_2)Al_4O_{20}(OH)_4(s) + 6C_2O_4^{2-}(aq) + 20H^+ \rightarrow 6AlC_2O_4^+(aq) + 6Si(OH)_4 + 2K^+$$

Reactions such as these largely determine the kinds and concentrations of solutes in surface water and groundwater. Acid hydrolysis, especially, is the predominant process that releases elements such as Na^+, K^+, and Ca^{2+} from silicate minerals.

15.7.3 Biological Aspects of Weathering

Organisms may play a strong role in weathering processes and formation of soil. Cavities on top of rock formations and boulders in some areas accumulate water, mineral debris, and organic debris. These cavities may serve as sites of miniature ecosystems, supporting initially cyanobacteria, green algae, fungi, bacteria, and insects. Organic acids released by the organisms and humic matter produced by degradation of vegetable matter in the cavities tend to dissolve the rock and enlarge the cavities. Small rock crystals are released that eventually weather to produce secondary minerals such as clays. Eventually, vascular plants begin to grow in the miniature ecosystems and embryonic soils develop.

15.8 GROUNDWATER IN THE GEOSPHERE

Groundwater (Figure 15.6) is a vital resource in its own right that plays a crucial role in geochemical processes, such as the formation of secondary minerals. The nature, quality, and mobility of groundwater are all strongly dependent upon the rock formations in which the water is held.

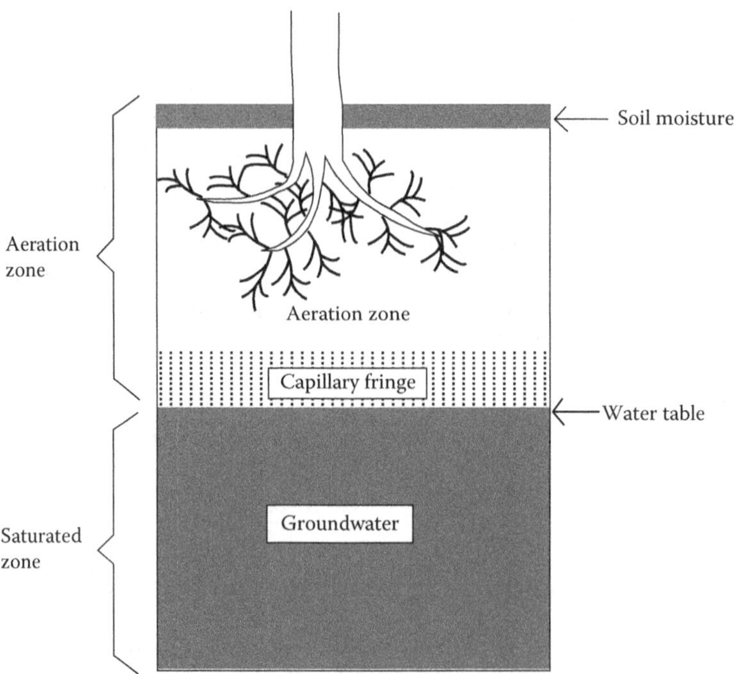

FIGURE 15.6 Some major features of the distribution of water underground.

Physically, an important characteristic of such formations is their *porosity*, which determines the percentage of rock volume available to contain water. Another important physical characteristic is *permeability*, which describes the ease of flow of the water through the rock. High permeability is usually associated with high porosity. However, clays tend to have low permeability even when a large percentage of the volume is filled with water.

Most groundwater originates as *meteoric* water from precipitation in the form of rain or snow. If water from this source is not lost by evaporation, transpiration, or to stream runoff, it may infiltrate into the ground. Initial amounts of water from precipitation onto dry soil are held very tightly as a film on the surfaces and in the micropores of soil particles in a *belt of soil moisture*. At intermediate levels, the soil particles are covered with films of water, but air is still present in larger voids in the soil. The region in which such water is held is called the *unsaturated zone* or *zone of aeration* and the water present in it is *vadose water*. At lower depths in the presence of adequate amounts of water, all voids are filled to produce a *zone of saturation*, the upper level of which is the *water table*. Water present in a zone of saturation is called *groundwater*. Because of its surface tension, water is drawn somewhat above the water table by capillary-sized passages in soil in a region called the capillary fringe.

The water table (Figure 15.7) is crucial for explaining and predicting the flow of wells and springs and the levels of streams and lakes. It is also an important factor in determining the extent to which pollutant and hazardous chemicals underground are likely to be transported by water. The water table can be mapped by observing the equilibrium level of water in wells, which is essentially the same as the top of the saturated zone. The water table is usually not level, but tends to follow the general contours of the surface topography. It also varies with differences in permeability and water infiltration. The water table is at surface level in the vicinity of swamps and frequently above the surface where lakes and streams are encountered. The water level in such bodies may be maintained by the water table. *Influent* streams or reservoirs are located above the water table; they lose water to the underlying aquifer and cause an upward bulge in the water table beneath the surface water.

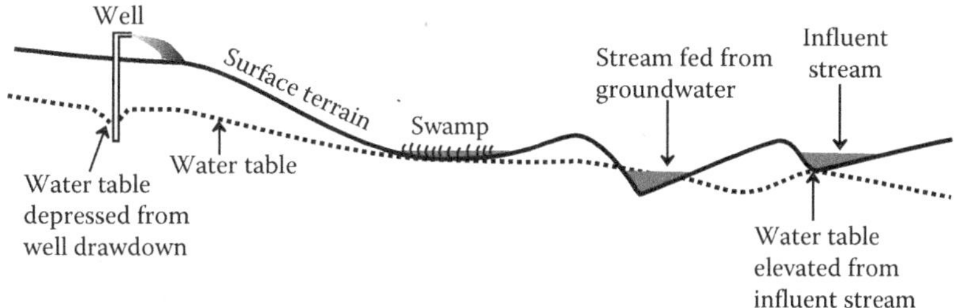

FIGURE 15.7 The water table and influences of surface features on it.

Groundwater *flow* is an important consideration in determining the accessibility of the water for use and transport of pollutants from underground waste sites. Various parts of a body of groundwater are in hydraulic contact so that a change in pressure at one point will tend to affect the pressure and level at another point. For example, infiltration from a heavy, localized rainfall may affect the water table at a point remote from the infiltration. Groundwater flow occurs as the result of the natural tendency of the water table to assume even levels by the action of gravity.

Groundwater flow is strongly influenced by rock permeability. Porous or extensively fractured rock is relatively highly *pervious*, meaning that water can migrate through the holes, fissures, and pores in such rock. Because water can be extracted from it, such a formation is called an *aquifer*. By contrast, an *aquiclude* is a rock formation that is too impermeable or unfractured to yield groundwater. Impervious rock in the unsaturated zone may retain water infiltrating from the surface to produce a *perched water table* that is above the main water table and from which water may be extracted. However, the amounts of water that can be extracted from such a formation are limited and the water is vulnerable to contamination.

15.8.1 Water Wells

Most groundwater is tapped for use by water wells drilled into the saturated zone. The use and misuse of water from this source have a number of environmental implications. In the United States, about two-third of the groundwater pumped is consumed for irrigation; lesser amounts of groundwater are used for industrial and municipal applications.

As water is withdrawn, the water table in the vicinity of the well is lowered. This *drawdown* of water creates a *zone of depression*. In extreme cases the groundwater is severely depleted and surface land levels can even subside (which is one reason why Venice, Italy is now very vulnerable to flooding). Heavy drawdown can result in infiltration of pollutants from sources such as septic tanks, municipal refuse sites, and hazardous waste dumps. When soluble iron(II) or manganese(II) is present in groundwater, exposure to air at the well wall can result in the formation of deposits of insoluble Fe(III) and Mn(IV) oxides produced by bacterially catalyzed processes:

$$4Fe^{2+}(aq) + O_2(aq) + 10H_2O \rightarrow 4Fe(OH)_3(s) + 8H^+ \quad (15.6)$$

$$2Mn^{2+}(aq) + O_2(aq) + (2x + 2)H_2O \rightarrow 2MnO_2{\cdot}xH_2O(s) + 4H^+ \quad (15.7)$$

Deposits of iron(III) and manganese(IV) that result from the processes outlined above line the surfaces from which water flows into the well with a coating that is relatively impermeable to water. The deposits fill the spaces that water must traverse to enter the well. As a result, they can seriously impede the flow of water into the well from the water-bearing aquifer. This creates major water source problems for municipalities using groundwater for water supply. As a result of this problem, chemical or mechanical cleaning, drilling of new wells, or even acquisition of new water sources may be required.

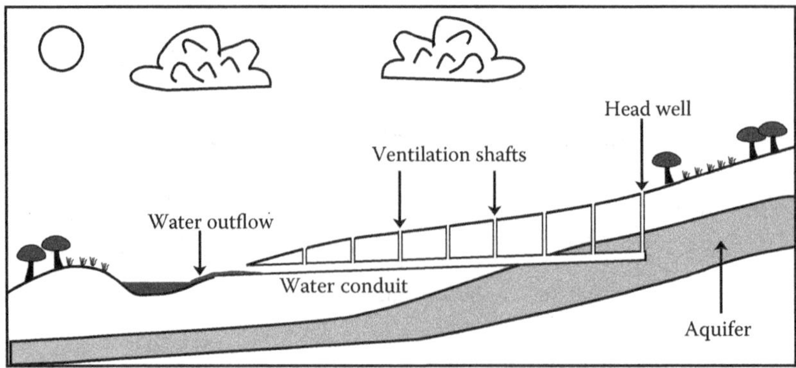

FIGURE 15.8 Qanats dating back 3000 years are used to transport water from aquifers at relatively high elevations to lower elevations where the water is used.

15.8.2 Qanats

An interesting ancient technology for utilizing groundwater is the *qanat* (Figure 15.8). Basically the qanat consists of a tunnel drilled into an aquifer and leading to an outflow located below the elevation of the aquifer. Typical lengths of the conduits in qanats are 10–16 km, although they have reached almost 30 km in length. Iran is the world's center of qanat technology dating back to ancient Persia some 3000 years ago. There are about 22,000 qanat units containing over 270,000 km of conduits in Iran that supply about 75% of the nation's water for irrigation and domestic use. Now the system in Iran is threatened by tube wells drilled into the aquifers that are lowering water tables and draining water from the qanats.

15.9 ENVIRONMENTAL ASPECTS OF THE GEOSPHERE

Most of the remainder of this chapter deals specifically with the environmental aspects of geology and human interactions with the geosphere. It discusses how natural geological phenomena affect the environment through occurrences such as volcanic eruptions that may blast so much particulate matter and acid gas into the atmosphere that it may have a temporary effect on global climate, or massive earthquakes that disrupt surface topography and disturb the flow and distribution of groundwater and surface water. Also discussed are human influences on the geosphere and the strong connection between the geosphere and the anthrosphere.[4]

Going back several billion years to its formation as a ball of dust particles collected from the universe and held together by gravitational forces, Earth has witnessed constant environmental change and disruption. During its earlier eons, Earth was a most inhospitable place for humans and, indeed, for any form of life. Heat generated by gravitational compression of primitive earth and by radioactive elements in its interior caused much of the mass of the planet to liquefy. Relatively high density iron sank into the core, and lighter minerals, primarily silicates, solidified and floated to the surface.

Although in the scale of a human lifetime, Earth changes almost imperceptibly; the planet is in fact in a state of constant change and turmoil. It is known that continents have formed, broken apart, and moved around. Rock formations produced in ancient oceans have been thrust up onto continental land and huge masses of volcanic rock exist where volcanic activity is now unknown. Today the angry bowels of Earth unleash enormous forces that push molten rock to the surface and move continents continuously as evidenced from volcanic activity, and from earthquakes resulting from the movement of great land masses relative to each other. The Earth's surface is constantly changing as new mountain ranges are heaved up and old ones are worn down.

Humans have learned to work with, against, and around natural earth processes and phenomena to exploit the Earth's resources and to make these processes and phenomena work for the benefit of humankind. Human efforts have been moderately successful in mitigating some of the major hazards posed by natural geospheric phenomena, although such endeavors often have had unforeseen detrimental consequences, sometimes many years after they were first applied. The survival of modern civilization and, indeed, of humankind will depend upon how intelligently humans work with Earth. That is why it is so important for humans to have a fundamental understanding of the geospheric environment.

An important consideration in human interaction with the geosphere is the application of engineering to geology. Engineering geology takes account of the geological characteristics of soil and rock in designing buildings, dams, highways, and other structures in a manner compatible with the geological strata on which they rest. Engineering geology must consider a large number of geological factors including type, strength, and fracture characteristics of rock, tendency for landslides to occur, susceptibility to settling, and likelihood of erosion. Engineering geology is an important consideration in land-use planning.

15.9.1 Natural Hazards

Earth presents a variety of natural hazards to the creatures that dwell on it. Some of these are the result of internal processes that arise from the movement of land masses relative to each other and from heat and intrusions of molten rock from below the surface. The most common such hazards are earthquakes and volcanoes. Whereas internal processes tend to force matter upward, often with detrimental effects, surface processes are those that generally result from the tendency of matter to seek lower levels. Such processes include erosion, landslides, avalanches, mudflows, and subsidence.

A number of natural hazards result from the interaction and conflict between solid earth and liquid and solid water. Perhaps the most obvious such hazard consists of floods when too much water falls as precipitation and seeks lower levels through stream flow. Wind can team with water to increase destructive effects, such as beach erosion and destruction of beachfront property resulting from wind-driven seawater. Ice, too, can have some major effects on solid earth. Evidence of such effects of the Ice Age times include massive glacial moraines left over from deposition of till from melting glaciers, and landscape features carved by advancing ice sheets.

15.9.2 Anthropogenic Hazards

All too often, attempts to control and reshape the geosphere to human demands have been detrimental to the geosphere and dangerous to human life and well-being. Such attempts may exacerbate damaging natural phenomena. A prime example of this interaction occurs when efforts are made to control the flow of rivers by straightening them and building levees. The initial results can be deceptively favorable in that a modified stream may exist for decades, flowing smoothly and staying within the confines imposed by humans. But eventually, the forces of nature are likely to overwhelm the efforts of humans to control them, such as when a record flood breaks levees and destroys structures constructed in flood-prone areas. Landslides of mounds of earthen material piled up from mining can be very destructive. Destruction of wetlands in an effort to provide additional farmland can have some detrimental effects on wildlife and on the overall health of ecosystems.

15.10 EARTHQUAKES

The loss of life and destruction of property by earthquakes makes them some of nature's more damaging natural phenomena.[5] The destructive effects of an earthquake are due to the release of energy.

The released energy moves from the quake's focus as seismic waves, discussed in Section 15.4. Literally millions of lives have been lost in past earthquakes, and damage from an earthquake in a developed urban area can easily run into billions of dollars. Earthquakes may cause catastrophic secondary effects, especially large, destructive ocean waves called tsunamis (discussed below).

Adding to the terror of earthquakes is their lack of predictability. An earthquake can strike at any time—during the calm of late night hours or in the middle of busy rush hour traffic. Although the exact prediction of earthquakes has so far eluded investigators, the locations where earthquakes are most likely to occur are much more well known. These are located in lines corresponding to boundaries along which tectonic plates collide and move relative to each other, building up stresses that are suddenly released when earthquakes occur. Such interplate boundaries are locations of preexisting faults and breaks. Occasionally, however, an earthquake will occur within a plate, made more massive and destructive because for it to occur the thick lithosphere has to be ruptured.

The scale of earthquakes can be estimated by the degree of motion that they cause and by their destructiveness. The former is termed the *magnitude* of an earthquake and is commonly expressed by the *Richter scale.* The Richter scale is open-ended, and each unit increase in the scale reflects a 10-fold increase in magnitude. Several hundred thousand earthquakes with magnitudes from two to three occur each year; they are detected by seismographs, but are not felt by humans. Minor earthquakes range from four to five on the Richter scale, and earthquakes cause damage at a magnitude greater than about five. Great earthquakes, which occur about once or twice a year, register over eight on the Richter scale.

The *intensity* of an earthquake is a subjective estimate of its potential destructive effect. On the Mercalli intensity scale, an intensity III earthquake feels like the passage of heavy vehicles; one with an intensity of VII causes difficulty in standing, damage to plaster, and dislodging of loose brick, whereas a quake with an intensity of XII causes virtually total destruction, throws objects upward, and shifts huge masses of earthen material. Intensity does not correlate exactly with magnitude.

Distance from the epicenter, the nature of underlying strata, and the types of structures affected may all result in variations in intensity from the same earthquake. In general, structures built on bedrock will survive with much less damage than those constructed on poorly consolidated material. Displacement of ground along a fault can be substantial, for example, up to 6 m along the San Andreas fault during the 1906 San Francisco earthquake. Such shifts can break pipelines and destroy roadways. Highly destructive surface waves can shake vulnerable structures apart.

The shaking and movement of ground are the most obvious means by which earthquakes cause damage. In addition to shaking it, earthquakes can cause the ground to rupture, subside, or rise. *Liquefaction* is an important phenomenon that occurs during earthquakes with ground that is poorly consolidated and in which the water table may be high. It results from separation of soil particles accompanied by water infiltration such that the ground behaves like a fluid.

Another devastating phenomenon consists of *tsunamis*, large ocean waves resulting from earthquake-induced movement of ocean floor. Tsunamis sweeping onshore at speeds up to 1000 km/h have destroyed many homes and taken many lives, often large distances from the epicenter of the earthquake itself. This effect occurs when a tsunami approaches land and forms huge breakers, some as high as 10–15 m, or even higher. On April 1, 1946, an earthquake off the coast of Alaska generated a Tsunami estimated to be more than 30 m high that killed five people on a nearby lighthouse. About 5 h later, a Tsunami generated by the same earthquake reached Hilo, Hawaii, and killed 159 people with a wave exceeding 15 m high. The March 27, 1964, Alaska earthquake generated a tsunami over 10 m high that hit a freighter docked at Valdez, tossing it around like matchwood. Miraculously, nobody on the freighter was killed, but 28 people on the dock died.

Literally millions of lives have been lost in past earthquakes, and damage from an earthquake in a developed urban area can easily run into billions of dollars. As examples, a massive earthquake in Egypt and Syria in 1201 AD took over 1 million lives, one in Tangshan, China, in 1976 killed about 650,000, and the 1989 Loma Prieta earthquake in California cost about $7 billion.

Significant progress has been made in designing structures that are earthquake resistant. As evidence of that, during a 1964 earthquake in Niigata, Japan, some buildings tipped over on their sides due to liquefaction of the underlying soil, but remained structurally intact! Other areas of endeavor that can lessen the impact of earthquakes are the identification of areas susceptible to earthquakes, discouraging development in such areas, and educating the public about earthquake hazards. Accurate prediction would be a tremendous help in lessening the effects of earthquakes, but so far has been generally unsuccessful. Most challenging of all is the possibility of preventing major earthquakes. One unlikely possibility would be to detonate nuclear explosives deep underground along a fault line to release stress before it builds up to an excessive level. Fluid injection to facilitate slippage along a fault has also been considered.

There is some evidence to suggest that human activities can contribute to the occurrence of earthquakes in rare cases. On May 12, 2008, the 7.9 magnitude Wenchuan earthquake in Sichuan Province, China, that left 80,000 people dead or missing may have been triggered by the mass of water contained in the Zipingpu reservoir completed in 2004. Containing about 300 million metric tons of water, this reservoir, which was built <2 km from a major fault line, has been estimated to have increased pressure on underlying strata as much as 25-fold. The Wenchuan earthquake originated only about 5 km from the reservoir and spread in a direction that seismologists believe is consistent with a quake triggered by the reservoir. As the reservoir was filling, from 2004 to 2005 the area registered about 730 minor earthquakes of a magnitude of three or less, likely caused in part by increasing pressure from the impounded water. In addition to the pressure from the mass of the water, water infiltration from the reservoir may have lubricated the underlying strata making them more prone to movement. Seismologists emphasize that the Wenchuan earthquake would have taken place eventually without the added pressure of the impounded water, but the construction may have hastened the quake by several hundred years.

15.11 VOLCANOES

Volcanic eruptions can be terrifying and extremely destructive events.[6] On May 18, 1980, Mount St. Helens, a volcano in Washington State, erupted, blowing out about 1 km^3 of material. This massive blast spread ash over half of the United States, causing about $1 billion in damages and killing an estimated 62 people, many of whom were never found. Many volcanic disasters have been recorded throughout history. Perhaps the best known of these is the 79 AD eruption of Mount Vesuvius, which buried the Roman city of Pompei with volcanic ash.

Temperatures of *lava*, molten rock flowing from a volcano, typically exceed 500°C and may get as high as 1400°C or more. Lava flows destroy everything in their paths, causing buildings and forests to burn and burying them under rock that cools and becomes solid. Often more dangerous than a lava flow are the *pyroclastics* produced by volcanoes and consisting of fragments of rock and lava. Some of these particles are large and potentially very damaging, but they tend to fall quite close to the vent. Ash and dust may be carried for large distances and, in extreme cases, as was the case in ancient Pompei, may bury large areas to some depth with devastating effects. The explosion of Tambora volcano in Indonesia in 1815 blew out about 30 km^3 of solid material. The ejection of so much solid into the atmosphere had such a devastating effect on global climate that the following year was known as "the year without a summer," causing widespread hardship and hunger because of global crop failures.

A special kind of particularly dangerous pyroclastic consists of *nuéeardente*. This term, French for "glowing cloud," refers to a dense mixture of hot toxic gases and fine ash particles reaching temperatures of 1000°C that can flow down the slopes of a volcano at speeds of up to 100 km/h. In 1902, a nuéeardente was produced by the eruption of Mont Pelée on Martinique in the Caribbean. Of as many as 40,000 people in the town of St. Pierre, the only survivor was a terrified prisoner shielded from the intense heat by the dungeon in which he was imprisoned.

One of the more spectacular and potentially damaging volcanic phenomena is a *phreatic eruption* that occurs when infiltrating water is superheated by hot magma and causes a volcano to literally explode. This happened in 1883 when uninhabited Krakatoa in Indonesia blew up with an energy release of the order of 100 megatons of TNT. Dust was blown 80 km into the stratosphere, and a perceptible climatic cooling was noted for the next 10 years. As is the case with earthquakes, volcanic eruptions may cause the devastating tsunamis. Krakatoa produced a tsunami 40 m high that killed 30,000–40,000 people on surrounding islands.

Some of the most damaging health and environmental effects of volcanic eruptions are caused by gases and particulate matter. Possible effects are discussed in Sections 14.8 and 15.18.

15.11.1 Mud Volcanoes

Mud volcanoes consist of eruptions of mud, water, and gas from underground reservoirs of mud pressurized by overlying rock formations. The largest mud volcano on Earth formed in 2008 in the Sidoario district of East Java in Indonesia. It appeared at the site of an oil and gas exploration well drilled into the mud formation at a depth of 3000 m and it is suspected that the well caused the volcano to form. Large quantities of petroleum-contaminated mud along with some methane and toxic hydrogen sulfide have been issuing from the volcano since it was formed. Although no humans have been injured, more than 12,000 had to be evacuated and 20 factories, 15 mosques, and 18 schools have been inundated. As the mud has been disgorged, an area of surrounding land surface has subsided, in places by as much as 100 m. More than 1100 mostly small mud volcanoes from natural sources have been observed worldwide. Some large ones in Azerbaijan emit large amounts of methane gas and have been burning for centuries. Worldwide, mud volcanoes emit 10–20 million metric tons of methane to the atmosphere each year.

15.12 SURFACE EARTH MOVEMENT

Mass movements are the result of gravity acting upon rock and soil on the Earth's surface. This produces a shearing stress on earthen materials located on slopes that can exceed the shear strength of the material and produce landslides and related phenomena involving the downward movement of geological materials. Such phenomena are affected by several factors, including the kinds and, therefore, strengths of materials, slope steepness, and degree of saturation with water. Usually, a specific event initiates mass movement. This can occur when excavation by humans makes the slopes steeper, by the action of torrential rains, or by earthquakes.

Illustrated in Figure 15.9, landslides refer to events in which large masses of rock and dirt move down the slope rapidly. Such events occur when material resting on a slope at an *angle of repose* is acted upon by gravity to produce a *shearing stress*. This stress may exceed the forces of friction or *shear strength*. Weathering, fracturing, water, and other factors may induce the formation of *slide planes* or *failure planes* such that a landslide results.

Loss of life and property from landslides can be substantial. In 1970, a devastating avalanche of soil, mud, and rocks initiated by an earthquake slid down Mount Huascaran in Peru killing an estimated 20,000 people. Sometimes the effects are indirect. In 1963, a total of 2600 people were killed near the Vaiont Dam in Italy when a sudden landslide filled the reservoir behind the dam with earthen material. Although the dam held, the displaced water spilled over its abutments as a wave 90 m high, wiping out structures and lives in its path.

Although often ignored by developers, the tendency toward landslides is predictable and can be used to determine areas in which homes and other structures should not be built. Slope stability maps based upon the degree of slope, the nature of underlying geological strata, climatic conditions, and other factors can be used to assess the risk of landslides. Evidence of a tendency for land to slide can be observed from effects on existing structures, such as walls that have lost their alignment, cracks in foundations, and poles that tilt. The likelihood of landslides can be minimized by moving material

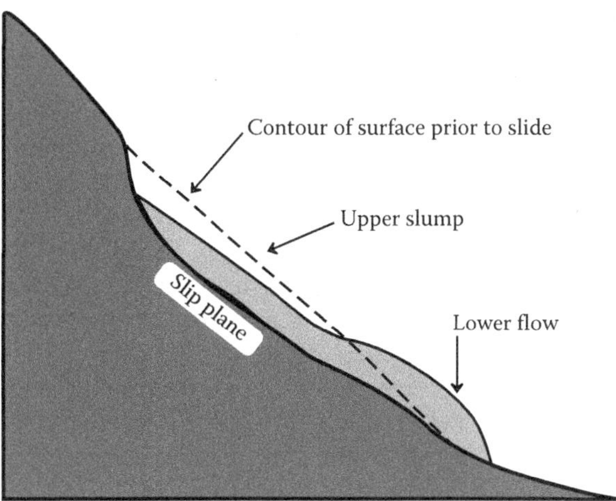

FIGURE 15.9 A landslide occurs when Earth moves along a slip plane. Typically, a landslide consists of an upper slump and lower flow. The latter serves to stabilize the slide, and when it is disturbed, such as by cutting through it to construct a road, Earth may slide further.

from the upper to the lower part of a slope, avoiding the loading of slopes, and avoiding measures that might change the degree and pathways of water infiltration into slope materials. In cases where the risk is not too severe, retaining walls may be constructed that reduce the effects of landslides.

Several measures can be used to warn of landslides. Simple visual observations of changes in the surface can be indicative of an impending landslide. More sophisticated measures include tilt meters and devices that sense vibrations accompanying the movement of earthen materials.

In addition to landslides, there are several other kinds of mass movements that have the potential to be damaging. *Rockfalls* occur when rocks fall down slopes so steep that at least part of the time the falling material is not in contact with the ground. The fallen material accumulates at the bottom of the fall as a pile of *talus*. A much less spectacular event is *creep*, in which movement is slow and gradual. The action of frost—frost heaving—is a common form of creep. Though usually not life-threatening, over a period of time creep may ruin foundations and cause misalignment of roads and railroads with significant property damage often being the result.

Sinkholes are a kind of Earth movement resulting when surface earth falls into an underground cavity. They rarely injure people but may cause spectacular property damage. Cavities that produce sinkholes may form by the action of water containing dissolved carbon dioxide on limestone (see Chapter 3, Reaction 3.7); loss of underground water during drought or from heavy pumping, thus removing support that previously kept soil and rock from collapsing; heavy underground water flow; and other factors that remove solid material from underground strata.

Special problems are presented by permanently frozen ground in arctic climates such as Alaska or Siberia. In such areas the ground may remain permanently frozen, thawing to only a shallow depth during the summer. This condition is called *permafrost*. Permafrost poses particular problems for construction, especially where the presence of a structure may result in thawing such that the structure rests in a pool of water-saturated muck on top of a slick surface of frozen water and soil. The construction and maintenance of highways, railroads, and pipelines, such as the Trans-Alaska pipeline in Alaska, can become quite difficult in the presence of permafrost.

Some types of soils, particularly the so-called expansive clays, expand and shrink markedly as they become saturated with water and dry out. Although essentially never life-threatening, the movement of structures and the damage caused to them by expansive clays can be very high. Aside from years when catastrophic floods and earthquakes occur, the monetary damage done by the action of expansive soil exceeds that of earthquakes, landslides, floods, and coastal erosion combined!

15.13 STREAM AND RIVER PHENOMENA

A *stream* consists of water flowing through a channel. The area of land from which water is drawn that flows into a stream is the stream's *drainage basin*. The sizes of streams are described by *discharge* defined as the volume of water flowing past a given point on the stream per unit time. Discharge and *gradient*, the steepness of the downward slope of a stream, determine the stream *velocity*.

Internal processes raise masses of land and whole mountain ranges, which in turn are shaped by the action of streams. Streams cut down mountain ranges, create valleys, form plains, and produce great deposits of sediment, thus playing a key role in shaping the geospheric environment. Streams spontaneously develop bends and curves by cutting away the outer parts of stream banks and depositing materials on the inner parts. These curved features of streams are known as *meanders*. Left undisturbed, a stream forms meanders across a valley in a constantly changing pattern. The cutting away of material by the stream and the deposition of sediment eventually form a generally flat area. During times of high stream flow, the stream leaves its banks, inundating parts or all of the valley, thus creating a *floodplain*.

A *flood* occurs when a stream develops a high flow such that it leaves its banks and spills out onto the floodplain. Floods are arguably the most common and damaging of surface phenomena in the geosphere. Though natural and in many respects beneficial occurrences, floods cause damage to structures located in their paths, and the severity of their effects is greatly increased by human activities.

A number of factors determine the occurrence and severity of floods. One of these is the tendency of particular geographic areas to receive large amounts of rain within short periods of time. One such area is located in the middle of the continental United States where warm, moisture-laden air from the Gulf of Mexico is carried northward during the spring months to collide with cold air from the north; the resultant cooling of the moist air can cause torrential rains to occur, resulting in severe flooding. In addition to season and geography, geological conditions have a strong effect on flooding potential. Rain falling on a steep surface tends to run off rapidly, creating flooding. A watershed can contain relatively massive quantities of rain if it consists of porous, permeable materials that allow a substantial rate of infiltration, assuming that it is not already saturated. Plants in a watershed tend to slow runoff and loosen soil, enabling additional infiltration. Through transpiration (see Chapter 16, Section 16.2), plants release moisture to the atmosphere quickly, enabling soil to absorb more moisture.

Several terms are used to describe flooding. When the *stage* of a stream, that is, the elevation of the water surface, exceeds the stream bank level, the stream is said to be at *flood stage*. The highest stage attained defines the flood *crest*. *Upstream* floods occur close to the inflow from the drainage basin, usually the result of intense rainfall. Whereas upstream floods usually affect smaller streams and watersheds, *downstream floods* occur on larger rivers that drain large areas. Widespread spring snowmelt and heavy, prolonged spring rains, often occurring together, cause downstream floods.

Floods are made more intense by higher fractions and higher rates of runoff, both of which may be aggravated by human activities. This can be understood by comparing a vegetated drainage basin to one that has been largely denuded of vegetation and paved over. In the former case, rainfall is retained by vegetation, such as grass cover. Thus the potential flood water is delayed, the time span over which it enters a stream is extended, and a higher proportion of the water infiltrates into the ground. In the latter case, less rainfall infiltrates, and the runoff tends to reach the stream quickly and to be discharged over a shorter time period, thus leading to more severe flooding. These factors are illustrated in Figure 15.10.

The conventional response to the threat of flooding is to control a river, particularly by the construction of raised banks called *levees*. In addition to raising the banks to contain a stream, the stream channel may be straightened and deepened to increase the volume and velocity of water flow, a process called *channelization*. Although effective for common floods, these measures may exacerbate extreme floods by confining and increasing the flow of water upstream such that the

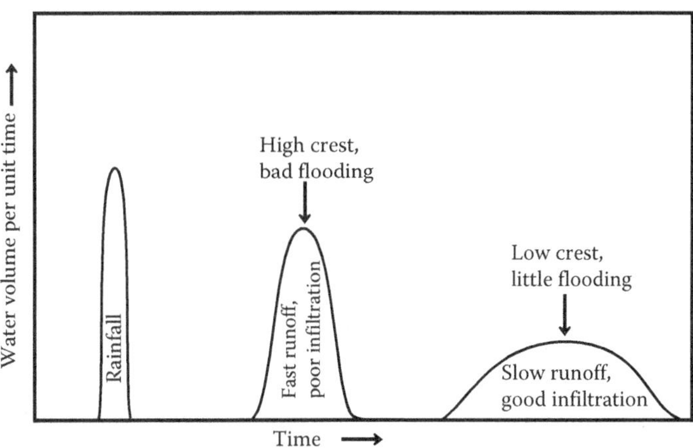

FIGURE 15.10 Influence of runoff on flooding.

capacity to handle water downstream is overwhelmed. Another solution is to construct dams to create reservoirs for flood control upstream. Usually such reservoirs are multipurpose facilities designed for water supply, recreation, and to control river flow for navigation in addition to flood control. The many reservoirs constructed for flood control in recent decades have been reasonably successful. There are, however, conflicts in the goals for their uses. Ideally, a reservoir for flood control should remain largely empty until needed to contain a large volume of floodwater, an approach that is obviously inconsistent with other uses. Another concern is that of exceeding the capacity of the reservoir, or dam failure, the latter of which can lead to catastrophic flooding.

15.14 PHENOMENA AT THE LAND/OCEAN INTERFACE

The coastal interface between land masses and the ocean is an important area of environmental activity. The land along this boundary is under constant attack from the waves and currents from the ocean, so that most coastal areas are always changing. The most common structure of the coast is shown in cross section in Figure 15.11. The beach, consisting of sediment, such as sand formed by wave action on coastal rock, is a sloping area that is periodically inundated by ocean waves. Extending from approximately the high tide mark to the dunes lining the landward edge of the beach is a relatively level area called the *berm*, which is usually not washed over by ocean water. The level of water to which the beach is subjected varies with the tides. Through wind action the

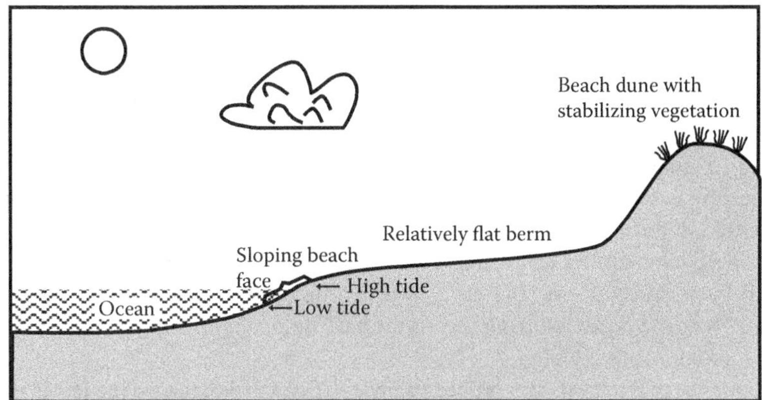

FIGURE 15.11 Cross section of the ocean/land interface along a beach.

surface of the water is in constant motion as undulations called *ocean waves*. As these waves reach the shallow water along the beach, they "touch bottom" and are transformed to *breakers*, characterized by crested tops. These breakers crashing onto a beach give it much of its charm, but can also be extremely destructive.

Coastlines exhibit a variety of features. Steep valleys carved by glacial activity, then filled with rising seawater, constitute the fjords along the coast of Norway. Valleys, formerly on land, now filled with seawater, constitute *drowned valleys*. *Estuaries* occur where tidal saltwater mixes with inflowing freshwater.

Erosion is a constant feature of a beachfront. Unconsolidated beach sand can be shifted readily—sometimes spectacularly through great distances over short periods of time—by wave action. Sand, pebbles, and rock in the form of rounded cobbles constantly wear against the coast by wave action, exerting a continuous abrasive action called *milling*. This action is augmented by the chemical weathering effects of seawater, in which the salt content may play a role.

Some of the more striking alterations to coastlines occur during storms, such as hurricanes and typhoons. The low pressure that accompanies severe storms tends to suck ocean water upward. This effect, usually combined with strong winds blowing onshore and coinciding with high tide, can cause ocean water to wash over the berm on a beach to attack dunes or cliffs inland. Such a *storm surge* can remove large quantities of beach, damage dune areas, and wash away structures unwisely constructed too close to the shore. A storm surge associated with a hurricane washed away most of the structures in Galveston, Texas, in 1900, claiming 6000 lives, and, despite the protection of a large floodwall constructed after the 1900 hurricane, Galveston was largely destroyed by Hurricane Ike in September 2008. A major concern with respect to coastal storms is loss of beaches, which are highly prized for their recreational value.

An especially vulnerable part of the coast consists of barrier islands—long, low strips of land roughly paralleling the coast some distance offshore. High storm surges may wash completely over barrier islands, partially destroying them and shifting them around. Many dwellings unwisely constructed on barrier islands, such as the outer banks of North Carolina, have been destroyed by storm surges during hurricanes.

15.14.1 The Threat of Rising Sea Levels

Large numbers of people live at a level near, or in some cases below, sea level. As a result, any significant temporary or permanent rise in sea level poses significant risks to lives and property. Such an event occurred on February 1, 1953 when high tides and strong winds combined to breach the system of dikes protecting much of the Netherlands from seawater. About one-sixth of the country was flooded as far inland as 64 km from the coast, killing about 2000 people and leaving approximately 100,000 without homes.

Although isolated instances of flooding by seawater caused by combinations of tidal and weather phenomena will continue to occur, a much more long-lasting threat is posed by long-term increases in sea level. These could result from global warming due to the greenhouse gas emissions discussed in Chapter 14. Several factors could raise ocean levels to destructive highs, also a result of greenhouse warming. Simple expansion of warmed oceanic water could raise sea levels by about 1/3 m over the next century. The melting of glaciers, such as those in the Alps, has probably raised ocean levels by about 5 cm during the last century, and the process is continuing. The greatest concern, however, is that global warming could cause the great West Antarctic ice sheet to melt, which would raise sea levels by as much as 6 m. In February and March 2002, a mass of ice larger than the state of Rhode Island became detached from the Antarctic Peninsula, raising fears that the Antarctic ice sheet may be melting relatively rapidly.

The measurement of sea levels has proven to be a difficult task because the levels of the surface of land keep changing. Land most recently covered with Ice Age glaciers in areas such as Scandinavia is still "springing back" from the immense mass of the glaciers, so that sea levels measured by gauges fixed on land actually appear to be dropping by several millimeters per year in such locations. An opposite situation exists on the east coast of North America where land was pushed outward and raised around the edge of the enormous sheet of ice that covered Canada and the northern United States about 20,000 years ago and is now settling back. Factors such as these illustrate the advantages of remarkably accurate satellite technology now used in the determination of sea levels.

15.15 PHENOMENA AT THE LAND/ATMOSPHERE INTERFACE

The interface between the atmosphere and land is a boundary of intense environmental activity. The combined effects of air and water tend to cause significant changes to the land materials at this interface. The top layer of exposed land is especially susceptible to physical and chemical weathering. Here, air laden with oxidant oxygen contacts rock, originally formed under reducing conditions, causing oxidation reactions to occur. Acid naturally present in rainwater as dissolved CO_2 or present as pollutant sulfuric, sulfurous, nitric, or hydrochloric acid can dissolve portions of some kinds of rocks. Organisms such as lichens, which consist of fungi and algae growing symbiotically on rock surfaces, drawing carbon dioxide, oxygen, or nitrogen from air can grow on rock surfaces at the boundary of the atmosphere and geosphere, causing additional weathering to take place.

One of the most significant agents affecting exposed geospheric solids at the atmosphere/geosphere boundary is wind. Consisting of air moving largely in a horizontal fashion, wind both erodes solids and acts as an agent to deposit solids on geospheric surfaces. The influence of wind is especially pronounced in dry areas. A major factor in wind erosion is wind *abrasions* in which solid particles of sand and rock carried by wind tend to wear away exposed rock and soil. Loose, unconsolidated sand and soil may be removed in large volumes by wind, a process called *deflation*.

The potential for wind to move matter is illustrated by the formation of large deposits of *loess*, consisting of finely divided soil carried by wind. Loess particles are typically several tens of micrometers in size, small enough to be carried great distances by wind. Especially common are loess deposits that originated with matter composed of rock ground to a fine flour by ice age glaciers. This material was first deposited in river valleys by flood waters issuing from melting glaciers and then blown some distance from the rivers by strong winds after drying out.

One of the more common geospheric features created by wind is a *dune*, consisting of a mound of debris, usually sand, dropped when wind slows down. When a dune begins to form, it forms an obstruction that slows wind even more, so that more sediment is dropped. The result is that in the presence of sediment-laden wind, dunes several meters or more high may form rapidly. In forming a dune, heavier, coarser particles settle first so that the matter in dunes is sorted according to size, just like sediments deposited by flowing streams. In areas where winds are prevalently from one direction, as is usually the case, dunes show a typical shape as illustrated in Figure 15.12. It is seen that the steeply sloping side, called the *slip face*, is downwind.

Some of the environmental effects of dunes result from their tendency to migrate with the prevailing winds. Migration occurs as matter is blown by the wind up the gently sloping face of the dune and falls down the slip face. Migrating sand dunes have inundated forest trees, and dust dunes in drought-stricken agricultural areas have filled road ditches, causing sharply increased maintenance costs.

15.16 EFFECTS OF ICE

The power of ice to alter the geosphere is amply demonstrated by the remains of past glacial activity from the Ice Age. Those large areas of the Earth's surface that were once covered with layers of glacial ice 1 or 2 km in thickness show evidence of how the ice carved the surface, left massive piles of rock and gravel, and left rich deposits of freshwater. The enormous weight of glaciers on the Earth's surface compressed it, and in places it is still springing back 10,000 or so years after the glaciers retreated. Today the influence of ice on the Earth's surface is much less than it was during

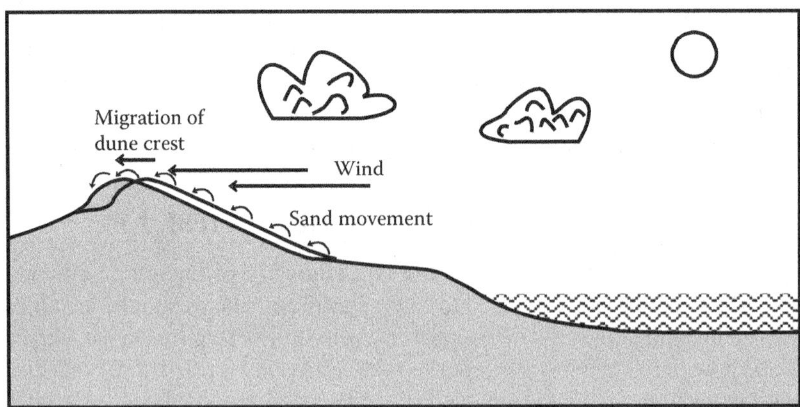

FIGURE 15.12 Shape and migration of a dune as determined by the prevailing wind direction.

the Ice Age, and there is substantial concern that melting of glaciers by greenhouse warming will raise sea levels so high that coastal areas will be inundated.

Glaciers form at sufficiently high latitudes and altitudes such that snow does not melt completely each summer. This occurs when snow becomes compacted over several years to several thousand years such that the frozen water turns to crystals of true ice. Huge masses of ice with areas of several thousand square kilometers or more, and often around 1 km thick, occur in polar regions and are called *continental glaciers.* Both Greenland and the Antarctic are covered by continental glaciers. *Alpine glaciers* occupy mountain valleys.

Glaciers on a slope flow as a consequence of their mass. This rate of flow is usually only a few meters per year, but may reach several kilometers per year. If a glacier flows into the sea, it may lose masses of ice as icebergs, a process called *calving.* Ice may also be lost by melting along the edges. The processes by which ice is lost are termed *ablation.*

Glacial ice affects the surface of the geosphere by both erosion and deposition. It is easy to imagine that a flowing mass of glacial ice is very efficient in scraping away the surface over which it flows, a process called *abrasion.* Adding to the erosive effect is the presence of rocks frozen into the glaciers, which can act like tools to carve the surface of the underlying rock and soil. Whereas abrasion tends to wear rock surfaces away producing a fine rock powder, larger bits of rock can be dislodged from the surface over which the glacier flows and be carried along with the glacial ice.

When glacial ice melts, the rock that has been incorporated into it is left behind. This material is called *till,* or if it has been carried for some distance by water running off the melting glacier it is called *outwash.* Piles of rock left by melting glaciers produce unique structures called *moraines.*

Although the effects of glaciers described above are the most spectacular manifestations of the action of ice on the geosphere, at a much smaller level ice can have some very substantial effects. Freezing and expansion of water in pores and small crevices in rock are a major contributor to physical weathering processes. Freeze/thaw cycles are also very destructive to some kinds of structures, such as stone buildings.

15.17 EFFECTS OF HUMAN ACTIVITIES

Human activities have profound effects on the geosphere. Such effects may be obvious and direct, such as strip mining, or rearranging vast areas for construction projects, such as roads and dams. Or the effects may be indirect, such as pumping so much water from underground aquifers that the ground subsides, or abusing soil such that it no longer supports plant life well and erodes. As the source of minerals and other resources used by humans, the geosphere is dug up, tunneled, stripped bare, rearranged, and subjected to many other kinds of indignities. The land is often severely disturbed, air can be polluted with dust particles during mining, and water may be polluted. Many of these effects, such as soil erosion caused by human activities, are addressed elsewhere in this book.

15.17.1 Extraction of Geospheric Resources: Surface Mining

Many human effects on the geosphere are associated with the extraction of resources from the Earth's crust. This is done in a number of ways, the most damaging of which can be surface mining. Surface mining is employed in the United States to extract virtually all of the rock and gravel that is mined, well over half of the coal, and numerous other resources. Properly done, with appropriate restoration practices, surface mining does minimal damage and may even be used to improve surface quality, such as by the construction of surface reservoirs where rock or gravel have been extracted. In earlier times before strict reclamation laws were in effect, surface mining, particularly of coal, left large areas of land scarred, devoid of vegetation, and subject to erosion.

Several approaches are employed in surface mining. Sand and gravel located under water are extracted by *dredging* with draglines or chain buckets attached to large conveyers. In most cases resources are covered with an *overburden* of earthen material that does not contain any of the resource that is being sought. This material must be removed as *spoil*. *Open-pit mining* is, as the name implies, a procedure in which gravel, building stone, iron ore, and other materials are simply dug from a big hole in the ground. Some of these pits, such as several from which copper ore has been taken in the United States, are truly enormous in size.

The most well-known (sometimes infamous) method of surface mining is *strip mining*, in which strips of overburden are removed by draglines and other heavy earth-moving equipment to expose seams of coal, phosphate rock, or other materials. Heavy equipment is used to remove a strip of overburden, and the exposed mineral resource is removed and hauled away. Overburden from a parallel strip is then removed and placed over the previously mined strip, and the procedure is repeated numerous times. Older practices left the replaced overburden as relatively steep erosion-prone banks. On highly sloping terrain, overburden is removed on progressively higher terraces and placed on the terrace immediately below.

15.17.2 Environmental Effects of Mining and Mineral Extraction

Some of the environmental effects of surface mining have been mentioned above. Although surface mining is most often considered for its environmental effects, subsurface mining may also have a number of effects, some of which are not immediately apparent and may be delayed for decades. Underground mines have a tendency to collapse, leading to severe subsidence. Mining disturbs groundwater aquifers. Water seeping through mines and mine tailings may become polluted. One of the more common and damaging effects of mining on water occurs when pyrite, FeS_2, commonly associated with coal, is exposed to air and becomes oxidized to sulfuric acid by bacterial action to produce acid mine water (see Section 6.14). Some of the more damaging environmental effects of mining are the result of the processing of mined materials. Usually, ore is only part, often a small part, of the material that must be excavated. Various *beneficiation* processes are employed to separate the useful fraction of ore, leaving a residue of *tailings*. A number of adverse effects can result from environmental exposure of tailings. For example, residues left from the beneficiation of coal are often enriched with pyrite, FeS_2, which is oxidized microbiologically and chemically to produce damaging acidic drainage (acid mine water). Uranium ore tailings unwisely used as fill material have contaminated buildings with radioactive radon gas.

15.18 AIR POLLUTION AND THE GEOSPHERE

The geosphere can be a significant source of air pollutants. Of these geospheric sources, volcanic activity is one of the most common. Volcanic eruptions, fumaroles, hot springs, and geysers can emit toxic and acidic gases, including carbon monoxide, hydrogen chloride, and hydrogen sulfide. Greenhouse gases that tend to increase global climatic warming—carbon dioxide and methane—can come from volcanic sources. Massive volcanic eruptions may inject huge amounts of particulate matter into the atmosphere. The incredibly enormous 1883 eruption of the East Indies volcano Krakatoa blew about 2.5 km^3 of solid matter into the atmosphere, some of which penetrated well into the stratosphere. This material stayed aloft long enough to circle Earth several times, causing red sunsets and a measurable lowering of temperature worldwide.

The 1982 eruption of the southern Mexico volcano El Chicón showed the importance of the type of particulate matter in determining effects on climate. The matter given off by this eruption was unusually rich in sulfur, so that an aerosol of sulfuric acid formed and persisted in the atmosphere for about 3 years, during which time the mean global temperature was lowered by several tenths of a degree due to the presence of atmospheric sulfuric acid. By way of contrast, the eruption of Mount St. Helens in Washington State in the United States 2 years earlier had little perceptible effect on climate, although the amount of material blasted into the atmosphere was about the same as that from El Chicón. The material from the Mount St Helens eruption had comparatively little sulfur in it, so the climatic effects were minimal.

Thermal smelting processes used to convert metal fractions in ore into usable forms have caused a number of severe air pollution problems that have affected the geosphere. Many metals are present in ores as sulfides, and smelting can release large quantities of sulfur dioxide, as well as particles that contain heavy metals such as arsenic, cadmium, or lead. The resulting acid and heavy metal pollution of surrounding land can cause severe damage to vegetation so that devastating erosion occurs. One such area is around a large nickel smelter in Sudbury, Ontario, Canada, where a large area of land has become denuded of vegetation. Similar dead zones have been produced by copper smelters in Tennessee and in eastern Europe, including the former Soviet Union.

Soil and its cultivation produce significant quantities of atmospheric emissions. Waterlogged soil, particularly that cultivated for rice, generates significant quantities of methane, a greenhouse gas. The microbial reduction of nitrate in soil releases nitrous oxide, N_2O, to the atmosphere. However, soil and rock can also remove atmospheric pollutants. It is believed that microorganisms in soil account for the loss from the atmosphere of some carbon monoxide, which some fungi and bacteria can metabolize. Carbonate rocks, such as calcium carbonate, $CaCO_3$, can neutralize acid from atmospheric sulfuric acid and acid gases.

As discussed in Section 9.6, masses of atmospheric air can become trapped and stagnant under conditions of a temperature inversion in which the vertical circulation of air is limited by the presence of a relatively warm layer of air overlaying a colder layer at ground level. The effects of inversions can be aggravated by topographical conditions that tend to limit circulation of air. Figure 15.13 shows such a condition in which surrounding mountain ridges limit horizontal air movement. Air pollutants may be forced up a mountain ridge from a polluted area to significantly higher altitudes than they would otherwise reach. Because of this "chimney effect," air pollutants may reach mountain pine forests that are particularly susceptible to damage from air pollutants such as ozone formed along with photochemical smog.

15.19 WATER POLLUTION AND THE GEOSPHERE

Water pollution is addressed in detail elsewhere in this book. Much water pollution arises from interactions of groundwater and surface water with the geosphere.[7] These aspects are addressed briefly here.

The relationship between water and the geosphere is twofold. The geosphere may be severely damaged by water pollution. This occurs, for example, when water pollutants produce contaminated sediments, such as those polluted by heavy metals or PCBs. In some cases the geosphere serves as a source of water pollutants. Examples include acid produced by exposed metal sulfides in the geosphere or synthetic chemicals improperly discarded in landfills and leaking into groundwater.

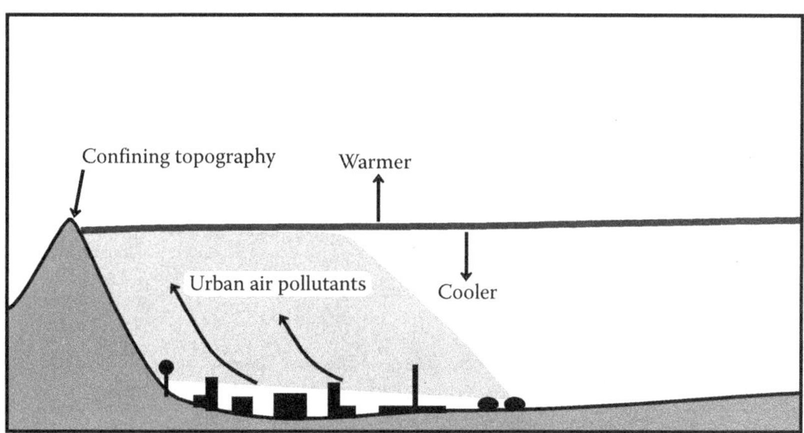

FIGURE 15.13 Topographical features, such as confining mountain ridges, may work with temperature inversions to increase the effects of air pollution.

The sources of water pollution are divided into two main categories. The first of these consists of *point sources*, which enter the environment at a single, readily identified entry point. An example of a point source would be a sewage water outflow. Point sources tend to be those directly identified as having their origins from human activities. *Nonpoint sources* of pollution are those from broader areas. Such a source is water contaminated by fertilizer from fertilized agricultural land, or water contaminated with excess alkali leached from alkaline soils. Nonpoint sources are relatively harder to identify and monitor. Pollutants associated with the geosphere are usually nonpoint sources.

An especially common and damaging geospheric source of water pollutants consists of sediments carried by water from land into the bottoms of bodies of water. Most such sediments originate from agricultural land that has been disturbed such that soil particles are eroded from land into water. The most common manifestation of sedimentary material in water is opacity, which seriously detracts from the esthetics of the water. Sedimentary material deposited in reservoirs or canals can clog them and eventually make them unsuitable for water supply, flood control, navigation, and recreation. Suspended sediment in water used as a water supply can clog filters and add significantly to the cost of treating the water. Sedimentary material can devastate wildlife habitats by reducing food supplies and ruining nesting sites. Turbidity in water can severely curtail photosynthesis, thus reducing primary productivity necessary to sustain the food chains of aquatic ecosystems.

15.20 WASTE DISPOSAL AND THE GEOSPHERE

The geosphere receives many kinds and large amounts of wastes. Its ability to cope with such wastes with minimal damage is one of its most important characteristics and is dependent upon the kinds of wastes disposed of on it. A variety of wastes, ranging from large quantities of relatively innocuous municipal refuse to much smaller quantities of potentially lethal radioactive wastes, are deposited on land or in landfills. These are addressed briefly in this section.

15.20.1 Municipal Refuse

The currently favored method for disposing of municipal solid wastes—household garbage—is in *sanitary landfills* (Figure 15.14) consisting of refuse piled on top of the ground or into a depression such as a valley, compacted, and covered at frequent intervals by soil. This approach permits frequent covering of the refuse so that loss of blowing trash, water contamination, and other undesirable effects are minimized. A completed landfill can be put to beneficial uses, such as a recreational area; because of settling, gas production, and other factors, landfill surfaces are generally not

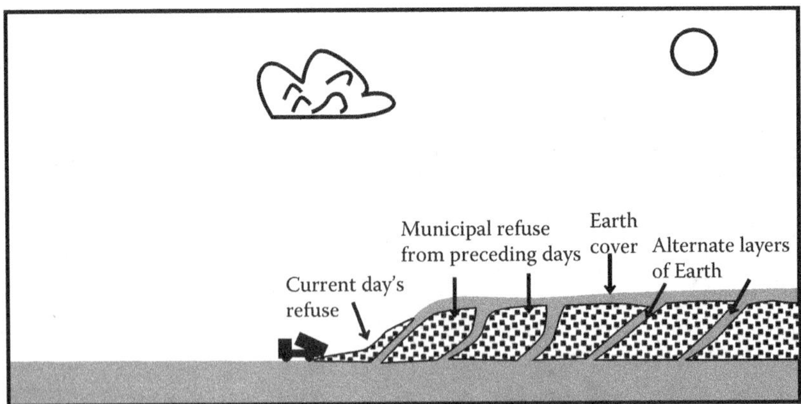

FIGURE 15.14 Structure of a sanitary landfill.

suitable for building construction. Modern sanitary landfills are much preferable to the open dump sites that were once the most common means of municipal refuse disposal.

Although municipal refuse is much less dangerous than hazardous chemical waste, it still poses some hazards. Despite prohibitions against the disposal of cleaners, solvents, lead storage batteries, and other potentially hazardous materials in landfills, materials that pose some environmental hazards do find their way into landfills and can contaminate their surroundings.

Landfills produce both gaseous and aqueous emissions. Biomass in landfills quickly depletes oxygen by aerobic biodegradation of microorganisms in the landfill,

$$\{CH_2O\}\text{ (biomass)} + O_2 \rightarrow CO_2 + H_2O \quad (15.8)$$

emitting carbon dioxide. Over a period of many decades the buried biodegradable materials undergo anaerobic biodegradation

$$2\{CH_2O\} \rightarrow CO_2 + CH_4 \quad (15.9)$$

releasing methane as well as carbon dioxide. Although often impractical and too expensive, it is desirable to reclaim the methane as fuel, and some large sanitary landfills are major sources of methane. Released methane is a greenhouse gas and can pose significant explosion hazards to structures built on landfills. Although produced in much smaller quantities than methane, hydrogen sulfide, H_2S, is also generated by anaerobic biodegradation. This gas is toxic and has a bad odor. In a properly designed sanitary landfill, hydrogen sulfide releases are small and the gas tends to oxidize before it reaches the atmosphere in significant quantities.

Water infiltrating into sanitary landfills dissolves materials from the disposed refuse and runs off as *leachate*. Contaminated leachate is the single greatest potential pollution problem with refuse disposal sites, so it is important to minimize its production by designing landfills in a way that keeps water infiltration as low as possible. The anaerobic degradation of biomass produces organic acids that give the leachate a tendency to dissolve acid-soluble solutes, such as heavy metals. Leachate can infiltrate into groundwater posing severe contamination problems. This is minimized by siting sanitary landfills over formations of poorly permeable clay or depositing layers of clay in the landfill before refuse is put in it. In addition, impermeable synthetic polymer liners may be placed in the bottom of the landfill. In areas of substantial rainfall, infiltration into the landfill exceeds its capacity to hold water so that leachate flows out. In order to prevent water pollution downstream, this leachate should be controlled and treated.

Hazardous chemical wastes are disposed of in the so-called *secure landfills*, which are designed to prevent leakage and geospheric contamination of toxic chemicals disposed of in them. Such a landfill is equipped with a variety of measures to prevent contamination of groundwater and the surrounding geosphere. The base of the landfill is made of compacted clay that is largely impermeable to leachate. An impermeable polymer liner is placed over the clay liner. The surface of the landfill is covered with material designed to reduce water infiltration, and the surface is designed with slopes that also minimize the amount of water running in. Elaborate drainage systems are installed to collect and treat leachate.

The most pressing matter pertaining to geospheric disposal of wastes involves radioactive wastes. Most of these wastes are *low-level* wastes, including discarded radioactive laboratory chemicals and pharmaceuticals, filters used in nuclear reactors, and ion-exchange resins used to remove small quantities of radionuclides from nuclear reactor cooler water. Disposed of in properly designed landfills, such wastes pose minimal hazards.

Of greater concern are the *high-level* radioactive wastes, primarily fission products of nuclear power reactors and by-products of nuclear weapons manufacture. Many of these wastes are currently stored as solutions in tanks, many of which have outlived their useful lifetimes and pose leakage hazards, at sites such as the federal nuclear facility at Hanford, Washington, where plutonium was generated in large quantities during post-World War II years. Eventually, such wastes must be placed in the geosphere such that they will pose no hazards. Numerous proposals have been advanced for their disposal, including disposal in salt formations, subduction zones in the seafloor, and ice sheets. The most promising sites appear to be those in poorly permeable formations of igneous rock. Among these are basalts, which are strong, glassy igneous types of rock found in the Columbia River plateau. Granite and pyroclastic welded tuffs fused by past high-temperature volcanic eruptions are other possibilities as sites for disposing of nuclear wastes and keeping them isolated for tens of thousands of years.

LITERATURE CITED

1. Wenk, H.-R. and A. Bulakh, *Minerals: Their Constitution and Origin*, Cambridge University Press, Cambridge, UK, 2004.
2. Jackson, K., *Plate Tectonics*, Lucent Books, San Diego, CA, 2005.
3. Keller, E. A. and R. H. Blodgett, *Natural Hazards: Earth's Processes as Hazards, Disasters, and Catastrophes*, 2nd ed., Pearson Prentice Hall, Upper Saddle River, NJ, 2008.
4. Keller, E. A., *Introduction to Environmental Geology*, 4th ed., Pearson Prentice Hall, Upper Saddle River, NJ, 2008.
5. Kusky, T., *Earthquakes: Plate Tectonics and Earthquake Hazards*, Facts on File, New York, 2008.
6. Kusky, T., *Volcanoes: Eruptions and Other Volcanic Hazards*, Facts on File, New York, 2008.
7. LaMoreaux, P. E., *Environmental Hydrogeology*, 2nd ed., Taylor & Francis, Boca Raton, FL, 2008.

SUPPLEMENTARY REFERENCES

Atkinson, D., *Weathering, Slopes and Landforms*, Hodder & Stoughton, London, 2004.

Bashkin, V. N., *Modern Biogeochemistry: Environmental Risk Assessment*, 2nd ed., Springer, Dordrecht, The Netherlands, 2006.

Bryant, E., *Natural Hazards*, 2nd ed., Cambridge University Press, Cambridge, UK, 2005.

Chamley, H., *Geosciences, Environment and Man*, Elsevier, New York, 2003.

Coch, N. K., *Geohazards: Natural and Human*, Prentice Hall, Upper Saddle River, NJ, 1995.

De Boer, J. Z. and D. T. Sanders, *Volcanoes in Human History: The Far Reaching Effects of Major Eruptions*, Princeton University Press, Princeton, NJ, 2001.

Donnelly, K. J., *Ice Ages of the Past and the Future*, The Rosen Publishing Group, New York, 2003.

Eby, G. N., *Principles of Environmental Geochemistry*, Thomson Brooks/Cole, Pacific Grove, CA, 2004.

Faure, G., *Principles and Applications of Geochemistry: A Comprehensive Textbook for Geology Students,* Prentice Hall, Upper Saddle River, NJ, 1998.

Gates, A. E. and D. Ritchie, *Encyclopedia of Earthquakes and Volcanoes*, 3rd ed., Facts on File, New York, 2007.

Holland, H. D. and K. K. Turekian, *Treatise on Geochemistry*, Elsevier/Pergamon, Amsterdam, 2004.

Hyndman, D. and D. Hyndman, *Natural Hazards and Disasters*, 2nd ed., Thomson Brooks/Cole, Belmont, CA, 2009.

Marti, J. and G. Ernst, Eds, *Volcanoes and the Environment*, Cambridge University Press, Cambridge, UK, 2005.

McCollum, S., *Volcanic Eruptions, Earthquakes, and Tsunamis*, Chelsea House, New York, 2007.

Montgomery, C. W., *Environmental Geology*, 8th ed., McGraw-Hill, Boston, 2008.

Murty, T. S., U. Aswathanarayana, and N. Nirupama, Eds, *The Indian Ocean Tsunami*, Taylor & Francis, London, 2007.

Nash, D. J. and S. J. McLaren, Eds, *Geochemical Sediments and Landscapes*, Blackwell Publishing, Malden, MA, 2007.

Pipkin, B. W., *Geology and the Environment*, 5th ed., Thomson Brooks/Cole, Belmont, CA, 2008.

Sammonds, P. R. and J. M. T. Thompson, Eds, *Advances in Earth Science: From Earthquakes to Global Warming*, Imperial College Press, London, 2007.

Savino, J. and M. D. Jones, *Supervolcano: The Catastrophic Event that Changed the Course of Human History (Could Yellowstone be Next?)*, New Page Books, Franklin Lakes, NJ, 2007.

Skinner, H. C. W. and A. R. Berger, Eds, *Geology and Health: Closing the Gap*, Oxford University Press, New York, 2003.

Stone, G. W., *Raging Forces: Life on a Violent Planet*, National Geographic, Washington, DC, 2007.

Van der Pluijm, Ben A. and S. Marshak, *Earth Structure: An Introduction to Structural Geology and Tectonics*, W.W. Norton, New York, 2004.

Waltham, T., *Foundations of Engineering Geology*, Taylor & Francis, London, 2009.

QUESTIONS AND PROBLEMS

1. Of the following, the one that is *not* a manifestation of desertification is (explain): (A) declining groundwater tables, (B) salinization of topsoil and water, (C) production of deposits of MnO_2 and $Fe_2O_3{\cdot}H_2O$ from anaerobic processes, (D) reduction of surface waters, (E) unnaturally high soil erosion.
2. Give an example of how each of the following chemical or biochemical phenomena in soils operates to reduce the harmful nature of pollutants (explain): (A) oxidation–reduction processes, (B) hydrolysis, (C) acid–base reactions, (D) precipitation, (E) sorption, (F) biochemical degradation.
3. Why do silicates and oxides predominate among the Earth's minerals?
4. Give the characteristic that the minerals with the following formulas have in common: NaCl, Na_2SO_4, $CaSO_4{\cdot}2H_2O$, $MgCl_2{\cdot}6H_2O$, $MgSO_4{\cdot}7H_2O$, $KMgClSO_4{\cdot}\frac{1}{4}H_2O$, $K_2MgCa_2(SO_4)_4{\cdot}2H_2O$.
5. Explain how the following are related: weathering, igneous rock, sedimentary rock, soil.
6. Match the following:

A. Metamorphic rock	1. Produced by the precipitation or coagulation of dissolved or colloidal weathering products
B. Chemical sedimentary rocks	2. Contain residues of plant and animal remains
C. Detrital rock	3. Formed from action of heat and pressure on sedimentary rock
D. Organic sedimentary rocks	4. Formed from solid particles eroded from igneous rocks as a consequence of weathering

7. From where does most flowing water that contains dissolved load originate? Why does such water tend to come from this source?
8. What role might be played by water pollutants in the production of dissolved load and in the precipitation of secondary minerals from it?
9. As defined in this chapter, are the ions involved in ion replacement the same as exchangeable cations? If not, why not?

10. Speculate regarding how water present in poorly consolidated soil might add to the harm caused by earthquakes.
11. In what sense may volcanoes contribute to air pollution? What possible effects may this have on climate?
12. Explain how excessive pumping of groundwater might adversely affect streams, particularly in regard to the flow of small streams.
13. Which three elements are most likely to undergo oxidation as part of the chemical weathering process? Give example reactions of each.
14. Match the following:

A. Groundwater	1. Water from precipitation in the form of rain or snow
B. Vadose water	2. Water present in a zone of saturation
C. Meteoric water	3. Water held in the unsaturated zone or zone of aeration
D. Water in capillary	4. Water drawn somewhat above the water table by surface fringe tension

15. Large areas of central Kansas have vast deposits of halite. What is halite? What does this observation say about the geological history of the area?
16. What is the distinction between weathering and erosion? Suggest ways in which air pollution may contribute to both phenomena.
17. One way in which coal and other fossil fuels may be used without contributing to higher levels of greenhouse gas carbon dioxide in the atmosphere is through carbon sequestration by pumping carbon dioxide into mineral strata. Explain with a chemical reaction how formations of limestone (calcium carbonate) might be used for this purpose. Suggest how this might cause problems on the surface.
18. Deposits of iron minerals are often found where groundwater flows to the surface. Use a chemical reaction to explain this observation.

16 Soil and Agricultural Environmental Chemistry

16.1 SOIL AND AGRICULTURE

Because of its role in supporting plant growth, soil is a crucial part of the geosphere. Like the very sparse layer of stratospheric ozone that is essential to protect terrestrial organisms from destructive solar ultraviolet radiation, the layer of soil on the earth's surface is extremely thin. If Earth were the size of a geography-class globe, the average thickness of productive soil on its surface would be less than that of a human cell! Because of its importance in sustainability and its fragile nature, soil and production of food and materials from it are discussed in some detail in this chapter.

Soil and agricultural practices are strongly tied with the environment and sustainability. Some of these considerations are addressed later in this chapter along with a discussion of soil erosion and conservation. Cultivation of land and agricultural practices along with other anthrospheric activities strongly influence the atmosphere, the hydrosphere, and the biosphere. Although this chapter deals primarily with soil, the topic of agriculture in general is introduced for perspective.

Although the most obvious use of soil is for plant growth leading to food production, it serves many functions in the maintenance of sustainability. It holds water, regulates water supplies, and serves as a medium to filter and conduct water from precipitation into groundwater aquifers. It serves to recycle raw materials and nutrients. It is a habitat for a large variety of organisms, especially fungi and bacteria. Soil interfaces with the anthrosphere as an engineering medium that is dug up, moved, and smoothed over to make roads, dams, and other engineering constructs.

The study of soil is called *pedology* or, more simply, *soil science.* To humans and most terrestrial organisms, soil is the most important part of the geosphere. Though only a tissue-thin layer compared to the earth's total diameter, soil is the medium that produces most of the food required by most living things. Good soil—and a climate conducive to its productivity—is the most valuable asset a society can have.

In addition to being the site of most food production, soil is the receptor of large quantities of pollutants, such as particulate matter from power plant smokestacks. Fertilizers, pesticides, and some other materials applied to soil often contribute to water and air pollution. Therefore, soil is a key component of environmental chemical cycles. It is a valuable part of the earth's natural capital.

Soil itself can become an air or water pollutant, especially when it has been abused by poor agricultural practices, deforestation, or desertification. Fine soil particles make up a large fraction of the matter in yellow clouds described as massive air pollution phenomena in Chapter 15. Furthermore, topsoil washed into streams and bodies of water by water erosion can be deposited as damaging sediments.

16.1.1 Agriculture

Agriculture, the production of food by growing crops and livestock, provides for the most basic of human needs. No other industry impacts as much as agriculture does on the environment. Agriculture is absolutely essential for the maintenance of the huge human populations now on Earth. The displacement of native plants, destruction of wildlife habitat, erosion, pesticide pollution, and other environmental aspects of agriculture have enormous potential for environmental damage. Survival of humankind on Earth demands environmentally friendly and sustainable agricultural practices. On the other hand, growth of domestic crops removes (at least temporarily) greenhouse gas carbon dioxide from the atmosphere and provides potential sources of renewable resources of energy and fiber that can substitute for petroleum-derived fuels and materials.

Agriculture can be divided into the two main categories of *crop farming*, in which plant photosynthesis is used to produce food and fiber, and *livestock farming*, in which domesticated animals are grown for meat, milk, and other animal products. Crop farming produces food consumed directly by humans, food for livestock, and fiber. Livestock farming involves the raising of animals for meat, dairy products, eggs, wool, and hides. Freshwater fish and even crayfish are raised on "fish farms." Beekeeping provides honey.

Agriculture is based on domestic plants engineered by early farmers from their wild plant ancestors. Without perhaps much of an awareness of what they were doing, early farmers selected plants with desired characteristics for the production of food. This selection of plants for domestic use brought about evolutionary change so profound that the products often barely resembled their wild ancestors. Plant breeding based on scientific principles of heredity is a very recent development dating from around 1900. One of the major objectives of plant breeding has been to increase yield. Yields of crops can also be increased by selecting for resistance to insects, drought, and cold. In some cases, the goal is to increase nutritional value, such as raising the content of essential amino acids.

The development of hybrids has vastly increased yields and other desired characteristics of a number of important crops. Basically, *hybrids* are the offspring of crosses between two different *true-breeding* strains. Often quite different from either parent strain, hybrids tend to exhibit "hybrid vigor" and to have significantly higher yields. The most success with hybrid crops has been obtained with corn (maize). Corn is one of the easiest plants to hybridize because of the physical separation of the male flowers, which grow as tassels on top of the corn plant, from female flowers, which are attached to incipient ears on the side of the plant. Despite past successes by more conventional means and some early disappointments with "genetic engineering," application of recombinant DNA technology (Section 16.13) will probably eventually overshadow all the advances ever made in plant breeding.

In addition to plant strains and varieties, numerous other factors are involved in crop production. Weather is an obvious factor and shortages of water, chronic in many areas of the world, are mitigated by irrigation. Here, automated techniques and computer control can play an important, more environmentally friendly role by minimizing the quantities of water required. The application of chemical fertilizer has vastly increased crop yields. The judicious application of pesticides, especially herbicides, but including insecticides and fungicides as well, has increased crop yields and reduced losses greatly. Use of herbicides has had an environmental benefit in reducing the degree of mechanical cultivation of soil required. Indeed, "no-till" and "low-till" agriculture (conservation tillage) are now widely practiced on a large scale.

The rearing of domestic animals may have significant environmental effects. Effluent from waste lagoons associated with concentrated livestock feeding operations can cause serious water pollution problems. Goats and sheep have destroyed pastureland in the Near East, Northern Africa, Portugal, and Spain. Of particular concern are the environmental effects of raising cattle. Significant amounts of forest land have been converted to marginal pasture land to raise beef. Production of one pound of beef requires about four times as much water and four times as much feed as does production of one pound of chicken. An interesting aspect of the problem is emission of greenhouse-gas methane by anaerobic bacteria in the digestive systems of cattle and other ruminant animals; cattle rank right behind wetlands and rice paddies as producers of atmospheric methane. However, because of the action of specialized bacteria in their stomachs, cattle and other ruminant animals are capable of converting otherwise unusable cellulose to food.

16.1.2 Pesticides and Agriculture

Pesticides, particularly insecticides and herbicides, are an integral part of modern agricultural production. In the United States, agricultural pesticides are regulated under the Federal Insecticide, Fungicide, and Rodenticide Act (FIFRA), first passed in 1947, revised in a major way in 1972, and subjected to several amendments since then. Pesticides are responsible for much of the high productivity of modern agriculture as well as some of the major pollution problems associated with agriculture.

An interesting development regarding the use of herbicides in the late 1990s was the production of transgenic crops resistant to specific herbicides. The Monsanto company pioneered this approach with the development of "Roundup Ready" crops that resist the herbicidal effects of Monsanto's flagship Roundup® herbicide (glyphosate). The seedlings of crops resistant to the herbicide are not harmed on exposure to it, whereas competing weeds are killed. Although sales of glyphosate have been greatly increased by Roundup Ready crops, especially soybeans, increased plantings of these crops has reduced the use of herbicides overall with a net overall benefit to the environment.

The use on transgenic crops of glyphosate, for which the structural formula is shown in Chapter 7, Figure 7.14, has made it a uniquely important product, the most widely produced pesticide in the world. Glyphosate binds strongly to soil colloids and is readily degraded by soil microorganisms. The properties of glyphosate make it difficult to measure in soil and water samples. The molecule is very polar and soluble in water, but not soluble in organic solvents commonly used to extract pollutants for analysis. It binds strongly to metal ions as well as organic, mineral, and clay solids making its isolation difficult. Because of glyphosate's structural similarity to naturally occurring amino acids and other plant biomolecules, there are numerous interferences in its determination.

16.2 NATURE AND COMPOSITION OF SOIL

Soil, a variable mixture of minerals, organic matter, and water capable of supporting plant life on the earth's surface, is the most fundamental requirement for agriculture. It is the final product of the weathering action of physical, chemical, and biological processes on rocks, which largely produces clay minerals. The organic portion of soil consists of plant biomass in various stages of decay. High populations of bacteria, fungi, and animals such as earthworms may be found in soil. Soil contains air spaces and generally has a loose texture (Figure 16.1).

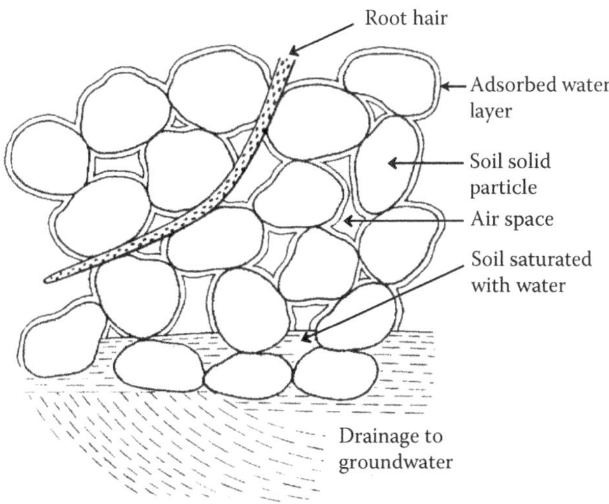

FIGURE 16.1 Fine structure of soil, showing solid, water, and air phases.

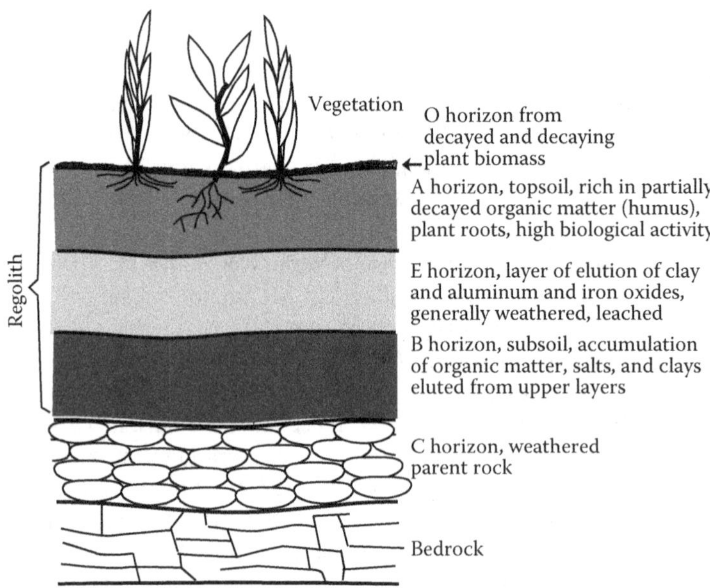

FIGURE 16.2 Soil profile showing soil horizons.

The solid fraction of typical productive soil is approximately 5% organic matter and 95% inorganic matter. Some soils, such as peat soils, may contain as much as 95% organic material. Other soils contain as little as 1% organic matter.

Typical soils exhibit distinctive layers called *horizons*, with increasing depth (Figure 16.2). Horizons form as a result of complex interactions among processes that occur during weathering. Rainwater percolating through soil carries dissolved and colloidal solids to lower horizons where they are deposited. Biological processes, such as bacterial decay of residual plant biomass, produce slightly acidic CO_2, organic acids, and complexing compounds that are carried by rainwater to lower horizons where they interact with clays and other minerals, altering the properties of the minerals. The top layer of soil, typically several inches in thickness, is known as the A horizon, or *topsoil*. This is the layer of maximum biological activity in the soil, contains most of the soil organic matter, and is essential for plant productivity. Various soils may have a variety of horizons, the main ones of which are described in Figure 16.2.

Soils exhibit a large variety of characteristics that are used for their classification for various purposes, including crop production, road construction, and waste disposal. Soil profiles are discussed above. The parent rocks from which soils are formed obviously play a strong role in determining the composition of soils. Other soil characteristics include strength, workability, soil particle size, permeability, and degree of maturity.

16.2.1 Water and Air in Soil

Large quantities of water are required for the production of most plant materials. For example, several hundred kilograms of water are required to produce 1 kg of dry hay. Water is part of the three-phase, solid–liquid–gas system making up soil. It is the basic transport medium for carrying essential plant nutrients from solid soil particles into plant roots and to the farthest reaches of the plant's leaf structure (Figure 16.3). The water in a plant evaporates into the atmosphere from the plant's leaves, a process called *transpiration*.

Normally, because of the small size of soil particles and the presence of small capillaries and pores in the soil, the water phase is not totally independent of soil solid matter. The availability of water to plants is governed by gradients arising from capillary and gravitational forces. The availability of

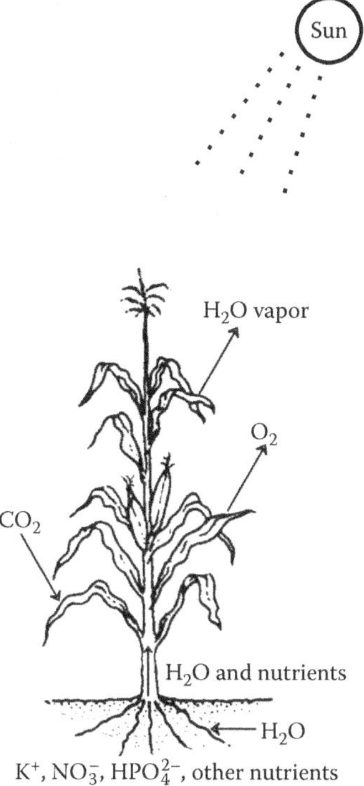

FIGURE 16.3 Plants transport water from the soil to the atmosphere by transpiration. Nutrients are also carried from the soil to the plant extremities by this process. Plants remove CO_2 from the atmosphere and add O_2 by photosynthesis. The reverse occurs during plant respiration.

nutrient solutes in water depends upon concentration gradients and electrical potential gradients. Water present in larger spaces of soil is relatively more available to plants and readily drains away. Water held in smaller pores or between the unit layers of clay particles is held much more strongly. Soils high in organic matter may hold appreciably more water than other soils, but it is relatively less available to plants because of physical and chemical sorption of the water by the organic matter.

There is a very strong interaction between clays and water in soil. Water is absorbed on the surfaces of clay particles. Because of the high surface/volume ratio of colloidal clay particles, a great deal of water may be bound in this manner. Water is also held between the unit layers of the expanding clays, such as the montmorillonite clays. As soil becomes waterlogged (water-saturated), it undergoes drastic changes in physical, chemical, and biological properties. Oxygen in such soil is rapidly used up by the respiration of microorganisms that degrade soil organic matter. In such soils, the bonds holding soil colloidal particles together are broken, which causes disruption of soil structure. Thus, the excess water in such soils is detrimental to plant growth, and the soil does not contain the air required by most plant roots. Most useful crops, with the notable exception of rice, cannot grow on waterlogged soils.

One of the most marked chemical effects of waterlogging is a reduction of p*E* by the action of organic reducing agents acting through bacterial catalysts. Thus, the redox condition of the soil becomes much more reducing, and the soil p*E* may drop from that of water in equilibrium with air (+13.6 at pH 7) to 1 or less. One of the more significant results of this change is the mobilization of iron and manganese as soluble iron(II) and manganese(II) through reduction of their insoluble higher oxides:

$$MnO_2 + 4H^+ + 2e^- \rightarrow Mn^{2+} + 2H_2O \quad (16.1)$$

$$Fe_2O_3 + 6H^+ + 2e^- \rightarrow 2Fe^{2+} + 3H_2O \quad (16.2)$$

Although soluble manganese is generally found in soil as Mn^{2+} ion, soluble iron(II) occurs frequently as negatively charged iron–organic chelates. Strong chelation of iron(II) by soil fulvic acids (Chapter 3) apparently enables reduction of iron(III) oxides at more positive p*E* values than would otherwise be possible. This causes an upward shift in the Fe(II)–$Fe(OH)_3$ boundary as shown in Figure 4.4.

Some soluble metal ions such as Fe^{2+} and Mn^{2+} are toxic to plants at high levels. Their oxidation to insoluble oxides may cause formation of deposits of Fe_2O_3 and MnO_2, which clog tile drains in fields.

Roughly 35% of the volume of typical soil is composed of air-filled pores. Whereas the normal dry atmosphere at sea level contains 21% O_2 and 0.04% CO_2 by volume, these percentages may be quite different in soil air because of the decay of organic matter:

$$\{CH_2O\} + O_2 \rightarrow CO_2 + H_2O \tag{16.3}$$

This process consumes oxygen and produces CO_2. As a result, the oxygen content of air in soil may be as low as 15% and the carbon dioxide content may be several percent higher. Thus, the decay of organic matter in soil increases the equilibrium level of dissolved CO_2 in groundwater. This lowers the pH and contributes to weathering of carbonate minerals, particularly calcium carbonate (see Reaction 3.7). As discussed in Section 16.3, CO_2 also shifts the equilibrium of the process by which roots absorb metal ions from soil.

16.2.2 The Inorganic Components of Soil

The weathering of parent rocks and minerals to form the inorganic soil components results ultimately in the formation of inorganic colloids. These colloids are repositories of water and plant nutrients, which may be made available to plants as needed. Inorganic soil colloids often absorb toxic substances in soil, thus playing a role in detoxification of substances that otherwise would harm plants. The abundance and nature of inorganic colloidal material in soil are obviously important factors in determining soil productivity.

The uptake of plant nutrients by roots often involves complex interactions with the water and inorganic phases. For example, a nutrient held by inorganic colloidal material has to traverse the mineral/water interface and then the water/root interface. This process is often strongly influenced by the ionic structure of soil inorganic matter.

As noted in Section 15.2, the most common elements on the earth's crust are oxygen, silicon, aluminum, iron, calcium, sodium, potassium, and magnesium. Therefore, minerals composed of these elements—particularly silicon and oxygen—constitute most of the mineral fraction of the soil. Common soil mineral constituents are finely divided quartz (SiO_2), orthoclase ($KAlSi_3O_8$), albite ($NaAlSi_3O_8$), epidote ($4CaO \cdot 3(AlFe)_2O_3 \cdot 6SiO_2 \cdot H_2O$), geothite ($FeO(OH)$), magnetite ($Fe_3O_4$), calcium and magnesium carbonates ($CaCO_3$, $CaCO_3 \cdot MgCO_3$), and oxides of manganese and titanium.

16.2.3 Organic Matter in Soil

Though typically comprising <5% of a productive soil, organic matter largely determines soil productivity. It serves as a source of food for microorganisms, undergoes chemical reactions such as ion exchange, and influences the physical properties of soil. Some organic compounds even contribute to the weathering of mineral matter, the process by which soil is formed. For example, $C_2O_4^{2-}$, oxalate ion, produced as a soil fungi metabolite, occurs in soil as the calcium salts whewellite and weddelite. Oxalate in soil water dissolves minerals, thus speeding the weathering process and increasing the availability of nutrient ion species. This weathering process involves oxalate complexation of iron or aluminum in minerals, represented by the reaction

$$3H^+ + M(OH)_3(s) + 2CaC_2O_4(s) \rightarrow M(C_2O_4)_2^-(aq) + 2Ca^{2+}(aq) + 3H_2O \tag{16.4}$$

in which M is Al or Fe. Some soil fungi produce citric acid and other chelating organic acids that react with silicate minerals and release potassium and other nutrient metal ions held by these minerals.

The strong chelating agent, 2-ketogluconic acid,

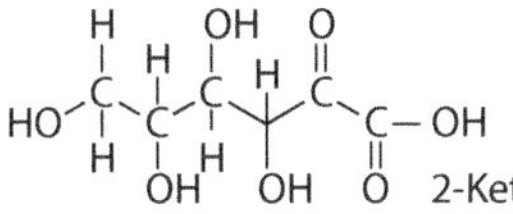

2-Ketogluconic acid

is produced by some soil bacteria. By solubilizing metal ions, it may contribute to the weathering of minerals. It may also be involved in the release of phosphate from insoluble phosphate compounds.

Biologically active components of the organic soil fraction include polysaccharides, amino sugars, nucleotides, and organic sulfur and phosphorus compounds. Humus, a water-insoluble material that biodegrades very slowly, makes up the bulk of soil organic matter. The organic compounds in soil are summarized in Table 16.1.

The accumulation of organic matter in soil is strongly influenced by temperature and by the availability of oxygen. Since the rate of biodegradation decreases with decreasing temperature, organic matter does not degrade rapidly in colder climates and tends to build up in soil. In water and in waterlogged soils, decaying vegetation does not have easy access to oxygen, and organic matter accumulates. The organic content may reach 90% in areas where plants grow and decay in soil saturated with water.

The presence of naturally occurring polynuclear aromatic (PAH) compounds is an interesting feature of soil organic matter. These compounds, some of which are carcinogenic, are discussed as air pollutants in Chapters 10 and 12. PAH compounds found in soil include fluoranthene, pyrene, and chrysene. PAH compounds in soil result from combustion from both natural sources (grass fires) or pollutant sources. Terpenes and plant pigments, such as β-carotene, also occur in soil organic matter.

TABLE 16.1
Major Classes of Organic Compounds in Soil

Compound Type	Composition	Significance
Humus	Degradation-resistant residue from plant decay, largely C, H, and O	Most abundant organic component, improves soil physical properties, exchanges nutrients, reservoir of fixed N
Fats, resins, and waxes	Lipids extractable by organic solvents	Generally, only several percent of soil organic matter, may adversely affect soil physical properties by repelling water, perhaps phytotoxic
Saccharides	Cellulose, starches, hemi cellulose, gums	Major food source for soil microorganisms, help stabilize soil aggregates
N-containing organics	Nitrogen bound to humus, amino acids, amino sugars, other compounds	Provide nitrogen for soil fertility
Phosphorus compounds	Phosphate esters, inositol phosphates (phytic acid), phospholipids	Sources of plant phosphate

16.2.4 Soil Humus

Of the organic components listed in Table 16.1, *soil humus* is by far the most significant.[1] Humus, composed of a base-soluble fraction called *humic* and *fulvic* acids (described in Section 3.17), and an insoluble fraction called *humin*, is the residue left when bacteria and fungi biodegrade plant material. The bulk of plant biomass consists of relatively degradable cellulose and degradation-resistant lignin, which is a polymeric substance with a higher carbon content than cellulose. Among lignin's prominent chemical components are aromatic rings connected by alkyl chains, methoxyl groups, and hydroxyl groups. These structural artifacts occur in soil humus and give it many of its characteristic properties.

The process by which humus is formed is called *humification*. Soil humus is similar to its lignin precursors, but has more carboxylic acid groups. Part of each molecule of humic substance is non-polar and hydrophobic, and part is polar and hydrophilic. Such molecules are called *amphiphiles*, and they form micelles (see Section 5.4 and Figure 5.4) in which the nonpolar parts compose the inside of small colloidal particles and the polar functional groups are on the outside. Amphiphilic humic substances probably also form bilayer surface coatings on mineral grains in soil. During the humification process, the nitrogen/carbon ratio of the organic matter increases as carbon is lost to CO_2 evolved during biodegradation and nitrogen fixed by nitrogen-fixing bacteria is incorporated into the humic residue.

Humic substances influence soil properties to a degree out of proportion to their small percentage in soil. They strongly bind metals and serve to hold micronutrient metal ions in soil. Because of their acid–base character, humic substances serve as buffers in soil. The water-holding capacity of soil is significantly increased by humic substances. These materials also stabilize aggregates of soil particles and increase the sorption of organic compounds by soil.

Humic materials in soil strongly sorb many solutes in soil water and have a particular affinity for heavy polyvalent cations. Soil humic substances may contain levels of uranium more than 10^4 times that of the water with which they are in equilibrium. Thus, water becomes depleted of its cations (or purified) in passing through humic-rich soils. Humic substances in soils also have a strong affinity for organic compounds with low water solubility such as DDT or atrazine, a herbicide widely used to kill weeds in corn fields (see structural formula in Chapter 7, Figure 7.12).

In some cases, there is a strong interaction between the organic and inorganic portions of soil. This is especially true of the strong complexes formed between clays and humic (fulvic) acid compounds. In many soils, 50–100% of soil carbon is complexed with clay. These complexes play a role in determining the physical properties of soil, soil fertility, and stabilization of soil organic matter. One of the mechanisms for the chemical binding between clay colloidal particles and humic organic particles is probably of the flocculation type (see Chapter 5) in which anionic organic molecules with carboxylic acid functional groups serve as bridges in combination with cations to bind clay colloidal particles together as a floc. Support is given to this hypothesis by the known ability of NH_4^+, Al^{3+}, Ca^{2+}, and Fe^{3+} cations to stimulate clay–organic complex formation. The synthesis, chemical reactions, and biodegradation of humic materials are affected by interaction with clays. The lower-molecular-mass fulvic acids may be bound to clay, occupying spaces in layers in the clay.

16.2.5 The Soil Solution

The *soil solution* is the aqueous portion of soil that contains dissolved matter from soil chemical and biochemical processes in soil and from exchange with the hydrosphere and biosphere. This medium transports chemical species to and from soil particles and provides intimate contact between the solutes and the soil particles. In addition to providing water for plant growth, it is an essential pathway for the exchange of plant nutrients between roots and solid soil.

A major challenge with respect to soil solution is obtaining a sample of it. The most straightforward means is collection of drainage water, but that may miss soil solution bound in capillaries and on surface films. Pressure/vacuum treatment, centrifugation, or displacement with another liquid may be employed.

Most of the solutes in soil are present as salts composed of H^+, Ca^{2+}, Mg^{2+}, K^+, and Na^+ cations (with lower levels of Fe^{2+}, Mn^{2+}, and Al^{3+}) and HCO_3^-, CO_3^{2-}, HSO_4^-, SO_4^{2-}, Cl^-, and F^- anions. The Fe^{2+}, Mn^{2+}, and Al^{3+} cations are generally present in hydrolyzed forms or bound with humic substances. Some anions become bound with H^+ (e.g., HCO_3^- formed from CO_3^{2-}). Multicharged ions tend to form ion pairs in solution such as $CaSO_4(aq)$.

16.3 ACID–BASE AND ION-EXCHANGE REACTIONS IN SOILS

One of the more important chemical functions of soils is the exchange of cations. As discussed in Chapter 5, the capacity of a sediment or soil to exchange cations is expressed as the CEC, the number of meq of monovalent cations that can be exchanged per 100 g of soil (on a dry-weight basis). The CEC should be looked upon as a conditional constant since it may vary with soil conditions such as p*E* and pH. Both the mineral and organic portions of soils exchange cations. Clay minerals exchange cations because of the presence of negatively charged sites on the mineral, resulting from the substitution of an atom of lower oxidation number for one of higher number, for example, magnesium for aluminum. Organic materials exchange cations because of the presence of the carboxylate group and other basic functional groups. Humus typically has a very high CEC. The CEC of peat may range from 300 to 400 meq/100 g. Values of CEC for soils with more typical levels of organic matter are around 10–30 meq/100 g.

Cation exchange in soil is the mechanism by which potassium, calcium, magnesium, and essential trace-level metals are made available to plants. When nutrient metal ions are taken up by plant roots, hydrogen ion is exchanged for the metal ions. This process, plus the leaching of calcium, magnesium, and other metal ions from the soil by water containing carbonic acid, tends to make the soil acidic:

$$\{Soil\}Ca^{2+} + 2CO_2 + 2H_2O \rightarrow \{Soil\}(H^+)_2 + Ca^{2+}(root) + 2HCO_3^- \tag{16.5}$$

Soil acts as a buffer and resists changes in pH. The buffering capacity depends upon the type of soil.

The oxidation of pyrite, FeS_2, in soil causes formation of acid-sulfate soils, sometimes called "cat clays":

$$FeS_2 + \tfrac{7}{2}O_2 + H_2O \rightarrow Fe^{2+} + 2H^+ + 2SO_4^{2-} \tag{16.6}$$

Cat clay soils may have pH values as low as 3.0. These soils, which are commonly found in Delaware, Florida, New Jersey, and North Carolina, are formed when neutral or basic marine sediments containing FeS_2 become acidic upon oxidation of pyrite when exposed to air. For example, soil reclaimed from marshlands and used for citrus groves has developed high acidity detrimental to plant growth. In addition, H_2S released by reaction of FeS_2 with acid is very toxic to citrus roots.

Soils are tested for potential acid-sulfate formation using a peroxide test. This test consists of oxidizing FeS_2 in the soil with 30% H_2O_2,

$$FeS_2 + \tfrac{15}{2}H_2O_2 \rightarrow Fe^{3+} + H^+ + 2SO_4^{2-} + 7H_2O \tag{16.7}$$

then testing for acidity and sulfate. Appreciable levels of sulfate and a pH below 3.0 indicate potential to form acid-sulfate soils. If the pH is above 3.0, either little FeS_2 is present or sufficient $CaCO_3$ is in the soil to neutralize the H_2SO_4 and acidic Fe^{3+}.

Pyrite-containing mine spoils (residue left over from mining) also form soils similar to acid-sulfate soils of marine origin.[2] In addition to producing acid, pyrite wastes can evolve phytotoxic H_2S.

A phytotoxic agent associated with acidic soils is aluminum as the Al^{3+} ion.[3] As the third most abundant element in the Earth's crust and a major component of clays, aluminum is present in most soils. It poses no problem in alkaline or neutral soils in which aluminum is bound as insoluble forms such as $Al(OH)_3$, but becomes toxic to plants in acidic soils that liberate soluble Al^{3+} ion. Aluminum toxicity to plants is a problem for the cultivation of approximately 8 billion acres of land worldwide that suffer from excess acidity, including about 86 million acres in the United States. It is believed that aluminum is toxic to plants because of its adverse effects on root tips. Barley is especially susceptible to the toxic effects of aluminum whereas wheat and corn tolerate the metal relatively well. These aluminum-tolerant plants apparently secrete an organic acid at their root tips that complexes Al^{3+} and prevents it from adversely affecting the roots. Through genetic engineering it may be possible to develop plant varieties that are aluminum-tolerant and make possible the cultivation of land that is otherwise unusable due to aluminum toxicity.

16.3.1 Adjustment of Soil Acidity

Most common plants grow best in soil with a pH near neutrality. If the soil becomes too acidic for optimum plant growth, it may be restored to productivity by liming, ordinarily through the addition of calcium carbonate:

$$\{Soil\}(H^+)_2 + CaCO_3 \rightarrow \{Soil\}Ca^{2+} + CO_2 + H_2O \tag{16.8}$$

In areas of low rainfall, soils may become too basic (alkaline) due to the presence of basic salts such as Na_2CO_3. Alkaline soils may be treated with aluminum or iron sulfate, which release acid on hydrolysis:

$$2Fe^{3+} + 3SO_4^{2-} + 6H_2O \rightarrow 2Fe(OH)_3(s) + 6H^+ + 3SO_4^{2-} \tag{16.9}$$

Sulfur added to soils is oxidized by bacterially mediated reactions to sulfuric acid:

$$S + \tfrac{3}{2}O_2 + H_2O \rightarrow 2H^+ + SO_4^{2-} \tag{16.10}$$

and sulfur is used, therefore, to acidify alkaline soils. The huge quantities of sulfur now being removed from fossil fuels to prevent air pollution by sulfur dioxide may make the treatment of alkaline soils by sulfur much more attractive economically.

16.3.2 Ion-Exchange Equilibria in Soil

Competition of different cations for cation-exchange sites on soil cation exchangers may be described semiquantitatively by exchange constants. For example, soil reclaimed from an area flooded with seawater will have most of its cation-exchange sites occupied by Na^+, and restoration of fertility requires binding of nutrient cations such as K^+:

$$\{Soil\}Na^+ + K^+ \rightleftarrows \{Soil\}K^+ + Na^+ \tag{16.11}$$

The exchange constant for this reaction is K_c,

$$K_c = \frac{N_K[Na^+]}{N_{Na}[K^+]} \tag{16.12}$$

which expresses the relative tendency of soil to retain K^+ and Na^+. In this equation, N_K and N_{Na} are the equivalent ionic fractions of potassium and sodium, respectively, bound to soil, and $[Na^+]$ and

$[K^+]$ are the concentrations of these ions in the surrounding soil water. For example, a soil with all cation-exchange sites occupied by Na^+ would have a value of 1.00 for N_{Na}; with one-half of the cation-exchange sites occupied by Na^+, N_{Na} is 0.5; and so on. The exchange of anions by soil is not so clearly defined as is the exchange of cations. In many cases, the exchange of anions does not involve a simple ion-exchange process. This is true of the strong retention of orthophosphate species by soil. At the other end of the scale, nitrate ion is very weakly retained by the soil.

Anion exchange may be visualized as occurring at the surfaces of oxides in the mineral portion of soil. A mechanism for the acquisition of surface charge by metal oxides is shown in Chapter 5, Figure 5.5, using MnO_2 as an example. At low pH, a metal oxide surface may have a net positive charge enabling it to hold anions, such as chloride, by electrostatic attraction as shown below where M represents a metal:

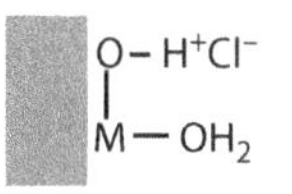

$$\begin{array}{l} O-H^+Cl^- \\ | \\ M-OH_2 \end{array}$$

At higher pH values, the metal oxide surface has a net negative charge due to the formation of OH^- ion on the surface caused by loss of H^+ from the water molecules bound to the surface:

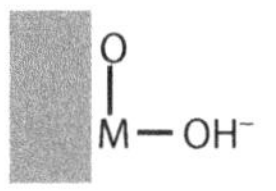

$$\begin{array}{l} O \\ | \\ M-OH^- \end{array}$$

In such cases, it is possible for anions such as HPO_4^{2-} to displace hydroxide ion and bond directly to the oxide surface:

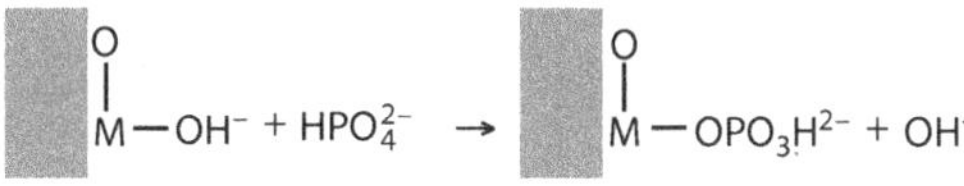

$$\begin{array}{l} O \\ | \\ M-OH^- \end{array} + HPO_4^{2-} \rightarrow \begin{array}{l} O \\ | \\ M-OPO_3H^{2-} \end{array} + OH^- \qquad (16.13)$$

16.4 MACRONUTRIENTS IN SOIL

One of the most important functions of soil in supporting plant growth is to provide essential plant nutrients—macronutrients and micronutrients. Macronutrients are those elements that occur in substantial levels in plant biomass and fluids. Micronutrients (Section 16.6) are elements that are essential only at very low levels and generally are required for the functioning of essential enzymes.

The elements generally recognized as essential macronutrients for plants are carbon, hydrogen, oxygen, nitrogen, phosphorus, potassium, calcium, magnesium, and sulfur. Carbon, hydrogen, and oxygen are obtained from the atmosphere. The other essential macronutrients must be obtained from soil. Of these, nitrogen, phosphorus, and potassium are the most likely to be lacking and are commonly added to soil as fertilizers. Because of their importance, these elements are discussed separately in Section 16.5.

Calcium-deficient soils are relatively uncommon. Application of lime, a process used to treat acid soils (see Section 16.3), provides a more than adequate calcium supply for plants. However, calcium uptake by plants and leaching by carbonic acid (Reaction 16.5) may produce a calcium deficiency in soil. Acid soils may still contain an appreciable level of calcium which, because of competition by hydrogen ion, is not available to plants. Treatment of acid soil to restore the pH to near neutrality generally remedies the calcium deficiency. In alkaline soils, the presence of high levels of sodium, magnesium, and potassium sometimes produces calcium deficiency because these ions compete with calcium for availability to plants.

Most of the 2.1% of magnesium in the earth's crust is rather strongly bound in minerals. Exchangeable magnesium held by ion-exchanging organic matter or clays is considered available to plants. The availability of magnesium to plants depends upon the calcium/magnesium ratio. If this ratio is too high, magnesium may not be available to plants and magnesium deficiency results. Similarly, excessive levels of potassium or sodium may cause magnesium deficiency.

Sulfur is assimilated by plants as the sulfate ion, SO_4^{2-}. Soils deficient in sulfur do not support plant growth well, largely because sulfur is a component of some essential amino acids and of thiamin and biotin. Sulfate ion is generally present in the soil as immobilized insoluble sulfate minerals or as soluble salts that are readily leached from the soil and lost as soil water runoff. Unlike the case of nutrient cations such as K^+, little sulfate is adsorbed to the soil (i.e., bound by ion-exchange binding), where it is resistant to leaching while still available for assimilation by plant roots.

Soil sulfur deficiencies have been found in a number of regions of the world. Whereas most fertilizers used to contain sulfur, its use in commercial fertilizers has declined. With continued use of sulfur-deficient fertilizers, it is possible that sulfur will become a limiting nutrient in more cases.

As noted in Section 16.3, the reaction of FeS_2 with acid in acid-sulfate soils may release H_2S, which is very toxic to plants and which also kills many beneficial microorganisms. Toxic hydrogen sulfide can also be produced by reduction of sulfate ion through microorganism-mediated reactions with organic matter. Production of hydrogen sulfide in flooded soils may be inhibited by treatment with oxidizing compounds, one of the most effective of which is KNO_3.

16.5 NITROGEN, PHOSPHORUS, AND POTASSIUM IN SOIL

Nitrogen, phosphorus, and potassium are plant nutrients that are obtained from soil. They are so important for crop productivity that they are commonly added to soil as fertilizers. The environmental chemistry of these elements is discussed here and their production as fertilizers in Section 16.7.

16.5.1 NITROGEN

Figure 16.4 summarizes the primary sinks and pathways of nitrogen in soil. In most soils, over 90% of the nitrogen content is organic. This organic nitrogen is primarily the product of the biodegradation of dead plants and animals. It is eventually hydrolyzed to NH_4^+, which can be oxidized to NO_3^- by the action of bacteria in the soil.

Nitrogen bound to soil humus is especially important in maintaining soil fertility. Unlike potassium or phosphate, nitrogen is not a significant product of mineral weathering. Nitrogen-fixing organisms ordinarily cannot supply sufficient nitrogen to meet peak demand. Inorganic nitrogen

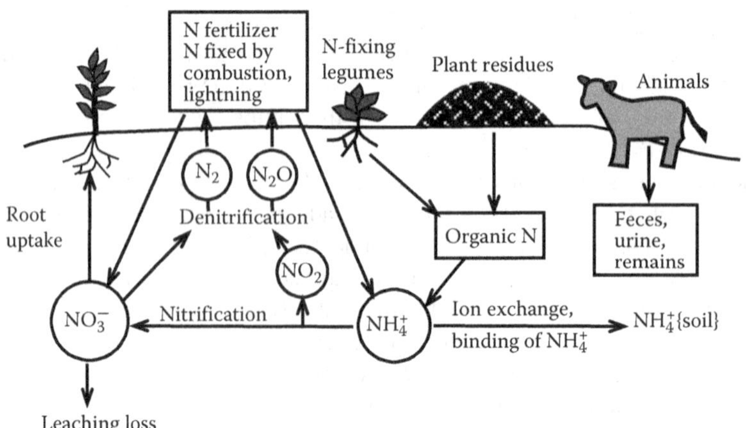

FIGURE 16.4 Nitrogen sinks and pathways in soil.

from fertilizers and rainwater is often largely lost by leaching. Soil humus, however, serves as a reservoir of nitrogen required by plants. It has the additional advantage that its rate of decay, hence its rate of nitrogen release to plants, roughly parallels plant growth—rapid during the warm growing season, slow during the winter months.

Nitrogen is an essential component of proteins and other constituents of living matter. Plants and cereals grown on nitrogen-rich soils not only provide higher yields, but are often substantially richer in protein and, therefore, more nutritious. Nitrogen is most generally available to plants as nitrate ion, NO_3^-. Some plants such as rice may utilize ammonium nitrogen; however, other plants are poisoned by this form of nitrogen. When nitrogen is applied to soils in the ammonium form, nitrifying bacteria perform an essential function in converting it to available nitrate ion.

Plants may absorb excessive amounts of nitrate nitrogen from soil. This phenomenon occurs particularly in heavily fertilized soils under drought conditions. Forage crops containing excessive amounts of nitrate can poison ruminant animals such as cattle or sheep. Plants having excessive levels of nitrate can endanger people when used for ensilage, an animal food consisting of finely chopped plant material such as partially matured whole corn plants, fermented in a structure called silo. Under the reducing conditions of fermentation, nitrate in ensilage may be reduced to toxic NO_2 gas, which can accumulate to high levels in enclosed silos. There have been many cases reported of persons being killed by accumulated NO_2 in silos.

Nitrogen fixation is the process by which atmospheric N_2 is converted to nitrogen compounds available to plants. Human activities are resulting in the fixation of a great deal more nitrogen than would otherwise be the case. Artificial sources now account for 30–40% of all nitrogen fixed. These include chemical fertilizer manufacture, nitrogen fixed during fuel combustion, combustion of nitrogen-containing fuels, and the increased cultivation of nitrogen-fixing legumes (see the following paragraph). Of some concern with this increased fixation of nitrogen is the possible effect upon the atmospheric ozone layer by N_2O released during denitrification of fixed nitrogen.

Before the widespread introduction of nitrogen fertilizers, soil nitrogen was provided primarily by legumes. These are plants such as soybeans, alfalfa, and clover, which contain on their root structures bacteria capable of fixing atmospheric nitrogen. Leguminous plants have a symbiotic (mutually advantageous) relationship with the bacteria that provide their nitrogen. Legumes may add significant quantities of nitrogen to soil, up to 10 pounds per acre per year, which is comparable to amounts commonly added as synthetic fertilizers. Soil fertility with respect to nitrogen may be maintained by rotating plantings of nitrogen-consuming plants with plantings of legumes, a fact recognized by agriculturists as far back as the Roman era.

The nitrogen-fixing bacteria in legumes exist in special structures on the roots called root nodules (see Figure 16.5). The rod-shaped bacteria that fix nitrogen are members of a special genus,

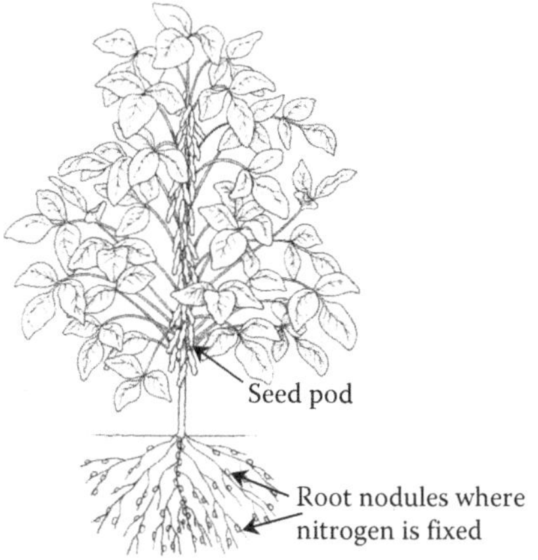

FIGURE 16.5 A soybean plant, showing root nodules where nitrogen is fixed.

Rhizobium. These bacteria may exist independently, but cannot fix nitrogen except in symbiotic combination with plants. Although all species of *Rhizobium* appear to be very similar, they exhibit a great deal of specificity in their choice of host plants. Curiously, legume root nodules also contain a form of hemoglobin, which must somehow be involved in the nitrogen-fixation process. Some studies suggest that some kinds of *Rhizobium* bacteria secrete substances near plant roots that promote plant growth by secreting or causing the plants to release substances that facilitate the uptake of nutrients or inhibit plant pathogens.[4]

Nitrate pollution of some surface waters and groundwater has become a major problem in some agricultural areas (see Chapter 7). Although fertilizers have been implicated in such pollution, there is evidence that feedlots are a major source of nitrate pollution. The growth of livestock populations and the concentration of livestock in feedlots have aggravated the problem. Such concentrations of cattle, coupled with the fact that a steer produces approximately 18 times as much waste material as a human, have resulted in high levels of water pollution in rural areas with small human populations. Streams and reservoirs in such areas frequently are just as polluted as those in densely populated and highly industrialized areas.

Nitrate in farm wells is a common and especially damaging manifestation of nitrogen pollution from feedlots because of the susceptibility of ruminant animals to nitrate poisoning. The stomach contents of ruminant animals such as cattle and sheep constitute a reducing medium (low p*E*) and contain bacteria capable of reducing nitrate ion to toxic nitrite ion:

$$NO_3^- + 2H^+ + 2e^- \rightarrow NO_2^- + H_2O \tag{16.14}$$

The origin of most nitrate produced from feedlot wastes is amino nitrogen present in nitrogen-containing waste products. Approximately, one-half of the nitrogen excreted by cattle is contained in the urine. Part of this nitrogen is proteinaceous and the other part is in the form of urea, NH_2CONH_2. As a first step in the degradation process, the amino nitrogen is probably hydrolyzed to ammonia or ammonium ion:

$$RNH_2 + H_2O \rightarrow R\text{–}OH + NH_3\ (NH_4^+) \tag{16.15}$$

This product is then oxidized through microorganism-catalyzed reactions to nitrate ion:

$$NH_3 + 2O_2 \rightarrow H^+ + NO_3^- + H_2O \tag{16.16}$$

Under some conditions, an appreciable amount of the nitrogen originating from the degradation of feedlot wastes is present as ammonium ion. Ammonium ion is rather strongly bound to soil (recall that soil is a generally good cation exchanger), and a small fraction is fixed as nonexchangeable ammonium ion in the crystal lattice of clay minerals. Because nitrate ion is not strongly bound to soil, it is readily carried through soil formations by water. Many factors, including soil type, moisture, and level of organic matter, affect the production of ammonia and nitrate ion originating from feedlot wastes, and a marked variation is found in the levels and distributions of these materials in feedlot areas.

16.5.2 Phosphorus

Although the percentage of phosphorus in plant material is relatively low, it is an essential component of plants. Phosphorus, like nitrogen, must be present in a simple inorganic form before it can be taken up by plants. In the case of phosphorus, the utilizable species is some form of orthophosphate ion. In the pH range that is present in most soils, $H_2PO_4^-$ and HPO_4^{2-} are the predominant orthophosphate species.

Orthophosphate is most available to plants at pH values near neutrality. It is believed that in relatively acidic soils, orthophosphate ions are precipitated or sorbed by species of Al(III) and

Fe(III). In alkaline soils, orthophosphate may react with calcium carbonate to form relatively insoluble hydroxyapatite:

$$3HPO_4^{2-} + 5CaCO_3(s) + 2H_2O \rightarrow Ca_5(PO_4)_3(OH)(s) + 5HCO_3^- + OH^- \quad (16.17)$$

In general, because of these reactions, little phosphorus applied as fertilizer leaches from the soil. This is important from the standpoint of both water pollution and utilization of phosphate fertilizers.

16.5.3 Potassium

Relatively high levels of potassium are utilized by growing plants. Potassium activates some enzymes and plays a role in the water balance in plants. It is also essential for some carbohydrate transformations. Crop yields are generally greatly reduced in potassium-deficient soils. The higher the productivity of the crop, the more the potassium removed from soil. When nitrogen fertilizers are added to soils to increase productivity, removal of potassium is enhanced. Therefore, potassium may become a limiting nutrient in soils heavily fertilized with other nutrients.

Potassium is one of the most abundant elements in the Earth's crust, of which it makes up 2.6%; however, much of this potassium is not easily available to plants. For example, some silicate minerals such as leucite, $K_2O{\cdot}Al_2O_3{\cdot}4SiO_2$, contain strongly bound potassium. Exchangeable potassium held by clay minerals is relatively more available to plants.

16.6 MICRONUTRIENTS IN SOIL

Boron, chlorine, copper, iron, manganese, molybdenum (for N-fixation), and zinc are considered essential plant micronutrients. These elements are needed by plants only at very low levels and frequently are toxic at higher levels. There is some chance that other elements will be added to this list as techniques for growing plants in environments free of specific elements improve. Most of these elements function as components of essential enzymes. Manganese, iron, chlorine, and zinc may be involved in photosynthesis. It is possible that sodium, silicon, nickel, and cobalt may also be essential nutrients for some plants.

Iron and manganese occur in a number of soil minerals. Sodium and chlorine (as chloride) occur naturally in soil and are transported as atmospheric particulate matter from marine sprays (see Chapter 10). Some of the other micronutrients and trace elements are found in primary (unweathered) minerals that occur in soil. Boron is substituted isomorphically for Si in some micas and is present in tourmaline, a mineral with the formula $NaMg_3Al_6B_3Si_6O_{27}(OH,F)_4$. Copper is isomorphically substituted for other elements in feldspars, amphiboles, olivines, pyroxenes, and micas; it also occurs as trace levels of copper sulfides in silicate minerals. Molybdenum occurs as molybdenite (MoS_2). Vanadium is isomorphically substituted for Fe or Al in oxides, pyroxenes, amphiboles, and micas. Zinc is present as the result of isomorphic substitution for Mg, Fe, and Mn in oxides, amphiboles, olivines, and pyroxenes and as trace zinc sulfide in silicates. Other trace elements that occur as specific minerals, sulfide inclusions, or by isomorphic substitution for other elements in minerals are chromium, cobalt, arsenic, selenium, nickel, lead, and cadmium.

The trace elements listed above may be coprecipitated with secondary minerals (see Section 15.2) that are involved in soil formation. Such secondary minerals include oxides of aluminum, iron, and manganese (precipitation of hydrated oxides of iron and manganese very efficiently removes many trace metal ions from solution); calcium and magnesium carbonates; smectites; vermiculites; and illites.

Micronutrient deficiencies have developed in some soils. In some cases, deficiencies are the result of soil factors including pH, p*E*, biological activity, CEC, and soil contents of organic matter and clay. Factors involving the plants and their root systems cause variations in micronutrient uptake. These include root and root hair morphology and surface area and association of roots with microorganisms. Also involved are root secretions including H^+, OH^-, HCO_3^- ions; citric, oxalic, and other acids; and enzymes, such as phosphatases. Major factors involved in micronutrient deficiency include loss of topsoil, leaching of nutrients from soil, liming of acid soils (Ca^{2+} from lime competes with micronutrient metal ions for root uptake), more intensive cultivation, and intense use of more purified fertilizers.

Some plants accumulate extremely high levels of specific trace metals.[5] Those accumulating more than 1.00 mg/g of dry weight are called *hyperaccumulators*. Nickel and copper both undergo hyperaccumulation in some plant species. As an example of a metal hyperaccumulator, *Aeolanthus biformifolius DeWild* growing in copper-rich regions of Shaba Province, Zaire, contains up to 1.3% copper (dry weight) and is known as a "copper flower."

The hyperaccumulation of metals by some plants has led to the idea of *phytoremediation* in which plants growing on contaminated ground accumulate metals, which are then removed with the plant biomass. Plants used for phytoremediation should be deep rooted and produce large quantities of biomass that has accumulated pollutants. A plant that is promising for its potential to perform phytoremediation is *Thlaspicaerulescans* (alpine pennycress), a member of the broccoli and cabbage family that can grow on soils with high levels of zinc and cadmium and accumulate these metals in the plant shoots and leaves. Phytoremediation of uranium-contaminated soils is facilitated by adding citrate to the soil to mobilize the uranium in a form that can be taken up by plant roots.

16.7 FERTILIZERS

Crop fertilizers contain nitrogen, phosphorus, and potassium as major components. Magnesium, sulfate, and micronutrients may also be added. Fertilizers are designated by numbers, such as 6-12-8, showing the respective percentages of nitrogen expressed as N (in this case 6%), phosphorus as P_2O_5 (12%), and potassium as K_2O (8%). Farm manure corresponds to an approximately 0.5-0.24-0.5 fertilizer. The organic fertilizers such as manure must undergo biodegradation to release the simple inorganic species (NO_3^-, $H_xPO_4^{x-3}$, K^+) assimilable by plants.

Most modern nitrogen fertilizers are made by the Haber process, in which N_2 and H_2 are combined over a catalyst at temperatures of approximately 500°C and pressures up to 1000 atm:

$$N_2 + 3H_2 \rightarrow 2NH_3 \tag{16.18}$$

The anhydrous ammonia product has a very high nitrogen content of 82%. It may be added directly to the soil, for which it has a strong affinity because of its water solubility and formation of ammonium ion:

$$NH_3(g)(\text{water}) \rightarrow NH_3(aq), \tag{16.19}$$

$$NH_3(aq) + H_2O \rightarrow NH_4^+ + OH^- \tag{16.20}$$

Special equipment is required, however, because of the toxicity of ammonia gas. Aqua ammonia, a 30% solution of NH_3 in water, may be used with much greater safety. It is sometimes added directly to irrigation water. It should be pointed out that ammonia vapor is toxic and NH_3 is reactive with some substances. Improperly discarded or stored ammonia can be a hazardous waste.

Ammonium nitrate, NH_4NO_3, is a common solid nitrogen fertilizer. It is made by oxidizing ammonia over a platinum catalyst, converting the nitric oxide product to nitric acid and reacting the nitric acid with ammonia. The molten ammonium nitrate product is forced through nozzles at the top of a *prilling tower* and solidifies to form small pellets while falling through the tower. The particles are coated with a water repellent. Ammonium nitrate contains 33.5% nitrogen. Although convenient to apply to soil, it requires considerable care during manufacture and storage because it is explosive. Ammonium nitrate also poses some hazards. It is mixed with fuel oil to form an explosive that serves as a substitute for dynamite in quarry blasting and construction. This mixture was used to devastating effect in the dastardly bombing of the Oklahoma City Federal Building in 1995.

Urea,

$$H_2N-\overset{\overset{\displaystyle O}{\|}}{C}-NH_2$$

is easier to manufacture and handle than ammonium nitrate. It is now the favored solid nitrogen-containing fertilizer. The overall reaction for urea synthesis is

$$CO_2 + 2NH_3 \rightarrow CO(NH_2)_2 + H_2O \tag{16.21}$$

involving a rather complicated process in which ammonium carbamate, chemical formula $NH_2CO_2NH_4$, is an intermediate.

Other compounds used as nitrogen fertilizers include sodium nitrate (obtained largely from Chilean deposits, see Section 15.2), calcium nitrate, potassium nitrate, and ammonium phosphates. Ammonium sulfate, a byproduct of coke ovens, used to be widely applied as fertilizer. The alkali metal nitrates tend to make soil alkaline whereas ammonium sulfate leaves an acidic residue.

Phosphate minerals are found in several states, including Idaho, Montana, Utah, Wyoming, North Carolina, South Carolina, Tennessee, and Florida. The principal mineral is fluorapatite, $Ca_5(PO_4)_3F$. The phosphate from fluorapatite is relatively unavailable to plants, and fluorapatite is frequently treated with phosphoric or sulfuric acids to produce superphosphates:

$$2Ca_5(PO_4)_3F(s) + 14H_3PO_4 + 10H_2O \rightarrow 2HF(g) + 10Ca(H_2PO_4)_2{\cdot}H_2O \tag{16.22}$$

$$2Ca_5(PO_4)_3F(s) + 7H_2SO_4 + 3H_2O \rightarrow 2HF(g) + 3Ca(H_2PO_4)_2{\cdot}H_2O + 7CaSO_4 \tag{16.23}$$

The superphosphate products are much more soluble than the parent phosphate minerals. The HF produced as a byproduct of superphosphate production can create air pollution problems.

Phosphate minerals are rich in trace elements required for plant growth, such as boron, copper, manganese, molybdenum, and zinc. Ironically, these elements are lost to a large extent when the phosphate minerals are processed to make fertilizer. Ammonium phosphates are excellent, highly soluble phosphate fertilizers. Liquid ammonium polyphosphate fertilizers consisting of ammonium salts of pyrophosphate, triphosphate, and small quantities of higher polymeric phosphate anions in aqueous solution work very well as phosphate fertilizers. The polyphosphates are believed to have the additional advantage of chelating iron and other micronutrient metal ions, thus making the metals more available to plants.

Potassium fertilizer components consist of potassium salts, generally KCl. Such salts are found as deposits in the ground or may be obtained from some brines. Very large deposits are found in Saskatchewan, Canada. These salts are all quite soluble in water. One problem encountered with potassium fertilizers is the luxury uptake of potassium by some crops, which absorb more potassium than is really needed for their maximum growth. In a crop where only the grain is harvested, leaving the rest of the plant in the field, luxury uptake does not create much of a problem because most of the potassium is returned to the soil with the dead plant. However, when hay or forage is harvested, potassium contained in the plant as a consequence of luxury uptake is lost from the soil.

16.7.1 Fertilizer Pollution

One of the more troublesome problems that has arisen from increased fertilizer use in recent times is water pollution from agricultural land runoff enriched in nitrogen, phosphorus, and potassium from fertilizers. On a local scale, lakes and reservoirs have become eutrophied as the result of fertilizer runoff. The nutrients in fertilizer cause excessive algal growth, the algal biomass decays and consumes oxygen, and bodies of water are seriously damaged by oxygen depletion. As a nonpoint source of pollution, fertilizer runoff presents challenging control problems.

One of the more spectacular manifestations of agricultural fertilizer pollution is the Gulf of Mexico dead zone, which develops and grows in size each summer in a region of the Gulf receiving Mississippi River runoff. It is attributed to runoff of water enriched in agricultural fertilizers, especially nitrogen, from the fertile Mississippi river watershed, which enables excessive growth of photosynthetic microorganisms (phytoplankton) in the Gulf. The biomass of these organisms along with the bodies of zooplankton and their waste products that feed on them settle to the bottom of the Gulf where microbial decay results in depletion of DO (hypoxia) that kills fish and other marine animal life. Aggravating the effect of low oxygen is the production in anoxic sediments of toxic, odorous hydrogen sulfide. The formation of this zone is intensified by the layer of less dense freshwater from the Mississippi river that tends to prevent mixing with the seawater that it overlays, preventing penetration of oxygen to bottom layers. In 2002 the dead zone of eutrophied, oxygen-depleted water grew to 8500 square miles, the largest ever recorded. Because of severe flooding in agricultural regions of the Mississippi River watershed that washed large amounts of fertilizer residues into the river, the Gulf dead zone was projected to reach a record area in 2008, but was reduced to 8000 square miles by the churning effect of Hurricane Dolly. Efforts have been made to persuade agricultural interests to use less fertilizer to reduce the area of the dead zone. More frequent, lighter applications of fertilizer should also reduce runoff.

16.8 POLLUTANTS FROM LIVESTOCK PRODUCTION

As it is currently practiced, livestock production generates significant amounts of environmental pollutants. Livestock manure has a very high BOD and can rapidly deplete water of oxygen when it gets into waterways. Decomposition of animal waste products produces inorganic nitrogen that can contaminate water with potentially toxic nitrate. Inorganic nitrogen and phosphorus released to water from the decomposition of livestock wastes can cause eutrophication of water. Nitrous oxide, N_2O, released to the atmosphere from livestock waste degradation can be an air pollutant. Methane generated in the anaerobic degradation of livestock wastes is a potent greenhouse gas.

Several organoarsenic compounds, especially roxarsone,

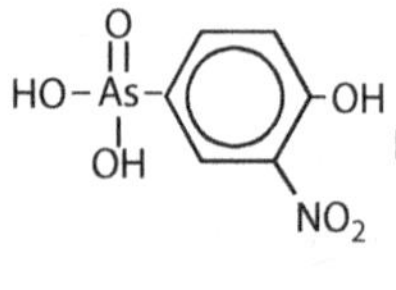

HO–As(=O)(OH)–C₆H₃(OH)(NO₂)

Roxarsone (3-nitro-4-hydroxy-phenylarsonic acid)

have been added to poultry feed to control disease (coccidiosis), enhance the efficiency of feed utilization, stimulate growth and egg production, and improve meat appearance. Roxarsone and its degradation products have been found in chicken litter and arsenic has been detected in soils fertilized with chicken litter. Although roxarsone has a relatively low toxicity, its inorganic biodegradation products are rather toxic. By 2008 arsenic compounds in chicken feed had been reduced significantly and some major producers had discontinued these additives.

16.9 PESTICIDES AND THEIR RESIDUES IN SOIL

The following are four major concerns regarding pesticides in soil that need to be considered with respect to their licensing and regulation:

- Carryover of pesticides and biologically active degradation products to crops grown in later seasons
- Biological effects on organisms in terrestrial and aquatic ecosystems including bioaccumulation and transfer through food chains

- Groundwater contamination
- Effects on soil fertility

Herbicides are the most common chemicals that influence soil and the organisms that it supports because to be effective they must generally come into direct contact with soil and persist long enough to be effective.

An important environmental characteristic of chemicals such as herbicides in soil is the formation of *bound residues*, which are parent compounds or metabolites in soil or in the organisms that inhabit soil and that are not extractable by common extraction procedures.[6] A common way of studying bound residues consists of spiking soil with compounds labeled with radioactive carbon-14 and subjecting the spiked soil to extraction. It is possible to measure the radiolabeled carbon in the extracts, in carbon dioxide respiration product, and in the soil residue. However, this kind of procedure does not necessarily reflect the bioavailability of a substance in soil and its related characteristics of biodegradation, bioaccumulation, or mineralization.

It is well established that as pesticides and other foreign compounds remain in soil for longer periods of time the following occur:

- The substances become increasingly resistant to extraction and desorption processes
- They become significantly less bioavailable to organisms
- Overall toxicity is decreased

These bioavailability and extractability effects are usually attributed to interaction of the contaminant compound with soil organic matter and are more pronounced for soils with high organic matter content. Molecules become trapped within micropores of the organic matter and are thus rendered less environmentally reactive. The apparent increase in binding with time may be due to slow migration of the molecules into more remote and smaller micropores and perhaps formation of covalent bonds to organic matter.

The bioavailability of foreign compounds in soil has traditionally been estimated by extraction of the materials from soil by various solvent compositions. However, it is generally believed that this overstates the environmental and biological availability of bound residues. As alternatives, tests involving microbial mineralization and uptake by earthworms using compounds radiolabeled with carbon-14 to follow the fate of residues have been used. The extent to which soil bacteria convert bound residues to inorganic carbon is a reasonably good measure of bioavailability. Uptake of bound residues by earthworms provides evidence of bioavailability to multicelled organisms capable of ingesting soil.

16.9.1 Soil Fumigants

Soil fumigants are volatile substances applied to soil to combat bacteria, fungi, nematodes, arthropods, and weeds, primarily on fields used to grow potatoes, tomatoes, strawberries carrots, and peppers. Because of fumigants' use on food crops and the potential for exposure of workers who tend and harvest these crops, the safety of these substances has been a matter of considerable concern. Structural formulas of the most common soil fumigants are shown in Figure 16.6. The most widely accepted soil fumigant has been methyl bromide, H_3CBr, but it was phased out of use in most industrialized countries including the United States in 2005 except for some very limited "critical use exemptions." Metam sodium is the most widely applied fumigant and the third most widely used pesticide in the United States, primarily on potato fields. In soil it breaks down to methyl isothiocyanate, which is the active agent (shown in Figure 16.6). In 2007 the U.S. EPA approved limited use of methyl iodide, H_3CI. Dimethyl disulfide is a relatively new fumigant that is commonly applied along with chloropicrin.

Although it has been assumed that because of their volatility soil fumigants do not persist in soil or cause groundwater pollution problems, studies with carbon-14 labeled fumigants has shown bound residues of these substances in soil. The levels of these residues in soil are strongly correlated with organic matter in the soil and they are apparently bound with fulvic acid (see discussion of

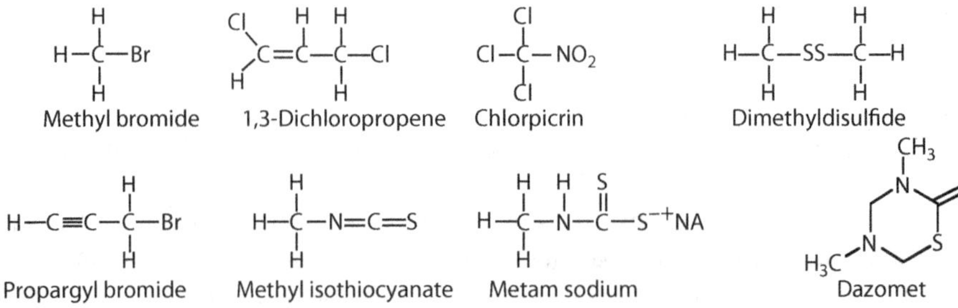

FIGURE 16.6 Soil fumigants. Methyl bromide was once the most commonly used soil fumigant, but it is now banned because of its potential to destroy stratospheric ozone.

humic substances in Section 3.17) in soil.[7] Soil fumigants are believed to act as toxicants to pests in soil by alkylating biological macromolecules including DNA and proteins. It has been suggested that the bound residues of soil fumigants form in soil as the result of alkylation of fulvic acid molecules, thus effectively removing and detoxifying the fumigants.

16.10 WASTES AND POLLUTANTS IN SOIL

Soil receives large quantities of waste products. Much of the sulfur dioxide emitted in the burning of sulfur-containing fuels ends up as soil sulfate. Atmospheric nitrogen oxides are converted to nitrates in the atmosphere, and the nitrates eventually are deposited on soil. Soil sorbs NO and NO_2, and these gases are oxidized to nitrate in the soil. Carbon monoxide is converted to CO_2 and possibly to biomass by soil bacteria and fungi. Particulate lead from automobile exhausts is found at elevated levels in soil along heavily traveled highways. Elevated levels of lead from lead mines and smelters are found on soil near such facilities.

Soil is the receptor of many hazardous wastes from landfill leachate, lagoons, and other sources (see Chapter 20). In some cases, land farming of degradable hazardous organic wastes is practiced as a means of disposal and degradation. The degradable material is worked into the soil, and soil microbial processes bring about its degradation. As discussed in Chapter 8, sewage and fertilizer-rich sewage sludge may be applied to soil.

The various constituents of soil have different affinities for organic contaminants. Natural organic matter, primarily humic substances, have a relatively high affinity for organic contaminants and heavy metal ions. Many soils contain elemental carbon, *black carbon*, material in the ash left over from the burning of crop residues, including sugarcane trash, wheat straw, and rice straw. This material is probably an important repository of organic contaminants in soil.

Volatile organic compounds (VOCs) such as benzene, toluene, xylenes, dichloromethane, trichloroethane, and trichloroethylene, may contaminate soil in industrialized and commercialized areas, particularly in countries in which enforcement of regulations is not very stringent. One of the more common sources of these contaminants is leaking underground storage tanks. Landfills built before current stringent regulations were enforced and improperly discarded solvents are also significant sources of soil VOCs.

Measurements of levels of PCBs in soils that have been archived for several decades provide interesting insight into the contamination of soil by pollutant chemicals and subsequent loss of these substances from soil.[8] Analyses of soils from the United Kingdom dating from the early 1940s to 1992 showed that the PCB levels increased sharply from the 1940s, reaching peak levels around 1970. Subsequently, levels fell sharply and now are back to early 1940s concentrations. This fall was accompanied by a shift in distribution to the more highly chlorinated PCBs, which was attributed by those doing the study to volatilization and long-range transport of the lighter PCBs away from the soil. These trends parallel levels of PCB manufacture and use in the United Kingdom from the early 1940s to the present. This is consistent with the observation that relatively high concentrations of PCBs have been observed in remote Arctic and sub-Arctic regions, attributed to condensation in colder climates of PCBs volatilized in warmer regions.

Some pollutant organic compounds are believed to become bound with humus during the humification process that occurs in soil. This largely immobilizes and detoxifies the compounds. Binding of pollutant compounds by humus is particularly likely to occur with compounds that have structural similarities to humic substances, such as phenolic and anilinic compounds illustrated by the following two examples:

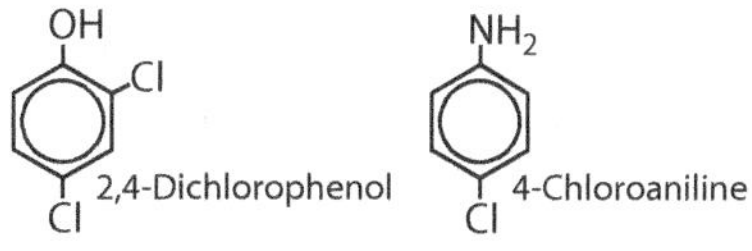

OH, Cl, Cl — 2,4-Dichlorophenol; NH_2, Cl — 4-Chloroaniline

Such compounds can become covalently bonded to humic substance molecules as bound residues.

Soil receives enormous quantities of pesticides as an inevitable result of their application to crops. The degradation and eventual fate of these pesticides on soil largely determines their ultimate environmental effects. Detailed knowledge of these effects are now required for licensing of a new pesticide (in the United States under the FIFRA). Among the factors to be considered are the sorption of the pesticide by soil; leaching of the pesticide into water, as related to its potential for water pollution; effects of the pesticide on microorganisms and animal life in the soil; and possible production of relatively more toxic degradation products.

Adsorption by soil is a key aspect of pesticide degradation and plays a strong role in the speed and degree of degradation. The degree of adsorption and the speed and extent of ultimate degradation are influenced by a number of other factors. Some of these, including solubility, volatility, charge, polarity, and molecular structure and size, are properties of the medium. Adsorption of a pesticide by soil components may have several effects. Under some circumstances, it retards degradation by separating the pesticide from the microbial enzymes that degrade it, whereas under other circumstances the reverse is true. Purely chemical degradation reactions may be catalyzed by adsorption. Loss of the pesticide by volatilization or leaching is diminished. The toxicity of a herbicide to plants may be reduced by sorption on soil. The forces holding a pesticide to soil particles may be of several types. Physical adsorption involves van der Waals forces arising from dipole–dipole interactions between the pesticide molecule and charged soil particles. Ion exchange is especially effective in holding cationic organic compounds, such as the herbicide paraquat,

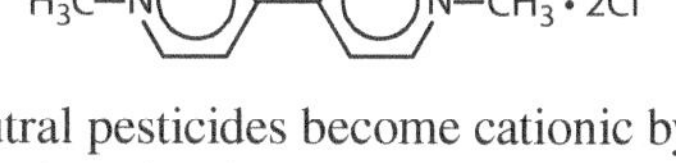

$$H_3C\overset{+}{-}N\langle\bigcirc\rangle\text{—}\langle\bigcirc\rangle N\overset{+}{-}CH_3 \cdot 2Cl^-$$

to anionic soil particles. Some neutral pesticides become cationic by protonation and are bound as the protonated positive form. Hydrogen bonding is another mechanism by which some pesticides are held to soil. In some cases, a pesticide may act as a ligand coordinating with metals in soil mineral matter.

The three primary ways in which pesticides are degraded in or on soil are *chemical degradation*, *photochemical reactions*, and, most important, *biodegradation*. Various combinations of these processes may operate in the degradation of a pesticide.

Chemical degradation of pesticides has been observed experimentally in soils and clays sterilized to remove all microbial activity. A number of purely chemical hydrolytic reactions of pesticides occur in soil. For example, clays have been shown to catalyze the hydrolysis of *o,o*-dimethyl-*o*-2,4,5-trichlorophenyl thiophosphate (also called trolene, ronnel, etrolene, or trichlorometafos), an effect attributed to –OH groups on the mineral surface:

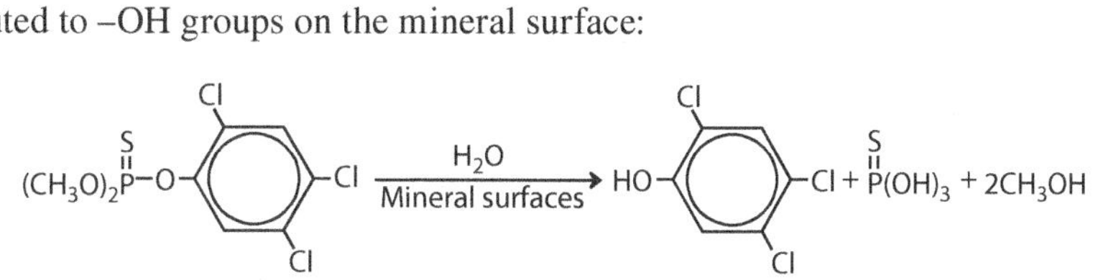

$$(CH_3O)_2\overset{S}{\overset{\|}{P}}\text{–O–}C_6H_2Cl_3 \xrightarrow[\text{Mineral surfaces}]{H_2O} \text{HO–}C_6H_2Cl_3 + \overset{S}{\overset{\|}{P}}(OH)_3 + 2CH_3OH \qquad (16.24)$$

A number of pesticides have been shown to undergo *photochemical reactions*, that is, chemical reactions brought about by the absorption of light (see Chapter 9). Frequently, isomers of the pesticides are formed as products. Many of the studies reported apply to pesticides in water or on thin films, and relatively less is known about the photochemical reactions of pesticides on soil and plant surfaces.

16.10.1 Biodegradation and the Rhizosphere

Although insects, earthworms, and plants may play roles in the *biodegradation* of pesticides and other pollutant organic chemicals, microorganisms have the most important role. Several examples of microorganism-mediated degradation of organic chemical species are given in Chapter 6.

The *rhizosphere*, the layer of soil in which plant roots are especially active, is a particularly important part of soil with respect to biodegradation of wastes. It is a zone of increased biomass and is strongly influenced by the plant root system and the microorganisms associated with plant roots. The rhizosphere may have more than 10 times the microbial biomass per unit volume compared to nonrhizospheric zones of soil. This population varies with soil characteristics, plant and root characteristics, moisture content, and exposure to oxygen. If this zone is exposed to pollutant compounds, microorganisms adapted to their biodegradation may also be present.

Plants and microorganisms exhibit a strong synergistic relationship in the rhizosphere, which benefits the plant and enables highly elevated populations of rhizospheric microorganisms to exist. Epidermal cells sloughed from the root as it grows and carbohydrates, amino acids, and root-growth-lubricant mucigel secreted from the roots all provide nutrients for microorganism growth. Root hairs provide a hospitable biological surface for colonization by microorganisms.

The biodegradation of a number of synthetic organic compounds has been demonstrated in the rhizosphere. Understandably, studies in this area have focused on herbicides and insecticides that are widely used on crops. Among the organic species for which enhanced biodegradation in the rhizosphere has been demonstrated are 2,4-D herbicide, parathion, carbofuran, atrazine, diazinon, volatile aromatic alkyl and aryl hydrocarbons, chlorocarbons, and surfactants.

16.11 SOIL LOSS AND DEGRADATION

Soil is a fragile resource that can be lost by erosion or become so degraded that it is no longer useful to support crops. Soil degradation is discussed here and measures to prevent and reverse it in Section 16.12.

The physical properties of soil and, hence, its susceptibility to erosion, are strongly affected by the cultivation practices to which the soil is subjected. *Desertification* refers to the process associated with drought and loss of fertility by which soil becomes unable to grow significant amounts of plant life. Desertification caused by human activities is a common problem globally, occurring in diverse locations such as Argentina, the Sahara, Uzbekistan, the U.S. Southwest, Syria, and Mali. It is a very old problem dating back many centuries to the introduction of domesticated grazing animals to areas where rainfall and groundcover were marginal. The most notable example is desertification aggravated by domesticated goats in the Sahara region. Desertification involves a number of interrelated factors, including erosion, climate variations, water availability, loss of fertility, loss of soil humus, and deterioration of soil chemical properties.

A related problem is *deforestation*, loss of forests. The problem is particularly acute in tropical regions, where the forests contain most of the existing plant and animal species. In addition to extinction of these species, deforestation can cause devastating deterioration of soil through erosion and loss of nutrients.

Soil erosion is the loss of soil by the action of both water and wind; water is the primary source of erosion. Millions of tons of topsoil are carried by the Mississippi River and swept from its mouth

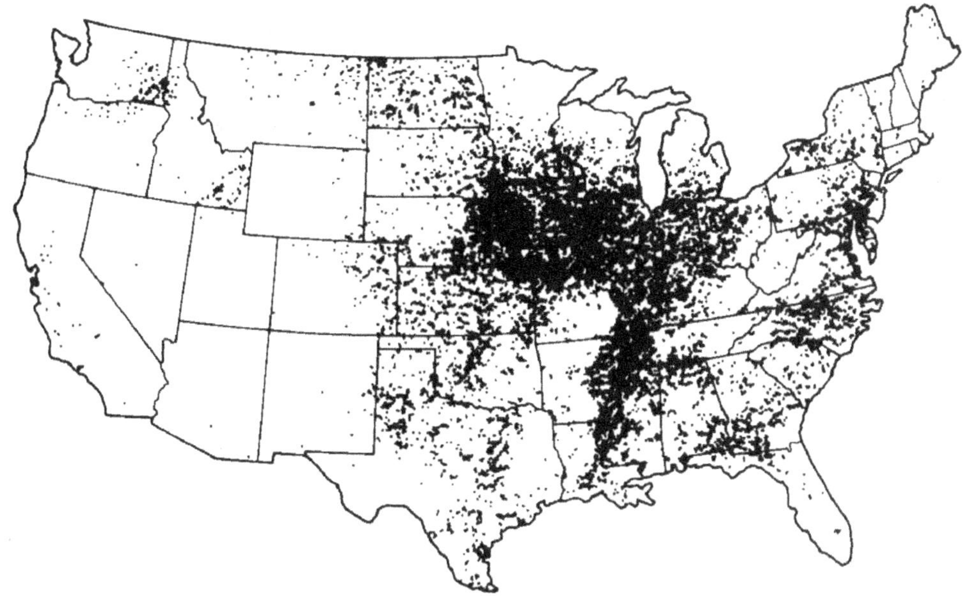

FIGURE 16.7 Pattern of soil erosion in the continental United States as of 1977. The dark areas indicate locations where the greatest erosion is occurring.

each year. About one-third of the U.S. topsoil has been lost since cultivation began in the continent. Figure 16.7 shows the pattern of soil erosion in the continental United States in 1977.

Wind erosion, such as occurs on the generally dry, high plains soils of eastern Colorado, poses another threat. After the Dust Bowl days of the 1930s, much of this land was allowed to revert to grassland, and the topsoil was held in place by the strong root systems of the grass cover. However, in an effort to grow more wheat and improve the sale value of the land, much of it was later returned to cultivation. Although freshly cultivated grassland may yield well for one or two years, the nutrients and soil moisture are rapidly exhausted and the land becomes very susceptible to wind erosion.

16.11.1 Soil Sustainability and Water Resources

The conservation of soil and the protection of water resources are strongly interrelated. Most freshwater falls initially on soil, and the condition of the soil largely determines the fate of the water and how much is retained in a usable condition. The land area upon which rainwater falls is called a *watershed*. In addition to collecting the water, the watershed determines the direction and rate of flow and the degree of water infiltration into groundwater aquifers (see the hydrologic cycle in Figure 3.1). Excessive rates of water flow prevent infiltration, lead to flash floods, and cause soil erosion. Measures taken to enhance the utility of land as a watershed also fortunately help prevent erosion. Some of these measures, which are discussed in Section 16.12, involve modification of the contour of the soil, particularly terracing, construction of waterways, and construction of water-retaining ponds. Waterways are planted with grass to prevent erosion, and water-retaining crops and bands of trees can be planted on the contour to achieve much the same goal. Reforestation and control of damaging grazing practices conserve both soil and water.

Covering soil with impermeable paving can prevent water infiltration to groundwater and increase the extent and rate of runoff. To help alleviate that problem Chicago has started an innovative "Green Alley" program in which alleys are paved with a porous concrete that allows water infiltration and eventual percolation into groundwater. This helps maintain groundwater levels and serves to reduce the burden on the stormwater collection system. Microorganisms that become embedded in the porous concrete serve to degrade organic matter including petroleum products that are carried by the infiltrating water.

16.12 SAVING THE LAND

With food as the most basic need that humans have, the sustainability of means to produce food is a top priority. The most basic part of food sustainability is the preservation of soil and its ability to support plant life, especially in the protection of soil from erosion. The preservation of soil from erosion is commonly termed *soil conservation*. There are a number of solutions to the soil erosion problem. Some are old, well-known agricultural practices, such as terracing, contour plowing (Figure 16.8), and periodically planting fields with cover crops, such as clover. For some crops conservation tillage (no-till agriculture) greatly reduces erosion. This practice consists of planting a crop among the residue of the previous year's crop, without plowing. Weeds are killed in the newly planted crop row by application of a herbicide prior to planting. The surface residue of plant material left on top of the soil prevents erosion.

Operators of at least one vinyard in the Napa Valley of California plant clover, oats, peas, and mustard as winter cover between the rows of vines. The clover and peas are legumes that fix atmospheric nitrogen and add to soil fertility. The crops are cut at the beginning of the summer growing season and the residues left on the soil to prevent weed growth, reduce erosion, and add fertility. One interesting aspect of the system is that perches are provided in the vinyards to attract hawks and owls that prey on rodents that live in the cut plant residues.

Another, more experimental, solution to the soil erosion problem is the cultivation of perennial plants, which develop large root systems and come up each spring after being harvested the previous fall. For example, a perennial corn plant has been developed by crossing corn with a distant, wild relative, teosinte, which grows in Central America. Unfortunately, the resulting plant does not give outstanding grain yields. It should be noted that an annual plant's ability to propagate depends upon producing large quantities of seeds, which is why plants harvested for their grain (seeds) are annual plants. In contrast, a perennial plant must develop a strong root system with bulbous growths called rhizomes, which store food for the coming year. However, it is possible that the application of genetic engineering may result in the development of perennial crops with good seed yields. The cultivation of such a crop would significantly reduce soil erosion.

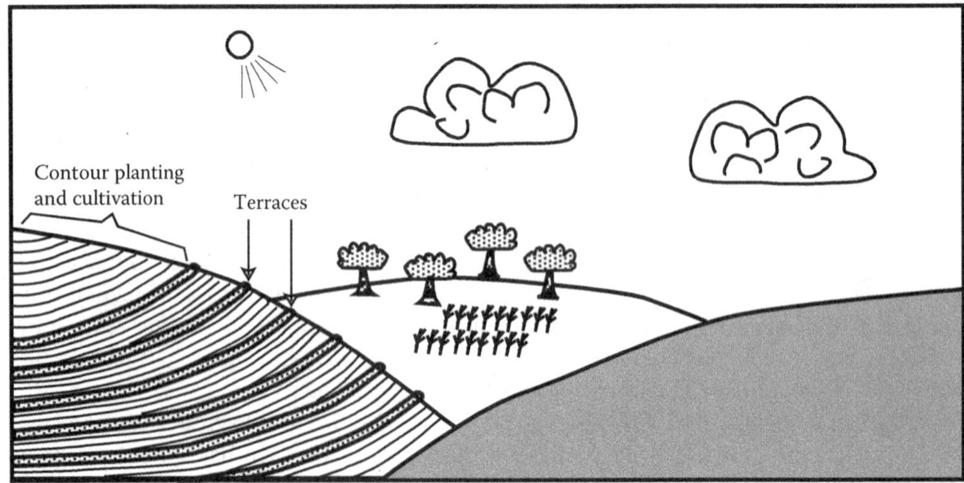

FIGURE 16.8 Construction of terraces on the contour of land and planting cops on the contour are practices that have been very effective in reducing soil erosion.

16.12.1 Agroforestry

The best known perennial plants are trees, which are very effective in stopping soil erosion. Wood from trees can be used as fuel, as raw material, for construction, and as food (glucose from wood cellulose, see below). There is a tremendous unrealized potential for an increase in the production of tree biomass. In the past, trees were often allowed to grow naturally with native varieties and without the benefit of any special agricultural practices, such as fertilization. The productivity of biomass from trees can be greatly increased with improved varieties, including those that are genetically engineered, and with improved cultivation and fertilization.

A promising alternative in sustainable agriculture is *agroforestry* in which crops are grown in strips between rows of trees as shown in Figure 16.9. The trees stabilize the soil, particularly on sloping terrain. By choosing trees with the capability to fix nitrogen, the system can be self-sufficient in this essential nutrient.

Cultivation called *alley cropping across the slope* (Figure 16.9) uses fast-growing, nitrogen-fixing trees to hold sloping soil in place. Spreading nutrient-rich tree prunings on the soil where crops are grown between crop seasons fertilizes the soil, adding organic matter and holding the soil in place. The trees potentially have economic value in providing wood for construction, firewood for cooking, fruit, and nuts. Genetically engineered trees may even provide high-value pharmaceuticals and specialty chemicals in the future. At the bottom of the slope, a buffer strip of grass can serve to filter nutrients and suspended soil from runoff from the fields. Potentially, rich topsoil collected by the buffer strip can be returned to higher levels to enrich the soil.

The most important use for wood is, of course, as lumber for construction. This use will remain important as higher energy costs increase the costs of other construction materials, such as steel, aluminum, and cement. Wood is about 50% cellulose, which can be hydrolyzed by enzyme processes to yield glucose sugar. The glucose can be used directly as food, fermented to ethyl alcohol for fuel (gasohol), or employed as a carbon and energy source for protein-producing yeasts. Given these and other potential uses, the future of trees as an environmentally desirable and profitable crop is very promising.

16.12.2 Soil Restoration

Soil can be impaired by loss of fertility, erosion, buildup of salinity, contamination by phytotoxins, such as zinc from sewage sludge, and other insults. Soil has a degree of resilience and can largely recover whenever the conditions leading to its degradation are removed. However, in many cases, more active measures called *soil restoration* are required to restore soil productivity, through the application of restoration ecology. Measures taken in soil restoration may include physical alteration of the soil to provide terraces and relatively flat areas not subject to erosion. Organic matter can be

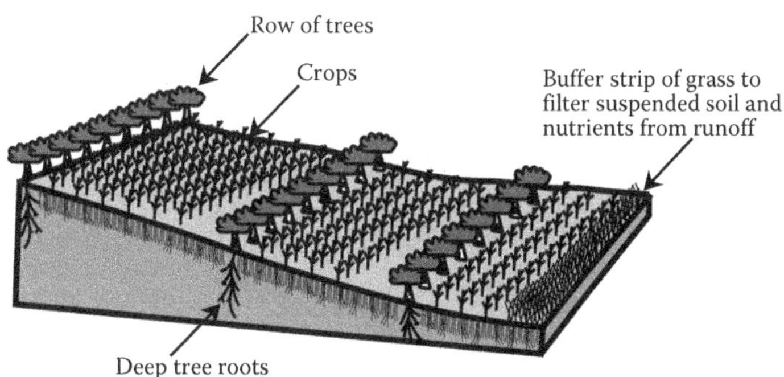

FIGURE 16.9 Alley cropping of crops between rows of trees running across sloping land can be an effective means of practicing agroforestry sustainably.

restored by planting crops the residues of which are cultivated into the soil for partially decayed biomass. Nutrients may be added and contaminants neutralized. Excess acid or base can be neutralized and salinity can be leached from the soil. Some toxic substances can be removed by plants in the process of phytoremediation (see Section 16.6). As the demand for food increases and damage to soil becomes more evident, soil restoration will become a very important endeavor.

16.13 GENETIC ENGINEERING AND AGRICULTURE

The nuclei of living cells contain the genetic instruction for cell reproduction. These instructions are in the form of a special material called deoxyribonucleic acid (DNA). In combination with proteins, DNA makes up the cell chromosomes. During the 1970s, the ability to manipulate DNA through genetic engineering became a reality, and since then it has become the basis of a major industry. Such manipulation falls into the category of recombinant DNA technology. Recombinant DNA gets its name from the fact that it contains DNA from two different organisms, recombined together. This technology promises some exciting developments in agriculture and, indeed, is expected to lead to a "second green revolution."

The first "green revolution" of the mid-1960s used conventional plant-breeding techniques of selective breeding, hybridization, cross-pollination, and back-crossing, to develop new strains of rice, wheat, and corn, which, when combined with chemical fertilizers, yielded spectacularly increased crop yields. For example, India's output of grain increased 50%. By working at the cellular level, however, it is now possible to greatly accelerate the process of plant breeding. Thus, plants may be developed that resist particular diseases, grow in seawater, or have much higher productivity. The possibility exists for developing entirely new kinds of plants.

One exciting possibility with genetic engineering is the development of plants other than legumes that fix their own nitrogen. For example, if nitrogen-fixing corn could be developed, the savings in fertilizer would be enormous. Furthermore, since the nitrogen is fixed in an organic form in plant root structures, there would be no pollutant runoff of chemical fertilizers.

Another promising possibility with genetic engineering is increased efficiency of photosynthesis. Plants utilize only about 1% of the sunlight striking their leaves, so there is appreciable room for improvement in that area.

Cell-culture techniques can be applied in which billions of cells are allowed to grow in a medium and develop mutants which, for example, might be resistant to particular viruses or herbicides or have other desirable qualities. If the cells with the desired qualities can be regenerated into whole plants, results can be obtained that might have taken decades using conventional plant-breeding techniques.

The United States has by far the largest land area devoted to genetically modified crops, 143 million acres of corn, soybean, cotton, canola, squash, papaya, and alfalfa in 2008. Ranking next were Argentina (47 million acres), Brazil (37 million acres), Canada (17 million acres), and India (15 million acres). China ranked sixth with 9 million acres of genetically modified crops but was planning a big increase in such crops in an effort to feed its huge population.

There has been strong resistance to transgenic crops and livestock in some quarters. This is especially true in Europe where restrictions on genetically modified foods and strict labeling requirements are in force. A particular concern with GM crops in Europe is the possibility of gene transfer to non-GM varieties such as might occur with cross-pollination of GM corn to other varieties of corn by means of corn pollen. Genes from GM crops could be transferred to wild plants that might act as particularly aggressive weeds.

Despite the enormous potential of the "green revolution," genetic engineering, and more intensive cultivation of land to produce food and fiber, these technologies cannot be relied upon to support an uncontrolled increase in world population, and may even simply postpone an inevitable day of reckoning with the consequences of population growth. Changes in climate resulting from global warming (greenhouse effect, Section 14.2), ozone depletion (by CFCs, Section 14.6), or natural disasters such as massive volcanic eruptions or collisions with large meteorites can, and almost certainly will, result in worldwide famine conditions in the future that no agricultural technology will be able to alleviate.

16.14 GREEN CHEMISTRY AND SUSTAINABLE AGRICULTURE

The practice of green chemistry may significantly enhance agricultural productivity and sustainability. To understand why this is so, it is useful to consider some of the problems that have arisen from spectacular advances in agricultural productivity with the application of advanced technologies over the last 100 years:

- Pesticides, herbicides, and fertilizers and their products have accumulated on agricultural lands and waters leading to adverse effects on wildlife, the environment, and potentially humans as well.
- Nontarget organisms have suffered and insect and weed pests have built up resistance to agents used in their eradication.
- Poorly trained and inadequately protected personnel in less developed countries have suffered adverse effects from modern agricultural products.
- Disposal problems have arisen with respect to obsolete pesticides.

The application of green chemistry to agriculture holds promise for preventing or alleviating problems such as these.

Agriculture is a science of living organisms applied to human needs for food and fiber production. So in attempting to find more sustainable and environment-friendly approaches to agriculture, it is reasonable to look to natural ecosystems that have evolved over the eons that enable various species of plants and animals to thrive. Such an approach is based upon *biomimetics* in which humans attempt to mimic natural life systems.

Pesticides that come from natural sources such as plants or bacteria are called *biopesticides*. These substances are usually more environmentally friendly than synthetic pesticides, although the blanket assumption that anything from a natural source is automatically safer than synthetic materials should not be made. Some substances, such as botulinum from *botulinus* bacteria or ricin from castor beans are among the most toxic substances known. In cases where genetic material is introduced into plants, there is the possibility that the material may get into wild species, such as weeds, with unintended consequences. There is also concern that proteins in biopesticides may be allergens. The three major categories of biopesticides are

- Microbial pesticides composed of microorganisms, such as insecticide-producing cells of *Bacillus thuringiensis* bacteria
- Plant-incorporated protectants, such as plants genetically engineered to make the insecticide produced by *Bacillus thuringiensis*
- Biochemical pesticides that control pests by nontoxic effects, such as sex attractants that confuse insects' reproductive activities

Advantages of biopesticides include a generally lower toxicity than conventional pesticides, high specificity for target pests, effectiveness in very small quantities, and rapid decomposition. Biopesticides are generally most effective when used in a program of integrated pest management.

A promising biomimetic approach is the use of naturally produced pesticides and pheromones in combating pests. Pheromones are substances produced by organisms, especially insects, in small quantities that are used for communication, especially with respect to reproduction. Because of their low toxicities, specificities for kinds of insects, and minute quantities required, pheromones are very environmentally friendly. Some insects can detect a single molecule of pheromone.

Some pheromones are relatively simple compounds. One example is methylsalicylate, which is released by some plants infested with insects such as aphids. This compound is an antiaphrodisiac released by male aphids during mating to make the females unattractive to other males, and release of this substance in the vicinity of aphid-infested crops interferes with aphid reproduction.

Another approach using pheromones is to place these materials in traps containing insecticides. The quantities of pheromones required are minute, as little as 50 mg per trap. For example, the diamondback moth, a cabbage pest, can be controlled with traps baited with its pheromone and containing permethrin insecticide.

Pheromone-based pest control has been found to be especially effective in control of moths, which are pests for a significant number of crops. One such pest is the gypsy moth, which is a serious problem for forest trees. Another is the coddling moth, which is the most serious pest in Washington State apple orchards.

One of the greatest deterrent to the widespread use of pheromones for insect control is cost. Recombinant DNA technologies are now being used to enable microorganisms such as yeasts to produce pheromones, and these technologies may significantly reduce costs of pheromones for pest control.

Efforts are under way to find natural products with insecticidal properties. The greatest success to date with natural products insecticides has been with the proteinaceous materials produced by strains of *Bacillus thuringiensis* (Bt). Consisting of a family of proteins that are selective for various insect pests, Bt destroys insect larvae by causing their digestive tracts to deteriorate, but it is not appreciably toxic to noninsect organisms and it breaks down in the environment. The genes for producing Bt have been spliced into corn and other crops by recombinant DNA techniques, and large fractions of the corn and cotton crops now grown in the United States and some other countries consist of these genetically modified strains.

Another pesticide from natural sources is Dow Agrosciences' Spinosad produced by the bacterium *Saccaropolyspora spinosa*. This substance is a neurotoxin to insect pests that afflict a number of economic pests including those that attack fruit, vegetables, trees, and cotton. Spinosad is nonvolatile, does not bioaccumulate, and has a low toxicity to mammals. Rohm and Haas have developed CONFIRM, a dicylhydrazine that mimics ecdysone, a hormone that initiates molting in insects. CONFIRM causes caterpillars to quit eating so that they die of dehydration and starvation. Another insecticide that operates by the same mechanism is azadirachtin, which is extracted from the kernels of the neem tree.

Development of a biofungicide from natural sources, AgraQuest's Serenade, was one of the winners of the 2003 Presidential Green Chemistry Challenge Awards. Serenade is a mixture of more than 30 lipopeptides such as agrastatin A (Figure 16.10) produced by a bacterium isolated from a California orchard and designated *Bacillus subtilis*, strain QST-713. Advantages of this fungicide include negligible toxicity to nontarget organisms and ready biodegradability. The active ingredients of the substance can be made by fermentation of natural materials, an inherently "green" process.

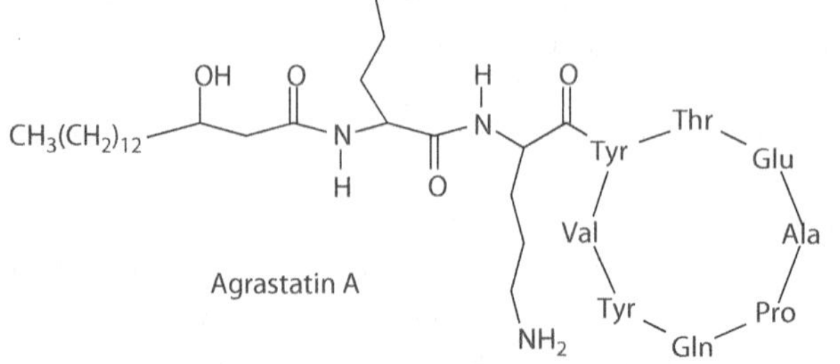

FIGURE 16.10 Agrastatin A, one of more than 30 lipopeptides contained in fungicidal serenade. The three-letter abbreviations in the ring structure represent amino acid residues in the peptide chain (see Figure 22.2).

Dow Agrosciences won the 2008 Presidential Green Chemistry Challenge Award for the development of Spinetoram, a chemical modification of spinosyns, which are oxic fermentation products of the soil bacterium, *Saccharopolyspora spinosa*. Winner of the 1999 Presidential Green Chemistry Award, Spinosad, described above, is a combination of spinosyns A and D, that is a very effective, safe, and environmentally benign insecticide for vegetables, but not very effective for insect control on tree fruits and tree nuts. Using relatively minor modifications of existing spinosyns, the developers of Spinetoram devised a material that retains the favorable environmental properties of Spinosad and is effective for the control of insect pests on fruit and nut trees. The mammalian acute oral toxicity of Spinetoram is only about 1/1000 that of azinphos-methyl and 1/44 that of phosmet, the two organophosphate insecticides that it is designed to replace.

Weeds are another major class of agricultural pest. Numerous herbicides have been developed to kill weeds. Probably one of the safest of these—and certainly one of the most widely used—is Monsanto's Roundup, glyophosate discussed in Section 16.1. A problem with Roundup and many other herbicides is that they may kill crops that they are used to protect. Soybeans, cotton, corn, and other crops have now been genetically engineered that are not harmed by Roundup, allowing its direct application to these crops to kill competing weeds, thus significantly reducing the amount of herbicide required.

Plants are susceptible to a variety of fungal, bacterial, and viral diseases. An interesting approach to combating such diseases is by activating plants' natural defense mechanisms. This has been done successfully with harpin, a protein produced by *Erwinia amylovora*, the bacterial agent of fire blight in apples and pears. *E. coli* bacteria have been genetically modified to produce harpin, which is sold as Eden Bioscience's Messenger biochemical pesticide and is reputed to enable plants to resist disease and generally thrive better.

Green chemistry can be applied to fertilizer utilization. A problem with urea, a favored fertilizer source of nutrient nitrogen is that it undergoes biologically mediated hydrolysis,

$$CO(NH_2)_2 + H_2O \rightarrow CO + 2NH_3 \tag{16.25}$$

leading to loss of as much as 30% of its nitrogen from evaporation of ammonia. Applied in relatively small quantities, *N*-(*n*-butyl)thiophosphorictriamide, a urease enzyme inhibitor, reduces the microbial hydrolysis of urea and increases the efficiency of urea fertilizer utilization. Thermal polyaspartate, a polyelectrolyte material formed by condensation and base treatment of the naturally occurring amino acid aspartic acid,

```
   O  H  H  O
   ‖  |  |  ‖
HO—C—C—C—C—OH
      |  |
      N  H   Aspartic acid
     / \
    H   H
```

and containing an abundance of carboxylate groups ($-CO_2^-$) groups has been found to be effective in stimulating plant uptake of fertilizer, thus reducing the amount of fertilizer required. This material has the advantage of coming from biological sources and being biodegradable.

16.15 AGRICULTURE AND HEALTH

Soil, the plants that grow on it, and the animals that consume these plants constitute an important connection between the geosphere and the biosphere and may influence health. An important connection is the incorporation into food of micronutrient elements essential for human health. One such nutrient (which is toxic at overdose levels) is selenium. It is definitely known that the health of animals is adversely affected in selenium-deficient areas as it is in areas of selenium excess. Human health might be similarly affected.

There are some striking geographic correlations with the occurrence of cancer, some of which may be due to soil type and its influence on food. A high incidence of stomach cancer occurs in areas with certain types of soils in the Netherlands, the United States, France, Wales, and Scandinavia. These soils are high in organic matter content, are acidic, and are frequently waterlogged.

One possible reason for the existence of "stomach cancer-producing soils" is the production of cancer-causing secondary metabolites by plants and microorganisms. Secondary metabolites are biochemical compounds that are of no apparent use to the organism producing them. It is believed that they are formed from the precursors of primary metabolites when the primary metabolites accumulate to excessive levels.

The role of soil in environmental health is not definitively known, and has not been as extensively studied as it deserves. The amount of research on the influence of soil in producing foods that are more nutritious and lower in content of naturally occurring toxic substances is quite small compared to research on higher soil productivity. It is hoped that the environmental health aspects of soil and its products will receive much greater emphasis in the future.

16.15.1 Food Contamination

Food is subject to contamination from faulty agricultural practices. The most common type of contamination is microbiological, often from bacteria transferred to food from human or animal wastes sources. In 2006, numerous cases including several fatalities due to hemolytic uremic syndrome were reported in the United States attributed to bagged spinach contaminated with *E. coli* O157:H7 bacteria. In 2008, the U.S. fresh tomato industry was devastated by erroneous reports of *Salmonella* contamination eventually attributed to peppers imported from Mexico. Some sources of peppers from Mexico as well as irrigation water used on the pepper crops were found to be contaminated with *Salmonella*.

The most common source of chemical contamination of food is from application of pesticides. Although large quantities of herbicides and soil fumigants are commonly applied to soil, insecticides are more likely to be found in food because to be effective they must usually be applied to plants. Sewage sludge applied to soil is a potential source of chemical contamination, particularly from phytotoxic zinc and heavy metals.

16.16 PROTECTING THE FOOD SUPPLY FROM ATTACK

As the basic provider of food that all humans need for their existence, the agricultural system is clearly of utmost importance and requires maximum protection from attack. In modern industrialized countries such as the United States, agriculture is highly specialized and now lacks the diversity which would otherwise protect it from, for example, infestation with pathogens to which widely grown varieties of crops are susceptible. Past disease outbreaks, such as the hoof and mouth disease infestation of livestock in England in 2001, serve as reminders of the vulnerability of the agricultural sector. Even a single incident, such as the finding of a cow with mad cow disease in Washington state in December 2003, can cause great anxiety and have significant economic repercussions.

Although chemical attack on agricultural systems is plausible, it is hard to imagine such an attack on a scale that would cause great damage. It is possible that crops could be sprayed with toxic chemicals before harvest, and, if these materials got into the food supply, could cause some anxiety and economic disruption. However, it would be difficult to mount such an attack in secrecy and on a scale required to do significant harm.

Infestation by crop-destroying insects could cause enormous loss and has done so throughout history. Huge swarms of locusts that devored all plant material in their paths are described in the Bible and still devastate agricultural areas. The boll weevil nearly wiped out the U.S. cotton crop during 1800s. Some of the most damaging insect infestations have come from exotic insect species introduced to areas where they previously were not found such as has been the case with beetle species that have killed millions of trees in U.S. forests.

Microbial diseases pose a significant threat to agriculture and could be used as attacking agents. Insofar as plants are concerned, the greatest microbial threats are from fungi. Fungi are notorious for forming huge numbers of very durable spores that could be readily spread. Microorganisms pose an even greater threat from exotic insect species introduced to areas where they previously were not found. This has threat to livestock. The outbreak of hoof and mouth disease in England that required destruction of tens of thousands of animals was mentioned above. Animals are subject to a number of viral and bacterial diseases. Perhaps the most notorious of these is anthrax, which can be spread to humans as well.

Perhaps the greatest terrorist threat to food supplies comes from potential contamination of edible products with pathogens. Human illnesses caused by bacterial contamination of food occur every year. Consumers were made ill and some died in the United States from *Salmonella*-contaminated peppers in 2008 and from peanuts contaminated by *Salmonella* from unsanitary handling conditions in 2008/2009. Terrorists could deliberately contaminate food with such pathogens, although it would be difficult to do so on a large enough scale to sicken significant numbers of people.

LITERATURE CITED

1. Stehouwer, R., Soil chemistry and the quality of humus, *Bio Cycle*, **45**, 41–46, 2004.
2. Martin, F., I. Garcia, M. Diez, M. Sierra, M. Simon, and C. Dorronsoro, Soil alteration by continued oxidation of pyrite tailings, *Applied Geochemistry*, **23**, 1152–1165, 2008.
3. Poschenrieder, C., B. Gunse, I. Corrales, and J. Barcelo, A glance into aluminum toxicity and resistance in plants, *Science of the Total Environment*, **400**, 356–368, 2008.
4. Hossain, S. A. M., Potential use of *Rhizobium* spp. to improve fitness of non-nitrogen-fixing plants, *Acta Agriculturae Scandinavica, Section B: Soil and Plant Science*, **58**, 352–358, 2008.
5. Memon, A. R. and P. Schroeder, Implications of metal accumulation mechanisms to phytoremediation, *Environmental Science and Pollution Research*, **16**, 162–175, 2009.
6. Gevao, B., K. C. Jones, K. T. Semple, A. Craven, and P. Burauel, Nonextractable pesticide residues in soil, *Environmental Science and Technology*, **37**, 138A–144A, 2003.
7. Xu, J. M., J. Gan, S. K. Papiernik, J. O. Becker, and S. R. Yates, Incorporation of fumigants into soil organic matter, *Environmental Science and Technology*, **37**, 1288–1291, 2003.
8. Alcock, R. E., A. E. Johnston, S. P. McGrath, M. L. Berrow, and K. C. Jones, Long-term changes in the polychlorinated biphenyl content of united kingdom soils, *Environmental Science and Technology*, **27**, 1918–1923, 1993.

SUPPLEMENTARY REFERENCES

Altieri, M. A., and C. I. Nicholls, *Biodiversity and Pest Management in Agroecosystems*, 2nd ed., Food Products Press, New York, 2004.

Baker, C. J. and K. E. Saxton, Eds, *No-Tillage Seeding in Conservation Agriculture*, 2nd ed., Cabi Publishing, Wallingford, UK, 2007.

Boardman, J. and J. Poesen, Eds, *Soil Erosion in Europe*, Wiley, Hoboken, NJ, 2006.

Brady, N. C. and R. R. Weil, *The Nature and Properties of Soils*, 15th ed., Pearson Education, Upper Saddle River, NJ, 2007.

Chesworth, W., Ed., *Encyclopedia of Soil Science*, Springer, Dordrecht, The Netherlands, 2008.

Clark, J. M. and H. Ohkawa, Eds, *Environmental Fate and Safety Management of Agrochemicals*, American Chemical Society, Washington, DC, 2005.

Coats, J. R. and H. Yamamoto, Eds, *Environmental Fate and Effects of Pesticides*, American Chemical Society, Washington, DC, 2003.

Conklin, A. R., *Introduction to Soil Chemistry: Analysis and Instrumentation*, Wiley, Hoboken, NJ, 2005.

Coyne, M. S. and J. A. Thompson, *Fundamental Soil Science*, Thomson Delmar Learning, Clifton Park, NY, 2006.

Eash, N. S., *Soil Science Simplified*, 5th ed., Blackwell Publishing, Ames, IA, 2008.

Essington, M. E., *Soil and Water Chemistry: An Integrative Approach*, Taylor & Francis/CRC Press, Boca Raton, FL, 2004.

Farndon, J., *Life in the Soil*, Blackbirch Press, San Diego, CA, 2004.

Ferry, N. and M. R. Angharad, Eds, *Environmental Impact of Genetically Modified Crops*, Gatehouse, Cambridge, MA, 2009.

Francis, C. A., R. P. Poincelot, and G. W. Bird, *Developing and Extending Sustainable Agriculture: A New Social Contract*, Haworth Food and Agricultural Products Press, New York, 2006.

Gan, J., Ed., *Synthetic Pyrethroids: Occurrence and Behavior in Aquatic Environments*, American Chemical Society, Washington, DC, 2008.

Gardiner, D. T. and R. W. Miller, *Soils in Our Environment*, 11th ed., Pearson/Prentice Hall, Upper Saddle River, NJ, 2008.

Gordon, S., Ed., *Critical Perspectives on Genetically Modified Crops and Food*, Rosen Publishing Group, New York, 2006.

Halford, N. G., Ed., *Plant Biotechnology: Current and Future Applications of Genetically Modified Crops*, Hoboken, NJ, 2006.

Horne, P. and J. Page, *Integrated Pest Management for Crops and Pastures*, Landlinks Press, Collingwood, Victoria, Australia, 2008.

Hunter, B. T., *Soil and Your Health: Healthy Soil is Vital to Your Health*, Basic Health Publications, North Bergen, NJ, 2004.

Lal, R., *Soil Degradation in the United States: Extent, Severity, and Trends*, CRC Press/Lewis Publishers, Boca Raton, FL, 2003.

Liang, G. H. and D. Z. Skinner, Eds, *Genetically Modified Crops: Their Development, Uses, and Risks*, Food Products Press, New York, 2004.

Maredia, K. M., D. Dakouo, and D. Mota-Sanchez, Eds, *Integrated Pest Management in the Global Arena*, CABI Publishers, New York, 2003.

Meiners, R. E. and B. Yandle, Eds, *Agricultural Policy and the Environment*, Rowan & Littlefield Publishers, Lanham, MD, 2003.

Morgan, R. P. C., *Soil Erosion and Conservation*, 3rd ed., Blackwell Publishing, Malden, MA, 2005.

Norris, R. F., P. Caswell-Chen, and M. Kogan, *Concepts in Integrated Pest Management*, Prentice Hall, Upper Saddle River, NJ, 2003.

Paul, E. A., *Soil Microbiology, Ecology, and Biochemistry*, 3rd ed., Academic Press, Boston, 2007.

Plaster, E. J., *Soil Science and Management*, 5th ed., Delmar Cengage Learning, Clifton Park, NJ, 2009.

Plimmer, J. R., Ed., *Encyclopedia of Agrochemicals*, Wiley-Interscience, New York, 2003.

Poincelot, R. P., *Sustainable Horticulture: Today and Tomorrow*, Prentice Hall, Upper Saddle River, NJ, 2004.

Reilly, J. M., Ed., *Agriculture: The Potential Consequences of Climate Variability and Change for the United States*, Cambridge University Press, New York, 2002.

Singer, M. J. and D. N. Munns, *Soils: An Introduction*, 6th ed., Pearson Prentice Hall, Upper Saddle River, NJ, 2006.

Sparks, D. L., *Environmental Soil Chemistry*, 2nd ed., Academic Press, Boston, 2003.

Tabatabai, M. A. and D. L. Sparks, Eds, *Chemical Processes in Soils*, Soil Science Society of America, Madison, WI, 2005.

Van Elsas, Jan Dirk, J. K. Jansson, and J. T. Trevors, Eds, *Modern Soil Microbiology*, 2nd ed., Taylor & Francis/CRC Press, Boca Raton, FL, 2007.

Wesseler, J. H. H., *Environmental Costs and Benefits of Transgenic Crops*, Springer, Dordrecht, The Netherlands, 2005.

White, R. E., *Principles and Practice of Soil Science: The Soil as a Natural Resource*, 4th ed., Blackwell Publishing, Malden, MA, 2006.

Wolf, B. and G. Snyder, *Sustainable Soil Science: Using Organic Matter to Improve Crop Production*, Food Products Press, New York, 2003.

QUESTIONS AND PROBLEMS

1. Give two examples of reactions involving manganese and iron compounds that may occur in waterlogged soil.
2. What temperature and moisture conditions favor the buildup of organic matter in soil?

3. "Cat clays" are soils containing a high level of iron pyrite, FeS_2. Hydrogen peroxide, H_2O_2, is added to such a soil, producing sulfate as a test for cat clays. Suggest the chemical reaction involved in this test.
4. What effect upon soil acidity would result from heavy fertilization with ammonium nitrate accompanied by exposure of the soil to air and the action of aerobic bacteria?
5. How many moles of H^+ ion are consumed when 200 kg of $NaNO_3$ undergo denitrification in soil?
6. What is the primary mechanism by which organic material in soil exchanges cations?
7. Prolonged waterlogging of soil does *not*: (a) increase NO_3^- production, (b) increase Mn^{2+} concentration, (c) increase Fe^{2+} concentration, (d) have harmful effects upon most plants, (e) increase production of NH_4^+ from NO_3^-.
8. Of the following phenomena, the one that eventually makes soil more basic is: (a) removal of metal cations by roots, (b) leaching of soil with CO_2-saturated water, (c) oxidation of soil pyrite, (d) fertilization with $(NH_4)_2SO_4$, (e) fertilization with KNO_3.
9. How many metric tons of farm manure are equivalent to 100 kg of 10-5-10 fertilizer?
10. How are the chelating agents that are produced from soil microorganisms involved in soil formation?
11. What specific compound is both a particular animal waste product and a major fertilizer?
12. What happens to the nitrogen/carbon ratio as organic matter degrades in soil?
13. To prepare a rich potting soil, a greenhouse operator mixed 75% "normal" soil with 25% peat. Estimate the CEC in meq/100 g of the product.
14. Explain why plants grown on either excessively acidic or excessively basic soils may suffer from calcium deficiency.
15. What are two mechanisms by which anions may be held by soil mineral matter?
16. What are the three major ways in which pesticides are degraded in or on soil?
17. Lime from lead mine tailings containing 0.5% lead was applied at a rate of 10 metric tons per acre of soil and worked in to a depth of 20 cm. The soil density was 2.0 g/cm. To what extent did this add to the burden of lead in the soil? There are 640 acres per square mile and 1609 m per mile.
18. Match the soil or soil-solution constituent in the left column with the soil condition described on the right, below:

1. High Mn^{2+} content in soil solution	A. "Cat clays" containing initially high levels of pyrite, FeS_2
2. Excess H^+	B. Soil in which biodegradation has not occurred to a great extent
3. High H^+ and SO_4^{2-} concentrations	C. Waterlogged soil
4. High organic content	D. Soil, the fertility of which can be improved by adding limestone

19. What are the processes occurring in soil that operate to reduce the harmful effects of pollutants?
20. Under what conditions do the reactions below occur in soil? Name two detrimental effects that can result from these reactions:

$$MnO_2 + 4H^+ + 2e^- \rightarrow Mn^{2+} + 2H_2O \quad \text{and} \quad Fe_2O_3 + 6H^+ + 2e^- \rightarrow 2Fe^{2+} + 3H_2O$$

21. What are four important effects of organic matter in soil?
22. How might irrigation water treated with fertilizer containing potassium and ammonia become depleted of these nutrients in passing through humus-rich soil?

17 Green Chemistry and Industrial Ecology

17.1 CHANGING THE BAD OLD WAYS

The chemical industry has come a long way in the approximately half century since the following was stated in *American Chemical Industry—A History*, by W. Haynes, Van Nostrand Publishers, 1954: "By sensible definition, any by-product of a chemical operation for which there is no profitable use is a waste. The most convenient, least expensive way of disposing of said waste—up the chimney or down the river—is the best." Fortunately, this barbaric attitude toward waste has been long regarded as completely wrong and unacceptable. Environmental chemistry is largely involved with the problems caused by improper discharges of pollutants from the anthrosphere to the other environmental spheres. This chapter deals largely with ways in which such problems can be avoided before environmental problems develop.

In recognition of the environmental effects of the chemical industry and related enterprises, many laws have been passed and implemented throughout the world to regulate chemical processes and products. These laws have emphasized dealing with environmental problems after they have occurred, a "command and control" approach. Compliance with environmental laws over the last several decades has involved the expenditure of more than a trillion dollars worldwide. These laws have unquestionably had very positive effects on environmental quality, have been effective in helping to save some species from extinction, and have improved human health and quality of life. However, necessary as it is, the regulatory approach to enhancing environmental quality has some definite deficiencies. Its effective implementation and maintenance have required legions of regulators and have resulted in the expenditure of vast sums for litigation that could be better used directly to enhance environmental quality. Especially from the viewpoint of those regulated, some regulations have appeared to be petty, poorly cost effective, and, in the worst cases, counterproductive.

A modern industrial society will always require regulations of various kinds to maintain environmental quality and even to ensure its continued existence. But, are there alternatives to some of the regulations? Most desirable are alternatives that help to ensure environmental quality by "natural," self-regulating means. In recent years, it has become increasingly evident that, at least to a certain extent, there are alternatives to a purely regulatory approach for the chemical industry and other enterprises that have potentially huge influences on the environment and sustainability.

One alternative to the regulatory approach to pollution control is provided in part by the practice of *industrial ecology*, which in its modern form can be traced to a 1989 article by Frosch and Gallopoulos.[1] Industrial ecology views industrial systems as interacting to mutual advantage in a way that minimizes environmental and sustainability impacts and that processes materials and energy with maximum efficiency and minimum waste in a manner analogous to the metabolism of matter and energy in natural ecosystems. Starting in the mid-1990s, *green chemistry* has developed as a dynamic and rapidly developing discipline dealing with the sustainable practice of chemistry. Green chemistry and industrial ecology are closely related and one cannot be practiced effectively without the other. This chapter addresses green chemistry and industrial ecology as essential disciplines in maintaining environmental quality.

17.2 GREEN CHEMISTRY

Green chemistry can be defined as *the sustainable practice of chemical science and technology within the framework of good practice of industrial ecology in a manner that is safe and nonpolluting and that consumes minimum amounts of materials and energy while producing little or no waste material and which minimizes the use and handling of hazardous substances and does not release such substances to the environment.*[2] The inclusion of industrial ecology in this definition carries with it a number of implications regarding minimum consumption of raw materials, maximum recycling of materials, minimum production of unusable by-products, and other environmentally friendly factors favorable to the maintenance of sustainability. Figure 17.1 illustrates the main aspects of green chemistry.

A key aspect of green chemistry is *sustainability*. Ideally, green chemistry is self-sustaining for several reasons. One of these is economic because green chemistry in its most developed form is less costly in strictly monetary terms than chemistry the way it has been traditionally practiced. Green chemistry is sustainable in terms of materials because of its minimal but highly efficient use of raw materials. And green chemistry is sustainable in terms of wastes, because it does not cause an intolerable accumulation of hazardous waste products.

In implementing the practice of green chemistry, two often-complementary approaches are the following.

- Use existing chemicals, but make them by environmentally benign syntheses.
- Substitute chemicals made by environmentally benign syntheses for existing chemicals.

Both approaches need to be used. And both approaches challenge the ingenuity of chemists and chemical engineers to come up with innovative solutions to environmental problems arising from the chemical industry.

17.2.1 Twelve Principles of Green Chemistry

Green chemistry is guided by *Twelve Principles of Green Chemistry*,[3] aspects of which are discussed in this chapter.

1. Design chemical products and processes to *prevent waste*. One of life's most common lessons is that it is better to not make a mess than to clean it up once made. Failure to follow

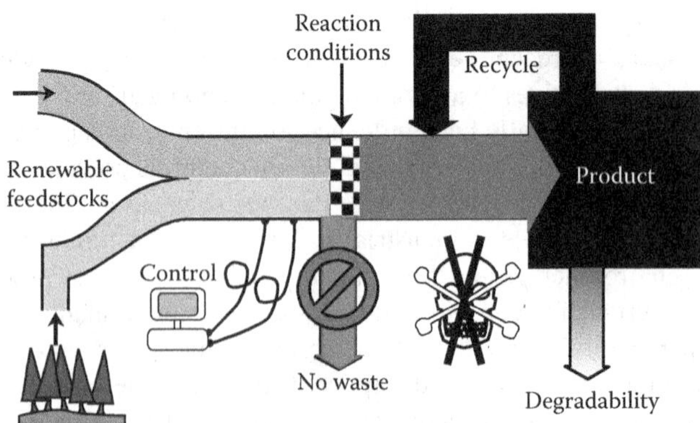

FIGURE 17.1 Illustration of the definition of green chemistry, which emphasizes renewable feedstocks, exacting control, mild reaction conditions, maximum recycling of materials, minimal wastes, and degradability of products that might enter the environment.

this simple rule has resulted in most of the troublesome chemical hazardous waste sites that are causing problems throughout the world today.

2. Design chemicals and products for *maximum safety* while still maintaining their effectiveness. The practice of green chemistry is making substantial progress in designing chemicals and new approaches to the use of chemicals such that effectiveness is retained and even enhanced while toxicity is reduced.
3. *Minimize hazards* in chemical synthesis using and generating substances with minimum toxicity and environmental hazard. Substances to be avoided when possible include toxic chemicals that pose health hazards to workers. They include substances that are likely to become air or water pollutants and harm the environment or organisms in the environment. In cases in which hazardous substances cannot be avoided, only the minimum amount needed should be used or produced and only at the time the substances are required. Here the connection between green chemistry and environmental chemistry is especially strong.
4. Use *renewable feedstocks* wherever possible. Raw materials extracted from Earth are depleting in that there is a finite supply that cannot be replenished after they are used. For these depleting feedstocks, recycling should be practiced to the maximum extent possible. Biomass feedstocks are highly favored in those applications for which they work.
5. *Use catalysts* for optimum conditions of chemical synthesis and minimum by-product. Reagents should be as selective as possible for their specific function.
6. *Avoid chemical derivatives* used as blocking agents and for other purposes in chemical synthesis to avoid generating excess by-products. In the synthesis of an organic compound it is often necessary to modify or protect groups on a molecule during the course of the synthesis. This often results in the generation of by-products not incorporated into the final product, such as occurs when a protecting group is bonded to a specific location on a molecule, then removed when protection of the group is no longer needed. Since these processes generate by-products that may require disposal, they should be avoided if possible.
7. *Maximize atom economy*: One of the most effective ways to prevent generation of wastes is to make sure that the maximum fraction of materials involved in making a product are incorporated into the final product. Therefore, the practice of green chemistry is largely about incorporation of all raw materials into the product, if at all possible. The degree to which this is accomplished is termed atom economy.
8. *Use safer reaction media*: Chemical synthesis as well as many manufacturing operations make use of auxiliary substances that are not part of the final product. In chemical synthesis, such a substance consists of solvents in which chemical reactions are carried out. Another example consists of separating agents that enable separation of product from other materials. Since these kinds of materials may end up as wastes or (in the case of some volatile, toxic solvents) pose health hazards, their use should be minimized and preferably avoided.
9. *Increase energy efficiency*: One way in which this can be done is to run reactions under mild temperature and pressure conditions, which also increases safety. Energy consumption poses economic and environmental costs in virtually all synthesis and manufacturing processes. In a broader sense, the extraction of energy, such as fossil fuels pumped from or dug out of the ground, has significant potential to damage the environment. One successful approach to energy use under mild conditions has been the application of biological processes, which, because of the conditions under which organisms grow, must occur at moderate temperatures and in the absence of toxic substances.
10. *Design for degradability* of chemicals and products to produce innocuous products upon degradation. This requires careful consideration of the downstream fates of products taking account of their environmental chemistry.

11. Use *in-process real-time monitoring and control* to reduce wastes and pollution, maximize safety, and minimize energy consumption. Exacting "real-time" control of chemical processes is essential for efficient, safe operation with minimum production of wastes. This goal has been made much more attainable by modern computerized controls.
12. *Minimize the potential for accidents* by designing processes that use materials unlikely to cause explosions, fires, and harmful releases. Accidents, such as spills, explosions, and fires, are a major hazard in the chemical industry. Not only are these incidents potentially dangerous in their own right, they tend to spread toxic substances into the environment and increase exposure of humans and other organisms to these substances.

Green chemistry is sustainable chemistry. There are several important respects in which green chemistry is sustainable.

- *Economic*: At a high level of sophistication green chemistry normally costs less in strictly economic terms (to say nothing of environmental costs) than chemistry as it is normally practiced.
- *Materials*: By efficiently using materials, maximum recycling, and minimum use of virgin raw materials, green chemistry is sustainable with respect to materials.
- *Waste*: By reducing insofar as possible, or even totally eliminating their production, green chemistry is sustainable with respect to wastes.

17.3 REDUCTION OF RISK: HAZARD AND EXPOSURE

A major goal in the manufacture and use of commercial products, and, indeed, in practically all areas of human endeavor, is the reduction of risk. Much of the design and practice of green chemistry is about risk reduction. There are two major aspects of risk—the *hazard* presented by a product or process and *exposure* of humans or other potential targets to those hazards:

$$\text{Risk} = F\{\text{hazard} \times \text{exposure}\} \tag{17.1}$$

This relationship simply states that risk is a function of hazard times exposure. It shows that risk can be reduced by a reduction of hazard, a reduction of exposure, and various combinations of both.

The command and control approach to reducing risk has concentrated upon reduction of exposure. Such efforts have used various kinds of controls and protective measures to limit exposure. The most common example of such a measure in the academic chemistry laboratory is the wearing of goggles to protect the eyes. Goggles will not by themselves prevent acid from splashing into the face of a student, but they do prevent the acid from contacting fragile eye tissue. Explosion shields will not prevent explosions, but they do retain glass fragments that might harm the chemist or others in the vicinity.

Reduction of exposure is unquestionably effective in preventing injury and harm. However, it does require constant vigilance and even nagging of personnel, as any laboratory instructor charged with making laboratory students wear their safety goggles at all times will attest. It does not protect the unprotected, such as a visitor who may walk bare faced into a chemical laboratory ignoring the warnings for required eye protection. On a larger scale, protective measures may be very effective for workers in a chemical manufacturing operation but useless to those outside the area or the environment beyond the plant walls who do not have protection. Protective measures are most effective against acute effects, but less so against long-term chronic exposures that may cause toxic responses over many years' period of time. Finally, the protective equipment can fail and there is always the possibility that humans will not use it properly.

Where feasible, hazard reduction is a much more certain way of reducing risk than is exposure reduction. The human factors that play so prominently in successfully limiting exposure and that require a conscious, constant effort are much less crucial when hazards have been reduced. Compare, for example, the use of a volatile, flammable, somewhat toxic organic solvent employed for cleaning and degreasing of machined metal parts with that of a water solution of a nontoxic cleaning agent used for the same purpose. To safely work around the solvent requires an unceasing effort and constant vigilance to avoid hazards such as formation of explosive mixtures with air, presence of ignition sources that could result in a fire, and excessive exposure by inhalation or absorption through skin that might cause peripheral neuropathy (a nerve disorder) in workers. Failure of protective measures can result in a bad accident or serious harm to worker health. The water-based cleaning solution, however, would not present any of these hazards so that failure of protective measures would not create a problem.

Normally, measures taken to reduce risk by reducing exposure have an economic cost that cannot be reclaimed in lower production costs or enhanced value of product. Of course, failure to reduce exposure can have direct, high economic costs in areas such as higher claims for worker compensation. In contrast, hazard reduction often has the potential to substantially reduce operating costs. Safer feedstocks are often less costly as raw materials. The elimination of costly control measures can lower costs overall. Again, to use the comparison of an organic solvent to a water-based cleaning solution, the organic solvent is almost certain to cost more than the aqueous solution containing relatively low concentrations of detergents and other additives. Although the organic solvent will at least require purification for recycle and perhaps even expensive disposal as a hazardous waste, the water solution may be purified by relatively simple processes, and perhaps even biological treatment, then safely discharged as wastewater to a municipal wastewater treatment facility. It should be kept in mind, however, that not all low-hazard materials are cheap, and may be significantly more expensive than their more hazardous alternatives. And, in some cases, nonhazardous alternatives simply do not exist.

Since hazardous substances manifest their hazards largely through their chemical reactions and characteristics, it is convenient to classify them chemically. Although the chemical variability of hazardous substances make such a classification system somewhat inexact, several categories can be defined based on chemical behavior. These are the following:

- *Combustible* and *flammable* substances, strong reducers that burn readily or violently in the presence of atmospheric oxygen
- *Oxidizers* that provide oxygen for the combustion of reducers
- *Reactive* substances that are likely to undergo rapid, violent reactions, often in an unpredictable manner
- *Corrosive* substances that are generally sources of H^+ ion or OH^- ion and that tend to react destructively with materials, particularly metals

Some hazardous substances fall into more than one of these groups, which increases the dangers that they pose.

Often the greatest concern with hazardous substances has to do with toxicity. Toxic substances are not so easy to classify in terms of chemical properties as are substances belonging to the classifications listed above. It is more appropriate to classify toxic substances on the basis of their biochemical properties. Of special use in making these classifications are *structure–activity relationships* that relate known structural features and functional groups to likely toxic effects.

Three kinds of hazardous substances stand out as candidates for reduction in the practice of green chemistry. The first of these consists of *heavy metals*, such as lead, mercury, or cadmium. As elements, these substances are indestructible. They have a wide range of adverse biological effects. Another category consists of *persistent, nonbiodegradable organic materials*, such as PCBs. Often not extremely toxic, these substances persist in the environment and exhibit a tendency to become magnified through biological food chains, adversely affecting organisms at or near the end of the food chain. The classic example of such compounds is insecticidal DDT, which caused reproductive problems for birds, such as falcons or eagles, at the top of the food chain. A third category of troublesome hazardous substances consists of *volatile organic compounds* (VOCs). These have been particularly prevalent in industrial settings because of their uses as solvents for organic reactions, vehicles in paints and coatings, and for cleaning parts. In the latter two applications, the most convenient means of dealing with these volatile materials was to allow them to evaporate, so that large quantities were simply discharged to the atmosphere.

17.3.1 The Risks of Not Taking Risks

There are limits to the reduction in risk beyond which efforts to do so become counterproductive. As in other areas of endeavor, there are circumstances in which there is no choice but to work with hazardous substances. Some things that are inherently dangerous are rendered safe by rigorous training, constant attention to potential hazards, and understanding of hazards and the best way to deal with them. Consider the analogy of commercial flight. When a large passenger aircraft lands, 100 ton of aluminum, steel, flammable fuel, and fragile human flesh traveling at a speed of twice the legal interstate speed limits for automobiles come into sudden contact with an unforgiving concrete runway. That procedure is inherently dangerous! But it is carried out hundreds of thousands of times per year throughout the world with but few injuries and fatalities, a tribute to the generally superb design, construction, and maintenance of aircraft and the excellent skills and training of aircrew. The same principles that make commercial air flight generally safe also apply to the handling of hazardous chemicals by properly trained personnel under carefully controlled conditions.

So, although much of this book is about risk reduction as it relates to chemistry, we must always be mindful of the risks of not taking risks. If we become so timid in all of our enterprises that we refuse to take risks, scientific and economic progress will stagnate. If we get to the point that no chemical can be made if its synthesis involves the use of a potentially toxic or otherwise hazardous substance, the progress of chemical science and the development of such beneficial products as new life-saving drugs or innovative chemicals for treating water pollutants will be held back. Many argue that nuclear power entails significant risks as an energy source and that development of this power source must therefore be stopped. But, balance that potential risk against the virtually certain risk of continuing to use fossil fuels that produce greenhouse gases that cause global climate warming, and a strong case can be made for continued development of nuclear energy with newer reactor designs that incorporate strong safety features. Another example is the use of thermal processes for treating hazardous wastes, somewhat risky because of the potential for the release of toxic substances or air pollutants, but still the best way to convert many kinds of hazardous wastes to innocuous materials.

17.4 WASTE PREVENTION AND GREEN CHEMISTRY

One of the basic goals of the practice of green chemistry is to reduce waste. Waste prevention is better than having to treat or clean up wastes. In the earlier years of chemical manufacture the direct costs associated with producing large quantities of wastes were very low because such wastes were simply discarded into waterways, onto the ground, or in the air as stack emissions. With the passage and enforcement of environmental laws after about 1970, costs for waste treatment increased steadily. One of the most commonly cited examples of costly waste remediation is the cleanup of PCBs in Hudson River (New York) sediment from PCBs that were discharged into the river as by-products of electrical equipment manufacture. General Electric has agreed to remove PCBs from sediments along 65 km of the river requiring the movement of about 2.5 million cubic meters of material at a cost now estimated to be around $700 million. The cleanup of pollutants including asbestos, dioxins, pesticide manufacture residues, perchlorate, and mercury are costing various concerns hundreds of millions of dollars. From a purely economic standpoint, therefore, a green chemistry approach that avoids these costs is very attractive, in addition to its large environmental benefits.

Although the costs of such things as engineering controls, regulatory compliance, personnel protection, wastewater treatment, and safe disposal of hazardous solid wastes have certainly been worthwhile for society and the environment, they have become a large fraction of the overall cost of doing business. Companies must now do *full cost accounting*, which includes costs of emissions, waste disposal, cleanup, and protection of personnel and the environment. In industrialized countries, costs of complying with environmental and occupational health regulations are of a magnitude similar to those of research and development for industry as a whole.

Considering the economic costs, alone, waste prevention must have a high priority in addition to its being dictated by environmental considerations. Therefore, a top priority in the practice of green chemistry is to avoid producing wastes. In the past, this largely meant producing wastes as usual, then processing them to reduce their hazards or to salvage something of value from them. With the proper practice of green chemistry, it means designing and operating integrated systems which, by their nature, do not produce wastes.

17.5 GREEN CHEMISTRY AND SYNTHETIC CHEMISTRY

Synthetic chemistry is the branch of chemical science involved with developing means of making new chemicals and developing improved ways of synthesizing existing chemicals. A key aspect of green chemistry is the involvement of synthetic chemists in the practice of environmental chemistry. Synthetic chemists, whose major objective has always been to make new substances and to make them cheaper and better, have come relatively late to the practice of environmental chemistry and green chemistry. Other areas of chemistry have been involved much longer in pollution prevention and environmental protection. From the beginning, analytical chemistry has been a key to discovering and monitoring the severity of pollution problems. Physical chemistry has played a strong role in explaining and modeling environmental chemical phenomena. The application of physical chemistry to atmospheric photochemical reactions has been especially useful in explaining and preventing harmful atmospheric chemical effects including photochemical smog formation and stratospheric ozone depletion. Other branches of chemistry have been instrumental in studying various environmental chemical phenomena. Now synthetic chemists, those who make chemicals and whose activities drive chemical processes, have become much more involved in making the manufacture, use, and ultimate disposal of chemicals as environmentally friendly as possible.

Before environmental and health and safety issues gained their current prominence, the economic aspects of chemical manufacture and distribution were relatively simple. The economic factors involved included costs of feedstock, energy requirements, and marketability of product. Now, however, costs must include those arising from regulatory compliance, liability, end-of-pipe waste treatment, and costs of waste disposal. By eliminating or greatly reducing the use of toxic or hazardous feedstocks, catalysts, and reaction media, and by avoiding the generation of dangerous intermediates and by-products, green chemistry eliminates or greatly diminishes the additional costs that have come to be associated with meeting environmental and safety requirements of conventional chemical manufacture.

As illustrated in Figure 17.2, there are two general and often complementary approaches to the implementation of green chemistry in chemical synthesis, both of which challenge the imaginations of chemists and chemical engineers. The first of these is to use existing feedstocks but make them by more environmentally benign processes. The second approach is to substitute other feedstocks that are made by environmentally benign approaches. In some cases, a combination of the two approaches is used.

17.5.1 YIELD AND ATOM ECONOMY

Traditionally, synthetic chemists have used *yield*, defined as a percentage of the degree to which a chemical reaction or synthesis goes to completion, to measure the success of a chemical synthesis. For example, if a chemical reaction shows that 100 g of product should be produced, but only 85 g is produced, the yield is 85%. A synthesis with a high yield may still generate significant quantities

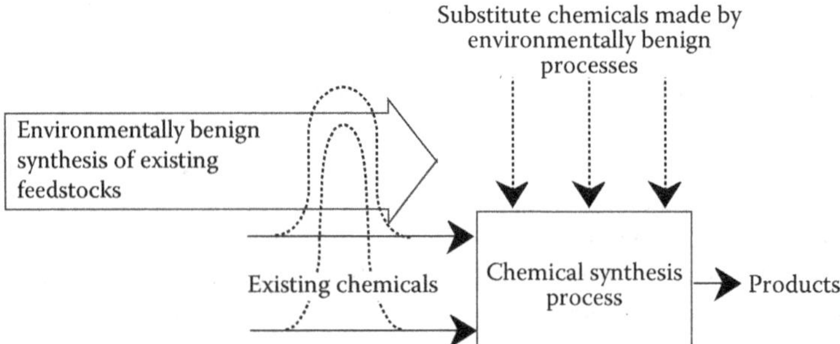

FIGURE 17.2 Two general approaches to the implementation of green chemistry. The dashed loops on the left represent alternative approaches to environmentally benign means of providing chemicals already used for chemical synthesis. A second approach, where applicable, is to substitute entirely different, environmentally safer raw materials.

of useless by-products if the reaction does so as part of the synthesis process. Instead of yield, green chemistry emphasizes *atom economy*, the fraction of reactant material that actually ends up in final product. Atom economy is expressed by the equation

$$\text{Atom economy} = \frac{\text{Molar mass desired product}}{\text{Total molar mass of materials generated}} \tag{17.2}$$

With 100% atom economy, all of the material that goes into the synthesis process is incorporated into the product. For efficient utilization of raw materials, a 100% atom economy process is most desirable. Figure 17.3 illustrates the concepts of yield and atom economy.

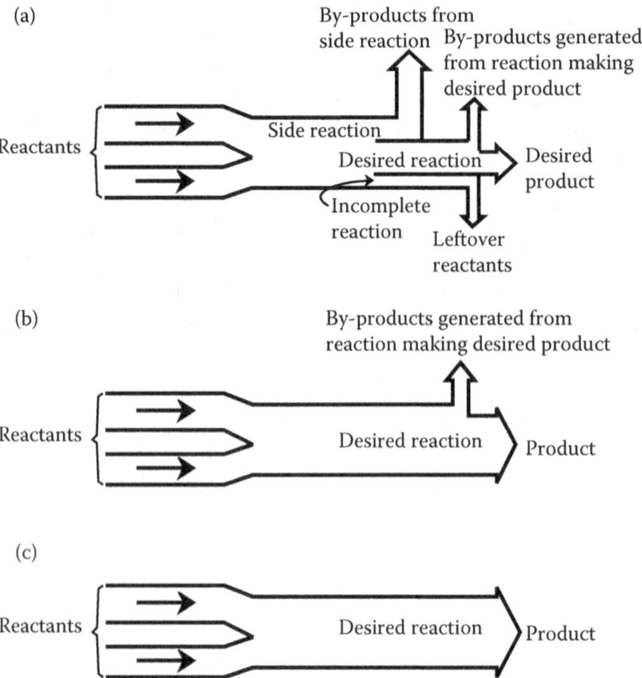

FIGURE 17.3 Illustration of percent yield and atom economy: (a) typical reaction with less than 100% yield and with by-products; (b) reaction with 100% yield, but with by-products inherent to the reaction; (c) reaction with 100% atom economy, no leftover reactants, no by-products.

Another quantitative measure of green chemistry is the E-factor based on waste produced relative to the amount of product.[4] In its simplest form the mathematical expression of the E-factor is the following:

$$\text{E-factor} = \frac{\text{Total mass of waste}}{\text{Total mass of product}} \tag{17.3}$$

17.6 FEEDSTOCKS

A crucial decision that must be made in implementing a process to manufacture a chemical or product is selection of *feedstock* from which the product is made. To the extent possible, the feedstock should place minimal demands on the earth's resources. It should be a relatively safe material, and its acquisition and refining should be safe. In some cases the processes and reagents required to isolate an otherwise safe feedstock make its acquisition hazardous, such as in the use of highly toxic cyanide to remove low levels of gold from ore. Wherever possible, a feedstock should be *renewable*; for example, feedstocks from biomass that can be grown repeatedly are often preferable to depletable petroleum feedstocks.

In a generalized sense, the overall process of obtaining a useful product from a feedstock, such as petroleum or biological materials, can be divided into the three categories shown in Figure 17.4. Both the petroleum and biological sources of potential feedstocks are relatively well known. For petroleum feedstocks, the technology for separation has been developed to a high degree of sophistication. The technologies for obtaining raw materials from plant sources, such as extraction of oils with solvents or separation of cellulose from lignin in wood, are also well advanced. Because of the high degree of development of the petrochemical industry, the science of converting petroleum feedstocks to desired products is very well developed, but is less so in the case of biological feedstocks.

17.6.1 Biological Feedstocks

With some billions of tons of carbon fixed as biomass each year, there exists an enormous potential for the use of biological materials as feedstocks. The most obvious such material consists of wood from trees. Large quantities of cellulose are generated each year in the production of crops such as corn and wheat. Although a fraction of crop by-product biomass should be returned to soil to maintain its condition as a growth medium for plants, there is a substantial excess that could be used as feedstock. Biological processes, particularly plant growth, produce a number of potentially useful biopolymers including, in addition to cellulose, hemicellulose, starch, lignin, lipids, and proteins. Plants are also useful sources of smaller molecules, including monosaccharides (glucose),

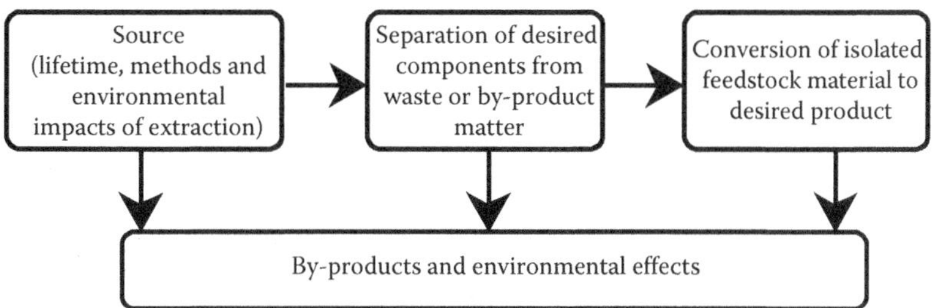

FIGURE 17.4 The three major steps in obtaining a feedstock and converting it to a useful product. Each of these steps has environmental implications and can benefit from the application of the principles of green chemistry.

disaccharides (sucrose), amino acids, waxes, fats, oils, and terpene hydrocarbons, including those used to make natural rubber. The potential of genetic engineering to produce plants that have high yields of feedstock chemicals could lead to the development of exciting new sources of biological feedstocks.

Potential feedstocks from biological sources tend to be more complex than those from petroleum. This offers the advantage of starting with a material in which much of the synthesis required to make a product has already been done by a living plant. Furthermore, many desired products have relatively high oxygen contents, and biological materials tend to contain bound oxygen. This can avoid the operation of converting a petroleum-based hydrocarbon feedstock to an oxygenated compound, which often requires severe conditions, hazardous oxidizing agents, and potentially troublesome catalysts. In some cases, however, the complexity of biological feedstocks is a disadvantage because it is usually more difficult to convert a complex molecule to a significantly different one than it is when starting with a relatively simple feedstock.

Biological feedstocks are often complex mixtures of materials that require separation, processing, and purification. Factories that perform these functions on biological feedstocks are *biorefineries*.[5] Biorefineries should be designed and operated in keeping with the best practice of green chemistry. This would include such things as using fermentation-produced ethanol or supercritical carbon dioxide (see Section 17.9) as solvents, the use of biocatalysts where possible, avoidance of severe conditions, and high efficiency of energy utilization.

Arguably the most important biological raw materials for chemical synthesis are carbohydrates produced by plants, including glucose, fructose, sucrose, starch, and

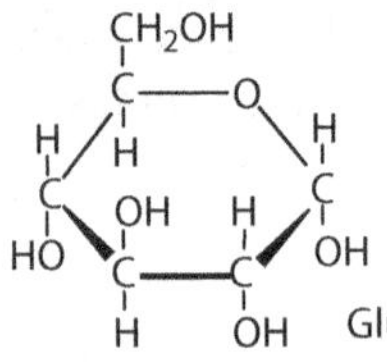

CH_2OH (structural formula)

Glucose

cellulose. Starch and cellulose are large-molecule polymers of glucose, the structural formulas of which are represented in Chapter 22, Figures 22.4 and 22.5. Carbohydrates, such as starch from corn grain and sucrose from sugar cane, are generated in high quantities by a number of plants. Furthermore, cellulosic wastes can be hydrolyzed to generate simple sugars that can be used in chemical synthesis. A major advantage of carbohydrates in chemical synthesis is their abundance of hydroxyl functional groups, as shown in the structural formula of glucose above. Such functionality provides sites for the attachment of other functionalities and for initiating chemical reactions leading to desired products. The inherent biodegradability of carbohydrates provides them with some environmental advantages as well.

The most abundant biomaterial produced by plants is cellulose. Cellulose currently has a number of uses as a feedstock, and these will grow in the future as plant biomass supplants diminishing petroleum feedstocks.

In addition to cellulose, a very abundant material in plant biomass is lignin, a complex binding material that serves to hold wood and other plant materials together. Unlike carbohydrate cellulose, which has a uniform structure and is composed entirely of glucose monomer units, lignin has a variable three-dimensional molecular structure, the most common feature of which is the phenylpropane unit (a benzene ring attached to a three-carbon chain) along with other hydrocarbon and oxygen-containing groups. The molecular masses of lignin may exceed 15,000. Lignin's variability severely limits its use as a chemical feedstock. A major detriment to the use of lignin in green chemistry is its resistance to any sort of biological processes that might be considered for its conversion to useful products. One of the more promising potential uses for abundant lignin resources is the production of phenolic compounds.

Described in Chapter 22, Section 22.5, lipids are substances that can be extracted from biological materials by organic solvents. Lipids and hydrocarbon terpenes produced by plants have a number of potential uses as feedstocks. Relatively high in hydrogen and carbon and low in oxygen contents, lipids are oils, greases, and waxes with properties similar to a number of materials in petroleum. Some lipids can be used to synthesize synthetic liquid fuels and can even be used directly as diesel fuel. An early diesel engine that won the Grand Prize at the Paris Exhibition of 1900 ran on peanut oil, chosen because the French government wanted to promote this material produced in abundance in some of its African colonies.

Oils and fats from biological sources are very important raw materials for the chemical industry.[6] These materials provide a variety of fatty acids including oleic acid from new strains of sunflower, linolenic acid from soybean and linseed, rinoleic acid from castor oil, erucic acid from rapeseed, and a variety of oils from newly available plant sources. For chemical synthesis, the fatty acids from natural sources offer advantages because of the carboxylic acid group and, for some fatty acids, the presence of reactive C=C double bonds in the carbon chain.

17.7 REAGENTS

Reagents are chemical agents that act upon feedstocks in synthesis. A reagent may be partly or even fully incorporated into a product, or it may act to produce a chemical change in the feedstock. Judicious selection of reagents for carrying out chemical processes can be a crucial factor in developing a successful green chemistry process. Use of a benign feedstock may be of relatively little use if large quantities of hazardous reagents are required for its processing.

Two important factors driving reagent selection are *product selectivity* and *product yield.* High product selectivity means higher conversion of raw material to the desired product with minimal by-product generation. High product yield means a high percentage of the desired product is obtained relative to the maximum yield calculated from stoichiometric considerations. Both high-product selectivity and high-product yield reduce the amounts of extraneous material that must be handled and separated.

In selecting safe reagents and feedstocks, consideration of structure–activity relationships can be very useful. It is known that certain structural features or functional groups tend to create particular kinds of hazards. For example, the presence of oxygen and nitrogen—particularly multiple nitrogen atoms—in close proximity in a molecule tends to make it reactive or even explosive. The presence of the N–N=O functional group means that a compound is an *N–nitroso* compound (*nitrosamine*), many of which are carcinogenic. The ability to donate methyl groups to biomolecules may make a compound mutagenic or carcinogenic; substitution of longer-chain hydrocarbon groups can reduce this hazard.

In evaluating the safety of reagents and choosing safer alternatives, special attention should be given to *functional groups* consisting of particular groupings of atoms. The carcinogenic potential of the N–N=O group was mentioned above. Aldehydes tend to be irritants to animals and are photochemically active, so that they can contribute to smog formation when released to the atmosphere. Whenever possible it is best to use alternative compounds when particular functional groups are likely to be a problem. It is also sometimes possible to mask functional groups to produce less dangerous forms, and then unmask them at the point in the synthesis where the functionality is needed.

Oxidation is one of the most common operations in chemical synthesis. The nature of oxidation often requires harsh conditions and harsh reagents, such as permanganate (MnO_4^-), chromium(VI) compounds (such as potassium dichromate, $K_2Cr_2O_7$), toxic and costly osmium tetroxide, or *m*-chloroperbenzoic acid. Therefore, one of the main objectives of green chemistry has been the development of more benign oxidizing agents and oxidation reactions.

Reduction is a common operation in chemical synthesis. As with oxidation, the reagents used for reduction tend to be reactive and difficult to handle. Two of the most common such reagents are lithium aluminum hydride, $LiAlH_4$, which is flammable and rather hazardous to use, and tributyltin hydride, which can release volatile, toxic tin-containing products. Both oxidizing and reducing agents used in chemical synthesis produce by-products that must be carefully disposed of at significant expense.

Alkylation consists of the attachment of alkyl groups, such as the methyl group, $-CH_3$, to an organic molecule. The attachment of an alkyl group to a nitrogen atom in an amine is used as a step in the synthesis of a variety of dyes, pharmaceuticals, pesticides, plant growth regulators, and other specialty chemicals. In such cases, alkylation is commonly carried out using alkyl halides or alkyl sulfates in the presence of a base as shown below for the attachment of a methyl group to N in aniline:

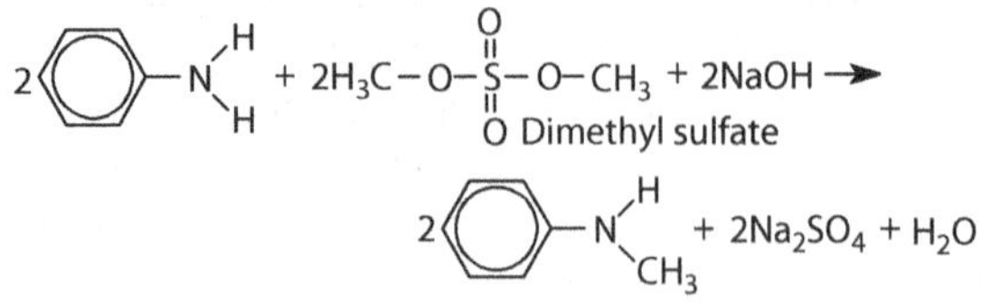

$$2\,C_6H_5{-}NH_2 + 2H_3C{-}O{-}S(=O)_2{-}O{-}CH_3 \text{ (Dimethyl sulfate)} + 2NaOH \rightarrow 2\,C_6H_5{-}NH{-}CH_3 + 2Na_2SO_4 + H_2O \quad (17.4)$$

This kind of reaction produces significant amounts of inorganic salt by-product, such as Na_2SO_4. Also, alkyl halides and alkyl sulfates pose toxicity concerns; dimethyl sulfate is a suspect human carcinogen.

Relatively nontoxic dialkyl carbonates are very effective alkylating agents that promise to provide a safer alternative to alkyl halides or sulfates for some kinds of alkylation reactions. The promise of dimethyl carbonate for methylation (alkylation in which the methyl group, $-CH_3$, is attached) has been enhanced by the straightforward synthesis of this compound from methanol and carbon monoxide in the presence of a copper salt:

$$2H{-}CH_2{-}OH + CO + \frac{1}{2}O_2 \rightarrow H_3C{-}O{-}C(=O){-}O{-}CH_3 \text{ (Dimethyl carbonate)} + H_2O \quad (17.5)$$

Dimethyl carbonate can be used for methylation of nitrogen in amine compounds at a temperature of 180°C under continuous-flow gas–liquid phase-transfer catalysis conditions (which involve transfer of organic ionic reactant species between water and an organic phase) as shown below for the methylation of aniline:

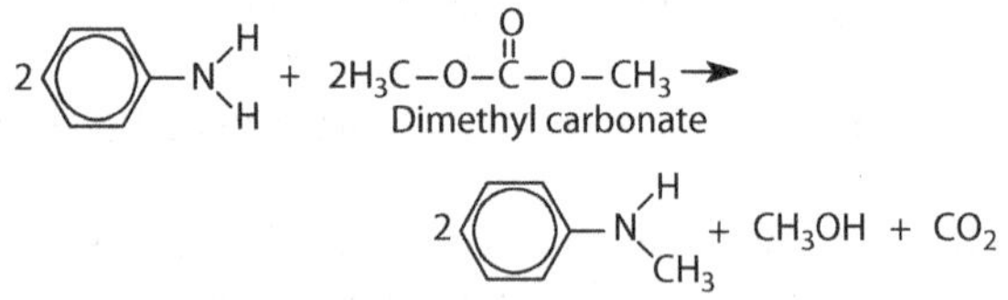

$$2\,C_6H_5{-}NH_2 + 2H_3C{-}O{-}C(=O){-}O{-}CH_3 \text{ (Dimethyl carbonate)} \rightarrow 2\,C_6H_5{-}NH{-}CH_3 + CH_3OH + CO_2 \quad (17.6)$$

The use of dimethyl carbonate as a methylating agent offers the twin advantages of up to 99% efficiency of conversion to product with selectivities of 99% or more for the monomethyl product. The by-products are innocuous carbon dioxide and methanol, which can be recirculated through the process for making dimethyl carbonate (Reaction 17.5). Dimethyl carbonate is useful, for example, in making monomethyl organonitrogen derivatives required for the synthesis of analgesics, such as ibuprofen.

17.8 STOICHIOMETRIC AND CATALYTIC REAGENTS

In discussions of green chemistry distinction is often made between stoichiometric reactions and catalytic reactions. A *stoichiometric reaction* written in a generalized form as

$$A + B + C \rightarrow \text{Product} + D + E \quad (17.7)$$

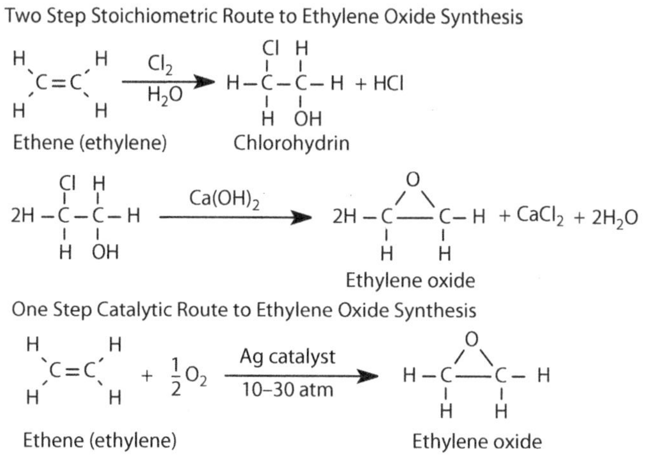

Two Step Stoichiometric Route to Ethylene Oxide Synthesis

$$H_2C{=}CH_2 \xrightarrow[H_2O]{Cl_2} H{-}CHCl{-}CH(OH){-}H + HCl$$

Ethene (ethylene) Chlorohydrin

$$2H{-}CHCl{-}CH(OH){-}H \xrightarrow{Ca(OH)_2} 2\,H_2C{-}O{-}CH_2 + CaCl_2 + 2H_2O$$

Ethylene oxide

One Step Catalytic Route to Ethylene Oxide Synthesis

$$H_2C{=}CH_2 + \frac{1}{2}O_2 \xrightarrow[10\text{–}30\ atm]{Ag\ catalyst} H_2C{-}O{-}CH_2$$

Ethene (ethylene) Ethylene oxide

FIGURE 17.5 The stoichiometric and catalytic routes to ethylene oxide synthesis.

is one in which two or more substances enter into a reaction to provide a desired product. If the reaction is simply that between A and B and if all these reagents are consumed to yield product, the reaction is 100% atom-efficient. Such an ideal is rarely realized in practice because (1) either A or B is a limiting reagent and one of the other reagents is left over, (2) only part of at least one of the reactants ends up in product, and (3) additional reagents are required to make the reaction proceed.

Where possible, it is preferable to replace stoichiometric reactions with *catalytic reactions* in which, ideally, a catalyst that is not itself consumed enables the reaction to occur rapidly, specifically, and with 100% atom efficiency. The distinction between stoichiometric and catalytic reactions can be illustrated with the example of the synthesis of ethylene oxide, widely used in the synthesis of ethylene glycol (antifreeze), ethylene glycol ester solvents, polyesters, ethanolamines, and other products, outlined in Figure 17.5. Although it gives relatively high yields, the stoichiometric chlorohydrin route to ethylene oxide requires hazardous and expensive chlorine gas as a reagent and produces 3.5 kg of waste calcium chloride per kilogram ethylene oxide product. The catalytic route gives an approximately 80% yield without generating by-products and the unreacted reagents can simply be recycled through the synthesis. By 1975, the catalytic route had supplanted the chlorhydrin synthesis pathway.

17.9 MEDIA AND SOLVENTS

Media is a term used to refer to the matrix in which or on which chemical processes occur. The type and strength of interaction between the media and reactants in a chemical process play a very important role in determining the type, degree, and rate of the process. Although media may include solids upon which reactions take place, by far the most common type of media consists of liquid *solvents* in which reagents are dissolved. Solvents interact to various extents and in various ways with the solutes dissolved in them. An important phenomenon is *solvation*, in which the solvent molecules interact with solute molecules. A common example of this is the attraction of polar water solvent molecules for dissolved cations and anions. This ability makes water an extraordinarily good solvent for ionic substances—acids, bases, and salts—that are commonly used in chemical reactions. The ability of water to form hydrogen bonds is particularly important in its ability to dissolve a wide range of biological materials capable of forming hydrogen bonds.

Beyond their uses as reaction media, solvents have other uses, particularly in separation, purification, and cleaning. Water is the most abundant and safest solvent and should be used wherever possible. In fact, one of the main objectives of the practice of green chemistry is to convert processes to the use of water solvent wherever possible. However, water is not a suitable solvent for a wide range of organic substances used industrially. Therefore, the use of organic solvents consisting of hydrocarbons and hydrocarbon derivatives, such as chlorinated hydrocarbons, is unavoidable in many cases. In addition to serving as reaction media, organic solvents are used as cleaners, degreasers, and as extractants to remove organic substances from solids. A major use of organic solvents is as the liquid *vehicle* to enable the application, spreading, and impregnation (such as of cloth) of dissolved or suspended dyes and other agents in formulations of paints, coatings, inks, and related materials.

By their nature, solvents cause more environmental and health problems than do other participants in the chemical synthesis process. Most solvents are volatile and tend to escape into the workplace and atmosphere. Hydrocarbon solvents are flammable and can form explosive mixtures with air. Released to the atmosphere, they can be instrumental in causing formation of photochemical smog (see Chapter 13). A number of adverse health effects are attributed to solvents. Carbon tetrachloride, CCl_4, causes lipid peroxidation in the body that can result in severe damage to the liver. Benzene causes blood disorders and is suspected of causing leukemia. Volatile C_5–C_7 alkanes damage nerves and can result in a condition known as peripheral neuropathy. The possibility of their being carcinogenic is always a consideration in dealing with solvents in the workplace.

A key consideration in selecting and using a solvent for the proper practice of green chemistry consists of its toxicological and environmental effects on biological systems. This obviously includes toxic effects to humans. Physical properties, such as volatility, density, and solubility, are important in estimating potential environmental and biological effects. Lipophilicity, the tendency to dissolve lipid tissue, is a measure of the ability of a solvent to penetrate skin and hence is an important factor in determining its biological effects. The environmental persistence and biodegradation of solvents should be considered. Special care must be taken to not use or at least prevent the release to the atmosphere of volatile solvents that may become involved in photochemical reactions leading to photochemical smog formation. The formerly widespread use of volatile chlorofluorocarbon (CFC) solvents as blowing agents to produce porous plastics and plastic foams resulted in widespread dissipation of these ozone destroying CFC vapors to the atmosphere (see Chapter 14).

Much of the progress that has been made toward the goal of green chemistry has come with replacement of potentially troublesome solvents with less dangerous ones. Some pertinent examples of such replacement are shown in Table 17.1. The best solvent to use, when possible, is water. It is discussed below.

17.9.1 Water, the Greenest Solvent

Although it does not truly dissolve hydrophobic organic substances, water may hold such substances in suspension as finely divided colloidal matter, thus in some cases enabling its use in place of organic solvents as a medium for organic reactions and for other applications. In addition to its not dissolving organic substances, water suffers from the disadvantage of reacting strongly with some reagents, such as $AlCl_3$ used in Friedel–Crafts reactions, strongly reducing $LiAlH_4$, and metallic sodium used in some applications. On the other hand, precisely because water is such a poor solvent for organic substances—the *hydrophobic effect*—some organic reactions proceed better in a water medium than in organic solvents. Water is an excellent solvent for some of the more hydrophilic biological molecules, such as glucose, that are gaining favor as reactants for green chemical processes.

Largely ignored during the development of organic synthesis, water is getting renewed attention as a medium for organic chemical reactions and processes. This is due largely to water being the ultimate green solvent with no detrimental environmental, safety (flammability), or toxicological aspects to its use. As the price for feedstocks for making organic solvents has increased, the fact that water is essentially free increases its attractiveness. Water is a good solvent for many biological materials that are gaining favor as green chemistry feedstocks, and, as discussed above, its repulsion of some organophilic reagents can be advantageous in some cases. Water-insoluble organic products are readily separated from water without having to distill off an organic solvent. The control of heat and

TABLE 17.1
Solvents for Which Substitutes Have Been Developed

Solvent	Disadvantages of the Solvent	Substitute	Characteristics of the Substitute
Benzene	Toxic, causing blood disorders and suspected of causing leukemia, metabolized to toxic phenol	Toluene	Much less toxic than benzene because of the presence of a metabolically oxidizable methyl substituent group; produces harmless hippuric acid metabolite
n-Hexane	Neurotoxic causing peripheral neuropathy manifested by mobility loss, reduced sensations in extremeties	2,5-Dimethyl-hexane	Lacks toxicity characteristics of *n*-hexane, significantly higher boiling point may be a disadvantage
Glycol ethers	Ethylene glycol monomethyl ether and ethylene glycol monoethyl ether have adverse reproductive and developmental effects in animals	1-Methoxy-2-propanol	Less toxic than the glycol ethers, but still effective as a solvent
Various organic solvents	Flammability, toxicity, poor biodegradability, tendency to contribute to photochemical smog	Supercritical fluid carbon dioxide	Widely available, good solvent for organic solutes, readily removed by evaporation, nonpolluting, except as a greenhouse gas if allowed to escape

temperature is an important aspect of many chemical processes. For heat and temperature control, water is the best solvent to use because of its very high heat capacity (see Chapter 3, Table 3.1).

17.9.2 Dense Phase Carbon Dioxide as a Solvent

Substances that are normally considered to be gases take on special properties when highly compressed. The general diagram shown in Figure 17.6 shows that, at temperatures exceeding a critical temperature, T_c, and pressures exceeding a critical pressure, P_c, the distinction between liquid and gas disappears and a substance becomes a *supercritical fluid*. The most widely studied supercritical fluid is that formed by carbon dioxide, for which T_c is 31.1°C and P_c is 73.8 atm. Supercritical fluids have many useful solvent properties. It has also been found, however, that highly compressed carbon dioxide below the critical point, where it is not supercritical, but may exist as a mixture of liquid and gas, has some excellent solvent properties as well. The term *dense phase fluid* is used to designate a highly compressed, dense substance that may be a supercritical fluid, highly compressed gas, or mixture of gas and liquid.

Dense phase fluids have a number of interesting solvent properties and have been extensively investigated as extractants and for chromatographic separations (supercritical fluid chromatography). An important characteristic of these fluids is their much lower viscosities than conventional liquids; that of supercritical CO_2 is only about 1/30 the viscosity of liquids normally used as solvents. This means that solutes diffuse much more readily in supercritical fluids, thus enabling them to react much faster. The wide latitude within which pressures and temperatures of dense phase fluids may be changed enables their properties to be varied widely.

Supercritical fluid carbon dioxide is an excellent solvent for organic solutes. This has led to its uses in place of organochlorine solvents for cleaning metal parts and in dry cleaning. A major advantage of supercritical fluid carbon dioxide in some applications is that it is readily evaporated

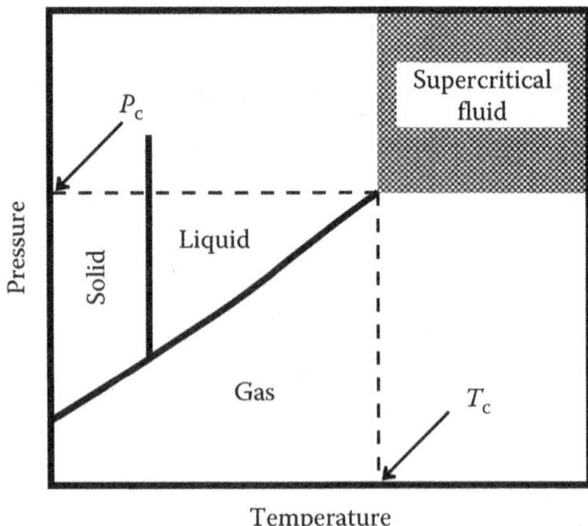

FIGURE 17.6 Temperature–pressure plot showing supercriticality.

from solutes by releasing pressure. This has led to interest in the solvent as a vehicle for paints and coatings. With the appropriate apparatus, the released carbon dioxide can be reclaimed and recompressed back to a supercritical state for recycling. This capability tends to overcome criticism of the use of carbon dioxide, which is, of course, a greenhouse gas when released to the atmosphere. Supercritical fluid carbon dioxide has a highly organophilic nature, which may be excessive for some applications with more polar or ionic solutes. Addition of polar cosolvents, such as methyl alcohol, can overcome this disadvantage. A further advantage of supercritical fluid carbon dioxide as a solvent is its ability to dissolve gases, enhanced by the very high pressure under which supercritical fluid carbon dioxide must be maintained. This enables reactions to occur efficiently with gaseous reactants in supercritical fluid carbon dioxide that would otherwise not be possible.

Supercritical fluids have some disadvantages. The major disadvantage of such solvents is the special apparatus required to maintain relatively severe supercritical conditions. However, supercritical fluid carbon dioxide's advantages of low cost, high abundance from a number of sources, nontoxic nature, nonflammability, and the fact that it is not classified as a volatile organic solvent are leading to increased uses of this solvent.

17.9.3 Gas-Expanded Solvents

Some of the handling and safety issues involved with the use of supercritical carbon dioxide can be overcome with *gas-expanded solvents* that are composed of subcritical pressurized gas (usually carbon dioxide) in contact with an organic solvent. The carbon dioxide dissolves in the organic liquid and expands the volume of the liquid several fold. As much as 80% of the organic solvent can be replaced by this approach. The properties of the solvent are altered significantly by the presence of carbon dioxide. An interesting possibility is that carbon dioxide, which is used in fire extinguishers, can render volatile hydrocarbon solvents much less flammable enabling their substitution for environmentally and toxicologically problematic chlorinated hydrocarbon liquids.

17.10 ENHANCING REACTIONS

One of the most important aspects of green chemistry is the enhancement of the speed and degree of completion of chemical reactions. The two major ways in which this can be done are by adding energy to the reaction system or by reducing the activation energy required to make the reaction go.

Catalysts are substances that enable reactions to proceed more rapidly (or at all). Catalysts can enable all kinds of reactions including oxidation, reduction, and a variety of organic reactions. The choice of catalysts is very important in the practice of green chemistry because they can add to the hazards of chemical processes and produce troublesome by-products and product contaminants. This can occur, for example, with homogeneous catalysts that are intimately mixed with the reagents involved in chemical synthesis. Catalysts that are most amenable to the practice of green chemistry are heterogeneous catalysts, such as molecular sieves, that can be kept entirely separate from products. Insofar as possible, such catalysts should be nontoxic.

In general, the greenest catalysts are enzymes that have developed in organisms to carry out biochemical processes.[7] Because they developed in organisms under conditions conducive to life, enzymatic reactions are by their nature carried out in water under ambient temperature and pressure conditions. They are generally very rapid and specific. Such reactions have been used for thousands of years in fermentation processes, such as the conversion of sucrose in cane sugar to glucose and fructose sugars by *invertase* enzymes and the subsequent fermentation of these two monosaccharides to ethanol by *zymase* enzyme in yeasts. In recent years, aided by recombinant DNA (genetic engineering) techniques, enzyme processes have been developed to carry out a wide variety of chemical processes. Among the functions that enzyme catalysts can perform are oxidation and reduction reactions, hydrolysis reactions, transfer of functional groups to or from organic molecules, and additions to groups across double bonds or the reverse.

The most basic way of introducing energy into a reaction system is by external heating. This can be done by simply heating a reaction vessel from the outside, with steam coils immersed in a reaction vessel, or by resistance-heated electrical coils immersed in the reaction system. The practice of green chemistry seeks to develop more elegant means of introducing energy into a reaction system as alternatives to external heating.

Microwaves can be used to add energy to reactions to enhance reaction rates. Microwaves are electromagnetic radiation with wavelengths of 1 cm to 1 m (frequency 30 GHz to 300 Hz). To avoid interference with microwave bands used in communication, industrial and household microwave generators commonly operate at 2.45 GHz. Microwaves are absorbed by polar molecules, such as those of water, causing rapid re-orientation of the molecules in a microwave field. The result is a high input of energy directly into substances subjected to microwaves thereby adding energy and speeding up reactions. Microwave energy can be put directly into relatively small volumes of reaction media, reducing material requirements and minimizing wastes. Microwaves can be used to enhance reactions in (1) water media, (2) polar organic solvents such as dimethylformamide, and (3) media-free reactions, such as mixed solid reactants.

Subjecting a reaction medium to ultrasound energy at frequencies between 20 and 100 kHz can introduce very high energy pulses into the medium. This approach to enhancing reactions is commonly called *sonochemistry*. Commonly, the ultrasound is produced by the piezoelectric effect through which crystals of substances such as ceramic impregnated barium titanate are subjected to rapidly reversing electrical fields converting the electrical energy to sound energy with an efficiency that can reach 95%. An advantage of sonochemistry is that it can introduce high energy into microscopic regions enabling reactions to occur without appreciably heating the reaction medium.

The passage of a direct current of electricity through a reaction medium can cause both reductions and oxidations to occur. Reduction, the addition of electrons, e^-, can occur at the relatively negatively charged cathode, and oxidation, the loss of electrons, at the relatively positively charged anode. A simple example of an electrochemical process used to manufacture industrial chemicals occurs when a direct current is passed through molten sodium chloride, NaCl to produce elemental sodium and elemental chlorine. At the cathode, sodium ion is reduced:

$$Na^+ + e^- \rightarrow Na(l) \tag{17.8}$$

and at the anode, chloride ion is oxidized to elemental chlorine gas:

$$2Cl^- \rightarrow Cl_2(g) + 2e^- \tag{17.9}$$

giving the following net reaction:

$$2Na^+ + 2Cl^- \rightarrow 2Na(l) + Cl_2(g) \tag{17.10}$$

The reaction utilizes electrical energy efficiently and occurs with 100% atom economy.

Electrochemical oxidation and reduction can be controlled by the potentials applied, by the media in which they occur, and by the electrodes used. In a sense, electrochemical processes use "matter-free" reagents; no other approach comes any closer to the attainment of ideal green chemistry.

Photochemical reactions use the energy of photons of light or ultraviolet radiation to cause reactions to occur. For electromagnetic radiation of frequency ν, the energy of a photon is given by the equation $E = h\nu$, where h is Planck's constant. Since a photon can be absorbed directly by a molecule or a functional group on a molecule, the application of electromagnetic radiation of the appropriate energy to a reaction medium can introduce a high amount of energy into a reactant species without significantly heating the medium. Photochemical energy can be used to cause synthesis reactions to occur more efficiently and with less production of waste by-products than nonphotochemical processes.[8,9] One example is the acylation of benzoquinone with an aldehyde to produce an acylhydroquinone, an intermediate used to make some specialty polymers:

$$\text{Benzoquinone} + \mathrm{H{-}C({=}O){-}R} \xrightarrow{h\nu} \text{An acylhydroquinone} \tag{17.11}$$

Benzoquinone

An acylhydroquinone
(R is a hydrocarbon group)

This reaction occurs with 100% atom economy. Unlike the standard Friedel–Crafts type of reaction, which utilizes the catalytic effect of Lewis acid-type acidic halides, particularly aluminum chloride, $AlCl_3$, the photochemical process does not require moisture- and air-sensitive, potentially reactive catalytic substances.

A reagent does not have to absorb a photon directly to undergo a photochemically induced reaction. In some cases photochemically reactive species may be added that absorb photons, then produce reactive excited species or free radicals that carry out additional reactions. An example of this is provided with hydrogen peroxide, which absorbs photons

$$H_2O_2 + h\nu \rightarrow HO\bullet + HO\bullet \tag{17.12}$$

to produce reactive hydroxyl radicals that react with a number of other species.

17.11 INDUSTRIAL ECOLOGY

Figure 17.7 illustrates the old way of industrial production that prevailed in the earlier days of the industrial revolution. Raw materials and energy were taken from nonrenewable sources, including wood from old growth timber, and products were made without consideration of wastes or pollutants. Air and water pollutants were simply released to the atmosphere and waterways, respectively, and by-product solid, semisolid, and liquid wastes were placed in the geosphere without consideration of their effects. After products were used, they were simply discarded to the environment without thought of environmental effects. The idea of sustainability was a foreign concept. As a result, air and water became polluted and the land was littered with waste dumps and discarded consumer products.

Some of the ill effects of reckless production without regard to sustainability were noted in Section 17.1. At the beginning of this chapter it was mentioned that a viable alternative to the old

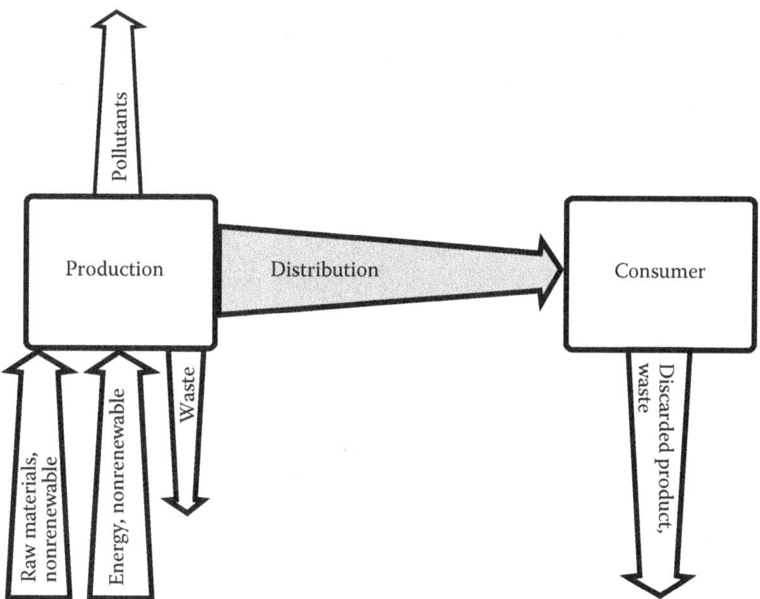

FIGURE 17.7 The old model of production without regard to consumption of nonrenewable energy and resources or consideration of pollutant release or waste disposal.

unsustainable ways is provided in part by the practice of industrial ecology, which emerged as a recognized discipline around 1990. *Industrial ecology* is a comprehensive approach to production, distribution, utilization, and termination of goods and services in a manner that maximizes mutually beneficial utilization of materials and energy among enterprises, thereby minimizing consumption of nonrenewable raw materials and energy while preventing the production of wastes and pollutants.[10] The practice of industrial ecology involves optimization of materials utilization starting with raw material and progressing through finished material, to component, to product, and finally to the final fate of obsolete product and its components. In addition to materials and resources, industrial ecology considers energy and capital.

Figure 17.8 shows industrial ecology operating in an idealized manner. This figure illustrates the minimal inputs of raw materials and energy, minimum production of wastes, and maximum circulation of materials within the systems that are characteristic of ideal systems of industrial ecology.

Industrial ecology is analogous to ecology in nature in which organisms create intricate interdependent webs whereby individual kinds of organisms utilize waste products from others for their own needs. This results in extremely high efficiencies of resource utilization in nature with essentially zero wastes. Organisms operate in natural ecosystems. Similarly, enterprises that practice industrial ecology operate in industrial ecosystems.

Several important analogies exist between industrial ecology and natural ecology. One of these is the analogy behind the evolution of organisms and their ecosystems through natural selection in which those organisms most suited to particular environments evolve through genetic processes. Similarly, systems of industrial ecology have evolved through natural selection processes. If a waste product becomes available, an enterprise is likely to develop to utilize the product. If a more efficient enterprise for utilizing the waste appears, it will become dominant. These kinds of practices have occurred ever since industries first emerged, centuries before industrial ecology was formally recognized. It should be kept in mind that with current understanding of the nature of industrial ecology, something like "intelligent design" has the potential to be applied to industrial ecosystems. Based on knowledge of industrial enterprises, their needs, and their markets, and utilizing the enormous computing powers now available to planners, systems of industrial ecology can be planned, constructed, and operated that are functional and efficient.

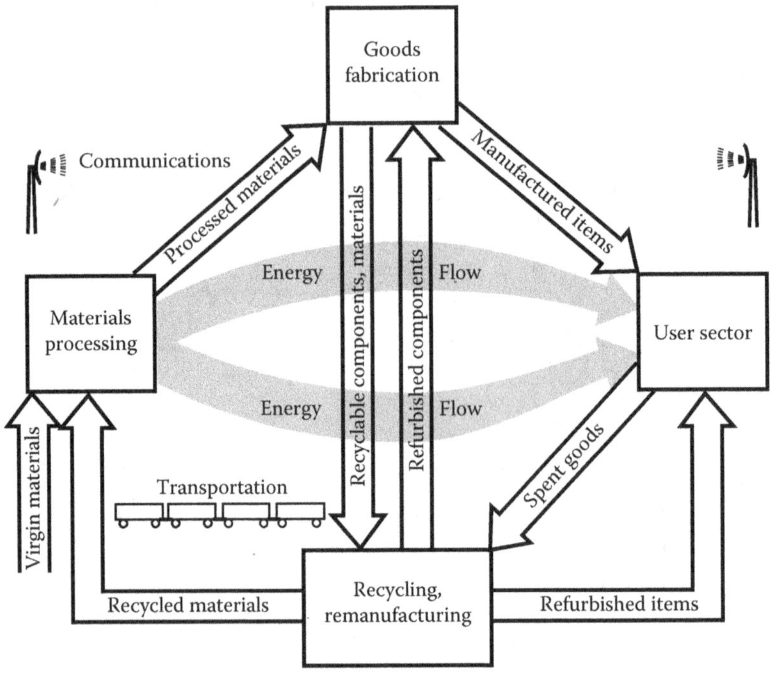

FIGURE 17.8 The major components of an industrial ecosystem showing maximum flow of materials within the system.

Just as natural ecosystems are not static and undergo ecological succession, such as grasslands → shrublands → forests, industrial ecosystems undergo a form of ecological succession. Rapid developments in manufacturing techniques, changes in sources of raw materials or energy, and shifting markets will inevitably change the mix of enterprises in an industrial ecosystem and the ways in which they interact. Therefore, a form of natural selection must operate to provide the most efficient possible systems of industrial ecology.

17.12 THE FIVE MAJOR COMPONENTS OF AN INDUSTRIAL ECOSYSTEM

Industrial ecosystems include all aspects of production, processing, and consumption. As shown in Figure 17.8, there are several distinct components of an industrial ecosystem, which can be categorized as the following: these are (1) a primary materials producer, (2) a source or sources of energy, (3) a materials processing and manufacturing sector, (4) a waste processing sector, and (5) a consumer sector. In such an idealized system operating with the best practice of industrial ecology, the flow of materials among the major hubs is very high. Each constituent of the system evolves in a manner that maximizes the efficiency with which the system utilizes materials and energy.

It is convenient to consider the *primary materials producers* and the *energy generators* together because both materials and energy are required in order for the industrial ecosystem to operate. The primary materials producer or producers may consist of one or several enterprises devoted to providing the basic materials that sustain the industrial ecosystem. Most generally, in any realistic industrial ecosystem, a significant fraction of the material processed by the system consists of virgin materials. In a number of cases, and increasingly so as pressures build to recycle materials, significant amounts of the materials come from recycling sources.

The processes that virgin materials entering the system are subjected to vary with the kind of material, but can generally be divided into several major steps. Typically, the first step is extraction, designed to remove the desired substance as completely as possible from the other substances with which it occurs. This stage of materials processing can produce large quantities of waste material requiring disposal, as is the case with some metal ores in which the metal makes up a small percentage of the ore that is mined. In other cases, such as corn grain providing the basis of a corn products industry, the "waste,"—in this specific example the cornstalks associated with the grain—can be left on the soil to add humus and improve soil quality. A concentration step may follow extraction to put the desired material into a purer form. After concentration, the material may be put through additional refining steps including separations. Following these steps, the material is usually subjected to additional processing and preparation leading to the finished materials. Throughout the various steps of extraction, concentration, separation, refining, processing, preparation, and finishing, various physical and chemical operations are used, and wastes requiring disposal may be produced. Recycled materials may be introduced at various parts of the process, although they are usually introduced into the system following the concentration step.

The extraction and preparation of energy sources can follow many of the steps outlined above for the extraction and preparation of materials. For example, the processes involved in extracting uranium from ore, enriching it in the fissionable uranium-235 isotope, and casting it into fuel rods for nuclear fission power production include all of those outlined above for materials. On the other hand, some rich sources of coal are essentially scooped from a coal seam and sent to a power plant for power generation with only minimal processing, such as sorting and grinding.

Recycled materials added to the system at the primary materials and energy production phase may be from both pre and postconsumer sources. As examples, recycled paper may be macerated and added at the pulping stage of paper manufacture. Recycled aluminum may be added at the molten metal stage of aluminum metal production.

Finished materials from primary materials producers are fabricated to make products in the *materials processing and manufacturing sector*, which is often a very complex system. For example, the manufacture of an automobile requires steel for the frame, plastic for various components, rubber in tires, lead in the battery, and copper in the wiring, along with a large number of other materials. Typically, the first step in materials manufacturing and processing is a forming operation. For example, sheet steel suitable for making automobile frames may be cut, pressed, and welded into the configuration needed to make a frame. At this step some wastes may be produced that require disposal. An example of such wastes consists of carbon fiber/epoxy composites left over from forming parts such as jet aircraft engine housings. Finished components from the forming step are fabricated into finished products that are ready for the consumer market.

The materials processing and manufacturing sector presents several opportunities for recycling. At this point it may be useful to define two different streams of recycled materials:

- *Process recycle streams* consisting of materials recycled in the manufacturing operation itself
- *External recycle streams* consisting of materials recycled from other manufacturers or from postconsumer products

Materials suitable for recycling can vary significantly. Generally, materials from the process recycle streams are most suitable for recycling because they are the same materials used in the manufacturing operation. Recycled materials from the outside, especially those from postconsumer sources, may vary greatly in their characteristics because of the lack of effective controls over recycled postconsumer materials. Therefore, manufacturers may be reluctant to use such substances.

In the *consumer sector*, products are sold or leased to the consumers who use them. The duration and intensity of use vary widely with the product; paper towels are used only once, whereas an automobile may be used thousands of times over many years. In all cases, however, the end of the useful lifetime of the product is reached and it is either (1) discarded or (2) recycled. The success of a total industrial ecology system may be measured largely by the degree to which recycling predominates over disposal.

Recycling has become so widely practiced that an entirely separate *waste processing sector* of an economic system may now be defined. This sector consists of enterprises that deal specifically with the collection, separation, and processing of recyclable materials and their distribution to end-users. Such operations may be entirely private or they may involve cooperative efforts with governmental sectors. They are often driven by laws and regulations that provide penalties against simply discarding used items and materials, as well as positive economic and regulatory incentives for their recycle.

17.13 INDUSTRIAL METABOLISM

Industrial metabolism refers to the processes to which materials and components are subjected in industrial ecosystems. It is analogous to the metabolic processes that occur with food and nutrients in biological systems. Like biological metabolism, industrial metabolism may be addressed at several levels. A level of industrial metabolism at which green chemistry, especially, comes into play is at the molecular level where substances are changed chemically to give desired materials or to generate energy. Industrial metabolism can be addressed within individual unit processes in a factory, at the factory level, at the industrial ecosystem level, and even globally.

A significant difference between industrial metabolism as it is now practiced and natural metabolic processes relates to the wastes that these systems generate. Natural ecosystems have developed such that true wastes are virtually nonexistent. For example, even those parts of plants that remain after biodegradation of plant materials form soil humus (see Chapter 16) that improves the conditions of soil on which plants grow. Anthropogenic industrial systems, however, have developed in ways that generate large quantities of wastes, where a waste may be defined as *dissipative use of natural resources*. Furthermore, human use of materials has a tendency to dilute and dissipate materials and disperse them to the environment. Materials may end up in a physical or chemical form from which reclamation becomes impractical because of the energy and effort required. A successful industrial ecosystem overcomes such tendencies.

Organisms performing their metabolic processes degrade materials to extract energy (catabolism) and synthesize new substances (anabolism). Industrial ecosystems perform analogous functions. The objective of industrial metabolism in a successful industrial ecosystem is to make desired goods with the least amount of by-product and waste. This can pose a significant challenge. For example, to produce lead from lead ore for automobile batteries requires mining large quantities of ore, extracting the relatively small fraction of the ore consisting of lead sulfide mineral, and roasting and reducing the mineral to get lead metal. The whole process generates large quantities of lead-contaminated tailings left over from mineral extraction and significant quantities of by-product sulfur dioxide, which must be reclaimed to make sulfuric acid and not released to the environment. The recycling pathway, by way of contrast, takes essentially pure lead from recycled batteries and simply melts it down to produce lead for new batteries; the advantages of recycling in this case are obvious. Industrial metabolic processes that emphasize recycling are desirable because recycling gives essentially constant reservoirs of materials in the recycling loop.

Living organisms have elaborate systems of control. Considering the metabolism that occurs in an entire natural ecosystem, it is seen to be *self-regulating*. If herbivores that consume plant biomass become too abundant and diminish the stock of the biomass, their numbers cannot be sustained, the population dies back, and their food source rebounds. The most successful ecosystems are those in which this self-regulating mechanism operates continuously without wide variations in populations. Industrial systems do not inherently operate in a self-regulating manner that is advantageous to their surroundings, or even to themselves in the long run. Examples of the failure of self-regulation of industrial systems abound in which enterprises have wastefully produced large quantities of goods of marginal value, running through limited resources in a short time, and dissipating materials to their surroundings, polluting the environment in the process. Despite these bad experiences, within a proper framework of laws and regulations designed to avoid wastes and excess, industrial ecosystems can be designed to operate in a self-regulating manner. Such self-regulation operates best under conditions of maximum recycling in which the system is not dependent on a depleting resource of raw materials or energy.

17.14 MATERIALS FLOW AND RECYCLING IN AN INDUSTRIAL ECOSYSTEM

In an industrial ecosystem, there are several points at which materials may be recycled and there are several points at which wastes are produced. The potential for the greatest production of waste lies in the earlier stages of the cycle in which large quantities of materials with essentially no use associated with the raw material, such as ore tailings, may require disposal. In many cases, little if anything of value can be obtained from such wastes and the best thing to do with them is to return them to their source (usually a mine), if possible. Another big source of potential wastes, and often the one that causes the most problems, consists of postconsumer wastes generated when a product's life cycle is finished. With a properly designed industrial ecology cycle, such wastes can be minimized and, ideally, totally eliminated.

In general, the amount of waste per unit output decreases in going through the industrial ecology cycle from virgin raw material to final consumer product. Also, the amount of energy expended in dealing with waste or in recycling decreases farther into the cycle. For example, waste iron from the milling and forming of automobile parts may be recycled from a manufacturer to the primary producer of iron as scrap steel. In order to be used, such steel must be remelted and run through the steel manufacturing process again, with a considerable consumption of energy. However, a postconsumer item, such as an engine block, may be refurbished and recycled to the market with relatively less expenditure of energy.

17.15 THE KALUNDBORG INDUSTRIAL ECOSYSTEM

The most often cited example of a functional industrial ecosystem is that of Kalundborg, Denmark. The various components of the Kalundborg industrial ecosystem are shown in Figure 17.9. To a

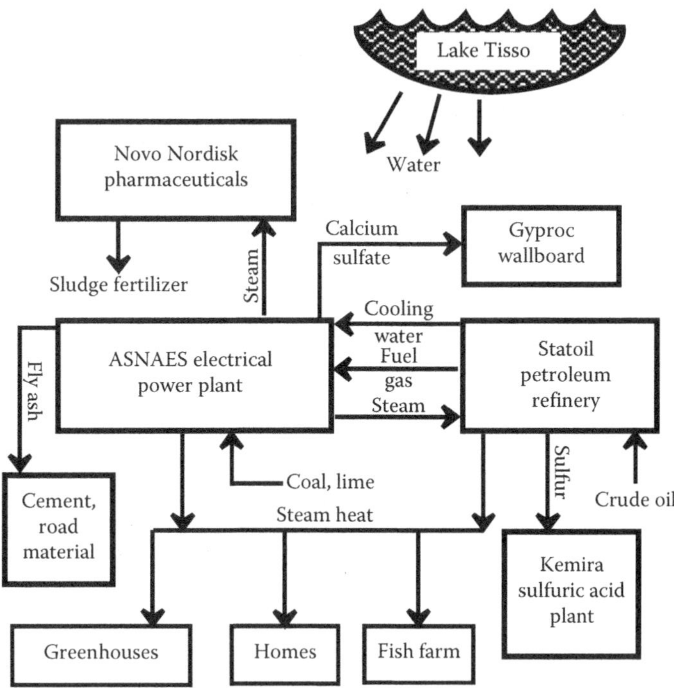

FIGURE 17.9 The Kalundborg, Denmark, industrial ecosystem, which is frequently cited as a successful example of the practice of industrial ecology.

degree, the Kalundborg system developed spontaneously, without being specifically planned as an industrial ecosystem. It is based on two major energy suppliers, the 1500-MW ASNAES coal-fired electrical power plant and the 4–5 million ton/yr Statoil petroleum refining complex, each the largest of its kind in Denmark. The electric power plant sells process steam to the oil refinery, from which it receives fuel gas and cooling water. Sulfur removed from the petroleum goes to the Kemira sulfuric acid plant. By-product heat from the two energy generators is used for district heating of homes and commercial establishments, as well as to heat greenhouses and a fish farming operation. Steam from the electrical power plant is used by the $2 billion/yr Novo Nordisk pharmaceutical plant, a firm that produces industrial enzymes and 40% of the world's supply of insulin. This plant generates a biological sludge that is used by area farms for fertilizer, applied both as a liquid sludge and in a dried form. Calcium sulfate produced as a by-product of sulfur removal by lime scrubbing from the electrical plant is used by the Gyproc company to make wallboard. The wallboard manufacturer also uses clean-burning gas from the petroleum refinery as fuel. Fly ash generated from coal combustion goes into cement and roadbed fill. Lake Tisso serves as a freshwater source. Other examples of efficient materials utilization associated with Kalundborg include use of sludge from the plant that treats water and wastes from the fish farm's processing plant for fertilizer, and blending of excess yeast from Novo Nordisk's insulin production as a supplement to swine feed.

The development of the Kalundborg complex occurred over a long period of time, beginning in the 1960s, and provides some guidelines for the way in which an industrial ecosystem can evolve naturally. The first of many synergistic (mutually advantageous) arrangements was cogeneration of usable steam along with electricity by the ASNAES electrical power plant. The steam was first sold to the Statoil petroleum refinery, then, as the advantages of large-scale, centralized production of steam became apparent, steam was also provided to homes, greenhouses, the pharmaceutical plant, and the fish farm. The need to produce electricity more cleanly than was possible simply by burning high-sulfur coal resulted in two more synergies. Operation of a lime-scrubbing unit for sulfur removal on the power plant stack produced large quantities of calcium sulfate, which found a market in the manufacture of gypsum wallboard. It was also found that a clean-burning gas by-product of the petroleum refining operation could be substituted in part for the coal burned in the power plant, further reducing pollution.

The implementation of the Kalundborg ecosystem occurred largely because of the close personal contact among the managers of the various facilities in a relatively close social and professional network over a long period of time. All the contracts have been based on sound business fundamentals and have been bilateral. Each company has acted upon its perceived self-interest, and there has been no master plan for the system as a whole. The regulatory agencies have been cooperative, but not coercive in promoting the system. The industries involved in the agreements have fit well, with the needs of one matching the capabilities of the other in each of the bilateral agreements. The physical distances involved have been small and manageable; it is not economically feasible to ship commodities such as steam or fertilizer sludges for long distances.

17.16 CONSIDERATION OF ENVIRONMENTAL IMPACTS IN INDUSTRIAL ECOLOGY

By its nature, industrial production has an impact upon the environment. Whenever raw materials are extracted, processed, used, and eventually discarded, some environmental impacts will occur. In designing an industrial ecological system, several major kinds of environmental impacts must be considered in order to minimize them and keep them within acceptable limits. These impacts and the measures taken to alleviate them are discussed below.

For most industrial processes, the first environmental impact is that of extracting raw materials. This can be a straightforward case of mineral extraction, or it can be less direct, such as utilization of biomass grown on forest or cropland. A basic decision, therefore, is the choice of the kind of material to be used. Wherever possible, materials should be chosen that are not likely to be in short supply in the foreseeable future. As an example, the silica used to make the lines employed for fiber-optics communication is in unlimited supply and a much better choice for communication conduits than copper wire made from limited resources of copper ore.

Industrial ecology systems should be designed to reduce or even totally eliminate air pollutant emissions. Among the most notable recent progress in that area has been the marked reduction and even total elimination of solvent vapor emissions (volatile organic carbon, VOC), particularly those from organochlorine solvents. Some progress in this area has been made with more effective trapping of solvent vapors. In other cases, the use of the solvents has been totally eliminated. This is the case for CFCs, which are no longer used in plastic foam blowing and parts cleaning because of their potential to affect stratospheric ozone. Other air pollutant emissions that should be eliminated are hydrocarbon vapors, including those of methane, CH_4, and oxides of nitrogen or sulfur.

Discharges of water pollutants should be entirely eliminated wherever possible. For many decades, efficient and effective water treatment systems have been employed that minimize water pollution. However, these are "end-of-pipe" measures, and it is much more desirable to design industrial systems such that potential water pollutants are not even generated.

Industrial ecology systems should be designed to prevent production of liquid wastes that may have to be sent to a waste processor. Such wastes fall into the two broad categories of water-based wastes and those contained in organic liquids. Under current conditions the largest single constituent of so-called "hazardous wastes" is water. Elimination of water from the waste stream automatically prevents pollution and reduces amounts of wastes requiring disposal. The solvents in organic wastes largely represent potentially recyclable or combustible constituents. A properly designed industrial ecosystem does not allow such wastes to be generated or to leave the factory site.

In addition to liquid wastes, many solid wastes must be considered in an industrial ecosystem. The most troublesome are toxic solids that must be placed in a secure hazardous waste landfill. The problem has become especially acute in some industrialized nations in which the availability of landfill space is severely limited. In a general sense, solid wastes are simply resources that have not been properly utilized. Closer cooperation among suppliers, manufacturers, consumers, regulators, and recyclers can minimize quantities and hazards of solid wastes.

Whenever energy is expended, there is a degree of environmental damage. Therefore, energy efficiency must have a high priority in a properly designed industrial ecosystem. Significant progress has been made in this area in recent decades, as much because of the high costs of energy as for environmental improvement. More efficient devices, such as high-efficiency electric motors, and approaches, such as cogeneration of electricity and heat, that make the best possible use of energy resources are highly favored. An important side benefit of more efficient energy utilization is the lowered emissions of air pollutants, including greenhouse gases.

17.17 LIFE CYCLES: EXPANDING AND CLOSING THE MATERIALS LOOP

In a general sense, the traditional view of product utilization is the one-way process of extraction → production → consumption → disposal shown in the upper portion of Figure 17.10. Materials that are extracted and refined are incorporated into the production of useful items, usually by processes that produce large quantities of waste by-products. After the products are worn out, they are discarded. This essentially one-way path results in a relatively large exploitation of resources, such as metal ores, and a constant accumulation of wastes. As shown at the bottom of Figure 17.10, however, the one-way path outlined above can become a cycle in which manufactured goods are used, and then recycled at the end of their life spans. As one aspect of such a cyclic system, it is often useful for manufacturers to assume responsibility for their products, to maintain "stewardship." Ideally, in such a system a product and/or the material in it would have a never-ending life cycle; when its useful lifetime is exhausted, it is either refurbished or converted into another product.

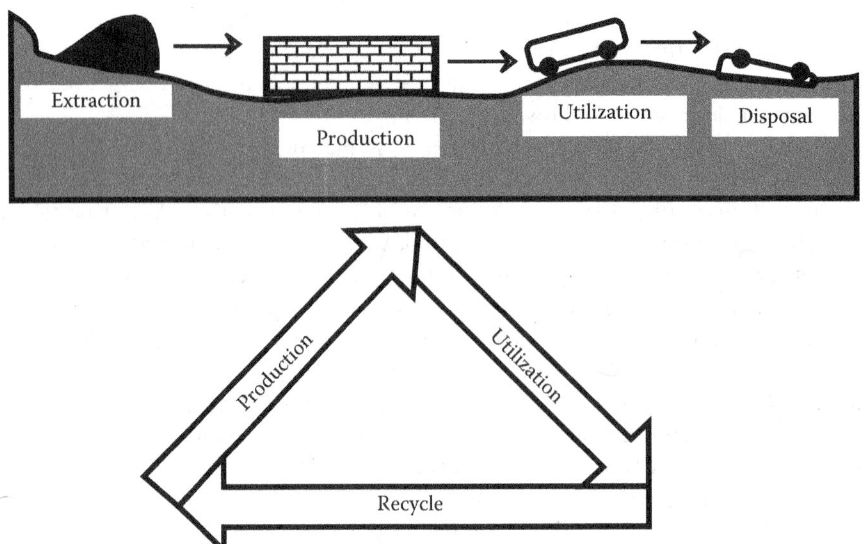

FIGURE 17.10 The one-way path of conventional utilization of resources to make manufactured goods followed by disposal of the materials and goods at the end consumes large quantities of materials and makes large quantities of wastes (top). In an ideal industrial ecosystem (bottom), the loop is closed and spent products are recycled to the production phase.

From the discussion above and in the rest of this chapter, it may be concluded that industrial ecology is all about *cyclization of materials*. This approach is summarized in a statement attributed to Kumar Patel of the University of California at Los Angles, "The goal is *cradle to reincarnation*, since if one is practicing industrial ecology correctly there is no grave." For the practice of industrial ecology to be as efficient as possible, cyclization of materials should occur at the highest possible level of material purity and stage of product development as discussed under *embedded utility* below.

In considering life cycles, it is important to note that commerce can be divided into the two broad categories of *products* and *services*. Although most commercial activity used to be concentrated on providing large quantities of goods and products, demand has been largely satisfied for some segments of the population, and the wealthier economies are moving more to a service-based system. Much of the commerce required for a modern society consists of a mixture of services and goods. The trend toward a service economy offers two major advantages with respect to waste minimization. Obviously, a pure service involves little material. Secondly, a service provider is in a much better position to control materials to ensure that they are recycled and to control wastes, ensuring their proper disposal. A commonly cited example is that of photocopy machines. They provide a service, and a heavily used copy machine requires frequent maintenance and cleaning. The parts of such a machine and the consumables, such as toner cartridges, consist of materials that eventually will have to be discarded or recycled. In this case it is often reasonable for the provider to lease the machine to users, taking responsibility for its maintenance and ultimate fate. The idea could even be expanded to include recycling of the paper processed by the copier, with the provider taking responsibility for recyclable paper printed by the machine.

In many cases, to be practical, recycling must be practiced on a larger scale than simply that of a single industry or product. For example, recycling plastics used in soft drink bottles to make new soft drink bottles is not allowed because of the possibilities for contamination. However, the plastics can be used as raw material for auto parts. Usually, different companies are involved in making auto parts and soft drink bottles.

17.17.1 Product Stewardship

The degree to which products are recycled is strongly affected by the custody of the products. For example, batteries containing cadmium or mercury pose significant pollution problems when they are purchased by the public; used in a variety of devices, such as calculators and cameras; then discarded through a number of channels, including municipal refuse. However, when such batteries are used within a single organization, it is possible to ensure that almost all of them are returned for recycling. In cases such as this, systems of stewardship can be devised in which marketers and manufacturers exercise a high degree of control of the product. This can be done through several means. One is for the manufacturer to retain ownership of the product, as is commonly practiced with photocopy machines. Another mechanism is one in which a significant part of the purchase price is refunded for trade-in of a spent item. This approach could work very well with batteries containing cadmium or mercury. The normal purchase price could be doubled, then discounted to half with the trade-in of a spent battery.

17.17.2 Embedded Utility

Figure 17.11 can be regarded as an "energy/materials pyramid" showing that the amounts of energy and materials involved decrease going from the raw material to the finished product. The implication of this diagram is that significantly less energy, and certainly no more materials, are involved when recycling is performed near the top of the materials flow chain rather than near the bottom.

In low-level recycling, a material or component is taken back to near the beginning of the steps through which it is made. For example, an automobile engine block might be melted down to produce molten metal from which new blocks are then cast. With high-level recycling, the item or material is recycled as close to the final product as possible. In the case of the automobile engine block, it may be cleaned, the cylinder walls rehoned, the flat surfaces replaned, and the block used as the platform for assembling a rebuilt engine. In this example and many others that can be cited, high-level recycling uses much less energy and materials and is inherently more efficient.

The greater usability and lower energy requirements for recycling products higher in the order of material flow are called *embedded utility*. One of the major objectives of a system of industrial

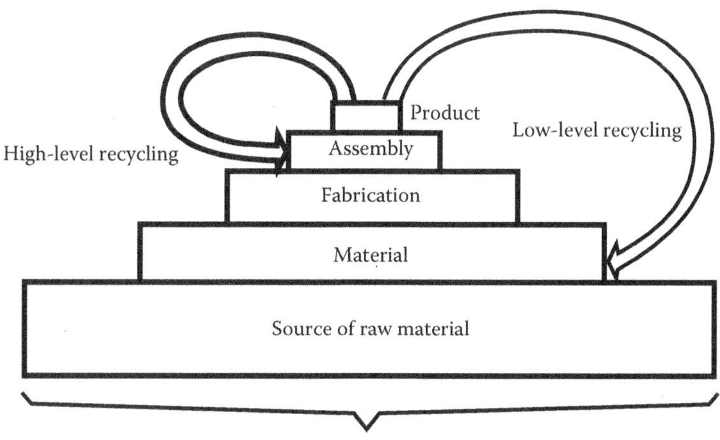

FIGURE 17.11 A material flow chain or energy/materials pyramid. Less energy and materials are involved when recycling is done near the end of the flow chain, thus retaining embedded utility.

ecology and, therefore, one of the main reasons for performing life-cycle assessments is to retain the embedded utility in products by measures such as recycling as near to the end of the material flow as possible, and replacing only those components of systems that are worn out or obsolete. An example of the latter occurred during the 1960s when efficient and highly reliable turboprop engines were retrofitted to still serviceable commercial aircraft airframes to replace relatively complex piston engines, thus extending the lifetime of the aircraft by a decade or more.

17.18 LIFE-CYCLE ASSESSMENT

From the beginning, industrial ecology must consider process/product design in the management of materials, including the ultimate fates of materials when they are discarded. The product and materials in it should be subjected to an entire *life-cycle assessment* or analysis. A life-cycle assessment applies to products, processes, and services through their entire life cycles from extraction of raw materials—through manufacturing, distribution, and use—to their final fates from the viewpoint of determining, quantifying, and ultimately minimizing their environmental impacts. It takes account of manufacturing, distribution, use, recycling, and disposal. Life-cycle assessment is particularly beneficial in determining the relative environmental merits of alternative products and services. At the consumer level, this could consist of an evaluation of paper versus styrofoam drinking cups. On an industrial scale, life-cycle assessment could involve evaluation of nuclear versus fossil-energy-based electrical power plants.

A basic step in life-cycle analysis is *inventory analysis*, which provides qualitative and quantitative information regarding consumption of material and energy resources (at the beginning of the cycle) and releases to the anthrosphere, hydrosphere, geosphere, and atmosphere (during or at the end of the cycle). It is based on various materials cycles and budgets, and it quantifies materials and energy required as input and the benefits and liabilities posed by products. The related area of *impact analysis* provides information about the kind and degree of environmental impacts resulting from a complete life cycle of a product or activity. Once the environmental and resource impacts have been evaluated, it is possible to do an *improvement analysis* to determine measures that can be taken to reduce impacts on the environment or resources.

In making a life-cycle analysis the following must be considered:

- If there is a choice, selection of the kinds of materials that will minimize waste
- Kinds of materials that can be reused or recycled
- Components that can be recycled
- Alternate pathways for the manufacturing process or for various parts of it

Although a complete life-cycle analysis is expensive and time consuming, it can yield significant returns in lowering environmental impacts, conserving resources, and reducing costs. This is especially true if the analysis is performed at an early stage in the development of a product or service. Improved computerized techniques are making significant advances in the ease and efficacy of life-cycle analyses. Until now, life-cycle assessments have been largely confined to simple materials and products such as reusable cloth versus disposable paper diapers. A major challenge now is to expand these efforts to more complex products and systems such as aircraft or electronics products.

17.18.1 Scoping in Life-Cycle Assessment

A crucial early step in life-cycle assessment is *scoping* the process by determining the boundaries of time, space, materials, processes, and products to be considered. Consider as an example the manufacture of parts that are rinsed with an organochloride solvent in which some solvent is lost by evaporation to the atmosphere, by staying on the parts, during the distillation and purification process by which the solvent is made suitable for recycling, and by disposal of waste solvent that cannot be repurified. The scope of the life-cycle assessment could be made very narrow by confining it to the process as it exists. An assessment could be made of the solvent losses, the impacts of these losses, and means for reducing the losses, such as reducing solvent emissions to the atmosphere by installation of activated carbon air filters or reducing losses during purification by employing more efficient distillation processes. A more broadly scoped life-cycle assessment would be to consider alternatives to the organochloride solvent. An even broader scope would consider whether the parts even need to be manufactured; Are there alternatives to their use?

17.19 CONSUMABLE, RECYCLABLE, AND SERVICE (DURABLE) PRODUCTS

In industrial ecology, most treatments of life-cycle analysis make the distinction between *consumable products*, which are essentially used up and dispersed to the environment during their life cycle and *service or durable products*, which essentially remain in their original form after use. Gasoline is clearly a consumable product, whereas the automobile in which it is burned is a service product. It is useful, however, to define a third category of products that clearly become "worn out" when employed for their intended purpose, but which remain largely undispersed to the environment. The motor oil used in an automobile is such a substance in that most of the original material remains after use. Such a category of material may be called a *recyclable commodity*.

17.19.1 Desirable Characteristics of Consumables

Consumable products include laundry detergents, hand soaps, cosmetics, windshield washer fluids, fertilizers, pesticides, laser printer toners, and all other materials that are impossible to reclaim after they are used. The environmental implications of the use of consumables are many and profound. In the late 1960s and early 1970s, for example, nondegradable surfactants in detergents caused severe foaming and esthetic problems at water treatment plants and sewage outflows, and the phosphate builders in the detergents promoted excessive algal growth in receiving waters resulting in eutrophication (see Chapter 7). Lead in consumable leaded gasoline was widely dispersed to the environment when the gasoline was burned. These problems have now been remedied with the adoption of phosphate-free detergents with biodegradable surfactants and mandatory use of unleaded gasoline.

Since they are destined to be dispersed into the environment, consumables should meet several "environmentally friendly" criteria, including the following:

- *Degradability*: This usually means biodegradability, such as that of household detergent constituents that occurs in waste treatment plants and in the environment. Chemical degradation may also occur.
- *Nonbioaccumulative*: Lipid-soluble, poorly biodegradable substances, such as DDT and PCBs, tend to accumulate in organisms and to be magnified through the food chain. This characteristic should be avoided in consumable substances.
- *Nontoxic*: To the extent possible, consumables should not be toxic in the concentrations that organisms are likely to be exposed to them. In addition to their not being acutely toxic, consumables should not be mutagenic, carcinogenic, or teratogenic (cause birth defects).

17.19.2 Desirable Characteristics of Recyclables

Recyclables is used here to describe materials that are not used up in the sense that laundry detergents or photocopier toners are consumed, but are not durable items. In this context, recyclables can be understood to be chemical substances and formulations. The HCFCs used as refrigerant fluids fall into this category, as does ethylene glycol mixed with water in automobile engine antifreeze/antiboil formulations (although rarely recycled in practice).

Insofar as possible, recyclables should be minimally hazardous with respect to toxicity, flammability, and other hazards. For example, both volatile hydrocarbon solvents and organochloride (chlorinated hydrocarbon) solvents are recyclable after use for parts degreasing and other applications requiring a good solvent for organic materials. The hydrocarbon solvents have relatively low toxicities, but may present flammability hazards during use and reclamation for recycling. The organochloride solvents are less flammable, but may pose a greater toxicity hazard. An example of such a solvent is carbon tetrachloride, which is so nonflammable that it was once used in fire extinguishers, but the current applications of which are highly constrained because of its high toxicity.

An obviously important characteristic of recyclables is that they should be designed and formulated to be amenable to recycling. In some cases there is little leeway in formulating potentially recyclable materials; motor oil, for example, must meet certain performance criteria, including the ability to lubricate, stand up to high temperatures, and other attributes, regardless of its ultimate fate. In other cases formulations can be modified to enhance recyclability. For example, the use of bleachable or removable ink in newspapers enhances the recyclability of the newsprint, enabling it to be restored to an acceptable level of brightness.

For some commodities, the potential for recycling is enormous. This can be exemplified by lubricating oils. The volume of motor oil sold in the United States each year for internal combustion engines is about 2.5 billion liters, a figure that is doubled if all lubricating oils are considered. A particularly important aspect of utilizing recyclables is their collection. In the case of motor oil, collection rates are low from consumers who change their own oil, and they are responsible for the dispersion of large amounts of waste oil to the environment.

17.19.3 Desirable Characteristics of Service Products

Since, in principle at least, service products are destined for recycling, they have comparatively lower constraints on materials and higher constraints on their ultimate disposal. A major impediment to the recycling of service products is the lack of convenient channels through which they can be put into the recycling loop. Television sets and major appliances such as washing machines or ovens have many recyclable components, but often end up in landfills and waste dumps simply because there is no handy means for getting them from the user and into the recycling loop. In such cases, government intervention may be necessary to provide appropriate channels. One partial remedy to the disposal/recycling problem consists of leasing arrangements or payment of deposits on items such as batteries to ensure their return to a recycler. The terms "de-shopping" or "reverse shopping" describe a process by which service commodities would be returned to a location such as a parking lot where they could be collected for recycling. According to this scenario, the analogy to a supermarket would be a facility in which service products are collected and disassembled for recycling.

Much can be done in the design of service products to facilitate their recycle. One of the main characteristics of recyclable service products must be ease of disassembly so that remanufacturable components and recyclable materials, such as copper wire, can be readily removed and separated for recycling.

17.20 DESIGN FOR ENVIRONMENT

Design for environment is the term given to the approach of designing and engineering products, processes, and facilities in a manner that minimizes their adverse environmental impacts and, where possible, maximizes their beneficial environmental effects. In modern industrial operations, design for environment is part of a larger scheme termed "design for *X*," where "*X*" can be any one of a number of characteristics such as assembly, manufacturability, reliability, and serviceability. In making such a design, numerous desired characteristics of the product must be considered, including ultimate use, properties, costs, and appearance. Design for environment requires that the designs of the product, the process by which it is made, and the facilities involved in making it conform to appropriate environmental goals and limitations imposed by the need to maintain environmental quality. It must also consider the ultimate fate of the product, particularly whether or not it can be recycled at the end of its normal life span.

17.20.1 Products, Processes, and Facilities

In discussing design for environment, the distinctions among products, processes, and facilities must be kept in clear perspective. *Products*—automobile tires, laundry detergents, and refrigerators—are items sold to consumers. *Processes* are the means of producing products and services. For example, tires are made by a process in which hydrocarbon monomers are polymerized to produce rubber that is molded in the shape of a tire with a carcass reinforced by synthetic fibers and steel wires. A *facility* is where processes are carried out to produce or deliver products or services. In cases where services are regarded as products, the distinction between products and processes becomes blurred. For example, a lawn care service delivers products in the forms of fertilizers, pesticides, and grass seeds, but also delivers pure services including mowing, edging, and sod aeration.

Although *products* tend to get the most public attention in consideration of environmental matters, *processes* often have more environmental impact. Successful process designs tend to stay in service for many years and to be used to make a wide range of products. While the product of a process may have minimal environmental impact, the process by which the product is made may have marked environmental effects. An example is the manufacture of paper. The environmental impact of paper as a product, even when improperly discarded, is not terribly great, whereas the process by which it is made involves harvesting wood from forests, high use of water, potential emission of a wide range of air pollutants, and other factors with profound environmental implications.

Processes develop symbiotic relationships when one provides a product or service utilized in another. An example of such a relationship is that between steel making and the process for the production of oxygen required in the basic oxygen process by which carbon and silicon impurities are oxidized from molten iron to produce steel. The long lifetimes and widespread applicability of popular processes make their design for environment of utmost importance.

The nature of a properly functioning system of industrial ecology is such that processes are even more interwoven than would otherwise be the case, because by-products from some processes are used by other processes. Therefore, the processes employed in such a system and the interrelationships and interdependencies among them are particularly important. A major change in one process may have a “domino effect” on the others.

17.20.2 Key Factors in Design for Environment

Two key choices that must be made in design for environment are those involving materials and energy. The choices of materials in an automobile illustrate some of the possible trade-offs. Steel as a component of automobile bodies requires relatively large amounts of energy and involves significant environmental disruption in the mining and processing of iron ore. Steel is a relatively heavy material, so more energy is involved in moving automobiles made of steel. However, steel is durable, has a high rate of recycling, and is produced initially from abundant sources of iron ore. Aluminum is much lighter than steel and quite durable. It is one of the most commonly recycled commodities. Good primary sources of aluminum, bauxite ores, are not as abundant as iron ores, and large amounts of energy are required in the primary production of aluminum. Plastics are another source of automotive components. The light weight of plastic reduces automotive fuel consumption, plastics with desired properties are readily made, and molding and shaping plastic parts is a straightforward process. However, plastic automobile components have a low rate of recycling.

Three related characteristics of a product that should be considered in design for environment are durability, repairability, and recyclability. *Durability* simply refers to how well the product lasts and resists breakdown in normal use. Some products are notable for their durability; ancient two-cylinder John Deere farm tractors from the 1930s and 1940s are legendary in farming circles for their durability, enhanced by the affection engendered in their owners who tend to preserve them. *Repairability* is a measure of how easy and inexpensive it is to repair a product. A product that can be repaired is less likely to be discarded when it ceases to function for some reason. *Recyclability* refers to the degree and ease with which a product or components of it may be recycled. An important aspect of recyclability is the ease with which a product can be disassembled into constituents consisting of a single material that can be recycled. It also considers whether the components are made of materials that can be recycled.

17.20.3 Hazardous Materials in Design for Environment

A key consideration in the practice of design for environment is the reduction of the dispersal of hazardous materials and pollutants. This can entail the reduction or elimination of hazardous materials in manufacture, an example of which was the replacement of stratospheric ozone-depleting CFCs in foam blowing of plastics. If appropriate substitutes can be found, somewhat toxic and persistent chlorinated solvents should not be used in manufacturing applications such as parts washing. The use of hazardous materials in the product—such as batteries containing toxic cadmium, mercury, and lead—should be eliminated or minimized. Pigments containing heavy metal cadmium or lead should not be used if there are any possible substitutes. The substitution of HCFCs and hydrofluorocarbons for ozone-depleting CFCs in products (refrigerators and air conditioners) is an example of a major reduction in environmentally damaging materials in products. The elimination of extremely persistent PCBs from electrical transformers removed a major hazardous waste problem due to the use of a common product.

17.21 INHERENT SAFETY

The use of nitroglycerin in the construction of the Central Pacific Railroad in California during the 1860s provides a lesson in safe handling of materials. On April 3, 1866, 70 crates of California-bound dangerously explosive nitroglycerin blew up aboard the steamship European being unloaded on the Caribbean coast of Panama killing 50 people and causing substantial damage. Only 2 weeks later two crates of nitroglycerin, which had been refused delivery because of their condition, exploded in an office of the Wells Fargo Company in San Francisco killing 15 people and causing massive destruction, although instantaneously resolving the question regarding what to do with the damaged goods. Only 2 days later six workers on the Central Pacific Railroad in the Sierra Nevada mountains were killed by an explosion of nitroglycerin that was being used in place of much less effective black gunpowder, for blasting on the railroad construction. As a consequence of these accidents, California legislators quickly passed legislation forbidding the transportation of nitroglycerin through San Francisco and Sacramento.

It appeared that the measures taken by the California authorities would prevent the use of the powerfully explosive nitroglycerin in the construction of the Central Pacific causing a huge delay of the project. However, James Howden, a British chemist, won contracts to make nitroglycerin on site for the massive Central Pacific construction project and he produced up to 100 lbs per day of the explosive as needed with no further injuries resulting from accidental nitroglycerin explosions. This was an early example of safe handling of material. The glycerin and sulfuric and nitric acids needed to make nitroglycerin could be transported to the site in relative safety and with no explosion hazard. The truly hazardous nitroglycerin was made only as needed so that at no time was there a large amount available in one place to cause an explosion. Finally, the nitroglycerin was only transported short distances greatly minimizing the possibility of transportation accidents with this notoriously sensitive explosive.

A chemical process is said to be inherently safe when permanent measures have been integrated into the process to reduce or eliminate specific hazards. Five approaches by which this may be done are the following.

1. Use only minimum quantities of hazardous substances. In the example of nitroglycerin discussed above relatively small quantities of the explosive were made at any one time.
2. Use a less hazardous substance. In 1867 Alfred Nobel invented dynamite, which consisted of nitroglycerin absorbed onto a carrier such as sawdust and which was much safer to handle than pure nitroglycerin but almost as effective as an explosive.
3. Use safer conditions. In a chemical process, this might consist of carrying out reactions at lower temperatures and pressures over a catalyst so that if a malfunction occurs, the results are much less catastrophic.
4. Simplify. As a general rule, simpler processes are safer processes. Added steps add possibilities for things to go wrong.
5. Maximize continuous steady-state operation with continuous processes that avoid problems likely to occur during startup and shutdown.

17.21.1 Increased Safety with Smaller Size

Safety can be increased by using "green reactors,"[11] often by the simple expedient of down-sizing operations and quantities of materials, a strategy of *minimization*. A common example of minimization is the substitution of small continuous flow reactors for large batch processes (Figure 17.12). In scaling up from batch laboratory processes to commercial production, large batch reactors often have been employed. Such a reactor containing large amounts of material can be very problematic if something goes wrong, such as an uncontrolled runaway reaction. Such reactors often have been used because of slow mixing and heating that required reactants to be in contact for long periods of time, whereas the actual reaction is often very quick once the reactants are in contact. It is often desirable to substitute a very small loop reactor with efficient mixing and rapid energy transfer so that the quantities of reactants in the reaction process are very much reduced (Figure 17.12). In such a case, only much smaller quantities are involved and if anything goes wrong the amounts of material that must be dealt with are much smaller.

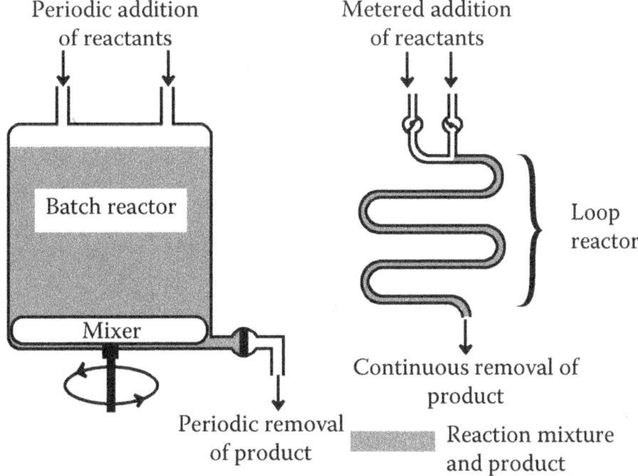

FIGURE 17.12 Batch reactor (left) and continuous flow reactor (right) for chemical synthesis. Note the much smaller volume (typically by a factor of 1/100) of reaction mixture and product in the continuous flow reactor.

17.22 INDUSTRIAL ECOLOGY AND ECOLOGICAL ENGINEERING

Ecological engineering seeks to integrate the anthrosphere and its activities with natural ecosystems to mutual advantage. Ecological engineering has many commonalities with industrial ecology. Ecological engineering combines systems ecology with engineering. It develops ecotechnologies through designing, constructing, and managing systems of natural ecology integrated with the anthrosphere.

The greatest success to date from the application of ecological engineering is the construction and operation of wetlands to treat wastewater. Constructed wetlands do not necessarily duplicate the living species and other aspects of natural wetlands and may even be located in areas where wetlands have not previously existed. They do provide the essential ingredients—water, sunlight, nutrients, and confined conditions—that enable wetland-based ecosystems to develop and thrive. Other endeavors involving ecological engineering include restoration ecology of areas damaged by development, phytoremediation using living plants to remove pollutants from polluted areas, stream restoration, and soil bioengineering that uses ecosystems and the plants in them to reduce soil erosion and to increase agricultural productivity in a sustainable manner.

The most ambitious project based on ecological engineering to date in the United States is the restoration of the Florida Everglades. This huge area is a vast wetland that has been drained and damaged by channelization projects carried out by the U.S. Army Corps of Engineers. Now a massive effort is underway to reverse the damage. The Everglades will never be what they were before human intervention, but the successful completion of the project will see the re-establishment of a functional wetlands system to the benefit of alligators, other wildlife, and eventually humans.

LITERATURE CITED

1. Frosch, R. A. and N. E. Gallopoulos, Strategies for manufacturing, *Scientific American*, **261**, 94–102, 1989.
2. Manahan, S. E., *Green Chemistry and the Ten Commandments of Sustainability*, 2nd ed., ChemChar Research, Columbia, MO, 2006.
3. Anastas, P. T. and J. C. Warner, *Green Chemistry: Theory and Practice*, Oxford University Press, New York, 2000.
4. Sheldon, R. A., E factors, green chemistry and catalysis: An odyssey, *Chemical Communications*, **29**, 3352–3365, 2008.
5. Clark, J. H., F. E. I. Deswarte, and T. J. Farmer, The integration of green chemistry into future biorefineries, *Biofuels, Bioproducts and Biorefining*, **3**, 72–90, 2009.
6. Metzger, J. O. and U. Bornscheuer, Lipids as renewable resources: Current state of chemical and biotechnological conversion and diversification, *Applied Microbiology and Biotechnology*, **71**, 13–22, 2006.
7. Ran, N., L. Zhao, Z. Chen, and J. Tao, Recent applications of biocatalysis in developing green chemistry for chemical synthesis at the industrial scale, *Green Chemistry*, **10**, 361–372, 2008.
8. Herrmann, J.-M., C. Duchamp, M. Karkmaz, B. T. Hoai, B. H. Lachheb, H., E. Puzenat, and C. Guillard, Environmental green chemistry as defined by photocatalysis, *Journal of Hazardous Materials*, **146**, 624–629, 2007.
9. Oelgemoller, M., C. Jung, and J. Mattay, Green photochemistry: Production of fine chemicals with sunlight, *Pure and Applied Chemistry*, **79**(11), 1939–1947, 2007.
10. Gallopoulos, N. E., Industrial ecology: An overview, *Progress in Industrial Ecology*, **3**, 10–27, 2006.
11. Doble, M., Green Reactors, *Chemical Engineering Progress*, **104**, 33–42, 2008.

SUPPLEMENTARY REFERENCES

Ahluwalia, V. K. and M. Kidwai, *New Trends in Green Chemistry*, Kluwer Academic Publishers, Boston, 2004.

Allen, D. T. and D. R. Shonnard, *Green Engineering: Environmentally Conscious Design of Chemical Processes*, Prentice Hall, Upper Saddle River, NH, 2002.

Anastas, P., Ed., *Handbook of Green Chemistry*, Wiley-VCH, New York, 2010.

Ayres, R. U. and L. W. Ayres, Eds, *A Handbook of Industrial Ecology*, Edward Elgar Publishing, Cheltenham, UK, 2002.

Ayres, R. U. and B. Warr, *The Economic Growth Engine: How Energy and Work Drive Material Prosperity*, Edward Elgar Publishing, Northampton, MA, 2009.

Barr-Kumar, R., *Green Architecture: Strategies for Sustainable Design*, Barr International, Washington, DC, 2003.

Beer, T. and A. Ismail-Zadeh, *Risk Science and Sustainability: Science for Reduction of Risk and Sustainable Development for Society*, Kluwer Academic Publishers, Boston, 2003.

Booth, D. E., *Hooked on Growth: Economic Addictions and the Environment*, Rowman & Littlefield Publishers, Lanham, MD, 2004.

Clark J. and D. MacQuarrie, *Handbook of Green Chemistry and Technology*, Blackwell Science, Malden, MA, 2002.

DeSimone, J. M., W. Tumas, Eds, *Green Chemistry Using Liquid and Supercritical Carbon Dioxide*, Oxford University Press, New York, 2003.

DeSimone, L. D. and F. Popoff, *Eco-efficiency: The Business Link to Sustainable Development*, MIT Press, Cambridge, MA, 1997.

Doble, M. and A. K. Kruthiventi, *Green Chemistry and Processes*, Elsevier, Amsterdam, 2007.

Doxsee, K. M. and J. E. Hutchison, *Green Organic Chemistry: Strategies, Tools, and Laboratory Experiments*, Thomson-Brooks/Cole, Monterey, CA, 2004.

Ehrenfeld, J. R., Industrial ecology: Coming of age, *Environmental Science and Technology*, **36**, 281A–285A, 2002.

El-Haggar, S. M., *Sustainable Industrial Design and Waste Management: Cradle-to-Cradle for Sustainable Development*, Elsevier Academic Press, Amsterdam, 2007.

Graedel, T. E. and B. R. Allenby, *Industrial Ecology*, 2nd ed., Prentice Hall, Upper Saddle River, NJ, 2003.

Graedel, T. E. and J. A. Howard-Grenville, *Greening the Industrial Facility: Perspectives, Approaches, and Tools*, Springer, Berlin, 2005.

Green, K. and S. Randles, Eds, *Industrial Ecology and Spaces of Innovation*, Edward Elgar Publishing, Northampton, MA, 2006.

Gupta, S. M. and A. J. D. Lambert, Eds, *Environment Conscious Manufacturing*, Taylor & Francis/CRC Press, Boca Raton, FL, 2008.

Hendrickson, C. T., L. B. Lave, and H. Scott Matthews, *Environmental Life Cycle Assessment of Goods and Services: An Input-Output Approach*, Resources for the Future, Washington, DC, 2006.

Horvath, I. T. and P. T. Anastas, Innovations and green chemistry, *Chemical Reviews*, **107**, 2169–2173, 2007.

Hunkeler, D., K. Lichtenvort, and G. Rebitzer, Eds, *Environmental Life Cycle Costing*, Taylor & Francis/CRC Press, Boca Raton, FL, 2008.

Islam, M. R., Ed., *Nature Science and Sustainable Technology*, Nova Science Publishers, New York, 2007.

Keeler, M. and B. Burke, *Fundamentals of Integrated Design for Sustainable Building*, Wiley, Hoboken, NJ, 2009.

Kronenberg, J., *Ecological Economics and Industrial Ecology: A Case Study of the Integrated Product Policy of the European Union*, Routledge, New York, 2007

Kruger, P., *Alternative Energy Resources: The Quest for Sustainable Energy*, Wiley, Hoboken, NJ, 2006.

Kutz, M., Ed., *Environmentally Conscious Transportation*, Wiley, Hoboken, NJ, 2008.

Lankey, R. L. and P. T. Anastas, Eds, *Advancing Sustainability through Green Chemistry and Engineering*, American Chemical Society, Washington, DC, 2002.

Levett, R., *A Better Choice of Choice: Quality of Life, Consumption and Economic Growth*, Fabian Society, London, 2003.

Li, C.-J. and B. M. Trost, Green chemistry for chemical synthesis, *Proceedings of the National Academy of Sciences of the United States of America*, **105**, 13197–13202, 2008.

Lifset, R. and T. E. Graedel, Industrial ecology: Goals and definitions, in *A Handbook of Industrial Ecology*, Robert U. Ayres and L. Ayres, Eds, Edward Elgar Publishing, Cheltenham, UK, 2002.

Lutz, W. and W. Sanderson, Eds, *The End of World Population Growth: Human Capital and Sustainable Development in the 21st Century*, Earthscan, Sterling, VA, 2004.

Madu, C. N., *Handbook of Environmentally Conscious Manufacturing*, Kluwer Academic Publishers, Boston, 2001.

McDonough, W. and M. Braungart, *Cradle to Cradle: Remaking The Way We Make Things,* North Point Press, New York, 2002.

Nelson, W. M., *Green Solvents for Chemistry: Perspectives and Practice*, Oxford University Press, New York, 2003.

Nelson, W. M., Ed., *Agricultural Applications in Green Chemistry*, Oxford University Press, New York, 2004.
OECD, *The Application of Biotechnology to Industrial Sustainability*, OECD, Paris, 2001.
Simpson, R. D., M. A. Toman, and R. U. Ayres, Eds, *Scarcity and Growth Revisited: Natural Resources and the Environment in the New Millennium*, Resources for the Future, Washington, DC, 2005.
Socolow, R., C. Andrews, F. Berkhout, and V. Thomas, Eds, *Industrial Ecology and Global Change*, Cambridge University Press, New York, 1994.
Sonnemann, G., F. Castells, and M. Schuhmacher, *Integrated Life-Cycle and Risk Assessment for Industrial Processes*, CRC Press/Lewis Publishers, Boca Raton, FL, 2003.
Tundo, P., A. Perosa, and F. Zecchini, *Methods and Reagents for Green Chemistry*, Wiley-Interscience, Hoboken, NJ, 2007.
Tundo, P. and V. Esposito, Eds, *Green Chemical Reactions*, Springer, Dordrecht, The Netherlands, 2008.
Udo de Haes, H. A., *Life-cycle Impact Assessment: Striving Towards Best Practice*, Society of Environment Toxicology and Chemistry, Pensacola, FL, 2002.
Van den Bergh, Jeroen C. J. M. and M. A. Janssen, Eds, *Economics of Industrial Ecology: Materials, Structural Change, and Spatial Scales*, MIT Press, Cambridge, MA, 2004.
Viegas, J., Ed., *Critical Perspectives on Planet Earth*, Rosen Publishing Group, New York, 2007.
Von Gleich, A., R. U. Ayres, and S. Gössling-Reisemann, Eds, *Sustainable Metals Management: Securing our Future—Steps Towards a Closed Loop Economy*, Springer, Dordrecht, The Netherlands, 2006.

QUESTIONS AND PROBLEMS

1. Although abundant, why is lignin not a good candidate as a raw material?
2. What structural features of the compound

```
    H      H      H
    |      |      |
 H—C——C——C—H
    |      |      |
 ONO2  ONO2  ONO2
```

   would make it hazardous? What is the nature of the hazard?
3. What is the potential use of dialkyl carbonates, such as dimethyl carbonate, in green chemical synthesis?
4. What are some uses of organic solvents other than for reaction media? What are some of the drawbacks of organic solvents for these uses?
5. In terms of interaction with reagents, what is the greatest disadvantage of water as a solvent? What is the greatest advantage of water as a solvent for a variety of solutes of biological origin?
6. What is a dense phase fluid? What form of dense phase carbon dioxide is produced at very high pressures and slightly elevated temperature?
7. What are the advantages of supercritical fluid carbon dioxide solvent? Why is carbon dioxide's volatility an advantage in some cases?
8. Discuss how electrons and photons can be regarded as catalysts. In what respects are they "massless" reagents?
9. Look up on the internet the Haber process for the synthesis of ammonia. Discuss and compare the conditions and relative advantages and disadvantages of the Haber process and its biological alternatives within the context of green chemistry.
10. Look up the Presidential Green Chemistry Award on the internet. From the information obtained, list several examples of green chemistry that have been put into practice on an industrial scale.
11. What is the relationship of industrial ecology to green chemistry? In which ways are industrial ecology and green chemistry complementary?
12. What is the meaning of "command and control"? What are its limitations in the control of pollution?
13. In what sense does the practice of green chemistry ensure environmental quality by "natural," self-regulating means?

14. What is the role of sustainability in the practice of green chemistry?
15. How is atom economy defined? In what sense is it a key aspect of the practice of green chemistry?
16. Look up the phenomenon of mineralization as it occurs in biological ecosystems. Name and describe a process analogous to mineralization that occurs in an industrial ecosystem.
17. How are the terms industrial metabolism, industrial ecosystem, and sustainable development related to industrial ecology?
18. From the definition of symbiosis, explain what is meant by industrial symbiosis. How is industrial symbiosis related to industrial ecology?
19. Justify or refute the statement that in an operational industrial ecosystem only energy is consumed.
20. In what sense is the consumer sector the most difficult part of an industrial ecosystem?
21. In what sense might a "moon station" or a colony on Mars advance the practice of industrial ecology?
22. Look up some information regarding the nature of electronic apparatus in the 1940s and 1950s. In what sense do modern solid-state electronic devices illustrate both dematerialization (use of less material) and material substitution (substitution of more readily available materials for those that are scarce)?
23. What are the enterprises that serve to underpin the Kalundborg industrial ecosystem? How might they compare with the basic enterprises of an industrial ecosystem consisting of rural counties in the state of Iowa?
24. How does "design for recycling" (DFR) relate to embedded utility?
25. Distinguish among consumable, durable (service), and recyclable products.
26. List some of the "environmentally friendly" criteria met by soap as a consumable commodity.
27. Consider a university as an industrial ecosystem in which the ultimate "consumer" is society that utilizes and benefits from educated graduates. Describe ways in which the university fits the model of an industrial ecosystem and ways in which it does not. Is there any recycling? Can you suggest ways in which a university might become a more efficient ecosystem?
28. Suppose that it is proposed to construct a huge system to divert a significant amount of water from near the mouth of the Mississippi River and pump it with power provided by giant wind farms in Texas across the southern United States and into southern California and northern Mexico. Suggest how such a project might constitute an industrial ecosystem and what it would include. Suggest advantages and possible disadvantages.
29. The Mississippi River water that would be used in the project suggested in the preceding question contains algal (plant) nutrients in the form of phosphates, inorganic nitrogen, and potassium that cause excessive plant growth (eutrophication) in large regions of the Gulf of Mexico near the mouth of the river. The water also contains relatively high levels of oxygen-demanding organics, silt, and some industrial chemicals which, along with eutrophication, result in the formation of a "dead zone" at certain times of the year in the Gulf of Mexico. Suggest how ecological engineering could be applied to the proposed water project to mitigate these water pollution problems and deliver a clean water product to the end users.
30. Globalization of economies is a contentious issue. Suggest how globalization may relate to the practice of industrial ecology. Suggest ways in which globalization may help and may hurt the proper practice of industrial ecology.

18 Resources and Sustainable Materials

18.1 WHERE TO GET THE STUFF WE NEED?

One of the greatest challenges facing humankind is posed by the demand for materials that humans need (or at least want) to meet their desires for higher material standards of living. The economic effects of this demand were illustrated painfully during the approximately 2005–2008 time period when demand for materials such as crude oil, aluminum, copper, lead, zinc, phosphate minerals (for fertilizer), and other commodities sent prices of these and many other materials soaring. This demand was fueled by a number of factors operating throughout the world including the evolving economies of highly populated China and India and buying binges of consumers in the United States made over-confident by rapidly increasing housing values, increasing prices of stock, and easy credit on credit cards. In early 2008, it began to look like the price of crude oil would soar far above \$150 per barrel, gasoline would surpass \$5.00 per gallon in the Unites States (still inexpensive by European standards), and grain for food and animal feed would continue to increase beyond record levels. The prices of metals had risen such that thieves were raiding unoccupied houses to steal aluminum and copper as some other thieves were even cutting catalytic converters from vehicles to get to their precious metal contents. Around mid-2008, a wrenching adjustment occurred as it became obvious that such price increases were unsustainable, prices of commodities such as crude oil dropped drastically, and home prices in the United States fell dramatically as countries throughout much of the world experienced the worst economic downturn (a very big recession or a mini-depression) since the Great Depression of the 1930s.

These events clearly illustrate the importance of materials for modern societies. The acquisition, utilization, and disposal of materials have enormous environmental consequences. Earth simply cannot support the current trajectory of material use. This is especially so as countries with large and growing populations aspire to the standard of living enjoyed by modern industrialized countries such as the United States, Canada, and Australia. Calculations of how many "Earths" would be required to satisfy material demands if all the world's people enjoyed a U.S. standard of living have shown that perhaps as many as 10 planets such as ours would be needed. It is obvious that materials are of the utmost importance in sustainability. This chapter is devoted to materials and the resources from which they are acquired. Energy is a resource of particular importance, so Chapter 19, "Sustainable Energy: The Key to Everything," is devoted to the topic of energy.

The materials needed for modern societies can be provided from either *extractive* (nonrenewable) or *renewable* sources. Extractive industries remove irreplaceable mineral resources from the earth's crust. The utilization of mineral resources is strongly tied with technology, energy, and the environment. Perturbations in one usually cause perturbations in the others. For example, reductions in automotive exhaust pollutant levels to curtail air pollution have required use of catalytic devices that contain platinum-group metals, a valuable and irreplaceable natural resource. Implementation of the best practices of industrial ecology and green chemistry will be required to improve environmental quality with reduced consumption of nonrenewable material resources.

In discussing nonrenewable sources of minerals as well as energy, it is useful to define two terms related to available quantities. The first of these is *resources*, defined as quantities that are estimated to be *ultimately* available. The second term is *reserves*, which refers to well-identified resources that can be profitably utilized with existing technology.

18.2 MINERALS IN THE GEOSPHERE

There are numerous kinds of mineral deposits that are used in various ways. These are, for the most part, sources of metals which occur in *batholiths* composed of masses of igneous rock that have been extruded in a solid or molten state into the surrounding rock strata. In addition to deposits formed directly from solidifying magma, associated deposits are produced by water interacting with magma. Hot aqueous solutions associated with magma can form rich *hydrothermal* deposits of minerals. Several important metals, including lead, zinc, and copper, are often associated with hydrothermal deposits.

Some useful mineral deposits are formed as *sedimentary deposits* along with the formation of sedimentary rocks. *Evaporites* are produced when seawater is evaporated. Common mineral evaporites are halite (NaCl), sodium carbonates, potassium chloride, gypsum ($CaSO_4 \cdot 2H_2O$), and magnesium salts. Many significant iron deposits consisting of hematite (Fe_2O_3) and magnetite (Fe_3O_4) were formed as sedimentary bands when earth's atmosphere was changed from reducing to oxidizing as photosynthetic organisms produced oxygen, precipitating the oxides from the oxidation of soluble Fe^{2+} ion.

Deposition of suspended rock solids by flowing water can cause segregation of the rocks according to differences in size and density. This can result in the formation of useful *placer* deposits that are enriched in desired minerals. Gravel, sand, and some other minerals, such as gold, often occur in placer deposits.

Some mineral deposits are formed by the enrichment of desired constituents when other fractions are weathered or leached away. The most common example of such a deposit is bauxite, Al_2O_3, remaining after silicates and other more soluble constituents have been dissolved from aluminum-rich minerals by the weathering action of water under the severe conditions of hot tropical climates with very high levels of rainfall. This kind of material is called a *laterite*.

18.2.1 Evaluation of Mineral Resources

World mineral resources are obviously critical to the well-being of modern civilization. In order to make its extraction worthwhile, a mineral must be enriched at a particular location in the earth's crust relative to the average crustal abundance. Normally applied to metals, such an enriched deposit is called an *ore*. The value of an ore is expressed in terms of a *concentration factor*:

$$\text{Concentration factor} = \frac{\text{Concentration of material in ore}}{\text{Average crustal concentration}} \tag{18.1}$$

Obviously, higher concentration factors are always desirable. Required concentration factors decrease with average crustal concentrations and with the value of the commodity extracted. A concentration factor of 4 might be adequate for iron, which makes up a relatively high percentage of the earth's crust. Concentration factors must be several hundred or even several thousand for less expensive metals that are not present at very high percentages in the earth's crust. However, for an extremely valuable metal, such as platinum, a relatively low concentration factor is acceptable because of the high financial return obtained from extracting the metal.

Acceptable concentration factors are a sensitive function of the price of a metal. Shifts in price can cause significant CCS (copper/chrome/arsenic). If the price of a metal increases by, for example, 50%, and the increase appears to be long term, it becomes profitable to mine deposits that had not been mined previously. The opposite can happen, as is often the case when richer sources are discovered or substitute materials are found.

In addition to large variations in the concentration factors of various ores, there are extremes in the geographic distribution of mineral resources. The United States is perhaps about average for all nations in terms of its mineral resources, possessing significant resources of copper, lead, iron, gold, and molybdenum, but virtually without resources of some important strategic metals, including chromium, tin, and platinum-group metals. For its size and population, South Africa is particularly blessed with some important metal mineral resources.

18.3 EXTRACTION AND MINING

Minerals are usually extracted from the earth's crust by various kinds of mining procedures, but other techniques may be employed as well. The raw materials so obtained include inorganic compounds such as phosphate rock, sources of metal such as lead sulfide ore, clay used for firebrick, and structural materials, such as sand and gravel.

Surface mining, which can consist of digging large holes in the ground, or strip mining, is used to extract minerals that occur near the surface. A common example of surface mining is quarrying of rock. Vast areas have been dug up to extract coal. Because of past mining practices, surface mining got a well-deserved bad name. With modern reclamation practices, however, topsoil is first removed and stored. After the mining is complete, the topsoil is spread on top of overburden that has been replaced such that the soil surface has gentle slopes and proper drainage. Topsoil spread over the top of the replaced spoil, often carefully terraced to prevent erosion, is seeded with indigenous grass and other plants, fertilized, and watered, if necessary, to provide vegetation. The end result of carefully done *mine reclamation* projects is a well-vegetated area suitable for wildlife habitat, recreation, forestry, grazing, and other beneficial purposes. Such a project can be regarded as an application of ecological engineering (see Section 17.22).

Water pollution is often a problem associated with mining. One of the most common problems is the formation of acid mine water (H_2SO_4) from microbial action on pyrite (FeS_2) exposed to the atmosphere in mining a number of kinds of mineral ores (see Section 6.14). Processes have been developed for the treatment of this acid rock drainage, such as those using sulfate-reducing bacteria in bioreactors.[1]

Extraction of minerals from placer deposits formed by deposition from water has obvious environmental implications. Mining of placer deposits can be accomplished by dredging from a boom-equipped barge. Another means that can be used is hydraulic mining with large streams of water. One interesting approach for more coherent deposits is to cut the ore with intense water jets, then suck up the resulting small particles with a pumping system. These techniques have a high potential to pollute water and disrupt waterways and are environmentally controversial.

For many minerals, underground mining is the only practical means of extraction. An underground mine can be very complex and sophisticated. The structure of the mine depends on the nature of the deposit. It is of course necessary to have a shaft that reaches to the ore deposit. Horizontal tunnels extend out into the deposit, and provision must be made for sumps to remove water and for ventilation. Factors that must be considered in designing an underground mine include the depth, shape, and orientation of the ore body, as well as the nature and strength of the rock in and around it; thickness of overburden; and depth below the surface.

Usually, significant amounts of processing are required before a mined product is used or even moved from the mine site. Such processing, and the by-products of it, can have significant environmental effects. Even rock to be used for aggregate and for road construction must be crushed and sized, a process that has the potential to emit air-polluting dust particles into the atmosphere. Crushing is also a necessary first step for further processing of ore. Some minerals occur to an extent of a few percent or even less in the rock taken from the mine and must be concentrated on site so that the residue does not have to be hauled far. These concentration processes along with roasting, extraction, and in some cases chemical leaching of the ore are categorized as *extractive metallurgy*.

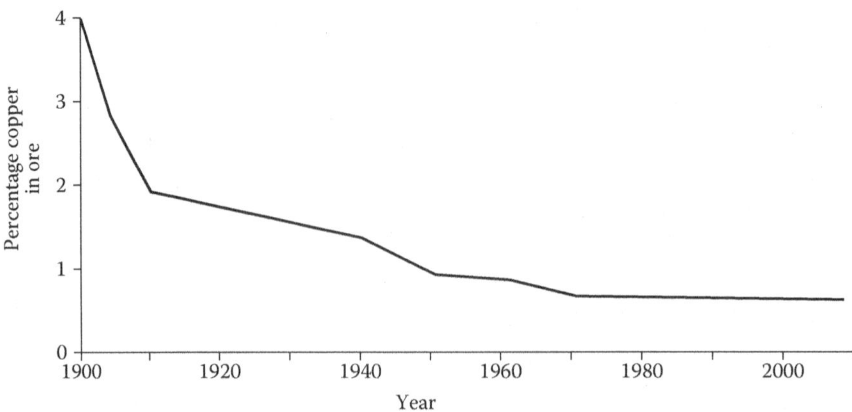

FIGURE 18.1 Average percentage of copper in ore that has been mined.

One of the more environmentally troublesome by-products of mineral refining consists of waste *tailings*. By the nature of the mineral processing operations employed, tailings are usually finely divided and, as a result, subject to chemical and biochemical weathering processes. Heavy metals associated with metal ores can be leached from tailings, producing water runoff contaminated with cadmium, lead, and other pollutants. Adding to the problem are some of the processes used to refine ore. Large quantities of cyanide solution are used in some processes to extract low levels of gold from ore, posing obvious toxicological hazards.

Environmental problems resulting from exploitation of extractive resources—including disturbance of land, air pollution from dust and smelter emissions, and water pollution from disrupted aquifers—are aggravated by the fact that the general trend in mining involves utilization of less rich ore. This is illustrated in Figure 18.1, showing the average percentage of copper in copper ore mined since 1900. The average percentage of copper in ore mined in 1900 was about 4%, but by 1982 it was about 0.6% in domestic ore, and 1.4% in richer foreign ore. Ore as low as 0.1% copper may eventually be processed. Increased demand for a particular metal, coupled with the necessity to utilize lower-grade ore, has a vicious multiplying effect on the amount of ore that must be mined and processed and accompanying environmental consequences.[2]

The proper practice of industrial ecology can be used to significantly reduce the effects of mining and mining by-products. One way in which this can be done is to entirely eliminate the need for mining, utilizing alternate sources of materials. An example of such utilization, widely hypothesized but not yet put into practice to a large extent, is the extraction of aluminum from coal ash. This would have the double advantage of minimizing amounts of waste ash and reducing the need to mine scarce aluminum ore.

18.4 METALS

The majority of elements are metals, most of which are of crucial importance as resources. The availability and annual usage of metals vary widely with the kind of metal. Some metals are abundant and widely used in structural applications; iron and aluminum are prime examples. Other metals, especially those of the platinum group (platinum, palladium, iridium, rhodium) are very precious and their use is confined to applications such as catalysts, filaments, or electrodes for which only small quantities are required. Some metals are considered to be "crucial" because of their applications for which no substitutes are available and shortages or uneven distribution in supply that occur. Such a metal is chromium, used to manufacture stainless steel (especially for parts exposed to high temperatures and corrosive gases), jet aircraft, automobiles, hospital equipment, and mining equipment. The platinum-group metals are used as catalysts in the chemical industry, in petroleum refining, and in automobile exhaust antipollution devices.

Metals exhibit a wide variety of properties and uses. They come from a number of different compounds; in some cases two or more compounds are significant mineral sources of the same metal. Usually these compounds are oxides or sulfides. However, other kinds of compounds and, in the cases of gold and platinum-group metals, the elemental (native) metals themselves serve as metal ore. Table 18.1 lists the important metals, their properties, major uses, and sources.

TABLE 18.1
Worldwide and Domestic Metal Resources

Metals	Properties[a]	Major Uses	Ores, Aspects of Resources[b]
Aluminum	mp 660°C, sg 2.70, malleable ductile, good electrical conductor	Metal products, including autos, aircraft, electrical equipment, electrical transmission lines	From bauxite ore containing 35–55% Al_2O_3, limited U.S. resources, ample world resources
Chromium	mp 1903°C, sg 7.14, hard, silvery color	Metal plating, stainless steel, wear-resistant and cutting tool alloys, chromium chemicals, including chromates	From chromite, an oxide mineral containing Cr, Mg, Fe; virtually none in the United States; most resources in South Africa, Zimbabwe, Russia
Cobalt	mp 1495°C, sg 8.71, bright, silvery color	Manufacture of hard, heat-resistant alloys, permanent magnet alloys, driers, pigments, glazes, animal feed additive	From a variety of minerals, such as linnaeite, Co_3S_4, and as a by-product of other metals; abundant global and U.S. resources
Copper	mp 1083°C, sg 8.96, ductile, malleable, excellent electrical conductor	Electrical conductors, alloys, chemicals, many uses	Occurs in low percentages as sulfides, oxides, and carbonates, some U.S. resources, marginally adequate world resources with increasing prices as supplies tighten
Gold	mp 1063°C, sg 19.3	Jewelry, basis of currency, electronics, increasing industrial uses	In various minerals at only around 10 ppm for ores currently processed in the United States; by-product of copper refining; <10% of the world resources in the United States
Iron	mp 1535°C, sg 7.86, silvery metal, in (rare) pure form	Most widely produced metal, usually as steel, a high-tensile strength material containing 0.3–1.7% C; made into many specialized alloys	Occurs as hematite (Fe_2O_3), goethite, ($Fe_2O_3 \cdot H_2O$), and magnetite (Fe_3O_4), abundant global and U.S. resources
Lead	mp 327°C, sg 11.35, silvery color	Fifth most widely used metal; storage batteries, chemicals, uses in gasoline, pigments, and ammunition largely eliminated for environmental reasons	Major source is galena, PbS; limited U.S. resources, marginally adequate world supplies, large fraction from recycled metal
Manganese	mp 1244°C, sg 7.3, hard, brittle, gray-white	Sulfur and oxygen scavenger in steel, manufacture of alloys, dry cells, gasoline additive, chemicals	Found in several oxide minerals, no U.S. production, adequate world resources in several countries, China is the largest producer
Mercury	mp −38°C, bp 357°C, sg 13.6	Instruments, electronic apparatus, electrodes, chemicals, uses curtailed because of toxicity	From cinnabar, HgS, no U.S. production except for recycled mercury, China is the major foreign producer, much mercury comes from recycling

continued

TABLE 18.1 (continued)
Worldwide and Domestic Metal Resources

Metals	Properties[a]	Major Uses	Ores, Aspects of Resources[b]
Molybdenum	mp 2620°C, sg 9.01, ductile, silvery-gray	Alloys, pigments, catalysts, chemicals, lubricants	Molybdenite (MoS_2) and wulfenite ($PbMoO_4$) are major ores. United States, Chile, and China are major producers, large global resources
Nickel	mp 1455°C	Stainless steel, specialty alloys, rechargeable batteries, coins, growing uses in high technology, catalysts (such as for hydrogenation of vegetable oils)	Found in ores associated with iron; Russia, Australia, and Canada are major producers; supplies are tight
Silver	mp 961°C, sg 10.5, shiny metal	Sterling ware, jewelry, bearings, dentistry, solders, electronics, decreasing use in photographic film with digital photographic equipment	Found with sulfide minerals, by-product of Cu, Pb, Zn, produced in a number of countries including the United States, in relatively short supply
Tin	mp 232°C, sg 7.31	Coatings, solders, bearing alloys, bronze, chemicals, organometallic biocides	Many forms associated with granitic rocks and chrysolites; no U.S. production; China, Indonesia, and Peru are major producers; much tin is recycled
Titanium	mp 1677°C, sg 4.5, silvery color	Strong, corrosion resistant, used in aircraft, valves, pumps, paint pigments. (TiO_2 in white pigments)	Commonly as TiO_2, ninth in elemental abundance, widely produced worldwide including the United States
Tungsten	mp 3380°C, sg 19.3, gray, heat resistant	Very strong, high melting point, used in alloys, drill bits, turbines, nuclear reactors, to make tungsten carbide	Found as tungstates, such as scheelite ($CaWO_4$); abundant resources in the United States and worldwide
Vanadium	mp 1917°C, sg 5.87, gray	Used to make strong steel alloys	In igneous rocks, primarily a by-product of other metals. China, Russia, and South Africa are main producers
Zinc	mp 420°C, sg 7.14, bluish-white	Widely used in alloys (brass), galvanized steel, paint pigments, chemicals; fourth in world metal production	Found in many ore minerals produced in numerous countries including the United States, in relatively short supply

[a] Abbreviations: bp, boiling point; mp, melting point; sg, specific gravity.

[b] Availability and levels of use depend on price, technology, recent discoveries, and other factors and are subject to fluctuation.

18.5 METAL RESOURCES AND INDUSTRIAL ECOLOGY

Metals come from two sources, the geosphere from which they are mined and the anthrosphere in which metals are recycled. In the case of relatively abundant metals that are inexpensive to extract from ores and that do not pose major environmental problems in their disposal, the geosphere is the major source (iron is such a metal). For metals in limited supply that should never be discarded because of adverse environmental effects, recycling predominates; lead is the prime example. The current anthrospheric stock of iron in the United States is now estimated at 3200 million metric tons such that a system of perfect recycling could entirely eliminate the need to mine additional iron ore.[3]

Considerations of industrial ecology are very important in extending and efficiently utilizing metal resources. More than any other kind of resource, metals lend themselves to recycling and to the practice of industrial ecology. This section briefly addresses the industrial ecology of metals.

18.5.1 Aluminum

Aluminum metal has a remarkably wide range of uses resulting from its properties of low density, high strength, ready workability, corrosion resistance, and high electrical conductivity. The use and disposal of aluminum present no environmental problems and it is one of the most readily recycled of all metals.

The environmental problems associated with aluminum result from the mining and processing of *bauxite* aluminum ore, which contains 40–60% alumina, Al_2O_3, associated with water molecules, the result of weathering away of more soluble minerals, particularly in high-rainfall regions of the tropics (see laterites in Section 18.2). Strip mining of bauxite from thin seams causes significant disturbance to the geosphere. The commonly used Bayer process for aluminum refining dissolves alumina, shown below as the hydroxide $Al(OH)_3$, from bauxite at high temperatures with sodium hydroxide as sodium aluminate:

$$Al(OH)_3 + NaOH \rightarrow NaAlO_2 + 2H_2O \tag{18.2}$$

leaving behind large quantities of caustic "red mud." This residue, which is rich in oxides of iron, silicon, and titanium, has virtually no uses and a high potential to produce pollution. Aluminum hydroxide is then precipitated in the pure form at lower temperatures and calcined at about 1200°C to produce pure anhydrous Al_2O_3. The anhydrous alumina is electrolyzed in molten cyrolite, Na_3AlF_6, at carbon electrodes to produce aluminum metal. All of these steps are very energy intensive, which makes recycling of aluminum metal particularly desirable.

An interesting possibility that could avoid many of the environmental problems associated with aluminum production is the use of coal fly ash as a source of the metal. Produced in large quantities as a by-product of electricity generation, fly ash is essentially free. Its anhydrous nature avoids the expense of removing water; it is a finely divided homogeneous material. Aluminum, along with iron, manganese, and titanium, can be extracted from coal fly ash with acid. If aluminum is extracted as the chloride salt, $AlCl_3$, it can be electrolyzed as the chloride by the ALCOA process. Although this process has not yet been proven to be competitive with the Bayer process, it may become so in the future.

Gallium is a metal that commonly occurs with aluminum ore and may be produced as a by-product of aluminum manufacture. Gallium combined with arsenic or with indium and arsenic is useful in semiconductor applications, including integrated circuits, photoelectric devices, and lasers. Although important, these applications require only miniscule amounts of gallium compared to major metals.

18.5.2 Chromium

Chromium is of crucial importance to industrialized societies because of its use in stainless steel and superalloys used in jet engines, nuclear power plants, chemical-resistant valves, and other applications in which a material that resists heat and chemical attack is required.

As noted in Table 18.1, supplies of chromium are poorly distributed around the world. To conserve chromium it is important that it be handled according to good practices of industrial ecology. Several measures may be taken in this respect. Chromium is almost impossible to recover from chrome-plated objects, and this use should be eliminated insofar as possible, as has been done with much of the decorative chrome-plated adornments formerly put on automobiles. Chromium(VI) (chromate) is a toxic form of the metal and its uses should be eliminated wherever possible. The use of chromium in leather tanning and miscellaneous chemical applications should be curtailed. Chromium was once widely used in the preparation of treated CCS (copper/chrome/arsenic) lumber, which resists fungal decay and termites, but this product is now banned because of its content of toxic arsenic.

18.5.3 Copper

Copper is a low-toxicity, corrosion-resistant metal widely used because of its workability (ductility and malleability), electrical conductivity, and ability to conduct heat. In addition to its use in electrical wire, where in some applications it is now challenged by aluminum, copper is also used in tubing, copper pipe, shims, gaskets, and other applications.

There are at least two major environmental problems associated with the extraction and refining of copper. The first of these is the dilute form in which copper ore now occurs (see Figure 18.1), such that in the United States 150–175 ton of inert material (not counting overburden removed in strip mining) must be processed and discarded to produce a ton of copper metal. The second problem is the occurrence of copper as the sulfide so that in the production of copper, large amounts of sulfur must be recovered as a by-product or, unfortunately in some less developed countries, released into the atmosphere as pollutant SO_2.

An advantage to copper for recycling is that it is used primarily as the metal, which represents "stored energy" in that it does not require energy for reduction to the metal. Recycling rates of scrap copper appear low in part because so much of the inventory of copper metal is tied up in long-lasting electrical wire, in structures, and other places where the lifetime of the metal is long. (This is in contrast to lead, where the main source of recycled metal is storage batteries, which last only 2–5 years.) An impediment to copper recycling is the difficulty of recovering copper components from circuits, plumbing, and other applications.

18.5.4 Cobalt

Cobalt is a "strategic" metal with very important applications in alloys, particularly in heat-resistant applications, such as jet engines. The major source of cobalt is as a by-product of copper refining, although it can also be obtained as a by-product of nickel and lead. As much as 50% of the cobalt in these sources is lost to tailings, slag, or other wastes, so there is a significant potential to improve the recovery of cobalt. Relatively low percentages of cobalt are recycled as scrap.

18.5.5 Lead

The industrial ecology of lead is very important because of the widespread use of this metal and its toxicity. Global fluxes of lead from the anthrosphere are shown in Figure 18.2.

Somewhat more than half of the lead processed by humans comes from the geosphere, mostly as lead ore, with the remainder coming from recycled lead. By far the greatest use of lead is in batteries, and the amount of battery lead recycled each year approaches that taken from the geosphere. A small fraction of lead is dissipated as wastes associated with mining, refining, and use of the metal. Former uses of lead other than in batteries, including pigments, solder, cable sheathing, formed products, and shot in ammunition, have been severely curtailed; recycling lead from such applications is difficult. Although much of the lead in batteries is recycled,[4] about one-third of it is lost and this represents another area of potential improvement in the utilization of lead.

18.5.6 Lithium

Lithium, atomic number 3, atomic mass 6.941, is emerging as a metal of particular importance with respect to energy sustainability. It is the first metal in the periodic table and the lightest with

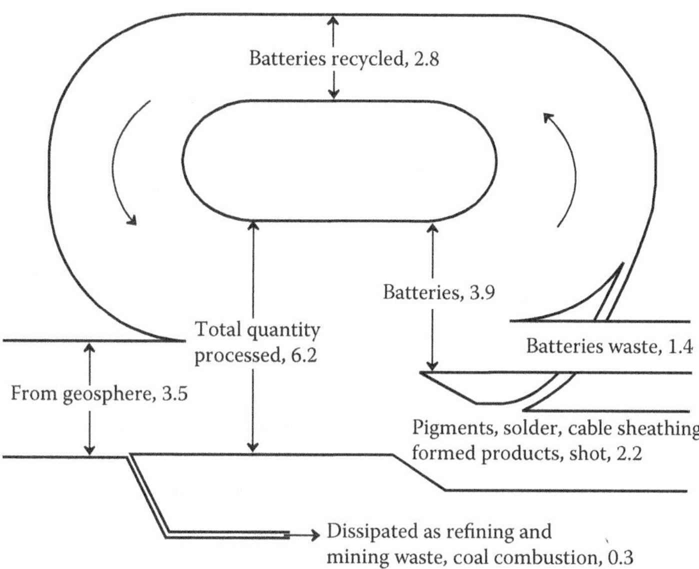

FIGURE 18.2 Flux of lead in the anthrosphere, globally, on an annual basis in millions of tons per year. Lead from the geosphere includes metal mined and a small quantity dissipated by coal combustion.

a density in the metallic form of only 0.531 g/cm^3. Until recently, the demand for lithium was relatively low with uses in glasses, enamels, specialty ceramics, lubricant greases, and as a pharmaceutical agent to treat some mental disorders. Lithium forms the Li^+ ion which, because of its low mass, is a very efficient charge carrier in powerful, compact lithium ion batteries, which operate by means of the movement of Li^+ ions between electrodes through an electrolyte and across a separator. Because of their abilities to store large amounts of electrical energy in relatively small rechargeable batteries, these devices have become very popular in portable electronic devices, especially laptop computers.

Although the popularity of lithium ion batteries in portable electronic devices has increased the demand for lithium, it is miniscule compared to the demand that will be created for this metal as lithium ion batteries are developed for use in electric and especially hybrid vehicles that have both electric motors and internal combustion engines. Known world resources of lithium are limited. As of 2009, these were estimated at 5.4 million tons in Bolivia, 3 million in Chile, 1.1 million in China, and 410,000 in the United States. The harvesting of lithium salts from the flat-salt desert of Uyuni in Bolivia has become an important source of income for the local people although meeting projected demand will require larger mechanized operations. The limited number of locations of known lithium deposits and political uncertainties regarding these sources have raised concerns regarding acquisition of sufficient supplies of this metal to meet projected demands. However, it is likely that there are significant lithium resources in various parts of the world that have not been found, largely because they have not been sought in the past due to limited demand for this element.

18.5.7 Potassium

Potassium deserves special mention as a metal because the potassium ion, K^+, is an essential element required for plant growth. It is mined as potassium minerals and applied to soil as plant fertilizer. Potassium minerals consist of potassium salts, generally KCl. Such salts are found as deposits in the ground or may be obtained from some brines. Very large deposits are found in Saskatchewan, Canada. All the salts used as potassium sources are quite soluble in water.

18.5.8 Zinc

Zinc is relatively abundant and not particularly toxic, so its industrial ecology is of less concern than that of toxic lead or scarce chromium. As with other metals, the mining and processing of zinc can pose some environmental concerns. Zinc occurs as ZnS (a mineral called sphalerite), and the sulfur must be reclaimed as SO_2 in the smelting of zinc. Zinc minerals often contain significant fractions of lead and copper, as well as significant amounts of toxic arsenic and cadmium. Zinc ore typically containing about 6% zinc is concentrated by a flotation process in which air is bubbled into a slurry of finely ground zinc ore in water treated with special chemicals and a froth concentrate of zinc sulfide containing up to 50% zinc is skimmed off the top. Up to 90% of the zinc in the ore is recovered.

Zinc is widely used as the metal and to make alloys, particularly brass, an alloy with copper that is widely recycled. Lesser amounts of zinc are used to make zinc chemicals. One of the larger uses for zinc is as a corrosion-resistant coating on steel. This application, refined to a high degree in the automotive industry in recent years, has significantly lengthened the life span of automotive bodies and frames. It is difficult to reclaim zinc from zinc plating. However, zinc is a volatile element and it can be recovered in baghouse dust from electric arc furnaces used to reprocess scrap steel.

The most important zinc compound is zinc oxide, ZnO. Formerly widely used as a paint pigment, this white substance is now employed as an accelerating and activating agent for hardening rubber products, particularly tires. The other two major compounds of zinc employed commercially are zinc chloride used in dry cells, as a disinfectant, and to vulcanize rubber, and zinc sulfide, used in zinc electroplating baths.

Two aspects of zinc may be addressed with respect to its industrial ecology. The first of these is that, although it is not very toxic to animals, zinc is phytotoxic (toxic to plants) and soil can be "poisoned" by exposure to zinc from zinc smelting or from application of zinc-rich sewage sludge. The second of these is that the recycling of zinc is complicated by its dispersal as a plating on other metals. However, means do exist to reclaim significant fractions of such zinc, such as from electric arc furnaces as mentioned above.

18.6 NONMETAL MINERAL RESOURCES

A number of minerals other than those used to produce metals are important resources. As with metals, the environmental aspects of mining many of these minerals are quite important. Typically, even the extraction of ordinary rock and gravel can have important environmental effects.

Clays are secondary minerals formed by weathering processes on parent minerals (see Chapter 15, Section 15.7). Clays have a variety of uses. About 70% of the clays used are miscellaneous clays of variable composition that have uses for a number of applications including filler (such as in paper), brick manufacture, tile manufacture, and portland cement production. Somewhat more than 10% of the clay used is fireclay, which has the characteristic of being able to withstand firing at high temperatures without warping. This clay is used to make a variety of refractories, pottery, sewer pipe, tile, and brick. Somewhat <10% of the clay that is used is kaolin, which has the general formula $Al_2(OH)_4Si_2O_5$. Kaolin is a white mineral that can be fired without losing shape or color. It is employed to make paper filler, refractories, pottery, dinnerware, and as a petroleum-cracking catalyst. About 7% of clay mined consists of bentonite and fuller's earth, a clay of variable composition used to make drilling muds, petroleum catalyst, carriers for pesticides, sealers, and clarifying oils. Very small quantities of a highly plastic clay called ball clay are used to make refractories, tile, and whiteware. U.S. production of clay is about 42 million metric tons per year, and global and domestic resources are abundant.

Fluorine compounds are widely used in industry. Large quantities of fluorspar, CaF_2, are required as a flux in steel manufacture. Synthetic and natural cryolite, Na_3AlF_6, is used as a solvent for aluminum oxide in the electrolytic preparation of aluminum metal. Sodium fluoride is added to water to help prevent tooth decay, a measure commonly called water fluoridation. World reserves of high-grade fluorspar are around 190 million metric tons, about 13% of which is in the United States. This is sufficient for several decades at projected rates of use. A great deal of by-product fluorine is recovered from the processing of fluorapatite, $Ca_5(PO_4)_3F$, used as a source of phosphorus (see below).

Micas are complex aluminum silicate minerals that are transparent, tough, flexible, and elastic. Muscovite, $K_2O \cdot 3Al_2O_3 \cdot 6SiO_2 \cdot 2H_2O$, is a common type of mica. Better grades of mica are cut into sheets and used in electronic apparatus, capacitors, generators, transformers, and motors. Finely divided mica is widely used in roofing, paint, welding rods, and many other applications. Sheet mica is imported into the United States, and finely divided "scrap" mica is recycled domestically. Shortages of this mineral are unlikely.

Pigments and *fillers* of various kinds are used in large quantities. The only naturally occurring pigments still in wide use are those containing iron. These minerals are colored by limonite, an amorphous brown-yellow compound with the formula $2Fe_2O_3 \cdot 3H_2O$, and hematite, composed of gray-black Fe_2O_3. Along with varying quantities of clay and manganese oxides, these compounds are found in ocher, sienna, and umber. Manufactured pigments include carbon black, titanium dioxide, and zinc pigments. About 1.5 million metric tons of carbon black, manufactured by the partial combustion of natural gas, are used in the United States each year, primarily as a reinforcing agent in tire rubber.

Over 7 million metric tons of minerals are used in the United States each year as fillers for paper, rubber, roofing, battery boxes, and many other products. Among the minerals used as fillers are carbon black, diatomite, barite, fuller's earth, kaolin (see clays, above), mica, limestone, pyrophyllite, and wollastonite ($CaSiO_3$).

Although sand and gravel are the cheapest of mineral commodities per ton, the average annual dollar value of these materials is greater than all but a few mineral products because of the huge quantities involved. In tonnage, sand and gravel production is by far the greatest of nonfuel minerals. Almost 1 billion tons of sand and gravel are employed in construction in the United States each year, largely to make concrete structures, road paving, and dams. Slightly more than that amount is used to manufacture portland cement and as construction fill. Although ordinary sand is predominantly silica, SiO_2, about 30 million tons of a more pure grade of silica are consumed in the United States each year to make glass, high-purity silica, silicon semiconductors, and abrasives.

At present, old river channels and glacial deposits are used as sources of sand and gravel. Many valuable deposits of sand and gravel are covered by construction and lost to development. Transportation and distance from source to use are especially crucial for this resource. Environmental problems involved with defacing land can be severe, although bodies of water used for fishing and other recreational activities frequently are formed by removal of sand and gravel.

18.7 PHOSPHATES

Phosphate minerals are of particular importance because of their essential use in the manufacture of fertilizers applied to land to increase crop productivity. In addition, phosphorus is used for supplementation of animal feeds, synthesis of detergent builders, and preparation of chemicals such as pesticides and medicines. The most common phosphate minerals are hydroxyapatite, $Ca_5(PO_4)_3(OH)$, and fluorapatite, $Ca_5(PO_4)_3F$. Ions of Na, Sr, Th, and U are found substituted for calcium in apatite minerals. Small amounts of PO_4^{3-} may be replaced by AsO_4^{3-} and the arsenic must be removed for food applications. Approximately 17% of world phosphate production is from igneous minerals, primarily fluorapatites. About three-fourth of world phosphate production is from sedimentary deposits, generally of marine origin. Vast deposits of phosphate, accounting for approximately 5% of world phosphate production, are derived from guano droppings of seabirds and bats. Current U.S. production of phosphate rock is around 40 million metric tons per year, most of it from Florida. Most of the available phosphate ore is of low grade so that its processing generates large quantities of by-product.[5]

Naturally occurring phosphate minerals are not sufficiently soluble to be used as phosphate fertilizer. For commercial phosphate fertilizer production, these minerals are treated with phosphoric or sulfuric acids to produce more soluble superphosphates as described in Chapter 16, Section 16.7. The hydrogen fluoride (HF) produced as a by-product of superphosphate production from fluorapatite can create air pollution problems, and the recovery of fluorides is an important aspect of the industrial ecology of phosphate production.

There are at least two major reasons why the industrial ecology of phosphorus is particularly important. The first of these is that current rates of phosphate use would exhaust known reserves of phosphate within two or three generations. Although additional sources of phosphorus will be found and exploited, it is clear that this essential mineral is in distressingly short supply relative to human consumption; phosphate shortages, along with sharply higher prices, will eventually cause a crisis in food production. The second significant aspect of the industrial ecology of phosphorus is the pollution of waterways by waste phosphate, a plant, and algal nutrient. This results in excessive growth of algae in the water, followed by decay of the plant biomass, consumption of dissolved oxygen, and an undesirable condition of eutrophication.

Excessive use of phosphate coupled with phosphate pollution suggests that phosphate wastes, such as from sewage treatment, should be substituted as sources of plant fertilizer. Several other partial solutions to the problem of phosphate shortages are the following.

- Development and implementation of methods of fertilizer application that maximize efficient utilization of phosphate
- Genetic engineering of plants that have minimal phosphate requirements and that utilize phosphorus with maximum efficiency
- Development of systems to maximize the utilization of phosphorus-rich animal wastes

18.8 SULFUR

Sulfur is an important nonmetal; its greatest single use is in the manufacture of sulfuric acid. However, the element is employed in a wide variety of other industrial and agricultural products. Current consumption of sulfur amounts to approximately 10 million metric tons per year in the United States. The four most important sources of sulfur are (in decreasing order) deposits of elemental sulfur, H_2S recovered from sour natural gas, organic sulfur recovered from petroleum, and pyrite (FeS_2). Recovery of sulfur from coal used as a fuel is a huge potential, largely untapped source of this important nonmetal.

The resource situation for sulfur differs from that of phosphorus in several significant respects. Although sulfur is an essential nutrient like phosphorus, most soils contain sufficient amounts of nutrient sulfur, and the major uses of sulfur are in the industrial sector. The sources of sulfur are varied and abundant and supply is no problem either in the United States or worldwide; sulfur recovery from fossil fuels as a pollution control measure could even result in surpluses of this element.

About 90% of the use of sulfur in the world is for the manufacture of sulfuric acid. Almost two-third of the sulfuric acid consumed is used to make phosphate fertilizers as discussed in Section 18.7, in which case the phosphorus ends up as waste "phospho-gypsum," $CaSO_4 \cdot xH_2O$. Other uses of sulfuric acid include lead storage batteries, steel pickling, petroleum refining, extraction of copper from copper ore, and the chemical industry.

The industrial ecology of sulfur needs to emphasize reduction of wastes and sulfur pollution, rather than supply of this element. Unlike many resources, such as most common metals, the uses of sulfur are for the most part dissipative, and the sulfur is "lost" to agricultural land, paper products, petroleum products, or other environmental sinks. There are two major environmental concerns with sulfur. One of these is the emission of sulfur into the atmosphere, which occurs mostly as pollutant sulfur dioxide and is largely manifested by production of acidic precipitation and dry deposition. The second major environmental concern with sulfur is that it is used mostly as sulfuric acid and is not incorporated into products, thus posing the potential to pollute water and create acidic wastes. Acid purification units are available to remove significant amounts of sulfuric acid from waste acid solutions for recycling.

18.8.1 Gypsum

Calcium sulfate in the form of the dihydrate $CaSO_4 \cdot 2H_2O$ is the mineral *gypsum*, one of the most common forms in which waste sulfur is produced. As noted, large quantities of this material are produced as a by-product of phosphate fertilizer manufacture. Another major source of gypsum is its production when lime is used to remove sulfur dioxide from power plant stack gas:

$$Ca(OH)_2 + SO_2 \rightarrow CaSO_3 + H_2O \tag{18.3}$$

to produce a calcium sulfite product that can be oxidized to calcium sulfate. About 100 million metric tons of gypsum are mined each year for a variety of uses, including production of portland cement, to make wallboard, as a soil conditioner to loosen tight clay soils, and numerous other applications.

Calcium sulfate from industrial or natural (gypsum) sources can be calcined at a very low temperature of only 159°C to produce $CaSO_4 \cdot \frac{1}{2}H_2O$, a material known as plaster of Paris, which was once commonly used for the manual application of plaster to walls. Plaster of Paris mixed with water forms a plastic material that sets up as the solid dihydrate:

$$CaSO_4 \cdot \tfrac{1}{2}H_2O + \tfrac{3}{2}H_2O \rightarrow CaSO_4 \cdot 2H_2O \tag{18.4}$$

Cast into sheets coated with paper, this material produces plasterboard commonly used for the interior walls of homes and other buildings. Historically, plaster of Paris was used for mortar and other structural applications, and it has the potential for similar applications today.

The very large quantities of gypsum that are mined suggest that by-product calcium sulfate, especially that produced with phosphate fertilizers and from flue gas desulfurization, should be a good candidate for reclamation through the practice of industrial ecology. The low temperature (see above) required to convert hydrated calcium sulfate to $CaSO_4 \cdot \frac{1}{2}H_2O$, which can be set up as a solid by mixing with water, suggests that the energy requirements for a gypsum-based by-products industry should be modest. Low-density gypsum blown as a foam and used as a filler in composites along with sturdy reinforcing materials should have good insulating, fire-resistant, and structural properties for building construction.

18.9 WOOD: A MAJOR RENEWABLE RESOURCE

Fortunately, one of the major natural resources in the world, wood, is a renewable resource. Production of wood and wood products is the fifth largest industry in the United States, and forests cover one-third of the U.S. surface area. Wood ranks first worldwide as a raw material for the manufacture of other products, including lumber, plywood, particle board, cellophane, rayon, paper, methanol, plastics, and turpentine.

Chemically, wood is a complicated substance consisting of long cells having thick walls composed largely of polysaccharides such as cellulose, a glucose polymer shown in Chapter 22, Figure 22.5. In addition to the solid wood fraction consisting of cellulose and lignin (see Chapter 17, Section 17.6), a wide variety of tannins, pigments, sugars, starch, cyclitols, gums, mucilages, pectins, galactans, terpenes, hydrocarbons, acids, esters, fats, fatty acids, aldehydes, resins, sterols, and waxes can be extracted from wood by water, alcohol-benzene, ether, and steam distillation.

Substantial amounts of methanol, a useful synthetic chemical and fuel once commonly called *wood alcohol*, are obtained from pyrolysis of wood.

A major use of wood is in paper manufacture. The widespread use of paper is a mark of an industrialized society. The manufacture of paper is a highly advanced technology. Paper consists essentially of cellulosic fibers tightly pressed together. To make paper, the lignin fraction must first be removed from the wood, leaving the cellulosic fraction. Both the sulfite and alkaline processes for accomplishing this separation have resulted in severe water and air pollution problems, now significantly alleviated through the application of advanced treatment technologies.

Wood fibers and particles can be used for making fiberboard, paper-base laminates (layers of paper held together by a resin and formed into the desired structures at high temperatures and pressures), particle board (consisting of wood particles bonded together by a phenol-formaldehyde or urea-formaldehyde resin), and nonwoven textile substitutes consisting of wood fibers held together by adhesives. Wood chemical by-products including glucose and cellulose are potential major products from the 60 million metric tons of wood wastes produced in the United States each year.

18.10 EXTENDING RESOURCES THROUGH THE PRACTICE OF INDUSTRIAL ECOLOGY

A tremendous potential exists for applying the practice of industrial ecology to lower the burden on virgin raw materials and sources of energy. These approaches include using less material (dematerialization), substitution of a relatively more abundant and safe material for one that is scarce and/or toxic, extracting useful materials from wastes (waste mining), and recycling materials and items. Properly applied, these measures can not only conserve increasingly scarce raw materials, but can also increase wealth as it is conventionally defined.

The greatest potential for extending material resources is by recycling through the practice of industrial ecology. One of the greatest savings in recycling is energy saved by the practice.

Materials vary in their amenability to recycling. Arguably the most recyclable materials are metals in a relatively pure form. Such metals are readily melted and recast into other useful components. Among the least recyclable materials are mixed polymers or composites, the individual constituents of which cannot be readily separated. The chemistry of some polymers is such that, once they are prepared from monomers, they are not readily broken down again and reformed to a useful form. This section briefly addresses the kinds of materials that are recycled or that are candidates for recycling in a functional system of industrial ecology.

An important aspect of industrial ecology applied to recycling materials consists of the separation processes that are employed to "unmix" materials for recycling at the end of a product cycle. An example of this is the separation of graphite carbon fibers from the epoxy resins used to bind them together in carbon fiber composites. The chemical industry provides many examples where separations are required. For example, the separation of toxic heavy metals from solutions or sludges can yield a metal product, leaving nontoxic water and other materials for safe disposal or reuse.

18.10.1 METALS

Pure metals are easily recycled, and the greatest challenge is to separate the metals into a pure state. One of the more difficult problems with metals recycling is the mixing of metals, such as problems that occur with metal alloys when a metal is plated onto another metal, or with components made of two or more metals in which it is hard to separate the metals. A common example of the complications from mixing metals is the contamination of iron with copper from copper wiring or other components made from copper. As an impurity, copper produces steel with inferior mechanical characteristics. Another problem can be the presence of toxic cadmium used as plating on steel parts.

Recycling metals can take advantage of the technology developed over many years of technology for the separation of metals that occur together in ores. Examples of by-product metals recovered during the refining of other metals are gallium from aluminum; arsenic from lead or copper; precious metal iridium, osmium, palladium, rhodium, and ruthenium from platinum; and cadmium, germanium, indium, and thorium from zinc.

18.10.2 Plastics and Rubber

Much attention has been given to the recycling of plastics in recent years. Compared to metals, plastics are much less recyclable because recycling is technically difficult and plastics are less valuable than metals. There are two general classes of plastics, a fact that has a strong influence on their recyclability. Thermoplastics are those that become fluid when heated and solid when cooled. Since they can be heated and reformed multiple times, thermoplastics are generally amenable to recycling. Recyclable thermoplastics include polyalkenes (low-density and high-density polyethylene and polypropylene); polyvinylchloride (PVC), used in large quantities to produce pipe, house siding, and other durable materials; polyethylene terephthalate; and polystyrene. Plastic packaging materials are commonly made from thermoplastics and are potentially recyclable. Fortunately, from the viewpoint of recycling, thermoplastics make up most of the quantities of plastics used.

Thermosetting plastics are those that form molecular cross linkages between their polymeric units when they are heated. These bonds set the shape of the plastic, which does not melt when it is heated. Therefore, thermosetting plastics cannot be simply remolded; they are not very amenable to recycling, and often burning them for their heat content is about the only use to which they may be put. An important class of thermosetting plastics consists of the epoxy resins, characterized by an oxygen atom bonded between adjacent carbons (1,2-epoxide or oxirane). Epoxies are widely used in composite materials combined with fibers of glass or graphite. Other thermosetting plastics include cross-linked phenolic polymers, some kinds of polyesters, and silicones. When recycling is contemplated, the best use for thermosetting plastics is for the fabrication of entire components that can be recycled.

Contaminants are an important consideration in recycling plastics. A typical kind of contaminant is paint used to color the plastic object. Adhesives and coatings of various kinds may also act as contaminants. Such materials may weaken the recycled material or decompose to produce gases when the plastic is heated for recycling. Toxic cadmium used to enable polymerization of plastics, a "tramp element" in recycling parlance, can hinder recycling of plastics and restrict the use of the recycled products.

Large quantities of recycled polyester plastics from bottles can be used to make carpets. As of 2008 Mohawk carpets was using 14,000 plastic drink bottles per minute, about one-third of those available for recycling in the United States, to make Everstrand carpet. It should be noted that the plastic used for bottled drinking water bottles, alone, in the United States requires around 175 million liters of petroleum each year for their manufacture generating almost 1.5 million metric tons per year of waste. Only about one in four of these plastic bottles are recycled.

Although natural rubber was widely recycled during World War II, the synthetic rubbers developed as substitutes during the war were much less amenable to recycling. Countries throughout the world produce millions of spent vehicle tires, which can cause severe problems in waste dumps and are commonly used as fuel. One approach that is proving successful is to freeze the shredded rubber from tires in liquid nitrogen and grind it to a very fine powder. This material is used as an additive for paints, coatings, and sealants to which it imparts some rubber properties including elasticity and impact resistance along with protection against ultraviolet radiation and ozone. The powdered rubber product is also added to rubber used in new tires.

18.10.3 Lubricating Oil

Lubricating oils are used in vast quantities and are prime candidates for recycling. The most simplistic means of recycling lubricating oil is to burn it, and large volumes of oil are burned for fuel. This is a very low level of recycling and will not be addressed further here.

For many years the main process for reclaiming waste lubricating oil used treatment with sulfuric acid followed by clay. This process generated large quantities of acid sludge and spent clay contaminated with oil. These undesirable by-products contributed substantial amounts of wastes to hazardous waste disposal sites. Current state-of-the-art practices of lubricating oil reclamation do not utilize large quantities of clay for cleanup, but instead use solvents, vacuum distillation, and catalytic hydrofinishing to produce a usable material from spent lubricating oil. The first step is dehydration to remove water and stripping to remove contaminant fuel (gasoline) fractions. If solvent treatment is used, the oil is then extracted with a solvent, such as isopropyl or butyl alcohols or methylethylketone. After treatment with a solvent, the waste oil is commonly centrifuged to remove impurities that are not soluble in the solvent. The solvent is then stripped from the oil. The next step is a vacuum distillation that removes a light fraction useful for fuel and leaves a heavy residue that can be used for fuel. The lubricating oil can then be subjected to hydrofinishing, chemical addition of H_2 over a catalyst, to produce a suitable lubricating oil product.

LITERATURE CITED

1. Bless, D., B. Park, S. Nordwick, M. Zaluski, H. Joyce, R. Hiebert, Randy, and C. Clavelot, Operational lessons learned during bioreactor demonstrations for acid rock drainage treatment, *Mine Water and the Environment*, **27**, 241–250, 2008.
2. Rosa, R. N. and R. N. Diogo, Exergy cost of mineral resources, *International Journal of Exergy*, **5**, 532–555, 2008.
3. Mueller, D. B., T. Wang, B. Duval, and T. E. Graedel, Exploring the engine of anthropogenic iron cycles, *Proceedings of the National Academy of Sciences of the United States of America*, **104**, 16111–16116, 2006.
4. Genaidy, A. M., R. Sequeira, T. Tolaymat, J. Kohler, and M. Rinder, An exploratory study of lead recovery in lead-acid battery lifecycle in U.S. market: An evidence-based approach, *Science of the Total Environment*, **407**, 7–22, 2008.
5. Negm, A. A. and A.-Z. M. Abouzeid, Utilization of solid wastes from phosphate processing plants, *Physicochemical Problems of Mineral Processing*, **42**, 5–16, 2008.

SUPPLEMENTARY REFERENCES

Aswathanarayana, U., *Mineral Resources Management and the Environment*, Taylor & Francis, London, 2003.

Clark, J. H. and F. Deswarte, Eds, *Introduction to Chemicals from Biomass*, Wiley, New York, 2008.

Dewulf, J. and H. Van Langenhove, Eds, *Renewables-Based Technology: Sustainability Assessment*, Wiley, Hoboken, NJ, 2006.

European Environment Agency, *Sustainable Use and Management of Natural Resources*, European Environment Agency, Copenhagen, 2005.

Grafton, R. Q., *The Economics of the Environment and Natural Resources*, Blackwell Publishing, Malden, MA, 2004.

Grant W. S., J. C. Hendee, and W. F. Sharpe, *Introduction to Forests and Renewable Resources*, 7th ed., McGraw-Hill, Boston, 2003.

Johnsen, O., *Minerals of the World*, Princeton University Press, Princeton, NJ, 2002.

Johnson, K. M., *USGS Mineral Resources Program–Supporting Stewardship of America's Natural Resources*, U.S. Department of the Interior Geological Survey, Reston, VA, 2006.

Journel, A. G. and P. C. Kyriakidis, *Evaluation of Mineral Reserves: A Simulation Approach*, Oxford University Press, Oxford, UK, 2004.

Kozlowski, R., G. Zaikov, and F. Pudel, Eds, *Renewable Resources: Obtaining, Processing, and Applying*, Nova Science Publishers, Hauppauge, NY, 2009.

Kozlowski, R., G. E. Zaikov, and F. Pudel, Eds, *Renewable Resources and Plant Biotechnology*, Nova Science Publishers, New York, 2006.

Lee, K. N., *The Compass and Gyroscope: Integrating Science and Politics for the Environment*, Island Press, Washington, DC, 1993.

Mineral Resources Forum, available at http://www.natural-resources.org/minerals/ (The Mineral Resources Forum website is an information resource dealing with minerals, metals, and sustainable development.)

Rajendran, S., K. Srinivasamoorthy, and S. Aravindan, Eds, *Mineral Exploration: Recent Strategies*, New India Publishing Agency, New Delhi, 2007.

Roonwal, G. S., K. Shahriar, and H. Ranjbar, Eds, *Mineral Resources and Development*, Daya Publishing House, Delhi, India, 2005.

Solid Waste Association of North America, *Successful Approaches to Recycling Urban Wood Waste*, U.S. Department of Agriculture, Forest Service, Forest Products Laboratory, Madison, WI, 2002.

USGS, available at http://minerals.usgs.gov/ (The Mineral Resources Program presented on this website provides and communicates current, impartial information on the occurrence, quality, quantity, and availability of mineral resources.)

Yu, L., *Biodegradable Polymer Blends and Composites from Renewable Resources*, Wiley, Hoboken, NJ, 2009.

QUESTIONS AND PROBLEMS

1. Lead, zinc, and copper are often associated with hydrothermal deposits and comonly occur as the metal sulfides. What does this suggest about the p*E* conditions (see Chapter 4) in the hydrothermal waters under which these deposits were formed?
2. The world's largest trona deposit is in the Green River Basin of Wyoming. Look up trona on the internet and suggest how this deposit was formed.
3. It is believed that the earth's oceans before the emergence of living organisms in them contained large amounts of dissolved Fe^{2+}. Describe with chemical equations how this iron was converted to deposits of Fe_2O_3, an important iron ore, by biochemical and chemical processes. Consider the solubility of $Fe(OH)_3$ and what happens to it when it is heated.
4. Using internet resources, look up the chemical reactions involved in extraction of gold from ore with cyanide. Why are the solutions maintained at a relatively high pH? What are the environmental implications of the process?
5. Calculate approximately how many tons of tailings were produced per ton of copper metal recovered from ore in 1900 compared to now. What are the environmental implications of these figures.
6. Although very rare in nature, why was gold one of the first metals discovered and used by humans? From what you can learn about the chemistry of copper and that of iron, suggest why copper metal was discovered and used before iron. Although iron almost never occurs in the elemental form in that part of the earth's crust accessible to humans, suggest how elemental iron was discovered. Where might elemental iron have been observed in nature in very rare and spotty locations around Earth?
7. What is the chemical reaction for the calcination of the aluminum compound precipitated from bauxite? In what two respects is the production of aluminum from ore energy-intensive? How does aluminum metal recycling reduce the consumption of energy in aluminum production?
8. What were the two applications of lead that led to its widespread dispersal in the environment until about the 1970s?
9. What environmental measures have the potential to lead to a surplus of sulfur?
10. What are the chemical processes for extracting cellulose from wood to make paper? Why are these environmentally problematic? Look up the chemical composition of cotton and suggest why it can be used to make an especially good (but expensive) grade of paper.
11. Two of the main approaches that employ the practice of industrial ecology for using smaller amounts of material and more readily available substances are dematerialization and

material substitution. Compare the nature of electronics and communications apparatus from the 1950s to the corresponding apparatus used today and show how both dematerialization and material substitution have led to much less use of materials in electronics and communications.

12. According to information from the U.S. National Renewable Energy Laboratory, "industrial biorefineries have been identified as the most promising route to the creation of a new domestic biobased industry." What is a biorefinery? How does it compare to a petroleum refinery? How can the widespread application of biorefineries aid sustainability?
13. Although alloys greatly increase the utility of metals, how is alloying potentially detrimental to metal recycling?
14. What are the two reasons that plastics are generally less recyclable than metals? Why are thermosetting plastics less recyclable than thermoplastics?
15. Why are present practices for recycling lubricating oils much more sustainable than those employed in decades past?

19 Sustainable Energy: The Key to Everything

19.1 ENERGY PROBLEM

It is a rather bold claim that sustainable energy is "the key to everything." But a convincing argument can be made that most problems in the physical environment can be solved to at least a large degree if enough energy is available, if it is inexpensive enough, and if it can be used without doing irreparable environmental harm. Consider the following environmental and sustainability problems that can be solved at least in part with sufficient sustainable energy:

- Water: With enough energy, wastewater can be reclaimed to drinking water standards by reverse osmosis and other energy-consuming technologies and seawater can be desalinated.
- Food: With enough energy, marginal land can be reclaimed by measures such as leveling, terracing, and rock removal and irrigation water can be pumped from long distances or desalinated to grow food. Greenhouses can be heated even during winter months to grow high-value specialty foods.
- Wastes: The disposal of hazardous organic wastes in landfills, though widely practiced, is not a good idea. With sufficient energy, such wastes can be converted to forms that cannot do harm.
- Transportation: With sufficient sustainable energy, transportation problems can be solved by technologies such as electrified railways.
- Fuels: Biomass sources of fixed carbon can be converted to hydrocarbon fuels for applications for which there are no viable alternatives (such as aircraft) without adding any net amounts of greenhouse gas carbon dioxide to the atmosphere.

The list above can be extended to many other areas and to a large number of sustainability and environmental problems. The great challenge is, of course, that systems of energy utilization have developed that are unsustainable. One of the most obvious sustainability challenges is that humankind is running out of the sources of energy around which its economic systems have become based. This is especially true of petroleum. Peak U.S. petroleum production was passed several years ago, and it is likely that peak world production will be reached within a few years of 2010. Exorbitantly high petroleum prices in the first part of 2008 followed by a precipitous fall in price as world economies collapsed at the end of the year emphasized the economic uncertainties of reliance on petroleum as an energy source, especially by nations that do not have domestic sources. There are still abundant resources of coal, but their utilization to provide for energy with current technology almost certainly will cause unacceptable global warming. Therefore, the great challenge facing humankind during the next several decades is to develop sources of energy that will meet energy needs and that will not ruin Earth and its climate.

There are energy alternatives to fossil fuels that can be developed, that are environmentally safe (or can be made so), and that, taken in total, can be adequate to supply energy needs. These include wind, solar, biomass, geothermal, and nuclear energy sources. Some other miscellaneous sources, such as tidal energy, may contribute as well. Fossil fuels will continue to be used and may contribute sustainably for decades with sequestration of greenhouse gas carbon dioxide. And, of course, energy conservation and greatly enhanced efficiency of energy use will make substantial contributions. This chapter discusses the energy alternatives listed above with emphasis on energy sustainability.

19.2 NATURE OF ENERGY

Energy is the capacity to do work (basically, to move matter around) or *heat* in the form of the movement of atoms and molecules. *Kinetic energy* is contained in moving objects. One such example is the energy contained in a rapidly spinning flywheel, a device of potentially great importance for energy storage to even out the energy flow from intermittent solar and wind sources. *Potential energy* is stored energy, such as in an elevated reservoir of water used as a means of storing hydroelectric energy for later use that can be run through a hydroelectric turbine to generate electricity as needed.

A very important form of potential energy is *chemical energy* stored in the bonds of molecules and released, usually as heat, when chemical reactions occur. For example, in the case of methane, CH_4, in natural gas, when the methane burns,

$$CH_4 + 2O_2 \rightarrow CO_2 + 2H_2O \tag{19.1}$$

the difference between the bond energies in the CO_2 and H_2O products and the CH_4 and O_2 reactants is released, primarily in the form of heat. If the heat is released by combustion of methane in a gas turbine, part of the heat energy can be converted to *mechanical energy* in the form of the rapidly spinning turbine and an electrical generator to which it is attached. The generator, in turn, converts the mechanical energy to *electrical energy.*

The standard unit of energy is the *joule*, abbreviated *J*. A total of 4.184 J of heat energy will raise the temperature of 1 g of liquid water by 1°C. This amount of heat is equal to 1 *calorie* of energy (1 cal = 4.184 J), the unit of energy formerly used in scientific work. A joule is a small unit, and the kilojoule, kJ, equal to 1000 J is widely used in describing chemical processes. The "calorie" commonly used to express the energy value of food (and its potential to produce fat) is actually a kilocalorie, kcal, equal to 1000 cal.

Power refers to energy generated, transmitted, or used per unit time. The unit of power is the *watt*, equal to an energy flux of 1 J/s). A compact fluorescent light bulb adequate to illuminate a desk area might have a rating of 21 W. A large power plant may put out electricity at a power level of 1000 *megawatts* (MW, where 1 MW is equal to 1 million watts). Power on a national or global scale is often expressed in *gigawatts*, each one of which is equal to a billion watts or even *terawatts*, where a terawatt is equal to a trillion watts.

The science that deals with energy in its various forms and with work is *thermodynamics*. There are some important laws of thermodynamics. The *first law of thermodynamics* states that energy is neither created nor destroyed. This law is also known as the *law of conservation of energy*. The first law of thermodynamics must always be kept in mind in the practice of green technology, the best practice of which requires the most efficient use of energy. Thermodynamics enables calculation of the amount of usable energy. As described by the laws of thermodynamics, only a relatively small amount of the potential energy in fuel can be converted to mechanical energy or electrical energy with the remaining energy from the combustion of the fuel dissipated as heat. With the application of green technology, much of this heat is salvaged for applications such as district heating of homes.

Although energy is neither created nor destroyed, the amount of useful energy that can be obtained from a system can be dissipated. Useful mechanical energy can be produced, for example, in a heat engine by harnessing part of the energy flow from a hot to a cooler part of the system (Section 19.4 and Equation 19.5). The amount of this useful energy is called *exergy* and it can go to zero in a system that has reached equilibrium.

19.3 SOURCES OF ENERGY USED IN THE ANTHROSPHERE

Before the 1800s, most of the energy used in the anthrosphere came from biomass produced during plant photosynthesis. Houses were heated with wood. Soil was cultivated and goods and people moved by the power of animals or by humans themselves, obtaining their energy from food biomass. Wind drove sailing ships and windmills and falling water moved waterwheels. These sources were renewable and sustainable, with solar energy captured by photosynthesis to generate biomass, wind produced by temperature and pressure differences in solar-heated masses of atmospheric air, and flowing water moved as part of the solar-driven hydrologic cycle.

Although coal from readily accessible deposits had long been used in small quantities for home heating, exploitation of this energy source grew rapidly after the invention of the steam engine around 1800. During the 1800s, coal became the predominant source of energy in the United States, England, Europe, and other countries that had readily accessible coal resources, a major shift from renewable biomass, wind, and water to a depletable resource that had to be dug from the ground. By 1900, petroleum had become a significant source of energy and by 1950 had surpassed coal as the United States' leading energy supply. Lagging behind petroleum, natural gas had become a significant energy supplier by 1950. By 1950, hydroelectric power was providing a large fraction of energy used in the anthrosphere and still is. By around 1975, nuclear energy was supplying significant amounts of electricity and has maintained an appreciable share worldwide until the present. Miscellaneous renewable sources including geothermal and, more recently, solar and wind energy are making increasing contributions to total energy supply. Biomass still contributes significantly to the total of the sources of energy used.

Figure 19.1 shows U.S. and world energy sources used annually as of the year 2000. The predominance of *fossil fuel* petroleum, natural gas, and coal are obvious. Estimates of the amounts of fossil fuels available differ. Those of the quantities of recoverable fossil fuels in the world before 1800 as estimated around 1970 are given in Figure 19.2. By far, the greatest recoverable fossil fuel is in the form of coal and lignite.

Although world coal resources are enormous and potentially can fill energy needs for a century or two, their utilization would become intolerable to earth's environment because of environmental disruption from mining and emissions of carbon dioxide long before coal resources were exhausted. Assuming only uranium-235 as a fission fuel source, total recoverable reserves of nuclear fuel are roughly about the same as fossil fuel reserves. These are several orders of magnitude higher if the use of breeder reactors that convert normally unfissionable uranium-238 to fissionable plutonium-239 is assumed. Extraction of only 2% of the deuterium present in the earth's oceans would yield

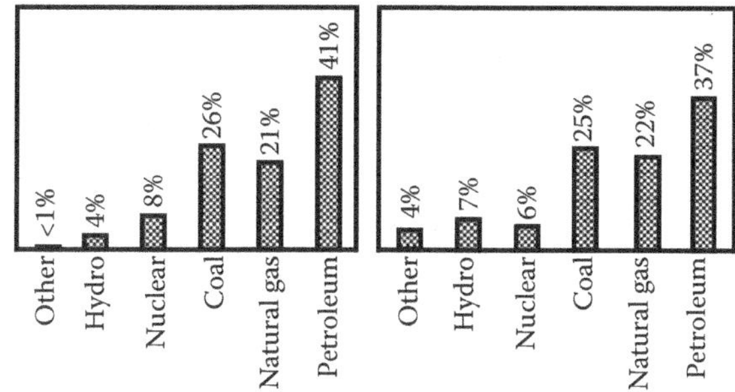

FIGURE 19.1 U.S. (left) and world (right) sources of energy. Percentages of total rounded to the nearest 1%.

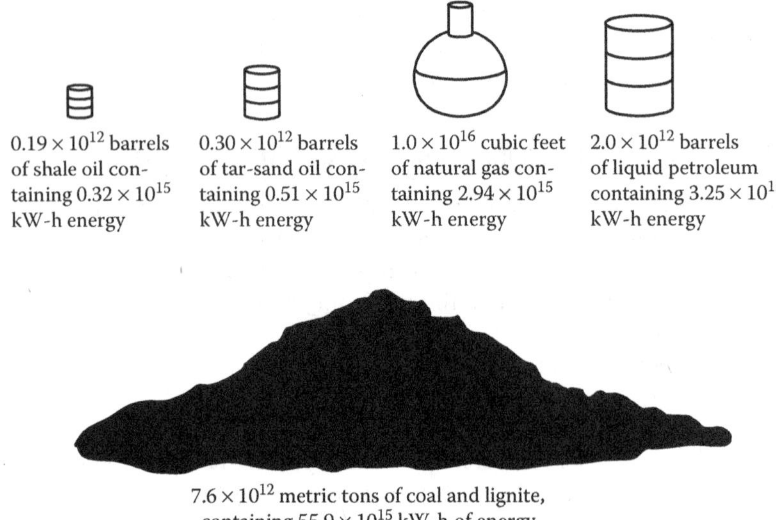

FIGURE 19.2 Original amounts of the world's recoverable fossil fuels (quantities in thermal kilowatt hours of energy based on data taken from M. K. Hubbert, "The energy resources of the earth," in *Energy and Power*, W. H. Freeman and Co., San Francisco, 1971).

about a billion times as much energy by controlled nuclear fusion as was originally present in fossil fuels! This prospect is tempered by the lack of success in developing a controlled nuclear fusion reactor. Geothermal power, currently utilized in northern California, Italy, Iceland, and New Zealand, has the potential for providing a significant percentage of energy worldwide. The same limited potential is characteristic of several renewable energy resources, including hydroelectric energy, tidal energy, and especially wind power. All of these will continue to contribute significant amounts of energy. Renewable, nonpolluting solar energy comes close to being an ideal energy source and it almost certainly has a bright future.

The contributions of different fossil fuels to greenhouse gas carbon dioxide emissions vary with the chemical nature of the fuel with greater contributions from those that have relatively less hydrogen. For example, the chemical reaction for the combustion of methane, CH_4,

$$CH_4 + 2O_2 \rightarrow CO_2 + 2H_2O + \text{energy} \tag{19.2}$$

shows that 2 molecules of H_2O are produced for each molecule of CO_2. Since the conversion of chemically bound hydrogen to H_2O produces a large amount of heat, a relatively smaller amount of CO_2 is released per unit of heat generated. Petroleum hydrocarbons, such as those in gasoline or diesel fuel contain essentially only 2 atoms of H per atom of C. The combustion of such a molecule showing the conversion of 1 atom of C to CO_2 is represented as

$$CH_2 + \tfrac{3}{2}O_2 \rightarrow CO_2 + H_2O + \text{energy} \tag{19.3}$$

demonstrating that only half as much hydrocarbon-bound H is burned per molecule of CO_2 produced, so significantly less energy is produced per C atom than in the combustion of natural gas. Coal is even worse. Coal is a black hydrocarbon with an approximate simple formula of $CH_{0.8}$, so the combustion of an atom of carbon in coal can be represented as the following:

$$CH_{0.8} + 1.2O_2 \rightarrow CO_2 + 0.4H_2O + \text{energy} \tag{19.4}$$

Much less hydrocarbon-bound hydrogen is available to burn per atom of C in coal compared to petroleum or, especially, natural gas, so the amount of carbon dioxide emitted to the atmosphere per unit energy produced from coal is higher than with petroleum and much higher than with natural gas.

The problem that industrialized societies have become dependent on unsustainable fossil energy is clear, the solution less so. Alternatives must be developed and the transition to them will not be easy. The alternatives are discussed later in this chapter.

19.4 ENERGY DEVICES AND CONVERSIONS

Energy occurs in many forms and its utilization requires conversion to other forms. Many devices exist for the utilization of energy and its conversion to other forms. The most common of these are shown in Figure 19.3. The types of energy available, the forms in which it is utilized, and the processes by which it is converted to other forms have a number of implications for green technology and sustainability. For example, the wind turbine shown in Figure 19.3(1), once in place, continues to pump electricity into the power grid with virtually no harm to the environment (although some people regard them as unsightly while others think they are picturesque), whereas the steam power plant shown in Figure 19.3(2) requires mining of depletable coal, combustion of the fossil fuel with its potential for air pollution, control of air pollutants, and means for cooling the steam exiting the turbine, with its potential for thermal pollution of waterways.

An important aspect of energy utilization is its conversion to usable forms. For example, gasoline burned in automobile engines comes originally from petroleum pumped from underground, the petroleum constituents are separated, molecules with properties suitable for engine fuel are produced chemically, the gasoline product is burned in an internal combustion engine converting chemical to mechanical energy, and the mechanical energy is transmitted to the wheels of the automobile in the form of kinetic energy that moves the automobile. Significantly, less than half of the energy in the gasoline is actually converted to mechanical energy of the automobile's motion; the rest is dissipated as waste heat through the engine's cooling system.

Figure 19.4 illustrates major forms of energy and conversions between them. A significant point of this illustration is the very large ranges of energy conversion efficiencies from just a few percent or less to almost 100%. These differences suggest areas in which improvements may be sought. One of the most striking efficiencies is the less than 0.5% conversion of light energy to chemical energy by photosynthesis. Despite such a low conversion efficiency, photosynthesis has generated the fossil fuels, from which industrialized societies now get their energy and provide a significant fraction of energy in areas where wood and agricultural wastes are used. Doubling photosynthesis efficiency with genetically engineered plants could be a major factor in making biomass a more desirable energy source. Replacement of woefully inefficient incandescent light bulbs with fluorescent bulbs that are 5–6 times more efficient in converting electrical energy to light can save large amounts of energy and will be mandated in the United States by legislation passed in 2007.

A particularly important energy conversion carried out in the anthrosphere is that of heat, such as from chemical combustion of fuel, to mechanical energy used to propel a vehicle or run an electrical generator. This occurs, for example, when gasoline in a gasoline engine burns, generating hot gases that move pistons in the engine connected to a crankshaft that converts the up-and-down movement of the piston to rotary motion that drives a vehicle's wheels. It also occurs when hot steam generated at high pressure in a boiler flows through a turbine connected directly to an electrical generator. A device, such as a steam turbine, in which heat energy is converted to mechanical energy is called a heat engine. Unfortunately, the laws of thermodynamics dictate that the conversion of heat to mechanical energy is always much less than 100% efficient. The efficiency of this conversion is given by the Carnot equation,

$$\text{Percent efficiency} = \frac{T_1 - T_2}{T_1} \times 100 \tag{19.5}$$

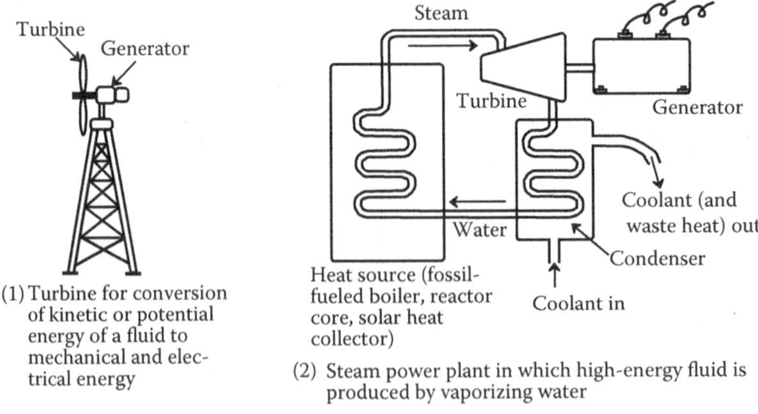

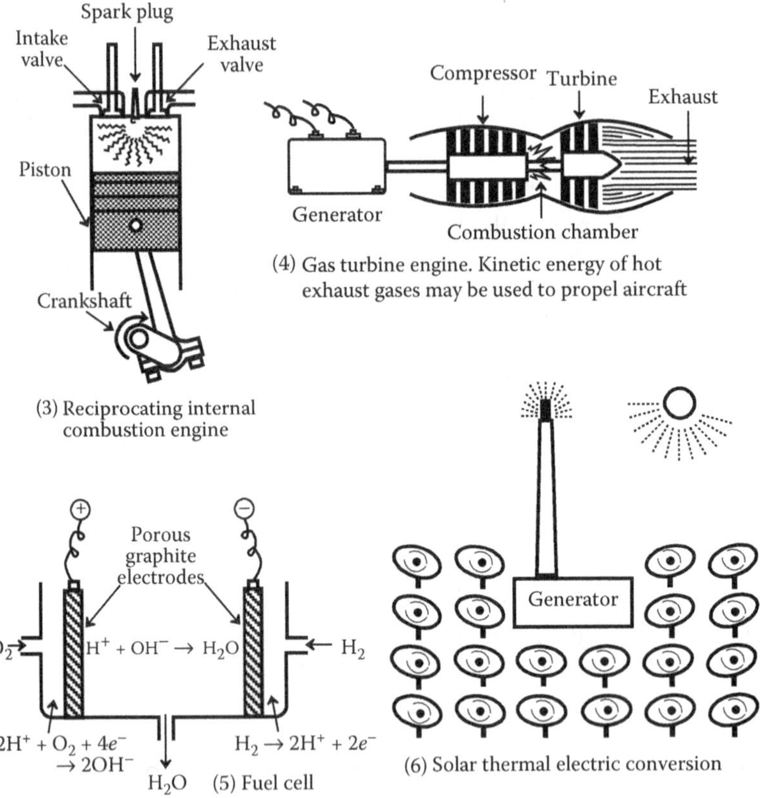

FIGURE 19.3 Examples of many devices for the collection of energy and its conversion to other forms.

in which T_1 is the inlet temperature (e.g., of steam into a steam turbine) and T_2 is the outlet temperature, both expressed in Kelvin (°C+273). Consider a steam turbine as shown in Figure 19.5. Substitution into the Carnot equation of 875 K for T_1 and 335 K for T_2 gives a maximum theoretical efficiency of 62%. However, it is not possible to introduce all the steam at the highest temperature and friction losses occur so that the energy conversion efficiency of most modern steam turbines is just below 50%. About 80% of the chemical energy released by combustion of fossil fuel in a boiler is actually transferred to water to produce steam so that the net efficiency for conversion of chemical energy in fossil fuels to mechanical energy to produce electricity is about 40%. The overall conversion of

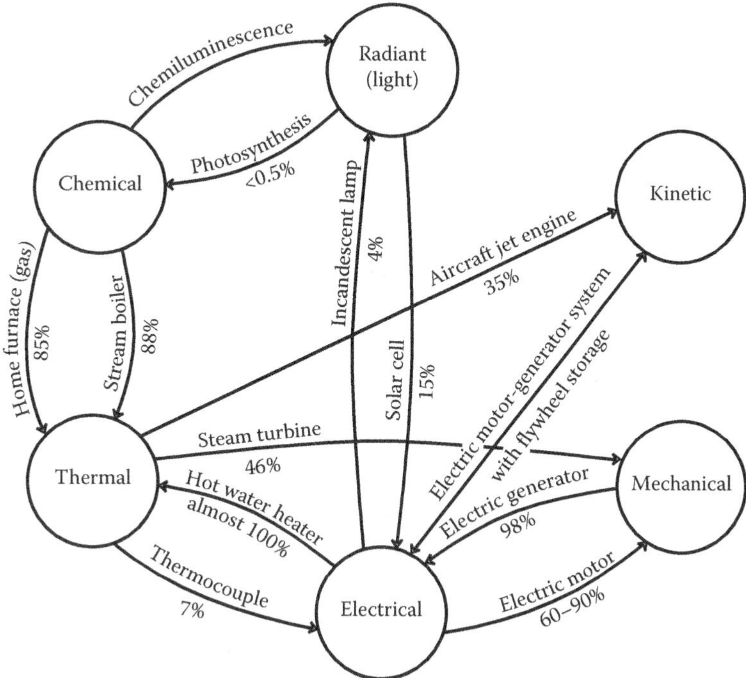

FIGURE 19.4 Major types of energy and conversions between them showing conversion efficiencies.

chemical energy to electricity is essentially the same because an electrical generator converts virtually all of the energy of a rotating turbine to electricity. Because nuclear reactor peak temperatures are limited for safety reasons, their conversion of nuclear energy to electricity is only about 30%.

A particularly important machine for converting chemical to mechanical energy is the internal combustion piston engine shown in Figure 19.6. Most internal combustion engines operate on a cycle of four strokes. In the first of these, the piston moves downward drawing air or an air/fuel mixture into the cylinder. Next, with both valves closed, the air or air/fuel mixture is compressed as the piston moves upward. With the piston near the top of the cylinder (a point at which fuel may be injected if only air is compressed), ignition occurs and the burning fuel creates a mass of highly pressurized combustion gas in the cylinder, which drives the piston down in the third stroke. The exhaust valve then opens and the exhaust gas is expelled during the exhaust stroke.

The efficiency of the internal combustion engine increases with the peak temperature reached by the burning fuel, which increases with the degree of compression during the compression stroke

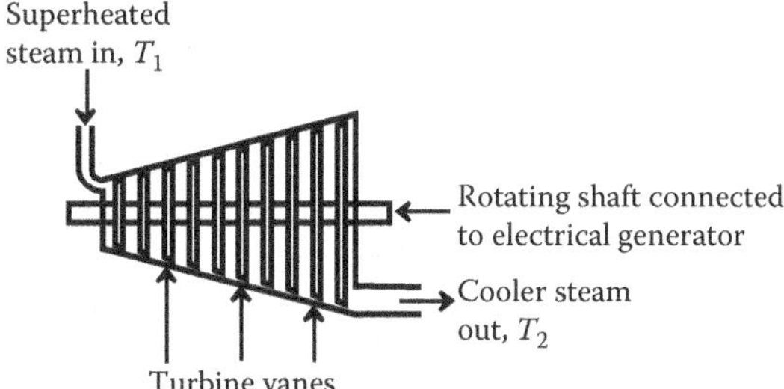

FIGURE 19.5 In a steam turbine, superheated steam impinges on vanes attached to a shaft to produce mechanical energy. For generation of electricity, the shaft is coupled to an electrical generator.

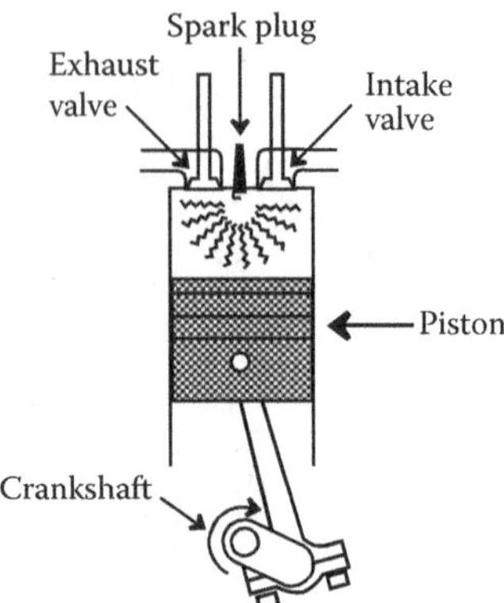

FIGURE 19.6 An internal combustion piston engine in which a very rapidly burning mixture of air and fuel drives a piston downward during the power stroke and this motion is converted to rotary mechanical motion by the crankshaft.

(around 20:1 for a modern diesel engine). This temperature is highest for the diesel engine in which the compression is so high that fuel injected into the combustion chamber ignites without a spark plug ignition source. Whereas a standard gasoline engine is typically about 25% efficient in converting chemical energy in fuel to mechanical energy, a diesel engine is typically 37% efficient, with some reaching higher values.

Although highly superior from the standpoint of efficiency, diesel engines do have some disadvantages with respect to emissions. The first of these is that the combustion zone is not homogeneous because the fuel is injected into the highly compressed air at the top of the compression stroke resulting in incomplete combustion and production of carbon particles; improperly adjusted diesel engines are a major source of particle air pollution in urban areas. In addition, because of their very high combustion temperatures and high ignition pressures, diesel engines tend to produce elevated levels of air pollutant nitrogen oxides. Recent advances in diesel engine design, computerized control, and exhaust pollutant control devices have greatly reduced diesel engine emissions.

19.4.1 Fuel Cells

Fuel cells are devices that convert the energy released by electrochemical reactions directly to electricity without going through a combustion process and electricity generator. Fuel cells are the primary means for utilizing hydrogen fuel and are becoming more common as electrical generators. A fuel cell has an anode at which elemental hydrogen is oxidized, releasing electrons to an external circuit, and a cathode at which elemental oxygen is reduced by electrons introduced from the external circuit, as shown by the half-reactions in Figure 19.7. The H^+ ions generated at the anode migrate to the cathode through a solid membrane permeable to protons. The net reaction is

$$2H_2 + O_2 \rightarrow 2H_2O + \text{electrical energy} \tag{19.6}$$

and the only product of the fuel cell reactions is water.

Although elemental hydrogen is the ultimate fuel for fuel cells, it may be produced by the chemical breakdown of hydrogen-rich fuels, such as methane, methanol, or even gasoline, a process that also generates carbon dioxide. Tubular-style solid-oxide fuel cells, such as those manufactured by Siemens Westinghouse, operate at an elevated temperature of about 1000°C and produce an exhaust that is hot enough to drive a turbine or even to cogenerate steam. Such systems may be able to develop overall efficiencies of up to 80%.

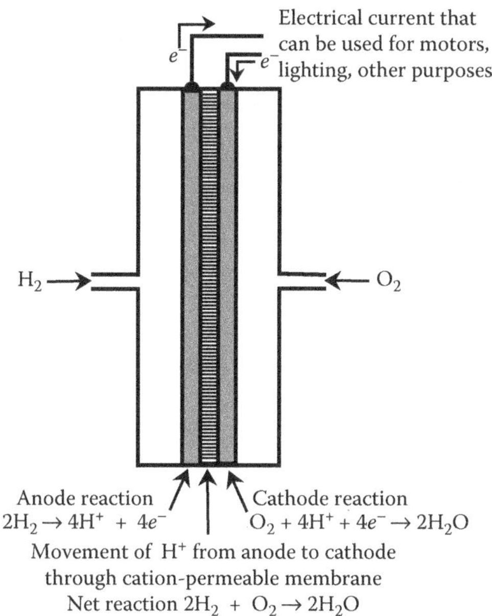

FIGURE 19.7 Cross-sectional diagram of a fuel cell in which elemental hydrogen can be reacted with elemental oxygen to produce electricity directly with water as the only chemical product.

19.5 GREEN TECHNOLOGY AND ENERGY CONVERSION EFFICIENCY

One of the best ways to conserve fuel resources is through increasing the efficiency of energy conversion including that of chemical to mechanical energy with the intermediate step of production of heat energy. Many advances have been made in this area since the late 1800s. Part of the increase in conversion of fuel energy to electricity going from around 4% conversion in 1900 to more than 40% at present resulted from increasing the input temperature (T_1 in the Carnot equation) in the heat engines driving electrical generators. Energy use efficiency increased by more than fourfold when picturesque steam engines were replaced by diesel/electric locomotives during the 1940s and 1950s. Substitution of diesel engines for gasoline engines in trucks and farm and construction equipment have resulted in gains in energy efficiency.

Much of the increased efficiency in fuel utilization has come from improved materials that allow higher operating temperatures. In addition to high-temperature-tolerant metals in engines, a contribution has been made by lubricating oils that do not break down at high temperatures. Much of the progress has been achieved with better engineering, now greatly aided by computerized design, evaluation, and manufacturing of engines. Engineers of a century ago had never heard of green technology, and probably would not have cared had they known about it. But they did understand costs of fuel (which, on the basis of constant value currency, were often higher then than they are now) and they welcomed the greater efficiencies they achieved on the basis of costs.

A key aspect of the most efficient conversion of chemical to mechanical energy in engines is the precise control of operational aspects such as ignition timing, valve timing, and fuel injection. In modern engines, key operation parameters are controlled by computers leading to optimum efficiency in engine operation.

As an inevitable consequence of the thermodynamics described by the Carnot equation, engines that convert heat to mechanical energy cannot utilize much of the heat, which is carried away by an engine cooling system. Typically, a small portion of this heat is used in automotive heaters on cold days. On a broader scale, such as municipal electrical systems, this heat can be used for heating buildings. Such efficiencies are discussed under "Combined Power Cycles" in Section 19.17.

19.6 ENERGY CONSERVATION AND RENEWABLE ENERGY SOURCES

Any consideration of energy needs and production must take energy conservation into consideration. This does not have to mean cold classrooms with thermostats set at 60°F in mid-winter, nor swelteringly hot homes with no air-conditioning, nor total reliance on the bicycle for transportation, although these and even more severe conditions are routine in many countries. The fact remains that the United States and several other industrialized nations have wasted energy at a deplorable rate. For example, U.S. energy consumption is higher per capita than that of some other countries that have equal, or significantly better, living standards. Obviously, a great deal of potential exists for energy conservation that will ease the energy problem.

Efficient use of energy can in fact correlate positively with higher economic standards. Figure 19.8 shows a plot of the ratio of energy use per unit of gross domestic product in developed industrialized nations and illustrates a steady and favorable decrease of energy required relative to economic output. Whereas in 2000, 1.7 barrels of oil equivalents were required per $1000 gross domestic product in developed nations, the corresponding figure for developing nations, which tend to lack advanced means of using energy efficiently, was 5.2 barrels, or 3 times as much. These figures indicate the substantial potential for decreased energy consumption by energy conscious development of the less industrially advanced nations as well as the additional conservation that can be achieved if citizens of industrialized nations can be persuaded to forego wasteful energy practices such as using excessively large and inefficient vehicles and living in overly large dwellings.

Transportation is the economic sector with the greatest potential for increased efficiencies. The private auto and airplane are only about one-third as efficient as buses or trains for transportation. Transportation of freight by a truck requires about 3800 Btu/ton-mile, compared to only 670 Btu/ton-mile for a train. Truck transport is terribly inefficient (as well as dangerous, labor-intensive, and environmentally disruptive) compared to rail transport. Major shifts in current modes of transportation in the United States will not come without anguish, but energy conservation dictates that they be made.

Figure 19.9 shows the trend in U.S. automobile fuel economy during recent decades. The gains through about 1990 were very impressive, then dropped off as less fuel efficient vehicles became more popular. If the same trends from this period would have been maintained, the U.S. automobile fleet would by now average close to 40 miles per gallon (MPG). Such a figure is readily achievable without seriously compromising safety or comfort and, as is obvious from the figure, with much lower emissions from pollutants today as compared to in 1970. In 2007, the U.S. Congress passed legislation mandating higher fuel economy standards for vehicles sold in the United States.

Household and commercial uses of energy are relatively efficient. Here again, appreciable savings can be made. The all-electric home requires much more energy (considering the percentage

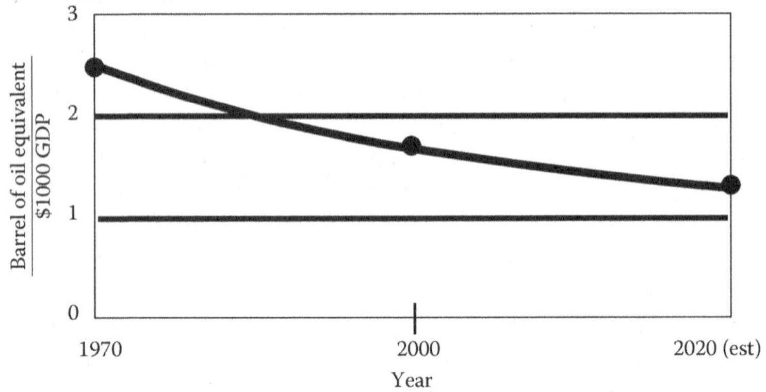

FIGURE 19.8 Plot of barrel of oil equivalent required per $1000 gross domestic product (GDP) as a function of year in industrialized nations.

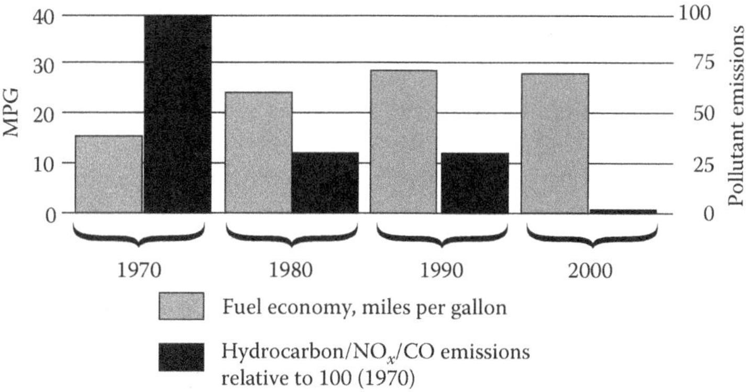

FIGURE 19.9 U.S. auto fleet fuel economy and emissions over three decades. Fuel economy has improved markedly while emissions have been greatly reduced.

wasted in generating electricity) than a home heated with fossil fuels. The sprawling ranch-house style home uses much more energy per person than does an apartment unit, row house, or even a home of comparable floor area built in a compact format (more like a square box). Improved insulation, sealing around the windows, and other measures can conserve a great deal of energy. Electricity generating plants centrally located in cities can provide waste heat for commercial and residential heating and cooling and, with proper pollution control, can use municipal refuse for part of their fuel, thus reducing quantities of solid wastes requiring disposal.

One of the greatest contributions to energy conservation and energy use efficiency in very recent years has been the *hybrid vehicle* that uses an internal combustion engine to produce electricity that is stored for propulsion of the vehicle in a nickel-metal-hydride battery (Figure 19.10). Although not greatly more efficient for prolonged driving at highway speeds, these vehicles have achieved improvements up to 50% in stop-and-go driving in traffic. For routine operation, the internal combustion engine supplies all the power needed plus additional power, if required, to run the generator to recharge a battery (larger than the battery in a conventional automobile but significantly smaller than the battery in an all-electric vehicle). When a surge of power is required, electricity from the storage battery drives the electric motor to produce the additional power. The braking system also generates electricity that is stored in the battery. When the vehicle is stopped, the internal combustion engine does not run, which also saves fuel.

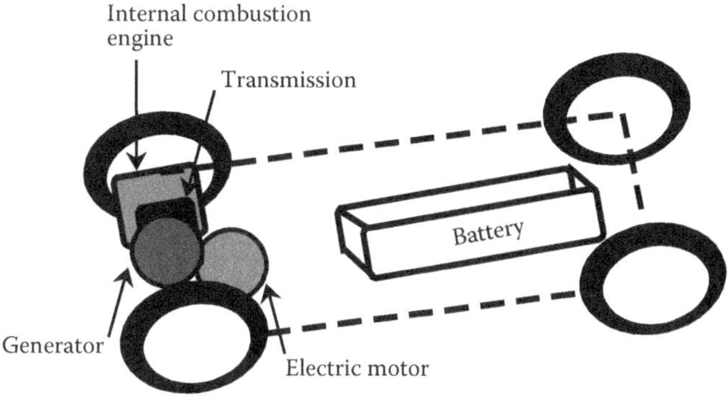

FIGURE 19.10 Major components of a hybrid automobile in which an internal combustion engine drives a generator to produce electricity to drive a vehicle. Electrical energy is stored in a battery. Electrical energy is also salvaged for storage from braking the vehicle.

Vastly reduced fuel consumption can be achieved by hybrid vehicles now under development that have batteries that can be charged by an external electrical source as well as the onboard engine. Such vehicles could give around 40–50 km driving range from an overnight charge alone, sufficient for the majority of routine driving. As the battery runs down, the internal combustion engine would recharge it as is the case with current hybrids. As of 2009, batteries were under development that could be recharged efficiently by both an external source of electricity and by an internal combustion engine coupled to a generator.

Although gasoline engines are now employed in hybrid vehicles, the ultimate in fuel economy could be achieved with an inherently more efficient diesel engine as the internal combustion engine component. By allowing the diesel engine to run at a generally steady rate, the output of exhaust pollutants, which are produced at higher levels by diesel engines as the engine speed is changed, could be greatly reduced. Furthermore, diesel engines idle with remarkably little fuel consumption, so that the diesel engine would not need to be turned off when the vehicle is stopped, thus staying hot and further reducing emissions when it is brought up to speed.

As scientists and engineers undertake the crucial task of developing alternative energy sources to replace dwindling petroleum and natural gas supplies, energy conservation must receive proper emphasis. In fact, zero energy-use growth, at least on a per capita basis, is a worthwhile and achievable goal. Such a policy would go a long way toward solving many environmental problems. With ingenuity, planning, and proper management, it could be achieved while increasing the standard of living and quality of life.

Closely related to energy conservation is the concept of *renewable energy* from sources that do not run out. Essentially all of these depend on energy from the sun including direct solar energy, wind driven by the solar heating of air masses, falling water from the solar-powered hydrologic cycle, and biomass formed from photosynthesis. For most of its lifetime on Earth, humankind has depended entirely on renewable sources of energy, and most countries are again emphasizing these sources.

Enlightened sustainable energy policies are being implemented in some developing countries. China implemented a new law on renewable energy at the beginning of 2006. This policy encourages renewable energy alternatives including wind power, biomass energy, and biomethane generation. Long known for its utilization of wastes (including even use of human wastes as fertilizer for growing vegetables), China has constructed many waste-to-methane generators in rural areas with 17 million families served by such facilities by 2006. Experimental biopower projects burning crop biomass by-products have been undertaken. As of 2006, China had 80 million square meters of solar collectors to heat water, equivalent to the energy from 10 million tons per year of coal. The total renewable energy capacity of China in 2006 was 7% of China's energy use, equivalent to 160 million tons of coal per year.

19.7 PETROLEUM AND NATURAL GAS

Liquid *petroleum* occurs in rock formations ranging in porosity from 10% to 30%. Up to half of the pore space is occupied by water. The oil in these formations must flow over long distances to an approximately 15-cm diameter well from which it is pumped. The rate of flow depends on the permeability of the rock formation, the viscosity of the oil, the driving pressure behind the oil, and other factors. Because of limitations in these factors, *primary recovery* of oil yields an average of about 30% of the oil in the formation, although it is sometimes as little as 15%. More oil can be obtained using *secondary recovery* techniques, which involve forcing water under pressure into the oil-bearing formation to drive the oil out. Primary and secondary recovery together typically extract somewhat less than 50% of the oil from a formation. Finally, *tertiary recovery* can be used to extract even more oil, normally through the injection of pressurized carbon dioxide, which forms a mobile solution with the oil and allows it to flow more easily to the well. Other chemicals, such as detergents, may be used to aid in tertiary recovery. Currently, about 300 billion barrels of U.S. oil are not available through primary recovery alone. A recovery efficiency of 60% through secondary or tertiary techniques could double the amount of available petroleum. Much of this would come from fields that have already been abandoned or essentially exhausted using primary recovery techniques.

The year 2008 was a very interesting one for world oil production and consumption. In July of that year, prices reached record levels of more than \$150/barrel and many experts forecast that prices would continue to increase with gasoline going well over the \$5/gallon reached in parts of the United States. With the world economic collapse that took place in the fall of 2008, petroleum prices fell precipitously, reaching levels of around \$40/barrel. The year also saw the first drop in world oil consumption since 1983 down to 85.8 million barrels per day and U.S. consumption dropped 6.3% to 19.4 million barrels per day.

Shale oil is a possible substitute for liquid petroleum. Shale oil is a pyrolysis product of oil shale, a rock containing organic carbon in a complex structure of biological origin from eons past called kerogen. Oil shale is believed to contain approximately 1.8 trillion barrels of shale oil that could be recovered from deposits in Colorado, Wyoming, and Utah. In the Colorado Piceance Creek basin alone, more than 100 billion barrels of oil could be recovered from prime shale deposits. However, the environmental implications of recovering shale oil by heating oil shale include production of vast amounts of greenhouse gas carbon dioxide. Furthermore, potential water pollution problems from water-soluble salt residues left over from the pyrolysis of oil shale make it unlikely that this resource will ever be developed on a large scale.

Natural gas, consisting almost entirely of methane, is a very attractive fuel that produces few pollutants and less carbon dioxide per unit energy than any other fossil fuel. In addition to its use as a fuel, natural gas can be converted to many other hydrocarbon materials. It can be used as a raw material for the Fischer–Tropsch synthesis of gasoline. As of the early 2000s, increased demand for natural gas had led to tight supplies in the United States. Production of natural gas from coal seams in Wyoming has required pumping saline, alkaline water from the seams, which has caused water pollution problems. By 2009, exploitation of natural gas in "tight" shale formations opened by fracturing with water had resulted in increased supply of natural gas in the United States and some other countries.

19.8 COAL

From Civil War times until World War II, *coal* was the dominant energy source behind industrial expansion in most nations. However, after World War II, the greater convenience of lower cost petroleum resulted in a decrease in the use of coal for energy in the United States and in a number of other countries. Annual coal production in the United States fell by about one-third, reaching a low of approximately 400 million tons in 1958, but since then coal production for electricity generation has reached about 1 billion tons per year in the United States. About one-third of the world's energy and around 50% of electrical energy is provided by coal.

The general term *coal* describes a large range of solid fossil fuels derived from partial degradation of plants. Coal is differentiated largely by *coal rank* based on the percentage of fixed carbon, percentage of volatile matter, and heating value. The approximate average empirical formula of coal is $CH_{0.8}$ and coal typically contains from 1 to several percent sulfur, nitrogen, and oxygen. Of these elements, sulfur bound to the organic coal molecule and mixed with coal as mineral pyrite, FeS_2, presents major environmental problems because of production of air pollutant sulfur dioxide during combustion. Much of the FeS_2 can be removed physically from coal prior to combustion and sulfur dioxide can be removed from stack gas by various scrubbing processes.

19.8.1 Coal Conversion

As shown in Figure 19.11, coal can be converted to gaseous, liquid, or low-sulfur, low-ash solid fuels such as coal char (coke) or solvent-refined coal (SRC). Coal conversion is an old idea; a house belonging to William Murdock at Redruth, Cornwall, England, was illuminated with coal

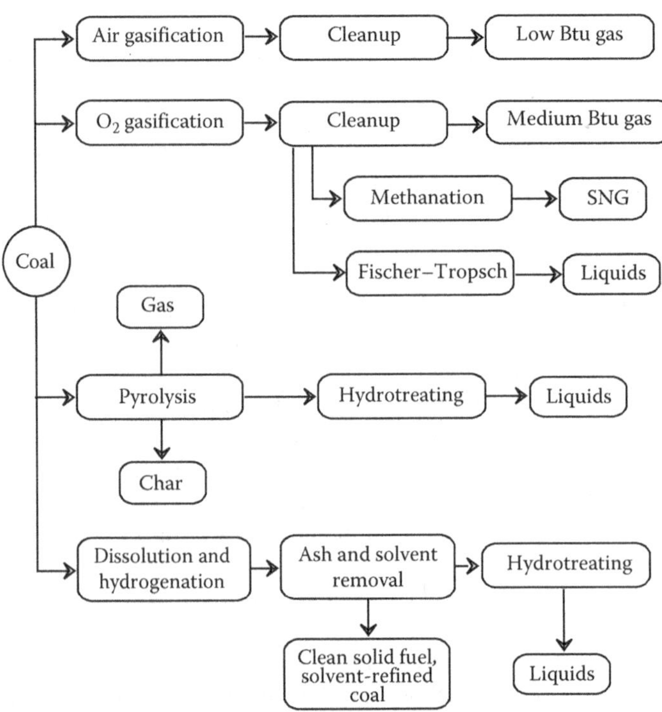

FIGURE 19.11 Routes to coal conversion. Btu refers to British thermal units, a measure of the heat energy that can be obtained from a fuel. Methanation means synthesis of CH_4 gas. Hydrogenation and hydrotreating refer to reaction with elemental H_2 gas.

gas in 1792. The first municipal coal-gas system was employed to light Pall Mall in London in 1807. The coal-gas industry began in the United States in 1816. The early coal-gas plants used coal pyrolysis (heating in the absence of air) to produce a hydrocarbon-rich product particularly useful for illumination. Later in the 1800s, the water-gas process was developed, in which steam was impinged on hot coal coke to produce a mixture consisting primarily of H_2 and CO. It was necessary to add volatile hydrocarbons to this "carbureted" water-gas to bring its illuminating power up to that of gas prepared by coal pyrolysis. The United States had 11,000 coal gasifiers operating in the 1920s. At the peak of its use in 1947, the water-gas method accounted for 57% of U.S.-manufactured gas. The gas was made in low-pressure, low-capacity gasifiers that by today's standards would be inefficient and environmentally unacceptable (many sites of these old plants have been designated as hazardous waste sites because of residues of coal tar and other wastes). This was definitely not a green technology because of the high toxicity of carbon monoxide in the gas product, and many people were killed by release of CO in their houses. During World War II, Germany developed a major synthetic petroleum industry based on coal, which reached a peak capacity of 100,000 barrels per day in 1944. A synthetic petroleum plant operating in Sasol, South Africa, reached a capacity of several tens of thousands of tons of coal per day in the 1970s and currently produces hydrocarbons and feedstocks equivalent to about 150,000 barrels of petroleum per day.

A number of environmental implications are involved in the widespread use of coal conversion. These include strip mining, water consumption in arid regions, lower overall energy conversion compared to direct coal combustion, and increased output of atmospheric carbon dioxide. These plus economic factors have prevented coal conversion from being practiced on a very large scale. However, coal conversion does enable relatively facile carbon sequestration (see Section 19.9), which could enable much more sustainable coal utilization.

19.9 CARBON SEQUESTRATION FOR FOSSIL FUEL UTILIZATION

Carbon sequestration, which prevents carbon dioxide generated by fossil fuels from entering the atmosphere, holds the promise of enabling utilization of fossil fuels without contributing to greenhouse warming. Basically, the various schemes that have been proposed entail capturing carbon dioxide from a product or waste stream and sequestering it in a place where it cannot enter the atmosphere. Several approaches have been suggested or tried for capturing carbon dioxide and there are several possibilities for sequestration. The term "carbon sequestration" is also used in a more general sense to apply to the removal of carbon dioxide from the atmosphere, especially by photosynthesis.

There are several possible sinks in which carbon dioxide can be sequestered. The largest of these, a natural sink for the gas, is the ocean. Earth's oceans have an almost inexhaustible capacity for carbon dioxide. However, lowering the average pH of the oceans by as little as 0.1 pH unit from acidic carbon dioxide could have a serious adverse effect on ocean life and productivity; there is some evidence to suggest that shells of some small sea creatures have become thinner because of higher levels of dissolved carbon dioxide in ocean water. Deep saline formations also have a very high capacity for carbon dioxide sequestration. Depleted oil and gas reservoirs and unmineable coal seams have much lower, but still significant, carbon dioxide capacities.

Geological carbon dioxide sequestration can be accomplished by injecting the gas into porous sedimentary formations at depths exceeding approximately 1000 m. Experience in the petroleum industry with underground disposal of carbon dioxide and injection of the gas into oil-bearing formations for petroleum recovery have provided the technology required for geological carbon dioxide sequestration. The carbon dioxide injected into sedimentary formations rises and is confined by poorly permeable caprock. Breaches in caprock, such as those from abandoned oil wells, can result in carbon dioxide release. Eventually, the carbon dioxide dissolves in the generally saline pore waters in the sedimentary formation into which it is injected. Chemical reactions in the water and with the surrounding geological strata can result in long-term stability of the carbon dioxide. It is much more effective and less expensive to force carbon dioxide into unsaturated saline groundwater where it is dissolved than to compress the gas into a pressurized gas form underground.[1]

The first commercial application of carbon dioxide sequestration has operated since 1996 in the North Sea about 240 km from the Norwegian coast in a region known as the Sleipner oil and gas field. The natural gas product from this field is about 9% carbon dioxide, a value that must be reduced to 2.5% for commercial distribution of the gas. Whereas all other gas-producing operations simply discharge the carbon dioxide removed to the atmosphere, at Sleipner it is pumped under pressure into a 200-m thick layer of sandstone, the Utsira formation, which is about 1000 m below the seabed. A mixture of carbon dioxide and hydrogen sulfide removed from sour natural gas that is abundant in Alberta, Canada, is now being sequestered underground.[2]

The easiest sources of carbon dioxide to capture are those from industrial processes that generate the gas in high concentrations. An example of such a process is the fermentation of carbohydrates to produce ethanol for fuel or other uses. This source provides much of the carbon dioxide that is used commercially at present. The largest anthropogenic source of carbon dioxide discharged to the atmosphere is generated in power plants fueled with fossil fuels. These sources present a substantial challenge for carbon dioxide removal because they are so dilute. A power plant fueled with carbon-rich coal produces an exhaust stream that is 13–15% carbon dioxide, whereas one burning hydrogen-rich methane produces only 3–5% carbon dioxide. A third possibility is to capture carbon dioxide released from the gasification of fossil fuels, particularly coal (see Section 19.8). Normally, gasification is performed using oxygen as an oxidant, and the initial product consists of carbon dioxide and combustible H_2 and CO gases. Carbon monoxide in the synthesis gas product can be reacted with steam,

$$CO + H_2O \rightarrow H_2 + CO_2 \tag{19.7}$$

to produce nonpolluting elemental hydrogen fuel and carbon dioxide.

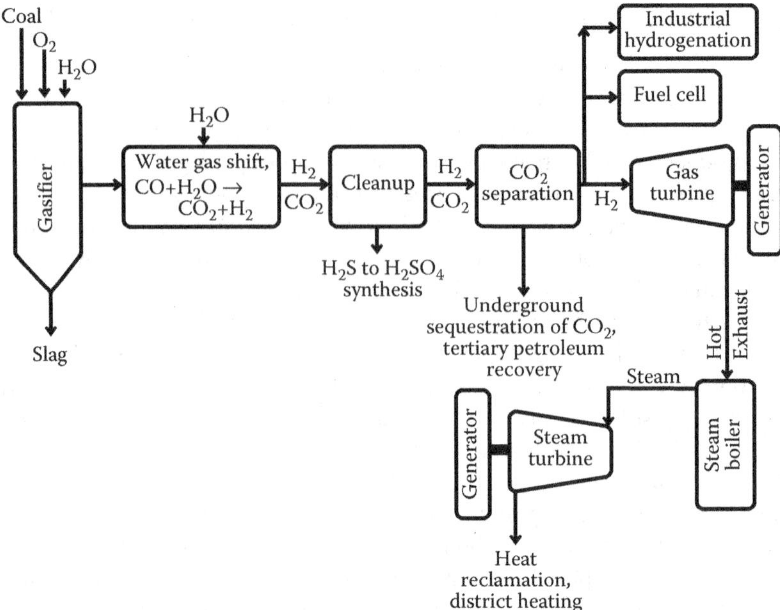

FIGURE 19.12 A coal-based carbon sequestration operation. Electricity is generated at generators connected both to the gas turbine and steam turbine. Reclaimed hydrogen sulfide is used to synthesize sulfuric acid. Hydrogen product can be employed for fuel cells and industrial hydrogenation as well as to power the gas turbine. Sequestered carbon dioxide can be used for petroleum recovery or simply disposed underground. Waste heat is reclaimed for district heating and industrial heating applications.

In late 2007, the U.S. Department of Energy announced funding for the construction of the first U.S. power plant with carbon dioxide sequestration, the FutureGen coal-powered power plant to be located at a site in Mattoon, Illinois. The project was abruptly canceled in early 2008 allegedly because of greatly increased costs, but a Government Accountability Office report released in March 2009 showed that the costs had been overestimated by $500 million because of an accounting cost mathematics error! The 275 MW power plant was to be large enough to supply electricity to 275,000 households and to sequester 1–2 million tons of carbon dioxide per year. Figure 19.12 is a schematic of a coal-fired power plant with carbon sequestration such as the FutureGen power plant. A gas turbine powered by hydrogen fuel produced by coal gasification generates electricity. Hot exhaust gas from the gas turbine is used to raise steam in a boiler and this steam powers a steam turbine that is also coupled to a generator. This combination results in very efficient electrical power generation. The production of intermediate hydrogen adds options for powering fuel cells or for using hydrogen in chemical synthesis or manufacture of synthetic hydrocarbon fuels. By-product hydrogen sulfide must be removed from the hydrogen product and can be used for making sulfuric acid, an important industrial chemical, or can be sequestered with the carbon dioxide. The carbon dioxide by-product is sequestered in mineral formations at depths up to around 2000 m. For power plants located near petroleum-bearing formations, carbon dioxide can be pumped into oil-bearing mineral formations for tertiary petroleum recovery.

19.10 INDUSTRIAL ECOLOGY FOR ENERGY AND CHEMICALS

An excellent example of a system of industrial ecology for sustainable utilization of fossil fuels is provided by the Great Plains Synfuel plant located near Beulah, North Dakota. This complex is designed to take advantage of abundant deposits of lignite, a relatively low-heating-value, high-water-content form of coal that occurs in abundance in North Dakota. A schematic diagram

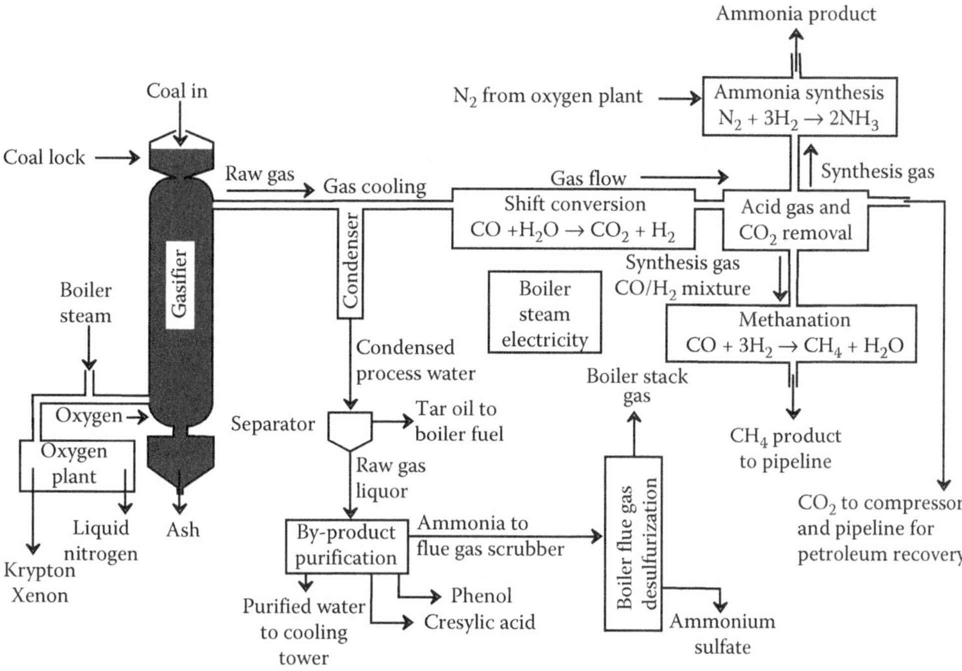

FIGURE 19.13 Schematic of the Great Plains lignite coal gasification complex.

of the system is shown in Figure 19.13. The heart of the plant consists of 14 Lurgi-type gasifiers that are 13 m high and 4 m in diameter processing 16,000 metric tons of lignite per day at temperatures up to 1200°C to produce a synthesis gas mixture of combustible H_2 and CO along with by-products CO_2, water, and smaller quantities of tars, oils, phenolic compounds, ammonia, and H_2S. Useful fuel hydrocarbons, phenol, cresols, ammonia, and sulfur compounds are extracted from the water, which is used for cooling water. The gas mixture is subjected to a shift reaction that increases the ratio of H_2 to CO and this mixture is reacted to produce methane (CH_4) that is fed into a natural gas pipeline. Part of the synthesis gas stream is diverted to an ammonia synthesis plant to make this valuable fertilizer product. A key feature of the plant is the extraction of CO_2 from the synthesis gas. The carbon dioxide is compressed and sent through a pipeline to Canada where it is pumped underground to depleted petroleum-bearing strata for secondary recovery of crude oil.

The Great Plains Synfuels plant provides an excellent example of a diversified industrial ecology complex. It utilizes an abundant resource (lignite coal) in a way that significantly reduces the greenhouse gas emissions and other potential environmental impacts from this resource. Rather than having to transport large quantities of lignite by rail to distant power plants, the energy content of the lignite is converted to methane, the most environmentally friendly fossil fuel, which is moved with minimum environmental disruption through pipelines. An even higher value product, fertilizer ammonia, is synthesized at the site using elemental hydrogen made from gasifying the lignite and elemental nitrogen isolated from air in the air liquefaction operation required to prepare elemental oxygen required for gasification. The carbon dioxide by-product is not released to the atmosphere where it could aggravate the global warming problem but is pumped to oilfields where it is pumped underground to increase crude oil recovery. Commercially valuable ammonium sulfate by-product, a useful source of nitrogen and sulfur soil nutrients, is recovered from the water released during lignite gasification. In an area where water shortages can occur, the relatively large amount of water released by lignite gasification is used for cooling and as boiler feedwater.

19.11 NUCLEAR ENERGY

The awesome power of the atom nucleus revealed at the end of World War II held out enormous promise for the production of abundant, cheap energy. This promise has never really come to full fruition, although nuclear energy currently provides a significant percentage of electric energy in many countries, and it may be the only source of electrical power that can meet world demand without unacceptable environmental degradation, particularly through the generation of greenhouse gases.

Nuclear fission for power production is carried out in nuclear power reactors in which the fission (splitting) of uranium-235 or plutonium nuclei occurs. Each such event generates two radioactive fission product atoms of roughly half the mass of the nucleus fissioned, an average of 2.5 neutrons, plus an enormous amount of energy compared to normal chemical reactions. The neutrons, initially released as fast-moving, highly energetic particles, are slowed to thermal energies in a moderator medium. For a reactor operating at a steady state, exactly one of the neutron products from each fission is used to induce another fission reaction in a chain reaction (Figure 19.14).

The energy from these nuclear reactions is used to heat water in the reactor core and produce steam to drive a steam turbine, as shown in Figure 19.15. As noted in Section 19.4, temperature limitations make nuclear power less efficient in converting heat to mechanical energy and, therefore, to electricity, than fossil energy conversion processes.

A limitation of fission reactors is the fact that only 0.71% of natural uranium is fissionable uranium-235. This situation could be improved by the development of *breeder reactors*, which convert uranium-238 (natural abundance 99.28%) to fissionable plutonium-239.

A major consideration in the widespread use of nuclear fission power is the production of large quantities of highly radioactive waste products. These remain lethal for thousands of years. They must either be stored in a safe place or be disposed of permanently in a safe manner. At the present time, spent fuel elements are being stored under water at the reactor sites. Under current regulations in most countries, the wastes from this fuel will eventually have to be buried. An alternative favored by many investigators is to process the material in the spent fuel elements to remove radioactive products from uranium fuel, isolate the relatively short-lived fission products that decay spontaneously within several hundred years, and bombard the longer-lived nuclear wastes with neutrons in nuclear reactors. The absorption of neutrons by the nuclei of the nuclear waste elements causes *transmutation* in which the elements are converted to other elements or fission products with shorter half lives resulting in relatively rapid production of stable isotopes. Radioactive waste elements for which transmutation is feasible include plutonium, americium, neptunium, curium, technetium-99, and iodine-129. Plutonium, americium, neptunium, and curium are heavy actinide elements that are fissionable and add fuel value in a nuclear reactor.

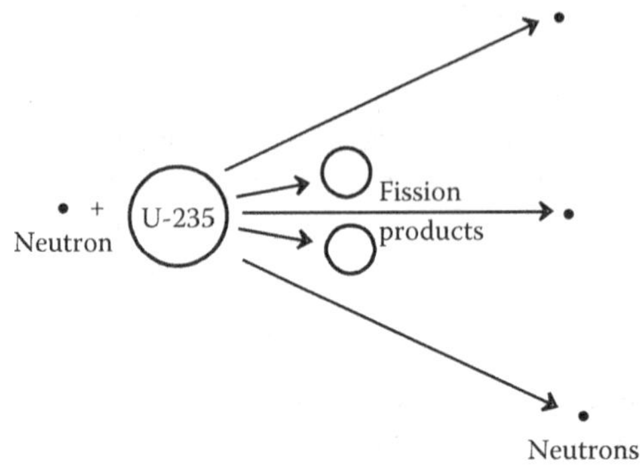

FIGURE 19.14 Fission of a uranium-235 nucleus.

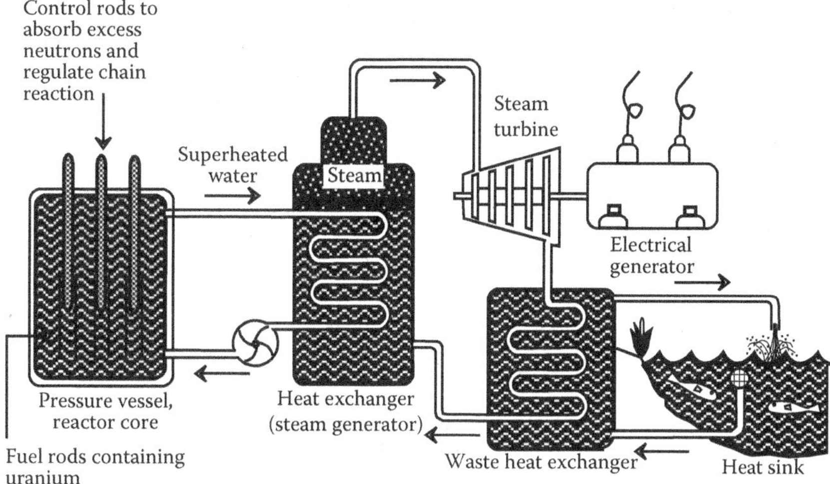

FIGURE 19.15 A typical nuclear fission power plant.

Another problem to be faced with nuclear fission reactors is their eventual decommissioning. There are three possible solutions. One is dismantling soon after shutdown, in which the fuel elements are removed, various components are flushed with cleaning fluids, and the reactor is cut up by remote control and buried. "Safe storage" involves letting the reactor stand 30–100 years to allow for radioactive decay, followed by dismantling. The third alternative is entombment, that is, encasing the reactor in a concrete structure.

The course of nuclear power development was altered drastically by two accidents. The first of these occurred on March 28, 1979, with a partial loss of coolant water from the Metropolitan Edison Company's nuclear reactor located on Three Mile Island in the Susquehanna River, 28 miles outside of Harrisburg, Pennsylvania. Known as the TMI incident, the result was a loss of control, overheating, and partial disintegration of the reactor core. Some radioactive xenon and krypton gases were released and some radioactive water was dumped into the Susquehanna River. Eventually the reactor building was sealed. A much worse accident occurred at Chernobyl in the Soviet Union in April 1986, when a reactor blew up and burned, spreading radioactive debris over a wide area and killing a number of people (officially 31, but certainly many more). Thousands of people were evacuated and the entire reactor structure had to be entombed in concrete and steel plate. Food was seriously contaminated as far away as northern Scandinavia.

As of 2006, 28 years had passed since a new nuclear electric power plant had been ordered in the United States, in large part because of the projected high costs of new nuclear plants. Although this tends to indicate hard times for the nuclear industry, pronouncements of its demise may be premature and by 2008, several U.S. utilities had begun the process of ordering new nuclear power plants and a significant number of such plants were under construction around the world. Properly designed nuclear fission reactors can generate large quantities of electricity reliably and safely and have done so for decades in U.S. naval submarines and carrier ships and in France, which gets about 80% of its electricity from nuclear sources. The single most important factor that may lead to the renaissance of nuclear energy is the threat to the atmosphere from greenhouse gases produced in large quantities by fossil fuels. It can be argued that nuclear energy is the only proven alternative that can provide the amounts of energy required within acceptable limits of cost, reliability, and environmental effects.

New designs for nuclear power plants can enable power reactors that are much safer and environmentally acceptable than those built with older technologies. The proposed new designs incorporate built-in passive safety features that work automatically in the event of problems that could lead to incidents such as TMI or Chernobyl with older reactors. These devices—which depend on passive phenomena such as gravity feeding of coolant, evaporation of water, or convection flow of fluids—give the reactor the desirable characteristics of *passive stability*. They have also enabled significant simplification of hardware, with only about half as many pumps, pipes, and heat exchangers as are contained in older power reactors.

19.11.1 Nuclear Fusion

The fusion of a deuterium nucleus and a tritium nucleus releases a lot of energy as shown below, where MeV stands for million electron volts, a unit of energy:

$$^{2}_{1}\mathrm{H} + {}^{3}_{1}\mathrm{H} \rightarrow {}^{4}_{2}\mathrm{He} + {}^{1}_{0}\mathrm{n} + 17.6\ \mathrm{MeV}\ \text{(energy released per fusion)} \quad (19.8)$$

This reaction is responsible for the enormous explosive power of the "hydrogen bomb." So far it has eluded efforts at containment for a practical continuous source of energy. And since physicists have been trying to make it work on a practical basis for the last approximately 50 years, it will probably never be done. (Within about 15 years after the discovery of the phenomenon of nuclear fission, it was being used in a power reactor to power a nuclear submarine.) However, the tantalizing possibility of using the essentially limitless supply of deuterium, an isotope of hydrogen, from earth's oceans for nuclear fusion still give some investigators the hope of a practical nuclear fusion reactor and some research continues toward that end.

Nuclear fusion was the subject of one of the greatest scientific embarrassments of modern times that occurred in 1989. This incident came about when investigators at the University of Utah announced that they had accomplished the so-called cold fusion of deuterium during the electrolysis of deuterium oxide (heavy water). This resulted in an astonishing flurry of activity as scientists throughout the world sought to repeat the results, whereas others ridiculed the idea. Unfortunately, for the dream of a cheap and abundant source of energy, the skeptics were right, and the whole story of cold fusion stands as a lesson in the (temporary) triumph of wishful technological thinking over scientific good sense.

19.12 GEOTHERMAL ENERGY

Underground heat in the form of steam, hot water, or hot rock used to produce steam has been employed as an energy resource for about a century and can be regarded as largely renewable. This energy was first harnessed for the generation of electricity at Larderello, Italy, in 1904, and has since been developed in Japan, Russia, New Zealand, the Philippines, and at the Geysers in northern California.

Underground dry steam is relatively rare, but is the most desirable from the standpoint of power generation. More commonly, energy reaches the surface as superheated water and steam. In some cases, the water is so pure that it can be used for irrigation and livestock; in other cases, it is loaded with corrosive, scale-forming salts. Utilization of the heat from contaminated geothermal water generally requires that the water be reinjected into the hot formation after heat removal to prevent contamination of surface water.

The utilization of hot rocks for energy requires fracturing of the hot formation, followed by injection of water and withdrawal of steam. This technology is still in the experimental state, but promises approximately 10 times as much energy production as steam and hot water sources.

Land subsidence and seismic effects, such as the mini-earthquakes that occur when water is pumped under extreme pressure into hot rock formations that fracture as a consequence, are environmental factors that may hinder the development of geothermal power. However, this energy source holds considerable promise, and its development continues.

19.13 THE SUN: AN IDEAL, RENEWABLE ENERGY SOURCE

Solar power is an ideal source of energy that is unlimited in supply, widely available, and inexpensive. It does not add to the earth's total heat burden or produce chemical air and water pollutants. On a global basis, utilization of only a small fraction of solar energy reaching the earth could provide for all energy needs. In the United States, for example, with conversion efficiencies ranging from 10% to 30%, it would only require collectors ranging in area from one-tenth down to one-thirtieth that of the state of Arizona to satisfy present U.S. energy needs. (This is still an enormous amount of land, and there are economic and environmental problems related to the use of even a fraction of this amount of land for solar energy collection. Certainly, many residents of Arizona would not be pleased at having so much of the state covered by solar collectors, and some environmental groups would protest the resultant shading of rattlesnake habitat.)

There are vast land areas available in the United States and throughout the world that receive exceptional levels of solar energy suitable for power generation. Factors involved in evaluating areas for solar energy generation include proximity to the Equator, relatively high altitude, and consistent absence of cloud cover. The terrestrial area that has optimum conditions is in the Sahara Desert of southeast Niger in Africa, which receives an average of 6.78 kWh/m^2 each day, which is close to the amount of energy required to heat water for a typical U.S. household.

Solar power cells (photovoltaic cells) for the direct conversion of sunlight to electricity have been developed and are widely used for energy in space vehicles. With present technology, however, they are relatively expensive in most places for large-scale generation of electricity, although the economic gap is narrowing. Most schemes for the utilization of solar power depend on the collection of thermal energy followed by conversion to electrical energy. The simplest example of such an approach involves focusing sunlight on a steam-generating boiler (see Illustration 6 in Figure 19.3). Parabolic reflectors can be used to focus sunlight on pipes containing heat-transporting fluids. Selective coatings on these pipes ensure that most of the incident energy is absorbed.

The most efficient and reliable type of solar-powered heat engine to generate electricity is the *Stirling engine*, which has achieved close to 30% overall efficiency in the conversion of solar energy to electricity. The Stirling engine uses hydrogen gas as a working fluid in a sealed system in which pistons connected by rods in cylinders pump the gas back and forth in the system. The gas is heated at a heater head kept at a relatively high temperature by sunlight focused from an array of reflective mirrors that concentrate sunlight on the heater head. The resultant pressure forces one of the pistons down in a power stroke. The gas is then transferred to a piston on the cooled side of the engine through a regenerator consisting of a material that captures some of the heat to be used in the next cycle, thereby greatly increasing efficiency. For a system planned for installation in California's Mojave desert,[3] each Stirling unit is rated at 25 kW power, running at 1800 rpm, and generating electricity at 480 V and 60 Hz. A major advantage of this system is its modular character that enables scaling from a single unit to a very large number generating hundreds of MW of energy in a single location. As of 2008, it appeared that the Stirling system would become a leading contender for the utilization of solar power in areas such as California's Mojave Desert and Africa's Sahara that receive intense, consistent sunlight.

The direct conversion of energy in sunlight to electricity is accomplished by special solar voltaic cells. Most types of photovoltaic cells depend on the electronic properties of silicon atoms containing low levels of other elements. In a typical photovoltaic cell, the cell consists of two layers of silicon, a donor layer that is doped with about 1 ppm of arsenic atoms and an acceptor layer that is doped with about 1 ppm of boron. Recall from Chapter 3 that Lewis symbols use dots to represent the outermost valence electrons of atoms, those that can be lost, gained, or shared in chemical bonds. Examination of the Lewis symbols of silicon, arsenic, and boron

$$\cdot\dot{\text{Si}}: \qquad \cdot\dot{\underset{\cdot}{\text{As}}}: \qquad \dot{\text{B}}:$$

shows that substitution of an arsenic atom with its 5 valence electrons for a silicon atom with its 4 valence electrons in the donor layer gives a site with an excess of 1 electron, whereas substitution of a boron atom with only 3 electrons for a silicon atom in the acceptor layer gives a site "hole" that is deficient in 1 electron. The surface of a donor layer in contact with an acceptor layer contains electrons that are attracted to the acceptor layer. When light shines on this area, the energy of the photons of light can push these electrons back onto the donor layer, from which they can go through an external circuit back to the acceptor layer, as shown in Figure 19.16. This flow of electrons constitutes an electrical current that can be used for energy.

Solar voltaic cells based on crystalline silicon have operated with a 30% efficiency for experimental cells and 15–20% for commercial units available in 2008, at a cost of around 20–25 cents/kWh, compared to 4–7 cents/kWh for fossil fuel fired power plants and 6–9 cents for those fired by biomass. Part of the high cost results from the fact that the silicon used in the cells must be cut as small wafers from silicon crystals for mounting on the cell surfaces. Significant advances in costs and technology are being made with thin-film photovoltaics, which use an amorphous silicon alloy. These cells are only about half as efficient as those made with crystalline silicon, but cost only about 25% as much. A newer approach to the design and construction of amorphous silicon film photovoltaic devices uses three layers of amorphous silicon to absorb, successively, short wavelength ("blue"), intermediate wavelength ("green"), and long wavelength ("red") light, as shown in Figure 19.17. Thin-film solar panels constructed with this approach have achieved solar-to-electricity energy conversion efficiencies just over 10%, lower than those using crystalline silicon, but higher than other amorphous film devices. The low cost and relatively high conversion efficiencies of these solar panels should enable production of electricity at only about twice the cost of conventional electrical power, which would be competitive in some situations.

Systems have been developed in which solar cells lining glass tubes are used to generate electricity. Such a configuration is well adapted to rooftops because it captures light from all angles including even light from below reflected from a white roof. It has been estimated that covering all flat rooftops in the United States with such collectors could provide about 150 GW of electricity, 15% of U.S. consumption.

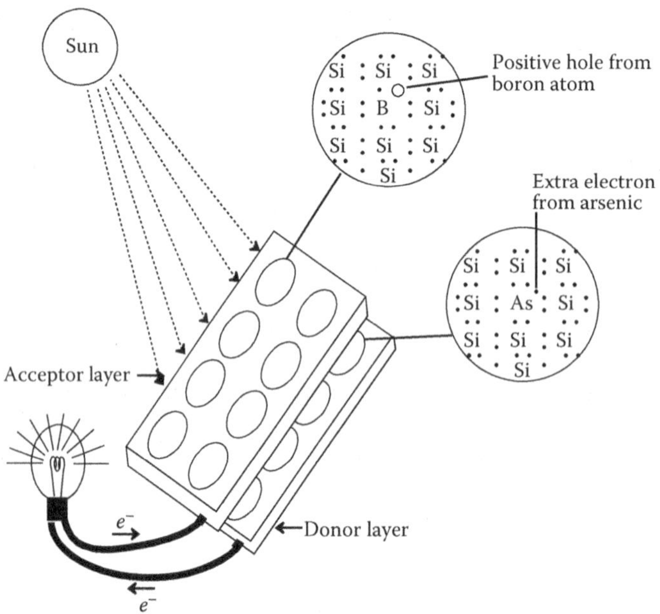

FIGURE 19.16 Operation of a photovoltaic cell.

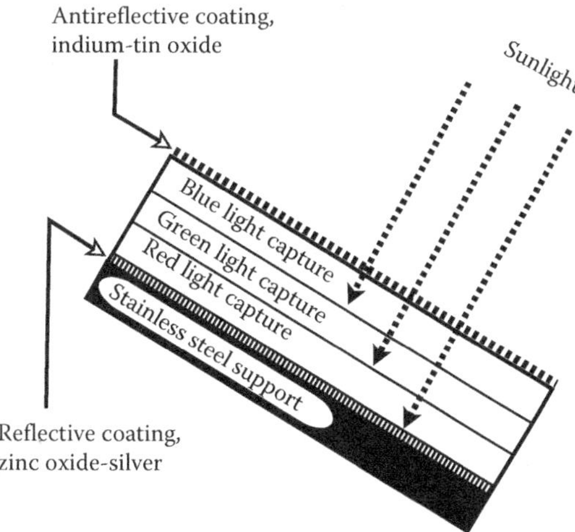

FIGURE 19.17 High-efficiency thin-film solar photovoltaic cell using amorphous silicon.

A major disadvantage of solar energy is its intermittent nature. However, flexibility inherent in an electric power grid would enable it to accept up to 15% of its total power input from solar energy units without special provision for energy storage. Existing hydroelectric facilities may be used for pumped-water energy storage in conjunction with solar electricity generation. Heat or cold can be stored in water, in a latent form in water (ice) or eutectic salts, or in beds of rock. Enormous amounts of heat can be stored in water as a supercritical fluid contained at high temperatures and very high pressures deep underground. Mechanical energy can be stored with compressed air or flywheels. Utilization of solar energy to produce elemental hydrogen as a means to store, transfer, and utilize energy, as discussed in Section 19.16, will probably come into widespread use.

No really insurmountable barriers exist to block the development of solar energy, such as might be the case with fusion power. In fact, the installation of solar space and water heaters became widespread in the late 1970s, and research on solar energy was well supported in the United States until after 1980, when it became fashionable to believe that free-market forces had solved the "energy crisis." At present, shortages of energy and concern over the effects of global warming due to the use of fossil fuels is leading to much increased interest in solar energy. With the installation of more heating devices and the probable development of some cheap, direct solar electrical generating capacity, it is likely that during the coming century, solar energy will be providing an appreciable percentage of energy needs in areas receiving abundant sunlight.

19.14 ENERGY FROM MOVING AIR AND MOVING WATER

19.14.1 Surprising Success of Wind Power

Wind power using huge turbines mounted on high towers and coupled to electrical generators is emerging at a somewhat surprising rate as a source of renewable energy. Although used for centuries with windmills that drove grain grinding and water pumping operations and during the early 1900s for small-scale electricity generation, especially in remote locations, modern large-scale wind-powered electrical generators emerged during the 1990s as economical means of generating electrical power. Wind power is completely renewable and nonpolluting. It is an indirect means of utilizing solar energy because winds are caused by the movement of air masses heated by the sun (Figure 19.18).

Wind power has become a major factor in world energy generation with an estimated capacity of almost 90,000 MW. (A megawatt is abbreviated as MW. Large amounts of power are often expressed

FIGURE 19.18 Wind-powered electrical generators mounted on towers are becoming increasingly common sights in the world in areas where consistent wind makes this nonpolluting source of renewable energy practical.

in gigawatts, where 1 GW = 1000 MW. A 1000-MW power plant is considered to be quite large and can provide electricity for about 250,000 average U.S. homes. Typical modern large wind turbines generate 1.5–4 MW each.) As of the beginning of 2008, the United States had just over 15,500 MW of wind power capacity installed with plants in 30 states. As of mid-2007, Texas was the leading state for wind energy production in the United States with an installed capacity of 3352 MW and an additional 1246 MW under construction. Long the leader in wind energy production, California is now second with about 2400 MW. As of 2008, the next five states were Iowa (1375 MW), Minnesota (1350 MW), Washington (1290 MW), Colorado (1065 MW), and Oklahoma (680 MW). Wind energy in the United States is now sufficient to power 3.9 million average households eliminating as much as 25 million tons of carbon dioxide emissions per year.

Wind power has become a major factor in world energy supply. The European Union (EU) countries have more wind power capacity than the rest of the world combined. By the beginning of 2008, EU countries had an installed wind power generation capacity estimated to be around 56,000 MW. At the beginning of 2008, the United States was the leading country in wind energy capacity followed by Germany, India, Spain, and China. Because of its aggressive program for wind power development, China was expected to advance to second within 2 or 3 years. As of 2008, Calgary, Canada, had in operation a new $140 million wind electrical power system that powers three-fourth of the city's municipal buildings and provides the energy for the municipal light rail transportation system. The system is equivalent to taking 30,000 automobiles off the road.

Wind power is especially attractive for some agricultural regions. One reason is that such regions often have low population densities so that there are fewer objections to wind power installations. A commercial wind turbine generating 1.8 MW is a formidable machine, typically mounted on towers around 80 m high to take advantage of higher, more consistent wind speeds at such heights, and may have blades 40 m long. However, the footprints of these structures are relatively small and do not occupy much farmland. The electricity is conveyed from the turbines in underground lines eliminating surface power lines. Adding to the potential attractiveness of wind power in agricultural regions is the fact that electricity generated by wind energy can be used to electrolyze water to produce elemental H_2 and O_2, an application not handicapped by wind's intermittent nature. The H_2 currently required to make NH_3 fertilizer is usually produced from steam reacting with methane (natural gas), a relatively expensive process, and its production from water by inexpensive wind energy could help keep the price of ammonia fertilizer at reasonable levels. Furthermore, both elemental H_2 and O_2 can be used to convert crop by-product biomass to hydrocarbon fuels, as discussed in Section 19.15.

Northern regions, including parts of Alaska, Canada, the Scandinavian countries, and Russia often have consistently strong wind conditions conducive to the generation of wind power. Isolation from other sources of energy makes wind power attractive for many of these regions. Severe climate conditions in these regions pose special challenges for wind generators. One problem can be the buildup of rime consisting of ice condensed directly on structures from supercooled fog in air. (In warmer regions, the remains of insects zapped by the rotating turbine blades have built up to the point of reducing the aerodynamic efficiency of the blades.)

Although wind turbines have been growing in size leading to much greater economy in power production, there is also a market for small wind turbines that can be used for an individual home or commercial building. Small wind turbines have been installed in a number of locations including Logan International Airport in Boston and the Brooklyn Naval yard. Although the cost of these turbines is much less than that of large commercial turbines, the cost of electricity produced by the small turbines is much higher.

19.14.2 Energy from Moving Water

Water flowing in contact with a device called a waterwheel is one of the oldest sources of power other than humans or animals. Grain mills driven by waterpower existed in ancient Greece and Rome, and large waterwheels developing up to 50 hp were constructed in the Middle Ages. In colonial North America, waterwheels drove grist mills and sawmills and were further applied to leather, textile, and machine shop operations. Because of problems with low water flow in the summer and ice formation in the winter, these operations were rapidly displaced when steam engines became available in the early 1800s.

With the development of electric power in the late 1800s, waterpower underwent a spectacular renaissance to drive electrical generators. The first practical hydroelectric plant went into operation on the Fox River near Appleton, Wisconsin, in 1882. Hydroelectric power grew rapidly as an energy source from that time and by 1980 accounted for about 25% of world electrical production and 5% of total world energy use. The potential to construct hydroelectric plants is favored by mountainous terrain with large river valleys and is distributed relatively evenly around the world. China has about one-tenth of the world's potential for hydroelectric power. About 99% of Norway's electric power is hydroelectric accounting for about 50% of that country's energy use.

The largest hydroelectric project in the world is the Three Gorges installation on the huge Yangtze River in China. Located at the end of a number of steep canyons, the dam spans 2.3 km across the river valley and reaches a height of 185 m. When filled, the reservoir formed by the dam will extend for 630 km with an average width of 1.3 km. When the dam was finished in 2009, it had 26 generating units each capable of generating 700 MW of hydroelectric power, a total capacity of 18.2 GW. After 2009, six more units are to be constructed in a subterranean powerhouse bringing the total generating capacity to 22.4 GW.

The sustainability and environmental acceptability of hydroelectric power present a mixed picture. In the modern age, construction of water impoundments tends to displace significant numbers of people; more than 1 million were displaced for the Chinese Three Gorges project. Altering the flow of rivers can change their aquatic ecology. Esthetics can be harmed by filling scenic river valleys with impounded water. In some cases in the United States, dams are being dismantled to restore river valleys to their former state. However, hydroelectric power prevents release of greenhouse gases from equivalent fossil energy-powered plants. Reservoirs can provide recreational facilities and serve as sources of fish and water used by municipalities, industries, and agriculture.

19.14.3 Energy from Moving Water without Dams

A promising approach to the utilization of energy from moving water without building dams is provided by *hydrokinetic* and *wave energy conversion* devices that tap the kinetic energy of moving water in river flows, tides, or ocean waves. These devices can be put in natural streams, tidal estuaries, ocean currents, and constructed waterways. One promising device consists of a turbine with large blades spaced relatively far apart and connected directly to a generator that can be placed in a water current, such as in a river. Such turbines can be attached directly to structures constructed on waterway beds or can be attached to bridge supports. Bidirectional turbines have been designed for use in tidal currents that flow back and forth.

19.15 BIOMASS ENERGY

Fossil fuels originally came from photosynthetic processes. Photosynthesis is a promising source of combustible chemicals to be used for energy production including transportation fuels[4] and could certainly produce all needed organic raw materials to substitute for petroleum in the current petrochemicals industry. It suffers from the disadvantage of being a very inefficient means of solar energy collection (a collection efficiency of only a fraction of a percent by photosynthesis is typical of most common plants). However, the overall energy conversion efficiency of several plants, such as sugarcane, is around 0.6%. Furthermore, some plants, such as *Euphorbia lathyrus* (gopher plant), a small bush growing wild in California, produce hydrocarbon emulsions directly. The fruit of the Philippine plant, *Pittsosporum reiniferum*, can be burned for illumination due to its high content of hydrocarbon terpenes, primarily α-pinene and myrcene. Conversion of agricultural plant residues to energy could be employed to provide much of the energy required for agricultural production. Indeed, until about 80 years ago, virtually all of the energy required in agriculture—hay and oats for horses, home-grown food for laborers, and wood for home heating—originated from plant materials produced on the land. (An interesting exercise is to calculate the number of horses required to provide the energy used for transportation at the present time in the Los Angeles basin. It can be shown that such a large number of horses would fill the entire basin with manure at a rate of several feet per day.)

Annual world production of biomass is estimated at 146 billion metric tons, mostly from uncontrolled plant growth. Many farm crops and trees can produce around 2 metric tons per acre per year of dry biomass, and some algae and grasses can produce significantly more. The heating value of this biomass is 5000–8000 Btu/lb about half of typical values for coal. However, biomass contains virtually no ash or sulfur, both problems with coal. Another sustainability advantage of biomass is that all of the carbon in it is taken from carbon dioxide in the atmosphere so that biomass combustion does not add any net quantities of carbon dioxide to the atmosphere. Indeed, use of biomass to produce hydrogen-rich methane or elemental hydrogen along with sequestration of by-product carbon dioxide would result in an overall loss of carbon dioxide from the atmosphere.

As it has been throughout history, biomass is significant as heating fuel, and in some parts of the world is the fuel most widely used for cooking. Scavenging wood for cooking fuel has been a major contributor to deforestation in some areas. About 15% of Finland's energy needs are provided by wood and wood products (including black liquor by-product from pulp and paper manufacture), about one-third of which is from solid wood. Despite the charm of a wood fire and the sometimes pleasant odor of wood smoke, air pollution from wood-burning stoves and furnaces is a significant problem in some areas. Currently, wood provides about 8% of world energy needs. This percentage could increase through the development of energy plantations consisting of trees grown solely for their energy content.

Biomass could be used to replace much of the 100 million metric tons of petroleum and natural gas currently consumed in the manufacture of primary chemicals in the world each year. Among the sources of biomass that could be used for chemical production are grains and sugar crops (for ethanol manufacture), oilseeds, animal by-products, manure, and sewage (the last two for methane generation). The biggest potential source of chemicals is the lignocellulose making up the bulk of most plant material (see below).

19.15.1 Ethanol Fuel

A major option for converting photosynthetically produced biochemical energy to forms suitable for internal combustion engines is the production of ethanol, C_2H_6O, by the fermentation of sugars from biomass. Suitably designed internal combustion engines can burn pure ethanol or a mixture of 85% ethanol and 15% gasoline called E85. More commonly, ethanol is blended in proportions of around 10% with gasoline to give *gasohol*, a fuel that can be used in existing internal combustion engines with little or no adjustment.

Gasohol boosts octane rating and reduces emissions of carbon monoxide. From a resource viewpoint, because of its photosynthetic origin, alcohol may be considered a renewable resource rather than a depletable fossil fuel. Ethanol is most commonly produced biochemically by the fermentation of carbohydrates. Brazil, a country that produces copious amounts of fermentable sugar from sugarcane, has been a leader in the manufacture of ethanol for fuel uses, with about 16 billion liters produced in 2006. All motor fuels in Brazil contain at least 24% ethanol and some fuel is essentially pure ethanol. A significant fraction of the gasoline consumed in the United States is supplemented with ethanol.

Although most of the ethanol that has been produced for fuel has been made from the fermentation of grain or sugar, there is legitimate concern that, considering the energy that goes into producing grain ethanol, there is no net energy gain. A potentially much more abundant and cheaper source of ethanol consists of biomass generated as a by-product of crop production including straw from wheat or rice production or cornstalks from growing corn. In the past, much rice straw from commercial production in the United States was simply burned to save the cost of cultivating it back into the soil. Straw cannot be fermented directly, but must be broken down to hexose and pentose sugars for fermentation. This has traditionally been done with acid treatment, which is expensive, although technologies exist for recycling acid. It is now generally agreed that production of ethanol from plant biomass by-products will require enzymatic hydrolysis with cellulase enzyme to produce the required sugars. The Canadian Iogen Corporation has a means for obtaining fermentable sugars from wheat straw and other plant materials and has attempted to develop a cost-effective commercial process.

19.15.2 Biodiesel Fuel

Biodiesel fuel is a growing source of renewable liquid hydrocarbon fuels. Unlike fuel ethanol, which must be transported in corrosion-resistant truck tanker trailers or railroad tankers, biodiesel fuel is readily moved through existing pipelines. Rudolf Diesel developed the high-compression, compression-ignited diesel engine in the late 1800s and first operated the engine in Augsburg, Germany, in 1893. He demonstrated his invention at the World Fair in Paris in 1900, receiving the "Grand Prix" (highest prize) for his invention. Interestingly, the fuel used in this and other demonstrations of Diesel's engine was peanut oil, and vegetable oils were the main source of fuel for diesel engines during the first two decades of their use.

Vegetable oils were eventually replaced by petroleum-based hydrocarbons. More recently, diesel fuels that are derivatives of the fatty acids in the oils have been synthesized from vegetable oil feedstocks. Vegetable oils from soybeans and other biological sources are used to make biodiesel fuel as discussed below.

Vegetable oils are fatty acid esters of glycerol, a three-carbon alcohol with 3-OH groups attached. To produce biodiesel fuel, the glycerol esters are hydrolyzed by strong base (NaOH) in the presence of methanol alcohol ($HOCH_3$) and the fatty acids are converted to their methyl esters, the molecules composing biodiesel fuel:

$$\underset{\text{Glycerol ester}}{\begin{array}{l} \mathrm{H{-}\overset{\displaystyle H}{\underset{|}{C}}{-}O{-}\overset{\displaystyle O}{\overset{\|}{C}}{-}R} \\ \mathrm{H{-}\underset{|}{C}{-}O{-}\overset{\displaystyle O}{\overset{\|}{C}}{-}R} \\ \mathrm{H{-}\underset{\displaystyle H}{\underset{|}{C}}{-}O{-}\overset{\displaystyle O}{\overset{\|}{C}}{-}R} \end{array}} + 3HOCH_3 \xrightarrow{NaOH} \underset{\text{Glycerol}}{\begin{array}{l} \mathrm{H{-}\overset{\displaystyle H}{\underset{|}{C}}{-}OH} \\ \mathrm{H{-}\underset{|}{C}{-}OH} \\ \mathrm{H{-}\underset{\displaystyle H}{C}{-}OH} \end{array}} + \underset{\text{Fatty acid methyl esters in biodiesel fuel}}{3H_3CO{-}\overset{\displaystyle O}{\overset{\|}{C}}{-}R} + H_2O \tag{19.9}$$

In this reaction, R stands for a long hydrocarbon chain in one of a number of fatty acids including stearic acid, linoleic acid, oleic acid, lauric acid, and behenic acid. For example, in stearic acid, R is a straight chain with 17 carbon atoms, $C_{17}H_{35}$.

Major oils that are used for biodiesel fuel production are rapeseed, sunflower, soybean, palm, coconut, and jatropha. Rapeseed, long grown for animal feed, is the largest source of oil for biodiesel manufacture and is widely produced in Europe, whereas soybean oil predominates in the United States. Both of these oils offer the advantage of providing a protein-rich animal feed after the oil has been squeezed from the seed. Palm oil, and to a lesser extent, coconut oil (from coconut trees) and jatropha (from *Jatropha curcus*, planted for hedges) are attractive because they are from perennial plants that thrive in the tropics. (In late 2008, Air New Zealand flew a Boeing 747 aircraft on which one engine was fueled with 50% jatropha-derived fuel and 50% conventional jet fuel; the aircraft had three other engines in case the experiment failed.) By 2009, the explosive growth of palm oil tree plantations in Malaysia and Indonesia was resulting in high levels of rain forest destruction.

An interesting possibility for biodiesel fuel production is algae that have an oil content exceeding 50%. Such algae can grow profusely in ponds fed with nutrient-rich effluent from wastewater treatment plants. Oil-producing algae have also been grown in the carbon dioxide-rich atmosphere provided by power plant stack gases. By growing algae in treated wastewater within a carbon-dioxide-rich atmosphere from power plants, nutrients can be removed from wastewater, thus reducing eutrophication in receiving waters, and some of the carbon dioxide can be removed from stack gas emissions.

The potential productivity of algae for biomass fuel production is spectacular. Whereas soybean and palm oil sources typically produce, respectively, up to 200 and 2500 L/acre annually of biodiesel fuel, algae have the potential to yield as much as 40,000 L/acre/year of the fuel. Furthermore, with adequate supplies of water, algae can be grown on desert lands that are unsuitable for growing soil-based crops. One experimental system grows algae in water that circulates through transparent plastic tubes. Some algae thrive in saltwater, which can come from the sea or underground formations bearing brackish water.

19.15.3 Unrealized Potential of Lignocellulose Fuels

Both grain- and sugar-based ethanol as well as vegetable oils used to make biodiesel fuel are not the best candidates for biomaterial fuels because they use only relatively small fractions of the plants consisting of the parts that have the most value for food, animal feed, and raw materials. A much more abundant source of fuel consists of the *lignocellulose* parts of plants, a material composed of large polymeric molecules with an approximate empirical formula of CH_2O that composes the structural members of plants including tree wood, stalks, straw, corncobs, and leaves. The three major components of lignocellulose and their approximate empirical formulas are cellulose/starch, $[C_6(H_2O)_5)]_n$; hemicellulose, $[C_5(H_2O)_4)]_n$; and lignin, $[C_{10}(H_{12}O)_3)]_n$.

Large amounts of crop by-product biomass are generated annually. Assuming conservatively that the amount of this material available in the United States is equal to the mass of corn grain produced, about 230 million tons of crop by-product biomass could be made available for fuel each year; the actual figure might be much higher.

The amount of biomass that could be generated from dedicated trees and grass is very high, an estimated 2240 million tons per year in the United States alone. A major advantage of this source is that it comes from perennial plants that can be grown on erodable land, much of which has been taken from agricultural production as the result of government programs. One of the plants that is remarkably productive of biomass consists of hybrid poplars, from the genus *Populis* that includes cottonwoods and aspens. These trees may grow more than 2 m/year and will establish new growth from the stumps of harvested trees, retaining their root systems that prevent soil erosion.

The grass most commonly considered for its biomass productivity is switchgrass. Native to North America, switchgrass is disease- and pest-resistant and requires little fertilizer. It tolerates both drought and flooding very well. Upland varieties of switchgrass grow up to 2 m tall in one growing season on well-drained soils. Lowland varieties can reach heights of 4 m and grow best on heavy soils in bottomlands. Improved varieties of switchgrass have been developed for animal forage and yield around 8 tons/acre of biomass each year. Average biomass yields of forests are only about half as high.

Another high-yielding grass native to swampy regions, such as the Florida everglades, is sawgrass (*Cladium jamaicense*), which gets its name from the saw-like serrations on its leaves. It is well adapted to cultivation in wet areas where other crops cannot be grown and has the additional advantage of providing good wildlife cover.

There are numerous pathways by which biomass and biowastes (including sewage and animal wastes) can be converted to energy, fuels, and feedstocks. These are the following:

- Direct combustion to produce heat and to raise steam to generate electricity
- Fermentation to produce ethanol (from sugar) or methane
- Pyrolysis to yield gaseous fuels (particularly methane), liquids (including hydrocarbons and oxygenated species), and solid carbon
- Thermochemical gasification to produce CO, H_2, CH_4, by-product liquids, and solid carbon
- Fischer–Tropsch synthesis of hydrocarbons from CO and H_2 derived from biomass
- Hydrogenation of oxygenated liquids from biomass to produce hydrocarbons
- Methyl esterification of oils to produce methyl ester biodiesel fuels

Sustainable preparation of biofuels from biomass makes it desirable to recover nutrients such as phosphorus and potassium from the biomass during processing.

Biomass can be used directly as a fuel with a heating value on a dry mass basis about half that of coal. It is extremely low in sulfur and ash and its mineral ash components do not contain toxic elements, such as the arsenic that occurs in some coals. The most direct way to use biomass fuel is direct combustion to produce heat. Biomass can be converted to other high-value fuels including hydrocarbons by gasification. One approach to biomass gasification begins with combustion of part of the biomass (represented by the formula $\{CH_2O\}$) with pure molecular oxygen oxidant (to avoid diluting the product gas with N_2 from air),

$$\{CH_2O\} + O_2 \rightarrow CO_2 + H_2O + \text{heat} \tag{19.10}$$

yielding heat required for the rest of the gasification process. Under the oxygen-deficient conditions through which gasification is carried out, part of the biomass is partially oxidized to combustible carbon monoxide, CO:

$$\{CH_2O\} + \tfrac{1}{2}O_2 \rightarrow CO + H_2O + \text{heat} \tag{19.11}$$

Part of the biomass is pyrolyzed by the heat produced by Reaction 19.10

$$\{CH_2O\} + heat \rightarrow C + H_2O \quad (19.12)$$

yielding hot carbon and combustible gaseous and liquid by-products. The hot carbon reacts with steam,

$$C + H_2O + heat \rightarrow CO + H_2 \quad (19.13)$$

yielding a *synthesis gas* mixture of CO and H_2. Biomass may also react as it is heated,

$$\{CH_2O\} + heat \rightarrow CO + H_2 \quad (19.14)$$

yielding synthesis gas. The carbon monoxide in synthesis gas can be subjected to the water-gas shift reaction,

$$CO + H_2O \rightarrow CO_2 + H_2 \quad (19.15)$$

giving elemental H_2 as the only gaseous product. The CO_2 generated in Reactions 19.10 and 19.15 can be separated and sequestered, such as by pumping into underground petroleum formations to enable recovery of petroleum, thereby preventing the release of this greenhouse-warming gas to the atmosphere.

Elemental hydrogen can be used as an end product of biomass gasification directly as a fuel in gas turbines and other heat engines or to generate electricity in fuel cells. Elemental hydrogen can be used to synthesize ammonia, NH_3, an important industrial chemical and fertilizer. A mixture of CO and H_2 can be reacted over a catalyst in a methanation reaction

$$CO + 3H_2 \rightarrow CH_4 + H_2O \quad (19.16)$$

to produce methane, which, made by this method, is called *synthetic natural gas* (SNG). With different proportions of hydrogen and CO reactants and a different catalyst, a mixture of CO and H_2 can react according to the Fischer–Tropsch reaction to yield a variety of hydrocarbons including gasoline, jet fuel, and diesel fuel as shown by the reaction below for the synthesis of octane, one of the hydrocarbons in gasoline:

$$8CO + 17H_2 \rightarrow C_8H_{18} + 8H_2O \quad (19.17)$$

A similar reaction can also be used to make ethanol and methanol, CH_3OH, which can be used as a fuel, gasoline additive, and to produce H_2 for fuel cells in vehicles.

An interesting possibility for increasing the amount of hydrocarbon fuel that can be obtained from biomass is to use H_2 and O_2 produced by the electrolysis of water with electricity generated by wind power,

$$2H_2O + \text{electrical energy} \rightarrow 2H_2(g) + O_2(g) \quad (19.18)$$

to react with biomass for gasification. The pure elemental oxygen can be used to produce energy from biomass as shown by Reaction 19.10 without diluting the gas product with elemental nitrogen, N_2, as would be the case with air oxidant, and enabling sequestration of the CO_2 product. The elemental hydrogen generated by electrolysis of water can be reacted directly with biomass. Ideally, direct hydrogenation of biomass would produce predominantly hydrocarbons, such as methane generated by the following reaction:

$$\{CH_2O\} + 2H_2 \rightarrow CH_4 + H_2O \quad (19.19)$$

More commonly, hydrogenation produces a mixture of many liquids, most of which contain oxygen in compounds such as ethers, alcohols, and ketones. For fuel use, these compounds must be further treated with hydrogen to remove oxygen and leave hydrocarbon liquids. Such a process as outlined above provides a means for using the energy originally generated by windpower to make a high-energy hydrocarbon fuel that can be used for transportation or other purposes.

19.15.4 Biogas

A significant source of clean-burning methane can be obtained from the anoxic (oxygen-free) bacterial fermentation of biomass of a variety of kinds. Representing biomass as $\{CH_2O\}$, the biochemical reaction is the following:

$$2\{CH_2O\} \rightarrow CH_4 + CO_2 \quad (19.20)$$

This reaction has long been used in the anoxic digesters of sewage treatment plants to reduce the amount of degradable organic matter in excess sewage sludge. Such a well-balanced plant makes enough methane to provide for all its energy needs. Large livestock feeding operations may have digesters to produce methane from livestock manure and other biological wastes associated with the feeding operation. Another source of methane generated by anoxic fermentation is obtained by burying collector pipes in mounds of municipal solid wastes and collecting the off-gas.

An outstanding example of "grass roots" energy sustainability is provided by the widespread use of biogas generators in rural China, an area chronically short of fuel. The Chinese have built millions of anoxic digesters located just below the ground surface in which all kinds of biomass including animal wastes, human excrement, and vegetable and crop wastes are degraded in the absence of air to produce combustible methane gas. The digesters are usually located in proximity to dwellings and to sources of waste, including latrines and pigsties. Covered by a thin layer of soil, these facilities take advantage of solar heating to accelerate fermentation. In addition to producing fuel, anoxic digestion alleviates waste disposal and health problems through the destruction of pathogens in the waste by the fermentation process, which often takes place at a somewhat elevated temperature that kills most pathogenic organisms. An additional advantage is that the fertilizer value of the wastes is conserved so that residues from the digesters can be placed on soil to serve as fertilizer for crops without much concern over spreading pathogens. The contents of both nitrogen and phosphorus in the digester residues are much higher than in biomass that has been treated by mulching in an air environment.

The gas from the Chinese biogas facilities is used largely for cooking and lamps. In larger installations, internal combustion engines have been adapted to run in part on biogas. In the case of gasoline engines, biogas is fed into carburetors through an adjustable feed valve and in the case of diesel engines, it is fed into the air/fuel intake manifold to supplement the minimal amount of diesel fuel injected into the combustion chamber to initiate combustion during engine operation.

A major methane-generating system in the United States is overseen by Central Vermont Public Service in Vermont, utilizing manure from several large dairy farms. The largest dairy in this complex is the Pleasant Valley Farm where 1500 milking cows produce over 40 million pounds of milk a year. As a profitable by-product of this operation, anoxic fermentation for 3 weeks of manure from the cows produces methane used to generate 3.5 million kWh of electricity a year, enough to power 500 homes. The farm went from spending about $200/day for electricity to making about $1200/day from renewable electricity sales in 2008. The system cost about $2 million. As of late 2008, Central Vermont Public Service had five farms with a total of about 5000 cows in its "Cow Power" renewable energy program.

19.16 HYDROGEN AS A MEANS TO STORE AND UTILIZE ENERGY

Hydrogen gas, H_2, is an ideal chemical fuel in some respects that may serve as a storage medium for solar energy. There are several ways to generate elemental hydrogen.[5] As noted in the preceding section, solar-generated electricity can be used to electrolyze water:

$$2H_2O + \text{electrical energy} \rightarrow 2H_2(g) + O_2(g) \quad (19.21)$$

The hydrogen fuel product, and even elemental oxygen produced as an electrolysis by-product, can be piped some distance and the hydrogen burned without pollution, or it may be used in a fuel cell (Figure 19.7). This may, in fact, make possible a "hydrogen economy." Disadvantages of using hydrogen as a fuel include its low heating value per unit volume and the wide range of explosive mixtures it forms with air. Although not yet economical, photochemical processes can be used to split water to H_2 and O_2 that can be used to power fuel cells.

Fuel-cell-powered vehicles are now practical in some applications and Honda began production of the first commercial fuel-cell automobile in 2008. One of the greatest barriers to their widespread adoption has been their inability to carry sufficient hydrogen for an acceptable range. Several solutions to this problem are now being investigated. One is the potentially problematic use of very cold liquid hydrogen as a fuel source. Another is the use of very high pressure containers composed of multilayer cylinders wrapped with carbon composite and filled with hydrogen at pressures up to 10,000 psi (about 670 × atmospheric pressure!), reputed to contain sufficient hydrogen to propel an automobile up to 300 miles. Other systems use catalysts to break down liquid fuels, such as methanol or gasoline, to generate hydrogen for fuel cells.

Although there has been much enthusiasm in some quarters for hydrogen fuel and predictions of a new hydrogen economy, some of the more enthusiastic arguments for hydrogen fuel may be too optimistic. The most important point is that, unlike fossil fuels, hydrogen is not a primary source of energy and has to be made by processes such as the electrolysis of water (Reaction 19.18) that use other sources of energy. Most of the 6 million tons of elemental hydrogen used in the United States each year is made from steam reforming of methane from natural gas:

$$CH_4 + H_2O \rightarrow 3H_2 + CO \quad (19.22)$$

The carbon monoxide product can be reacted with steam as noted in the preceding section,

$$CO + H_2O \rightarrow CO_2 + H_2 \quad (19.23)$$

to produce additional H_2 and the CO_2 product of this reaction can be sequestered as discussed in Section 19.9.

In principle, the process described above and the utilization of elemental hydrogen in fuel cells can provide a transport fuel that is pollution-free. However, methane gas is easier to store than elemental hydrogen and the modern internal combustion engine with associated emissions control equipment is virtually pollution-free. So the intermediate production of elemental hydrogen is unlikely to be the greenest approach. Production of elemental hydrogen by electrolysis of water using electricity from renewable sources, such as photovoltaics and windpower, is essentially nonpolluting, but it requires balancing the relatively inefficient electrolysis process against the value of the electricity that it consumes.

19.17 COMBINED POWER CYCLES

Combined power cycles enable much more efficient utilization of combustible fuels by first using the heat of combustion in a turbine coupled to an electrical generator, raising steam in a boiler

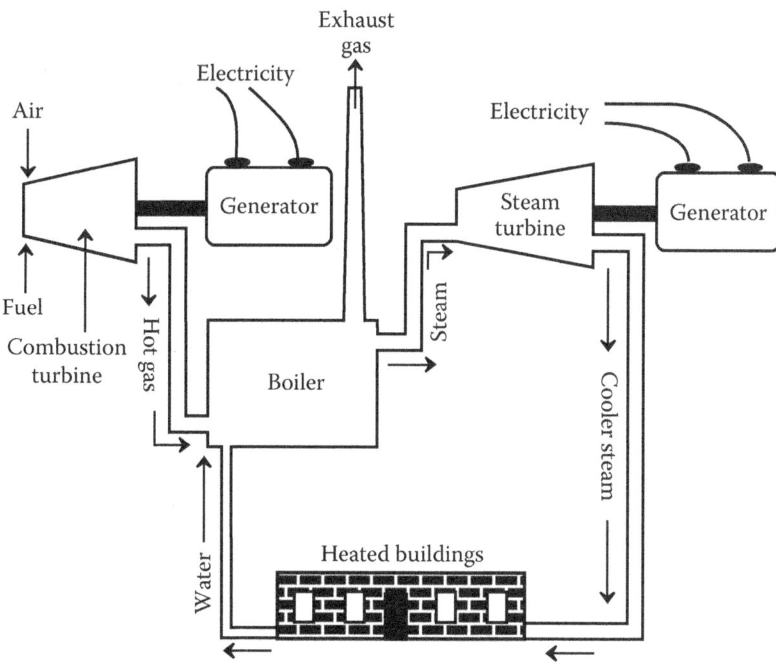

FIGURE 19.19 A combined power cycle in which combustible gas or oil is first used to fire a gas turbine connected to an electrical generator. The hot gases from this turbine are fed to a boiler to raise steam, which drives a steam turbine, also connected to a generator. The still hot exhaust steam from the steam turbine is used for process heat or conveyed to commercial or residential buildings for heating. The water condensed from the steam in this final application is returned to the power plant to generate more steam, thus conserving water and avoiding the necessity to treat more water to the high purity standards required by the boiler.

with the hot exhaust gas from this turbine, using the steam to power a second turbine linked to a generator, and finally using the steam and hot water from the steam turbine for applications such as processing in the chemical industry, heating commercial buildings, or heating homes. A schematic diagram of a combined power cycle system is illustrated in Figure 19.19. The water condensed from the steam used for heating is pure and is recycled to the boiler, thus minimizing the amount of makeup boiler feedwater, which requires expensive treatment to make it suitable for use in boilers. The use of steam leaving a steam turbine for heating, a concept known as *district heating*, is commonly practiced in Europe (and many university campuses in the United States) and can save large amounts of fuel otherwise required for heating. Such a system as the one described is in keeping with the best practice of industrial ecology and should be employed whenever it is practical to do so.

19.18 A SYSTEM OF INDUSTRIAL ECOLOGY FOR METHANE PRODUCTION

Figure 19.20 illustrates an industrial ecosystem based on the production of methane fuel from renewable biomass and wind energy sources. Methane is the ideal clean-burning fuel for virtually all applications that currently use fossil fuels except for aircraft. With hybrid automobile technology, methane can be used as motor fuel giving an acceptable range between refueling. With previously untapped sources coming on line in the United States and a number of other countries, methane is relatively available and inexpensive. Pipeline systems are in place for its distribution. And, when methane sources are eventually exhausted, the gas can be synthesized from biomass and elemental hydrogen produced from the electrolysis of water.

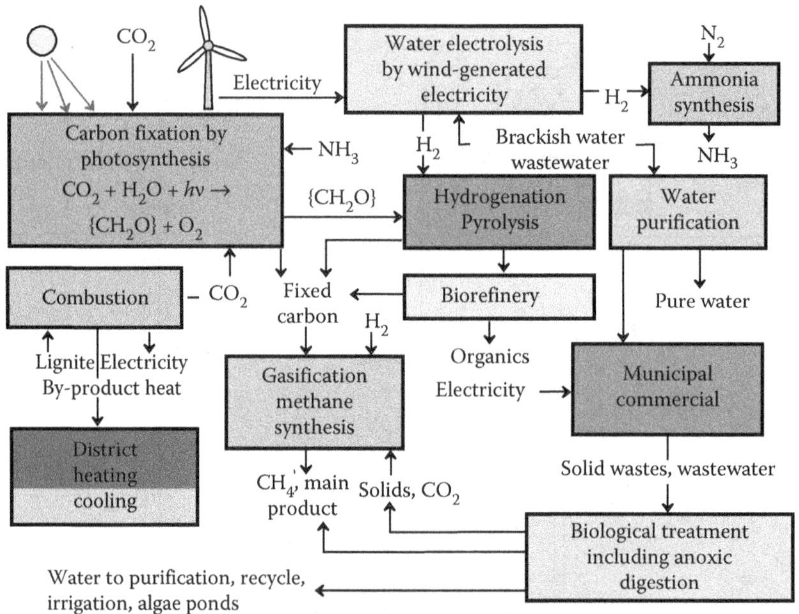

FIGURE 19.20 An industrial ecosystem based on the synthesis of methane and other fuels from biomass using wind-generated electrical power as the major energy source driving the system.

A system for producing methane renewably is outlined in Figure 19.20. Biomass is used as a source of fixed carbon that is reacted with elemental hydrogen made by electrolyzing water using renewable wind energy. Another source of methane in the system is anoxic fermentation of wastes, including sewage. Instead of municipal refuse in the community being disposed in landfill, it is first subjected to anoxic fermentation to generate methane after which the residues are thermochemically gasified to provide feedstock for chemical synthesis. Other aspects of the system are largely self-explanatory from the figure.

LITERATURE CITED

1. Leonenko, Y. and D. Keith, Reservoir engineering to accelerate the dissolution of CO_2 stored in aquifers, *Environmental Science and Technology*, **42**, 2742–2747, 2008.
2. Bachu, S. and W. D. Gunter, Acid-gas injection in the Alberta Basin, Canada: A CO_2-storage experience, *Geological Society Special Publication*, **233**, 225–234, 2004.
3. For an explanation of the planned system see: http://www.stirlingenergy.com/
4. Larson, E. D., A review of life-cycle analysis studies on liquid biofuel systems for the transport sector, *Energy for Sustainable Development*, **10**, 109–126, 2006.
5. Holladay, J. D., J. Hu, D. L. King, and Y. Wang, An overview of hydrogen production technologies, *Catalysis Today*, **139**, 244–260, 2009.

SUPPLEMENTARY REFERENCES

Amos, S., *Energy: A Historical Perspective and 21st Century Forecast*, The American Association of Petroleum Geologists, Tulsa, OK, 2005.
Brebbia, C. A. and V. Popov, Eds, *Energy and Sustainability*, WIT Press, Southampton, UK, 2007.
Brenes, M. D., Ed., *Biomass and Bioenergy: New Research*, Nova Science Publishers, New York, 2006.
Coley, D. A., *Energy and Climate Change: Creating a Sustainable Future*, Wiley, Hoboken, NJ, 2008.
Dickson, M. and M. Fanelli, Eds, *Geothermal Energy: Utilization and Technology*, Earthscan, Sterling, VA, 2005.

Fay, J. A. and D. S. Golomb, *Energy and the Environment*, Oxford University Press, New York, 2002.

Goetzberger, A. and V. U. Hoffmann, *Photovoltaic Solar Energy Generation*, Springer, Berlin, 2005.

Harvey, L. and D. Danny, Ed., *A Handbook on Low-Energy Buildings and District-Energy System: Fundamentals, Techniques and Examples*, Earthscan, Sterling, VA, 2006.

Hau, E., *Windturbines: Fundamentals, Technologies, Application, and Economics*, 2nd ed., Springer, Berlin, 2006.

Herbst, A. M. and G. W. Hopley, *Nuclear Energy Now: Why the Time Has Come for the World's Most Misunderstood Energy Source*, Wiley, Hoboken, NJ, 2007.

Hofman, K. A., Ed., *Energy Efficiency, Recovery and Storage*, Nova Science Publishers, New York, 2007.

Infield, D. and L. Freris, *Renewable Energy in Power Systems*, Wiley, Chichester, UK, 2008.

Jamasb, T., W. J. Nuttall, and M. G. Pollitt, Eds, *Future Electricity Technologies and Systems*, Cambridge University Press, Cambridge, UK, 2006.

Kanninen, B., Ed., *Atomic Energy*, Green Haven Press, San Diego, 2006.

Krauter, S. C. W., *Solar Electric Power Generation—Photovoltaic Energy Systems*, Springer, New York, 2006.

Kruger, P., *Alternative Energy Resources: The Quest for Sustainable Energy*, Wiley, Hoboken, NJ, 2006.

Larminie, J. and A. Dicks, *Fuel Cell Systems Explained*, 2nd ed., Wiley, New York, 2003.

Manwell, J. F., J. G. McGowan, and A. L. Rogers, *Wind Energy Explained: Theory, Design and Application*, Wiley, New York, 2002.

McGowan, T., *Biomass and Alternate Fuel Systems: An Engineering and Economic Guide*, Hoboken, NJ, 2009.

Nag, A., Ed., *Biofuels Refining and Performance*, McGraw-Hill, New York, 2008.

Nakaya, A., *Energy Alternatives*, ReferencePoint Press, San Diego, CA, 2008.

Nutall, W. J., *Nuclear Renaissance: Technologies and Policies for the Future of Nuclear Power*, Bristol, Philadelphia, 2005.

O'Hayre, R., *Fuel Cell Fundamentals*, Wiley, Hoboken, NJ, 2006.

Patel, M. R., *Wind and Solar Power Systems: Design, Analysis, and Operation*, 2nd ed., Taylor & Francis, London, 2006.

Peinke, J., *Wind Energy*, Springer, New York, 2006.

Romm, J. J., *The Hype about Hydrogen: Fact and Fiction in the Race to Save the Climate*, Island Press, Washington, 2004.

Savage, L., Ed., *Geothermal Power*, Greenhaven Press, Detroit, 2007.

Silveira, S., Ed., *Bioenergy, Realizing the Potential*, Elsevier, Amsterdam, 2005.

Soetaert, W. and E. Vandamme, *Biofuels*, Wiley, Hoboken, NJ, 2008.

Sørensen, B., *Renewable Energy Conversion, Transmission, and Storage*, Elsevier/Academic Press, Amsterdam, 2007.

Suppes, G. J. and T. S. Storvick, *Sustainable Nuclear Power*, Elsevier/Academic Press, Amsterdam, 2007.

Tabak, J., *Solar and Geothermal Energy*, Facts On File, New York, 2009.

Weiss, C. and W. B. Bonvillian, *Structuring an Energy Technology Revolution*, MIT Press, Cambridge, MA, 2009.

Worldwatch Institute, *Biofuels for Transport: Global Potential and Implications for Energy and Agriculture*, Earthscan, Sterling, VA, 2007.

QUESTIONS AND PROBLEMS

1. The tail of a firefly glows, although it is not hot. Explain the kind of energy transformation that is most likely involved in the firefly's tail producing light.
2. What is the standard unit of energy? What unit did it replace? What is the relationship between these two units?
3. Which law states that energy is neither created nor destroyed?
4. What is the special significance of 1340 W?
5. What is the reaction in nature by which solar energy is converted to chemical energy?
6. In what respects is wind both one of the oldest, as well as one of the newest, sources of energy?
7. What are two major problems with reliance on coal and petroleum for energy?
8. Why does natural gas contribute less to greenhouse warming than does petroleum and much less than coal?
9. How might coal be utilized for energy without producing greenhouse gas carbon dioxide?

10. What is a large limiting factor in growing biomass for fuel, and in what respect does this limit hold hope for the eventual use of biomass fuel?
11. What relationship describes the limit to which heat energy can be converted to mechanical energy?
12. Why does a diesel-powered vehicle have significantly better fuel economy than a gasoline-powered vehicle of similar size?
13. Why is a nuclear power plant less efficient in converting heat energy to electricity than is a fossil-fueled power plant?
14. Instead of having a spark plug that ignites the fuel, a diesel engine has a glow plug that operates only during engine startup. Explain the operation of the glow plug.
15. Cite two examples of vastly increased efficiency of energy utilization that took place during the 1900s.
16. Describe a combined power cycle. How may it be tied with district heating?
17. What are three reactions used in biomass gasification?
18. What is a major proposed use of liquid methanol as a fuel for the future?
19. Describe a direct and an indirect way to produce electricity from solar energy.
20. What is the distinction between donor and acceptor layers in photovoltaic cells?
21. Using internet resources for information, list some possible means for storing energy generated from solar radiation?
22. What are the advantages of *Pittsosporum reiniferum* and *Euphorbia lathyrus* for the production of biomass energy?
23. Corn produces biomass in large quantities during its growing season. What are two potential sources of biomass fuel from corn, one that depends on the corn grain and the other that does not?
24. Does the use of biomass for fuel contribute to greenhouse gas carbon dioxide? Explain.
25. What fermentation process is used to generate a fuel from wastes, such as animal wastes?
26. What are two potential pollution problems that accompany the use of geothermal energy to generate electricity?
27. What basic phenomenon is responsible for nuclear energy? What keeps the process going?
28. What is the biggest problem with nuclear energy? Why is it not such a bad idea to store spent nuclear fuel at a reactor site for a number of years before moving it?
29. What is meant by passive stability in nuclear reactor design?
30. What is the status of thermonuclear fusion for power production?
31. Arrange the following energy conversion processes in order from the least to the most efficient: (a) electric hot water heater, (b) photosynthesis, (c) solar cell, (d) electric generator, (e) aircraft jet engine.
32. Considering the Carnot equation and common means for energy conversion, what might be the role of improved materials (metal alloys and ceramics) in increasing energy conversion efficiency?
33. As it is now used, what is the principle or basis for the production of energy from uranium by nuclear fission? Is this process actually used for energy production? What are some of its environmental disadvantages? What is one major advantage?
34. What would be at least two highly desirable features of nuclear fusion power if it could ever be achieved in a controllable fashion on a large scale?
35. Justify describing the sun as "an ideal energy source." What are two big disadvantages of solar energy?
36. What are some of the greater implications of the use of biomass for energy? How might such widespread use affect greenhouse warming? How might it affect agricultural production of food?

20 Nature, Sources, and Environmental Chemistry of Hazardous Wastes

20.1 INTRODUCTION

A *hazardous substance* is a material that may pose a danger to living organisms, materials, structures, or the environment by explosion or fire hazards, corrosion, toxicity to organisms, or other detrimental effects. What, then, is a hazardous waste? A simple definition of a *hazardous waste* is that it is a hazardous substance that has been discarded, abandoned, neglected, released or designated as a waste material, or one that may interact with other substances to be hazardous. The definition of hazardous waste is addressed in greater detail in Section 20.2, but in a simple sense it is a material that has been left where it should not be and that may cause harm to one if one encounters it.

20.1.1 History of Hazardous Substances

Humans have always been exposed to hazardous substances, going back to prehistoric times when they inhaled noxious volcanic gases or succumbed to carbon monoxide from inadequately vented fires in cave dwellings sealed too well against Ice-Age cold. Slaves in Ancient Greece developed lung disease from weaving mineral asbestos fibers into cloth to make it more degradation resistant. Some archaeological and historical studies have concluded that lead wine containers were a leading cause of lead poisoning in the more affluent ruling class of the Roman Empire, leading to erratic behavior such as fixation on spectacular sporting events, chronic unmanageable budget deficits, speculative purchases of overvalued stock, illicit trysts in governmental offices, and ill-conceived, overly ambitious military ventures in remote foreign lands. Alchemists who worked during the Middle Ages often suffered debilitating injuries and illnesses resulting from the hazards of their explosive and toxic chemicals. During the 1700s, runoff from mine spoils piles began to create serious contamination problems in Europe. As the production of dyes and other organic chemicals developed from the coal tar industry in Germany during the 1800s, pollution and poisoning from coal tar by-products were observed. By around 1900, the quantity and variety of chemical wastes produced each year were increasing sharply with the addition of wastes such as spent steel and iron pickling liquor, lead battery wastes, chromic wastes, petroleum refinery wastes, radium wastes, and fluoride wastes from aluminum ore refining. As the century progressed into the World War II era, the wastes and hazardous by-products of manufacturing increased markedly from sources such as chlorinated solvents manufacture, pesticides synthesis, polymers manufacture, plastics, paints, and wood preservatives.

The Love Canal affair of the 1970s and 1980s brought hazardous wastes to public attention as a major political issue in the United States. Starting around 1940, this site in Niagara Falls, New York, had received about 20,000 metric tons of chemical wastes containing at least 80 different chemicals.

By 1994, state and federal governments had spent well over $100 million to clean up the site and relocate residents.

Other areas containing hazardous wastes that received attention included an industrial site in Woburn, Massachusetts, that had been contaminated by wastes from tanneries, glue-making factories, and chemical companies dating back to about 1850; the Stringfellow Acid Pits near Riverside, California; the Valley of the Drums in Kentucky; and Times Beach, Missouri, an entire town that was abandoned because of contamination by TCDD (dioxin).

The problem of hazardous wastes is truly international in scope. As a result of the problem of dumping such wastes in developing countries, the 1989 Basel Convention on the Control of Transboundary Movement of Hazardous Wastes and their Disposal was held in Basel, Switzerland in 1989, and has since been signed by more than 100 countries. This treaty defines a long List A of hazardous wastes, a List B of nonhazardous wastes, and a List C of as yet unclassified materials. An example of a material on List C is PVC-coated wire, which is harmless, itself, but may release dioxins or heavy metals when thermally treated.

Actions subsequent to the Basel Convention have dealt with specific kinds of wastes. One of these consists of persistent organic pollutants (POP) addressed by the Stockholm Convention on Persistent Organic Pollutants of 2001. POP are organic compounds or mixtures including pesticides, industrial products, and manufacturing by-products that persist in the environment because of their resistance to physical, chemical, and biological processes. Those targeted for elimination in their production and use are aldrin, chlordane, dieldrin, endrin, heptachlor, hexachlorobenzene (HCB), mirex, toxaphene, and PCBs. The production and use of DDT are to be restricted. Targeted for minimization and ultimate elimination are manufacturing by-product PCBs, dibenzo-*p*-dioxins, and dibenzofurans.

Rapidly developing industrial economies normally have growing problems with hazardous wastes. During the 1990s and early 2000s, this was particularly true of the rapidly growing economies of highly populous China and India. By 2005, China was reported to be generating about 900 million metric tons of industrial solid wastes each year of which 10.6 million metric tons were classified as hazardous waste.[1]

20.1.2 Legislation

Governments in a number of nations have passed legislation to deal with hazardous substances and wastes. In the United States such legislation has included the following:

- Toxic Substances Control Act (TSCA) of 1976
- Resource Conservation and Recovery Act (RCRA) of 1976 [amended and reauthorized by the Hazardous and Solid Wastes Amendments Act (HSWA) of 1984]
- Comprehensive Environmental Response, Compensation, and Liability Act (CERCLA) of 1980

RCRA legislation charged the U.S. EPA with protecting human health and the environment from improper management and disposal of hazardous wastes by issuing and enforcing regulations pertaining to such wastes. RCRA requires that hazardous wastes and their characteristics be listed and controlled from the time of their origin until their proper disposal or destruction. Regulations pertaining to firms generating and transporting hazardous wastes require that they keep detailed records, including reports on their activities and manifests to ensure proper tracking of hazardous wastes through transportation systems. Approved containers and labels must be used, and wastes can only be delivered to facilities approved for treatment, storage, and disposal. In 2007, almost 47 million tons of RCRA-regulated wastes were reported from 16,349 generators in the United States. About 3000 facilities are involved in the treatment, storage, or disposal of RCRA wastes.

CERCLA (Superfund) legislation deals with actual or potential releases of hazardous materials that have the potential to endanger people or the surrounding environment at uncontrolled or abandoned hazardous waste sites in the United States. The act requires responsible parties or the government to clean up waste sites. Among CERCLA's major purposes are the following:

- Site identification
- Evaluation of danger from waste sites
- Evaluation of damages to natural resources
- Monitoring of release of hazardous substances from sites
- Removal or cleanup of wastes by responsible parties or government

CERCLA was extended for five years by the passage of the Superfund Amendments and Reauthorization Act (SARA) of 1986, legislation with greatly increased scope and additional funding. Actually longer than CERCLA, SARA encouraged the development of alternatives to land disposal that favor permanent solutions reducing volume, mobility, and toxicity of wastes; increased emphasis upon public health, research, training, and state and citizen involvement; and establishment of a new program for leaking underground (petroleum) storage tanks. After 1986, few new legislative initiatives dealing with hazardous wastes were forthcoming in the United States.

During the earlier years of Superfund, it was funded by taxes levied on corporations that used, disposed of, or profited from toxic and hazardous chemicals. Since 1995, Congress has refused to renew these fees and a $3.8 billion surplus in the fund was gradually used up, running dry in 2003. Funding from general revenues and from fines levied against parties found responsible for improper waste disposal has kept the Superfund cleanup program going, but at a decreased rate. A total of $1.8 billion was spent on Superfund projects in the United States in 1999 and that figure dropped to $1.3 billion in 2007. Since 1980, the Superfund program has placed around 1600 waste sites on the National Priorities list designated for priority cleanup. Of these, somewhat more than 300 have been completely cleaned up and delisted and the majority of cleanup has been completed at somewhat more than 700 of the sites. Work was completed at a rate of about 75 sites per year during the 1990s, but has now dropped to about half that level.

20.2 CLASSIFICATION OF HAZARDOUS SUBSTANCES AND WASTES

Many specific chemicals in widespread use are hazardous because of their chemical reactivities, fire hazards, toxicities, and other properties. There are numerous kinds of hazardous substances, usually consisting of mixtures of specific chemicals. These include such things as explosives; flammable liquids; flammable solids, such as magnesium metal and sodium hydride; oxidizing materials; such as peroxides; corrosive materials, such as strong acids; etiologic agents that cause disease; and radioactive materials.

20.2.1 Characteristics and Listed Wastes

For regulatory and legal purposes in the United States, hazardous substances are listed specifically and are defined according to general characteristics. Under the authority of the RCRA, the U.S. EEPA defines hazardous substances in terms of the following characteristics:

- *Ignitability*, characteristic of substances that are liquids, the vapors of which are likely to ignite in the presence of ignition sources; nonliquids that may catch fire from friction or contact with water and which burn vigorously or persistently; ignitable compressed gases; and oxidizers

- *Corrosivity*, characteristic of substances that exhibit extremes of acidity or basicity or a tendency to corrode steel
- *Reactivity*, characteristic of substances that have a tendency to undergo violent chemical change (e.g., explosives, pyrophoric materials, water-reactive substances, or cyanide- or sulfide-bearing wastes)
- *Toxicity*, defined in terms of a standard extraction procedure followed by chemical analysis for specific substances

In addition to classification by characteristics, EPA designates more than 450 *listed wastes*, which are specific substances or classes of substances known to be hazardous. Each such substance is assigned an EPA *hazardous waste number* in the format of a letter followed by three numerals, where a different letter is assigned to substances from each of the four following lists:

- *F-type wastes from nonspecific sources*: For example, quenching wastewater treatment sludges from metal heat treating operations where cyanides are used in the process (F012).
- *K-type wastes from specific sources*: For example, heavy ends from the distillation of ethylene dichloride in ethylene dichloride production (K019).
- *P-type acute hazardous wastes*: Wastes that have been found to be fatal to humans in low doses, or capable of causing or significantly contributing to an increase in serious irreversible or incapacitating reversible illness. These are mostly specific chemical species such as fluorine (P056) or 3-chloropropane nitrile (P027).
- *U-type miscellaneous hazardous wastes*: These are predominantly specific compounds such as calcium chromate (U032) or phthalic anhydride (U190).

Compared to RCRA, CERCLA gives a rather broad definition of hazardous substances that includes the following:

- Any element, compound, mixture, solution, or substance, the release of which may substantially endanger public health, public welfare, or the environment
- Any element, compound, mixture, solution, or substance in reportable quantities designated by CERCLA Section 102
- Certain substances or toxic pollutants designated by the Federal Water Pollution Control Act.
- Any hazardous air pollutant listed under Section 112 of the Clean Air Act
- Any imminently hazardous chemical substance or mixture that has been the subject of government action under Section 7 of the TSCA
- With the exception of those suspended by Congress under the Solid Waste Disposal Act, any hazardous waste listed or having characteristics identified by RCRA § 3001

20.2.2 Hazardous Wastes

Three basic approaches to defining hazardous wastes are (1) a qualitative description by origin, type, and constituents; (2) classification by characteristics largely based upon testing procedures; and (3) by means of concentrations of specific hazardous substances. Wastes may be classified by general type such as "spent halogenated solvents" or by industrial sources such as "pickling liquor from steel manufacturing."

20.2.2.1 Hazardous Wastes and Air and Water Pollution Control

Somewhat paradoxically, measures taken to reduce air and water pollution (Figure 20.1) have had a tendency to increase production of hazardous wastes. Most water treatment processes yield sludges

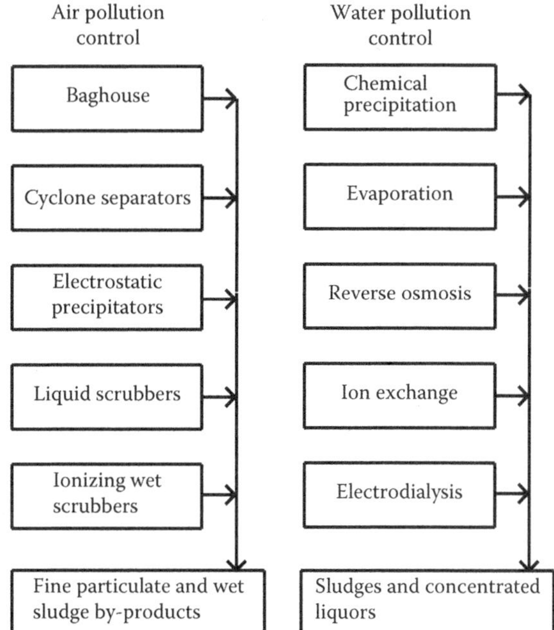

FIGURE 20.1 Potential contributions of air and water pollution control measures to hazardous wastes production.

or concentrated liquors that require stabilization and disposal. Air scrubbing processes likewise produce sludges. Baghouses and precipitators used to control air pollution all yield significant quantities of solids, some of which are hazardous.

20.3 SOURCES OF WASTES

Quantities of hazardous wastes produced each year are not known with certainty and depend upon the definitions used for such materials. In 1988, the figure for RCRA-regulated wastes in the United States was placed at 290 million tons. However, most of this material was water, with only a few million tons consisting of solids. Some high-water-content wastes are generated directly by processes that require large quantities of water in waste treatment, and other aqueous wastes are produced by mixing hazardous wastes with wastewater.

Some wastes that might exhibit a degree of hazard are exempt from RCRA regulation by legislation. These exempt wastes include the following:

- Fuel ash and scrubber sludge from power generation by utilities
- Oil and gas drilling muds
- By-product brine from petroleum production
- Cement kiln dust
- Waste and sludge from phosphate mining and beneficiation
- Mining wastes from uranium and other minerals
- Household wastes

20.3.1 Types of Hazardous Wastes

In terms of quantity by mass, the greatest quantities are those from categories designated by hazardous waste numbers preceded by F and K, respectively. The former are those from nonspecific sources and include the following examples:

- F001—the spent halogenated solvents used in degreasing: tetrachloroethylene, trichloroethylene, methylene chloride, 1,1,1-trichloroethane, carbon tetrachloride, and the chlorinated fluorocarbons; and sludges from the recovery of these solvents in degreasing operations
- F004—the spent nonhalogenated solvents: cresols, cresylic acid, and nitro-benzene; and still bottoms from the recovery of these solvents
- F007—spent plating-bath solutions from electroplating operations
- F010—quenching-bath sludge from oil baths from metal heat treating operations

The "K-type" hazardous wastes are those from specific sources produced by industries such as the manufacture of inorganic pigments, organic chemicals, pesticides, explosives, iron and steel, and nonferrous metals, and from processes such as petroleum refining or wood preservation; some examples are given below:

- K001—bottoms sediment sludge from the treatment of wastewaters from wood-preserving processes that use creosote and/or pentachlorophenol
- K002—wastewater treatment sludge from the production of chrome yellow and orange pigments
- K020—heavy ends (residue) from the distillation of vinyl chloride in vinyl chloride monomer production
- K043—2,6-dichlorophenol waste from the production of 2,4-D
- K047—pink/red water from TNT operations
- K049—slop oil emulsion solids from the petroleum refining industry
- K060—ammonia lime still sludge from coking operations
- K067—electrolytic anode slimes/sludges from primary zinc production

The remainder of wastes consists of reactive wastes, corrosive wastes, toxic wastes, ignitable wastes, and "P" wastes (discarded commercial chemical products, off-specification species, containers, and spill residues), "U" wastes, and unspecified types.

20.3.2 Hazardous Waste Generators

Several hundred thousand companies generate hazardous wastes in the United States, but most of these generators produce only small quantities. Hazardous waste generators are unevenly distributed geographically across the continental United States, with a relatively large number located in the industrialized upper Midwest, including the states of Illinois, Indiana, Ohio, Michigan, and Wisconsin.

Industry types of hazardous waste generators can be divided among the seven following major categories, each containing of the order of 10–20% of hazardous waste generators: chemicals and allied products manufacture, petroleum-related industries, fabricated metals, metal-related products, electrical equipment manufacture, "all other manufacturing," and nonmanufacturing and nonspecified generators. About 10% of the generators produce more than 95% of all hazardous wastes. Whereas, as noted above, the number of hazardous waste generators is distributed relatively evenly among several major types of industries, 70–85% of the *quantities* of hazardous wastes are generated by the chemical and petroleum industries. Of the remainder, about three-fourth comes from metal-related industries and about one-fourth from all other industries.

Household hazardous wastes are potential contaminants of municipal refuse that are supposed to be collected separately by municipalities. The problem of household hazardous wastes is complicated by the literally millions of individual household sources and the variability in the diligence with which citizens collect and segregate the wastes. A variety of substances occur in household hazardous wastes including insecticides, cleaners, oils, batteries, and fluorescent light bulbs (which contain small amounts of mercury). One of the most common components of these wastes consists of leftover paint. Household paint wastes have been the subject of a study in Denmark that found that most of the paint waste was composed of water-based paint.[2] The content of heavy metals in the paint waste was found to be less than that of ordinary household waste and it was concluded that it would be safe to discard water-based paint with ordinary household refuse.

20.4 FLAMMABLE AND COMBUSTIBLE SUBSTANCES

Most chemicals that are likely to burn accidentally are liquids. Liquids form *vapors*, which are usually more dense than air and thus tend to settle. The tendency of a liquid to ignite is measured by a test in which the liquid is heated and periodically exposed to a flame until the mixture of vapor and air ignites at the liquid's surface. The temperature at which ignition occurs under these conditions is called the *flash point*.

With these definitions in mind, it is possible to divide ignitable materials into four major classes. A *flammable solid* is one that can ignite from friction or from heat remaining from its manufacture, or which may cause a serious hazard if ignited. Explosive materials are not included in this classification. A *flammable liquid* is one having a flash point below 60.5°C (141°F). A *combustible liquid* has a flash point in excess of 60.5°C, but below 93.3°C (200°F). Whereas gases are substances that exist entirely in the gaseous phase at 0°C and 1 atm pressure, a *flammable compressed gas* meets specified criteria for lower flammability limit (LFL), flammability range (see below), and flame projection.

In considering the ignition of vapors, two important concepts are those of flammability limit and flammability range. Values of the vapor/air ratio below which ignition cannot occur because of insufficient fuel define the LFL. Similarly, values of the vapor/air ratio above which ignition cannot occur because of insufficient air define the *upper flammability limit* (UFL). The difference between upper and LFLs at a specified temperature is the *flammability range*. Table 20.1 gives some examples of these values for common liquid chemicals. The percentage of flammable substance for best combustion (most explosive mixture) is labeled "optimal." In the case of acetone, for example, the optimal flammable mixture is 5.0% acetone.

TABLE 20.1
Flammabilities of Some Common Organic Liquids

		Volume Percent in Air	
Liquid	**Flash Point (°C)[a]**	**LFL[b]**	**UFL[b]**
Diethyl ether	–43	1.9	36
Pentane	–40	1.5	7.8
Acetone	–20	2.6	13
Toluene	4	1.27	7.1
Methanol	12	60	37
Gasoline (2,2,4-tri-methylpentane)	—	1.4	7.6
Naphthalene	157	0.9	5.9

[a] Closed-cup flash point test.
[b] LFL, lower flammability limit; UFL, upper flammability limit at 25°C.

One of the more disastrous problems that can occur with flammable liquids is a boiling liquid expanding vapor explosion (BLEVE). These are caused by rapid pressure buildup in closed containers of flammable liquids heated by an external source. The explosion occurs when the pressure buildup is sufficient to break the container walls.

20.4.1 Combustion of Finely Divided Particles

Finely divided particles of combustible materials are somewhat analogous to vapors with respect to flammability. One such example is a spray or mist of hydrocarbon liquid in which oxygen has the opportunity for intimate contact with the liquid particles causing the liquid to ignite at a temperature below its flash point.

Dust explosions can occur with a large variety of solids that have been ground to a finely divided state. Many metal dusts, particularly those of magnesium and its alloys, zirconium, titanium, and aluminum, can burn explosively in air. In the case of aluminum, for example, the reaction is the following:

$$4\text{Al(powder)} + 3\text{O}_2\text{(from air)} \rightarrow 2\text{Al}_2\text{O}_3. \tag{20.1}$$

Coal dusts and grain dusts have caused many fatal fires and explosions in coal mines and grain elevators, respectively. Dusts of polymers such as cellulose acetate, polyethylene, and polystyrene can also be explosive.

20.4.2 Oxidizers

Combustible substances are reducing agents that react with *oxidizers* (oxidizing agents or oxidants) to produce heat. Diatomic oxygen, O_2, from air is the most common oxidizer. Many oxidizers are chemical compounds that contain oxygen in their formulas. The halogens (periodic table group 7A) and many of their compounds are oxidizers. Some examples of oxidizers are given in Table 20.2.

An example of a reaction of an oxidizer is that of concentrated HNO_3 with copper metal, which gives toxic NO_2 gas as a product:

$$4\text{HNO}_3 + \text{Cu} \rightarrow \text{Cu(NO}_3)_2 + 2\text{H}_2\text{O} + 2\text{NO}_2. \tag{20.2}$$

TABLE 20.2
Examples of Some Oxidizers

Name	Formula	State of Matter
Ammonium nitrate	NH_4NO_3	Solid
Ammonium perchlorate	NH_4ClO_4	Solid
Bromine	Br_2	Liquid
Chlorine	Cl_2	Gas (stored as liquid)
Fluorine	F_2	Gas
Hydrogen peroxide	H_2O_2	Solution in water
Nitric acid	HNO_3	Concentrated solution
Nitrous oxide	N_2O	Gas (stored as liquid)
Ozone	O_3	Gas
Perchloric acid	$HClO_4$	Concentrated solution
Potassium permanganate	$KMnO_4$	Solid
Sodium dichromate	$Na_2Cr_2O_7$	Solid

20.4.3 Spontaneous Ignition

Substances that catch fire spontaneously in air without an ignition source are called *pyrophoric*. These include several elements—white phosphorus, the alkali metals (group 1A), and powdered forms of magnesium, calcium, cobalt, manganese, iron, zirconium, and aluminum. Also included are some organometallic compounds, such as ethyllithium (LiC_2H_5) and phenyllithium (LiC_6H_5), and some metal carbonyl compounds, such as iron pentacarbonyl ($Fe(CO)_5$). Another major class of pyrophoric compounds consists of metal and metalloid hydrides, including lithium hydride, LiH; pentaborane, B_5H_9; and arsine, AsH_3. Moisture in air is often a factor in spontaneous ignition. For example, lithium hydride undergoes the following reaction with water from moist air:

$$LiH + H_2O \rightarrow LiOH + H_2 + \text{heat} \tag{20.3}$$

The heat generated from this reaction can be sufficient to ignite the hydride so that it burns in air:

$$2LiH + O_2 \rightarrow Li_2O + H_2O \tag{20.4}$$

Some compounds with organometallic character are also pyrophoric. An example of such a compound is diethylethoxyaluminum:

H_5C_2
\
 $Al\text{-}OC_2H_5$ Diethylethoxyaluminum
/
H_5C_2

Many mixtures of oxidizers and oxidizable chemicals catch fire spontaneously and are called *hypergolic mixtures*. Nitric acid and phenol form such a mixture.

20.4.4 Toxic Products of Combustion

Some of the greater dangers of fires are from toxic products and by-products of combustion. The most obvious of these is carbon monoxide, CO, which can cause serious illness or death because it forms carboxyhemoglobin with hemoglobin in the blood so that the blood no longer carries oxygen to body tissues. Toxic SO_2, P_4O_{10}, and HCl are formed by the combustion of sulfur, phosphorus, and organochlorine compounds, respectively. A large number of noxious organic compounds such as aldehydes are generated as by-products of combustion. In addition to forming carbon monoxide, combustion under oxygen-deficient conditions produces PAHs consisting of fused ring structures. Some of these compounds, such as benzo[*a*]pyrene, below, are precarcinogens that are acted upon by enzymes in the body to yield cancer-producing metabolites.

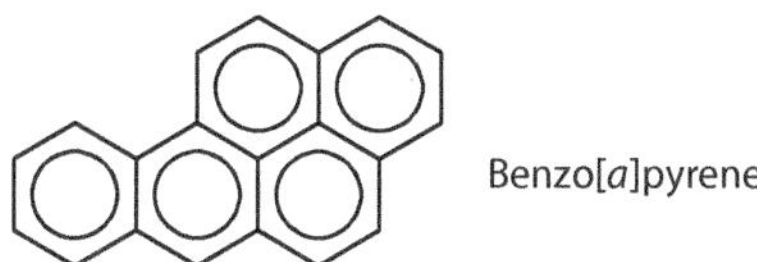

Benzo[*a*]pyrene

20.5 REACTIVE SUBSTANCES

Reactive substances are those that tend to undergo rapid or violent reactions under certain conditions. Such substances include those that react violently or form potentially explosive mixtures with water. An example is sodium metal, which reacts strongly with water as follows:

$$2Na + 2H_2O \rightarrow 2NaOH + H_2 + \text{heat} \tag{20.5}$$

This reaction usually generates enough heat to ignite the sodium and hydrogen. Explosives constitute another class of reactive substances. For regulatory purposes, substances are also classified as reactive that produce toxic gases or vapors when they react with water, acid, or base. Hydrogen sulfide or hydrogen cyanide is the most common toxic substances released in this manner.

Heat and temperature are usually very important factors in reactivity. Many reactions require energy of activation to get them started. The rates of most reactions tend to increase sharply with increasing temperature and most chemical reactions give off heat. Therefore, once a reaction is started in a reactive mixture lacking an effective means of heat dissipation, the rate may increase exponentially with time, leading to an uncontrollable event. Other factors that may affect reaction rate include physical form of reactants (e.g., a finely divided metal powder that reacts explosively with oxygen, whereas a single mass of metal barely reacts), rate and degree of mixing of reactants, degree of dilution with nonreactive media (solvent), presence of a catalyst, and pressure.

Some chemical compounds are self-reactive in that they contain oxidant and reductant in the same compound. Nitroglycerin, a strong explosive with the formula $C_3H_5(ONO_2)_3$, decomposes spontaneously to CO_2, H_2O, O_2, and N_2 with a rapid release of a very high amount of energy. Pure nitroglycerin has such a high inherent instability that only a slight blow may be sufficient to detonate it. TNT is also an explosive with a high degree of reactivity. However, it is inherently relatively stable in that some sort of detonating device is required to cause it to explode.

20.5.1 Chemical Structure and Reactivity

As shown in Table 20.3, some chemical structures are associated with high reactivity. High reactivity in some organic compounds results from unsaturated bonds in the carbon skeleton, particularly where multiple bonds are adjacent (allenes, C=C=C) or separated by only one carbon–carbon single bond (dienes, C=C—C=C). Some organic structures involving oxygen are very reactive. Examples are oxiranes, such as ethylene oxide,

```
       O
      / \
  H-C-----C-H
    |     |
    H     H   Ethylene oxide
```

hydroperoxides (ROOH), and peroxides (ROOR′), where R and R′ stand for hydrocarbon moieties such as the methyl group, $—CH_3$. Many organic compounds containing nitrogen along with carbon and hydrogen are very reactive. Included are triazenes containing a functionality with three nitrogen atoms (R—N=N—N), some azo compounds (R—N=N—R′), and some nitriles in which a nitrogen atom is triply bonded to a carbon atom:

$$R—C\equiv N \quad \text{Nitrile}$$

Functional groups containing both oxygen and nitrogen tend to impart reactivity to an organic compound. Examples of such functional groups are alkyl nitrates ($R—ONO_2$), alkyl nitrites (R—O—N=O), nitroso compounds (R—N=O), and nitro compounds ($R—NO_2$).

Many different classes of inorganic compounds are reactive. These include some of the halogen compounds of nitrogen (shock-sensitive nitrogen triiodide, NI_3, is an outstanding example), compounds with metal–nitrogen bonds (NaN_3), halogen oxides (ClO_2), and compounds with oxyanions of the halogens. An example of the last group of compounds is ammonium perchlorate, NH_4ClO_4, which was involved in a series of massive explosions that destroyed 8 million lb of the compound, injured 300 workers, caused $75 million in damage, and demolished a 40 million lb per year U.S. rocket fuel plant near Henderson, Nevada, in 1988. (By late 1989, a new $92 million plant for the manufacture of ammonium perchlorate had been constructed near Cedar City in a remote region of southwest Utah. Prudently, the buildings at the new plant were placed at large distances from each other!)

TABLE 20.3
Examples of Reactive Compounds and Structures

Organic	
Allenes	C=C=C
Dienes	C=C−C=C
Azo compounds	C−N=N−C
Triazenes	C−N=N−N
Hydroperoxides	R−OOH
Peroxides	R−OO−R'
Oxides	−C(−O−)C− (epoxide ring)
Alkyl nitrates	$R-O-NO_2$
Nitro compounds	$R-NO_2$
Rings that are not 6-membered	(five-membered ring)
Heterogeneous rings	(six-membered ring containing Y)
Inorganic	
Nitrous oxide	N_2O
Nitrogen halides	NCl_3, NI_3
Interhalogen compounds	BrCl
Halogen oxides	ClO_2
Halogen azides	ClN_3
Hypohalites	NaClO

Explosives such as nitroglycerin or TNT are single compounds containing both oxidizing and reducing functions in the same molecule. Such substances are commonly called *redox compounds*. Some redox compounds have even more oxygen than is needed for a complete reaction and are said to have a positive balance of oxygen, some have exactly the stoichiometric quantity of oxygen required (zero balance, maximum energy release), and others have a negative balance and require oxygen from outside sources to completely oxidize all components. TNT has a substantial negative balance of oxygen; ammonium dichromate [$(NH_4)_2Cr_2O_7$] has a zero balance, reacting with exact stoichiometry to H_2O, N_2, and Cr_2O_3; and treacherously explosive nitroglycerin has a positive balance as shown by the following reaction:

$$4C_3H_5N_3O_9 \rightarrow 12CO_2 + 10H_2O + 6N_2 + O_2. \quad (20.6)$$

20.6 CORROSIVE SUBSTANCES

Corrosive substances are regarded as those that dissolve metals or cause oxidized material to form on the surface of metals—rusted iron is a prime example—and, more broadly, cause deterioration of materials, including living tissue, that they contact. Most corrosives belong to at least one of the four following chemical classes: (1) strong acids, (2) strong bases, (3) oxidants, and (4) dehydrating agents. Table 20.4 lists some of the major corrosive substances and their effects.

TABLE 20.4
Examples of Some Corrosive Substances

Name and Formula	Properties and Effects
Nitric acid, HNO_3	Strong acid and strong oxidizer, corrodes metal, reacts with protein in tissue to form yellow xanthoproteic acid, lesions are slow to heal
Hydrochloric acid, HCl	Strong acid, corrodes metals, gives off HCl gas vapor, which can damage respiratory tract tissue
Hydrofluoric acid, HF	Corrodes metals, dissolves glass, causes particularly bad burns to flesh
Alkali metal hydroxides, NaOH and KOH	Strong bases, corrode zinc, lead, and aluminum, substances that dissolve tissue and cause severe burns
Hydrogen peroxide, H_2O_2	Oxidizer, all but very dilute solutions cause severe burns
Interhalogen compounds such as ClF, BrF_3	Powerful corrosive irritants that acidify, oxidize, and dehydrate tissue
Halogen oxides such as OF_2, Cl_2O, Cl_2O_7	Powerful corrosive irritants that acidify, oxidize, and dehydrate tissue
Elemental fluorine, chlorine, bromine (F_2, Cl_2, Br_2)	Very corrosive to mucous membranes and moist tissue, strong irritants

20.6.1 Sulfuric Acid

Sulfuric acid is a prime example of a corrosive substance. As well as being a strong acid, concentrated sulfuric acid is also a dehydrating agent and oxidant. The tremendous affinity of H_2SO_4 for water is illustrated by the heat generated when water and concentrated sulfuric acid are mixed. If this is done incorrectly by adding water to the acid, localized boiling and spattering can occur that result in personal injury. The major destructive effect of sulfuric acid on skin tissue is removal of water with accompanying release of heat. Sulfuric acid decomposes carbohydrates by removal of water. In contact with sugar, for example, concentrated sulfuric acid reacts to leave a charred mass. The reaction is

$$C_{12}H_{22}O_{11} \xrightarrow{H_2SO_4} 11H_2O\ (H_2SO_4) + 12C + \text{heat}. \tag{20.7}$$

Some dehydration reactions of sulfuric acid can be very vigorous. For example, the reaction with perchloric acid produces unstable Cl_2O_7, and a violent explosion can result. Concentrated sulfuric acid produces dangerous or toxic products with a number of other substances, such as toxic carbon monoxide (CO) from reaction with oxalic acid, $H_2C_2O_4$; toxic bromine and sulfur dioxide (Br_2, SO_2) from reaction with sodium bromide, NaBr; and toxic, unstable chlorine dioxide (ClO_2) from reaction with sodium chlorate, $NaClO_3$.

Contact with sulfuric acid causes severe tissue destruction resulting in a severe burn that may be difficult to heal. Inhalation of sulfuric acid fumes or mists damages tissues in the upper respiratory tract and eyes. Long-term exposure to sulfuric acid fumes or mists has caused erosion of teeth!

20.7 TOXIC SUBSTANCES

Toxicity is of the utmost concern in dealing with hazardous substances. This includes both long-term chronic effects from continual or periodic exposures to low levels of toxicants and acute effects from a single large exposure. Toxic substances are covered in greater detail in Chapters 23 and 24.

20.7.1 Toxicity Characteristic Leaching Procedure

For regulatory and remediation purposes, a standard test is needed to measure the likelihood of toxic substances getting into the environment and causing harm to organisms. The U.S. EPA specifies a test called the *Toxicity Characteristic Leaching Procedure (TCLP)* designed to determine the toxicity hazard of wastes.[3] The test was designed to estimate the availability to organisms of both inorganic and organic species in hazardous materials present as liquids, solids, or multiple phase mixtures and does not test for the direct toxic effects of wastes. Basically, the procedure consists of leaching a material with a solvent designed to mimic leachate generated in a municipal waste disposal site, followed by chemical analysis of the leachate.

20.8 PHYSICAL FORMS AND SEGREGATION OF WASTES

Three major categories of wastes based upon their physical forms are *organic materials*, *aqueous wastes*, and *sludges*. These forms largely determine the course of action taken in treating and disposing of the wastes. The *level of segregation*, a concept illustrated in Figure 20.2, is very important in treating, storing, and disposing of different kinds of wastes. It is relatively easy to deal with wastes that are not mixed with other kinds of wastes, that is, those that are highly segregated. For example, spent hydrocarbon solvents can be used as fuel in boilers. However, if these solvents are mixed with spent organochlorine solvents, the production of contaminant hydrogen chloride during combustion may prevent fuel use and require disposal in special hazardous waste incinerators. Further mixing with inorganic sludges adds mineral matter and water. These impurities complicate the treatment processes required by producing mineral ash in incineration or lowering the heating value of the material incinerated because of the presence of water. Among the most difficult types of wastes to handle and treat are those with the least segregation, of which a "worst-case scenario" would be "dilute sludge consisting of mixed organic and inorganic wastes," as shown in Figure 20.2.

An important factor related to segregation and mixing of wastes is the possibility of *waste incompatibility*. It is crucial that wastes not be mixed that will react together adversely or where one waste may exacerbate problems with another. For example, mixing acid with metal sulfide wastes can result in release of toxic H_2S gas. Chelating agents such as EDTA mixed with heavy-metal salts can cause the heavy metals to become mobile as anionic chelates.

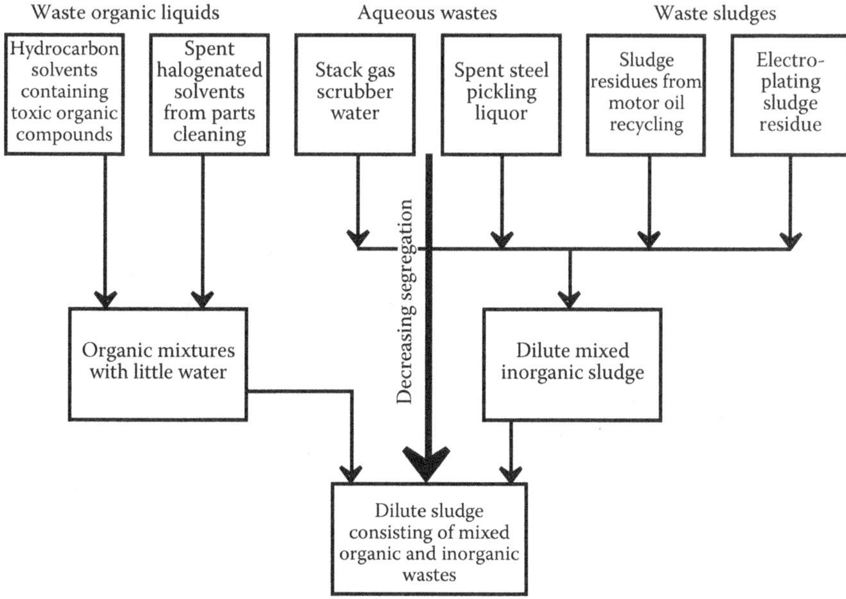

FIGURE 20.2 Illustration of waste segregation.

Concentration of wastes is an important factor in their management. A waste that has been concentrated or, preferably, never diluted is generally much easier and more economical to handle than one that is dispersed in a large quantity of water or soil. Dealing with hazardous wastes is greatly facilitated when the original quantities of wastes are minimized and the wastes remain separated and concentrated insofar as possible.

20.9 ENVIRONMENTAL CHEMISTRY OF HAZARDOUS WASTES

The properties of hazardous materials, their production, and what makes a hazardous substance a hazardous waste were discussed in preceding parts of this chapter. Hazardous materials normally cause problems when they enter the environment and have detrimental effects on organisms or other parts of the environment. Therefore, the present chapter deals with the environmental chemistry of hazardous materials. In discussing the environmental chemistry of hazardous materials, it is convenient to consider the following five aspects based upon the definition of environmental chemistry:

- Origins
- Transport
- Reactions
- Effects
- Fates

It is also useful to consider the five environmental spheres as defined and outlined in Chapter 1:

- Anthrosphere
- Geosphere
- Hydrosphere
- Atmosphere
- Biosphere

Hazardous materials almost always originate in the anthrosphere, are often discarded into the geosphere, and are frequently transported through the hydrosphere or the atmosphere. The greatest concern for their effects is usually on the biosphere, particularly human beings. Figure 20.3 summarizes these relationships.

There are a variety of ways in which hazardous wastes get into the environment. Although now much more controlled by pollution prevention laws, hazardous substances have been deliberately added to the environment by humans. Wastewater containing a variety of toxic substances has

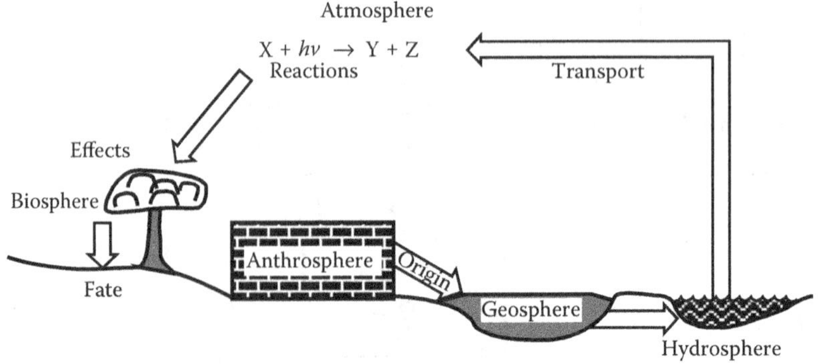

FIGURE 20.3 Scheme of interactions of hazardous wastes in the environment.

been discharged in large quantities into waterways. Hazardous gases and particulate matter have been discharged into the atmosphere through stacks from power plants, incinerators, and a variety of industrial operations. Hazardous wastes have been deliberately spread on soil or placed in landfills in the geosphere. Evaporation and wind erosion may move hazardous materials from wastes dumps into the atmosphere, or they may be leached from waste dumps into groundwater or surface waters. Underground storage tanks or pipelines have leaked a variety of materials into soil. Accidents, fires, and explosions may distribute dangerous materials into the environment. Another source of such materials consists of improperly operated waste treatment or storage facilities.

20.10 PHYSICAL AND CHEMICAL PROPERTIES OF HAZARDOUS WASTES

Having considered the generation of hazardous wastes from the anthrosphere, the next thing to consider is their properties, which determine movement and other kinds of behaviors. These properties can be generally divided into physical and chemical properties.

The behavior of waste substances in the atmosphere is largely determined by their volatilities. In addition, their solubilities in water determine the degree to which they are likely to be removed with precipitation. Water solubility is the most important physical property in the hydrosphere. The movement of substances through the action of water in the geosphere is largely determined by the degree of sorption to soil, mineral strata, and sediments.

Volatility is a function of the vapor pressure of a compound. Vapor pressures at a particular temperature can vary over many orders of magnitude. Of common organic liquids, diethyl ether has one of the highest vapor pressures, whereas those of PCBs are very low. When a volatile liquid is present in soil or in water, its water solubility also determines how well it evaporates. For example, although methanol boils at a lower temperature than benzene, the much lower solubility of benzene in water means that it has the greater tendency to go from the hydrosphere or geosphere into the atmosphere.

The environmental movement, effects, and fates of hazardous waste compounds are strongly related to their chemical properties. For example, a toxic heavy-metal cationic species, such as Pb^{2+} ion, may be strongly held by negatively charged soil solids. If the lead is chelated by the chelating EDTA anion, represented Y^{4-}, it becomes much more mobile as PbY^{2-}, an anionic form. Chelation of cobalt(II) and cobalt(III) with NTA anion (Section 3.14) makes the metal much more mobile in mineral strata. Oxidation state can be very important in the movement of hazardous substances. The reduced states of iron and manganese, Fe^{2+} and Mn^{2+}, respectively, are water soluble and relatively mobile in the hydrosphere and geosphere. However, in their common oxidized states, Fe(III) and Mn(IV), these elements are present as insoluble $Fe_2O_3 \cdot xH_2O$ and MnO_2, which have virtually no tendency to move. Furthermore, these iron and manganese oxides will sequester heavy-metal ions, such as Pb^{2+} and Cd^{2+}, preventing their movement in the soluble form.

The major properties of hazardous substances and their surroundings that determine the environmental transport of such substances are the following:

- Physical properties of the substances, including vapor pressure and solubility
- Physical properties of the surrounding matrix
- Physical conditions to which wastes are subjected. Higher temperatures and erosive wind conditions enable volatile substances to move more readily
- Chemical and biochemical properties of wastes. Substances that are less chemically reactive and less biodegradable will tend to move farther before breaking down

20.11 TRANSPORT, EFFECTS, AND FATES OF HAZARDOUS WASTES

The transport of hazardous wastes is largely a function of their physical properties, the physical properties of their surrounding matrix, the physical conditions to which they are subjected, and chemical factors. Highly volatile wastes are obviously more likely to be transported through the atmosphere and more soluble ones to be carried by water. Wastes will move farther faster in porous, sandy formations than in denser soils. Volatile wastes are more mobile under hot, windy conditions, and soluble ones during periods of heavy rainfall. Wastes that are more chemically and biochemically reactive will not move so far as less reactive wastes before breaking down.

20.11.1 Physical Properties of Wastes

The major physical properties of wastes that determine their amenability to transport are volatility, solubility, and the degree to which they are sorbed to solids, including soil and sediments.

The distribution of hazardous waste compounds between the atmosphere and the geosphere or hydrosphere is largely a function of compound volatility. Usually, in the hydrosphere, and often in soil, hazardous waste compounds are dissolved in water; therefore, the tendency of water to hold the compound is a factor in its mobility. For example, although ethyl alcohol has a higher vapor pressure and lower boiling temperature (77.8°C) than toluene (110.6°C), vapor of the latter compound is more readily evolved from soil because of its limited solubility in water compared to ethanol, which is totally miscible with water.

20.11.2 Chemical Factors

As an illustration of chemical factors involved in transport of wastes, consider cationic inorganic species consisting of common metal ions. These inorganic species can be divided into three groups based upon their retention by clay minerals. Elements that tend to be highly retained by clay include cadmium, mercury, lead, and zinc. Potassium, magnesium, iron, silicon, and NH_4^+ ions are moderately retained by clay, whereas sodium, chloride, calcium, manganese, and boron ions are poorly retained. The retention of the last three elements is probably biased in that they are leached from clay, so that negative retention (elution) is often observed. It should be noted, however, that the retention of iron and manganese is a strong function of oxidation state in that the reduced forms of Mn and Fe are relatively poorly retained, whereas the oxidized forms of $Fe_2O_3 \cdot xH_2O$ and MnO_2 are very insoluble and stay on soil as solids.

20.11.3 Effects of Hazardous Wastes

The effects of hazardous wastes in the environment may be divided among effects on organisms, effects on materials, and effects on the environment. These are addressed briefly here and in greater detail in later sections.

The ultimate concern with wastes has to do with their toxic effects on animals, plants, and microbes. Virtually all hazardous waste substances are poisonous to a degree, some extremely so. The toxicity of a waste is a function of many factors, including the chemical nature of the waste, the matrix in which it is contained, circumstances of exposure, the species exposed, manner of exposure, degree of exposure, and time of exposure. The toxicities of hazardous wastes are discussed in more detail in Chapters 23 and 24.

As defined in Section 20.6, many hazardous wastes are *corrosive* to materials, usually because of extremes of pH or because of dissolved salt content. Oxidant wastes can cause combustible substances to burn uncontrollably. Highly reactive wastes can explode, causing damage to materials and structures. Contamination by wastes, such as by toxic pesticides in grain, can result in substances becoming unfit for use.

In addition to their toxic effects in the biosphere, hazardous wastes can damage air, water, and soil. Wastes that get into air can cause deterioration of air quality, either directly or by the formation of secondary pollutants. Hazardous waste compounds dissolved in, suspended in, or floating as surface films on the surface of water can render it unfit for use and for sustenance of aquatic organisms.

Soil exposed to hazardous wastes can be severely damaged by alteration of its physical and chemical properties and ability to support plants. For example, soil exposed to concentrated brines from petroleum production may become unable to support plant growth so that the soil becomes extremely susceptible to erosion.

20.11.4 Fates of Hazardous Wastes

The fates of hazardous waste substances are addressed in more detail in subsequent sections. As with all environmental pollutants, such substances eventually reach a state of physical and chemical stability, although that may take many centuries to occur. In some cases, the fate of a hazardous waste material is a simple function of its physical properties and surroundings.

The fate of a hazardous waste substance in water is a function of the substance's solubility, density, biodegradability, and chemical reactivity. Dense, water-immiscible liquids may simply sink to the bottoms of bodies of water or aquifers and accumulate there as "blobs" of liquid. Biodegradable substances are broken down by bacteria, a process for which the availability of oxygen is an important variable. Substances that readily undergo bioaccumulation are taken up by organisms, exchangeable cationic materials become bound to sediments, and organophilic materials may be sorbed by organic matter in sediments.

The fates of hazardous waste substances in the atmosphere are often determined by photochemical reactions. Ultimately, such substances may be converted to nonvolatile, insoluble matter and precipitate from the atmosphere onto soil or plants.

20.12 HAZARDOUS WASTES AND THE ANTHROSPHERE

As the part of the environment where humans process substances, the anthrosphere is the source of most hazardous wastes. These materials may come from manufacturing, transportation activities, agriculture, and any one of a number of activities in the anthrosphere. Hazardous wastes may be in any physical form and may include liquids, such as spent halogenated solvents used in degreasing parts; semisolid sludges, such as those generated from the gravitation separation of oil/water/solids mixtures in petroleum refining; and solids, such as baghouse dusts from the production of pesticides.

Releases of hazardous wastes from the anthrosphere commonly occur through incidents such as spills of liquids, accidental discharge of gases or vapors, fires, and explosions. RCRA regulations designed to minimize such accidental releases from the anthrosphere and to deal with them when they occur are contained in 40 CFR 265.31 (Title 40 of the Code of Federal Regulations, Part 265.31). Under these regulations, hazardous waste generators are required to have specified equipment, trained personnel, and procedures that protect human health in the event of a release, and that facilitate remediation if a release occurs. An effective means of communication for summoning help and giving emergency instruction must be available. Also required are firefighting capabilities including fire extinguishers and adequate water. To deal with spills, a facility is required to have on hand absorbents, such as granular vermiculite clay, or absorbents in the form of pillows or pads. Neutralizing agents for corrosive substances that may be used should be available as well.

As noted above, hazardous wastes originate in the anthrosphere. However, to a large extent, they move, have effects, and end up in the anthrosphere as well. Large quantities of hazardous substances are moved by truck, rail, ship, and pipeline. Spills and releases from such movement, ranging from minor leaks from small containers to catastrophic releases of petroleum from wrecked tanker ships, are a common occurrence. Much effort in the area of environmental protection can be profitably devoted to minimizing and increasing the safety of the transport of hazardous substances through the anthrosphere.

In the United States, the transportation of hazardous substances is regulated through the U.S. Department of Transportation (DOT). One of the ways in which this is done is through the *manifest* system of documentation designed to accomplish the following goals:

- Acts as a tracking device to establish responsibility for the generation, movement, treatment, and disposal of the waste.
- By requiring the manifest to accompany the waste, such as during truck transport, it provides information regarding appropriate actions to take during emergencies such as collisions, spills, fires, or explosions.
- Acts as the basic documentation for recordkeeping and reporting.

Many of the adverse effects of hazardous substances occur in the anthrosphere. One of the main examples of such effects occurs as corrosion of materials that are strongly acidic or basic or that otherwise attack materials. Fire and explosion of hazardous materials can cause severe damage to anthrospheric infrastructure.

The fate of hazardous materials is often in the anthrosphere. One of the main examples of a material dispersed in the anthrosphere consists of lead-based anticorrosive paints that have been spread on steel structural members.

20.13 HAZARDOUS WASTES IN THE GEOSPHERE

The sources, transports, interactions, and fates of contaminant hazardous wastes in the geosphere involve a complex scheme, some aspects of which are illustrated in Figure 20.4. As illustrated in the figure, there are numerous vectors by which hazardous wastes can get into groundwater. Leachate from a landfill can move as a waste plume carried along by groundwater, in severe cases, draining into a stream or into an aquifer where it may contaminate well water. Sewers and pipelines may leak hazardous substances into the geosphere. Such substances seep from waste lagoons into geological strata, eventually contaminating groundwater. Wastes leaching from sites where they have been spread on land for disposal or as a means of treatment can contaminate the geosphere and groundwater. In some cases, wastes are pumped into deep wells as a means of disposal.

The movement of hazardous waste constituents in the geosphere is largely by the action of flowing water in a waste plume as shown in Figure 20.4. The speed and degree of waste flow depend upon numerous factors. Hydrologic factors such as water gradient and permeability of the solid formations through which the waste plume moves are important. The rate of flow is usually rather slow, typically several centimeters per day. An important aspect of the movement of wastes through the geosphere is *attenuation* by the mineral strata. This occurs because waste compounds are sorbed

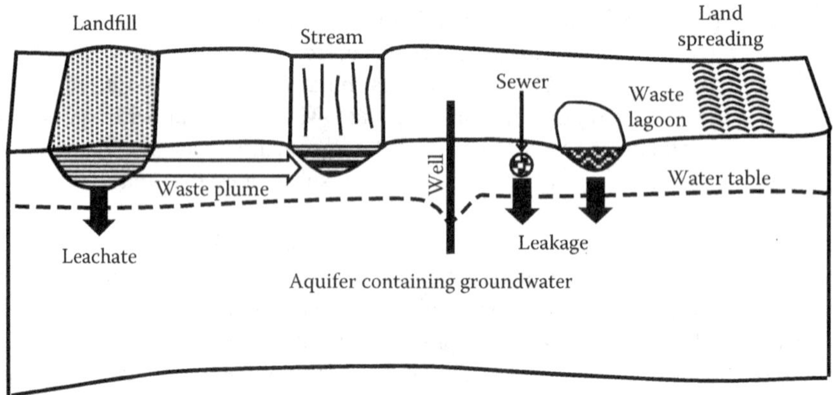

FIGURE 20.4 Sources and movement of hazardous wastes in the geosphere.

to solids by various mechanisms. A measure of the attenuation can be expressed by a *distribution coefficient*, K_d,

$$K_d = \frac{C_s}{C_w} \tag{20.8}$$

where C_s and C_w are the equilibrium concentrations of the constituent on solids and in solution, respectively. This relationship assumes relatively ideal behavior of the hazardous substance that is partitioned between water and solids (the sorbate). A more empirical expression is based on the Freundlich equation:

$$C_s = K_F C_{eq}^{1/n} \tag{20.9}$$

where K_F and $1/n$ are empirical constants.

Several important properties of the solid determine the degree of sorption. One obvious factor is surface area. The chemical nature of the surface is also important. Among the important chemical factors are presence of sorptive clays, hydrous metal oxides, and humus (particularly important for the sorption of organic substances).

In general, sorption of hazardous waste solutes is higher above the water table in the unsaturated zone of soil. This region tends to have a higher surface area and favor aerobic biodegradation processes.

The chemical nature of the leachate is important in sorptive processes of hazardous substances in the geosphere. Organic solvents or detergents in leachates will solubilize organic materials, preventing their retention by solids. Acidic leachates tend to dissolve metal oxides:

$$M(OH)_2(s) + 2H^+ \rightarrow M^{2+} + 2H_2O \tag{20.10}$$

thus preventing sorption of metals in insoluble forms. This is a reason that leachates from municipal landfills, which contain weak organic acids, are particularly prone to transport metals. Solubilization by acids is particularly important in the movement of heavy-metal ions.

Heavy metals are among the most dangerous hazardous waste constituents that are transported through the geosphere. Many factors affect their movement and attenuation. The temperature, pH, and reducing nature (as expressed by the negative log of the electron activity, p*E*) of the solvent medium are important. The nature of the solids, especially the inorganic and organic chemical functional groups on the surface, the CEC, and the surface area of the solids largely determine the attenuation of heavy-metal ions. In addition to being sorbed and undergoing ion exchange with geospheric solids, heavy metals may undergo oxidation–reduction processes, precipitate as slightly soluble solids (especially sulfides), and in some cases, such as occurs with mercury, undergo microbial methylation reactions that produce mobile organometallic species.

The importance of chelating agents interacting with metals and increasing their mobilities has been illustrated by the effects of chelating EDTA on the mobility of radioactive heavy metals, especially ^{60}Co. The EDTA and other chelating agents, such as diethylenetriaminepentaacetic acid (DTPA) and nitrilotriacetic acid (NTA), were used to dissolve metals in the decontamination of radioactive facilities and were codisposed with radioactive materials at Oak Ridge National Laboratory (Tennessee) during the period 1951–1965. Unexpectedly high rates of radioactive metal mobility were observed, which was attributed to the formation of anionic species such as $^{60}CoT^-$ (where T^{3-} is the chelating NTA anion). Whereas unchelated cationic metal species are strongly retained in soil by precipitation reactions and cation exchange processes

$$Co^{2+} + 2OH^- \rightarrow Co(OH)_2(s) \tag{9.10}$$

$$2Soil\}^-H^+ + Co^{2+} \rightarrow (Soil\}^-)_2Co^{2+} + 2H^+ \tag{9.11}$$

anion bonding processes are very weak, so that the chelated anionic metal species are not strongly bound. Naturally occurring humic acid chelating agents may also be involved in the subsurface movement of radioactive metals. It is now generally accepted that poorly biodegradable, strong chelating agents, of which EDTA is the prime example, will facilitate transport of metal radionuclides, whereas oxalate and citrate will not do so because they form relatively weak complexes and are readily biodegraded. Biodegradation of chelating agents tends to be a slow process under subsurface conditions.

Soil can be severely damaged by hazardous waste substances. Such materials may alter the physical and chemical properties of soil and thus its ability to support plants. Some of the more catastrophic incidents in which soil has been damaged by exposure to hazardous materials have arisen from soil contamination from SO_2 emitted from copper or lead smelters, or from brines from petroleum production. Both of these contaminants stop the growth of plants and, without the binding effects of viable plant root systems, topsoil is rapidly lost by erosion.

20.14 HAZARDOUS WASTES IN THE HYDROSPHERE

Hazardous waste substances may enter the hydrosphere as leachate from waste landfills, drainage from waste ponds, seepage from sewer lines, or runoff from soil. Deliberate release into waterways also occurs, and is a particular problem in countries with lax environmental enforcement. There are, therefore, numerous ways by which hazardous materials may enter the hydrosphere.

For the most part, the hydrosphere is a dynamic, moving system, so that it provides perhaps the most important variety of pathways for moving hazardous waste species in the environment. Once in the hydrosphere, hazardous waste species can undergo a number of processes by which they are degraded, retained, and transformed. These include the common chemical processes of precipitation–dissolution, acid–base reactions, hydrolysis, and oxidation–reduction reactions. Also included are a wide variety of biochemical processes which, in most cases, reduce hazards, but in some cases, such as the biomethylation of mercury, greatly increase the risks posed by hazardous wastes.

The unique properties of water have a strong influence on the environmental chemistry of hazardous wastes in the hydrosphere. Aquatic systems are subject to constant change. Water moves with groundwater flow, stream flow, and convection currents. Bodies of water become stratified so that low-oxygen reducing conditions may prevail in the bottom regions of a body of water and there is a constant interaction of the hydrosphere with the other environmental spheres. There is a continuing exchange of materials between water and the other environmental spheres. Organisms in water may have a strong influence on even poorly biodegradable hazardous waste species through bioaccumulation mechanisms.

Figure 20.5 shows some of the pertinent aspects of hazardous waste materials in bodies of water, with emphasis upon the strong role played by sediments. An interesting kind of hazardous waste

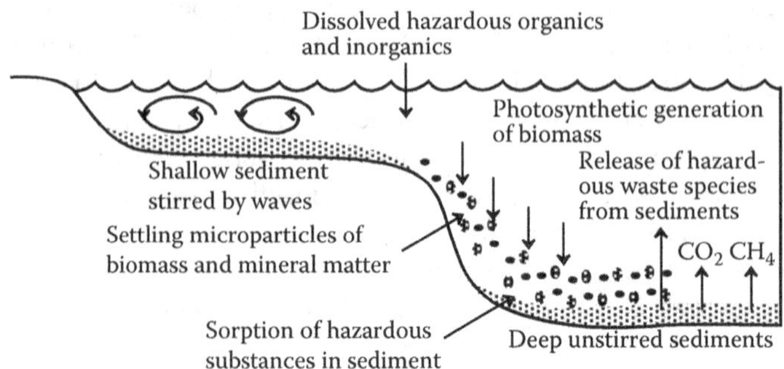

FIGURE 20.5 Aspects of hazardous wastes in surface water in the hydrosphere. The deep unstirred sediments are anaerobic and the site of hydrolysis reactions and reductive processes that may act on hazardous waste constituents sorbed to the sediment.

material that may accumulate in sediments consists of dense, water-immiscible liquids that may sink to the bottom of bodies of water or aquifers and remain there as "blobs" of liquid. Hundreds of tons of PCB wastes have accumulated in sediments in the Hudson River in New York State and are the subject of a heated debate regarding how to remediate the problem.

Hazardous waste species undergo a number of physical, chemical, and biochemical processes in the hydrosphere, which strongly influence their effects and fates. The major ones of these are listed below:

- *Hydrolysis reactions* are those in which a molecule is cleaved with addition of a molecule of H_2O. An example of a hydrolysis reaction is the hydrolysis of dibutyl phthalate, Hazardous Waste Number U069:

$$C_6H_4(C(=O)-O-C_4H_9)_2 + 2H_2O \longrightarrow C_6H_4(C(=O)-OH)_2 + 2HOC_4H_9$$

- *Precipitation reactions*, such as the formation of insoluble lead sulfide from soluble lead(II) ion in the anaerobic regions of a body of water:

$$Pb^{2+} + HS^- \rightarrow PbS(s) + H^+$$

An important part of the precipitation process is normally *aggregation* of the colloidal particles first formed to produce a cohesive mass. Precipitates are often relatively complicated species, such as the basic salt of lead carbonate, $2PbCO_3 \cdot Pb(OH)_2$. Heavy metals, a common ingredient of hazardous waste species precipitated in the hydrosphere, tend to form hydroxides, carbonates, and sulfates with the OH^-, HCO_3^-, and SO_4^{2-} ions that commonly are present in water, and sulfides are likely to be formed in bottom regions of bodies of water where sulfide is generated by anaerobic bacteria. Heavy metals are often coprecipitated as a minor constituent of some other compound, or are sorbed by the surface of another solid.

- *Oxidation–reduction reactions* commonly occur with hazardous waste materials in the hydrosphere, generally mediated by microorganisms. An example of such a process is the oxidation of ammonia to toxic nitrite ion mediated by Nitrosomonas bacteria:

$$NH_3 + \tfrac{3}{2}O_2 \rightarrow H^+ + NO_2^-(s) + H_2O$$

- *Biochemical processes*, which often involve hydrolysis and oxidation–reduction reactions. Organic acids and chelating agents, such as citrate, produced by bacterial action may solubilize heavy-metal ions. Bacteria also produce methylated forms of metals, particularly mercury and arsenic.
- *Photolysis reactions* and miscellaneous chemical phenomena. Photolysis of hazardous waste compounds in the hydrosphere commonly occurs on surface films exposed to sunlight on the top of water.

Hazardous waste compounds have a number of effects on the hydrosphere. Perhaps the most serious of these is the contamination of groundwater, which, in some cases, can be almost irreversible. Waste compounds accumulate in sediments, such as river or estuary sediments.[4] Hazardous waste compounds dissolved in, suspended in, or floating as surface films on the surface of water can render it unfit for use and for sustenance of aquatic organisms.

Many factors determine the fate of a hazardous waste substance in water. Among these are the substance's solubility, density, biodegradability, and chemical reactivity. As discussed above and in Section 20.16, biodegradation largely determines the fates of hazardous waste substances in the hydrosphere. In addition to biodegradation, some substances are concentrated in organisms by bioaccumulation processes and may become deposited in sediments as a result. Organophilic materials may be sorbed by organic matter in sediments. Cation-exchanging sediments have the ability to bind cationic species, including cationic metal ions and organics that form cations.

20.15 HAZARDOUS WASTES IN THE ATMOSPHERE

Hazardous waste chemicals can enter the atmosphere by evaporation from hazardous waste sites, by wind erosion, or by direct release. Hazardous waste chemicals usually are not evolved in large enough quantities to produce secondary air pollutants, such as photochemical smog, formed by reactions of air pollutants in the atmosphere. Therefore, species from hazardous waste sources are usually of most concern in the atmosphere as primary pollutants emitted in localized areas at a hazardous waste site. Plausible examples of primary air pollutant hazardous waste chemicals include corrosive acid gases, particularly HCl; toxic organic vapors, such as vinyl chloride (U043); and toxic inorganic gases, such as HCN potentially released by the accidental mixing of waste cyanides with strong acids:

$$H_2SO_4 + 2NaCN \rightarrow Na_2SO_4 + 2HCN(g) \tag{20.11}$$

Primary air pollutants such as these are almost always of concern only adjacent to the site or to workers involved in site remediation. One such substance that has been responsible for fatal poisonings at hazardous waste sites, usually tanks that are undergoing cleanup or demolition, is highly toxic hydrogen sulfide gas, H_2S.

An important characteristic of a hazardous waste material that enters the atmosphere is its *pollution potential.* This refers to the degree of environmental threat posed by the substance acting as a primary pollutant or to its potential to cause harm from secondary pollutants.

Another characteristic of a hazardous waste material that determines its threat to the atmosphere is its *residence time*, which can be expressed by an estimated atmospheric half-life, $t_{1/2}$. Among the factors that go into estimating atmospheric half-lives are water solubilities, rainfall levels, and atmospheric mixing rates.

Hazardous waste compounds in the atmosphere that have significant water solubilities are commonly removed from the atmosphere by *dissolution* in water. The water may be in the form of very small cloud or fog particles or it may be present as rain droplets.

Some hazardous waste species in the atmosphere are removed by *adsorption onto aerosol particles.* Typically, the adsorption process is rapid so that the lifetime of the species is that of the aerosol particles (typically a few days). Adsorption onto solid particles is the most common removal mechanism for highly nonvolatile constituents such as benzo[*a*]pyrene.

Dry deposition is the name given to the process by which hazardous waste species are removed from the atmosphere by impingement onto soil, water, or plants on the earth's surface. These rates are dependent upon the type of substance, the nature of the surface that they contact, and weather conditions.

A significant number of hazardous waste substances leave the atmosphere much more rapidly than predicted by dissolution, adsorption onto particles, and dry deposition, meaning that chemical processes must be involved. The most important of these are photochemical reactions, commonly involving hydroxyl radical, $HO^{\bullet}$. Other reactive atmospheric species that may act to remove hazardous waste compounds are ozone (O_3), atomic oxygen (O), peroxyl radicals ($HOO^{\bullet}$), alkylperoxyl radicals ($ROO^{\bullet}$), and NO_3. Although its concentration in the troposphere is relatively low, $HO^{\bullet}$ is so reactive that it tends to predominate in the chemical processes that remove hazardous waste species from air. Hydroxyl radical undergoes *abstraction reactions* that remove H atoms from organic compounds:

$$R\text{—}H + HO^{\bullet} \rightarrow R^{\bullet} + H_2O \tag{20.12}$$

and may react with those containing unsaturated bonds by addition as illustrated by the following reaction:

$$\mathrm{R(H)C{=}CH_2} + \mathrm{HO\cdot} \longrightarrow \mathrm{H{-}\dot{C}(R){-}C(H)(H){-}OH} \quad (20.13)$$

The free radical products are very reactive. They react further to form oxygenated species, such as aldehydes, ketones, and dehalogenated organics, eventually leading to the formation of particles or water-soluble materials that are readily scavenged from the atmosphere.

Direct photodissociation of hazardous waste compounds in the atmosphere may occur by the action of the shorter wavelength light that reaches to the troposphere and is absorbed by a molecule with a light-absorbing group called a *chromophore*:

$$\mathrm{R{-}X} + h\nu \rightarrow \mathrm{R^\bullet} + \mathrm{X^\bullet} \quad (20.14)$$

Among the factors involved in assessing the effectiveness of direct absorption of light to remove species from the atmosphere are light intensity, quantum yields (chemical reactions per quantum absorbed), and atmospheric mixing. The requirement of a suitable chromophore limits direct photolysis as a removal mechanism for most compounds other than conjugated alkenes, carbonyl compounds, some halides, and some nitrogen compounds, particularly nitro compounds, all of which commonly occur in hazardous wastes.

20.16 HAZARDOUS WASTES IN THE BIOSPHERE

Microorganisms, bacteria, fungi, and, to a certain extent, protozoa may act metabolically on hazardous wastes substances in the environment. Therefore, the ecotoxicology of hazardous wastes, that is, their toxic effects as they influence organisms in ecosystems, is very important. Most of these substances are *anthropogenic* (made by human activities), and most are classified as *xenobiotic* molecules that are foreign to living systems. Although by their nature xenobiotic compounds are degradation resistant, almost all classes of them—nonhalogenated alkanes, halogenated alkanes (trichloroethane, dichloromethane), nonhalogenated aryl compounds (benzene, naphthalene, and benzo[*a*]pyrene), halogenated aryl compounds (HCB, pentachlorophenol), phenols (phenol, cresols), PCBs, phthalate esters, and pesticides (chlordane, parathion)—can be at least partially degraded by various microorganisms.

Bioaccumulation occurs in which wastes are concentrated in the tissue of organisms. It is an important mechanism by which wastes enter food chains. *Biodegradation* occurs when wastes are converted by biological processes to generally simpler molecules; the complete conversion to simple inorganic species, such as CO_2, NH_3, SO_4^{2-}, and $H_2PO_4^-/HPO_4^{2-}$, is called *mineralization*. The production of a less toxic product by biochemical processes is called *detoxification*. An example is the bioconversion of highly toxic organophosphate paraoxon insecticide to *p*-nitrophenol, which is only about 1/200 as toxic:

$$\mathrm{(H_5C_2O)_2P({=}O){-}O{-}C_6H_4{-}NO_2} \xrightarrow[\text{Enzymes}]{\mathrm{H_2O,\{O\}}} \mathrm{HO{-}C_6H_4{-}NO_2} + \text{Other products} \quad (20.15)$$

20.16.1 Microbial Metabolism in Waste Degradation

Two major divisions of biochemical metabolism that operate on hazardous waste species are *oxic (aerobic) processes* that use molecular O_2 as an oxygen source and *anoxic (anaerobic) processes*, which make use of another oxidant such as sulfate, SO_4^{2-}, which is reduced to H_2S. (This has the benefit of precipitating insoluble metal sulfides in the presence of hazardous waste heavy metals.) Because molecular oxygen does not penetrate to such depths, anaerobic processes predominate in the deep sediments as shown in Figure 20.5.

For the most part, anthropogenic compounds resist biodegradation much more strongly than do naturally occurring compounds. Given the nature of xenobiotic substances, there are very few enzyme systems in microorganisms that act specifically on these substances, especially in making an initial attack on the molecule. Therefore, most xenobiotic compounds are acted upon by a process called *cometabolism*, which occurs concurrently with normal metabolic processes. An interesting example of cometabolism is provided by the white rot fungus, *Phanerochaete chrysosporium*, which has been promoted for the treatment of hazardous organochlorides such as PCBs, DDT, and chlorodioxins. This fungus uses dead wood as a carbon source and has an enzyme system that breaks down wood lignin, a degradation-resistant biopolymer that binds the cellulose in wood. Under appropriate conditions, this enzyme system attacks organochloride compounds and enables their mineralization.

The susceptibility of a xenobiotic hazardous waste compound to biodegradation depends upon its physical and chemical characteristics. Important physical characteristics include water solubility, hydrophobicity (aversion to water), volatility, and lipophilicity (affinity for lipids). In organic compounds, certain structural groups—branched carbon chains, ether linkages, meta-substituted benzene rings, chlorine, amines, methoxy groups, sulfonates, and nitro groups—impart particular resistance to biodegradation.

Microorganisms vary in their ability to degrade hazardous waste compounds; virtually never does a single microorganism have the ability to completely mineralize a waste compound. Abundant aerobic bacteria of the *Pseudomonas* family are particularly adept at degrading synthetic compounds such as biphenyl, naphthalene, DDT, and many other compounds. *Actinomycetes*, microorganisms that are morphologically similar to both bacteria and fungi, degrade a variety of organic compounds including degradation-resistant alkanes and lignocellulose, as well as pyridines, phenols, nonchlorinated aryls, and chlorinated aryls.

Fungi are particularly noted for their ability to attack long-chain and complex hydrocarbons, and are more successful than bacteria in the initial attack on PCB compounds. The potential of the white rot fungus, *Phanerochaete chrysosporium*, to degrade biodegradation-resistant compounds, especially organochloride species, was previously noted.

20.16.2 Ecotoxicology of Hazardous Wastes

Biologically, the greatest concern with wastes has to do with their toxic effects on animals, plants, and microbes. Virtually all hazardous waste substances are poisonous to a degree, some extremely so. Toxicities vary markedly with the physical and chemical nature of the waste, the matrix in which it is contained, the type and condition of the species exposed, and the manner, degree, and time of exposure.

Ecotoxicology, the study of how chemicals affect organisms in the environment, is an important part of the risk assessment of hazardous wastes.[5] *Eco-epidemiology* is a more broadly based area that seeks to evaluate over a relatively long term how the health of a biological community is affected by the physical and chemical nature of the environment with respect to foreign chemical substances in it. A key aspect of the ecotoxicology of hazardous wastes is the determination of the concentration at which chemicals begin to have significant effects on organisms in the environment. The ecotoxicities of products likely to be released to the environment are evaluated starting with a quantitative (QSAR) approach (a largely computational exercise) and progressing through acute, chronic, and model ecosystem testing (a tiered effects process) in order to evaluate the likely effects of compounds upon survival, growth, and reproduction of organisms. The expense and complexity of testing increase with higher levels in the tiered approach and later tests may not be required if those earlier in the sequence show that there is unlikely to be a problem. Only the most widely used compounds are subject to expensive chronic toxicity or model ecosystem tests.

20.17 HAZARDOUS SUBSTANCES IN TERRORISM

A wide variety of substances used in everyday commerce or made specifically for their destructive or toxic effects have a high potential for use in terrorist attacks. Some of these substances are discussed here. Toxic substances with potential for use in terrorist incidents are discussed further in Chapter 24.

Substances used to attack people can be distributed through a number of conduits including air, food, drinking water, and pharmaceuticals. They are especially effective against concentrated groups of people, such as those gathered in shopping malls, theaters, or transportation facilities. In part because they are readily diluted to nonhazardous levels by wind and air currents, chemical agents are of less concern for attacks on dispersed populations out of doors than are biological agents or nuclear weapons. The quantities of chemical agents required for effective attack vary widely from just a few grams of nerve gas to hundreds of kilograms of ammonia or chlorine. Because of chemophobic concerns of the general public, even a relatively harmless chemical attack is likely to cause widespread panic to exposed individuals. One can imagine the consternation that could be caused by an ominous looking, but largely harmless, fog of ammonium chloride produced by simultaneous release of ammonia and hydrogen chloride gas in a crowded area.

Hazardous chemicals that have the potential for use in terrorist attacks include the following:

1. Toxic and otherwise hazardous industrial chemicals
2. Military poisons
3. Combustible substances and oxidizers
4. Corrosive substances
5. Highly reactive substances including explosives

The use of relatively large quantities of hazardous industrial chemicals is a characteristic of all modern industrialized societies. The potential hazards posed by such chemicals are exemplified by the catastrophic Bhopal, India, incident of 1984 that killed 3500 people when methyl isocyanate was accidentally released. Elemental chlorine and phosgene are widely used and were among the first military poisons employed in World War I. A more recent example of large-scale poisoning by a hazardous chemical occurred in the Chuandongbei natural gas field of southwestern China in December 2003, when a drill penetrated a highly pressurized pocket of natural gas containing a high level of toxic hydrogen sulfide, H_2S. Toxic gas spewing from the well killed at least 191 people including a high proportion of the population of Xiaoyang village, the community closest to the well. Approximately 600 people in the town of Zhonghe were treated for poisoning and thousands of people were evacuated. To prevent additional poisoning, the gas was ignited, creating a huge fire and converting the hydrogen sulfide to noxious, but much less toxic sulfur dioxide.

Industrial chemicals are very well controlled inside plant walls and incidents of harm from this source of chemicals are relatively rare. A greater hazard is posed by the transport of such chemicals by rail, truck, or other means. Major transportation accidents in which hazardous chemicals are released occur relatively frequently. Rail cars and trucks transporting such chemicals are vulnerable to terrorist attack and even hijacking that could result in exposure of human populations. In this respect, the practices of industrial ecology and green chemistry offer enormous potential for reducing the hazards. In many cases, hazardous chemical intermediates can be produced on site as needed and by just-in-time production, thereby eliminating the need to transport large quantities of such chemicals and preventing their accumulation in quantities that might pose hazards.

Substances that burn and oxidizers required for combustion are in widespread use and must be transported from sources to users. The potential of flammable substances to cause death and

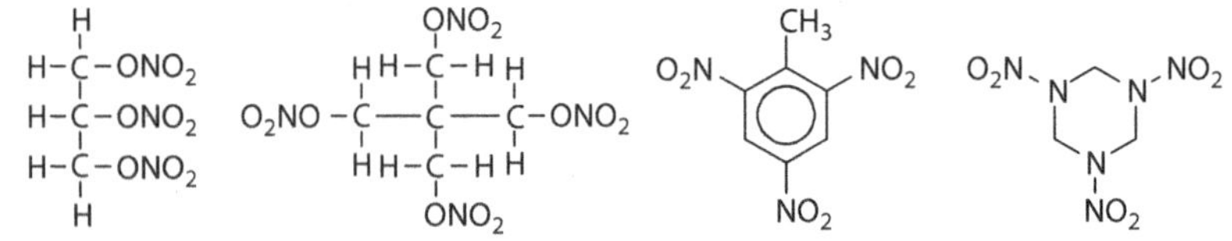

Nitroglycerin | Pentaerythritol tetranitrate (PETN) | 2,4,6-Trinitrotoluene (TNT) | 1,3,5-Trinitro-1,3,5-triazacyclohexane (RDX)

FIGURE 20.6 Common military explosives.

destruction was tragically illustrated by the 2001 attack on the New York World Trade Center in which the destructive agent was flammable jet fuel. Flammable substances include gases and volatile liquids that can spread some distance from their source through conduits such as a subway tunnels and sewers. Pipelines through which large quantities of flammable substances are transported are vulnerable to terrorist attack. Mixed with air, flammable vapors can cause devastating explosions.

Oxidizers can greatly accelerate the burning of flammable materials. This was illustrated tragically in the 1997 crash of a ValueJet airliner in the Florida Everglades in which sodium chlorate used for aircraft oxygen generators was placed in the luggage hold and provided an oxidizer that burned tires and other flammable materials resulting in the crash.

Corrosive substances that attack materials and flesh can be used in terrorist attacks. Acidic, dehydrating sulfuric acid has been used to blind people in criminal attacks. Acids and other corrosive substances can be used to attack infrastructure, such as communications equipment.

Explosives are the substances most widely used in terrorist attacks. Blasts from relatively small amounts of explosives can bring down aircraft. Much larger quantities have been used in some of the more prominent recent terrorist attacks including the 1995 bombing of the Murrah Federal Building in Oklahoma City and the bombing of the U.S. Embassy in Kenya.

Many kinds of explosives have been developed that can be used for illegal purposes. These include gunpowder, nitroglycerin (the explosive in dynamite), 2,4,6-TNT, 1,3,5-trinitro-1,3,5-triazacyclohexane (RDX), and pentaerythritol tetranitrate (PETN). Structural formulas of several dangerous explosives are shown in Figure 20.6. Explosives can be made from readily available materials, especially a mixture of ammonium nitrate fertilizer with fuel oil, which was used in the destruction of the Murrah Federal Building in Oklahoma City.

20.17.1 Detection of Hazardous Substances

The detection of explosives and other hazardous substances is extremely important in combating terrorist threats. Metal detectors and x-ray imaging, which have been the standard means for finding weapons and bombs on airline passengers and in their luggage, are of limited use in detecting hazardous substances. Stationary ion mobility spectrometers and chemiluminescence sensors can be used to detect residues of explosives such as RDX, PETN, or TNT. Typically, sampling is done by swabbing luggage; careful cleaning of luggage can reduce chances of detection. Nuclear quadrupole resonance (NQR), which produces a signal from the nuclei of atoms of ^{14}N, which constitute 99.6% of naturally occurring nitrogen, is an especially promising explosives detection technique. The information it provides correlates with the functional groups in specific kinds of explosives enabling their detection. Other technologies that hold promise for explosives detection include x-ray diffraction, microwave/millimeter wave scanners, and NQR.

NQR is an especially promising explosives detection technique because of its specificity for the detection of nitrogen found in all common explosives and because of its ability to detect explosives in containers and even land mines. It produces a signal from the nuclei of atoms of ^{14}N, which constitute 99.6% of naturally occurring nitrogen. A pulse of radiofrequency radiation from a transmitter coil excites the ^{14}N spins to higher quantized energy levels and measurement of the particular precession frequency that is followed as the nuclear spins return to their equilibrium positions provides the identities and abundance of nitrogen-containing functional groups. This information correlates with specific kinds of explosives enabling their detection.

One of the most sensitive means for detecting hazardous chemical substances that might be involved in attacks is "canine olfactory detection." It consists of the use of dogs to sniff odors of substances at very low levels. This detection system makes use of the dogs' approximately 220 million mucus-coated olfactory receptors (about 40 times as many as humans) by which they detect odors and is more sophisticated than any yet developed by humans. (However, for various reasons, including the desire to get rewards, dogs sometime exhibit the temperamental and unpredictable nature often attributed to modern computerized instruments and some humans. According to one authoritative source quoting an expert in canine olfaction, "Dogs lie. We know they do.")[6]

20.17.2 Removing Hazardous Agents

An important aspect of resisting the effects of attacks by some hazardous substances is their removal, such as from contaminated air, or decontamination, such as from contaminated surfaces. Filters of various kinds can be employed to remove particles from both air and water. Adsorbents including activated carbon and molecular sieves can be used to sequester molecular contaminants. Liquids and solutions that can absorb contaminants or even react with them chemically can be used in scrubbers and packed sorbent reaction beds.

A common problem in decontamination is that the materials used for this purpose may be incompatible with contaminated apparatus such as the electronic equipment. Although water sprayed as a fine mist into an atmosphere contaminated with chlorine, hydrogen fluoride, hydrogen sulfide, and other water-soluble materials can be an effective removal agent, it is hardly compatible with sensitive computer equipment. Reactive foams might be somewhat more preferable in some cases.

Robots offer some significant potential for counterterrorism such as in assessing attack sites, performing some types of search and rescue operations, inspecting containers, performing routine surveillance, and doing some of the labor-intensive and tedious work of decontamination. In this respect, robots are most likely to function as assistants to humans rather than substitutes for them. Although the idea of robots functioning entirely on their own in complex tasks such as decontamination remains a distant dream, continued advances in computerized control and artificial intelligence will undoubtedly increase the role of robots in counterterrorism activities of the future.

LITERATURE CITED

1. Duan, H., Q. Huang, Q. Wang, B. Zhou, and J. Li, Hazardous waste generation and management in China: A review, *Journal of Hazardous Materials*, **158**, 221–227, 2008.
2. Fjelsted, L. and T. H. Christensen, Household hazardous waste: Composition of paint waste, *Waste Management and Research*, **25**, 502–509, 2007.
3. Toxicity Characteristic Leaching Procedure, Test Method 1311 in *Test Methods for Evaluating Solid Waste, Physical/Chemical Methods*, EPA Publication SW-846, 3rd ed., (November, 1986, last amended 2008), as amended by Updates I, II, IIA, IIB, III, IIIA, IIIB, and IV in pdf format from http://www.epa.gov/SW-846
4. Magdaleno, A., A. Mendelson, A. F. d. Iorio, A. Rendina, and J. Moretton, Genotoxicity of leachates from highly polluted lowland river sediments destined for disposal in landfill, *Waste Management*, **28**, 2134–2139, 2008.
5. Wilke, B.-M., F. Riepert, C. Koch, and T. Kuene, Ecotoxicological characterization of hazardous wastes, *Ecotoxicology and Environmental Safety*, **70**, 283–293, 2008.
6. Derr, M., With Dog Detectives, Mistakes Can Happen, *New York Times*, December 24, 2002, p. D1.

SUPPLEMENTARY REFERENCES

ASTM Committee D-18 on Soil and Rock, *ASTM Standards Relating to Environmental Site Characterization*, 2nd ed., ASTM International, West Conshohocken, PA, 2002.

Barth, R. C., P. D. George, and R. H. Hill, *Environmental Health And Safety For Hazardous Waste Sites*, American Industrial Hygiene Association, Fairfax, VA, 2002.

Cabaniss, A. D., Ed., *Handbook on Household Hazardous Waste*, Government Institute/Scarecrow Press, Lanham, MD, 2008.

Carson, P. A. and C. Mumford, *Hazardous Chemicals Handbook*, 2nd ed., Butterworth-Heinemann, Boston, 2002.

Cheremisinoff, N. P., *Handbook of Industrial Toxicology and Hazardous Materials*, Marcel Dekker, New York, 1999.

Cheremisinoff, N. P., *Industrial Solvents Handbook*, 2nd ed., Marcel Dekker, New York, 2003.

Chermisinoff, D., *Fire and Explosion Hazards Handbook of Industrial Chemicals*, Jaico Publishing House, Mumbai, India, 2005.

Davletshina, T. A. and N. P. Cheremisinoff, *Fire and Explosion Hazards Handbook of Industrial Chemicals*, Noyes Publications, Westwood, NJ, 1998.

Gallant, B., *Hazardous Waste Operations and Emergency Response Manual*, Wiley-Interscience, Hoboken, NJ, 2006.

Garrett, T. L., Ed., *The RCRA Practice Manual*, 2nd ed., American Bar Association, Section of Environment, Energy, and Resources, Chicago, 2004.

Hackman, C. L., E. Ellsworth Hackman, III, and M. E. Hackman, *Hazardous Waste Operations and Emergency Response Manual and Desk Reference*, McGraw-Hill, New York, 2002.

Hawley, C., *Hazardous Materials Air Monitoring and Detection Devices*, 2nd ed., Thomson/Delmar Learning, Clifton Park, NY, 2007.

Hawley, C., *Hazardous Materials Handbook: Awareness and Operations Levels*, Delmar Cengage, Clifton Park, NY, 2008.

Hawley, C., *Hazardous Materials Incidents*, 3rd ed., Thomson/Delmar Learning, Clifton Park, NY, 2008.

Hocking Martin B. B., *Handbook of Chemical Technology and Pollution Control*, 3rd ed., Elsevier, Boston, 2006.

LaGrega, Michael D., P. L. Buckingham, and J. C. Evans, *Hazardous Waste Management*, 2nd ed., McGraw-Hill, Boston, MA, 2001.

McGowan, K., *Hazardous Waste*, Lucent books, San Diego, CA, 2001.

Meyer, E., *Chemistry of Hazardous Materials*, Prentice-Hall, Upper Saddle River, NJ, 1998.

Office of Technology Assessment, *Technologies and Management Strategies for Hazardous Waste Control*, University Press of the Pacific, Honolulu, 2005.

Pohanish, R. P., *Sittig's Handbook of Toxic and Hazardous Chemicals and Carcinogens*, 5th ed., Noyes Publications, Westwood, NJ, 2007.

Sara, M. N., *Site Assessment and Remediation Handbook,* 2nd ed., CRC Press/Lewis Publishers, Boca Raton, FL, 2003.

Spencer, A. B. and G. R. Colonna, *NFPA Pocket Guide to Hazardous Materials*, National Fire Protection Association, Quincy, MA, 2003.

Streissguth, T., *Nuclear and Toxic Waste*, Greenhaven Press, San Diego, CA, 2001.

Teets, J. W. and D. Reis, *RCRA: Resource Conservation and Recovery Act*, American Bar Association, Section of Environment, Energy, and Resources, Chicago, 2004.

Urben, P. G., Ed., *Bretherick's Handbook of Reactive Chemical Hazards*, 7th ed., Elsevier, Amsterdam, 2007.

Wang, L. K., *Handbook of Industrial and Hazardous Wastes Treatment*, 2nd ed., Marcel Dekker, New York, 2004.

QUESTIONS AND PROBLEMS

1. Match each of the following kinds of hazardous substances on the left with a specific example of each from the right, below:

1. Explosives	a. Oleum, sulfuric acid, caustic soda
2. Compressed gases	b. White phosphorus
3. Radioactive materials	c. NH_4ClO_4

4. Flammable solids	d. Hydrogen, sulfur dioxide
5. Oxidizing materials	e. Nitroglycerin
6. Corrosive materials	f. Plutonium, cobalt-60

2. Of the following, the property that is *not* a member of the same group as the other properties listed is: (A) substances that are liquids whose vapors are likely to ignite in the presence of ignition sources, (B) nonliquids that may catch fire from friction or contact with water and which burn vigorously or persistently, (C) ignitable compressed gases, (D) oxidizers, (E) substances that exhibit extremes of acidity or basicity.
3. In what respects may it be said that measures taken to alleviate air and water pollution tend to aggravate hazardous waste problems?
4. Why is attenuation of metals likely to be very poor in acidic leachate? Why is attenuation of anionic species in soil less than that of cationic species?
5. Discuss the significance of LFL, UFL, and flammability range in determining the flammability hazards of organic liquids.
6. Concentrated HNO_3 and its reaction products pose several kinds of hazards. What are these?
7. What are substances called that catch fire spontaneously in air without an ignition source?
8. Name four or five hazardous products of combustion and specify the hazards posed by these materials.
9. What kind of property tends to be imparted to a functional group of an organic compound containing both oxygen and nitrogen?
10. Match the corrosive substance from the column on the left, below, with one of its major properties from the right column:

1. Alkali metal hydroxides	A. Oxidizer
2. Hydrogen peroxide	B. Strong bases
3. Hydrofluoric acid, HF	C. Dissolves glass
4. Nitric acid, HNO_3	D. Reacts with tissue to form yellow xanthoproteic acid

11. Rank the following wastes in increasing order of segregation: (A) mixed halogenated and hydrocarbon solvents containing little water, (B) spent steel pickling liquor, (C) dilute sludge consisting of mixed organic and inorganic wastes, (D) spent hydrocarbon solvents free of halogenated materials, (E) dilute mixed inorganic sludge.
12. Inorganic species may be divided into three major groups based upon their retention by clays. What are the elements commonly listed in these groups? What is the chemical basis for this division? How might anions (Cl^-, NO_3^-) be classified?
13. In what form would a large quantity of hazardous waste PCB likely be found in the hydrosphere?
14. The TCLP was originally devised to mimic a "mismanagement scenario" in which hazardous wastes were disposed along with biodegradable organic municipal refuse. Discuss how this procedure reflects the conditions that might arise from circumstances in which hazardous wastes and actively decaying municipal refuse were disposed together.
15. What are three major properties of wastes that determine their amenability to transport?
16. List and discuss the significance of major sources for the origin of hazardous wastes, that is, their main modes of entry into the environment. What are the relative dangers posed by each of these? Which part of the environment would each be most likely to contaminate?
17. What is the influence of organic solvents in leachates upon attenuation of organic hazardous waste constituents?
18. What features or characteristics should a compound possess in order for direct photolysis to be a significant factor in its removal from the atmosphere?

19. Describe the particular danger posed by codisposal of strong chelating agents with radionuclide wastes. What may be said about the chemical nature of the latter with regard to this danger?
20. Describe a beneficial effect that might result from the precipitation of either $Fe_2O_3 \cdot xH_2O$ or $MnO_2 \cdot xH_2O$ from hazardous wastes in water.
21. Why are secondary air pollutants from hazardous waste sites usually of only limited concern as compared to primary air pollutants? What is the distinction between the two?
22. Match the following physical, chemical, and biochemical processes dealing with the transformations and ultimate fates of hazardous chemical species in the hydrosphere on the left with the description of the process on the right, below:

1. Precipitation reactions	A. Molecule is cleaved with the addition of H_2O
2. Biochemical processes	B. Often involve hydrolysis and oxidation–reduction
3. Oxidation–reduction	C. By sediments and by suspended matter
4. Hydrolysis reactions	D. Generally mediated by microorganisms
5. Sorption	E. Generally accompanied by aggregation of colloidal particles suspended in water

23. As applied to hazardous wastes in the biosphere, distinguish among biodegradation, biotransformation, detoxification, and mineralization.
24. What is the potential role of *Phanerochaete chrysosporium* (white rot fungus) in treatment of hazardous waste compounds? For which kinds of compounds might it be most useful?
25. Which part of the hydrosphere is most subject to long-term, largely irreversible contamination from the improper disposal of hazardous wastes in the environment?
26. Several physical and chemical characteristics are involved in determining the amenability of a hazardous waste compound to biodegradation. These include hydrophobicity, solubility, volatility, and affinity for lipids. Suggest and discuss ways in which each one of these factors might affect biodegradability.
27. List and discuss some of the important processes determining the transformations and ultimate fates of hazardous chemical species in the hydrosphere.

21 Industrial Ecology for Waste Minimization, Utilization, and Treatment

21.1 INTRODUCTION

Chapter 20 dealt with the nature and production of hazardous wastes. Clearly, there have been severe hazardous waste problems in the United States and throughout the world. Since the 1970s, much has been done to reduce and clean up hazardous wastes. Legislation dealing with wastes has been passed, regulations have been put forth and refined, and numerous waste sites have been characterized and treated. Many of the financial resources expended on hazardous wastes have been devoted to litigation in attempts to establish the identities and contributions of various parties to waste problems. This chapter discusses how environmental chemistry, industrial ecology, and green chemistry can be applied to hazardous waste management to develop measures by which chemical wastes can be minimized, recycled, treated, and disposed. In descending order of desirability, hazardous waste management attempts to accomplish the following:

- Do not produce it
- If making it cannot be avoided, produce only minimum quantities
- Recycle it
- If it is produced and cannot be recycled, treat it, preferably in a way that makes it nonhazardous
- If it cannot be rendered nonhazardous, dispose of it in a safe manner
- Once it is disposed, monitor it for leaching and other adverse effects

The *effectiveness* of a hazardous waste management system is a measure of how well it reduces the quantities and hazards of wastes. As shown in Figure 21.1, the best management option consists of measures that prevent generation of wastes. Next in order of desirability is recovery and recycling of waste constituents. Next is destruction and treatment with conversion to nonhazardous waste forms. The least desirable option is disposal of hazardous materials in storage or landfill.

21.2 WASTE REDUCTION AND MINIMIZATION

During recent years, substantial efforts have been made in reducing the quantities of wastes and, therefore, the burden of dealing with wastes. Much of this effort has been the result of legislation and regulations restricting wastes, along with the resulting concerns over possible legal actions and lawsuits. In many cases—and ideally in all—minimizing the quantities of wastes produced is simply good business. Wastes are materials, materials have value and, therefore, all materials should be used for some beneficial purpose and not discarded as wastes, usually at a high cost for waste disposal.

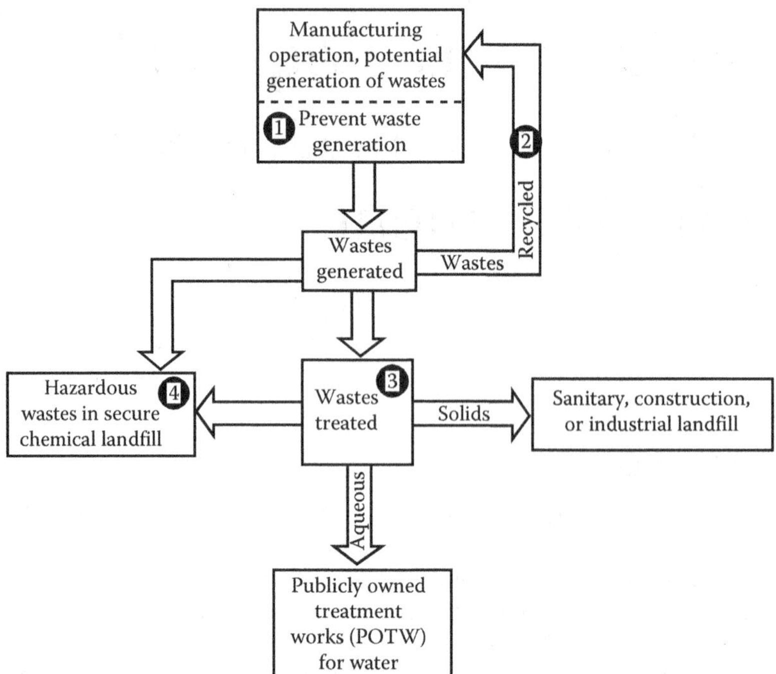

FIGURE 21.1 Order of effectiveness of waste treatment management options. The darkened circles indicate the degree of effectiveness from the most desirable (1) to the least (4).

Industrial ecology is all about the efficient use of materials. Therefore, by its nature, a system of industrial ecology is also a system of waste reduction and minimization.[1] In reducing quantities of wastes, it is important to take the broadest possible view. This is because dealing with one waste problem in isolation may simply create another. Early efforts to control air and water pollution resulted in problems from hazardous wastes isolated from industrial operations. A key aspect of industrial ecology is its approach based upon industrial systems as a whole, making such systems the best means of dealing with wastes by avoiding their production.

Many hazardous waste problems can be avoided at early stages by *waste reduction* (cutting down quantities of wastes from their sources) and *waste minimization* (utilization of treatment processes that reduce the quantities of wastes requiring ultimate disposal).[2]

Ways in which quantities of wastes can be reduced include source reduction, waste separation and concentration, resource recovery, and waste recycling. The most effective approaches to minimizing wastes center around careful control of manufacturing processes, taking into consideration discharges and the potential for waste minimization at every step of manufacturing. Viewing the process as a whole (as outlined for a generalized chemical manufacturing process in Figure 21.2) often enables crucial identification of the source of a waste, such as a raw material impurity, catalyst, or process solvent. Once a source is identified, it is much easier to take measures to eliminate or reduce the waste. The most effective approach to minimizing wastes is to emphasize waste minimization as an integral part of plant design.

Modifications of the manufacturing process can yield substantial waste reduction. Some such modifications are of a chemical nature. Changes in chemical reaction conditions can minimize production of by-product hazardous substances. In some cases, potentially hazardous catalysts, such as those formulated from toxic substances, can be replaced by catalysts that are nonhazardous or that can be recycled rather than discarded. Wastes can be minimized by volume reduction, for example, through dewatering and drying sludge.

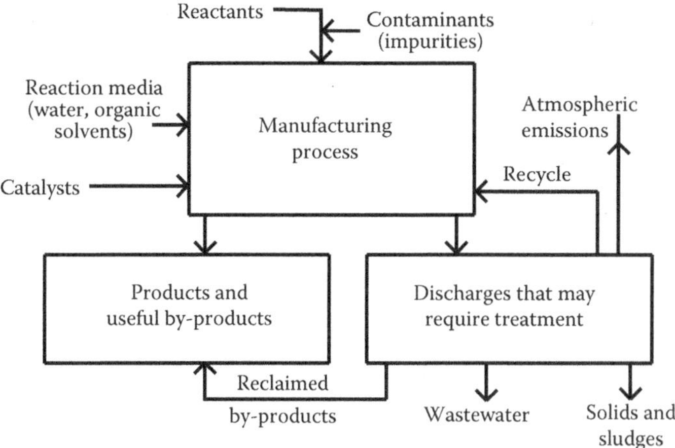

FIGURE 21.2 Chemical manufacturing process from the viewpoint of discharges and waste minimization.

Many kinds of waste streams are candidates for minimization. As examples, such waste streams identified at U.S. Government federal facilities have included solvents used for cleaning and degreasing, spent motor oil from gasoline and diesel engines, leftover and waste paint thinners, antifreeze/antiboil engine cooling formulations, batteries, inks, exposed photographic film, and pathology wastes. The sources of the wastes are as varied as the waste streams themselves. Motor pool maintenance garages generate used motor oil and spent coolants. Hospitals, clinics, and medical laboratories generate pathology wastes. Aircraft maintenance depots where aircraft and their parts are cleaned, chemically stripped of paint and coatings, repainted, and electroplated generate large quantities of effluents including organic materials. Other facilities generating wastes include equipment and arms maintenance facilities, photo developing and printing laboratories, paint shops, and arts and crafts shops.

A crucial part of the process for reducing and minimizing wastes is the development of a material balance, which is an integral part of the practice of industrial ecology. Such a balance addresses various aspects of waste streams, including sources, identification, and quantities of wastes and methods and costs of handling, treatment, recycling, and disposal. Priority waste streams can then be subjected to detailed process investigations to obtain the information needed to reduce wastes.

There are encouraging signs of progress in the area of waste minimization. All major companies have initiated programs to minimize quantities of wastes produced. Typically, more than 97% of oil-based petroleum refinery waste sludges that used to be discarded to landfill are now subjected to coking to yield useful hydrocarbon liquids and gases and coke, a solid carbon material with commercial value. Similar successes have been achieved with what were once waste products in a number of industries.

21.3 RECYCLING

Wherever possible, recycling and reuse should be accomplished on-site because it avoids having to move wastes and because a process that produces recyclable materials is often the most likely to have use for them. The four broad areas in which something of value may be obtained from wastes are the following:

- Direct recycle as raw material to the generator, as with the return to feedstock of raw materials not completely consumed in a synthesis process
- Transfer as a raw material to another process; a substance that is a waste product from one process may serve as a raw material for another, sometimes in an entirely different industry

- Utilization for pollution control or waste treatment, such as use of waste alkali to neutralize waste acid
- Recovery of energy, for example, from the incineration of combustible hazardous wastes

21.3.1 Examples of Recycling

Recycling of scrap industrial impurities and products occurs on a large scale with a number of different materials. Most of these materials are not hazardous but, as with most large-scale industrial operations, their recycling may involve the use or production of hazardous substances. Some of the more important examples are the following:

- *Ferrous metals* composed primarily of iron and used largely as feedstock for electric-arc furnaces
- *Nonferrous metals*, including aluminum (which ranks next to iron in terms of quantities recycled), copper and copper alloys, zinc, lead, cadmium, tin, silver, and mercury
- *Metal compounds*, such as metal salts
- *Inorganic substances* including alkaline compounds (such as sodium hydroxide used to remove sulfur compounds from petroleum products), acids (steel pickling liquor where impurities permit reuse), and salts (e.g., ammonium sulfate from coal coking used as fertilizer)
- *Glass*, which makes up about 10% of municipal refuse
- *Paper*, commonly recycled from municipal refuse
- *Plastic*, consisting of a variety of moldable polymeric materials and composing a major constituent of municipal wastes
- *Rubber*
- *Organic substances*, especially solvents and oils, such as hydraulic and lubricating oils
- *Catalysts* from chemical synthesis or petroleum processing
- Materials with *agricultural uses*, such as waste lime or phosphate-containing sludges used to treat and fertilize acidic soils

21.3.2 Waste Oil Utilization and Recovery

Waste oil generated from lubricants and hydraulic fluids is one of the more commonly recycled materials. Annual production of waste oil in the United States is of the order of 4 billion liters. Around half of this amount is burned as fuel and lesser quantities are recycled or disposed as waste. The collection, recycling, treatment, and disposal of waste oil are all complicated by the fact that it comes from diverse, widely dispersed sources and contains several classes of potentially hazardous contaminants. These are divided between organic constituents (PAHs, chlorinated hydrocarbons) and inorganic constituents (aluminum, chromium, and iron from wear of metal parts; barium and zinc from oil additives; lead from leaded gasoline).

21.3.2.1 Recycling Waste Oil

The processes used to convert waste oil to a feedstock hydrocarbon liquid for lubricant formulation are illustrated in Figure 21.3. The first of these uses distillation to remove water and light ends that have come from condensation and contaminant fuel. The second, or processing, step may be a vacuum distillation in which the three products are oil for further processing, a fuel oil cut, and a heavy residue. The processing step may also employ treatment with a mixture of solvents including isopropyl and butyl alcohols and methylethyl ketone to dissolve the oil and leave contaminants as a sludge; or contact with sulfuric acid to remove inorganic contaminants followed by treatment with clay to take out acid and contaminants that cause odor and color. The third step shown in Figure 21.3 employs vacuum distillation to separate lubricating oil stocks from a fuel fraction and heavy residue. This phase of treatment may also involve hydrofinishing, treatment with clay, and filtration.

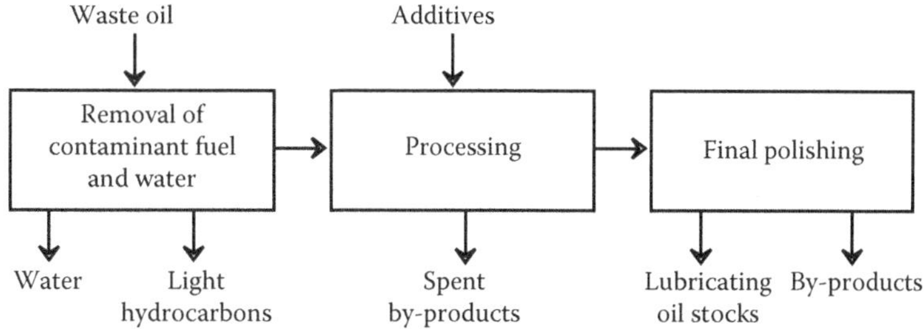

FIGURE 21.3 Major steps in reprocessing waste oil.

21.3.2.2 Waste Oil Fuel

For economic reasons, waste oil that is to be used for fuel is given minimal treatment of a physical nature, including settling, removal of water, and filtration. Metals in waste fuel oil become highly concentrated in its fly ash, which may be hazardous.

21.3.3 Waste Solvent Recovery and Recycle

The recovery and recycling of waste solvents has some similarities to the recycling of waste oil and is also an important enterprise. Among the many solvents listed as hazardous wastes and recoverable from wastes are dichloromethane, tetrachloroethylene, trichloroethylene, 1,1,1-trichloroethane, benzene, liquid alkanes, 2-nitropropane, methylisobutyl ketone, and cyclohexanone. For reasons of both economics and pollution control, many industrial processes that use solvents are equipped for solvent recycle. The basic scheme for solvent reclamation and reuse is shown in Figure 21.4.

Because of their impact on material use and environmental effects, solvents are given a high priority in the practice of green chemistry. A number of operations are used in solvent recovery and purification. Entrained solids are removed by settling, filtration, or centrifugation. Drying agents may be used to remove water from solvents, and various adsorption techniques and chemical treatment may be required to free the solvent from specific impurities. Fractional distillation, often requiring several distillation steps, is the most important operation in solvent purification and recycling. It is used to separate solvents from impurities, water, and other solvents.

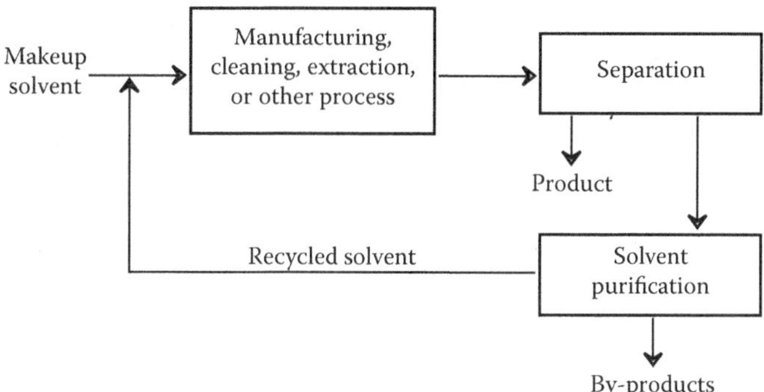

FIGURE 21.4 Overall process for recycling solvents.

21.3.4 Recovery of Water from Wastewater

It is often desirable to reclaim water from wastewater. This is especially true in regions where water is in short supply. Even where water is abundant, water recycling is desirable to minimize the amount of water that is discharged.

A little more than half of the water used in the United States is consumed by agriculture, primarily for irrigation. Steam-generating power plants consume about one-fourth of the water, and other uses, including manufacturing and domestic uses, account for the remainder.

The three major manufacturing consumers of water are chemicals and allied products, paper and allied products, and primary metals. These industries use water for cooling, processing, and boilers. Their potential for water reuse is high and their total consumption of water is projected to drop in future years as recycling becomes more common.

The degree of treatment required for reuse of wastewater depends upon its application. Water used for industrial quenching and washing usually requires the least treatment, and wastewater from some other processes may be suitable for these purposes without additional treatment. At the other end of the scale, boiler makeup water, potable (drinking) water, water used to directly recharge aquifers, and water that people will directly contact (in boating, water skiing, and similar activities) must be of very high quality.

The treatment processes applied to wastewater for reuse and recycle depend upon both the characteristics of the wastewater and its intended uses. Solids can be removed by sedimentation and filtration. BOD is reduced by biological treatment, including trickling filters and activated sludge treatment. For uses conducive to the growth of nuisance algae, nutrients may have to be removed. The easiest of these to handle is nutrient phosphate, which can be precipitated with lime. Nitrogen can be removed by denitrification processes.

Two of the major problems with industrial water recycling are heavy metals and dissolved toxic organic species. Heavy metals may be removed by ion exchange or precipitation by base or sulfide. The organic species are usually removed with activated carbon filtration. Some organic species are biologically degraded by bacteria in biological wastewater treatment.

One of the greater sources of potentially hazardous wastewater arises from oil/water separators at wash racks where manufactured parts and materials are washed. Because of the use of surfactants and solvents in the washwater, the separated water tends to contain emulsified oil that is incompletely separated in an oil/water separator. In addition, the sludge that accumulates at the bottom of the separator may contain hazardous constituents, including heavy metals and some hazardous organic constituents. Several measures that include the incorporation of good industrial ecology practice may be taken to eliminate these problems. One such measure is to eliminate the use of surfactants and solvents that tend to contaminate the water, and to use surfactants and solvents that are more amenable to separation and treatment. Another useful measure is to treat the water to remove harmful constituents and recycle it. This not only conserves water and reduces disposal costs, but also enables recycling of surfactants and other additives.

The ultimate water quality is achieved by processes that remove solutes from water, leaving pure H_2O. A combination of activated carbon treatment to remove organics, cation exchange to remove dissolved cations, and anion exchange for dissolved anions can provide very high quality water from wastewater. Reverse osmosis (see Chapter 8) can accomplish the same objective. However, these processes generate spent activated carbon, ion-exchange resins that require regeneration, and concentrated brines (from reverse osmosis) that require disposal, all of which have the potential to end up as hazardous wastes.

21.4 PHYSICAL METHODS OF WASTE TREATMENT

This section addresses predominantly physical methods for waste treatment and the following section addresses methods that utilize chemical processes. It should be kept in mind that most waste

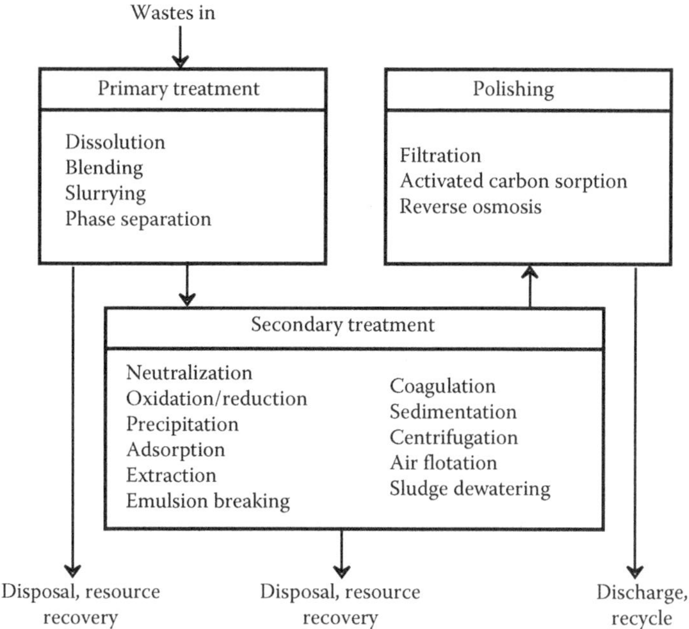

FIGURE 21.5 Major phases of waste treatment.

treatment measures have both physical and chemical aspects. The appropriate treatment technology for hazardous wastes obviously depends upon the nature of the wastes, which may vary in physical form, density, chemical composition, solubility in water, solubility in organic solvents, volatility, and other characteristics.

As shown in Figure 21.5, waste treatment may occur at three major levels—*primary*, *secondary*, and *polishing*—somewhat analogous to the treatment of wastewater (see Chapter 8). Primary treatment is usually preparation for further treatment, although it can result in the removal of by-products and reduction of the quantity and hazard of the waste. Secondary treatment detoxifies, destroys, and removes hazardous constituents. Polishing usually refers to treatment of water removed from wastes before discharge. The term may also apply to the treatment of other products so that they may be safely discharged or recycled.

21.4.1 Methods of Physical Treatment

Knowledge of the physical behavior of wastes has been used to develop various unit operations for waste treatment that are based upon physical properties. These operations include the following:

- Phase separation:
  - Filtration
  - Sedimentation
  - Flotation
- Phase transfer:
  - Extraction
  - Sorption
- Phase transition:
  - Distillation
  - Evaporation
  - Physical precipitation

- Membrane separations:
  - Reverse osmosis
  - Hyper- and ultrafiltration

21.4.1.1 Phase Separations

The most straightforward means of physical treatment involves separation of components of a mixture that are already in two different phases. *Sedimentation* and *decanting* are easily accomplished with simple equipment. In many cases, the separation must be aided by mechanical means, particularly *filtration* or *centrifugation*. *Flotation* is used to bring suspended organic matter or finely divided particles to the surface of a suspension. In the process of *dissolved air flotation* (DAF), air is dissolved in the suspending medium under pressure and comes out of solution when the pressure is released as minute air bubbles attached to suspended particles, which causes the particles to float to the surface.

An important and often difficult waste treatment step is *emulsion breaking* in which colloidal-sized *emulsions* are caused to aggregate and settle from suspension. Agitation, heat, acid, and the addition of *coagulants* consisting of organic polyelectrolytes, or inorganic substances such as an aluminum salt, may be used to cause the particles to stick together and settle out.

21.4.1.2 Phase Transition

A second major class of physical separation is that of *phase transition* in which a material changes from one physical phase to another. It is best exemplified by *distillation*, which is used in treating and recycling solvents, waste oil, aqueous phenolic wastes, xylene contaminated with paraffin from histological laboratories, and mixtures of ethylbenzene and styrene. Distillation produces *distillation bottoms* (still bottoms), which are often hazardous and polluting. These consist of unevaporated solids, semisolid tars, and sludges from distillation. Specific examples are distillation bottoms from the production of acetaldehyde from ethylene (hazardous waste number K009) and still bottoms from toluene reclamation distillation in the production of disulfoton (K036). The landfill disposal of these and other hazardous distillation bottoms, once widely practiced, is now severely limited.

Evaporation is usually employed to remove water from an aqueous waste to concentrate it. A special case of this technique is *thin-film evaporation* in which volatile constituents are removed by heating a thin layer of liquid or sludge waste spread on a heated surface.

Drying is removal of solvent or water from a solid or semisolid (sludge) or the removal of solvent from a liquid or suspension—is a very important operation because water is often the major constituent of waste products, such as sludges obtained from emulsion breaking. In *freeze drying*, the solvent, usually water, is sublimed from a frozen material. Hazardous waste solids and sludges are dried to reduce the quantity of waste, to remove solvent or water that might interfere with subsequent treatment processes, and to remove hazardous volatile constituents. Dewatering can often be improved with addition of a filter aid, such as diatomaceous Earth, during the filtration step.

Stripping is a means of separating volatile components from less volatile ones in a liquid mixture by the partitioning of the more volatile materials to a gas phase of air or steam (steam stripping). The gas phase is introduced into the aqueous solution or suspension containing the waste in a stripping tower that is equipped with trays or packed to provide maximum turbulence and contact between the liquid and gas phases. The two major products are condensed vapor and a stripped bottoms residue. Examples of two volatile components that can be removed from water by air stripping are benzene and dichloromethane. Air stripping can also be used to remove ammonia from water that has been treated with a base to convert ammonium ion to volatile ammonia.

Physical precipitation is used here as a term to describe processes in which a solid forms from a solute in solution as a result of a physical change in the solution, as compared to chemical precipitation (see Section 21.5) in which a chemical reaction in solution produces an insoluble material. Cooling the solution, evaporation of solvent, or alteration of solvent composition may cause physical precipitation. The most common type of physical precipitation by solvent alteration occurs when a water-miscible organic solvent is added to an aqueous solution so that the solubility of a salt is lowered below its concentration in the solution.

21.4.1.3 Phase Transfer

Phase transfer consists of the transfer of a solute in a mixture from one phase to another. An important type of phase transfer process is *solvent extraction* in which a substance is transferred from solution in one solvent (usually water) to another (usually an organic solvent) without any chemical change. Solvents may be used for *leaching* substances from solids or sludges. Solvent extraction and the major terms applicable to it are summarized in Figure 21.6. The same terms and general principles apply to leaching. A major application of solvent extraction to waste treatment has been in the removal of phenol from by-product water produced in coal coking, petroleum refining, and chemical syntheses that involve phenol.

One of the more promising approaches to solvent extraction and leaching of hazardous wastes is the use of *supercritical fluids*, most commonly CO_2, as extraction solvents. A supercritical fluid is one that has characteristics of both liquid and gas and consists of a substance above its critical temperature and pressure (31.1°C and 73.8 atm, respectively, for CO_2). After a substance has been extracted from a waste into a supercritical fluid at high pressure, the pressure can be released, resulting in separation of the substance extracted. The fluid can then be compressed again and recirculated through the extraction system. Some possibilities for treatment of hazardous wastes by extraction with supercritical CO_2 include removal of organic contaminants from wastewater, extraction of organohalide pesticides from soil, extraction of oil from emulsions used in aluminum and steel processing, and regeneration of spent activated carbon. Waste oils contaminated with PCBs, metals, and water can be purified using supercritical ethane.

Transfer of a substance from a solution to a solid phase is called *sorption*. The most important sorbent is *activated carbon*, which, in some cases, is used for complete treatment of hazardous wastes as well as pretreatment of waste streams going into processes such as reverse osmosis to improve treatment efficiency and reduce fouling. Effluents from other treatment processes, such as biological treatment of degradable organic solutes in water, can be polished with activated carbon. Activated carbon sorption is most effective for removing from water those hazardous waste materials that are poorly water-soluble and that have high molar masses, such as xylene, naphthalene (U165), chlorinated hydrocarbons, phenol (U188), and aniline. Activated carbon does not work well for organic compounds that are highly water-soluble or polar.

Solids other than activated carbon can be used for sorption of contaminants from liquid wastes. These include synthetic resins composed of organic polymers and mineral substances. Of the latter, clay is employed to remove impurities from waste lubricating oils in some oil recycling processes.

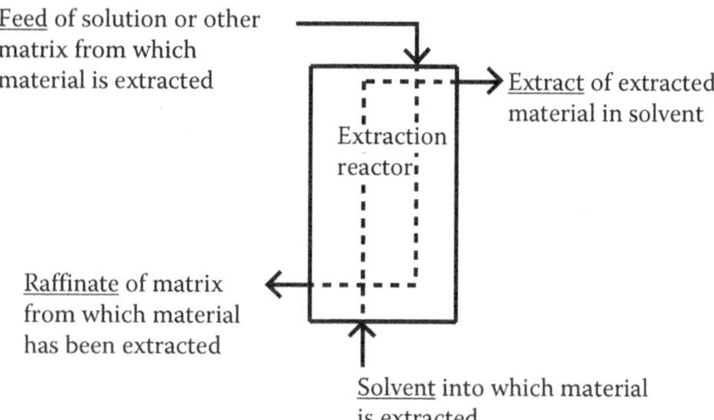

FIGURE 21.6 Outline of solvent extraction/leaching process with important terms underlined.

21.4.1.4 Molecular Separation

A third major class of physical separation is *molecular separation*, often based upon *membrane processes* in which dissolved contaminants or solvent pass through a size-selective membrane under pressure. The products are a relatively pure solvent phase (usually water) and a concentrate enriched in the solute impurities. The common membrane processes are discussed in Section 8.6 and summarized in Table 8.1. Ultrafiltration and hyperfiltration are especially useful for concentrating suspended oil, grease, and fine solids in water. They also serve to concentrate solutions of large organic molecules and heavy metal ion complexes.

21.5 CHEMICAL TREATMENT: AN OVERVIEW

The applicability of chemical treatment to wastes depends upon the chemical properties of the waste constituents, particularly acid–base, oxidation–reduction, precipitation, and complexation behavior; reactivity; flammability/combustibility; corrosivity; and compatibility with other wastes. The major unit operations for chemical waste treatment are the following:

- Acid/base neutralization
- Chemical extraction and leaching
- Ion exchange
- Chemical precipitation
- Oxidation
- Reduction

Some of the more sophisticated means available for treatment of wastes have been developed for pesticide disposal.

21.5.1 Acid–Base Neutralization

Waste acids and bases are treated by *neutralization*:

$$H^+ + OH^- \rightarrow H_2O \tag{21.1}$$

Although simple in principle, neutralization can present some problems in practice. These include evolution of volatile contaminants, mobilization of soluble substances, excessive heat generated by the neutralization reaction, and corrosion to apparatus. By adding too much or too little of the neutralizing agent, it is possible to get a product that is too acidic or basic. Lime, $Ca(OH)_2$, is widely used as a base for treating acidic wastes. Because of lime's limited solubility, solutions of excess lime do not reach extremely high pH values. Sulfuric acid, H_2SO_4, is a relatively inexpensive acid for treating alkaline wastes. However, addition of too much sulfuric acid can produce highly acidic products; for some applications, acetic acid, CH_3COOH, is preferable. As noted above, acetic acid is a weak acid and an excess of it does little harm. It is also a natural product and biodegradable.

Neutralization, or pH adjustment, is often required prior to the application of other waste treatment processes. Processes that may require neutralization include oxidation/reduction, activated carbon sorption, wet air oxidation, stripping, and ion exchange. Microorganisms usually require a pH in the range of 6–9, so neutralization may be required prior to biochemical treatment.

21.5.1.1 Acid Recovery

Rather than neutralizing acid, in some cases, it may be recovered for its economic value. Acid recovery is in keeping with the best practice of green chemistry. An example of acid recovery is the use of trioctyl phosphate to recover acetic acid and nitric acid from a waste acid mixture of these acids and H_3PO_4 produced in some processes such as etching in the production of liquid crystals.

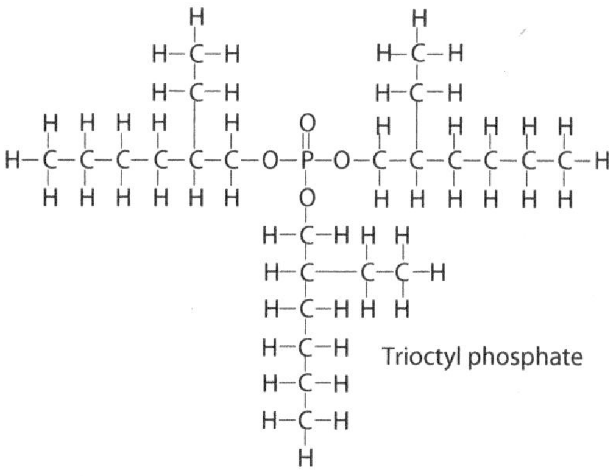

```
                     H                 H
                   H-C-H             H-C-H
                   H-C-H             H-C-H
   H  H  H  H    |  H     O     H  |  H  H  H  H
 H-C--C--C--C--C--C--O-P-O--C--C--C--C--C--C-H
   H  H  H  H  H  H     O     H  H  H  H  H  H
                      H-C-H H  H
                      H-C----C--C-H
                      H-C-H H  H
                      H-C-H      Trioctyl phosphate
                      H-C-H
                      H-C-H
                        H
```

21.5.2 Chemical Precipitation

Chemical precipitation is used in hazardous waste treatment primarily for the removal of heavy metal ions from water as shown below for the chemical precipitation of cadmium:

$$Cd^{2+}(aq) + HS^{-}(aq) \longrightarrow CdS(s) + H^{+}(aq) \tag{21.2}$$

21.5.2.1 Precipitation of Metals

The most widely used means of precipitating metal ions is by the formation of hydroxides such as chromium(III) hydroxide:

$$Cr^{3+} + 3OH^{-} \longrightarrow Cr(OH)_3 \tag{21.3}$$

The source of hydroxide ion is a base (alkali), such as lime [$Ca(OH)_2$], sodium hydroxide (NaOH), or sodium carbonate (Na_2CO_3). Most metal ions tend to produce basic salt precipitates, such as basic copper(II) sulfate, $CuSO_4 \cdot 3Cu(OH)_2$, which is formed as a solid when hydroxide is added to a solution containing Cu^{2+} and SO_4^{2-} ions. The solubilities of many heavy metal hydroxides reach a minimum value, often at a pH in the range of 9–11, then increase with increasing pH values due to the formation of soluble hydroxo complexes, as illustrated by the following reaction:

$$Zn(OH)_2(s) + 2OH^{-}(aq) \longrightarrow Zn(OH)_4^{2-}(aq) \tag{21.4}$$

The chemical precipitation method that is used most is precipitation of metals as hydroxides and basic salts with lime. Sodium carbonate can be used to precipitate hydroxides [$Fe(OH)_3 \cdot xH_2O$], carbonates ($CdCO_3$), or basic carbonate salts [$2PbCO_3 \cdot Pb(OH)_2$]. The carbonate anion produces hydroxide by virtue of its hydrolysis reaction with water:

$$CO_3^{2-} + H_2O \longrightarrow HCO_3^{-} + OH^{-} \tag{21.5}$$

Carbonate, alone, does not give as high a pH as do alkali metal hydroxides, which may have to be used to precipitate metals that form hydroxides only at relatively high pH values.

The solubilities of some heavy metal sulfides are extremely low, so precipitation by H_2S or other sulfides (see Reaction 21.2) can be a very effective means of treatment. Hydrogen sulfide is a toxic gas that is itself considered to be a hazardous waste (U135). Iron(II) sulfide (ferrous sulfide) can be used as a safe source of sulfide ion to produce sulfide precipitates with other metals that are less soluble than FeS. However, toxic H_2S can be produced when metal sulfide wastes contact acid:

$$MS + 2H^{+} \longrightarrow M^{2+} + H_2S \tag{21.6}$$

Some metals can be precipitated from solution in the elemental metal form by the action of a reducing agent such as sodium borohydride:

$$4Cu^{2+} + NaBH_4 + 2H_2O \rightarrow 4Cu + NaBO_2 + 8H^+ \quad (21.7)$$

or with more active metals in a process called *cementation*:

$$Cd^{2+} + Zn \rightarrow Cd + Zn^{2+} \quad (21.8)$$

Regardless of the method used to precipitate a metal, the form of the metal in the waste solution can be an important consideration. Chelated metals can be especially difficult to remove.

21.5.2.2 Coprecipitation of Metals

In some cases, advantage may be taken of the phenomenon of coprecipitation to remove metals from wastes. A good example of this application is the coprecipitation of lead from battery industry wastewater with iron(III) hydroxide. Raising the pH of such a wastewater consisting of dilute sulfuric acid and contaminated with Pb^{2+} ion precipitates lead as several species, including $PbSO_4$, $Pb(OH)_2$, and $Pb(OH)_2 \cdot 2PbCO_3$. In the presence of iron(III), gelatinous $Fe(OH)_3$ forms, which coprecipitates the lead, leading to significantly greater lead removal, typically to below 0.2 ppm.

21.5.3 Oxidation–Reduction

As shown by the reactions in Table 21.1, *oxidation* and *reduction* can be used for the treatment and removal of a variety of inorganic and organic wastes. Some waste oxidants can be used to treat oxidizable wastes in water and cyanides. Ozone, O_3, is a strong oxidant that can be generated on-site by an electrical discharge through dry air or oxygen. Ozone employed as an oxidant gas at levels of 1–2 wt% in air and 2–5 wt% in oxygen has been used to treat a large variety of oxidizable contaminants, effluents, and wastes including wastewater and sludges containing oxidizable constituents. It is also used to partially oxidize organic waste solutes in water to make them more amenable to biological treatment.

Fenton oxidation is also used to oxidize organic substances in water, often as a final polishing step prior to release of the treated wastewater. The Fenton process entails adding iron(II) salts, such

TABLE 21.1
Oxidation–Reduction Reactions Used to Treat Wastes

Waste Substance	Reaction with Oxidant or Reductant
Oxidation of Organics	
Organic matter, $\{CH_2O\}$	$\{CH_2O\} + 2\{O\} \rightarrow CO_2 + H_2O$
Aldehyde	$CH_3CHO + \{O\} \rightarrow CH_3COOH$ (acid)
Oxidation of Inorganics	
Cyanide	$2CN^- + 5OCl^- + H_2O \rightarrow N_2 + 2HCO_3^- + 5Cl^-$
Iron(II)	$4Fe^{2+} + O_2 + 10H_2O \rightarrow 4Fe(OH)_3 + 8H^+$
Sulfur dioxide	$2SO_2 + O_2 + 2H_2O \rightarrow 2H_2SO_4$
Reduction of Inorganics	
Chromate	$2CrO_4^{2-} + 3SO_2 + 4H^+ \rightarrow Cr_2(SO_4)_3 + 2H_2O$
Permanganate	$MnO_4^- + 3Fe^{2+} + 7H_2O \rightarrow MnO_2(s) + 3Fe(OH)_3(s) + 5H^+$

as $FeSO_4$, acidification, addition of hydrogen peroxide, and acid neutralization. Hydrogen peroxide reacts with iron(II),

$$Fe(II) + H_2O_2 \rightarrow Fe(III) + OH^- + HO^\bullet \quad (4.77)$$

to generate highly reactive hydroxyl radical, HO•, which oxidizes even refractory organic materials. A disadvantage of Fenton treatment is the iron-containing sludge that is generated when the acidic reaction mixture is neutralized.

21.5.4 Electrolysis

As shown in Figure 21.7, *electrolysis* is a process in which one species in solution (usually a metal ion) is reduced by electrons at the *cathode* and another gives up electrons to the *anode* and is oxidized there. In hazardous waste applications, electrolysis is most widely used in the recovery of cadmium, copper, gold, lead, silver, and zinc. Metal recovery by electrolysis is made more difficult by the presence of cyanide ion, which stabilizes metals in solution as the cyanide complexes, such as $Ni(CN)_4^{2-}$.

Electrolytic removal of contaminants from solution can be by direct electrodeposition, particularly of reduced metals, and as the result of secondary reactions of electrolytically generated precipitating agents. A specific example of both is the electrolytic removal of both cadmium and nickel from wastewater contaminated by nickel/cadmium battery manufactured using fibrous carbon electrodes. At the cathode, cadmium is removed directly by reduction to the metal:

$$Cd^{2+} + 2e^- \rightarrow Cd \quad (21.9)$$

At relatively high cathodic potentials, hydroxide is formed by the electrolytic reduction of water:

$$2H_2O + 2e^- \rightarrow 2OH^- + H_2 \quad (21.10)$$

or by the reduction of molecular oxygen, if it is present:

$$2H_2O + O_2 + 4e^- \rightarrow 4OH^- \quad (21.11)$$

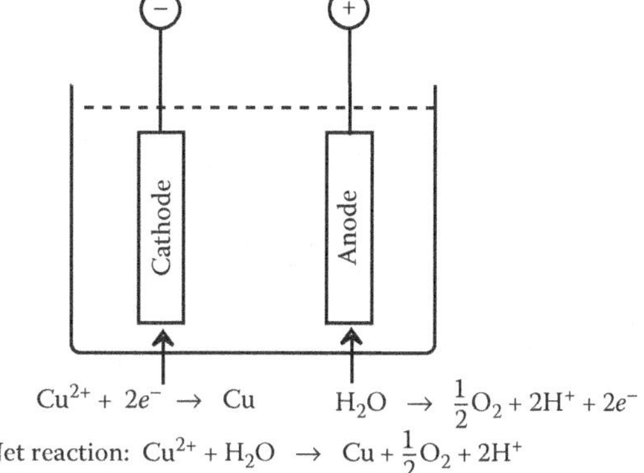

FIGURE 21.7 Electrolysis of copper solution.

If the localized pH at the cathode surface becomes sufficiently high, cadmium can be precipitated and removed as colloidal $Cd(OH)_2$. The direct electrodeposition of nickel is too slow to be significant, but it is precipitated as solid $Ni(OH)_2$ at pH values above 7.5 from OH^- generated at the cathode.

Cyanide, which is often present as an ingredient of electroplating baths with metals such as cadmium and nickel, can be removed by oxidation with electrolytically generated elemental chlorine at the anode. Chlorine is generated by the anodic oxidation of added chloride ion:

$$2Cl^- \rightarrow Cl_2 + 2e^- \tag{21.12}$$

The electrolytically generated chlorine then breaks down cyanide by a series of reactions for which the overall reaction is the following:

$$2CN^- + 5Cl_2 + 8OH^- \rightarrow 10Cl^- + N_2 + 2CO_2 + 4H_2O \tag{21.13}$$

21.5.5 Hydrolysis

One of the ways to dispose of chemicals that are reactive with water is to allow them to react with water under controlled conditions, a process called *hydrolysis*. Inorganic chemicals that can be treated by hydrolysis include metals that react with water; metal carbides, hydrides, amides, alkoxides, and halides; and nonmetal oxyhalides and sulfides. Examples of the treatment of these classes of inorganic species are given in Table 21.2.

Organic chemicals may also be treated by hydrolysis. For example, toxic acetic anhydride is hydrolyzed to relatively safe acetic acid:

$$\underset{\substack{\text{Acetic anhydride}\\\text{(an acid anhydride)}}}{CH_3-C(=O)-O-C(=O)-CH_3} + H_2O \rightarrow 2CH_3-C(=O)-OH \tag{21.14}$$

21.5.6 Chemical Extraction and Leaching

Chemical extraction or *leaching* in hazardous waste treatment is the removal of a hazardous constituent by chemical reaction with an extractant in solution. Poorly soluble heavy metal salts can be extracted by reaction of the salt anions with H^+ as illustrated by the following:

$$PbCO_3 + H^+ \rightarrow Pb^{2+} + HCO_3^- \tag{21.15}$$

TABLE 21.2
Inorganic Chemicals That May Be Treated by Hydrolysis

Class of Chemical	Reaction with Oxidant or Reductant
Active metals (calcium)	$Ca + 2H_2O \rightarrow H_2 + Ca(OH)_2$
Hydrides (sodium aluminum hydride)	$NaAlH_4 + 4H_2O \rightarrow 4H_2 + NaOH + Al(OH)_3$
Carbides (calcium carbide)	$CaC_2 + 2H_2O \rightarrow Ca(OH)_2 + C_2H_2$
Amides (sodium amide)	$NaNH_2 + H_2O \rightarrow NaOH + NH_3$
Halides (silicon tetrachloride)	$SiCl_4 + 2H_2O \rightarrow SiO_2 + 4HCl$
Alkoxides (sodium ethoxide)	$NaOC_2H_5 + H_2O \rightarrow NaOH + C_2H_5OH$

Acids also dissolve basic organic compounds such as amines and aniline. Extraction with acids should be avoided if cyanides or sulfides are present to prevent formation of toxic hydrogen cyanide or hydrogen sulfide. Nontoxic weak acids are usually the safest to use. These include acetic acid, CH_3COOH, and the acid salt, NaH_2PO_4.

Chelating agents, such as dissolved EDTA (HY^{3-}), dissolve insoluble metal salts by forming soluble species with metal ions:

$$FeS + HY^{3-} \rightarrow FeY^{2-} + HS^{-} \qquad (21.16)$$

Heavy metal ions in soil contaminated by hazardous wastes may be present in a coprecipitated form with insoluble iron(III) and manganese(IV) oxides, Fe_2O_3 and MnO_2, respectively. These oxides can be dissolved by reducing agents, such as solutions of sodium dithionate/citrate or hydroxylamine. This results in the production of soluble Fe^{2+} and Mn^{2+} and the release of heavy metal ions, such as Cd^{2+} or Ni^{2+}, which are removed with the water.

21.5.7 Ion Exchange

Ion exchange is a means of removing cations or anions from solution onto a solid resin, which can be regenerated by treatment with acids, bases, or salts. The greatest use of ion exchange in hazardous waste treatment is for the removal of low levels of heavy metal ions from wastewater:

$$2H^{+-}\{CatExchr\} + Cd^{2+} \rightarrow Cd^{2+-}\{CatExchr\}_2 + 2H^{+} \qquad (21.17)$$

Ion exchange is employed in the metal plating industry to purify rinsewater and spent plating bath solutions. Cation exchangers are used to remove cationic metal species, such as Cu^{2+}, from such solutions. Anion exchangers remove anionic cyanide metal complexes [e.g., $Ni(CN)_4^{2-}$] and chromium(VI) species, such as CrO_4^{2-}. Radionuclides may be removed from radioactive wastes and mixed waste by ion-exchange resins.

21.6 GREEN WASTE TREATMENT BY PHOTOLYSIS AND SONOLYSIS

Photolysis and sonolysis are discussed together because of some similarities in their modes of action and because they are both potentially green chemistry waste treatment processes that can be carried out without added reagents or with minimal quantities of relatively harmless reactants. Both provide means of introducing high energy into the treatment system (see Section 17.10); photolysis does so through the action of photons of ultraviolet light or even x-rays or gamma rays, whereas sonolysis does so by "phonons" of high frequency ultrasound. Both generate chemically reactive intermediate species that break down waste molecules.

Photolytic reactions were discussed in Chapter 9. *Photolysis* can be used to destroy a number of kinds of hazardous wastes. In such applications, it is most useful in breaking chemical bonds in refractory organic compounds. TCDD (see Section 7.11), one of the most troublesome and refractory of wastes, can be treated by ultraviolet light in the presence of hydrogen atom donors {H} resulting in reactions such as the following:

Cl Cl O Cl Cl O + $h\nu$ {H} $\longrightarrow$ H Cl O Cl Cl O (21.18)

As photolysis proceeds, the H–C bonds are broken, the C–O bonds are broken, and the final product is a harmless organic polymer.

An initial photolysis reaction can result in the generation of reactive intermediates that participate in *chain reactions* that lead to the destruction of a compound. By far, the most important reactive intermediate is free radical $HO^{\bullet}$. Solid titanium dioxide, TiO_2, can be used as a photocatalyst leading to chain reactions involving active radicals. When the surface of TiO_2 is irradiated with ultraviolet radiation, "holes" (h) are generated at sites where excited electrons (e^-) are produced:

$$TiO_2 + h\nu \rightarrow TiO_2(h + e^-) \qquad (21.19)$$

The electrons can react with dissolved O_2 to produce reactive O_2^-:

$$TiO_2(e^-) + O_2 \rightarrow O_2^- \qquad (21.20)$$

The surface holes may take electrons from dissolved hydroxyl ion to produce reactive hydroxyl radicals on the TiO_2 surface:

$$TiO_2(h) + OH^- \rightarrow HO^{\bullet} \qquad (21.21)$$

The hydroxyl radicals may then initiate reactions leading to the destruction of waste compounds or the TiO_2(h) sites may initiate reactions directly with waste molecules by abstracting electrons from them.

In the example just cited, solid TiO_2 acts as a *sensitizer* to absorb radiation and generate reactive species that destroy wastes. Other sensitizers may be added to solution to generate hydroxyl radical for waste treatment. The addition of a chemical oxidant, such as potassium peroxydisulfate, $K_2S_2O_8$, enhances destruction by oxidizing active photolytic products. One of the most common processes that entails generation of active hydroxyl radical is the *Fenton reaction* involving iron(II) and hydrogen peroxide (see Reaction 4.77). Sunlight impinging on water containing iron can generate both Fe(II) and H_2O_2 with the potential to use for waste treatment.

Hazardous waste substances other than TCDD that have been destroyed by photolysis are herbicides including atrazine, 2,4-D, 2-(2,4-dichlorophenoxy) propionic acid; 2,4-dichlorophenol; 2,4,6-trichlorophenol; TNT; and PCBs.

Sonolysis in water with ultrasound (typically 660 kHz frequency) generates reactive hydroxyl radical by the following reaction:

$$HOH \xrightarrow{\text{Ultrasound}} HO\bullet + H\bullet \qquad (21.22)$$

The hydroxyl radical undergoes reactions with waste compounds leading to their destruction. The waste solution may be saturated with O_2 gas to facilitate waste oxidation. Sonolysis may be combined with photolysis for more effective waste destruction.

21.7 THERMAL TREATMENT METHODS

Thermal treatment of hazardous wastes can be used to accomplish most of the common objectives of waste treatment—volume reduction; removal of volatile, combustible, mobile organic matter; and destruction of toxic and pathogenic materials. The most widely applied means of thermal treatment of hazardous wastes is *incineration*. Incineration utilizes high temperatures, an oxidizing atmosphere, and often turbulent combustion conditions to destroy wastes. Methods other than incineration that make use of high temperatures to destroy or neutralize hazardous wastes are discussed briefly at the end of this section.

21.7.1 Incineration

Hazardous waste incineration will be defined here as a process that involves exposure of the waste materials to oxidizing conditions at a high temperature, usually in excess of 900°C. Normally, the heat required for incineration comes from the oxidation of organically bound carbon and hydrogen contained in the waste material or in supplemental fuel:

$$\text{C(organic)} + O_2 \rightarrow CO_2 + \text{heat} \tag{21.23}$$

$$\text{4H(organic)} + O_2 \rightarrow 2H_2O + \text{heat} \tag{21.24}$$

These reactions destroy organic matter and generate heat required for endothermic reactions, such as the breaking of C–Cl bonds in organochlorine compounds.

21.7.1.1 Incinerable Wastes

Ideally, incinerable wastes are predominantly organic materials that will burn with a heating value of at least 5000 Btu/lb and preferably over 8000 Btu/lb. Such heating values are readily attained with wastes having high contents of the most commonly incinerated waste organic substances, including methanol, acetonitrile, toluene, ethanol, amyl acetate, acetone, xylene, methyl ethyl ketone, adipic acid, and ethyl acetate. In some cases, however, it is desirable to incinerate wastes that will not burn alone and which require *supplemental fuel*, such as methane and petroleum liquids. Examples of such wastes are nonflammable organochlorine wastes, some aqueous wastes, or soil in which the elimination of a particularly troublesome contaminant is worth the expense and trouble of incinerating it. Inorganic matter, water, and organic hetero element contents of liquid wastes are important in determining their incinerability.

21.7.2 Hazardous Waste Fuel

Many industrial wastes, including hazardous wastes, are burned as *hazardous waste fuel* for energy recovery in industrial furnaces and boilers and in incinerators for nonhazardous wastes, such as sewage sludge incinerators. This process is called *coincineration*, and more combustible wastes are utilized by it than are burned solely for the purpose of waste destruction. In addition to heat recovery from combustible wastes, it is a major advantage to use an existing on-site facility for waste disposal rather than a separate hazardous waste incinerator. One of the greatest applications of coincineration is use of hazardous waste liquids containing organochlorine solvents and other pollutants as fuel for the preparation of portland cement (see below).

21.7.3 Incineration Systems

The four major components of hazardous waste incineration systems are shown in Figure 21.8. *Waste preparation* for liquid wastes may require filtration, settling to remove solid material and water, blending to obtain the optimum incinerable mixture, or heating to decrease viscosity. Solids may require shredding and screening. Atomization is commonly used to feed liquid wastes. Several mechanical devices, such as rams and augers, are used to introduce solids into the incinerator.

The most common kinds of *combustion chambers* are liquid injection, fixed hearth, rotary kiln, and fluidized bed. These types are discussed in more detail later in this section.

Often the most complex part of a hazardous waste incineration system is the *air pollution control system*, which involves several operations. The most common operations in air pollution control from hazardous waste incinerators are combustion gas cooling, heat recovery, quenching, particulate matter removal, acid gas removal, and treatment and handling of by-product solids, sludges, and liquids.

Hot ash is often quenched in water. Prior to disposal it may require dewatering and chemical stabilization. A major consideration with hazardous waste incinerators and the types of wastes

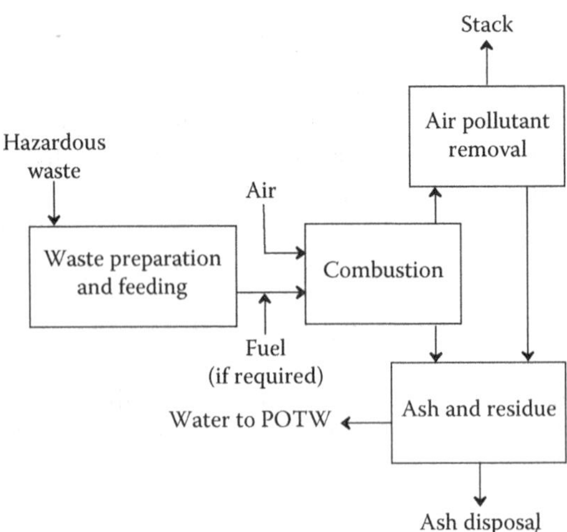

FIGURE 21.8 Major components of a hazardous waste incinerator system.

that are incinerated is the disposal problem posed by the ash, especially with respect to potential leaching of heavy metals.

21.7.4 Types of Incinerators

Hazardous waste incinerators may be divided among the following, based upon type of combustion chamber:

- *Rotary kiln* (about 40% of U.S. hazardous waste incinerator capacity) in which the primary combustion chamber is a rotating cylinder lined with refractory materials, and an afterburner downstream from the kiln to complete destruction of the wastes.
- *Liquid injection incinerators* (also about 40% of U.S. hazardous waste incinerator capacity) that burn pumpable liquid wastes dispersed as small droplets.
- *Fixed-hearth incinerators* with single or multiple hearths upon which combustion of liquid or solid wastes occurs.
- *Fluidized-bed incinerators* in which combustion of wastes is carried out on a bed of granular solid (such as limestone) held in a suspended (fluid-like) state by injection of air to remove pollutant acid gas and ash products.
- *Advanced design incinerators* including *plasma incinerators* that make use of an extremely hot plasma of ionized gas injected through an electrical arc; *electric reactors* that use resistance-heated incinerator walls at around 2200°C to heat and pyrolyze wastes by radiative heat transfer; *infrared systems* that generate intense infrared radiation by passing electricity through silicon carbide resistance heating elements; *molten salt combustion* that uses a bed of molten sodium carbonate at about 900°C to destroy the wastes and retain gaseous pollutants; and *molten glass processes* that use a pool of molten glass to transfer heat to the waste and to retain products in a poorly leachable glass form.

21.7.5 Combustion Conditions

The key to effective incineration of hazardous wastes lies in the combustion conditions. These require (1) sufficient free oxygen in the combustion zone to ensure combustion of the wastes; (2) turbulence for thorough mixing of waste, oxidant, and (in cases where the waste does not have sufficient fuel content to be burned alone) supplemental fuel; (3) high combustion temperatures above about 900°C to ensure that thermally resistant compounds do react; and (4) sufficient residence time (at least 2 s) to allow reactions to occur.

21.7.6 Effectiveness of Incineration

EPA standards for hazardous waste incineration are based upon the effectiveness of destruction of the *principal organic hazardous constituents* (POHCs). Measurement of these compounds before and after incineration gives the *destruction removal efficiency* (DRE) according to the formula

$$DRE = \frac{W_{in} - W_{out}}{W_{in}} \times 100 \quad (21.25)$$

where W_{in} and W_{out} are the mass flow rates of the POHC input and output (at the stack downstream from emission controls), respectively. U.S. EPA regulations call for destruction of 99.99% of POHCs and 99.9999% ("six nines") destruction of 2,3,7,8-tetrachlorodibenzo-*p*-dioxin, commonly called TCDD or "dioxin."

21.7.7 Wet Air Oxidation

Organic compounds and oxidizable inorganic species can be oxidized by oxygen in aqueous solution. The source of oxygen usually is air. Rather extreme conditions of temperature and pressure are required, with a temperature range of 175–327°C and a pressure range of 300–3000 psig (2070–20,700 kPa). The high pressures allow a high concentration of oxygen to be dissolved in the water and the high temperatures enable the reaction to occur.

Wet air oxidation has been applied to the destruction of cyanides in electroplating wastewaters. The oxidation reaction for sodium cyanide is the following:

$$2Na^+ + 2CN^- + O_2 + 4H_2O \rightarrow 2Na^+ + 2HCO_3^- + 2NH_3 \quad (21.26)$$

Organic wastes can be oxidized in supercritical water, taking advantage of the ability of supercritical fluids to dissolve organic compounds (see Section 21.4). Wastes are contacted with water and the mixture raised to a temperature and pressure required for supercritical conditions for water. Oxygen is then pumped in sufficient to oxidize the wastes. The process produces only small quantities of CO and no SO_2 or NO_x. It can be used to degrade PCBs, dioxins, organochlorine insecticides, benzene, urea, and numerous other materials.

21.7.8 UV-Enhanced Wet Oxidation

Hydrogen peroxide (H_2O_2) can be used as an oxidant in solution assisted by ultraviolet radiation ($h\nu$). For the oxidation of organic species represented in general as $\{CH_2O\}$, the overall reaction is

$$2H_2O_2 + \{CH_2O\} + h\nu \rightarrow CO_2 + 3H_2O \quad (21.27)$$

The ultraviolet radiation breaks chemical bonds and serves to form reactive oxidant species, such as $HO^{\bullet}$.

21.7.9 Destruction of Hazardous Wastes in Cement Manufacture

One of the most effective means of destroying hazardous wastes is to use them as supplemental fuel in the manufacture of portland cement. In this application, cement manufacture is an excellent example of the practice of industrial ecology, destroying a waste and utilizing it as fuel.[3] Portland cement is made by heating limestone, clay, and sand to a very high temperature such that the mineral mixture fuses, after which it is ground to the fine powder that composes cement. When a cement kiln is cofired with hazardous wastes, the very high temperatures, strongly oxidizing conditions, and relatively long residence times are ideal for destroying organic wastes. Furthermore, the basic solids in the charge that is fired to make cement act to sequester acid gases, such as HCl generated in the combustion of organohalide compounds. Collection of solids from the cement kiln flue prevents release of potential pollutant ash materials.

21.8 BIODEGRADATION OF WASTES

Biodegradation of wastes is a term that describes the conversion of wastes by enzymatic biological processes to simple inorganic molecules (mineralization) and, to a certain extent, to biological materials. Usually, the products of biodegradation are molecular forms that tend to occur in nature and that are in greater thermodynamic equilibrium with their surroundings than are the starting materials. *Detoxification* refers to the biological conversion of a toxic substance to a less toxic species. Microbial bacteria and fungi possessing enzyme systems required for biodegradation of wastes are usually best obtained from populations of indigenous microorganisms at a hazardous waste site where they have developed the ability to degrade particular kinds of molecules. Biological treatment offers a number of significant advantages and has considerable potential for the degradation of hazardous wastes, even *in situ*.

Under the label of *bioremediation*, the use of microbial processes to destroy hazardous wastes has been the subject of intense investigation for many years.[4] Doubts still exist about claims for its effectiveness in a number of applications. It must be kept in mind, however, that there are many factors that can cause biodegradation to fail as a treatment process. Often physical conditions are such that mixing of wastes, nutrients, and electron acceptor species (such as oxygen) is too slow to permit biodegradation to occur at a useful rate. Low temperatures may make reactions too slow to be useful. Toxicants, such as heavy metals, may inhibit biological activity, and some metabolites produced by the microorganisms may be toxic to them.

21.8.1 Biodegradability

The *biodegradability* of a compound is influenced by its physical characteristics, such as solubility in water and vapor pressure, and by its chemical properties, including molar mass, molecular structure, and presence of various kinds of functional groups, some of which provide a "biochemical handle" for the initiation of biodegradation. With the appropriate organisms and under the right conditions, even substances such as phenol that are considered to be biocidal to most microorganisms can undergo biodegradation.

Recalcitrant or *biorefractory* substances are those that resist biodegradation and tend to persist and accumulate in the environment. Such materials are not necessarily toxic to organisms, but simply resist their metabolic attack. However, even some compounds regarded as biorefractory may be degraded by microorganisms that have had the opportunity to adapt to their biodegradation; for example, DDT is degraded by properly acclimated *Pseudomonas*. Chemical pretreatment, especially by partial oxidation, can make some kinds of recalcitrant wastes much more biodegradable.

Properties of hazardous wastes and their media can be changed to increase biodegradability. This can be accomplished by adjustment of conditions to optimum temperature, pH (usually in the range of 6–9), stirring, oxygen level, and material load and by addition of nutrients. Biodegradation can be aided by removal of toxic organic and inorganic substances, such as heavy metal ions.

21.8.2 Aerobic Treatment

Aerobic waste treatment processes utilize aerobic bacteria and fungi that require molecular oxygen, O_2. These processes are often favored by microorganisms, in part, because of the high energy yield obtained when molecular oxygen reacts with organic matter. Aerobic waste treatment is well adapted to the use of an activated sludge process. It can be applied to hazardous wastes such as chemical process wastes and landfill leachates. Some systems use powdered activated carbon as an additive to absorb organic wastes that are not biodegraded by microorganisms in the system.

Contaminated soils can be mixed with water and treated in a bioreactor to eliminate biodegradable contaminants in the soil. It is possible, in principle, to treat contaminated soils biologically in place by pumping oxygenated, nutrient-enriched water through the soil in a recirculating system.

21.8.3 Anaerobic Treatment

Anaerobic waste treatment in which microorganisms degrade wastes in the absence of oxygen can be practiced on a variety of organic hazardous wastes. Compared to the aerated activated sludge process, anaerobic digestion requires less energy; yields less sludge by-product; generates hydrogen sulfide (H_2S), which precipitates toxic heavy metal ions; and produces methane gas, CH_4, which can be used as an energy source.

The overall process for anaerobic digestion is a fermentation process in which organic matter is both oxidized and reduced. The simplified reaction for the anaerobic fermentation of a hypothetical organic substance, "$\{CH_2O\}$," is the following:

$$2\{CH_2O\} \rightarrow CO_2 + CH_4 \tag{21.28}$$

In practice, the microbial processes involved are quite complex. Most of the wastes for which anaerobic digestion is suitable consist of oxygenated compounds, such as acetaldehyde or methylethyl ketone.

21.8.4 Reductive Dehalogenation

Reductive dehalogenation is a mechanism by which halogen atoms are removed from organohalide compounds by anaerobic bacteria. It is an important means of detoxifying alkyl halides (particularly solvents), aryl halides, and organochlorine pesticides, all of which are important hazardous waste compounds, and which were discarded in large quantities in some of the older waste disposal dumps. Reductive dehalogenation is the main means by which some of the more highly halogenated waste compounds are biodegraded; such compounds include tetrachloroethene, HCB, pentachlorophenol, and the more highly chlorinated PCB congeners (see dehalorespiration in Section 6.13).

The two general processes by which reductive dehalogenation occurs are *hydrogenolysis*, as shown by the example in Reaction 21.29,

$$C_6H_5Cl + 2H \rightarrow C_6H_6 + HCl \tag{21.29}$$

and *vicinal reduction*:

$$H_3C{-}CHCl{-}CH_2Cl + 2e^- \rightarrow H_3C{-}CH{=}CH_2 + 2Cl^- \tag{21.30}$$

Vicinal reduction removes two adjacent halogen atoms and works only on alkyl halides, not aryl halides. Both processes produce innocuous inorganic halide (Cl^-).

21.9 PHYTOREMEDIATION

Phytoremediation uses various kinds of plants including trees to treat soil and water contaminated with organic contaminants, heavy metals, and radionuclides. Although phytoremediation is generally regarded as using plants to remove pollutants from soil, it also includes growing plants on contaminated soil, often under conditions that are not normally conducive to plant growth. A cover of plants—phytostabilization—significantly reduces loss of pollutants from water or wind erosion and provides a soil medium conducive to pollutant biodegradation. Enhanced biodegradation of pollutants by microorganisms in the rhizosphere, where plant roots grow, is discussed in Section 16.10.

One way in which plants act to decontaminate soil is by absorbing contaminants through their roots and transporting them to the rest of the plant. The plant biomass may then be burned or otherwise processed to remove the pollutants; use of plants to "mine" valuable metals, such as nickel, has even been seriously suggested. Beyond simple uptake of pollutants, growing plants have active enzyme systems, such as nitroreductase, dioxygenases, and laccases that can degrade organic pollutants. Plants that absorb volatile organic pollutants, such as benzene, toluene, ethylbenzene, xylene, and MTBE have an undesirable tendency to release these materials to the atmosphere.

Fast-growing, deep-rooted tree varieties are favorites for phytoremediation. Willows and poplars, including hybrid varieties that grow with spectacular rapidity, have been favored. Domestic crops including alfalfa, clover, corn, rye, sorghum, and soy have been used on soil contaminated with petroleum hydrocarbons. Tobacco plants bred for phytoremediation have been used on soils contaminated with cadmium, copper, and zinc.

21.10 LAND TREATMENT AND COMPOSTING

21.10.1 Land Treatment

Soil may be viewed as a natural filter for wastes. Soil has physical, chemical, and biological characteristics that can enable waste detoxification, biodegradation, chemical decomposition, and physical and chemical fixation. Therefore, *land treatment* of wastes may be accomplished by mixing the wastes with soil under appropriate conditions. In many cases, the reclamation of contaminated soil to restore it to productivity is an important activity.

Soil is a natural medium for a number of living organisms that may have an effect upon biodegradation of hazardous wastes. Of these, the most important are bacteria, including those from the genera *Agrobacterium*, *Arthrobacteri*, *Bacillus*, *Flavobacterium*, and *Pseudomonas*. *Actinomycetes* and fungi are important organisms in decay of vegetable matter and may be involved in biodegradation of wastes.

Microorganisms useful for land treatment are usually present in sufficient numbers to provide the inoculum required for their growth. The growth of these indigenous microorganisms may be stimulated by adding nutrients and an electron acceptor to act as an oxidant (for aerobic degradation) accompanied by mixing. The most commonly added nutrients are nitrogen and phosphorus. Oxygen can be added by pumping air underground or by treatment with hydrogen peroxide, H_2O_2. In some cases, such as for treatment of hydrocarbons on or near the soil surface, simple tillage provides both oxygen and the mixing required for optimum microbial growth.

Wastes that are amenable to land treatment are biodegradable organic substances. However, in soil contaminated with hazardous wastes, bacterial cultures may develop that are effective in degrading normally recalcitrant compounds through acclimation over a long period of time. Land treatment is most used for petroleum refining wastes and is applicable to the treatment of fuels and wastes from leaking underground storage tanks. It can also be applied to biodegradable organic chemical wastes, including some organohalide compounds. Land treatment is not suitable for the treatment of wastes containing acids, bases, toxic inorganic compounds, salts, heavy metals, and organic compounds that are excessively soluble, volatile, or flammable.

21.10.2 Composting

Composting of hazardous wastes is the biodegradation of solid or solidified materials in a medium other than soil. Bulking material, such as plant residue, paper, municipal refuse, or sawdust may be added to retain water and enable air to penetrate to the waste material. Successful composting of hazardous waste depends upon a number of factors, including those discussed above under land treatment. The first of these is the selection of the appropriate microorganism or *inoculum*. Once a successful composting operation is underway, a good inoculum is maintained by recirculating spent compost to each new batch. Other parameters that must be controlled include oxygen supply, moisture content (which should be maintained at a minimum of about 40%), pH (usually around neutral), and temperature. The composting process generates heat, so if the mass of the compost pile is sufficiently high, it can be self-heating under most conditions. Some wastes are deficient in nutrients, such as nitrogen, which must be supplied from commercial sources or from other wastes.

21.11 PREPARATION OF WASTES FOR DISPOSAL

Immobilization, stabilization, fixation, and solidification are terms that describe sometimes overlapping techniques whereby hazardous wastes are placed in a form suitable for long-term disposal. These aspects of hazardous waste management are addressed below.

21.11.1 Immobilization

Immobilization includes physical and chemical processes that reduce surface areas of wastes to minimize leaching. It isolates the wastes from their environment, especially groundwater, so that they have the least possible tendency to migrate. This is accomplished by physically isolating the waste, reducing its solubility, and decreasing its surface area. Immobilization usually improves the handling and physical characteristics of wastes.

21.11.2 Stabilization

Stabilization means the conversion of a waste from its original form to a physically and chemically more stable material that is less likely to cause problems during handling and disposal, and less likely to be mobile after disposal. Stabilization may include solidification (see below) and chemical reactions that generate products that are less volatile, soluble, and reactive. Stabilization is required for land disposal of wastes. *Fixation* is a stabilization process that binds a hazardous waste in a less mobile and less toxic form.

21.11.3 Solidification

Solidification may involve chemical reaction with a solidification agent, mechanical isolation in a protective binding matrix, or a combination of chemical and physical processes. It can be accomplished by evaporation of water from aqueous wastes or sludges, sorption onto solid material, reaction with cement, reaction with silicates, encapsulation, or imbedding in polymers or thermoplastics.

In many solidification processes, such as reaction with portland cement, water is an important ingredient of the hydrated solid matrix. Therefore, the solid should not be heated excessively or exposed to extremely dry conditions, which could result in diminished structural integrity from loss of water. In some cases, however, heating a solidified waste is an essential part of the overall solidification procedure. For example, an iron hydroxide matrix can be converted to highly insoluble, refractory iron oxide by heating. Heating can convert organic constituents of solidified wastes to inert carbon. Heating is an integral part of the process of vitrification (see below).

21.11.3.1 Sorption to a Solid Matrix Material

Hazardous waste liquids, emulsions, sludges, and free liquids in contact with sludges may be solidified and stabilized by fixing onto solid *sorbents*, including activated carbon (for organics), fly ash, kiln dust, clays, vermiculite, and various proprietary materials. Sorption may convert liquids and semisolids to dry solids, improve waste handling, and reduce solubility of waste constituents, and can be used to improve waste compatibility with substances such as portland cement used for solidification and setting. Specific sorbents may also be used to stabilize pH and p*E* (a measure of the tendency of a medium to be oxidizing or reducing, see Chapter 4).

The action of sorbents can include simple mechanical retention of wastes, physical sorption, and chemical reactions. It is important to match the sorbent to the waste. A substance with a strong affinity for water should be employed for wastes containing excess water and one with a strong affinity for organic materials should be used for wastes with excess organic solvents.

21.11.3.2 Thermoplastics and Organic Polymers

Hazardous waste materials may be mixed with hot *thermoplastic* liquids and solidified in the cooled thermoplastic matrix, which is rigid but deformable. The thermoplastic material most used for this purpose is asphalt bitumen. Other thermoplastics, such as paraffin and polyethylene, have also been used to immobilize hazardous wastes.

Among the wastes that can be immobilized with thermoplastics are those containing heavy metals, such as electroplating wastes. Organic thermoplastics repel water and reduce the tendency toward leaching in contact with groundwater. Compared to cement, thermoplastics add relatively less material to the waste.

A technique similar to that described above uses *organic polymers* produced in contact with solid wastes to imbed the wastes in a polymer matrix. Three kinds of polymers that have been used for this purpose include polybutadiene, urea-formaldehyde, and vinyl ester-styrene polymers. This procedure is more complicated than is the use of thermoplastics but, in favorable cases, yields a product in which the waste is held more strongly.

21.11.3.3 Vitrification

Vitrification or *glassification* consists of imbedding wastes in a glass material. In this application, glass may be regarded as a high-melting-temperature inorganic thermoplastic. Molten glass can be used or glass can be synthesized in contact with the waste by mixing and heating with glass constituents—silicon dioxide (SiO_2), sodium carbonate (Na_2CO_3), and calcium oxide (CaO). Other constituents may include boron oxide, B_2O_3, which yields a borosilicate glass that is especially resistant to changes in temperature and chemical attack. In some cases, glass is used in conjunction with thermal waste destruction processes, serving to immobilize hazardous waste ash constituents. Some wastes are detrimental to the quality of the glass. Aluminum oxide, for example, may prevent glass from fusing.

Vitrification is relatively complicated and expensive, the latter because of the energy consumed in fusing glass. Despite these disadvantages, it is the favored immobilization technique for some special wastes and has been promoted for solidification of radionuclear wastes because glass is chemically inert and resistant to leaching. However, high levels of radioactivity can cause deterioration of glass and lower its resistance to leaching.

21.11.3.4 Solidification with Cement

Portland cement is widely used for solidification of hazardous wastes including wastes containing both metals and organic contaminants. In this application, portland cement provides a solid matrix for isolation of the wastes, chemically binds water from sludge wastes, and may react chemically with wastes (e.g., the calcium and base in portland cement react chemically with inorganic arsenic wastes to reduce their solubilities). However, most wastes are held physically in the rigid portland cement matrix and are subject to leaching.

As a solidification matrix, portland cement is most applicable to inorganic sludges containing heavy metal ions that form insoluble hydroxides and carbonates in the basic carbonate medium provided by the cement. The success of solidification with portland cement strongly depends upon whether or not the waste adversely affects the strength and stability of the concrete product. A number of substances—organic matter such as petroleum or coal; some silts and clays; sodium salts of arsenate, borate, phosphate, iodate, and sulfide; and salts of copper, lead, magnesium, tin, and zinc—are incompatible with portland cement because they interfere with its set and cure, producing a mechanically weak product and resulting in deterioration of the cement matrix with time. However, a reasonably good disposal form can be obtained by absorbing organic wastes with a solid material, which in turn is set in portland cement.

21.11.3.5 Solidification with Silicate Materials

Water-insoluble *silicates* (pozzolanic substances), containing oxyanionic silicon such as SiO_3^{2-}, are used for waste solidification. There are a number of such substances, some of which are waste products, including fly ash, flue dust, clay, calcium silicates, and ground-up slag from blast furnaces. Soluble silicates, such as sodium silicate, may also be used. Silicate solidification usually requires a setting agent, which may be portland cement (see above), gypsum (hydrated $CaSO_4$), lime, or compounds of aluminum, magnesium, or iron. The product may vary from a granular material to a concrete-like solid. In some cases, the product is improved by additives, such as emulsifiers, surfactants, activators, calcium chloride, clays, carbon, zeolites, and various proprietary materials.

Success has been reported for the solidification of both inorganic wastes and organic wastes (including oily sludges) with silicates. The advantages and disadvantages of silicate solidification are similar to those of portland cement discussed above. One consideration that is especially applicable to fly ash is the presence in some silicate materials of leachable hazardous substances, which may include arsenic and selenium.

21.11.3.6 Encapsulation

As the name implies, *encapsulation* is used to coat wastes with an impervious material so that they do not come in contact with their surroundings. For example, a water-soluble waste salt encapsulated in asphalt would not dissolve so long as the asphalt layer remains intact. A common means of encapsulation uses heated, molten thermoplastics, asphalt, and waxes that solidify when cooled. A more sophisticated approach to encapsulation is to form polymeric resins from monomeric substances in the presence of the waste.

21.11.4 Chemical Fixation

Chemical fixation is a process that binds a hazardous waste substance in a less mobile, less toxic form by a chemical reaction that alters the waste chemically. Physical and chemical fixations often occur together, and sometimes it is a little difficult to distinguish between them. Polymeric inorganic silicates containing some calcium and often some aluminum are the inorganic materials most widely used as a fixation matrix. Many kinds of heavy metals are chemically bound in such a matrix, as well as being held physically by it. Similarly, some organic wastes are bound by reactions with matrix constituents.

21.12 ULTIMATE DISPOSAL OF WASTES

Regardless of the destruction, treatment, and immobilization techniques used, there will always remain, from hazardous wastes, some material that has to be put somewhere. This section briefly addresses the ultimate disposal of ash, salts, liquids, solidified liquids, and other residues that must be placed where their potential to do harm is minimized.

21.12.1 Disposal Aboveground

In some important respects, disposal aboveground, essentially in a pile designed to prevent erosion and water infiltration, is the best way to store solid wastes. Perhaps its most important advantage is that it avoids infiltration by groundwater that can result in leaching and groundwater contamination common to storage in pits and landfills. In a properly designed aboveground disposal facility, any leachate that is produced drains quickly by gravity to the leachate collection system where it can be detected and treated.

Aboveground disposal can be accomplished with a storage mound deposited on a layer of compacted clay covered with impermeable membrane liners laid somewhat above the original soil surface and shaped to allow leachate flow and collection. The slopes around the edges of the storage mound should be sufficiently great to allow good drainage of precipitation, but gentle enough to deter erosion.

21.12.2 Landfill

Landfill historically has been the most common way of disposing of solid hazardous wastes and some liquids, although it is being severely limited in many nations by new regulations and high land costs. Landfill involves disposal that is at least partially underground in excavated cells, quarries, or natural depressions. Usually, fill is continued above ground to utilize space most efficiently and provide a grade for drainage of precipitation.

The greatest environmental concern with landfill of hazardous wastes is the generation of leachate from infiltrating surface water and groundwater with resultant contamination of groundwater supplies. Modern hazardous waste landfills provide elaborate systems to contain, collect, and control such leachate.

There are several components in a modern landfill. A landfill should be placed on a compacted, low-permeability medium, preferably clay, which is covered by a flexible-membrane liner consisting of watertight impermeable material. This liner is covered with granular material in which is installed a secondary drainage system. Next is another flexible-membrane liner, above which is installed a primary drainage system for the removal of leachate. This drainage system is covered with a layer of granular filter medium, upon which the wastes are placed. In the landfill, wastes of different kinds are separated by berms consisting of clay or soil covered with liner material. When the fill is complete, the waste is capped to prevent surface water infiltration and covered with compacted soil. In addition to leachate collection, provision may be made for a system to treat evolved gases, particularly when methane-generating biodegradable materials are disposed in the landfill.

The flexible-membrane liner made of rubber (including chlorosulfonated polyethylene) or plastic (including chlorinated polyethylene, high-density polyethylene, and PVC) is a key component of state-of-the-art landfills. It controls seepage out of, and infiltration into, the landfill. Obviously, liners have to meet stringent standards to serve their intended purpose. In addition to being impermeable, the liner material must be strongly resistant to biodegradation, chemical attack, and tearing.

Capping is done to cover the wastes, prevent infiltration of excessive amounts of surface water, and prevent release of wastes to overlying soil and the atmosphere. Caps come in a variety of forms and are often multilayered. Some of the problems that may occur with caps are settling, erosion, ponding, damage by rodents, and penetration by plant roots.

21.12.3 Surface Impoundment of Liquids

Many liquid hazardous wastes, slurries, and sludges are placed in *surface impoundments*, which usually serve for treatment and often are designed to be filled in eventually as landfill disposal sites. Most liquid hazardous wastes and a significant fraction of solids are placed in surface impoundments in some stage of treatment, storage, or disposal.

A surface impoundment may consist of an excavated "pit," a structure formed with dikes, or a combination thereof. The construction is similar to that discussed above for landfills in that the bottom and walls should be impermeable to liquids and provision must be made for leachate collection. The chemical and mechanical challenges to liner materials in surface impoundments are severe so that proper geological siting and construction with floors and walls composed of low-permeability soil and clay are important in preventing pollution from these installations.

21.12.4 Deep-Well Disposal of Liquids

Deep-well disposal of liquids consists of their injection under pressure to underground strata isolated by impermeable rock strata from aquifers. Early experience with this method was gained in the petroleum industry where disposal is required for large quantities of saline wastewater coproduced with crude oil. The method was later extended to the chemical industry for the disposal of brines, acids, heavy metal solutions, organic liquids, and other liquids.

A number of factors must be considered in deep-well disposal. Wastes are injected into a region of elevated temperature and pressure, which may cause chemical reactions to occur involving the waste constituents and the mineral strata. Oils, solids, and gases in the liquid wastes can cause problems such as clogging. Corrosion may be severe. Microorganisms may have some effects. Most problems from these causes can be mitigated by proper waste pretreatment.

The most serious consideration involving deep-well disposal is the potential contamination of groundwater. Although injection is made into permeable saltwater aquifers presumably isolated from aquifers that contain potable water, contamination may occur. Major routes of contamination include fractures, faults, and other wells. The disposal well itself can act as a route for contamination if it is not properly constructed and cased (lined with pipe) or if it is damaged.

21.13 LEACHATE AND GAS EMISSIONS

21.13.1 Leachate

The production of contaminated leachate is a possibility with most disposal sites.[5] Therefore, new hazardous waste landfills require leachate collection/treatment systems as discussed in Section 21.12 and many older sites are required to have such systems retrofitted to them.

Chemical and biochemical processes have the potential to cause some problems for leachate collection systems. One such problem is clogging by insoluble manganese(IV) and iron(III) hydrated oxides upon exposure to air as described for water wells in Section 15.8.

Leachate consists of water that has become contaminated by wastes as it passes through a waste disposal site. It contains waste constituents that are soluble, not retained by soil, and not degraded chemically or biochemically. Some potentially harmful leachate constituents are products of chemical or biochemical transformations of wastes.

The best approach to leachate management is to prevent its production by limiting infiltration of water into the site. Rates of leachate production may be very low when sites are selected, designed, and constructed with minimal production of leachate as a major objective. A well maintained, low-permeability cap over the landfill is very important for leachate minimization.

21.13.2 Hazardous Waste Leachate Treatment

The first step in treating leachate is to characterize it fully, particularly with a thorough chemical analysis of possible waste constituents and their chemical and metabolic products. The biodegradability of leachate constituents should also be determined. The options available for the treatment of hazardous waste leachate are generally the physical, chemical, and biochemical processes used to treat industrial wastewaters as discussed earlier in this chapter and in Chapter 8.

21.13.3 Gas Emissions

In the presence of biodegradable wastes, methane and carbon dioxide gases are produced in landfills by anaerobic degradation (see Reaction 21.28). Gases may also be produced by chemical processes with improperly pretreated wastes, as would occur in the hydrolysis of calcium carbide to produce acetylene:

$$CaC_2 + 2H_2O \rightarrow C_2H_2 + Ca(OH)_2 \tag{21.31}$$

Odorous and toxic hydrogen sulfide, H_2S, may be generated by the chemical reaction of sulfides with acids or by the biochemical reduction of sulfate by anaerobic bacteria (*Desulfovibrio*) in the presence of biodegradable organic matter:

$$SO_4^{2-} + 2\{CH_2O\} + 2H^+ \xrightarrow[\text{bacteria}]{\text{Anaerobic}} H_2S + 2CO_2 + 2H_2O \tag{21.32}$$

Gases such as these may be toxic, combustible, or explosive. Furthermore, gases permeating through landfilled hazardous waste may carry along waste vapors, such as those of volatile aryl compounds and low-molar-mass halogenated hydrocarbons. Of these, the ones of most concern are benzene, 1,2-dibromoethane, 1,2-dichloroethane, carbon tetrachloride, chloroform, dichloromethane, tetrachloroethane, 1,1,1-trichloroethane, trichloroethylene, and vinyl chloride. Because of the hazards from these and other volatile species, it is important to minimize production of gases and, if significant amounts of gases are produced, they should be vented or treated by activated carbon sorption or flaring.

21.14 *In Situ* Treatment

In situ treatment refers to waste treatment processes that can be applied to wastes in a disposal site by direct application of treatment processes and reagents to the wastes. Where possible, *in situ* treatment is desirable for waste site remediation.

21.14.1 *In Situ* Immobilization

In situ immobilization is used to convert wastes to insoluble forms that will not leach from the disposal site. Heavy metal contaminants including lead, cadmium, zinc, and mercury can be immobilized by chemical precipitation as the sulfides by treatment with gaseous H_2S or alkaline Na_2S solution. Disadvantages include the high toxicity of H_2S and the contamination potential of soluble sulfide. Although precipitated metal sulfides should remain as solids in the anaerobic conditions of a landfill, unintentional exposure to air can result in oxidation of the sulfide and remobilization of the metals as soluble sulfate salts.

Oxidation and reduction reactions can be used to immobilize heavy metals *in situ*. Oxidation of soluble Fe^{2+} and Mn^{2+} to their insoluble hydrous oxides, $Fe_2O_3 \cdot xH_2O$ and $MnO_2 \cdot xH_2O$, respectively, can precipitate these metal ions and coprecipitate other heavy metal ions. However, subsurface reducing conditions could later result in reformation of soluble reduced species. Reduction can be used *in situ* to convert soluble, toxic chromate to insoluble chromium(III) compounds.

Chelation may convert metal ions to less mobile forms, although with most agents chelation has the opposite effect. The humin fraction of soil humic substances immobilizes metal ions.

21.14.2 Vapor Extraction

Many important wastes have relatively high vapor pressures and can be removed by vapor extraction. This technique works for wastes in soil above the level of groundwater, that is, in the vadose zone. Simple in concept, vapor extraction involves pumping air into injection wells in soil and withdrawing it, along with volatile components that it has picked up, through extraction wells. The substances vaporized from the soil are removed by activated carbon or by other means. In some cases, the air is pumped through an engine (which can be used to run the air pumps) and organic vapors are destroyed by conditions in the engine's combustion chambers. Pumping air is relatively efficient compared to groundwater pumping because of the much higher flow rates of air through soil compared to water. Vapor extraction is most applicable to the removal of VOCs such as chloromethanes, chloroethanes, chloroethylenes (such as trichloroethylene), benzene, toluene, and xylene.

21.14.3 Solidification *In Situ*

In situ solidification can be used as a remedial measure at hazardous waste sites. One approach is to inject soluble silicates followed by reagents that cause them to solidify. For example, injection of soluble sodium silicate followed by calcium chloride or lime forms solid calcium silicate.

21.14.4 Detoxification *In Situ*

When only one or a limited number of harmful constituents is present in a waste disposal site, it may be practical to consider detoxification *in situ*. This approach is most practical for organic contaminants including pesticides (organophosphate esters and carbamates), amides, and esters. Among the chemical and biochemical processes that can detoxify such materials are chemical and enzymatic oxidation, reduction, and hydrolysis. Chemical oxidants that have been proposed for this purpose include hydrogen peroxide, ozone, and hypochlorite.

Enzyme extracts collected from microbial cultures and purified have been considered for *in situ* detoxification. One cell-free enzyme that has been used for detoxification of organophosphate insecticides is parathion hydrolase. The hostile environment of a chemical waste landfill, including the presence of enzyme-inhibiting heavy metal ions, is detrimental to many biochemical approaches to *in situ* treatment. Furthermore, most sites contain a mixture of hazardous constituents, which might require several different enzymes for their detoxification.

21.14.5 Permeable Bed Treatment

Some groundwater plumes contaminated by dissolved wastes can be treated by *subsurface permeable reactive barriers* consisting of a permeable bed of material placed in a trench through which the groundwater must flow. Limestone in a permeable bed neutralizes acid and precipitates some heavy metal hydroxides or carbonates. Synthetic ion-exchange resins can be used in a permeable bed to retain heavy metals and even some anionic species, although competition with ionic species present naturally in the groundwater can cause some problems with their use. Activated carbon in a permeable bed will remove some organics, especially less soluble, higher-molar-mass organic compounds.

Granular *zerovalent iron* [iron(0)] offers some advantages as a permeable bed medium.[6] This material acts as a reducing agent and can dechlorinate organochlorine compounds, which are resistant to biodegradation. In addition to treating groundwaters contaminated with organochlorine compounds, permeable barriers containing zerovalent iron have been used to treat groundwaters polluted with inorganics, metals, munition wastes, radionuclides, and various other pollutants.

Permeable bed treatment requires relatively large quantities of reagent, which argues against the use of activated carbon and ion-exchange resins. In such an application, it is unlikely that either of these materials could be reclaimed and regenerated as is done when they are used in columns to treat wastewater. Furthermore, ions taken up by ion exchangers and organic species retained by activated carbon may be released at a later time, causing subsequent problems. Finally, a permeable bed that has been truly effective in collecting waste materials may, itself, be considered a hazardous waste requiring special treatment and disposal.

21.14.6 *In Situ* Thermal Processes

Heating of wastes *in situ* can be used to remove or destroy some kinds of hazardous substances. Steam injection, radio frequency, and microwave heating have been proposed for this purpose. Volatile wastes brought to the surface by heating can be collected and held as condensed liquids or by activated carbon.

One approach to immobilizing wastes *in situ* is high temperature vitrification using electrical heating. This process involves pouring conducting graphite on the surface between two electrodes and passing an electric current between the electrodes. In principle, the graphite becomes very hot and "melts" into the soil leaving a glassy slag in its path. Volatile species evolved are collected and, if the operation is successful, a nonleachable slag is left in place. It is easy to imagine problems that might occur, including difficulties in getting a uniform melt, problems from groundwater infiltration, and very high consumption of electricity.

21.14.7 Soil Washing and Flushing

Extraction with water containing various additives can be used to cleanse soil contaminated with hazardous wastes. When the soil is left in place and the water pumped into and out of it, the process is called *flushing*; when soil is removed and contacted with liquid the process is referred to as *washing*. Here, washing is used as a term applied to both processes.

The composition of the fluid used for soil washing depends upon the contaminants to be removed. The washing medium may consist of pure water or it may contain acids (to leach out metals or neutralize alkaline soil contaminants), bases (to neutralize contaminant acids), chelating agents (to solubilize heavy metals), surfactants (to enhance the removal of organic contaminants from soil and improve the ability of the water to emulsify insoluble organic species), or reducing agents (to reduce oxidized species). Soil contaminants may dissolve, form emulsions, or react chemically. Heavy metal salts; lighter aromatic hydrocarbons, such as toluene and xylenes; lighter organohalides, such as trichloro- or tetrachloroethylene; and light-to-medium molar mass aldehydes and ketones can be removed from soil by washing.

LITERATURE CITED

1. Wang, L. K. and D. Aulenbach, Implementation of industrial ecology for industrial hazardous waste management, *Hazardous Industrial Waste Treatment*, L. K. Wang, Ed., Taylor & Francis/CRC Press, Boca Raton, FL, 2007, pp. 1–13.
2. Hernandez, E. A. and V. Uddameri, Hazardous waste assessment, management, and minimization, *Water Environment Research*, **80**, 1648–1653, 2008.
3. Reijnders, L., The cement industry as a scavenger in industrial ecology and the management of hazardous substances, *Journal of Industrial Ecology*, **11**, 15–25, 2007.

4. Ong, S. K., Y. S. Rao, B. Alok, C. Pascale, R. D. Tyagi, and I. Lo, Eds, *Natural Processes and Systems for Hazardous Waste Treatment*, Society of Civil Engineers, Reston, VA, 2008.
5. Dhage, S. S., S. A. Patkie, and S. S. Tipnis, Pollution hazards of leachates from solid waste dump sites, *Pollution Research*, **24**, 135–142, 2005.
6. Noubactep, C., M. Guenther, D. Peter, S. Martin, and B. J. Merkel, Testing the suitability of zerovalent iron materials for reactive walls, *Environmental Chemistry*, **2**, 71–76, 2005.

SUPPLEMENTARY REFERENCES

ASTM Committee D-18 on Soil and Rock, *ASTM Standards Relating to Environmental Site Characterization*, 2nd ed., ASTM International, West Conshohocken, PA, 2002.

Barth, R. C., P. D. George, and R. H. Hill, *Environmental Health and Safety for Hazardous Waste Sites*, American Industrial Hygiene Association, Fairfax, VA, 2002.

Blackman, W. C., *Basic Hazardous Waste Management*, CRC Press/Lewis Publishers, Boca Raton, FL, 2001.

Cabaniss, A. D., Ed., *Handbook on Household Hazardous Waste*, Government Institute/Scarecrow Press, Lanham, MD, 2008.

Carson, P. A. and C. Mumford, *Hazardous Chemicals Handbook*, 2nd ed., Butterworth-Heinemann, Boston, 2002.

Chermisinoff, N. P. and T. A. Dalvetshina, *Fire and Explosion Hazards Handbook of Industrial Chemicals*, Jaico Publishing House, Mumbai, India, 2005.

Cheremisinoff, N. P., *Industrial Solvents Handbook*, 2nd ed., Marcel Dekker, New York, 2003.

Gallant, B., *Hazardous Waste Operations and Emergency Response Manual*, Wiley-Interscience, Hoboken, NJ, 2006.

Garrett, T. L., Ed., *The RCRA Practice Manual*, 2nd ed., American Bar Association, Section of Environment, Energy, and Resources, Chicago, 2004.

Hackman, C. L., E. E. Hackman, III, and M. E. Hackman, *Hazardous Waste Operations and Emergency Response Manual and Desk Reference*, McGraw-Hill, New York, 2002.

Hawley, C., *Hazardous Materials Air Monitoring and Detection Devices*, 2nd ed., Thomson/Delmar Learning, Clifton Park, NY, 2007.

Hawley, C., *Hazardous Materials Handbook: Awareness and Operations Levels*, Delmar Cengage, Clifton Park, NY, 2008.

Hawley, C., *Hazardous Materials Incidents*, 3rd ed., Thomson/Delmar Learning, Clifton Park, NY, 2008.

Hocking M. B. B., *Handbook of Chemical Technology and Pollution Control*, 3rd ed., Elsevier, Boston, 2006.

LaGrega, M. D., P. L. Buckingham, and J. C. Evans, *Hazardous Waste Management*, 2nd ed., McGraw-Hill, Boston, MA, 2001.

McGowan, K., *Hazardous Waste*, Lucent books, San Diego, CA, 2001.

Meyer, E., *Chemistry of Hazardous Materials*, Prentice-Hall, Upper Saddle River, NJ, 1998.

Nemerow, N. L., Ed., *Industrial Waste Treatment*, Elsevier/Butterworth-Heinemann, Amsterdam, 2007.

Nemerow, N. L. and F. J. Agardy, *Strategies of Industrial and Hazardous Waste Management*, Van Nostrand Reinhold, New York, 1998.

Network Environmental Systems Staff, *Hazardous Waste Compliance Manual*, 2nd ed., Network Environmental Systems, Folsom, CA, 2004.

Office of Technology Assessment, *Technologies and Management Strategies for Hazardous Waste Control*, University Press of the Pacific, Honolulu, 2005.

Pohanish, R. P., *Sittig's Handbook of Toxic and Hazardous Chemicals and Carcinogens*, 5th ed., Noyes Publications, Westwood, NJ, 2007.

Sara, M. N., *Site Assessment and Remediation Handbook*, 2nd ed., CRC Press/Lewis Publishers, Boca Raton, FL, 2003.

Sheha, R. R. and H. H. Someda, Eds, *Hazardous Waste: Classifications and Treatment Technologies*, Nova Science Publishers, New York, 2008.

Spencer, A. B. and G. R. Colonna, *NFPA Pocket Guide to Hazardous Materials*, National Fire Protection Association, Quincy, MA, 2003.

Tang, W. Z., *Physicochemical Treatment of Hazardous Wastes*, CRC Press, Boca Raton, FL, 2003.

Tedder, D. W. and F. G. Pohland, Eds, *Emerging Technologies in Hazardous Waste Management 8*, Kluwer Academic, New York, 2000.

Teets, J. W. and D. Reis, *RCRA: Resource Conservation and Recovery Act*, American Bar Association, Section of Environment, Energy, and Resources, Chicago, 2004.

Urben, P. G., Ed., *Bretherick's Handbook of Reactive Chemical Hazards*, 7th ed., Elsevier, Amsterdam, 2007.

Wang, L. K., N. K. Shammas, and Y.-T. Hung, Eds, *Advances in Hazardous Industrial Waste Treatment*, CRC/Taylor & Francis, Boca Raton, FL, 2008.

Wang, L. K., *Handbook of Industrial and Hazardous Wastes Treatment*, 2nd ed., Marcel Dekker, New York, 2004.

Woodside, G., *Hazardous Materials and Hazardous Waste Management*, Wiley, New York, 1999.

QUESTIONS AND PROBLEMS

1. Place the following in descending order of desirability for dealing with wastes and discuss your rationale for doing so (explain): (A) reducing the volume of remaining wastes by measures such as incineration, (B) placing the residual material in landfills, properly protected from leaching or release by other pathways, (C) treating residual material as much as possible to render it nonleachable and innocuous, (D) reduction of wastes at the source, (E) recycling as much waste as is practical.
2. Match the waste recycling process or industry from the column on the left with the kind of material that can be recycled from the list on the right, below:

A. Recycle as raw material to the generator	1. Waste alkali
B. Utilization for pollution control or waste treatment	2. Hydraulic and lubricating oils
C. Energy production	3. Incinerable materials
D. Materials with agricultural uses	4. Incompletely consumed feedstock material
E. Organic substances	5. Waste lime or phosphate-containing sludge

3. What material is recycled using hydrofinishing, treatment with clay, and filtration?
4. What is the "most important operation in solvent purification and recycle" that is used to separate solvents from impurities, water, and other solvents?
5. DAF is used in the secondary treatment of wastes. What is the principle of this technique? For what kinds of hazardous waste substances is it most applicable?
6. Match the process or industry from the column on the left with its "phase of waste treatment" from the list on the right, below:

A. Activated carbon sorption	1. Primary treatment
B. Precipitation	2. Secondary treatment
C. Reverse osmosis	3. Polishing
D. Emulsion breaking	
E. Slurrying	

7. Distillation is used in treating and recycling a variety of wastes, including solvents, waste oil, aqueous phenolic wastes, and mixtures of ethylbenzene and styrene. What is the major hazardous waste problem that arises from the use of distillation for waste treatment?
8. Supercritical fluid technology has a great deal of potential for the treatment of hazardous wastes. What are the principles involved with the use of supercritical fluids for waste treatment? Why is this technique especially advantageous? Which substance is most likely to be used as a supercritical fluid in this application? For which kinds of wastes are supercritical fluids most useful?
9. What are some advantages of using acetic acid, compared, for example, to sulfuric acid, as a neutralizing agent for treating waste alkaline materials?

10. Designate which of the following would be *least likely* to be produced by, or used as a reagent for the removal of heavy metals by their precipitation from solution (explain): (A) Na_2CO_3, (B) CdS, (C) $Cr(OH)_3$, (D) KNO_3, (E) $Ca(OH)_2$.
11. Both $NaBH_4$ and Zn are used to remove metals from solution. How do these substances remove metals? What are the forms of the metal products?
12. Of the following, thermal treatment of wastes is *not* useful for (explain): (A) volume reduction, (B) destruction of heavy metals, (C) removal of volatile, combustible, mobile organic matter, (D) destruction of pathogenic materials, (E) destruction of toxic substances.
13. From the following, choose the waste liquid that is least amenable to incineration and explain why it is not readily incinerated (explain): (A) methanol, (B) tetrachloroethylene, (C) acetonitrile, (D) toluene, (E) ethanol, (F) acetone.
14. Name and give the advantages of the process that is used to destroy more hazardous wastes by thermal means than are burned solely for the purpose of waste destruction.
15. What is the major advantage of fluidized-bed incinerators from the standpoint of controlling pollutant by-products?
16. Explain the best way to obtain microorganisms to be used in the treatment of hazardous wastes by biodegradation?
17. What are the principles of composting? How is it used to treat hazardous wastes?
18. How is portland cement used in the treatment of hazardous wastes for disposal? What might be some disadvantages of such a use?
19. What are the advantages of aboveground disposal of hazardous wastes as opposed to burying wastes in landfills?
20. Describe and explain the best approach to managing leachate from hazardous waste disposal sites.
21. An incinerator is operated primarily to destroy chlorophenols as the POHCs fed along with other less hazardous constituents at an average rate of 10 kg/h. The exhaust gas coming from the incinerator stack at a rate of 10 m^3/min contains 1 $\mu g/m^3$ of chlorophenols. What is the DRE of the incinerator for the POHC?
22. Can phytoremediation be described as a bioremediation process? Is it a biodegradation process? If not, how can it be used to treat hazardous wastes?

22 Environmental Biochemistry

22.1 BIOCHEMISTRY

The effects of pollutants and potentially harmful chemicals on living organisms are of particular importance in environmental chemistry. These effects are addressed under the topic of "Toxicological Chemistry" in Chapter 23, and for specific substances in Chapter 24. This chapter is designed to provide the fundamental background in biochemistry required to understand toxicological chemistry.

Most people have had the experience of looking through a microscope at a single cell. It may have been an ameba, alive and oozing about like a blob of jelly on the microscope slide, or a cell of bacteria, stained with a dye to make it show up more plainly. Or, it may have been a beautiful cell of algae with its bright green chlorophyll. Even the simplest of these cells is capable of carrying out a thousand or more chemical reactions. These life processes fall under the heading of *biochemistry*, that branch of chemistry that deals with the chemical properties, composition, and biologically-mediated processes of complex substances in living systems.[1]

Biochemical phenomena that occur in living organisms are extremely sophisticated. In the human body, complex metabolic processes break down a variety of food materials to simpler chemicals, yielding energy and the raw materials to build body constituents such as muscle, blood, and brain tissue. Impressive as this may be, consider a humble microscopic cell of photosynthetic cyanobacteria only about a micrometer in size, which requires only a few simple inorganic chemicals and sunlight for its existence. This cell uses sunlight energy to convert carbon from CO_2, hydrogen and oxygen from H_2O, nitrogen from NO_3^-, sulfur from SO_4^{2-}, and phosphorus from inorganic phosphate into all the proteins, nucleic acids, carbohydrates, and other materials that it requires to exist and reproduce. Such a simple cell accomplishes what could not be done by humans in even a vast chemical factory costing billions of dollars.

Ultimately, most environmental pollutants and hazardous substances are of concern because of their effects upon living organisms. The study of the adverse effects of substances on life processes requires some basic knowledge of biochemistry. Biochemistry is discussed in this chapter, with emphasis upon aspects that are especially pertinent to environmentally hazardous and toxic substances, including cell membranes, DNA, and enzymes.

Biochemical processes not only are profoundly influenced by chemical species in the environment, they largely determine the nature of these species, their degradation, and even their syntheses, particularly in the aquatic and soil environments. The study of such phenomena forms the basis of *environmental biochemistry.*[2]

22.1.1 Biomolecules

The biomolecules that constitute matter in living organisms often are polymers with molecular masses of the order of a million or even larger. As discussed later in this chapter, these biomolecules may be divided into the categories of carbohydrates, proteins, lipids, and nucleic acids. Proteins and nucleic acids consist of macromolecules, lipids are usually relatively small molecules, and carbohydrates range from small sugar molecules to high molar mass macromolecules such as those in cellulose.

The behavior of a substance in a biological system depends to a large extent upon whether the substance is hydrophilic ("water-loving") or hydrophobic ("water-hating"). Some important toxic substances are hydrophobic, a characteristic that enables them to traverse cell membranes readily. Part of the detoxification process carried on by living organisms is to render such molecules hydrophilic, therefore water-soluble and readily eliminated from the body.

22.2 BIOCHEMISTRY AND THE CELL

The focal point of biochemistry and biochemical aspects of toxicants is the *cell*, the basic building block of living systems where most life processes are carried out. Bacteria, yeasts, and some algae consist of single cells. However, most living things are made up of many cells. In a more complicated organism, the cells have different functions. Liver cells, muscle cells, brain cells, and skin cells in the human body are quite different from each other and do different things. Cells are divided into two major categories depending upon whether or not they have a nucleus: *eukaryotic* cells have a nucleus and *prokaryotic* cells do not. Prokaryotic cells are found predominately in single-celled organisms such as bacteria. Eukaryotic cells occur in multicellular plants and animals—higher life forms.

22.2.1 Major Cell Features

Figure 22.1 shows the major features of the *eukaryotic cell*, which is the basic structure in which biochemical processes occur in multicellular organisms. These features are the following:

- *Cell membrane*, which encloses the cell and regulates the passage of ions, nutrients, lipid-soluble ("fat-soluble") substances, metabolic products, toxicants, and toxicant metabolites into and out of the cell interior because of its varying *permeability* for different substances. The cell membrane protects the contents of the cell from undesirable outside influences. Cell membranes are composed in part of phospholipids that are arranged with their hydrophilic ("water-seeking") heads on the cell membrane surfaces and their hydrophobic ("water-repelling") tails inside the membrane. Cell membranes contain bodies of proteins that are involved in the transport of some substances through the membrane. One reason the cell membrane is very important in toxicology and environmental biochemistry is because it regulates the passage of toxicants and their products into and out of the cell interior. Furthermore, when its membrane is damaged by toxic substances, a cell may not function properly and the organism may be harmed.

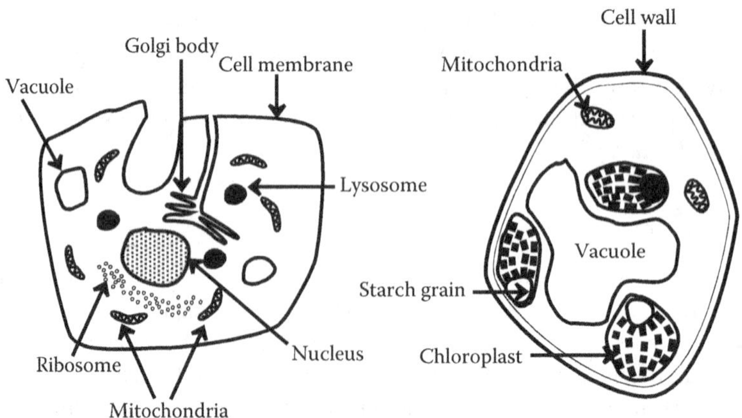

FIGURE 22.1 Some major features of the eukaryotic cell in animals (*left*) and plants (*right*).

- *Cell nucleus*, which acts as a sort of "control center" of the cell. It contains the genetic directions the cell needs to reproduce itself. The key substance in the nucleus is *deoxyribonucleic acid* (DNA). *Chromosomes* in the cell nucleus are made up of combinations of DNA and proteins. Each chromosome stores a separate quantity of genetic information. Human cells contain 46 chromosomes. When DNA in the nucleus is damaged by foreign substances, various toxic effects, including mutations, cancer, birth defects, and defective immune system function may occur.
- *Cytoplasm*, which fills the interior of the cell not occupied by the nucleus. Cytoplasm is further divided into a water-soluble proteinaceous filler called *cytosol*, in which are suspended bodies called *cellular organelles*, such as mitochondria or, in photosynthetic organisms, chloroplasts.
- *Mitochondria*, "powerhouses" which mediate energy conversion and utilization in the cell. Mitochondria are sites in which food materials—carbohydrates, proteins, and fats—are broken down to yield carbon dioxide, water, and energy, which is then used by the cell. The best example of this is the oxidation of the sugar glucose, $C_6H_{12}O_6$:

$$C_6H_{12}O_6 + 6O_2 \rightarrow 6CO_2 + 6H_2O + \text{energy}$$

  This kind of process is called *cellular respiration.*
- *Ribosomes*, which participate in protein synthesis.
- *Endoplasmic reticulum*, which is involved in the metabolism of some toxicants by enzymatic processes.
- *Lysosome*, a type of organelle that contains potent substances capable of digesting liquid food material. Such material enters the cell through a "dent" in the cell wall, which eventually becomes surrounded by cell material. This surrounded material is called a *food vacuole*. The vacuole merges with a lysosome and the substances in the lysosome bring about the digestion of the food material. The digestion process consists largely of *hydrolysis reactions* in which large, complicated food molecules are broken down into smaller units by the addition of water.
- *Golgi bodies*, that occur in some types of cells. These are flattened bodies of material that serve to hold and release substances produced by the cells.
- *Cell walls* of plant cells. These are strong structures that provide stiffness and strength. Cell walls are composed mostly of cellulose, which will be discussed later in this chapter.
- *Vacuoles* inside plant cells that often contain materials dissolved in water.
- *Chloroplasts* in plant cells that are involved in photosynthesis (the chemical process which uses energy from sunlight to convert carbon dioxide and water to organic matter). Photosynthesis occurs in these bodies. Food produced by photosynthesis is stored in the chloroplasts in the form of *starch grains*.

22.3 PROTEINS

Proteins are nitrogen-containing organic compounds which are the basic units of live systems. Cytoplasm, the jelly-like liquid filling the interior of cells, is largely protein. Enzymes, which act as catalysts of life reactions, are proteins; they are discussed later in the chapter. Proteins are made up of *amino acids* joined together in huge chains. Amino acids are organic compounds which contain the carboxylic acid group, $-CO_2H$, and the amino group, $-NH_2$. They are sort of a hybrid of carboxylic acids and amines. Proteins are polymers or *macromolecules* of amino acids containing from approximately 40 to several thousand amino acid groups joined by peptide linkages. Smaller molecule amino acid polymers, containing only about 10–40 amino acids per molecule, are called *polypeptides.* A portion of the amino acid left after the elimination of H_2O during polymerization is called a *residue*. The amino acid sequence of these residues is designated by a series of three-letter abbreviations for the amino acid.

Natural amino acids all have the following chemical group:

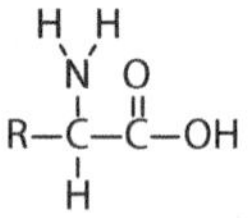

```
   H H
    \ /
    N  O
    |  ||
 R—C—C—OH
    |
    H
```

In this structure, the $-NH_2$ group is always bonded to the carbon next to the $-CO_2H$ group. This is called the "alpha" location, so natural amino acids are α-amino acids. Other groups, designated as "R," are attached to the basic α-amino acid structure. The R groups may be as simple as an atom of H found in glycine,

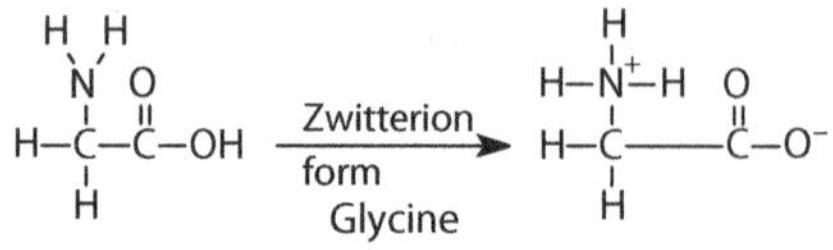

```
   H H                                H
    \ /                               |+
    N  O                          H—N—H  O
    |  ||        Zwitterion           |      ||
 H—C—C—OH   ——————————>     H—C———C—O⁻
    |            form                 |
    H           Glycine               H
```

or, they may be as complicated as the structure,

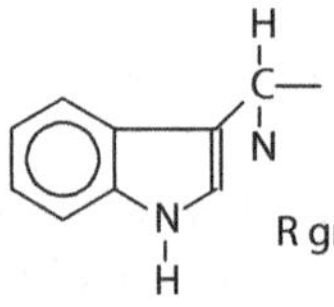

```
            H
            |
            C—
            |
            N
   (indole ring)
        N
        |
        H          R group found in tryptophan
```

found in tryptophan. There are 20 common amino acids in proteins, examples of which are shown in Figure 22.2. The amino acids are shown with uncharged $-NH_2$ and $-CO_2H$ groups. Actually, these functional groups exist in the charged *zwitterion* form as shown for glycine, above.

Amino acids in proteins are joined together by a specific bond called the *peptide linkage*. The formation of peptide linkages is a condensation process involving the loss of water. Consider as an example the condensation of alanine, leucine, and tyrosine, as shown in Figure 22.3. When these three amino acids join together, two water molecules are eliminated. The product is a tripeptide since there are three amino acids involved. The amino acids in proteins are linked as shown for this tripeptide, except that many more monomeric amino acid groups are involved.

Proteins may be divided into several major types that have widely varying functions. These are given in Table 22.1.

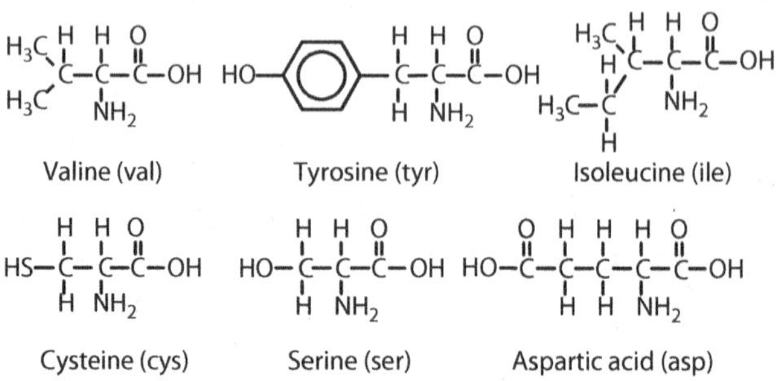

```
H3C H H O                  H H O            H3C H H O
   \| | ||                 | | ||              \| | ||
    C—C—C—OH  HO—(C6H4)—C—C—C—OH            H C—C—C—OH
   /  |                    | |                 /   |
H3C   NH2                  H NH2          H3C—C    NH2
                                               |
                                               H
 Valine (val)           Tyrosine (tyr)      Isoleucine (ile)

    H H O               H H O              O H H H O
    | | ||              | | ||             || | | | ||
HS—C—C—C—OH        HO—C—C—C—OH        HO—C—C—C—C—OH
    | |                 | |                   | | |
    H NH2               H NH2                 H H NH2
 Cysteine (cys)       Serine (ser)       Aspartic acid (asp)
```

FIGURE 22.2 Examples of naturally occurring amino acids.

FIGURE 22.3 Condensation of alanine, leucine, and tyrosine to form a tripeptide consisting of three amino acids joined by peptide linkages (outlined by dashed lines).

22.3.1 Protein Structure

The order of amino acids in protein molecules, and the resulting three-dimensional structures that form, provide an enormous variety of possibilities for *protein structure*. This is what makes life so diverse. Proteins have primary, secondary, tertiary, and quaternary structures. The structures of protein molecules determine the behavior of proteins in crucial areas such as the processes by which the body's immune system recognizes substances that are foreign to the body. Proteinaceous enzymes depend upon their structures for the very specific functions of the enzymes.

The order of amino acids in the protein molecule determines its *primary structure*. *Secondary protein structures* result from the folding of polypeptide protein chains to produce a maximum number of hydrogen bonds between them:

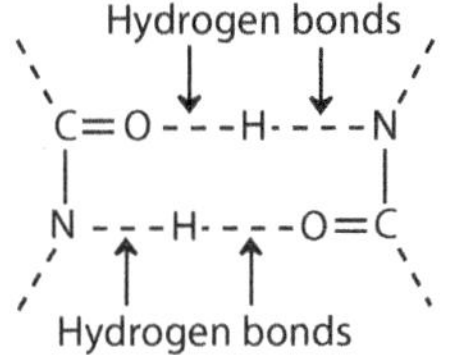

Illustration of hydrogen bonds between N and O atoms in peptide linkages, which constitute protein secondary structures

TABLE 22.1
Major Types of Proteins

Type of Protein	Example	Function and Characteristics
Nutrient	Casein (milk protein)	Food source. People must have an adequate supply of nutrient protein with the right balance of amino acids for adequate nutrition
Storage	Ferritin	Storage of iron in animal tissues
Structural	Collagen (tendons) keratin (hair)	Structural and protective components in organisms
Contractile	Actin, myosin in	Strong, fibrous proteins that can contract muscle tissue and cause movement to occur
Transport	Hemoglobin	Transport inorganic and organic species across cell membranes, in blood, between organs
Defense	—	Antibodies against foreign agents such as viruses produced by the immune system
Regulatory	Insulin, human growth hormone	Regulate biochemical processes such as sugar metabolism or growth by binding to sites inside cells or on cell membranes
Enzymes	Acetylcholinesterase	Catalysts of biochemical reactions (see Section 22.6)

The nature of the R groups on the amino acids determines the secondary structure. Small R groups enable protein molecules to be hydrogen-bonded together in a parallel arrangement. With larger R groups, the molecules tend to take a spiral form. Such a spiral is known as an *α-helix.*

Tertiary structures are formed by the twisting of α-helices into specific shapes. They are produced and held in place by the interactions of amino side chains on the amino acid residues constituting the protein macromolecules. Tertiary protein structure is very important in the processes by which enzymes identify specific proteins and other molecules upon which they act. It is also involved with the action of antibodies in blood which recognize foreign proteins by their shape and react to them. This is basically what happens in the case of immunity to a disease where antibodies in blood recognize specific proteins from viruses or bacteria and reject them.

Two or more protein molecules consisting of separate polypeptide chains may be further attracted to each other to produce a *quaternary structure.*

Some proteins are *fibrous proteins*, which occur in skin, hair, wool, feathers, silk, and tendons. The molecules in these proteins are long and threadlike and are laid out parallel in bundles. Fibrous proteins are quite tough and they do not dissolve in water.

Aside from fibrous protein, the other major type of protein form is the *globular protein.* These proteins are in the shape of balls and oblongs. Globular proteins are relatively soluble in water. A typical globular protein is hemoglobin, the oxygen-carrying protein in red blood cells. Enzymes are generally globular proteins.

22.3.2 Denaturation of Proteins

Secondary, tertiary, and quaternary protein structures are easily changed by a process called *denaturation*. These changes can be quite damaging. Heating, exposure to acids or bases, and even violent physical action can cause denaturation to occur. The albumin protein in egg white is denatured by heating so that it forms a semisolid mass. Almost the same thing is accomplished by the violent physical action of an egg beater in the preparation of meringue. Heavy metal poisons such as lead and cadmium change the structures of proteins by binding to functional groups on the protein surface.

22.4 CARBOHYDRATES

Carbohydrates have the approximate simple formula CH_2O and include a diverse range of substances composed of simple sugars such as glucose:

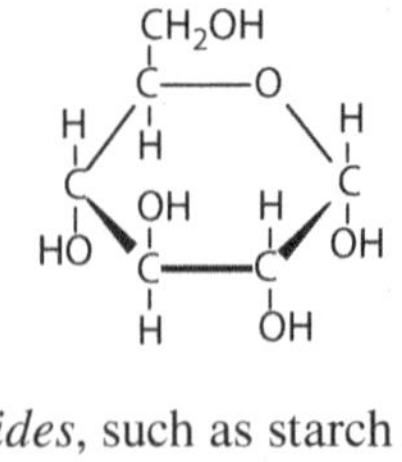

Glucose molecule

High-molar-mass *polysaccharides*, such as starch and glycogen ("animal starch"), are biopolymers of simple sugars.

When photosynthesis occurs in a plant cell, the energy from sunlight is converted to chemical energy in a carbohydrate. This carbohydrate may be transferred to some other part of the plant for use as an energy source. It may be converted to a water-insoluble carbohydrate for storage until it is needed for energy. Or it may be transformed to cell wall material and become part of the structure of the plant. If the plant is eaten by an animal, the carbohydrate is used for energy by the animal.

The simplest carbohydrates are the *monosaccharides*, also called *simple sugars.* Because they have six carbon atoms, simple sugars are sometimes called hexoses. Glucose (formula shown above) is the most common simple sugar involved in cell processes. Other simple sugars with the same

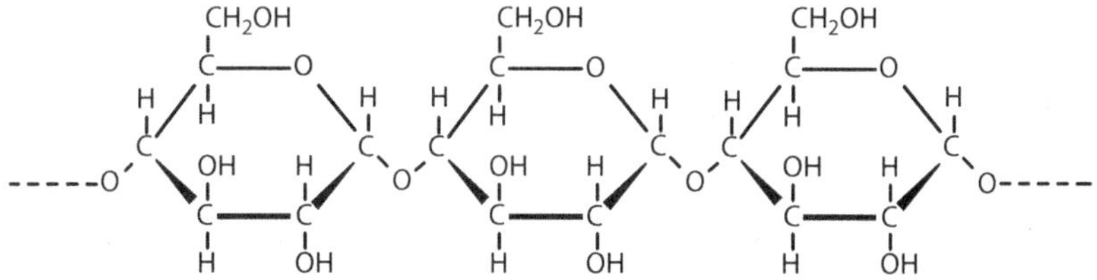

FIGURE 22.4 Part of a starch molecule showing units of $C_6H_{10}O_5$ condensed together.

formula but somewhat different structures are fructose, mannose, and galactose. These must be changed to glucose before they can be used in a cell. Because of its use for energy in body processes, glucose is found in the blood. Normal levels are from 65 to 110 mg glucose per 100 mL of blood. Higher levels may indicate diabetes.

Units of two monosaccharides make up several very important sugars known as *disaccharides*. When two molecules of monosaccharides join together to form a disaccharide,

$$C_6H_{12}O_6 + C_6H_{12}O_6 \rightarrow C_{12}H_{22}O_{11} + H_2O \tag{22.1}$$

a molecule of water is lost. Recall that proteins are also formed from smaller amino acid molecules by condensation reactions involving the loss of water molecules. Disaccharides include sucrose (cane sugar used as a sweetener), lactose (milk sugar), and maltose (a product of the breakdown of starch).

Polysaccharides consist of many simple sugar units hooked together. One of the most important polysaccharides is *starch*, which is produced by plants for food storage. Animals produce a related material called *glycogen*. The chemical formula of starch is $(C_6H_{10}O_5)_n$, where n may represent a number as high as several hundreds. What this means is that the very large starch molecule consists of many units of $C_6H_{10}O_5$ joined together. For example, if n is 100, there are 6 times 100 carbon atoms, 10 times 100 hydrogen atoms, and 5 times 100 oxygen atoms in the molecule. Its chemical formula is $C_{600}H_{1000}O_{500}$. The atoms in a starch molecule are actually present as linked rings represented by the structure shown in Figure 22.4. Starch occurs in many foods, such as bread and cereals. It is readily digested by animals, including humans.

Cellulose is a polysaccharide which is also made up of $C_6H_{10}O_5$ units. Molecules of cellulose are huge, with molecular masses of around 400,000. The cellulose structure (Figure 22.5) is similar to that of starch. Cellulose is produced by plants and forms the structural material of plant cell walls. Wood is about 60% cellulose and cotton contains over 90% of this material. Fibers of cellulose are extracted from wood and pressed together to make paper.

Humans and most other animals cannot digest cellulose because they lack the enzyme needed to hydrolyze the oxygen linkages between the glucose molecules. Ruminant animals (cattle, sheep, goats, moose) have bacteria in their stomachs that break down cellulose into products which can be used by the animal. Chemical processes are available to convert cellulose to simple sugars by the reaction

$$\underset{\text{Cellulose}}{(C_6H_{10}O_5)_n} + nH_2O \rightarrow \underset{\text{Glucose}}{nC_6H_{12}O_6} \tag{22.2}$$

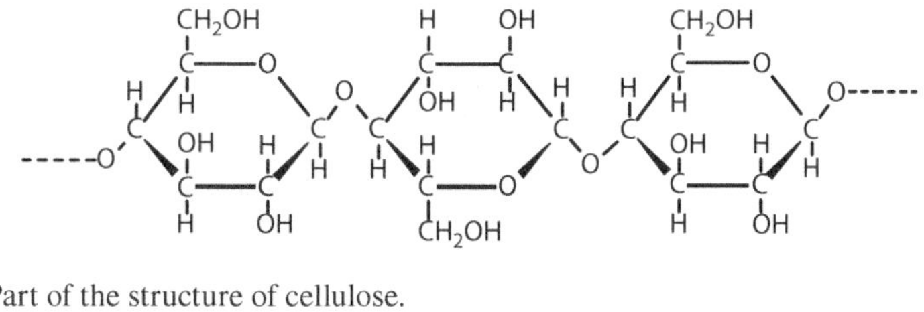

FIGURE 22.5 Part of the structure of cellulose.

where *n* may be 2000–3000. This involves breaking the linkages between units of $C_6H_{10}O_5$ by adding a molecule of H_2O at each linkage, a hydrolysis reaction. Large amounts of cellulose from wood, sugarcane, and agricultural products go to waste each year. The hydrolysis of cellulose enables these products to be converted to sugars, which can be fed to animals.

Carbohydrate groups are attached to protein molecules in a special class of materials called *glycoproteins*. Collagen is a crucial glycoprotein that provides structural integrity to body parts. It is a major constituent of skin, bones, tendons, and cartilage.

22.5 LIPIDS

Lipids are substances that can be extracted from plant or animal matter by organic solvents, such as chloroform, diethyl ether, or toluene (Figure 22.6). Whereas carbohydrates and proteins are characterized predominately by the monomers (monosaccharides and amino acids) from which they are composed, lipids are defined essentially by their physical characteristic of organophilicity. The most common lipids are fats and oils composed of *triglycerides* formed from the alcohol glycerol, $CH_2(OH)CH(OH)CH_2OH$, and a long-chain fatty acid such as stearic acid, $CH_3(CH_2)_{16}COOH$, as shown in

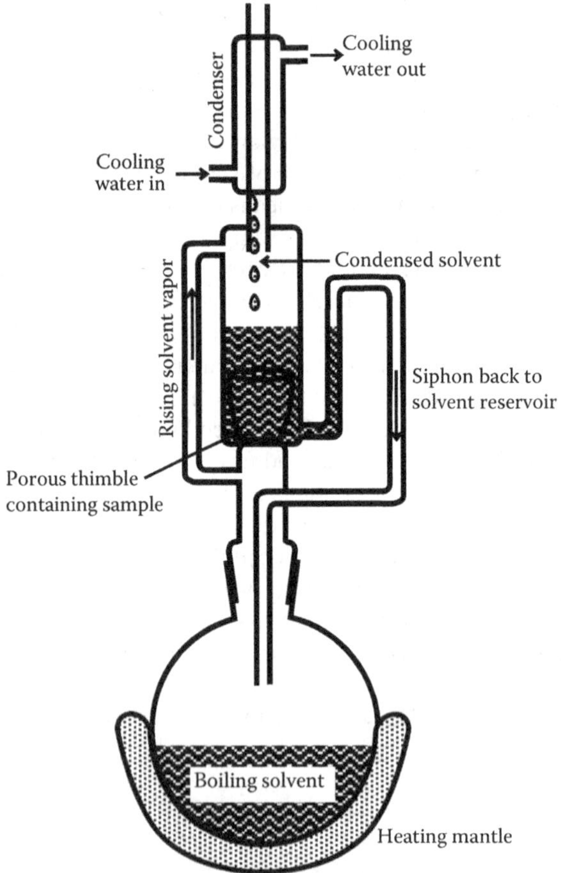

FIGURE 22.6 Lipids are extracted from some biological materials with a soxhlet extractor (above). The solvent is vaporized in the distillation flask by the heating mantle, rises through one of the exterior tubes to the condenser, and is cooled to form a liquid. The liquid drops onto the porous thimble containing the sample. Siphon action periodically drains the solvent back into the distillation flask. The extracted lipid collects as a solution in the solvent in the flask.

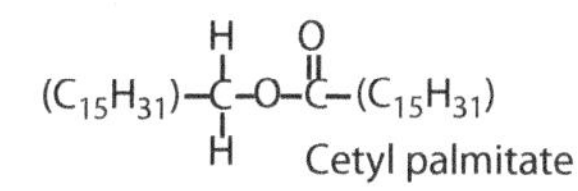

$$(C_{15}H_{31})-\overset{\displaystyle H}{\underset{\displaystyle H}{\overset{|}{\underset{|}{C}}}}-O-\overset{\displaystyle O}{\overset{\|}{C}}-(C_{15}H_{31})$$

Cetyl palmitate

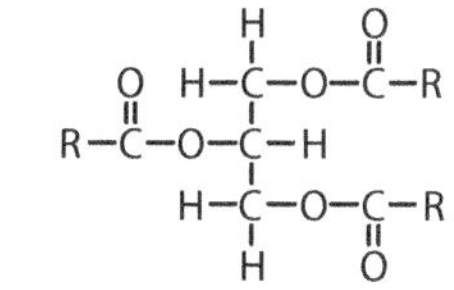

$$\begin{array}{ccccc} & & H & & O \\ & & | & & \| \\ O & H- & C & -O- & C-R \\ \| & & | & & \\ R-C-O- & & C & -H & \\ & & | & & \\ & H- & C & -O- & C-R \\ & & | & & \| \\ & & H & & O \end{array}$$

FIGURE 22.7 General formula of triglycerides, which make up fats and oils. The R group is from a fatty acid and is a hydrocarbon chain, such as $-(CH_2)_{16}CH_3$.

Figure 22.7. Numerous other biological materials, including waxes, cholesterol, and some vitamins and hormones, are classified as lipids. Common foods such as butter and salad oils are lipids. The longer chain fatty acids such as stearic acid are also organic-soluble and are classified as lipids.

Lipids are toxicologically important for several reasons. Some toxic substances interfere with lipid metabolism, leading to detrimental accumulation of lipids. Many toxic organic compounds are poorly soluble in water but are lipid-soluble, so that bodies of lipids in organisms serve to dissolve and store toxicants.

An important class of lipids consists of *phosphoglycerides* (glycerophosphatides), which may be regarded as triglycerides in which one of the acids bonded to glycerol is orthophosphoric acid. These lipids are especially important because they are essential constituents of cell membranes. These membranes consist of bilayers in which the hydrophilic phosphate ends of the molecules are on the outside of the membrane and the hydrophobic "tails" of the molecules are on the inside.

Waxes are also esters of fatty acids. However, the alcohol in a wax is not glycerol, but is often a very long chain alcohol. For example, one of the main compounds in beeswax is myricyl palmitate in which the alcohol portion of the ester has a very large hydrocarbon chain:

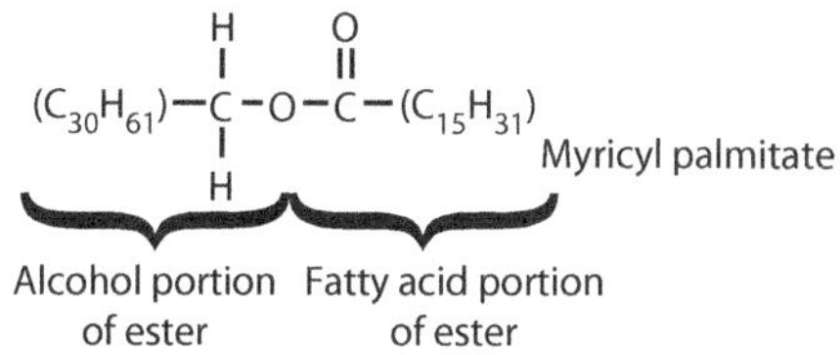

$$\underbrace{(C_{30}H_{61})-\overset{\displaystyle H}{\underset{\displaystyle H}{\overset{|}{\underset{|}{C}}}}}_{\text{Alcohol portion of ester}}\underbrace{-O-\overset{\displaystyle O}{\overset{\|}{C}}-(C_{15}H_{31})}_{\text{Fatty acid portion of ester}}$$

Myricyl palmitate

Waxes are produced by both plants and animals, largely as protective coatings. Waxes are found in a number of common products. Lanolin is one of these. It is the "grease" in sheep's wool. When mixed with oils and water, it forms stable colloidal emulsions consisting of extremely small oil droplets suspended in water. This makes lanolin useful for skin creams and pharmaceutical ointments. Carnauba wax occurs as a coating on the leaves of some Brazilian palm trees. Spermaceti wax is composed largely of cetyl palmitate (below), extracted from the blubber of the sperm whale. It is very useful in some cosmetics and pharmaceutical preparations.

Steroids are lipids found in living systems which all have the ring system shown in Figure 22.8 for cholesterol. Steroids occur in bile salts, which are produced by the liver and then secreted into the intestines. Their breakdown products contribute to the characteristic color of feces. Bile salts act upon fats in the intestine. They suspend very tiny fat droplets in the form of colloidal emulsions. This enables the fats to be broken down chemically and digested.

Some steroids are *hormones*. Hormones act as "messengers" from one part of the body to another. As such, they start and stop a number of body functions. Male and female sex hormones (estrogens) are examples of steroid hormones. Hormones are given off by glands in the body called *endocrine glands*. The locations of important endocrine glands are shown in Figure 22.9.

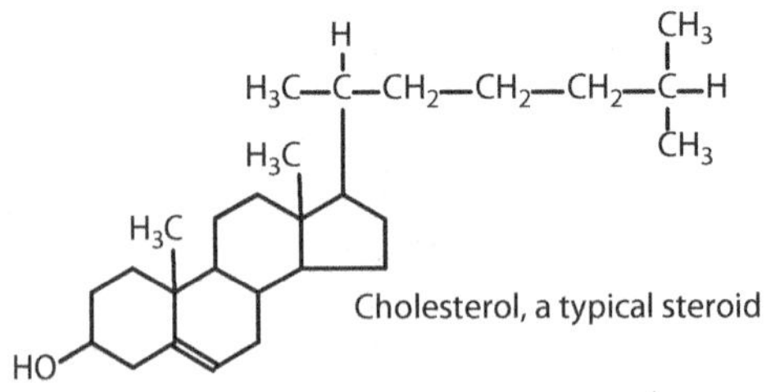

Cholesterol, a typical steroid

FIGURE 22.8 Steroids are characterized by the ring structure shown above for cholesterol.

22.6 ENZYMES

Catalysts are substances that speed up a chemical reaction without themselves being consumed in the reaction. The most sophisticated catalysts of all are those found in living systems. They bring about reactions that could not be performed at all, or only with great difficulty, outside a living organism. These catalysts are called *enzymes*. In addition to speeding up reactions by as much as ten- to a hundred million-fold, enzymes are extremely selective in the reactions they promote.

Enzymes are proteinaceous substances with highly specific structures that interact with particular substances or classes of substances called *substrates*. Enzymes act as catalysts to enable biochemical reactions to occur, after which they are regenerated intact to take part in additional reactions. The extremely high specificity with which enzymes interact with substrates results from their "lock and key" action based upon the unique shapes of enzymes as illustrated in Figure 22.10. This illustration shows that an enzyme "recognizes" a particular substrate by its molecular structure and binds to it to produce an *enzyme–substrate complex*. This complex then breaks apart to form one or more products different from the original substrate, regenerating the unchanged enzyme, which is then available to catalyze additional reactions. The basic process for an enzyme reaction is, therefore,

$$\text{enzyme} + \text{substrate} \rightleftarrows \text{enzyme} - \text{substrate complex} \rightleftarrows \text{enzyme} + \text{product} \quad (22.3)$$

Several important things should be noted about this reaction. As shown in Figure 22.10, an enzyme acts upon a specific substrate to form an enzyme–substrate complex because of the fit between their structures. As a result, something happens to the substrate molecule. For example, it might be split in two at a particular location. Then the enzyme–substrate complex comes apart, yielding the enzyme and products. The enzyme is not changed in the reaction and is now free to

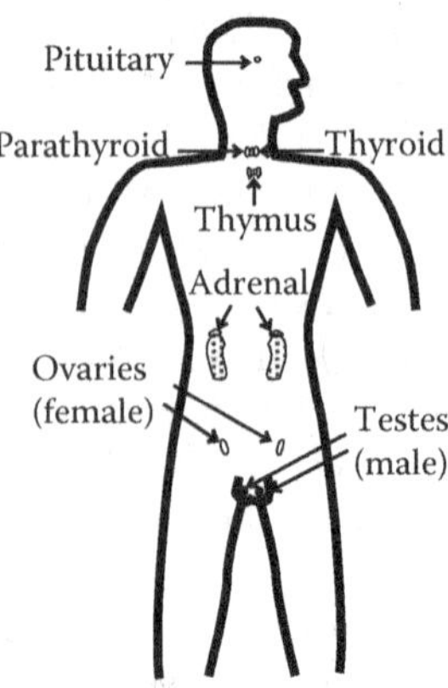

FIGURE 22.9 Locations of important endocrine glands.

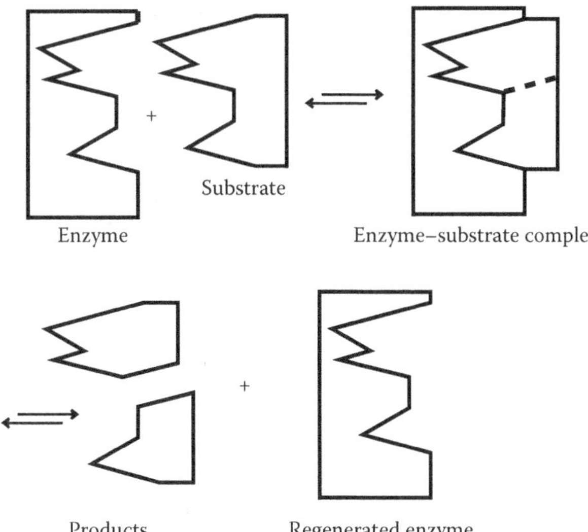

FIGURE 22.10 Representation of the "lock-and-key" mode of enzyme action which enables the very high specificity of enzyme-catalyzed reactions.

react again. Note that the arrows in the formula for enzyme reaction point both ways. This means that the reaction is *reversible*. An enzyme–substrate complex can simply go back to the enzyme and the substrate. The products of an enzymatic reaction can react with the enzyme to form the enzyme–substrate complex again. It, in turn, may again form the enzyme and the substrate. Therefore, the same enzyme may act to cause a reaction to go either way.

Some enzymes cannot function by themselves. In order to work, they must first be attached to *coenzymes*. Coenzymes normally are not protein materials. Some of the vitamins are important coenzymes.

Enzymes are named based on what they do. For example, the enzyme given off by the stomach, which splits proteins as part of the digestion process, is called *gastric proteinase*. The "gastric" part of the name refers to the enzyme's origin in the stomach. The "proteinase" denotes that it splits up protein molecules. The common name for this enzyme is pepsin. Similarly, the enzyme produced by the pancreas that breaks down fats (lipids) is called *pancreatic lipase*. Its common name is steapsin. In general, lipase enzymes cause lipid triglycerides to dissociate and form glycerol and fatty acids.

The enzymes mentioned above are *hydrolyzing enzymes*, which bring about the breakdown of high-molecular-mass biological compounds and add water. This is one of the most important types of the reactions involved in digestion. The three main classes of energy-yielding foods that animals eat are carbohydrates, proteins, and fats. Recall that the higher carbohydrates humans eat are largely disaccharides (sucrose, or table sugar) and polysaccharides (starch). These are formed by the joining together of units of simple sugars, $C_6H_{12}O_6$, with the elimination of an H_2O molecule at the linkage where they join. Proteins are formed by the condensation of amino acids, again with the elimination of a water molecule at each linkage. Fats are esters that are produced when glycerol and fatty acids link together. A water molecule is lost for each of these linkages when a protein, fat, or carbohydrate is synthesized. In order for these substances to be used as a food source, the reverse process must occur to break down large, complicated molecules of protein, fat, or carbohydrate to simple, soluble substances which can penetrate a cell membrane and take part in chemical processes in the cell. This reverse process is accomplished by hydrolyzing enzymes.

Biological compounds with long chains of carbon atoms are broken down into molecules with shorter chains by the breaking of carbon–carbon bonds. This commonly occurs by the elimination of CO_2 from carboxylic acids. For example, *pyruvate decarboxylase* enzyme acts upon pyruvic acid,

$$\underset{\text{Pyruvic acid}}{\mathrm{H{-}\overset{H}{\underset{H}{C}}{-}\overset{O}{\overset{\|}{C}}{-}\overset{O}{\overset{\|}{C}}{-}OH}} \xrightarrow[\text{decarboxylase}]{\text{Pyruvate}} \underset{\text{Acetaldehyde}}{\mathrm{H{-}\overset{H}{\underset{H}{C}}{-}\overset{O}{\overset{\|}{C}}{-}H}} + CO_2 \tag{22.4}$$

to split off CO_2 and produce a compound with one less carbon. It is by such carbon-by-carbon break down reactions that long chain compounds are eventually degraded to CO_2 in the body, or that long-chain hydrocarbons undergo biodegradation by the action of microorganisms in the water and soil environments.

Oxidation and reduction are the major reactions for the exchange of energy in living systems. Cellular respiration is an oxidation reaction in which a carbohydrate, $C_6H_{12}O_6$, is broken down to carbon dioxide and water with the release of energy:

$$C_6H_{12}O_6 + 6O_2 \rightarrow 6CO_2 + 6H_2O + \text{energy} \tag{22.5}$$

Actually, such an overall reaction occurs in living systems by a complicated series of individual steps. Some of these steps involve oxidation. The enzymes that bring about oxidation in the presence of free O_2 are called *oxidases.* In general, biological oxidation–reduction reactions are catalyzed by *oxidoreductase enzymes.*

In addition to the types of enzymes discussed above, there are many enzymes that perform miscellaneous duties in living systems. Typical of these are *isomerases*, which form isomers of particular compounds. For example, of the several simple sugars with the formula $C_6H_{12}O_6$, only glucose can be used directly for cell processes. The other isomers are converted to glucose by the action of isomerases. *Transferase enzymes* move chemical groups from one molecule to another, *lyase enzymes* remove chemical groups without hydrolysis and participate in the formation of C=C bonds or addition of species to such bonds, and *ligase enzymes* work in conjunction with adenosine triphosphate (ATP), (a high-energy molecule that plays a crucial role in energy-yielding, glucose-oxidizing metabolic processes), to link molecules together with the formation of bonds such as carbon–carbon or carbon–sulfur bonds.

Enzyme action may be affected by many different things. Enzymes require a certain hydrogen ion concentration (pH) to function best. For example, gastric proteinase requires the acid environment of the stomach to work well. When it passes into the much less acidic intestines, it stops working. This prevents damage to the intestine walls, which would occur if the enzyme tried to digest them. Temperature is critical. Not surprisingly, the enzymes in the human body work best at around 37°C (98.6°F), which is the normal body temperature. Heating these enzymes to around 60°C permanently destroys them. Some bacteria that thrive in hot springs have enzymes that work best at temperatures as high as that of boiling water. Other "cold-seeking" bacteria have enzymes adapted to near the freezing point of water.

One of the greatest concerns regarding the effects of surroundings upon enzymes is the influence of toxic substances. A major mechanism of toxicity is the alteration or destruction of enzymes by toxic agents—cyanide, heavy metals, or organic compounds such as insecticidal parathion. An enzyme that has been destroyed obviously cannot perform its designated function, whereas one that has been altered either may not function at all or may act improperly. The detrimental effects of toxicants on enzymes are discussed in more detail in Chapter 23.

22.7 NUCLEIC ACIDS

The structural formulas of the monomeric constituents of nucleic acids are given in Figure 22.11. These are pyrimidine or purine nitrogen-containing bases, two sugars, and phosphate. DNA

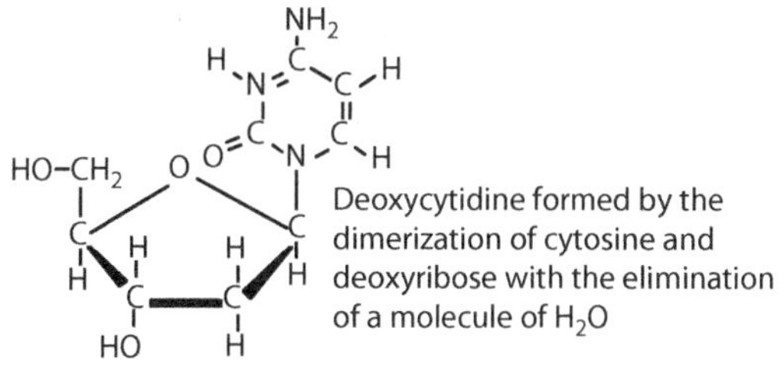

Deoxycytidine formed by the dimerization of cytosine and deoxyribose with the elimination of a molecule of H_2O

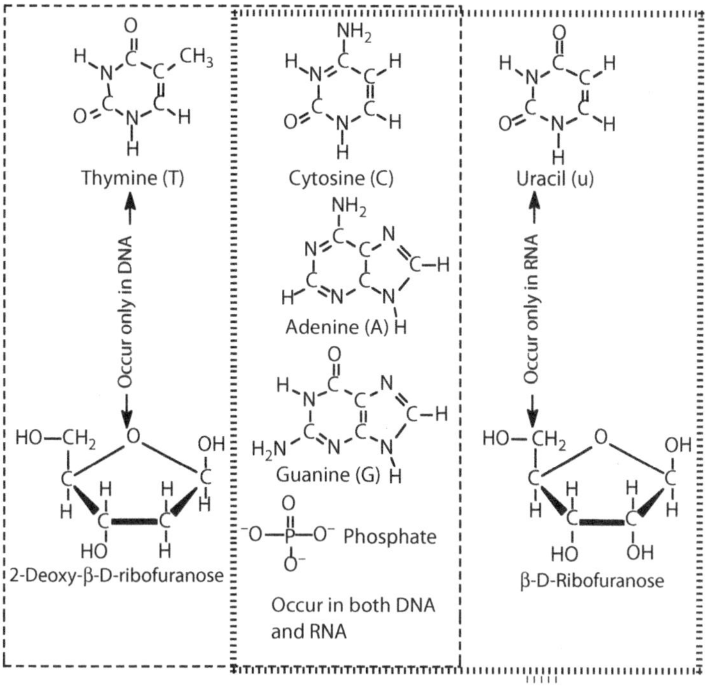

FIGURE 22.11 Constituents of DNA (enclosed by ----) and of RNA (enclosed by ''''').

molecules are made up of the nitrogen-containing bases adenine, guanine, cytosine, and thymine; phosphoric acid (H_3PO_4); and the simple sugar 2-deoxy-β-D-ribofuranose (commonly called deoxyribose). RNA molecules are composed of the nitrogen-containing bases adenine, guanine, cytosine, and uracil; phosphoric acid (H_3PO_4); and the simple sugar β-D-ribofuranose (ribose).

The formation of nucleic acid polymers from their monomeric constituents may be viewed as the following steps:

- Monosaccharide (simple sugar) + cyclic nitrogenous base yields a *nucleoside*.
- Nucleoside + phosphate yields *phosphate ester nucleotide*.

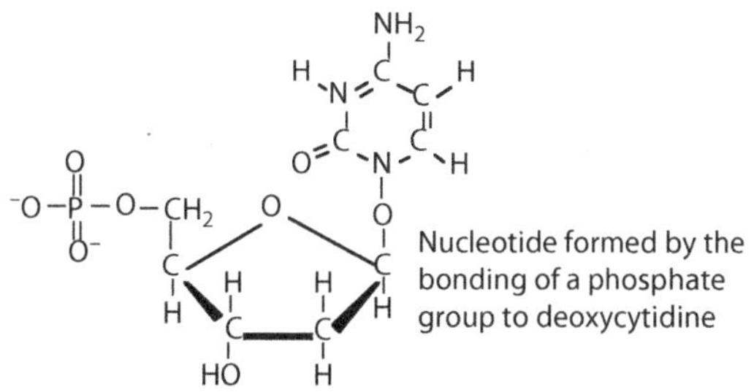

Nucleotide formed by the bonding of a phosphate group to deoxycytidine

- Polymerized nucleotide yields *nucleic acid* as shown by the structure below. In the nucleic acid, the phosphate negative charges are neutralized by metal cations (such as Mg^{2+}) or positively-charged proteins (histones).

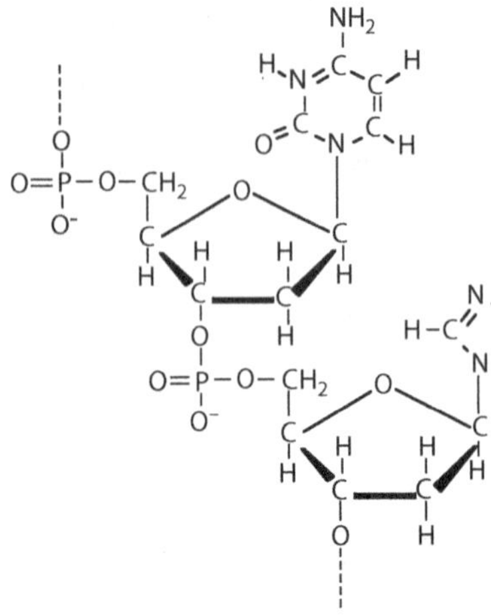

Segment of the DNA polymer showing linkage of two nucleotides

Molecules of DNA are huge with molecular masses >1 billion. Molecules of RNA are also quite large. The structure of DNA is that of the famed "double helix" (Figure 22.12). It was figured out in 1953 by an American scientist, James D. Watson, and Francis Crick, a British scientist. They received the Nobel Prize for this scientific milestone in 1962. This model visualizes DNA as a so-called double α-helix structure of oppositely wound polymeric strands held together by hydrogen bonds between opposing pyrimidine and purine groups. As a result, DNA has both a primary and a secondary structure; the former is due to the sequence of nucleotides in the individual strands of DNA and the latter results from the α-helix interaction of the two strands. In the

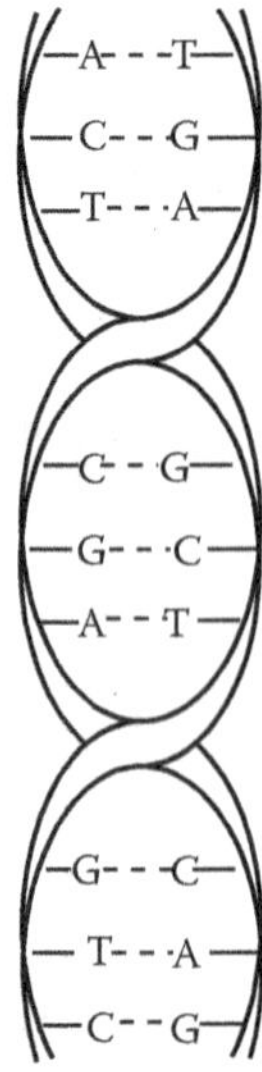

FIGURE 22.12 Representation of the double helix structure of DNA showing the allowed base pairs held together by hydrogen bonding between the phosphate/sugar polymer "backbones" of the two strands of DNA. The letters stand for adenine (A), cytosine (C), guanine (G), and thymine (T). The dashed lines, ---, represent hydrogen bonds.

secondary structure of DNA, only cytosine can be opposite guanine and only thymine can be opposite adenine and vice versa. Basically, the structure of DNA is that of two spiral ribbons "counter-wound" around each other as illustrated in Figure 22.12. The two strands of DNA are *complementary*. This means that a particular portion of one strand fits like a key in a lock with the corresponding portion of another strand. If the two strands are pulled apart, each manufactures a new complementary strand, so that two copies of the original double helix result. This occurs during cell reproduction.

The molecule of DNA is sort of a coded message. This "message," the genetic information contained in, and transmitted by nucleic acids, depends upon the sequence of bases from which they are composed. It is somewhat like the message sent by telegraph, which consists only of dots, dashes, and spaces in between. The key aspect of DNA structure that enables storage and replication of this information is the famed double helix structure of DNA mentioned above.

Portions of the DNA double helix may unravel, and one of the strands of DNA may produce a strand of RNA. This substance then goes from the cell nucleus out into the cell and regulates the synthesis of new protein. In this way, DNA regulates the function of the cell and acts to control life processes.

22.7.1 Nucleic Acids in Protein Synthesis

Whenever a new cell is formed, the DNA in its nucleus must be accurately reproduced from the parent cell. Life processes are absolutely dependent upon accurate protein synthesis as regulated by cell DNA. The DNA in a single cell must be capable of directing the synthesis of up to 3000 or even more different proteins. The directions for the synthesis of a single protein are contained in a segment of DNA called a *gene*. The process of transmitting information from DNA to a newly-synthesized protein involves the following steps:

- The DNA undergoes *replication*. This process involves separation of a segment of the double helix into separate single strands which then replicate such that guanine is opposite cytosine (and vice versa) and adenine is opposite thymine (and vice versa). This process continues until a complete copy of the DNA molecule has been produced.
- The newly replicated DNA produces *messenger RNA* (*m-RNA*), a complement of the single strand of DNA, by a process called *transcription*.
- A new protein is synthesized using m-RNA as a template to determine the order of amino acids in a process called *translation*.

22.7.2 Modified DNA

DNA molecules may be modified by the unintentional addition or deletion of nucleotides or by substituting one nucleotide for another. The result is a *mutation* that is transmittable to offspring. Mutations can be induced by chemical substances. This is a major concern from a toxicological viewpoint because of the detrimental effects of many mutations and because substances that cause mutations often cause cancer as well. DNA malfunction may result in birth defects and the failure to control cell reproduction results in cancer. Radiation from x-rays and radioactivity also disrupts DNA and may cause mutation.

22.8 RECOMBINANT DNA AND GENETIC ENGINEERING

As noted above, segments of DNA contain information for the specific syntheses of particular proteins. Within the last two decades, it has become possible to transfer this information between organisms by means of *recombinant DNA technology*, which has resulted in a new industry based on *genetic engineering*. Most often the recipient organisms are bacteria, which can be reproduced (cloned) over many orders of magnitude from a cell that has acquired the desired qualities. Therefore, to synthesize a particular substance such as human insulin or growth hormone, the required genetic information can be transferred from a human source to bacterial cells, which then produce the substance as part of their metabolic processes.

The first step in recombinant DNA gene manipulation is to lyze or "open up" a cell that has the genetic material needed and to remove this material from the cell. Through enzyme action, the sought-after genes are cut from the donor DNA chain. These are next spliced into small DNA molecules. These molecules, called *cloning vehicles*, are capable of penetrating the host cell and becoming incorporated into its genetic material. The modified host cell is then reproduced many times and carries out the desired biosynthesis.

Early concerns about the potential of genetic engineering to produce "monster organisms" or new and horrible diseases have been largely allayed, although caution is still required with this technology. In the environmental area, genetic engineering offers some hope for the production of bacteria engineered to safely destroy troublesome wastes and to produce biological substitutes for environmentally damaging synthetic pesticides.

Numerous possibilities exist for combining biology with chemistry to produce chemical feedstocks and products of various kinds. An example is the production of polylactic acid using lactic acid produced enzymatically with corn and polymerized by standard chemical processes. Much attention has been focused on the development of enzymes to perform a variety of chemical conversions. Another important area that uses transgenic organisms is the breeding of plants that produce natural insecticides, specifically the insecticide from *Bacillus thuringiensis*.

22.9 METABOLIC PROCESSES

Biochemical processes that involve the alteration of biomolecules fall under the category of *metabolism*. Metabolic processes may be divided into the two major categories of *anabolism* (synthesis) and *catabolism* (degradation of substances). An organism may use metabolic processes to yield energy or to modify the constituents of biomolecules.

22.9.1 Energy-Yielding Processes

Organisms can gain energy by one of three major processes, which are listed as follows:

- *Respiration*, in which organic compounds undergo catabolism that requires molecular oxygen (*aerobic respiration*) or that occurs in the absence of molecular oxygen (*anaerobic respiration*). Aerobic respiration uses the *Krebs cycle* to obtain energy from the following reaction:

$$C_6H_{12}O_6 + 6O_2 \rightarrow 6CO_2 + 6H_2O + \text{energy}$$

  About half of the energy released is converted to short-term stored chemical energy, particularly through the synthesis of ATP nucleotide. For longer-term energy storage, glycogen or starch polysaccharides are synthesized, and for still longer-term energy storage, lipids (fats) are generated and retained by the organism.
- *Fermentation*, which differs from respiration in not having an electron transport chain. Yeasts produce ethanol from sugars by fermentation:

$$C_6H_{12}O_6 \rightarrow 2CO_2 + 2C_2H_5OH$$

- *Photosynthesis*, in which light energy captured by plant and algal chloroplasts is used to synthesize sugars from carbon dioxide and water:

$$6CO_2 + 6H_2O + h\nu \rightarrow C_6H_{12}O_6 + 6O_2$$

Plants cannot always get the energy that they need from sunlight. During the dark, they must use stored food. Plant cells, like animal cells, contain mitochondria in which stored food is converted to energy by cellular respiration.

Plant cells, which use sunlight as a source of energy and CO_2 as a source of carbon, are said to be *autotrophic*. In contrast, animal cells must depend upon organic material manufactured by plants for their food. These are called *heterotrophic* cells. They act as "middlemen" in the chemical reaction between oxygen and food material using the energy from the reaction to carry out their life processes.

22.10 METABOLISM OF XENOBIOTIC COMPOUNDS

When toxicants or their metabolic precursors (*protoxicants*) enter a living organism, they may undergo several processes, including those that may make them more toxic or that detoxify them. Chapter 23 discusses the metabolic processes that toxicants undergo and the mechanisms by which they may cause damage to an organism. Emphasis is placed on *xenobiotic compounds*, which are those that are normally foreign to living organisms; on chemical aspects; and on processes that lead to products that can be eliminated from the organism. Of particular importance is *intermediary xenobiotic metabolism* which results in the formation of somewhat transient species that are different from both those ingested and the ultimate product that is excreted. These species may have significant toxicological effects. Xenobiotic compounds in general are acted upon by enzymes that function on a material that is in the body naturally—an *endogenous substrate*. For example, flavin-containing mono-oxygenase enzyme acts upon endogenous cysteamine to convert it to cystamine, but also functions to oxidize xenobiotic nitrogen and sulfur compounds.

Biotransformation refers to changes in xenobiotic compounds as a result of enzyme action. Reactions that are not mediated by enzymes may also be important in some cases. As examples of nonenzymatic transformations, some xenobiotic compounds bond with endogenous biochemical species without an enzyme catalyst, undergo hydrolysis in body fluid media, or undergo oxidation/reduction processes. The metabolic Phase I and Phase II reactions of xenobiotics discussed here and in Chapter 23 are enzymatic, however.

The likelihood that a xenobiotic species will undergo enzymatic metabolism in the body depends upon the chemical nature of the species. Compounds with a high degree of polarity, such as relatively ionizable carboxylic acids, are less likely to enter the body system and, when they do, tend to be quickly excreted. Therefore, such compounds are unavailable, or only available for a short time, for enzymatic metabolism. Volatile compounds, such as dichloromethane or diethyl ether, are expelled so quickly from the lungs that enzymatic metabolism is less likely. This leaves *nonpolar lipophilic compounds*, those that are relatively less soluble in aqueous biological fluids and more attracted to lipid species, as the most likely candidates for enzymatic metabolic reactions. Of these, the ones that are resistant to enzymatic attack (e.g., PCBs) tend to bioaccumulate in lipid tissue.

Xenobiotic species may be metabolized in many body tissues and organs. As part of the body's defense against the entry of xenobiotic species, the most prominent sites of xenobiotic metabolism are those associated with entry into the body, such as the skin and lungs. The gut wall through which xenobiotic species enter the body from the gastrointestinal tract is also a site of significant xenobiotic compound metabolism. The liver is of particular significance because materials entering systemic circulation from the gastrointestinal tract must first traverse the liver.

22.10.1 Phase I and Phase II Reactions

The processes that most xenobiotics undergo in the body can be divided into the two categories of Phase I reactions and Phase II reactions. A *Phase I reaction* introduces reactive, polar functional groups into lipophilic ("fat-seeking") toxicant molecules. In their unmodified forms, such toxicant molecules tend to pass through lipid-containing cell membranes and may be bound to lipoproteins, in which form they are transported through the body. Because of the functional group present, the product of a Phase I reaction is usually more water-soluble than the parent xenobiotic species, and more importantly, possesses a "chemical handle" to which a substrate material in the body may become attached so that the toxicant can be eliminated from the body. The binding of such a substrate is a *Phase II reaction*, and it produces a *conjugation product* that is amenable to excretion from the body.

In general, the changes in structure and properties of a compound that result from a Phase I reaction are relatively mild. Phase II processes, however, usually produce species that are much different from the parent compounds. It should be emphasized that not all xenobiotic compounds undergo both Phase I and Phase II reactions. Such a compound may undergo only a Phase I reaction and be excreted directly from the body. Or a compound that already possesses an appropriate functional group capable of conjugation may undergo a Phase II reaction without a preceding Phase I reaction. Phase I and Phase II reactions are discussed in more detail as they relate to toxicological chemistry in Chapters 23 and 24.

LITERATURE CITED

1. Voet, D., J. G. Voet, and C. W. Pratt, *Fundamentals of Biochemistry*, Wiley, New York, 2008.
2. Stanley E. M., *Toxicological Chemistry and Biochemistry*, 3rd ed., CRC Press/Lewis Publishers, Boca Raton, FL, 2002.

SUPPLEMENTARY REFERENCES

Berg, J. M., J. L. Tymoczko, and L. Stryer, *Biochemistry*, 6th ed., W.H. Freeman, New York, 2007.

Bettelheim, F. A., W. H. Brown, and J. March, *Introduction to General, Organic & Biochemistry*, 8th ed., Thomson-Brooks/Cole, Belmont, CA, 2007.

Bowsher, C., M. Steer, and A. Tobin, *Plant Biochemistry*, Garland Science, New York, 2008.

Campbell, M. K. and S. O. Farrell, *Biochemistry*, 5th ed., Thomson-Brooks/Cole, Belmont, CA, 2006.

Champe, P. C., R. A. Harvey, and D. R. Ferrier, *Biochemistry*, 4th ed., Lippincott Williams & Wilkins, Baltimore, 2008.

Chesworth, J. M., T. Stuchbury, and J. R. Scaife, *An Introduction to Agricultural Biochemistry*, Chapman and Hall, London, 1998.

Elliott, W. H. and D. C. Elliott, *Biochemistry and Molecular Biology*, 4th ed., Oxford University Press, New York, 2009.

Garrett, R. H. and C. M. Grisham, *Biochemistry*, 4th ed., Thomson-Brooks/Cole, Belmont, CA, 2009.

Horton, H. R., *Principles of Biochemistry*, 4th ed., Prentice Hall, Upper Saddle River, NJ, 2006.

Kuchel, P. W., K.E. Cullen, and G. B. Ralston, *Schaum's Easy Outline of Biochemistry*, McGraw-Hill, Boston, 2002.

McKee, T. and J. R. McKee, *Biochemistry: The Molecular Basis of Life*, 4th ed., Oxford University Press, New York, 2009.

Moore, J., T. and R. Langley, *Biochemistry for Dummies*, Wiley Publishing, Inc., Indianapolis, IN, 2008.

Swanson, T. A., S. I. Kim, and M. J. Glucksman, *Biochemistry and Molecular Biology*, 4th ed., Lippincott Williams & Wilkins, Philadelphia, 2007.

Tymoczko, J. L., J. M. Berg, and L. Stryer, *Biochemistry: A Short Course*, W.H. Freeman and Co., New York, 2009.

Voet, D. and J. G. Voet, *Fundamentals of Biochemistry: Life at the Molecular Level*, 3rd ed., Wiley, Hoboken, NJ, 2008.

QUESTIONS AND PROBLEMS

1. What is the toxicological importance of lipids? How are lipids related to hydrophobic pollutants and toxicants?
2. What is the function of a hydrolase enzyme?

3. Match the cell structure on the left with its function on the right, below:

1. Mitochondria	a. Toxicant metabolism
2. Endoplasmic reticulum	b. Fills the cell
3. Cell membrane	c. Deoxyribonucleic acid
4. Cytoplasm	d. Mediate energy conversion and utilization
5. Cell nucleus	e. Encloses the cell and regulates the passage of materials into and out of the cell interior

4. The formula of simple sugars is $C_6H_{12}O_6$. The simple formula of higher carbohydrates is $C_6H_{10}O_5$. Of course, many of these units are required to make a molecule of starch or cellulose. If higher carbohydrates are formed by joining together the molecules of simple sugars, why is there a difference in the ratios of C, H, and O atoms in the higher carbohydrates as compared to the simple sugars?
5. Why does wood contain so much cellulose?
6. What would be the chemical formula of a trisaccharide made by the bonding together of three simple sugar molecules?
7. The general formula of cellulose may be represented as $(C_6H_{10}O_5)_x$. If the molar mass of a molecule of cellulose is 400,000, what is the estimated value of x?
8. During 1 month a factory for the production of simple sugars, $C_6H_{12}O_6$, by the hydrolysis of cellulose processes 1 million pounds of cellulose. The percentage of cellulose that undergoes the hydrolysis reaction is 40%. How many pounds of water are consumed in the hydrolysis of cellulose each month?
9. What is the structure of the largest group of atoms common to all amino acid molecules?
10. Glycine and phenylalanine can join together to form two different dipeptides. What are the structures of these two dipeptides?
11. One of the ways in which two parallel protein chains are joined together, or cross linked, is by way of an –S–S– link. Which amino acid, might be the most likely to be involved in such a link? Explain your choice.
12. Fungi, which break down wood, straw, and other plant material, have what are called "exoenzymes." Fungi have no teeth and cannot break up plant material physically by force. Knowing this, what do you suppose an exoenzyme is? Explain how you think it might operate in the process by which fungi break down something as tough as wood.
13. Many fatty acids of lower molecular weight have a bad odor. Speculate as to the reasons that rancid butter has a bad odor. What chemical compound is produced that has a bad odor? What sort of chemical reaction is involved in its production?
14. The long-chain alcohol with 10 carbons is called decanol. What do you think would be the formula of decyl stearate? To what class of compounds would it belong?
15. Write an equation for the chemical reaction between sodium hydroxide and cetyl stearate. What are the products?
16. What type of endocrine gland is found only in females? What type of these glands is found only in males?
17. The action of bile salts is a little like that of soap. What function do bile salts perform in the intestine? Look up the action of soaps in Chapter 5, and explain how you think bile salts may function somewhat like soap.
18. If the structure of an enzyme is illustrated as,

how should the structure of its substrate be represented?

19. Look up the structures of ribose and deoxyribose. Explain where the "deoxy" came from in the name deoxyribose.
20. In what respect are an enzyme and its substrate like two opposite strands of DNA?
21. For what discovery are Watson and Crick noted?
22. Why does an enzyme no longer function if it is denatured?

23 Toxicological Chemistry

23.1 INTRODUCTION TO TOXICOLOGY AND TOXICOLOGICAL CHEMISTRY

Ultimately, most pollutants and hazardous substances are of concern because of their toxic effects. The general aspects of these effects are addressed in this chapter under the heading toxicological chemistry; the toxicological chemistry of specific classes of chemical substances is addressed in this chapter. In order to understand toxicological chemistry, it is essential to have some understanding of biochemistry, the science that deals with chemical processes and materials in living systems. Biochemistry was summarized in Chapter 22.

23.1.1 Toxicology

A *poison*, or *toxicant*, is a substance that is harmful to living organisms because of its detrimental effects on tissues, organs, or biological processes. Common end point effects of toxic substances are destruction of cells, mutation of DNA leading to cancer, and disruption of the signaling mechanisms by which cell development and function are controlled. Most toxicants are foreign to the bodies of individuals that are affected (see xenobiotics in Section 23.5) and tend to have an affinity for lipids. Therefore, they have a tendency to cross the lipid membranes of cells and to build up to toxic levels. In many cases toxicants undergo metabolism to produce an active species that causes poisoning.

Toxicology is the science of poisons. Whether a substance is poisonous depends upon the type of organism exposed, the amount of the substance, and the route of exposure. In the case of human exposure, the degree of harm done by a poison can depend strongly upon whether the exposure is to the skin, by inhalation, or through ingestion.

Toxicants to which subjects are exposed to in the environment or occupationally may be in several different physical forms. This may be illustrated for toxicants that are inhaled. *Gases* are substances such as carbon monoxide in air that are normally in the gaseous state under ambient conditions of temperature and pressure. *Vapors* are gas-phase materials that have evaporated or sublimed from liquids or solids. *Dusts* are respirable solid particles produced by grinding bulk solids, whereas *fumes* are solid particles from the condensation of vapors, often metals or metal oxides. *Mists* are liquid droplets.

Often a toxic substance is in solution or mixed with other substances. A substance with which the toxicant is associated (the solvent in which it is dissolved or the solid medium in which it is dispersed) is called the *matrix*. The matrix may have a strong effect upon the toxicity of the toxicant.

There are numerous variables related to the ways in which organisms are exposed to toxic substances. One of the most crucial of these, *dose*, is discussed in Section 23.2. Another important factor is the *toxicant concentration*, which may range from the pure substance (100%) down to a very dilute solution of a highly potent poison. Both the *duration* of exposure per exposure incident and the *frequency* of exposure are important. The *rate* of exposure and the total time period over which the organism is exposed are both important situational variables. The exposure *site* and *route* also affect toxicity.

It is possible to classify exposures on the basis of acute versus chronic and local versus systemic exposure, giving four general categories. *Acute local* exposure occurs at a specific location over a time period of a few seconds to a few hours and may affect the exposure site, particularly the skin, eyes, or mucous membranes. The same parts of the body can be affected by *chronic local* exposure, for which the time span may be as long as several years. *Acute systemic* exposure is a brief exposure or exposure to a single dose and occurs with toxicants that can enter the body, such as by inhalation or ingestion, and affect organs, such as the liver, that are remote from the entry site. *Chronic systemic* exposure differs in that the exposure occurs over a prolonged time period.

In discussing exposure sites for toxicants it is useful to consider the major routes and sites of exposure, distribution, and elimination of toxicants in the body as shown in Figure 23.1. The major routes of accidental or intentional exposure to toxicants by humans and other animals are the skin (percutaneous or dermal route), the lungs (inhalation, respiration, pulmonary route), and the mouth (oral route); minor routes of exposure are rectal, vaginal, and parenteral (intravenous or intramuscular, a common means for the administration of drugs or toxic substances in test subjects). Of these, the dermal route is the most difficult to quantify. It is particularly important in children, whose activities bring them into contact with contaminated dirt, pesticides, household chemicals, and other environmental pollutants. Furthermore, the skin of children is relatively more permeable to toxic substances, increasing their risk of dermal exposure.

The way that a toxic substance is introduced into the complex system of an organism is strongly dependent upon the physical and chemical properties of the substance. The pulmonary system is most likely to take in toxic gases or very fine, respirable solid or liquid particles. In other than a respirable form, a solid usually enters the body orally. Absorption through the skin is most likely for liquids, solutes in solution, and semisolids, such as sludges.

The defensive barriers that a toxicant may encounter vary with the route of exposure. Toxic substances ingested orally are absorbed through the intestinal epithelium, which has detoxification systems that help reduce the effects of the substances. Toxic elemental mercury is absorbed through the alveoli in the lungs much more readily than through the skin or gastrointestinal tract. Most test exposures to animals are through ingestion or gavage (introduction into the stomach through a tube). Pulmonary exposure is often favored with subjects that may exhibit refractory behavior when

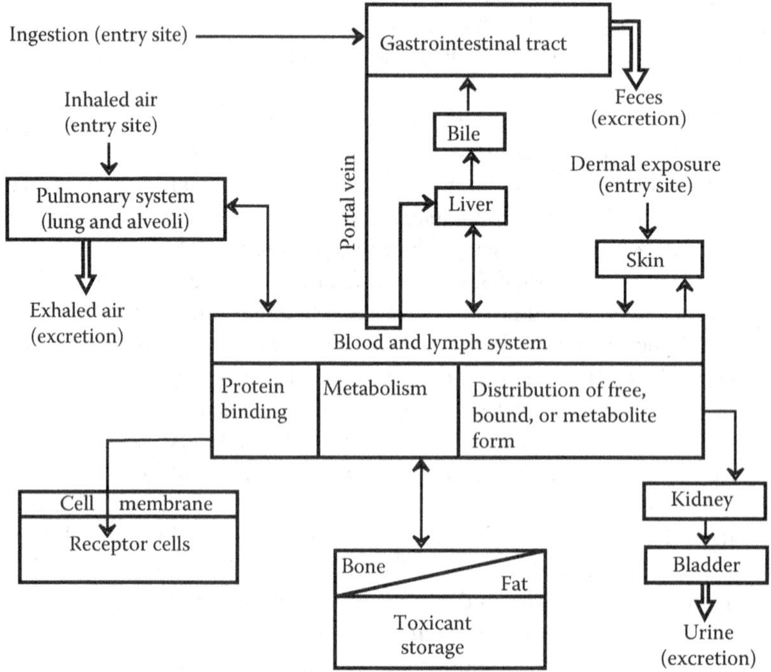

FIGURE 23.1 Major sites of exposure, metabolism, and storage, routes of distribution and elimination of toxic substances in the body.

noxious chemicals are administered by means requiring a degree of cooperation from the subject. Intravenous injection may be chosen for deliberate exposure when it is necessary to know the concentration and effect of a xenobiotic substance in the blood. However, pathways used experimentally that are almost certain not to be significant in accidental exposures can give misleading results when they avoid the body's natural defense mechanisms.

An interesting historical example of the importance of the route of exposure to toxicants is provided by cancer caused by contact of coal tar with skin. The major barrier to dermal absorption of toxicants is the *stratum corneum*, or horny layer. The permeability of skin is inversely proportional to the thickness of this layer, which varies by location on the body in the order soles and palms > abdomen, back, legs, arms > genital (perineal) area. Evidence of the susceptibility of the genital area to absorption of toxic substances is to be found in accounts of the high incidence of cancer of the scrotum among chimney sweeps in London described by Sir Percival Pott, Surgeon General of Britain during the reign of King George III. The cancer-causing agent was coal tar condensed in chimneys. This material was more readily absorbed through the skin in the genital areas than elsewhere, leading to a high incidence of scrotal cancer. (The chimney sweeps' conditions were aggravated by their lack of appreciation of basic hygienic practices, such as bathing and regular changes of underclothing.)

Organisms can serve as indicators of various kinds of pollutants. In this application, organisms are known as *biomonitors*. For example, higher plants, fungi, lichens, and mosses can be important biomonitors for heavy metal pollutants in the environment.

23.1.2 Synergism, Potentiation, and Antagonism

The biological effects of two or more toxic substances can be different in kind and degree from those of one of the substances alone. One of the ways in which this can occur is when one substance affects the way in which another undergoes any of the steps in the kinetic phase as discussed in Section 23.7 and illustrated in Figure 23.9. Chemical interaction between substances may affect their toxicities. Both substances may act upon the same physiologic function, or two substances may compete for binding to the same receptor (molecule or other entity acted upon by a toxicant). When both substances have the same physiologic function, their effects may be simply *additive* or they may be *synergistic* (the total effect is greater than the sum of the effects of each separately). *Potentiation* occurs when an inactive substance enhances the action of an active one, and *antagonism* when an active substance decreases the effect of another active one.

23.2 DOSE–RESPONSE RELATIONSHIPS

Toxicants have widely varying effects upon organisms. Quantitatively, these variations include minimum levels at which the onset of an effect is observed, the sensitivity of the organism to small increments of toxicant, and levels at which the ultimate effect (particularly death) occurs in most exposed organisms. Some essential substances, such as nutrient minerals, have optimum ranges above and below which detrimental effects are observed (see Section 23.5 and Figure 23.4).

Factors such as those just outlined are taken into account by the *dose–response* relationship, which is one of the key concepts of toxicology. *Dose* is the amount, usually per unit body mass, of a toxicant to which an organism is exposed. *Response* is the effect upon an organism resulting from exposure to a toxicant. In order to define a dose–response relationship, it is necessary to specify a particular response, such as death of the organism, as well as the conditions under which the response is obtained, such as the length of time from administration of the dose. Consider a specific response of a population of the same kinds of organisms. At relatively low doses, none of the organisms exhibits the response (e.g., all live), whereas at higher doses all of the organisms exhibit the response (e.g., all die). In between, there is a range of doses over which some of the organisms respond in the specified manner and others do not, thereby defining a dose–response curve. Dose–response relationships differ among different kinds and strains of organisms, types of tissues, and populations of cells.

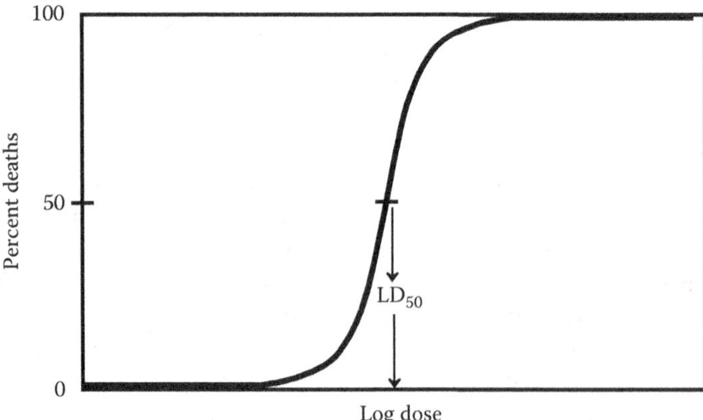

FIGURE 23.2 Illustration of a dose–response curve in which the response is the death of the organism. The cumulative percentage of deaths of organisms is plotted on the *Y* axis.

Figure 23.2 shows a generalized dose–response curve. Such a plot may be obtained, for example, by administering different doses of a poison in a uniform manner to a homogeneous population of test animals and plotting the cumulative percentage of deaths as a function of the log of the dose. The dose corresponding to the mid-point (inflection point) of the resulting S-shaped curve is the statistical estimate of the dose that would kill 50% of the subjects and is designated as LD_{50}. The estimated doses at which 5% (LD_5) and 95% (LD_{95}) of the test subjects die are obtained from the graph by reading the dose levels for 5% and 95% fatalities, respectively. A relatively small difference between LD_5 and LD_{95} is reflected by a steeper S-shaped curve, and vice versa. Statistically, 68% of all values on a dose–response curve fall within ±1 standard deviation of the mean at LD_{50} and encompass the range from LD_{16} to LD_{84}.

23.3 RELATIVE TOXICITIES

Table 23.1 illustrates standard *toxicity ratings* that are used to describe estimated toxicities of various substances to humans. In terms of fatal doses to an adult human of average size, a "taste" of a supertoxic substance (just a few drops or less) is fatal. A teaspoonful of a very toxic substance could have the same effect. However, as much as a quart of a slightly toxic substance might be required to kill an adult human.

When there is a substantial difference between LD_{50} values of two different substances, the one with the lower value is said to be the more *potent*. Such a comparison must assume that the dose–response curves for the two substances being compared have similar slopes.

23.3.1 Nonlethal Effects

So far, toxicities have been described primarily in terms of the ultimate effect—deaths of organisms or lethality. This is obviously an irreversible consequence of exposure. In many, and perhaps most, cases, sublethal and reversible effects are of greater importance. This is obviously true of drugs, where death from exposure to a registered therapeutic agent is rare, but other effects, both detrimental and beneficial, are usually observed. By their very nature, drugs alter biological processes; therefore, the potential for harm is almost always present. The major consideration in establishing drug dose is to find a dose that has an adequate therapeutic effect without undesirable side effects. A dose–response curve can be established for a drug that progresses from noneffective levels through effective, harmful, and even lethal levels. A low slope for this curve indicates a wide range of effective dose and a wide margin of safety (see Figure 23.3). This term applies to other substances, such as pesticides, for which it is desirable to have a large difference between the dose that kills a target species and that which harms a desirable species.

TABLE 23.1
Toxicity Scale with Example Substances[a]

Substance	Approximate LD_{50}	Toxicity Rating
	-10^5	1. Practically nontoxic $>1.5 \times 10^4$ mg/kg
DEHP[b] →	–	
Ethanol →	-10^4	2. Slightly toxic, 5×10^3 to 1.5×10^4 mg/kg
Sodium chloride →	–	
Malathion →	-10^3	
Chlordane →	–	3. Moderately toxic, 500 to 5000 mg/kg
Heptachlor →	-10^2	
	–	4. Very toxic, 50 to 500 mg/kg
Parathion →	–10	
	–	
TEPP[c] →	–1	5. Extremely toxic, 5 to 50 mg/kg
Tetrodotoxin[d] →	-10^{-1}	
	–	
	-10^{-2}	
	–	6. Supertoxic, <5 mg/kg
TCDD[e] →	-10^{-3}	
	–	
	-10^{-4}	
	–	
Botulinus toxin →	-10^{-5}	

[a] Doses are in units of milligrams of toxicant per kilogram of body mass. Toxicity ratings on the right are given as numbers ranging from 1 (practically nontoxic) through 6 (supertoxic) along with estimated lethal oral doses for humans in mg/kg. Estimated LD_{50} values for substances on the left have been measured in test animals, usually rats, and apply to oral doses.
[b] Bis(2-ethylhexyl)phthalate.
[c] Tetraethylpyrophosphate.
[d] Toxin from puffer fish.
[e] TCDD, commonly called "dioxin."

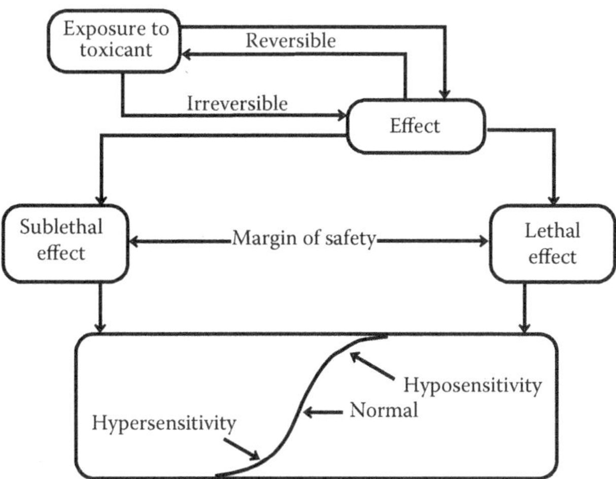

FIGURE 23.3 Effects of and responses to toxic substances.

23.4 REVERSIBILITY AND SENSITIVITY

Sublethal doses of most toxic substances are eventually eliminated from an organism's system. If there is no lasting effect from the exposure, it is said to be *reversible*. However, if the effect is permanent, it is termed *irreversible*. Irreversible effects of exposure remain after the toxic substance is eliminated from the organism. Figure 23.3 illustrates these two kinds of effects. For various chemicals and different subjects, toxic effects may range from the totally reversible to the totally irreversible.

23.4.1 HYPERSENSITIVITY AND HYPOSENSITIVITY

Examination of the dose–response curve shown in Figure 23.2 reveals that some subjects are very sensitive to a particular poison (e.g., those killed at a dose corresponding to LD_5), whereas others are very resistant to the same substance (e.g., those surviving a dose corresponding to LD_{95}). These two kinds of responses illustrate *hypersensitivity* and *hyposensitivity*, respectively; subjects in the mid-range of the dose–response curve are termed *normals*. These variations in response tend to complicate toxicology in that there is no specific dose guaranteed to yield a particular response, even in a homogeneous population.

In some cases hypersensitivity is induced. After one or more doses of a chemical, a subject may develop an extreme reaction to it. This occurs with penicillin, for example, in cases where people develop such a severe allergic response to the antibiotic that exposure is fatal if countermeasures are not taken.

23.5 XENOBIOTIC AND ENDOGENOUS SUBSTANCES

Xenobiotic substances are those that are foreign to a living system, whereas those that occur naturally in a biologic system are termed *endogenous*. Xenobiotic substances that cause harm in organisms are commonly metabolized by them. The levels of endogenous substances must usually fall within a particular concentration range in order for metabolic processes to occur normally. Levels below a normal range may result in a deficiency response or even death, and the same effects may occur above the normal range. This kind of response is illustrated in Figure 23.4.

Examples of endogenous substances in organisms include various hormones, glucose (blood sugar), and some essential metal ions, including Ca^{2+}, K^+, and Na^+. The optimum level of calcium in human blood serum occurs over a rather narrow range of 9–9.5 milligrams per deciliter (mg/dL). Below these values a deficiency response known as hypocalcemia occurs, manifested by muscle cramping. At serum levels above about 10.5 mg/dL hypercalcemia occurs, the major effect of which is kidney malfunction.

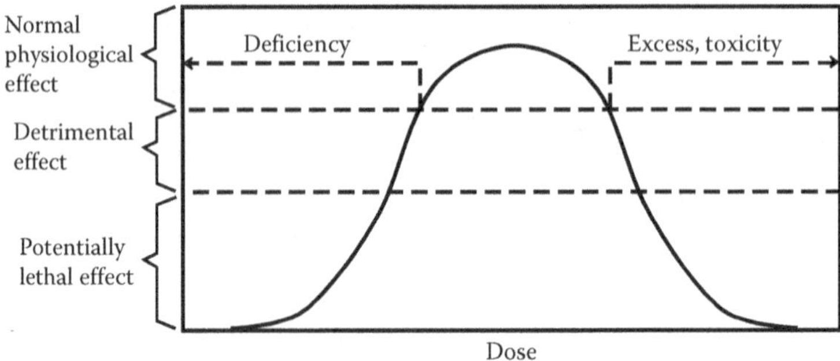

FIGURE 23.4 Biological effect of an endogenous substance in an organism showing optimum level, deficiency, and excess.

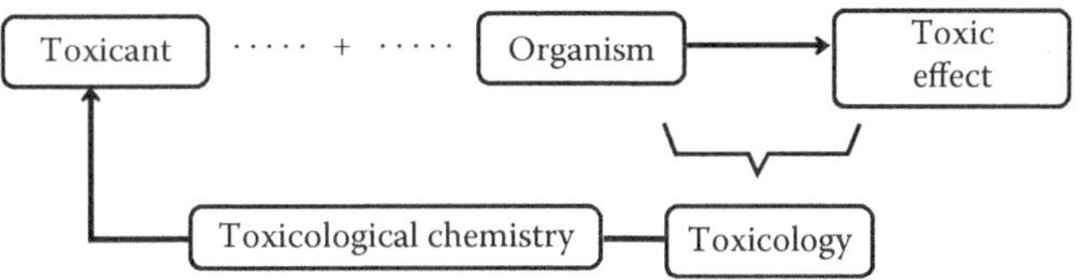

FIGURE 23.5 Toxicology is the science of poisons. Toxicological chemistry relates toxicology to the chemical nature of toxicants.

23.6 TOXICOLOGICAL CHEMISTRY

23.6.1 Toxicological Chemistry Defined

Toxicological chemistry is the science that deals with the chemical nature and reactions of toxic substances, including their origins, uses, and chemical aspects of exposure, fates, and disposal.[1] Toxicological chemistry addresses the relationships between the chemical properties and molecular structures of molecules and their toxicological effects. Figure 23.5 outlines the terms discussed above and the relationships among them.

Much of what is known about xenobiotic substances in living systems is based upon intensive research upon pharmaceutical compounds in organisms. *Pharmacodynamics* deals with what a drug does to a body including the dose–response relationship, sites and mechanisms of pharmaceutical actions, therapeutic effects, and side effects. What the body does to a drug is addressed by *pharmacokinetics*, including uptake, distribution, metabolism, retention, and excretion.

23.6.2 Toxicants in the Body

The processes by which organisms metabolize xenobiotic species are enzyme-catalyzed Phase I and Phase II reactions, which are described briefly here. The substances that undergo these reactions are treated theoretically by the science of *quantitative structure-activity relationships* (QSAR), which relates the chemical nature of substances to their biochemical reactions.

23.6.2.1 Phase I Reactions

Lipophilic xenobiotic species in the body tend to undergo *Phase I reactions* that make them more water-soluble and reactive by the attachment of polar functional groups, such as –OH (Figure 23.6). Most Phase I processes are "microsomal mixed-function oxidase" reactions catalyzed by the cytochrome P-450 enzyme system, associated with the *endoplasmic reticulum* of the cell and occurring most abundantly in the liver of vertebrates.[2] An example of a Phase I reaction of a xenobiotic is epoxide formation of chloroprene, used to make solvent-resistant synthetic rubbers:

Chloroprene —{O}, P-450→ epoxide —H_2O→ diol; Chloroprene —{O} P450→ epoxide —Rearrange→ products (23.1)

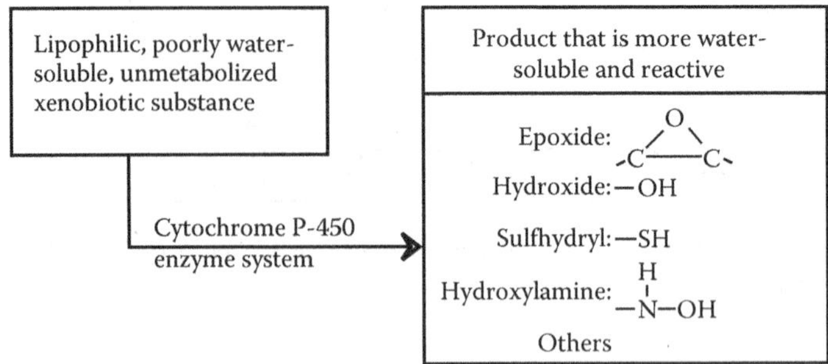

FIGURE 23.6 Illustration of Phase I reactions.

23.6.2.2 Phase II Reactions

A *Phase II reaction* occurs when an endogenous species is attached by enzyme action to a polar functional group which often, though not always, is the result of a Phase I reaction on a xenobiotic species. Phase II reactions are called *conjugation reactions* in which enzymes attach *conjugating agents* to xenobiotics, their Phase I reaction products, and nonxenobiotic compounds (Figure 23.7). The *conjugation product* of such a reaction is usually less toxic than the original xenobiotic compound, less lipid-soluble, more water-soluble, and more readily eliminated from the body. The major conjugating agents and the enzymes that catalyze their Phase II reactions are glucuronide (UDP glucuronyltransferase enzyme), glutathione (glutathionetransferase enzyme), sulfate (sulfotransferase enzyme), and acetyl (acetylation by acetyltransferase enzymes). The most abundant conjugation products are glucuronides. A glucuronide conjugate is illustrated in Figure 23.8, where –X–R represents a xenobiotic species conjugated to glucuronide, and R is an organic moiety.

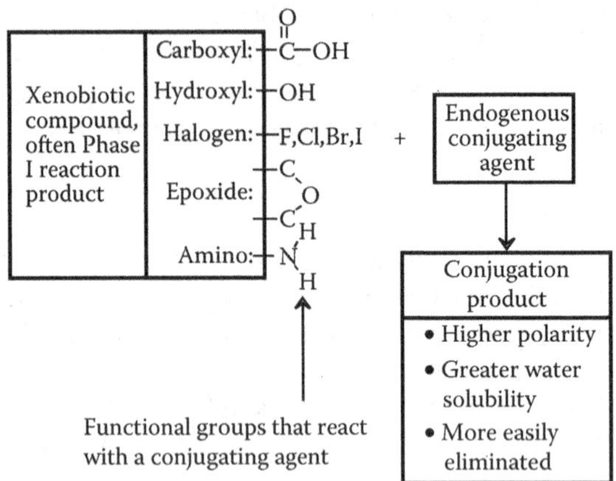

FIGURE 23.7 Illustration of Phase II reactions.

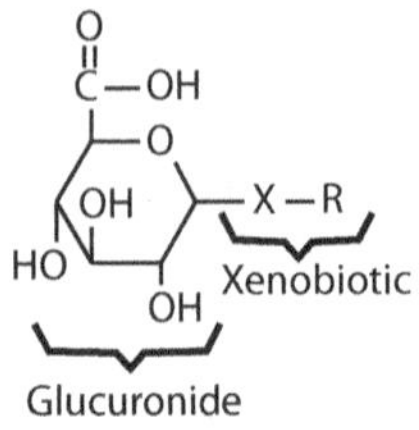

Xenobiotic
Glucuronide

FIGURE 23.8 Glucuronide conjugate formed from a xenobiotic, HX–R.

For example, if the xenobiotic compound conjugated is phenol, HXR is HOC_6H_5, X is the O atom, and R represents the phenyl group, C_6H_5.

23.7 KINETIC PHASE AND DYNAMIC PHASE

23.7.1 Kinetic Phase

The major routes and sites of absorption, metabolism, binding, and excretion of toxic substances in the body are illustrated in Figure 23.1. Toxicants in the body are metabolized, transported, and excreted; they have adverse biochemical effects; and they cause manifestations of poisoning. It is convenient to divide these processes into two major phases, a kinetic phase and a dynamic phase.

In the *kinetic phase*, a toxicant or the metabolic precursor of a toxic substance (*protoxicant*) may undergo absorption, metabolism, temporary storage, distribution, and excretion, as illustrated in Figure 23.9. A toxicant that is absorbed may be passed through the kinetic phase unchanged as an *active parent compound*, metabolized to a *detoxified metabolite* that is excreted, or converted to a toxic *active metabolite*. These processes occur through Phase I and Phase II reactions discussed above.

23.7.2 Dynamic Phase

In the *dynamic phase* (Figure 23.10), a toxicant or toxic metabolite interacts with cells, tissues, or organs in the body to cause some toxic response. The three major subdivisions of the dynamic phase are the following:

- *Primary reaction* with a receptor or target organ
- A *biochemical response*
- *Observable effects*

23.7.2.1 Primary Reaction in the Dynamic Phase

A toxicant or an active metabolite reacts with a receptor. The process leading to a toxic response is initiated when such a reaction occurs. A typical example is when benzene epoxide, the initial product of the Phase I reaction of benzene (see Figure 24.1), forms an adduct with a nucleic acid unit in

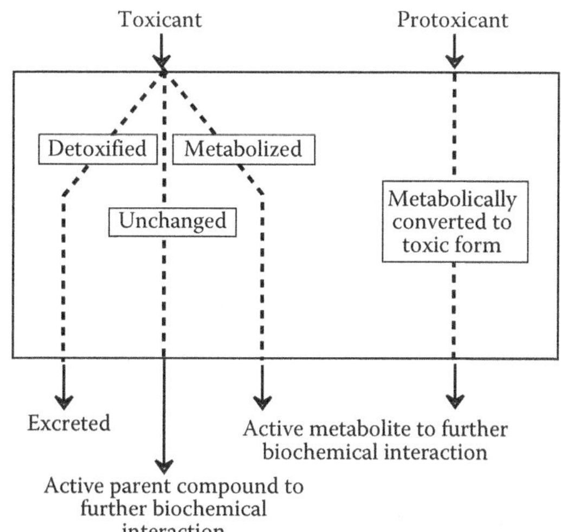

FIGURE 23.9 Processes involving toxicants or protoxicants in the kinetic phase.

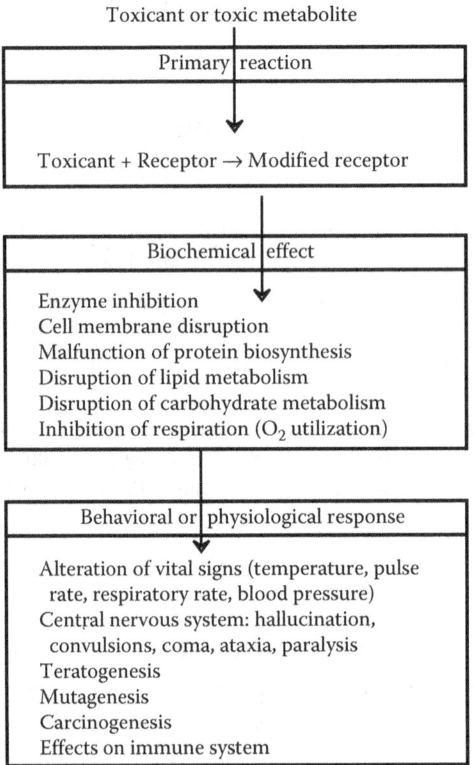

FIGURE 23.10 Dynamic phase of toxicant action.

DNA (receptor) resulting in alteration of the DNA. (Many species that cause a toxic response are reactive intermediates, such as benzene epoxide, which have a brief lifetime but a strong tendency to undergo reactions leading to a toxic response while they are around.) DNA adduct formation with benzene epoxide is an *irreversible* reaction between a toxicant and a receptor. A *reversible* reaction that can result in a toxic response is illustrated by the binding between carbon monoxide and oxygen-transporting hemoglobin (Hb) in blood:

$$O_2Hb + CO \rightleftarrows COHb + O_2 \quad (23.2)$$

23.7.2.2 Biochemical Effects in the Dynamic Phase

The binding of a toxicant to a receptor may result in some kind of biochemical effect. The major ones are the following:

- Impairment of enzyme function by binding to the enzyme, coenzymes, metal activators of enzymes, or enzyme substrates
- Alteration of cell membrane or carriers in cell membranes
- Interference with carbohydrate metabolism
- Interference with lipid metabolism resulting in excess lipid accumulation ("fatty liver")
- Interference with respiration, the overall process by which electrons are transferred to molecular oxygen in the biological oxidation of energy-yielding substrates
- Stopping or interfering with protein biosynthesis by their action on DNA
- Interference with regulatory processes mediated by hormones or enzymes

23.7.2.3 Responses to Toxicants

Among the more immediate and readily observed manifestations of poisoning are alterations in the *vital signs* of *temperature*, *pulse rate*, *respiratory rate*, and *blood pressure*. Poisoning by some substances may cause an abnormal skin color (jaundiced, yellow skin from CCl_4 poisoning) or excessively moist or dry skin. Toxic levels of some materials or their metabolites cause the body to have unnatural *odors*, such as the bitter almond odor of HCN in tissues of victims of cyanide poisoning. Symptoms of poisoning manifested in the eye include *miosis* (excessive or prolonged contraction of the eye pupil), *mydriasis* (excessive pupil dilation), *conjunctivitis* (inflammation of the mucus membrane that covers the front part of the eyeball and the inner lining of the eyelids), and *nystagmus* (involuntary movement of the eyeballs). Some poisons cause a moist condition of the mouth, whereas others cause a dry mouth. Gastrointestinal tract effects including pain, vomiting, or paralytic ileus (stoppage of the normal peristalsis movement of the intestines) occur as a result of poisoning by a number of toxic substances.

Central nervous system poisoning may be manifested by *convulsions*, *paralysis*, *hallucinations*, and *ataxia* (lack of coordination of voluntary movements of the body), as well as abnormal behavior, including agitation, hyperactivity, disorientation, and delirium. Severe poisoning by some substances, including organophosphates and carbamates, causes *coma*, the term used to describe a lowered level of consciousness.

Prominent among the more chronic responses to toxicant exposure are mutations, cancer, and birth defects and effects on the immune system. Other observable effects, some of which may occur soon after exposure, include gastrointestinal illness, cardiovascular disease, hepatic (liver) disease, renal (kidney) malfunction, neurologic symptoms (central and peripheral nervous systems), and skin abnormalities (rash, dermatitis).

Often the effects of toxicant exposure are subclinical in nature. The most common of these are some kinds of damage to the immune system, chromosomal abnormalities, modification of functions of liver enzymes, and slowing of conduction of nerve impulses.

23.8 TERATOGENESIS, MUTAGENESIS, CARCINOGENESIS, AND EFFECTS ON THE IMMUNE AND REPRODUCTIVE SYSTEMS

23.8.1 Teratogenesis

Teratogens are chemical species that cause birth defects. These usually arise from damage to embryonic or fetal cells. However, mutations in germ cells (egg or sperm cells) may cause birth defects, such as Down's syndrome.

The biochemical mechanisms of teratogenesis are varied. They include enzyme inhibition by xenobiotics; deprivation of the fetus of essential substrates, such as vitamins; interference with energy supply; or alteration of the permeability of the placental membrane.

Fetuses exposed to toxic substances *in utero* are the most vulnerable to the effects of xenobiotics. Such toxicants can traverse the placental barrier and enter the fetal blood stream, the detoxification enzyme systems of fetuses are not as effective as they become after birth, and their organ systems are immature and subject to damage during development. The result of such exposures can be birth defects, retardation of interuterine growth, and later development of afflictions including diabetes and coronary artery disease.[3]

23.8.2 Mutagenesis

Mutagens alter DNA to produce inheritable traits. Although mutation is a natural process that occurs even in the absence of xenobiotic substances, most mutations are harmful. The mechanisms of mutagenicity are similar to those of carcinogenicity, and mutagens often cause birth defects as well. Therefore, mutagenic hazardous substances are of major toxicological concern.

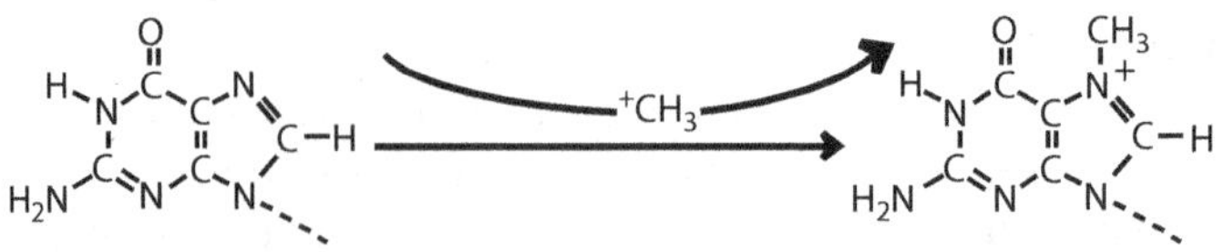

Guanine bound to DNA → ($^+CH_3$) Methylated guanine in DNA

FIGURE 23.11 Alkylation of guanine in DNA.

23.8.2.1 Biochemistry of Mutagenesis

To understand the biochemistry of mutagenesis, it is important to recall from Chapter 22 that DNA contains the nitrogenous bases adenine, guanine, cytosine, and thymine. The order in which these bases occur in DNA determines the nature and structure of newly produced RNA, a substance generated as a step in the synthesis of new proteins and enzymes in cells. Exchange, addition, or deletion of any of the nitrogenous bases in DNA alters the nature of RNA produced and can change vital life processes, such as the synthesis of an important enzyme. This phenomenon, which can be caused by xenobiotic compounds, is a mutation that can be passed on to progeny, usually with detrimental results.

There are several ways in which xenobiotic species may cause mutations. It is beyond the scope of this work to discuss these mechanisms in detail. For the most part, however, mutations due to xenobiotic substances are the result of chemical alterations of DNA, such as those discussed in the two examples below.

Nitrous acid, HNO_2, is an example of a chemical mutagen that is often used to cause mutations in bacteria. To understand the mutagenic activity of nitrous acid it should be noted that three of the nitrogenous bases—adenine, guanine, and cytosine—contain the amino group, $-NH_2$. The action of nitrous acid is to replace amino groups with a hydroxy group. When this occurs, the DNA may not function in the intended manner, causing a mutation to occur.

Alkylation, the attachment of a small alkyl group, such as $-CH_3$ or $-C_2H_5$, to an N atom on one of the nitrogenous bases in DNA, is one of the most common mechanisms leading to mutation. The methylation of "7" nitrogen in guanine in DNA to form N-methylguanine is shown in Figure 23.11. O-alkylation may also occur by attachment of a methyl or other alkyl group to the oxygen atom in guanine.

A number of mutagenic substances act as alkylating agents. Prominent among these are the compounds shown in Figure 23.12.

Alkylation occurs by way of generation of positively charged electrophilic species that bond to electron-rich nitrogen or oxygen atoms on the nitrogenous bases in DNA. The generation of such species usually occurs by way of biochemical and chemical processes. For example, dimethylnitrosamine (structural formula in Figure 23.12) is activated by oxidation through cellular NADPH to produce the following highly reactive intermediate:

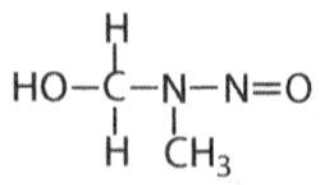

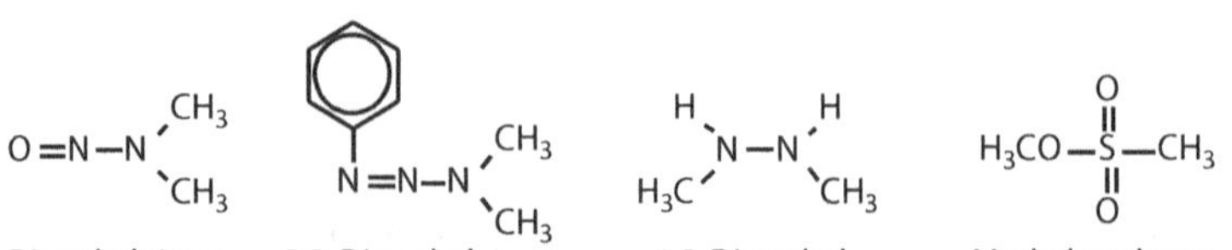

$$HO-CH_2-N(CH_3)-N=O$$

$O=N-N(CH_3)_2$ Dimethylnitrosamine; $C_6H_5-N=N-N(CH_3)_2$ 3,3-Dimethyl-1-phenyltriazine; $H_3C(H)N-N(H)CH_3$ 1,2-Dimethylhydrazine; $H_3CO-S(=O)_2-CH_3$ Methylmethanesulfonate

FIGURE 23.12 Examples of simple alkylating agents capable of causing mutations.

This product undergoes several nonenzymatic transitions, losing formaldehyde and generating a methyl carbocation, $^{+}CH_3$, that can methylate nitrogenous bases on DNA:

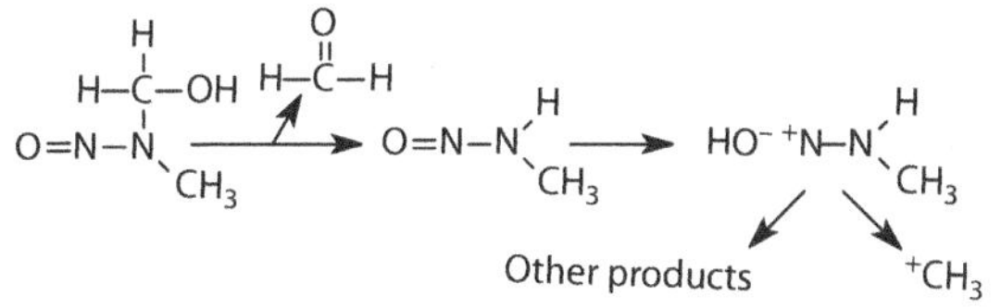

$$\mathrm{O{=}N{-}N(CH_2OH)(CH_3)} \xrightarrow{-\,\mathrm{H_2C{=}O}} \mathrm{O{=}N{-}N(H)(CH_3)} \longrightarrow \mathrm{HO^{-}\;{}^{+}N{-}N(H)(CH_3)} \longrightarrow \text{Other products} + {}^{+}\mathrm{CH_3} \tag{23.3}$$

One of the more notable mutagens is tris(2,3-dibromopropyl)phosphate, commonly called "tris," that was used as a flame retardant in children's sleepwear. Tris was found to be mutagenic in experimental animals and metabolites of it were found in children wearing the treated sleepwear. This strongly suggested that tris is absorbed through the skin and its uses were discontinued.

23.8.3 Carcinogenesis

Cancer is a condition characterized by the uncontrolled replication and growth of the body's own (somatic) cells. *Carcinogenic agents* may be categorized as follows:

- Chemical agents, such as nitrosamines and PAHs
- Biological agents, such as hepadnaviruses or retroviruses
- Ionizing radiation, such as x-rays
- Genetic factors, such as selective breeding

Clearly, in some cases, cancer is the result of the action of synthetic and naturally occurring chemicals. The role of xenobiotic chemicals in causing cancer is called *chemical carcinogenesis*. It is often regarded as the single most important facet of toxicology and is clearly the one that receives the most publicity.

Chemical carcinogenesis has a long history. As noted earlier in this chapter, in 1775 Sir Percival Pott, Surgeon General serving under King George III of England, observed that chimney sweeps in London had a very high incidence of cancer of the scrotum, which he related to their exposure to soot and tar from the burning of bituminous coal. Around 1900 a German surgeon, Ludwig Rehn, reported elevated incidences of bladder cancer in dye workers exposed to chemicals extracted from coal tar; 2-naphthylamine

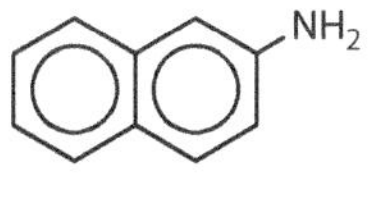

NH_2 2-Naphthylamine

was shown to be largely responsible. Other historical examples of carcinogenesis include observations of cancer from tobacco juice (1915), oral exposure to radium from painting luminescent watch dials (1929), tobacco smoke (1939), and asbestos (1960).

An important and unresolved question regarding carcinogens is the existence of thresholds above which carcinogenesis occurs and below which there is no effect.[4] Rodent tests for carcinogens are normally conducted at doses 1000–10,000 times those to which humans are exposed and the probabilities of substances causing cancer in humans are extrapolated linearly to the low doses at which humans are exposed under the assumption that there is no threshold below which the carcinogen has no effect. Some authorities believe that such extrapolations are unrealistic and overstate cancer risk in part because they do not take into account mechanisms possessed by humans to combat genotoxic and carcinogenic insults leading to cancer. Such mechanisms include detoxication and error-free DNA repair.

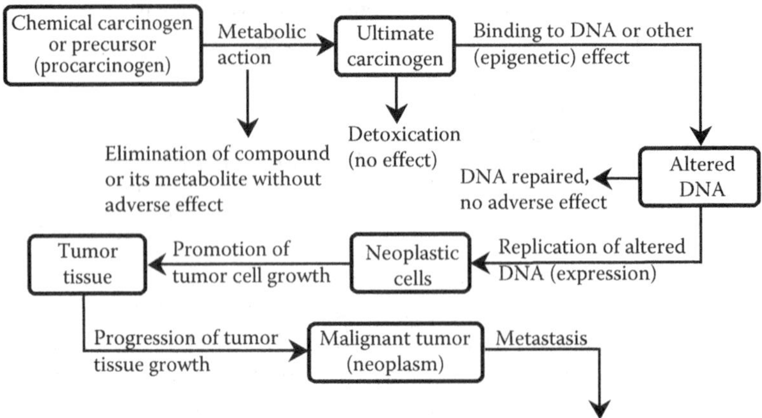

FIGURE 23.13 Outline of the process by which a carcinogen or procarcinogen may cause cancer.

23.8.3.1 Biochemistry of Carcinogenesis

Large expenditures of time and money on the subject in recent years have yielded a much better understanding of the biochemical bases of chemical carcinogenesis. The overall processes for the induction of cancer may be quite complex, involving numerous steps. However, it is generally recognized that there are two major steps in carcinogenesis: an *initiation stage* followed by a *promotional stage*. These steps are further subdivided as shown in Figure 23.13.

Initiation of carcinogenesis may occur by the reaction of a *DNA-reactive species* with DNA, or by the action of an *epigenetic carcinogen* that does not react with DNA and is carcinogenic by some other mechanism. Most DNA-reactive species are *genotoxic carcinogens* because they are also mutagens. These substances react irreversibly with DNA. They are either electrophilic or, more commonly, metabolically activated to form electrophilic species, as is the case with electrophilic $^{+}CH_3$ generated from dimethylnitrosamine, discussed under mutagenesis above. Most cancer-causing substances require metabolic activation to produce electrophilic species capable of forming adducts with DNA thereby causing gene mutations. The substances that are activated are called *procarcinogens*. The metabolic species actually responsible for carcinogenesis is termed an *ultimate carcinogen*. Some species that are intermediate metabolites between precarcinogens and ultimate carcinogens are called *proximate carcinogens*. Carcinogens that do not require biochemical activation are categorized as *primary* or *direct-acting carcinogens*. Some procarcinogens and primary carcinogens are shown in Figure 23.14.

Most substances classified as epigenetic carcinogens are *promoters* that act after initiation. Manifestations of promotion include increased numbers of tumor cells and decreased length of time for tumors to develop (shortened latency period). Promoters do not initiate cancer, are not electrophilic, and do not bind with DNA. The classic example of a promoter is decanoyl phorbol acetate or phorbol myristate acetate, which is extracted from croton oil.

23.8.3.2 Alkylating Agents in Carcinogenesis

Chemical carcinogens usually have the ability to form covalent bonds with macromolecular life molecules. Such covalent bonds can form with proteins, peptides, RNA, and DNA. Although most binding is with other kinds of molecules, which are more abundant, the DNA adducts are the significant ones in initiating cancer. Prominent among the species that bond to DNA in carcinogenesis are the alkylating agents that attach alkyl groups—methyl (CH_3) or ethyl (C_2H_5)—to DNA. A similar type of compound, *arylating agents*, act to attach aryl moieties, such as the phenyl group,

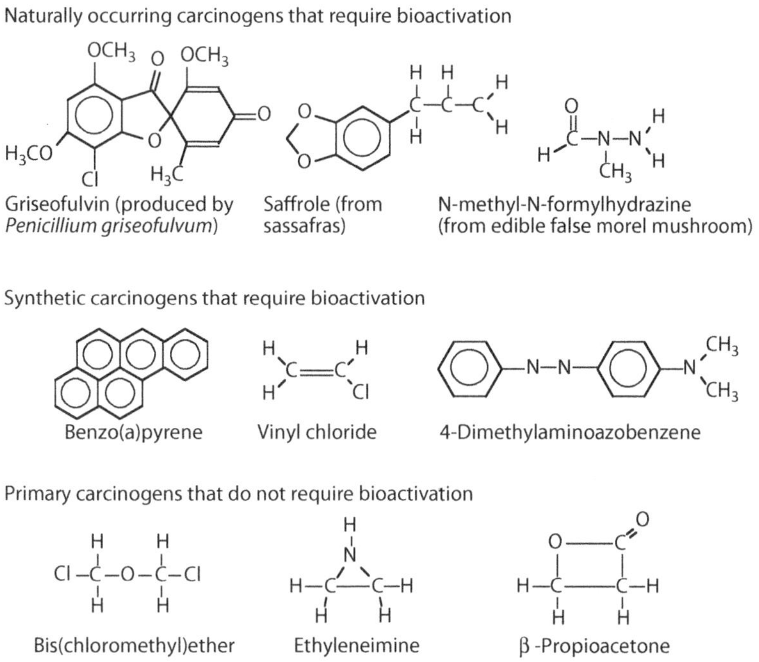

FIGURE 23.14 Examples of the major classes of naturally occurring and synthetic carcinogens, some of which require bioactivation, and others that act directly.

to DNA. As shown by the examples in Figure 23.15, the alkyl and aryl groups become attached to N and O atoms in the nitrogenous bases that compose DNA. This alteration in DNA can trigger initiation of the sequence of events that results in the growth and replication of neoplastic (cancerous) cells. The reactive species that donate alkyl groups in alkylation are usually formed by metabolic activation through the action of enzymes. This process was shown for conversion of dimethylnitrosamine to a methylating metabolic intermediate in the discussion of mutagenesis earlier in this section.

23.8.4 Testing for Carcinogens

Only a few chemicals have definitely been established as human carcinogens. A well documented example is vinyl chloride, $CH_2{=}CHCl$, which is known to have caused a rare form of liver cancer (angiosarcoma) in individuals who cleaned autoclaves in the PVC fabrication industry. In some cases chemicals are known to be carcinogens from epidemiological studies of exposed humans. Animals are used to test for carcinogenicity, and the results can be extrapolated, although with much uncertainty, to humans.

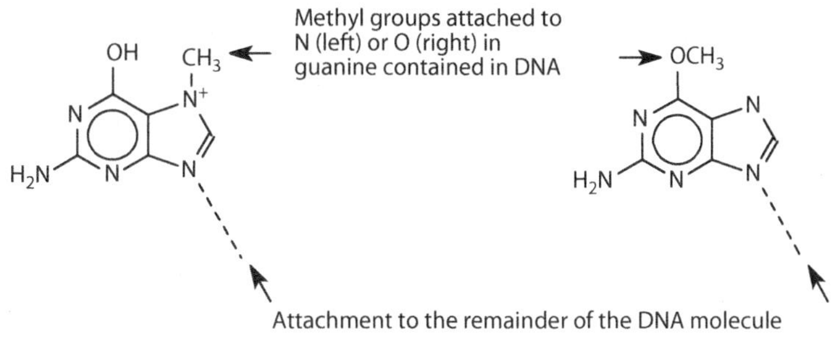

FIGURE 23.15 Alkylated (methylated) forms of the nitrogenous base guanine.

23.8.4.1 Bruce Ames Test

Mutagenicity used to infer carcinogenicity is the basis of the *Bruce Ames* test, in which observations are made of the reversion of mutant histidine-requiring *Salmonella* bacteria back to a form that can synthesize its own histidine. The test makes use of enzymes in homogenized liver tissue to convert potential procarcinogens to ultimate carcinogens. Histidine-requiring *Salmonella* bacteria are inoculated onto a medium that does not contain histidine, and those that mutate back to a form that can synthesize histidine establish visible colonies that are assayed to indicate mutagenicity.

According to Bruce Ames, the pioneer developer of the test that bears his name, animal tests for carcinogens that make use of massive doses of chemicals have a misleading tendency to give results that cannot be accurately extrapolated to assess cancer risks from smaller doses of chemicals. This is because the huge doses of chemicals used kill large numbers of cells, which the organism's body attempts to replace with new cells. Rapidly dividing cells greatly increase the likelihood of mutations that result in cancer simply as the result of rapid cell proliferation, not genotoxicity.

23.8.5 Immune System Response

The *immune system* acts as the body's natural defense system to protect it from xenobiotic chemicals; infectious agents, such as viruses or bacteria; and neoplastic cells, which give rise to cancerous tissue. Adverse effects on the body's immune system are being increasingly recognized as important consequences of exposure to hazardous substances. Toxicants can cause *immunosuppression*, which is the impairment of the body's natural defense mechanisms. Xenobiotics can also cause the immune system to lose its ability to control cell proliferation, resulting in leukemia or lymphoma.

Another major toxic response of the immune system is *allergy* or *hypersensitivity*. This kind of condition results when the immune system overreacts to the presence of a foreign agent or its metabolites in a self-destructive manner. Among the xenobiotic materials that can cause such reactions are beryllium, chromium, nickel, formaldehyde, some kinds of pesticides, resins, and plasticizers.

23.8.6 Endocrine Disruption

Some pollutants are of particular concern because of their potential to disrupt the crucial endocrine gland activities that regulate the metabolism and reproductive functions of organisms.[5] Because they live in water, fish, frogs, and reptiles such as alligators are particularly susceptible to such substances present as water pollutants. Exposed fish have exhibited reproductive dysfunction, alterations in secondary sex characteristics, and abnormal serum steroid levels. Substances that exhibit hormone-like activity in test subjects are called *hormonally active agents.* Of particular concern are the substances that act like estrogen, the female sex hormone, that survive wastewater treatment and get into waterways receiving treated wastewater. Examples of such compounds are shown in Figure 23.16. Among such substances are estrogen, an endogenous sex hormone; 17a-ethinylestradiol, an ingredient of oral contraceptives, and chemicals from industrial and consumer sources, such as the last two examples in Figure 23.16. Estrogenic substances from artificial sources are called *xenoestrogens* and include antioxidants, bisphenol A, dioxins, PCBs, phytoestrogens (from plants), some pesticides (chlordecone, dieldrin, DDT and its metabolites, methoxychlor, toxaphene), preservatives, and phthalic esters (butylbenzyl phthalate).

23.9 HEALTH HAZARDS

In recent years attention in toxicology has shifted away from readily recognized, usually severe, acute maladies that developed on a short time scale as a result of brief, intense exposure to toxicants, toward delayed, chronic, often less severe illnesses caused by long-term exposure to low levels of toxicants. Although the total impact of the latter kinds of health effects may be substantial, their assessment is very difficult because of factors such as uncertainties in exposure, low occurrence above background levels of disease, and long latency periods.

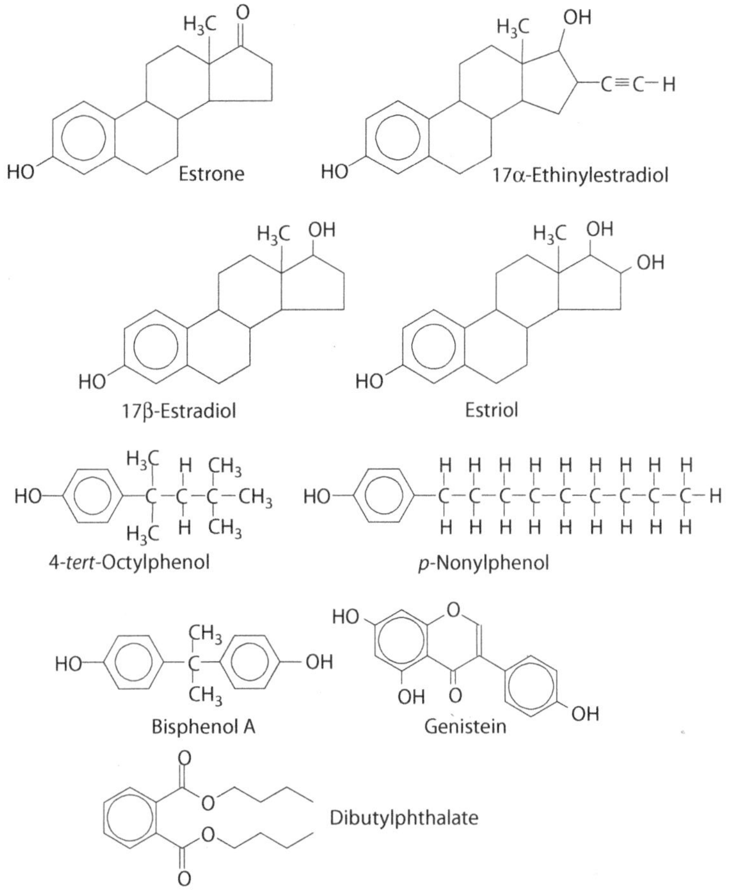

FIGURE 23.16 Endocrine disruptors that may be discharged into the environment and affect organisms. Estrone (a natural estrogen), 17α-ethinylestradiol (a constituent of oral contraceptives), 17β-estradiol, and estriol are steroid estrogens. Both 4-*tert*-octylphenol and *p*-nonylphenol are breakdown products of nonionic surfactants, bisphenol A is an ingredient of epoxy resins and other polymers, genistein is synthesized by trees to make the wood disease-resistant and is present in pulp mill discharges, and dibutylphthalate is a plasticizer in plastics.

23.9.1 Assessment of Potential Exposure

A critical step in assessing exposure to toxic substances, such as those from hazardous waste sites is evaluation of potentially exposed populations. The most direct approach to this is to determine chemicals or their metabolic products in organisms. For inorganic species this is most readily done for heavy metals, radionuclides, and some minerals, such as asbestos. Symptoms associated with exposure to particular chemicals may also be evaluated. Examples of such effects include readily apparent effects, for example, skin rashes, or subclinical effects, such as chromosomal damage.

23.9.2 Epidemiological Evidence

Epidemiological studies applied to toxic environmental pollutants, such as those from hazardous wastes, attempt to correlate observations of particular illnesses with probable exposure to such wastes. There are two major approaches to such studies. One approach is to look for diseases known to be caused by particular agents in areas where exposure is likely from such agents in hazardous wastes. A second approach is to look for *clusters* consisting of an abnormally large number of cases of a particular disease in a limited geographic area, then attempt to locate sources of exposure to hazardous wastes that may be responsible. The most common types of maladies observed in clusters are spontaneous abortions, birth defects, and particular types of cancer.

Epidemiologic studies are complicated by long latency periods from exposure to onset of disease, lack of specificity in the correlation between exposure to a particular waste and the occurrence of a disease, and background levels of a disease in the absence of exposure to a hazardous waste capable of causing the disease.

23.9.3 Estimation of Health Effects Risks

An important part of estimating the risks of adverse health effects from exposure to toxicants involves extrapolation from experimentally observable data. Usually the end result needed is an estimate of a low occurrence of a disease in humans after a long latency period resulting from low-level exposure to a toxicant for a long period of time. The data available are almost always taken from animals exposed to high levels of the substance for a relatively short period of time. Extrapolation is then made using linear or curvilinear projections to estimate the risk to human populations. There are, of course, very substantial uncertainties in this kind of approach.

23.9.4 Risk Assessment

Toxicological considerations are very important in estimating potential dangers of pollutants and hazardous waste chemicals. One of the major ways in which toxicology interfaces with the area of hazardous wastes is in *health risk assessment*, providing guidance for risk management, cleanup, or regulation needed at a hazardous waste site based upon knowledge about the site and the chemical and toxicological properties of wastes in it. Risk assessment includes the factors of site characteristics; substances present, including indicator species; potential receptors; potential exposure pathways; and uncertainty analysis. It may be divided into the following major components:

- Identification of hazard
- Exposure assessment
- Dose–response assessment
- Risk characterization

LITERATURE CITED

1. Manahan, S. E., *Toxicological Chemistry and Biochemistry*, 3rd ed., Lewis Publishers/CRC Press, Boca Raton, FL, 2002.
2. Myasoedova, K. N, New findings in studies of cytochromes P450, *Biochemistry* (Moscow), **73**, 965–969 (2008).
3. Barr, D. B., A. Bishop, and L. L. Needham, Concentrations of xenobiotic chemicals in the maternal-fetal unit, *Reproductive Toxicology*, **23**, 260–266 (2007).
4. Nohmi, T., N. Toyoda-Hokaiwado, M. Yamada, K. Masumura, M. Honma, and S. Fukushima, Meeting report on the international symposium on genotoxic and carcinogenic thresholds, *Genes and Environment*, **30**, 101–107 (2008).
5. Norris, D. O. and J. A. Carr, *Endocrine Disruption: Biological Basis for Health Effects in Wildlife and Humans*, Oxford University Press, New York, 2006.

SUPPLEMENTARY REFERENCES

Baselt, R. C., *Disposition of Toxic Drugs and Chemicals in Man*, 6th ed., Biomedical Publications, Foster City, CA, 2002.

Benigni, R., Ed., *Quantitative Structure-Activity Relationship (QSAR) Models of Mutagens and Carcinogens*, CRC Press, Boca Raton, FL, 2003.

Bingham, E., B. Cohrssen, and C. H. Powell, *Patty's Toxicology*, 5th ed., Wiley, New York, 2001.
Boelsterli, U. A., *Mechanistic Toxicology: The Molecular Basis of How Chemicals Disrupt Biological* Targets, 2nd ed., CRC Press, Boca Raton, FL, 2007.
Dart, R. C., *Medical Toxicology*, 3rd ed., Lippincott, Williams & Wilkins, Philadelphia, 2003.
Fenton, J., *Toxicology: A Case-Oriented Approach*, CRC Press, Boca Raton, FL, 2002.
Greenberg, M. I., R. J. Hamilton, S. D. Phillips, and G. McCluskey, Eds, *Occupational, Industrial, and Environmental Toxicology*, 2nd ed., Mosby, St. Louis, 2003.
Hoffman, D. J., B. A. Rattner, G. A. Burton, Jr., and J. Cairns, Jr., *Handbook of Ecotoxicology*, 2nd ed., Lewis Publishers/CRC Press, Boca Raton, FL, 2002.
Ioannides, C., Ed., *Cytochromes P450: Role in the Metabolism and Toxicity of Drugs and Other Xenobiotics*, RSC Publications, Cambridge, UK, 2008.
Joshi, B. D., P. C. Joshi, and N. Joshi, Eds, *Environmental Pollution and Toxicology*, A.P.H. Publishing Corporation, New Delhi, India, 2008.
Klaassen, C. D., Ed., *Casarett and Doull's Toxicology: The Basic Science of Poisons*, 7th ed., McGraw-Hill Medical, New York, 2008.
Landis, W. G. and M.-H. Yu, *Introduction to Environmental Toxicology: Impacts of Chemicals upon Ecological Systems*, 3rd ed., CRC Press/Lewis Publishers, Boca Raton, FL, 2004.
Leikin, J. B. and F. P. Paloucek, *Poisoning and Toxicology Handbook*, 4th ed., Taylor & Francis/CRC Press, Boca Raton, FL, 2008.
Leonard, B., Ed., *Report on Carcinogens: Carcinogen Profiles*, 10th ed., Collingdale, PA, 2002.
Lippmann, M., Ed., *Environmental Toxicants: Human Exposures and their Health Effects*, 3rd ed., Wiley, New York, 2009.
Newman, M. C. and W. H. Clements, *Ecotoxicology: A Comprehensive Treatment*, Taylor & Francis/CRC Press, Boca Raton, FL 2008.
Nichol, J., *Bites and Stings. The World of Venomous Animals*, Facts on File, New York, 1989.
Parvez, S. H., Ed., *Molecular Responses to Xenobiotics*, Elsevier, Amsterdam, 2001.
Pohanish, R. P. and M. Sittig, *Sittig's Handbook of Toxic and Hazardous Chemicals and Carcinogens*, Knovel Corporation, Norwich, NY, 2002.
Public Health Service, National Toxicology Program, *11th Report on Carcinogens*, U.S. Department of Health and Human Services, Washington, DC, 2008, available at http://ntp.niehs.nih.gov/ntp/roc/toc11.html
Romano, J. A., B. J. Lukey, and H. Salem, Eds, *Chemical Warfare Agents: Chemistry, Pharmacology, Toxicology, and Therapeutics*, 2nd ed., Taylor & Francis/CRC Press, Boca Raton, FL, 2008.
Rosenstock, L., *Textbook of Clinical Occupational and Environmental Medicine*, Saunders Health Sciences Division, Philadelphia, 2004.
Saferstein, R., *Forensic Science: From the Crime Scene to the Crime Lab*, Pearson Prentice Hall, Upper Saddle River, NJ, 2009.
Santos, E. B., Ed., *Ecotoxicology Research Developments*, Nova Science Publishers, New York, 2009.
Smart, R. C. and E. Hodgson, Eds, *Molecular and Biochemical Toxicology*, 4th ed., Wiley, Hoboken, NJ, 2008.
Stine, K. E. and T. M. Brown, *Principles of Toxicology*, 2nd ed., Taylor & Francis/CRC Press, Boca Raton, FL 2006.
Timbrell, J. A., *Principles of Biochemical Toxicology*, 4th ed., Informa Healthcare, New York, 2009.
Ullmann's Industrial Toxicology, Wiley-VCH, New York, 2005.
Walker, C. H., *Principles of Ecotoxicology*, 3rd ed., Taylor & Francis/CRC Press, Boca Raton, FL, 2006.
Ware, G., *Reviews of Environmental Contamination and Toxicology*, Springer, New York, 2007 (published annually).
Wexler, P., Ed., *Encyclopedia of Toxicology*, 2nd ed., Elsevier, New York, 2005.
Williamson, J. A., P. J. Fenner, J. W. Burnett, and J. F. Rifkin, Eds, *Venomous and Poisonous Marine Animals: A Medical and Biological Handbook*, University of New South Wales Press, Sydney, Australia, 1996.
Wilson, S. H. and W. A. Suk, *Biomarkers of Environmentally Associated Disease*, Taylor & Francis/CRC Press Boca Raton, FL, 2002.
Zelikoff, J. and P. L. Thomas, Eds, *Immunotoxicology of Environmental and Occupational Metals*, Taylor & Francis, London, 1998.

QUESTIONS AND PROBLEMS

1. How are conjugating agents and Phase II reactions involved with some toxicants?
2. What is the toxicological importance of proteins, particularly as related to protein structure?

3. What is the toxicological importance of lipids? How are lipids related to hydrophobic pollutants and toxicants?
4. What are Phase I reactions? What enzyme system carries them out? Where is this enzyme system located in the cell?
5. Name and describe the science that deals with the chemical nature and reactions of toxic substances, including their origins, uses, and chemical aspects of exposure, fates, and disposal.
6. What is a dose–response curve?
7. What is meant by a toxicity rating of 6?
8. What are the three major subdivisions of the *dynamic phase* of toxicity, and what happens in each?
9. Characterize the toxic effect of carbon monoxide in the body. Is its effect reversible or irreversible? Does it act on an enzyme system?
10. Of the following, choose the one that is *not* a biochemical effect of a toxic substance (explain): (A) impairment of enzyme function by binding to the enzyme, (B) alteration of cell membrane or carriers in cell membranes, (C) change in vital signs, (D) interference with lipid metabolism, (E) interference with respiration.
11. Distinguish among teratogenesis, mutagenesis, carcinogenesis, and immune system effects. Are there ways in which they are related?
12. As far as environmental toxicants are concerned, compare the relative importance of acute and chronic toxic effects and discuss the difficulties and uncertainties involved in studying each.
13. What are some of the factors that complicate epidemiologic studies of toxicants?
14. Alkylating agents do not or are not (explain): (A) formed by metabolic activation, (B) attach groups such as CH_3 to DNA, (C) include some species that cause cancer, (D) alter DNA, (E) noted for being electron-pair donors or nucleophiles.
15. Of the following, if any, the *untrue* statement regarding Phase I reactions is (explain): (A) they tend to introduce reactive, polar functional groups onto lipophilic ("fat-seeking") toxicant molecules, (B) the product of a Phase I reaction is usually more water-soluble than the parent xenobiotic species, (C) the product of a Phase I reaction possesses a "chemical handle" to which a substrate material in the body may become attached so that the toxicant can be eliminated from the body, (D) Phase I reactions are generally conjugation reactions through which an endogenous conjugating agent is attached, (E) Phase I reactions are catalyzed by enzymes.

24 Toxicological Chemistry of Chemical Substances

24.1 INTRODUCTION

Toxicological chemistry, defined and discussed in Chapter 23, centers on the relationship between the chemical nature of toxicants and their toxicological effects. This chapter discusses this relationship with regard to some of the major pollutants and hazardous substances. The first section deals with toxicological aspects of elements (particularly heavy metals), the presence of which in a compound frequently means that the compound is toxic. It also discusses the toxicities of some commonly used elemental forms, such as the chemically uncombined elemental halogens. The following section discusses the toxicological chemistry of inorganic compounds, many of which are produced from industrial processes. It also provides a brief discussion of organometallic compounds. The next section deals with the toxicology of organic compounds. The toxicological properties of hydrocarbons and oxygen-containing organic compounds are discussed along with other organic substances containing functional groups, such as alcohols and ketones. This section also discusses the toxicities of organic nitrogen, halide, sulfur, and phosphorus compounds, some of which are used as pesticides or military poisons. Finally, toxic natural products are discussed.

24.1.1 ATSDR Toxicological Profiles

A very useful source of information about the toxicological chemistry of various kinds of toxic substances is published by the U.S. Department of Health and Human Services, Public Health Service Agency for Toxic Substances, and Disease Registry ATSDR's Toxicological Profiles. The detailed documents pertaining to these substances, which are listed in Table 24.1, can be accessed through the website cited in the table.

24.2 TOXIC ELEMENTS AND ELEMENTAL FORMS

24.2.1 Ozone

Ozone (O_3, see Chapters 9, 13, and 14) has several toxic effects. Air containing 1 ppm by volume ozone has a distinct odor. Inhalation of ozone at this level causes severe irritation and headache. Ozone irritates the eyes, upper respiratory system, and lungs. Inhalation of ozone can sometimes cause fatal pulmonary edema. *Pulmonary* refers to lungs and *edema* to an accumulation of fluid in tissue spaces; therefore, *pulmonary edema* is an abnormal accumulation of fluid in lung tissue. Chromosomal damage has been observed in subjects exposed to ozone.

Ozone generates free radicals in tissue. These reactive species can cause lipid peroxidation, oxidation of sulfhydryl (–SH) groups, and other destructive oxidation processes. Compounds that protect organisms from the effects of ozone include radical scavengers, antioxidants, and compounds containing sulfhydryl groups.

TABLE 24.1
Materials Listed by ATSDR[a]

Acetone
Acrolein
Acrylonitrile
Aldrin/dieldrin
Aluminum
Americium
Ammonia
Aniline
Antimony
Arsenic
Asbestos
Atrazine
Barium
Benzene
Benzidine
2,3-Benzofuran
Beryllium
Bis(2-chloroethyl)ether
Bis(chloromethyl)ether
Blister agents(Lewisite)
Boron
Bromodichloromethane
Bromoform and dibromochloromethane
Bromomethane
1,3-Butadiene
2-Butanone
2-Butoxyethanol
Cadmium
Calcium or sodium hypochlorite
Carbon disulfide
Carbon Tetrachloride
Cesium
Chlordane
Chlorfenvinphos
Clofenvinfos
Chlorinated dibenzo-*p*-dioxins (CDDs)
Chlorodibenzofurans (CDFs), dibenzofuranos
Chloroethane
Chloroform
Chloromethane
Chlorophenols
Chlorpyrifos
Chromium
Cobalt
Copper
Crankcase oil, used
Creosote
Cresols
Crotonaldehyde
Cyanide
DDT, DDE, DDD
DEHP, di(2-ethyl-hexyl)phthalate
Diazinon
Diborane
1,2-Dibromo-3-chloropropane
1,2-Dibromoethane
Dichlorobenzenes
Dichlorobenzidine
1,1-Dichloroethane
1,2-Dichloroethane
1,1-Dichloroethene
1,2-Dichloroethene
1,2-Dichloropropane
1,3-Dichloropropene
Dichlorvos
Diethyl phthalate
Diisopropyl methylphosphonate (DIMP)
Di-*n*-butylphtalate
Di-1,3-dinitrobenzene and 1,3,5-trinitrobenzene
Dinitrocresols
Dinitrophenols, 2,4- and 2,6-dinitrotoluene (HCCPD)
Di-*n*-octylphthalate (DNOP)
Dioxins (CDDs)
Dibenzo-*p*-dioxinas
Diphenylhydrazine
Disulfoton
Endosulfan
Endrin
Ethion
Ethylbenzene
Ethylene glycol
Ethylene oxide
Fluorides, hydrogen fluoride, and fluorine
Fuel oils
Gasoline, automotive
Heptachlor
Heptachlor epoxide
Hexachlorobutadiene
Hexachlorocyclohexane
Hexachlorocyclopentadiene
Hexachloroethane
Hexamethylene diisocyanate (HDI)
Hydrogen peroxide
Hydrogen sulfide
Iodine
Ionizing radiation
Isophorone
Jet fuels
Kerosene (fuel oils)
Lead
Malathion
Manganese
4,4′-Methylenebis (2-chloroaniline)
Mercury
Methoxychlor
Methyl isocyanate
Methyl mercaptan
Methyl parathion
Methyl *t*-butyl ether (MTBE)
Methylene chloride
Methylenedianiline
Mirex and chlordecone
Naphthalene, 1-methylnapthalene, 2-methylnapthalene
Nerve agents (GA, GB, GD, VX)
Nickel
Nitrobenzene
Nitrogen oxides
Nitrophenols
n-Nitrosodimethylamine
n-Nitrosodi-*n*-propylamine
Nitrosodiphenylamine
Otto fuel II
Pentachlorophenol
Perchlorates
Phenol
Phosgene
Phosgene oxime
Phosphine
Phosphorus
Plutonium
Polybrominated biphenyls (PBBs)
Polybrominated diphenyl ethers (PBDEs)
Polychlorinated biphenyls (PCBs)
Polycyclic aromatic hydrocarbons (PAHs)
Propylene glycol
Pyrethroids
Pyridine
Radium
RDX
Selenium
Selenium hexafluoride

continued

TABLE 24.1 (continued)
Materials Listed by ATSDR[a]

Silver	Tetrachloroethylene	Trichloropropane
Sodium hydroxide	Thallium	Trinitrotoluene (TNT)
Stoddard solvent	Thorium	2,4,6-TNT
Strontium	Tin	Tungsten
Styrene	Titanium	Uranium
Sulfur dioxide	Toluene	Vanadium
Sulfur trioxide	Total petroleum hydrocarbons	Vinyl acetate
Sulfuric acid	Toxaphene	Vinyl chloride
Synthetic vitreous fibers	Trichloroethane	Xylene
1,1,2,2-Tetrachloroethane	Trichloroethylene (TCE)	Zinc

[a] U.S. Agency for Toxic Substances and Disease Registry: http://www.atsdr.cdc.gov/toxfaq.html#bookmark05

24.2.2 White Phosphorus

Elemental white phosphorus can enter the body by inhalation, by skin contact, or orally. It is a systemic poison, that is, one that is transported through the body to sites remote from its entry site. White phosphorus causes anemia, gastrointestinal system dysfunction, bone brittleness, and eye damage. Exposure also causes *phossy jaw*, a condition in which the jawbone deteriorates and becomes fractured.

24.2.3 Elemental Halogens

Elemental *fluorine* (F_2) is a pale yellow, highly reactive gas that is a strong oxidant. It is a toxic irritant and attacks skin, eye tissue, and the mucous membranes of the nose and respiratory tract. *Chlorine* (Cl_2) gas reacts in water to produce a strongly oxidizing solution. This reaction is responsible for some of the damage caused to the moist tissue lining the respiratory tract when the tissue is exposed to chlorine. The respiratory tract is rapidly irritated by exposure to 10–20 ppm of chlorine gas in air, causing acute discomfort that warns of the presence of the toxicant. Even brief exposure to 1000 ppm of Cl_2 can be fatal.

Bromine (Br_2) is a volatile, dark red liquid that is toxic when inhaled or ingested. Like chlorine and fluorine, it is strongly irritating to the mucous tissue of the respiratory tract and eyes and may cause pulmonary edema. The toxicological hazard of bromine is limited somewhat because its irritating odor elicits a withdrawal response.

Elemental solid *iodine* (I_2) is irritating to the lungs much like Cl_2 or Br_2. However, the relatively low vapor pressure of iodine limits exposure to I_2 vapor.

24.2.4 Heavy Metals

Heavy metals (Section 7.3) are toxic in their chemically combined forms and some, notably mercury, are toxic in the elemental form. The toxic properties of some of the most hazardous heavy metals and metalloids are discussed here.

Although not truly a *heavy* metal, *beryllium* (atomic mass 9.01) is one of the more hazardous toxic elements. Its most serious toxic effect is berylliosis, a condition manifested by lung fibrosis and pneumonitis, which may develop after a latency period of 5–20 years. Beryllium is a hypersensitizing agent and exposure to it causes skin granulomas and ulcerated skin. Beryllium was used in the nuclear weapons program in the United States, and it is believed that 500–1000 cases of beryllium poisoning have occurred or will occur in the future as a result of exposure to workers. In July

1999, the U.S. Department of Energy acknowledged these cases of beryllium poisoning and proposed legislation to compensate the victims. This resulted in the Energy Employees Occupational Illness Compensation Program of 2000 under which qualified workers with beryllium disease can receive a lump sum payment of $150,000 as reimbursement for disability and medical care resulting from the disease.[1]

Cadmium adversely affects several important enzymes; it can also cause painful osteomalacia (bone disease) and kidney damage. Inhalation of cadmium oxide dusts and fumes results in cadmium pneumonitis characterized by edema and pulmonary epithelium necrosis (death of tissue lining lungs).

Lead, widely distributed as metallic lead, inorganic compounds, and organometallic compounds, has a number of toxic effects, including inhibition of the synthesis of hemoglobin. It also adversely affects the central and peripheral nervous systems and the kidneys. Its toxicological effects have been widely studied.

Arsenic is a metalloid that forms a number of toxic compounds. The toxic +3 oxide, As_2O_3, is absorbed through the lungs and intestines. Biochemically, arsenic acts to coagulate proteins, forms complexes with coenzymes, and inhibits the production of adenosine triphosphate (ATP) in essential metabolic processes involving the utilization of energy. Arsenic is the toxic agent in one of the greatest environmental catastrophes of the last century, the result of its ingestion through well water in Bangladesh is discussed in Chapter 7, Section 7.4.

An interesting historical footnote with respect to arsenic poisoning is that of the *arsenic eaters* who were smugglers traversing the Alps of southern Austria as recently as the early 1900s. By carrying heavy loads on their backs through the mountains, they were able to evade tolls levied by local governmental bodies in the region. Starting with doses of arsenic trioxide of 10 mg twice a week, these individuals gradually raised their intake to more than 450 mg per week, which is almost 10 times the lower fatal dose for nonacclimated individuals. This enabled their bodies to switch from oxic sugar metabolism to the anoxic lactic acid cycle for their energy needs so that they could endure the vigorous exercise of carrying heavy loads of smuggled goods at high altitudes! One can only wonder how many potential arsenic eaters died in attempts to acclimate their bodies to this deadly poison.

Elemental *mercury* vapor can enter the body through inhalation and be carried by the bloodstream to the brain where it penetrates the blood–brain barrier. It disrupts metabolic processes in the brain causing tremor and psychopathological symptoms such as shyness, insomnia, depression, and irritability. Divalent ionic mercury, Hg^{2+}, damages the kidney. Organometallic mercury compounds are also very toxic, dimethylmercury, $Hg(CH_3)_2$, spectacularly so, killing most of the early investigators who first synthesized it.

24.3 TOXIC INORGANIC COMPOUNDS

24.3.1 CYANIDE

Both *hydrogen cyanide* (HCN) and *cyanide salts* (which contain CN^- ion) are rapidly acting poisons; a dose of only 60–90 mg is sufficient to kill a human. Metabolically, cyanide bonds to iron(III) in iron-containing ferricytochrome oxidase enzyme (see enzymes, Section 22.6), preventing its reduction to iron(II) in the oxidative phosphorylation process by which the body utilizes O_2. This prevents utilization of oxygen in cells, so that metabolic processes cease.

24.3.2 CARBON MONOXIDE

Carbon monoxide, CO, is a common cause of accidental poisonings. At CO levels in air of 10 ppm, impairment of judgment and visual perception occur; exposure to 100 ppm causes dizziness, headache, and weariness; loss of consciousness occurs at 250 ppm; and inhalation of 1000 ppm results in rapid death. Chronic long-term exposures to low levels of carbon monoxide are suspected of causing disorders of the respiratory system and the heart.

After entering the blood stream through the lungs, carbon monoxide reacts with hemoglobin (Hb) to convert oxyhemoglobin (O_2Hb) to carboxyhemoglobin (COHb):

$$O_2Hb + CO \rightleftarrows COHb + O_2 \quad (24.1)$$

In this case, hemoglobin is the receptor (Section 23.7) acted on by the carbon monoxide toxicant. Carboxyhemoglobin is much more stable than oxyhemoglobin so that its formation prevents hemoglobin from carrying oxygen to body tissues. The equilibrium constant expression for the preceding reaction is

$$\frac{[COHb]}{[O_2Hb]} = M \times \frac{P_{CO}}{P_{CO_2}}. \quad (24.2)$$

where [COHb] and [O_2Hb] are the equilibrium concentrations of carboxyhemoglobin and oxyhemoglobin, respectively, in blood, *P* denotes the partial pressures of CO and O_2 in air, and *M* is the Haldane constant for which values of 210–220 are commonly used for adult human hemoglobin.

24.3.3 Nitrogen Oxides

The two most common toxic oxides of nitrogen are NO and NO_2, of which the latter is regarded as the more toxic. Nitrogen dioxide causes severe irritation of the innermost parts of the lungs resulting in pulmonary edema. In cases of severe exposures, fatal *bronchiolitis fibrosa obliterans* may develop approximately 3 weeks after exposure to NO_2. Fatalities may result from even brief periods of inhalation of air containing 200–700 ppm of NO_2. Biochemically, NO_2 disrupts lactic dehydrogenase and some other enzyme systems, possibly acting much like ozone, a stronger oxidant mentioned in Section 24.2. Free radicals, particularly $HO^{\bullet}$, are likely formed in the body by the action of nitrogen dioxide and the compound probably causes *lipid peroxidation* in which the C=C double bonds in unsaturated body lipids are attacked by free radicals and undergo chain reactions in the presence of O_2, resulting in their oxidative destruction.

Nitrous oxide, N_2O, is used as an oxidant gas and in dental surgery as a general anesthetic. This gas was once known as "laughing gas," and was used in the late 1800s as a "recreational gas" at parties held by some of our not-so-staid Victorian ancestors. Nitrous oxide is a central nervous system depressant and can act as an asphyxiant.

24.3.4 Hydrogen Halides

Hydrogen halides (general formula HX, where X is F, Cl, Br, or I) are relatively toxic gases. The most widely used of these gases are HF and HCl; their toxicities are discussed here.

24.3.4.1 Hydrogen Fluoride

Hydrogen fluoride (HF, mp –83.1°C, bp 19.5°C) is used as a clear, colorless liquid or gas or as a 30–60% aqueous solution of *hydrofluoric acid*, both referred to as HF here. Both are extreme irritants to any part of the body that they contact, causing ulcers in affected areas of the upper respiratory tract. Lesions caused by contact with HF heal poorly, and tend to develop gangrene.

Fluoride ion, F^-, is toxic in soluble fluoride salts, such as NaF, causing *fluorosis*, a condition characterized by bone abnormalities and mottled, soft teeth. Livestock are especially susceptible to poisoning from fluoride fallout on grazing land; severely afflicted animals become lame and even die. Industrial pollution has been a common source of toxic levels of fluoride. However, about 1 ppm of fluoride used in some drinking water supplies prevents tooth decay.

24.3.4.2 Hydrogen Chloride

Gaseous *hydrogen chloride* and its aqueous solution, called *hydrochloric acid*, both denoted as HCl, are much less toxic than HF. Hydrochloric acid is a natural physiological fluid present as a dilute solution in the stomachs of humans and other animals. However, the inhalation of HCl vapor can cause spasms of the larynx as well as pulmonary edema and even death at high levels. The high affinity of hydrogen chloride vapor for water tends to dehydrate the eyes and respiratory tract tissue.

24.3.5 Interhalogen Compounds and Halogen Oxides

Interhalogen compounds, including ClF, BrCl, and BrF_3, are extremely reactive and are potent oxidants. They react with water to produce hydrohalic acid solutions (HF, HCl) and nascent oxygen {O}. Too reactive to enter biological systems in their original chemical state, interhalogen compounds tend to be powerful corrosive irritants that acidify, oxidize, and dehydrate tissue, much like those of the elemental forms of the elements from which they are composed. Because of these effects, the skin, eyes, and mucous membranes of the mouth, throat, and pulmonary systems are especially susceptible to attack.

Major halogen oxides, including fluorine monoxide (OF_2), chlorine monoxide (Cl_2O), chlorine dioxide (ClO_2), chlorine heptoxide (Cl_2O_7), and bromine monoxide (Br_2O), tend to be unstable, highly reactive, and toxic compounds that pose hazards similar to those of the interhalogen compounds discussed above. Chlorine dioxide, the most commonly used halogen oxide, is employed for odor control and bleaching wood pulp. As a substitute for chlorine in water disinfection, it produces fewer undesirable chemical by-products, particularly trihalomethanes.

The most important of the oxyacids and their salts formed by halogens are hypochlorous acid, HOCl, and hypochlorites, such as NaOCl, used for bleaching and disinfection. The hypochlorites irritate the eyes, skin, and mucous membrane tissue because they react to produce active (nascent) oxygen ({O}) and acid as shown by the reaction below:

$$HClO \rightarrow H^+ + Cl^- + \{O\} \tag{24.3}$$

24.3.6 Inorganic Compounds of Silicon

Silica (SiO_2, quartz) occurs in a variety of types of rocks such as sand, sandstone, and diatomaceous earth. *Silicosis* resulting from human exposure to silica dust from construction materials, sandblasting, and other sources has been a common occupational disease. A type of pulmonary fibrosis that causes lung nodules and makes victims more susceptible to pneumonia and other lung diseases, silicosis is one of the most common disabling conditions resulting from industrial exposure to hazardous substances. It can cause death from insufficient oxygen or from heart failure in severe cases.

Silane, SiH_4, and disilane, H_3SiSiH_3, are examples of inorganic *silanes*, which have H–Si bonds. Numerous organic ("organometallic") silanes exist in which alkyl moieties are substituted for H. Little information is available regarding the toxicities of silanes.

Silicon tetrachloride, $SiCl_4$, is the only industrially significant compound of the *silicon tetrahalides*, a group of compounds with the general formula SiX_4, where X is a halogen. The two commercially produced *silicon halohydrides*, general formula $H_{4-x}SiX_x$, are dichlorosilane (SiH_2Cl_2) and trichlorosilane ($SiHCl_3$). These compounds are used as intermediates in the synthesis of organosilicon compounds and in the production of high-purity silicon for semiconductors. Silicon tetrachloride and trichlorosilane, fuming liquids that react with water to give off HCl vapor have suffocating odors and are irritants to the eyes, nasal, and lung tissue.

24.3.7 Asbestos

Asbestos is the name given to a group of fibrous silicate minerals, typically those of the serpentine group, for which the approximate chemical formula is $Mg_3(Si_2O_5)(OH)_4$. Asbestos has been widely used in structural materials, brake linings, insulation, and pipe manufacture. Inhalation of asbestos may cause asbestosis (a pneumonia condition), mesothelioma (tumor of the mesothelial tissue lining the chest cavity adjacent to the lungs), and bronchogenic carcinoma (cancer originating with the air passages in the lungs) so that uses of asbestos have been severely curtailed and widespread programs have been undertaken to remove the material from buildings.

24.3.8 Inorganic Phosphorus Compounds

Phosphine (PH_3), a colorless gas that undergoes autoignition at 100°C, is a potential hazard in industrial processes and in the laboratory. Symptoms of potentially fatal poisoning from phosphine gas include pulmonary tract irritation, depression of the central nervous system, fatigue, vomiting, and difficult, painful breathing.

Tetraphosphorus decoxide, P_4O_{10}, is produced as a fluffy white powder from the combustion of elemental phosphorus and reacts with water from air to form syrupy orthophosphoric acid. Because of the formation of acid by this reaction and its dehydrating action, P_4O_{10} is a corrosive irritant to the skin, eyes, and mucous membranes.

The most important of the *phosphorus halides*, general formulas PX_3 and PX_5, is phosphorus pentachloride used as a catalyst in organic synthesis, as a chlorinating agent, and as a raw material to make phosphorus oxychloride ($POCl_3$). Because they react violently with water to produce the corresponding hydrogen halides and oxo phosphorus acids,

$$PCl_5 + 4H_2O \rightarrow H_3PO_4 + 5HCl \tag{24.4}$$

the phosphorus halides are strong irritants to the eyes, skin, and mucous membranes.

The major *phosphorus oxyhalide* in commercial use is phosphorus oxychloride ($POCl_3$), a faintly yellow fuming liquid. Reacting with water to form toxic vapors of hydrochloric acid and phosphorus acid (H_3PO_3), phosphorus oxyhalide is a strong irritant to the eyes, skin, and mucous membranes.

24.3.9 Inorganic Compounds of Sulfur

A colorless gas with a foul, rotten egg odor, *hydrogen sulfide* is very toxic. In some cases, inhalation of H_2S kills faster than even hydrogen cyanide; rapid death ensues from exposure to air containing more than about 1000 ppm H_2S due to asphyxiation from respiratory system paralysis. Lower doses cause symptoms that include headache, dizziness, and excitement due to damage to the central nervous system. General debility is one of the numerous effects of chronic H_2S poisoning.

Sulfur dioxide, SO_2, dissolves in water to produce sulfurous acid, H_2SO_3; hydrogen sulfite ion, HSO_3^-; and sulfite ion, SO_3^{2-}. Because of its water solubility, sulfur dioxide is largely removed in the upper respiratory tract. It is an irritant to the eyes, skin, mucous membranes, and respiratory tract. Some individuals are hypersensitive to sodium sulfite (Na_2SO_3), which has been used as a chemical food preservative. Because of threats to hypersensitive individuals, these uses were severely restricted in the United States in early 1990.

Number one in synthetic chemical production, *sulfuric acid* (H_2SO_4) is a severely corrosive poison and dehydrating agent in the concentrated liquid form; it readily penetrates the skin to reach subcutaneous tissue causing tissue necrosis with effects resembling those of severe thermal burns. Sulfuric acid fumes and mists irritate the eyes and respiratory tract tissue, and industrial exposure has even caused tooth erosion in workers.

TABLE 24.2
Inorganic Sulfur Compounds

Compound Name	Formula	Properties
Sulfur		
Monofluoride	S_2F_2	Colorless gas, mp −104°C, bp 99°C, toxicity similar to HF
Tetrafluoride	SF_4	Gas, bp −40°C, mp −124°C, powerful irritant
Hexafluoride	SF_6	Colorless gas, mp −51°C, remarkably low toxicity when pure, but may be contaminated with toxic lower fluorides
Monochloride	S_2Cl_2	Oily, fuming orange liquid, mp −80°C, bp 138°C, strong irritant to the eyes, skin, and lungs
Tetrachloride	SCl_4	Brownish/yellow liquid/gas, mp −30°C, decomposes above 0°C, irritant
Trioxide	SO_3	Solid anhydride of sulfuric acid reacts with moisture or steam to produce sulfuric acid
Sulfuryl chloride	SO_2Cl_2	Colorless liquid, mp −54°C, bp 69°C, used for organic synthesis, corrosive toxic irritant
Thionyl chloride	$SOCl_2$	Colorless-to-orange fuming liquid, mp −105°C, bp 79°C, toxic corrosive irritant
Carbon oxysulfide	COS	Volatile liquid by-product of natural gas or petroleum refining, toxic narcotic
Carbon disulfide	CS_2	Colorless liquid, industrial chemical, narcotic and central nervous system anesthetic

The more important halides, oxides, and oxyhalides of sulfur are listed in Table 24.2. The major toxic effects of these compounds are given in the table.

24.3.10 Perchlorate

Perchlorate anion, ClO_4^-, is chemically unreactive and was long regarded as toxicologically insignificant, but it is now known that it inhibits the action of iodide, I^-, required for proper functioning of the thyroid. Physiologically, ClO_4^- competes with I^- and diminishes iodide uptake by the thyroid to the extent that it has even been used to treat cases of hyperthyroidism. The reduction of I^- uptake by ClO_4^- competition inhibits the production of thyroid hormones. This is of particular concern because of the role of thyroid hormones in brain development before and after birth. Iodide carriers that are inhibited by perchlorate have been studied in frogs.[2] It is suspected that water contaminant ClO_4^- can alter sex ratios and inhibit metamorphosis, such as change from tadpoles to adult frogs, in amphibians.

24.3.11 Organometallic Compounds

As discussed elsewhere in this book, a variety of toxic organometallic compounds are found in the environment. The toxicological properties of some organometallic compounds—pharmaceutical organoarsenicals, organomercury fungicides, and tetraethyllead antiknock gasoline additives—that were widely used in the past are well known. It is important to consider the potential toxicities of relatively newer organometallic compounds used in semiconductors as catalysts and for chemical synthesis.

Organometallic compounds often behave in the body in ways totally unlike the inorganic forms of the metals that they contain. This is due in large part to the fact that, compared to inorganic forms, organometallic compounds have an organic nature and higher lipid solubility.

24.3.11.1 Organolead Compounds

Perhaps the most notable toxic organometallic compound is tetraethyllead, $Pb(C_2H_5)_4$, a colorless, oily liquid that was widely used as a gasoline additive to boost octane rating. Tetraethyllead has a strong affinity for lipids and can enter the body by all three common routes of inhalation, ingestion, and absorption through the skin. Acting differently from inorganic compounds in the body, it affects the central nervous system with symptoms such as fatigue, weakness, restlessness, ataxia, psychosis, and convulsions. Recovery from severe lead poisoning tends to be slow. In cases of fatal tetraethyllead poisoning, death has occurred as soon as 1 or 2 days after exposure.

24.3.11.2 Organotin Compounds

The greatest number of organometallic compounds in commercial use are those of tin—tributyltin chloride and related tributyltin (TBT) compounds. These compounds have bactericidal, fungicidal, and insecticidal properties. They have particular environmental significance because of their widespread applications as industrial biocides, now increasingly limited because of their environmental and toxicological effects. Organotin compounds are readily absorbed through the skin, sometimes causing a skin rash. They probably bind with sulfur groups on proteins and appear to interfere with mitochondrial function.

24.3.11.3 Carbonyls

Metal carbonyls, regarded as extremely hazardous because of their toxicities include nickel tetracarbonyl [$Ni(CO)_4$], cobalt carbonyl, and iron pentacarbonyl. Some of the hazardous carbonyls are volatile and readily taken into the body through the respiratory tract or through the skin. The carbonyls affect tissue directly and they break down to toxic carbon monoxide and products of the metal, which have additional toxic effects.

24.3.11.4 Reaction Products of Organometallic Compounds

An example of the production of a toxic substance from the burning of an organometallic compound is provided by the combustion of diethylzinc:

$$Zn(C_2H_5)_2 + 7O_2 \rightarrow ZnO(s) + 5H_2O(g) + 4CO_2(g) \tag{24.5}$$

Zinc oxide is used as a healing agent and food additive. However, inhalation of zinc oxide fume particles produced by the combustion of zinc organometallic compounds causes zinc *metal fume fever*. This is an uncomfortable condition characterized by elevated temperature and "chills."

24.4 TOXICOLOGY OF ORGANIC COMPOUNDS

24.4.1 Alkane Hydrocarbons

Gaseous methane, ethane, propane, *n*-butane, and isobutane (both C_4H_{10}) are regarded as *simple asphyxiants* that form mixtures with air containing insufficient oxygen to support respiration. The most common toxicological occupational problem associated with the use of hydrocarbon liquids in the workplace is dermatitis, caused by dissolution of the fat portions of the skin and characterized by inflamed, dry, scaly skin. Inhalation of volatile liquid 5–8 carbon *n*-alkanes and branched-chain alkanes may cause central nervous system depression manifested by dizziness and loss of coordination. Exposure to *n*-hexane and cyclohexane results in loss of myelin (a fatty substance constituting a sheath around certain nerve fibers) and degeneration of axons (part of a nerve cell through which nerve impulses are transferred out of the cell). This has resulted in multiple disorders of the nervous system (*polyneuropathy*) including muscle weakness and impaired sensory function of the hands and feet. In the body, *n*-hexane is metabolized to 2,5-hexanedione, a Phase I oxidation product that can be observed in urine of exposed individuals and used as a biological monitor of exposure to *n*-hexane.

2,5-hexanedione

24.4.2 Alkene and Alkyne Hydrocarbons

Ethylene, a widely used colorless gas with a somewhat sweet odor, acts as a simple asphyxiant and anesthetic to animals and is phytotoxic (toxic to plants). The toxicological properties of propylene (C_3H_6) are very similar to those of ethylene. Colorless, odorless, gaseous 1,3-butadiene is an irritant to the eyes and respiratory system mucous membranes; at higher levels it can cause unconsciousness and even death. It causes cancer in animals and is rated as a probable human carcinogen. Acetylene, H–C≡C–H, is a colorless gas with a garlic odor. It acts as an asphyxiant and narcotic, causing headache, dizziness, and gastric disturbances. Some of these effects may be due to the presence of impurities in the commercial product.

24.4.3 Benzene and Aromatic Hydrocarbons

Inhaled benzene is readily absorbed by blood, from which it is strongly taken up by fatty tissues. For the nonmetabolized compound, the process is reversible and benzene is excreted through the lungs. As shown in Figure 24.1, benzene is converted to phenol by a Phase I oxidation reaction (see Section 23.6) in the liver. The reactive and short-lived benzene epoxide intermediate known to occur in this reaction is probably responsible for much of the unique toxicity of benzene, which involves damage to bone marrow. In addition to phenol, several other oxygenated derivatives of benzene are produced when it is metabolized, as is *trans,trans*-muconic acid, produced by cleavage of the benzene ring.

Benzene is a skin irritant, and progressively higher local exposures can cause skin redness (erythema), burning sensations, fluid accumulation (edema), and blistering. Inhalation of air containing about 7 g/m^3 of benzene causes acute poisoning within an hour because of a narcotic effect upon the central nervous system manifested progressively by excitation, depression, respiratory system failure, and death. Inhalation of air containing more than about 60 g/m^3 of benzene can be fatal within a few minutes.

Long-term exposures to lower levels of benzene cause nonspecific symptoms, including fatigue, headache, and appetite loss. Chronic benzene poisoning causes blood abnormalities, including a lowered white cell count, an abnormal increase in blood lymphocytes (colorless corpuscles introduced to the blood from the lymph glands), anemia, a decrease in the number of blood platelets required for clotting (thrombocytopenia), and damage to bone marrow. It is thought that preleukemia, leukemia, or cancer may result.

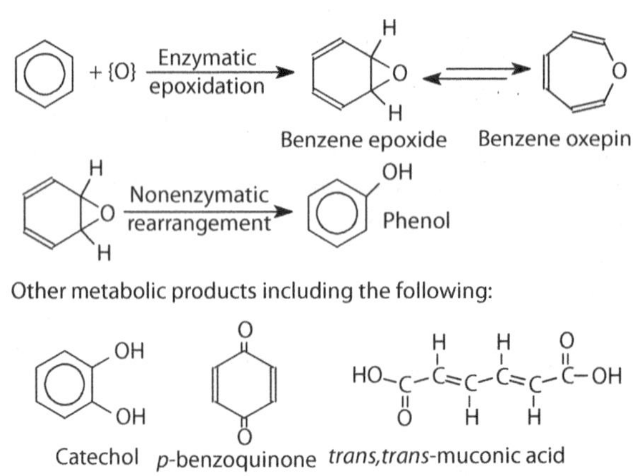

FIGURE 24.1 Conversion of benzene to phenol in the body.

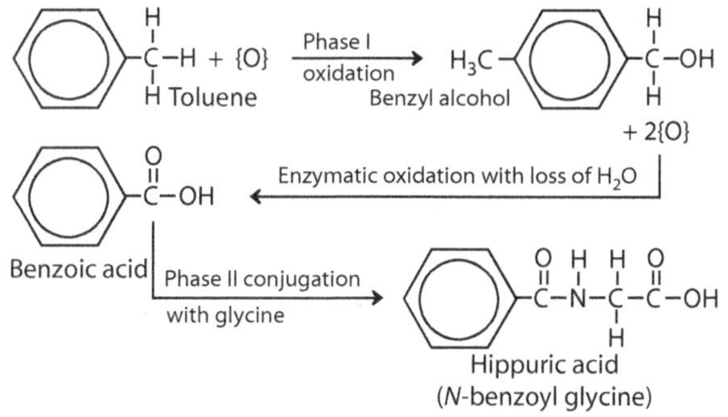

Toluene + {O} → (Phase I oxidation) → Benzyl alcohol + 2{O} → (Enzymatic oxidation with loss of H_2O) → Benzoic acid → (Phase II conjugation with glycine) → Hippuric acid (*N*-benzoyl glycine)

FIGURE 24.2 Metabolic oxidation of toluene with conjugation to hippuric acid, which is excreted with urine.

24.4.3.1 Toluene

Toluene, a colorless liquid boiling at 101.4°C, is classified as moderately toxic through inhalation or ingestion; it has a low toxicity by dermal exposure. Toluene can be tolerated without noticeable ill effects in ambient air up to 200 ppm. Exposure to 500 ppm may cause headache, nausea, lassitude, and impaired coordination without detectable physiological effects. Massive exposure to toluene has a narcotic effect, which can lead to coma. Because it possesses an aliphatic side chain that can be oxidized enzymatically to products that are readily excreted from the body (see the metabolic reaction scheme in Figure 24.2), toluene is much less toxic than benzene.

24.4.3.2 Naphthalene

As is the case with benzene, *naphthalene* undergoes a Phase I oxidation reaction that places an epoxide group on the aromatic ring. This process is followed by Phase II conjugation reactions to yield products that can be eliminated from the body in the urine.

Exposure to naphthalene can cause anemia and marked reductions in red cell count, hemoglobin, and hematocrit in individuals exhibiting genetic susceptibility to these conditions. Naphthalene causes skin irritation or severe dermatitis in sensitized individuals. Headaches, confusion, and vomiting may result from inhalation or ingestion of naphthalene. Death from kidney failure occurs in severe instances of poisoning.

24.4.3.3 Polycyclic Aromatic Hydrocarbons

Benzo[*a*]pyrene (see Section 10.4) is the most studied of the PAHs. Some metabolites of PAH compounds, particularly the 7,8-diol-9,10-epoxide of benzo[*a*]pyrene shown in Figure 24.3, are known to cause cancer. There are two stereoisomers of this metabolite, both of which are known to be potent mutagens and presumably can cause cancer. They act by forming adducts with DNA.

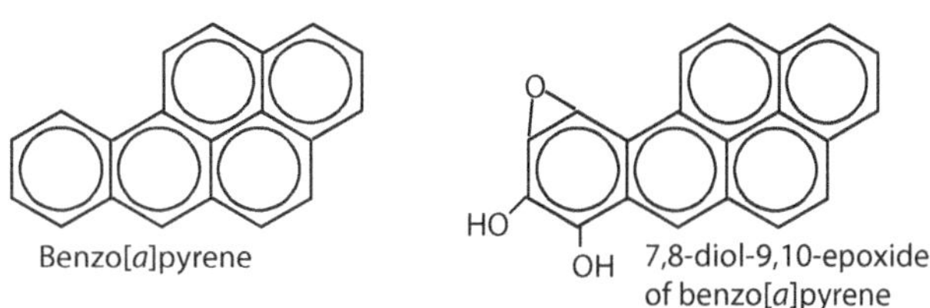

FIGURE 24.3 Benzo[*a*]pyrene and its carcinogenic metabolic product.

24.4.4 Oxygen-Containing Organic Compounds

24.4.4.1 Oxides

Hydrocarbon *oxides* such as ethylene oxide and propylene oxide,

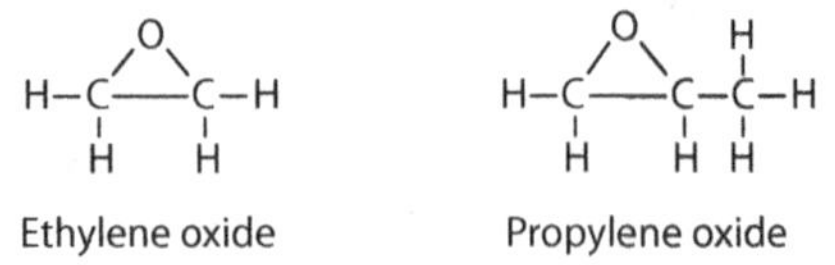

Ethylene oxide   Propylene oxide

which are characterized by an *epoxide* functional group bridging oxygen between two adjacent C atoms, are widely used and toxic. Ethylene oxide, a gaseous colorless, sweet-smelling, flammable, explosive gas used as a chemical intermediate, sterilant, and fumigant, is moderately to highly toxic, a mutagen, and carcinogenic to experimental animals. Inhalation of relatively low levels of this gas results in respiratory tract irritation, headache, drowsiness, and dyspnea, whereas exposure to higher levels causes cyanosis, pulmonary edema, kidney damage, peripheral nerve damage, and even death. Propylene oxide is a colorless, reactive, volatile liquid (bp 34°C) with uses similar to those of ethylene oxide and similar, though less severe, toxic effects. The toxicity of 1,2,3,4-butadiene epoxide, the oxidation product of 1,3-butadiene, is notable in that it is a direct-acting (primary) carcinogen.

24.4.4.2 Alcohols

Human exposure to the three light alcohols shown in Figure 24.4 is common because they are widely used industrially and in consumer products.

Methanol, which has caused many fatalities when ingested accidentally or consumed as a substitute for beverage ethanol, is metabolically oxidized to formaldehyde and formic acid. In addition to causing acidosis, these products affect the central nervous system and the optic nerve. Acute exposure to lethal doses causes an initially mild inebriation, followed in about 10–20 h by unconsciousness, cardiac depression, and death. Sublethal exposures can cause blindness from deterioration of the optic nerve and retinal ganglion cells. Inhalation of methanol fumes may result in chronic, low-level exposure.

Ethanol is usually ingested through the gastrointestinal tract, but can be absorbed as vapor by the alveoli of the lungs. Ethanol is oxidized metabolically more rapidly than methanol, first to acetaldehyde (discussed later in this section), then to CO_2. Ethanol has numerous acute effects resulting from central nervous system depression. These range from decreased inhibitions and slowed reaction times at 0.05% blood ethanol, through intoxication, stupor, and—at more than 0.5% blood ethanol—death. Ethanol also has a number of chronic effects, of which the addictive condition of alcoholism and cirrhosis of the liver are the most prominent.

Despite its widespread use in automobile cooling systems, exposure to ethylene glycol is limited by its low vapor pressure. However, inhalation of droplets of ethylene glycol can be very dangerous. In the body, ethylene glycol initially stimulates the central nervous system, then depresses it. Glycolic acid, chemical formula $HOCH_2CO_2H$, formed as an intermediate metabolite in the metabolism of ethylene glycol, may cause acidemia, and oxalic acid produced by further oxidation may precipitate in the kidneys as solid calcium oxalate, CaC_2O_4, causing clogging.

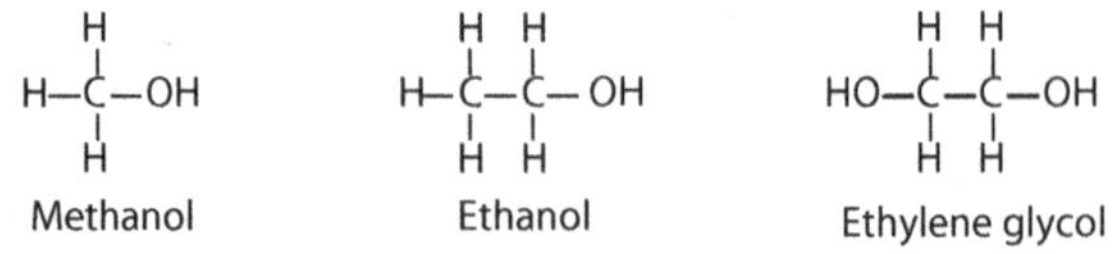

Methanol   Ethanol   Ethylene glycol

FIGURE 24.4 *Alcohols* such as these three compounds are oxygenated compounds in which the hydroxyl functional group is attached to an alkyl or alkenyl hydrocarbon skeleton.

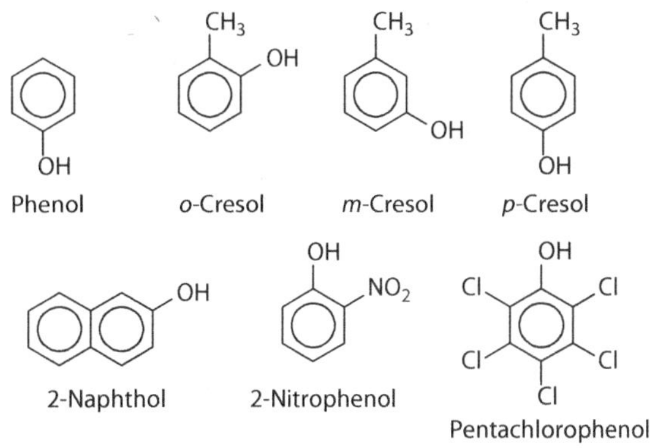

FIGURE 24.5 Some phenols and phenolic compounds.

Of the higher alcohols, 1-butanol is an irritant, but its toxicity is limited by its low vapor pressure. Unsaturated (alkenyl) allyl alcohol, $CH_2{=}CHCH_2OH$, has a pungent odor and is strongly irritating to the eyes, mouth, and lungs.

24.4.5 Phenols

Figure 24.5 shows some of the more important phenolic compounds, aryl analogs of alcohols which have properties much different from those of the aliphatic and olefinic alcohols. Nitro groups ($-NO_2$) and halogen atoms (particularly Cl) bonded to the aromatic rings strongly affect the chemical and toxicological behavior of phenolic compounds.

Although the first antiseptic to be used on wounds and in surgery, phenol is a protoplasmic poison that damages all kinds of cells and is alleged to have caused "an astonishing number of poisonings" since it came into general use.[3] The acute toxicological effects of phenol are largely upon the central nervous system and death can occur as soon as one and half hour after exposure. Acute poisoning by phenol can cause severe gastrointestinal disturbances, kidney malfunction, circulatory system failure, lung edema, and convulsions. Fatal doses of phenol may be absorbed through the skin. Key organs damaged by chronic phenol exposure include the spleen, pancreas, and kidneys. The toxic effects of other phenols resemble those of phenol.

24.4.5.1 Aldehydes and Ketones

Some of the more common carbonyl compounds (aldehydes and ketones) are shown in Figure 24.6. Because of their widespread use, these compounds are toxicologically significant.

Formaldehyde is uniquely important because of its widespread use and toxicity. In the pure form, formaldehyde is a colorless gas with a pungent, suffocating odor. It is commonly encountered as *formalin*, a 37–50% aqueous solution of formaldehyde containing some methanol. Exposure to inhaled formaldehyde via the respiratory tract is usually due to molecular formaldehyde vapor,

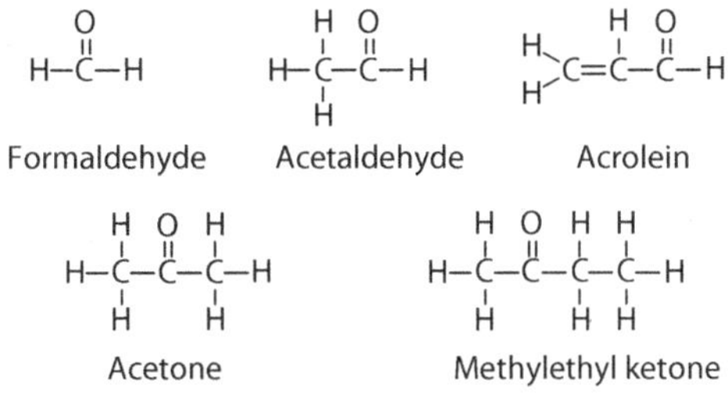

FIGURE 24.6 Commercially and toxicologically significant aldehydes and ketones.

whereas exposure by other routes is usually due to formalin. Prolonged, continuous exposure to formaldehyde can cause hypersensitivity. A severe irritant to the mucous membrane linings of both the respiratory and alimentary tracts, formaldehyde reacts strongly with functional groups in molecules. Formaldehyde has been shown to be a lung carcinogen in experimental animals. The toxicity of formaldehyde is largely due to its metabolic oxidation product, formic acid (see Section 24.4.5.2).

The lower aldehydes are relatively water-soluble and intensely irritating. These compounds attack exposed moist tissue, particularly the eyes and mucous membranes of the upper respiratory tract. (Some of the irritating properties of photochemical smog, Chapter 13, are due to the presence of aldehydes.) However, aldehydes that are relatively less soluble can penetrate further into the respiratory tract and affect the lungs. Colorless, liquid acetaldehyde is relatively less toxic than acrolein and acts as an irritant and systemically as a narcotic to the central nervous system. Extremely irritating, lachrimating acrolein vapor has a choking odor and inhalation of it can cause severe damage to respiratory tract membranes. Tissue exposed to acrolein may undergo severe necrosis, and direct contact with the eye can be especially hazardous.

The ketones shown in Figure 24.6 are relatively less toxic than the aldehydes. Pleasant-smelling acetone can act as a narcotic; it causes dermatitis by dissolving fats from skin. Not many toxic effects have been attributed to methyl ethyl ketone. It is suspected of having caused neuropathic disorders in shoe factory workers.

24.4.5.2 Carboxylic Acids

Formic acid, HCO_2H, is a relatively strong acid that is corrosive to tissue. In Europe, decalcifier formulations for removing mineral scale that contain about 75% formic acid are sold; and children ingesting these solutions have suffered corrosive lesions to the mouth and esophageal tissue. Although acetic acid as a 4–6% solution in vinegar is an ingredient of many foods, pure acetic acid (glacial acetic acid) is extremely corrosive to the tissue that it contacts. Ingestion of or skin contact with acrylic acid can cause severe damage to tissues.

24.4.5.3 Ethers

The common ethers have relatively low toxicities because of the low reactivity of the C–O–C functional group which has very strong carbon–oxygen bonds. Exposure to volatile diethyl ether is usually by inhalation and about 80% of this compound that gets into the body is eliminated unmetabolized as the vapor through the lungs. Diethyl ether depresses the central nervous system and is a depressant widely used as an anesthetic for surgery. Low doses of diethyl ether cause drowsiness, intoxication, and stupor, whereas higher exposures cause unconsciousness and even death.

24.4.5.4 Acid Anhydrides

Strong-smelling, intensely lachrimating *acetic anhydride*,

$$\mathrm{H{-}\overset{H}{\underset{H}{C}}{-}\overset{O}{\overset{\|}{C}}{-}O{-}\overset{O}{\overset{\|}{C}}{-}\overset{H}{\underset{H}{C}}{-}H}\quad \text{Acetic anhydride}$$

is a systemic poison. It is very corrosive to the skin, eyes, and upper respiratory tract, causing blisters and burns that heal only slowly. Levels in the air should not exceed 0.04 mg/m^3 and adverse effects to the eyes can occur at about 0.4 mg/m^3.

24.4.5.5 Esters

Many esters (Figure 24.7) have relatively high volatilities so that the pulmonary system is a major route of exposure. Because of their generally good solvent properties, esters penetrate tissues and tend to dissolve body lipids. For example, vinyl acetate acts as a skin defatting agent. Because they hydrolyze in water, ester toxicities tend to be the same as the toxicities of the acids and alcohols

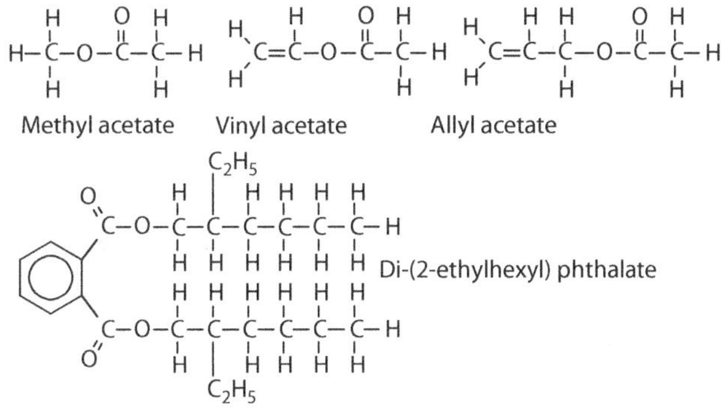

FIGURE 24.7 Examples of esters.

from which they were formed. Many volatile esters exhibit asphyxiant and narcotic action. Whereas many of the naturally occurring esters have insignificant toxicities at low doses, allyl acetate and some of the other synthetic esters are relatively toxic.

Insofar as potential health effects are concerned, di-(2-ethylhexyl) phthalate (DEHP) is arguably the ester of most concern. This is because of the use of this ester at levels of around 30% as a plasticizer to impart flexibility to PVC plastic. As a consequence of the widespread use of DEHP-containing PVC plastics, DEHP has become a ubiquitous contaminant found in water, sediment, food, and biological samples. There has been concern about its use in medical applications such as in bags used to hold intravenous solutions administered to medical patients. Although the acute toxic effects of DEHP are low, widespread direct exposure of humans is worrisome.

24.4.6 Organonitrogen Compounds

Organonitrogen compounds constitute a large group of compounds with diverse toxicities. Examples of several of the kinds of organonitrogen compounds discussed here are given in Figure 24.8.

24.4.6.1 Aliphatic Amines

The lower amines, such as the methylamines, are rapidly and easily taken into the body by all common exposure routes. They are basic and react with water in tissue,

$$R_3N + H_2O \rightarrow R_3NH^+ + OH^- \qquad (24.6)$$

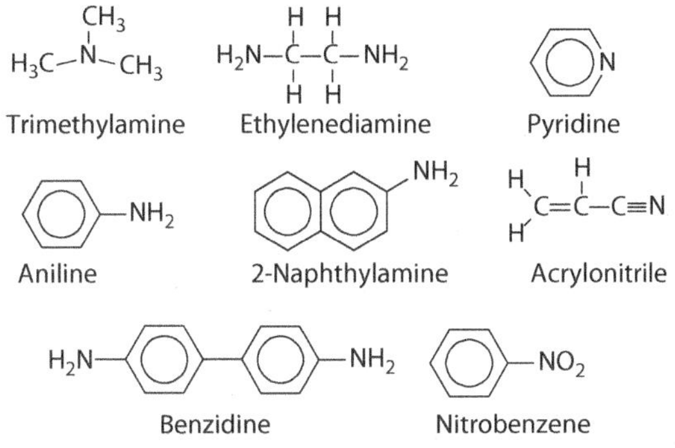

FIGURE 24.8 Some toxicologically significant organonitrogen compounds.

thereby raising the pH of the tissue to harmful levels, acting as corrosive poisons (especially to sensitive eye tissue), and causing tissue necrosis at the point of contact. Among the systemic effects of amines are necrosis of the liver and kidneys, lung hemorrhage and edema, and sensitization of the immune system. The lower amines are among the more toxic substances in routine, large-scale use.

Ethylenediamine is the most common of the *alkyl polyamines*, compounds in which two or more amino groups are bonded to alkane moieties. Its toxicity rating is only 3, but it is a strong skin sensitizer and can damage eye tissue.

24.4.6.2 Carbocyclic Aromatic Amines

Aniline is a widely used industrial chemical and is the simplest of the *carbocyclic aromatic amines*, a class of compounds in which at least one substituent group is an aromatic hydrocarbon ring bonded directly to the amino group. There are numerous compounds with many industrial uses in this class of amines. Some of the carbocyclic aromatic amines have been shown to cause cancer in the human bladder, ureter, and pelvis, and are suspected of being lung, liver, and prostate carcinogens. A very toxic colorless liquid with an oily consistency and distinct odor, aniline readily enters the body by inhalation, ingestion, and through the skin. Metabolically, aniline converts iron(II) in hemoglobin to iron(III). This causes a condition called *methemoglobinemia*, characterized by cyanosis and a brown-black color of the blood, in which the hemoglobin can no longer transport oxygen in the body. This condition is not reversed by oxygen therapy.

Both *1-naphthylamine* (a-naphthylamine) and *2-naphthylamine* (b-naphthylamine) are proven human bladder carcinogens. In addition to being a proven human carcinogen, *benzidine*, 4,4′-diaminobiphenyl, is highly toxic and has systemic effects that include blood hemolysis, bone marrow depression, and kidney and liver damage. It can be taken into the body orally, by inhalation into the lungs, and by skin sorption.

24.4.6.3 Pyridine

Pyridine, a colorless liquid with a sharp, penetrating, "terrible" odor, is an aromatic amine in which an N atom is part of a 6-membered ring. This widely used industrial chemical is only moderately toxic with a toxicity rating of 3. Symptoms of pyridine poisoning include anorexia, nausea, fatigue, and, in cases of chronic poisoning, mental depression. In a few rare cases, pyridine poisoning has been fatal.

24.4.6.4 Acrylamide: Toxic Potato Chips?

In 2002, Swedish investigators reported finding potentially harmful amounts of acrylamide in fried and baked foods, but not in boiled foods. Acrylamide is known to be a neurotoxin and is a carcinogen to rats and possibly to humans as well. The formation of acrylamide has been shown to occur when glucose and the amino acid asparagine are heated together to temperatures exceeding 120°C. The reaction sequence below, known as the Maillard reaction, results in the formation of acrylamide from glucose and asparagine:

$$\text{Glucose} + \underset{\text{Asparagine}}{HO{-}C(=O){-}CH(NH_2){-}CH_2{-}C(=O){-}NH_2} \xrightarrow[-CO_2]{-H_2O} \text{(glucose ring)}{-}N{=}CH{-}CH_2{-}C(=O){-}NH_2 \xrightarrow{\text{Several steps}} \underset{\text{Acrylamide}}{H_2C{=}CH{-}C(=O){-}NH_2} \quad (24.7)$$

24.4.6.5 Nitriles

Nitriles contain the –C≡N functional group. Colorless, liquid *acetonitrile*, CH_3CN, is widely used in the chemical industry. With a toxicity rating of 3–4, acetonitrile is considered relatively safe, although it has caused human deaths, perhaps by metabolic release of cyanide. *Acrylonitrile*, a colorless liquid with a peach-seed (cyanide) odor, is highly reactive because it contains both nitrile and C=C groups. Ingested, absorbed through the skin, or inhaled as vapor, acrylonitrile metabolizes to release deadly HCN, which it resembles toxicologically.

24.4.6.6 Nitro Compounds

The simplest of the *nitro compounds, nitromethane* H_3CNO_2, is an oily liquid that causes anorexia, diarrhea, nausea, and vomiting and damages the kidneys and liver. *Nitrobenzene*, a pale yellow, oily liquid with an odor of bitter almonds or shoe polish, can enter the body by all routes. It has a toxic action much like that of aniline, converting hemoglobin to methemoglobin, which cannot carry oxygen to body tissue. Nitrobenzene poisoning is manifested by cyanosis.

24.4.6.7 Nitrosamines

N-nitroso compounds (*nitrosamines*) contain the N–N=O functional group and have been found in a variety of materials to which humans may be exposed, including beer, whiskey, and cutting oils used in machining. Cancer may result from exposure to a single large dose or from chronic exposure to relatively small doses of some nitrosamines. Once widely used as an industrial solvent and known to cause liver damage and jaundice in exposed workers, dimethylnitrosamine was shown to be carcinogenic from studies starting in the 1950s.

CH_3
|
N–N=O
|
CH_3

Dimethylnitrosamine
(N-nitrosodimethylamine)

24.4.6.8 Isocyanates and Methyl Isocyanate

Compounds with the general formula R–N=C=O, *isocyanates* are widely used industrial chemicals noted for the high chemical and metabolic reactivity of their characteristic functional group. *Methyl isocyanate*, H_3C–N=C=O, was the toxic agent involved in the catastrophic industrial poisoning in Bhopal, India, on December 2, 1984, the worst industrial accident in history. In this incident, several tons of methyl isocyanate were released, killing 2000 people and affecting about 100,000. The lungs of victims were attacked; survivors suffered long-term shortness of breath and weakness from lung damage, as well as numerous other toxic effects including nausea and bodily pain.

24.4.6.9 Organonitrogen Pesticides

Pesticidal *carbamates* are characterized by the structural skeleton of carbamic acid outlined by the dashed box in the structural formula of carbaryl in Figure 24.9. Widely used on lawns and gardens, insecticidal *carbaryl* has a low toxicity to mammals. Highly water-soluble *carbofuran* is a systemic insecticide in that it is taken up by the roots and leaves of plants; insects that feed on the leaves are poisoned. The toxic effects on animals exposed to carbamates are due to the fact that they inhibit acetylcholinesterase directly without the need to first undergo biotransformation. This effect is relatively reversible because of metabolic hydrolysis of the carbamate ester.

Reputed to have "been responsible for hundreds of human deaths,"[4] herbicidal *paraquat* has a toxicity rating of 5. Dangerous or even fatal acute exposures can occur by inhalation of spray, skin contact, and ingestion. Paraquat is a systemic poison that affects enzyme activity and is devastating to a number of organs. Pulmonary fibrosis results in animals that have inhaled paraquat aerosols,

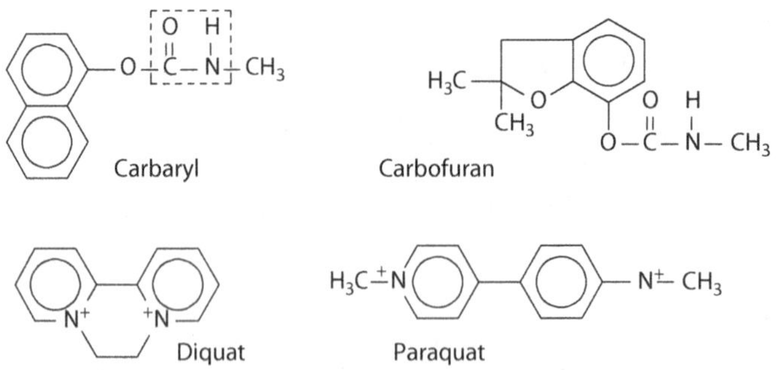

Carbaryl Carbofuran

Diquat Paraquat

FIGURE 24.9 Examples of organonitrogen pesticides.

and the lungs are also adversely affected by nonpulmonary exposure. Acute exposure may cause variations in the levels of catecholamine, glucose, and insulin. The most prominent initial symptom of poisoning is vomiting, followed within a few days by dyspnea, cyanosis, and evidence of impairment of the kidneys, liver, and heart. Pulmonary fibrosis, often accompanied by pulmonary edema and hemorrhaging, is observed in fatal cases.

Arguably the most widely encountered organonitrogen herbicide is *atrazine*, the structural formula of which is shown in Figure 7.12.[5] This compound is frequently encountered in drinking water supplies, especially in agricultural areas.

24.4.7 Organohalide Compounds

Because of their persistence and tendency to accumulate in lipid tissues, the ecotoxicology of organohalide compounds is quite important. Their occurrence in the environment has become less prevalent as their manufacture has been curtailed for environmental reasons. An interesting aspect of these compounds is their occurrence in human mothers' milk, which can be analyzed for environmental exposure to them.[6]

24.4.7.1 Alkyl Halides

The toxicities of alkyl halides, such as carbon tetrachloride, CCl_4, vary a great deal with the compound. Most of these compounds cause depression of the central nervous system, and individual compounds exhibit specific toxic effects.

Carbon tetrachloride was once widely available in general merchandise stores as a product to remove grease stains from clothing. It was also sold sealed in baseball-sized spherical glass containers that could be hurled at kitchen fires to extinguish the flames. During many years of its use as a consumer product, carbon tetrachloride compiled a grim record of toxic effects which led the U.S. Food and Drug Administration (FDA) to prohibit its household use in 1970. It is a systemic poison that affects the nervous system when inhaled, and the gastrointestinal tract, liver, and kidneys when ingested. The biochemical mechanism of carbon tetrachloride toxicity involves reactive radical species, including the following,

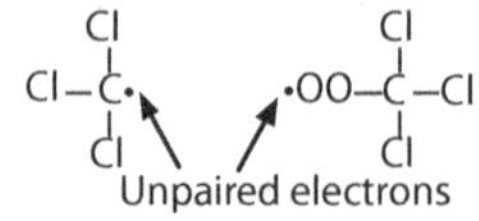

$$Cl_3C\cdot \qquad \cdot OOCCl_3$$

Unpaired electrons

that react with biomolecules, such as proteins and DNA. The most damaging such reaction occurs in the liver as *lipid peroxidation*, consisting of the attack of free radicals on unsaturated lipid molecules, followed by oxidation of the lipids through a free radical mechanism.

24.4.7.2 Alkenyl Halides

The most significant *alkenyl* or *olefinic organohalides* are the lighter chlorinated compounds, such as vinyl chloride and tetrachloroethylene:

$$\underset{\text{Vinyl chloride}}{H_2C{=}CHCl} \qquad \underset{\text{Tetrachloroethylene}}{Cl_2C{=}CCl_2}$$

Because of their widespread use and disposal in the environment, the numerous acute and chronic toxic effects of the alkenyl halides are of considerable concern.

The central nervous system, respiratory system, liver, and blood and lymph systems are all affected by vinyl chloride exposure, which has been widespread because of this compound's use in polyvinylchloride manufacture. Most notably, vinyl chloride is carcinogenic, causing a rare angiosarcoma of the liver. This deadly form of cancer has been observed in workers chronically exposed to vinyl chloride while cleaning autoclaves in the polyvinylchloride fabrication industry. The alkenyl organohalide, 1,1-dichloroethylene, is a suspect human carcinogen based upon animal studies and its structural similarity to vinyl chloride. The toxicities of both 1,2-dichloroethylene isomers are relatively low. These compounds act in different ways in that the *cis* isomer is an irritant and narcotic, whereas the *trans* isomer affects both the central nervous system and the gastrointestinal tract, causing weakness, tremors, cramps, and nausea. A suspect human carcinogen, trichloroethylene has caused liver carcinoma in experimental animals and is known to affect numerous body organs. Like other organohalide solvents, trichloroethylene causes skin dermatitis from dissolution of skin lipids, and it can affect the central nervous and respiratory systems, liver, kidneys, and heart. Symptoms of exposure include disturbed vision, headaches, nausea, cardiac arrhythmias, and burning/tingling sensations in the nerves (paresthesia).

Tetrachloroethylene damages the liver, kidneys, and central nervous system. It is a suspect human carcinogen. It is metabolically oxidized to tetrachloroethylene oxide,

$$Cl_2C{=}CCl_2 \xrightarrow[\text{P450}]{\{O\}} \underset{\text{Tetrachloroethylene oxide}}{Cl_2C\overset{O}{\frown}CCl_2} \tag{24.8}$$

which is probably the reactive intermediate reponsible for its toxicity. The tetrachloroethylene oxide intermediate is further metabolized to trichloroacetic acid, Cl_3CO_2H, and other products.

24.4.7.3 Aryl Halides

Individuals exposed to irritant monochlorobenzene by inhalation or skin contact suffer symptoms to the respiratory system, liver, skin, and eyes. Ingestion of this compound causes effects similar to those of toxic aniline, including incoordination, pallor, cyanosis, and eventual collapse.

The dichlorobenzenes are irritants that affect the same organs as monochlorobenzene. Some animal tests have suggested that 1,2-dichlorobenzene is a potential cancer-causing substance. *Para*-dichlorobenzene (1,4-dichlorobenzene), a chemical used in air fresheners and mothballs, has been known to cause profuse rhinitis (running nose), nausea, jaundice, liver cirrhosis, and weight loss associated with anorexia. It is not known to be a carcinogen. Its major urinary metabolite is 2,5-dichlorophenol, which is eliminated principally as the glucuronide or sulfate.

Because of their once widespread use in electrical equipment, as hydraulic fluids, and in many other applications, PCBs (see Section 7.12) became widespread, extremely persistent environmental pollutants. PCBs have a strong tendency to undergo bioaccumulation in lipid tissue. Polybrominated biphenyl analogs (PBBs) were much less widely used and distributed. However, PBBs were involved in one major incident that resulted in catastrophic agricultural losses when livestock feed contaminated with PBB flame retardant caused massive livestock poisoning in Michigan in 1973.

The greatest current concerns with brominated compounds are centered around brominated fire retardants. These are primarily pentabromo, octabromo, and decabromodiphenyl ethers (see below):

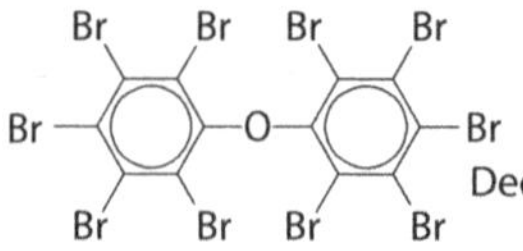

Decabromodiphenyl ether

They are very persistent pollutants and, although their toxicities to humans are not well established, environmental exposure to them is a matter of some concern. Unfortunately, effective substitutes for these compounds as fire retardants are lacking. Tightened standards for polybrominated compounds by the EU are causing concerns among electrical and electronic manufacturers who must comply with more stringent fireproofing standards in the face of expected phaseout of polybrominated fire retardants.

24.4.8 Organohalide Pesticides

Exhibiting a wide range of kind and degree of toxic effects, many organohalide insecticides (see Section 7.11) affect the central nervous system, causing tremor, irregular eye jerking, changes in personality, and loss of memory. Such symptoms are characteristic of acute DDT poisoning. However, the acute toxicity of DDT to humans is very low and, when used for the control of typhus and malaria in World War II, it was applied directly to people. The chlorinated cyclodiene insecticides—aldrin, dieldrin, endrin, chlordane, heptachlor, endosulfan, and isodrin—act on the brain, releasing betaine esters and causing headaches, dizziness, nausea, vomiting, jerking muscles, and convulsions. Dieldrin, chlordane, and heptachlor have caused liver cancer in test animals, and some chlorinated cyclodiene insecticides are teratogenic or fetotoxic. Because of these effects, aldrin, dieldrin, heptachlor, and chlordane have been prohibited from use in the United States.

The major *chlorophenoxy* herbicides are 2,4-dichlorophenoxyacetic acid (2,4-D), 2,4,5-trichlorophenoxyacetic acid (2,4,5-T or Agent Orange, now banned), and Silvex. Large doses of 2,4-dichlorophenoxyacetic acid have been shown to cause nerve damage, (peripheral neuropathy), convulsions, and brain damage. With a toxicity somewhat less than that of 2,4-D, Silvex is largely excreted unchanged in the urine. The toxic effects of 2,4,5-T (used as a herbicidal warfare chemical called "Agent Orange") have resulted from the presence of TCDD (commonly known as "dioxin," discussed below), a manufacturing by-product. Autopsied carcasses of sheep poisoned by this herbicide have exhibited nephritis, hepatitis, and enteritis.

24.4.8.1 TCDD

Polychlorinated dibenzodioxins are compounds which have the same basic structure as that of TCDD,

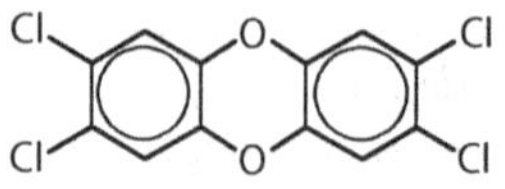

TCDD (2,3,7,8-tetrachloro-dibenzo-p-dioxin)

but have different numbers and locations of chlorine atoms on the ring structure. Extremely toxic to some animals, the toxicity of TCDD to humans is rather uncertain; it is known to cause a skin condition called chloracne. TCDD has been a manufacturing by-product of some commercial products (see the discussion of 2,4,5-T, above), a contaminant identified in some municipal incineration emissions, and a widespread environmental pollutant from improper waste disposal. This compound has been released in a number of industrial accidents, the most massive of which exposed several tens of thousands of people to a cloud of chemical emissions spread over an approximately 3-square-mile area at the Givaudan-La Roche Icmesa manufacturing plant near Seveso, Italy, in 1976. On an encouraging note from a toxicological perspective, no abnormal occurrences of major malformations were found in a study of several thousand children born in the area within 6 years after the release.

24.4.8.2 Chlorinated Phenols

The chlorinated phenols used in largest quantities have been *pentachlorophenol* (Chapter 7) and the trichlorophenol isomers used as wood preservatives. Although exposure to these compounds has been correlated with liver malfunction and dermatitis, contaminant polychlorinated dibenzodioxins may have caused some of the observed effects.

24.4.9 Organosulfur Compounds

Despite the high toxicity of H_2S, not all organosulfur compounds are particularly toxic. Their hazards are often reduced by their strong, offensive odors that warn of their presence.

Inhalation of even very low concentrations of the alkyl *thiols*, such as methanethiol, H_3CSH, can cause nausea and headaches; higher levels can cause increased pulse rate, cold hands and feet, and cyanosis. In extreme cases, unconsciousness, coma, and death occur. Like H_2S, the alkyl thiols are precursors to cytochrome oxidase poisons.

An oily, water-soluble liquid, *methylsulfuric acid* is a strong irritant to the skin, eyes, and mucous tissue. Colorless, odorless *dimethyl sulfate* is highly toxic and is a primary carcinogen that does not require bioactivation to cause cancer. Skin or mucous membranes exposed to dimethyl sulfate develop conjunctivitis and inflammation of nasal tissue and respiratory tract mucous membranes following an initial latent period during which few symptoms are observed. Damage to the liver and kidneys, pulmonary edema, cloudiness of the cornea, and death within 3–4 days can result from heavier exposures.

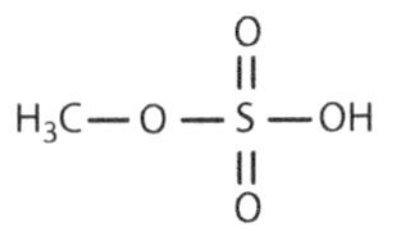

$$H_3C-O-\overset{\overset{\displaystyle O}{\|}}{\underset{\underset{\displaystyle O}{\|}}{S}}-OH \qquad\qquad H_3C-O-\overset{\overset{\displaystyle O}{\|}}{\underset{\underset{\displaystyle O}{\|}}{S}}-OCH_3$$

Methylsulfuric acid 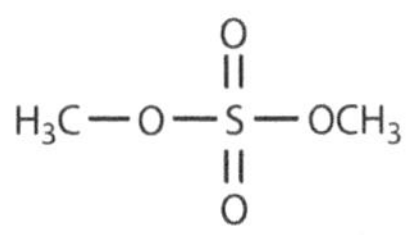   Dimethylsulfate

24.4.9.1 Sulfur Mustards

A typical example of deadly *sulfur mustards*, compounds used as military poisons or "poison gases," is mustard oil [bis(2-chloroethyl)sulfide]:

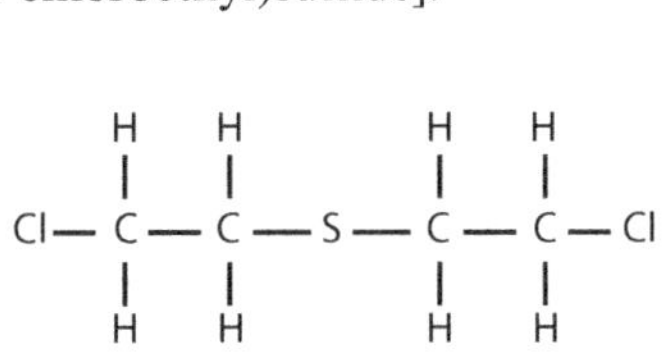

$$Cl-\overset{\overset{\displaystyle H}{|}}{\underset{\underset{\displaystyle H}{|}}{C}}-\overset{\overset{\displaystyle H}{|}}{\underset{\underset{\displaystyle H}{|}}{C}}-S-\overset{\overset{\displaystyle H}{|}}{\underset{\underset{\displaystyle H}{|}}{C}}-\overset{\overset{\displaystyle H}{|}}{\underset{\underset{\displaystyle H}{|}}{C}}-Cl \quad \text{Mustard oil}$$

An experimental mutagen and primary carcinogen, mustard oil produces vapors that penetrate deep within tissue, resulting in destruction and damage at some depth from the point of contact; penetration is very rapid, so that efforts to remove the toxic agent from the exposed area are ineffective after 30 min. This military "blistering gas" poison causes tissue to become severely inflamed with lesions that often become infected. These lesions in the lung can cause death.

24.4.10 Organophosphorus Compounds

Organophosphorus compounds have varying degrees of toxicity. Some of these compounds, such as the "nerve gases" produced as industrial poisons, are deadly in minute quantities. The toxicities of major classes of organophosphate compounds are discussed in this section.

24.4.10.1 Organophosphate Esters

Some organophosphate esters are shown in Figure 24.10. *Trimethyl phosphate* is probably moderately toxic when ingested or absorbed through the skin, whereas moderately toxic *triethyl phosphate*, $(C_2H_5O)_3PO$, damages nerves and inhibits acetylcholinesterase. Notoriously toxic *tri-o-cresyl phosphate* (TOCP) apparently is metabolized to products that inhibit acetylcholinesterase. Exposure to TOCP causes degeneration of the neurons in the body's central and peripheral nervous systems with early symptoms of nausea, vomiting, and diarrhea accompanied by severe abdominal pain. About 1–3 weeks after these symptoms have subsided, peripheral paralysis develops manifested by "wrist drop" and "foot drop," followed by slow recovery, which may be complete or leave a permanent partial paralysis, whereas the toxic effects of tetraethyl pyrophosphate (TEPP) or paraoxon develop much more rapidly. Humans poisoned by parathion exhibit skin twitching and respiratory distress. In fatal cases, respiratory failure occurs due to central nervous system paralysis.

Briefly used in Germany as a substitute for insecticidal nicotine, TEPP is a very potent acetylcholinesterase inhibitor. With a toxicity rating of 6 (supertoxic), TEPP is deadly to humans and other mammals.

24.4.10.2 Phosphorothionate and Phosphorodithioate Ester Insecticides

Because esters containing the P=S (thiono) group are resistant to nonenzymatic hydrolysis and are not as effective as P=O compounds in inhibiting acetylcholinesterase, they exhibit higher insect:mammal toxicity ratios than their nonsulfur analogs. Therefore, *phosphorothionate* and *phosphorodithioate* esters (Figure 24.11) have been widely used as insecticides. The insecticidal activity of these compounds requires metabolic conversion of P=S to P=O (oxidative desulfuration). Environmentally, organophosphate insecticides are superior to many of the organochlorine insecticides because the organophosphates readily undergo biodegradation and do not bioaccumulate.

The first commercially successful phosphorothionate/phosphorodithioate ester insecticide was *parathion*, *O,O*-diethyl-*O*-*p*-nitrophenylphosphorothionate, first licensed for use in 1944. This insecticide, which inhibits acetylcholinesterase, has a toxicity rating of 6 (supertoxic). Since its use began, several hundred people have been killed by parathion, including 17 of 79 people exposed to

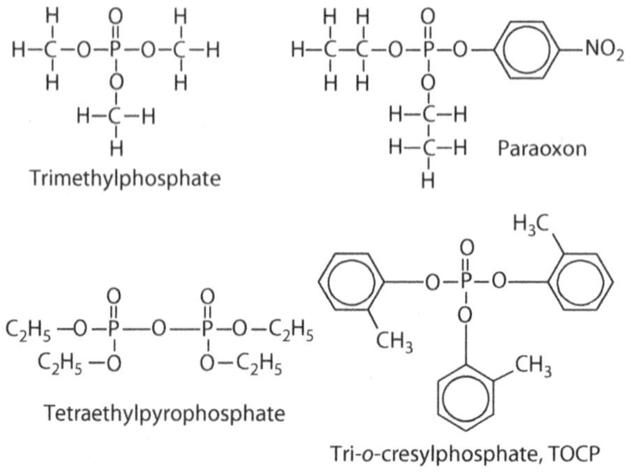

Trimethylphosphate

Paraoxon

Tetraethylpyrophosphate

Tri-*o*-cresylphosphate, TOCP

FIGURE 24.10 Some organophosphate esters.

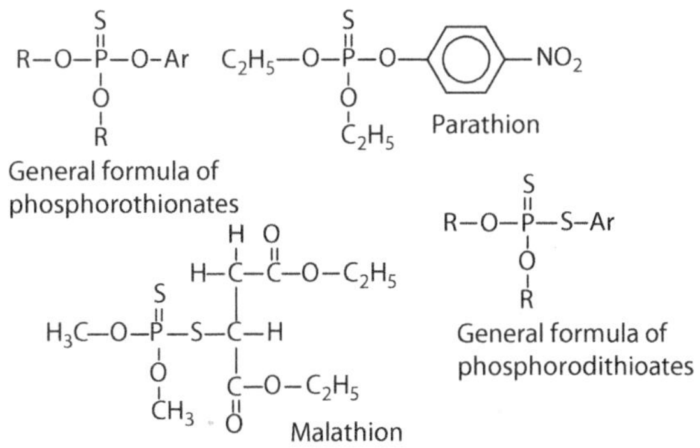

FIGURE 24.11 Phosphorothionate and phosphorodithioate ester insecticides. Malathion contains hydrolyzable carboxyester linkages.

contaminated flour in Jamaica in 1976. As little as 120 mg of parathion has been known to kill an adult human, and a dose of 2 mg has been fatal to a child. Most accidental poisonings have occurred by absorption through the skin. Methylparathion (a closely related compound with methyl groups instead of ethyl groups) is regarded as extremely toxic, and it is no longer used. In order for parathion to have a toxic effect, it must be converted metabolically to paraoxon (Figure 24.10), which is a potent inhibitor of acetylcholinesterase. Because of the time required for this conversion, symptoms develop several hours after exposure.

Malathion is the best known of the phosphorodithioate insecticides. It has a relatively high insect:mammal toxicity ratio because of its two carboxyester linkages which are hydrolyzable by carboxylase enzymes (possessed by mammals, but not insects) to relatively nontoxic products. For example, although malathion is a very effective insecticide, its LD_{50} for adult male rats is about 100 times that of parathion.

24.4.10.3 Organophosphorus Military Poisons

Powerful inhibitors of acetylcholinesterase enzyme, organophosphorus "nerve gas" military poisons include *Sarin* and *VX*, for which structural formulas are shown below. (The possibility that military poisons such as these might be used in war was a major concern during the 1991 Mid-East conflict, which, fortunately, ended without their being employed. The existence of military poison nerve gas was one of the justifications for the 2003 Iraqi war, but these "weapons of mass destruction" were not found.) A systemic poison to the central nervous system that is readily absorbed as a liquid through the skin, Sarin may be lethal at doses as low as about 0.01 mg/kg; a single drop can kill a human.

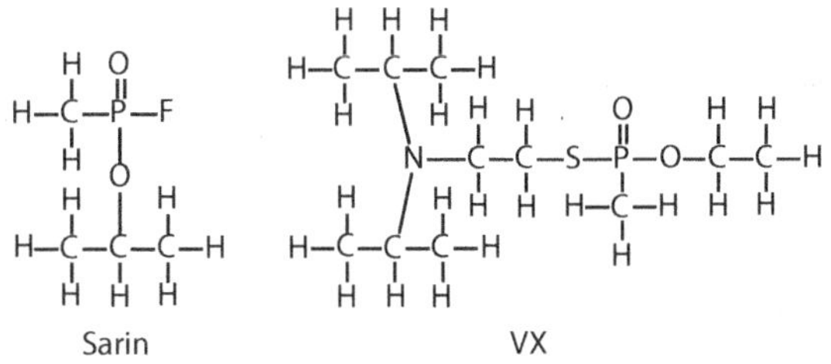

24.5 TOXIC NATURAL PRODUCTS

Nature produces a wide variety of toxic substances. People can ingest toxic natural products in a variety of forms and food can be a source of mutagens that can be cancer-suspect agents. One of the most toxic substances known is botulinum toxin produced by anaerobic *Clostridium botulinum* bacteria. (Despite its extreme toxicity, botulinum toxin has a number of medicinal uses primarily as a muscle relaxant; it is the active ingredient in Botox used to treat wrinkles.) A number of organisms wage forms of chemical warfare. Venomous serpents produce venoms of various kinds to kill prey and for defense.

Food poisoning of various kinds can be caused by ingestion of natural products. One of the most severe forms of food poisoning is paralytic shellfish poisoning in which various varieties of shellfish accumulate toxins from single-celled dinoflagellates. These toxins block neuronal transmission and can be fatal.

Mycotoxins are toxic secondary metabolites from fungi that have a wide range of structures and a variety of toxic effects. Human and animal exposure to mycotoxins usually result from ingestion of food upon which fungal molds have grown. Among the many kinds of molds that produce mycotoxins are *Aspergillus flavus*, *Fusarium*, *Trichoderma*, *Aspergillus*, and *Penicillium*. Perhaps the most well-known mycotoxins are the *aflatoxins*, such as aflatoxin B_1 produced by *Aspergillus*:

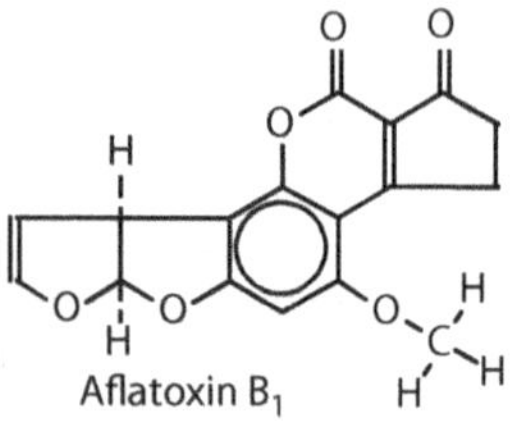

Aflatoxin B_1

Molds that produce mycotoxins grow on a variety of food products including corn, cereal grains, rice, apples, peanuts, and milk. Other mycotoxins include ergot alkaloids, ochratoxins, fumonisins, trichothecenes, tremorgenic toxins, satratoxin, zearalenone, and vomitotoxin.

LITERATURE CITED

1. NIOSH website: http://www.cdc.gov/niosh/ocas/ocaseeoi.html
2. Carra, D. L., J. A. Carrb, R. E. Willisa, and T. A. Pressley, "A perchlorate sensitive iodide transporter in frogs," *General and Comparative Endocrinology*, **156**, 9–14, 2008.
3. Gosselin, R. E., R. P. Smith, and H. C. Hodge, Eds, "Phenol," in *Clinical Toxicology of Commercial Products*, 5th ed., Williams and Wilkins, Baltimore/London, pp. III-344–III-348, 1984.
4. Gosselin, R. E., R. P. Smith, and H. C. Hodge, Eds, *Clinical Toxicology of Commercial Products*, 5th ed., Williams and Wilkins, Baltimore/London, pp. III-328–III-336, 1984.
5. Benotti, M. J., R. A. Trenholm, B. J. Vanderford, J. D. Holady, B. D. Stanford, and S. A. Snyder, Shane "Pharmaceuticals and endocrine disrupting compounds in U.S. drinking water," *Environmental Science and Technology*, **43**, 597–603, 2009.
6. Zietz, B. P., M. Hoopmann, M. Funcke, R. Huppmann, R. Suchenwirth, and E. Gierden, "Long-term biomonitoring of polychlorinated biphenyls and organochlorine pesticides in human milk from mothers living in Northern Germany," *International Journal of Hygiene and Environmental Health*, **211**, 624–638, 2008.

SUPPLEMENTARY REFERENCES

Baselt, R. C., *Disposition of Toxic Drugs and Chemicals in Man*, 6th ed., Biomedical Publications, Foster City, CA, 2002.

Benigni, R., Ed., *Quantitative Structure–Activity Relationship (QSAR) Models of Mutagens and Carcinogens*, CRC Press, Boca Raton, FL, 2003.

Bingham, E., B. Cohrssen, and C. H. Powell, *Patty's Toxicology*, 5th ed., Wiley, New York, 2001.

Carey, J., Ed., *Ecotoxicological Risk Assessment of the Chlorinated Organic Chemicals*, SETAC Press, Pensacola, FL, 1998.

Cooper, A. R., L. Overholt, H. Tillquist, and D. Jamison, *Cooper's Toxic Exposures Desk Reference with CD-ROM*, CRC Press/Lewis Publishers, Boca Raton, FL, 1997.

Dart, R. C., *Medical Toxicology*, 3rd ed., Lippincott, Williams & Wilkins, Philadelphia, 2003.

Ellenhorn, M. J. and S. S. Ellenhorn, *Ellenhorn's Medical Toxicology: Diagnosis and Treatment of Human Poisoning*, 2nd ed., Williams & Wilkins, Baltimore, MD, 1997.

Fenton, J., *Toxicology: A Case-Oriented Approach*, CRC Press, Boca Raton, FL, 2002.

Greenberg, M. I., R. J. Hamilton, and S. D. Phillips, Eds, *Occupational, Industrial, and Environmental Toxicology*, Mosby, St. Louis, 1997.

Hall, S. K., J. Chakraborty, and R. J. Ruch, Eds, *Chemical Exposure and Toxic Responses*, CRC Press/Lewis Publishers, Boca Raton, FL, 1997.

Hodgson, E. and R. C. Smart, Eds, *Introduction to Biochemical Toxicology*, 3rd ed., Wiley-Interscience, New York, 2001.

Jameson, C. W., Ed., *Report on Carcinogens: Carcinogen Profiles*, 11th ed., National Toxicology Program, Research Triangle Park, NC, 2005.

Johnson, B. L., *Impact of Hazardous Waste on Human Health: Hazard, Health Effects, Equity and Communication Issues*, Ann Arbor Press, Chelsea, MI, 1999.

Joshi, B. D., P. C. Joshi, and N. Joshi, Eds, *Environmental Pollution and Toxicology*, A.P.H. Publishing Corporation, New Delhi, India, 2008.

Klaassen, C. D., Ed., *Casarett and Doull's Toxicology: The Basic Science of Poisons*, 7th ed., McGraw-Hill Medical, New York, 2008.

Leikin, J. B. and F. P. Paloucek, *Poisoning and Toxicology Handbook*, 4th ed., Taylor & Francis/CRC Press, Boca Raton, FL, 2008.

Lippmann, M., Ed., *Environmental Toxicants: Human Exposures and their Health Effects*, 3rd ed., Wiley, New York, 2009.

Nichol, J., *Bites and Stings. The World of Venomous Animals*, Facts on File, New York, 1989.

Patnaik, P., *A Comprehensive Guide to the Hazardous Properties of Chemical Substances*, 2nd ed., Wiley, New York, 1999.

Pohanish, R. P. and M. Sittig, *Sittig's Handbook of Toxic and Hazardous Chemicals and Carcinogens*, Knovel Corporation, Norwich, NY, 2002.

Public Health Service, National Toxicology Program, *11th Report on Carcinogens*, U.S. Department of Health and Human Services, Washington, DC, 2008, available from the following website: http://ntp.niehs.nih.gov/ntp/roc/toc11.html

Romano, J. A., B. J. Lukey, and H. Salem, Eds, *Chemical Warfare Agents: Chemistry, Pharmacology, Toxicology, and Therapeutics*, 2nd ed., Taylor & Francis/CRC Press, Boca Raton, FL, 2008.

Rosenstock, L., *Textbook of Clinical Occupational and Environmental Medicine*, Saunders Health Sciences Division, Philadelphia, 2004.

Smart, R. C. and E. Hodgson, Eds, *Molecular and Biochemical Toxicology*, 4th ed., Wiley, Hoboken, NJ, 2008.

Ullmann's Industrial Toxicology, Wiley-VCH, New York, 2005.

Ware, G., *Reviews of Environmental Contamination and Toxicology*, Vol. 190, Springer, New York, 2007.

Wexler, P., Ed., *Encyclopedia of Toxicology*, 2nd ed., Elsevier, New York, 2005.

Williamson, J. A., P. J. Fenner, J. W. Burnett, and J. F. Rifkin, Eds, *Venomous and Poisonous Marine Animals: A Medical and Biological Handbook*, University of New South Wales Press, Sydney, Australia, 1996.

Wilson, S. H. and W. A. Suk, *Biomarkers of Environmentally Associated Disease*, Taylor & Francis/CRC Press, Boca Raton, FL, 2002.

Zelikoff, J. and P. L. Thomas, Eds, *Immunotoxicology of Environmental and Occupational Metals*, Taylor & Francis, London, 1998.

QUESTIONS AND PROBLEMS

1. List and discuss two elements that are invariably toxic in their elemental forms. For another element, list and discuss two elemental forms, one of which is quite toxic and the other of which is essential for the body. In what sense is even the toxic form of this element "essential for life?"
2. What is a toxic substance that bonds to iron(III) in iron-containing ferricytochrome oxidase enzyme, preventing its reduction to iron(II) in the oxidative phosphorylation process by which the body utilizes O_2?

3. What are interhalogen compounds, and which elemental forms do their toxic effects most closely resemble?
4. Name and describe the three health conditions that may be caused by inhalation of asbestos.
5. Why might tetraethyl lead be classified as "the most notable toxic organometallic compound"?
6. What is the most common toxic effect commonly attributed to low-molar-mass alkanes?
7. Information about the toxicities of many substances to humans is lacking because of limited data on direct human exposure. (Volunteers to study human health effects of toxicants are in notably short supply.) However, there is a great deal of information available about human exposure to phenol and the adverse effects of such exposure. Explain.
8. Comment on the toxicity of the compound below:

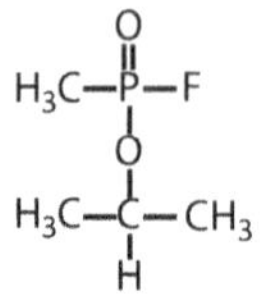

$$H_3C-\overset{\overset{O}{\|}}{P}-F$$
$$|$$
$$O$$
$$|$$
$$H_3C-\underset{H}{\underset{|}{C}}-CH_3$$

9. What are neuropathic disorders? Why are organic solvents frequently the cause of such disorders?
10. What is a major metabolic effect of aniline? What is this effect called? How is it manifested?
11. What are the organic compounds characterized by the N−N=O functional group? What is their major effect on health?
12. What structural group is characteristic of carbamates? For what purpose are these compounds commonly used? What are their major advantages in such an application?
13. What is lipid peroxidation? Which common toxic substance is known to cause lipid peroxidation?
14. Biochemically, what do organophosphate esters such as parathion do that could classify them as "nerve poisons"?
15. Although benzene and toluene have a number of chemical similarities, their metabolisms and toxic effects are quite different. Explain.
16. Match each compound in the figure below with its description from the following and explain your choices: (A) By-product of herbicide manufacture, (B) produced by fungi, (C) carcinogen, (D) although potentially a toxic pollutant, there are no good substitutes for its use.

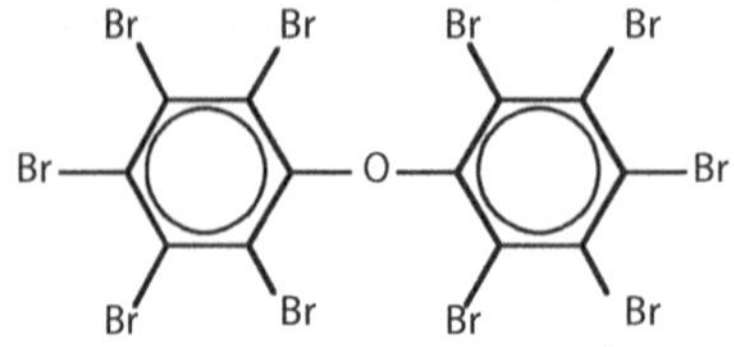

Br Br Br Br
Br O Br
Br Br Br Br
(1)

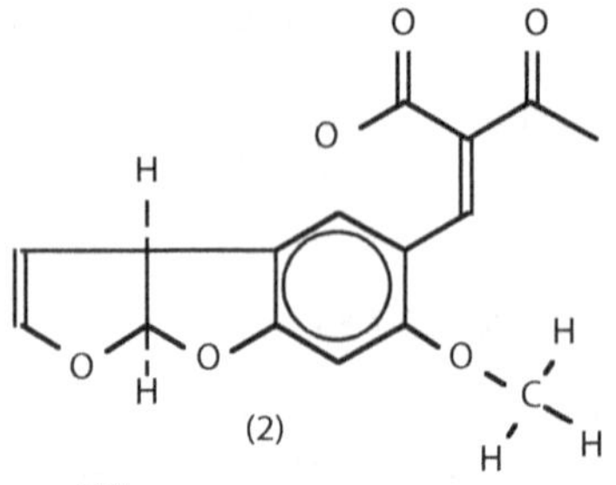

H O O
O
O O O
H H
C
H H
(2)

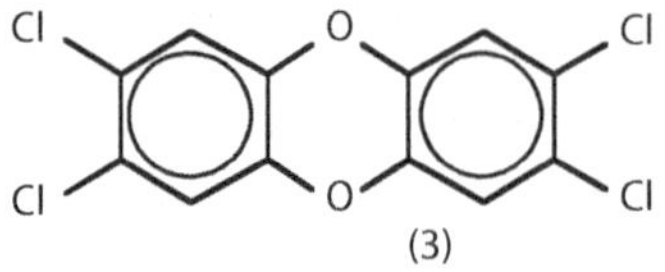

Cl O Cl
Cl O Cl
(3)

CH_3
N—N=O
CH_3
(4)

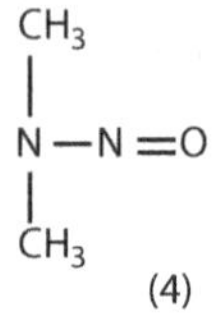

17. Consider four substances mentioned in this chapter, "A, B, C, and D," that are toxic because they interfere with oxygen transport or utilization in the body. A does not affect hemoglobin, but it can result in asphyxiation. B and C prevent hemoglobin from transporting O_2. Toxic substance D prevents the body from utilizing O_2 in metabolic processes. Because D has an affinity for a metal ion in a particular oxidation state, C can actually act as an antidote to poisoning from D. Give plausible identities of these four species and explain your choice in detail.
18. Match each toxic substance from the list on the left, below, with its effect or characteristic from the list on the right and explain each choice:

A. Methanol	1. Inhibits acetylcholine esterase
B. Parathion	2. Carcinogen
C. Phthalate esters	3. Affects optic nerve leading to blindness
D. Dimethylnitrosamine	4. Very widespread

25 Chemical Analysis of Water and Wastewater

25.1 GENERAL ASPECTS OF ENVIRONMENTAL CHEMICAL ANALYSIS

Scientists' understanding of the environment can only be as good as their knowledge of the identities and quantities of pollutants and other chemical species in water, air, soil, and biological systems. Therefore, proven, state-of-the-art techniques of chemical analysis, properly employed, are essential to environmental chemistry. The present time is a very exciting period in the evolution of analytical chemistry, characterized by the development of new and improved analysis techniques that enable detection of much lower levels of chemical species and a vastly increased data throughput. These developments pose some challenges. Because of the lower detection limits of some instruments, it is now possible to see quantities of pollutants that would have escaped detection previously, resulting in difficult questions regarding the setting of maximum allowable limits of various pollutants. The increased output of data from automated instruments has in many cases overwhelmed human capacity to assimilate and understand it.

Challenging problems still remain in developing and utilizing techniques of environmental chemical analysis. Not the least of these problems is knowing which species should be measured, or even whether or not an analysis should be performed at all. The quality and choice of analyses is much more important than the number of analyses performed. Indeed, a persuasive argument can be made that, given modern capabilities in analytical chemistry, too many analyses of environmental samples are performed, whereas fewer, more carefully planned analyses would yield more useful information.

In addition to a discussion of water analysis, this chapter covers some of the general aspects of environmental chemical analysis and the major techniques that are used to determine a wide range of analytes (species measured). Many techniques are common to water, air, soil, and biological sample analyses, and reference is made to them in chapters that follow.

25.1.1 Error and Quality Control

A crucial aspect of any chemical analysis is the validity and quality of the data that it produces. All measurements are subject to error, which may be *systematic* (of the same magnitude and same direction) or *random* (varying in both magnitude and direction). Systematic errors cause the measured values to vary consistently from the true values; this variation is known as the *bias*. The degree to which a measured value comes close to the actual value of an analytical measurement is called the *accuracy* of the measurement, reflecting both systematic and random errors. It is essential for the analyst to determine these error components in the measurement of environmental samples, including water samples. The identification and control of systematic and random errors falls in the category of *quality control (QC)* procedures. It is beyond the scope of this chapter to go into any detail on these crucial procedures for which the reader is referred to a work on standard methods for the analysis of water.[1]

In order for the results from a laboratory to be meaningful, the laboratory needs a *quality assurance plan* specifying measures taken to produce data of known quality. An important aspect of such a plan is the use of laboratory control standards consisting of samples with very accurately known analyte levels in a carefully controlled matrix. Such standard reference materials are available in the United States from the National Institute of Standards and Technology (NIST) for many kinds of samples.

Many environmental analytes are present at very low levels, which challenge the ability of the method used to detect and accurately quantify them. (Typically, pharmaceuticals and their metabolites are present at nanogram to low-picogram per liter levels in wastewater.) Therefore, the *detection limit* of a method of analysis is quite important. Defining detection limit has long been a controversial topic in chemical analysis. Every analytical method has a certain degree of noise. The detection limit is an expression of the lowest concentration of analytes that can be measured above the noise level with a specified degree of confidence in an analytical procedure. In the detection of analytes, two kinds of errors can be defined. A Type I error occurs when the measurement finds an analyte present when it is actually absent. A Type II error occurs when the measurement finds an analyte absent when it is actually present.

Detection limits can be further categorized into several different subcategories. The *instrument detection limit (IDL)* is the analyte concentration capable of producing a signal 3 times the standard deviation of the noise. The *lower level of detection (LLD)* is the quantity of analyte that will produce a measurable signal 99% of the time; it is about 2 times the IDL. The *method detection limit* (*MDL*) is measured like the LLD, except that the analyte is taken through the whole analytical procedure, including steps such as extraction and sample cleanup; it is about 4 times the IDL. Finally, the *practical quantitation limit (PQL)*, which is about 20 times the IDL, is the lowest level achievable in routine analysis in laboratories.

25.1.2 Water Analysis Methods

Methods of analysis are published for a huge number of water constituents and contaminants. It is not possible to cover them in a comprehensive manner in one short chapter. For analytical procedures, the reader is referred to sources of methods. The most comprehensive of these is the classic *Standard Methods for the Examination of Water and Wastewater*.[2] The U.S. EPA publishes methods for water analysis that are listed in an index of methods[3] and available from the National Technical Information Service.[4] Another useful source of methods is available on a CD ROM from Genium Publishing Corp.[5] Current issues in water analysis are reviewed periodically in the journal *Analytical Chemistry*.[6]

25.2 CLASSICAL METHODS

Before sophisticated instrumentation became available, most important water quality parameters and some air pollutant analyses were done by *classical methods*, which require only chemicals, balances to measure masses, burettes, volumetric flasks and pipettes to measure volumes, and other simple laboratory glassware. The two major classical methods are *volumetric analysis*, in which volumes of reagents are measured, and *gravimetric analysis*, in which masses are measured. Some of these methods are still used today, and many have been adapted to instrumental and automated procedures.

The most common classical methods for pollutant analysis are titrations, largely used for water analysis. Some of the titration procedures used are discussed in this section.

Acidity (see Section 3.7) is determined simply by titrating hydrogen ion with base. Titration to the methyl orange endpoint (pH 4.5) yields the "free acidity" due to strong acids (HCl and H_2SO_4). Carbon dioxide does not, of course, appear in this category. Titration to the phenolphthalein endpoint (pH 8.3) yields total acidity and accounts for all acids except those weaker than HCO_3^-.

Alkalinity may be determined by titration with H_2SO_4 to pH 8.3 to neutralize bases as strong as, or stronger than, carbonate ion,

$$CO_3^{2-} + H^+ \rightarrow HCO_3^- \tag{25.1}$$

or by titration to pH 4.5 to neutralize bases weaker than CO_3^{2-}, but as strong as, or stronger than, HCO_3^-:

$$HCO_3^- + H^+ \rightarrow H_2O + CO_2(g) \tag{25.2}$$

Titration to the lower pH yields total alkalinity.

The calcium and magnesium ions responsible for water hardness are readily titrated at pH 10 with a solution of EDTA, a chelating agent discussed in Sections 3.10 and 3.13. The titration reaction is

$$Ca^{2+}(\text{or } Mg^{2+}) + H_2Y^{2-} \rightarrow CaY^{2-}(\text{or } MgY^{2-}) + 2H^+ \tag{25.3}$$

where H_2Y^{2-} is the partially ionized EDTA chelating agent. Eriochrome Black T is used as an indicator, and it requires the presence of magnesium, with which it forms a wine red complex.

Several oxidation–reduction titrations can be used for environmental chemical analysis. Oxygen is determined in water by the Winkler titration. The first reaction in the Winkler method is the oxidation of manganese(II) to manganese(IV) by the oxygen analyte in a basic medium; this reaction is followed by acidification of the brown hydrated MnO_2 in the presence of I^- ion to release free I_2, and then titration of the liberated iodine with standard thiosulfate, using starch as an endpoint indicator:

$$Mn^{2+} + 2OH^- + \frac{1}{2}O_2 \rightarrow MnO_2(s) + H_2O \tag{25.4}$$

$$MnO_2(s) + 2I^- + 4H^+ \rightarrow Mn^{2+} + I_2 + 2H_2O \tag{25.5}$$

$$I_2 + 2S_2O_3^{2-} \rightarrow S_4O_6^{2-} + 2I^- \tag{25.6}$$

A back-calculation from the amount of thiosulfate required yields the original quantity of dissolved oxygen (DO) present. Biochemical oxygen demand (BOD) (see Section 7.9), is determined by adding a microbial "seed" to the diluted sample, saturating with air, incubating for five days, and determining the oxygen remaining. The results are calculated to show BOD as mg/L of O_2. A BOD of 80 mg/L, for example, means that biodegradation of the organic matter in a liter of the sample would consume 80 mg of oxygen.

25.3 SPECTROPHOTOMETRIC METHODS

25.3.1 Absorption Spectrophotometry

Absorption spectrophotometry of light-absorbing species in solution, historically called colorimetry when visible light is absorbed, is still used for the analysis of some water and air pollutants. Basically, absorption spectrophotometry consists of measuring the percent transmittance (%T) of monochromatic light passing through a light-absorbing solution as compared to the amount of light passing through a blank solution containing everything in the medium but the sought-for constituent (100%). The absorbance (A) is defined as

$$A = \log\frac{100}{\%T} \tag{25.7}$$

The relationship between A and the concentration (C) of the absorbing substance is given by Beer's law:

$$A = abC \tag{25.8}$$

where a is the absorptivity, a wavelength-dependent parameter characteristic of the absorbing substance; b is the path length of the light through the absorbing solution; and C is the concentration of the absorbing substance. A linear relationship between A and C at constant path length indicates adherence to Beer's law. In many cases, analyses may be performed even when Beer's law is not obeyed, if a suitable calibration curve is prepared. A color-developing step is usually required in which the sought-for substance reacts to form a colored species, and in some cases a colored species is extracted into a nonaqueous solvent to provide a more intense color and a more concentrated solution.

A number of solution spectrophotometric methods have been used for the determination of water and air pollutants. Some of these are summarized in Table 25.1.

25.3.2 Atomic Absorption and Emission Analyses

Atomic absorption analysis is commonly used for the determination of metals in environmental samples. This technique is based on the absorption of monochromatic light by a cloud of atoms of the analyte metal. The monochromatic light can be produced by a source composed of the same atoms as those being analyzed. The source produces intense electromagnetic radiation with a wavelength exactly the same as that absorbed by the atoms, resulting in extremely high selectivity. The basic components of an atomic absorption instrument are shown in Figure 25.1. The key element is the hollow cathode lamp in which atoms of the analyte metal are energized such that they become electronically excited and emit radiation with a very narrow wavelength band characteristic of the

TABLE 25.1
Solution Spectrophotometric (Colorimetric) Methods of Analysis

Analyte	Reagent and Method
Arsenic	Reaction of arsine, AsH_3, with silver diethylthiocarbamate in pyridine, forming a red complex
Boron	Reaction with curcumin, forming red rosocyanine
Bromide	Reaction of hypobromite with phenol red to form bromphenol blue-type indicator
Chlorine	Development of color with orthotolidine
Cyanide	Formation of a blue dye from the reaction of cyanogen chloride, CNCl, with pyridine-pyrazolone reagent, measured at 620 nm
Fluoride	Decolorization of a zirconium-dye colloidal precipitate ("lake") by the formation of colorless zirconium fluoride and free dye
Nitrate and nitrite	Nitrate is reduced to nitrite, which is diazotized with sulfanilamide and coupled with *N*-(l-naphthyl)-ethylenediamine dihydrochloride to produce a highly colored azo dye measured at 540 nm
Nitrogen, Kjeldahl method	Digestion in sulfuric acid to NH_4^+ followed by treatment with alkaline phenol reagent and sodium hypochlorite to form blue indo-phenate phenol measured at 630 nm
Phenols	Reaction with 4-aminoantipyrine at pH 10 in the presence of potassium ferricyanide, forming an antipyrine dye which is extracted into pyridine and measured at 460 nm
Phosphate	Reaction with molybdate ion to form a phosphomolybdate, which is selectively reduced to intensely colored molybdenum blue
Selenium	Reaction with diaminobenzidine, forming colored species absorbing at 420 nm
Silica	Formation of molybdosilicic acid with molybdate, followed by reduction to a heteropoly blue measured at 650 or 815 nm
Sulfide	Formation of methylene blue
Surfactants	Reaction with methylene blue to form blue salt
Tannin and lignin	Blue color from tungstophosphoric and molybdophosphoric acids

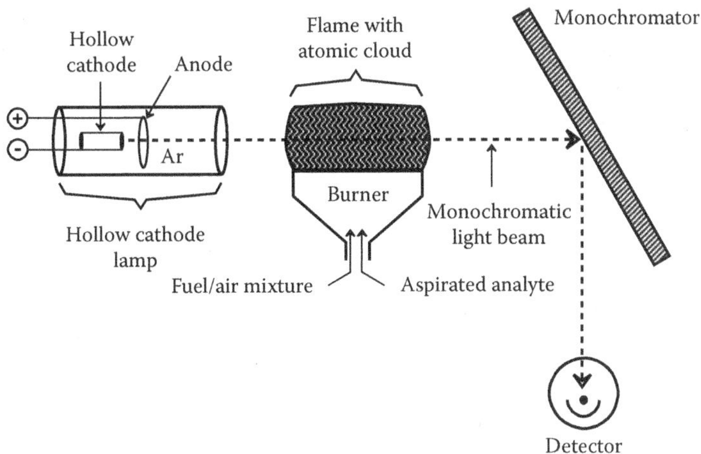

FIGURE 25.1 Basic components of a flame atomic absorption spectrophotometer.

metal. This radiation is guided by the appropriate optics through a flame into which the sample is aspirated. In the flame, most metallic compounds are decomposed, and the metal is reduced to the elemental state, forming a cloud of atoms. These atoms absorb a fraction of radiation in the flame. The fraction of radiation absorbed increases with the concentration of the sought-for element in the sample according to the Beer's law relationship (Equation 25.8). The attenuated light beam next goes to a monochromator to eliminate extraneous light resulting from the flame, and then to a detector.

Atomizers other than a flame can be used. The most common of these is the graphite furnace, an electrothermal atomization device that consists of a hollow graphite cylinder placed so that the light beam passes through it. A small sample of up to 100 μL is inserted through a hole on the upper side of the horizontally oriented tube. An electric current is passed through the tube to heat it—gradually at first to dry the sample, then rapidly to vaporize and excite the metal analyte. The absorption of metal atoms in the hollow portion of the tube is measured and recorded as a spike-shaped signal. A diagram of a graphite furnace with a typical output signal is shown in Figure 25.2. The major advantage of the graphite furnace is that it gives detection limits up to 1000 times lower than those of conventional flame devices.

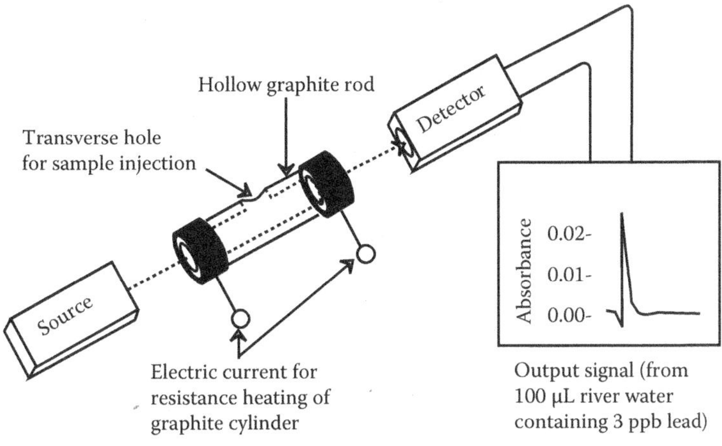

FIGURE 25.2 Graphite furnace for atomic absorption analysis and typical output signal.

A special technique for the flameless atomic absorption analysis of mercury involves room temperature reduction of mercury to the elemental state by tin(II) chloride in solution, followed by sweeping the mercury into an absorption cell with air. Nanogram (10^{-9} g) quantities of mercury can be determined by measuring mercury absorption at 253.7 nm.

25.3.3 Atomic Emission Techniques

Metals may be determined in water, atmospheric particulate matter, and biological samples very well by observing the spectral lines emitted when they are heated to a very high temperature. An especially useful atomic emission technique is inductively coupled with plasma atomic emission spectroscopy (ICP/AES). The "flame" in which analyte atoms are excited in plasma emission consists of an incandescent plasma (ionized gas) of argon heated inductively by radiofrequency energy at 4–50 MHz and 2–5 kW (Figure 25.3). The energy is transferred to a stream of argon through an induction coil, producing temperatures up to 10,000 K. The sample atoms are subjected to temperatures around 7000 K, twice those of the hottest conventional flames (e.g., nitrous oxide–acetylene operates at 3200 K). Since emission of light increases exponentially with temperature, lower detection limits are obtained. Furthermore, the technique enables emission analysis of some of the environmentally important metalloids such as arsenic, boron, and selenium. Interfering chemical reactions and interactions in the plasma are minimized as compared to flames. Of greatest significance, however, is the capability of analyzing as many as 30 elements simultaneously, enabling a true multielement analysis technique. Plasma atomization combined with mass spectrometric measurement of analyte elements (ICP/MS) has developed as an especially powerful means for multielement analysis that can even be used for some nonmetals.

25.4 ELECTROCHEMICAL METHODS OF ANALYSIS

Potentiometric, voltammetric, and amperometric water analyses utilize electrochemical sensors. Potentiometry is based on the general principle that the relationship between the potential of a

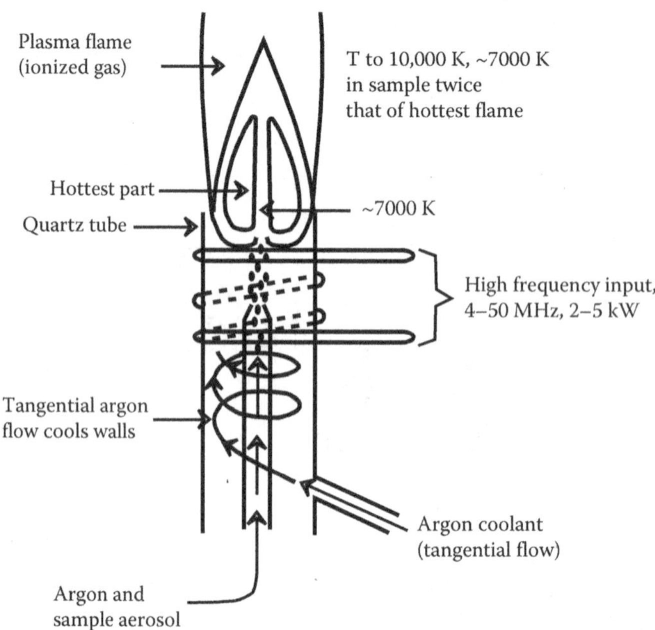

FIGURE 25.3 Schematic diagram showing inductively coupled plasma used for optical emission spectroscopy.

measuring electrode and that of a reference electrode is a function of the log of the activity of an ion in solution. For a measuring electrode responding selectively to a particular ion, this relationship is given by the Nernst equation,

$$E = E^0 + \frac{2.303RT}{zF}\log(a_z) \tag{25.9}$$

where E is the measured potential; E^0 is the standard electrode potential; R is the gas constant; T is the absolute temperature; z is the signed charge (+ for cations, − for anions); F is the Faraday constant; and a is the activity of the ion being measured. At a given temperature, the quantity $2.303RT/F$ has a constant value; at 25°C, it is 0.0592 V (59.2 mV). At constant ionic strength, the activity, a, is directly proportional to concentration, and the Nernst equation for electrodes responding to Cd^{2+} and F^- may be written as

$$E\ (\text{in mV}) = E^0 + \frac{59.2}{2}\log[Cd^{2+}] \tag{25.10}$$

$$E = E^0 - 59.2\log[F^-] \tag{25.11}$$

Electrodes that respond more or less selectively to various ions are called *ion-selective electrodes*. Generally, the potential-developing component is a membrane of some kind that allows for selective exchange of the sought-for ion. The glass electrode used for the measurement of hydrogen-ion activity and pH is the oldest and most widely used ion-selective electrode. The potential is developed at a glass membrane that selectively exchanges hydrogen ion in preference to other cations, giving a Nernstian response to hydrogen ion activity, a_{H^+}:

$$E = E^0 + 59.2\log(a_{H^+}),\quad E = E^0 - 59.2\ \text{pH} \tag{25.12}$$

The pH meters used with glass electrodes are calibrated directly in pH units using standard buffer solutions.

Of the ion-selective electrodes other than glass electrodes, the fluoride electrode is the most successful. It is well behaved, relatively free of interferences, and has an adequately low detection limit and a long range of linear response. Like all ion-selective electrodes, its electrical output is in the form of a potential signal that is proportional to log of concentration. Because of the logarithmic response of potential to concentration, small errors in E lead to relatively high concentration errors.

Voltammetric techniques, the measurement of current resulting from potential applied to a microelectrode, have found some applications in water analysis. One such technique is differential-pulse polarography, in which the potential is applied to the microelectrode in the form of small pulses superimposed on a linearly increasing potential. The current is read near the end of the voltage pulse and compared to the current just before the pulse was applied. It has the advantage of minimizing the capacitive current from charging the microelectrode surface, which sometimes obscures the current due to the reduction or oxidation of the species being analyzed. Anodic-stripping voltammetry involves deposition of metals on an electrode surface over a period of several minutes followed by stripping them off very rapidly using a linear anodic sweep. The electrodeposition concentrates the metals on the electrode surface, and increased sensitivity results. An even better technique is to strip the metals off using a differential-pulse signal. A differential-pulse anodic-stripping voltammogram of copper, lead, cadmium, and zinc in tap water is shown in Figure 25.4.

25.5 CHROMATOGRAPHY

First described in the literature in the early 1950s, gas chromatography (GC) has played an essential role in the analysis of organic materials. GC is both a qualitative and a quantitative technique; for

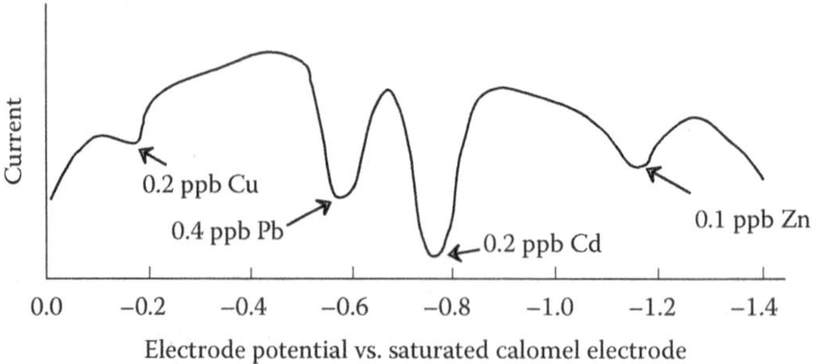

FIGURE 25.4 Differential-pulse anodic-stripping voltammogram of tap water at a mercury-plated, wax-impregnated graphite electrode.

some analytical applications of environmental importance, it is remarkably sensitive and selective. GC is based on the principle that when a mixture of volatile materials transported by a carrier gas is passed through a column containing an adsorbent solid phase or, more commonly, an absorbing liquid phase coated on a solid material, each volatile component will be partitioned between the carrier gas and the solid or liquid. The length of time required for the volatile component to traverse the column is proportional to the degree to which it is retained by the nongaseous phase. Since different components may be retained to different degrees, they will emerge from the end of the column at different times. If a suitable detector is available, the time at which the component emerges from the column and the quantity of the component are both measured. A recorder trace of the detector response appears as peaks of different sizes, depending on the quantity of material producing the detector response. Both quantitative and (within limits) qualitative analyses of the sought-for substances are obtained.

The essential features of a gas chromatograph are shown schematically in Figure 25.5. The carrier gas generally is argon, helium, hydrogen, or nitrogen. The sample is injected as a single compact plug into the carrier gas stream immediately ahead of the column entrance. If the sample is liquid, the injection chamber is heated to vaporize the liquid rapidly. The separation column may consist of a metal or glass tube packed with an inert solid of high surface area covered with a liquid phase, or it may consist of an active solid, which enables the separation to occur. More commonly, capillary columns are employed, which consist of very small diameter, very long tubes in which the liquid phase is coated on the inside of the column.

The component that primarily determines the sensitivity of gas chromatographic analysis and, for some classes of compounds, the selectivity as well, is the detector. One such device is the thermal conductivity detector, which responds to changes in the thermal conductivity of gases passing over it. The electron-capture detector, which is especially useful for halogenated hydrocarbons and phosphorus compounds, operates through the capture of electrons emitted by a β-particle source. The flame-ionization gas chromatographic detector is very sensitive for the detection of organic compounds. It is based on the phenomenon by which organic compounds form highly conducting fragments, such as C^+, in a flame. Application of a potential gradient across the flame results in a small current that may be readily measured. The mass spectrometer, described in Section 25.6, may be used as a detector for a GC. A combined GC/MS instrument is an especially powerful analytical tool for organic compounds.

Chromatographic analysis requires that a compound exhibit at least a few mm of vapor pressure at the highest temperature at which it is stable. In many cases, organic compounds that cannot be chromatographed directly may be converted to derivatives that are amenable to gas chromatographic analysis. It is seldom possible to analyze organic compounds in water by direct injection of the water into the GC; higher concentration is usually required. Two techniques commonly employed to remove volatile compounds from water and concentrate them are extraction with solvents and

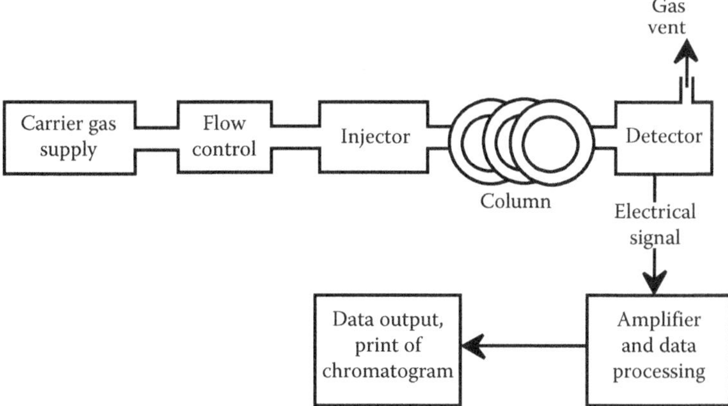

FIGURE 25.5 Schematic diagram of the essential features of a GC.

purging volatile compounds with a gas, such as helium; concentrating the purged gases on a short column; and driving them off by heat into the chromatograph.

25.5.1 High-Performance Liquid Chromatography

A liquid mobile phase used with very small column-packing particles enables high-resolution chromatographic separation of materials in the liquid phase. Very high pressures up to several thousand psi are required to get a reasonable flow rate in such systems. Analysis using such devices is called *high-performance liquid chromatography (HPLC)* and offers an enormous advantage in that the materials analyzed need not be changed to the vapor phase, a step that often requires preparation of a volatile derivative or results in decomposition of the sample. The basic features of a high-performance liquid chromatograph are the same as those of a gas chromatograph shown in Figure 25.5, except that a solvent reservoir and high-pressure pump are substituted for the carrier gas source and regulator. A hypothetical HPLC chromatogram is shown in Figure 25.6. Refractive index and ultraviolet detectors are both used for the detection of peaks coming from the liquid chromatograph column. Fluorescence detection can be especially sensitive for some classes of compounds.

Mass spectrometric detection of HPLC effluents has led to the development of LC/MS analysis. This powerful tool enables measurement of highly polar water contaminants without the necessity of forming volatile derivatives of analytes to be separated by GC and can be used to measure analytes down to the nanogram per liter levels. LC/MS is especially useful for the determination of a variety of emerging contaminants including pharmaceuticals and their metabolites, hormones, and endocrine-disrupting species.

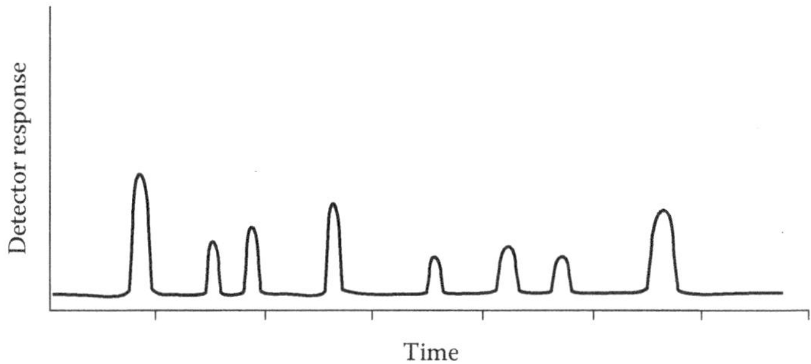

FIGURE 25.6 Hypothetical HPLC chromatogram.

TABLE 25.2
Chromatography-Based EPA Methods for Organic Compounds in Water

	Method Number			
Class of Compounds	**GC**	**GC/MS**	**HPLC**	**Example Analytes**
Purgeable halocarbons	601			Carbon tetrachloride
Purgeable aromatics	602			Toluene
Acrolein and acrylonitrile	603			Acrolein
Phenols	604			Phenol and chlorophenols
Benzidines			605	Benzidine
Phthalate esters	606			Bis(2-ethylhexylphthalate)
Nitrosamines	607			*N*-nitroso-*N*-dimethylamine
Organochlorine pesticides and PCBs	608			Heptachlor, PCB 1016
Nitroaromatics and isophorone	609			Nitrobenzene
PAHs	610		610	Benzo[*a*]pyrene
Haloethers	611			Bis(2-chloroethyl) ether
Chlorinated hydrocarbons	612			1,3-Dichlorobenzene
2,3,7,8-TCDD		613		2,3,7,8-TCDD
Organophosphorus pesticides	614			Malathion
Chlorinated herbicides	615			Dinoseb
Triazine pesticides	619			Atrazine
Purgeable organics		624		Ethylbenzene
Base/neutrals and acids		625		More than 70 organic compounds
Dinitro aromatic pesticides		646		Basalin (Fluchloralin)
Volatile organic compounds		1624		Vinyl chloride

25.5.2 Chromatographic Analysis of Water Pollutants

The U.S. EPA has developed a number of chromatography-based standard methods for determining water pollutants. Some of these methods use the purge-and-trap technique, bubbling gas through a column of water to purge volatile organics from the water followed by trapping the organics on solid sorbents, whereas others use solvent extraction to isolate and concentrate the organics. These methods are summarized in Table 25.2.

25.5.3 Ion Chromatography

The liquid chromatographic determination of ions, particularly anions, by *ion chromatography* has enabled the measurement of species that used to be very troublesome for water chemists. The development of this technique has been facilitated by special detection techniques using suppressors to enable detection of analyte ions in the chromatographic effluent. Ion chromatography has been developed for the determination of most of the common anions, including arsenate, arsenite, borate, carbonate, chlorate, chlorite, cyanide, the halides, hypochlorite, hypophosphite, nitrate, nitrite, phosphate, phosphite, pyrophosphate, selenate, selenite, sulfate, sulfite, sulfide, trimetaphosphate, and tripolyphosphate. Cations, including the common metal ions, can also be determined by ion chromatography.

25.6 MASS SPECTROMETRY

Mass spectrometry (MS) is particularly useful for the identification of specific organic pollutants. It depends on the production of ions by an electrical discharge or chemical process, followed by

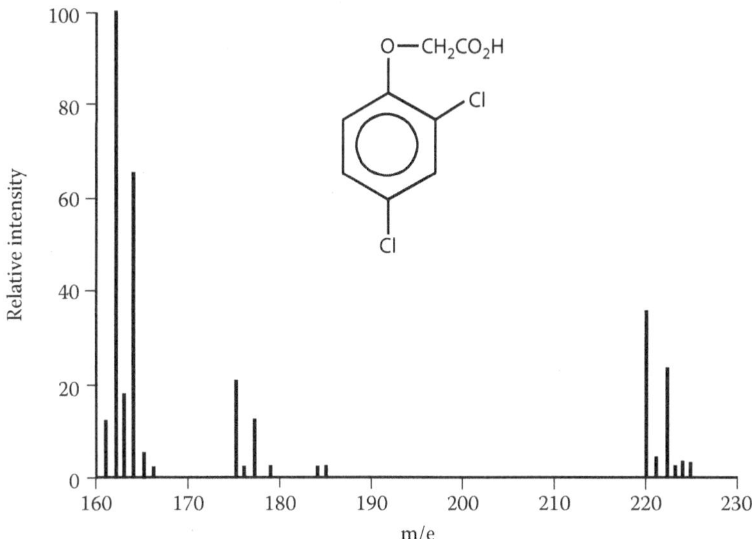

FIGURE 25.7 Partial mass spectrum of the herbicide 2,4-dichlorophenoxyacetic acid (2,4-D), a common water pollutant.

separation based on the charge-to-mass ratio and measurement of the ions produced. The output of a mass spectrometer is a mass spectrum, such as the one shown in Figure 25.7. A mass spectrum is characteristic of a compound and serves to identify it. Computerized data banks for mass spectra have been established and are stored in computers interfaced with mass spectrometers. Identification of a mass spectrum depends on the purity of the compound from which the spectrum is taken. Prior separation by GC with continual sampling of the column effluent by a mass spectrometer, commonly called GC-MS, is particularly effective in the analysis of organic pollutants.

25.7 ANALYSIS OF WATER SAMPLES

The preceding sections of this chapter have covered the major kinds of analysis techniques that are used on water. In this section, several specific aspects of water analysis are addressed.

25.7.1 Physical Properties Measured in Water

The commonly determined physical properties of water are color, residue (solids), odor, temperature, specific conductance, and turbidity. Most of these terms are self-explanatory and will not be discussed in detail. All of these properties either influence or reflect the chemistry of the water. Solids, for example, arise from chemical substances either suspended or dissolved in the water and are classified physically as total, filterable, nonfilterable, or volatile. Specific conductance is a measure of the degree to which water conducts alternating current and reflects, therefore, the total concentration of dissolved ionic material. By necessity, some physical properties must be measured in the water without sampling (see discussion of water sampling below).

25.7.2 Water Sampling

It is beyond the scope of this text to describe water sampling procedures in detail. It must be emphasized, however, that the acquisition of meaningful data demands that correct sampling and storage procedures be used. These procedures may be quite different for various species in water.

In general, separate samples must be collected for chemical and biological analysis because the sampling and preservation techniques differ significantly. Usually, the shorter the time interval between sample collection and analysis, the more accurate the analysis will be. Indeed, some analyses must be performed in the field within minutes of sample collection. Others, such as the determination of temperature, must be performed on the body of water itself. Within a few minutes after collection, water pH may change, dissolved gases (oxygen, carbon dioxide, hydrogen sulfide, and chlorine) may be lost, or other gases (oxygen and carbon dioxide) may be absorbed from the atmosphere. Therefore, analyses of temperature, pH, and dissolved gases should always be performed in the field. Furthermore, precipitation of calcium carbonate accompanies changes in the pH-alkalinity-calcium carbonate relationship following sample collection. Analysis of a sample after standing may thus give erroneously low values for calcium and total hardness.

Oxidation–reduction reactions may cause substantial errors in analysis. For example, soluble iron(II) and manganese(II) are oxidized to insoluble iron(III) and manganese(IV) compounds when an anaerobic water sample is exposed to atmospheric oxygen. Microbial activity may decrease phenol or biochemical oxygen demand (BOD) values, change the nitrate–nitrite–ammonia balance, or alter the relative proportions of sulfate and sulfide. Iodide and cyanide are frequently oxidized. Chromium(VI) in solution may be reduced to insoluble chromium(III). Sodium, silicate, and boron are leached from glass container walls.

Samples can be divided into two major categories. *Grab samples* are taken at a single time and in a single place. Therefore, they are very specific with respect to time and location. *Composite samples* are collected over an extended time and may encompass different locations as well. In principle, the average results from a large number of grab samples give the same information as a composite sample. A composite sample has the advantage of providing an overall picture from only one analysis. On the other hand, it may miss extreme concentrations and important variations that occur over time and space.

25.7.2.1 Extractors

The ease and effectiveness of various kinds of solid-phase devices for water sampling is steadily increasing their use in water analysis. Based on size and physical configuration, several categories of such devices are available. One of these is the conventional solid-phase extractor (SPE) containing an extracting solid in a column. Activated carbon has been used for decades for this purpose, but synthetic materials, such as those composed of long hydrocarbon chains (C18) bound to solids have been found to be quite useful. A typical procedure uses a polymer-divinylstyrene extraction column to remove pesticides from water. The pesticide analytes are eluted from the SPE with ethyl acetate and measured by GC.

Solid-phase microextraction (SPME) devices constitute a second kind of SPE. These make use of very small diameter tubes in which analytes are bonded directly to the extractor walls, and then eluted directly into a chromatograph. The use of SPME devices for the determination of haloethers in water has been described.

A third kind of device that is simple, convenient, and available for a number of classes of compounds consists of disks composed of substances that bind with and remove analytes from water when the water is filtered through them. As an example, solid-phase extraction disks can be used to remove and concentrate radionuclides from water, including ^{99}Tc, ^{137}Cs, ^{90}Sr, and ^{238}Pu. Organic materials sampled from water with such disks include haloacetic acids and acidic and neutral herbicides.

Liquid–liquid extraction of organic-soluble analytes has long been used to extract these materials from water and isolate them in a solvent that can be reduced to a small volume in which analytes can be further concentrated by a Kuderna–Danish evaporator or nitrogen blowing. Relatively high consumption of solvent (creating a disposal problem) and emulsion formation can be problems with liquid–liquid extraction. As an alternative for volatile and semivolatile analytes, membrane extraction can be used. This procedure entails running the water sample over one side of a membrane through which the analytes migrate to another solvent on the other side. Water can even be used as a collector solvent. In the case of ionizable organics, for example, the pH on the sample side can be adjusted such that the analyte species are neutral, in which form they can migrate through an organophilic membrane, and be collected in an aqueous solution that is at a pH adjusted such that the analyte species are ionized and hence cannot permeate back through the membrane. A procedure exists for the determination of semivolatile compounds in water in which the extracting solvent traverses the interior of hollow fibers and collects analytes from the water sample flowing over the outer surface of the fibers. By pressurizing the solvent, a significant fraction of it is forced through the membrane into the water thereby concentrating the organic-soluble analytes in the solvent phase.

25.7.3 Water Sample Preservation

It is not possible to completely protect a water sample from changes in composition. However, various additives and treatment techniques can be employed to minimize sample deterioration. These methods are summarized in Table 25.3.

The most general method of sample preservation is refrigeration to 4°C. Freezing normally should be avoided because physical changes including precipitate formation and gas loss may adversely affect sample composition. Acidification is commonly applied to metal samples to prevent their precipitation, and it also slows microbial action. In the case of metals, the samples should be filtered before adding acid to enable determination of dissolved metals. Sample holding times vary from zero for parameters such as temperature or DO measured by a probe to 6 months for metals. Many different kinds of samples, including those to be analyzed for acidity, alkalinity, and various forms of nitrogen or phosphorus, should not be held for more than 24 h. Details on water sample preservation are to be found in standard references on water analysis listed at the end of this chapter.

25.7.4 Total Organic Carbon in Water

It is important to measure dissolved organic carbon (Chapter 7) in water because it exerts an oxygen demand in water, often is in the form of toxic substances, and is a general indicator of water pollution. The measurement of total organic carbon, TOC, in water is performed by methods that, for the most part, totally oxidize the dissolved organic material to produce carbon dioxide. The amount of carbon dioxide evolved is taken as a measure of TOC.

TOC can be determined by a technique that uses a dissolved oxidizing agent, commonly potassium peroxydisulfate, $K_2S_2O_8$, promoted by ultraviolet light to oxidize the organic carbon to CO_2. Phosphoric acid is added to the sample, which is sparged with nitrogen to drive off inorganic CO_2. After sparging, the sample is pumped to a chamber containing a lamp emitting ultraviolet radiation of 184.9 nm. This radiation produces reactive free radical species such as the hydroxyl radical, $HO^\bullet$,

TABLE 25.3
Preservatives and Preservation Methods Used with Water Samples

Preservative or Technique Used	Effect on Sample	Type of Samples for Which the Method is Employed
Nitric acid	Keeps metals in solution	Metal-containing samples
Sulfuric acid	Bactericide	Biodegradable samples containing organic carbon, oil, or grease
	Formation of sulfates with volatile bases	Samples containing amines or ammonia
Sodium hydroxide	Formation of sodium salts from volatile acids or cyanides	Samples containing volatile organic acids
Chemical	Fix a particular reaction constituent	Samples to be analyzed for DO using the Winkler method

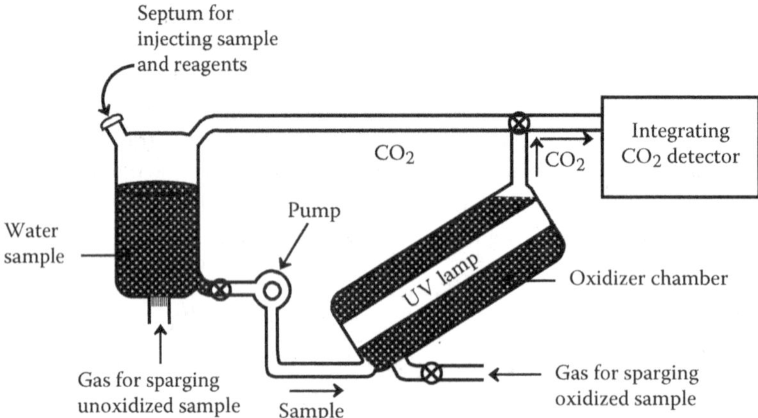

FIGURE 25.8 TOC analyzer employing UV-promoted sample oxidation.

(see Chapters 9, 12, and 13). These active species bring about the rapid oxidation of dissolved organic compounds as shown in the following general reaction:

$$\text{Organics} + HO^{\bullet} \rightarrow CO_2 + H_2O \qquad (25.13)$$

After the completion of oxidation, the CO_2 is sparged from the system and measured with a gas chromatographic detector or by absorption in ultrapure water followed by a conductivity measurement. Figure 25.8 is a schematic of a TOC analyzer.

25.7.5 Measurement of Radioactivity in Water

There are several potential sources of radioactive materials that may contaminate water (see Section 7.13). Radioactive contamination of water is normally detected by measurements of gross β and gross α activity, a procedure that is simpler than detecting individual isotopes. The measurement is made from a sample formed by evaporating water to a very thin layer on a small pan, which is then inserted inside an internal proportional counter. This setup is necessary because β particles can penetrate only very thin detector windows, and α particles have essentially no penetrating power. More detailed information can be obtained for radionuclides that emit γ rays by the use of γ spectrum analysis. This technique employs solid-state detectors to resolve rather closely spaced γ peaks in the sample's spectra. In conjunction with multichannel spectrometric data analysis, it is possible to determine a number of radionuclides in the same sample without chemical separation. This method requires minimal sample preparation.

25.7.6 Biological Toxins

Toxic substances produced by microorganisms are of some concern in water. Photosynthetic cyanobacteria and some kinds of algae growing in water produce potentially troublesome toxic substances. An HPLC/MS method of analysis for such toxins has been described.[7]

25.7.7 Summary of Water Analysis Procedures

The main chemical parameters commonly determined in water are summarized in Table 25.4. In addition to these, a number of other solutes, especially specific organic pollutants, may be determined in connection with specific health hazards or incidents of pollution.

TABLE 25.4
Chemical Parameters Commonly Determined in Water

Chemical Species	Significance in Water	Methods of Analysis
Acidity	Indicative of industrial pollution or acid mine drainage	Titration
Alkalinity	Water treatment, buffering, algal productivity	Titration
Aluminum	Water treatment, buffering	AA[a], ICP[b]
Ammonia	Algal productivity, pollutant	Spectrophotometry
Arsenic	Toxic pollutant	Spectrophotometry, AA, ICP
Barium	Toxic pollutant	AA, ICP
Beryllium	Toxic pollutant	AA, ICP, fluorimetry
Boron	Toxic to plants	Spectrophotometry, ICP
Bromide	Seawater intrusion, industrial waste	Spectrophotometry, potentiometry, ion chromatography
Cadmium	Toxic pollutant	AA, ICP
Calcium	Hardness, productivity, treatment	AA, ICP, titration
Carbon dioxide	Bacterial action, corrosion	Titration, calculation
Chloride	Saline water contamination	Titration, electrochemical, ion chromatography
Chlorine	Water treatment	Spectrophotometry
Chromium	Toxic pollutant (hexavalent Cr)	AA, ICP, colorimetry
Copper	Plant growth	AA, ICP
Cyanide	Toxic pollutant	Spectrophotometry, potentiometry, ion chromatography
Fluoride	Water treatment, toxic at high levels	Spectrophotometry, potentiometry, ion chromatography
Hardness	Water quality, water treatment	AA, titration
Iodide	Seawater intrusion, industrial waste	Catalytic effect, potentiometry ion chromatography
Iron	Water quality, water treatment	AA, ICP, colorimetry
Lead	Toxic pollutant	AA, ICP, voltammetry
Lithium	May indicate some pollution	AA, ICP, flame photometry
Magnesium	Hardness	AA, ICP
Manganese	Water quality (staining)	AA, ICP
Mercury	Toxic pollutant	Flameless atomic absorption
Methane	Anaerobic bacterial action	Combustible-gas indicator
Nitrate	Algal productivity, toxicity	Spectrophotometry, ion chromatography
Nitrite	Toxic pollutant	Spectrophotometry, ion chromatography
Nitrogen (albuminoid)	Proteinaceous material	Spectrophotometry
(organic)	Organic pollution indicator	Spectrophotometry
Oil and grease	Industrial pollution	Gravimetry
Organic carbon	Organic pollution indicator	Oxidation-CO_2 measurement
Organic contaminants	Organic pollution indicator	Activated carbon adsorption
Oxygen	Water quality	Titration, electrochemical
Oxygen demand		
(biochemical)	Water quality and pollution	Microbiological titration
(chemical)	Water quality and pollution	Chemical oxidation-titration
Ozone	Water treatment	Titration
Pesticides	Water pollution	GC
pH	Water quality and pollution	Potentiometry
Phenols	Water pollution	Distillation-colorimetry
Phosphate	Productivity, pollution	Spectrophotometry

continued

TABLE 25.4 (continued)
Chemical Parameters Commonly Determined in Water

Chemical Species	Significance in Water	Methods of Analysis
Phosphorus	Water quality and pollution (hydrolyzable)	Spectrophotometry
Potassium	Productivity, pollution	AA, ICP, flame photometry
Selenium	Toxic pollutant	Spectrophotometry, ICP, neutron activation
Silica	Water quality	Spectrophotometry, ICP
Silver	Water pollution	AA, ICP
Sodium	Water quality, saltwater intrusion	AA, ICP, flame photometry
Strontium	Water quality	AA, ICP, flame photometry
Sulfate	Water quality, water pollution	Ion chromatography
Sulfide	Water quality, water pollution	Spectrophotometry, titration, chromatography
Sulfite	Water pollution, oxygen scavenger	Titration, ion chromatography
Surfactants	Water pollution	Spectrophotometry
Tannin, lignin	Water quality, water pollution	Spectrophotometry
Vanadium	Water quality, water pollution	ICP
Zinc	Water quality, water pollution	AA, ICP

[a] AA denotes atomic absorption.

[b] ICP stands for inductively coupled plasma techniques in which the atoms in the atomized plasma are detected by atomic emission or by MS.

25.8 AUTOMATED WATER ANALYSES

Huge numbers of water analyses must often be performed in order to get meaningful results and for reasons of economics. This has resulted in the development of a number of automated procedures in which the samples are introduced through a sampler and the analyses performed and results posted without manual manipulation of reagents and apparatus. Such procedures have been developed and instruments marketed for the determination of a number of analytes, including alkalinity, sulfate, ammonia, nitrate/nitrite, and metals. Colorimetric procedures are popular for such automated analytical instruments, using simple, rugged colorimeters for absorbance measurements. Figure 25.9 shows an automated analytical system for the determination of alkalinity. The reagents and sample liquids are transported through the analyzer by a peristaltic pump consisting basically of rollers moving over flexible tubing. By using different sizes of tubing, the flow rates of the reagents are proportioned. Air bubbles are introduced into the liquid stream to aid mixing and to separate one sample from another. Mixing of the sample and various reagents is accomplished in mixing coils. Since many color-developing reactions are not rapid, a delay coil is provided that allows the color to develop before reaching the colorimeter. Bubbles are removed from the liquid stream by a debubbler prior to introduction into the flow cell for colorimetric analysis.

25.9 SPECIATION

Speciation refers to the determination of specific species of elements rather than simply analysis of the elements. It applies especially to metals and metalloids, which may be relatively easy to measure as the element, but much more difficult to determine as specific species of the element. For example, it is a relatively straightforward process to measure total tin in water, but much more challenging (and important) to determine organotin species that might be present as pollutants. Arsenic provides an important example of significant elemental species in water. Inorganic arsenic can be present as either As(III) or As(V), of which the former is generally more toxic. Organoarsenic compounds can

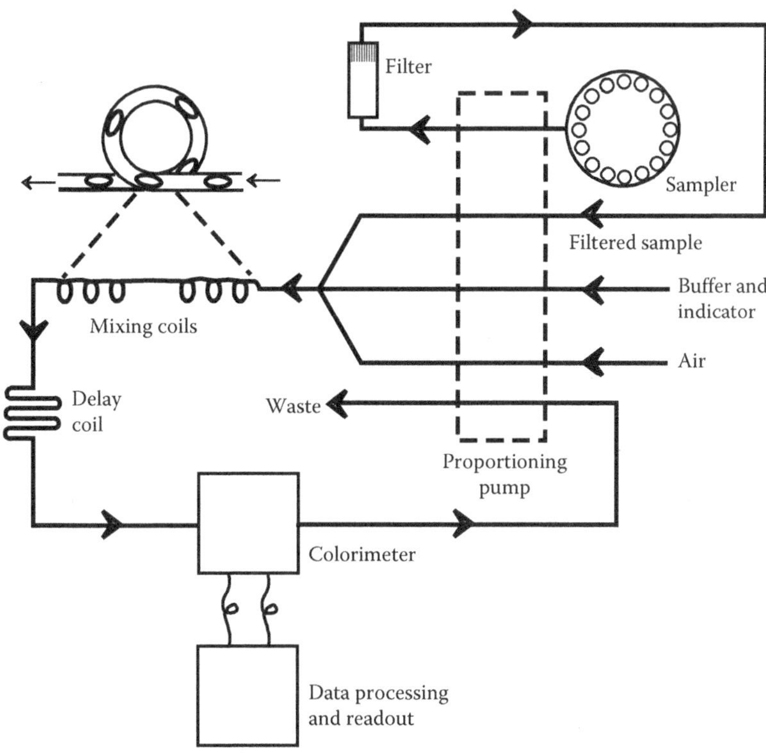

FIGURE 25.9 Automated analyzer system for the determination of total alkalinity in water. Addition of a water sample to a methyl orange solution buffered to pH 3.1 causes a loss of color in proportion to the alkalinity in the sample.

be produced microbially and include monomethylarsonate and dimethylarsinate. Organoarsenic compounds are also used as additives to livestock feed and get into water supplies as runoff from feeding operations; roxarsone is one such compound.

The most generally applicable means of measuring elemental species is separation by chromatography followed by measurement of elemental levels in the effluent. Specifically for arsenic, separation by anion exchange chromatography followed by measurement of arsenic with inductively coupled plasma-MS is a good technique.

25.10 EMERGING CONTAMINANTS IN WATER ANALYSIS

Emerging contaminants in water analysis refers to species that have been recognized as significant water contaminants relatively recently, the long-term effects of which are not very well known. Approximately 3000 different substances are now used as pharmaceutical ingredients including antibiotics, painkillers, antidiabetics, contraceptives, antidepressants, impotence drugs, and α blockers and these substances and their metabolites get into water at very low levels. Major classes of emerging contaminants include the following:

- Pharmaceuticals and their metabolites
- Endocrine disruptors
- Polybrominated compounds
- Microbial toxins
- Personal care products
- Disinfection products

- Organometallic compounds
- Nanomaterials

Some specific substances that have been recognized as emerging contaminants relatively recently include the following: benzotriazoles, complexing agents used as anticorrosives; naphthenic acids, which are toxic endocrine disruptors remaining from the extraction of crude oil from tar sands, particularly in Alberta, Canada; ethylenedibromide, formerly added to leaded gasoline but still a problem in some groundwaters; and dioxane, widely produced as a solvent stabilizer.

Emerging contaminants in drinking water supplies are generally indicative of contamination by incompletely treated wastewater. Pharmaceuticals and their metabolites have the potential to cause direct health effects. Endocrine disruptors may affect health as well as sexual development in aquatic organisms. The organometallic compounds of most concern are the organotin compounds, which have industrial uses and have been used as biocides. Microbial toxins include substances produced by photosynthetic bacteria and algae growing in water supplies. An example consists of microcystins, cyclic peptides produced by cyanobacteria.

A variety of analytical methods are used to determine concentrations of emerging contaminants in water. Especially useful are chromatographic methods coupled with MS or tandem MS (two mass spectrometers in tandem). Some of the specific emerging contaminants that can be determined by solid-phase extraction of water contaminants measured by liquid chromatography/tandem mass spectrometry (LC/MS/MS) using electrospray ionization include meprobamate, ethynylestradiol, progesterone, caffeine, testosterone, androstenedione, acetaminophen, erythromycin, TCEP, hydrocodone, oxybenzone, DEET, carbamazepine, diazepam, dilantin, sulfamethoxazole, trimethoprim, atrazine, triclosan, pentoxifylline, diclofenac, ibuprofen, naproxen, gemfibrozil, fluoxetine, and iopromide. Including pharmaceuticals, steroids, personal care products, and pesticides, this list gives an idea of compounds included among emerging contaminants.

25.11 CHIRAL CONTAMINANTS

Chiral molecules are those in which four different groups are arranged in three dimensions around an atom, usually of carbon, that constitutes a chiral center (see alanine below) such that the molecule cannot be directly superimposed over its mirror image. Two chiral molecules of the same compound, called *enantiomers*, have identical melting points, boiling points, and solubilities as well as generally identical chemical properties. They differ, often markedly, in their biochemical properties because of the exact fit required between enzyme active sites and the substrate molecules upon which the enzymes act. Therefore, chiral isomers may exhibit completely different environmental and toxicological behaviors. For example, one enantiomer of a herbicide may undergo biodegradation readily whereas the other is resistant to it. Biochemical differences between enantiomers may be especially pronounced for pesticides. For example, one enantiomer of a herbicide may kill weeds very effectively whereas the other has no effect. An example is the *R* enantiomer of herbicidal mecoprop (Figure 7.14), which is now marketed as a herbicide in place of the racemic mixture with the *S* enantiomer.

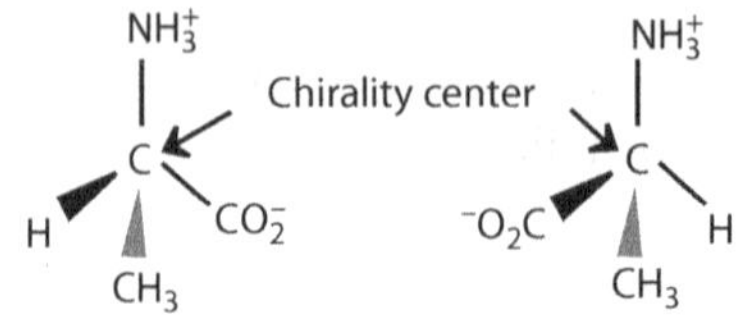

NH_3^+ | Chirality center | NH_3^+

C, H, CO_2^-, CH_3 ^-O_2C, C, H, CH_3

The two enantiomers of the amino acid alanine

Although important, the ability to determine specific enantiomers of contaminants in water is relatively new. Most commonly, the compounds are separated by LC and measured by MS. This requires a separation medium that is selective for particular enantiomers. Cyclodextrins are most commonly used for this purpose.

LITERATURE CITED

1. "Data Quality," Section 1030 in Eaton, A. D., L. S. Clesceri, E. W. Rice, A. E. Greenberg, and M. A. H. Franson, Eds, *Standard Methods for the Examination of Water and Wastewater*, 21st ed., American Public Health Association, Washington, DC, pp. 1-13–1-22, 2005.
2. Eaton, A. D., L. S. Clesceri, E. W. Rice, A. E. Greenberg, and M. A. H. Franson, Eds, *Standard Methods for the Examination of Water and Wastewater*, 21st ed., American Public Health Association, Washington, DC, 2005.
3. *Index to EPA Test Methods*, April 2003 revised, http://www.epa.gov/ne/info/testmethods/
4. *Methods and Guidance for the Analysis of Water*, Version 2.0 on CD ROM including EPA Series, 500, 600, and 1600 Methods, National Technical Information Service (NTIS), U.S. Department of Commerce, Springfield, VA, 2003.
5. *Understanding Environmental Methods*, Version 5.0 on CD ROM, Genium Publishing Corp., Amsterdam, NY, 2009.
6. Richardson, S. D., "Water analysis: Emerging contaminants and current issues," *Analytical Chemistry*, **79**, 4295–4324, 2007.
7. Hedmana, C. J., W. R. Kricka, D. A. Karner Perkinsa, E. A. Harrahyb, and W. C. Sonzognia, "New measurements of cyanobacterial toxins in natural waters using high performance liquid chromatography coupled to tandem mass spectrometry," *Journal of Environmental Quality*, **37**, 1817–1824, 2008.

SUPPLEMENTARY REFERENCES

AWWA Staff, *Simplified Procedures for Water Examination*, 5th ed., American Water Works Association, Denver, CO, 2002.

AWWA Staff, *Basic Laboratory Procedures for Wastewater Examination*, 4th ed., Water Environment Federation, Alexandria, VA, 2002.

Crompton, T. R., *Determination of Anions in Natural and Treated Waters*, Taylor & Francis, New York, 2002.

Crompton, T. R., *Chromatography of Natural and Treated Waters*, Taylor & Francis, New York, 2005.

Crompton, T. R., *Determination of Metals in Natural and Treated Waters*, Taylor & Francis, New York, 2002.

Dieken, F. P., *Methods Manual for Chemical Analysis of Water and Wastes*, Alberta Environmental Centre, Vergeville, Alberta, Canada, 1996.

Eaton, A. D., L. S. Clesceri, E. W. Rice, A. E. Greenberg, and M. A. H. Franson, Eds, *Standard Methods for the Examination of Water and Wastewater*, 21st ed., American Public Health Association, Washington, DC, 2005.

Kolle, W., *Wasseranalysen*, Wiley, New York, 2002.

Lewinsky, A. A., Ed., *Hazardous Materials and Wastewater: Treatment, Removal and Analysis*, Nova Science Publishers, New York, 2007.

Meyers, R. A., Ed., *The Encyclopedia of Environmental Analysis and Remediation*, Wiley, New York, 1998.

Nollet, L. M. L., Ed., *Handbook of Water Analysis*, 2nd ed., Taylor & Francis/CRC Press, Boca Raton, FL, 2007.

Patnaik, P., *Handbook of Environmental Analysis: Chemical Pollutants in Air, Water, Soil, and Solid Wastes*, CRC Press/Lewis Publishers, Boca Raton, FL, 1997.

Quevauviller, P. and C. Thompson, *Analytical Methods for Drinking Water: Advances in Sampling and Analysis*, Wiley, Hoboken, NJ, 2006.

Rump, H. H., *Laboratory Manual for the Examination of Water*, 3rd ed., Wiley, New York, 2001.

Tomar, M., *Laboratory Manual for the Quality Assessment of Water and Wastewater*, CRC Press/Lewis Publishers, Boca Raton, FL, 1999.

www.epa.gov/safewater/methods/methods.html links to U.S. EPA and non-EPA methods of analysis for drinking water.

www.epa.gov/safewater/methods/sourcalt.html links to methods developed by U.S. EPA's Office of Ground Water and Drinking Water.

www.epa.gov/nerlcwww links to methods for the determination of microorganism in water.

www.epa.gov/nerlcwww/ordmeth.htm links to drinking water and marine water methods recently developed by U.S. EPA's Office of Research and Development.

www.epa.gov/safewater/dwinfo local drinking water.

QUESTIONS AND PROBLEMS

1. A soluble water pollutant forms ions in solution and absorbs light at 535 nm. What are the two physical properties of water influenced by the presence of this pollutant?
2. A sample was taken from the bottom of a deep, stagnant lake. Upon standing, bubbles evolved from the sample; the pH went up; and a white precipitate formed. From these observations, what may be said about the dissolved CO_2 and hardness in the water?
3. For which of the following analytes may nitric acid be used as a water sample preservative: H_2S; CO_2; metals; coliform bacteria; and cyanide?
4. In the form of what compound is oxygen fixed in the Winkler analysis of O_2?
5. Of the following analytical techniques, the water analysis techniques that would best distinguish between the hydrated $Ag(H_2O)_6^+$ ion and the complex $Ag(NH_3)_2^+$ ion by direct measurement of the uncomplexed ion are (a) neutron-activation analysis, (b) atomic absorption, (c) inductively coupled plasma atomic emission spectroscopy, (d) potentiometry, and (e) flame emission.
6. A water sample was run through the colorimetric procedure for the analysis of nitrate, giving 55.0% transmittance. A sample containing 1.00 ppm nitrate run through the exactly identical procedure gave 24.6% transmittance. What was the concentration of nitrate in the first sample?
7. What is the molar concentration of HCl in a water sample containing HCl as the only contaminant and having a pH of 3.80?
8. A 200-mL sample of water required 25.12 mL of 0.0200 N standard H_2SO_4 for titration to the methyl orange endpoint, pH 4.5. What was the total alkalinity of the original sample?
9. Analysis of a lead-containing sample by graphite-furnace atomic absorption analysis gave a peak of 0.075 absorbance units when 50 μL of pure sample was injected. Lead was added to the sample such that the added concentration of lead was 6.0 μg/L. Injection of 50 μL of "spiked" sample gave an absorbance of 0.115 absorbance units. What was the concentration of lead in the original sample?
10. In a 2.63×10^{-4} M standard fluoride solution, a fluoride electrode read −0.100 V versus a reference electrode, and it read −0.118 V in an appropriately processed fluoride sample. What was the concentration of fluoride in the sample?
11. The activity of iodine-131 ($t_{1/2} = 8$ days) in a water sample 24 days after collection was 520 pCi/L. What was the activity on the day of collection?
12. Neutron irradiation of exactly 2.00 mL of a standard solution containing 1.00 mg/L of unknown heavy metal "X" for exactly 30 s gave an activity of 1257 counts per minute, when counted exactly 33.5 min after the irradiation, measured for a radionuclide product of "X" having a half-life of 33.5 min. Irradiation of an unknown water sample under identical conditions (2.00 mL, 30.0 s, and the same neutron flux) gave 1813 counts per minute when counted 67.0 min after irradiation. What was the concentration of "X" in the unknown sample?
13. Why is magnesium-EDTA chelate added to a magnesium-free water sample before it is to be titrated with EDTA for Ca^{2+}?
14. For what type of sample is the flame-ionization detector most useful?
15. Manganese from a standard solution was oxidized to MnO_4^- and diluted such that the final solution contained 1.00 mg/L of Mn. This solution had an absorbance of 0.316. A 10.00 mL wastewater sample was treated to develop the MnO_4 color and diluted to 250.0 mL. The diluted sample had an absorbance of 0.296. What was the concentration of Mn in the original wastewater sample?

26 Analysis of Wastes and Solids

26.1 INTRODUCTION

The analysis of hazardous wastes of various kinds for a variety of potentially dangerous substances is one of the most important aspects of hazardous waste management.[1] These analyses are performed for a number of reasons including tracing the sources of wastes, assessing the hazards posed by the wastes to surroundings and to waste remediation personnel, and determining the best means of waste treatment. This chapter is a brief overview of several of the main considerations applied to the analysis of wastes. Here, wastes are broadly defined to include all kinds of solids, semisolids, sludges, liquids, contaminated soils, sediments, and other kinds of materials that are either wastes themselves or are contaminated by wastes.

For the most part, the substances determined as part of waste analysis, the *analytes*, are measured by techniques that are used for the determination of the same analytes in water (see methods described in Chapter 25) and, to a lesser extent, in air. However, the preparation techniques that must be employed for waste analysis are usually more complex than those used for the same analytes in water. That is because the matrices in which the waste analytes are contained are usually relatively complicated, which makes it difficult to recover all the analytes from the waste and which introduces interfering substances. As a result, the lower limits at which substances can be measured in wastes (the practical quantitation limit, see Section 25.1) are usually much higher than in water.

There are several distinct steps in the analysis of a waste. Compared to water, wastes are often highly heterogeneous, which makes the first step, collection of representative samples, difficult. Whereas water samples can often be introduced into an analytical instrument with minimal preparation, the processing of hazardous wastes to get a sample that can be introduced into an instrument is often relatively complicated. Such processing can consist of dilution of oily samples with an organic solvent, extraction of organic analytes into an organic solvent, extraction with supercritical or subcritical hot liquids, evolution and collection of volatile organic analytes, or digestion of solids with strong acids and oxidants to extract metals for atomic spectrometric analysis. The products of these processes must often be subjected to relatively involved sample cleanup procedures to remove contaminants that might interfere with the analysis or damage the analytical instrument.

Over a number of years, the U.S. EPA has developed specialized methods for the characterization of wastes. These methods are given in the publication entitled *Test Methods for Evaluating Solid Waste, Physical/Chemical Methods*, which is periodically updated to keep it current.[2] Because of the difficult and exacting nature of many of the procedures in this work and because of the hazards associated with the use of reagents such as strong acids and oxidants used for sample digestion and solvents employed to extract organic analytes, anyone attempting analyses of hazardous waste materials should use this resource and follow procedures carefully with special attention to precautions. "SW-846," as it is commonly known, is available in a convenient CD/ROM form as part of a comprehensive summary of environmental analytical methods.[3] ASTM International, originally known as the American Society for Testing and Materials publishes comprehensive works on the analysis of environmental samples including solid wastes.[4]

26.2 SAMPLE DIGESTION FOR ELEMENTAL ANALYSIS

In order to analyze a solid waste sample by flame atomic absorption spectroscopy, graphite furnace absorption spectroscopy, inductively coupled argon plasma spectroscopy, or inductively coupled argon plasma mass spectrometry (MS), the sample must first be digested to get the analyte metals in solution. Digestion dissolves only those fractions of metals that can be put into solution under relatively extreme conditions and therefore enables measurement of available metals. It should be noted that sample digestion procedures generally use highly corrosive, dangerous reagents that are strong acids and strong oxidants. Therefore, digestion should be carried out only by carefully trained personnel using the proper equipment, including fume hoods and adequate personnel protection.

EPA Method 3050 is a procedure for acid digestion of sediments, sludges, and soils. A sample of up to 2 g is treated with a mixture of nitric acid and hydrogen peroxide; the sample is then refluxed with either conc. HNO_3 or conc. HCl, then refluxed with dilute HCl, filtered, and the filtrate analyzed for metals.

Microwave heating can be used to assist the digestion of samples. The procedure for the digestion of aqueous liquids consists of mixing a 45-mL sample with 5 mL of concentrated nitric acid, placing it in a fluorocarbon (Teflon) digestion vessel, and heating for 20 min. After digestion is complete, the sample is cooled, solids are separated by filtration or centrifugation, and the liquid remaining is analyzed by the appropriate atomic spectrometric technique.

Method 3052 is a procedure for microwave-assisted acid digestion of siliceous and organically based matrices. It can be used on a variety of kinds of samples including biological tissues, oils, oil-contaminated soils, sediments, sludges, and soil. This method is not appropriate for analyses of leachable metals, but is used for the measurement of total metals. A sample of up to 0.5 g is digested with microwave heating for 15 min in an appropriate acid mixture held in a fluorocarbon polymer container. Commonly, the reagents employed are a mixture of 9 mL of conc. nitric acid and 3 mL of hydrofluoric acid, although other acid mixtures employing reagents such as conc. HCl and hydrogen peroxide may be used. The sample is heated in the microwave oven to 180°C and held at that temperature for at least 9.5 min. After heating, the residual solids are filtered off and the filtrate is analyzed for metals.

Many kinds of hazardous waste samples contain metals dissolved or suspended in viscous petroleum products, including oils, oil sludges, tars, waxes, paints, paint sludges, and other hydrocarbon materials. Method 3031 can be used to dissolve these metals—including antimony, arsenic, barium, beryllium cadmium, chromium, cobalt, copper, lead molybdenum, nickel, selenium, silver, thallium, vanadium, and zinc—in a form suitable for atomic spectrometric analysis. The procedure involves mixing 0.5 g of sample with 0.5 g of finely ground $KMnO_4$ and 1.0 mL of conc. H_2SO_4, which causes a strongly exothermic reaction to occur as the hydrocarbon matrix is oxidized. After the reaction has subsided, 2 mL of conc. HNO_3 and 2 mL of conc. HCl are added, the sample is filtered with filter paper, the filter paper is digested with conc. HCl, and the sample is diluted and analyzed for metals.

26.3 ANALYTE ISOLATION FOR ORGANICS ANALYSIS

The determination of organic analytes requires that they be isolated from the sample matrix. Since organic analytes are generally soluble in organic solvents, they can usually be extracted from samples with a suitable solvent. Although extraction works well for nonvolatile and semivolatile analytes, it is not so suitable for VOCs, which are readily vaporized during sample processing. The volatile materials are commonly isolated by techniques that take advantage of their high vapor pressures.

26.3.1 Solvent Extraction

Method 3500 is a procedure for extracting nonvolatile or semivolatile compounds from a liquid or solid sample. The sample is extracted with an appropriate solvent, dried, and concentrated by evaporating solvent in a Kuderna–Danish apparatus prior to further processing for analysis.

A number of methods more complicated than Method 3500 have been devised for extracting nonvolatile and semivolatile analytes from waste samples. Method 3540 uses extraction with a Soxhlet extractor. This device, illustrated in Chapter 22, Figure 22.6, for the extraction of lipids from biological tissue, provides for recirculation of continuously redistilled fresh solvent over a sample of soils, sludges, and wastes. The sample is first mixed with anhydrous Na_2SO_4 to dry it, then placed inside an extraction thimble in the Soxhlet apparatus, which redistills a relatively small volume of extraction solvent over the sample. After extraction, the sample may be dried, concentrated, and exchanged into another solvent prior to analysis.

Method 3545 uses pressurized fluid extraction at 100°C and a pressure up to 2000 psi to remove organophilic analyte species from solid samples including soils, clays, sediments, sludges, and waste solids. Used for the extraction of semivolatile organic compounds, organophosphorus pesticides, organochlorine pesticides, chlorinated herbicides, and PCBs, it requires less solvent and takes less time than the Soxhlet extraction described above. Before extraction, the finely ground 10–30 g sample should be dried to prevent residual water from interfering with the extraction. It may be air-dried or dried with anhydrous Na_2SO_4 or pelletized diatomaceous Earth. An extraction time of 5 min is typically used.

Method 3550 uses sonication with ultrasound to expedite the extraction of nonvolatile and semivolatile organic compounds from solids including soils, sludges, and wastes. The procedure calls for subjecting the finely divided dried sample mixed with solvent to ultrasound for a brief period of time. Low concentration samples may be subjected to multiple extractions with additional fresh solvent.

26.3.2 Supercritical Fluid Extraction

Although the requirement for specialized high-pressure equipment has limited its application, extraction with supercritical carbon dioxide maintained at temperatures and pressures above the critical point where separate liquid and vapor phases do not exist is a very effective means of extracting some organic analytes from solid waste samples. A typical example of this technique is the extraction of estrogenic bisphenol A from sewage sludge for analysis.

Method 3561 is used to extract PAHs such as acenaphthene, benzo(a)pyrene, fluorene, and pyrene from solid samples with supercritical carbon dioxide. The actual procedure is relatively complicated and is divided into three steps. The first step extracts the more volatile compounds with pure supercritical carbon dioxide at moderately low density and temperature. Less volatile PAHs are removed in the second step using supercritical carbon dioxide and methanol modifier. A third step using pure carbon dioxide is employed to purge the modifier from the system.

26.3.3 Pressurized Liquid Extraction and Subcritical Water Extraction

Pressurized liquid extraction and subcritical water extraction offer some of the advantages of supercritical fluid extraction but under generally less severe conditions. Compared to conventional solvent extraction, both of these techniques are faster and consume less solvent.

Pressurized liquid extraction, also called *accelerated solvent extraction*, uses relatively small volumes (60 mL or less) of organic solvents heated to 50°C to 200°C under sufficient pressure to maintain the liquid state. This enables extraction of solid samples within 30 min. Typically, polyhalogenated dibenzo-*p*-dioxins and benzo-*p*-furans have been extracted from solid samples within 15 min with a 1/1 mixture of *n*-hexane and acetone at 130°C.[4] Compound degradation was minimal.

Subcritical water extraction, or *pressurized hot water extraction* uses water heated to 100°C to 374°C pressurized to maintain the liquid state. The dielectric constant of water under these conditions changes with temperature, which can be adjusted to change the extraction characteristics and make it an effective extractant for organophilic analytes. Specifically, as the temperature of subcritical liquid water is raised, its dielectric constant and polarity decrease making it a more "organic-like" extractant. Because it is more polar than organic solvents, water can also be more effective than common organic solvents in extracting polar analytes. Because of these characteristics and the environmentally friendly character of water, hot water extraction is finding greater use in the analysis of solid waste materials and sludges.

A typical application of subcritical water extraction is the extraction of brominated flame retardants from sediment samples. The technique is reputed to be more effective than conventional Soxhlet extraction with solvents.

26.4 SAMPLE CLEANUP

Processing of most waste, soil, and sediment samples results in the extraction of extraneous substances that can cause the observation of extraneous peaks, be detrimental to peak resolution and column efficiency, and be damaging to expensive columns and detectors. In order to obtain meaningful analytical results from the often complex samples of solid wastes, proper extraction and processing prior to analysis are critical. *Sample cleanup* refers to a variety of measures that can be taken to remove these constituents from sample extracts by a number of procedures including distillation, partitioning with immiscible solvents, adsorption chromatography, gel permeation chromatography, or chemical destruction of interfering substances with acid, alkali, or oxidizing agents; two or more of these techniques may be used in combination. The most widely applicable cleanup technique is gel permeation chromatography, which can be used to separate substances with high molecular masses from the analytes of interest. Treatment by adsorption column chromatography with alumina, Florisil, or silica gel can be used to isolate a relatively narrow polarity range of analytes away from interfering substances. Acid–base partitioning can be used in the determination of materials such as chlorophenoxy herbicides and phenols to separate acidic, basic, and neutral organics. Table 26.1 shows the uses of the main sample cleanup techniques.

Alumina column cleanup makes use of highly porous granular aluminum oxide. Available in acidic, neutral, and basic pH ranges, this solid is packed into a column topped with a water-absorbing substance over which the sample is eluted with a suitable solvent, which leaves interferences on the column. After elution, the sample is concentrated, exchanged with another solvent if necessary, then analyzed. Florisil is an acidic magnesium silicate and a registered trade name of Floridin Co. It is used in a column cleanup procedure in a manner similar to alumina. Silica gel is a weakly acidic amorphous silicon oxide. It can be activated by heating for several hours at 150°C to 160°C and used for the separation of

TABLE 26.1
Sample Cleanup Techniques and Their Applications

Number	Type	Applications
3610	Alumina column	Phthalate esters, nitrosamines
3611	Alumina column cleanup and separation of petroleum wastes	PAHs, petroleum wastes
3620	Florisil column	Phthalate esters, nitrosamines, organochlorine pesticides, PCBs, chlorinated hydrocarbons, organophosphorus pesticides
3630	Silica gel	PAHs
3630(b)	Silica gel	Phenols
3640	Gel permeation chromatography	Phenols, phthalate esters, nitrosamines, organochlorine pesticides, PCBs, nitroaromatics, cyclic ketones, polycyclic aromatic hydrocarbons, chlorinated hydrocarbons, organophosphorus pesticides, priority pollutant semivolatiles
3650	Acid–base liquid-liquid partition	Phenols, priority pollutant semivolatiles
3660	Sulfur cleanup	Organochlorine pesticides, PCBs, priority pollutant semivolatiles

hydrocarbons. Deactivated silica gel containing 10–20% water acts as an adsorbent for compounds with ionic and nonionic functionalities such as dyes, alkali metal cations, terpenoids, and plasticizers. It is used in a column as described for alumina above. Gel-permeation chromatography separates solutes by size carried over a hydrophobic gel by organic solvents. A gel must be chosen that will separate the appropriate size range of analytes and interferences. The gel is preswelled before loading onto a column and flushed extensively with solvent before the sample is introduced for separation.

26.5 SAMPLE PREPARATION FOR VOCs

Many different VOCs are found at waste sites and in various kinds of waste solid and sludge samples. These include benzene, bromomethane, chloroform, 1,4-dichlorobenzene, dichloromethane, styrene, toluene, vinyl chloride, and the xylene isomers. Various methods are employed to isolate and concentrate VOCs from the solid, sludge, or liquid matrices in which they are contained.

Method 5021, "volatile organic compounds in soils and other solid matrices using equilibrium headspace analysis" is used to isolate VOCs from soil, sediment, or solid waste samples for determination by gas chromatography (GC) or GC/MS. This procedure makes use of a special glass headspace vial that contains at least 2 g of sample. A matrix-modifying preservative solution, internal standards, and surrogate compounds may be added to the sample and mixed thoroughly by rotating the vial. The sample is heated to 85°C for approximately 1 h prior to injection into the gas chromatograph. For injection, helium is forced under pressure into the vial and a sample of the gas in the headspace above the solid sample is forced out of the vial and into the gas chromatograph system for quantitation.

Method 5030 is a purge-and-trap procedure that can be used to collect for gas chromatographic analysis poorly water-soluble compounds that have boiling points below 200°C, which includes a wide variety of compounds commonly occurring in hazardous wastes. For samples in water, helium is bubbled through the sample and the volatile analyte compounds are absorbed on a sorbent column. Solid samples can be dispersed in methanol and the methanol added to water for the purging step. After the purging is complete, the sorbent column is heated and flushed with carrier gas to sweep the sample compounds into the gas chromatograph for qualitative and quantitative analysis of the VOCs present.

Method 5035 is a closed-system, purge-and-trap and extraction method applicable to the determination of VOCs in soil, sediment, and waste samples. For the determination of very low levels of VOCs in soil, a sealed sample vial is used that remains sealed throughout the sample processing operations. An approximately 5-g sample is weighed into a sample vial containing a stirring bar and sodium bisulfate preservative solution, and the vial is sealed and transported to the laboratory as soon as possible. For analysis, reagent water, surrogates, and internal standards are injected into the vial without opening it. The contents of the vial are purged with helium gas into an appropriate sampling trap, then flushed into the gas chromatograph for measurement.

Method 5031 is used to isolate volatile, nonpurgeable, water-soluble compounds by azeotropic distillation. The analyte compounds for which it is suitable include acetone, acetonitrile, acrylonitrile, allyl alcohol, 1-butanol, *t*-butyl alcohol, crotonaldehyde, 1,4-dioxane, ethanol, ethyl acetate, ethylene oxide, isobutyl alcohol, methanol, methylethyl ketone, methylisobutyl ketone, *n*-nitroso-di-*n*-butylamine, paraldehyde, 2-pentanone, 2-picoline, 1-propanol, 2-propanol, propionitrile, pyridine, and *o*-toluidine. The technique takes advantage of the formation of an azeotrope liquid mixture of water and analytes that boil at a constant temperature and give off vapors of a constant composition. For separation of the analyte species, a 1-L sample is adjusted to pH 7 with a buffer, and a small sample of condensate enriched in analyte species is collected. The organics are measured in the azeotrope solution distilled from the sample.

Method 5032 employs vacuum distillation to isolate volatile organic analytes from liquid, solid, and oily waste matrices, and even animal tissues. Examples of compounds isolated for analysis by this procedure include acetone, benzene, carbon disulfide, chloroform, ethanol, styrene, tetrachloroethene, vinyl chloride, and the *o*-, *m*-, *p*-xylene isomers. Such compounds should be no more than minimally soluble in water and should boil below 180°C. The sample in water is distilled under a vacuum and the water condensed in a chilled condenser. Volatile analyte constituents not condensed with the water vapor are swept into a cryogenic trap maintained at liquid nitrogen temperature of −196°C for collection. For analysis, the contents of the trap are evaporated at an elevated temperature and swept into the chromatograph.

Method 3585, waste dilution for volatile organics, is employed to place a nonaqueous waste sample of volatile organics into the appropriate form for injection into a gas chromatograph. It is applicable to samples containing analytes at levels of 1 mg/kg or higher. The procedure calls for placing a 1 g oil-phase sample into a vial marked for a volume of 10 mL, diluting with *n*-hexadecane or other appropriate solvent, sealing the vial, and mixing to dissolve the sample. Because the diluted sample usually contains residual materials from the sample with a tendency to foul the gas chromatograph, the sample is injected through a replaceable direct injection liner containing Pyrex glass wool.

26.6 BIOASSAY AND IMMUNOASSAY SCREENING OF WASTES

Several important biologically based methods are used to characterize wastes and particularly to assess the dangers that they pose to the biosphere.[5] Acute toxicity tests that cover a relatively short period of the life cycle of organisms can be useful in finding materials that may pose immediate hazards. Long-lasting effects are assayed by chronic toxicity tests that measure such things as fertilization, growth, and reproduction. These tests normally have been conducted over the full life cycle of the organism or shortened to 30 days, an "early-stage test." The U.S. EPA has developed a 7-day chronic toxicity test that applies to early life-cycle stages of organisms, a time period during which the organisms are most sensitive to toxicants.

Immunoassay has emerged as a useful technique for screening wastes for specific kinds of pollutants.[6] Commercial immunoassay techniques have been developed that permit very rapid analyses of large numbers of samples. A variety of immunoassay techniques have been developed. These techniques all use biologically produced antibodies that bind specifically to analytes or classes of analytes. This binding is combined with chemical processes that enable detection through a signal-producing species (reporter reagent) such as enzymes, chromophores, fluorophores, and luminescent compounds. The reporter reagent binds with the antibody. When an analyte is added to the antibody to displace the reagent, the concentration of displaced reagent is proportional to the level of analyte displacing it from the antibody. Detection of the displaced reporter reagent enables quantification of the analyte.

Immunoassay techniques are divided into the two major categories: heterogeneous and homogeneous; the former requires a separation (washing) step whereas the latter does not require such a step. Typically, when heterogeneous procedures are used, the antibody is immobilized on a solid support on the inner surface of a disposable test tube. The sample is contacted with the antibody displacing reporter reagent, which is removed by washing. The amount of reagent displaced, commonly measured spectrophotometrically, is proportional to the amount of analyte added. Very widely used enzyme immunoassays make use of reporter reagent molecules bound with enzymes, and kits are available for enzyme-linked immunosorbent assays (ELISA) of a number of organic species likely to be found in hazardous wastes.

Immunoassay techniques have been approved for the determination of numerous analytes commonly found in hazardous wastes. Where the EPA method numbers are given in parentheses, these include pentachlorophenol (4010), 2,4-D (4015), PCBs (4020), petroleum hydrocarbons (4030), PAHs (4035), toxaphene (4040), chlordane (4041), DDT (4042), TNT explosives in soil (4050), and hexahydro-1,3,5-trinitro-1,3,5-triazine (RDX) in soil (4051). ELISA have been reported for monitoring pentachlorophenol, BTEX (benzene, toluene, ethylbenzene, and *o*-, *m*-, and *p*-xylene) in industrial effluents.

26.7 DETERMINATION OF CHELATING AGENTS

Strong chelating agents in wastes have been found to play an important role in the mobility of heavy metals and metal radionuclides at waste disposal sites, with their potential to contaminate groundwater. Therefore, the determination of chelating agents, such as EDTA and N-(2-hydroxyethyl) ethylenediaminetriacetic acid (HEDTA), is an important analytical procedure for wastes. Most chelating agents of concern at waste sites are polar and nonvolatile, which prevents their determination by direct gas chromatographic methods. One of the more satisfactory methods for their determination makes use of derivatization to produce volatile species suitable for gas chromatographic analysis. A method has been described for the determination of chelating agents in hazardous wastes from radioactive wastes using derivatization followed by gas chromatographic/mass spectrometric (GC/MS) analysis.[7] The study involved wastes contained in a potentially leaking double-shell storage tank at the U.S. Department of Energy Hanford site. Treatment with BF_3 and methanol produced volatile derivatives that were measured by GC/MS. In addition to EDTA and HEDTA, the study showed the presence of NTA and citrate chelating agents, and chelating nitrosoiminiodiacetate was produced as an artifact of the analytical procedure.

26.8 TOXICITY CHARACTERISTIC LEACHING PROCEDURE

The TCLP is specified to determine the potential toxicity hazard of various kinds of wastes.[8] The test was designed to estimate the availability to organisms of both inorganic and organic species in

TABLE 26.2
Contaminants Determined in TCLP

EPA Hazardous Waste Number	Contaminant	Regulatory Level (mg/L)	EPA Hazardous Waste Number	Contaminant	Regulatory Level (mg/L)
Heavy Metals (Metalloids)					
D004	Arsenic	5.0	D032	Hexachlorobenzene	0.13[b]
D005	Barium	100.0	D033	Hexachlorobutadiene	0.5
D006	Cadmium	1.0	D034	Hexachloroethane	3.0
D007	Chromium	5.0	D035	Methylethyl ketone	200.0
D008	Lead	5.0	D036	Nitrobenzene	2.0[b]
D009	Mercury	0.2	D037	Pentachlorophenol	100.0
D010	Selenium	1.0	D038	Pyridine	5.0[b]
D011	Silver	5.0	D039	Tetrachloroethylene	0.7
Organics			D040	Trichloroethylene	0.5
D018	Benzene	0.5	D041	2,4,5-Trichlorophenol	400.0
D019	Carbon tetrachloride	0.5	D042	2,4,6-Trichlorophenol	2.0
D021	Chlorobenzene	100.0	D043	Vinyl chloride	0.2
D022	Chloroform	6.0	**Pesticides**		
D023	*o*-cresol	200.0[a]	D012	Endrin	0.02
D024	*m*-cresol	200.0[a]	D013	Lindane	0.4
D025	*p*-cresol	200.0[a]	D014	Methoxychlor	10.0
D026	Cresol	200.0[a]	D015	Toxaphene	0.5
D027	1,4-Dichlorobenzene	7.5	D016	2,4-D	10.0
D028	1,2-Dichloroethane	0.5	D017	2,4,5-TP (Silvex)	1.0
D029	1,1-Dichloroethylene	0.7	D020	Chlordane	0.03
D030	2,4-Dinitrotoluene	0.13[b]	D031	Heptachlor (and its epoxide)	0.008

[a] If *o*-, *m*-, and *p*-cresol concentrations cannot be differentiated, the total cresol (D026) concentration is used. The regulatory level of total cresol is 200 mg/L.

[b] Quantitation limit is greater than the calculated regulatory level. The quantitation limit therefore becomes the regulatory level.

hazardous materials present as liquids, solids, or multiple phase mixtures by producing a leachate, the TCLP extract, which is analyzed for the specific toxicants listed in Table 26.2.

The procedure for conducting the TCLP is rather involved. The procedure need not be run at all if a total analysis of the sample reveals that none of the pollutants specified in the procedure could exceed regulatory levels. At the opposite end of the scale, analysis of any of the liquid fractions of the sample showing that any regulated species would exceed regulatory levels even after the dilutions involved in the TCLP measurement have been carried out designate the sample as hazardous, and the TCLP measurement is not required.

In conducting the TCLP test, if the waste is a liquid containing less than 0.5% solids, it is filtered through a 0.6–0.8 μm glass fiber filter and the filtrate is designated as the TCLP extract. At levels of the solids exceeding 0.5%, any liquid present is filtered off for separate analysis and the solid is extracted to provide a TCLP extract (after size reduction, if the particles exceed certain size limitations). The choice of the extraction fluid is determined by the pH of the aqueous solution produced by shaking a mixture of 5 g of solids and 96.5 mL of water. If the pH is <5.0, a pH 4.93 acetic acid/sodium acetate buffer is used for extraction; otherwise, the extraction fluid used is a pH of 2.88 ± 0.05 solution of dilute acetic acid. Extractions are carried out in a sealed container rotated end-over-end for 18 h. The liquid portion is then separated and analyzed for the specific substances given in Table 26.2. If values exceed the regulatory limits, the waste is designated as "toxic."

LITERATURE CITED

1. Taricksa, J. R., Y. T. Hung, and K. H. Li, On-site monitoring and analysis of industrial pollutants, Chapter 4 in, *Hazardous Industrial Waste Treatment*, Lawrence K. Wang, Yung-Tse Hung, Howard H. Lo, and Constantine Yapijakis, Eds, pp. 133–136, Taylor & Francis/CRC Press, Boca Raton, FL, 2007.
2. *Test Methods for Evaluating Solid Waste, Physical/Chemical Methods*, EPA Publication SW-846, 3rd ed., (1986), as amended by Updates I (1992), II (1993, 1994, 1995) and III (1996), U.S. Government Printing Office, Washington, DC.
3. *Understanding Environmental Methods*, Version 5.0 on CD ROM, Genium Publishing Corporation, Amsterdam, NY, 2007.
4. *ASTM Book of Standards Volume 11.04: Water and Environmental Technology: Environmental Assessment; Hazardous Substances and Oil Spill Responses; Waste Management,* ASTM International, West Conshohocken, PA, 2008.
5. Selivanovskaya, S. Y. and V. Z. Latypova, Bioassay of industrial waste pollutants, Chapter 2 in, *Hazardous Industrial Waste Treatment*, L. K. Wang, Y.-T. Hung, H. H. Lo, and C. Yapijakis, Eds, pp. 15–62, Taylor & Francis/CRC Press, Boca Raton, FL, 2007.
6. Castillo, M., A. Oubiña, and D. Barcelo, Evaluation of ELISA kits followed by liquid chromatography-atmospheric pressure chemical ionization-mass spectrometry for the determination of organic pollutants in industrial effluents, *Environmental Science and Technology*, **32**, 2180–2184, 1998.
7. Grant, K. E., G. M. Mong, R. B. Lucke, and J. A. Campbell, Quantitative determination of chelators and their degradation products in mixed hazardous wastes from tank 241-SY-101 using derivatization GC/MS, *Journal of Radioanalytical and Nuclear Chemistry*, **221**, 383–402, 1996.
8. Toxicity Characteristic Leaching Procedure, Test Method 1311 in *Test Methods for Evaluating Solid Waste, Physical/Chemical Methods*, EPA Publication SW-846, 4th ed., U.S. Government Printing Office, Washington, DC, 2008.

SUPPLEMENTARY REFERENCES

Dieken, F. P., *Methods Manual for Chemical Analysis of Water and Wastes*, Alberta Environmental Centre, Vergeville, Alberta, Canada, 1996.

Franchetti, M. J., *Solid Waste Analysis and Minimization*, McGraw-Hill, New York, 2009.

Gavasci, R., F. Lombardi, A. Polettini, and P. Sirini, Leaching tests on solidified products, *Journal of Solid Waste Technology Management*, **25**, 14–20, 1998.

Margesin, R. and F. Schinner, Eds, *Manual for Soil Analysis: Monitoring and Assessing Soil Bioremediation,* Springer, Berlin, 2005.

Morales-Munoz, S., J. L. Luque-Garcia, and M. D. Luque De Castro, Approaches for accelerating sample preparation in environmental analysis, *Critical Reviews in Environmental Science and Technology*, **33**, 391–421, 2003.

Pansu, M. and J. Gautheyrou, *Handbook of Soil Analysis: Mineralogical, Organic and Inorganic Methods*, Springer, Berlin, 2006.

Patnaik, P., Ed., *Handbook of Environmental Analysis: Chemical Pollutants in Air, Water, Soil, and Solid Wastes*, CRC Press, Boca Raton, FL, 1997.

Que H. S. S., *Hazardous Waste Analysis*, Government Institutes, Rockville, MD, 1999.

Rayment, G. E., R. Sadler, N. Craig, B. Noller, and B. Chiswell, *Chemical Analysis of Contaminated Land*, CRC Press, Boca Raton, FL, 2003.

Reeve, *Introduction to Environmental Analysis*, Wiley, New York, 2002.

Ulery, A. L. and L. R. Drees, Eds, *Methods of Soil Analysis. Part 5, Mineralogical Methods*, Soil Science Society of America, Madison, WI, 2008.

Williford, C. W., Jr. and R. M. Bricka, Extraction of TNT from aggregate soil fractions, *Journal of Hazardous Materials*, **66**, 1–13, 1999.

QUESTIONS AND PROBLEMS

The use of internet resources is assumed in answering any of the questions. These would include such things as constants and conversion factors as well as additional information needed to complete an answer.

1. Explain the uses of microwave in hazardous waste analysis. How is ultrasound employed in hazardous waste analysis?
2. Does sample digestion necessarily give an analysis leading to total metals? Why might it not be advantageous to measure total metals in a sample?
3. What is the distinction between a Kuderna–Danish apparatus and a Soxhlet apparatus?
4. How is anhydrous Na_2SO_4 used in organics analysis?
5. How does the purge-and-trap procedure differ from azeotropic distillation? For what kinds of compounds would the two procedures be employed?
6. What is the purpose of sample cleanup? Why is cleanup more commonly applied to samples to be analyzed for organic contaminants than for metals?
7. For what purpose is a cryogenic trap used in organics analysis? What advantage might it have over the solid sorbent traps used in conventional purge-and-trap analysis?
8. What do benzene, bromomethane, chloroform, 1,4-dichlorobenzene, dichloromethane, styrene, toluene, vinyl chloride, and *o*-xylene have in common?
9. What is the principle of immunoassay? What makes it specific for compounds or narrow classes of compounds? Why might it be especially suitable as a survey technique for hazardous waste sites? What is ELISA?
10. In what sense is the TCLP a measure of available toxicants?
11. Under what circumstances is it unnecessary to run the TCLP in evaluating the toxicity hazard of a waste material?
12. Which is the most widely applicable sample cleanup technique? What are three other kinds of materials used in column cleanup of samples?
13. Look up the conditions under which water is a supercritical fluid. What are the advantages of using subcritical water as an extractant when it is effective in that application?
14. The TCLP does not necessarily remove all analytes from a sample. Does this mean that the test is of limited validity? Explain why it might be useful.

27 Analysis of the Atmosphere and Air Pollutants

27.1 ATMOSPHERIC MONITORING

Good analytical methodology, particularly that applicable to automated analysis and continuous monitoring, is essential to the study and alleviation of air pollution. The atmosphere is a particularly difficult analytical system because of the very low levels of substances to be analyzed; sharp variations in pollutant level with time and location; differences in temperature and humidity; and difficulties encountered in reaching desired sampling points, particularly those substantially above the earth's surface. Furthermore, although improved techniques for the analysis of air pollutants are continually being developed, a need still exists for new analytical methodology and the improvement of existing methodology.

Much of the earlier data on air pollutant levels were unreliable as a result of inadequate analysis and sampling methods. An atmospheric pollutant analysis method does not have to give the actual value to be useful. One which gives a relative value may still be helpful in establishing trends in pollutants levels, determining pollutant effects, and locating pollution sources. Such methods may continue to be used while others are being developed.

Atmospheric analysis is a dynamic area that continues to benefit from advances in instrumentation and techniques. Standard methods of atmospheric chemical analysis have been summarized.[1,2]

27.1.1 Air Pollutants Measured

The air pollutants generally measured may be placed in several different categories. In the United States, one such category contains materials for which ambient (surrounding atmosphere) standards have been set by the EPA. These are sulfur dioxide, carbon monoxide, nitrogen dioxide, nonmethane hydrocarbons, and particulate matter. The standards are categorized as primary and secondary. *Primary standards* are those defining the level of air quality necessary to protect public health. *Secondary standards* are designed to provide protection against known or expected adverse effects of air pollutants, particularly upon materials, vegetation, and animals. Another group of air pollutants to be measured consists of those known to be specifically hazardous to human health, such as asbestos, beryllium, and mercury. Still another category of air pollutants contains those regulated in new installations of selected stationary sources, such as coal-cleaning plants, cotton gins, lime plants, and paper mills. Some pollutants in this category are visible emissions, acid (H_2SO_4) mist, particulate matter, nitrogen oxides, and sulfur oxides. These substances often must be monitored in the stack to ensure that emission standards are being met. A fourth category consists of the emissions of mobile sources (motor vehicles)—hydrocarbons, CO, and NO_x. A fifth group consists of miscellaneous elements and compounds, such as certain heavy metals, fluoride, chlorine, phosphorus, PAHs, PCBs, odorous compounds, reactive organic compounds, and radionuclides. Much of the remainder of this chapter is devoted to a discussion of the analytical methods for most of the species mentioned above.

The units in which air pollutants and air-quality parameters are expressed include for gases and vapors, μg/m^3 (alternatively, ppm by volume); for mass of particulate matter, μg/m^3; for particulate matter count, number per cubic meter; for visibility, km; for instantaneous light transmission, percentage of light transmitted; for emission and sampling rates, m^3/min; for pressure, mm Hg; and for temperature, °C. Air volumes should be converted to conditions of 10°C and 760 mm Hg (1 atm), assuming ideal gas behavior.

27.2 SAMPLING

The ideal atmospheric analysis techniques are those that work successfully without sampling, such as long-path laser resonance absorption monitoring. For many analyses, however, various types of sampling are required. In some very sophisticated monitoring systems, samples are collected and analyzed automatically and the results are transmitted to a central receiving station. Often, however, a batch sample is collected for later chemical analysis.

The analytical result from a sample can only be as good as the method employed to obtain that sample. A number of factors enter into obtaining a good sample. The size of the sample required (total volume of air sampled) decreases with increasing concentration of pollutant and increasing sensitivity of the analytical method. Often a sample of 10 or more cubic meters is required. The sampling rate is determined by the equipment used and generally ranges from approximately 0.003 to 3.0 m^3/min. The duration of sampling time influences the result obtained, as shown in Figure 27.1. The actual concentration of the pollutant is shown by the solid line. A sample collected over an 8-h period has the concentration shown by the dashed line, whereas samples taken over 1-h intervals exhibit the concentration levels shown by the dotted line. Sampling techniques are discussed briefly for specific kinds of analytes later in this chapter.

Methods for sampling atmospheric particles are discussed in Section 27.9. The main methods for particle sampling are simple sedimentation, filtration, and impingement of an airstream onto a surface that collects particles.

Sampling for vapors and gases may range from methods intended to collect only one specific pollutant to those designed to collect all pollutants. The most fundamental method of sampling is *whole-air sampling* in which a volume of air is collected in a bag or steel or glass container. Loss of analyte by adsorption onto container walls can be a problem with this technique. Essentially, all pollutants may be removed from an air sample cryogenically by freezing or by liquifying the air in collectors maintained at a low temperature. Analytes can be collected on solid sorbents over which air is filtered or by bubbling through liquids. Collection methods can be *dynamic* in which air is pumped through a sampling medium or *passive* in which a sampling medium, such as a semipermeable bag filled with sorbent material, is simply exposed to air over a period of time (see below).

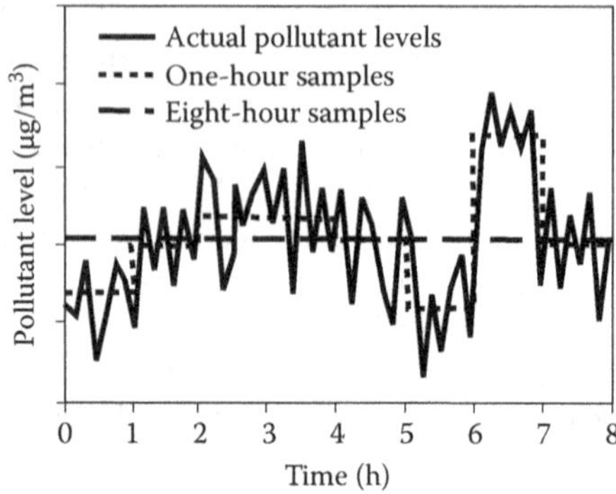

FIGURE 27.1 Effect of duration of sampling upon observed values of air pollutant levels.

Also used in water analysis, *solid-phase microextraction* combines sampling and preconcentration in which VOCs are collected from air by pumping over a small amount of fiber holding an extraction medium. The analyte can then be evolved from the extraction device directly into a gas chromatograph for measurement.

Denuders are among the more useful sampling devices for certain kinds of air pollutants. Denuders solve a major sampling problem by enabling collection of gas-phase pollutants free of contamination by particles. Otherwise, for some pollutants such as acids, it is not possible to distinguish relative amounts of analytes in the gas phase from those in particulate form.

Diffusion denuders run a laminar flow airstream through a tube, the walls of which are covered with a sorptive or reactive collecting medium for the analytes of interest. The diffusion coefficients of small particles are only about 10^{-4} those of gases, so that the particles go through the tube, and the gases diffuse to the walls and are collected. Open-bore denuders consist of tubes with coated walls. A more efficient device is the *annular denuder* composed of concentric tubes separated by 1–2 mm of annular space. *Thermodenuders* use heat to drive off analytes collected by the device. They can be used for semicontinuous analysis by employing a collection/analysis cycle involving alternate collection and thermal desorption. *Diffusion scrubbers* are denuders in which walls are composed of membranes such that gas from the sample goes through the wall and is collected in a collecting liquid.

An annular denuder coated with citric acid can be used to collect gaseous ammonia in air. Gaseous HCl, HNO_2, and SO_2 can be collected on the wall of a denuder coated with sodium carbonate.

Passive air sampling devices can be used for semiquantitative analysis of air contaminants. POP, such as PCBs may be collected from ambient air by devices consisting of triolein lipid material held inside tubes of permeable membrane material suspended in contact with the air. Such samplers may be left in place for many days or weeks and yield an integrated value of air contaminants.

Somewhat related to passive air sampling devices in principle are *surrogate sampling* media consisting of materials that are in contact with the air naturally for prolonged periods of time. Pine needles and tree bark have been used as surrogate sampling media. An interesting surrogate sampling medium is butter from cows' milk. This medium is especially useful for lipophilic POP, such as PCBs. These materials deposit from air onto forage eaten by cows and become concentrated in the milk fat and hence in butter.

Chemisorption reactions in which analytes react chemically with substances on a solid surface can be used to capture some air pollutants for analysis. One such process uses 2,4-dinitrophenylhydrazine contained in a cartridge through which air is pumped to sample the important class of carbonyl compounds (aldehydes and ketones) in the atmosphere. The adsorbent reacts with carbonyl compounds to form hydrazone compounds as shown by the reaction of acetaldehyde below and the hydrazone products are eluted with acetonitrile and measured by high-performance liquid chromatography using spectrophotometric detection.

$$(O_2N)_2C_6H_3\text{–NH–NH}_2 + H\text{–C(=O)–CH}_3 \longrightarrow (O_2N)_2C_6H_3\text{–NH–N=CH–CH}_3 + H_2O \qquad (27.1)$$

2,4-Dinitrophenylhydrazine   Ethanal-2,4-dinitrophenylhydrazone

27.3 METHODS OF ANALYSIS

A very large number of different analytical techniques are used for atmospheric pollutant analysis. Some of these, the uses of which are not confined to atmospheric analysis, were discussed in Chapters 25 and 26. Techniques confined largely to atmospheric analysis are discussed in the remainder of this chapter. A summary of some of the main instrumental techniques for air monitoring is presented in Table 27.1.

TABLE 27.1
Major Instrumental Techniques Used for Air Pollutant Analysis

Pollutant	Method	Potential Interferences
SO_2 (total S)	Flame photometric (FPD)	H_2S, CO
SO_2	GC (FPD)	H_2S, CO
SO_2	Spectrophotometric (pararosaniline wet chemical)	H_2S, HCl, NH_3, NO_2, O_3
SO_2	Electrochemical	H_2S, HCl, NH_3, NO, NO_2, O_3, C_2H_4
SO_2	Conductivity	HCl, NH_3, NO_2
SO_2	Gas-phase spectrophotometric	NO, NO_2, O_3
O_3	Chemiluminescent	H_2S
O_3	Electrochemical	NH_3, NO_2, SO_2
O_3	Spectrophotometric (potassium iodide reaction, wet chemical)	NH_3, NO_2, NO, SO_2
O_3	Gas-phase spectrophotometric	NO_2, NO, SO_2
CO	Infrared	CO_2 (at high levels)
CO	GC (with flame ionization detector)	—
CO	Electrochemical	NO, C_2H_4
CO	Catalytic combustion-thermal detection	NH_3
CO	Infrared fluorescence	—
NO_2	Chemiluminescent	NH_3, NO, NO_2, SO_2
NO_2	Spectrophotometric (azo-dye reaction, wet chemical)	NO, SO_2, NO_2, O_3
NO_2	Electrochemical	HCl, NH_3, NO, NO_2, SO_2, O_3, CO
NO_2	Gas-phase spectrophotometric	NH_3, NO, NO_2, SO_2, CO
NO_2	Conductivity	HCl, NH_3, NO, NO_2, SO_2

The U.S. EPA specifies reference methods of analysis for selected air pollutants to determine compliance with the primary and secondary national ambient air quality standards for those pollutants. These methods are published annually in the *Code of Federal Regulations*.[3] They are not necessarily state-of-the art, and in some cases are outdated and cumbersome. However, for regulatory and legal purposes they provide reliable measurements proven to be valid.

27.4 DETERMINATION OF SULFUR DIOXIDE

The reference method for the analysis of sulfur dioxide is the spectrophotometric pararosaniline method developed by West and Gaeke.[4] It is applicable to the analysis of 0.005–5 ppm SO_2 in ambient air. The method makes use of a collecting solution of 0.04 M potassium tetrachloromercurate to collect sulfur dioxide according to the following reaction:

$$HgCl_4^{2-} + SO_2 + H_2O \rightarrow HgCl_2SO_3^{2-} + 2H^+ + 2Cl^- \quad (27.2)$$

Typically, sampling involves scrubbing 30 L of air through 10 mL of scrubbing solution with a collection efficiency of around 95%. The $HgCl_2SO_3^{2-}$ complex stabilizes readily oxidized sulfur dioxide against the oxidizing agents of oxidants such as ozone and nitrogen oxides. For analysis, sulfur dioxide in the scrubbing medium is reacted with formaldehyde:

$$HCHO + SO_2 + H_2O \rightarrow HOCH_2SO_3H \quad (27.3)$$

The adduct formed is then reacted with uncolored organic pararosaniline hydrochloride to produce a red-violet dye. Although NO_2 at levels above about 2 ppm interferes, the interference may be eliminated by reducing the NO_2 to N_2 gas with sulfamic acid, H_2NSO_3H. The level of SO_2 measured

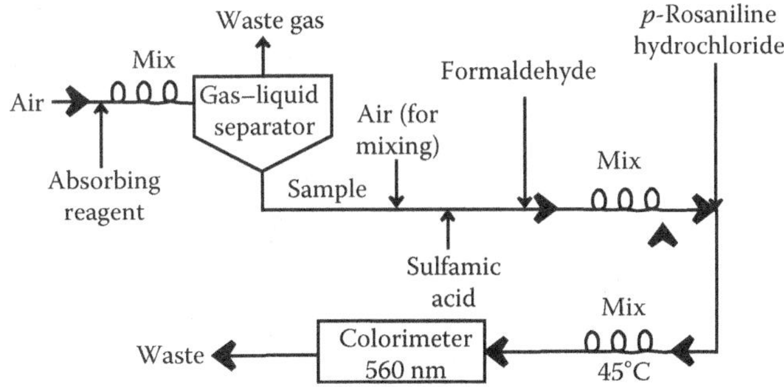

FIGURE 27.2 Block diagram of an automated system for the determination of sulfur dioxide by the pararosaniline method.

spectrophotometrically is corrected to standard ambient conditions of 25°C and atmospheric pressure (101 kPa). The lower limit of detection of SO_2 in 10 mL of the collecting solution is 0.75 μg. This represents a concentration of 25 μg SO_2/m^3 (0.01 ppm) in an air sample of 30 standard liters (short-term sampling), and a concentration of 13 μg SO_2/m^3 (0.005 ppm) in an air sample of 288 standard liters (long-term sampling).

Performed manually, the West–Gaeke method for sulfur dioxide analysis is cumbersome and complicated. However, the method has been refined to the point that it can be done automatically with continuous-monitoring equipment. A block diagram of such an analyzer is shown in Figure 27.2.

Conductimetry (*amperometry*) was used as the basis for a commercial continuous sulfur dioxide analyzer as early as 1929. Generally, the sulfur dioxide is collected in a solution of hydrogen peroxide, which oxidizes the SO_2 to H_2SO_4, and the resulting increased electrical conductance of the sulfuric acid solution is measured. Sulfur dioxide can be determined by ion chromatography by bubbling SO_2 through hydrogen peroxide solution to produce SO_4^{2-}, followed by analysis of the sulfate by ion chromatography, a method that separates ions on a chromatography column and detects them very sensitively by conductivity measurement. Flame photometry, sometimes in combination with GC, is used for the detection of sulfur dioxide and other gaseous sulfur compounds. The gas is burned in a hydrogen flame, and the sulfur emission line at 394 nm is measured.

Several direct spectrophotometric methods are used for sulfur dioxide measurement, including nondispersive infrared absorption, Fourier transform infrared (FTIR) analysis, ultraviolet absorption, molecular resonance fluorescence, and second-derivative spectrophotometry. The principles of these methods are the same for any gas measured.

27.5 NITROGEN OXIDES

Several methods have been used to determine nitrogen oxides. As noted in Table 27.1, these methods include electrochemical methods, direct measurements of nitrogen oxides spectrophotometrically in the gas phase, and a wet chemical method based on formation of an azo dye. However, gas-phase chemiluminescence is the favored method of NO_x analysis.[5] The general phenomenon of chemiluminescence was defined in Section 9.8. It results from the emission of light from electronically excited species formed by a chemical reaction. In the case of NO, ozone is used to bring about the reaction, producing electronically excited nitrogen dioxide:

$$NO + O_3 \rightarrow NO_2^* + O_2 \tag{27.4}$$

The species loses energy and returns to the ground state through emission of light in the 600–3000 nm range. The emitted light is measured by a photomultiplier; its intensity is proportional to the concentration of NO. A schematic diagram of the device used is shown in Figure 27.3.

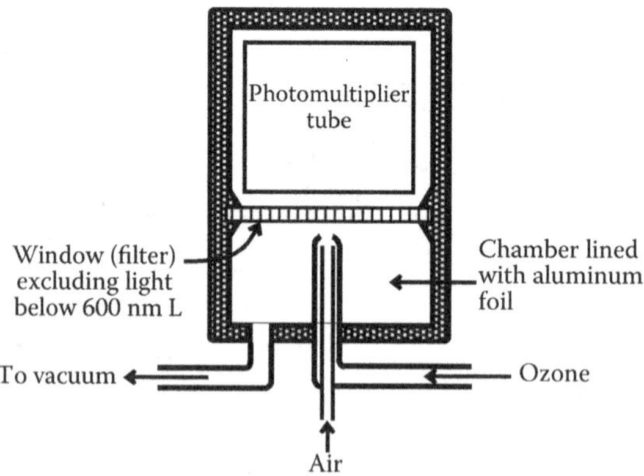

FIGURE 27.3 Chemiluminescence detector for NO_x.

Since the chemiluminescence detector system depends upon the reaction of O_3 with NO, it is necessary to convert NO_2 to NO in the sample prior to analysis. This is accomplished by passing the air sample over a thermal converter, which brings about the desired conversion. Analysis of such a sample gives NO_x, the sum of NO and NO_2. Chemiluminescence analysis of a sample that has not been passed over the thermal converter gives NO. The difference between these two results is NO_2.

Other nitrogen compounds besides NO and NO_2 undergo chemiluminescence by reacting with O_3, and these may interfere with the analysis if present at an excessive level. Particulate matter also causes interference which may be overcome by employing a membrane filter on the air inlet.

This analysis technique is illustrative of chemiluminescence analysis in general. Chemiluminescence is an inherently desirable technique for the analysis of atmospheric pollutants because it can avoid wet chemistry, is basically simple, and lends itself well to continuous monitoring and instrumental methods. Another chemiluminescence method, that employed for the analysis of ozone, is described in the next section.

27.6 ANALYSIS OF OXIDANTS

The atmospheric oxidants that are commonly analyzed include ozone, hydrogen peroxide, organic peroxides, and chlorine. The classic manual method for the analysis of oxidants is based upon their oxidation of I^- ion followed by spectrophotometric measurement of the product. The sample is collected in 1% KI buffered at pH 6.8. Oxidants react with iodide ion as shown by the following reaction of ozone:

$$O_3 + 2H^+ + 3I^- \rightarrow I_3^- + O_2 + H_2O \tag{27.5}$$

The absorbance of the colored I_3^- product is measured spectrophotometrically at 352 nm. Generally, the level of oxidant is expressed in terms of ozone, although it should be noted that not all oxidants—PAN, for example—react with the same efficiency as O_3. Oxidation of I^- may be used to determine oxidants in a concentration range of several hundredths of a ppm to approximately 10 ppm. Nitrogen dioxide gives a limited response to the method, and reducing substances interfere seriously.

Now the favored method for oxidant analysis uses chemiluminescence.[6] The chemiluminescent reaction is that between ozone and ethylene. Chemiluminescent radiation from this reaction is emitted over a range of 300–6000 nm, with a maximum at 435 nm. The intensity of emitted light is directly proportional to the level of ozone. Ozone concentrations ranging from 0.003 to 30 ppm may be measured. Ozone for calibrating the instrument is generated photochemically from the absorption of ultraviolet radiation by oxygen.

27.7 ANALYSIS OF CARBON MONOXIDE

Carbon monoxide can be measured in the atmosphere by absorption of infrared radiation at 4.7 μm. Observation of the absorbance of solar radiation at this wavelength through the atmosphere can be used to calculate total column abundance and column concentrations of carbon monoxide. Carbon monoxide is analyzed in the atmosphere by nondispersive infrared spectrometry.[7] This technique depends upon the fact that carbon monoxide absorbs infrared radiation strongly at certain wavelengths. Therefore, when such radiation is passed through a long (typically 100 cm) cell containing trace levels of carbon monoxide, more of the infrared radiant energy is absorbed.

A nondispersive infrared spectrometer differs from standard infrared spectrometers in that the infrared radiation from the source is not dispersed according to wavelength by a prism or grating. The nondispersive infrared spectrometer is made very specific for a given compound, or type of compound, by using the sought-for material as part of the detector, or by placing it in a filter cell in the optical path. A diagram of a nondispersive infrared spectrometer selective for CO is shown in Figure 27.4. Radiation from an infrared source is "chopped" by a rotating device so that it alternately passes through a sample cell and a reference cell. In this particular instrument, both beams of light fall on a detector which is filled with CO gas and separated into two compartments by a flexible diaphragm. The relative amounts of infrared radiation absorbed by the CO in the two sections of the detector depend upon the level of CO in the sample. The difference in the amount of infrared radiation absorbed in the two-compartments causes slight differences in heating, so that the diaphragm bulges slightly toward one side. Very slight movement of the diaphragm can be detected and recorded. By means of this device, carbon monoxide can be measured from 0 to 150 ppm, with a relative accuracy of ±5% in the optimum concentration range.

Flame-ionization GC detection can also be used for the analysis of carbon monoxide. This detector system was described in Chapter 25, Section 25.5. It is selective for hydrocarbons, and conversion of CO to methane in the sample is required. This is accomplished by reaction with hydrogen over a nickel catalyst at 360°C:

$$CO + 3H_2 \rightarrow CH_4 + H_2O \tag{27.6}$$

A major advantage of this approach is that the same basic instrumentation may be used to measure hydrocarbons.

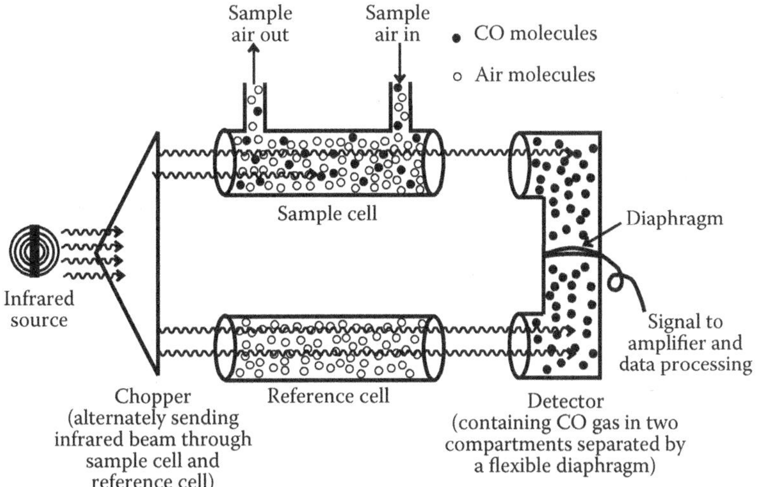

FIGURE 27.4 Nondispersive infrared spectrometer for the determination of carbon monoxide in the atmosphere.

Carbon monoxide may also be analyzed by measuring the heat produced by its catalytic oxidation to CO_2 over a catalyst consisting of a mixture of MnO_2 and CuO. Differences in temperature between a cell in which the oxidation is occurring and a reference cell through which part of the sample is flowing are measured by thermistors. A vanadium oxide catalyst can be used for the oxidation of hydrocarbons, enabling their simultaneous analysis.

27.8 DETERMINATION OF HYDROCARBONS AND ORGANICS

Monitoring of hydrocarbons in atmospheric samples takes advantage of the very high sensitivity of the hydrogen flame ionization detector to measure this class of compounds. Known quantities of air are run through the flame ionization detector 4 to 12 times per hour to provide a measure of total hydrocarbon content. A separate portion of each sample goes into a stripper column to remove water, carbon dioxide, and nonmethane hydrocarbons. Methane and carbon monoxide, which are not retained by the stripper column, are separated by a chromatographic column, passed through a catalytic reduction tube, and then to a flame ionization detector. Eluting first, methane is not changed by the reduction tube and is detected as methane by the detector. The carbon monoxide is reduced to methane, as shown by Reaction 27.6 in the preceding section, then detected as the methane product by the flame ionization detector. Concentrations of nonmethane hydrocarbons are given by subtracting the methane concentrations from the total hydrocarbons.

Using the method described above, total hydrocarbons can be determined in a range of 0–13 mg/ m^3, corresponding to 0–10 ppm. Methane can be measured over a range of 0–6.5 mg/m^3 (0–10 ppm).

27.8.1 Determination of Specific Organics in the Atmosphere

The method for hydrocarbons described above gives total hydrocarbons and methanes. In some cases it is important to have a method to determine individual organics because of their toxicities, ability to form photochemical smog, as indicators of photochemical smog, and as means of tracing sources of pollution. Numerous techniques have been published for the determination of organic compounds in the atmosphere. For example, whole-air samples can be collected in Tedlar bags, concentrated cryogenically at −180°C, then thermally desorbed and measured with high-resolution capillary column GC.

The U.S. EPA has published several procedures for determining the identities and concentrations of key organic air pollutants classified as air toxics.[8] These include VOCs, such as benzene, vinyl chloride, and chloroform; organochlorine pesticides and PCBs; aldehydes and ketones; phosgene; N-nitroso compounds; phenols, dioxin; nonmethane hydrocarbons, and PAHs. Sampling methods for various ones of these analytes include collection in canisters or bags, adsorption onto various solid sorbents, collection in liquids, and collection and reactions in reagent solutions. GC predominates as an analysis technique with detection by MS, flame ionization and electron capture. High-performance liquid chromatography is also used for the determination of some analytes.

27.9 ANALYSIS OF PARTICULATE MATTER

Particles are almost always removed from air or gas (such as exhaust flue gas) prior to analysis. The two main approaches to particle isolation are filtration and removal by methods that cause the gas stream to undergo a sharp bend such that particles are collected on a surface.

27.9.1 Filtration

The method commonly used for determining the quantity of total suspended particulate matter in the atmosphere draws air over filters that remove the particles.[9] A *Hi-Vol sampler* for particles is essentially a glorified vacuum cleaner that draws air through a filter. The samplers are usually placed under a shelter which excludes precipitation and particles larger than about 0.1 mm in diameter, favoring collection of particles up to 25–50 μm diameter. These devices efficiently collect particles from a large volume of air, typically 2000 m^3, and typically over a 24-h period (Figure 27.5).

The filters used in a Hi-Vol sampler are usually composed of glass fibers and have a collection efficiency of at least 99% for particles with 0.3 μm diameter. Particles with diameters exceeding 100 μm remain on the filter surface, whereas particles with diameters down to approximately 0.1 μm are collected on the glass fibers in the filters. Efficient collection is achieved by using very small diameter fibers (<1 μm) for the filter material.

The technique described here is most useful for determining total levels of particulate matter. Prior to taking the sample, the filter is maintained at 15–35°C at 50% relative humidity for 24 h, then weighed. After sampling for 24 h, the filter is removed and equilibrated for 24 h under the same conditions used prior to its installation on the sampler. The filter is then weighed and the quantity of particulate matter per unit volume of air is calculated.

The range over which particulate matter can be measured is approximately 2–750 μg/m^3, where volume is expressed at 25°C and 1 atm (760 mm Hg, 101 kPa) pressure. The lower limit is determined by limitations in measuring mass, and the upper limit by limited flow rate when the filter becomes clogged.

Size separation of particles can be achieved by filtration through successively smaller filters in a *stacked filter unit.* Another approach uses the *virtual impactor*, a combination of an air filter and an impactor (discussed below). In the virtual impactor, the gas stream being sampled is forced to make a sharp bend. Particles larger than about 2.5 μm do not make the bend and are collected on a filter. The remaining gas stream is then filtered to remove smaller particles. A similar approach is the basis for a reference method for the collection of particulate matter with an aerodynamic diameter less than or equal to a nominal 10 μm, PM_{10}.[10] As discussed in Chapter 10, small particulate matter is of special concern because of its respirability and other properties. This has led to the definition of another category of particulate matter, $PM_{2.5}$, consisting of particles having an aerodynamic diameter less than or equal to a nominal 2.5 μm. To determine $PM_{2.5}$, ambient air is drawn into an inertial particle size separator and the $PM_{2.5}$ particles are collected on a polytetrafluoroethylene (PTFE) filter for weighing over a period of about 24 h.[11] The lower detection limit of the method is about 2 μg/m^3.

Results obtained by the analysis of particulate matter collected by filters should be treated with some caution. A number of reactions may occur on the filter and during the process of removing the sample from the filter. This can cause serious misinterpretation of data. For example, volatile particulate matter may be lost from the filter. Furthermore, because of chemical reactions on the filter, the material analyzed may not be the material that was collected. *Artifact particulate matter* forms

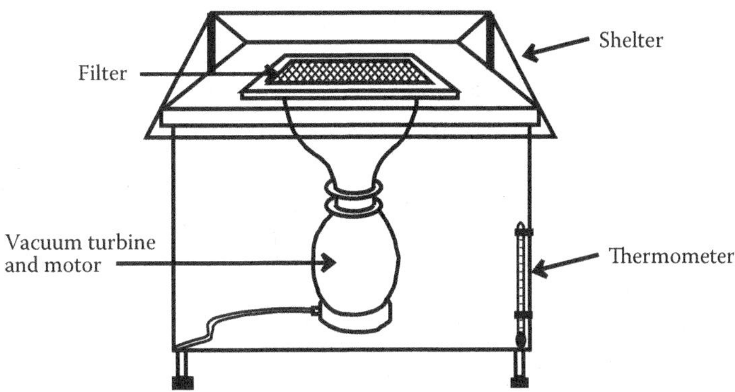

FIGURE 27.5 Hi-Vol sampler for the collection of particulate matter from the atmosphere for analysis.

from the oxidation of acid gases on alkaline glass fibers. This is a major problem with sulfur dioxide retained and oxidized to sulfate, especially on alkaline filters. Artifact particulate matter gives an exaggerated value of particulate matter concentration.

27.9.2 Collection by Impactors

Impactors cause a relatively high velocity gas stream to undergo a sharp bend such that particles are collected on a surface impacted by the stream. The device may be called a dry or wet impactor depending upon whether collecting surface is dry or wet; wet surfaces aid particle retention. Size segregation can be achieved with an impactor because larger particles are preferentially impacted, and smaller particles continue in the gas stream. The cascade impactor, accomplishes size separation by directing the gas stream onto a series of collection slides through successively smaller orifices, which yield successively higher gas velocities. Particles may break up into smaller pieces from the impact of impingement; therefore, in some cases impingers yield erroneously high values for levels of smaller particles.

One of the major difficulties in particle analysis is the lack of suitable filter material. Different filter materials serve very well for specific applications, but none is satisfactory for all applications. Fiber filters composed of polystyrene are very good for elemental analysis because of low background levels of inorganic materials. However, they are not useful for organic analysis. Glass-fiber filters have good weighing qualities and are therefore very useful for determining total particle concentration; however, metals, silicates, sulfates, and other species are readily leached from the fine glass fibers, introducing error into analysis for inorganic pollutant analysis.

27.9.3 Particle Analysis

A number of chemical analysis techniques can be used to characterize atmospheric particulate pollutants.[12] These include atomic absorption, inductively coupled plasma techniques, x-ray fluorescence, neutron activation analysis, and ion-selective electrodes for fluoride analysis. Chemical microscopy is an extremely useful technique for the characterization of atmospheric particles. Either visible or electron microscopy may be employed. Particle morphology and shape tell the experienced microscopist a great deal about the material being examined. Reflection, refraction, microchemical tests, and other techniques may be employed to further characterize the materials being examined. Microscopy may be used for determining levels of specific kinds of particles and for determining particle size.

27.9.4 X-Ray Fluorescence

X-ray fluorescence is another multielement analysis technique that can be applied to a wide variety of environmental samples. It is especially useful for the characterization of atmospheric particulate matter, but it can be applied to some water and soil samples as well. This technique is based upon measurement of x-rays emitted when electrons fall back into inner shell vacancies created by bombardment with energetic x-rays, gamma radiation, or protons. The emitted x-rays have an energy characteristic of the particular atom. The wavelength (energy) of the emitted radiation yields a qualitative analysis of the elements, and the intensity of radiation from a particular element provides a quantitative analysis. A schematic diagram of a wavelength-dispersive x-ray fluorescence spectrophotometer is shown in Figure 27.6. An excitation source, normally an x-ray tube emitting "white" x-rays (a continuum), produces a primary beam of energetic radiation which excites fluorescent x-rays in the sample. A radioactive source emitting gamma rays or protons from an accelerator may also be used for excitation. For best results, the sample should be mounted as a thin layer, which means that segments of air filters containing fine particulate matter make ideal samples. The fluorescent x-rays are passed through a collimator to select a parallel secondary beam, which is dispersed according to wavelength by diffraction with a

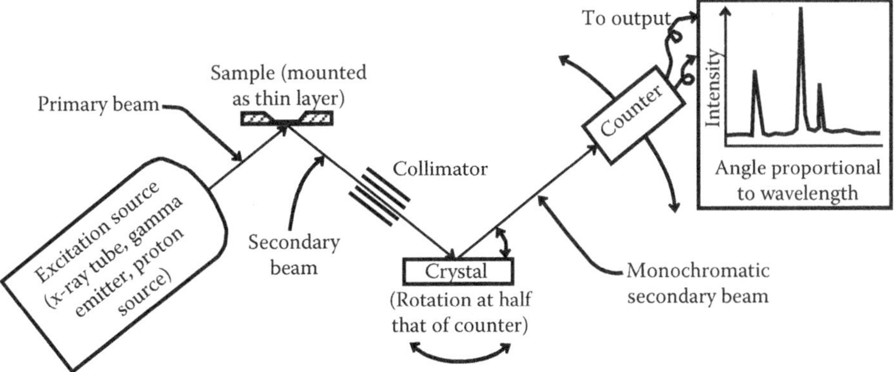

FIGURE 27.6 Wavelength-dispersive x-ray fluorescence spectrophotometer.

crystal monochromator. The monochromatic x-rays in the secondary beam are counted by a detector which rotates at a degree twice that of the crystal to scan the spectrum of emitted radiation.

Energy-selective detectors of the Si(Li) semiconductor type enable measurement of fluorescent x-rays of different energies without the need for wavelength dispersion. Instead, the energies of a number of lines falling on a detector simultaneously are distinguished electronically. An energy-dispersive x-ray fluorescence spectrum from an atmospheric particulate sample is shown in Figure 27.7. A significant advantage of x-ray fluorescence multielement analysis is that sensitivities and detection limits do not vary greatly across the periodic table as they do with methods such as neutron activation analysis or atomic absorption. Proton-excited x-ray emission is particularly sensitive.

27.9.5 Determination of Lead in Particulate Matter

Because of its toxicity, widespread industrial use, and former use in gasoline, lead is one of the most important contaminants in atmospheric particulate matter. The U.S. EPA specifies a method for the determination of lead in atmospheric particulate matter.[13] Lead collected from the atmosphere is

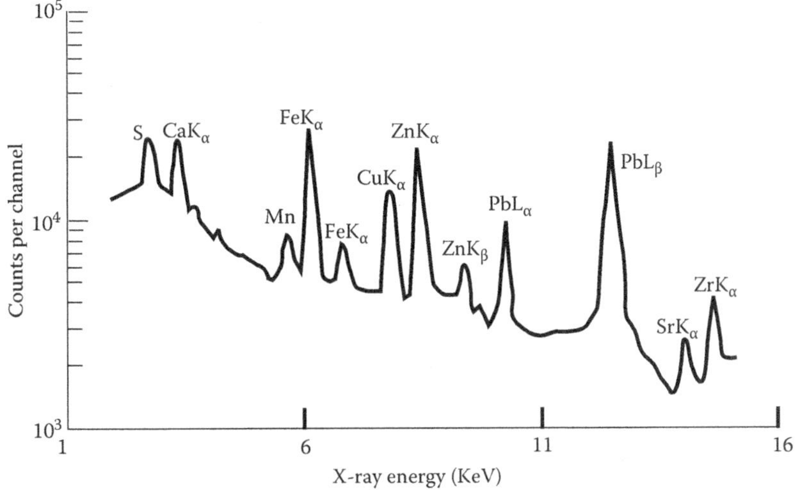

FIGURE 27.7 Energy-dispersive x-ray fluorescence spectrum from an atmospheric particle sample.

extracted from particulate matter with heated nitric acid or by a mixture of nitric and hydrochloric acid facilitated by ultrasonication. The lead in the extract is then measured by flame atomic absorption spectrometry. The lower detection limit is typically 0.07 μg Pb/m^3, using a collection volume of 2400 m^3 of air.

27.10 DIRECT SPECTROPHOTOMETRIC ANALYSIS OF GASEOUS AIR POLLUTANTS

From the foregoing discussion, it is obvious that measurement techniques that depend upon the use of chemical reagents, particularly liquids, are cumbersome and complicated. It is a tribute to the ingenuity of instrument designers that such techniques are being applied successfully to atmospheric pollutant monitoring. Direct spectrophotometric techniques are much more desirable when they are available and when they are capable of accurate analysis at the low levels required. One such technique, nondispersive infrared spectrophotometry, was described in Section 27.7 for the analysis of carbon monoxide. Three other direct spectrophotometric methods are FTIR spectroscopy, tunable diode laser spectroscopy, and differential optical absorption spectroscopy. These techniques may be used for point air monitoring, in which a sample is monitored at a given point, generally by measurement in a long absorption cell. In-stack monitoring may be performed to measure effluents. A final possibility is the collection of long-line data (sometimes using sunlight as a radiation source), yielding concentrations in units of concentration-length (ppm-m). If the path length is known, the concentration may be calculated. This approach is particularly useful for measuring concentrations in stack plumes.

The low levels of typical air constituents require long path lengths, sometimes up to several kilometers, for spectroscopic measurements. These may be achieved by locating the radiation source some distance from the detector, by the use of a distant retroreflector to reflect the radiation back to the vicinity of the source, or by cells in which a beam is reflected multiple times to achieve a long path length.

A typical open-path FTIR system for remote monitoring of air pollutants uses a single unit (telescope) that functions as both a transmitter and receiver of infrared radiation (Figure 27.8). The radiation is generated by a silicon carbide glower, modulated by a Michaelson inteferometer, and transmitted to a retroreflector, which reflects it back to the telescope where its intensity is measured. The modulated infrared signal, called an inteferogram, is processed by a mathematical algorithm, the Fourier transform, to give a spectrum of the absorbing substances. This spectrum is fitted mathematically to spectra of the absorbing species to give their concentrations.

Dispersive absorption spectrometers are basically standard spectrometers with a monochromator for selection of the wavelength to be measured. They are used to measure air pollutants by determining absorption at a specified part of the spectrum of the sought-for material. Of course, other gases or particulate matter that absorb or scatter light at the chosen wavelength interfere.

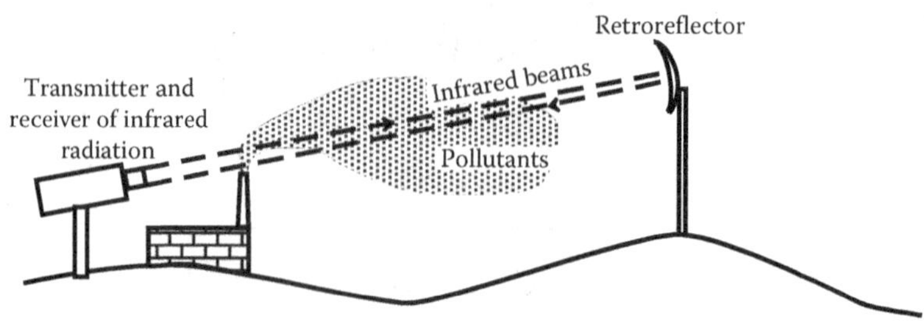

FIGURE 27.8 FTIR system for remote sensing of air pollutants.

These instruments are generally applied to in-stack monitoring. Sensitivity is increased by using long path lengths or by pressurizing the cell.

Second-derivative spectroscopy is a useful technique for trace gas analysis. Basically, this technique varies the wavelength by a small value around a specified nominal wavelength. The second derivative of light intensity versus wavelength is obtained. In conventional absorption spectrophotometry, a decrease in light intensity as the light passes through a sample indicates the presence of at least one substance—and possibly many—absorbing at that wavelength. Second-derivative spectroscopy, however, provides information regarding the change in intensity with wavelength, thereby indicating the presence of specific absorption lines or bands which may be superimposed on a relatively high background of absorption. Much higher specificity is obtained. The spectra obtained by second-derivative spectrometry in the ultraviolet region show a great deal of structure and are quite characteristic of the compounds being observed.

Lidar, which stands for light detection and ranging (analogous to radar, radio detection and ranging), is finding numerous applications in atmospheric monitoring. Lidar systems send short pulses of light or infrared radiation into the atmosphere and collect radiation scattered back from molecules or particles in the atmosphere. Computer analysis of the signal enables measurement of species in the atmosphere.

Differential absorption lidar uses two different laser wavelengths, one of which is absorbed by the analyte molecules and the other of which is not. The intensity difference of the two return signals provides a measure of the analyte concentration.

Target velocities can be measured by *Doppler lidar*. Target molecules moving toward the detector exhibit slightly reduced wavelengths in the return signal (blue shift) and those moving away give slightly longer wavelengths (red shift). This technique is very useful for the analysis of aerosol particles and, since these are carried at the essentially the same velocity as the wind, it provides a remote measurement of wind velocity.

LITERATURE CITED

1. *ASTM Book of Standards Volume 11.07: Atmospheric Analysis*, ASTM International, West Conshohocken, PA, 2008.
2. Garcia-Jares, C., J. Regueiro, R. Barro, T. Dagnac, and M. Llompart, Analysis of industrial contaminants in indoor air. Part 2. Emergent contaminants and pesticides. Departamento de Quimica Analitica, Nutricion y Bromatologia, Instituto de Investigacion y Analisis Alimentarios, Universidad de Santiago de Compostela, Santiago de Compostela, Spain. *Journal of Chromatography, A*, **1216** (3), 567–597, 2009.
3. 40 *Code of Federal Regulations*, Part 50, Office of the Federal Register, National Archives and Records Administration, Washington, DC, July 1, annually.
4. Reference Method for the Determination of Sulfur Dioxide in the Atmosphere (Pararosaniline Method), 40 *Code of Federal Regulations*, Part 50, Appendix A, 2009.
5. Measurement Principle and Calibration Procedure for the Measurement of Nitrogen Dioxide in the Atmosphere (Gas-Phase Chemiluminescence), 40 *Code of Federal Regulations*, Part 50, Appendix F, 2009.
6. Measurement Principle and Calibration Procedure for the Measurement of Ozone in the Atmosphere, 40 *Code of Federal Regulations*, Part 50, Appendix D, 2009.
7. Measurement Principle and Calibration Procedure for the Measurement of Carbon Monoxide in the Atmosphere (Nondispersive Infrared Photometry), 40 *Code of Federal Regulations*, Part 50, Appendix C, 2009.
8. U.S. Environmental Protection Agency, *Air Toxic Methods*, available at http://www.epa.gov/ttnamti1/airtox.html, updated annually.
9. Reference Method for the Determination of Suspended Particulate Matter in the Atmosphere (High-Volume Method), 40 *Code of Federal Regulations*, Part 50, Appendix B, 2009.
10. Reference Method for the Determination of Particulate Matter as PM_{10} in the Atmosphere (High-Volume Method), 40 *Code of Federal Regulations*, Part 50, Appendix J, 2009.
11. Reference Method for the Determination of Fine Particulate Matter as $PM_{2.5}$ in the Atmosphere, 40 *Code of Federal Regulations*, Part 50, Appendix L, 2009.

12. Murphy, D. M., Atmospheric science: Something in the air (a review of techniques for the analysis of particles in the air), *Science*, **307**, 1888–1890, 2005.
13. Reference Method for the Determination of Lead in Suspended Particulate Matter Collected from Ambient Air, 40 *Code of Federal Regulations*, Part 50, Appendix G, 2009.

SUPPLEMENTARY REFERENCES

Guzzi, R., Ed., *Exploring the Atmosphere by Remote Sensing Techniques*, Springer-Verlag, Berlin, 2003.

Hoffman, T., Ed., *Atmospheric Analytical Chemistry*, Springer, Berlin, 2006.

Keith L. H. and M. M. Walker, Eds, *Handbook of Air Toxics: Sampling, Analysis, and Properties*, CRC Press/Lewis Publishers, Boca Raton, FL, 1995.

Matson, P. A. and R. C. Harriss, Eds, *Biogenic Trace Gases: Measuring Emissions from Soil and Water*, Blackwell Science, Cambridge, MA, 1995.

Meier, A., *Determination of Atmospheric Trace Gas Amounts and Corresponding Natural Isotopic Ratios by Means of Ground-Based FTIR Spectroscopy in the High Arctic*, Alfred-Wegener-Institut für Polar und Meeresforschung, Bremen, Germany, 1997.

Michulec, M., W. Wardencki, M. Partyka, and J. Namiesnik, Analytical techniques used in monitoring of atmospheric air pollutants, *Critical Reviews in Analytical Chemistry*, **35**, 117–133, 2005.

Optical Society of America, *Laser Applications to Chemical and Environmental Analysis*, Optical Society of America, Washington, DC, 1998.

Parlar, H., Ed., *Essential Air Monitoring Methods from the MAK-Collection for Occupational Health and Safety*, Wiley-VCH, Weinheim, Germany, 2006.

Spurny, K. R., Ed., *Analytical Chemistry of Aerosols*, CRC Press/Lewis Publishers, Boca Raton, FL, 1999.

Willeke, K. and P. A. Baron, Eds, *Aerosol Measurement: Principles, Techniques, and Applications*, Van Nostrand Reinhold, New York, 1993.

Winegar, E. D. and L. H. Keith, Eds, *Sampling and Analysis of Airborne Pollutants*, CRC Press/Lewis Publishers, Boca Raton, FL, 1993.

QUESTIONS AND PROBLEMS

1. What device is employed to make a nondispersive infrared analyzer selective for the compound being determined?
2. Suggest how MS would be most useful in air pollutant analysis.
3. What is a required characteristic of the absorption cell used for the direct spectrophotometric measurement of gaseous pollutants in the atmosphere, and how may this characteristic be achieved in a cell of manageable dimensions?
4. If 0.250 g of particulate matter is the minimum quantity required for accurate weighing on a Hi-Vol sampler filter, how long must such a sampler be operated at a flow rate of 2.00 m^3/min to collect a sufficiently large sample in an atmosphere containing 5 μg/m^3 of particulate matter?
5. The atmosphere around a chemical plant is suspected of containing a number of heavy metals in the form of particulate matter. After a review of Chapter 10, suggest several methods that would be useful for a qualitative and roughly quantitative analysis of the metals in the particulate matter.
6. Assume that the signal from a chemiluminescence analyzer for NO is proportional to NO concentration. For the same rate of air flow, an instrument gave a signal of 135 μamp for an air sample that had been passed over a thermal converter and 49 μamp with the converter out of the stream. A standard sample containing 0.233 ppm NO gave a signal of 80 μamp. What was the level of NO_2 in the atmospheric sample?
7. Using internet resources, look up the function of permeation tubes in providing standards for atmospheric analysis. A permeation tube containing NO_2 lost 208 mg of the gas in 124 min at 20°C. What flow rate of air at 20°C should be used with the tube to prepare a standard atmospheric sample containing exactly 1.00 ppm by volume of NO_2?

8. What solution should be used in the "wet impinger" designed to collect samples for the determination of metals in the atmosphere?
9. An atmosphere contains 0.10 ppm by volume of SO_2 at 25°C and 1.00 atm pressure. What volume of air would have to be sampled to collect 1.00 mg of SO_2 in tetrachloromercurate solution?
10. Assume that 20% of the surface of a membrane filter used to collect particulate matter consists of circular openings with a uniform diameter of 0.45 μm. How many openings are on the surface of a filter with a diameter of 5.0 cm?
11. Some atmospheric pollutant analysis methods have been used in the past that later have been shown not to give the "true" value. In what respects may such methods still be useful?
12. How may ion chromatography be used for the analysis of nonionic gases?
13. A total of 40 L of air saturated with water vapor at atmospheric pressure and 25°C was bubbled through 500 mL of an oxidizing solution, converting all sulfur dioxide in the air to sulfate. The concentration of sulfate was determined by ion chromatography to be 2×10^{-3} mol/L. Calculate the concentration by volume of SO_2 in the atmosphere on a dry air basis.
14. When large numbers of analyses are performed, an important consideration is the degree to which an analytical procedure is "green." Compare the analyses for atmospheric sulfur dioxide, nitrogen dioxide, ozone, and carbon monoxide described in References given in this chapter and from the internet with respect to the degree to which they are "green."

28 Analysis of Biological Materials and Xenobiotics

28.1 INTRODUCTION

As defined in Chapter 23, Section 23.5, a xenobiotic species is one that is foreign to living systems. Common examples include heavy metals, such as lead, that serve no physiologic function, and synthetic organic compounds that are not natural. Exposure of organisms to xenobiotic materials is a very important consideration in environmental and toxicological chemistry. Therefore, the determination of exposure by various analytical techniques is one of the more crucial aspects of environmental chemistry.

This chapter deals with the determination of xenobiotic substances in biological materials. Although such substances can be measured in a variety of tissues, the greatest concern is their presence in human tissues and other samples of human origin. Therefore, the methods described in this chapter apply primarily to exposed human subjects. They are essentially identical to methods used on other animals and, in fact, most were developed through animal studies. Significantly, different techniques may be required for plant or microbiological samples.

The measurement of xenobiotic substances and their metabolites in blood, urine, breath, and other samples of biological origin to determine exposure to toxic substances is called *biological monitoring*. Comparison of the levels of analytes measured with the degree and type of exposure to foreign substances is a crucial aspect of toxicological chemistry. It is an area in which rapid advances are being made. A complete reference on standardized methods for laboratory analysis practiced by clinical toxicologists confirming to standards stipulated by the International Standards Organization (ISO) was published in 2009.[1] Several books on biological monitoring such as those by Angerer, Draper, Baselt, and Kneip and coauthors are listed at the end of this chapter under "supplementary references."

The two main approaches to workplace monitoring of toxic chemicals are workplace monitoring, using samplers that sample xenobiotic substances from workplace air, and biological monitoring. Although the analyses are generally much more difficult, biological monitoring is a much better indicator of exposure because it measures exposure through all routes—oral and dermal as well as inhalation—and it gives an integrated value of exposure. Furthermore, biological monitoring is very useful in determining the effectiveness of measures taken to prevent exposure, such as protective clothing and hygienic measures.

28.2 INDICATORS OF EXPOSURE TO XENOBIOTICS

Biological markers or *biomarkers* refer to evidence used to monitor exposure, effects, and recovery from effects of xenobiotic substances. They (1) provide confirmation of exposure, (2) enable assessment of ill effects, and (3) provide evidence of an individual's response to exposure. Effects on the liver are especially useful as biomarkers because of the liver's key role in xenobiotic metabolism and its response to virtually all toxicants. In dealing with toxic substances, the liver often suffers damage, including cell death, lesions, fat abnormalities (fatty liver), impairment of bile flow, and vascular disorders. Prominent among the indicators of liver damage from toxicants are proteins indicating such damage that occur in blood (serum and plasma) and urine.

The two major considerations in determining exposure to xenobiotics are the type of sample and the type of analyte. Both of these are influenced by what happens to a xenobiotic material when it gets into the body. For some exposures, the entry site composes the sample. This is the case, for example, in exposure to asbestos fibers in the air, which is manifested by lesions to the lung. More commonly, the analyte may appear at some distance from the site of exposure, such as lead in bone that was originally taken in by the respiratory route. In other cases, the original xenobiotic is not even present in the analyte. An example of this is methemoglobin in blood, the result of exposure to aniline absorbed through the skin.

The two major kinds of samples analyzed for xenobiotics exposure are blood and urine. Both these samples are analyzed for *systemic xenobiotics*, which are those that are transported in the body and metabolized in various tissues. Xenobiotic substances and their metabolites and adducts are absorbed into the body and transported through the bloodstream. Therefore, blood is of unique importance as a sample for biological monitoring. Blood is not a simple sample to process, and subjects often object to the process of taking it. Upon collection, blood may be treated with an anticoagulant, usually a salt of EDTA, and processed for analysis as whole blood. It may also be allowed to clot and be centrifuged to remove solids; the liquid remaining is blood serum.

Recall from Chapter 22 that as the result of Phase I and Phase II reactions, xenobiotics tend to be converted to more polar and water-soluble metabolites. These are eliminated with the urine, making urine a good sample to analyze as evidence of exposure to xenobiotic substances. Urine has the advantage of being a simpler matrix than blood and one that subjects more readily give for analysis. Other kinds of samples that may be analyzed include breath (for volatile xenobiotics and volatile metabolites), hair or nails (for trace elements, such as selenium), adipose tissue (fat), and milk (obviously limited to lactating females). Various kinds of organ tissue can be analyzed in cadavers, which can be useful in trying to determine the cause of death by poisoning.

The choice of the analyte actually measured varies with the xenobiotic substance to which the subject has been exposed. Therefore, it is convenient to divide xenobiotic analysis on the basis of the type of chemical species determined. The most straightforward analyte is, of course, the xenobiotic itself. This applies to elemental xenobiotics, especially metals, which are almost always determined in the elemental form. In a few cases, organic xenobiotics can also be determined as the parent compound. However, organic xenobiotics are commonly metabolized to other products by Phase I and Phase II reactions. Usually, the Phase I reaction product is measured, often after it is hydrolyzed from the Phase II conjugate, using enzymes or acid hydrolysis procedures. Thus, for example, *trans,trans*-muconic acid can be measured as evidence of exposure to the parent compound benzene. In other cases, a Phase II reaction product is measured; for example, hippuric acid is determined as evidence of exposure to toluene. Some xenobiotics or their metabolites form adducts with endogenous materials in the body, which are then measured as evidence of exposure. A simple example is the adduct formed between carbon monoxide and hemoglobin, namely carboxyhemoglobin. More complicated examples are the adducts formed by the carcinogenic Phase I reaction products of PAHs with DNA or hemoglobin. Another class of analytes consists of endogenous substances produced on exposure to a xenobiotic material. Methemoglobin formed as a result of exposure to nitrobenzene, aniline, and related compounds is an example of such a substance that does not contain any of the original xenobiotic material but is formed as the result of exposure to a toxic substance. Another class of substance causes measurable alterations in enzyme activity. The most common example of this is the inhibition of acetylcholinesterase enzyme by organophosphates or carbamate insecticides.

28.3 DETERMINATION OF METALS

28.3.1 Direct Analysis of Metals

Several biologically important metals can be determined directly in body fluids, especially urine, by atomic absorption. In the simplest cases, the urine is diluted with water or with acid and a portion analyzed directly by graphite furnace atomic absorption, taking advantage of the very high sensitivity of that technique for some metals. Metals that can be determined directly in urine by this approach include chromium, copper, lead, lithium, and zinc. Very low levels of metals can be measured using a graphite furnace atomic absorption technique, and Zeeman background correction with a graphite furnace enables measurement of metals in samples that contain enough biological material to cause significant amounts of "smoke" during the atomization process, so that ashing the samples is less necessary.

A variety of metals may be determined in blood and serum simply diluted with appropriate reagents such as ammonia, TritonX-100 surfactant, and EDTA using inductively coupled plasma atomization with mass spectrometric detection. This procedure enables analyses giving adequate detection limits for the measurement of cadmium, cobalt, copper, lead, rubidium, and zinc.

28.3.2 Metals in Wet-Ashed Blood and Urine

Several toxicologically important metals are readily determined from wet-ashed blood or urine using atomic spectroscopic techniques. The ashing procedure may vary, but always entails heating the sample with strong acid and oxidant to dryness and redissolving the residue in acid. A typical procedure is digestion of blood or urine for cadmium analysis, which consists of mixing the sample with a comparable volume of concentrated nitric acid, heating to a reduced volume, adding 30% hydrogen peroxide oxidant, heating to dryness, and dissolving in nitric acid prior to measurement by atomic absorption or emission. Mixtures of nitric, sulfuric, and perchloric acid are effective although they are somewhat hazardous media for digesting blood, urine, or tissue. Wet ashing followed by atomic absorption analysis can be used for the determination in blood or urine of cadmium, chromium, copper, lead, manganese, and zinc, among other metals. Although atomic absorption, especially highly sensitive graphite furnace atomic absorption, has long been favored for measuring metals in biological samples, the multielement capability and other advantages of inductively coupled plasma atomic spectroscopy has led to its use for determining metals in blood and urine samples.

28.3.3 Extraction of Metals for Atomic Absorption Analysis

A number of procedures for the determination of metals and biological samples call for the extraction of the metal with an organic chelating agent in order to remove interferences and concentrate the metal to enable detection of low levels. The urine or blood sample may be first subjected to wet ashing to enable extraction of the metal. Beryllium from an acid-digested blood or urine sample may be extracted by acetylacetone into methylisobutyl ketone prior to atomic absorption analysis. Almost all of the common metals can be determined by this approach using appropriate extractants.

The availability of strongly chelating extractant reagents for a number of metals has led to the development of procedures in which the metal is extracted from minimally treated blood or urine, then quantified by atomic absorption analysis. The metals for which such extractions can be used include cobalt, lead, and thallium extracted into organic solvent as the dithiocarbamate chelate, and nickel extracted into methylisobutyl ketone as a chelate formed with ammonium pyrrolidinedithiocarbamate.

Several metals or metalloids are converted to a volatile form, usually hydrides, for analysis. Arsenic, antimony, and selenium can be reduced to AsH_3, SbH_3, and H_2Se, respectively, which can be determined by atomic absorption or other means. Mercury is reduced to a volatile metal, which is evolved from solution and measured by cold vapor atomic absorption.

28.4 DETERMINATION OF NONMETALS AND INORGANIC COMPOUNDS

Several nonmetals and nonmetallic species are determined in biological samples. One important example is fluoride, which occurs in biological fluids as the fluoride ion, F^-. In some cases of occupational exposure or exposure through food or drinking water, excessive levels of fluoride in the body can be a health concern. Fluoride is readily determined potentiometrically with a fluoride ion-selective electrode. The sample is diluted with an appropriate buffer and the potential of the fluoride electrode measured very accurately versus a reference electrode, with the concentration calculated from a calibration plot. Even more accurate values can be obtained by the use of standard addition in which the potential of the electrode system in a known volume of sample is read, a measured amount of standard fluoride is added, and the shift in potential is used to calculate the unknown concentration of fluoride.

Another nonmetal for which a method of determining biological exposure would be useful is white phosphorus, the most common and relatively toxic elemental form. Unfortunately, there is no chemical method suitable for the determination of exposure to white phosphorus that would distinguish such exposure from relatively high background levels of organic and inorganic phosphorus in body fluids and tissues.

Toxic cyanide can be isolated in a special device called a Conway microdiffusion cell by treatment with acid, followed by collection of the weakly acidic HCN gas that is evolved in a base solution. The cyanide released can be measured spectrophotometrically by the formation of a colored species.

Carbon monoxide is readily determined in blood by virtue of the colored carboxyhemoglobin that it forms with hemoglobin. The procedure consists of measuring the absorbances at wavelengths of 414, 421, and 428 nm of the blood sample, a sample through which oxygen has been bubbled to change all the hemoglobin to the oxyhemoglobin form, and a sample through which carbon monoxide has been bubbled to change all the hemoglobin to carboxyhemoglobin. With the appropriate calculations, a percentage conversion to carboxyhemoglobin can be obtained.

28.5 DETERMINATION OF PARENT ORGANIC COMPOUNDS

A number of organic compounds can be measured as the unmetabolized compound in blood, urine, and breath. In some cases, the sample can be injected along with its water content directly into a gas chromatograph. Direct injection is used for the measurement of acetone, *n*-butanol, dimethylformamide, cyclopropane, halothane, methoxyflurane, diethyl ether, isopropanol, methanol, methyl *n*-butyl ketone, methyl chloride, methylethyl ketone, toluene, trichloroethane, and trichloroethylene.

For the determination of volatile compounds in blood or urine, a straightforward approach is to liberate the analyte at an elevated temperature, allowing the volatile compound to accumulate in *headspace* above the sample followed by direct injection of the headspace gas into a gas chromatograph. A reagent such as perchloric acid may be added to deproteinize the blood or urine sample and facilitate release of the volatile xenobiotic compound. Among the compounds determined by this approach are acetaldehyde, dichloromethane, chloroform, carbon tetrachloride, benzene, trichloroethylene, toluene, cyclohexane, and ethylene oxide. The use of multiple detectors for the gas chromatographic determination of analytes in headspace increases the versatility of this technique and enables the determination of a variety of physiologically important volatile organic compounds.

Purge-and-trap techniques are those in which volatile analytes are evolved from blood or urine in a gas stream and collected on a trap for subsequent chromatographic analysis. Such techniques employing gas chromatographic separation and mass spectrometric detection can be used to determine a number of volatile organic compounds in blood, for example.

28.6 MEASUREMENT OF PHASE I AND PHASE II REACTION PRODUCTS

28.6.1 Phase I Reaction Products

For a number of organic compounds, the most accurate indication of exposure is to be obtained by determining their Phase I reaction products. This is because many compounds are metabolized in the body and do not show up as the parent compound. And those fractions of volatile organic compounds that are not metabolized may be readily eliminated with expired air from the lungs and may thus be missed. In cases where a significant fraction of the xenobiotic compound has undergone a Phase II reaction, the Phase I product may be regenerated by acid hydrolysis.

One of the compounds commonly determined as its Phase I metabolite is benzene, which undergoes the following reactions in the body (see Chapter 24, Section 24.4):

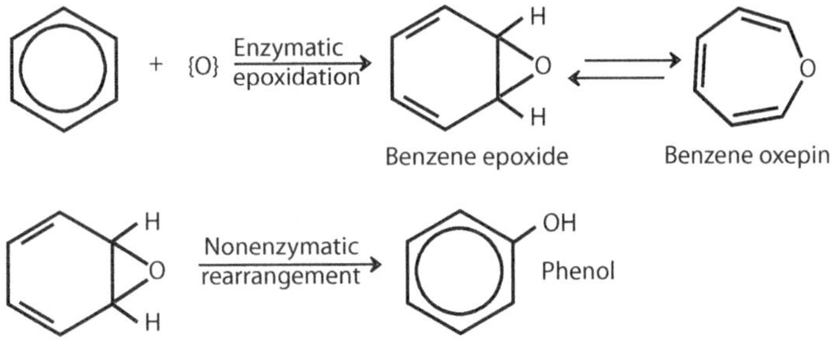

+ {O} Enzymatic epoxidation → Benzene epoxide ⇌ Benzene oxepin (28.1)

Nonenzymatic rearrangement → Phenol

Therefore, exposure to benzene can be determined by analysis of urine for phenol. Although a very sensitive colorimetric method for phenol involving diazotized *p*-nitroaniline has long been available, gas chromatographic analysis is now favored. The urine sample is treated with perchloric acid to hydrolyze phenol conjugates and the phenol is extracted into diisopropyl ether for chromatographic analysis. Two other metabolic products of benzene, *trans,trans*-muconic acid, and S-phenyl mercapturic acid are now commonly measured as more specific biomarkers of benzene exposure.

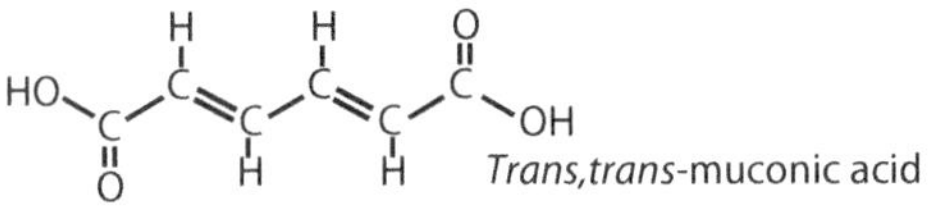

Trans,trans-muconic acid

Insecticidal carbaryl undergoes the metabolic reaction shown in Reaction 28.2. Therefore, the analysis of 1-naphthol in urine indicates exposure to carbaryl. The 1-naphthol that is conjugated by a Phase II reaction is liberated by acid hydrolysis, and is then determined spectrophotometrically or by chromatography

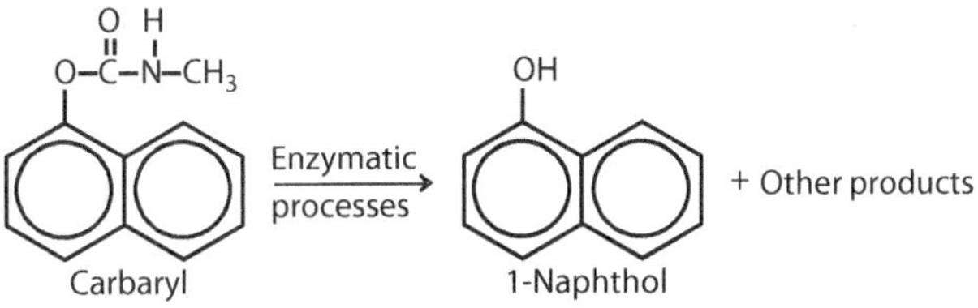

Carbaryl Enzymatic processes → 1-Naphthol + Other products (28.2)

In addition to the examples discussed above, a number of other xenobiotics are measured by their Phase I reaction products. These compounds and their metabolites are listed in Table 28.1. These methods are used for determining metabolites in urine. Normally the urine sample is acidified to release the Phase I metabolites from the Phase II conjugates that they might have formed and, except where direct sample injection is employed, the analyte is collected as vapor or extracted into an organic solvent. In some cases, the analyte is reacted with a reagent that produces a volatile derivative that is readily separated and detected by gas chromatography.

TABLE 28.1
Phase I Reaction Products of Xenobiotics Determined

Parent Compound	Metabolite	Method of Analysis
Cyclohexane	Cyclohexanol	Extraction of acidified, hydrolyzed urine with dichloromethane followed by gas chromatography
Diazinone	Organic phosphates	Colorimetric determination of phosphates
p-Dichlorobenzene	2,5-Dichlorophenol	Extraction into benzene, gas chromatographic analysis
Dimethylformamide	Methylformamide	Gas chromatography with direct sample introduction
Dioxane	β-hydroxyethoxy-acetic acid	Formation of volatile methyl ester, gas chromatography
Ethybenzene	Mandelic acid and related aryl acids	Extraction of acids, formation of volatile derivatives, gas chromatography
Ethylene glycol Monomethyl ether	Methoxyacetic acid	Extracted with dichloromethane, converted to volatile methyl derivative, gas chromatography
Formaldehyde	Formic acid	Gas chromatography of volatile formic acid derivative
Hexane	2,5-Hexanedione	Gas chromatography after extraction with dichloromethane
n-heptane	2-Heptanone, valerolactone, 2,5-heptanedione	Measurement in urine by GC/MS
Isopropanol	Acetone	Gas chromatography following extraction with methylethyl ketone
Malathion	Organic phosphates	Colorimetric determination of phosphates
Methanol	Formic acid	Gas chromatography of volatile formic acid derivative
Methyl bromide	Bromide ion	Formation of volatile organobromine compounds, gas chromatography
Nitrobenzene	*p*-Nitrophenol	Gas chromatography of volatile derivative
Parathion	*p*-Nitrophenol	Gas chromatography of volatile derivative
Polycyclic aromatic hydrocarbons	1-Hydroxypyrene	HPLC of urine
Styrene	Mandelic acid	Extraction of acids, formation of volatile derivatives, gas chromatography
Tetrachloroethylene	Trichloroacetic acid	Extracted into Pyridine and measured colorimetrically
Trichloroethane	Trichloroacetic acid	Extracted into Pyridine and measured colorimetrically
Trichloroethylene	Trichloroacetic acid	Extracted into Pyridine and measured colorimetrically

28.6.2 Phase II Reaction Products

Hippuric acids, which are formed as Phase II metabolic products from toluene, xylenes, benzoic acid, ethylbenzene, and closely related compounds, can be determined as biological markers of exposure. The formation of hippuric acid from toluene is shown in Chapter 23, Figure 23.2, and the formation of 4-methylhippuric acid from *p*-xylene is shown below:

$$H_3C\text{-}C_6H_4\text{-}CH_3 + \{O\} \xrightarrow[\text{Oxidation}]{\text{Phase I}} H_3C\text{-}C_6H_4\text{-}CH_2\text{-}OH$$

$$H_3C\text{-}C_6H_4\text{-}CH_2\text{-}OH + 2\{O\} \xrightarrow{\text{Enzymatic oxidation with loss of } H_2O} H_3C\text{-}C_6H_4\text{-}C(=O)\text{-}OH \quad (28.3)$$

$$H_3C\text{-}C_6H_4\text{-}C(=O)\text{-}OH \xrightarrow[\text{with glycine}]{\text{Phase II conjugation}} H_3C\text{-}C_6H_4\text{-}C(=O)\text{-}N(H)\text{-}CH_2\text{-}C(=O)\text{-}OH$$

Other metabolites that may be formed from aryl solvent precursors include mandelic acid and phenylgloxylic acid.

Exposure to toluene can be detected by extracting hippuric acid from acidified urine into diethyl ether/isopropanol and direct ultraviolet absorbance measurement of the extracted acid at 230 nm. Potentially, methylhippuric acid can be measured in urine as evidence of exposure to xylenes. However, when the analysis is designed to detect the xylenes, ethylbenzene, and related compounds, several metabolites related to hippuric acid may be formed and the ultraviolet spectrometric method lacks specificity. To improve specificity, the various acids produced from these compounds can be extracted from acidified urine into ethyl acetate, derivatized to produce volatile species, and quantified by gas chromatography.

A disadvantage in measuring toluene exposure by hippuric acid is the production of this metabolite from natural sources other than toluene, and the determination of tolulylmercapturic acid is now favored as a biomarker of toluene exposure. An interesting sidelight is that dietary habits can cause uncertainties in the measurement of xenobiotic metabolites. An example of this is the measurement of exposure of workers to 3-chloropropene by the production of allylmercapturic acid. This metabolite is also produced by garlic, and garlic consumption by workers can be a confounding factor in the method. Thiocyanate monitored as evidence of exposure to cyanide is increased markedly by the consumption of cooked cassava!

28.6.3 Mercapturates

Mercapturates have proven to be very useful Phase II reaction products for measuring exposure to xenobiotics, especially because of the sensitive determination of these substances by HPLC separation, and fluorescence detection of their *o*-phthaldialdehyde derivatives. In addition to toluene mentioned above, the xenobiotics for which mercapturates may be monitored include styrene, structurally similar to toluene; acrylonitrile; allyl chloride; atrazine; butadiene; and epichlorohydrin.

The formation of mercapturates or mercapturic acid derivatives by metabolism of xenobiotics is the result of Phase II conjugation by glutathione. *Glutathione* (commonly abbreviated GSH) is a crucial conjugating agent in the body. This compound is a tripeptide, meaning that it is composed of three amino acids linked together. These amino acids and their abbreviations are glutamic acid (Glu), cysteine (Cys), and glycine (Gly). The formula of GSH may be represented as illustrated in Figure 28.1, where the SH is shown specifically because of its crucial role in forming the covalent link to a xenobiotic compound. GSH conjugate may be excreted directly, although this is rare. More commonly, the GSH conjugate undergoes further biochemical reactions that produce mercapturic acids (compounds with N-acetylcysteine attached) or other species. The specific mercapturic acids can be monitored as biological markers of exposure to the xenobiotic species that result in their formation. The overall process for the production of mercapturic acids as applied to a generic xenobiotic species, HX-R (see previous discussion), is illustrated in Figure 28.1.

An interesting application of mercapturate analysis for ingestion of specific compounds is the measurement of S-allylmercapturic acid (structural formula below) in urine as a biomarker of ingestion of garlic for health reasons. The precursor to this compound is γ-glutamyl-S-allyl-L-cysteine that occurs naturally in garlic.

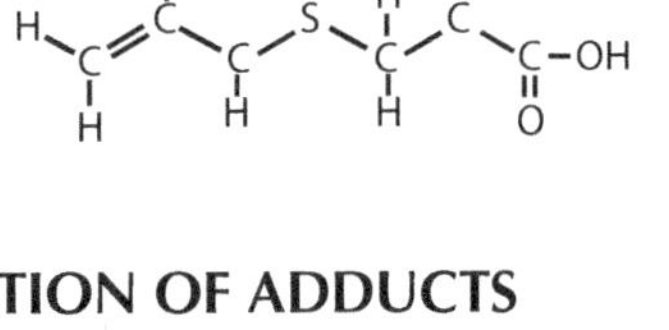

S-allylmercapturic acid

28.7 DETERMINATION OF ADDUCTS

Determination of adducts is often a useful and elegant means of measuring exposure to xenobiotics. Adducts, as the name implies, are substances produced when xenobiotic substances add to endogenous

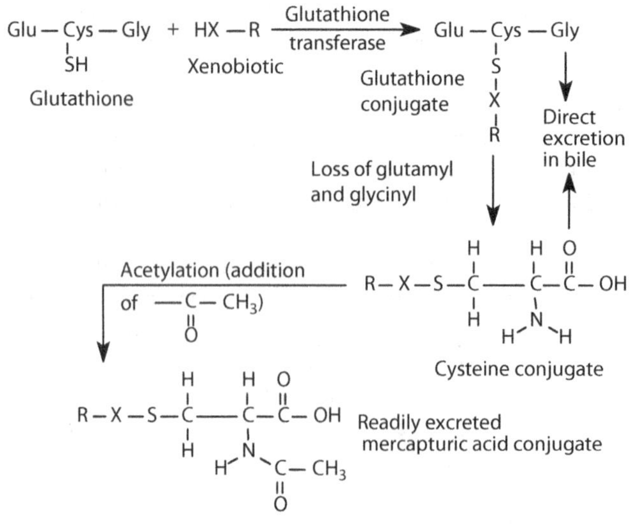

FIGURE 28.1 GSH conjugate of a xenobiotic species (HX-R) followed by formation of GSH and cysteine conjugate intermediates (which may be excreted in bile) and acetylation to form readily excreted mercapturic acid conjugate.

chemical species. The measurement of carbon monoxide from its hemoglobin adduct was discussed in Section 28.4. In general, adducts are produced when a relatively simple xenobiotic molecule adds to a large macromolecular biomolecule that is naturally present in the body. The fact that adduct formation is a mode of toxic action, such as that occurs in the methylation of DNA during carcinogenesis (Chapter 23, Section 23.8), makes adduct measurement as a means of biological monitoring even more pertinent.

Adducts to hemoglobin are perhaps the most useful means of biological monitoring by adduct formation. Hemoglobin is, of course, present in blood, which is the most accurate type of sample for biological monitoring. Adducts to blood plasma albumin are also useful monitors and have been applied to the determination of exposure to toluene diisocyanate, benzo(a)pyrene, styrene, styrene oxide, and aflatoxin B1.

The DNA adduct of styrene oxide has been measured to indicate exposure to carcinogenic styrene oxide. Another example of how DNA adducts can be measured as evidence of xenobiotic exposure is provided by the tobacco-specific carcinogen 4-(methylnitrosamino)-1-(3-pyridyl)-1-butanone, which is activated by metabolic processes and adds to deoxyguanosine in DNA to give the guanine adduct, O^6-[1-oxo-1-(3-pyridyl)but-4-yl]dGuo, as shown in Figure 28.2. Measurement of this adduct by liquid chromatography/electrospray ionization-mass spectrometry (LC/ESI-MS) or LC/ESI-tandem mass spectrometry (LC/ESI-MS/MS) can provide evidence of exposure to the carcinogen.

One disadvantage of biological monitoring by adduct formation can be the relatively complicated procedures and expensive, specialized instruments required. Lysing red blood cells may be required to release the hemoglobin adducts, derivatization may be necessary, and the measurements of the final analyte species can require relatively sophisticated instrumental techniques. Despite these complexities, the measurement of hemoglobin adducts is emerging as a method of choice for a number of xenobiotics including acrylamide, acrylonitrile, 1,3-butadiene, 3,3′-dichlorobenzidine, ethylene oxide, and hexahydrophthalic anhydride.

28.8 THE PROMISE OF IMMUNOLOGICAL METHODS

As discussed in Chapter 26, Section 26.6, immunoassay methods offer distinct advantages in specificity, selectivity, simplicity, and costs. Although used in simple test kits for blood glucose and

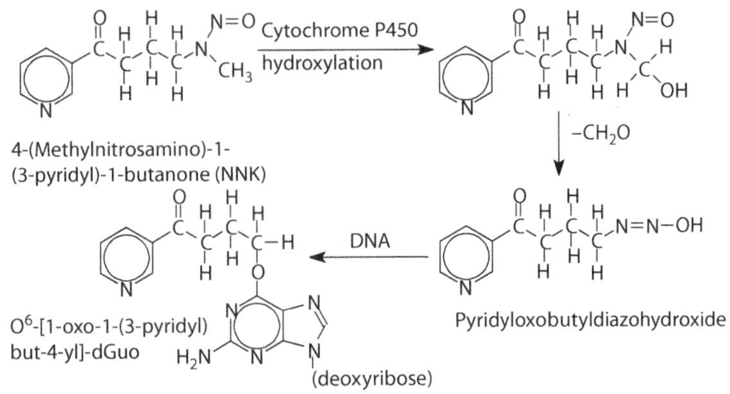

FIGURE 28.2 Formation of a DNA adduct from 4-(methylnitrosamino)-1-(3-pyridyl)-1-butanone, a tobacco-specific carcinogen.

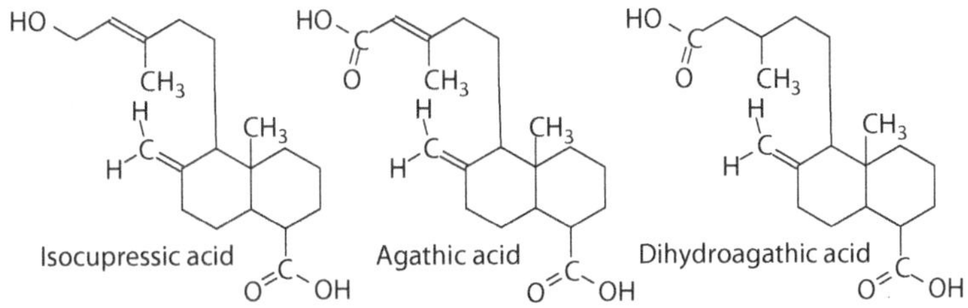

FIGURE 28.3 Isocupressic acid involved in causing abortions in pregnant cows and two of its metabolites.

pregnancy testing, immunoassay methods have been limited in biological monitoring of xenobiotics, in part because of interferences in complex biological systems. Because of their inherent advantages, however, it can be anticipated that immunoassays will grow in importance for biological monitoring of xenobiotics. As an example of such an application, PCBs have been measured in blood plasma by immunoassays. Methods have been developed for the immunoassay of cyanobacterial toxins (saxitoxins, microcystins, and nodularins) in biological samples.

Calves are aborted from pregnant cows that ingest isocupressic acid by eating vegetation from lodgepole pine, ponderosa pine, common juniper, and Monterey cypress. As evidence of exposure to this toxic agent, an immunoassay has been developed that responds to the abortifacient agent as well as its major metabolites, agathic acid, dihydroagathic acid, and tetrahydroagathic acid (Figure 28.3).[2] The method can be used to determine these compounds in both blood serum and urine of cattle.

In addition to immunoassay measurement of xenobiotics and their metabolites, immunological techniques can be used for the separation of analytes from complex biological samples employing immobilized antibodies. This approach has been used to isolate aflatoxicol from urine and enable its determination along with aflatoxins B1, B2, G1, G2, M1, and Q1 using high-performance liquid chromatography and postcolumn derivatization/fluorescence detection.[3] A monoclonal antibody reactive with S-phenylmercapturic acid, an important Phase II reaction product of benzene resulting from GSH conjugation, has been generated from an appropriate hapten–protein conjugate. The immobilized antibody has been used in a column to enrich S-phenylmercapturic acid from the urine of workers exposed to benzene.[4] Significantly, more such applications can be anticipated in future years.

LITERATURE CITED

1. Külpmann, Ed., *Wolf Rüdiger, Clinical Toxicological Analysis: Procedures, Results, Interpretation*, Wiley, New York, 2009.
2. Lee, S. T., D. R. Gardner, M. Garrosian, K. E. Panter, A. N. Serrequi, T. K. Schoch, and B. L. Stegelmeier, Development of enzyme-linked immunosorbent assays for isocupressic acid and serum metabolites of isocupressic acid, *Journal of Agricultural and Food Chemistry*, **51**, 3228–3233, 2003.
3. Kussak, A., B. Andersson, K. Andersson, and C.-A. Nilsson, Determination of aflatoxicol in human urine by immunoaffinity column clean-up and liquid chromatography, *Chemosphere*, **36**, 1841–1848, 1998.
4. Ball, L., A. S. Wright, N. J. Van Sittert, and P. Aston, Immunoenrichment of urinary s-phenylmercapturic acid, *Biomarkers*, **2**, 29–33, 1997.

SUPPLEMENTARY REFERENCES

Angerer, J. K. and K.-H. Schaller, *Analyses of Hazardous Substances in Biological Materials*, Vol. 1, VCH, Weinheim, Germany, 1985.

Angerer, J. K. and K.-H. Schaller, *Analyses of Hazardous Substances in Biological Materials*, Vol. 2, VCH, Weinheim, Germany, 1988.

Angerer, J. K. and K.-H. Schaller, *Analyses of Hazardous Substances in Biological Materials*, Vol. 3, VCH, Weinheim, Germany, 1991.

Angerer, J. K. and K.-H. Schaller, *Analyses of Hazardous Substances in Biological Materials*, Vol. 4, VCH, Weinheim, Germany, 1994.

Angerer, J. K. and K.-H. Schaller, *Analyses of Hazardous Substances in Biological Materials*, Vol. 5, Wiley, New York, 1996.

Angerer, J. K. and K.-H. Schaller, *Analyses of Hazardous Substances in Biological Materials*, Vol. 6, Wiley, New York, 1999.

Angerer, J. K. and K.-H. Schaller, *Analyses of Hazardous Substances in Biological Materials*, Vol. 7, Wiley, New York, 2001.

Angerer, J. K. and K.-H. Schaller, *Analyses of Hazardous Substances in Biological Materials*, Vol. 8, Wiley, New York, 2003.

Angerer, J. K. and M. Muller, Eds, *Analyses of Hazardous Substances in Biological Materials*, Vol. 9, *Markers of Susceptibility*, Wiley, New York, 2004.

Baselt, R. C., *Disposition of Toxic Drugs and Chemicals in Man,* 7th ed., Biomedical Publications, Foster City, CA, 2004.

Committee on National Monitoring of Human Tissues, Board on Environmental Studies and Toxoicology, Commission on Life Sciences, *Monitoring Human Tissues for Toxic Substances*, National Academy Press, Washington, DC, 1991.

Conti, M. E., Ed., *Biological Monitoring: Theory and Applications: Bioindicators and Biomarkers for Environmental Quality and Human Exposure Assessment*, WIT, Southampton, UK, 2008.

Hee, S. Q., *Biological Monitoring: An Introduction*, Van Nostrand Reinhold, New York, 1993.

Kneip, T. J. and J. V. Crable, *Methods for Biological Monitoring*, American Public Health Association, Washington, DC, 1988.

Lauwerys, R. R. and P. Hoet, Industrial Chemical Exposure: Guidelines for Biological *Monitoring*, 3rd ed., Taylor & Francis/CRC Press, Boca Raton, FL, 2001.

Mendelsohn, M. L., J. P. Peeters, and M. J. Normandy, Eds, *Biomarkers and Occupational Health: Progress and Perspectives*, Joseph Henry Press, Washington, DC, 1995.

Richardson, M., Ed., *Environmental Xenobiotics*, Taylor & Francis, London, 1996.

Saleh, M. A., J. N. Blancato, and C. H. Nauman, *Biomarkers of Human Exposure to Pesticides*, American Chemical Society, Washington, DC, 1994.

Travis, C. C., Ed., *Use of Biomarkers in Assessing Health and Environmental Impacts of Chemical Pollutants*, Plenum Press, New York, 1993.

Whitacre, D., Ed., *Reviews of Environmental Contamination and Toxicology*, Springer, New York, 2009.

Williams, W. P., *Human Exposure to Pollutants: Report on the Pilot Phase of the Human Exposure Assessment Locations Programme*, United Nations Environment Programme, New York, 1992.

World Health Organization, *Biological Monitoring of Chemical Exposure in the Workplace*, World Health Organization, Geneva, Switzerland, 1996.

QUESTIONS AND PROBLEMS

1. Personnel monitoring in the workplace is commonly practiced with vapor samplers that workers carry around. How does this differ from biological monitoring? In what respects is biological monitoring superior?
2. Why is blood arguably the best kind of sample for biological monitoring? What are some of the disadvantages of blood in terms of sampling and sample processing? What are some disadvantages of blood as a matrix for analysis? What are the advantages of urine? Discuss why urine might be the kind of sample most likely to show metabolites and least likely to show parent species.
3. Distinguish among the following kinds of analytes measured for biological monitoring: parent compound, Phase I reaction product, Phase II reaction product, adducts.
4. What is wet ashing? For what kinds of analytes is wet ashing of blood commonly performed? What kinds of reagents are used for wet ashing, and what are some of the special safety precautions that should be taken with the use of these kinds of reagents for wet ashing?
5. What species is commonly measured potentiometrically in biological monitoring?
6. Compare the analysis of Phase I and Phase II metabolic products for biological monitoring. How are Phase II products converted back to Phase I metabolites for analysis?
7. Which biomolecule is most commonly involved in the formation of adducts for biological monitoring? What is the problem with measuring adducts for biological monitoring?
8. What are the two general uses of immunology in biological monitoring? What is the disadvantage of immunological techniques? Discuss the likelihood that immunological techniques will find increasing use in the future as a means of biological monitoring.
9. The determination of DNA adducts is a favored means of measuring exposure to carcinogens. Based on what is known about the mechanism of carcinogenicity, why would this method be favored? What might be some limitations of measuring DNA adducts as evidence of exposure to carcinogens?
10. How are mercapturic acid conjugates formed? What special role do they play in biological monitoring? What advantage do they afford in terms of measurement?
11. For what kinds of xenobiotics are trichloroacetic acid measured? Suggest the pathways by which these compounds might form trichloroacetic acid metabolically.
12. Match each xenobiotic species from the column on the left with the analyte that is measured in its biological monitoring from the column on the right.

1. Methanol	a. Mandelic acid
2. Malathion	b. A diketone
3. Styrene	c. Organic phosphates
4. Nitrobenzene	d. Formic acid
5. *n*-Heptane	e. *p*-Nitrophenol

13. From the examples given in Section 28.8, it appears that immunological methods for xenobiotic analysis are favored for moderately complex molecules of biological origin. Suggest why this may be so. What do you think could be the major advantages and disadvantages of immunological methods?
14. The response of a fluoride ion-selective electrode obeys a Nernstian equation

$$E = E_a - 59.2 \log[F^-]$$

where E is the measured potential of the electrode versus a reference electrode in millivolts (mv), E_a is a constant for a given electrode system, and $[F^-]$ is the concentration of fluoride ions in mol/L (M). An analyst placed a fluoride electrode and a reference electrode in 100 mL of a urine sample and set E to zero. Addition of 100 mL of a 1.00×10^{-4} M standard F^- solution shifted the reading of E to -15.0 mv. What was the concentration of F^- in the sample?

Index

B

D

N

O

P

Q

R

S

T

U

V

W

X

Y

Z